AF826025

Lipid Technologies and Applications

Lipid Technologies and Applications

edited by

664.3
LIP

Frank D. Gunstone
Mylnefield Research Services, Ltd.
Scottish Crop Research Institute
Invergowrie
Dundee, Scotland

Fred B. Padley
Consultant
Loders Croklaan, b.v.
A Division of Quest International
Wormerveer, The Netherlands

MARCEL DEKKER

MARCEL DEKKER, INC. NEW YORK • BASEL • HONG KONG

6199

The Library of Congress Cataloging-in-Publication Data

Lipid technologies and applications / edited by Frank D. Gunstone, Fred B. Padley.
p. cm.
Includes index.
ISBN 0-8247-9838-4 (alk. paper)
1. Lipids. I. Gunstone, F. D. II. Padley, F. B. (Fred B.)
TP453.L56L558 1997
664'.3—dc21

97-8301
CIP

THE UNIVERSITY OF READING LIBRARY

The publisher offers discounts on this book when ordered in bulk quantities. For more information, write to Special Sales/Professional Marketing at the address below.

This book is printed on acid-free paper.

Copyright © 1997 by MARCEL DEKKER, INC. All Rights Reserved.

Neither this book nor any part may be reproduced or transmitted in any form or by any means, electronic or mechanical, including photocopying, microfilming, and recording, or by any information storage and retrieval system, without permission in writing from the publisher.

MARCEL DEKKER, INC.
270 Madison Avenue, New York, New York 10016
http://www.dekker.com

FOOD.

Current printing (last digit):
10 9 8 7 6 5 4 3 2 1

PRINTED IN THE UNITED STATES OF AMERICA

Preface

This book is concerned with the wide-ranging use of oils and fats. Annual production of such materials from vegetable and animal sources will be around 90–100 million tonnes by the end of this century. A significant part of these products is used in the country of origin, while roughly two-thirds is exported or imported in the form of seed or extracted oil. The use of such materials is more widespread than is generally realized.

Most fat and oil products are used for human food, and some, such as milk, butter, olive oil, and fish oil, have been part of the human diet for centuries. Others, such as spreading fats, baking fats, frying oils, salad oils, ice cream, chocolate, and vanaspati, are largely products of the 20th century. About 80% of these products is consumed in these forms and 6% is used in animal feed.

Nonfood uses account for the remaining 14%. These uses are many and varied, and fat-based products find their way into many areas of human life. Most nonfood uses of fatty acids and their derivatives are based on their amphiphilic nature (i.e., their ability to be involved with both polar and nonpolar materials at the same time). Soap for cleansing and oils for their emollient properties and as carriers of pigments and perfumes are examples of personal care products that have been used for centuries. These are now joined by other fat-based products for personal care, domestic and industrial cleaning, coatings, and lubricants, etc.

Today these materials have added significance because they are produced from renewable resources in contrast to nonrenewable resources such as oil, gas, and coal and they are readily biodegradable, so that when they have served their purpose, they are easily converted to simpler molecules for reuse in natural processes.

Biotechnology plays an increasingly important role in the production and usage of lipids. This is already apparent in two ways. The first way is in opportunities to improve and modify the yields and the characteristics of oils, particularly from vegetable and microbiological sources. Second, procedures are being developed to modify natural fats using enzymes as catalysts as an alternative to those that are now extensively employed often under harsh temperature and pressure conditions.

This book is divided into six parts, covering the nature of lipids and their major sources, while the second part examines the processes by which these lipids are recovered from natural sources, how they are refined, and how they are modified to broaden their range of uses: the third, fourth, and fifth parts focus on the major food uses, and the sixth on the major nonfood uses. Each part comprises several chapters and the authors of each chapter were asked to prepare authoritative accounts of their topics: these accounts cover the present state of the topic, the required starting materials, the volume of the products obtained, and indications of future trends. We would like to express our gratitude to all the contributors to this book.

Frank D. Gunstone
Fred B. Padley

Contents

Contributors

K. T. Achaya CSIR Emeritus Scientist (retired), Bangalore, India

David A. Allen Technical Consultant, Partnership Solutions, Liverpool, England

John Bentley Consultant, Buckinghamshire, England

Guido Bognolo ICI Performance Chemicals, Everberg, Belgium

Iain J. Campbell Unilever Research Laboratory, Sharnbrook, Bedfordshire, England

Eric G. Carter Pharma Business Nutrition Department, Pharmacia & Upjohn Inc., Clayton, North Carolina

Henri-Jean Caupin Engineering Polymers Division, Elf Atochem S-A, Paris, France

Y. M. Choo Palm Oil Research Institute of Malaysia (PORIM), Kajan, Selangor Darul Ehsan, Malaysia

Keith Coupland Department of Research and Development, Croda Universal Ltd., Hull, England

John W. Finley* Nabisco Foods, East Hanover, New Jersey

Eric Flack† Danisco Ingredients (UK) Limited, Bury St. Edmunds, Suffolk, England

Phil C. Garnsworthy Department of Agriculture and Horticulture, University of Nottingham, Sutton Bonington Campus, Loughborough, England

H. Douglas Goff Department of Food Science, University of Guelph, Guelph, Ontario, Canada

---

**Current affiliation:* Monsanto Company, Chesterfield, Missouri

†*Current affiliation:* Consultant, Greenewood, The Park, Great Barton, Bury St. Edmunds, Suffolk, England

Frank D. Gunstone Mylnefield Research Services, Ltd., Scottish Crop Research Institute, Invergowrie, Dundee, Scotland

Michael I. Gurr Maypole Scientific Services, St. Mary's, Isles of Scilly, England

Martin H. Hilder Manufacturing Application Group, Oil and Dairy Based Foods, Unilever Research Laboratory, Vlaardingen, The Netherlands

Alan D. James Akzo Nobel Chemicals Ltd., Littleborough, Lancashire, England

Malcolm G. Jones Water Structuring, Foams and Emulsions, Unilever Research Laboratory, Sharnbrook, Bedfordshire, England

L. P. Klemann Nabisco, Inc., East Hanover, New Jersey

Wicher T. Koetsier Unichema International, Emmerich, Germany

John M. Krochta Department of Food Science and Technology, University of California–Davis, Davis, California

Niels Krog Danisco Ingredients, Brabrand, Denmark

Robert T. Lyons Pharmaceutical Product Development, Pharmacia & Upjohn, Inc., Clayton, North Carolina

A. N. Ma Palm Oil Research Institute of Malaysia (PORIM), Kajang, Selangor Darul Ehsan, Malaysia

A. R. Macrae Unilever Research Laboratory, Sharnbrook, Bedfordshire, England

Theo Mang Fuchs Petrolub AG Oil + Chemie, Mannheim, Germany

Keith Masters NIRO A/S, Soeborg, Denmark

A. McDonald* Nabisco, Inc., East Hanover, New Jersey

Timothy L. Mounts Food Quality and Safety Research, National Center for Agricultural Utilization Research, Agricultural Research Service, U.S. Department of Agriculture, Peoria, Illinois

Julie A. Nichols Product Applications Laboratory, Croda Universal Ltd., Hull, England

A. S. H. Ong Malaysian Palm Oil Promotion Council, Kuala Lumpur, Malaysia

Fred B. Padley Bedford, Bedfordshire, England, Consultant, Loders Croklaan bv., a Division of Quest International, Wormerveer, The Netherlands

Maurice R. Porter Maurice R. Porter & Associates, Sully, Vale of Glamorgan, South Wales

A. Rozendaal Fat Technology and Oil Processing Unit, Unilever Research Laboratory, Vlaardingen, The Netherlands

Michael Schneider Department of Research & Development, Lucas Meyer GmbH & Co., Hamburg, Germany

Thomas H. Shellhammer Department of Food Science and Technology, University of California–Davis, Davis, California

Ralph E. Timms Consultant, Oils and Fats, The Cottages, Swinderby, Lincoln, England

**Current affiliation:* Consultant, Chicago, Illinois

Maurice A. Williams Anderson International Corp., Cleveland, Ohio

Julian Wiseman Department of Agriculture and Horticulture, The University of Nottingham, Sutton Bonington Campus, Loughborough, England

1

Fatty Acids and Lipids Structure

Frank D. Gunstone
Mylnefield Research Services Ltd., Scottish Crop Research Institute, Invergowrie, Dundee, Scotland

I. INTRODUCTION

It is almost obligatory to start any book on lipids with an account of the chemical structure of the lipids and their component acids, and this book is not different in this respect. The following material is based on recently published fuller accounts [1,2], where further details may be sought if required.

There is no generally accepted definition of the class of natural products designated as lipids, but I adhere to the view that lipids are composed of fatty acids or closely related components such as the corresponding alcohols and the sphingosine bases. Expressed slightly differently, they represent those products of the acetate-malonate scheme of biosynthesis derived through the reductive pathway.

II. FATTY ACID NOMENCLATURE

Fatty acids are designated in several different ways. Despite the alternative descriptions set out here, many trivial names are still widely used. These names were often given before the chemical structure of the acid was elucidated and were frequently chosen to indicate the source of the acid. Examples include palmitic (from palm oil), oleic (from olive oil, *Olea europea*), linoleic and linolenic (from linseed oil), ricinoleic (from castor oil, *Ricinus communis*), and arachidic acid (from groundnut oil, *Arachis hypogea*). A trivial name may be assigned or may continue to be used because the systematic name is cumbersome, as for example with α-eleostearic acid, which is simpler than 9*c*11*t*13*t*-octadecatrienoic acid. Trivial names are easy to use, but they are not in themselves indicative of structure.

Systematic names are based on internationally accepted rules agreed to by organic chemists and biochemists. Those who know the rules can interconvert systematic names and structures. As a simple example, oleic acid is *cis*-9-octadecenoic or *Z*-9-octadecenoic acid. This is a carboxylic acid (oic) with 18 carbon atoms (octadec) and one olefinic center (en) that lies between carbons 9 and 10 (counting from the carboxyl end) and has the *cis* (*Z*) configuration, i.e.,

$$CH_3(CH_2)_7CH{=}CH\ (CH_2)_7COOH$$

which may also be represented by the line drawing

Representations like this are increasingly popular. They are more useful when the number of double bonds and/or other functional groups is larger and the saturated sections of the molecule are short. The structure shown below is not easily recognizable as stearic acid (octadecanoic) because the number of carbon atoms represented in this structure is not immediately apparent and has to be carefully counted. This difficulty is overcome in the shorter formulation shown to the right.

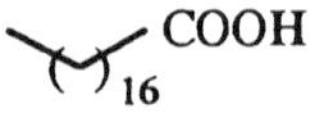

The line forms are useful because of their immediate visual impact; they are not convenient for tabulated data or for insertion into lines of text. Because the word form can be complex and clumsy, systematic or trivial names are sometimes abbreviated to two or three capital letters as in the following examples:

Gamma-linolenic acid	GLA
Arachidonic acid	AA
Eicosapentaenoic acid	EPA
Docosahexaenoic acid	DHA

Another way of designating fatty acids involves the use of numbers such as 18:2. This symbol describes an acid such as linoleic with 18 carbon atoms (assumed to be straight-chain) and two unsaturated centers (assumed to be *cis*-olefinic). Since there are many isomeric compounds that could be represented by this symbol, additional descriptors may be added thus:

18:2 (9,12), 18:2 (9*c*,12*c*), 18:2 (9Z,12*Z*), 18:2 (*n*-6)

All these refer to the same acid. The first indicates the position of the two unsaturated centers in the C_{18} chain with reference to COOH = 1. The second and third formulations confirm the *cis* or *Z* configuration of the double bonds. The symbol Δ is sometimes added to show that numbering is with respect to the acid function. The fourth designation introduces a further concept in fatty acid nomenclature.

It is sometimes useful to designate double bond positions with respect to the CH_3 end group, and this is done with symbols such as ω6 or *n*-6, which indicate that the first double bond is on carbon 6, counting from the methyl group. In the absence of other information it is assumed that all the double bonds are methylene-interrupted and have the *cis* (*Z*) configuration. Symbols such as *c*, *t*, and *e* are used to show *cis*, *trans*, and ethylenic unsaturation (where configuration is not known or does not apply), and *a* (acetylenic) or *y* (ynoic) is used to show a triple bond.

III. FATTY ACIDS—MAIN STRUCTURAL FEATURES

The number of known natural fatty acids exceeds 1000, though only a relatively small number, perhaps 20–50, are of common concern. Based on a survey of these 1000 structures, and noting particularly the structures of those acids produced most commonly in nature, it is possible to make four general statements. Each of these is generally true, but there are exceptions to all four. The exceptions are frequently trivial, but sometimes they are significant. Though originally based on a survey of chemical structures, it is now clear that these statements reflect the underlying biosynthetic pathways by which the acids are produced in nature.

1. Natural fatty acids, both saturated and unsaturated, are straight-chain compounds with an even number of carbon atoms in their molecules. This is true for the great majority of structures and for the more abundant acids. Chain lengths range from two to over 80 carbon atoms, though they are most commonly between C_{12} and C_{22}. Despite the validity of this statement, acids with an odd number of carbon atoms (e.g., heptadecanoic, C_{17}) occur, as do those with branched chains (e.g., isopalmitic, anteisononadecanoic) or with carbocyclic units (e.g., sterculic, chaulmoogric).

2. Acids with one unsaturated center are usually olefinic compounds with *cis* (*Z*) configuration and with the double bond in one of a limited number of preferred positions. This is most commonly Δ9 (i.e., nine carbon atoms from the carboxyl group as in oleic) or *n*-9 (i.e., nine carbon atoms from the methyl group as in oleic or erucic acid). But double bonds occur in other positions (e.g., petroselinic, 6*c*-18:1) or have *trans* configuration (e.g., elaidic 9*t*-18:1) or can be replaced by an acetylenic unit (e.g., tariric 6*a*-18:1).

3. Polyunsaturated acids are mainly polyolefinic with a methylene-interrupted arrangement of double bonds having *cis* (*Z*) configuration. That is, *cis* double bonds are separated from each other by one CH_2 group as in arachidonic acid:

The 1,4 pattern of unsaturation is characteristic of natural fatty acids and differs from that in acyclic isoprenoids, which is usually 1,3 (conjugated) or 1,5. Polyunsaturated acids occur in biosynthetically related families. The most important are the *n*-6 acids based on linoleic acid and the *n*-3 acids based on α-linolenic acid. In contrast to this very common pattern of unsaturation, some acids have conjugated unsaturation, which may be *cis* or *trans* (e.g., eleostearic, calendic, parinaric acids); some have mixed en/yne unsaturation, which may be conjugated (e.g., isanic) or nonconjugated (e.g., crepenynic); and some have nonconjugated unsaturation, which is not entirely methylene-interrupted. These are known as non-methylene-interrupted polyenes (e.g., columbinic, pinolenic).

4. Fatty acids rarely have functional groups apart from the carboxyl group and the various types of unsaturation already discussed. Nevertheless, acids are known that also contain a fluoro, hydroxy, keto, or epoxy group. Two important examples are ricinoleic (12-hydroxyoleic) and vernolic (12,13-epoxyoleic) acid.

These generalizations have a biosynthetic basis, and even some of the exceptions are accommodated in the general biosynthetic scheme with only minor modifications.

Based on the annual production of commercial vegetable oilseeds it has been estimated that eight acids account for about 97% of the total production: lauric (4%), myristic (2%), palmitic (11%), stearic (4%), oleic (34%), linoleic (34%), α-linolenic (5%), and erucic (3%). The level of linolenic acid would rise if all green tissue were taken into account. The major acids in animal fats and in fish oils are myristic, palmitic, palmitoleic, stearic, oleic, eicosenoic, arachidonic, EPA, docosenoic, and DHA.

Table 1 Names and Selected Physical Properties of Some Alkanoic Acids

			Acid		Methyl ester		
Chain length	Systematic name	Trivial name	mp (°C)	bp (°C)[a]	mp (°C)	bp (°C)[a]	Mol wt
4	Butanoic	Butyric	−5.3	164	—	103	88.1
6	Hexanoic	Caproic	−3.2	206	−69.6	151	116.2
8	Octanoic	Caprylic	16.5	240	−36.7	195	144.2
10	Decanoic	Capric	31.6	271	−12.8	228	172.3
12	Dodecanoic	Lauric	44.8	130[1]	5.1	262	200.3
14	Tetradecanoic	Myristic	54.4	149[1]	19.1	114[1]	228.4
16	Hexadecanoic	Palmitic	62.9	167[1]	30.7	136[1]	256.4
18	Octadecanoic	Stearic	70.1	184[1]	37.8	156[1]	284.5
20	Eicosanoic	Arachidic	76.1	204[1]	46.4	188[2]	312.5
22	Docosanoic	Behenic	80.0	—	51.8	206[2]	340.6
24	Tetracosanoic	Lignoceric	84.2	—	57.4	222[2]	368.6

[a] bp at 760 mm or at 1 or 2 mm as indicated by superscript.
Source: Adapted from Ref. 1, p. 1.

IV. SATURATED ACIDS

Table 1 gives the systematic and trivial names and selected physical properties of some C_4–C_{24} acids.

A. Short- and Medium-Chain Saturated Acids (C_4–C_{14})

Members of the C_4–C_{14} group of acids occur in milk fats and in the lauric group of vegetable oils. Cow's milk fat, for example, contains butanoic (butyric, C_4) acid at a level of about 4% (by weight). This may not seem very much, but because of the low molecular weight of the C_4 acid compared with the more common C_{18} acids it represents about 8.5% on a molar basis, which means that it could be present in up to 25% of milk fat glycerides. Also present are lower levels of the C_6–C_{12} acids. The short-chain acids are retained in butter and in other products made from milk fat.

Some seed oils contain very high levels of lauric acid (C_{12}, about 50%) with significant levels of caprylic (C_8), capric (C_{10}), and myristic (C_{14}) acids also. The best known of these are coconut oil and palmkernel oil. At the present time there is interest in other sources of these medium-chain acids. Among these are the cuphea oils, which show great promise but still need to be fully domesticated before they produce an easily cultivated crop. A transgenic rapeseed has also been developed that produces a lauric-rich oil.

B. Palmitic and Stearic Acid

Palmitic acid (16:0) is the most widely occurring saturated acid. It is present in fish oils (10–30%), in milk and body fats of land animals (up to 30%), and in virtually all vegetable fats at levels of between 5 and 50%. Useful sources of palmitic acid include cottonseed oil (15–30%), palm oil (30–60%), Chinese vegetable tallow (60–70%), lard (20–30%), and tallow from sheep and cattle (25–35%).

Despite the popular impression, stearic acid (18:0) is much less common than palmitic acid. It is, however, a major component of the tallows of ruminant animals (5–40%) and a significant component in a number of vegetable tallows (solid fats of vegetable origin) including cocoa butter (30–35%), Illipe or Borneo tallow (40%), and shea butter (~45%). Stearic acid is also easily made by hydrogenation of readily available oleic, linoleic, and linolenic acids.

Since many population groups in the world have scruples about animal products, vegetable sources of palmitic acid and stearic acid are preferred. These are used in food and non-food products (surfactants, cosmetics, personal hygiene products) to obtain products that find general acceptance.

C. Long-Chain Acids

Saturated acids of chain length greater than 18 carbon atoms are present at low levels in a few seed oils and at higher levels only in a few uncommon sources. The C_{20}–C_{30} members are often present in waxes. A convenient source of some of these acids is groundnut oil, which contains 5–8% long-chain acids, including arachidic (20:0), behenic (22:0), and lignoceric (24:0) acids. Rarer but richer sources include rambutan tallow (*Nephelium lappaceum*, ~35% 20:0), kusum (*Schleichera trifuga*, 20–30% 20:0), *Lophira alata* (15–30% 22:0), and *L. procera* (~20% 22:0) seed fats. Long-chain acids can also be made from more readily available shorter chain acids (C_{12}–C_{18}) by appropriate chain-extension procedures. These reaction sequences are not confined to adding one or two carbon atoms at a time, as methods exist for adding five or six carbon atoms in one cycle of reactions.

V. MONOENE ACIDS

Over 100 monoene acids have been described. These fall almost entirely in the range C_{10}–C_{30}, with C_{16}, C_{18}, and C_{22} members the most common. Most have *cis* (*Z*) configuration, and the most common are either Δ9 or *n*-9, i.e., the double bond is nine carbon atoms from the carboxyl group (Δ9) or the methyl group (*n*-9).

9-Hexadecenoic acid (palmitoleic, zoomaric) is a minor component (<1%) of many seed oils and animal fats. It is more significant in fish oils (up to 10%) and is a major component of a few seed oils such as macadamia oil (around 20%).

9-Octadecenoic acid (oleic) is the most widely distributed and the most extensively produced of all fatty acids. It is the prototype of all monoene acids and serves as the biological precursor of other *n*-9 monoene acids and of the *n*-9 family of polyene acids. Olive oil (60–80%) and almond oil (60–70%) are rich sources of this acid, and it is also present at high levels in several nut oils (55–75%). It is a significant component in low-erucic rapeseed oil, groundnut oil, and palm oil. High-oleic varieties of sunflower oil (80% and above) and safflower oil (70–75%) are now available, and the former is commonly used as a source of oleic-rich glycerides. Oleic acid is a major component in most animal fats, where it is often accompanied by low levels of other 18:1 isomers. This may or may not matter depending on the end use of the acid.

Other octadecenoic acids include petroselinic (6*c*), elaidic (9*t*), and vaccenic (11*c* and 11*t*). Petroselinic acid is a major component (35–85%) of many seed oils of the families Umbelliferae, Araliaceae, and Garryaceae. Examples include carrot, caraway, parsnip, and parsley. Attempts are being made to develop coriander as a new crop containing about 80% petroselinic acid.

Eicosenoic acid (20:1) occurs in fish oils and also accompanies erucic acid (22:1) in vegetable oils rich in the C_{22} acid. It is present mainly as the Δ9 (*n*-11) and/or Δ11 (*n*-9) isomers.

The most important docosenoic acid (22:1) is erucic acid (Δ13*c*, *n*-9), present at high levels in seed oils of the Cruciferae. This acid has been bred out of rapeseed grown for food use, but there is a considerable demand for erucic acid in the form of its amide. Erucic acid is available on a commercial scale from high-erucic rapeseed oil (45–50%), mustard seed oil (~60%), and the seed oil of *Crambe abyssinica* (~60%). The major docosenoic acid in fish oils is the Δ11 isomer (cetoleic).

Jojoba oil (from *Simmondsia chinensis*) is unusual in that its major lipids are ester waxes (esters of long-chain alcohols and long-chain acids) rather than glycerol esters. Most of the acids and alcohols are C_{20} and C_{22} compounds with one double bond. In meadowfoam oil (*Limnanthes alba*) more than 95% of the fatty acids are C_{20} and C_{22} compounds. These are mainly 20:1 (5*c*, 63%) and 22:1 (5*c*, 3% and 13*c*, 10%) along with 22:2 (5*c*13*c*, 18%).

The major C_{24} monoene acid [nervonic (Δ15*c*, *n*-9)] is a significant component of many sphingolipids. Honesty seed oil (*Lunaria biennis*), with C_{22} (43%) and C_{24} (25%) monoene acids, is probably the most convenient source of nervonic acid.

VI. METHYLENE-INTERRUPTED POLYENE ACIDS

The most important polyene acids have a methylene-interrupted pattern of unsaturation with two to six double bonds and *cis* (Z) configuration. They are grouped in families and are biosynthetically related to other members of the same family. The two major groups are the *n*-6 acids based on linoleic acid and the *n*-3 acids based on α-linolenic acid. Two minor groups are based on oleic acid (*n*-9 family) and on 9-hexadecenoic acid (*n*-7 family). The common and important C_{18} members are present in most vegetable oils, whereas the C_{16}, C_{20}, and C_{22} acids are present in fish oils and in the lipids of other animals. Some of these acids cannot be made by animals and are essential dietary requirements of vegetable origin. Structures and names are given in Table 2, and some comments on the more important *n*-6 and *n*-3 acids follow.

Linoleic acid (18:2, *n*-6) is the most common of all the polyene acids and serves as a prototype for other acids in this category. It is present in all vegetable fats and is a major component in several. Among the more common sources are soybean (45–60%), low-erucic rapeseed (10–30%), groundnut (15–45%), rice bran (35–40%), poppy (~70%), safflower (55–80%), sunflower (20–75%), corn (maize) (35–60%), cotton (35–60%), sesame (~45%), and tall oil fatty acids (40–50%).

γ-Linolenic acid (18:3, *n*-6, GLA) is not a common acid, but recent interest in its use as a dietary supplement has directed attention to its three common sources: evening primrose oil (8–12%), borage oil (20–25%), and blackcurrant seed oil (15–17%).

Table 2 Methylene-Interrupted Polyene Acids

Trivial name	Structure	Double bond position[a]
Linoleic	18:2 (*n*-6)	9,12
γ-Linolenic	18:3 (*n*-6)	6,9,12
α-Linolenic	18:3 (*n*-3)	9,12,15
Stearidonic	18:4 (*n*-3)	6,9,12,15
Dihomo-γ-linolenic	20:3 (*n*-6)	8,11,14
Mead's acid	20:3 (*n*-9)	5,8,11
Arachidonic	20:4 (*n*-6)	5,8,11,14
Eicosapentaenoic (EPA)	20:5 (*n*-3)	5,8,11,14,17
Docosapentaenoic (DPA)	22:5 (*n*-3)	7,10,13,16,19
Docosahexaenoic (DHA)	22:6 (*n*-3)	4,7,10,13,16,19

[a] All double bonds have *cis* configuration.

Arachidonic acid (20:4, *n*-6, AA) is of considerable importance as the precursor of many important C_{20} metabolites such as the prostaglandins, thromboxanes, and leukotrienes (members of the eicosanoid cascade). This acid is a minor component of many fish oils and attains higher levels in animal phospholipids such as those from egg and liver. Though rare in the plant kingdom, it has been identified in mosses, ferns, and some algae and fungi.

α-Linolenic acid (18:3, *n*-3) is an important component of lipids in leaves, stems, and roots. It is a minor but significant component of soybean oil (4–11%) and of rapeseed oil (5–15%) and a major component of linseed oil (50–60%) and perilla oil (60–70%). Eicosapentaenoic acid (20:5, EPA) and docosahexaenoic acid (22:6, DHA) are also important *n*-3 acids. The former is a precursor of a series of eicosanoids, and the latter is present at high levels in the lipids of sperm and the retina. The major sources of these two acids are the fish oils, up to 30% of which may be the two acids combined. One or both are also produced by some microorganisms.

Mead's acid (20:3, *n*-9) is of interest in that it becomes significant in animal lipids only during periods of essential fatty acid deficiency.

VII. OTHER POLYENE, ACETYLENIC, AND ALLENIC ACIDS

The best known conjugated polyene acids have 18 carbon atoms and three or four double bonds per molecule. Two Δ9,11,13,15-tetraenes are derived biologically from α-linolenic acid (Δ9,12,15), and the Δ9,11,13- and the Δ8,10,12-trienes are derived from linoleic acid (Δ9,12). These are listed in Table 3 with structures, trivial names, and major sources.

Linoleic acid furnishes a mixture of conjugated dienes (9,11 and 10,12 isomers) in a number of enzymatic and nonenzymatic processes. This mixture is reported to display anticarcinogenic properties, with the 9*c*11*t* acid being the active isomer. The mechanism of this activity is not known, and the suggestion that it acts as an antioxidant has been questioned. Conjugated linoleic acid (CLA) is present at low levels in many dietary sources. The highest levels

Table 3 Natural Octadecatrienoic and Octadecatetraenoic Acids with Conjugated Unsaturation

Configuration	Trivial name	Typical source (seed oils)
Trienes (18:3)		
8*c*10*t*12*c*	Jacaric	*Jacaranda mimosifolia*
8*t*10*t*12*c*	Calendic	*Calendula officinalis*
8*t*10*t*12*t*	—	*C. officinalis*
9*c*11*t*13*c*	Catalpic	*Catalpa ovata*
9*c*11*t*13*t*	α-Eleostearic[a,b]	*Aleurites fordii*[d]
9*t*11*t*13*t*	β-Eleostearic	*A. fordii*[d]
9*t*11*t*13*c*	Punicic	*Punica granatum*
Tetraenes (18:4)		
9*c*11*t*13*t*15*c*	α-Parinaric[c]	*Impatiens balsamina*
9*t*11*t*13*t*15*t*	β-Parinaric	—

[a] Also kamlolenic, 18-hydroxy-α-eleostearic (*Mallotus philippinensis*).
[b] Also licanic, 4-oxo-α-eleostearic (*Licania rigida*).
[c] Also chrysobalanic, 4-oxo-α-parinaric (*Chrysobalanus icaco*).
[d] Tung oil.
Source: Adapted from Ref. 1, p. 7.

appear in ruminant and dairy fats. For example, lamb fat contains 5.6 mg of CLA per gram of fat, over 90% of it the 9*c*11*t* isomer.

Some polyene acids have incomplete methylene-interrupted unsaturation. These are found in some seed oils, some mycobacteria, and some marine lipids, especially sponges. The additional double bond is frequently Δ3 or Δ5. Examples include a diene acid (20:2 5*c*13*c*) in meadowfoam oil, pinolenic acid (18:3 5*c*9*c*12*c*) in tall oil, columbinic acid (18:3 5*t*9*c*12*c*) in aquilegia (columbine) oil, and long-chain trienes such as 26:3 (5,9,19), 28:3 (5,9,19), and 30:3 (5,9,23) in sponge lipids (demospongic acids).

Acetylenic acids occur only rarely in seed oils and in mosses. Structures have been identified that contain acetylenic unsaturation only (e.g., tariric 18:1 6*a*); mixed olefinic and acetylenic unsaturation, which may be conjugated (ximenynic or santalbic, 18:2 9*c*11*t*) or nonconjugated (crepenynic, 18:2 9*c*12*a*); and yet others that also contain a hydroxyl group (helenynolic, 9-OH 18:2 10*t*12*c*).

A few allenic acids are also known and are probably produced by rearrangement of acetylenic precursors. The allenic acids are optically active by virtue of the chiral allenic group and include laballenic (18:2 5*e*6*e*) and lamenallenic (18:3 5*e*6*e*16*t*) acids.

VIII. ACIDS WITH *TRANS* OLEFINIC UNSATURATION

Polyene acids having *trans* olefinic groups in conjugated and non-methylene-interrupted systems have already been discussed. Natural acids with *trans* unsaturation are very uncommon. Nevertheless they are a significant component (about 5%) of human dietary intake. They arise from two major sources: (1) from those cooking and spreading fats that have been produced by partial hydrogenation with a heterogeneous catalyst and (2) from dairy and other ruminant fats. These last result from biohydrogenation of linoleic and linolenic acid in the animal's diet. For example, butterfat contains ten 18:1 *trans* acids of which about 75% is *trans*-vaccenic acid (Δ11*t*). The *trans* acids in human milk fat (2–4%) probably arise from these dietary sources. Dietary *trans* acids are also present in some refined oils exposed to high temperatures (230–250°C) during the refining process. The major acids of this type (present, e.g., in refined soybean and rapeseed oil) are the 9*c*12*c*15*t* and 9*t*12*c*15*c* isomers from linolenic acid. Linoleic acid undergoes a similar reaction less readily to give the 9*c*12*t* and 9*t*12*c* diene acids.

IX. BRANCHED-CHAIN ACIDS

There are many branched-chain acids, though these seldom occur at a significant level. Examples include the iso acids (mainly with an even number of carbon atoms), the anteiso acids (mainly with an uneven number of carbon atoms) found in wool wax and other animal fats, and the phytol-based acids present at low levels in fish oils.

$$\underset{\text{iso acids}}{CH_3\overset{\overset{\displaystyle CH_3}{|}}{C}H(CH_2)_nCOOH} \qquad \underset{\text{anteiso acids}}{CH_3CH_2\overset{\overset{\displaystyle CH_3}{|}}{C}H(CH_2)_nCOOH}$$

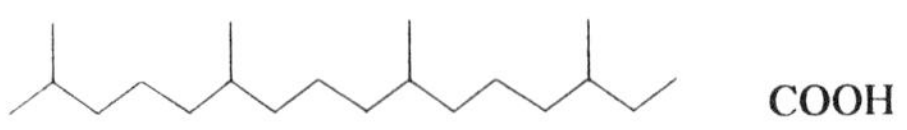
phytanic acid (C_{20})

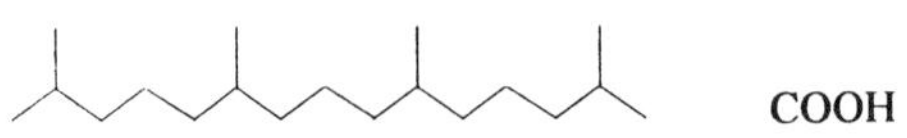
pristanic acid (C_{19})

X. CYCLIC ACIDS

The most common cyclic acids contain a cyclopropane, cyclopropene, or cyclopentene unit. Cyclopropane acids occur in bacterial membrane phospholipids and are mainly C_{17} or C_{19} (lactobacillic) acids. The cyclopropane unit, like a *cis* double bond, introduces a discontinuity in the molecule and increases fluidity in the membrane.

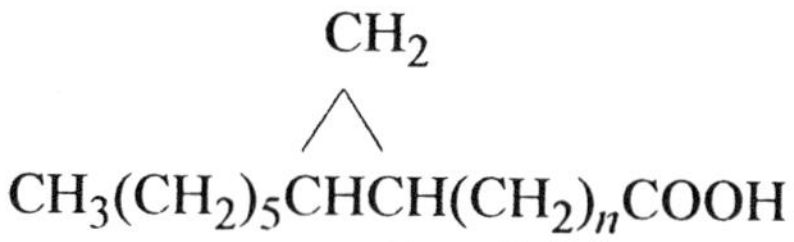

$n = 7$ or 9

The best known cyclopropene acids are malvalic and sterculic, which are present at high levels in sterculia oils and at lower levels in kapok seed oil (~12%) and in cottonseed oil (~1%). These acids are highly reactive and are destroyed during refining and hydrogenation of the oils. They have attracted interest because they inhibit the biodesaturation of stearic to oleic acid. 2-Hydroxysterculic acid has also been identified. This compound is probably an intermediate in the bioconversion of sterculic to malvalic acid by α-oxidation.

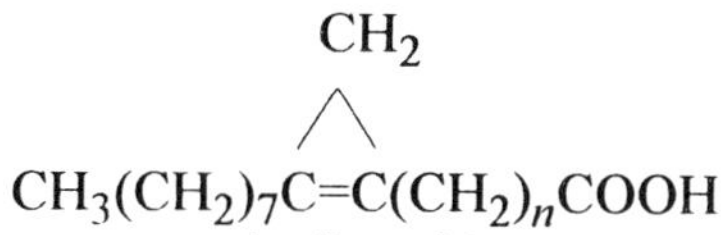

malvalic acid $n = 6$
sterculic acid $n = 7$

Seed fats of the Flacourtiaceae are unique in containing several cyclopentene acids. They range from C_6 to C_{20} acids and may also be unsaturated in the side chain. These acids are optically active by virtue of the chiral center in the cyclopentene unit.

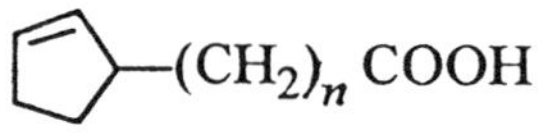

($n = 0$-14)

XI. OXYGENATED ACIDS

The most common of the oxygenated acids contain hydroxy, epoxy, or furanoid units. Other systems not included here include keto (oxo) and methoxy groups.

A. Hydroxy Acids

The best known natural hydroxy acid is ricinoleic (12-hydroxyoleic) acid. This is the major acid (~90%) in castor oil (*Ricinus communis* seed oil). An isomer (9-OH 18:1 12*c*) occurs in *Strophanthus* and *Wightia* seed oils. Other acids of this type include densipolic, lesquerolic, and auricolic (Table 4). Lesquerolic acid is the C_{20} homologue of ricinoleic acid, and attempts are being made to develop *Lesquerella fendleri* as a new crop.

Castor oil and ricinoleic acid are important materials used in cosmetics, in lubricants both before and after hydrogenation, and as a drying oil after dehydration (dehydrated castor oil, DCO). Also important are the products of pyrolysis (heptanal and 10-undecenoic acid and materials derived from these) and of alkali fusion (2-octanol, 2-octanone, 10-hydroxydecanoic acid, and the C_{10} dibasic acid, sebacic acid). Of less importance are some hydroxy acids based on 8-, 9-, or 13-hydroxystearic acid with conjugated unsaturation. Some typical examples are isanolic (8-OH 18:3 9*a*11*a*17*e*), helenynolic (9-OH 18:2 10*t*12*a*), dimorphecolic (9-OH 18:2 10*t*12*c*), and coriolic (13-OH 18:2 9*c*11*t*).

Table 4 Mid-Chain Hydroxy Acids Without Conjugated Unsaturation

Chain length and unsaturation	Position of hydroxyl group	Trivial name	Source (seed oils)
16:1 9*c*	12	—	*Lesquerella densipila*
18:1 9*c*	12*R*	Ricinoleic	*Ricinus communis*
18:1 12*c*	9*S*	Isoricinoleic	*Strophanthus* and *Wightia* spp
18:2 9*c*15*c*	12*R*	Densipolic	*L. densipila*
20:1 11*c*	14*R*	Lesquerolic	*L. densipila*
20:2 11*c*17*c*	14	Auricolic	*L. auriculata*

Source: Adapted from Ref. 1, p. 15.

B. Epoxy Acids

Several natural epoxy acids are known. Vernolic (*cis*-12,13-epoxyoleic) was first discovered in 1954 and is still the best known in this group. It occurs at high levels in the seed oils of *Vernonia anthelmintica* (70–75%), *V. galamensis* (73–78%), *Cephalocroton cordofanus* (60–65%), *Euphorbia lagascae* (60–65%), Stokes aster (65–80%), and *Erlanga tomentosa* (50–55%), and attempts are being made to develop *V. galamensis* and *E. lagascae* as commercial crops.

C. Furanoid Acids

Natural furanoid acids with the structures shown have been identified. They are present at low levels, often as a complete series, in fish oils. The proportion of these increases significantly during prolonged fasting. The short-chain urofuranic acids that have been identified in animal blood and urine are probably breakdown products of furanoid acids believed to come from dietary vegetable sources.

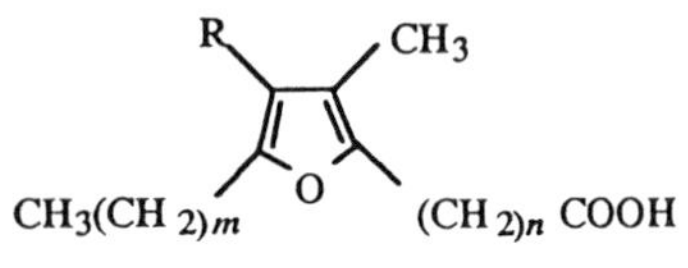

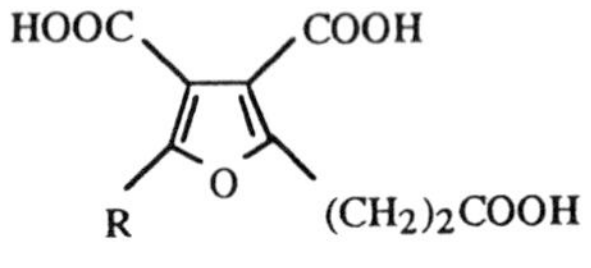

$R = CH_3$ or H

$m = 2$ or 4 $n = 6,8$, or 10

furanoid acids

$R = CH_3(CH_2)_4$ or $CH_3(CH_2)_2$

or CH_3CH_2CHOH or CH_3CH_2CO

urofuranic acids

XII. LIPID STRUCTURE

The fatty acids discussed in the previous section seldom occur in their free state but usually as esters and sometimes as amides (Table 5). Esters contain acid and alcohol components, and the most common alcohol present in lipids is glycerol (propane-1,2,3-triol). Important glycerolipids include the acylglycerols, the phospholipids, and the glycosylglycerides. Other alcohols that occur less commonly in lipids in an acylated form include carbohydrates (sugar esters), sterols (sterol esters), and long-chain alcohols (wax esters). The amides are derivatives of long-chain acids with amines such as the sphingosine bases or some amino acids.

Before detailing these it is useful to explain some terms that occur frequently. Simple or neutral lipids are triacylglycerols, some ether lipids, sterol esters, and wax esters. Complex

Table 5 Types of Lipids

Esters	
Of glycerol	Acylglycerols (glycerides)
	Phosphoglycerides (PA, PC, PE, PS, PI, PG)[a]
	Glycosylglycerides (MGDG, DGDG)[a]
Of other alcohols	
diols	—
sugars	Sugar esters
long-chain alcohols[b]	Wax esters
sterols	Sterol esters
Amides	
Of long-chain bases (sphingosine, etc)	Ceramides, cerebrosides, gangliosides
Of taurine	—
Of serine	—

[a]PA, phosphatidic acids; PC, phosphatidylcholines; PE, phosphatidylethanolamines; PS, phosphatidylserines; PI, phosphatidylinositols; PG, phosphatidylglycerols; MGDG, monogalactosyldiacylglycerols; DGDG, digalactosyldiacylglycerols.
[b]Also form ether lipids.

or polar lipids include phosphoglycerides, glycosyldiacylglycerols, and sphingolipids. Phospholipids are derivatives of phosphoric (or less commonly phosphonic) acid, glycolipids contain one or more carbohydrate units, and sulfolipids contain sulfur.

Glycerol contains a prochiral carbon atom carrying two CH_2OH groups. When these differ, as when they are acylated with different fatty acids, then the molecule is chiral and exists in two enantiomeric forms. To designate the stereochemistry of glycerol-containing components, the carbon atoms are stereospecifically numbered (*sn*). When the glycerol molecule is represented by a Fischer projection with the secondary hydroxyl to the left of the central (prochiral) carbon atom, the carbon atoms are numbered 1, 2, and 3 from top to bottom. Molecules that are stereospecifically numbered in this fashion have the prefix *sn*- immediately preceding the term glycerol. The prefix *rac*- in front of the full name shows that the compound is an equal mixture of both enantiomers. When *x*- is used, the configuration is unknown or un- specified.

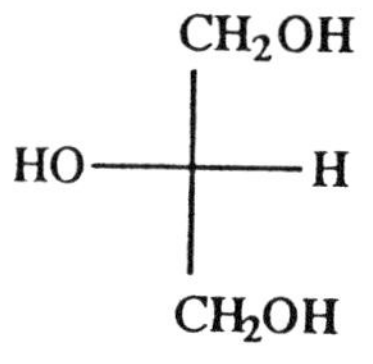

sn-1	α
sn-2	β
sn-3	α or α′

Any glycerol lipid will be chiral when the substituents at the *sn*-1 and *sn*-3 positions are different. Mirror image molecules or enantiomers possess opposite but equal rotations. However, if both substituents are long-chain acyl groups, then the optical rotations are very small and may be difficult to observe and measure.

XIII. ACYLGLYCEROLS

The major reserve lipids are triacylglycerols (formerly known as triglycerides). Mono- and diacylglycerols may also be present as minor components. However, these are important intermediates in the biosynthesis and catabolism of triacylglycerols and other classes of lipids. They are also prepared on a large scale for use as surface-active agents.

Monoacylglycerols (monoglycerides) are fatty acid monoesters of glycerol and exist in two isomeric forms, the α- and β-monoglycerides, whose structures are illustrated.

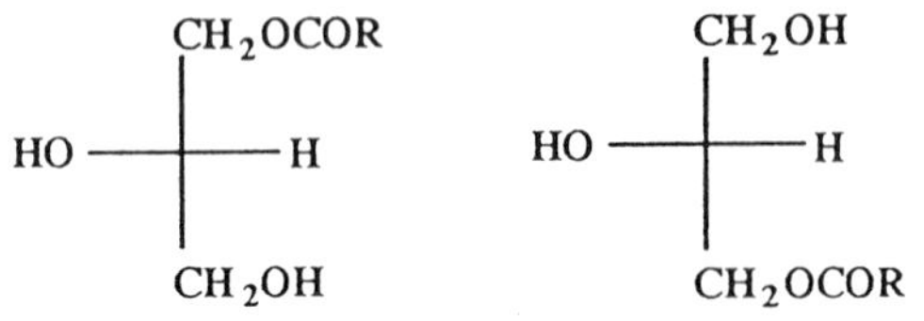

1- and 3- monoacyl-*sn*-glycerols
α-monoglyceride

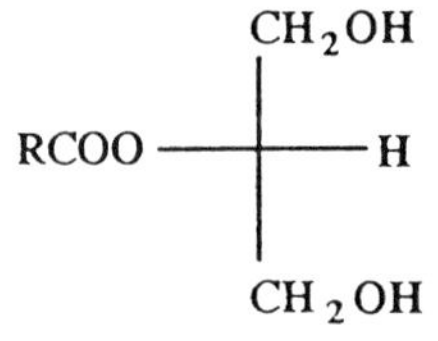

2-monoacyl-*sn*-glycerol
β-monoglyceride

Pure isomers readily change to a 90:10 mixture of the α and β isomers. This rearrangement is promoted by acid or alkali.

Diacylglycerols (diglycerides) are fatty acid diesters of glycerol, and they also exist in two isomeric forms, as illustrated. They readily form an equilibrium mixture, with the 1,3-diacylglycerol being the more stable.

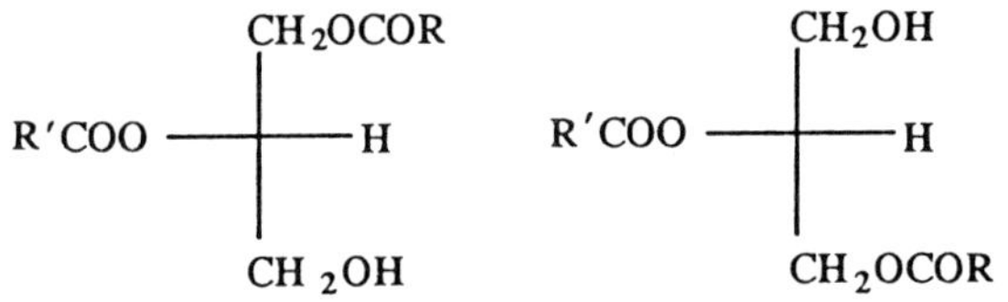

1,2- and 2,3-diacyl-*sn*-glycerols
αβ-diglycerides

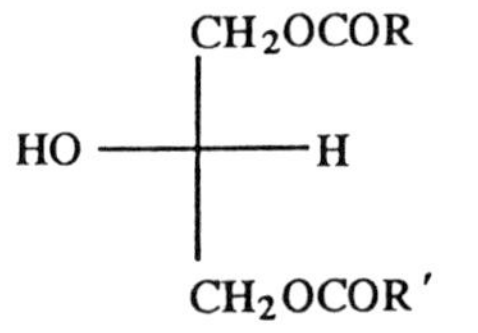

1,3-diacyl-*sn*-glycerol
αα'-monoglyceride

Triacylglycerols (triglycerides) are fully acylated derivatives of glycerol:

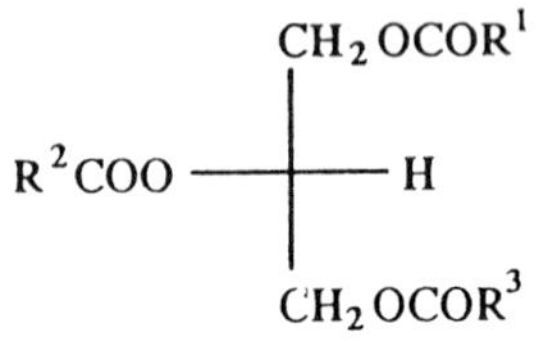

It is unusual for a natural triacylglycerol to have only one acid, though this does happen, e.g., triolein in olive oil and tripalmitin in palm oil. More usually, two or three different acids are present. The number of possible triacylglycerols rises very quickly with the number of

Table 6 Triacyl Glycerols Containing Only Palmitic (P) and Oleic (O) Acids

PPP	
POP	[OPP / PPO][a]
OPO	[POO / OOP][a]
OOO	

[a]Unsymmetrical triacylglycerols in enantiomeric pairs.

Table 7 Relation Between Number of Acids and Maximum Number of Triacylglycerols That Can Be Formed from Them

	Number of acids			
	5	10	20	n
Triacylglycerols				
All isomers distinguished	125	1000	8000	n^3
Excluding optical isomers	75	400	4200	$(n^3+n^2)/2$
No isomers distinguished	35	220	1540	$(n^3+3n^2+2n)/6$

acids present in the fatty acid pool. With only two acids there are eight triacylglycerols (including enantiomers, Table 6). With three acids this number is 27, and it rises very rapidly thereafter (Table 7). The triacylglycerol composition of some natural oils is discussed in Chapter 2.

On hydrolysis with aqueous acid or alkali, the triacylglycerols are split into glycerol and a mixture of fatty acids. Enzymatic hydrolysis usually occurs with some degree of regiospecificity. Most commonly this involves deacylation of the α positions (*sn*-1 and *sn*-3), with no reaction occurring at the β position, so the products are fatty acids from the α positions and 2-monoacylglycerols:

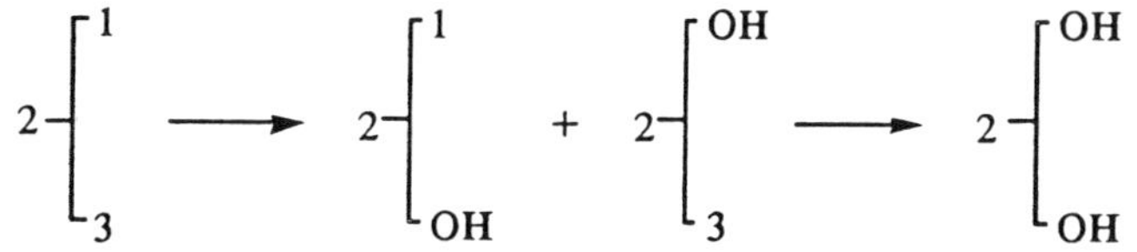

(1, 2, and 3 represent the mixed fatty acids present at each of these positions.)

XIV. GLYCOSYLDIACYLGLYCEROLS

The glycosyldiacylglycerols form a class of glycolipids that are present in plants (especially in the chloroplast) and in bacteria. Diacylglycerols are linked to one or more sugar units (especially galactose) at the *sn*-3 position. They are usually rich in polyunsaturated C_{18} acids. Typical examples are the monogalactosyldiacylglycerols (MGDG) and the digalactosyldiacylglycerols (DGDG).

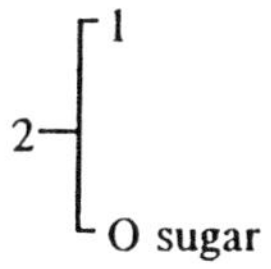

sugar unit	
galactose	MGDG
galactose-galactose	DGDG

XV. PHOSPHOGLYCERIDES

There are several types of phosphoglycerides (Table 8). These are based on the phosphatidic acids, which are themselves diacyl derivatives of 3-glycerophosphoric acid. The phosphatidic acids are monoesters of the tribasic phosphoric acid. The remaining phospholipids are diesters; i.e., a second hydroxy compound is associated with the molecule (Table 8).

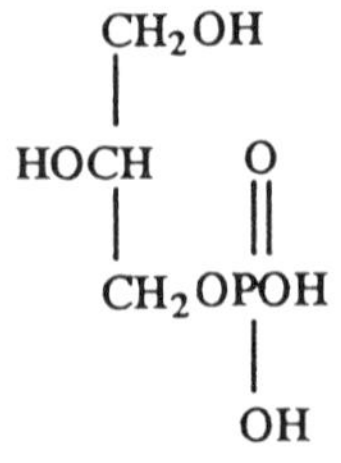

3-glycerophosphoric acid

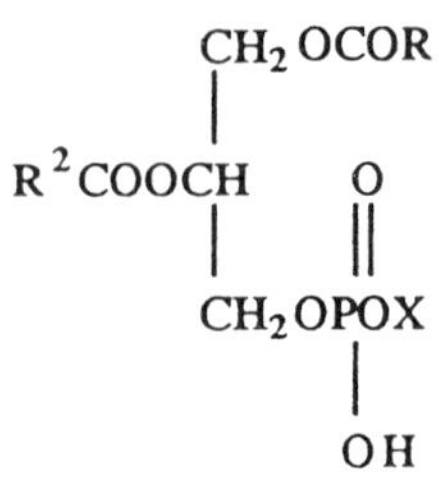

phosphatidic acids (PA, X = H)

The composition of some natural phospholipids is discussed in Chapter 2. Many phospholipids contain saturated acids at the *sn*-1 position and an unsaturated fatty acid at the *sn*-2 position, but this is not an invariable rule. The phosphoglycerides, with two lipophilic chains and a hydrophobic head group, are important structural lipids. They are essential parts of lipid bilayers that make up cell membranes.

Each phosphoglyceride has four ester groups, and their hydrolysis is illustrated in Table 9 for phosphatidylcholines. Complete hydrolysis occurs with aqueous acid, and the products are glycerol, fatty acids, phosphoric acid, and choline. With alkali the acyl groups are easily hydrolyzed, but the phosphate ester bonds react more slowly so that hydrolysis furnishes fatty acids and glycerophosphorylcholine (GPC), which is then slowly hydrolyzed to choline and glycerophosphoric acid. This last is a mixture of α and β isomers because of the migration of phosphoric acid among the hydroxyl groups.

Enzymatic hydrolysis is more selective and gives rise to a range of interesting products. Phospholipase A1 causes deacylation only at the *sn*-1 position, liberating the fatty acids from this position and leaving a lysophosphatidylcholine. Phospholipase A2 behaves similarly at the *sn*-2 position. Phospholipase B is now recognized as a mixture of the A1 and A2 enzymes and removes fatty acids from both positions, leaving glycerophosphorylcholine. Phospholipase C reacts at one of the phosphate ester bonds to give 1,2-diacylglycerols and phosphorylcholine, whereas phospholipase D cleaves the other phosphate ester bond to give choline and phosphatidic acids (Table 9).

These stereospecific processes have been exploited in a number of synthetic procedures. Starting with natural phosphatidylcholines, for example, it is possible to remove the fatty acids from the *sn*-1 and/or *sn*-2 positions and replace them with an acid of choice, thereby retaining the stereochemistry of the natural product. Also, the phosphatidylcholines can be converted to phosphatidic acids by reaction with phospholipase D and then be esterified as required to produce other phosphoglycerides.

Table 8 The Major Glycerophospholipids

Name	Abbreviation	X[a]
Phosphatidylcholines	PC	$CH_2CH_2\overset{+}{N}Me_3$
Phosphatidylethanolamines[b]	PE	$CH_2CH_2\overset{+}{N}H_3$
Phosphatidylserines	PS	$CH_2CH(\overset{+}{N}H_3)COO^-$
Phosphatidylinositols	PI	$C_6H_{11}O_5$
Phosphatidylglycerols	PG	$CH_2CH(OH)CH_2OH$

[a] See the structure of the phosphatidic acids (X = H) in the text.
[b] Also occur as esters of phosphonic acid (H_3PO_3).

Table 9 Enzymatic Hydrolysis of Phosphatidylcholines[a]

Phospholipase	Lipolysis products[b]	Source
A1	Fatty acids (*sn*-1), lyso PC	Snake venom
A2	Fatty acids (*sn*-2), lyso PC	Snake venom
B	Fatty acids (*sn*-1 and 2), GPC	—
C	DAG, phosphorylcholine	Bacteria (e.g., *Clostridia*)
D	PA, choline	Most plant tissues

[a]These enzymes are also effective with phosphatidylethanolamines.
[b]PC, phosphatidylcholines; GPC, glycerophosphorylcholines; DAG, diacylglycerols; PA, phosphatidic acids.

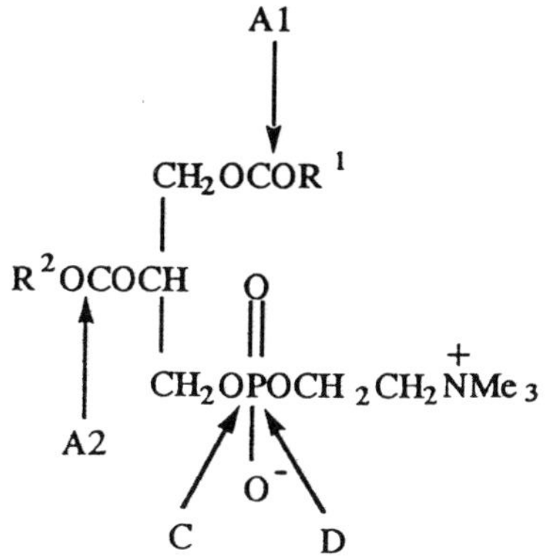

XVI. ETHER LIPIDS

Four types of monoether lipids have been recognized depending on whether they are related to triacylglycerols or phosphoglycerides and on whether the ether group does or does not contain a Δ1 double bond (*trans*). The ether group is generally in the *sn*-1 position, but some diethers (*sn*-1 and *sn*-2) are also known.

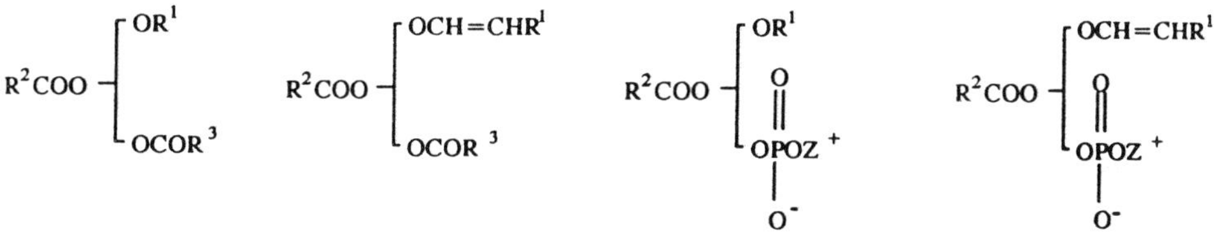

1 2 3 4

Compounds of type **1** occur in some marine oils. When they are hydrolyzed they yield fatty acids (2 mol) and glycerol ethers. The alkyl chain is commonly 16:0, 18:0, or 18:1, and the glycerol ethers (which are diols) derived from these are known as chimyl alcohol (16:0), batyl alcohol (18:0), and selachyl alcohol (9*c*-18:1). The vinyl ether group present in compounds of types **2** and **4** resists reaction with alkali and with lithium aluminum hydride but is hydrolyzed under acidic conditions to give an aldehyde.

$$\mathbf{2} \rightarrow \text{glycerol} + \text{fatty acids} + [R^1CH{=}CHOH] \rightleftharpoons R^1CH_2CHO$$

Ethanolamine derivatives of type **4** occur in animal tissue and were originally called plasmalogens to reflect the production of aldehydes. Their biochemical significance is not understood.

Compounds designated platelet-activating factor (PAF) have biological activity (aggregation, inflammatory, edemic) at very low concentrations and are specifically bound to receptors in platelets. They are 1-*O*-alkyl-2-acetyl-*sn*-glycerol-3-phosphocholines (**5**).

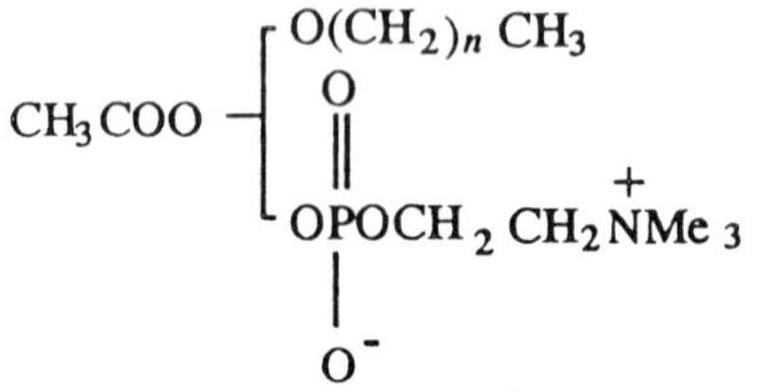

XVII. ACYL DERIVATIVES OF OTHER ALCOHOLS

All the lipids discussed so far have been esters of glycerol. There remain to be detailed some esters of fatty acids with other alcohols. Ester waxes are derivatives of long-chain acids and long-chain alcohols such as occur in fish (sperm whale oil, orange roughy oil), animals (beeswax, wool wax), and plants (carnauba, jojoba). Esters (and ethers) based on ethanediol and propane-1,3-diol have also been identified. The latter may be widely distributed at low levels but are difficult to detect in the presence of glycerol esters. Sterols, sugars, and other natural alcohols also occur as esters with long-chain alcohols. This acylation leads to a reduction in water solubility and an increase in fat solubility.

Cutins and suberins are plant polymers produced from mono-, di-, and trihydroxy C_{16} and C_{18} acids.

XVIII. SPHINGOLIPIDS

The component units of sphingolipids are fatty acids, often including 2-hydroxy acids, bound as amides to long-chain amines. These amines also have two or three hydroxyl groups, one of which is linked to one or more sugar units or to a phosphoester unit thus:

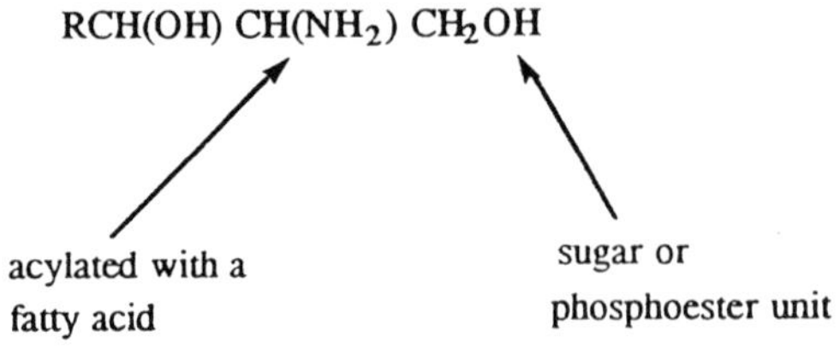

The long-chain bases (sphingoids) are mainly C_{18} or C_{20} compounds, and the structures of two such bases are illustrated here.

sphingosine (4*t*-sphingenine)	*t(E)* *(R)* *(S)* $CH_3(CH_2)_{12}CH{=}CHCH(OH)CH(NH_2)CH_2OH$
phytosphingosine	*(R)* *(S)* *(S)* $CH_3(CH_2)_{13}CH(OH)CH(OH)CH(NH_2)CH_2OH$

Their *N*-acyl derivatives are known generally as ceramides. The structures and occurrence of typical glycosphingolipids (gangliosides, cerebrosides) and a sphingomyelin are given in Table 10.

Table 10 Structure and Occurrence of Some Sphingolipids

$$\begin{array}{l} RCH{=}CHCH(OH)\underset{\displaystyle |\atop \displaystyle NHR^1}{C}HCH_2OR^2 \end{array}$$

Name	R^1	R^2	Occurrence
Sphingenine	H	H	—
Ceramides	COR′	H	Skin
Cerebrosides	COR′	glu or gal	Brain, spleen, ganglion cells
Gangliosides	COR′	glu-gal-sialic acid	Myelin sheath of brain and nerve
Sphingomyelin	COR′	$PO_3^-CH_2CH_2\overset{+}{N}Me_3$	Animal membranes

There are several diseases in which sphingolipids accumulate in various organs and tissues through enzymatic defects in sphingolipid metabolism. These are known generally as sphingolipidoses and include diseases associated with the names Tay-Sachs, Fabry, Gaucher, Farber, Niemann-Pick, and Krabbe.

Though structurally different, the phosphoglycerides and sphingolipids resemble each other in that both contain two long chains and a polar head group and are conveniently located in lipid bilayers.

REFERENCES

1. F. D. Gunstone, J. L. Harwood, and F. B. Padley (Eds.), *Lipid Handbook*, 2nd ed., Chapman and Hall, London, 1994.
2. F. D. Gunstone, *Fatty Acid and Lipid Chemistry*, Blackie, London, 1996.

2

Major Sources of Lipids

Frank D. Gunstone
Mylnefield Research Services Ltd., Scottish Crop Research Institute, Invergowrie, Dundee, Scotland

I. INTRODUCTION

This chapter is devoted to the major and some minor sources of lipids. The publications of Mielke GmbH of Hamburg (*Oil World*, etc.) provide figures for the production, imports and exports, and stocks of 17 major oil and fat sources [1]. These include three animal fats (butterfat, lard, and tallow/grease), three industrial oils (castor, fish, and linseed), and 11 vegetable oils (coconut, corn, cotton, groundnut, olive, palm, palmkernel, rapeseed, sesame, soybean, and sunflower). These are discussed here along with some other significant sources such as rice bran oil, tall oil fatty acids, and others. Sections II–XI are concerned with general matters and Section XII contains an account of individual sources.

It is appropriate to note that among vegetable oils, which together make up about 75% of the annual world production, some such as soy, cotton, corn, and rice bran oils are by-products, and the supply of such oils is controlled by the demand for the more important component, protein meal for animal feed from soybeans, fiber from cotton, or cereal from corn and rice. A second group, such as coconut, palm, palmkernel, and olive oils are produced from tree crops. The trees have to be planted and come to maturity, after which they generally produce annual crops for 20 years or more. Finally, some of the oils (groundnut, rapeseed, sesame, and sunflower) come from annual crops the availability of which can vary from year to year as they are planted annually according to their perceived profitability.

The major emphasis in this chapter is on fatty acids and triacylglycerols because these are the major components in each lipid source, but reference is also made, where relevant, to phospholipids, tocopherols, and other minor components.

II. NAMES

Table 1 lists the English, French, and botanical names of many lipid sources of vegetable origin. The original reference [2] contains some additional entries and also gives Russian names.

Table 1 Botanical Name of Plant and Name of Oil[a]

Plant	Oil	
	English	French
Aleurites fordii	Tung nut oil	Huile de tung
Aleurites moluccana, A. triloba	Candlenut oil, artist's oil, lumbang oil	Huile de bancoul
Arachis hypogea	Arachis seed oil, groundnut oil, peanut oil	Huile d'arachide
Attalea speciosa, Orbignya speciosa	Babassu oil	Huile de babassu
Borago officinalis	Borage seed oil	Huile de bourâche
Brassica juncea	Brown mustardseed oil	Huile de moutarde brune
	Indian mustardseed oil	Huile de moutarde d'Inde
Brassica napus	Colza oil, rapeseed oil	Huile de colzo
Brassica nigra	Black mustard seed oil	Huile de moutarde noire
Brassica rapa	Turnip seed oil	Huile de navette
Butyrospermum paradonum, B. parkii	Shea butter, shea nut oil	Amande de karité
Camelina sativa	Cameline seed oil, camelina seed oil, gold of pleasure	Huile de cameline
Camelina sinensis	Teaseed oil	Huile de thé
Cannabis sativa	Cannabis oil, hempseed oil	Huile de chenevis
Carthamus tinctorus	Safflower oil, safflower seed oil	Huile de carthame
Ceiba pentandra	Kapok seed oil	Huile de kapok
Cocos nucifera	Coconut oil, copra oil	Huile de coprah
		Huile de coco
Corylus aveliana	Hazelnut oil	Huile de noisette
Crambe abyssinica	Sea kale oil, crambe oil	Huile de crambe
Cucurbita maxima	Pumpkin seed oil, squash seed oil, vegetable marrow seed oil	Huile de cucurbitacéas (citrouille, courge, patiron)
Cuphea spp.	Cuphea seed oil	Huile de cuphéa
Elaies guinensis	Palm oil	Huile de palme
	Palmkernel oil	Huile de palmiste
Euphorbia lathyris	Spurge seed oil	Huile d'épurge
Fagus sylvatica	Beechnut oil	Huile de faîne
Glycine max	Soyabean oil, soybean oil	Huile de soja
Gossypium spp.	Cottonseed oil	Huile de coton
Guizotia abyssinica	Nigerseed oil	Huile de niger
Helianthus annus	Sunflower seed oil, sunflower oil	Huile de tournesol
Hevea brasiliensis	Rubberseed oil	Huile d'hévéa
Hibiscus esculentus	Okra seed oil	Huile de gombo
Jatropha curcas	Jatropha seed oil, physicnut oil	Huile de jatropha (pignon d'Inde, pulghère, purghère)
Juglans regia	Walnut oil	Huile de noix
Licania rigida	Oiticica oil	Huile d'oiticica
Limnanthes alba	Meadowfoam seed oil	—
Linum usitatissimum	Linseed oil	Huile de lin
Madhuca longifolia, Bassia longifolia	Indian illipe butter, mowrah butter	Beurre d'illipé
Oenothera biennis, O. lamarkiana	Evening primrose oil	Huile d'onagre

Table 1 Continued

Plant	Oil	
	English	French
Olea europeae	Olive oil	Huile d'olive
Orbignya oleifera	Babassu nut oil	Huile de Babasso
Oryza sativa	Rice bran oil	Huile de riz
Papaver somniferum	Poppyseed oil	Huile d'oeillette
Perilla frutescens	Perilla seed oil	Huile de périlla
Persea americana	Avocado pear oil, avocado kernel oil	Huile d'avocat
Prunus armaniaca, Armeniaca vulgaris	Apricot kernel oil	Huile de noyaux d'apricòt
Prunus domestica	Plum kernel oil	Huile de noyaux prune
Prunus dulcis, Prunus amygdalus	Almond oil	Huile d'amande
Prunus persica	Peach kernel oil	Huile de noyaux de pêche
Ribes nigrum	Blackcurrant pip oil	Huile de pépins de cassis
Ricinus communis	Castor oil, castor seed oil	Huile de ricin
Sapium sebiferum	Chinese vegetable tallow, stillingia oil	Huile de stillingia
Sesamum indicum	Sesame seed oil, sesame oil	Huile de sésame
Shorea macrophylla, Shorea stenoptera	Borneo oil, Borneo tallow, illipe butter	Suif de Borneo, beurre d'illipé de Borneo
Shorea robusta	Sal fat	Graisse de sal
Simmondsia chinensis	Jojoba seed oil	Huile de jojoba
Sinapsis alba	White mustard oil, white mustardseed oil	Huile de moutarde blanche
Theobroma cacao	Cocoa butter, theobroma oil	Beurre de cacao
Triticum aestivum	Wheatgerm oil	Huile de germe de blé
Vernonia spp.	Vernonia seed oil	Huile de vernonia
Vitellaria paradoxa	Shea butter	Beurre de karité
Vitis vinifera	Grapeseed oil	Huile de pépins de raisin
Zea mays	Maize oil, corn oil	Huile de germe de mais

[a] The original publication also gives the name of each oil in Russian.
Source: Ref. 2.

III. ANNUAL PRODUCTION AND FORECASTS

Table 2 contains production data and forecasts for the period 1986–2012 for 17 major oils and fats obtained from a recent Mielke publication [3]. These are average annual figures for the five-year period 1986–1990 and predictions for the quinquennia 1993–1997 and 2008–2012. Over these three time periods annual production is expected to rise from 76 to 91 to 132 million tonnes. Five of these oils—palm, palmkernel, soybean, rapeseed, and sunflower—predominate and together represent 53,59, and 65% of the total supply in the three time periods. Although the others rise in annual production, all (except corn oil) are expected to decline in terms of market share. Palm oil, which has long held second place to soybean oil, is predicted to become the dominant oil before the end of the first decade in the next century, and the two products of the oil palm will together represent one-fourth of the annual produc-

Table 2 Production and Forecasts of Average Annual Production (million tonnes) for 17 Major Oils during Three Quinquennia Between 1986 and 2012[a]

	1986–1990	1993–1997	2008–2012
Palm	9.1(12.0)	15.3(16.9)	29.8(22.6)
Soybean	15.3(20.2)	17.9(19.8)	25.1(18.9)
Rape	7.5 (9.9)	10.1(11.2)	15.6(11.8)
Sunflower	7.3 (9.7)	8.4 (9.3)	12.0 (9.1)
Tallow/grease	6.6 (8.7)	7.0 (7.7)	8.1 (6.1)
Lard	5.4 (7.1)	5.7 (6.3)	7.7 (5.8)
Butterfat	6.4 (8.5)	6.0 (6.6)	6.9 (5.2)
Cottonseed	3.6 (4.8)	4.2 (4.6)	5.9 (4.5)
Groundnut	3.7 (4.9)	4.2 (4.6)	5.7 (4.3)
Palmkernel	1.2 (1.6)	2.0 (2.2)	3.5 (2.6)
Coconut	3.0 (4.0)	3.0 (3.3)	3.3 (2.5)
Corn	1.3 (1.7)	1.7 (1.9)	2.7 (2.0)
Olive	1.9 (2.5)	2.0 (2.2)	2.1 (1.6)
Fish	1.5 (2.0)	1.2 (1.3)	1.3 (1.0)
Sesame	0.66(0.9)	0.71(0.8)	0.94(0.7)
Linseed	0.72(1.0)	0.64(0.7)	0.88(0.7)
Castor	0.38(0.5)	0.45(0.5)	0.55(0.4)
Total	75.6	90.5	132.1

[a]The figures in parentheses represent percentage of the total.
Source: Ref. 3.

tion in the 2008–2012 quinquennium. Most of this will be produced in two countries in southeast Asia—Indonesia and Malaysia.

IV. GEOGRAPHICAL AREAS OF PRODUCTION

Information summarized in Table 3 shows the major geographical sources of soybeans, rapeseed, sunflower seed, cottonseed, groundnuts, palm oil, and copra (the source of coconut oil)

Table 3 Major Geographical Areas of Production of Seven Vegetable Oils/Fats Based on Average Annual Figures for 1989–1992

	Total tonnes[a] $\times 10^6$	Major geographical sources (%)
Copra	4.9	Philippines (41), Indonesia (25), India (8)
Cottonseed	33.5	China (24), USA (16), FSU (13), India (13), Pakistan (10), Brazil (4)
Groundnut	16.4	India (34), China (25), USA (9)
Palm oil	11.6	Malaysia (55), Indonesia (23), Nigeria (5)
Rapeseed	25.2	China (26), India (20), Canada (14), EU (23), Poland (5)
Soybeans	105.9	USA (50), Brazil (17), Argentina (11), China (10)
Sunflower seed	22.3	FSU (29), Argentina (17), EU (18), China (6), USA (5), Eastern Europe (10)

[a]These are not figures for oil (except for palm) but for beans, seeds, etc.
Source: Ref. 4.

Table 4A Typical Fatty Acid Composition of Some Major Oils and Fats

	Fatty acid			
Oil or fat	16:0	18:1	18:2	Other (including, %)
Coconut[a]				
Corn	13	31	52	4
Cotton	24	19	53	4
Groundnut	13	37	41	9 (C_{20}–C_{24}, 7)
Olive	10	78	7	5
Palm	44	40	10	6
Palmkernel[a]				
Rape	4	56	26	14 (18:3, 10)
Sesame	9	38	45	8 (18:0, 6)
Soybean	11	22	53	14 (18:3, 8)
Sunflower	6	18	69	7 (18:0, 6)
Castor[b]				
Fish oils[c]				
Linseed	6	17	14	63 (18:3, 60)
Butter[d]				
Lard[d]				
Tallow[e]				

[a]See Table 11.
[b]See Section XII.A
[c]See Table 31.
[d]See Table 25.
[e]See Table 27.
Source: Ref. 5.

[4]. These figures show the importance of the United States, Argentina, India, China, and the countries of southeast Asia as sources of production. Export figures are somewhat different because highly populated countries such as India and China consume much of their indigenous oils. European Community countries growing rape and sunflower are not the same because these are grown mainly in northern and southern regions, respectively.

V. FATTY ACID COMPOSITION

The fatty acid composition of most of the oils and fats discussed in this chapter are summarized in Tables 4A–4C. Most vegetable oils contain high levels of oleic and/or linoleic acid along with palmitic acid as the major saturated acid. Sometimes the oleic acid content is very high, sometimes linoleic acid is high, and sometimes both have moderate values. For example, oleic acid exceeds 50% in olive oil (78%), in rapeseed (56%), and in the high-oleic varieties of safflower oil (Section XIII.M) and sunflower oil (Section XIII.P). Linoleic acid exceeds 50% in sunflower (69%), soybean (53%), cotton (53%), and corn (52%), and both acids exceed 30% in corn (31 and 52%), groundnut (37 and 41%), and sesame (38 and 45%). These examples are taken from the oils listed in Table 4A. Other examples are given in Table 4C.

Occasionally other acids beyond these three are present at significant levels. Examples include medium-chain acids (C_{10}–C_{14} and especially C_{12}) in coconut and palmkernel oils, stearic acid in cocoa butter and some similar hard fats, α-linolenic acid in linseed, rapeseed,

Table 4B Typical Fatty Acid Composition of Some Major and Minor Oils and Fats

	Fatty acid				
Oil or fat	16:0	18:0	18:1	18:2	Other (including, %)
Cocoa butter	24	34	37	3	2
Crambe	2	1	16	8	73 (18:3, 7; 22:1, 56)
Mustard	3	1	23	9	64 (18:3, 10; 20:1, 8; 22:1, 43)
Rape (high-erucic)	3	1	16	14	66 (18:3, 10; 20:1, 6; 22:1, 49)
Rice bran	16	2	42	37	3
Safflower					
High-oleic	6	2	74	16	2
High-linoleic	7	3	14	75	1
Sunflower					
High-oleic	4	5	81	8	2
High-linoleic	6	5	20	68	1
Tall oil					
American	1	2	46	36	15 (conjugated, 9; pinolenic, 2)
Scandinavian	1	2	30	45	22 (conjugated, 5; pinolenic, 9)
Tung oil	3	2	11	15	69 (eleostearic, 69)
Vernonia seed oil	3	3	5	14	75 (vernolic, 75)
Wheat germ	14	1	22	58	5 (18:3, 5)

Source: Ref. 5.

and soybean, and ricinoleic acid in castor oil. Several other examples are found throughout this chapter.

The more variable fatty acid composition in butter, lard, and tallow, and in a range of fish oils discussed in Sections XII.V–XII.X.

VI. TRIACYLGLYCEROLS

Since seed oils and animal fats consist mainly of triacylglycerols (>98%), the composition of these materials can be fully defined only in terms of the individual glycerol esters present. Such information is, however, not normally provided with traded samples, and there are several reasons for this.

1. The potential number of triacylglycerols is quite large. In the oversimplified case of an oil or fat with three major acids (such as palmitic, oleic, and linoleic), ignoring the minor acids, there could be 27 (3^3) different glycerol esters. If stereoisomers are ignored, this number falls to 18, and if all isomers are ignored it reduces to 10 species (PPP, POP, PLP, POO, PLL, POL, OOO, OOL, OLL, LLL). In a more realistic case with five different acids these numbers would be 125, 75, and 35, and the situation is even more complex with materials such as milk fats and fish oils containing many more acids. When it becomes possible to analyze natural material to this level of sophistication it will be difficult to understand the significance of the results because of the large number of different triacylglycerols, many of them at quite low levels.

2. The changes in fatty acid composition that occur in natural material and are recognized in Codex specifications are multiplied in the triacylglycerol composition. Small differences in fatty acid composition lead to larger changes in triacylglycerol composition. This implies that to define a fat or oil either there will have to be a wide range of permitted values for the major triacylglycerols or all samples will have to be analyzed.

Table 4C Fatty Acid Composition of Some Minor Oils and Fats

		Major component acid			
Common name	Botanical name	16:0	18:1	18:2	Other
Aceituno	*Simarauba glauca*	12	58	2	18:0, 28
Almond	*Prunus amygdalus*	7	61	30	
Apricot	*Prunus armeniaes*	5	60	32	
Argane	*Argania spinosa*	14	47	32	18:0, 6
Avocado	*Parsea americana*	12	71	14	
Babassu	*Orbignya martiana, O. oleifera*	9	13	3	12:0, 45; 14:0, 17
Brazil nut	*Berthrolletia myrtaceae*	14	29	47	18:0, 9
Buffalo gourd	*Cucurbita foetidissima*	9	25	64	
Candlenut	*Aleurites moluccana, A. trisperma*	6–8	17–25	38–45	18:3, 25–30
Cashew	*Anacardium occidentale*	11	60	17	
Cherry	*Prunus avium*	6	35	45	α-Elst, 10[b]
Chinese vegetable tallow[a]	*Sapium sebiferum*	62	27	1	
Gold of pleasure	*Camelina sativa*	4–6	15–20	16–24	20:1, 10–16; 18:3, 30–40
Grapeseed	*Vitis vinifera*	7	16	72	
Hazelnut (filbert)	*Corylus avellana*	7	69	21	
Hemp	*Cannabis sativa*	6	12	55	18:3, 25
Illipe butter (Borneo tallow)	*Shorea stenoptera*	18	35	1	18:0, 46
Kapok	*Ceiba pentandra* + others	16	21	37	mal, ster, 13[b]
Macadamia	*Macadamia tetraphylla*	8	56	3	16:1, 22
Mango	*Mangifera indica*	5–8	40–50	2–4	18:0, 40–45
Mango olein	—	7–10	48–53	4–8	18:0, 30–34
Meadowfoam	*Limnanthes alba*	—	1–3	1–3	20:1, 60–65; 22:1, 11–14; 22:2, 17–22
Mowrah	*Madhuca longifolia*	28	14	49	
Nectarine	—	6	66	27	
Oiticica	*Licania rigida*	7	6	—	Licanic, 78[b]
Passionflower	*Passiflora edulis*	8	13	77	
Peach	*Prunus persicae*	8	59	33	
Pecan	*Carya illipensis*	6	67	22	
Pistachio	*Pistachio vera*	9	69	18	
Plum	*Prunus domestica*	6	62	30	
Poppy	*Papaver somniferum*	10	11	72	18.3, 5
Pumpkin	*Cucurbita maxima*	11	47	34	
Safflower (HO)	*Carthamus tinctorus*	6	74	16	
Sal	*Shorea robusta*	4–7	37–43	0–4	18:0, 40–46
Sal olein	—	6–9	48–52	2–6	18:0, 31–35
Shea	*Butyrospermum paradoxum*	3–4	44–46	6–8	18:0, 42–45
Stillingia[a]	*Stillingia sebiferum*	5	13	23	18:3, 47; HO 8:2, 5; 10:2, 5
Sunflower (HO)	*Helianthus annus*	4	81	8	18:0, 5
Tea	*Thea sansanqua*	16	59	22	
Tobacco	*Nicotiana tabacum*	8	16	71	
Ucuhuba	*Virola sufinam*	7	8	2	12:0, 13; 14:0, 69
Walnut	*Juglans regia*	8	18	58	18:3, 8

[a] Stillingia oil and chinese vegetable tallow come from the same source.

[b] Eleostearic, 9*c*11*t*13*c*-18:3; malvalic, C_{18}-cyclopropene acid; sterculic, C_{19}-cyclopropene acid; licanic, 4-oxo-9*c*11*t*13*c*-18:3.

Source: Refs. 5 and 7.

3. A third problem is that there is still no simple way of measuring triacylglycerol composition and such analyses are usually a matter for the research laboratory rather than the quality control laboratory. This, however, is changing in the study of some of the harder fats. The several ways of studying this problem generally give partial information, and it is not easy to combine the results obtained by different procedures. The most important and widely used methods of analysis are as follows.

a. Separation of glycerol esters according to their levels of unsaturation by silver ion chromatography. Originally achieved by thin layer chromatography, this is now generally effected by HPLC using a silver ion column. The results are given in terms of triacylglycerols but not usually for individual molecular species. For example, a figure assigned to LLO represents the sum of three isomers depending on whether the oleic chain is in the *sn*-1, -2, or -3 position. Similarly, the value for LOP represents the six esters containing these three acids.
b. Separation according to partition number by reverse-phase HPLC. Again, the triacylglycerols identified are usually not individual molecular species but mixtures as described in (a).
c. Separation according to the number of carbon atoms in the three acyl chains by gas chromatography. This has improved since its first introduction, and a lot of useful information can be obtained in this way. Smaller separation by unsaturation is added to the major separation by carbon number.
d. Analysis by enzymatic procedures or by HPLC of mono- or diacylglycerols resulting from reaction with ethylmagnesium bromide. These are separated either on a chiral column or after reaction with a chiral reagent. These procedures can give an indication of the composition of fatty acids in each of the *sn*-1, -2, and -3 positions. Triacylglycerol composition can be calculated from the results only if it is assumed that the acids in each position are randomly associated with each other. This is described as a 1-random 2-random 3-random arrangement. It could be argued that the procedures that give information about the fatty acids in each of the *sn* positions are developments of fatty acid composition rather than triacylglycerol composition. Sometimes they are combined with other separation procedures carried out in a preparative manner.

Some results obtained by these methods are included with the discussion of individual oils in Section XIII.

VII. SAPONIFICATION VALUE, IODINE VALUE, AND TOTAL UNSAPONIFIABLE MATERIAL

Many oil and fat specifications still include a reference to classical values such as the saponification value, the iodine value, and the content of unsaponifiable material.

The saponification value (SV, the number of milligrams of potassium hydroxide required to saponify 1 g of oil) is less immediately indicative of fatty acid composition than the saponification equivalent (SE, the weight of oil saponified by 1 g-mol of potassium hydroxide). The latter is a measure of the average chain length of all the acyl chains. The high level of C_{18} acids in most oils gives a saponification equivalent around 291–297 (sap value 189–193). This figure is raised by the presence of unsaponifiable matter (nonacidic) and by the presence of acids with molecular weight greater than 297 as glycerol esters (e.g., ricinoleic acid, vernolic acid, C_{20}–C_{24} acids). Conversely, the saponification equivalent is reduced by the presence of palmitic acid at higher levels than usual (e.g., palm oil, cottonseed oil) and for the

Table 5 Saponification Value (SV), Saponification Equivalent (SE), Iodine Value (IV), and Content of Unsaponifiable Material in Some Vegetable Oils

Oil	SV[a]	SE[a]	IV	Unsaponifiable (%)
Castor	176–187	300–319	81–91	1.0
Cocoa butter	190–200	281–295	33–40	0.2–0.6
Coconut	248–265	212–226	6–11	1.5 max
Corn (maize)	187–195	288–300	103–128	2.8 max
Cottonseed	189–198	283–297	99–119	1.5 max
Groundnut	187–196	286–300	80–106	1.0 max
Linseed	189–195	288–297	177 min	1.5 max
Olive	184–196	286–305	75–94	1.5
Palm	190–209	268–295	50–55	1.2 max
Palmkernel	230–254	221–244	13–23	1.0 max
Rapeseed (low erucic)	188–193	291–298	110–126	2.0 max
Rice bran	181–189	297–310	99–108	3–5
Safflower	186–198	283–302	135–150	1.5 max
Sesame	187–195	288–300	104–120	2.0 max
Soy	189–195	288–297	120–143	1.5 max
Sunflower	188–194	289–298	110–143	1.5 max
Tung	189–195	288–297	160–175	0.5
Vernonia	165–210	267–340	104–108	6.7
Wheat germ	184–185	303–305	120–130	2–6
Trilaurate		213.0	0	—
Tripalmitate		269.1	0	—
Tristearate		297.2	0	—
Trioleate		295.2	86.0	—
Trilinoleate		293.1	173.2	—
Trilinolenate		291.1	261.5	—

[a] SE = 56108/SV.
Source: Ref. 5.

lauric oils (coconut oil, palmkernel oil). These factors have the reverse effect on the saponification value, as the two are related by the expression SE = 56,108/SV.

Iodine values are a measure of average unsaturation. Most oils rich in palmitic, oleic, and linoleic acid have iodine values in the range 90–130 depending on the relative levels of saturated, monoene, and diene acids. The iodine value is higher in oils rich in triene acids (linseed, tung) and lower when there is a high level of saturated acids such as palmitic or stearic (palm, cocoa butter) or of the medium-chain acids (coconut, palmkernel).

Vegetable oils generally contain up to 1.5% unsaponifiable material (sterols, 4-methylsterols, triterpene alcohols, tocopherols, hydrocarbon, etc.; see Section IX) though some higher values are apparent in Table 5.

VIII. PHOSPHOLIPIDS

Phospholipids of various types (Chapter 1) are present as minor components (0.5–3.0%) in most crude oils, from which they are largely removed during refining. They may be recovered in a by-product, generally called lecithin, which is a mixture of phospholipids and triacylglyc-

Table 6 Composition of Phospholipids from Selected Natural Sources

Source	Total phospholipid in oil (%)	Percent of total					
		PC	PE	PI	Sum PI/PA	PA	Other
Cottonseed	1.0	23	14	13		9	41
Groundnut	0.4	23	8	17		3	49
Palm	—	36	24	22		9	9
Rapeseed	2.5	48	9	20		—	23
Soybean	3.2	39	23	20		5	13
Lecithin, liquid		43	26	—	19	—	12
Lecithin, granular		32	18	—	32	—	18
Sunflower seed	1.5	49	21	28		2	—
Egg yolk[a]		25	5	tr		—	70

[a]Composition of total egg yolk lipid (including neutral lipids 62% and sterols 5%).
Source: Ref. 5.

erols. The material recovered from soybean oil is an important commercial product. It provides a source of mixed and individual phospholipids at various levels of purity. Lecithin is also available from rape, corn, and sunflower, and egg yolk is another important source of phospholipids. These are important surface-active compounds used extensively in the food, pharmaceutical, and cosmetic industries.

The major phospholipids in crude lecithin are usually phosphatidylcholines (PC), phosphatidylethanolomines (PE), phosphatidylinositols (PI), and phosphatidic acids (PA) accompanied by other phospholipids at lower levels. The proportion of these vary with the source and with the method of recovery, and some typical figures are given in Table 6. One commercial supplier reports his soybean lecithin to contain PC (20–23%), PE (16–21%), and PI (12–18%), with the balance being triacylglycerols and minor phospholipids. The component acids vary with the phospholipid source and the phospholipid class, and typical results are given in Table 7. The fatty acid composition of individual phospholipid types can be manipulated by enzymatic processes.

Table 7 Fatty Acid Composition of Some Phospholipids

Fatty acid	PC				PE	
	Soy	Rape	Sunflower	Egg	Soy	Egg
16:0	15	6	11	38	37	15
18:0	3	1	4	9	2	28
18:1	7	56	12	32	10	21
18:2	68	31	72	12	50	9
18:3	6	4	tr	—	1	—
20:4 (*n*-6)	—	—	—	2	—	10
22:6 (*n*-3)	—	—	—	5	—	13
Other	1	2	1	2	tr	4

Source: Ref. 6.

Table 8 Sterol Composition (Percentage of Total Sterol) of Some Vegetable Oils

	Palm-kernel	Coconut	Cotton-seed	Soy-bean	Corn	Ground-nut	Palm	Sun-flower	Rape
Total									
Range[a]	0.8–1.2	0.5–1.1	2.7–5.9	1.8–4.1	7.9–22.1	0.9–2.9	0.4–0.5	2.4–4.5	4.8–11.3
Average[a]	1.0	0.8	4.5	3.2	13.8	1.6	0.4	3.4	7.5
Cholesterol	2	2	1	1	tr	2	4	1	1
Brassicasterol	tr	1	tr	tr	tr	tr	ND	tr	10
Campesterol	10	9	8	20	21	17	23	10	34
Stigmasterol	14	12	1	18	6	8	11	9	tr
Sitosterol	67	46	84	54	63	58	57	60	49
Δ^5-Avenasterol	6	26	3	2	6	12	3	3	5
Δ^7-Stigmasterol	tr	2	1	3	2	2	1	10	1
Δ^7-Avenasterol	tr	1	1	1	1	1	1	5	tr
Other[b]	1	1	1	1	1	tr	—	2	tr

tr = trace; ND = not detected.
[a]In grams per kilogram.
Source: Refs. 5 and 8.

IX. STEROL COMPOSITION

Most vegetable oils contain 1000–5000 ppm (0.1–0.5%) of combined sterols present both as free sterols and as esters with long-chain acids. Higher levels are apparent in rapeseed oil (range 5000–11,000 ppm, mean value about 7500) and in corn oil (range 8000–22,000, mean value 14,000). Sitosterol is generally the major phytosterol, representing 50–80% of total sterol; campesterol, stigmasterol, and Δ^5-avenasterol frequently attain significant levels. Brassicasterol is significant only in seed oils of brassica species.

Cholesterol is generally considered to be a zoosterol and is not present in plant systems at any significant level. The normal value of 10–20 ppm is to be compared with the much higher values reported for animal fats (up to 1000 ppm), fish oils (up to 7000 ppm), dairy fats (2000–3000 ppm), and egg yolks (12,500 ppm ≡ 1.25%). Details are given in Table 8 [5,8].

X. TOCOPHEROLS AND VITAMIN E

Though present at still lower levels, the tocopherols and tocotrienols attract attention because of their vitamin E and antioxidant activities. The tocopherols are a series of benzopyranols with one, two, or three methyl groups attached to the phenolic ring. The molecules also have a C_{16} side chain on the pyran ring; this is saturated in the tocopherols and has three double bonds in the tocotrienols. There are four tocopherols and four tocotrienols designated α, β, γ, δ. Some typical contents are given in Table 9.

The tocopherols show different vitamin E activity in the order α (1.0) > β (0.5) > γ (0.1) > δ (0.03). The vitamin E activity of some oils, expressed as α-tocopherol equivalents, is listed in Table 9 in the extreme right column. These depend particularly on the level of α-tocopherol. Antioxidant activity is in the reverse order ($\delta > \gamma > \beta > \alpha$) so that the various oils do not follow a similar sequence of vitamin E and antioxidant activity.

Walnut oil and wheatgerm are good sources of tocopherols. Others with high vitamin E activity include sunflower, cottonseed, safflower, and palm oils. Antioxidant activity is high in soybean oil, corn oil, and walnut oil. Some oils have high oxidative stability despite

Table 9 Tocopherols and Vitamin E Activity in Oils and Fats[a]

	Tocopherols				Tocotrienols		Vit E activity as α-tocopherol equivalents
	α	β	γ	δ	α	β	
Coconut[b]	5	—	—	6	5	tr	7
Corn	112	50	602	19	—	—	198
Cottonseed	389	—	387	—	—	—	428
Groundnut	130	—	214	21	—	—	152
Olive	119	—	7	—	—	—	120
Palm[c]	256	—	316	70	146	3	335
Palmkernel	62	—	—	—	—	—	62
Rape	210	1	42	tr	tr	—	215
Safflower	342	—	71	—	—	—	349
Sesame	136	—	290	—	—	—	165
Soybean	75	15	797	266	2	tr	171
Sunflower	487	—	51	8	—	—	492
Walnut	563	—	595	450	—	—	636
Wheatgerm	1330	710	260	271	26	18	1736
Cod liver	220	—	—	—	—	—	220
Herring	92	—	—	—	—	—	92
Menhaden	75	—	—	—	—	—	75
Lard	12	—	7	—	7	—	15
Tallow	27	—	—	—	—	—	27

[a]All figures in milligrams per kilogram (i.e., ppm).
[b]Also γT3 19.
[c]Also γT3 286.
Source: Refs. 5 and 8.

their low α-tocopherol levels because other antioxidants (usually phenolic compounds) are also present. Oxidative stability will also be influenced by the level of unsaturation in the oil.

The tocopherols are partially removed during oil refining. This reduces oxidative stability but can furnish a concentrate that can be added back or added to other oils or fatty foods. Much natural antioxidant comes from a by-product of soybean refining. Concentrates of tocopherols from palm oil are also produced for use as vitamins (Palm Vit E).

XII. WAXES

Waxes include a variety of long-chain compounds occurring together in plants and animals and finding use in the food, pharmaceutical, and cosmetic industries. They are generally water-resistant materials made up of mixtures of hydrocarbons, ketones, acids, alcohols, and their esters and are used for the protection of surfaces of shoes, automobiles, furniture, paper, etc. They differ from the long-chain acids in triacylglycerols in that they are usually of higher molecular weight (up to C_{60} and beyond) and are frequently branched with one or more methyl groups, and although they may be unsaturated they generally do not contain the methylene-interrupted polyunsaturation characteristic of the common polyene acids [5,9]. The more important commercial materials are discussed in Section XII.U.

XIII. INDIVIDUAL SOURCES

A. Castor Oil

The castor oil plant (*Ricinus communis*), which grows in tropical, subtropical, and even some temperate regions, is cultivated mainly in India, Brazil, and China. The oil makes up about one-half of the shelled bean. The residual meal contains several poisonous and allergenic materials and can be used for animal feeds only after detoxification. (It is used mainly for ruminants.) The oil differs from other commercially available oils by reason of the presence of the hydroxy acid ricinoleic acid (Chapter 1, Section XI.A, page 9) at the 83–90% level. Compared with other vegetable oils, castor oil is more viscous, less soluble in hexane, and more soluble in alcohol as a consequence of the presence of the hydroxy acid. Ricinoleic acid and the esters are most commonly used after chemical modification.

1. Pyrolysis of castor methyl esters at 450–500° gives methyl 10-undecenoate (C_{11}) and heptanoic acid. The C_{11} compound is used to produce perfumes and polymers.
2. Alkali fusion at 250–275° converts the C_{18} hydroxy acid to C_8 (2-octanol or 2-octanone) and C_{10} (sebacic acid or 10-hydroxydecanoic acid) compounds.
3. The surface-active properties of ricinoleic acid or its esters can be enhanced by reaction of the midchain hydroxyl group. For example, sulfated castor oil (Turkey Red oil) is much used in the textile industry. It can also be ethoxylated.
4. Hydrogenation affords 12-hydroxystearic acid, which is used in the paint industry to introduce thixotropy and as a component of many greases in the form of its lithium salt.
5. Dehydration of castor oil by heating at 230–280° in the presence of an acidic catalyst gives isomers of octadecadienoic acid (mainly the $\Delta 9,11$ and $\Delta 9,12$ dienes). Dehydrated castor oil is used in paints and varnishes as a drying oil.

Unlike some oils containing hydroxy acids, the hydroxyl groups in castor oil are not acylated and remain free. It was shown some time ago [10] that the major triacylglycerols contain two or three ricinoleic acid groups: RRR (68%), RRO (8%), RRL (8%), RRSt (5%), and others (11%). Newer information has been obtained by mass spectrometry [11].

Seed oils containing other hydroxy acids are being developed as potential new oilseed crops. *Lesquerella fendleri* seeds yield an oil (25%) containing about 55% lesquerolic acid. This is the C_{20} homologue of ricinoleic acid (14-hydroxy-11-*cis*-eicosenoic acid) and should show properties similar to those of the C_{18} acid. Dimorphotheca oil contains dimorphecolic acid (9-hydroxy-10-*trans*,12-*cis*-octadecadienoic acid), which is readily dehydrated to give a C_{18} acid with a conjugated triene system.

B. Cocoa Butter

The cocoa bean (*Theobroma cacao*) is the source of both cocoa powder and cocoa butter, which are important ingredients of chocolate. The value of cocoa butter in this connection depends mainly on its melting behavior, which is related to the fatty acid and triacylglycerol composition. Cocoa butter is a solid fat that melts in the mouth, producing a desirable cooling sensation on the tongue. This is a consequence of the high level (64–77%) of symmetrical disaturated oleo glycerol esters (SOS). The content of unsymmetrical esters (SSO) is below 1%. The major glycerol esters are POP (18–23%), POSt (36–41%), and StOSt (23–31%) (P = palmitic, O = oleic, St = stearic, S = saturated acids). The crystalline forms and melting behavior of this fat have been studied intensively [12–14].

Because cocoa butter commands a good price that tends to fluctuate depending on the vagaries of cocoa production, there is an interest in less expensive fats from other sources that show similar melting properties [15]. No less than five different approaches to this problem are being examined.

1. Other natural fats with similar fatty acid and triacylglycerol composition are known (Table 10), and some of them are commercially available though at lower levels than cocoa butter [5]. Supplies are sometimes unreliable.
2. Fractions of oils with high levels of SOS can be isolated, usually by crystallization. For example, double fractionation of palm oil produces a product (palm midfraction, PMF) containing palmitic (48–54%), stearic (5%), oleic (35–38%), and linoleic (4–7%) acids.
3. Attempts have been made to produce a fat with the required properties by growing yeast in the presence of stearic acid or one of its simple alkyl esters or, alternatively, in the presence of sterculic acid, which inhibits the desaturation of stearic to oleic acid.
4. A concentrate of StOSt that is a valuable component of chocolate fat has been obtained by regiospecific lipolysis of high-oleic sunflower oil containing triolein with stearic acid using *Mucor miehei* as a source of lipase.

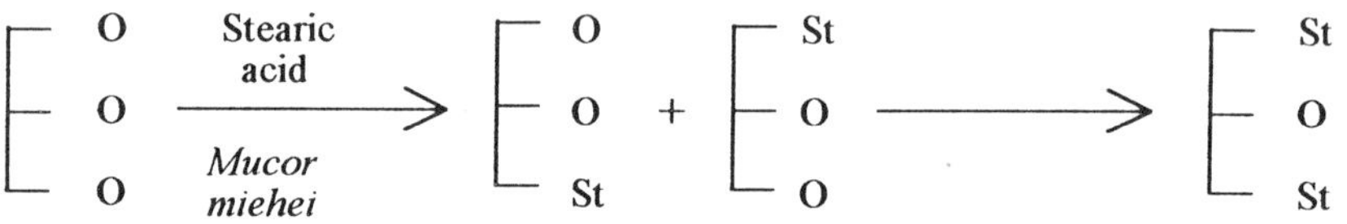

5. By genetic engineering it is possible to produce a rapeseed oil in which the level of stearic acid has been raised to about 40%. A high-stearate (15–30%) soybean oil has also been developed.

C. Coconut Oil

The coconut palm (*Cocos nucifera*) grows between the latitudes 20° north and 20° south of the equator, in the Philippines and in Indonesia. Inside the hard shell of the coconut is a thick endosperm layer. When dried and separated from the shell this is known as copra and is the source of coconut oil (65–70%). This oil is used for food and non-food uses in roughly equal amounts. For example, alcohols can be produced from coconut oil by hydrogenolysis. Coconut oil is an important lauric oil characterized by high levels of medium-chain acids, especially lauric acid (C_{12}) at about 50%, along with lower levels of the C_{14}, C_{10}, C_8, and C_6

Table 10 Natural Oils Containing High Levels of SOS Glycerides (wt %)

Source	Oil	mp (°C)	16:0	18:0	18:1	Other	SOS
Shorea robusta	Sal oil	30–35	5	44	42	9	64
S. stenoptera	Illipe butter	34–38	19–20	40–42	35–38	1	79
Vateria indica	Malabar tallow	34–36	12	41	45	2	74
Garcinia indica	Kokum butter	38–42	4	53	40	3	74
Madhuca butry	Phulwara butter	40–42	60	4	36	—	69
Madhuca latifolia,	Mowrah butter	—	24	19	43	14	—
M. longifolia		—	28	14	49	9	—
Butyrospermum parkii	Shea butter	—	3	44	46	6	—

Source: Ref. 5.

Table 11 Component Acids (wt %) and Major Triacylglycerols (>5%, mol %) in Coconut and Palmkernel Oils

Component acids

Acid	6:0	8:0	10:0	12:0	14:0	16:0	18:0	18:1	18:2
Coconut	0.5	8	7	47	18	9	3	6	2
Palmkernel	0.5	2.5	4	49	16	9	2	14	2

Major triacylglycerols

Saturated acyl groups	Number of isomers[a]	Coconut	Palmkernel
12, 12, 8	3	11.9	6.4
12, 12, 10	3	5.7	4.7
12, 12, 12	1	10.6	19.8
12, 12, 14	3	10.8	14.1
14, 12, 8	6	9.1	3.6
Other	—	51.9	51.4

[a]Not including stereoisomers.

saturated acids. Hydrolytic rancidity furnishes free acids that give the oil a soapy off-flavor. The acids can be further oxidized and decarboxylated to give odd-chain methyl ketones, which are the source of ketonic rancidity.

$$RCH_2CH_2COOH \rightarrow RCOCH_2COOH \rightarrow RCOCH_3$$

Triacylglycerol analysis by gas chromatography shows that the major glycerol esters have two or three lauric acid chains (Table 11).

D. Corn Oil (Maize Oil)

Corn (maize, *Zea mays*) is grown mainly for use as animal feed, with an increasing amount being subjected to wet milling to produce corn starch, sugar, and syrup. The oil, recovered mainly from corn germ, has good oxidative stability in spite of its high level of linoleic acid and is used as a salad oil, a frying oil, and in margarine (after partial hydrogenation). The levels of palmitic, oleic, and linoleic acid are in the range 9–17, 20–42, and 39–63 respectively (Codex values). Triacylglycerol composition has been examined by stereospecific analysis [16], silver ion chromatography [5], and gas chromatography [5]. These studies indicate that the major triacylglycerols include LLL (15%), LLO (21%), LLS (17%), LOO (14%), LOS (17%), LSS (5%), OOO (6%), and OOS (4%) or, alternatively, that the TAG molecules are mainly C_{52} (26%) or C_{54} compounds (66%).

E. Cottonseed Oil

Cotton (*Gossypium* spp) is a subtropical species now grown also in temperate zones. It is produced mainly in China but also in the United States, the former Soviet Union, India, Pakistan, and Brazil (Table 3). It is grown mainly for its fiber, but the seed is a useful by-product, yielding both oil (~30%) and protein (~30%). The seed represents only about 11–12% of the total value of the crop. A significant amount is fed to animals as whole seed. The oil is used as a salad or cooking oil after winterization (i.e., removal of the more solid triacylglycerols). In the early decades of this century cottonseed oil was the major oil in the United States, but it is now third after soybean and corn. On the world scale it ranks fifth among the vegetable oils after the four market leaders (soy, palm, rape, and sunflower).

Cottonseed oil imparts a desirable nutty flavor to crisps/chips. It also tends to form β' crystals. This is beneficial because it extends plastic range, aids aeration, and helps to contribute to uniform appearance.

Cottonseeds contain gossypol (up to 0.4%), which is considered to be undesirable. It concentrates in the meal rather than in the oil, from which it is finally removed during alkali refining. Because of the gossypol the meal is suitable only for ruminants. Glandless cotton contains lower levels of gossypol.

Cottonseed oil is rich in palmitic acid (22–26%), oleic acid (15–20%), and linoleic acid (49–58%). It also contains up to 10% C_{20}–C_{24} acids (mainly saturated) and about 1% sterculic and malvalic acids in the crude oil. The cyclopropene acids are undesirable components, but they are largely removed during refining, particularly deodorization, and also during hydrogenation. They are not considered to present any health hazard in cottonseed oil.

The major triacylglycerols include the isomers represented by PLL, PLP, LLL, POL, LOL, and POP, with palmitic acid present mainly in the *sn*-1 and -3 positions and hardly at all in the *sn*-2 position. The disaturated triacylglycerols (POP and PLP) are largely removed during winterization. Bland et al. [18] give a detailed triacylglycerol analysis indicating the presence of PLL (26%), LLL (16%), POL (14%), OLL (13%), and PLP (9%), with seven others making up the balance (22%).

F. Evening Primrose Oil (Also Borage and Blackcurrant Seed Oils)

Evening primrose (*Oenothera biennis*, *O. lamarkiana*) oil is one of a small number of oils containing γ-linoleic acid (6,9,12-18:3). This relatively rare triene acid of the *n*-6 polyene family has attracted attention because of claims that it is beneficial in the treatment of a number of diseases such as atopic eczema, premenstrual syndrome, multiple sclerosis, arthritis, and several others [19]. These claims are not universally accepted, but the oil is widely used as a health supplement. γ-Linolenic acid is the first intermediate in the bioconversion of linoleic to arachidonic acid, and the first step of Δ6 desaturation is known to be rate-limiting.

Evening primrose oil is also very rich in linoleic acid (70–75%), so that with the γ-linoleic acid (8–10%) the content of *n*-6 essential fatty acids is very high. Other commercial sources of γ-linolenic acid include the oils from borage (*Borago officinalis*) and from blackcurrant. Both these contain higher levels of γ-linolenic acid than evening primrose oil but they are used less extensively. Borage oil contains less linoleic acid and also contains C_{20}, C_{22}, and C_{24} monoene acids (total ~8%). Blackcurrant seed oil contains significant levels of *n*-3 poly-

Table 12A Fatty Acid Composition of Oils Containing γ-Linolenic Acid

	Fatty acid (%)					
Oil	16:0	18:0	18:1	18:2	18:3 (*n*-6)	Other
Blackcurrant	7	2	11	47	17	16[a]
Borage	10	4	16	38	23	9[b]
Evening primrose	6	2	9	72	10	1

[a]Including 18:3 (*n*-3), 13% and 18:4 (*n*-3), 3%.
[b]Including 20:1, 4.5%; 22:1; 2.5%; and 24:1, 1.5%.

Table 12B The Major Triacylglycerols (>1 wt %) in Borage and Evening Primrose Determined by Reverse-Phase HPLC[a]

TAG	Borage	Evening Primrose
LGG	9.8	0.7
LLG	15.4	17.6
PGG	2.4	—
LLL	10.1	54.3
OLG	12.3	1.6
PLG	9.5	1.2
OLL	8.2	13.7
LGD	2.5	—
PLL, OOG	8.9	7.9
SLG, POG	5.0	—
PPG	1.5	—
LLD	4.1	—
OOL	4.1	0.8
POL, SLL	0.3	2.2
Other	5.9	0

[a]Each TAG entry includes all possible isomers. Further information is also available in Ref. 48.
Source: Ref. 47.

ene acids (18:3 and 18:4). Analytical data are presented in Tables 12A and 12B [20]. Higher levels of GLA are present in a variety of upgraded oils.

G. Groundnut Oil (Monkey Nut, Peanut, Arachis Oil)

The groundnut (*Arachis hypogea*) is an annual legume grown in tropical and semitropical regions, especially in China and India, and in the United States at a lower level. The oil shows excellent oxidative stability and is considered to have a desirable flavor. Its major component acids are palmitic (8–14%), oleic (36–67%), and linoleic (14–43%) with typical values of 13, 37, and 44%, respectively. The oil also contains some 6–8% (total) of C_{20}, C_{22}, and C_{24} saturated and monoene acids.

The major triacylglycerols have been examined by silver ion chromatography [5] and by HPLC [17, 21] with the results detailed in Table 13.

Table 13 Triacylglycerol Composition of Groundnut Oil

LLL	LLO	LLS	LOO	LOS	LSS	OOO	OOS	OSS	Other	Ref.
4	20	12	17	22	6	6	10	3	—	5
6	26	8	21	13	2	5	16	1	2	21

Fatty acid composition [L (40,42), O (39, 43), S (21,15)] calculated from the above figures show that the oil samples do not have the same fatty acid composition and therefore cannot be expected to have the same TAG composition.
Source: Refs. 5 and 21.

Table 14 Triacylglycerol Composition of Linseed Oil

LnLnLn	LnLnL	LnLnO	LnLO	LnOO	LnLnS	LnLS	LnOS	Other	Ref.
35	14	19	6	6	5	7	4	4	23
24	14	15	5	5	12	5	7	13	5

Source: Refs. 5 and 23.

H. Linseed Oil

The flax plant (*Linus usitatissimum*) is grown either for its oil (linseed oil) or for its fiber (flax, linen). The oil variety is grown mainly in Argentina, India, the former Soviet Union, the United States, and Canada, with Argentina and Canada the main exporting countries.

The seed content is 35–44% oil that is characterized by a high iodine value (>177) related to the high level of α-linolenic acid (9,12,15–18:3), which is usually 50–60%. The meal remaining after removal of the oil is a valuable animal feed for cattle, pigs, chickens, and fish. The whole seed finds limited use in baked goods. The triacylglycerol composition of this oil is summarized in Table 14. Its stereospecific analysis is given in Refs. 5 and 22.

Because of its high unsaturation, the oil is used mainly in paints and varnishes, in the production of linoleum, and as a sealant for concrete. Production of linseed oil has changed little over the last 50 years because it has been replaced in some of its traditional uses by products of the petrochemical industry. However, there is now evidence of renewed demand for linseed oil in some of these products. For example, linoleum is being promoted on the basis of its durability and its nonflammability and on environmental grounds in respect of both its natural constituents (linseed oil, cork, woodflour, pine resin) and its ease and safety of disposal. It is especially acceptable in high traffic public areas such as shops, schools, hospitals, and airports.

Plant breeders in Australia have developed a variety of linseed with low levels of linolenic acid (~2%) and high levels of linoleic acid (~70%). Called linola, it yields a linoleic-rich oil (like sunflower) and can be grown in the same temperate zones as rapeseed (canola). It is already being grown in Australia and Canada and most recently in Europe. Table 15 compares the fatty acid composition of linseed and linola oils.

I. Olive Oil

Olive oil is obtained from the mesocarp of olives (*Olea europeae*). Commercial growth of the olive tree is confined almost entirely (~97%) to the Mediterranean countries of Italy, Greece, Spain, Turkey, and Tunisia. The ripe mesocarp is 15–40% oil that is especially rich in oleic

Table 15 Fatty Acid Composition of Linseed Oil and of Linola

	Fatty acid				
Oil	16:0	18:0	18:1	18:2	18:3
Linseed	6	3	17	14	60
Linola					
Glenelg	8	5	21	64	2
Croxton	8	3	18	69	2

acid. Codex ranges are palmitic 8–20%, oleic 55–83%, and linoleic 4–21%. Christie et al. [24,30] has given the fatty acid composition of olive oil at each of the *sn*-1, *sn*-2, and *sn*-3 positions. In a study using silver ion HPLC it was concluded that the major triacylglycerols are SSO (3%), SOO (31%), OOO (46%), SOL (6%), and OOL (14%). In another report, 99 olive oil samples were investigated by reverse-phase HPLC [23]. Six dominant glycerol esters make up 82–93% of the whole oil; the median values are LOO, 11%; OOO, 43%; POP, 3%; POL, 4%; POO 22%; and StOO, 5%.

In addition to the usual range of minor components, olive oil contains the C_{30} hydrocarbon squalene at a higher level (150–700 mg/100 g) than most other vegetable oils (5–50 mg/100 g).

The oil is available in several grades and commands a good price. For this reason the more expensive grades are subject to adulteration with poorer grades of olive oil and/or with other oils, and considerable effort is devoted to developing analytical procedures to detect this. The oil is used mainly for human consumption as a salad or cooking oil.

J. Palm and Palmkernel Oils

The oil palm (*Elais guinensis*) grows in tropical regions of Asia, Africa, and America and predominantly in Malaysia and Indonesia. In both of these latter countries there has been a considerable increase in production during the past 15 years. This increase is expected to continue for some years to come, and palm oil will replace soybean oil as the largest produced oil; also, Indonesia is expected to overtake Malaysia as the leading producer. For good production the tree requires a humid tropical climate with rainfall (~2000 mm) spread through the year. In terms of oil per hectare, the oil palm outproduces all other oil crops, yielding about 5–7 tonnes/hectare on well-managed plantations.

Palm trees begin to produce fruits after 3–4 years, peak after about 10 years, and may have a production life of up to 30 years. In later years oil yields decline and the increasing height of the tree may present harvesting problems. The fruit bunches (4–20 kg) contain 200–2000 individual fruits and yield palm oil (20–24%) and palmkernel oil (2–4%). Breeding objectives include still higher average yields, plants of lower height, higher ratios of kernel to mesocarp, and a modified fatty acid composition of the oil.

The oil palm produces two distinct oils: palm oil from the fleshy mesocarp and palmkernel oil from the fruit kernel. The usage of both oils is extended by fractionation. Palm oil gives a more valuable olein (an excellent frying oil) and a less valuable stearin (used in part as a replacement for tallow in the oleochemical industry). With palmkernel oil the stearin is the more valuable fraction [26]. The oils and their fractions can be further modified by blending with other oils, by partial hydrogenation, or by interesterification. Further fractionation of the palm oil fractions yields an intermediate fraction (PMF, palm midfraction) that can be used as a cocoa butter extender. Palm oil is characterized by high levels of carotene and of tocopherols and tocotrienols (vitamin E), which can also be isolated or concentrated as products of considerable value. Palm oil fatty acid distillate (PAFD), a by-product of the principal refining procedure, is an important component of animal feed.

Palm oil is widely used as a food oil with limited non-food uses also. It differs from other commercially available oils in its fatty acid and triacylglycerol composition. It contains almost equal amounts of saturated acids (mainly palmitic with some stearic) and unsaturated acids (mainly oleic with some linoleic acid). The proportion of palmitic and oleic acids in the major triacylglycerols leads to stability of the β' crystals. These are very desirable in the production of margarines and shortenings, especially those with high levels of unsaturated acids. Some information on triacylglycerol composition is given in Table 16.

Table 16 Triacylglycerol Composition of Palm Oil

(i) Major triacylglycerols (%)

PPP	POP	PLP	POSt	POO, PLSt	PLO	OOO	Other	Ref.
7	33	7	6	23	7	3	14	27
5	43	8	4	29	8	2	1	28
7	31	9	6	23	9	4	11	29

(ii) Distribution between *sn*-1, -2, and -3 positions[a] (%)

	16:0	18:0	18:1	18:2
TAG	48.4	3.7	36.3	10.0
sn-1	60.1	3.4	26.8	9.3
sn-2	13.3	0.2	67.9	17.5
sn-3	71.9	7.6	14.4	3.2

[a] 14:0 and 16:1 are also cited in the original paper [30].
Source: Refs. 27–30.

The triacylglycerol composition of palm oil has been examined by HPLC and GLC procedures and also by stereospecific analysis. As expected from the high levels of palmitic and oleic acids in this oil, the triacylglycerols containing these two acids (POP and POO) are the dominant species.

Palmkernel oil, like coconut oil, is an important lauric oil with significant food and non-food uses. Some information on triacylglycerol composition is listed in Table 11 along with that for coconut oil.

K. Rape and Other Brassica Species (Also Crambe and Mustard)

Several species of Brassica are available in commercial quantities including rape [*Brassica napus*, *B. rapa* (or *campestris*), and *B. juncea*] and mustard (*B. alba* and other species). Typically these oils were rich in erucic acid as is Crambe oil (*Crambe abyssinica* and *C. hispanica*), but some have been modified by classical or modern methods of seed breeding to give a different fatty acid composition.

Today erucic acid is available from high-erucic rapeseed oil (about 50%) or from Crambe oil (about 60%). At present there is a demand for about 20,000 tonnes of erucic acid based compounds derived from about 56,000 tonnes of erucic-rich oil. It is used mainly as the amide (slip agent for polyethylene film), alcohol (emollient), esters (lubricants), and various nitrogen derivatives (personal care products). There is also a demand for behenyl alcohol (pour point depressant) and behenyl glycerol (food emulsifier). Behenic acid (22:0) is the reduced form of erucic acid [31].

Low-erucic rapeseed is now the third largest source of oil (after soy and palm) and also of meal (after soy and cotton). The seed contains over 40% oil, which represents about 80% of the seed's monetary value. The residual meal is 33–44% high quality protein. Seed breeders have developed seeds that produce oil low in erucic acid (<5% or <2%) and meal low in the undesirable sulfur-containing glucosinolates. This low-erucic oil (known as LEAR or canola) finds many food uses and some non-food uses (biodiesel, lubricants). The crude oil is rich in phospholipids (~3.5%), though these are reduced to 10–300 ppm (phosphorus) after refining

Table 17 Stereospecific Analysis of High-Erucic Rapeseed Oil

	16:0	18:0	18:1	18:2	18:3	20:1	22:1
TAG	3.9	1.3	18.6	16.1	8.7	10.4	38.0
sn-1	5.6	1.9	16.5	1.2	3.3	17.4	51.1
sn-2	1.6	0.4	33.4	39.3	21.3	1.9	0.8
sn-3	4.9	1.7	5.6	7.4	1.3	12.0	62.3

Source: Ref. 32.

and are themselves a useful by-product. This oil grows in cooler agricultural regions including China, northern Europe, and Canada as well as on the Indian subcontinent. In common with other *Brassica* seeds, rapeseed oil contains brassicasterol at levels (~600 ppm) not attained in other species.

Plants of the *Brassica* species lend themselves to genetic engineering, and considerable progress has been made in changing their fatty acid composition. Oils with lauric acid (40%) or stearic acid (40%) are already undergoing extensive field trials, and other possibilities under active investigation include oils with ricinoleic acid, erucic acid at the 90% level, oleic acid at a high level, vernolic acid, and γ-linolenic acid and oils like jojoba oil that produce wax esters rather than glycerol esters. The high-oleic oil (>80%) is associated with the desired reduction in linolenic acid from 8–10% to less than 3%. Other breeding objectives include increasing seed yield, winter hardiness, and resistance to frost, disease, insects, lodging, and shattering and an improvement in meal quality.

Bergqvist and Kaufmann [28] used HPLC to demonstrate that the major triacylglycerols in low-erucic rapeseed oil are LLL (5%), LLO and LnOO (19%), LOO (27%), and OOO (41%). Takagi and Ando [32] reported a stereospecific analysis of the high-erucic oil (Table 17), and similar data were obtained by tandem mass spectrometry [33].

L. Rice Bran Oil

Rice (*Oryza sativa*) is an important cereal with an annual production of about 500 million tonnes. To produce white rice the hull is removed and the bran layer is abraded, giving about 8–10% of the rice grain. This contains the testa, cross cells, aleurone cells, part of the aleurone layer, and the germ and includes almost all of the oil and the majority of the protein, ash, vitamins, and fibre of the rice coreopsis. Less than 40% of rice is treated in this way. This indicates a potential of 15 million tonnes of bran and over 3 million tonnes of rice bran oil. In fact, present production is only about 0.45 million tonnes per annum, and not all of this is food grade [34].

Lipases liberated from the testa and the cross cells promote rapid hydrolysis of the oil, and it should therefore be extracted within a few hours of milling. The more free acid has to be removed, the more oil will be lost at the same time. Attempts have been made to upgrade rice bran oil with 30% free acid by reaction with glycerol and lipozyme followed by neutralization.

Refined rice bran oil is an excellent salad and frying oil with high oxidative stability. It also finds several non-food uses. The oil lowers serum cholesterol by reducing LDL and VLDL with no change in HLDL. This effect seems not to be related to fatty acid or triacylglycerol composition but to the unsaponifiable fraction and probably to the oryzanols (1.5–2.0% of oil), which are ferulic acid esters of sterols and triterpene alcohols.

The major acids in rice bran oil are palmitic (12–18%, typically 16%), oleic (40–50%, typically 42%), and linoleic (29–42%, typically 37%). The oil contains phospholipids (~5%), a wax that can be removed and finds commercial use, and unsaponifiable material (3–5%).

This last contains sterols (~1.8%), 4-methylsterols (~0.4%), triterpene alcohols (~1.2%), and less polar compounds (~0.8%) such as tocopherols, tocotrienols, and squalene.

M. Safflower Seed Oil

Safflower seed oil is considered a minor oil, for its annual production was about 250,000 tonnes in 1982–1983. Safflower (*Carthamus tinctorius*) is grown mainly in India and to a lesser extent in Mexico and the United States. The plant was grown for the dye present in its leaves long before there was any interest in the seed oil. It is normally a linoleic-rich oil (~75% linoleic acid) with L_3 (47%), L_2O (19%), and L_2S (18%) (L = linoleic, O = oleic, S = saturated acid) as the major triacylglycerols. An oleic-rich safflower oil (~74% oleic acid) has also been developed. Christie et al. [30] conducted stereospecific analysis with the results shown in Table 18.

N. Sesame Oil

Sesame (*Sesamum indicum*), the source of another minor oil (annual production about 660,000 tonnes), is grown mainly in India and China but also in Burma, Sudan, and Mexico. The seed contains 40–60% oil with almost equal levels of oleic (range 33–54%, average 40%) and linoleic acid (range 39–59%, average 46%). The oil contains sesamin (0.5–1.1%) and sesamolin (0.3–0.6%) and has very high oxidative stability. The view that this is due to the production of the phenol sesamol from sesamolin during refining has recently been questioned. Early work [5] indicates that the major triacylglycerols are LOO (22%), LLO (20%), LOS (17%), LLL (19%), LLS (9%), OOO (8%), OOS (8%), and other (6%).

O. Soybean Oil

Soybeans (*Glycine max*) have been grown for 4000–5000 years and have become an important oil crop during the last 50 years even though the oil itself is a by-product in the production of a valuable protein meal used worldwide as an animal feed. Soy is now the prime source of oil, though it is likely to be overtaken by the oil palm in the next decade.

About half the world crop is grown in the United States, followed by Brazil (~17%), Argentina (~10%), and China (~10%). The balance is produced in over 30 countries, and Italy, Paraguay, India, and Indonesia have recorded significant increases in the last 10–15

Table 18 Stereospecific Analysis of Safflower Seed Oil[a]

	16:0	18:0	18:1	18:2
TAG	7.9	2.1	13.7	76.3
Method 1				
sn-1	14.2	2.9	14.4	68.6
sn-2	2.9	0.2	15.3	81.5
sn-3	6.8	3.1	11.4	78.7
Method 2				
sn-1	16.5	2.7	14.0	66.8
sn-2	0.6	0.4	15.7	83.3
sn-3	9.1	3.0	11.0	76.9

[a]Two sets of results obtained in slightly different ways.
Source: Ref. 30.

Table 19 Stereospecific Analysis of Soybean TAG

Fatty acid	HPLC method			Enzyme method		
	sn-1	*sn*-2	*sn*-3	*sn*-1	*sn*-2	*sn*-3
16:0	16.7	2.2	16.1	16.9	1.4	16.6
18:0	5.4	0.3	4.6	5.2	0.3	4.7
18:1	24.3	23.4	24.6	22.3	23.0	26.9
18:2	46.4	68.1	47.0	47.4	69.2	45.2
18:3	6.4	5.7	7.0	7.5	5.7	5.9
Other	0.8	0.3	0.7	0.7	0.4	0.7

Source: Ref. 32.

years. The product is imported as whole beans, as oil, or as meal, and crushing is carried out in Japan and the EU countries as well as in the producer countries.

The bean contains high quality protein (38–42%) and oil (18–22%), which is usually recovered by hexane extraction. The meal represents 60–65% of the total value of the beans, and oil, though important, is a by-product.

The oil is used mainly in food—often after partial hydrogenation—as salad oil, cooking oil, and frying oil and in margarines and shortening. Non-food uses are based mainly on the high levels of unsaturation and include coatings, epoxidized oil, and dimer acids. Valuable by-products of the refining process include lecithin (a mixture of phospholipids), which is widely used because of its surface activity, and vitamin E (tocopherol) used as an antioxidant. The oil also contains sterols, triterpene alcohols, chlorophyll, and hydrocarbons as minor components.

Soybean oil is a linoleic-rich oil containing 6–10% α-linolenic acid. This is considered to be undesirable for most food purposes and has been reduced to 3–4% or even to 2% by seed breeding methods. The oil with only 3–4% linolenic acid has enhanced oxidative stability and improved flavor characteristics. A high stearate (15–30%) oil has also been developed in the hope that it can be used in margarine and shortenings without partial hydrogenation.

Stereospecific analysis has given the results in Table 19, and an older silver ion chromatographic procedure indicates the glyceride composition given in Table 20.

P. Sunflower Seed Oil

Sunflowers (*Helianthus annus*) grow well in temperate zones. They are grown extensively in the former Soviet Union (29.0%), Argentina (16.6%), EU (17.7% especially France, Spain, and Italy), China (5.7%), the United States (5.2%), and eastern Europe (9.8%).

The oil represents 80% of the total value of the seed. It is usually 60–75% linoleic acid and >90% oleic and linoleic acids. In this form it is widely used as a salad oil, as a cooking

Table 20 Major Triacylglycerols ($\geq$5%) in Soybean Oil by Silver Ion TLC

LeLL	7	LLO	16	LOS	12
LeLO	5	LLS	13	OOS	5
LLL	15	LOO	8	Other	19

Source: Ref. 5.

Table 21 Stereospecific Analysis of Sunflower Seed Oil[a]

	16:0	18:0	18:1	18:2
TAG	7.2	4.6	21.9	66.3
Method 1				
sn-1	10.6	3.3	16.6	69.5
sn-2	1.3	1.1	21.5	76.0
sn-3	9.7	9.2	27.6	53.5
Method 2				
sn-1	11.2	4.0	17.7	67.0
sn-2	0.7	0.4	20.4	78.5
sn-3	10.4	9.9	28.7	51.0

[a]Two sets of results obtained in slightly different ways.
Source: Ref. 30.

oil, and in the production of margarine and shortening. Its value in these food uses is enhanced by the virtual absence of linolenic acid. For the same reason it produces nonyellowing paints. Its triacylglycerol composition is given in Tables 21 and 22.

High-oleic varieties (Sunola, Highsun) with about 85% oleic acid have been developed. Their oil has increased oxidative stability and finds use as a good source of oleic acid in enzymatically modified products. Recent developments of sunflower seed and its oil are discussed by Haumann [35] and by Morrison et al. [36].

Q. Tall Oil Fatty Acids

The term tall oil comes from the word *tallolja* (the Swedish word for pine oil), which is a mixture of fatty acids with some neutral material (~9%). Tall oil fatty acids are by-products of the wood pulp industry in which pine wood chips are digested under pressure with an aqueous mixture of sodium hydroxide and sodium sulfide, which converts the acids to their sodium salts. Tall oil is produced mainly in North America (~250,000 tonnes) and Scandinavia (~90,000 tonnes), but the products from these two sources differ in composition because of the different wood species that are pulped (Table 23). The crude acid extract is distilled to give fatty acids (with <2% of resin acids) and resin acids (with <3% of fatty acids). The fatty acid fraction contains little or no saturated acid (<3%). It is mainly a mixture of oleic and linoleic acids (75–80%) with variable amounts of pinolenic acid (5*c*9*c*12*c*-18:3) and conjugated diene acids. Also present are C_{20} homologues of pinolenic acid (5,11,14- and 7,11,14-20:3) and cyclic acids, probably derived from pinolenic and related acids [37]. An example is cyclopinolenic acid, whose structure is shown here.

Table 22 Major Triacylglycerols (≥5%) in Sunflower Seed Oil by Ag^+ TLC

LLL	14	LOO	19
LLO	39	LOS	11
LLS	14	Other	3

Source: Ref. 5.

Table 23 Fatty Acid Composition of Tall Oil

						Other[a]			
	16:0	18:0	18:1	18:2	18:3	A	B	C	D
American	0.5	2	46	36	0	2	9	0.5–2	0.5–3
Scandinavian	0.5	2	30	45	1	9	5	0.5–2	0.5–3

[a] A, pinolenic acid; B, conjugated acids; C, resin acids; D, unsaponifiable.
Source: Ref. 37.

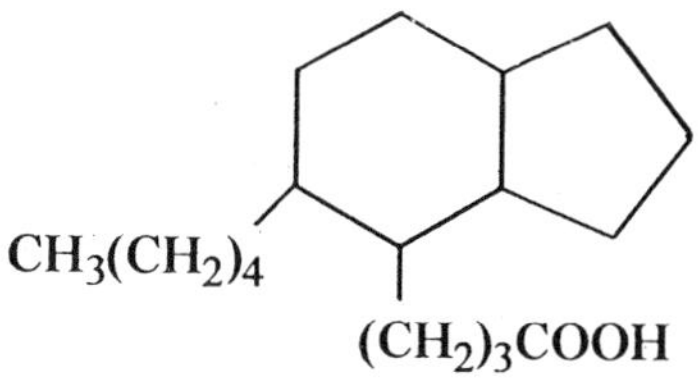

cyclopinolenic acid

Tall oil fatty acids contain some sulfur compounds that interfere with catalytic processes. They are therefore not usually converted to alcohols or to nitrogen-containing compounds but are used to produce dimer acids, alkyds, and coatings, detergents, and lubricants. New uses being investigated include employment as solvents, in inks, and as biodiesel oils.

R. Tung Oil (China Wood Oil)

Tung oil is a minor seed oil (~50,000 tonnes in 1990) of the tung tree (*Aleurites fordii* and *A. montana*) grown in China and the United States. It is characterized by a high content of α-eleostearic acid (~70%, 9*c*,11*t*,13*t*-18:3), which contains conjugated triene unsaturation. It dries (hardens through oxidative polymerization) more readily than linseed oil, which contains α-linolenic acid (a nonconjugated C_{18} triene acid and an isomer of eleostearic acid), and is used in enamels and varnishes. The acid and its glycerol ester are easily isomerized to the all-*trans* isomer (β-eleostearic acid), which has a higher melting point.

Oiticica oil and kamala oil contain other conjugated triene acids, and these and some others of this type are discussed in Chapter 1, Section VII, page 7.

S. Wheatgerm Oil

Wheatgerm oil is extracted from the bran and germ of the wheat grain (*Triticum gestivum*) (see discussion of rice bran oil). The oil is rich in linoleic acid (60–65%), but its value lies more in the high level of tocopherol [38]. It is one of the richest sources of vitamin E (see Table 9).

T. Vernonia Oil

Vernonia galamensis seed oil and some other seed oils (see Chapter 1, Section XI.B, page 10.) are characterized by their high levels of vernolic acid (*cis*-12,13-epoxyoleic acid), in the range 50–80% [5]. This acid with multiple functionality is of considerable potential though it has not yet been commercialized.

U. Waxes

Waxes are lipidlike materials of animal or vegetable origin. (Hydrocarbon waxes from mineral sources are not included in this account.) The true ester waxes are esters of long-chain acids and long-chain alcohols, and these are often, but not always, the major compounds of the natural waxes. Also present are free acids, free alcohols, hydrocarbons, and other minor compounds.

The major waxes available commercially include beeswax and wool wax from animal sources and carnauba (*Copernicia cerifera*), candelilla (*Euphorbia antisiphilitica*, *E. cerifera*, and *Pedilanthus pavenis*), ouricouri (*Syagrus coronata*, *Cocos coronata*, and *Attalea excelsa*), and jojoba oil (*Simmondsia chinensis*) from plant sources. Some indication of the composition and properties of these waxes is given in Table 24. Fuller details are to be found in a review by Hamilton [39], and Barnett [49] reported on lanolin and its many derivatives.

V. Butter (Cow's Milk) Fat

Practically all butter (~7 million tonnes/annum) is made from cow's milk fat, though some other milk fats such as those from goats and camels are in local demand. Butter is produced throughout the world and is used almost entirely for food purposes—mainly as a spread but also for baking and frying. Butter fat can be fractionated to give harder and softer fractions that find wide food uses [40,41]. Butterfat changes in composition between summer and winter, presumably reflecting different dietary intakes. It can also be modified by manipulation of the animal's diet [42].

Whole milk lipids are mainly triacylglycerols (97–98%) with smaller amounts of free acids (0.1–0.4), cholesterol (0.2–0.4%), diacylglycerols (0.3–0.6%), and phospholipids (0.2–1.0%). The latter are mainly phosphatidylcholines (35%), phosphatidylethanolamines (32%), and sphingomyelin (25%) along with lower levels of phosphatidylserine (3%) and phosphatidylinositol (5%). Butterfat is very complex in its fatty acid and triacylglycerol composition. It contains over 15 major acids, but over 500 components have been identified. These include, in addition to the usual range of the C_{16} and C_{18} acids, short- and medium-chain acids (C_4–C_{14}), a range of *trans* monoenes that are mainly 18:1 isomers with *trans*-vaccenic acid predo-

Table 24 Composition and Properties of Some Commercial Waxes

	Beeswax	Woolwax	Carnauba	Candelilla	Ouricouri	Jojoba
Composition						
Ester waxes	70–80[a]		84–85	28–29		~100
Acids	12–15		3–4	7–9		—
Alcohols	—		2–3	12–14[b]		
Hydrocarbons	10–15		2–3	50–51		—
Property						
mp (°C)	62–65	35–42	78–85	67–69	82–84	~7
Acid value	17–36	7–15	3–10	13–18	8–20	2
Sap. value	90–149	100–110	79–95	35–87	70–100	92
Ester value	64–84	85–100	—	—	75–85	—
Iodine value	7–16	15–30	7–14	14–27	6–8	82

[a] Mainly C_{40}–C_{50}.
[b] Including sterols.
Source: Ref. 39.

Table 25 Major Fatty Acids (wt %) in Summer and Winter Milk Fats

Fatty acid	June	December	Average
4:0	4.2	3.5	3.6
6:0	2.5	2.2	2.2
8:0	2.3	1.1	1.2
10:0	2.2	2.6	2.8
12:0	2.4	2.8	2.8
14:0	9.0	10.6	10.1
15:0	1.3	1.1	1.1
16:0	22.1	26.0	25.0
17:0	0.7	1.1	1.0
18:0	14.3	11.6	12.1
18:1	30.4	24.8	27.1
18:2	1.2	2.8	2.4
18:3 and 20.0	2.6	2.3	2.1
Other	4.8	7.5	6.5

Source: Ref. 43.

minating (11*t*-18:1), hydroxy acids that give rise to the γ- and δ-lactones (important flavor components), and mono- and multimethyl branched acids. Quantitative information on the major acids and triacylglycerols is given in Tables 25 and 26.

W. Lard and Tallow

Lard and tallow [45] are by-products of the rendering industry. They are rich in palmitic, stearic, and oleic acids but contain many minor components such as odd-chain acids, branched-chain acids, and *trans* acids. They contain high levels of cholesterol (lard 0.37–0.42%, beef tallow 0.08–0.14%, and mutton tallow 0.23–0.31%) and lower levels of natural antioxidants than are usually found in vegetable oils. In the tallows, unsaturated acids concentrate in the *sn*-2 positions and saturated acids in the *sn*-1 and *sn*-3 positions. This distribution pattern is different in lard, where, unusually, palmitic acid predominates in the *sn*-2 position. Tables 27–30 present data on the composition of lard and tallow.

X. Fish Oil

The total catch of fish has risen from around 20,000 tonnes in 1950 to almost 100,000 tonnes at the present time. This includes shellfish and fish from both marine and freshwater sources.

Table 26 Triacylglycerols (%) by Carbon Number in Australian Milk Fat

26	0.6	36	10.5	46	6.4
28	0.5	38	13.1	48	8.4
30	1.0	40	10.5	50	11.7
32	2.3	42	6.3	52	11.3
34	5.5	44	5.6	54	6.0

Source: Ref. 44.

Table 27 Major Fatty Acids in Lard and Beef and Mutton Tallows

	14:0	16:0	16:1	18:0	18:1	18:2
Typical values						
Lard	2.0	27.1	4.0	11.0	44.4	11.4
Beef	2.5	27.0	10.8	7.4	47.5	1.7
Mutton	5.6	27.0	1.6	31.7	31.7	1.6
Range						
Lard	0.5–2.5	20–32	1.7–5.0	5–24	35–62	3–16
Beef	1.4–6.3	20–37	6.7–8.8	6–40	26–50	0.5–5
Mutton	5.5	25.8	1.5	30.5	30	1.4

Source: Ref. 5.

Table 28 Major Triacylglycerols in Lard

PPSt	2.0	POO	5.2	OOO	11.7
StPSt	2.0	StOO	6.1	OPL	7.2
PPO	7.9	OPO	18.4	Other	24.6
StPO	12.8	StPL	2.1		

Source: Ref. 5.

Table 29 Major Triacylglycerols in Beef Tallow

PPSt	3.6	PPO	3.9	OPO	2.1
PStSt	3.2	StPO	3.6	*trans*	2.3[b]
POP	5.8	*trans*	4.5[a]	OOO	4.9
POSt	10.5	POO	12.1	other	32.8
StOSt	4.8	StOO	5.9		

[a]One double bond.
[b]Two double bonds.
Source: Ref. 5.

Table 30 Stereospecific Analysis of Tallow

	TAG	*sn*-1	*sn*-2	*sn*-3
14:0	3.4	0.2	4.3	2.9
16:0	29.5	45.7	21.0	25.5
18:0	26.0	33.1	12.5	31.2
18:1	33.1	16.6	52.0	33.7
Other	8.0	4.4	10.2	6.7

Source: Ref. 30.

The amount of fish produced by marine and freshwater aquacultures, now about 14,000 tonnes, has doubled during the last ten years. About one-third of the total catch, especially the small shoaling pelagic species, is used for industrial purposes, i.e., for direct conversion to fish oil and fish meal. Fish meal is the cheapest source of animal protein and is in high demand [50].

It has been reported that 1 tonne of anchovies will produce about 210 kg of meal (still containing about 10% oil) and 30 kg of oil. The same tonnage of anchovy is the source of about 11 kg of EPA and DHA, of which about 7.5 kg is in the oil and 3.5 kg in the meal. Of all the fish oil produced, about two-thirds is used in food products and one-third for animal feed. This latter proportion is rising, particularly to feed farmed fish.

Fish oil and fish meal are the most convenient sources of *n*-3 acids. The meal is used to feed poultry, pigs, and fish, and the *n*-3 acids are passed into the human food chain through these products. Traditionally fish oil has been used only after partial hydrogenation. This is necessary if the oil is to be used as a component of fat spreads. Partial hydrogenation gives a product of increased oxidative stability with the required melting behavior but with a reduction in nutritional value. The *n*-3 polyene acids are converted to acids of lower unsaturation with some double bonds of *trans* configuration. It has been demonstrated that with appropriate refining procedures the fish oils (without hydrogenation) can be incorporated into a range of fat-based products. Small amounts are also incorporated into infant formulas and supplied as health foods in capsule form. Products with enhanced levels of EPA and/or DHA are being developed.

Fish oils make only a minor contribution (~2%) to the total production of oils and fats. They come mainly from fish such as herring, capelin, sand eel, menhaden, sardine, and anchovy.

Fish oils generally contain a wide range of fatty acids ranging in chain length from C_{14} to C_{26} with zero to six double bonds. Their major acids are generally saturated (14:0, 16:0, and 18:0), monounsaturated (16:1, 18:1, 20:1, and 22:1), and members of the *n*-3 polyene family (18:4, 20:5, and 22:6). There are also many other acids at low levels, including branched-chain and cyclic acids. Their typical fatty acid composition is given in Table 31.

ABBREVIATIONS

D	Docosenoic acid
EU	European Union countries [12]
FSU	Former Soviet Union
G	γ-Linolenic acid or acyl group
GLA	γ-Linolenic acid or acyl group
L	Linoleic acid or acyl group

Table 31 Component Acids of Fish Oils (Typical results, wt %)

	14:0	16:0	16:1	18:0	18:1	18:2[a]	18:4[b]	20:1	20:5	22:1	22:5[b]	22:6[b]
Herring[c]	9	15	7	1	16	1	—	16	3	23	—	3
Menhaden	8	22	11	3	21	2	—	2	14	2	—	10
Capelin	8	9	16	1	8	1	1	17	9	20	1	3
Anchovy	8	20	9	3	15	1	2	3	18	2	1	11
Sardine[d]	8	16	9	4	11	1	2	3	17	4	3	13
Cod (liver)	4	14	12	3	22	1	1	12	7	11	trace	7

[a] *n*-6.
[b] *n*-3.
[c] Nova Scotia.
[d] California.
Source: Refs. 5 and 46.

Ln	α-Linolenic acid or acyl group
O	Oleic acid or acyl group
P	Palmitic acid or acyl group
PC	Phosphatidylcholines
PE	Phosphatidylethanolamines
PI	Phosphatidylinositides
R	Ricinoleic acid or acyl group
S	Saturated acid or acyl group
St	Stearic acid or acyl group
TAG	Triacylglycerols

REFERENCES

1. T. Mielke (Ed.), *Oil World Annual 1991* and later volumes, ISTA Mielke GmbH, Hamburg, Germany, 1991.
2. BSi Standards, *Nomenclature for Oilseeds*, BS 6406: 1993, ISO 5507: 1992, (1993).
3. T. Mielke, reported by editorial staff in *INFORM 5*: 715 (1994).
4. B. Howard, *Oils and Oilseeds to 1996*, The Economist Intelligence Unit, Special Report No. M.703, London, 1993.
5. F. D. Gunstone, J. L. Harwood, and F. B. Padley (Eds.), *The Lipid Handbook*, 2nd ed., Chapman and Hall, London, 1994.
6. F. D. Gunstone, *Fatty Acid and Lipid Chemistry*, Blackie, London, 1996.
7. B. S. Kamel and Y. Kakuda (Eds.), *Technological Advances in Improved and Alternative Sources of Lipids*, Blackie, London, 1994.
8. J. B. Rossell and J. L. R. Pritchard (Eds.), *Analysis of Oilseeds, Fats, and Fatty Foods*, Elsevier, London, 1991.
9. T. J. Walton, Waxes, cutin and suberin, in *Methods of Plant Biochemistry*, Vol. 4 (J. L. Harwood and J. R. Bowyer, Eds.), Academic, London, 1990, p. 105.
10. K. T. Achaya, B. M. Craig, and C. G. Youngs, The component fatty acids and glycerides of castor oil, *J. Am. Oil Chem. Soc. 41*: 783 (1964).
11. L. R. Hogge, D. C. Taylor, D. W. Reed, and E. W. Underhill, Characterization of castor bean neutral lipids by mass spectrometry/mass spectrometry, *J. Am. Oil Chem. Soc. 68*: 863 (1991).
12. V. K. S. Shukla, Cocoa butter, properties and quality, *Lipid Technol. 7*: 54 (1995).
13. V. K. S. Shukla, Confectionery fats, in *Developments in Oils and Fats* (R. J. Hamilton, Ed.) Blackie, London, 1995, p. 66.
14. I. Hachiya, T. Koyana, and K. Sato, Fat polymorphism and chocolate crystallisation, *Lipid Technol. 2*: 34 (1990).
15. J.B. Rossell, Vegetable fats for chocolate, couvertures and coatings, *Lipid Technol. 4*: 106 (1992).
16. I. A. de la Roche, E. J. Weber, and D. E. Alexander, Effects of fatty acid concentration and positional specificity on maize triglyceride structure, *Lipids 6*: 531 (1971).
17. G. Sempore and J. Bezard, Determination of molecular species of oil triacylglycerols by reversed-phase and chiral-phase high performance liquid chromatography, *J. Am. Oil Chem. Soc. 68*: 702 (1991).
18. J. M. Bland, E. J. Conkerton, and G. Abraham, Triacylglyceride composition of cottonseed oil by HPLC and GC, *J. Am. Oil Chem. Soc. 68*: 840 (1991).
19. D. F. Horrobin, Nutritional and medical importance of gamma-linolenic acid, *Prog. Lipid Res. 31*: 163 (1992).
20. F. D. Gunstone, Gamma linolenic acid—occurrence and physical and chemical properties, *Prog. Lipid Res. 31*: 145 (1992).

21. J. A. Singleton and H. E. Pattee, Characterisation of peanut oil triacylglycerols by HPLC, GLC, and EIMS, *J. Am. Oil Chem. Soc. 64*: 534 (1987).
22. T. Takagi, Chromatographic resolution of chiral lipid derivatives, *Prog. Lipid Res. 29*: 277 (1990).
23. M. H. J. Bergqvist and P. Kaufmann, A multivariate optimization of triacylglycerol analysis by high-performance liquid chromatography, *Lipids 28*: 667 (1993).
24. F. Santinelli, P. Damiani, and W. W. Christie, The triacylglycerol structure of olive oil determined by silver ion high-performance liquid chromatography in combination with stereospecific analysis, *J. Am. Oil Chem. Soc. 69*: 552 (1992). (See also Ref. 30.)
25. R. V. Flor, L. T. Hecking, and B. D. Martin, Development of high performance liquid chromatography criteria for determination of grades of commercial olive oils. Part 1. The normal ranges for the triacylglycerols, *J. Am. Oil Chem. Soc. 70*: 199 (1993).
26. W. Hamm, Trends in edible oil fractionation, *Trends Food Sci. Technol. 6*: 121 (1995).
27. S. Takano and Y. Kondoh, Triglyceride analyses by combined argentation/nonaqueous reversed phase high performance liquid chromatography, *J. Am. Oil Chem. Soc. 64*: 380 (1987).
28. M. H. Bergqvist and P. Kaufmann, A multivariate optimization of triacylglycerol analysis by high-performance liquid chromatography, *Lipids 28*: 667 (1993).
29. G. J. Sassano and B. S. F. Jeffrey, Gas chromatography of triacylglycerols in palm oil fractions with medium-polarity wide-bore columns, *J. Am. Oil Chem. Soc. 70*: 1111 (1993).
30. W. W. Christie, B. Nikolova-Damyanova, P. Laakso, and B. Herslof, Stereospecific analysis of triacyl-*sn*-glycerols via resolution of diastereomeric diacylglycerol derivatives by high performance liquid chromatography on silica, *J. Am. Oil Chem. Soc. 68*: 695 (1991).
31. C. Leonard, Sources and commercial applications of high-erucic vegetable oils, *Lipid Technol. 6*: 79 (1994).
32. T. Takagi and Y. Ando, Stereospecific analysis of triacylglycerols by chiral high-performance liquid chromatography, *Lipids 26*: 542 (1991).
33. H. Kallio and G. Currie, Analysis of low erucic acid turnip rapeseed oil (*Brassica campestris*) by negative ion chemical ionization tandem mass spectrometry. A method giving information on the fatty acid composition in positions *sn*-2 and *sn*-1/3 of triacylglycerols, *Lipids 28*: 207 (1993).
34. R. N. Sayre and R. M. Saunders, Rice bran and rice bran oil, *Lipid Technol. 2*: 72 (1990).
35. B. F. Haumann, Modified oil may be key to sunflower's future, *INFORM 5*: 1198 (1994).
36. W. H. Morrison, R. J. Hamilton, and C. Kalu, Sunflower seed oil, in *Developments in Oils and Fats* (R. J. Hamilton, Ed.), Blackie, London, 1995, p. 132.
37. A. Hase and S. Pajakkala, Tall oil a fatty acid source, *Lipid Technol. 6*: 110 (1994).
38. P. J. Barnes, Lipid composition of wheat germ and wheat germ oil, *Fette Seifen Anstrichm. 84*: 256 (1982).
39. R. J. Hamilton, Commercial waxes: their composition and applications, in *Waxes: Chemistry, Molecular Biology and Functions* (R. J. Hamilton, Ed.), The Oily Press, Dundee, Scotland, 1995, p. 257.
40. L. Eyres, Milk fat product development, *Lipid Technol. 1*: 12 (1959).
41. W. de Greyt and A. Huyghebaert, Food and non-food applications of milk fat, *Lipid Technol. 5*: 138 (1993).
42. W. Charteris, Monobutter: a fridge-spreadable butter, *Lipid Technol. 3*: 90 (1991).
43. R. G. Jensen, G. W. Gander, and J. Sampugna, Fatty acid composition of the lipids from pooled, raw milk, *J. Dairy Sci. 45*: 329 (1962).
44. R. E. Timms, Detection and quantification of non-milk fat in mixtures of milk fat and non-milk fat, *J. Dairy Res. 47*: 295 (1980).
45. R. R. Grummer, Production and consumption of animal by-products from the rendering industry, *Lipid Technol. 5*: 114 (1993).
46. The British Nutrition Foundation Report, *Unsaturated Fatty Acids: Nutritional and Physiological Significance*, Chapman and Hall, London, 1992.
47. Y. S. Huang, X. Lin, P. R. Redden, and D. F. Horrobin, *In vitro* hydrolysis of natural and

synthetic γ-linoleic acid containing triacylglycerols by pancreatic lipase, *J. Am. Oil Chem. Soc. 72*: 625 (1995).
48. P. Manninen, P. Laakso, and M. Kallio, Separation of γ- and α-linolenic-acid-containing triacylglycerols by capillary supercritical fluid chromatography, *Lipids 30*: 665 (1995).
49. G. Barnett, Lanolin and derivatives, *Cosmet. Toiletries 101*: 21 (1986).
50. R. J. Hamilton and R. D. Rice (Eds.), *Fish Oil, Technology, Nutrition, and Marketing*, P. J. Barnes and Associates, High Wycombe, UK, 1995.

3

Phospholipids

Michael Schneider
Lucas Meyer GmbH & Co., Hamburg, Germany

I. TERMINOLOGY

In the scientific literature, rational nomenclature according to IUPAC or abbreviated IUPAC nomenclature (e.g., 1,2-diacyl-*sn*-glycero-3-phosphocholine or phosphatidylcholine) [1] and trivial terms such as lecithin (for PC) and cephalin (for PE or PI/PA) still coexist. In industrial and commercial understanding, the term lecithin denotes a complex mixture of neutral lipids (predominantly triglycerides), polar lipids (phospho- and glycolipids), and carbohydrates. This blend is described as a generally permitted food additive "lecithin" in Europe under E 322 [2] and in the United States under the Code of Federal Regulations [3]. The two descriptions differ to a minor extent in their specification details but not in principle. As this book is about industrial application and technology, industrial terminology will be used throughout this chapter.

II. OCCURRENCE

Two main classes of phospholipids occur naturally in amounts that have led to industrial applications: phosphoglycerides and phosphosphingolipids. Table 1 gives the most commonly used terms and abbreviations for the phospholipids found in materials used for commercial purposes.

As phospholipids are integral structure elements of all kinds of membranes in every living organism, they can be obtained from all types of biomass. Commercial sources, however, are predominantly vegetable oil seeds (soybeans, rapeseed, sunflower seed). For pharmaceutical, cosmetic, and some dietary applications, egg yolk, milk, and brain have gained some importance.

Vegetable materials usually contain only small amounts of phospholipids, ranging from 0.3 to 2.5wt% maximum (on dry basis) [5–7]. Animal materials have substantially higher phospholipid contents: egg, 14% [8]; brain, 6% [9]; and whole milk, 2% [10].

Table 1 Phospholipid Classes

Class	Abbreviation
Phosphoglycerides	
1. Ester phospholipids[a]	
Phosphatidylcholine	PC
Phosphatidylethanolamine	PE
Phosphatidylserine	PS
Phosphatidylinositol	PI
Phosphatidylglycerol	PG
Diphosphatidylglycerol	DPG
Phosphatidic acid	PA
N-Acyl-phosphatidylethanolamine	NAPE
2. Ether phospholipids	
Choline plasmalogen	
Ethanolamine plasmalogen	
Phosphosphingolipids	
Ceramide phosphocholine (sphingomyelin)	SPM
Glycosylated inositolphosphoceramides	
Phytoglycolipids	PGL

[a]Along with the listed diester species, small amounts of monoacyl types, the so-called lysophospholipids, are found in all natural sources.
Source: Ref. 4.

III. COMPOSITION

The various methods of obtaining commercial phospholipid products are described later. All production methods are based on complex mixtures of various species of phospholipids that are by-products of vegetable oil processing or of extraction from wet or dried animal sub-

Table 2 Phospholipid Composition [wt%] as Extracted from Commercial Raw Material

	Soy[a]	Rapeseed[a]	Sunflower seed[a]	Corn[a]	Egg	Milk[b]	Brain[a] (bovine)
PC	21.9	24.6	25.4	30.4	74	27.0	32.4
PE	13.6	22.1	11.0	3.2	19.1	36.4	23.5
PI	12.0	14.7	19.4	16.3	0.4		4.3
PS	—		0.8	1.0	—		11.0
PG/DPG	2.3		1.2	1.4	0.6		2.1
PA	5.8		3.3	9.4	—		0.9
NAPE	2.8		1.0	2.6	—		
SPM	—				2.5	29.0	20.4
Lyso-PLs	2.9	19.4		5.4	2.9		3.0
Others	3.6	19.2			0.5	7.6	
Ref.	11	12	13	14	11	10	11

[a]Figures are based on the acetone-insoluble constituents of commercial lecithins, which also contain certain amounts of glycolipids.
[b]The data consider only phosphorus-containing polar lipids.

Table 3 Fatty Acid Composition [wt%] of Phospholipid Mixtures

	Soy	Rapeseed	Sunflower seed	Corn	Egg	Brain
16:0	21.4	18.3	14.8	22.8	30	1
18:0	3.8	0.6	3.0	1.5	16	41
18:1	12.0	21.3	13.3	26.5	29	31
18:2	57.0	47.9	68.7	48.5	14	—
18:3	5.8	7.4		0.7	1	—
20:4					5	3
22:6					3	8
Others					2	16
Ref.	14	15	13	14	11	11

stances. In addition, vegetable extraction products contain smaller or larger amounts of phosphorus-free polar lipids (glycolipids) and sometimes even carbohydrates. The phospholipid composition is a "fingerprint" typical of the raw material being extracted, (Table 2), and so also is the composition of the fatty acids linked to the phospholipid backbone (Table 3). The data in Table 3 reflect the composition of the natural phospholipid mixtures. The fatty acid composition of isolated phospholipids can differ considerably from the listed data. For more details, see Ref. 15.

IV. OBTAINING PHOSPHOLIPID PRODUCTS

Methods for obtaining phospholipids on an industrial scale are completely different for vegetable-derived products than for animal-derived products. Vegetable phospholipids are manufactured exclusively from by-products of the vegetable oil refining process. For animal phospholipids, specific extraction processes have been developed and are applied commercially.

A. Vegetable Phospholipids

Today the exclusive sources of vegetable phospholipids are commercial lecithins obtained in the first step of vegetable oil refining during the degumming process. To stabilize vegetable oils against sedimentation and off-taste development and also to enable further refining steps, phospho- and glycolipids must be removed. During the standard batch degumming process [16], 2% of water by volume is added to crude oil heated to approximately 70°C and agitated for ½–1 h. The addition of water causes the polar lipids contained in the oil to hydrate, leaving them oil-insoluble. Centrifugation of the hydrated gums leads to a mixture of water, phospho- and glycolipids, and also some triglycerides, free carbohydrates, traces of sterols, free fatty acids, and carotenoids. Careful drying of the wet gums immediately after removal from the crude oil leads to crude vegetable lecithin.

The literature reports dozens of variations of the degumming process [16], all with the aim of reducing the amount of phosphorus remaining in the oil to a minimum. Degumming conditions have an influence on crude vegetable lecithin composition and quality.

Table 4 shows a typical composition of soybean lecithin [14]. Traditionally, crude vegetable lecithins are the starting materials of choice for further fractionation and purification processes to obtain phospholipid compositions suitable for various industrial applications.

Table 4 Typical Composition of Crude Soybean Lecithin (%)

Triglycerides	34.2
Diglycerides	0.4
Free fatty acids	0.4
Sterols	0.8
Glycolipids	6.5
Carbohydrates	6.7
Phospholipids	50.0

B. Animal Phospholipids

In contrast to vegetable products, there is no comparable industrial animal oil processing technology, and therefore phospholipids have to be specifically extracted using various technologies. Since egg phospholipids are the sole products of commercial interest, the processing alternatives are those shown in Fig. 1.

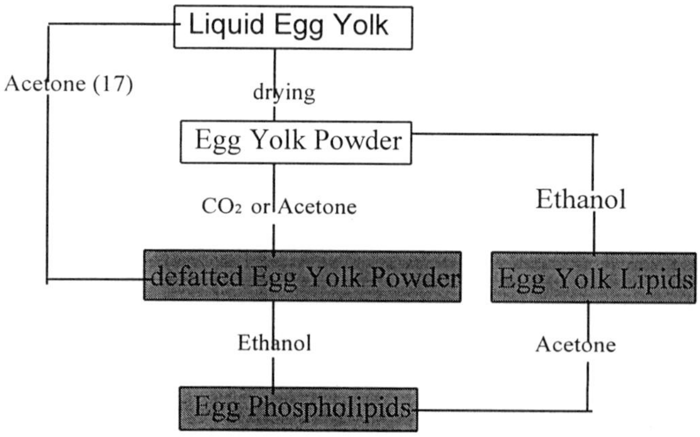

Fig. 1 Egg yolk phospholipid processing principles.

Methods for milk phospholipid extraction are numerous [17–22]. Some of them describe the use of solvent mixtures to use partitioning [18,19] or a combination of extraction and adsorption chromatography [21] or extraction with chloroform–methanol mixtures [22]. Ultrafiltration processes have also been described.

Brain phospholipid extraction is usually based on modifications of the Folch procedure [23,24]. Removal of nonlipidic impurities is a lengthy process that requires further solvent extraction or gel chromatography. A modification of the extraction process that does not require extraction or adsorption steps is described in Ref. 9.

V. INDUSTRIAL PHOSPHOLIPID PROCESSING

As this book is about industrial applications and technologies, only processes used on a commercial scale are described here.

Industrial processing can be split into

1. Solvent fractionation
2. Gas extraction
3. Chromatographic purification
4. Chemical modification
5. Enzymatic modification

A. Solvent Fractionation

Commercial solvent technologies are limited to a small selection of solvents that are appropriate to the image and use of a natural functional ingredient. These solvents are acetone, ethanol, compressed gases, and others (e.g., hexane, methanol, isopropanol). For many industrial applications, crude phospholipid products obtained from vegetable oil refining (crude lecithins) can be used directly. During the last 30–40 years, however, solvent fractionation has widened the range of commercial phospholipid products.

Table 5 gives an overview of today's industrial consumption of some phospholipid products. (Explanations of compositional details are given in the sections describing process details.)

Table 5 Commercial Phospholipid Products and Their Predominant Use

Product	Application[a]
Soybean lecithin	AF/FO/TI/CO/PH/DI
Deoiled soybean lecithin	AF/FO/TI/CO/PH/DI
Ethanol-soluble fractions	FO/CO/PH/DI
Ethanol-insoluble fractions	AF/FO
Deoiled soybean lecithin fractions (PC 50–80%)	FO/CO/PH/DI
Purified soybean PC (>90%)	PH/CO
Sunflower seed lecithin	AF/FO
Rapeseed lecithin	AF/FO/TI
Corn lecithin	AF/TI
Egg phospholipids (deoiled)	FO/DI/CO
Egg lipid mixes (PL-containing)	DI/PH
Milk phospholipids	CO/PH
Brain phospholipids	PH/DI

[a] AF, Animal feed; CO, cosmetics; DI, dietetic; FO, Food; PH, pharmaceutical; TI, technical industry.

B. Acetone Deoiling

Vegetable lecithins contain approximately 30–40% neutral lipids, predominantly triglycerides. To improve the handling of the highly viscous crude lecithins and to improve dispersibility properties, industry makes use of the fact that polar (phospho- and glyco-) lipids are almost insoluble in acetone whereas neutral lipids dissolve.

Acetone extraction leads to powdered or granulated products that contain 2–3% residual neutral lipids. The products are slightly hygroscopic but flow freely after addition of flow agents such as tricalcium phosphate or silica.

Because of the removal of natural tocopherols (~1000 ppm in crude soybean lecithin), the shelf life of deoiled products is usually limited to approximately 12 months. Table 6 gives the typical composition of deoiled soybean lecithin.

Table 6 Typical Composition of a Commercially Deoiled Soybean Lecithin

Component	Percent content	
Phospholipids		80
Phosphatidylcholine	24	
Phosphatidylethanolamine	22	
Phosphatidylinositol	14	
Phosphatidic Acid	7	
N-Acyl-phosphatidylethanolamine	6	
Lysophospholipids	3	
Glycolipids		12
Neutral lipids		2.5
Triglycerides	2	
Free fatty acids	0.5	
Sterols	tr	
Moisture		1

C. Compressed Gas Deoiling

Since the introduction of supercritical gas extraction on an industrial scale, which was first used for decaffeination of coffee and tea [25], a number of publications and patents have described the deoiling of lecithin [26–28]. Supercritical carbon dioxide at temperatures of ~40°C and at a pressure of 300–700 bar shows solvent characteristics like those of liquid acetone—it dissolves neutral lipids but leaves behind the polar lipids. Advantages of such a process are the mild temperature treatment, no oxidation during drying, no solvent residues, and no flammability risk.

However, owing to the powder structure of deoiled lecithin, only batch processes have been described so far—extracting the crude lecithin either in bulk or after nebulizing [29]. The major economic drawbacks of carbon dioxide extraction—batch processing and poor oil-carrying capacity—have been overcome with the development of countercurrent extraction processes using propane–carbon dioxide mixtures at 80 bar [30]. In cases where slightly reduced selectivity in splitting between neutral and polar lipids can be accepted, compressed propane at ~40 bar is an excellent alternative [31]. However, at present there is no commercial vegetable phospholipid product on the market that is being processed by compressed gas technology.

Processing methods used for vegetable lipid mixtures have also been applied to egg lipid blends [32]. Following the trend toward low cholesterol, low fat products, one can remove fat and cholesterol (neutral lipids) from dried egg yolk using either compressed carbon dioxide or propane or mixtures of the two. In the United States there is only one consumer product that has been prepared in this manner.

D. Alcohol Fractionation

Large quantities of vegetable lecithin "fractions" are produced mainly with ethanol or ethanol–water systems. Commercial interest in these fractions is due to the difference in their technological functionality [soluble fraction oil-in-water (o/w) emulsifier; insoluble fraction water-in-oil (w/o) emulsifier]. The interest of the industry in PC-enriched products for dietary purposes is also a driving force.

Table 7 shows phosphatidylcholine concentrations after fractionation of both a crude and a deoiled soybean lecithin [33].

Table 7 Alcohol Fractionation of Soybean Lecithin

	Triglycerides (%)	Phosphatidylcholine (%)	Yield (%)
Crude lecithin	39	15	
Soluble fraction			
Methanol	25	39	23
Ethanol (100%)	60	22	26
Ethanol (95%)	48	33	15
Ethanol (90%)	23	46	12
Isopropanol	54	19	49
Insoluble fraction			
Ethanol (95%)	38	7	85
Deoiled lecithin	2.5	24	
Soluble fraction			
Ethanol (100%)	2.5	52	28
Insoluble fraction			
Ethanol (100%)	2.5	12	72

What is described for soybean lecithin can in principle be applied also to other crude vegetable phospholipid products (lecithins). However, although the results are similar, the other lecithins are not used commercially.

E. Chromatographic Purification

A review of chromatographic purification is given in Ref. 33. Generally speaking, chromatographic separation is the method of choice for the preparation of polar lipids of high purity. It is almost impossible to give a brief account of the numerous methods published in the scientific and patent literature. Most of them are used for the quantitative analysis of tissue extracts and less for the production of commercial quantities of glycolipids or phospholipids. In this chapter the focus is on a few short descriptions only.

The best results in separation are usually achieved by combining different separation techniques—for example, solvent treatment of the crude lipid mixture and adsorption processes. Among the long list of adsorbents [aluminum oxide, silicic acid, Florisil (magnesium silicate), DEAE (diethylaminoethyl) cellulose, and ion-exchange resins], alumina and silica play an outstanding role. Both materials are used especially for low and medium pressure chromatography to produce large quantities of soybean and egg PC products.

A typical alumina separation is as follows. Deoiled soybean lecithin is fractionated first by treatment with ethanol to produce a soluble part with about 50% PC. This solution is then passed down a column filled with aluminum oxide. Nearly all of the PC-accompanying phospholipids are bound to the alumina. The effluent consists of almost pure choline phospholipids. One drawback of this method is the impossibility of removing lysophosphatidylcholine (LPC) from the material. In fact, additional LPC is formed during the chromatographic process.

Attempts to purify egg phospholipids show that in an ethanolic solution sphingomyelin has the same elution behavior as PC and is therefore included in the PC/LPC effluent.

The separation of highly purified PC as well as other phospholipids is possible using stepwise elution with a chloroform–methanol–water mixture of various compositions.

Table 8 lists some chromatographic systems used for purification of phospholipids.

In comparison with alumina separation, silicic acid chromatography seems to be more flexible with respect to solvent systems, separation characteristics, and reconditioning of the

Table 8 Chromatographic Systems for Phospholipid Purification

Adsorbent	Solvent system
Aluminum oxide	Ethanol
	Chloroform–methanol
Silica gel	Hexane
	Ethanol
	Hexane–isopropanol–water
	Chloroform–methanol
	Acetone–chloroform–methanol
	Chloroform–acetone
Florisil	Chloroform–methanol
Amberlite	Ethanol
	Ethyl acetate–ethanol
DEAE-cellulose	Chloroform–methanol–acetic acid

Source: Ref. 33.

column. Silica is also used for selective elution processes to obtain the different polar lipid classes of a given lecithin. The preferred solvent systems again are chloroform–methanol and methanol–acetone mixtures. Separation is mainly done stepwise but in some cases by gradient elution. For more details about the wide range of possibilities for chromatographic separation, refer to the literature on this subject [33].

As already mentioned, chromatography on an industrial scale today is mainly focused on the preparation of PC of natural origin (egg, soy), which is of great commercial interest and available in ton quantities. Preparation of the natural phospholipids such as PS and PI is of growing interest. The separation technique of choice is large-scale HPLC, which is also used in the production of semisynthetic and synthetic phospholipids.

Pure PC from a chromatographic process is still a mixture of various molecular species differing in fatty acid composition. This is demonstrated impressively in Table 9, which gives an example of the long list of molecular species of egg PC. However, so far the literature has given little indication of the possibility of separating phospholipids on the basis of the degree of unsaturation. One possibility for analytical applications is to use argentation chromatography, which uses silver nitrate impregnated silica as an adsorbent. The higher the degree of unsaturation, the more firmly the phospholipids are bound to the stationary phase and thus separated from each other. A recent development in this area is the use of reverse-phase HPLC. However, it is not known whether these processes are already used commercially. It seems to make more sense to synthesize the appropriate molecule.

Table 9 Molecular Species of Egg Yolk Phosphatidylcholines

Molecular species	Percent	Molecular species	Percent
16:0, 16:1	0.98	18:0, 18:2	11.22
16:0, 18:1	38.20	18:1, 18:2	1.50
18:0, 18:1	9.30	16:0, 20:4	2.52
16:1, 18:1	0.92	18:0, 20:4	3.36
18:1, 18:1	2.96	16:0, 22:6	1.84
16:0, 18:2	21.78	18:0, 22:6	1.60

Source: Ref. 34.

F. Modification of Phospholipids

The molecular structure of phospholipids can be changed by either enzymatic or chemical means. The aim of all these processes is to obtain tailormade technological and/or physiological properties that are somewhat different from those of the natural substances themselves.

1. Enzymatic Modification of Commercial Phospholipid Products

There are different hydrolytic enzyme systems known that are able to modify the phospholipid structure. They are all ester hydrolases and react site-specifically as follows:

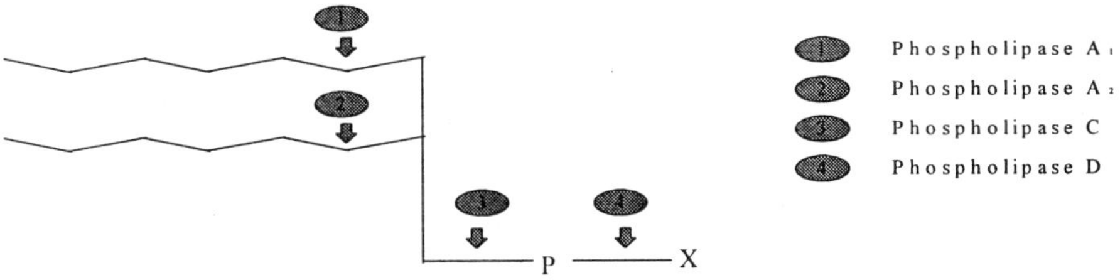

It is typical of hydrolytic enzymes that their catalytic influence can be used for hydrolysis or for transesterification. Commercial use to a large extent is known only with phospholipase A_2 from porcine pancreas (reaction 1) [35,36].

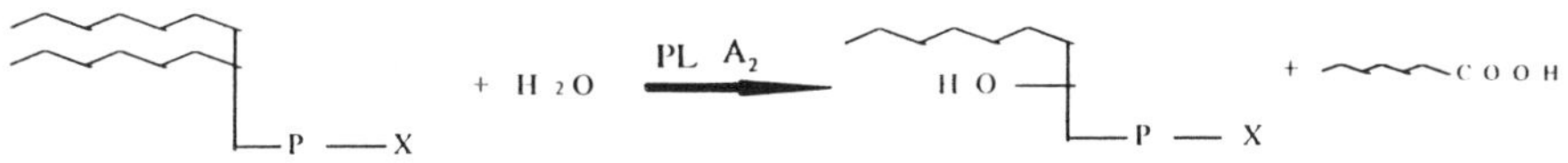

Partially hydrolyzed vegetable phospholipid products (lecithins) show clearly improved emulsifying properties up to a certain level of lysophospholipids. Above this level of conversion, water dispersibility increases but not the emulsifying capacity. On the other hand, with increased hydrolysis the interaction with proteins in complex food systems seems to increase.

For another hydrolytic process using enzyme technology, cabbage-derived phospholipase D is used (reaction 2). The reaction increases the phosphatidic acid (PA) content in lecithin products [37]. The resulting products are reported to have improved emulsion properties and the ability to mask the bitter taste of certain pharmaceuticals [38].

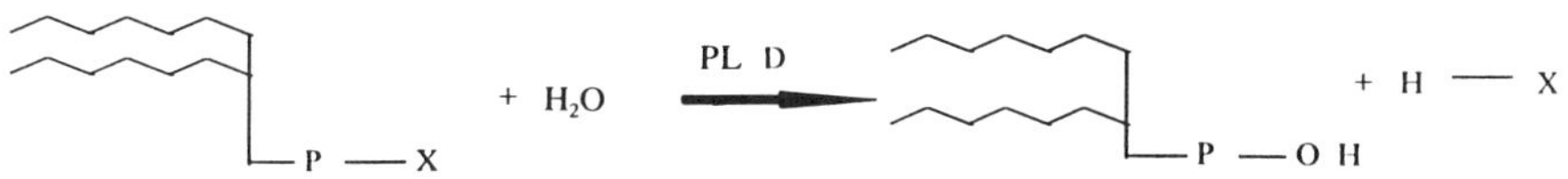

Whereas enzymatic transesterification of natural phospholipids with phospholipase A_2 (insertion of fatty acids with special pharmacological effects) is only of academic interest today [39], the phospholipase D catalyzed formation of phosphatidylglycerol is already a commercial process in Japan. The resulting products are reported to have improved emulsion capacity and temperature resistance. In Europe commercial interest is restricted to hydrolyzed but not transesterified products as the European food regulation (E 322) only permits the "hydrolyzed substances" [2].

Phospholipase A_1 and phospholipase C reacted products are not on the market because of the limited availability of phospholipase A_1 (and fatty acid migration from the 2- to the 1-position after hydrolysis) and a lack of interest in diglycerides formed by a phospholipase C reaction.

2. *Chemical Modification*

There are various ways to chemically modify phospholipid molecules, but only few of them are commercialized. The reason is that none of the resulting products has food grade status, except for the products of hydroxylation (allowed in the United States for baking). However, despite that problem, substantial development and application work has been done on chemically modified phospholipid products.

Hydroxylation. An effective way to improve emulsion properties of vegetable lecithins for o/w systems is hydroxylation [40]. Hydrogen peroxide reacts with the double bonds of unsaturated phospholipid fatty acids under the catalytic action of small amounts of organic acids of low molecular weight (e.g., lactic acid) to form dihydroxy fatty acid derivatives (reaction 3).

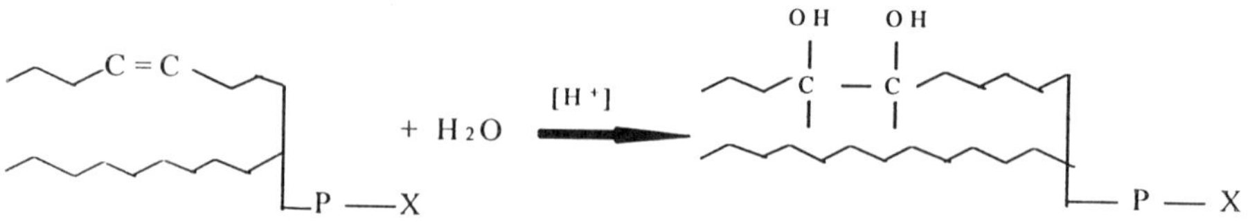

Acetylation. Another way to improve the hydrophilicity of vegetable lecithin products is acetylation. During the reaction of vegetable phospholipid mixtures with moderate amounts of acetic anhydride, acetylation of the free amino group of PE takes place (reaction 4). [41,42].

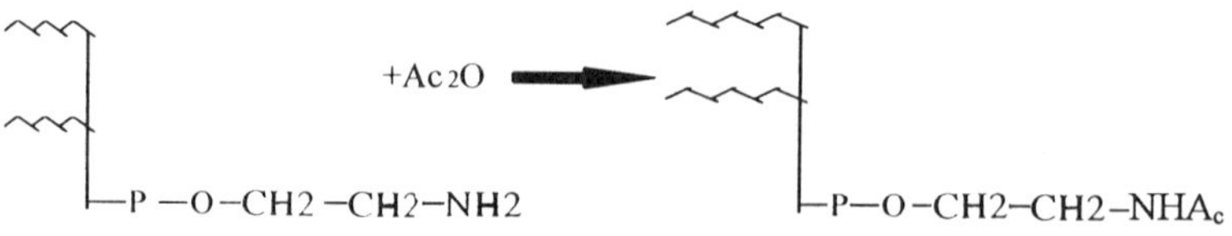

Acetylated phospholipid mixtures are excellent o/w emulsifiers and also exhibit good thermal stability. The latter is explained by the fact that there is no primary amino group left for Maillard-type darkening reactions with the carbohydrate contents of vegetable lecithin products.

Hydrogenation. All natural phospholipid mixtures (lecithins) contain more or less high levels of unsaturated fatty acids. This makes all products susceptible to oxidative processes, which lead finally to unpleasant taste and smell (rancidity).

During catalytic hydrogenation of natural phospholipids, the unsaturated fatty acids become saturated. As a result the products are more stable against oxidation but less soluble in fats and oils.

As a side effect of hydrogenation, most of the natural pigments are destroyed, turning the resulting products into white to off-white free-flowing powders.

The hydrogenation process for phospholipids requires quite drastic conditions regarding hydrogen pressure as well as type and quantity of catalyst. Only palladium and platinum catalysts are resistant enough to the poisoning effect of phospholipids to allow reasonable conversion rates [43–45].

Because of the lack of food grade status and the high costs, the process is limited to application in pharmaceuticals and cosmetic products.

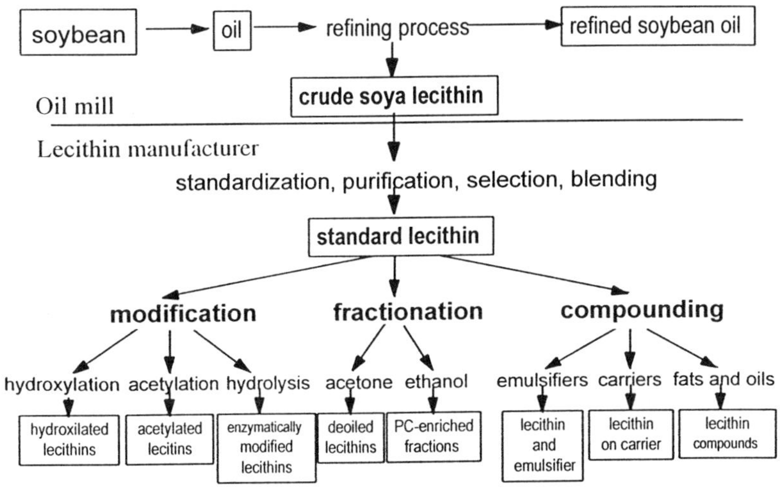

Fig. 2 Industrial processing of soybean lecithin (abbreviated).

To summarize this section on the industrial processing of natural phospholipid products, Fig. 2 presents a visual impression of industrial processing of the most commonly used soybean products.

VI. APPLICATIONS

Phospholipid applications follow two main principles. They can be used either for their technological functionality or for physiological benefits. The number of application areas has rapidly increased over the last decades not only because of their functional properties but also because of new scientific evidence for certain therapeutic activity.

Because of the vast range of applications this review concentrates on the industrially interesting ones and describes application principles in a very pragmatic way. For more scientific background, reference is made to the already existing literature [15,46–50].

A. Phospholipids as Functional Ingredients

Phospholipids are amphiphilic molecules of moderate hydrophilicity. Even though the HLB (hydrophilic-lipophilic balance) considerations do not fit perfectly for zwitterionic molecules, application technologists very often raise questions regarding indications of HLB values. Table 10 gives some guiding values based on numerous emulsion tests.

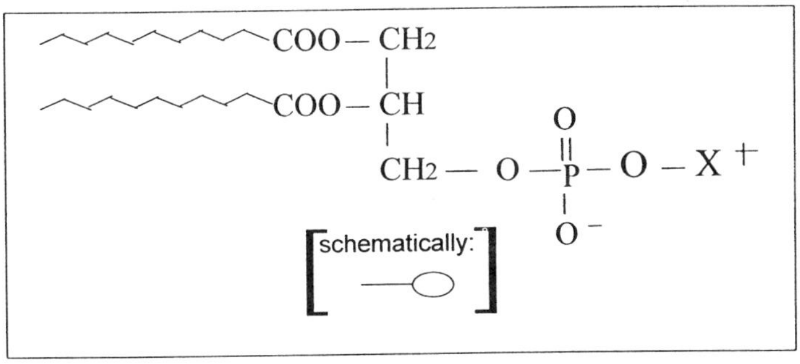

Table 10 Indication of HLB values for Commercial Phospholipid Products

Product	HLB
Crude soy lecithin	3
Deoiled soy lecithin	4
Deoiled egg lecithin	5
Hydrolyzed soy lecithin	7
Acetylated soy lecithin	8
Hydroxylated soy lecithin	8
Acetylated and hydrolyzed soy lecithin	9–10
Purified soy PC	7

Phospholipid products in crude or refined or modified form are used for wetting, emulsification, or dispersion of products from the technical industry, in feed and food products, and in cosmetics and pharmaceuticals. Application principles can be reduced to a very few behavior patterns of phospholipids as illustrated in Fig. 3 [51].

1. *Industrial Products*

Paints and Varnishes. Predominantly crude or modified crude vegetable lecithin products are used as dispersing aids for both organic and inorganic pigments. Applications range from printing inks to dispersion paints and varnishes. As phospholipids are quite lipophilic, they are used predominantly in systems based on organic solvents. Recent developments toward highly hydrophilic lecithins modified by acetylation, hydrolysis, and/or hydroxylation lead to products that seem to have attractive properties for water-based systems.

Another interesting application is in the manufacture of magnetic tapes. Here lecithins are used as dispersants for the magnetic metal oxide pigments [52,53].

Other technical applications are found in the use of phospholipid-containing formulas for plant pesticides [54,55] or fertilizers. For this area specific phospholipid concentrates loaded with active ingredients that can be easily diluted with water are of interest. The attraction of the formula is that in final dilution the product is a liposome preparation [56]. Liposomal pesticide or fertilizer preparations are reported to have improved functionality and contribute to reducing the use of toxic ingredients and to reducing environmental pollution [57,58]. (The liposome principle is described in detail in the section on cosmetics.)

A nice example of the combined action for both technological and physiological principles of phospholipids is the subject of a patent for the removal of petroleum spills [59]. In this application lecithin serves not only as an emulsifier/dispersant for a concentrated all-natural preparation but also as a nutritional cocktail for soil microorganisms, which are stimulated to "digest" the spilled oil.

2. *Feed Products*

The attraction of phospholipid products in feed preparations is, as just mentioned, the combination of the emulsifier function with nutritional benefits. Lecithins contribute to the animal's energy supply, they are sources of essential fatty acids and micronutrients such as choline and inositol, and they support fat metabolism.

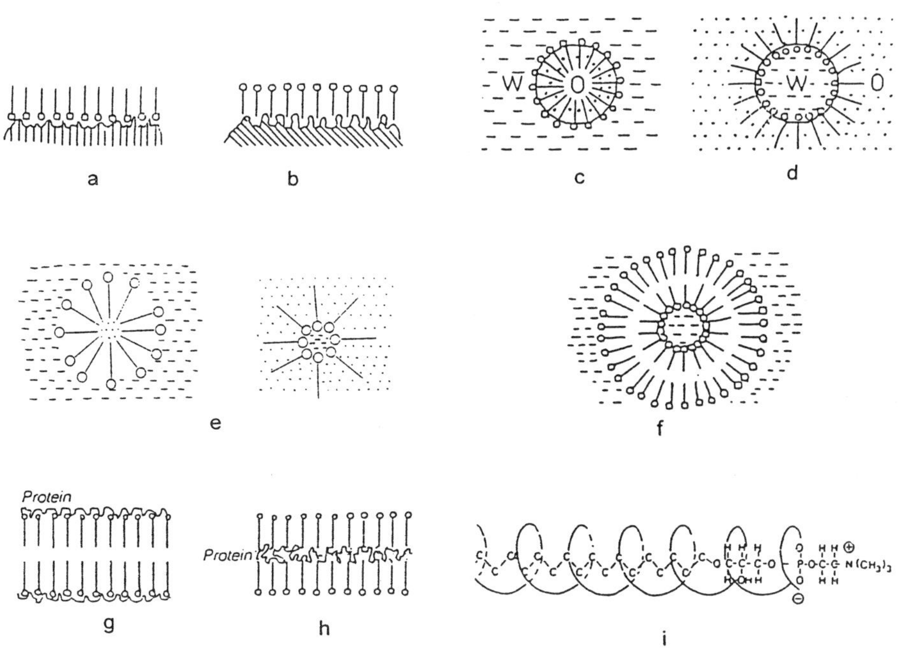

Fig. 3 Simplified schematic physicochemical action principles of phospholipids.
(a,b) Coating of hydrophilic (a) and lipophilic (b) surfaces (wetting, dispersing);
(c,d) stabilization of o/w (c) and w/o (d) emulsions;
(e) micellar organization (solubilization);
(f) liposome (entrapment of hydrophilic and lipophilic substances);
(g,h) association of phospholipids with proteins;
(i) inclusion of lysophospholipids in amylose.

The annual world consumption of lecithins for cow's milk replacers (refattened skim milk products) is 40,000 metric tons. They are used in combination with synthetic emulsifiers (ethoxylated products, sugar esters, polyglycerol derivatives, etc.).

Additionally it should be mentioned that deoiled powdered lecithins are gaining importance as extrusion aids in the manufacture of feed pellets, especially for fish farming (better oil uptake, better sinkability, improved extruder capacity).

More information on the use of phospholipids in feed is given in the subsection on animal nutrition, Section VI.B.5.

3. *Food Applications*

The industrial production of vegetable phospholipid products (lecithins) as by-products of the oil refining process started in the mid-1930s. The availability of a cheap and versatile natural emulsifier has led to a rapid increase in the use of lecithin, especially in the food industry. Approximately 60–65% of the annual production of lecithins is still going into various food products.

As there are so many different kinds of applications, it is impossible to be complete and go deeply into details in this chapter. The manufacturers of various phospholipid products offer excellent documentation and practical assistance in solving application problems.

Until recently, the application of lecithin to food products has been a predominantly empirical exercise that required long years of practical and pragmatic experience. Because of the complexity of the various products offered and the complicated physicochemical behavior of phospholipids and glycolipids, only a few basic principles have been discovered over the last 60 years of industrial application.

Food Emulsions. It has been demonstrated that purified phospholipids (commercially available material is predominantly phosphatidylcholine from soybeans and eggs) are poor emulsifiers, especially for o/w systems [60]. Of absolute importance to emulsion quality and type of emulsion (o/w or w/o) is the detailed composition of the phospholipid mixture. The more acidic types of phospholipids like PI and PA play an important role. At the pH value of food products they contribute to a highly negative zeta potential of the dispersed phase, resulting in improved physicochemical stability [61].

OIL-IN-WATER EMULSIONS. Mayonnaise, salad dressings, and similar sauces require protein in addition to phospholipids for stability. Natural vegetable lecithins are not very efficient emulsifiers for these types of emulsions as they are usually prepared at a low pH value (close to the isoelectric point of phospholipids, between pH 3 and 4) or they exhibit high ionic strength due to salt content or they are subject to extensive heat treatment during sterilization. All of these "stressful" conditions have a negative effect on emulsion stability. However, recent findings describe the specific functionality of hydrolyzed phospholipids (lysophospholipids) in the presence of proteins. Obviously a kind of lipoprotein structure formed during food preparation shows a good stabilizing effect against various kinds of physicochemical stress [62,63].

This principle of combining proteins with lysophospholipids may become a universal concept for complex food preparations that undergo stress treatment during manufacture such as sausages, heat-sterilized emulsions, mayonnaise, dressings, sauces, even bread (frozen doughs) and low fat spreads.

WATER-IN-OIL EMULSIONS (MARGARINE, SPREADS). Traditionally a typical margarine is 80% fat; "light" products such as low fat spreads can have fat contents of 40% or less.

One major disadvantage of margarine as a butter substitute is its poor spattering performance and the sedimentation and burning of the bulk protein in a hot pan. Combinations of monodiglycerides and lecithin or hydrolyzed lecithins have more or less solved the problem. Low fat spreads, however, require ethanol-fractionated lecithins with higher than natural PC contents and a specific PC/PE ratio of approximately 4:1, quite often in combination with gelatin or other stabilizers [64].

It is interesting, but not yet understandable, that deoiled lecithins—hydrolyzed or slightly PC-enriched—seem to perform better in low fat spreads than oil-containing liquid products. Sometimes the formula does not even require thickeners and stabilizers.

Separation Agents. Lecithin–oil combinations are widely used as pan release agents or separating agents in various food applications. However, lecithin can develop color problems when heated to temperatures of more than 80°C because of Maillard-type reactions between the primary amino group of PE and free carbohydrates contained in vegetable lecithins. Two strategies have been developed to overcome this problem: (1) ethanol fractionation to increase PC and decrease PE, one of the reaction partners for the browning products, or (2) acetylation as another way of reducing the percentage of primary amines. Ethanol-fractionated substances show an additional benefit for the conventional release agents, as these products dissolve clearly in vegetable oils and do not tend to sediment.

Confectionery. Lecithin has become a traditional ingredient in all types of chocolate masses. Its function is to cut down on the requirement for cocoa butter, the most expensive ingredient in chocolate products.

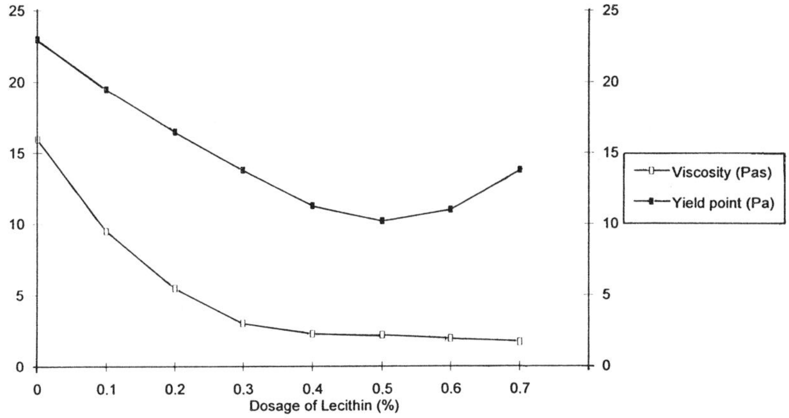

Fig. 4 Chocolate viscosity and yield point. (■) Viscosity (Pa-s); (□) yield point (Pa).

Even though the mode of action is not yet fully understood, lecithin acts because of its surface-active properties at the interface between sugar and cocoa particles and fat. Covering the surface of the hydrophilic ingredients, it reduces internal friction, i.e., it reduces the viscosity and yield point of liquid chocolate masses. The flow characteristics of these masses are dependent on many different factors including fat content, moisture, conching time, tempering, particle size and size distribution, and quality of cocoa butter. Accordingly, the optimal lecithin dosage can vary as much as the type of lecithin, whereas the latter is determined primarily by the type of chocolate.

Modern lecithin technology offers a range of custom products and fractions that are specially designed for chocolate bars, hollow articles, coating, and dietetic chocolate. Common to all types of products and all lecithins is the optimal dosage range of 0.3–0.7% (see Fig. 4). Above a critical quantity (obviously dependent on the grade of dispersity of the mass), viscosity again increases. The best time to add lecithin is shortly (30 min) before the end of conching in order to achieve homogeneous distribution and yet not hinder moisture evaporation and optimal flavor development.

Over recent years a range of phospholipid products (from soybeans) have been developed. These have specific properties and influence either viscosity or yield point or both, as indicated in Table 11.

Baking. Another industrially important yet diversified phospholipid application area is the baking industry [63]. The main uses are in flour treatment (to standardize baking properties, predominantly used in Germany and Belgium) and in bread improvers. Lecithins show their effects predominantly in all kinds of wheat-based bakery goods. They do not function nearly as well in rye products. The reason is again the specific tendency of phospholipids to interact with proteins—in the case of wheat products, with gluten. Van der Waals forces bind phospholipids to the protein network, increasing its elasticity. Gases developed during baking are kept inside the pores and are better enabled to expand the structure (i.e., increase the volume).

For two main reasons enzymatically hydrolyzed phospholipids are again of special interest. (1) They can be inserted in the helix structure of the amylose part of wheat starch, thus stabilizing it against reorganization during storage (retrogradation). The result is an improved shelf life. (2) They have a strong tendency to interact with hydrophobic parts of gluten, making the bread suspension/dispersion more stress-tolerant, i.e., heat-tolerant, contributing

Table 11 Lecithin Types and Properties in Chocolate

Product	Properties	
	Viscosity	Yield point
Standard soybean lecithin	↓	↓
PC-enriched soybean lecithin	↓ ↓ ↓	↑
Combination of soybean lecithin and PGPR [a]	↓	↓ ↓ ↓
Ammonium phosphatides [b]	↓	↓ ↓
Combination of lecithin and ammonium phosphatides	↓ ↓	↓ ↓

[a] PGPR = polyglycerol polyricinoleate.
[b] Ammonium phosphatides are based on partially hydrogenated rapeseed monodiglycerides, which have first been reacted with phosphorus pentoxide and then neutralized with ammonia. The products are of food grade status and registered under E 476 as food additives.

to an even distribution and a stable protein network. The result is greater volume with small, even pores.

Commercial bread improvers can contain various ingredients besides lecithin (emulsifiers like data ester, malt flour, enzymes, ascorbic acid, hydrocolloids, sugar, enzymes, etc.). In recent years there has been a strong tendency to use all-natural bread improvers that do not contain synthetic emulsifiers and even to avoid enzymes because of an ongoing discussion of allergies.

Instant Food Products. Modern food technology and development concentrate on convenience foods, which are easy to use, cheap to transport, and have long shelf life [64]. Instant products, i.e., dried products that have to be reconstituted by the consumer by dispersing or dissolving them in water, milk, etc., are probably the first generation of large-scale convenience foods. Typical instant products include dry milk products (whole, skim, refatted), cocoa and chocolate drinks, protein powders, and soups and sauces.

During reconstitution of the powdered products, two problems can occur: poor wetting because of fat content and/or particle structure and rapid gelling at the powder surface, leading to an impermeable gel layer surrounding the product. Two strategies have been developed to overcome or minimize these problems:

1. Agglomeration. Capillary porous structure forces the liquid to penetrate the particles.
2. Lecithination. Added phospholipid products mask free surface fat.

Phospholipid products can be included during manufacture of the dried product (i.e., during spray drying or blending) or can be applied onto the surface of the final product by spraying either a water dispersion of a deoiled lecithin powder or a low viscosity oily solution.

Ideal phospholipid blends are either of natural soybean origin and composition (in liquid or deoiled form) or moderately PC enriched products. The latter have the advantage of having much less influence on taste because of relatively low dosage levels.

4. *Cosmetics*

Phospholipids are well recognized natural ingredients in cosmetics [65] as they combine a long list of positive properties that are of special attraction to the cosmetic formulator. They are natural nonirritating emulsifiers and emollients that contain essential fatty acids and have pH-balancing, film-forming, and moisture retention properties. These qualities have made vegetable and egg phospholipid products traditional cosmetics ingredients. However, at the same time they suffer some disadvantages. They are yellowish to brown in color and oxidation-sensitive, are poor emulsifiers by themselves, and they have a strong smell. Most of these disadvantages (color, oxidation, smell) can be overcome by hydrogenation [66]. But the use of hydrogenated products is still limited, probably because of their limited emulsification power and high prices.

The discovery of their membrane-forming properties and the development of simple and cheap liposome manufacturing methods have led to an almost revolutionary use of phospholipids in cosmetics.

Liposomes. Liposomes are tiny globular vesicles. An aqueous cave is entrapped inside a structure of one to several double layers of phospholipids. In so-called multilamellar liposomes each double layer is separated from the other by an aqueous compartment.

In cosmetics, liposomes can be used for moisturizing as they carry a lot of water inside the onionlike structure (up to 15 mL of water per gram of phospholipid!). But they can also be considered a microencapsulation system for both water-soluble substances (in the aqueous compartments) and lipophilic ingredients (inside the lipophilic domain of the membranes). Liposomes not only protect sensitive ingredients, they also provide a long-lasting effect on the skin (multilamellar structures) and their solubilizing effect contributes to solving formulation problems.

Historically, the liposome manufacturing processes have been very complicated, requiring expensive equipment and sometimes toxic solvents, but a recently introduced method [67] has dramatically simplified even large-scale production. The so-called pro-liposome method also offers many formulation alternatives [68,69].

THE PRO-LIPOSOME PRINCIPLE. The pro-liposome method is based on the idea that addition of water to a pro-liposome mixture leads to the spontaneous formation of liposomes. A pro-liposome mixture consists of a carefully selected mixture of phospholipids solubilized in a hydrophilic medium (usually but not necessarily an ethanol–water mixture). The phospholipids are in stacked bilayer form. The addition of water under moderate mechanical agitation then triggers the formation of closed vesicles (liposomes). Active loading of water-soluble ingredients can be done by the addition of a concentrated aqueous solution to the pro-liposome mixture prior to final dilution. Lipophilic substances are ideally admixed with the pro-liposome preparation in ethanolic solution. For details of the process see Ref. 56.

5. *Pharmaceuticals*

Phospholipids are useful tools for the formulation of pharmaceutical products. Unlike food and feed applications, for pharmaceuticals there is a preference for egg-derived phospholipids. To my mind the only reason for this tendency is that egg phospholipids have been used in intravenous emulsions as safe ingredients since the 1960s. It is definitely not because of superior technological functionality.

Finished product categories where phospholipids have found application include total parenteral nutrition (TPN) (emulsifier), fat emulsions as drug carriers (emulsifier), aerosols (suspending aid), dermatological creams (emulsifier), liposomes (altered bioavailability), mixed micelles (solubilization), suppositories (for improved absorption), and antibiotics (for improved dispersibility). As can be seen from this list, phospholipids are efficiently used to solve formulation problems with lipophilic ingredients (oils, lipophilic drugs). As most newly developed drug substances are lipophilic, the perfectly biocompatible and nontoxic natural phospholipids (and blends) have a bright future as formulation tools.

Typical products used for pharmaceutical application are

Fractionated, sometimes chromatographically purified, deoiled soybean phospholipids with PC contents between 50 and 95%
Purified, deoiled, sometimes chromatographically purified, egg phospholipids with PC contents between 70 and 95%
Hydrogenated egg or soy products
Enzymatically hydrolyzed products (lysophopholipid-enriched)
Enzymatically converted products (e.g., phosphatidylglycerol, PG)

6. *Phospholipids as Antioxidants*

Whether phospholipids exhibit a direct primary antioxidative effect is not completely clear yet. There are published reports available on investigations in which purified phospholipids have been used. Whenever liquid soy lecithins or soy lecithin fractions have been tested, the effect observed can probably be explained by the natural content of tocopherols (~1000 ppm) in crude products. The stabilizing effect almost completely disappeared when deoiled lecithins were studied (after removal of neutral lipids including tocopherols with acetone) [70].

However, the synergistic effect of lecithins and phospholipids, especially in combination with ascorbyl palmitate and tocopherol, is unquestionable [71]. Speculation about the mechanism of antioxidative effects leads toward the metal-binding characteristics or synergistic effect of phosphatidylethanolamine [72] or even toward the prevention of oxidation because of the formation of Maillard-type products [73]. Various commercial products are offered on the market.

Care has to be taken that phospholipid-containing antioxidant cocktails are only used up to a maximum of 80°C. Above that temperature, the natural phospholipids start to degrade.

B. Physiology of Phospholipids

Soon after the isolation of phospholipids from animal and vegetable sources and the identification of similar molecules in human tissue, the pharmaceutical industry started to offer phospholipid preparations with various therapeutic indications. Since the 1950s, scientific investigators have discovered both the biological role of phospholipids and their therapeutic potential. Phospholipids are highly specialized lipids that are key constituents of cell membranes and are essential for the growth, maturation, maintenance, and functional capacity of all cells of the animal and human body. In addition to their roles in cell membranes, phospholipids are essential elements of the lipoproteins in blood circulation and of the surfactants that wet internal body surfaces, and they are used in the bile as emulsifiers.

The outer cell membrane can be considered the cell's gatekeeper, whereas the internal membranes are the metabolic work centers. Phospholipids create the functional microenvironment for the membrane proteins.

The crucial role of phospholipid-containing structures in human health has stimulated many research activities and clinical investigations.

Before the therapeutic ideas are described, a few words should be given to the metabolism of phospholipids after oral intake.

The average intake of phospholipids with normal food is 1.7–2.2 g/day [74]. These phospholipids, like all orally administered phospholipids, are hydrolyzed by pancreatic phospholipase A_2 after micellarization with bile fluids. The resulting 1-acyl-lysophospholipids can be taken up by the mucosa enterocytes.

Inside the mucosal cell, reesterification to intact phospholipids occurs using the pool of fatty acids present in the cells. Finally, phospholipids are released into the chyle, where they are transported as part of the chylomicron organization [75–77].

This brief outline of absorption seems to be true for about 50–60% of the orally taken phospholipids, the rest are split into smaller molecules, which are metabolized in the usual way.

Nutritional aspects of phospholipid supplementation in humans cover the following main areas: lipid (especially cholesterol) metabolism, liver therapy, neurological disorders, and memory improvement. In this overview only the major applications can be described, as there is a vast literature on the potential therapeutic effects of phospholipids. The purpose of this section is not to describe the biochemical background of their therapeutic activity but to focus on a description of the effect of industrial phospholipid products.

1. *Lipid Metabolism*

As coronary heart disease is the major cause of death in most societies, interest in using a natural supplement like soy phospholipids (most clinical studies have been carried out with soy phospholipid preparations) for preventive and probably even therapeutic reasons is high. Purified soy PC and even mixed soy phospholipids have a clear effect on lowering elevated serum lipid levels. An overview reporting on 27 clinical trials [78] shows that there is indeed such an effect. Two of the most striking sets of study results are summarized in Tables 12 and 13.

The most dramatic reduction in both cholesterol and triglycerides was seen when a fine (liposomal) aqueous dispersion of mixed soy phospholipids was applied (9 g/day). Whether dispersity may have had an influence on the effect has not yet been investigated. But it is obvious that results after administration of bulk phospholipids (e.g., in granular form) are not as good. However, the evaluation of trial results is very complicated, as study designs have been too different to allow direct comparison.

The HDL/LDL balancing effect of soybean phospholipid preparations is obvious, in contrast with vegetable triglycerides (soybean oil), which reduce both HDL and LDL cholesterol to a moderate extent. Phospholipids not only lower total cholesterol but at the same time raise HDL (the lipoprotein that washes out cholesterol) and lower LDL (the one that delivers it to the cells), thus having a twofold positive effect [81].

Finally the existing clinical data have led to the publication of a monograph description by the German health authorities [82]. This monograph grants the natural mixture of soybean phospholipids drug status for the indication of moderate disorders of lipid metabolism. The therapeutic dose is based on 3.5 g PC per day.

2. *Liver Therapy*

Various experimental and clinical results support the assumption that soybean phosphatidylcholine has protective and even curative and regenerative effects on liver cell membranes. In various animal models the hepatoprotective effect of PC against various toxic substances (ethanol, chlorinated hydrocarbons, drugs, etc.) has been impressively demonstrated. Also, more than 100 clinical studies on animals and humans with some type of liver disease have been conducted with PC. The main diseases treated were cirrhosis, hepatitis (acute and chronic), hepatic coma, fatty liver, and toxic liver damage from xenobiotics and from mushroom poisoning. Treatment lasted for weeks to months, and dosages ranged up to 3 g PC/day (orally).

Table 12 Cholesterol-Lowering Effect of a Lecithin Solution[a]

	Total cholesterol (mg/dL)		Triglycerides (mg/dL)		HDL (mg/dL)		LDL (mg/dL)	
	Before	After 4 wk	Before	After 4 wk	Before	After 4 wk	Before	After 4 wk
Lecithin	307 ± 32	211 ± 39	223 ± 39	156 ± 27	41 ± 6	54 ± 9	204 ± 31	162 ± 28
Placebo	322 ± 39	290 ± 35	201 ± 42	214 ± 34	38 ± 6	43 ± 7	192 ± 39	186 ± 32

[a]In a controlled study, 240 patients were given either 5.4 g deoiled lecithin per day in a 9% solution containing 16% ethanol or a placebo.
Source: Ref. 79.

Besides excellent tolerance, clinical improvements included [83–85] improvement of specific symptoms, better biochemical and histological outcomes, accelerated improvement of clinical findings, and shortened duration of hospitalization. These impressive findings pushed the German health authorities to publish another monograph on phospholipids [86]. Purified soybean PC (73–79%) has been given drug status as a liver therapeutic substance (dosage 1.5 g/day).

3. *Neurology*

The assumption that phosphatidylcholine may have a positive effect on degenerative neurological disorders goes back to the discovery that choline is able to pass the blood/brain barrier and to stimulate acetylcholine synthesis [87].

Administration of choline salts raised blood choline levels only for short times and led to a fishy body odor because of amine formation from choline. PC administration, however, efficiently raised choline levels for a much longer period without having the mentioned side effect [88]. Since then intensive clinical activities have begun to investigate PC effects.

The two major known neurodegenerative diseases, Alzheimer's and Parkinson's, feature significant losses of the neurons that produce acetylcholine. Dietary PC is a convenient precursor to make choline available for enhanced acetylcholine synthesis. But the results with both Alzheimer's and Parkinson's disease patients are not very convincing. For patients with fully developed symptoms, even extremely high doses did not lead to significant improvements of the symptoms [89,90].

Slightly better results could be achieved in CNS-related movement disorders like tardive dyskinesia, Friedreich's ataxia, Tourette's disease and Huntington's disease. Improvements were observed in the reduction of abnormal movements and in speaking ability [91,92].

Table 13 Lipid-Lowering Effect of Phosphatidylcholine[a]

	Total cholesterol (mg/dL)		Triglycerides (mg/dL)		HDL (mg/dL)		LDL (mg/dL)	
	Before	After 6 wk	Before	After 6 wk	Before	After 6 wk	Before	After 6 wk
PC	309	251	354	249	32	44	195	130
Placebo	285	310	327	415	38	41	168	182

[a]A group of 28 patients were given 1.8g PC per day in capsules or a placebo.
Source: Ref. 80.

In summary, phosphatidylcholine preparations for single therapeutic action are not sufficient but can assist ethical medication.

4. *Memory*

Moderate to no improvement of cognitive abilities is reported from the use of conventional types and forms of vegetable phospholipids. The situation after administration of bovine brain derived phosphatidylserine is different. Significant improvements could be observed, especially in all kinds of age-related cognitive decline. So far 15 clinical studies have been done, and all of them demonstrate impressively that PS seems to be a very effective nootropic substance. The mechanism for the PS activity is not fully understood yet. A good survey is given in Ref. 93. The therapeutic activity seems not to be linked to the origin of the PS. (Different origins would mean different fatty acid compositions.)

Since animal brain is not a very attractive raw material for therapeutic and dietary substances like PS, a "vegetable" PS has been produced by enzymatic transphosphatidylation:

$$\text{PC, PE, ...} + \text{Serine} \xrightarrow{\text{PL D}} \text{PS} + \text{Choline, Ethanolamine, ...}$$

This product has been introduced into the market in Israel and the United States. A clinical study showed that there seems to be no difference in therapeutic efficacy compared to the animal-derived product [94].

As mentioned earlier, there are thousands of articles describing other effects of phospholipids on human health (especially worthy of mention is the work of Zeisel and coworkers [95–98]). These findings have not yet led to major commercial and industrial applications, but it is certain that more products with novel therapeutic concepts will soon appear on the market.

5. *Phospholipids in Animal Nutrition*

Besides their use as emulsifiers in various feed preparations (see Section VI.A.2), phospholipid products have gained a substantial market as nutritional ingredients in feed formulations. Here they serve two functions: improvement of absorption of lipophilic nutrients (fat, vitamins) and supply of nutritional substances (essential fatty acids, choline, inositol, etc.).

For terrestrial animals the main function of phospholipids is the improvement of fat digestibility. The biosurfactant property of phospholipids enhances emulsification of dietary lipids in the small intestine. Since in young animals the digestive system is still immature, the quantity of secreted bile is most probably insufficient to emulsify the dietary fat, particularly if the levels of fat are high. Feed supplemented with lecithin therefore improves the emulsification of dietary lipids and contributes to better absorption of the fatty acids and to better utilization of added fat as the most efficient energy source.

Phosphatidylcholine is very involved in the metabolic transport of fatty acids, which can be transported from the liver only in the form of triglycerides within the lipoproteins. An essential constituent of those lipoproteins is phosphatidylcholine. When adequate supplies of phosphatidylcholine are not available, the liver is unable to export triglycerides and becomes infiltrated with fat (fatty liver syndrome). There is an absolute dependence on PC; no other phospholipid will suffice. As a result, fat digestibility is improved (equal to an increase in feed conversion) and fewer problems with diarrhea are observed in farm animals.

In the following sections the nutritional benefits of lecithin and phospholipid for growing terrestrial animals, fish, and crustaceans are described in more detail.

Table 14 Effect of Lecithin on the Development of Early Weaned Pigs[a]

	Control group	Group I	Group II	Group III
Deoiled lecithin (%)	—	1.2	—	1.2
Soybean oil (%)	—	—	6	6
Daily gain (g)	372	391	381	390
Feed conversion	1.66	1.67	1.78	1.75

[a] 3–8 weeks of age, test period 5 weeks.
Source: Ref. 99.

Piglets. A good quality starter feed for piglets has to meet three criteria: It must meet all of the nutrient requirements of the young pig, it must be palatable, and it must be easy to digest. Trials have demonstrated that piglet diets containing additional phospholipids have a positive effect on weight gain, feed efficiency, and good health (immunology). Table 14 presents the results of one study. Figure 5 shows the improvement in fat digestibility.

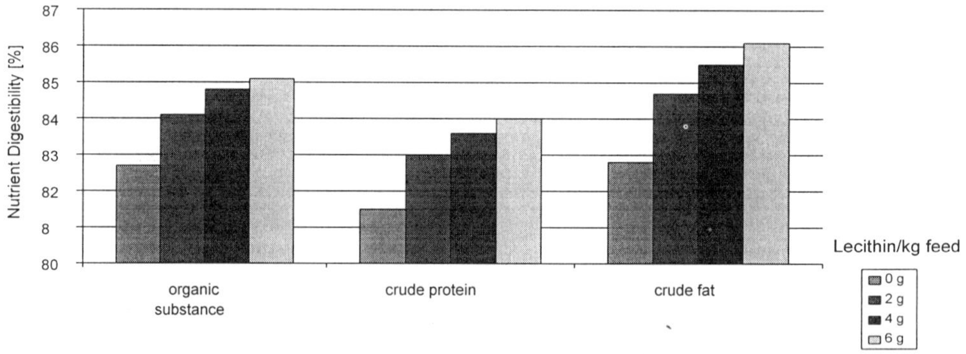

Fig. 5 Improvement in digestibility of fat and other feed nutrients in dependence on different dosages of deoiled lecithin (From Ref. 100.)

Crustaceans. It is known that phospholipids (especially PC + PI) are needed not only to promote the health of crustaceans but also for best performance [101,102]. Unlike terrestrial animals, crustaceans are incapable of synthesizing phospholipids. Crustacean diets, therefore, have to be fortified with phospholipids. The fortification of feed with phospholipids contributes to the inositol and choline needs of the animals due to the presence of PI and PC in the phospholipids. The requirements of *Penaeus japonicus* larvae for cholesterol and soybean phospholipids were studied [103]. The results obtained indicated that the survival rates of larvae increased as levels of soybean PC rose from 0 to 3.0% and those of cholesterol rose from 0 to 1.0% of the diets. However, the supplementation of higher levels of PC (6.0%) and cholesterol (5.0%) did not result in further improvement of survival rates. (See Table 15.)

In another trial, the addition of 3.0% deoiled lecithin to the diet of prawns significantly improved the weight gain and feed conversion rate (Fig. 6). The whole body of the prawns receiving a lecithin-deficient diet contained a lower concentration of phospholipids than the body of those fed lecithin-supplemented diets.

Fish. A critical situation is the nutritional requirement of larvae at the time of transition from endogenous to exogenous feeding. The fish egg contains levels of nutrients that must

Table 15 Survival Rate and Growth After Feeding Increasing Levels of Soybean PC and Cholesterol to *Penaeus japonicus* Larvae for 7 Days

Cholesterol in the diet (%)	Soybean PC in the diet [%] 0	1.0	3.0	6.0
	Survival rate (%)			
0	13	25	32	26
0.5	21	32	64	59
1.0	25	69	72	80
5.0	35	42	55	68
	Growth index			
0	4.0	5.0	5.5	5.5
0.5	4.5	6.7	7.0	6.9
1.0	5.0	7.0	7.0	6.9
5.0	5.0	6.2	6.5	6.3

meet the demands of both energy and growth of the embryo and the resulting fry. Throughout the period of embryogenesis, PC is the only lipid that declines in absolute terms [106].

From the results of this trial, it was concluded that PC has a range of biochemical and physiological functions during embryogenesis in many fish species. PC may also have a role in the transport of yolk components to the developing embryo. In addition, PC is undoubtedly an important source of long-chain (*n*-3) PUFAs for developing egg and larvae. Further nutritional attributes of PC include the provision of phosphorus and choline. The presence of PC in the diet of larval fish may also aid in the emulsification and absorption of dietary neutral lipid.

Many research workers have demonstrated the need for dietary phospholipids in juvenile salmonids [107,108]. But there is evidence that phospholipids are effective in larger salmonids also. In a field trial, growing rainbow trout were fed a commercial feed supplemented with 1.6% deoiled lecithin but in the absence of choline. The fish had an initial mean weight of

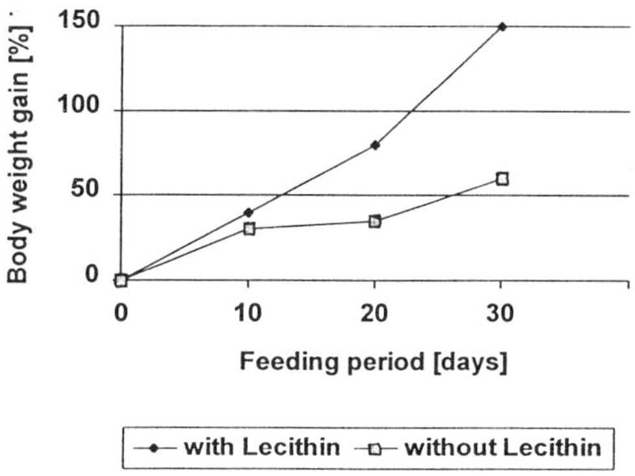

Fig. 6 Weight gain (%) of prawns fed diets with and without lecithin (From Ref. 105.)

Table 16 Effect of Deoiled Lecithin on the Performance of Growing Rainbow Trout (*Salmo gardneri*)

	Control	Trial	Change against control (%)
Deoiled lecithin (%)	—	1.4	
Choline chloride, 50%	0.5	—	
Number of trout	100	100	
Trial period (days)	99	99	
Initial weight (g)	270	262	−3.0
Final weight (g)	504	528	+4.8
Weight gain (g)	234	266	+13.7
Daily weight gain (g)	2.39	2.70	+13.0
Feed conversion	5.18	4.87	−6.0
Losses (%)	8.7	1.2	−86.2

Source: Ref. 109.

325 g. In the 61 day trial period, the lecithin-fed trout had 14.6% higher weight gain than that of the control group, which received a feed containing 0.5% choline chloride [109]. Results are summarized in Table 16.

To conclude this section about phospholipids in animal nutrition, it has to be mentioned that besides their biochemical and physiological properties, phospholipids also exhibit technological functions that are of industrial importance in the manufacture of feed products. These functions enable them to serve as emulsifiers, wetting and dispersing agents, extrusion aids, release agents, and antioxidants.

REFERENCES

1. IUPAC-IUB Committee on Biochemical Nomenclature Recommendation, *Lipids 12*: 455 (1977).
2. E 322 - Lecithine, *Amtsblatt Eur. Gem. L223*: 30 (1978).
3. Lecithin, *Fed. Reg.* Vol. 49, §184.1400 (1984).
4. O. W. Thiele (Ed.), *Lipide, Isoprenoide mit Steroiden* Georg Thieme, Stuttgart, 1979.
5. E. Schulze and E. Steiger, Über den Lecithingehalt der Pflanzensamen, *Z. Physiol. Chem. 13*: 365 (1889).
6. E. Schulze, Landwirtsch. Versuchsstationen, *Landwirtschaft 49*: 203 (1898).
7. E. Schulze, Über pflanzliche Phosphatide, *Chem. Z. 32*: 981 (1908).
8. D. A. Gornall and A. Kuksis, Molecular species of glycero phosphatides and triglycerides of egg yolk lipoproteins, *Can. J. Biochem. 49*: 51 (1971).
9. F. Phillips and O. S. Privett, A simplified procedure for the quantitative extraction of lipids from brain tissue, *Lipids 14*: 590 (1979).
10. O. S. Privett et al., in *Dairy Lipids and Lipid Metabolism* (H. F. Brink and D. Kritchevsky, Eds.) Avi, Westport, CT, 1968, p. 99.
11. Lucas Meyer, Hamburg, Germany, data based on ^{31}P-NMR.
12. L. A. Appelqvist and R. Ohlson (Eds.), *Rapeseed: Cultivation, Composition, Processing and Utilization,* Elsevier, Amsterdam, 1972.
13. S. S. Grewal, P. S. Sukhija, and L. S. Bhatia, Polar lipid composition during sunflower *Helianthus annuus* seed development, *Biochem. Physiol. Pflanzen 173*: 11 (1978).
14. E. J. Weber, Compositions of commercial corn and soybean lecithins, *J. Am. Oil Chem. Soc. 58*: 898 (1981).

15. B. F. Szuhaj and G. R. List (Eds.), *Lecithins*, American Oil Chemist's Society, Champaign, 1985.
16. F. J. Flider, The manufacture of soybean Lecithins, in *Lecithins* (B. F. Szuhaj and G. R. List, Eds.), American Oil Chemist's Society, Champaign, 1985, p. 21.
17. G. Rouser, A. J. Bauman, N. Nicolaides, and D. Heller, Paper chromatography of lipids: methods, applications and interpretations, *J. Am. Oil Chem. Soc. 38*: 565 (1961).
18. T. Majonnier and H. C. Troy (Eds.), *The Technical Control of Dairy Products*, Majonnier Bros. Co., Chicago, 1922.
19. J. C. Crane and B. E. Horrall, Phospholipids in dairy products. II. Determination of phospholipids and lecithin in lipids extracted from dairy products, *J. Dairy Sci. 26*: 935 (1943).
20. P. Walstra and J. J. De Graaf, Note on the determination of the phospholipid content of milk products, *Neth. Milk Dairy J. 16*: 283 (1962).
21. A. H. Duthie and S. Patton, New method for extraction of milk phospholipids, *J. Dairy Sci. 48*: 170 (1965).
22. J. Cerbulis, Distribution of lipids in various fractions of cow milk, *J. Agric. Food Chem. 15*: 784 (1967).
23. J. Folch, I. Ascoli, M. Lees, J. A. Meath, and F. N. Le Baron, Preparation of lipid extract from brain tissue, *J. Biol. Chem. 191*: 833 (1951).
24. J. Folch, M. Lees, and G. H. Stanley, A simple method for the isolation and purification of total lipids from animal tissues, *J. Biol. Chem. 226*: 497 (1957).
25. K. Zosel, Verfahren zur Entcoffeinierung von Rohkaffee, DP 2005293 (1970).
26. Kali-Chemi Pharma, Verfahren zur Gewinnung von Lecithin, DP 3011185 (1980).
27. Lucas Meyer, Verfahren zur Gewinnung von Lecithin, DP 3229041 (1990).
28. E. Stahl, Extraction of natural substances by dense gases, *Fette, Seifen, Anstrichm. 84*: 444 (1982).
29. E. Stahl and K.-W. Quirin, Deoiling of crude lecithin by jet extraction with dense carbon dioxide, *Fette, Seifen, Anstrichm. 87*: 219 (1985).
30. Lucas Meyer, Process for the preparation of lecithin, by preference the continuous preparation of pure lecithin, Eur. Patent 0156374 (1985).
31. S. Peter, Process for removing oil from raw lecithin, PCT 94/01004 (1994).
32. SKW Trostberg, Verfahren zur Herstellung von pulverförmigen Eiprodukten mit einem reduzierten Gehalt an Fetten und Cholesterinderivaten, DP 4236474 (1992).
33. M. Schneider, Fractionation and purification of lecithin, in *Lecithins, Sources, Manufacture & Uses* (B. F. Szuhaj, Ed.), Americal Oil Chemists Society, 1989, p. 109.
34. A. Kuksis and L. Marai, Determination of the complete structure of natural lecithins, *Lipids 2*: 217 (1967).
35. Unilever, Use of self-releasing binders on an isocyanate-/sulfonic acid basis for producing shaped articles, DP 1900859 (1969).
36. Lucas Meyer, Verfahren zur Herstellung eines hydrolisierten Lecithins, Eur. Patent Appl. 0260573 (1987).
37. M. Heller, N. Mozes, and I. Peri, Interactions of phospholipase D with 1,2 diacyl-*sn*-glycero-3-phosphocholine and dodecylsulfate, *Lipids 11*: 604 (1976).
38. Kao Corporation, Method for suppressing bitter taste of oral preparation, Eur. Patent Appl. 0631787 (1994).
39. C. S. Chen and C. J. Sih, General aspects and optimization of enantioselective biocatalysis in organic solvents by the use of lipases, *Angew. Chem. Int. Ed. Eng. 28*: 695 (1989).
40. General Mills, Journal box structure, U.S. Patent 2,445,948 (1944).
41. Central Soya, Method of preparing heat resistant lecithin release agent, U.S. Patent 4,479,977 (1984).
42. Central Soya Co., Process of phosphatide preparation, U.S. Patent 3,301,881 (1963).
43. Armour & Co, Verfahren zur Herstellung von stabilisierten Emulgatoren aus Soyabohnenphosphatid, DP 1002607 (1953).
44. American Lecithin Co., Hydrogenation of phosphatides, U.S. Patent 2870179 (1954).
45. D. V. Sokolski and T. Omaraliev, Hydrogenation of the lecithin fraction of soybean oil phosphatides, *Chem. Abstr. 68*: 96992 (1968).

46. G. Cevc (Ed.), *Phospholipid Handbook*, Marcel Dekker, New York, 1993.
47. G. B. Ansell (Ed.), *Form and Function of Phospholipids*, Elsevier, Amsterdam, 1973.
48. D. M. Small (Ed.), *Handbook of Lipid Research*, Plenum, New York, 1980.
49. G. Cevc and F. Paltauf (Eds.), *Phospholipids: Characterisation, Metabolism and Novel Biological Applications*, American Oil Chemist's Society, 1995.
50. B. F. Szuhaj (Ed.), *Lecithins, Sources, Manufacture & Uses*, 1989.
51. H. Pardun, *Die Pflanzenlecithine*, Verlag Chem. Ind., 1988.
52. Victor Co. of Japan Ltd., Magnetisches Aufzeichnungsmaterial, DP 2910487 (1978).
53. International Business Machines Corp., Magnetic inks containing lecithin as a surfactant, French Patent 1548700 (1966).
54. A. Nattermann, Aqueous phospholipid solutions containing a solubilizing agent, U.S. Patent 4,874,553 (1987).
55. Newman Agrochemicals, Agrochemical formulation, PCT 89/08628 (1988).
56. Phares Pharmaceutical, Liposome forming composition, Eur. Patent 0158441 (1985).
57. Kabi Pharmacia, Lipid formulation system, PCT 92/05771 (1992).
58. Nova Pharmaceuticals, Lipospheres for controlled delivery of substances, PCT 91/07171 (1991).
59. Cognis GmbH, Verbesserte Nährstoffgemische für die Bioremediation verschmutzter Böden und Gewässer, DP 4218243 (1992).
60. L. Rydhag and I. Wilton, The function of phospholipids of soybean lecithin in emulsions, *J. Am. Oil Chem. Soc. 58*: 830 (1981).
61. L. Rydhag, Importance of the phase-behavior of phospholipids for emulsion stability, *Fette, Seifen, Anstrichm 81*: 168 (1979).
62. Nestlé, Zusammensetzung für Wärmestabilisierung von Proteinen und Produkten daraus, Eur. Patent 0628256 (1993).
63. Unilever N. V., Hitzesterilisierbare Wasser- und Öl-Emulsionen, Eur. Patent 0328789 (1988).
64. Unilever N. V, Antispritzmittel für Wasser enthaltende Speisefette, DP 1492952 (1963).
65. Y. Pomeranz, Lecithin in baking, in *Lecithins* (B. F. Szuhaj and G. R. List, Eds.), American Oil Chemist's Society, Champaign, 1985, p. 289.
66. *Spray Tower Philosophy*, Lucas Meyer Publ. 5, 1988.
67. *Lecithins in Cosmetics*, Lucas Meyer Publ. 9, 1991.
68. *Liposomes—The Pro-Liposome Approach*, Lucas Meyer Publ. 14, 1993.
69. S. Perrett, M. Golding, and W. P. Williams, A simple method for the preparation of liposomes for pharmaceutical applications: characterization of the liposomes, *J. Pharm. Pharmacol. 43*: 154 (1991).
70. G. Mieth and F. Linow, Zur fettstabilisierenden Wirkung von Phosphatiden; 2. (Mitt.) Antioxidative Wirkung ausgewählter Phosphatide, *Nahrung 19*: 577 (1975).
71. G. Pongracz, Antioxidant mixtures for use in food, *Int. J. Vitamin Nutr. Res. 43*: 517 (1973).
72. D. H. Hildebrand, J. Terao, and M. Kito, Phospholipids plus tocopherols increase soybean oil stability, *J. Am. Oil Chem. Soc. 61*: 552 (1984).
73. C. K. Park and D. H. Kim, Relationship between fluorescence and antioxidant activity of ethanol extracts of a Maillard browning mixture, *J. Am. Oil Chem. Soc. 60*: 98 (1983).
74. B. Borgström, On the binding of bile salt to pancreatic lipase, in *Lipid Absorption: Biochemical and Clinical Aspects* (K. Rommel et al., Eds.), MTP Press, 1976.
75. A. Nilsson and B. Borgström, Absorption and metabolism of lecithin and lysolecithin by intestinal slices, *Biochem. Biophys. Acta 137*: 240 (1967).
76. R. O. Scow, Y. Stein, and O. Stein, Incorporation of dietary lecithin and lysolecithin into lymph chylomicrons in the rat, *J. Biol. Chem. 242*: 4919 (1967).
77. D. Lekim and H. Betzing, Intestinal absorption of polyunsaturated phosphatidylcholine in the rat, *Z. Physiol. Chem. 357*: 1321 (1976).
78. J. T. Knuiman, A. C. Beynan, and M. B. Katan, Lecithin intake and serum-cholesterol, *J. Am. Clin. Nutr. 49*: 266 (1989).
79. D. Kupke, U. Schmidt, and H. J. Reimann, The influence of lecithin on hyperlipemia—a report on prevention and treatment of hyperlipoproteinemia, *Med. Welt 38*: 1244 (1987).

80. A. K., Horsch, I. Majolk, C. C. Heuck, and E. Göpfert, Einfluss von Polyenylphosphatidylcholin (PPC) auf die Serumlipide bei Patienten mit Hyperlipoproteinaemie, *VASA 15*: 251 (1986).
81. J. N. Hawthorne, M. Hoccom, and J. E. O'Mullane, The nutritional value of linoleic acid given either as corn oil or as soya lecithin, *Clin. Dietol. 8*: 253 (1981).
82. BGA Monography, Lecithinum ex Soja (Sojalecithin), *Bundesanzeiger*, May 5, 1988, p. 2018.
83. K. J. Gundermann, Biological activity of polyunsaturated phosphatidylcholine in different diseases, in *Phospholipids: Characterisation, Metabolism and Novel Biological Applications* (G. Cevc and F. Paltauf, eds.), American Oil Chemist's Society, Champaign, 1995, p. 208.
84. C. S. Lieber, L. M. DeCarli, K. M. Mak, C. I. Kim, and M. A. Leo, Attenuation of alcohol-induced hepatic fibrosis by polyunsaturated lecithin, *Hepatology 12*: 1390 (1990).
85. C. S. Lieber, A. J. Robins, J. Li, L. M. Decarli, K. M. Mak, M. Ki, J. M. Fasulo, and M. A. Leo, Phosphatidylcholine protects against fibrosis and cirrhosis in the baboon, *Gastroenterology 106*: 152 (1994).
86. BGA Monography, Phospholipide aus Sojabohnen mit 73–79% 3-*sn*-Phosphatidylcholin, *Bundesanzeiger*, July 19, 1994, p. 7361.
87. E. L. Cohen and R. J. Wurtman, Brain acetylcholine: control by dietary choline, *Science 191*: 561 (1976).
88. R. J. Wurtman, M. J. Hirsch, and J. H. Growdon, Lecithin consumption raises serum-free-choline levels, *Lancet 2*: 68 (1977).
89. P. Etienne, S. Gauthier, D. Dastoor, B. Collier, and J. Rainer, Alzheimer's disease: clinical effect of lecithin treatment, in *Nutrition and the Brain* (A. Barbeau et al., Eds.), Raven Press, New York, 1979, p. 389.
90. P. Etienne, D. Dastoor, A. Gauthier, R. Ludwick, and B. Collier, Lecithin in the treatment of Alzheimer's disease, *Aging 19*: 369 (1982).
91. J. H. Growdon, A. J. Gelenber, J. Doller, M. J. Hirsch, and R. H. Wurtmann, Lecithin can suppress tardive dyskinesia, *N. Engl. J. Med. 298*: 1029 (1978).
92. A. Barbeau, Phosphatidylcholine lecithin in neurologic disorders, *Neurology 28*: 358 (1978).
93. P. Kidd, *Phosphatidylserine*, Lucas Meyer, Hamburg, Germany, 1995.
94. J. Gindin, M. Novikov, D. Kedar, A. Walter-Gunzburg, A. Nacer, O. Karta, E. Zurand, and S. Levi, Kaplan Hospital, Rehovot, Israel, unpublished results.
95. S. H. Zeisel and D. J. Canty, Choline phospholipids molecular mechanisms for human diseases, *J. Nutr. Biochem. 4*: 258 (1993).
96. S. H. Zeisel, K. A. da Costa, P. D. Franklin, E. A. Alexander, J. T. Lamont, N. F. Sheard, and A. Beiser, Choline, an essential nutrient for humans, *FASEB J. 5*: 2093 (1991).
97. S. H. Zeisel, Lecithin in health and disease, in *Lecithins* (B. F. Szuhaj and G. R. List, Eds.), American Oil Chemist's Society, Champaign, 1985, p. 323.
98. S. H. Zeisel, Lecithin in health and human nutrition, in *Lecithins, Sources, Manufacture & Uses* (B. F. Szuhaj, Ed.), American Oil Chemist's Society, Champaign, 1989, p. 225.
99. Lucas Meyer, Hamburg, Germany, unpublished results.
100. K. D. Günther, Rein-Lecithin als Wirkstoff in der Ferkelernährung, *Kraftfutter 4/94*: 129 (1994).
101. L. R. D'Abramo, C. L. Bordner, and D. E. Conklin, Relationship between dietary phosphatidylcholine and serum cholesterol in the lobster Humarus sp., *Marine Biol. 67*: 231 (1982).
102. A. Kanazawa, S. Teshima, and M. Sakamoto, Effects of dietary lipids, fatty acids and phospholipids on growth and survival of prawn (Penaeus japonicus) larvae, *Aquaculture 50*: 39 (1985).
103. J. W. Hertrampf, Feeding crustaceans with lecithin, *Feed Mag. 11*: 12 (1989).
104. S. Teshima, A. Kanazawa, H. Sasada, and M. Kawasaki, Requirements of the larval prawn, Penaeus japonicus, for cholesterol and soybean phospholipids, *Mem. Fac. Fish Kagoshima Univ. 31*: 193 (1982).
105. S. Teshima and A. Kanazawa, Role of dietary phospholipids in the transport of [^{14}C]-tripalmitin in prawn, *Bull. Jpn. Soc. Sci. Fish 52*: 519 (1986).
106. A. J. Fraser, C. Gamble, and B. Sargent, Changes in lipid content, lipid class composition and fatty acid composition of developing eggs and unfed larvae of cod (*Gadus morhua*), *Marine Biol. 99*: 307 (1988).

107. H. A. Poston, Effect of body size on growth survival and chemical composition of Atlantic salmon fed soy lecithin and choline, *Prog. Fish Cult. 52*: 218 (1990).
108. J. W. Hertrampf, Zur Phospholipid-Versorgung von Salmoniden, *Mühle Mischfuttertechnik 128*: 463 (1991).
109. H. N. Allwörden, Versuchsbericht des Institutes für Humanernährung und Lebensmittelkunde der Univ. Kiel, Germany, University of Kiel, personal communication, 1990.

4

Lipids and Nutrition

Michael I. Gurr
Maypole Scientific Services, St. Mary's, Isles of Scilly, England

I. INTRODUCTION

During the past decade or more there has been intense interest among nutritionists, health professionals, and consumers in the subject of dietary fat. Virtually all dietary recommendations in industrialized countries have focused primarily on reducing the amount of fat in the diet, and some have emphasized the need to modify its composition. This theme has been a regular feature of national daily newspapers, magazines, and radio and television programs.

There are several reasons. Many medical reports have highlighted the high prevalence of cardiovascular disease (CVD) in industrialized countries, and most have associated this with high consumption of fat, especially saturated fatty acids (SFA). Thus, the most recent such report from the U.K. Department of Health [1] clearly indicates that it considers that the incidence and mortality from CVD in the United Kingdom could be substantially reduced if the British population were to reduce its consumption of fat and SFA. This echoes the views of many similar national and international reports. A major health initiative in England (The Health of the Nation) aims to reduce average intakes of fat by 12% and of saturated fatty acids by 35% by the year 2005 [2].

High fat intakes or the consumption of an inappropriate balance of the different fatty acids have also been suggested as being connected with several other chronic diseases, but, apart from heart disease, most attention had been given to obesity. In the United Kingdom, which is not atypical of many industrialized countries, there is the paradoxical situation that whereas total energy and total fat intakes have been steadily declining, the proportion of the population deemed to be overweight has been rising. Professional views about the contribution of excessive intakes of fat (as distinct from carbohydrates or total energy) have undergone many changes during this century. Currently, the consensus seems to be that dietary fat may make a specific contribution to the development of obesity. The Health of the Nation aims to reduce the number of people between the ages 16 and 64 who are obese by one-fourth in males and at least one-third in females by the year 2005.

The widespread desire to reduce fat in the diet as painlessly as possible has given rise to remarkable growth in the availability of low or reduced fat products and a continuing quest for fat substitutes that will deliver the properties of fat but not its energy content. Nevertheless fat has its important place in the diet and should not be regarded as a dietary constituent that is entirely dispensable.

II. LIPIDS IN THE BODY

Table 1 indicates the main types of lipids in the human body and their functions. Briefly, the properties of biological membranes are strongly influenced by the degree of "fluidity" of the lipid bilayer, which is in turn dependent on the nature of the phospholipid fatty acids, the ratio of cholesterol to phospholipids, and interactions between the phospholipid polar headgroups and membrane proteins [3].

Fluidity can be influenced by dietary lipid intakes but seems to be regulated within certain limits by subtle changes in the proportions of phospholipids, cholesterol, and fatty acids. The degree of fluidity of the membrane is important for the proper functioning of membrane proteins that act as enzymes, transporters of small molecules across the membrane, or as receptors for substances such as hormones, antigens, and nutrients. For this reason, the composition of structural lipids in specific tissues is fairly conservative, in contrast to that of storage lipids, which may vary widely.

The body's energy reserves are mainly in the form of triacylglycerols in adipose tissue. Milk fat triacylglycerols provide a source of energy for the newborn baby. The fatty acid composition of human storage fat is strongly influenced by the fatty acid composition of the dietary fat.

Table 1 Types of Lipids in the Human Body and Their Functions

Structural Lipids. These lipids are present as a bilayer in biological membranes, as a layer on the surface of the skin, and in the lung surfactant lipid. Because of their specialized functions, their composition is conservative. Although fatty acid composition can be influenced by diet, homeostatic mechanisms operate to maintain physical properties within a limited range. The main components are the following.

Phospholipids. Major components. The "fluidity" of membranes is regulated by the degree of unsaturation of the fatty acids and interactions between the polar headgroups and the membrane proteins. Fatty acids may be highly saturated, as in the lung surfactant, or highly unsaturated, as in mitochondrial membranes. The phospholipid fatty acids provide a store of long-chain PUFA for eicosanoid formation.

Glycolipids. Minor components except in neural membranes.

Cholesterol. Major component. Regulates "fluidity" of bilayer. Membrane function is absolutely dependent on cholesterol. Even minor modifications to the structure of the steroid molecule lead to loss of function. Also provides precursor for synthesis of steroid hormones (adrenal gland), bile acids (liver), and vitamin D (skin).

Membranes also have many minor components, such as vitamin E as part of the lipid bilayer.

Storage Lipids. Triacylglycerol fatty acids provide long-term storage of fuel that can be mobilized during food shortage, for growth, and during vigorous exercise. The main storage depot is the adipose tissue. Milk fat provides a store of energy for growth of infants. Storage fatty acid composition is strongly influenced by the fatty acid composition of the diet. Small amounts of other lipids are dissolved in the storage triacylglycerols: cholesterol, cholesteryl ester, fat-soluble vitamins, and adventitious fat-soluble compounds derived from the diet.

Lipids in Transit. Fatty acids transit between adipose stores and structural sites via the blood esterified in lipoprotein lipids or as nonesterified fatty acids carried on blood albumin.

Lipids in both structural and storage pools are continually being replaced as fat stores become used as body fuels and as structural lipids become precursors for a variety of metabolic end products (see Sections IV and V).

III. LIPIDS IN FOOD

A. Sources of Dietary Lipids

Lipids in the human diet derive from the storage and structural lipids of plants and animals. More than 90% of dietary lipids are triacylglycerols originating from the adipose tissue or milk of animals or from plant seed oils, mainly in the form of manufactured products. While much of this is "visible fat" (the fat on a joint of meat, the cream on milk, spreading fats, or cooking oils), there is also much fat in the diet that is "hidden," such as the fat incorporated in baked foods, pastries, and confectionery goods and that present in processed meats, cheeses, and desserts. The phospholipid fatty acids of the structural lipids tend to be forgotten because they are also hidden but, although quantitatively in small amounts, they are nevertheless nutritionally very important. Table 2 indicates current sources of fat in the U.K. diet [4].

B. Composition of Food Lipids

Figure 1 illustrates the major classes of fatty acids provided by different types of lipids used in human foods. Detailed information on fatty acid compositions of individual food lipids can be found in Refs. 3 and 5–7.

Seed oils and animal storage fats provide the bulk of the lipids consumed in foods. However, some foods, although contributing little total fat to the diet, are nevertheless important sources of specific essential fatty acids. Note in particular the contrast between the *n*-6 PUFA-rich lipids of animal structural lipids and the *n*-3 PUFA-rich lipids of plant leaves. Human milk fat (whose composition is not illustrated in this figure) has a composition very different from that of cow's milk, a fact that is of great importance in human nutrition. The reader is referred to Jensen et al. [8] for further details of the composition of human milk fat and to

Table 2 Contribution of Fat to the U.K. Diet by Various Foods

	1981	1983	1985	1987	1993	1994
Fat in grams per person per day						
All dairy products	32	30	27	24	19	18
All meat	28	26	25	25	21	19
Margarine and other fats	26	26	24	26	18	20
Biscuits, cakes, pastries	6	7	6	7	7	7
All other foods	12	12	14	14	19	18
All foods	104	101	96	96	84	82
Fat as percent of dietary energy						
Total fat	42	43	43	42	41	39
Saturated fatty acids	19	19	18	17	16	15
Ratio of polyunsaturated to saturated fatty acids (P/S)	0.25	0.29	0.32	0.37	0.43	0.44
Energy intake						
Kilocalories	2210	2144	2016	2040	1930	1884
Megajoules	9.23	8.96	8.42	8.52	8.10	7.87

Source: Ministry of Agriculture, Fisheries and Food: National Food Survey for 1993, HMSO, London, 1994.

STRUCTURAL FATS	Meat	Leaf
SMCFA	-	-
12-16SFA	++	+
>16SFA	+	-
18MUFA	++	+
>18MUFA	-	-
t-MUFA	-	-
n-6PUFA	++	++
n-3PUFA	-	++++

LAURIC OILS	
SMCFA	+
12-16SFA	++++
>16SFA	-
18MUFA	+
>18MUFA	-
t-MUFA	-
n-6PUFA	-
n-3PUFA	-

HIGH PUFA SEED OILS	Soy	Sun
SMCFA	-	-
12-16SFA	+	+
>16SFA	-	-
18MUFA	++	+++
>18MUFA	-	-
t-MUFA	-	-
n-6PUFA	++++	++++
n-3PUFA	+	-

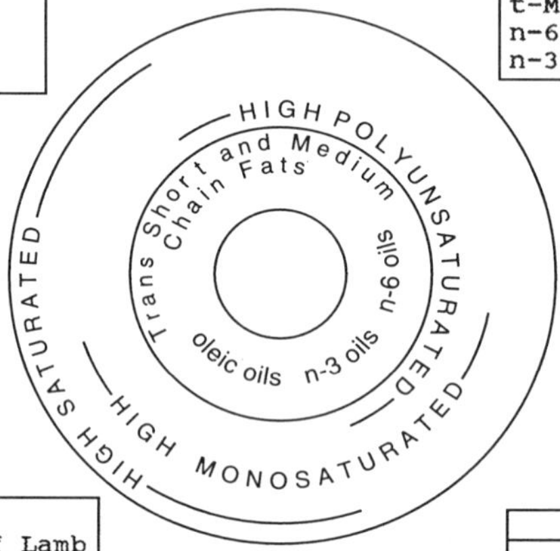

DAIRY	
SCMCFA	+
12-16SFA	+++
>16SFA	+
18MUFA	++
>18MUFA	-
t-MUFA	+
n-6PUFA	-
n-3PUFA	-

FISH OILS	Cod	Herring
SMCFA	-	-
12-16SFA	++	++
>16SFA	-	-
18MUFA	+	+
>18MUFA	++	+++
t-MUFA	-	-
n-6PUFA	-	-
n-3PUFA	+++	-

ANIMAL FATS	Lard	Beef	Lamb
SMCFA	-	-	-
12-16SFA	++	++	++
>16SFA	+	+	+
18MUFA	+++	+++	+++
>18MUFA	-	-	-
t-MUFA	-	+	+
n-6PUFA	+	-	+
n-3PUFA	-	-	-

PALM	
SMCFA	-
12-16SFA	+++
>16SFA	-
18MUFA	+++
>18MUFA	-
t-MUFA	-
n-6PUFA	+
n-3PUFA	-

HIGH MUFA OILS	Rape	Olive
SMCFA	-	-
12-16SFA	+	+
>16SFA	-	-
18MUFA	++++	++++
>18MUFA	-	-
t-MUFA	-	-
n-6PUFA	++	+
n-3PUFA	+	-

Fig. 1 At-a-glance indicator of the principal fatty acid classes in natural fats used in foods. The fats and oils illustrated are natural fats and oils used as ingredients, not manufactured fats. SMCFA, short- and medium-chain fatty acids, carbon chain lengths 4–10; 12–16SFA, saturated fatty acids with chain lengths between 12 and 16 carbon atoms; >16SFA, saturated fatty acids with chain lengths greater than 16 carbons, mainly stearic acid; 18MUFA, monounsaturated fatty acids with chain length of 18 carbons, almost entirely oleic acid; >18MUFA, monounsaturated fatty acids with chain lengths greater than 18 carbons, mainly 20 and 22 carbons; *t*MUFA, *trans*-unsaturated fatty acids, mainly 18-carbon monounsaturated; *n*-6PUFA, polyunsaturated fatty acids of the *n*-6 family, mainly linoleic and arachidonic acids; *n*-3PUFA, polyunsaturated fatty acids of the *n*-3 family, mainly α-linolenic acid in plant fats and eicosapentaenoic and docosahexaenoic acids in fish oils.

Goedhart and Bindels [9] for a detailed discussion of the contribution of milk fat to the nutrition of infants. Important differences in fatty acid composition lie mainly in the content of the long-chain PUFA (20 carbon atoms and above) of both the *n*-3 and *n*-6 families in human milk, which are important in the development of the infant's nervous system. Apart from differences in the fatty acid composition of the two milks, differences in the structure of the triacylglycerols are important to the digestion of milk fat. Human milk triacylglycerols resemble pig storage fat in that a high proportion (70%) of palmitic acid is esterified at the *sn*-2 position of glycerol, while the primary positions contain a high proportion of unsaturated fatty acids. Cow's milk triacylglycerols, however, have a much smaller proportion of 16:0

(40%) at position 2 and relatively large proportions of saturated fatty acids at positions 1 (16:0, 18:0) and 3 (4:0, 6:0) [6].

Animal fats provide on average 400 mg cholesterol per day in the U.K. diet and small amounts of fat-soluble vitamins: retinol (vitamin A), vitamin D, and tocopherol (vitamin E) (Section V.E). Seed oils and plant leaves also supply carotenoids, tocopherols, tocotrienols, and plant sterols, although the latter are not absorbed by the human gut.

IV. METABOLISM OF LIPIDS

The following is a brief account as a background to understanding the nutritional roles of lipids. A detailed account can be found in Refs. 3 and 10.

A. Digestion and Absorption

Lipid digestion in adults takes place mainly in the upper part of the small intestine. Hydrolysis of triacylglycerols catalyzed by the enzyme pancreatic lipase generates fatty acids and monoacylglycerols. These are absorbed by the enterocytes lining the small intestine and reconverted into triacylglycerols in the enterocytes. They are packaged, together with other fat-soluble substances absorbed from the diet, with a stabilizing coat of proteins and phospholipids, as chylomicrons (Fig. 2).

The chylomicrons, irrespective of the fat eaten, consist mainly of triacylglycerols esterified with fatty acids of chain length 14 carbon atoms or longer. Short-chain fatty acids (from milk and dairy products) and medium-chain fatty acids (from milk and coconut oil) are selectively transported into the portal blood as nonesterified fatty acids and are not incorporated into chylomicrons. The short- and medium-chain fatty acids are rapidly metabolized in the liver. Normally, they do not contribute to plasma lipids nor are they deposited in adipose tissue unless consumed in very large quantities [11].

B. Transport

Chylomicrons contain triacylglycerols, cholesterol, cholesteryl esters, phospholipids, and other minor lipids in different proportions as well as specific proteins known as apoproteins, designated apo A, B, C, D, and E. The latter not only help to stabilize the particles but also provide a means by which the different lipoproteins are recognized by receptors on the surface of cells, thus determining their ultimate fate.

Chylomicrons, which are the main carriers of dietary lipids, are cleared rapidly from the bloodstream as a result of hydrolysis by the enzyme lipoprotein lipase (LPL) at the surface of adipose tissue (Fig. 2). After a meal, the secretion of insulin activates adipose tissue LPL, releasing fatty acids from chylomicron lipids for storage in adipose tissue. When stored fatty acids are needed as fuel, insulin levels fall, LPL is "switched off," and a lipase in adipose tissue is activated, releasing fatty acids into the blood. LPL in muscle then becomes active, enabling that tissue to use fatty acids as a fuel.

Chylomicrons are not totally degraded by lipoprotein lipase. Smaller, denser lipoprotein particles called chylomicron remnants remain and are further metabolized in the liver to low density lipoprotein (LDL), which is depleted in triacylglycerols and enriched in cholesterol. The uptake of cholesterol by tissues can occur by nonspecific diffusion across the cell membrane or by interaction with specific LDL receptors on the cell surface. After binding to the receptor, the lipoprotein–receptor complex is taken into the cell, where the lipoprotein is broken down. The resulting free cholesterol interacts with intracellular membranes to inhibit a key enzyme in cholesterol biosynthesis. The intracellular production of cholesterol is thus

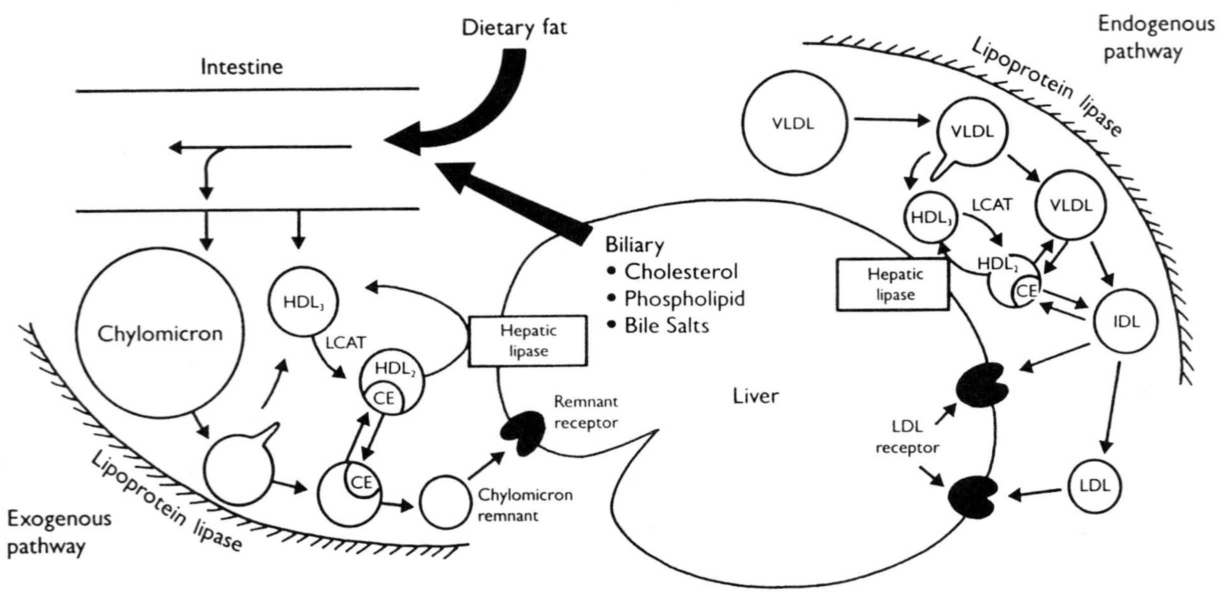

Fig. 2 The metabolism of lipoproteins. The endogenous pathway is concerned with the transport and metabolism of fat coming from the diet. The fats are packaged as chylomicrons, which circulate and are removed from the plasma by the enzyme lipoprotein lipase, mainly in the adipose tissue. The chylomicrons are not entirely consumed by this enzyme but are degraded to smaller particles called remnants, which are removed by the liver. Parts of the chylomicrons also go into making HDL, which, in conjunction with the enzyme LCAT, remove excess cholesterol from membranes and other lipoprotein particles, converting it into cholesteryl esters. This process involves the interconversion of two forms of HDL—HDL_2 and HDL_3—and cholesteryl esters are taken to the liver for further processing. The endogenous pathway is concerned with the transport and metabolism of fats made in the body itself. The products are the VLDL, which are metabolized in a manner analogous to the chylomicrons. Their remnants are usually called intermediate density lipoproteins (IDL) and are further metabolized to LDL, which are taken up by the specific LDL receptor. (From Ref. 86.)

reduced and does not become reactivated until circulating cholesterol concentrations (reflecting, in part, dietary intake of cholesterol) begin to fall. When the concentration of cholesterol in plasma is low, the number of receptors for LDL is low and the rate of endogenous cholesterol synthesis is high. When the concentration of cholesterol in plasma rises, the number of receptors increases, bringing cholesterol into the cell and shutting down the cell's own production. This elegant mechanism enables the body to maintain its supply of cholesterol within well-controlled limits depending on the supply of dietary cholesterol. This is necessary because cholesterol is a vital compound, yet in large concentrations it may have undesirable metabolic effects.

The liver is the main site for making new fat from carbohydrate or recycling old fat molecules, which it exports as very low density lipoprotein (VLDL) into the blood. This circulates and is metabolized in a very similar manner to that described for the chylomicrons.

High density lipoprotein (HDL) plays a vital role as a scavenger for cholesterol. Cholesterol is present in plasma lipoproteins mainly as cholesteryl esters. A blood enzyme converts unesterified cholesterol in HDL into cholesteryl esters. The HDL, which is now depleted of unesterified cholesterol, interacts with membranes, blood vessel walls, or other lipoprotein particles that may have an overabundance of cholesterol, removes the unesterified cholesterol, and transports it to the liver, where it is channeled into other metabolic pathways. The HDL, now depleted of cholesterol, is returned to plasma to continue the scavenging cycle.

C. Biosynthesis

Human beings living in developed countries generally consume diets that contain a high proportion of energy as fat. Under these conditions, the enzymes of fat biosynthesis are "switched off" or working at a low level and the needs for both structural and storage fats are satisfied from dietary intake. However, the phenomenon of turnover—the building up and breaking down of lipids—proceeds continuously in all tissues, and so a low level of enzymatic activity may be present at all times. Current opinion is that fatty acid synthesis from carbohydrate is an unimportant process in human beings, at least in adipose tissue, even when there is little fat in the diet.

High rates of fatty acid biosynthesis can occur in the mammary gland during lactation. The medium-chain fatty acids, caprylic (octanoic) and capric (decanoic) acid, are produced specifically in the mammary gland and in no other tissue. They can therefore act as markers of endogenous fatty acid biosynthesis. When the amounts of fatty acids in the human diet are very low, there is a marked elevation of the milk medium-chain fatty acids compared with concentrations in the milk of women eating a high fat diet. When the diet contains an appreciable amount of fat, this activity is suppressed, and much of the milk fat is derived from circulating lipoproteins that are broken down by mammary gland lipoprotein lipase activated by the hormone prolactin.

Diet may influence the regulation of lipid metabolism in several ways. One is to increase or restrict the availability of molecules that are substrates for the enzymes in a metabolic pathway. Another is the obligatory supply from the diet of a coenzyme, a small molecule that must be attached to the enzyme before the enzyme can act as a catalyst. The enzymes of fatty acid biosynthesis, for example, require pantothenic acid and biotin, both B-group vitamins, as coenzymes. Deficiency of these vitamins, therefore, leads to defects in lipid biosynthesis, which may have profound effects on health. An important effect of diet in the regulation of lipid metabolism is to bring about changes in the concentrations of specific circulating hormones, insulin being among the most important, that induce or repress the biosynthesis of some important enzymes of lipid metabolism. Currently there is much research into the ability of different fatty acids and other lipids to regulate the expression of various genes [10].

D. Essential Fatty Acids

1. *Essential Fatty Acid Deficiency*

In 1929, Burr and Burr [12] described how acute deficiency states could be produced in rats by feeding fat-free diets and how these deficiencies could be eliminated by adding only certain specific fatty acids to the diet. It is now recognized that the two primary essential fatty acids (EFA) for humans are linoleic and α-linolenic acids, because the body is unable to synthesize them.

Well-documented EFA deficiency in humans is rare but was first seen in children given fat-free diets when they developed skin conditions similar to those produced in rats. The skin abnormalities and other signs of EFA deficiency disappeared when linoleic acid was added to the diet [13]. More recently, it has become apparent that EFA deficiency is secondary to a number of disease conditions, so that part of the disease at least is amenable to dietary treatment [14]. Individuals who are unable to absorb long-chain fatty acids, either because of gastrointestinal disease or as a result of surgery [15], are as close to being deficient in essential fatty acids as any group yet studied. Other vulnerable groups are hospital patients on intravenous alimentation with glucose and amino acid mixtures [16].

2. *EFA Metabolism*

Animal tissues contain enzymes called desaturases that insert double bonds into saturated fatty acids, normally beginning at position 9 [3]. The enzyme bringing about this transformation is 9-desaturase. Enzymes are also present that catalyze the insertion of more double bonds to produce PUFA, but in animal tissues this can occur only at positions between the first double bond and the carboxyl group. The enzyme responsible for this second desaturation is 6-desaturase, which inserts a double bond beginning at position 6. Thus, oleic acid is converted into *cis*, *cis*-6,9-octadecadienoic acid (*n*-9), whereas linoleic acid (*n*-6) is converted into all-*cis*-6,9,12-octadecatrienoic acid (*n*-6), and α-linolenic acid (*n*-3) is converted into all-*cis*-

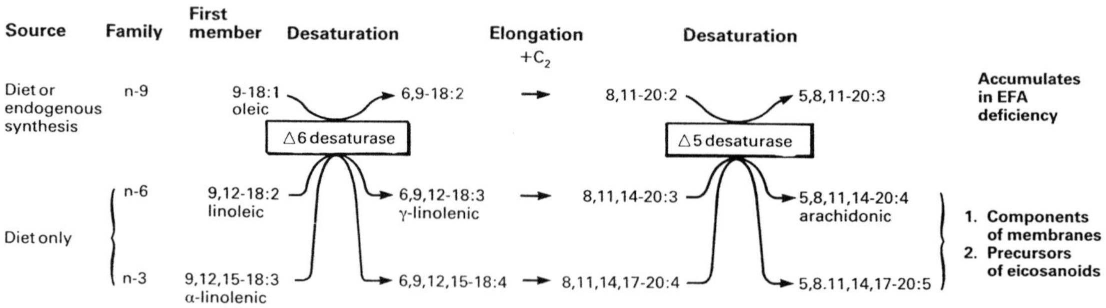

Fig. 3 The metabolism of three families of unsaturated fatty acids. The first member of the *n*-9 family, oleic acid, can be taken in from the diet or can be formed in body tissues, whereas linoleic and α-linolenic acids, the first members of the *n*-6 and *n*-3 families, respectively, can only come from the diet. The first step in their metabolism is the introduction of a new double bond at position 6 by the enzyme 6-desaturase. All three fatty acids compete for this enzyme. There follows a series of alternate competitive elongation and desaturation steps. The products formed from oleic acid accumulate only when linoleic acid is absent from the diet or present in only a very small amount. They are, therefore, markers of essential fatty acid deficiency. Arachidonic acid (*n*-6 family) is the major long-chain polyunsaturated fatty acid in human membranes of most tissues; DHA (*n*-3 family) is a characteristic component of nervous tissue. (From Ref. 86.)

6,9,12,15-octadecatetraenoic acid (*n*-3) (Fig. 3). Oleic, linoleic, and α-linolenic acids are the simplest members of the three major families of polyunsaturated fatty acids designated as *n*-9, *n*-6, and *n*-3, respectively.

An elongation by two carbon atoms and another double bond insertion at position 5 by the 5-desaturase then follow. Although the next higher PUFA contains a double bond in position 4, there is no evidence for the existence of a 4-desaturase. Instead there are two sequential chain elongations after the 5-desaturation followed by a further 6-desaturation. In the *n*-3 family this results in the formation of 24:6 (*n*-3), and retroconversion (chain shortening) of this metabolite results in the formation of docosahexaenoic acid [DHA, 22:6 (*n*-3)].

The long-chain PUFA products of EFA metabolism are further metabolized to biologically active compounds called eicosanoids (Section V.A.6).

E. Breakdown of Tissue Lipids

1. Lipolysis

All tissues contain lipases, which are enzymes that catalyze the cleavage of ester bonds in lipids [3]. As a result of the concerted activity of lipases and acyltransferases (enzymes that catalyze esterification) there is a continual turnover, or replacement, of all parts of lipid molecules in cells that allows fine control of their metabolism. Especially important is the membrane phospholipase that liberates essential fatty acids for the biosynthesis of eicosanoids.

2. Oxidation of Fatty Acids for Cellular Energy

The process known as beta oxidation is a controlled step-by-step breakdown of fatty acids to yield metabolic energy [3]. Beta oxidation is catalyzed by a group of enzymes in a specialized compartment of the cell called the mitochondrion. The complete oxidation of one molecule of the 16-carbon fatty acid palmitic acid yields eight molecules of acetic acid, which are further degraded to give carbon dioxide and water. During this process, the substance adenosine triphosphate (ATP) is generated. ATP is the most important source of chemical energy in the cell and is required to provide the power for many biosynthetic reactions (including the biosynthesis of lipids).

3. Peroxidation

Peroxidation of lipids leads to food spoilage and to a reduction in nutritive value [3,5,17]. It may also occur in the living body. When it does so in an uncoordinated way as a result of tissue disorganization or local lack of antioxidant defense, it may lead to free radical damage to tissues and quite possibly may contribute to disease. However, carefully regulated lipid peroxidation is a mechanism by which essential fatty acids (EFA) are converted into eicosanoids, which are intimately involved in metabolic regulation in all parts of the body [3] (Section V.A.6).

V. NUTRITIONAL CONTRIBUTIONS OF VARIOUS FOOD LIPIDS

A. Triacylglycerols

1. Palatability of Food: Flavor, Aroma, and Texture

Food is nutritious only insofar as it is good to eat, and the presence of fat contributes substantially to the palatability of food [5].

2. *Role in Storage and Provision of Metabolic Energy*

The energy value of triacylglycerols to the body is about 38 kJ (9 kcal) per gram, compared with about 17 kJ for proteins and 16 kJ for carbohydrates. As storage fuels, triacylglycerols have the advantage that they can be stored in anhydrous form, representing more energy for less bulk than complex polysaccharides such as food starches or body glycogen, which are highly hydrated.

Although the gross energies of fatty acids with different degrees of unsaturation (as measured by bomb calorimetry) differ slightly, these differences are insignificant nutritionally. The useful energy available to the body (metabolizable energy) depends mainly on the digestibility of the fat and on the chain length of the fatty acids. Most common food fats are efficiently digested by most people with insignificant differences in their digestible energies. However, saturated (and to a lesser extent unsaturated) fatty acids with chain lengths of more than 18 carbons are less well digested and absorbed, the efficiency falling as chain length increases. The energy values of medium- and short-chain fatty acids are considerably lower than those of the longer chain acids (Section VI.G.3).

3. *Supply of Structural Lipids*

Dietary fats also provide a supply of fatty acids for the body's structural lipids. These will be considered under the appropriate fatty acid classes.

4. *Saturated Fatty Acids*

Saturated fatty acids (SFA) are major components of the phospholipids in the membrane lipid bilayer, accounting for 15–50% of membrane fatty acids. The presence of saturated fatty acids is important in achieving the appropriate degree of fluidity for each particular membrane. The phospholipids of the lung surfactant lipid are esterified almost entirely with palmitic acid. Without this, the lungs would collapse during expiration. Thus although saturated fatty acids are not dietary essentials, they can be regarded as essential components of membrane structures.

Nonspecialists normally think of saturated fatty acids in terms of their alleged capacity to raise blood cholesterol, since this topic has received much publicity in relation to dietary fats and health (see Section VII). Different SFA are by no means equivalent in their influence on blood cholesterol, however [18].

Short- and medium-chain fatty acids (4:0, 6:0, 8:0, 10:0), particularly abundant in coconut and milk fats, have no effect on plasma cholesterol. Bonanome and Grundy [19] confirmed that stearic acid (18:0) has no effect on plasma cholesterol, an observation originally made in the 1960s.

Of the many SFA in the human diet, only three are now thought to have a marked influence on plasma cholesterol: lauric (12:0), myristic (14:0), and palmitic (16:0), and the main lipoprotein fraction affected is LDL. Hayes and Khosla [20] propose that the different hypercholesterolemic fatty acids have different thresholds at which they exert an effect on plasma cholesterol, which are dependent on the concentrations of linoleic acid (18:2, *n*-6) and cholesterol in the diet and the initial plasma LDL concentration. At a level of 18:2 in excess of about 6.5% of dietary energy, the LDL receptors are fully active and are able to remove LDL particles from plasma and maintain a relatively low plasma concentration of LDL. At lower dietary 18:2 inclusions, 14:0 overrides the effects of 18:2 and reduces the activity of the receptors. Palmitic acid is less potent, and more of this fatty acid is required to reduce receptor activity and only at very low intakes of 18:2.

5. *Monounsaturated Acids*

Most naturally occurring fatty acids of dietary significance have double bonds in the *cis* geometrical configuration. Oleic acid (*cis*-9-18:1) is the only monounsaturated fatty acid (MUFA) of quantitative nutritional significance. It is present widely in foods and is particularly abundant in olive and rapeseed oils. The cholesterolemic effects of MUFA have been reevaluated during the last decade. Although not all experiments have been well controlled [21], most found that MUFA, when substituted for SFA, lowered plasma total cholesterol as effectively as *n*-6 PUFA [18]. The lowering was almost entirely associated with the LDL fraction. When substituted for carbohydrates, MUFA resulted in a similarly low plasma LDL cholesterol but did not elicit the rise in VLDL (and, therefore, triacylglycerols) often seen with high carbohydrate diets. There is also a suggestion that the inclusion of MUFA in diets also helps to maintain a high concentration of HDL in contrast with high carbohydrate diets. They also have the advantage over PUFA that they are less prone to oxidation.

While there is little evidence that the very long chain MUFA (erucic, cetoleic) found in older varieties of rapeseed oil or in some fish oils (e.g., herring) influence plasma cholesterol, concern has been expressed that they may result in lipidosis and finally necrosis of heart muscle. However, there is little evidence of harm from the quantities normally ingested in practical human diets [22,23].

A small proportion of dietary MUFA has the double bond in the *trans* geometrical configuration. Although some *trans* double bonds occur naturally, many are introduced as a result of industrial hydrogenation. Their nutritional properties are discussed in Section VI.D.

6. *Polyunsaturated Fatty Acids*

Roles of Essential Fatty Acids The essentiality of linoleic and α-linolenic acids is dependent on their metabolism to the longer chain *n*-3 and *n*-6 PUFA, which have at least two vital functions [3,10].

First, they play a role in stabilizing biological membranes by creating physical properties that are optimal for the transport of substances across the membrane and for the biochemical reactions occurring in the membrane. For example, the membranes of liver mitochondria of EFA-deficient animals have smaller proportions of linoleic and arachidonic acids and larger proportions of oleic and all-*cis*-5,8,11-eicosatrienoic acid than those of healthy animals and are less efficient at oxidizing fatty acids and synthesizing ATP, which is a vital source of chemical energy in cells.

In contrast with tissues like liver, kidney, and muscle, whose PUFA tend to be predominantly of the *n*-6 family, nervous tissue and the retina of the eye have larger proportions of the longer chain acids with five and six double bonds, predominantly of the *n*-3 family. Their main precursor is α-linolenic acid, although the preformed long-chain PUFA may be obtained in the diet, mainly from fish oils. Fatty acids of the *n*-3 family may have a specific role in vision and brain function, and there is evidence to suggest that when the ratio of *n*-3 to *n*-6 fatty acids is modified experimentally in rat tissues some changes do occur in the electroretinograms and in behavioral responses. Although such a clear role has not been established in humans, the circumstantial evidence for the essentiality of α-linolenic acid is compelling [13,24].

Recent research suggests that the capacity of human tissues to convert 18:3 (*n*-3) into 22:6 (*n*-3) is limited, and if this is so, then a dietary supply of preformed 22:6 (*n*-3) may be quite important. The competition between *n*-6 and *n*-3 described earlier may enhance the importance of dietary 22: 6 (*n*-3) because diets in which the *n*-6 acids overwhelm the *n*-3 family (as tends to be the case in most industrialized countries) may cause a suppression of DHA formation from 18:3 (*n*-3) and its replacement by 22:5 (*n*-6) in membranes that have a specific need for 22:6 (*n*-3) [10].

A second major role for EFA is related to their metabolic conversion into a whole range of oxygenated compounds, eicosanoids, which exert a range of profound physiological activities at concentrations down to 10^{-9} g per gram of tissue. There are two basic metabolic pathways (Fig. 4), one regulated by an enzyme called cyclooxygenase, producing prostaglandins, thromboxanes, and prostacyclins; the other by an enzyme called lipoxygenase, producing leukotrienes.

The main sources of fatty acids for the production of eicosanoids are the membrane phospholipids, from which they are released by the action of phospholipase A. Phosphatidylinositol, which contains a high concentration of arachidonic acid in position 2, provides a major store of eicosanoid precursors. A crucial point about eicosanoids is that they are so potent in their biological activities that they must be produced close to the site where they are required and the excess broken down as soon as their job is done by enzymes that convert them into inactive metabolites.

Another feature of eicosanoids is that they may have different, sometimes opposing, activities when they are produced in different tissues. Broadly, eicosanoids are involved in what biologists call tissue homeostasis. This means that certain physiological functions are kept constantly within predetermined limits so that they do not under- or overperform. A good example is the tendency of a type of blood cell called platelets to aggregate together to plug

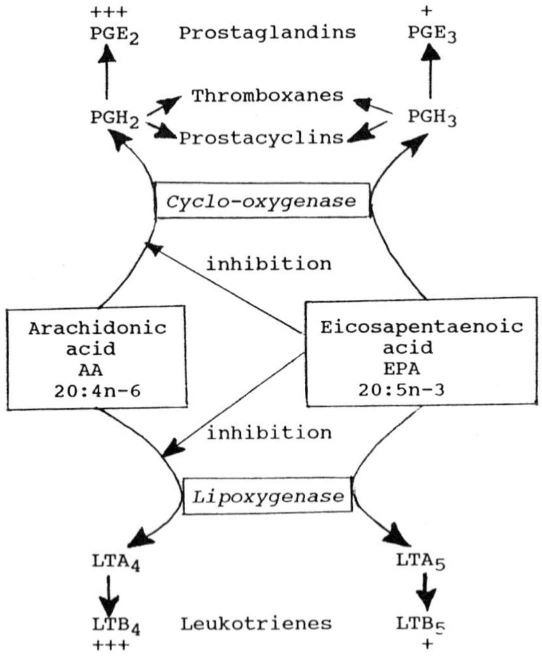

+++ signifies strongly inflammatory
\+ signifies weakly inflammatory

Fig. 4 Competition between the *n*-3 and *n*-6 polyunsaturated fatty acid pathways for production of eicosanoids. (From Ref. 87.)

a wound. If they do not aggregate sufficiently, the wound will not be healed; if they aggregate too much they may precipate a potentially fatal thrombosis. Thromboxanes, which are produced by the platelets themselves, stimulate the platelets to aggregate, whereas prostacyclins, produced by the walls of the arteries, inhibit the aggregation. It is the balance of these two opposing activities that keeps the platelet aggregation system functioning properly. Many more examples of these homeostatic mechanisms regulated by eicosanoids are being discovered. Other eicosanoid activities are those involved in regulating kidney function, inflammatory reactions, reactions of the cellular immune system, secretion of hormones such as insulin, reproduction, bone calcification, muscle contraction including the constriction and dilation of blood vessels, electrical activity of the heart, and various aspects of visual function [10].

Requirements for Essential Fatty Acids. It seems possible that there are different levels of requirements for EFA for their different functions. For example, there may be a relatively low level of requirement (as for most vitamins) to prevent overt signs of EFA deficiency such as skin conditions and growth retardation. This may be around 1% of dietary energy as linoleic acid and about one-fifth of this amount as α-linolenic acid [24]. At a higher level, there may be a larger but quantitatively less easily defined quantity required for such functions as maintaining an optimal balance of membrane fatty acids, preserving a reservoir for eicosanoid formation, and maintaining optimal concentrations of the different plasma lipoproteins. The absolute amounts required at this level are likely to be influenced by the amounts and types of other fatty acids in the diet because of the competition for metabolic pathways between EFA and non-EFA as well as competition between the EFA families. Because primary EFA may be only slowly converted into their long-chain PUFA derivatives, there may sometimes be a dietary requirement for the latter.

n-*6 Family.* In addition to its requirement as an EFA, linoleic acid results in a lowering of plasma cholesterol, and this may be important in reducing the risk of coronary heart disease (CHD) [21].

γ-Linolenic acid (GLA), the first metabolite of linoleic acid, formed from it by 6-desaturation, has entered the folklore as having health-giving properties. In the human body, GLA is merely an intermediate in the pathway from linoleic to arachidonic acid and does not accumulate in mammalian cells.

The rate limitation of the 6-desaturase in the *n*-6 pathway is important in understanding the claims for health benefits of GLA. 6-Desaturase activity decreases with age and is also depressed by the so-called stress hormones, adrenalin and cortisol, by alcohol, by fatty acids of the *n*-3 family, and in the condition of diabetes mellitus. Bioassays based on water permeability of skin show that GLA has higher EFA potency than its precursor, LA. A commonly used biochemical index of EFA deficiency has been the so-called triene/tetraene ratio. When the dietary supply of linoleic acid is limited, oleic acid becomes the prime substrate for 6-desaturase (Fig. 3). The major product of this (*n*-9) pathway is the trienoic acid 5,8,11-20:3, and its ratio to AA increases in EFA deficiency and decreases when LA is added back to the diet. The ratio was reduced far more rapidly when GLA was substituted for LA [25].

Reports on the blood cholesterol lowering potency of GLA are inconsistent. One publication claimed a large cholesterol-lowering effect that was disproportionate to the degree of unsaturation compared with linoleic acid, while others have found no differences between the effects of linoleic acid and GLA.

The literature is replete with claims for the therapeutic benefits of GLA. Disorders for which its efficacy has been tested in controlled clinical trials include atopic dermatitis, rheumatoid arthritis, diabetic neuropathy, multiple sclerosis, ulcerative colitis, hypertension, premenstrual syndrome, schizophrenia, hyperactivity, and several cancers. In almost all categories the results have been inconsistent.

The main sources of GLA are evening primrose, borage, and blackcurrant oils. There are preliminary indications that evening primrose oil may be more effective in some of the physiological effects described above than the other oils in which GLA occurs. One possible explanation is that GLA is present in evening primrose oil almost entirely as molecular species of triacylglycerol in which one GLA is combined with two LA molecules: dilinoleoyl-monogammalinolenoyl-glycerol (DLMG) [26]. Another possibility is that minor components of evening primrose oil, not GLA, are responsible for some of the effects.

n-3 Family. The parent, and simplest, member of the *n*-3 family is α-linolenic acid, all-*cis*-9,12,15-octadecatrienoic acid (α-18:3). It is synthesized only in higher plants, algae, and phytoplankton and is present universally in chloroplast membranes, where it is present almost exclusively as a constituent of galactosyl-diacylglycerols [3]. Some seed oils, however, contain α-18:3 in their triacylglycerols (e.g., linseed, rapeseed, and soybean oils, Fig. 1).

When α-18:3 is ingested by human beings, some is incorporated unchanged into lipids, a little is oxidized, and some is metabolized by a series of desaturations and elongations to longer chain PUFA. The predominant end products are eicosapentaenoic acid (all-*cis*-5,8,11,14,17-20:5, EPA) and docosahexaenoic acid (all-*cis*-4,7,10,13,16,19-22:6, DHA). However, the extent to which α-linolenic acid is converted into longer chain derivatives is limited, and the largest sources of these long-chain PUFA are the oils of fish that eat phytoplankton containing a high proportion of α-18: 3 in their lipids [6]. These fatty acids are present in triacylglycerols that are stored in the animal's liver (e.g., cod) or in the flesh (e.g., mackerel, herring).

Inclusion in the diet of four or more grams per day of *n*-3 polyunsaturated fatty acids from fish oils maintains low blood triacylglycerol concentrations in people with normal levels and significantly reduces their concentration in people with hypertriglyceridemia [27]. Unlike the *n*-6 family of PUFA, however, there is no consistent effect of *n*-3 PUFA on blood cholesterol. In some individuals, *n*-3 PUFA have even been known to raise blood total cholesterol. There are small and inconsistent effects on HDL cholesterol. The triacylglycerol lowering effect is due entirely to changes in plasma VLDL concentrations brought about by a reduction in VLDL triacylglycerol biosynthesis in the liver and to some extent by reduced clearance and catabolism of VLDL [27].

The conversion of long-chain *n*-3 PUFA into eicosanoids may explain a number of physiological effects of eating fish oils, including effects on blood clotting and on immune function. Thus, consumption of about 200–300 g/day of fatty fish results in prolonged bleeding times and reduced tendency for platelet aggregation. The effects of consumption on the order of 30 g/day or less that may be practical or acceptable in most western countries are less conclusive. Nevertheless, in one study, consumption of at least two portions of fatty fish per week (200–400 g) resulted in a significant decrease in mortality from coronary heart disease [28].

Inclusion of long-chain *n*-3 polyunsaturated fatty acids characteristic of many fish oils in the diet is quickly reflected in elevated concentrations of these fatty acids in plasma and in the membranes of red and white blood cells. The biochemical effect of an increase in the *n*-3/*n*-6 PUFA ratio in cell membranes is to alter the composition of the eicosanoids formed from these PUFA through the action of the enzymes cyclooxygenase and lipoxygenase. The strongly proinflammatory eicosanoids from arachidonic acid are reduced and replaced by the weakly inflammatory eicosanoids from eicosapentaenoic and docosahexaenoic acids [10]. This knowledge has formed the basis for dietary treatments for inflammatory diseases, of which rheumatoid arthritis has been most actively researched. Controlled dietary trials have demonstrated modest but potentially beneficial improvements in the severity of rheumatoid arthritis, and longer term studies are now needed to establish whether dietary supplements of *n*-3 PUFA should be standard recommended treatment [10,27,29].

7. *Distribution of Fatty Acids in Triacylglycerols*

The way in which fatty acids are distributed in a triacylglycerol may also influence plasma cholesterol irrespective of the overall composition of the fatty acids, and this may explain some apparent anomalies. Thus linoleic acid is more *hypo*cholesterolemic [30] and SFA more *hyper*cholesterolemic [31] when esterified in position 2 than in positions 1 or 3. Butterfat is much less *hyper*cholesterolemic when the positions of its fatty acids are randomized by interesterification [32].

B. Phospholipids

Phospholipids, as intact molecules, have not been regarded as essential dietary components because they are readily synthesized in the body. However, some revision of this view may be necessary. The most abundant phospholipid, phosphatidylcholine, provides a source of choline in the diet. Choline, in the form of acetylcholine, is a neurotransmitter involved in the transmission of nerve impulses. The only known pathway for the formation of choline in the human body is by methylation of another phospholipid, phosphatidylethanolamine. Therefore, a dietary source of choline as phosphatidylcholine or of ethanolamine as phosphatidylethanolamine should probably be regarded as essential.

Phospholipids are also carriers of PUFA, but because phospholipid fatty acids contribute only a small proportion of the normal dietary intake of fatty acids, it is debatable whether their nutritional contribution is significant. An exception may be the supply of arachidonic acid, present in the phospholipids of meats.

C. Glycolipids

The daily intake of glycolipids is very small. Their contribution of α-linolenic acid is important. There is little evidence that any of the sugar moieties supplied by glycolipids has dietary significance.

D. Sterols

1. *Cholesterol*

The influence of dietary cholesterol on plasma cholesterol is still a subject of debate. In the United States, dietary guidelines clearly emphasize the importance of dietary cholesterol as a determinant of plasma cholesterol, whereas in Europe the scientific evidence has generally been interpreted to indicate that it plays a minor role if any. McNamara [33] analyzed and reviewed 68 human intervention studies that examined the effect of dietary cholesterol on plasma cholesterol and concluded that there is an effect but it is small compared with the influence of SFA. A mean change of 100 mg/day dietary cholesterol causes a change of only 2.3 mg/dL plasma cholesterol. Somewhat larger reductions could be achieved by relatively modest reductions in body fat.

2. *Plant Sterols*

The amounts of plant sterols consumed in omniverous diets in western countries are very small but will be significantly higher in vegetarians. They inhibit the absorption of cholesterol, so that endogenous synthesis of cholesterol may be expected to be higher in vegetarians than in omnivores.

E. Fat-Soluble Vitamins

Dietary fat is also important insofar as it contributes several fat-soluble vitamins: A, D, E, and K. Requirements for the fat-soluble vitamins are of a different order of magnitude from those of the essential fatty acids, being measured in micrograms rather than grams per day. Some fat is necessary in the diet to improve the absorption and utilization of fat-soluble vitamins. However, there is little evidence that, within the normal range of fat intakes, the amount of dietary fat significantly affects the utilization of fat-soluble vitamins.

1. Vitamin A (Retinol) and Carotenoids

Retinol is found only in animal fats. Dark green-leaved vegetables and vegetable oils like palm oil contain a precursor (provitamin A), β-carotene, that is converted into retinol in the body. Milk fat is relatively rich in retinol equivalents. Fish liver oils are by far the most concentrated source of retinol, but mammalian storage fats such as lard contain none. The British diet provides on average about twice the "reference nutrient intake" for vitamin A. Reference nutrient intake (RNI) is defined by the U.K. Department of Health [24] as an intake likely to meet the needs of 97% of the population. The vitamin A RNI for adults is currently 600–700 μg/day. Two-thirds of average U.K. intake comes from retinol itself, and about one-third from carotene. About 21% of dietary retinol equivalents in the United Kingdom comes from milk and milk products [4].

Vitamin A, if taken in excess, will accumulate in the liver and eventually destroy it.

Vitamin A is required for vision and in the maintenance of cellular differentiation. There is mounting evidence that β-carotene may have specialized roles (e.g., as an antioxidant) that are independent of its function as a vitamin A precursor.

2. Vitamin D

Vitamin D is the name for a family of sterols with antirachitic properties. The only one of significance is the one now called vitamin D_3 or cholecalciferol. It is present in only a few foods, the richest sources being the liver oils of fish.

Cholecalciferol is produced by the ultraviolet irradiation of 7-dehydrocholesterol, a sterol derived from cholesterol and widely distributed in animal fats including the skin surface lipids.

Many people get little or no vitamin D from their diet but obtain all they require from the action of the ultraviolet rays in sunlight on 7-dehydrocholesterol in the skin. For this reason, many nutritionists have questioned whether vitamin D should be considered a vitamin at all. However, at least two groups of people may have a special need to obtain vitamin D from the diet. In the first group are infants, children, and pregnant and lactating women, whose requirements are particularly high. The vitamin D content of cow's milk and even human breast milk is quite low, but manufactured infant feeds are fortified with this vitamin. In the second group are people who are little exposed to sunlight, such as the housebound elderly and people living in northern latitudes or those who wear enveloping clothes. Dark-skinned immigrants to northern European countries are especially vulnerable. Infants and children who obtain too little vitamin D develop rickets, with deformed bones that are too weak to support their weight. The reason vitamin D preparations are provided for children and pregnant women and margarine is fortified in the United Kingdom and some other countries is that these degenerative changes soon become permanent if supplementation is not begun early enough.

Dietary vitamin D is absorbed along with the bulk of the dietary fat and carried into the blood in chylomicrons. In the liver, it is converted into 25-hydroxycholecalciferol, which is further converted into 1,25-dihydroxycholecalciferol in the kidneys. The latter can be regarded as a hormone that is primarily involved in the regulation of calcium and phosphorus metabolism.

Like vitamin A, vitamin D is toxic if consumed in excessive quantities.

3. Vitamin E

Vitamin E activity is shared by a family of compounds known as tocopherols, the most active of which is α-tocopherol. The richest sources are vegetable oils, cereal products, and eggs. α-Tocopherol is present in the lipid bilayers of biological membranes and may play a structural role there. It is known to be a powerful antioxidant and prevents the peroxidation of unsaturated fatty acids. The products of peroxidation of unsaturated fatty acids can cause damage to cells if the oxidative process is not kept in check, and such damage appears to be exacerbated in animals given diets deficient in vitamin E. These observations have led to the postulate that vitamin E plays a vital role in the prevention of the oxidation of membrane fatty acids in living cells. It is a generally held view that intake should be considered in relation to the PUFA content of the diet rather than in absolute amounts. A ratio of vitamin E to linoleic acid of about 0.6 mg/g is generally recommended. In general vegetable oils containing high concentrations of PUFA are sufficiently rich in vitamin E to give adequate protection. However, with the growing interest in fish oils, which are even more highly unsaturated, the question of vitamin E adequacy may well have to be reviewed. More unsaturated relatives of the tocopherols, the tocotrienols, share some vitamin E activities with the tocopherols but may also exhibit some independent biological activities.

VI. EFFECTS OF PROCESSING ON THE NUTRITIONAL CONTRIBUTIONS OF FATS

A. Mixing or Homogenization

Mixing and homogenization are normally mild processes, but sometimes they can disrupt food structure, releasing enzymes that degrade fats. Two classes of enzymes are of particular importance. Lipases hydrolyze the complex lipids in food (mainly either triacylglycerols or phosphoglycerides), leading to release of free fatty acids (FFA) [3]. When free fatty acid concentrations reach a certain threshold, off-flavors and unpleasant aromas may develop, but the FFA do not usually cause nutritional problems.

The other type of enzyme is an oxygenase or peroxidase that catalyzes the oxidative deterioration of fatty acids once they have been released by lipases. Some of these enzymes (such as the lipoxygenase of plants) are involved in the formation of pleasant flavors characteristic of particular plant foods (e.g., the characteristic smell and taste of freshly cut cucumber), but oxidation of larger quantities of fatty acids may lead to off-flavors and toxic compounds. These usually make the food unpalatable long before they reach concentrations that cause nutritional or toxicological problems. Thus combined effects of lipase and oxygenase action may sometimes be seen after milk homogenization, with accompanying rancidity problems. The enzymes are located in the milk fat globule membrane and cause the release and oxidation of unsaturated fatty acids present in the membrane phospholipids [34].

B. Heating

The functions of heating, which may be done industrially or in the home, are to make food palatable and also to make it safe by killing microorganisms. However, heating may reduce heat-labile and oxygen-sensitive nutrients, so there is a need to achieve a balance between microbiological safety and nutritive value.

Some changes occur during heating in which oxygen is not involved. These may occur, for example, in the bulk of a deep fat fryer where oxygen has little access. Combination of radicals under these circumstances gives rise to polymeric products formed from the triacylglycerol molecules in the heated fat. The first substances to be formed are cyclic monomers of triacylglycerols followed by formation of polymeric triacylglycerols.

Fats heated under normal household conditions (at, say, 180°C) generate 10–20% of polymeric material, but at this level the functional properties of the oil are not noticeably impaired, and feeding experiments with rats have not demonstrated any harmful effects. None is apparently deposited in adipose tissue [35].

The most important changes during heating involve reaction with oxygen from the air. Oxidation can take place when the fat is stored in the presence of oxygen at room or even refrigerator temperature but is greatly speeded up by heating and in the presence of trace metal catalysts (e.g., iron or copper).

The major pathway is attack by oxygen at the double bonds of unsaturated fatty acids with the initial formation of a peroxide [3,5,36]. The more highly unsaturated the fatty acid, the greater its susceptibility to oxidation.

Free radicals produced during lipid peroxidation can reduce concentrations of essential fatty acids (EFA), vitamin A, and vitamin E [36]. They may also cause damage to proteins, including enzymes, and DNA and may also generate carcinogens. Lipid peroxidation may be held in check by the presence of lipid-soluble antioxidants: α-tocopherol (vitamin E) and carotenes (both natural) and butylated hydroxytoluenene or butylated hydroxyanisole (synthetic).

Cholesterol can also autoxidize, especially when present in dried and/or heated foods (e.g., dried egg powder). Products include 25-dihydroxycholesterol, 7-ketocholesterol, 7-hydroxycholesterol, 5- and 7-hydroperoxides, and cholesterol-3,5,6-triol. These compounds have been shown to cause arterial disease in rabbits at levels at which cholesterol itself is ineffective [37].

C. Irradiation

Irradiation is also a process that may give rise to free radical formation and, therefore, lipid deterioration. This process is useful for controlling insect infestation of foods, reducing total microbial load, and reducing the numbers of specific pathogens in foods. It is not useful for all foods, and in many cases it can cause deterioration in palatability through poor flavor and texture.

Vitamins E, K, and ascorbic acid are particularly susceptible to irradiation damage. Surprisingly, carotene has not been shown to be much affected.

Some highly unsaturated oils may suffer extensive damage due to a combination of free radical formation and reduction of antioxidant nutrients that would normally reduce the extent of free radical damage. Thus, these two types of damage may reinforce one another. When mixtures of unsaturated fatty acids and proteins have been irradiated prior to animal feeding, the quality of the protein was found to be reduced, probably due to protein damage by peroxidized lipids. In general, however, irradiation does very little direct nutritional damage.

D. Hydrogenation

Vegetable oils and, in some countries such as the United Kingdom, fish oils are often hydrogenated industrially to provide fats of more desirable physical properties, texture, and keeping quality; these can be used in, for example, the manufacture of margarines, cooking fats, and shortenings. During this process, there is a decrease in total unsaturation and an increase in the concentration of saturated fatty acids, and some new isomeric fatty acids are usually produced. Double bonds are shifted along the hydrocarbon chain, and some geometrical isomerization from *cis* to *trans* occurs, but the amounts of isomers vary according to the processing conditions and the starting material [38]. Hydrogenated soybean oil or fish oil may contain as much as 50% by weight of the fatty acids as acids containing one or more *trans* sites. The final concentration, however, in a blended margarine or cooking fat will be much lower.

Monoenoic acids represent the major fraction, with *trans* unsaturation between positions 6 and 14. The most abundant isomers are those with unsaturation at positions 9, 10, and 11 [39,40]. A small proportion of isomers with diunsaturation (*cis,trans* and *trans,cis*) is also found, but improvements in catalysts have reduced the amounts of *trans,trans* isomers in modern hydrogenated fats almost to zero. A range of positional *cis* isomers are also produced during hydrogenation, but their biological properties have been largely unresearched.

In ruminant fats, the proportion of the *trans* group of acids is usually around 5% of the fatty acids present. The isomers present are qualitatively similar to those in industrially hydrogenated fats, but the quantities are different; the isomer with a *trans* bond in position 11 predominates [40].

Average consumption of *trans* fatty acids in the United Kingdom in the early 1990s is estimated to have been 4–6 g per person per day, representing about 5.5% of dietary fat intake or 2% of dietary energy. It has been estimated, however, that some individuals may consume up to 30 g of *trans* fatty acids per day and others as little as 1–2 g. A recorded decrease in the United Kingdom in recent years may have come about partly as a result of different (and probably better) methods of estimating intakes and partly because of a true decrease in consumption. Intakes of ruminant fats have tended to decrease, and the *trans* fatty acids content of hydrogenated fats has also fallen as a result of changes in raw materials and processing methods. In the United States, average current intakes are based on estimates from limited data but appear to be higher than in the United Kingdom (8–13 g per person per day).

The digestion and absorption of isomeric fatty acids, their transport around the body, their storage as body fat, or their utilization directly as a source of energy by biological oxidation are all similar to those processes for the more common fatty acid components of fat in the diet.

A number of studies suggest that *trans* fatty acids are treated more like saturated than unsaturated fatty acids by the biochemical systems responsible for the biosynthesis of membrane components. This is probably because the presence of a *trans* unsaturated bond produces an overall molecular shape at this site in the molecule similar to that of a saturated fatty acid.

In general, the proportion of *trans* fatty acids incorporated into fats at the various sites in the body is dependent on the proportion present in the diet. As with all fatty acids, however, once incorporated the *trans* fatty acids are subject to turnover. There is no evidence from animal studies or the limited data from human experiments that *trans* fatty acids accumulate specifically at any site in the body.

If the level of *trans* fatty acids in the diet is disproportionately high, the metabolism of EFA may be impaired and signs of essential fatty acid deficiency may develop. This may arise from either inhibition of or competition for desaturases. However, similar competition occurs with *cis*-MUFA, so the effect is not a unique property of *trans* fatty acids. Overt signs of EFA deficiency as a result of *trans* fatty acid consumption have not been recorded in humans. Nevertheless, some scientists believe that metabolic changes may occur that are characteristic of the early stages of EFA deficiency yet produce no overt clinical signs. Such changes, they argue, could involve subtle impairment of membrane function or of physiological functions such as the integrity of the vasculature. Such ideas need further research, but until such effects have been clearly demonstrated in practice it must be concluded that there is no sound evidence for specific adverse effects of *trans* fatty acids on essential fatty acid metabolism.

There is preliminary evidence that *trans* fatty acids from a mother's diet cross the placenta during pregnancy and interfere with EFA metabolism in the newborn. There appears to be an inverse association between the mother's *trans* fatty acid intake and the baby's birthweight [41], and this needs further research.

Taken together, several well-conducted human studies [38,42,43] have shown that dietary 18-carbon fatty acids with a single *trans* double bond increase the concentration of LDL cholesterol in the blood compared with a similar amount of dietary 18-carbon *cis*-monounsaturated fatty acid. This LDL cholesterol raising effect was directly related to the proportion of *trans* fatty acids in the dietary fat and was a little lower than that of the 14- and 16-carbon saturated fatty acids. Unlike the saturated fatty acids, however, which tend to raise HDL cholesterol slightly, the *trans* fatty acids significantly lowered HDL cholesterol. There appears to be a dose-related effect, at least at intakes between 3 and 11% of total energy. Average intakes in the United Kingdom are estimated to be about 2% of energy, and it is unlikely that such levels of consumption contribute more than a small part of the risk of CVD in that country. More information is required on the physiological effects of *trans* fatty acids from hydrogenated fish oils. There is evidence from at least two studies that *trans* acids in the diet lead to increases in the concentration of plasma lipoprotein (a) [Lp(a)] (an independent risk factor for CVD), especially in those individuals who already have high baseline plasma concentrations of this lipoprotein [44,45].

There is no firm evidence about the influence of *trans* fatty acids on thrombosis.

A large epidemiological study of 85,000 American nurses [46] reported that women consuming the highest levels of *trans* fatty acids had a 50% greater risk of CVD than those consuming the lowest amounts. The association was significant only for *trans* fatty acids from hydrogenated vegetable oils, not from ruminant fats. In view of this apparent anomaly of high risk from industrially produced *trans* fatty acids and no risk from ruminant *trans* fatty acids, it is prudent to consider whether the *trans* fatty acids content of hydrogenated fats might have been acting as a marker for socioeconomic status or for other aspects of an unhealthy lifestyle. Two recent studies in which the concentration of *trans* fatty acids in adipose tissue was used as an index of long-term intakes of *trans* fatty acids found no association between *trans* fatty acid intakes and coronary heart disease [47,48].

E. Interesterification

The process of interesterification, which involves treatment of a fat with a catalyst at an elevated temperature, leads to randomization of fatty acids on triacylglycerol molecules as distinct from their natural stereospecific distributions. Interesterification causes changes in the physical properties of the fats, such as crystal structure and melting point, and may influence their absorption and metabolism. Thus butterfat in which the fatty acids have been randomized by interesterification has minimal effect on plasma cholesterol [32], whereas natural butterfat tends to raise plasma cholesterol [49].

In experiments with laboratory animals designed to induce atherosclerosis by including high levels of fat in their diets, natural peanut oil was found to be strongly atherogenic, but the same did not apply to randomized peanut oil [50]. The reasons for these findings have not been established.

F. Fractionation

1. Medium-Chain Triacylglycerols

The most abundant sources of medium-chain triacylglycerols (MCT) used in the preparation of human foods are coconut and palmkernel oils, which contain 15% and 8% respectively of 8:0 + 10:0. The most abundant acid in these oils is lauric acid (12:0), which accounts for nearly 50% of the total fatty acids. Seed oils of *Cuphea* species also contain medium-chain fatty acids and could become another potential food source [6].

Medium-chain fatty acids are rapidly absorbed in conditions in which the absorption of long-chain fatty acids may be impaired. Failure to digest lipids may occur because a disease has resulted in insufficient lipase or bile salt; malabsorption is generally due to damage to the mucosal surfaces of the small intestine. Absorption may also "back up" due to congenital failure to produce specific apoproteins that are obligatory components of plasma lipoproteins. Patients with malabsorption are at risk of deficiencies in energy, fat-soluble vitamins, and essential fatty acids but can usually absorb medium-chain fatty acids [51].

For such patients, MCT, which can be purchased as cooking oils or fat spreads, provide useful sources of dietary fats, although of course they lack the essential fatty acids. MCT have also found applications as a source of energy in the enteral feeding of preterm babies.

An alternative application for MCT that has been much discussed is in reduced calorie diets and diets that will not elevate plasma cholesterol, because medium-chain fatty acids are not deposited in adipose tissue and do not contribute to plasma lipoproteins (Section IV.B). However, to my knowledge, convincing evidence for their long-term effectiveness in humans is yet to be published. Their effectiveness would need to be substantial in view of the cost of the product and competition from other low calorie fat substitutes now being developed.

A recent development in fractionation technology has been the production of spreadable butters by removal of higher melting triacylglycerols.

2. Reduced Cholesterol Foods

In reduced cholesterol foods, most or part of the cholesterol has been removed by treatment with cholesterol-degrading enzymes or by differential extraction, for example by supercritical carbon dioxide extraction. This technology has been applied to eggs, fish, beef, and milk fat [52]. There is little scientific justification for these products, since there is only a weak relationship between cholesterol in the diet and cholesterol in the blood. Their production is motivated by perceived marketing opportunities. It has been argued that in the long term, promotion of reduced cholesterol dairy products will do little for the credibility of the dairy industry [52,53].

G. Reduced Fat Products, Structured Fats, and Fat Substitutes

1. Reduced Fat Spreads

The aqueous phase in a margarine contributes about 15% of the weight of the product. It is now possible with skillful use of modern emulsifier technology to increase this to about 75%

or even more to give low fat spreads that may have applications in energy-reduced diets. These are often blends of vegetable and dairy fats to give a relatively high polyunsaturated fatty acid content yet retain a butterlike flavor.

In regard to "slimming," the concept of the low fat spread is that for the same bulk of spread the energy content can be reduced by half, and since fats have over twice as much energy per gram as carbohydrates and proteins, this can be a big potential saving. However, they will only be effective in human diets if there is no tendency to compensate for the low energy density of the food by automatically adjusting total food intake over the medium or even the long term. This has been the subject of considerable research interest, but the results are not yet entirely clear on the matter. One much quoted publication suggested that if the energy content of one meal is reduced, energy intake in later meals is increased subconsciously to maintain overall energy consumption [54].

2. *Undigested Fat Analogues*

In a low fat spread, the energy content of the actual fat phase is not reduced and the products have limitations when it comes to being used for cooking. Therefore, the search has been on for substances with fat-like physical properties yet without any energy value.

Several types of compounds have been tested including polyglycerol fatty acid esters and "retrofats," which are esters of fatty alcohols and citric acid [55]. The most recent and most promising types are the sucrose polyesters, and one company has exhaustively tested a product known widely as olestra [56]. Like all the compounds mentioned, olestra is not digested by enzymes in the small intestine nor is it absorbed intact. It therefore passes down the gut and contributes no energy to the body.

Nutritionists have been concerned that in clinical trials olestra caused abdominal cramping and loose stools in some individuals. More importantly, olestra may inhibit the absorption of fat-soluble vitamins (A,D,E, and K) and carotenoids from foods eaten at the same time as the olestra-containing products. For these and other reasons, it has taken a considerable time for the safety of olestra in foods to be evaluated. In January 1996 the US Food and Drug Administration approved the use of olestra in certain savory snack foods provided that the olestra-containing products have vitamins A,D,E, and K added to them and are specifically labeled. It is likely to be several years before the consumer acceptability of these products and their efficacy in reducing overall fat intakes can be properly assessed.

3. *Digested Reduced Energy Fats*

Another approach has been to devise digestible fats (thereby avoiding problems, real or imagined, of having undigested fat in the gut) that have intrinsically lower metabolizable energies than "normal" fats, contradicting the golden rule that all fats provide 9 kcal (38 kJ) per gram. It has long been recognized that the shorter the chain length of a fatty acid (and therefore the lower its molecular mass), the lower its gross energy per gram. Palmitic acid (16:0) consists of 75% carbon, whereas carbon constitutes 67% of capric acid (8:0) and 55% of butyric (4:0) acid, and the heats of combustion decrease accordingly. The metabolizable energy (ME) per gram also decreases with diminishing molecular mass. During beta oxidation, less ATP is produced per gram of short-chain than of long-chain fatty acids. Furthermore, the contribution of glycerol (with an ME value of 4.3 kcal/g) to the metabolizable energy of triacylglycerols becomes relatively more significant as the proportion of short-chain fatty acids increases and the molecular mass of the triacylglycerol decreases.

These theoretical considerations led to the logical development of mixed acid triacylglycerols containing short- and long-chain fatty acids. In 1991, Peters and colleagues [57] described Caprenin, a mixed acid glyceride of one long-chain acid (behenic acid, 22:0) and two medium-chain acids [capric (10:0) and caprylic (8:0)], and in three papers [58–60] published during 1994 another triacylglycerol having similar attributes and named SALATRIM was described. The authors cleverly combined the advantages of the reduced energy value of the short-chain fatty acids with the poorer absorption of stearic acid (18:0) to produce a triacylglycerol with significantly reduced energy value (5 kcal/g; 21 kJ/g) compared with conventional fats (9 kcal/g; 38 kJ/g).

An important point emerging from the study was that the actual absorption of 18:0 decreased as the proportion of 18:0 to short-chain fatty acids increased and was also lower when it was esterified in positions 1 or 3 than in position 2. In a human clinical trial in which 17 subjects received a diet containing 22% of energy from SALATRIM for 7 days, a metabolizable energy value of SALATRIM of 4.9 kcal/g and absorption of stearic acid ranging from 63 to 70% were reported [60].

SALATRIM is produced by interesterifying triacylglycerols containing short-chain fatty acids (e.g., tributyrin) with long-chain saturated triacylglycerols (e.g., hydrogenated rapeseed oil). It is therefore simple to produce by methods that are standard in the fats and oils processing industry. This material has the potential to reduce substantially the energy value of foods into which it is incorporated while still retaining desirable textural properties. Whether it turns out to be a useful aid to weight maintenance remains to be seen.

VIII. HOW MUCH OF A HEALTH PROBLEM IS FAT?

The intense consumer interest in fat is mainly due to much publicized associations between high fat intakes and ill health. It would be impossible to adequately review here the evidence for and against these associations. Instead, I briefly summarize what seems to be the current consensus among nutritionists, indicate where there have been recent changes in views and where current evidence is weak, and discuss priorities for future research. The main focus is on obesity and coronary heart disease, but other health aspects are also discussed briefly.

A. Obesity

1. *Body Fat Distribution*

Arguably, obesity is one of the major health problems in industrialized countries and one that seems resistant to all public health attempts to control it. Opinions differ widely, however, on how seriously to treat the problem. Few studies in which the degree of overweight or obesity has been tested as a risk factor for CHD have given clear-cut answers. Recent research may indicate why this has been so. The index of overweight frequently used in epidemiological or experimental studies as a measure of obesity, namely body mass index ($BMI = kg/m^2$), is an extremely rough guide because it does not indicate how the fat is distributed in the body. Fat distribution is now recognized as being far more important than the mere fact of being overweight. Broadly, people divide into two main types: those who have the bulk of their fat around the waist (mainly, but not exclusively, men) and those whose fat is mainly on the hips, thighs, and lower limbs (more characteristic of women). A simple measurement of fat distribution now in common use is the waist-to-hip girth ratio.

It is now recognized that central obesity is a significant risk factor for CHD whereas lower body or overall obesity have virtually no relationship. Central obesity and the common disorders glucose intolerance [and its main clinical manifestation insulin-independent (maturity onset) diabetes mellitus], hypertriacylglycerolemia, and hypertension form what is sometimes dubbed the "deadly quartet" [61]. The common link between the four is thought to be hyperinsulinemia, i.e., habitually raised blood insulin concentrations. Central obesity is characterized by extremely enlarged fat cells in the abdominal adipose tissue. The large fat cells are insulin-resistant and need more insulin to sustain their metabolism. This contributes to the raised blood insulin. They also have a greater tendency to break down their triacylglycerols, liberating free fatty acids into the plasma. High concentrations of fatty acids in the blood are thought to reduce the insulin sensitivity of muscle as well as interfering with the removal of insulin by the liver. The excess of insulin and fatty acids causes the liver to turn the fatty acid into VLDL, which explains the hyperlipidemia. Finally, high insulin levels are thought to act upon the kidneys and the vascular system in such a way as to produce high blood pressure, although the latter mechanisms are more obscure than those affecting blood lipids.

2. *Is a Fat Calorie More Fattening Than a Carbohydrate Calorie?*

It might seem that there is an obvious link between fat intake and obesity because (1) fat has, weight for weight, over twice as much energy as carbohydrates and (2) fat contributes to palatability and might therefore be expected to encourage overeating.

Many have argued that what matters in regard to the potential for gaining excessive weight is total energy intake rather than whether the calories are contributed by fat or carbohydrate. In other words, they have assumed that the "fattening potential" of 1 cal of fat in excess of requirements is equal to the fattening value of 1 cal of carbohydrate. In recent years, this view has been challenged [62,63].

It is clear that the body weight of adults (even if it is above what might be regarded as ideal) remains relatively stable over long periods of time despite considerable short-term fluctuations. The relative proportions of the major body constituents—protein, carbohydrate, and fat—remain relatively constant during these periods even though the food we eat every day may vary widely in composition. It is apparent that there is a balance between the accretion of body constituents and the oxidation of protein, carbohydrate, and fat that acts to correct any major perturbations from what the body regards as normal.

One thing is certain: The body cannot thwart the laws of thermodynamics. Body energy stores can remain stable only when energy intake is equivalent to energy expenditure over an extended period. There must be some mechanism, therefore, by which the body is able to adjust either its intake of energy as food or its expenditure of energy by oxidizing the components of that food, or both.

Body components that may be regulated are the energy stores: glycogen (small: about 1.5–2 kg) or fat in adipose tissue (large: potentially two orders of magnitude larger than glycogen stores). Historically, one school of thought held that energy intake was regulated by signals originating from body fat stores, working through detectors in the brain. This was called the lipostatic theory. When Tremblay and colleagues [64] induced fat loss in obese subjects by means of increased physical activity, they found that there was a point beyond which they could lose no further fat. This coincided with an inability of their fat cells to increase the rate at which they hydrolyzed triacylglycerols compared with the fat cells of lean subjects. These observations are consistent with some kind of lipostatic control mechanism.

Alternatively, since variations in glucose metabolism are known to affect food intake, another school of thought proposed the presence of brain receptors that were sensitive to differences in glucose concentration between the venous and arterial blood and stimulated according to the rate of glucose utilization. In support of this idea, there is recent evidence that depriving cells of glucose increases food intake and also that carbohydrate intakes are more stable than fat intakes in human beings [65].

These ideas, developed in the 1950s and 1960s, put emphasis on controls over food intake. In the 1970s and 1980s, the pendulum swung more toward concepts that individuals who are prone to obesity would be characterized by defects in energy expenditure that would favor a positive energy balance, whereas individuals who remained lean could adapt their energy expenditure according to fluctuations in food intake or to changes in body stores. Although such adaptive changes in energy expenditure have been well characterized in animals, there is little evidence for their importance in humans. Basic components of the current concept are as follows:

Equivalence of total energy intake and expenditure is not sufficient in itself to maintain energy balance. There must also be an equilibrium between the intake and oxidation of each macronutrient: protein, carbohydrate, and fat.

The rate of oxidation of carbohydrate matches carbohydrate intake quickly enough to prevent excessive exhaustion or accumulation of carbohydrate reserves.

The same is not true of fat oxidation. It appears that there is little or no ability to increase fat oxidation in response to excessive fat intakes in the short term, and since the fat stores are virtually unlimited, fat storage increases more readily in response to excessive consumption of fat than of carbohydrate. In the longer term, fat oxidation adapts so that the oxidation of the fuel mix is appropriate to the new body composition, bringing about a new equilibrium between intake and oxidation of fuels and resulting in a new, higher, stable weight.

Physical activity may either increase or attenuate food intake in people consuming high fat diets, depending on each individual's inherited ability to increase fat oxidation in response to exercise.

High fat diets are associated with appreciable short-term increases in energy intake. When given foods with a high ratio of fat to carbohydrate, people tend to consume sufficient carbohydrate to fulfill the expected daily carbohydrate need. This leads to an increase in energy from fat. There is some tendency for total food intake to be reduced (suggesting that some kind of compensation mechanism was at work), but, because of the high energy density of the foods, the body was not able to fully compensate, and energy intake increased. Some epidemiological studies seem to confirm that there is a positive statistical association between percent body fat and proportion of fat in the diet [65].

In summary, the answer to the question "Is a fat calorie more fattening than a carbohydrate calorie?" appears to be yes. The reasons are complex, but the main cause is the body's inability to increase fat oxidation rapidly in response to increases in fat consumption. Instead, fat is put into stores that are, to all intents and purposes, limitless. By contrast, when carbohydrate intake exceeds its limited storage capacity, the excess is readily oxidized. In the longer term, fat oxidation adjusts to the higher body fat stores, bringing intake and oxidation into equilibrium and resulting in a new stable weight. Whereas a decade or so ago the different views about the regulation of body weight seemed to be in conflict, the various observations described here can now be integrated into a model that goes some way to demonstrating how the body manages to maintain constancy of composition for relatively long periods despite large fluctuations in intake and in physical activity.

B. Cardiovascular Disease

It is recognized that many inherited and environmental factors interact to affect predisposition to the development of cardiovascular disease (CVD). Of the environmental factors, smoking and diet have been given most prominence, and of the various dietary components, most emphasis has been placed on fat. The consensus among nutritionists, epidemiologists, and medical practitioners is that SFA are the principal dietary components contributing to CVD through their influence on blood cholesterol, particularly LDL cholesterol. These particles are held to be a prerequisite for the development of atherosclerotic plaque [1]. Although the adherents to this hypothesis would be the first to accept the gross oversimplicity of this view, the unbiased reader of the U.K. Department of Health's most recent report, "Nutritional Aspects of Cardiovascular Disease," could surely be left in little doubt that those responsible for public health in the United Kingdom and elsewhere regard reduction in dietary SFA as the main plank in the strategy to reduce CVD.

I [66] and others [67–69] have argued strongly that the scientific evidence on which this strategy is based is weak in the extreme. A summary of the arguments is presented in Table 3.

A considerable weakness in the concept of SFA involvement in CVD as illustrated by the U.K. Department of Health report [1] is the focus on total cholesterol and LDL in the *fasting state*. In practice, most human beings in developed countries spend most of their time in the postprandial state [70]. Zilversmit [71] proposed that the most atherogenic particles were not LDL but the remnants of chylomicrons and VLDL after part of their triacylglycerol had been removed. This idea has now been extended and refined, and current views of the role of lipoproteins in CVD are summarized in the concept of the atherogenic lipoprotein phenotype (ALP) [70,72].

Briefly, the ALP describes a collection of abnormalities in lipoprotein metabolism that predispose apparently healthy people to increased CVD risk. Its characteristics are a moderately raised concentration of triacylglycerol, low concentration of HDL, and a predominance of unusually small, dense LDL and HDL particles in the postabsorptive state. The ALP is also associated with the condition of insulin resistance, in which tissues need more of the hormone than normal to accomplish a particular metabolic task. This causes the pancreas to secrete more insulin and results in habitually elevated blood insulin concentrations. There is a strong genetic basis for the ALP, which may mainly affect the enzyme lipoprotein lipase, which removes triacylglycerols from circulating chylomicrons immediately after a meal. However, it is quite clear that development of the condition involves diet, smoking, and physical activity interacting with the predisposing genes. It is quite possible that certain fatty acids directly influence the expression of the genes involved in the ALP, and this is currently the subject of vigorous research.

Table 3 Deficiencies in the Lipid Hypothesis

1. There is insufficient correspondence in vascular pathology (a) between animals and humans and (b) between general atherosclerosis and lesions seen in familial hypercholesterolemia.
2. International epidemiology is flawed by confounding factors and selection biases.
3. Within countries, epidemiology gives little support for a relationship between dietary fat and CVD. Fat intake does not explain differences in total CVD or differences in incidence of CVD according to region, sex, or social class.
4. Trends in CVD mortality are not coincident with changes in the amount and type of fat eaten.
5. The hypothesis cannot explain the greater risk in men compared with women, enhanced risk in postmenopausal women, or consistent recent decreases in mortality in women compared with more erratic changes in men in many countries. There have been no corresponding decreases in CVD incidence.
6. Plasma total cholesterol is a weak predictor of CVD compared with hemostatic factors. Less than 50% of CVD risk is accounted for by known "risk factors."
7. Extrapolations from drug trials in high risk individuals to the general population are unwarranted.
8. Intervention trials have not demonstrated a major impact of dietary fat modification on CVD and none on total mortality. Only the most stringent diets achieve useful plasma cholesterol reductions.

One reason this condition has been overlooked in the obsession with total and LDL cholesterol is the technical difficulties in routinely measuring lipoprotein subclasses, and another is that values for blood TAG, total cholesterol, and LDL that are characteristic of an ALP usually fall below the levels that clinicians regard as appropriate for taking therapeutic action after lipid screening. A likely outcome of the appreciation of the importance of this syndrome will be a swing away from the focus on dietary SFA and *n*-6 PUFA toward the *n*-3 PUFA. The latter are more effective in controlling levels of triacylglycerol-rich lipoproteins and in reversing the effects of insulin insensitivity, which lie at the heart of the ALP.

3. *Fat and Thrombosis*

The culmination of the series of events that lead to death from CVD is an arterial occlusion or thrombosis. By contrast with atherosclerosis, little research has been conducted on dietary influences on thrombosis. This has been mainly due to lack of adequate techniques for assessing the process of thrombus formation in the living human being. It is extremely complex and involves three main types of events: the aggregation of cellular elements in the blood (the platelets), the interactions of a sequence of blood clotting factors known collectively as the coagulation cascade, and a mechanism for the dissolution of clots once formed.

Experimental studies linking dietary lipids with thrombosis are still in their infancy and have been concerned mainly with (1) effects on platelet aggregation and (2) effects on different components of the coagulation cascade. Platelet studies rely on tests of function in vitro whose relationship with events in vivo is uncertain. These have been inconsistent, although some studies have shown increased potential for aggregation in diets with a high intake of SFA [73]. In studies on the components of the coagulation cascade, reduction of fat intake as a proportion of dietary energy reduced the plasma concentration of factor VII_c, but the composition of the fatty acids had no influence [73,74]. Most studies have found the concentration of plasma fibrinogen (a powerful predictor of CVD) to be uninfluenced by dietary fat. The few studies so far reported on the effects of dietary lipids on the fibrinolytic system (which dissolves clots) are conflicting [74].

Future research seems likely to be concentrated on investigating the role of total fat, SFA, and the *n*-3 PUFA on thrombus formation. Present indications are that total fat and SFA may be prothrombotic and the *n*-3 fatty acids antithrombotic. The balance of fatty acids in the diet will be of primary practical concern. Another important area of study will be to more clearly distinguish the dietary influences on the events leading to a stroke from those leading to a heart attack. Although they are both cardiovascular diseases and have ostensibly similar natural histories, heart attacks and stroke clearly respond quite differently to different risk factors, including diet [1].

C. Cancer

Cancer is a disease of cell growth regulation in which nearly all control has been lost over cell growth and the cells grow autonomously. It is generally thought to develop in three stages: *initiation*, in which some agent such as a chemical carcinogen causes a permanent alteration of the DNA of a cell; *promotion*, in which a "promoter" increases the chance of misinformation inserted into the genetic code being expressed; and *propagation*, which involves any factor that stimulates uncontrolled growth.

Doll and Peto [75] estimated that about one-third of cancer deaths might be due to diet, although their best guess ranged from 10 to 70%. Diet may influence cancer in three ways: (1) through provision of carcinogens in foods, (2) by influencing the conversion of harmless components in the gut into carcinogens by the gut microflora, or (3) as a result of the effect of nutrient balance (fat, proteins, carbohydrates, minerals, vitamins) on the initiation, promotion, or propagation of tumors. Information gained so far implies that energy, fat, and some proteins may contribute to cancer, while fiber, vitamins A, C, and E, the carotenoids, calcium, and some proteins may be protective.

Animal experients indicate that high fat diets result in increased tumor incidence but that this may be largely an effect of overall energy intake. Indeed, the most effective way to prevent or reduce cancer is to restrict energy intake. There must be a minimal dietary level of linoleic acid (*n*-6) for promotion to occur, but only in the context of a high fat diet. PUFA of the *n*-3 family seems to have a protective effect, but the evidence is so far insubstantial.

Evidence for associations between fat intake and human cancer comes from international epidemiological comparisons. The strongest relationships are with breast and colon cancer. Theoretically, the effects of fat on breast cancer could be mediated via an effect on hormone balance or through effects on the immune system via the control of eicosanoid synthesis, while effects on colon cancer might be mediated via effects on the secretion of bile salts that can be metabolized to carcinogens in the colon by the gut microflora. International epidemiological evidence is flawed by unreliable data on food intakes and by confounding factors such as smoking, reproductive practices, body weight, physical activity, and especially energy intakes, which have rarely been properly controlled. An anomaly is that fat intakes also correlate with intakes of fat-soluble vitamins, the latter being generally regarded as protective. Few within-country epidemiological studies, prospective studies, or case control studies have provided any strong evidence for or against a role for fat in most cancers. The report by the U.K. Department of Health [24] on dietary reference values was unable to make any specific recommendations about fat in relation to cancer.

The results of several studies undertaken principally to investigate risk factors for coronary heart disease have implied that whereas high plasma cholesterol concentrations may be regarded as increasing the overall risk of mortality from cardiovascular disease, very low plasma cholesterol concentrations may be associated with increased risk of cancer [76,77]. Firm conclusions about the true importance of low plasma cholesterol as a risk factor for cancer cannot yet be drawn, since several studies have been unable to demonstrate this association and no plausible biological mechanism can yet be advanced to explain it.

Despite the overwhelming emphasis on the adverse effects of dietary fat on cancer development, there has been growing interest in a naturally occurring fat-soluble cancer inhibitor: conjugated linoleic acid (CLA) [78]. CLA is mixture of isomers of linoleic acid (*cis*-9, *cis*-12-octadecadienoic acid) the main component being *cis*-9, *trans*-11-octadecadienoic acid. These are formed during the process of biohydrogenation of PUFA in the rumen of cows and subsequently find their way into milk and beef tissues. Evidence for involvement in cancer development has so far been confined to small laboratory animals and tests in vitro. Much further research is needed on the effects of CLA on human cancers before any firm conclusions can be drawn.

D. Inflammatory Disease

There is accumulating evidence that nutrition and particularly fat in the diet may in some way influence the functioning of the immune system. Certainly protein energy malnutrition markedly depresses the body's ability to mount an immune response, and deficiencies of specific nutrients, including zinc, vitamin A, and EFA, are also associated with immune malfunction. Likewise, overnutrition also seems to have an adverse effect on immune function. In particular, excessive intakes of *n*-6 PUFA seem to be immunosuppressive, and this phenomenon has been exploited in dietary support during clinical treatments designed to suppress the rejection of organ transplants and in the management of multiple sclerosis, a disease in which autoimmunity is thought to play a part [79]. The mechanism probably involves the regulation by dietary EFA of the amounts and types of eicosanoids formed, since these compounds are known to be involved in the regulation of cell-mediated immune function [80].

Rheumatoid arthritis (RA) is an inflammatory disease in which an inappropriate immune response plays a part. The idea that the severity of RA is related to diet, in particular SFA, has received widespread publicity, but the scientific basis for these beliefs is not substantial. A recent review [81] reflects the paucity of good scientific research and demonstrates overwhelming emphasis on the effects of the *n*-3 PUFA rather than on SFA.

E. Depression

The finding that some intervention trials that involved lowering blood cholesterol appeared to have resulted in more deaths from suicides and other so-called violent causes led to much speculation about whether changes in the cholesterol content of neural membranes might lead to functional changes resulting in altered states of mind [82]. In a recent review, Hibbeln and Salem [83] argue that focusing on cholesterol might be misleading. They suggest that in the cholesterol lowering studies other dietary changes, particularly increases in the *n*-6/*n*-3 PUFA ratio, might be more important. In a wide-ranging review they present a thought-provoking case for the aggravation of depressive conditions by a relative deficiency in dietary, and hence membrane, *n*-3 PUFA.

F. Postscript

This chapter has reviewed the many and varied roles of lipids in the body and in the diet. In assessing the roles of dietary lipids in health and disease, it is important to put fat in perspective with other dietary components and diet, in turn, in perspective with other environmental factors. It has been asserted frequently and sometimes stridently that western diets have changed drastically during the last century compared with diets to which, it is assumed, the body became accustomed during the evolution of the human race, and that this is a major cause of the many ills affecting developed nations [84,85]. Such arguments have to be treated with caution for a number of reasons.

First, the human body is very adaptable. Homeostatic mechanisms ensure that its cells, tissues, and organs adapt to changes in nutrient intakes by subtle changes in the amounts and activities of enzymes in metabolic pathways, by changes in membrane composition, and by changes in the numbers and activities of membrane receptors and other regulatory proteins. These changes take place over months, weeks, hours, or even minutes, certainly not centuries.

Second, the notion that a single "primitive" diet is appropriate for all human beings fits uneasily with the knowledge that diets, apparently consistent with good health and well-being, vary enormously throughout the world's communities, including severalfold differences between fat intakes expressed as a proportion of total energy as well as very large differences in the types of fats eaten.

Third, life expectancy and general health have steadily improved in the industrialized countries, a fact often forgotten or dismissed in the enthusiasm for branding dietary fat as a major killer.

Although a reduction in fat intake may be a distinct advantage for many people, especially those who find difficulty in maintaining energy balance, there is a danger that in this obsession with dietary fat other perhaps more important lifestyle factors are forgotten, including smoking, too little exercise, and raised blood pressure. The role of fat must be seen in perspective, and it should not be concluded that widespread changes in dietary fat consumption will lead inevitably to freedom from the so-called diseases of affluence. The evidence does not point that way. The key to a good diet is variety.

ABBREVIATIONS

AA	Arachidonic acid
ALP	Atherogenic lipoprotein phenotype
Apo-	Apoprotein
ATP	Adenosine triphosphate
BMI	Body mass index
CHD	Coronary heart disease
CLA	Conjugated linoleic acid
CVD	Cardiovascular disease
DHA	Docosahexaenoic acid
DLMG	Dilinoleoyl-monogammalinolenoyl glycerol
DNA	Deoxyribonucleic acid
EFA	Essential fatty acid(s)
EPA	Eicosapentaenoic acid
GLA	Gamma-linolenic acid
HDL	High density lipoprotein(s)
LA	Linoleic acid
LCAT	Lecithin-cholesterol acyltransferase
LDL	Low density lipoprotein
LPL	Lipoprotein lipase
LTA	Leukotriene A
LTB	Leukotriene B
MCT	Medium-chain triacylglycerol(s)
ME	Metabolizable energy
MUFA	Monounsaturated fatty acid(s)
PGE	Prostaglandin E
PGH	Prostaglandin H
PUFA	Polyunsaturated fatty acid(s)
RA	Rheumatoid arthritis
RNI	Reference nutrient intake
SFA	Saturated fatty acid(s)
SMCFA	Short- and medium-chain fatty acid(s)
VLDL	Very low density lipoprotein(s)

REFERENCES

1. Department of Health, Nutritional aspects of cardiovascular disease, *Report on Health and Social Subjects*, No. 46, Her Majesty's Stationery Office, London, 1994.
2. Department of Health, *The Health of the Nation: A Strategy for Health in England*, HMSO, London, 1992.
3. M. I. Gurr and J. L. Harwood, *Lipid Biochemistry: An Introduction*, 4th ed., Chapman and Hall, London, 1991.
4. Ministry of Agriculture, Fisheries and Food, *Household Food Consumption and Expenditure*, Annu. Rep. Nat. Food Survey Committee, HMSO, London, 1993.
5. M. I. Gurr, *Role of Fat in Food and Nutrition*, 2nd ed. Elsevier, London, 1992.

6. F. D. Gunstone, J. L. Harwood, and F. B. Padley, *The Lipid Handbook*, 2nd ed., Chapman and Hall, London, 1994.
7. B. Holland, A. A. Welch, I. D. Unwin, D. H. Buss, A. A. Paul, and D. A. T. Southgate, *McCance and Widdowson's The Composition of Foods*, 5th ed., Roy. Soc. Chem., Cambridge, UK, 1991.
8. R. G. Jensen, A. M. Ferris, C. J. Lammi-Keefe, and R. A. Henderson, Lipids of bovine and human milk: a comparison, *J. Dairy Sci. 73*: 223–240 (1990).
9. A. C. Goedhart and J. C. Bindels, The composition of human milk as a model for the design of infant formulas: recent findings and possible application, *Nutr. Res. Rev. 7*: 1–23 (1994).
10. British Nutrition Foundation, *Report of the Task Force on Unsaturated Fatty Acids—Nutritional and Physiological Significance*, British Nutrition Foundation and Chapman and Hall, London, 1992.
11. I. A. Prior, F. Davidson, C. E. Salmond, and Z. Czochanska, *Am. J. Clin. Nutr. 34*: 1552–1561 (1981).
12. G. O. Burr and M. M. Burr, A new deficiency disease produced by the rigid exclusion of fat from the diet, *J. Biol. Chem. 82*: 345–367 (1929).
13. L. Soderhjelm, H. F. Wiese, and R. T. Holman, The role of polyunsaturated fatty acids in human nutrition and metabolism, *Prog. Chem. Fats Other Lipids 9*: 555–585 (1970).
14. R. T. Holman and S. B. Johnson, Essential fatty acid deficiencies in man, in *Dietary Fats and Health* (E. G. Perkins and W. J. Visek, Eds.), Am. Oil Chem. Soc., Champaign, IL, 1983.
15. M. Press, H. Kikuchi, T. Shimoyama, and G. R. Thompson, Diagnosis and treatment of essential fatty acid deficiency in man, *Br. Med. J. 2*: 247–250 (1974).
16. R. T. Holman, Biological activities of and requirements for polyunsaturated fatty acids, *Prog. Chem. Fats Other Lipids 9*: 607–682 (1970).
17. J. C. Allen and R. J. Hamilton (Eds.), *Rancidity in Foods*, 2nd ed., Appl. Sci London, 1989.
18. S. M. Grundy and M. A. Denke, Dietary influences on serum lipids and lipoproteins, *J. Lipid Res. 31*: 1149–1172 (1990).
19. A. Bonanome and S. M. Grundy, Effect of dietary stearic acid on plasma cholesterol and lipoprotein levels, *N. Engl. J. Med. 318*: 1244–1248 (1988).
20. K. C. Hayes and P. Khosla, Dietary fatty acid thresholds and cholesterolemia, *FASEB J. 6*: 2600–2607 (1992).
21. M. I. Gurr, N. Borlak, and S. Ganatra, Dietary fat and plasma lipids; *Nutr. Res. Rev. 2*: 63–86 (1989).
22. R. O. Vles, Nutritional aspects of rapeseed oil, in *The Role of Fats in Human Nutrition* (A. J. Vergroesen, Ed.), Academic, London, 1975.
23. K. R. Norum, E. N. Christiansen, B. O. Cristophersen, and J. Bremer, Metabolic and nutritional aspects of long-chain fatty acids of marine origin, in *The Role of Fats in Human Nutrition*, 2nd ed. (A. J. Vergroesen and M. A. Crawford, Eds.), Academic, London, 1989.
24. Department of Health, Dietary Reference Values for Food Energy and Nutrients for the United Kingdom, *Report on Health and Social Subjects*, No. 41, HMSO, London, 1991.
25. A. G. Hassam, The role of evening primrose oil in nutrition and disease, in *The Role of Fats in Human Nutrition* (F. B. Padley and J. Podmore, Eds.), Ellis Horwood, Chichester, 1985, pp. 84–100.
26. D. F. Horrobin, Natural does not equal safe, *Pharm. Technol. Eur.*, December 1994, pp. 14–15.
27. R. S. Lees and M. Karel (Eds.), *Omega-3 Fatty Acids in Health and Disease*, Marcel Dekker, New York, 1990.
28. M. L. Burr, A. M. Fehily, J. F. Gilbert, S. Rogers, R. M. Holliday, P. M. Sweetnam, P. C. Elwood, and N. M. Deadman, Effects of changes in fat, fish and fibre intakes on death and myocardial infarction: Diet and Reinfarction Trial (DART), *Lancet ii*: 757–761 (1989).
29. L. G. Cleland, J. K. French, W. H. Betts, G. A. Murphy, and M. J. Elliott, Clinical and biochemical effects of dietary fish oil supplements in rheumatoid arthritis, *J. Rheumatol. 15*: 1471–1475 (1988).
30. I. Tamamoto, M. Sugano, and M. Wada, Hypocholesterolaemic effect of animal and plant fats in rats, *Atherosclerosis 13*: 171–184 (1971).

31. R. B. McGandy, D. M. Hegsted, and M. L. Myers, Use of semisynthetic fats in determining effects of specific dietary fatty acids on serum lipids in man, *Am. J. Clin. Nutr. 23*: 1288–1298 (1970).
32. A. Christophe, F. Matthys, R. Geers, and G. Verdonk, Nutritional studies with randomized butter: cholesterolaemic effects of butter oil and randomized butter oil in man, *Arch. Int. Physiol. Biochem. 86*: 413–415 (1978).
33. D. J. McNamara, Relationship between blood and dietary cholesterol, in *Meat and Health: Advances in Meat Research*, Vol. 6 (A. M. Pearson and T. R. Dutson, Eds.), Elsevier, London, 1990, pp. 63–87.
34. E. Renner, *Milk and Milk Products in Human Nutrition*, Volkswirtschaftlicher Verlag, Munich, 1983.
35. G. Billek, Heated fats in the diet, in *The Role of Fats in Human Nutrition* (F. B. Padley and J. Podmore, Eds.), Ellis Horwood, Chichester, 1985, pp. 163–172.
36. M. I. Gurr, Lipids: products of industrial hydrogenation, oxidation and heating, in *Nutritional and Toxicological Aspects of Food Processing*, (R. Walker and E. Quattrucci, Eds.), Taylor and Francis, London, 1988, pp. 149–155.
37. C. B. Taylor, S. K. Peng, N. T. Werthesen, P. Tham, and K. T. Lee, Spontaneously occurring angiotoxic derivatives of cholesterol, *Am. J. Clin. Nutr. 32*: 40–57 (1979).
38. British Nutrition Foundation, *Report of the Task Force on Trans Fatty Acids*, Br. Nutr. Foundation, London, 1995.
39. P. W. Parodi, Composition and structure of some consumer-available edible fats, *J. Am. Oil Chem. Soc. 53*: 530–534 (1976).
40. P. W. Parodi, Distribution of isomeric octadecenoic fatty acids in milk fat, *J. Dairy Sci. 59*: 1870–1873 (1976).
41. B. Koletsko, *Trans* fatty acids may impair biosynthesis of polyunsaturates and growth in man, *Acta Paediatr. 81*: 302–306 (1992).
42. R. P. Mensink and M. B. Katan, Effects of dietary trans fatty acids on high density and low density lipoprotein levels in healthy subjects, *N. Engl. J. Med. 323*: 439–444 (1990).
43. J. T. Judd, B. A. Clevidence, R. A. Muesing, J. Wittes, M. E. Sunkin, and J. J. Podczasy, Dietary *trans* fatty acids: effects on plasma lipids and lipoproteins of healthy men and women, *Am. J. Clin. Nutr. 59*: 861–868 (1994).
44. R. P. Mensink, P. L. Zock, M. B. Katan, and G. H. Hornstra, Effects of dietary *cis* and *trans* fatty acids on serum lipoprotein(a) levels in humans, *J. Lipid Res. 33*: 1493–1501 (1992).
45. P. J. Nestel, M. Noakes, B. Belling, R. McArthur, P. Clifton, E. Janus, and M. Abbey, Plasma lipoprotein lipid and Lp (a) changes with substitution of elaidic acid for oleic acid in the diet, *J. Lipid Res. 33*: 1029–1036 (1992).
46. W. C. Willett, M. J. Stampfer, J. E. Manson, G. A. Colditz, F. E. Speizer, B. A. Rosner, L. A. Sampson, and C. H. Hennekens, Intake of *trans* fatty acids and risk of coronary heart disease among women, *Lancet 341*: 581–585 (1993).
47. A. Aro, A. F. M. Kardinaal, I. Salminen, J. D. Kark, R. A. Riemersma, M. Delgado-Rodriguez, J. Gomez-Aracena, J. K. Huttunen, L. Kohlmeier, B. C. Martin, J. M. Martin-Moreno, V. P. Mazaev, J. Ringstad, M. Thamm, P. van't Veer, and F. J. Kok, Adipose tissue isomeric *trans* fatty acids and risk of myocardial infarction in nine countries: the EURAMIC study, *Lancet 345*: 273–278 (1995).
48. T. L. Roberts, D. A. Wood, R. A. Riemersma, P. J. Gallagher, and F. C. Lampe, *Trans* isomers of oleic and linoleic acids in adipose tissue and sudden cardiac death, *Lancet 345*: 278–282 (1995).
49. P. M. Kris-Etherton, J. Derr, D. C. Mitchell, V. A. Mustad, M. E. Russell, E. T. McDonell, D. Salabsky, and T. A. Pearson, The role of fatty acid saturation on plasma lipids, lipoproteins and apoproteins. I. Effects of whole food diets high in cocoa butter, olive oil, soya bean oil, dairy butter and milk chocolate on the plasma lipids of young men, *Metabolism 42*: 121–129 (1993).
50. D. Kritchevsky, S. A. Tepper, D. Vesselinovitch, and R. W. Wissler, Cholesterol vehicle in experimental atherosclerosis. 13. Randomized peanut oil, *Atherosclerosis 17*: 225–243 (1973).
51. K. Sickinger, Clinical aspects and therapy of fat malassimilation with particular reference to the

use of medium-chain triglycerides, in *The Role of Fat in Human Nutrition* (A. J. Vergroesen, Ed.), Academic, London, 1975, pp. 115–209.
52. R. Sieber, Cholesterol removal from animal food—can it be justified? *Lebensm. Wiss. Technol. 26*: 375–387 (1993).
53. E. Renner and M. I. Gurr, Do we need cholesterol reduced dairy products? *Dairy Ind. Int. 56*: 34–35 (1991).
54. R. W. Foltin, M. W. Fischman, T. H. Moran, B. J. Rolls, and T. H. Kelly, Caloric compensation for lunches varying in fat and carbohydrate content by humans in a residential laboratory, *Am. J. Clin. Nutr. 52*: 969–980 (1990).
55. V. K. Babayan, Modification of food to control fat intake, *J. Am. Oil Chem. Soc. 51*: 260–264 (1974).
56. K. W. Miller and G. S. Allgood, Nutritional assessment of Olestra, a non-calorific fat substitute, *Int. J. Food Sci. Nutr. 44*(Suppl. 1): S77–S82 (1993).
57. J. C. Peters, B. N. Holcombe, L. K. Hiller, and D. R. Webb, Caprenin. 3. Absorption and caloric value in adult humans, *J. Am. Coll. Toxicol. 10*: 357–367 (1991).
58. J. R. Hayes, D. H. Pence, S. Scheinbach, R. P. D'Amelia, L. P. Klemann, N. H. Wilson, and J. W. Finley, Review of triacylglycerol digestion, absorption and metabolism with respect to SALATRIM triacylglycerols, *J. Agric. Food Chem. 42*: 474–483 (1994).
59. L. P. Klemann, J. W. Finley, and G. A. Leveille, Estimation of the absorption coefficient of stearic acid in SALATRIM fats, *J. Agric. Food Chem. 42*: 484–488 (1994).
60. J. W. Finley, L. P. Klemann, G. A. Leveille, M. S. Otterburn, and C. G. Walchak, Caloric availability of SALATRIM in rats and humans, *J. Agric. Food Chem. 42*: 495–499 (1994).
61. N. M. Kaplan, The deadly quartet. Upper body obesity, glucose tolerance, hypertriglyceridaemia and hypertension, *Arch. Int. Med. 149*: 1514–1520 (1989).
62. A. Tremblay, Human obesity: a defect in lipid oxidation or in thermogenesis? *Int. J. Obesity 16*: 953–957 (1992).
63. J. P. Flatt, Importance of nutrient balance in body weight regulation, *Diabetes/Metab. Rev. 4*: 571–581 (1988).
64. A. Tremblay, J.-P. Després, and C. Bouchard, Adipose tissue characteristics of ex-obese long-distance runners, *Int. J. Obesity 8*: 641–648 (1984).
65. A. Tremblay, G. Plourde, J.-P. Després, and C. Bouchard, Impact of dietary fat content and fat oxidation on energy intake in humans, *Am. J. Clin. Nutr. 49*: 799–805 (1989).
66. M. I. Gurr, Dietary lipids and coronary heart disease: old evidence, new perspective, *Prog. Lipid Res. 31*: 195–243 (1992).
67. R. H. Rosenman, The questionable roles of the diet and serum cholesterol in the incidence of ischemic heart disease and its 20th century changes, *Homeostasis 34*: 1–44 (1993).
68. U. Ravnskov, Cholesterol lowering trials in coronary heart disease: frequency of citation and outcome. *Br. Med. J. 305*: 15–19 (1992).
69. W. E. Stehbens, The quality of epidemiological data in coronary heart disease and atherosclerosis, *J. Clin. Epidemiol. 46*: 1337–1346 (1993).
70. S. Sethi, M. J. Gibney, and C. M. Williams, Postprandial lipoprotein metabolism, *Nutr. Res. Rev. 6*: 161–183 (1993).
71. D. B. Zilversmit, Atherogenesis: a postprandial phenomenon, *Circulation 60*: 473–485 (1979).
72. B. A. Griffin and A. Zampelas, Influence of dietary fatty acids on the atherogenic lipoprotein phenotype, *Nutr. Res. Rev. 8*: 1–26 (1995).
73. S. Renaud, F. Godsey, E. Dumont, C. Thevenon, E. Ortchanian, and J. L. Martin, Influence of long term diet modification on platelet function and composition in Moselle farmers, *Am. J. Clin. Nutr. 43*: 136–150 (1986).
74. P. Marckmann, B. Sandstrom, and J. Jespersen, Effects of total fat content and fatty acid composition in diet on factor VII coagulant activity and blood lipids, *Atherosclerosis 80*: 227–233 (1990).
75. R. Doll and R. Peto, The causes of cancer, *J. Natl. Cancer Inst. 66*: 1191–1308 (1981).
76. C. G. Isles, D. J. Hole, C. R. Gillis, V. M. Hawthorne, and A. F. Lever, Plasma cholesterol,

coronary heart disease and cancer in the Renfrew and Paisley survey, *Br. Med. J. 298*: 920–924 (1989).

77. S. B. Hulley, J. M. B. Walsh, and T. B. Newman, Health policy on blood cholesterol: time to change direction, *Circulation 86*: 1026–1029 (1992).
78. P. W. Parodi, Conjugated linoleic acid: an anticarcinogenic fatty acid present in milk fat, *Aus. J. Dairy Technol. 49*: 93–97 (1994).
79. M. I. Gurr, The role of lipids in the regulation of the immune system, *Prog. Lipid Res. 22*: 257–287 (1983).
80. P. V. Johnson, Dietary lipids and immunity, *Adv. Lipid Res. 21*: 103–141 (1985).
81. D. M. Kremer and D. R. Robinson, Studies of dietary supplementation with omega-3 fatty acids in patients with rheumatoid arthritis, in *Health Effects of Omega-3 Polyunsaturated Fatty Acids in Seafoods* (*World Rev. Nutr. Diet.*, Vol. 66) (A. P. Simopoulos, R. R. Kifer, R. E. Martin, and S. M. Barlow, Eds.), Karger, Basel, 1991, pp. 367–382.
82. H. Engelberg, Low serum cholesterol and suicide, *Lancet 339*: 727–729 (1992).
83. J. R. Hibbeln and N. Salem, Dietary polyunsaturated fatty acids and depression: when cholesterol does not satisfy, *Am. J. Clin. Nutr. 62*: 1–9 (1995).
84. World Health Organization Study Group, Diet, Nutrition and the Prevention of Chronic Diseases, Tech. Rep. Ser. 797, WHO, Geneva, 1990.
85. S. B. Eaton and D. A. Nelson, Calcium in evolutionary perspective, *Am. J. Clin. Nutr. 54*: 281S–287S (1991).
86. M. I. Gurr, Fats in *Human Nutrition and Dietetics* (J. S. Garrow and W. P. T. James, Eds.), Churchill Livingstone, Edinburgh, 1993, Chap. 6.
87. M. I. Gurr, Polyunsaturated fatty acids of the *n*-3 family: influence on inflammatory disease, *Lipid Technol. 5*: 65–68 (1993).

5

Extraction of Lipids from Natural Sources

Maurice A. Williams
Anderson International Corp., Cleveland, Ohio

I. INTRODUCTION

Lipids are removed from lipid-containing materials by two methods: mechanical crushing and solvent extraction. Small volume processors prefer mechanical crushing to avoid having to build a costly solvent extraction system. High volume processors prefer solvent extraction to avoid having multiple mechanical screw presses operating in parallel. Usually volumes less than 100 Mg/day are mechanically crushed, and volumes above 300 Mg/day are solvent extracted. Volumes range from a few megagrams per day for a small crushing mill to 3000 or more for a large solvent extraction plant.

In mechanical crushing, the oilseed is cleaned, cracked, flaked, sometimes cooked, and conditioned to optimum moisture and temperature and then enters a mechanical screw press, which generates enough pressure to cause oil to flow out of the oilseed. Crushing can reduce the oil content of most oilseeds to less than 5%. In solvent extraction, the prepared oilseed passes through a percolation vessel on a moving screen (or perforated plate or some similar device) that transports the material through the vessel. Hot solvent, sprayed on top, percolates through the material, extracting the oil. The solvent is then removed from the oil and the solid residue. Solvent extraction can reduce oil levels below 1%.

II. PREPARATION

A. Cleaning

Oilseeds come into the mills along with stems, leaves, stones, sand or dirt, and sometimes weed seeds. Stems, dirt, and stones are abrasive and contain no oil. Weed seeds contain different kinds of oil. The first preparation step is to clean the incoming oilseeds. Stems and stones are usually larger than the oilseeds and can be screened out through revolving reels. Weed seeds are usually smaller and can be removed along with sand and dirt through vibrating screens. Magnets remove tramp iron.

Table 1 Preparation of Some Common Oilseeds

Oilseed	Particle Size	Cook at		Dry to	
		% H_2O	°C	% H_2O	°C
Copra	3 mm	—	—	2–2.5	127
Corn germ	Whole germ	—	—	2	115
Cottonseed	0.25 mm flakes	12–14	88	2.6–2.8	115–129
Linseed	0.25 mm flakes	11	88	2.5	115
Palm kernels	0.38 mm	—	—	1.5–2.0	115
Peanut	3 mm	9	88	2.5–3.0	121
Rapeseed	3 mm flakes	12	88	3	115
Safflower	3 mm	12	88	3	71
Sesame	Whole seed	—	—	2.5–3.0	127
Soybean	1.5 mm	—	—	2–2.25	132
Sunflower	0.25 mm flakes	8–9	88	3	110
Tung	3 mm	—	—	3	121

B. Dehulling

Hulls are abrasive. They usually contain less than 1% oil, and they add to the bulk entering the screw press. Removing hulls reduces wear on the screw press, increases oil yield (because hulls absorb oil within the press), and increases capacity (because the screw press presses only meats rather than whole oilseed). There are several different dehulling procedures. Bar hullers or disk hullers are used for medium-sized seeds such as sunflower seed or cottonseed. Hot aspiration systems are used for soybeans. Cracking machines are used for larger oilseeds. Seeds enveloped in thick, hard shells are dehulled by various devices that crack the hulls, then separate the hulls by screening or aspiration or by flotation and sometimes even by hand [1]. The objective of all procedures is to remove as much hull as possible with minimal loss of meats. A special case is cottonseed, which contains lint, the short ends of cotton fibers, still attached to the hull.

C. Flaking

Oilseed meats are usually flattened into thin flakes. This helps to more readily expose the oil to solvent than would prevail with whole or cracked meats. It also reduces power consumption of a screw press, which has to pulverize the seeds within the press to release the oil. Flaking also ensures a more uniform cook (if the oilseed requires cooking) because all oilseed particles are uniformly thick.

Two types of flakers are used. One type employs two heavy steel rolls in parallel, almost touching each other and revolving in opposite directions so that the two rolls pull the meats down between them. Mechanical jackscrews or hydraulic cylinders apply pressure against the bearing blocks supporting the rolls so that the rolls do not spread apart when the meats pass between them. Sometimes one roll revolves faster than the other. This helps the smearing action of the rolls.

The second type stacks flaking rolls atop each other, usually five high. This has the advantage that the weight of the upper rolls provides the pressure. The meats are progressively flattened into thinner and thinner flakes as they pass from roll to roll in their descent down the stack.

Five-high rolls are well accepted for flaking oilseeds in mechanical crushing plants. They are not widely used to prepare oilseeds for solvent extraction where large unbroken flakes are desired. A five-high roll, for example, flakes the seed four times and produces more broken flakes and fines than a single-pass flaker would.

D. Cooking and Drying

Cooking is excellent preparation for most oilseeds and is necessary for those too soft to withstand the pressures generated within a screw press. Cooking hardens those oilseeds, similar to the way cooking hardens eggs. Cooking also ruptures oil cells, thereby releasing the oil into a more fluid state. The temperature of cooking reduces oil viscosity, allowing it to flow more readily through the slotted walls of the screw press. Cooking inactivates troublesome enzymes that adversely affect oil and meal quality. Cooking also sterilizes the oilseed by destroying bacteria, molds, fungi, etc. that infest the incoming oilseed.

Cooking occurs when oilseeds are held at elevated temperature, preferably around 98°C, and at 10–12% moisture for at least 15 min without loss of moisture or drop in temperature. When the chemical and physical reactions caused by cooking are completed, the oilseed is dried to a lower moisture more compatible to the screw press. Sometimes separate vessels are used for cooking and drying. Sometimes the same vessel is sectioned into different compartments, one or two for cooking, the rest for drying. Sometimes cooking is done by extrusion followed by a separate drying step. Sometimes extrusion cooking is done in a high shear, low moisture extruder, eliminating the need for a dryer.

The moisture–temperature conditions of cooking must prevail throughout the entire cook cycle. If moisture is lost as the temperature is raised, the cooking suffers and may not occur. A cooker vessel should have a sealed inlet and a sealed outlet to prevent moisture within it from escaping. Feed screws pushing flakes against a hinged baffle plate as they fall into the cooker serve as inlet seals. Similar screws or alternative devices (described below) serve as discharge seals. Live steam (sometimes a mixture of live steam and water) is sprayed into the cooker chamber to elevate the oilseed moisture and provide a humid atmosphere within the cooker. Steam jackets on the shell provide heat.

Dryers, on the other hand, drive moisture out of the oilseed, reducing the moisture to around 2–3% for more efficient pressing. Dryers are vented, and usually a draft of air is induced to draw the water vapors away from the flakes. As flake moisture drops, the remaining moisture is more tenaciously gripped by the solids. If temperature were not raised, drying would stop.

Temperature, therefore, is raised. There are usually several drying vessels in series (or separate chambers in series) each heated through a steam jacket. By the time the flakes exit the last chamber (ready to enter the screw press) they should be at 2.5–3% moisture. They will have to be heated to 115–126°C to reach this moisture.

Once the material reaches the desired final moisture and temperature it should immediately enter the screw press. Long delays or, worse than that, pneumatic conveying of the dried product can seriously upset the critical moisture–temperature conditions required for optimum screw pressing. Preparation is summarized in Table 1.

1. Stack Cookers

Two types of jacketed cookers are used for oilseeds: stack cookers and horizontal cookers. A stack cooker looks like a cylindrical silo standing upright. The interior is divided into several horizontal sections, each with a steam-jacketed bottom. Down through the center is a rotating shaft with horizontal arms that sweep across the bottom of each stack. Flakes enter the top stack through an inlet seal (as described earlier).

Built into each stack bottom is a trapdoor that is pushed open every time the sweep arm passes over it. The magnitude by which each trapdoor is opened can be adjusted by external linkage. The flakes fall from stack to stack through these trapdoors. Each stack can be vented to a common header to accommodate drying. The vents on the lower stacks are usually open; those for the top stacks are closed to prevent moisture loss during cooking. Live steam (sometimes mixed with water) is injected into the top stack through a hollow passage in the sweep arm.

2. *Horizontal Cookers*

Horizontal cookers are arranged in clusters of separate vessels, one for cooking, three or more for drying. Each vessel is steam-jacketed and contains a revolving shaft with protruding stirring paddles. Flakes enter the cooker through an inlet seal, migrate through the vessel as the shaft tumbles them against the walls, and find a level within the cooker dependent on angle of repose and dams that are inserted into the discharge end of the cooker.

The flakes spill over the dams into a closed chamber above a discharge screw that conveys the flakes against a hinged plate (flake seal) as they fall away from the cooker. Live steam and water are injected through the inlet face plate of the cooker near the bottom so that the steam immediately contacts the flakes.

In comparison to a horizontal cooker, a stack cooker has an advantage in that it is more compact than a cluster of horizontal vessels. A disadvantage is the time required for the incoming material to reach cooking temperature. The incoming material falls on top of a thick layer of material that is slowly agitated beneath its surface. It takes several minutes before any of the incoming material contacts the steam-heated bottom. This allows oil-damaging enzymes in the incoming material several minutes to start their action before the material reaches inactivation temperature.

In a horizontal cooker, the incoming material immediately mixes with a large volume of tumbling material already heated by the jackets. Therefore the incoming material reaches enzyme inactivation temperature faster than it would in a stack cooker, and there is less damage done by the enzymes. However, if enzymatic damage is a serious problem, as with myrosinase in rapeseed [2], extrusion cooking (to be described below) can bring the oilseed to inactivation temperature within a few seconds.

E. Detoxification

Several toxins are troublesome in oilseeds. Cottonseed contains gossypol. Castor seed contains three toxins, CB-1A being the most difficult to inactivate. Seeds like peanuts and cottonseed that are permitted to mold can contain aflatoxin, a metabolic waste product of several aspergilli molds.

Gossypol is a deep red oil-soluble compound, unhealthy for chickens, turkeys, and baby pigs. Gossypol also darkens cottonseed oil if allowed to dissolve into the oil. Cooking, when done properly, can bind gossypol to the cottonseed protein [3], ensuring a light-colored oil. Some bound gossypol in the meal is released within the animal's digestive tract, however [4]. Most cottonseed meal is sold as an ingredient in ruminant feeds because, with their multichambered stomachs, ruminants are not harmed by gossypol.

The toxin CB-1A in castor can be inactivated by procedures Rhee developed at Texas A&M University [5,6] with chemical additives, such as $Ca(OH)_2$ or NaOH and NaOCl and using an extruder as a high pressure reactor vessel [7]. Aflatoxin is a metabolic waste product of several species of molds, including ordinary bread mold, *Aspergillus flavus*. These molds occur worldwide and are potential problems with any oilseed that molds before processing, especially with peanuts, corn, cottonseed, and copra. Methods have been developed for detoxifying aflatoxin-contaminated meals with ammonia. The ammonia reacts with the aflatoxin, breaking it into smaller nontoxic residues. Ammoniation can reduce aflatoxin levels from 300–1000 ppb to as low as 1–3 ppb, but it also reduces the meal's nutritional value [8].

Since prevention is a less expensive procedure, there is interest in reasonable precautions that could prevent the mold from growing in the first place [9,10]. Horne et al. [9] outline prevention methods on the farm, during harvest, and during storage that could minimize mold growth. Hron et al. [11] and Lusas et al. [12] developed procedures for extracting aflatoxin with alcohols. Those procedures will be discussed in more detail in Section V.B. Rhee [5] has adapted the castor meal deallergenation process to destroy aflatoxin using NaOH and NaOCl.

F. Extrusion Preparation

1. Background

In 1954, Zies and Baer developed an expanding extruder using some of the technology they used to build screw presses [13] (Fig. 1). This machine first served to cook and puff cereal grains for the pet food industry [14] and later found use in other industries [15].

In the 1960s, realizing that the expansion and extrusion also inactivated enzymes, Baer, Williams, and Zies developed a process for extrusion cooking of rice bran to inactivate lipase and agglomerate the bran into spongelike collets prior to solvent extraction [16]. In the 1970s, this expander was also used to transform flaked soybean and other oilseeds into porous collets [17].

An expander (or expanding extruder) consists of a rapidly rotating wormshaft within a cylindrical barrel. Oilseed enters one end of the barrel and is forced out through a die plate at the discharge end. Worm flighting is interrupted rather than continuous. The interruptions mesh with stationary pins protruding from the barrel wall. Flights that rotate between stationary pins impart a highly turbulent mastication to the oilseed. Injected steam elevates the moisture and temperature and softens the material.

Fig. 1 An expanding extruder. (Courtesy of Anderson International Corp.)

The oilseed, at 10–13% moisture and 110–115°C, is subjected to 1379–4137 kPa pressure for 10–30 s before discharge from the expander. When the oilseed emerges from the expander, some absorbed moisture flashes. This inflates the particles with internal pores and surface cracks, giving the particles a porous, spongelike internal structure.

2. *Cooking by Extrusion*

Since expanders cook within 20 s, they counteract the activity of troublesome enzymes, such as lipase in rice bran [18] and urease in soybean [19]. The short time between enzyme activation by air and moisture and inactivation by heat destroys the enzyme before it has time to cause damage. An expander is even more effective doing this than the horizontal atmospheric pressure cookers described earlier.

Canola, for example, contains enzymes that release phosphorus compounds into the oil. If flaked canola is processed through a stack cooker, the enzymes, which are triggered into action when flaking exposes them to air, have several minutes to release phosphatides into the oil before the slow-acting stack cooker brings the canola to a high enough temperature to inactivate the enzymes. An expander brings the canola to full temperature in 20 s. This results in a significant reduction of phosphorus compounds in the finished oil, 48 ppm compared to 350 ppm when using stack cookers. Reduced chlorophyll, free fatty acid, peroxide value, and greenish color are added benefits.

Zhang et al. [20] found a similar reduction of phosphorus in soybean oil. Conventionally prepared soybean yields degummed oil with 184 ppm phosphorus. Extrusion-prepared soybean yielded degummed oil with 67 ppm phosphorus.

In this application, the expander replaces the cooking tray in a stack cooker and cooks the oilseed at 88–105°C. If the oilseed is going to a solvent extractor, it would continue to a prepress or collet expander (same as flakes from a cooker). If the oilseed is going to a full press, then drying to optimum crushing moisture (2–3%) is still required.

3. *Extrusion before Crushing*

Expanding extruders are also used to prepare oilseeds for mechanical crushing. Nelson et al. [21] passed soybean through a high shear extruder and converted the bean into a frothy meal of almost fluidlike consistency. Fluidization was caused by moisture boiling through the freshly liberated oil. The soybean, at 10% moisture and ambient temperature, was cracked before entry into the extruder. High shear developed by the extruder pulverized the soybean and ruptured the oil cells. Heat generated by shaft friction raised the temperature to 135°C. This caused the moisture to flash from 10–14% down to 6–7% as the material flowed out of the extruder.

Oilseed expanders fitted with adjustable jaw chokes to replace the die plates (Fig. 2) and operated under low moisture, high shear conditions can transform cracked or uncracked oilseeds, with or without preheating, into frothy, semifluid extrudates that flash to 5–7% moisture en route to the screw press. A rotating cone point inserted between two laterally positioned stationary jaws generates the shear [22,23]. Both the proximity of the cone point to the jaws and the opening between the jaws are adjustable.

4. *Extrusion Before Solvent Extraction*

The first oilseed application for extrusion was rice bran [16,18]. Rice bran comes into a solvent extraction plant as a fine powder, which makes rice bran very difficult to extract. The powder also contains an enzyme, lipase, that splits triglycerides into free fatty acids. Acti-

Fig. 2 An oilseed extruder with an adjustable-jaw choke. (Courtesy of Anderson International Corp.)

vated when the bran is removed from the rice, lipase will raise the free fatty acid level approximately three to seven percentage points every day [24]. The cooking conditions within an expander inactivate lipase and convert the bran into porous, spongelike particles (collets) that permit rapid flow of solvent. The same procedures also transform flaked soybean and flaked cottonseed into porous collets that handle better in an extractor. Farnsworth et al. [25] describe early extrusion research with cottonseed.

Flaked oilseeds extract well if the flakes are good quality, but if flakes are thick or crumbly or if the feed to the flaker already contains fines, then extraction suffers. Extrusion is beneficial because it converts poor quality flakes into easily extracted collets. Even compared to good flakes, extruded collets offer great advantages because collets are larger, heavier, stronger, and more porous than flakes. This permits faster solvent flow (because of larger particles), greater extractor capacity (because the heavier, more porous collets occupy less space), and better drainage (because the collets are strong enough not to crumble as easily as flakes do) [26].

Extrusion converts the oilseed's protein into a gluelike condition that binds all the oilseed particles into a densely compacted elastic mass. When the material exits the expanding extruder into atmospheric conditions, the mass inflates with internal pores as the superheated moisture flashes into water vapor [27].

Some oilseeds are mechanically crushed and then solvent-extracted. The partially deoiled "cake" from the screw press is sent to a solvent extractor. Expanding extruders can transform the presscake into porous collets. Steam injected into the expander raises cake moisture 2–4 percentage points. This must be done before the cake has cooled and hardened. Once cool and hard, the denatured protein in the cake can no longer be transformed into a gelatinous, inflatable condition. If oilseeds of high oil content are prepressed to 15–30% oil, they extrude similarly to flaked soybean (at 18% oil) and flaked cottonseed meats (at 30–33% oil).

Slotted-Wall Expanders. Oilseeds containing more than 30–33% oil cannot be extruded through a closed-wall expander because the oil accumulates within the expander and stops steady-state operation. Anderson International (Cleveland, OH) developed a slotted-wall expander that can make collets from high oil materials by allowing the liberated oil to escape through the slotted-wall drainage cage [28] (Fig. 3). Field trials with canola [29] and other oilseeds [30] showed that this Hivex expander could extrude full-fat oilseeds at high oil levels and produce collets at 20–30% oil. Preparation for most oilseeds is to crack or flake the seeds and heat to approximately 60–70°C while reducing the moisture to approximately 8% and being careful not to denature the protein.

III. MECHANICAL EXTRACTION OR CRUSHING

A. Background

Most mechanical crushing of lipid-containing materials is done with screw presses. A screw press is constructed of a screwlike wormshaft that rotates within a slotted-wall, cylindrical

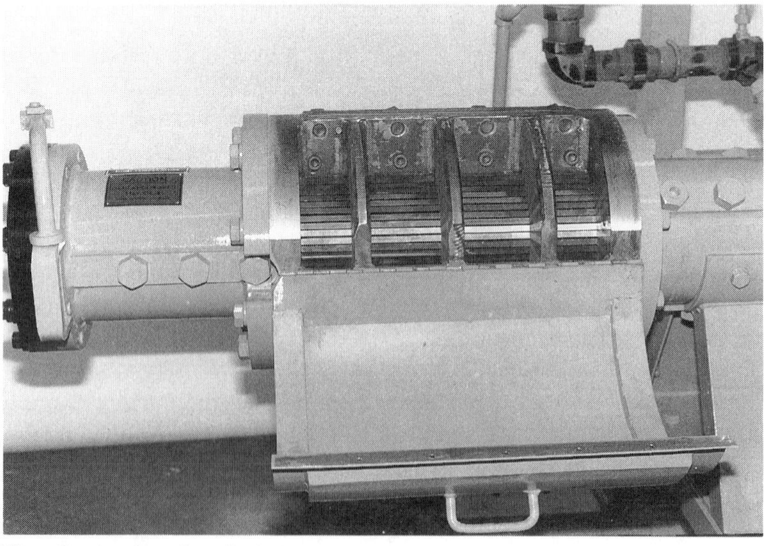

Fig. 3 A slotted-wall drainage cage on an oilseed extruder. (Courtesy of Anderson International Corp.)

housing. It accepts a continuous stream of lipid-bearing material, propels it through the housing, and compresses it under high pressure to squeeze out the oil. The first workable screw press was developed by Valerius D. Anderson [31] in 1900 and improved in 1906. His improved screw press had a barrel made of two longitudinal halves (each an assembly of bars and spacer shims) that were clamped together with heavy clamping bolts. Inserted between the barrel halves were two lug bars called knife bars, 180° apart. The lugs protruded from the barrel walls and meshed with interruptions in the wormshaft flighting. Figure 4 shows a modern press in cross section.

Incoming seeds are cleaned, dehulled, cracked or flaked, and sometimes cooked, as described earlier [32]. The final step is to reduce the oilseed's moisture to a level more compatible for screwpressing: 2–3% for full pressing, 4–6% for prepressing. If the moisture is not reduced, the solid material is too soft, and some of it flows out with the oil. If the moisture is too low, the material will overheat from friction generated by the wormshaft. Moisture is also important in the flaking and dehulling operations. Good crushing requires proper moisture control throughout the process.

In addition to oilseed preparation, three mechanical aspects of a screw press also influence its performance: the wormshaft, the barrels (or drainage cages), and the choke mechanism. The wormshaft revolves within the barrel, propelling the oilseed toward the discharge end, where it encounters the choking mechanism. Shaft speed, screw geometry, and applied horsepower influence capacity and pressure profile along the barrel. Choke position also influences the pressure profile, especially the final pressure before discharge. Drainage cage spacing determines the ease with which oil rather than solids flows through the walls.

B. Full Presses Versus Prepresses

Screw presses serve two applications for oilseeds: fullpressing and prepressing. Fullpressing requires shaft designs that generate pressure—estimated at 96,500 kPa—to squeeze out as

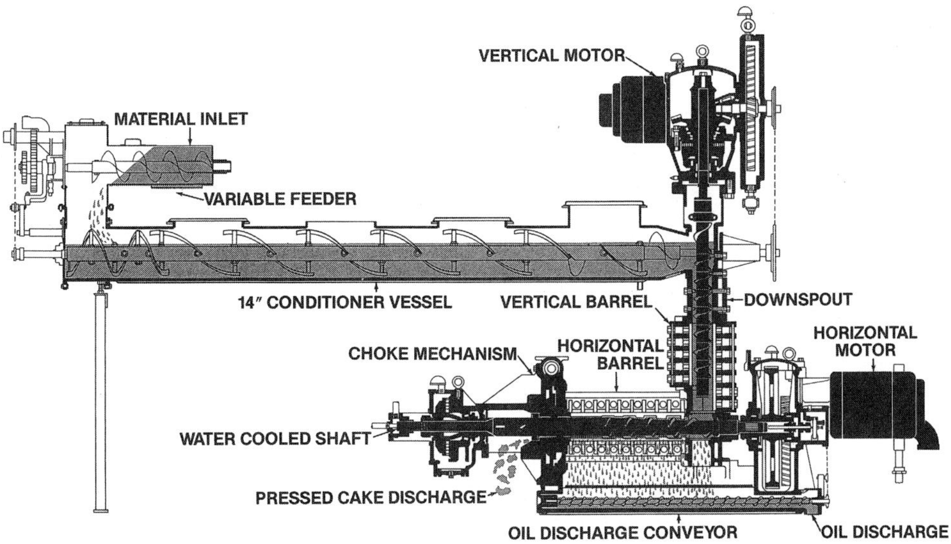

Fig. 4 Sectional view of a screw press showing knife bars and wormshaft flighting. (Courtesy of Anderson International Corp.)

much oil as possible, preferably to 3–5% residual. Animal materials can be reduced to 7–10% residual.

Prepressing is used to partially deoil the seed in preparation for solvent extraction. A prepress (Fig. 5) generates less pressure than a full press, consumes less kilowatts per kilogram of oilseed, and runs at higher capacity. The wormshaft revolves faster, inlet moisture is higher, and less choking (to apply backpressure) is needed. More oil, however, is left in the solids: 15–30%, compared to 3–5% for fullpressing.

C. The Wormshaft

The wormshaft picks up the feed entering the screw press and transports it along the barrel, subjecting it to greater and greater pressure by means of a steadily increasing hub diameter and/or shorter flight pitch. Ideally, wormshaft geometry should increase compaction while at the same time compensating for loss of volume as oil escapes through the barrels [33]. This is not easy because different oil-bearing materials generate different resistances to flow, which influences the pressure. Even the same material with different preparation can vary in its resistance to flow.

Hull content, for example, increases resistance to flow. Materials high in hull content do not easily slide through the screw press's channel (annular space between the shaft and the barrel). Shafts designed for these materials have uniform channel depth and uniform pitch. They rely on the material's resistance to flow to generate pressure. An adjustable choking device at the end of the shaft can increase the material's resistance to flow, thereby increasing pressure.

Softer materials flow more easily. There is not enough drag for a uniform channel depth, uniform pitch wormshaft to develop sufficient pressure. Shafts for soft materials rely on decreasing channel depth and decreasing pitch to generate pressure and also on "cones" or funnel-shaped segments inserted between the worm segments. Channel depth decreases rapidly as the oilseed flows across the cone and can be very thin at the cone's discharge end,

Fig. 5 A prepress for partially deoiling lipid-containing materials. (Courtesy of Anderson International Corp.)

3.17 mm in some cases. Cones present obstacles for the propelling action of worms. Worms preceding the cones have to generate higher pressure and therefore grind out more oil.

Some screw presses use two shafts, one that, in effect, prepresses, followed by a second one that fullpresses (shown in Fig. 4). The first shaft is placed upright above the main (horizontal) shaft and pushes feed down into the main shaft. The vertical shaft often has a drainage cage.

D. The Choking Mechanism

Most presses have a choking mechanism at the discharge end. Choking provides a means to adjust the press's internal pressure to compensate for differences in feed quality or preparation and to provide a means to compensate for wormshaft and barrel bar wear. Presses designed without chokes are not so tolerant of wear. Chokes also allow for starting up and shutting down the press under conditions of low internal pressure (choke wide open). Once the press is running, the choke is closed to increase pressure and bring down the residual oil. At shutdown, just before the feed flow is stopped, the choke is fully opened to ensure that all solidly pressed material can be discharged out of the press.

There are different types of choking mechanisms. One type resembles an iris diaphragm. A cluster of movable jaws is inserted into or retracted from the channel by means of a large ring supporting the opposite ends of the jaws and allowing them to insert or retract depending on which direction the ring is rotated. Rotation is accomplished by a worm gear operated manually or through a two-directional motor.

Other mechanisms resemble tapered cones that are mounted on the discharge end of the shaft and are slid toward the endplate of the barrel (or away from it) to restrict or clear the channel and thereby influence pressure.

E. Cooling the Press

A screw press generates frictional heat as the wormshaft slides against the oilseed. The heat could in time get the shaft and barrels hot enough to overheat the oil and scorch the solids. If this is of concern, then cooling is required.

Some screw presses are cooled by recirculating product oil over the drainage cages [34,35]. Product oil, cooled in a water-cooled heat exchanger to 48.9°C, flows at approximately 150 L/min over the barrel surfaces. This dilutes the hot freshly pressed oil with a larger volume of cool oil, flushes the barrels to rinse away any solids that linger on the barrel, absorbs heat from the screw press, keeping temperature within acceptable limits, and also blankets the drainage surfaces with flowing oil, thereby preventing the fresh oil from lingering on the hot barrel, exposed to the oxidizing effects of air. Another method of cooling screw presses is to water-cool the barrel housings. The freshly pressed hot oil then cools as it drips off the water-cooled barrels.

Wormshafts are also water-cooled. Flowing through a rotary seal, water enters through a tube inserted into a cylindrical opening bored longitudinally through the center of the shaft. Water flows through the tube toward one end of the shaft, then through the bore back toward the opposite end, and finally out through the rotary seal.

F. Slotted Barrels

The drainage cage is assembled of rectangular bars fitted into a supporting frame (Fig. 6). Each bar is separated from the others by spacing clips, which are made in different thicknesses

Fig. 6 A drainage cage showing the bars. (Courtesy of Anderson International Corp.)

to allow for different gaps between the bars. High pressure full presses use narrow spacers, 0.13–0.76 mm. Lower pressure prepresses, running at higher capacities and liberating a greater volume of oil, generally use thick spacers, 0.5–1.5 mm.

F. Removing Solids from the Oil

Freshly pressed oil contains 2–7% oil-free solids, sometimes more. The solids are usually removed in a two-step procedure. Larger particles are removed with a vibrating screen or a screening tank. A screening tank resembles a large reservoir for freshly pressed oil. Solids that settle to the bottom are captured by a slow-moving drag mechanism that lifts them up a shrouded passage on one end of the tank and slowly drags them over a drainage screen. Excess oil drains back into the tank; the solids continue to the conveyor, taking fresh material to the screw press. The second clarification step is a plate-and-frame filter press.

Unlike the first step, solids from a filter press do not flow continuously. They are accumulated within the filter press for several hours as the oil passes through the filtering cloths. Then the press is opened, and all the accumulated solids are discharged at once. To prevent disturbing the steady-state operation of the screw press, the discharged solids from the filter press are dropped into a hopper equipped with a variable-speed screw that meters the solids into the fresh material. The screw press, therefore, receives a uniform blend of fresh material, drained solids, and filter presscake.

IV. SPECIAL CASES

A. Olives

Olive oil production is concentrated in the Mediterranean region, which is the source of 98% of the olive oil marketed throughout the world [36]. They are processed within 3 days at low temperature, no higher than 40°C [37, p. 345] to avoid any deterioration of oil quality.

The olives are washed, and the pulp is ground into a thick paste (called malaxer) that is processed by one of three methods to obtain the oil. One method is pressing in batch hydraulic presses. The pastelike malaxer is spread onto pressing bags and stacked onto the lower pressure plate of a vertical hydraulic plate press. When pressure is applied, an emulsion of oil and water flows from the press. The emulsion is centrifuged, and the oil is filtered using diatomaceous earth [36,38]. Oil yields can be increased by adding enzymes to the malaxer [39].

The second method centrifuges the malaxer itself. A third method processes the malaxer in an extraction vessel equipped with stainless steel blades that are inserted into the malaxer. Because of differences in surface tension, oil rather than water clings to the blades and migrates along the blade's surface to drip off and flow out of the extractor [37, p. 350].

B. Palm Fruit Versus Palm Kernels

Palm fruit oil accounted for 16% of the world's vegetable oil production (in 1994) [40], second only to soybean oil [41]. The fruits grow in large grapelike clusters weighing 10–20 kg and containing more than 1500 fruits [42]. The fruits arrive at the oil mill containing 20–25% moisture [37, p. 219]. To avoid spoilage, the fruit is processed as quickly as possible. The first step is sterilization. The fruits arrive in narrow-gauge railway wagons, which are rolled into horizontal sterilization chambers.

After sterilization, the fruit clusters are removed from the wagons and fed into a stripper that strips the stalks from the individual fruits. The fruits are washed to remove sand and dirt, then pass into a digester, which breaks the structure of the skin and pulp but leaves the kernels intact.

The digested material is then pressed in continuous, high volume, low pressure screw presses, which press out a liquor consisting of approximately 53% oil, 40% water, and 7% finely divided solids [42]. The liquor goes to a continuous settling tank, where it separates into three phases: a top phase of clear oil containing small amounts of moisture and solids, a bottom watery phase containing most of the solids, and an intermediate phase not yet separated. Continuous streams flow from the top and the bottom. The top phase (the oil) is centrifuged to remove solids and undissolved water, then passes through an evaporative dehydrator to remove the dissolved water.

Screw press solids (a mixture of pulp solids, skin residue, and kernels) contain about 16% moisture. The kernels are freed of pulp in a pneumatic separator and are then conditioned to 10–12% moisture. This shrinks the meats, loosening the hulls. The kernels then pass through a centrifugal nutcracker [37, p. 219], which discharges a mixture of loose meats and hulls.

Hulls are removed by slurrying the mixture with water and passing it through a liquid cyclone. The meats pass out the bottom; the hulls pass out the top with the water. Another method mixes the cracked kernels with a clay (or salt) aqueous slurry at 1.18–1.20 specific gravity. The meats float; the hulls sink. The meats are dried to below 8% moisture and sent for crushing in high pressure oilseed presses as described earlier.

C. Animal Materials

Lipids are obtained from both animal and vegetable sources. These source materials vary considerably. Most vegetable lipids come from oilseeds. Most animal lipids come from fatty tissues from animal bodies if the fat is destined for human consumption or, if the fat is not destined for human consumption, from all lipid-bearing portions of animal bodies. Animal materials are prepared for lipid extraction much differently than oilseeds because they contain considerably more moisture and, on a moisture-free basis, contain more lipid, and the lipid (called fat rather than oil) is usually solid at room temperature.

Leaf fat from hogs has the highest fat content (92–95%). Fatty tissue from cattle contains 60–80% fat. These and similar materials are usually processed under hygienic conditions to produce edible-grade fats such as lard. All unused portions of animals from slaughterhouses, livestock that died before slaughter, dead farm animals, fish harvested for production of fish lipids, fish residues from fish processing plants, recycled restaurant grease—any source of readily available fatty animal tissue—is processed under less hygienic conditions to produce inedible-grade animal fats such as tallow, grease, and fish oil. These products find a ready market in industries that produce soaps, lubricants, and animal feeds.

The separation of fat from animal residues (called rendering) involves various methods to liberate the fat from the solids, remove most of the moisture, then press or solvent extract the residue to obtain the fat [1]. To obtain edible fats where color, flavor, and keeping qualities are important and the amount of solids in the raw material is small, the fatty tissue is mixed with water and agitated while being heated in order to release the fat from the tissue. The low fat solids residue, called tankage, is dried and marketed as an ingredient for animal feeds.

Another method of rendering (dry rendering) heats the material without addition of water, and the material's own moisture is boiled out. The resultant low moisture residue, containing all the fat and solids, is drained of free fat and fullpressed or solvent-extracted as oilseeds are. Dry rendering is the more economical of the two methods. It is done batchwise under atmospheric conditions, batchwise under pressure, and in continuous systems operating under atmospheric conditions, under pressure, and under vacuum [43,44].

Another method mixes the ground raw material with recycled fat to form a slurry of approximately 68% fat, 10% solids, and 22% moisture. The slurry is pumped through an evaporator, where the water is removed under vacuum, then through a centrifuge to remove most of the fat. The solid residue, at approximately 35% fat, is pressed to 10% fat in screw presses [45,46].

Fish can be rendered in the same equipment used for other animal materials, but fish is processed separately. High moisture, strong odor, and difficult processing (compared to other materials) encourage procedures and equipment modified specifically for fish.

Bernadini [37, p. 371] describes a rendering system that continuously feeds whole fish into horizontal cylindrical steam-heated cookers. Twenty minutes at 90–95°C sterilizes the fish and causes them to disintegrate into an aqueous mass. The cooker discharges into a continuously operated, low pressure screw press (similar to those used for palm fruit), which separates the mass into an aqueous phase containing most of the oil, some suspended solids, and most of the soluble solids, and a solid phase containing the larger solids and some oil and having about 50% moisture content.

The liquid phase is separated into oil and water fractions. The water fraction is concentrated by evaporation, and the solids-rich "stick water" is added to the presscake, which is thermally dried to 10% moisture. The finished fish meal has about 10% oil.

There has also been research into wet rendering with the assistance of enzymes that promote fat separation by dissolving the connective tissue. In 1982, Freeman [47] described the use of natural enzymes to improve the rendering of catfish viscera.

V. SOLVENT EXTRACTION

A. Traditional Solvents

In the early days of solvent extraction, many solvents were investigated, but by 1947 the most commonly used solvent was hexane. For a review of other solvents, see Johnson and Lusas [48], Lusas et al. [49], and Hron et al. (covering biorenewable solvents) [50].

Hexane is a clear hydrocarbon that boils at 68.9°C. It is derived from petroleum, is fully miscible with oil and immiscible with water, and does not impart an objectionable odor to the oil or meal. As long as it is in abundant supply, it will remain the solvent of choice. But hexane is flammable, which presents a safety hazard. Plants using hexane are advised to avoid conditions that could generate spark-causing static electricity or could permit solvent vapors to leak outside the vessels and accumulate in the plant. Advisory groups like The American National Fire Protection Association (NFPA) issue manuals that contain recommended safety precautions [51].

B. Alternative Solvents

Halogenated solvents have long been considered and sometimes have been used because of their nonflammability. Johnson et al. [52], extracting cottonseed with dichloromethane, reported a significant reduction of gossypol and aflatoxin in the meal.

Acetone can extract oil, gossypol, and aflatoxin. It is, however, miscible with water and must be protected against absorbing too much water from the incoming material. The U.S. Department of Agriculture sponsored research into the use of an azeotropic mixture of acetone, hexane, and water (ratios of 53:44:3) to extract gossypol along with the oil from cottonseed [53]. The gossypol was removed from the oil by conventional alkali refining. The resultant meal produced superior growth in chicks, showing that the gossypol was gone, but this three-component solvent was never accepted by the oilseed industry because of difficulty in evaporating an azeotropic mixture and because the acetone caused an off-odor in the meal [54].

Another alternative solvent is liquid CO_2 at (or above) its critical temperature and pressure. The critical point for carbon dioxide is a pressure of 7387 kPa and a temperature of 31°C. Extraction with supercritical CO_2 usually takes place at pressures of 20,670–103,352 kPa and temperatures of 50–80°C, depending on the product being extracted. Extractions are performed by circulating supercritical CO_2 through a pressurized cell filled with the oilseed. List and Friedrich [55] reported that almost all of the oil could be extracted from 0.10 mm soy flakes, but if the flakes are thicker (0.38 and 0.81 mm), only 87% or 67%, respectively, of the oil is extracted. The extraction vessels are operated batchwise rather than continuously.

Feidrich and coworkers [56,57] reported producing a better quality soybean oil using supercritical CO_2 than is obtained with hexane. Other oilseeds studied for supercritical fluids extraction are cottonseed [58], wet-milled and dry-milled corn germ [59], and rice bran [60]. Such extraction is feasible, but the required high operating pressure makes the equipment expensive. Supercritical fluid extraction is presently used on some high value materials such as coffee, hops, and natural flavor extracts.

Environmental concerns and the possible short supply of petroleum-based solvents have spurred research on extraction processes using ethanol and isopropanol. These solvents are attractive because they can be produced from starchy cereal grains and they can extract some undesirable constituents such as gossypol in cottonseed and aflatoxins.

Karnofsky and others worked with alcohol extraction [61,62] in the 1980s. Hron and co-workers [11,63–66] worked with ethanol and developed a benchtop process for extracting cottonseed oil along with gossypol and any aflatoxin in the cottonseed using 95% ethanol [11]. Abraham et al. [64] tempered cottonseed meats to 14% H_2O, flaked them to 0.229 mm, and dried them to 2% moisture (to ensure that the ethanol would not extract moisture from the incoming flakes). Unheated ethanol extracts gossypol in a first extraction step. The ambient temperature mixture then passes through an absorber to remove gossypol and returns to the first-stage extractor.

In the second extraction step, heated ethanol extracts oil and aflatoxin. The resultant oil–alcohol mixture is chilled to 4°C to reduce the oil's solubility in the alcohol. Centrifuging separates the oil and alcohol into fractions that are sent to reverse osmosis membranes and absorbers to remove aflatoxin and any remaining gossypol. The crude oil–alcohol is then miscella-refined. Hron et al. [65] report that aqueous 95% ethanol converts free gossypol into bound gossypol even at ambient temperature. Acidulating the aqueous ethanol with a tribasic acid and heating the cottonseed improves the extraction [66].

Lusas et al. [12] chose isopropanol because of its greater oil solubility, lower heat of vaporization, and freedom from scrutiny by alcohol taxing agencies. They chose 93–96% isopropanol to ensure a high enough oil-holding limit of the alcohol. Concentrated isopropanol can hold 48% oil in solution with the alcohol. A constant boiling azeotrope can hold only 12–14% oil.

Lusas et al. developed a pilot process that consists of adjusting cottonseed meats to 12% moisture, flaking them cold to 0.38 mm, preconditioning them in an expanding extruder at 88–93°C, holding the extruded meats for 30 min, and then passing them through a second, collet-forming expanding extruder at 104–110°C. The collets are then extracted with 95% isopropanol at 77°C in a simulated nine-stage countercurrent extraction. Solvent-to-meal ratio is 1:1 by weight. Free gossypol in the meal was reduced to approximately 200 ppm. Chill separating at 5°C removes most of the alcohol. The oil is then desolventized in a thin film evaporator. Subsequent caustic refining and bleaching produced oils of a quality that compares favorably with hexane-extracted oils.

The isopropanol is then reconcentrated by perevaporation (a membrane separation procedure in which the mixture is vaporized and the vapors are separated by the membrane). Vacuum draws water vapor, the smaller molecule, through the membrane. The water vapor is then condensed and discarded [67].

C. Extractors

1. *Background*

The extractor most commonly used today is a percolation extractor. A percolation extractor carries the lipid-bearing material through a vaportight chamber where solvent rains down through the material, dissolving out the oil, similar to the way a coffee percolator works. The material is transported through the extraction chamber on a woven wire belt (or by similar means), and the solvent makes multiple passes through the bed of material. Collection sumps under the material and overhead spray heads separate the flow of solvent and the lipid-bearing material into "stages." The flow is countercurrent: fresh solvent encounters previously extracted material, and fresh lipid-bearing material encounters solvent already containing some oil. This countercurrent flow permits high levels of oil removal using as little solvent as practical.

In most percolation extractors, incoming material is mixed with oil-laden solvent as the material is conveyed into the extractor. This extracts some readily available oil but also carries out loose particles of solids (fines). The oil-laden solvent (miscella) from this first stage is pumped through the second-stage compartment to allow the already formed bed of material to filter out the fines. The oil–solvent mixture then leaves the extractor as full miscella. Lean miscella from stage 3 skips stage 2 to mix with incoming feed in stage 1 then goes on to stage 2 as described above. The rest of the stages operate in true countercurrent fashion. Stage 3 gets miscella from stage 4, stage 4 gets miscella from stage 5, stage 5 gets miscella from stage 6, and stage 6, usually the last stage, gets washed with fresh solvent. Then the solids free drain for several minutes before discharge from the extraction chamber.

There are several major types of percolation extractors, classified mainly by their means of transporting solids: perforated belt, rectangular loop, circular basket (rotary or stationary), and sliding-bed [51].

2. *Perforated Belt Extractors*

Manufactured by DeSmet (Edegem, Belgium), the perforated belt extractor (Fig. 7) employs a flexible perforated belt to carry the lipid-bearing material through the extraction chamber. The belt, made of wire mesh or wedge-bar sections, forms a looped endless conveyor and carries the material as one continuous bed. Bed depth is adjustable and is usually 0.5–2 m. Belt speed determines residence time.

The belt travels on a slight incline through the extractor. This reinforces countercurrent flow by causing miscella washes to seep through the solids in a direction slanted toward the feed end. Rakes scrape ridges into the bed's surface after each miscella spray to promote more rapid penetration of miscella.

The miscella from each stage is pumped back onto the stage it came from. Miscella advances from stage to stage by overflowing the sumps under the belt.

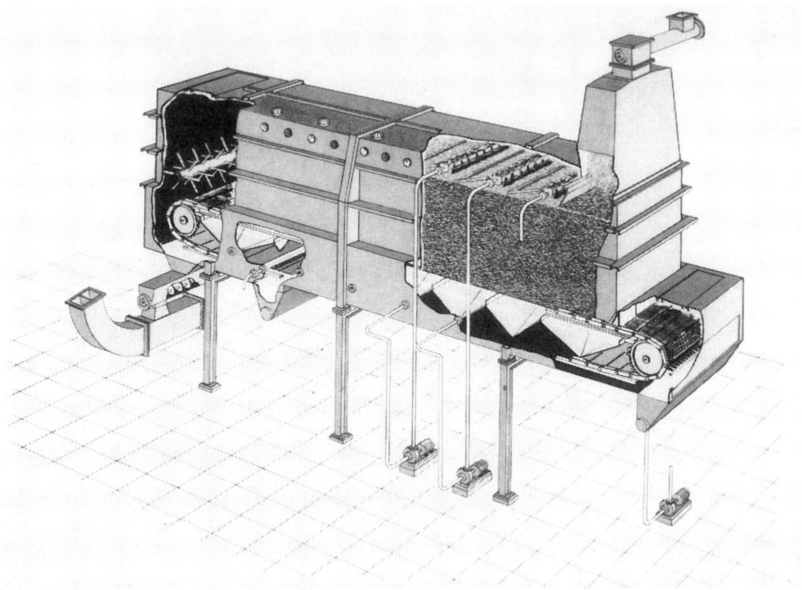

Fig. 7 Perforated belt extractor. (Courtesy of Extraction DeSmet SA/NV.)

3. Rectangular Loop Extractor

Manufactured by Crown Iron Works Co. (Minneapolis, MN) the rectangular loop extractor employs an en masse drag conveyor that moves the solid material through a closed-loop housing of rectangular cross section (Fig. 8). Fresh material flows into the drag conveyor just before the conveyor starts its downward pass to the lower level and is deposited as a shallow bed (usually less than 0.77 m deep) that travels a distance that is usually 50 times the bed depth.

Rich miscella is pumped into the material during the downward pass. The mixture is then pulled horizontally along a slotted vee-bar floor, through which the miscella drains into the first-stage sump. After the solids pass through stages 3, 4, and 5, they are lifted to the upper level and are turned upside down as the drag conveyor loops over the upper vee-bar flooring. The solids then pass through stages 6 and 7.

Miscella from each sump can be pumped back to the stage it came from or to the next stage. If the miscella is returned to the stage it came from, it can still move to the next stage through overflow weirs built into the sumps.

Crown later reversed the drag's direction of travel. Fresh material in the Model III enters at the beginning of horizontal travel on the upper level, travels through several stages on that level, then loops around to the lower level (turned over in transit) for the rest of the extraction. The spent material drops out on the lower level, allowing the drag conveyor to be empty as it rises to the upper level. Having the conveyor empty at this point reduces horsepower consumption of the extractor drive.

Fig. 8 Rectangular loop extractor. (Courtesy of Crown Iron Works Co.)

4. Circular Basket Extractors

There are two varieties of circular extractors. They look similar, but the parts that move in one do not move in the other. A "rotary" circular extractor is built around a slowly revolving circular disk divided into pie-shaped compartments that hold the material to be extracted. Incoming material falls into each compartment as that compartment passes under a stationary inlet spout. The compartments ("cells") then rotate under a circle of stationary spray heads that spray miscella in a countercurrent fashion into the cells. The miscella flowing out of the cells gathers in a circle of stationary sumps beneath the rotating disk. In this fashion, the lipid-bearing material passes under a series of spray heads, then under a spray of fresh solvent, then travels over a final drainage sump, and finally moves over a discharge hopper where the floor of the cell swings open, and the extracted, drained, but solvent-laden solids (marc) drop into a hopper. Continued rotation shuts the door and moves the now empty cell under the flake inlet. Called a rotary extractor because the cells rotate, this type of extractor is supplied by Davy International (Pittsburgh, PA) (Rotocel) (Fig. 9) and by Krupp (Hamburg, Germany) (Carrousel).

The other type of circular basket extractor is made of a circular disk of cells with wedge-shaped compartments, but the disk does not rotate. An apparatus containing the flake inlet spout and the fresh solvent and miscella spray heads revolves above the stationary circular disk. Another apparatus containing the drainage sumps and marc hopper (connected to a common drive with the top apparatus) revolves below the stationary disk. Termed a stationary basket extractor, this variety is supplied by French Oil Mill Machinery Co. (Piqua, OH).

All versions of circular extractors handle deep beds of solids, sometimes 2–3 m deep. There are several mechanical variations of both versions. In one variation, the individual cells have hinged bottoms that swing open to empty the cell. In another variation (termed the sliding plate), the solids slide against a common one-piece bottom that has an opening through which the solids fall when the extraction cycle is completed.

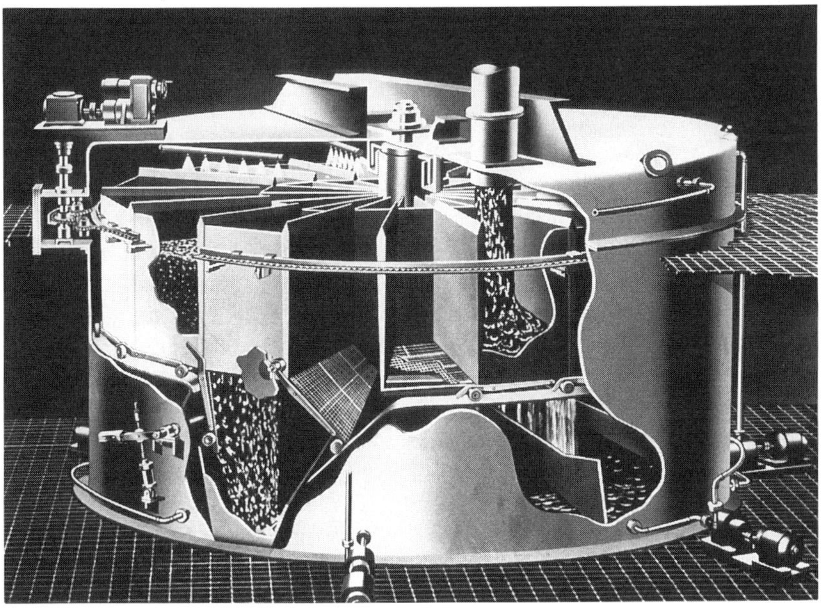

Fig. 9 A Rotocel circular basket extractor. (Courtesy of Davy International Corp.)

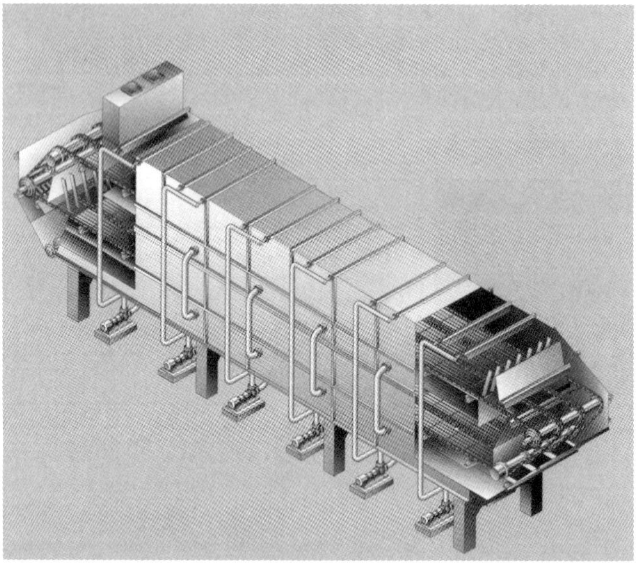

Fig. 10 Sliding-bed extractor. (Courtesy of Krupp Machienentechnik GmbH.)

5. *Sliding-Bed Extractors*

In a sliding-bed percolation extractor, a chain-and-sprocket assembly with drag bars slides (or drags) the material along two stationary perforated steel plates, one above the other. The solids enter on the upper plate, pass under spray heads that spray miscella onto them in a countercurrent flow pattern, and then drop to the lower plate, where they form a fresh bed of material to be slid under the rest of the miscella spray heads. The fresh hexane rinse is on the lower plate.

Manufactured by Krupp, the sliding-bed extractor is shown in Fig. 10. How deep the bed of material is on the plates is an easily adjusted variable. Bed depth for most materials is 0.5–1.3 m.

D. Solvent Recovery

Solvent is removed from the miscella in a two-step process: evaporation followed by steam-stripping. Evaporation is usually done in a double-stage evaporator. In the first stage, hot solvent vapors that were boiled out of the solids condense against the first-stage evaporator and boil solvent from the miscella. In the evaporator's second stage, jacket steam provides heat. The oil leaves containing approximately 5% solvent. It then enters a stripping column where the last traces of solvent are stripped out by live steam injected countercurrent to the oil flow.

Steam and solvent vapors (for this example, hexane) from the stripper are condensed, and the liquid water and hexane go to a solvent–water separator. The lighter hexane rises, spills over a weir at the top, and goes to the solvent work tank. The heavier water sinks, then flows up a siphon tube (pushed upward by the combined weight of water and solvent above it), and goes to a water stripper that strips out the last traces of solvent. The solvent-free effluent water then leaves the system.

Solvent is removed from the solvent-laden solids in desolventizers similar to the cooker vessels discussed earlier. The solvent is driven out thermally with heat supplied by steam jackets and by injected live steam, which strips the solvent in the same way live steam strips solvent from the oil. The heat and moisture (from the steam) also partially cook the solids and are used to "toast" or inactivate urease in soymeal. The cook conditions also lower the nitrogen solubility index (NSI) of the protein.

High NSI materials are desired for manufacturing meat analogues from vegetable proteins. If a processor wants to make high NSI products, a flash desolventizer is used. These vessels desolventize the marc in a circulating stream of preheated solvent vapor followed by a low temperature, low moisture finishing step that removes the last traces of solvent without denaturing the protein [68]. Some plants divert part of their marc to a flash desolventizer if they have a market for the higher value, high NSI flours. The rest of their marc (soymeal, for example) goes to a desolventizer/toaster to serve the animal feed market.

Besides oil and solids, there are two additional streams flowing out of a solvent extraction system: effluent air and effluent water. Air enters with the incoming material and through leaks in the vessels that operate under vacuum. Air leaves the system through a water-cooled vent condenser followed by a refrigerated vent condenser, a charcoal absorber, or a mineral oil scrubber and then passes through a flame arrester before release to the atmosphere.

Effluent water (in a hexane extraction system) passes through a steam-heated "wastewater stripper," where live steam strips out the last traces of hexane. All recovered hexane vapors are condensed and sent to the solvent–water separator. The hexane is recycled to the extractor. The water is discharged from the system.

REFERENCES

1. F. Norris, Extraction of fats and oils, in *Bailey's Industrial Oil and Fat Products* (Daniel Swern, Ed.), 4th ed., Vol. 2, Wiley, New York, 1982, p. 175.
2. J. R. Reynolds and C. G. Youngs, Effect of seed preparation on efficiency and oil quality in filtration extraction of rapeseed, *J. Am. Oil Chem. Soc. 41*: 63 (1964).
3. J. W. Dunning and R. J. Terstage, Effect of overcooking of cottonseed meats on quality of meals, *J. Am. Oil Chem. Soc. 29*: 153 (1952).
4. M. C. Calhoun, J. E. Huston, C. B. Calk, B. C. Baldwin, Jr., and S. W. Kuhlmann, *Cattle Research with Gossypol Containing Feeds*, National Cottonseed Products Assoc., Memphis, TN, 1991, pp. 39–51.
5. K. C. Rhee, Detoxification and deallergenation of oilseed meals, in *Proceedings of the World Conference on Oilseed and Vegetable Oil Processing Technology* (T. H. Applewhite, Ed.), AOCS Press, Champaign, IL, 1993, pp. 346–351.
6. The Processing of Castor Meal for Detoxification and Deallergenation, The International Castor Oil Association, Tech. Bull. 1-1989, Ridgewood, NJ, 1989, pp. 58–62.
7. J. Horton and M. A. Williams, A cooker-extruder for deallergenation of castor bean meal, *J. Am. Oil Chem. Soc. 66*: 227 (1989).
8. P. J. Coulty, Aflatoxin in cottonseed products—what can be done?, *Oil Mill Gaz. 96*: 26 (1991).
9. C. W. Horne, L. L. Boleman, C. G. Coffman, J. H. Denton, D. B. Laehorn, A. B. Wagner, Jr., M. L. Walker, R. E. Withers, Jr., and C. G. Woefel, *Mycotoxins in Feed and Food-Producing Crops*, Texas Agric. Extension Service, Texas A&M Univ., College Station, TX, Bull. B-1278, 12-88 (1988) Revision, pp. 17–24.
10. L. Lee, Aflatoxin, *J. Am. Oil Chem. Soc. 66*: 1398 (1989).
11. R. J. Hron, Sr., M. S. Kuk, G. Abraham, and P. J. Wan, Ethanol extraction of oil, gossypol and aflatoxin from cottonseed, *J. Am. Oil. Chem. Soc. 71*: 417 (1994).
12. E. W. Lusas, L. R. Watkins, and S. S. Koseoglu, Isopropyl alcohol to be tested as a solvent, *Int. News Fats, Oils Relat. Mater. (INFORM) 2*: 970 1991).
13. C. W. Zies, Apparatus for the preparation of food compounds, U.S. Patent 3,108,530 (to International Basic Economy Corp) (Oct. 29, 1963).

14. M. A. Williams, R. E. Horn, and R. P. Rugala, Extrusion, extrusion, extrusion, *Food Eng. 49*: 99 (1977) (Part 1) and *49*: 87 (1977) (Part 2).
15. M. A. Williams, Preparation of oilseeds to improve extraction of fats, *Extrusion Commun. 6*: 12 (1993).
16. S. Baer, M. A. Williams, and C. W. Zies, Pre-treatment of oleaginous plant materials, U.S. Patent 3,255,220 (to International Basic Economy Corp.) (June 7, 1966).
17. M. A. Williams, Extrusion preparation for oil extraction, *INFORM 6*: 289 (1995).
18. M. A. Williams and S. Baer, The expansion and extraction of rice bran, *J. Am. Oil Chem. Soc. 42*: 151 (1965).
19. M. A. Williams, Extruded starter pig feeds, *Feed Manage. 42*: 20 (1991).
20. F. Zhang, S. S. Koseoglu, and K. C. Rhee, Effect of expander process on the phospholipids in soybean oil, *J. Am. Oil Chem. Soc. 71*: 1145 (1994).
21. A. I. Nelson, W. B. Wijeratne, S. W. Yeh, T. M. Wei, and L. S. Wei, Dry extrusion as an aid to mechanical expelling of oil from soybeans, *J. Am Oil. Chem. Soc. 64*: 1341 (1987).
22. M. A. Williams, Adjustable-jaw expanders, *Feed Manage.* 44: 25 (1993).
23. M. A. Williams, Advancement of processing hardware, *Oil Mill Gaz. 10*: 20 (1995).
24. M. A. Williams, Extrusion of rice bran, in *The Proceedings of the World Congress on Vegetable Protein Utilization in Human Foods and Animal Feedstuffs* (T. H. Applewhite, Ed.), AOCS Press, Champaign, IL, 1989, pp. 100–102.
25. J. T. Farnsworth, L. A. Johnson, J. P. Wagner, L. R. Watkins, and E. W. Lusas, Enhancing direct solvent extraction of oilseeds by extrusion preparation, *Oil Mill Gaz. 91*: 30 (November 1986).
26. E. W. Lusas and L. R. Watkins, Oilseeds: extrusion for solvent extraction, *J. Am. Oil Chem. Soc. 65*: 1109 (1988).
27. M. A. Williams, Extrusion preparation for solvent extraction, *Oil Mill Gaz. 91*: 24 (November 1986).
28. M. A. Williams, Apparatus and method for the continuous extrusion and partial deliquefaction of oleaginous materials, U.S. Pat. 4,901,635 (to Anderson International Corp.) (Feb. 20, 1990).
29. M. A. Williams, Expanders for high-oil seeds, *Oil Mill Gaz. 95*: 10 (1989).
30. M. A. Williams, Using expanders to improve extractability, *INFORM 1*: 959 (1990).
31. V. D. Anderson, Press, U.S. Pat. 647,354 (Apr. 10, 1900).
32. J. W. Dunning, Unit operations in a mechanical extraction mill, *J. Am. Oil Chem. Soc. 33*: 462 (1956).
33. M. A. Williams, Placing expanders and screw presses under automatic controls: some entry level systems, *Oil Mill Gaz. 96*: 38 (June 1991).
34. R. T. Anderson, Oil settling tanks, U.S. Patent 2,365,442 (to V. D. Anderson Co.) (Dec. 19, 1942).
35. R. T. Anderson, Screening and conveying means for the solids of oil separating tanks, U.S. Patent 2,420,927 (to the V. D. Anderson Co.) (May 20, 1947).
36. S. Latta, Gourmet oils in the 1990s, *INFORM 2*: 99 (1991).
37. E. Bernadini, *The New Oil and Fat Technology*, Tecnologie s.r.i., Rome, 1973.
38. FAO, *Olive Oil Technology*, Food and Agriculture Organization of the United Nations, Rome, 1975.
39. F. M. Christensen, Extraction by aqueous enzymatic processes, *INFORM 2*: 984 (1991).
40. Y. Basiron and R. Abdullah, The palm oil situation in south east Asia, *INFORM 6*: 891 (1995).
41. Anon., Palm oil update, *INFORM 6*: 884 (1995).
42. K. G. Berger, Production of palm oil from fruit, *J. Am. Oil Chem. Soc. 60*: 158A (1983).
43. W. H. Prokop, Rendering systems for processing animal by-product materials, *J. Am. Oil Chem. Soc. 62*: 805 (1985).
44. S. Ottewell, Rendering meat safe?, *Chem. Eng. 477*: 15 (1990).
45. C. Greenfield, Process of dehydration of fatty materials, U.S. Patent 2,651,647 (Sept. 8, 1953).
46. C. Greenfield, Dehydration of fluid fatty mixtures above normal coagulation temperature, U.S. Patent 3,076,715 (Feb. 5, 1963).

47. D. W. Freeman, in *Proceedings of the 11th Annual Meeting, Inland Commercial Fisheries Association* (D. Yocum, Ed.), Arkansas Game and Fish Commission, Little Rock, AR, 1982, pp. 1–10.
48. L. A. Johnson and E. W. Lusas, Comparison of alternative solvents for oils extraction, *J. Am. Oil Chem. Soc. 60*: 229 (1983).
49. E. W. Lusas, L. R. Watkins, and K. C. Rhee, Separation of fats and oils by solvent extraction: non-traditional methods, in *Proceedings of the World Conference on Edible Fats and Oils Processing: Basic Principles and Modern Practices* (D. R. Erickson, Ed.), AOCS Press, Champaign, IL, 1990, pp. 56–78.
50. R. J. Hron, Sr., S. P. Koltun, and A. V. Graci, Jr., Biorenewable solvents for vegetable oil extraction, *J. Am Oil Chem. Soc. 59*: 674A (1982).
51. NFPA, *NFPA 36, Solvent Extraction Plants*, 1993 ed., National Fire Protection Association, Quincy, ME.
52. L. A. Johnson, J. T. Farnsworth, N. Z. Sadek, N. Chamkasem, E. W. Lusas, and B. L. Reid, Pilot plant studies on extracting cottonseed with methylene chloride, *J. Am. Oil Chem. Soc. 63*: 647 (1986).
53. G. E. Mann, F. L. Carter, V. L. Frampton, A. B. Watts, and C. Johnson, Evaluation of cottonseed meals prepared by extraction with acetone–water mixtures, *J. Am. Oil Chem. Soc. 39*: 86 (1962).
54. R. J. Hron, Sr. and M. S. Kuk, Acetone extracted cottonseed meals without catty odors, *J. Food Sci. 54*: 1088 (1989).
55. G. R. List and J. P. Friedrich, Processing characteristics and oxidative stability of soybean oil extracted with supercritical carbon dioxide at 50°C and 8,000 psi, *J. Am. Oil Chem. Soc. 62*: 82 (1985).
56. J. P. Friedrich, G. R. List, and A. J. Heakin, Petroleum-free extraction of oil from soybeans with supercritical CO_2, *J. Am. Oil Chem. Soc. 59*: 288 (1982).
57. J. P. Friedrich and E. H. Pryde, Supercritical CO_2 extraction of lipid-bearing materials and characteristics of the products, *J. Am. Oil Chem. Soc. 61*: 223 (1984).
58. G. R. List, J. P. Friedrich, and J. Pominski, Characterization and processing of cottonseed oil obtained by extraction with supercritical carbon dioxide, *J. Am. Oil Chem. Soc. 61*: 1847 (1984).
59. G. R. List, J. P. Friedrich, and D. D. Christianson, Properties and processing of corn oils obtained by extraction with supercritical carbon dioxide, *J. Am. Oil Chem. Soc. 61*: 1849 (1984).
60. W. Zhao, A. Shishikura, K. Fujimoto, K. Arai, and S. Saito, Fractional extraction of rice bran oil with supercritical carbon dioxide, *Agr. Biol. Chem. 51*: 1773 (1987).
61. G. B. Karnofsky and R. J. Hansotte, Process for extracting oleaginous seed materials particularly cottonseed with aqueous alcohols, U.S. Pat. 4,359,417 (to Dravo Corp.) (Nov. 16, 1982).
62. G. Karnofsky, Ethanol and Isopropanol as solvents for full-fat cottonseed extraction, *Oil Mill Gaz. 86*: 34 (1981).
63. R. J. Hron, Sr. and S. P. Koltun, An aqueous ethanol extraction process for cottonseed oil, *J. Am. Oil Chem. Soc. 61*: 1457 (1984).
64. G. Abraham, K. M. Decossas, R. J. Hron, Sr., and M. S. Kuk, Process engineering economic evaluation of the ethanol extraction of cottonseed: preliminary analysis, *J. Am. Oil Chem. Soc. 68*: 418 (1991).
65. R. J. Hron, Sr., G. Abraham, M. S. Kuk, and G. S. Fisher, Acidic ethanol extraction of cottonseed, *J. Am. Oil Chem. Soc. 69*: 951 (1992).
66. G. Abraham, R. J. Hron, Sr., M. S. Kuk, and P. J. Wan, Water accumulation in the alcohol extraction of cottonseed, *J. Am. Oil Chem. Soc. 70*: 207 (1993).
67. E. W. Lusas, L. R. Watkins, S. S. Koseoglu, K. C. Rhee, E. Hernandez, M. N. Riaz, W. H. Johnson, Jr., and S. C. Doty, New isopropanol system shows promise, *INFORM 5*: 1245 (1994).
68. E. D. Milligan and J. F. Suriano, System for production of high and low protein dispersibility index edible extracted soybean flakes, *J. Am. Oil Chem. Soc. 51*: 158 (1974).

6

Refining

David A. Allen
Partnership Solutions, Liverpool, England

I. INTRODUCTION

The large majority of oils and fats are not fully suitable for human consumption immediately after extraction from seeds, nuts, fruits, fish, animals, etc. For markets in developed countries, their tastes and odors are generally too strong and rather unpleasant. In addition, the fats and oils may contain components that make them unsuitable for particular applications, in terms of either quality or performance. The best known exception to this general statement is olive oil. Cold pressed and (extra) virgin oils are much prized for their flavor and aroma. Only the lower grades of olive oil are likely to be fully refined prior to use.

The processes that together are referred to as refining or purification are chosen to effect removal of a large proportion of the non-triacylglycerol components from oils and fats. Ideally the antioxidants naturally present in many crude oils and fats would not be removed by the refining processes. Unfortunately, the process conditions necessary to remove undesirable materials are also effective in removing some proportion of the natural antioxidants and vitamins.

A significant number of undesirable materials may be present in crude oils and fats and should be either completely or almost completely removed by the refining processes. Table 1 provides a list of these non-triacylglycerol impurities. Note, however, that it is unlikely that any one crude oil or fat would contain all of the components listed.

In spite of the large number of possible impurities in crude oils and fats, trading specifications are generally focused principally, or even exclusively, on those impurities that are certain to affect yield, i.e., that are commercially based rather than technically sophisticated specifications. The principal safeguard to the refiner is found in the Good Merchantable Quality (GMQ) or Fair Average Quality (FAQ) clause of the contract. Refiners will usually carry out more extensive analysis of a crude oil shipment than is required simply to confirm that the oil meets the commercial specification. Over the years a refiner builds up his own picture of what is typical and acceptable in terms of non-triacylglycerol content for a range of crude oils and fats. He may be able to relate certain analytical results to refinery yield performance or refined oil quality parameters. Thus if, or when, he receives a parcel of oil that does not

Table 1 Non-Triacylglycerol Components of Fats and Oils

Free fatty acids
Partial glycerols (mono- and diacylglycerols)
Phosphoacylglycerols
Pigments
Sterols
Tocopherols
Oxidation products
Trace elements (copper, iron, sulfur)
Protein fragments
Waxes
Hydrocarbons
Moisture
Dirt
Pesticide residues
Polycyclic aromatic hydrocarbons

meet the requirements of GMQ, he is able to recognize this quickly and provide convincing documentary evidence to support his claim that the oil is not GMQ/FAQ.

II. PROCESSING ALTERNATIVES

As already indicated, the complete refining or purification process comprises a series of discrete processing steps. There are two major alternative processing sequences, which are known in the industry as "chemical" and "physical" refining (Fig. 1). These terms are drawn from the different methods of free fatty acid removal—either neutralization by caustic soda (or other alkali) solution, which is a chemical reaction, or distillation, which is a physical process.

The physical refining route can offer important advantages to the refiner. These are becoming ever more significant as concerns for the environment lead to increasingly stringent limits on the quality, and quantity, of water discharged from refineries. Such limits result from the fact that the soap produced by alkali neutralization of free fatty acids is usually acidified to convert the soap to a mixture of free fatty acid and neutral oil (known as acid oil). As a result of the neutralization and subsequent acidification processes, a considerable quantity of wastewater is generated. This can be relatively high in fatty matter and suspended solids as well as containing sodium sulfate in solution. All three parameters (fatty matter, suspended solids, and sodium sulfate) are likely to be limited by the water authorities with heavy fines for exceeding prescribed limits.

There is no "neutralization" (refining) step in physical refining (see Fig. 1). Thus the resultant need to acidify soapstock and treat wastewater is avoided. Furthermore, the yield loss on physical refining is lower than that achieved by chemically refining the same oil. Because neutralization and water washing steps are not required for physical refining, there is a saving in capital investment. This is substantial because of the high cost of the sophisticated continuous separators or centrifuges used for soapstock and washwater separations.

In an ideal world, therefore, a refiner would choose to operate only a physical refining process sequence. Unfortunately, unless oil quality objectives are compromised, certain oils are not naturally suited to physical refining. This is principally because of the relatively high contents of phospholipids and gums. To make more oils suitable for physical refining, certain major companies have developed processes that can reduce the phosphatide content of some

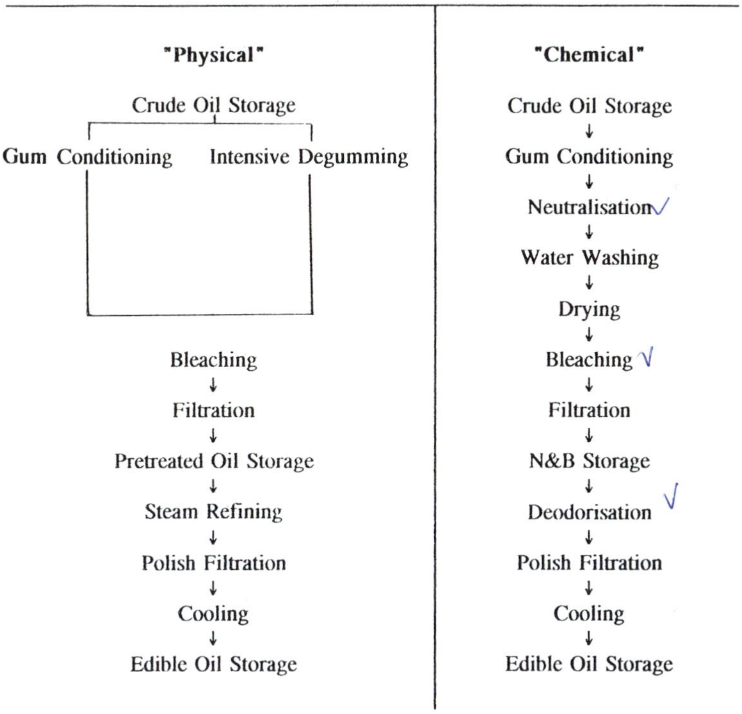

Fig. 1 Refining process alternatives.

types of crude oils to a sufficient extent that they will physically refine satisfactorily. These processes are summarized in Section IV.A. At present the major processes are covered by patents, but they can be licensed from the patent holders.

III. CHEMICAL REFINING

A. Degumming

Soft or liquid oils are usually degummed with water at the completion of the extraction processes. As a result, the remaining phosphatide levels are much lower, but the majority of these phosphatides are nonhydratable, that is, they are not affected by contact with water alone. The use of a more aggressive medium, such as citric or phosphoric acid, will hydrate the phosphatides, permitting separation from the oil either at a discrete degumming stage or together with the soapstock produced at the neutralization stage.

The equipment used to bring about intimate contact between the crude oil itself and a small quantity of concentrated acid varies according to the type of oil being processed and the level of phosphatides present. Crude oil fed to the pretreatment process is normally heated to a temperature in the range 60–85°C. Specialist mixers are used to disperse the small quantity (0.05–0.2%, based on the weight or throughput of crude oil) of concentrated (85%) phosphoric acid in the bulk of the crude oil and to provide a short residence time for the hydration reaction to take place. Mixer types commonly employed include disk and knife mixers. Not only do they offer different intensities of mixing but they also offer quite different residence times, varying from a few seconds to 1–2 mins, respectively [1]. Often refineries are equipped with both disk and knife mixers, which are selected on the basis of the type and quality of oil being processed.

Hydration of phosphatides is a slow process compared with the rapid reaction of an alkali solution with free fatty acid to produce soap. Thus additional reaction time is often provided in the form of an in-line reaction vessel, which may be unagitated or fitted with a gentle low shear agitator. Typical additional residence time could be 5–9 mins [1].

If gums are to be recovered separately for use as commercial lecithin, citric acid should be used in preference to phosphoric acid. The latter tends to char the lecithin, leading to dark-colored product.

The above descriptions refer to continuously operating plants. Degumming pretreatment can also be carried out in batch plants fitted with slowly rotating (approximately 30 rpm) gate agitators. The crude oil is heated to 55–85°C, and the acid solution is added under agitation. Water (0.5–2%) is then added. After a few hours of agitation, the hydrated gums will have flocculated. They may then be separated from the oil using a conventional "refining" (disk-type) centrifuge in preference to gravity settlement, which is both slow and of poorer yield.

The main reason for carrying out the pretreatment step to remove gums, both hydratable and nonhydratable, is that certain phosphatides, most notably the calcium and magnesium salts of phosphatidic and lysophatidic acids [2], are strong emulsifying agents. If these phosphatides were still present in the oil during alkali neutralization and subsequent soapstock separation, the yield of neutral oil would be adversely affected by emulsification losses.

In addition to hydrating nonhydratable phosphatides, phosphoric acid has the effect of reducing the chlorophyll content of the oil. This effect can be very important when processing rapeseed or canola oil produced from unripened or damaged seeds. In this case the dosage of concentrated phosphoric acid needs to be near or above the top of the usage range quoted above.

B. Neutralization

The primary objective of the neutralization (or refining) step is the removal of almost all the free fatty acids present in the crude oil. As already stated, an alkali solution [usually sodium hydroxide (caustic soda)] reacts readily with free fatty acid to produce soap and water. Whereas the free fatty acids were fully dissolved in the oil, soap is almost completely oil-insoluble, making subsequent separation relatively straightforward.

As soon as the oil–phosphatide–phosphoric acid mixture from the degumming stage comes into contact with caustic soda, any excess phosphoric acid is immediately neutralized, and no further phosphatide hydration takes place. To allow for neutralization of excess phosphoric acid, the quantity of caustic soda dosed into the oil stream is increased to compensate for the total addition of phosphoric acid.

From a production viewpoint, it is important to achieve the desired reduction in free fatty acid content and the lowest practical level of yield loss. Caustic soda will react with triacylglycerols as well as with free fatty acids. Thus there is inevitably some loss of neutral oil by saponification. Neutral oil losses can also occur because of emulsification into the water phase and occlusion of droplets into the soapstock itself.

Appropriate selection of the following process variables will enable the refiner to achieve the desired reduction in free fatty acid content with an acceptable minimum loss of neutral oil.

1. The type of alkali used—sodium hydroxide (caustic soda), sodium carbonate (soda ash), sodium silicate, ammonium hydroxide
2. The strength of the alkali solution
3. The excess of alkali solution above the stoichiometric quantity required to neutralize the free fatty acid and phosphoric acid (calculated from chemical equations)
4. The temperature at which reaction is carried out

5. The type and degree of agitation during and after alkali addition
6. The time between alkali addition and soapstock separation

Caustic soda is the most widely used alkali for refining. It has the effect of thoroughly cleansing the oil and effecting some decolorization. It also saponifies some neutral oil, especially when used in concentrated solution or large excess. To reduce saponification losses, sodium carbonate is sometimes used in place of or in combination with caustic soda. Sodium carbonate is a much milder alkali, resulting in less unwanted saponification but also less decolorization.

Pardun [3] first reported the possibility of using ammonium hydroxide for refining soybean oil. He claimed that environmental benefits were achieved by avoiding the need for splitting the soapstock with mineral acids. Ammonia may be boiled off the soapstock without the need for acid treatment.

The stronger the alkali solution used, the greater the removal of gums and pigments as well as free fatty acids, but the higher the losses due to saponification. The soapstock formed is also thicker, making it more difficult to discharge from the centrifuge and increasing the risk of neutral oil being lost by occlusion or entrapment in the soap itself. Lower strength alkali solutions overcome these problems but can lead to increased oil losses due to emulsification.

The higher the excess of alkali solution used, the greater the saponification loss but the greater the cleansing effect on the oil. Low levels of excess alkali reduce the saponification loss but increase the risk of emulsification leading to losses. Choice of alkali type, strength, and excess is generally based on experience with the type of oil being processed. Minor adjustments are made depending on how well a particular parcel of oil is processing.

Combining a long contact time with high temperature and vigorous agitation leads to increases in saponification losses. This has the advantage of maximizing pigment removal and improving phase separation.

Modern continuous refineries (see Figs. 2 and 3) are equipped to offer the refiner a range of mixing time and intensity options and a range of contact times as well as ease of selection and control of throughput rate and process temperature(s). This gives the refiner the ability to handle a number of different oil types and a range of oil qualities using the same equipment.

The primary separator is used to separate the soapstock, plus gums and dirt, from neutralized oil. Almost invariably new refinery installations are fitted with self-cleaning separators for this duty. These units are designed to permit the centrifuge bowl to be separated briefly into two parts, leaving a small space through which viscous sludges are discharged. The very high centrifugal forces generated at the inner surface of the centrifuge bowl are sufficient to propel even thick sludges through the self-cleaning space during a "bowl shoot." The refinery operator can set the frequency and duration of these bowl shoots, depending on the quality of the oil being processed and the time since the centrifuge was last cleaned manually.

Over time, sludge material also builds up between the conical "disks," which are located over the main rotating shaft in the center of the separator. This buildup restricts the flow of oil and soapstock across the faces of the disks and vertically between disks. The process operator will find that normal adjustment of the backpressure control valve is insufficient to ensure satisfactory separation between the water and oil phases. There is then little alternative but to stop production and either clean the centrifuge in place (cleaning in place, CIP) or strip it down and clean each disk manually.

Many continuous refining plants are equipped with three separators. In this case the first two (following the direction of oil flow) are identical, permitting use as either primary soapstock separators or washwater separators. When one primary separator is off-line for cleaning,

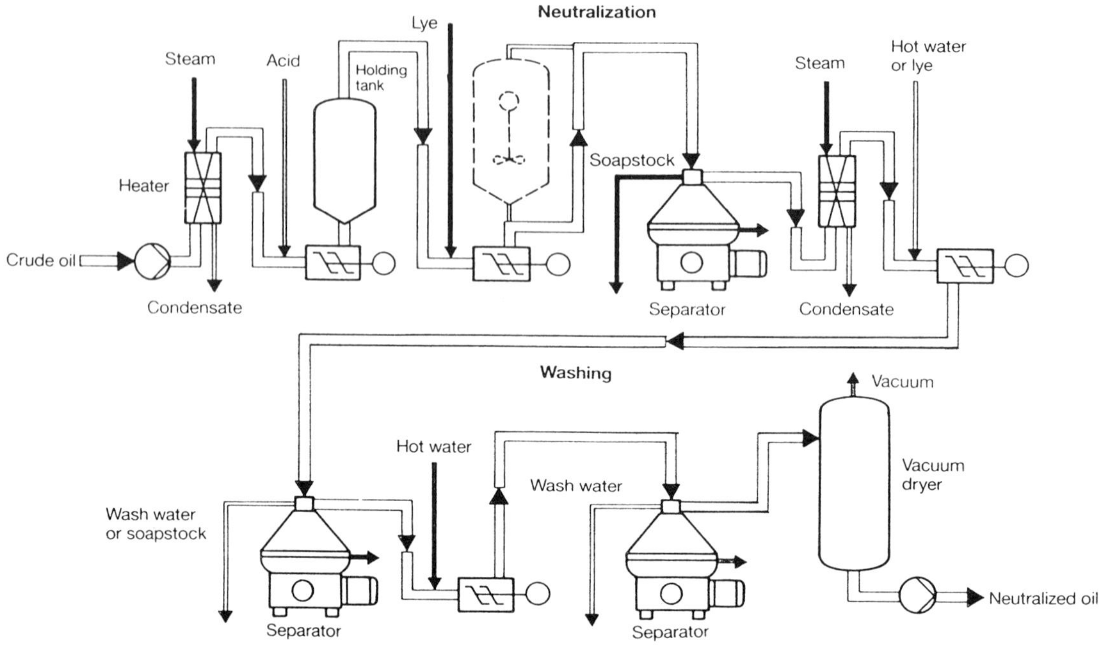

Fig. 2 Flow diagram of a modern continuous refining (neutralization) plant. (Courtesy of Westfalia Separator Limited.)

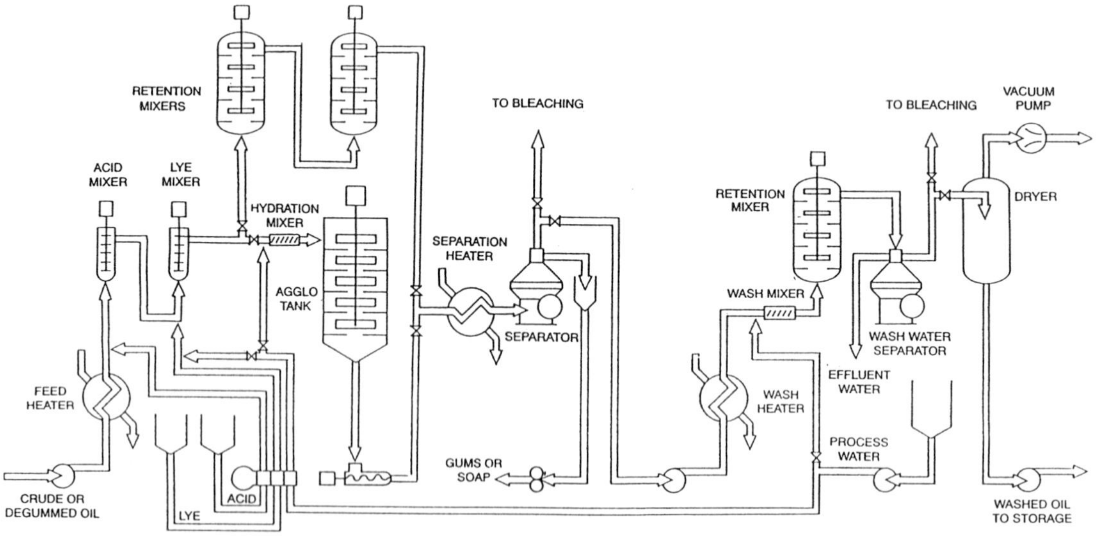

Fig. 3 Possible mixer combinations for a multipurpose refining plant. (Courtesy of RBD Technologies.)

the refinery can continue operating at a reduced throughput by using the second identical separator with just one water wash separator. By this means, lost production is minimized, but quality is maintained.

C. Washing and Drying

Neutral refined oil leaving the primary separator typically contains 500–1000 mg soap/kg oil. This soap must be removed from the oil before further processing. Hot soft water, at a higher temperature than that of the oil, is added at a rate equivalent to 10–20% of the oil flowrate. After only brief contact with the oil, soapy water is separated from the oil in the second centrifugal separator. In many continuous plants, there is further washwater addition followed by another centrifugal separation. This third separator is used only for washing. It therefore requires less sophisticated internal fittings than the so-called primary separators, making it lower in cost. To reduce water consumption, many plants operate countercurrent washing, whereby fresh water is used for the second wash stage. The dilute soapy water produced is used for the first wash stage.

After two water washes, the oil should be almost completely soap-free, containing less than 50 mg/kg. This level may be further reduced if citric or phosphoric acid is added either to the water used for the final wash or to the oil prior to drying. Such acid treatment eliminates the soap almost completely. Soap removal is important because during bleaching, soap will preferentially adsorb onto the bleaching clay, effectively deactivating a portion of the surface for color pigment adsorption.

After water washing, the refined oil is dried to a moisture content of approximately 0.1% prior to bleaching or intermediate storage. Drying is typically carried out under a moderate vacuum (20–50 torr) and at an oil temperature of 90–100°C.

The viscous soapstock from the primary separator is usually mixed with the washwater from the wash separators prior to acidification with a concentrated mineral acid, most commonly sulfuric acid. The acid reacts with the soap to convert it back to a mixture of free fatty acid and neutral oil. This commodity is known as acid oil and is sold as a by-product either for use in animal feed or for use in the oleo chemical industry. Acid oil is separated from the water phase, usually by gravity settlement of the water. The water phase, comprising principally sodium sulfate solution and excess sulfuric acid, is adjusted to a neutral pH, using caustic soda solution, prior to being subjected to effluent treatment to reduce the contents of suspended solids and separable grease and oil to permitted levels prior to discharge to, for example, a public effluent system.

D. Refining with Dilute Alkali

An alternative to the processes outlined above that use strong sodium hydroxide solutions in the range 16–24°Baumé (3.1–5.3 Normal), is the Zenith process [4]. A schematic for this process is shown in Fig. 4. This process sequence involves three semicontinuous steps, although a fully continuous version is now available. The P-unit (see Fig. 4) is a vacuum vessel fabricated from acid-resistant steel and divided into three or four trays, depending on the desired reaction time and the required oil throughput. Concentrated phosphoric acid is automatically dosed from the holding vessel on the top of the P-unit. The top tray provides steam heating to ensure that the oil is deaerated and held at optimum reaction temperature. All trays are provided with agitators mounted on a common drive shaft. Valves at the bottom of each tray operate to an automatic cycle. The quantity of phosphoric acid added ranges from 0.03 to 0.1% of oil weight for low phosphatide oils such as coconut oil, palmkernel oil, tallow, lard, and partially refined hydrogenated oils. Between 0.1 and 0.2% of phosphoric acid is

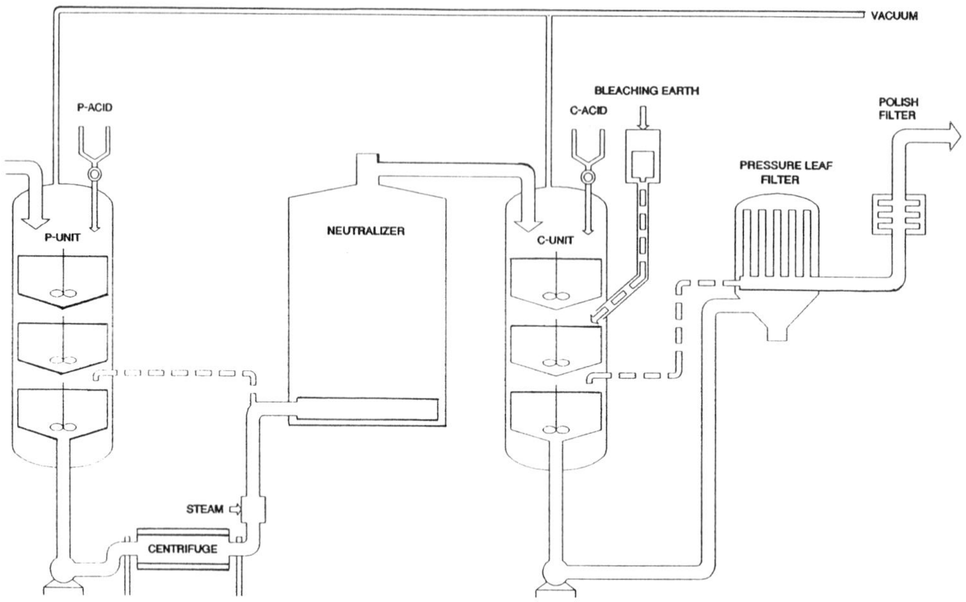

Fig. 4 Flow diagram of the Zenith dilute alkali refining process and associated bleaching plant. (Courtesy of Campro International Inc.)

required for the liquid seed oils. Up to 0.4% may be required for oils containing more than 200 mg/kg of phosphorus or high chlorophyll levels.

After the P-unit, the oil is centrifuged to separate the dirt and denatured phosphatides, in the form of a sludge. The oil then enters the bottom of the neutralizer vessel, in which it is transformed into a series of very fine droplets (approx 10^4 m^{-3}), which rise upward through the dilute alkali solution filling the neutralizer. During their upward passage, the free fatty acids in the oil droplets are neutralized by the alkali and the resulting soap dissolves in the alkali solution. Any water-soluble impurities also transfer into the aqueous solution.

Neutralized oil accumulates on the top surface of the alkali solution before passing continuously to the C-unit. The strength of the alkali solution is initially approximately 2°Baumé (0.35 Normal), except for coconut oil refining, when the solution is stronger, 4.6°Baumé (0.8 Normal). As oil is refined, the alkali solution becomes weaker. When it reaches 0.1 Normal it is replaced. The cleaning of the oil prior to neutralization coupled with the use of very dilute alkali solutions results in relatively very low saponification of triacylglycerols. Thus, yields of refined oil are close to theoretical.

In the C-unit, or bleacher, the refined oil is first treated with citric acid solution to split any residual soap and to inactivate any oxidation catalysts. Bleaching clay is then added to decolorize the oil. The quantity of bleaching clay required is likely to be approximately the same as that required for bleaching refined, washed oil from strong alkali treatment.

Finally, bleaching clay is separated from the oil on a pressure leaf filter, followed by a polish or guard filter to ensure that only clean oil goes into storage or for further processing.

In the Zenith process there is no need to water wash the neutralized oil because soap levels are already low (50–100 mg/kg). Spent alkali–soapstock solution is acidulated with mineral acid to produce an "acid oil" by-product. Because weak alkali is used, there is little decolorization of the oil.

E. Batch Refining

Batch refining is carried out in open or, preferably, closed tanks. These are usually vertical cylindrical vessels fitted with conical bottoms. The conical shape assists in removing only the aqueous phase at the completion of the various process steps.

Batch refining kettles are fitted with variable-speed agitators and steam/cooling coils for temperature adjustment. Capacities can vary greatly but are generally in the range of 5–25 tonnes.

At the outset of processing, oil is charged to the vessel and heated to the target temperature for refining, 80–95°C. Caustic soda solution of the chosen strength is sprinkled slowly onto the surface of the crude oil, which is subjected to gentle agitation. Soapstock begins to form almost immediately. After a short reaction period following the completion of caustic soda addition, the agitator is stopped, and the heavier aqueous phase, including the soap, is allowed to settle under gravity. The settling rate will be affected by factors such as the size of the soapstock particles, the size of oil and water droplets, and the viscosities of the oil and aqueous phases. Where relatively thick soapstocks are formed, e.g., from saturated fats and from cottonseed and sunflower oils, the use of a dilute caustic soda is preferred in order to reduce soapstock viscosity.

When the aqueous phase has fully settled, it is run off from a valve at the lowest point of the conical bottom of the tank. It is common practice to have a sight glass between the valve outlet and the inlet to the discharge pipe. If this is properly illuminated, the operator can readily determine when discharge of the aqueous phase is complete and the oil phase is starting to discharge.

The hot neutralized oil is washed two to four times with hot water to ensure that soapstock has been removed insofar as possible. Note, however, that residual soapstock levels from batch refining are significantly higher than those achieved by continuous centrifugal separation of washwater from neutral oil. This residual soap will be adsorbed onto the clay used in the bleaching step that usually follows refining.

It is common practice to use the same vessel for bleaching as for neutralization and water washing. If this is done, a vacuum pressure is produced within the vessel and the oil temperature is raised to dry the oil to an acceptably low moisture level prior to bleaching.

Occasionally, problems of either slow settling of soapstock or emulsion formation are encountered in batch refining. These can often be overcome by using a dilute solution of common salt for the first water wash.

Batch refining is a relatively slow process. It may be possible to refine only one batch of oil during a production shift. This is in marked contrast with the production rates of 10–15 t/h achieved in many continuous refining plants.

F. Bleaching

The process is known as bleaching because when first introduced its primary purpose was seen as the removal of color pigments from the oil. More recently it has been concluded that bleaching clays also remove or at least reduce the contents of other impurities such as soap, trace metals, phosphatides, and sulfur compounds. Primary oxidation products, hydroperoxides, are broken down on the surface of the bleaching clay, and secondary oxidation products, carbonyl compounds, are adsorbed onto the surface. Decolorization still occurs during bleaching, but its effect is often less significant than that which takes place at the higher temperatures of deodorization. Polycyclic aromatic hydrocarbons containing five or more benzene rings in their structure are removed during bleaching only if activated carbon is used in addition to bleaching clay.

Bleaching clays are naturally occurring bentonite and montmorillonite clays. Usually these materials are activated using a mineral acid such as sulfuric acid. The acid removes some minerals from the clay, resulting in larger volume micropores and smaller particle size [5]. The principal benefit of acid activation is that it permits reduced usage of clay to meet a particular bleached oil specification. A slight negative is that acid-activated clay retains more oil per unit weight than natural clay. The lower usage level, however, more than offsets this effect in most cases. Mag [6] reviews the theory of adsorption bleaching, including the mechanisms of acid activation of bleaching clay.

As mentioned above, activated carbon can also be used for bleaching. It is generally more expensive than activated bleaching clay. Activated carbon retains more oil than bleaching clay. Its very fine particle size makes it difficult to handle and to filter. Certain grades of activated carbon are used specifically to remove the so-called heavy polycyclic aromatic hydrocarbons, which will not be stripped from the oil during subsequent deodorization. Preferably, the oil is treated with bleaching clay before the activated carbon is added. This avoids using the active sites on the carbon to adsorb the easy-to-remove components like soap and color pigments. It has been stated that a combination of 10–20 parts of bleaching clay to one of carbon can be more effective than bleaching clay alone [7]. Where removal of polycyclic aromatic hydrocarbons (PAH) is a priority, a much higher proportion of activated carbon will be required. Coconut oil, some fish oils, and some rapeseed oils may require activated carbon treatment to reduce the content of heavy PAH [8].

Bleaching clays are expected to have good adsorption properties for the components to be removed from the oil, not too high a retention of oil, and good filterability. The latter factor is important in ensuring that clear, particle-free oil is produced soon after a clean filter is brought on-line and oil flowrates do not diminish rapidly as filter cake builds up on the filter medium.

The precise mechanism of the adsorption process is not yet fully understood [9]. Bleaching is carried out at elevated temperature (90–120°C), almost always under a vacuum pressure to exclude oxygen. The use of vacuum pressure tends to remove some of the residual moisture from the oil, following the drying stage at the completion of neutralization and water washing. It also removes some "free" water from the bleaching clay itself. This may not be desirable because experience has shown that bleaching clay effectiveness can sometimes be enhanced by adding small amounts of moisture to the oil prior to adding bleaching clay [10].

Temperature and time are the major process parameters affecting the bleaching achieved with a particular dosage of an activated bleaching clay. Bleaching times of 10–30 min are generally required to ensure that adsorption reaches or approaches equilibrium. Shorter times may be sufficient in certain circumstances, but commercial equipment is usually sized to provide residence times in this range at the design throughput rates for oil.

A small rise in FFA may be observed during bleaching. This is likely to be due to a combination of mineral acid leaching out from the activated bleaching clay and some hydrolysis of the oil.

At the completion of bleaching, the peroxide value of the oil is generally zero or very low, unless levels of bleaching clay below about 0.5% have been used or the unbleached oil had a peroxide value above about 10 meq/kg. Anisidine values are not reduced to the same extent during bleaching unless uneconomic quantities of bleaching clay are employed to achieve this particular result.

Many refineries have noted what is generally known as the press effect. This results in a slightly lower color in the bleached oil after filtration than immediately prior to filtration. If oil is truly in equilibrium with the bleaching clay prior to filtration, then there will be no press effect [12]. The existence of a press effect in many operating situations indicates that oil and clay are not in equilibrium at the end of the bleaching period. Some refineries are able to use the press effect by shutting off the flow of clay to the bleacher when the filter is almost full and then running for a period of time, during which only the press effect reduces the oil color.

1. Equipment for Bleaching

As with refining and water washing, bleaching may be carried out as a batch process. It is more usual, however, to have a semicontinuous or fully continuous bleaching plant. Often this takes oil directly from the outlet of the dryer, which marks the completion of the water washing process.

Batch bleaching can be carried out in the same vessel as neutralization and washing. It can also be carried out in a separate but similar vessel. As mentioned above, bleaching usually takes place under vacuum. Batch bleaching has the advantage of providing easy segregation between different oils in a multi-oil refinery. It also has the advantage of a relatively low capital cost. Its major disadvantages are the range of bleaching times that result from a long filtration time and the fact that the bleacher often has to be returned to atmospheric pressure before filtration can proceed, thus exposing the oil to oxygen.

Semicontinuous and fully continuous bleaching processes are carried out in virtually identical equipment. The major difference lies in the stepwise sequencing of oil through a series of compartments in the semicontinuous process compared with continuous flow along a predetermined path in the continuous process. With semicontinuous bleaching, neutralized oil is pumped through a heat exchanger to a deaeration unit and into the top tray within the main bleacher vessel. Here the temperature is further increased to the desired bleaching temperature. Bleaching clay and/or activated carbon is added directly to the agitated oil in the top tray. All the trays within the bleacher vessel are held under the same vacuum pressure (25–50 torr absolute pressure).

After a predetermined time interval, the oil is discharged by gravity into the tray below, and so on until it reaches the bottom of the bleacher. This region acts as a feed tank for the filters.

Because of the segregated tray arrangement and careful sequencing of the drop valves, changes from one oil to another can be made in this plant with only traces of cross contamination.

In many plants, bleaching clay is added directly to the oil. This is generally preferred, because Brimberg [10] showed that a bleaching effect is then achieved more rapidly than with alternative means of clay addition. Alternatives include dispersing the clay in a small side stream of oil before adding the resultant slurry back to the main stream and making a slurry of bleaching clay in oil that has already been bleached, then adding that to the main oil stream.

Currently, the driving force behind the introduction of new and improved bleaching clays and bleaching processes is the desire to reduce the amount of absorbent required for effective bleaching. Silica hydrogels (see Section III.G) are one example of new bleaching media. One example of a process designed to achieve a reduction in bleaching clay usage is shown in Fig. 5. This provides an alternative to the usual agitated vessel (often divided into a number of sequential compartments) that provides the contact time between bleaching clay and oil necessary for effective bleaching. In the design shown, bleaching clay is dosed into neutralized refined oil in a small agitated tank. The resultant slurry is then drawn into a tall slim vessel within which it is initially heated to bleaching temperature while being agitated with steam. Heated slurry overflows to fall through a vacuum space to the main residence time volume,

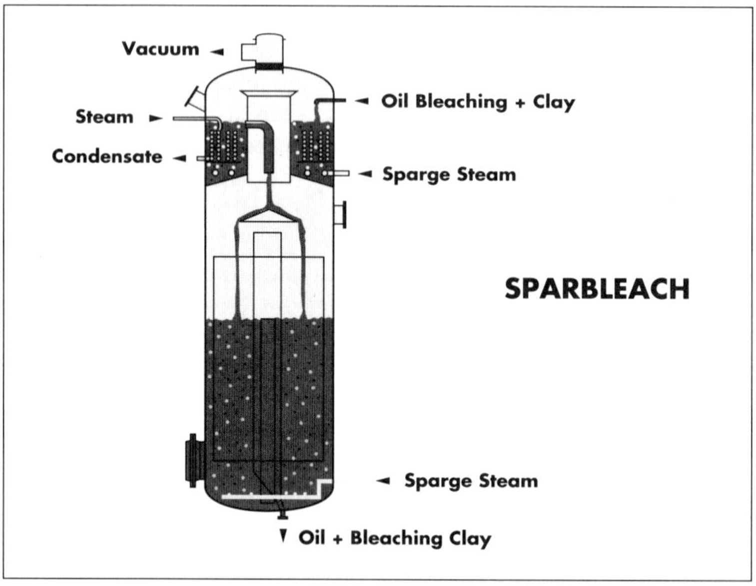

Fig. 5 Diagram of the reactor for the Sparbleach process. (Courtesy of Extraction de Smet NV/SA.)

where it is again steam sparged until it flows from the vessel for filtration. As well as a saving in bleaching clay consumption, savings in maintenance costs are also claimed because there is no mechanical agitation.

2. *Filtration of Bleaching Clay*

At the completion of bleaching, spent clay is filtered from the oil stream. A number of different filter designs are in use. These include plate-and-frame filter presses, both manual and with fully automatic cleaning; chamber presses, which may be fitted with diaphragms to permit oil to be squeezed out of the filter cake; and filters that use a woven stainless steel mesh as the filter medium. These filter "leaves" may be mounted horizontally in a vertical cylindrical tank and cleaned by spinning the central mounting shaft. Alternatively, they may be mounted in a vertical or horizontal cylindrical tank and cleaned by vibration of the leaves themselves. These filters may discharge spent filter cake either internally through a dump valve or externally into a receiving tray or screw conveyor.

All these filters are capable of fully automatic cleaning, which may be started after a set time period, at a preset differential pressure, or directly by a plant operator. Each filter type requires the oil to be recycled for several minutes after a clean filter is brought into operation to allow sufficient filter cake to build up on the filter mesh for the filtrate to be sparkling clear. It is common practice to position a "guard" filter immediately downstream from the main filter to trap any particles that have remained in the oil. This is often a bag filter with relatively little holding capacity for bleaching clay. Thus it blocks quickly in the event of a significant malfunction of the main filter, which serves to alert plant operators to the need to take appropriate action.

During the cleaning sequence, the filter cake is blown with steam or, preferably, nitrogen to reduce the quantity of oil remaining in the filter cake and therefore lost to the process. Typically, the oil content of spent filter cake is around 20–25%, but it may be as high as 40% depending on the characteristics of the clay, the blowing medium, and the type of filter.

Oil recovered by blowing the filter cake is of a poorer quality than the remainder of the production. It should be returned to crude or rebleached and not simply mixed with the filtered oil.

Spent bleaching clay can present a fire hazard, especially where high iodine value oils have been processed. Spontaneous combustion usually starts in a smoldering pocket. Plant operators need to be alert for early signs of trouble and be prepared to take immediate action.

3. *Spent Bleaching Clay*

Spent clay is a waste product. Fahn and Zschau [12] present an overview of possible disposal methods. More recently there have been attempts to regenerate bleaching clays for reuse [13], but commercial processes seem to be several years away. Some work has been reported [14] on regenerating silica–alumina mixtures. Early work suggests that regeneration may not be possible if these materials are contaminated with phosphatides or soap. This is counter to a major function of these materials. Thus any progress toward successful regenerability is likely to be slow.

4. *Bleaching Plant Operation*

Most bleaching processes are controlled on the basis of the Lovibond red color of the bleached, filtered oil. For certain applications, e.g., soy oil for bottling, chlorophyll content is also measured and bleaching clay addition is increased or decreased as appropriate. Most plants operate with a standard time for filters to remain on-line prior to cleaning. Alternatives are to use the differential pressure across the filter unit to indicate that the filter is ready for cleaning or to monitor filtration rate, which decreases as the filter cake thickness increases.

Whichever method is adopted, it is crucially important to clean a filter before filter cake thicknesses are such that there is "bridging" between adjacent filter leaves. If bridging occurs, the filter will not discharge the filter cake correctly when subjected to the normal cleaning sequence. It will be necessary to dismantle the filter elements and clean them manually, entailing additional cost and downtime.

Care must be taken when withdrawing filter leaves for cleaning and when replacing clean leaves. The woven stainless steel mesh that provides the filter medium is relatively easily damaged. Obviously, a tear in a filter surface will result in bleaching clay passing through the filter and rapidly filling the guard filter.

Although the filters described are cleaned automatically in place, a proportion of the bleaching clay is driven into the apertures of the mesh and is not removed by vibration or spinning. When a number of the filter elements are fouled in this way, they require additional cleaning. This is often undertaken away from the filter body either by treatment with hot caustic soda in a specialist cleaning tank or by jet washing with hot water.

Because of the need for periodic extra cleaning of filter elements, a spare set or half-set is maintained in a clean condition for use when elements are removed for external cleaning. This provision keeps the downtime of the filter unit to an absolute minimum.

G. Silica Refining

As mentioned in the previous section, silica hydrogel may be used for bleaching. It was initially introduced to oils and fats processing as an alternative to bleaching clay. Silica hydrogel has a much higher capacity than bleaching clays for soaps, phosphatides, and, in particular, trace metals [15–19]. Unfortunately, it does not have any significant capacity for color pigments including chlorophyll. Thus, when used as a bleaching medium, silica hydrogel is normally used in combination with bleaching clay.

Preferably the silica hydrogel is added to the refined and washed oil several minutes before the bleaching clay is added. This staged addition maximizes the benefits of silica hydrogel by adsorbing all the soap and most of the phosphatides present, leaving the subsequently added bleaching clay to adsorb color pigments, trace metals, and oxidation products. When using this two-step addition, savings on total bleaching media have been achieved [20]. The use of a combination of Trisyl (W. R. Grace) silica hydrogel with a bleaching clay was found to improve phosphatide removal and color stability in physically refined palm oil [21].

A reduction in the amount of solid adsorbent used increases the time "bleaching" filters remain on-line and reduces the quantity of neutral oil lost with the adsorbent. It also reduces the amount of solid waste that must be disposed of. Financial savings are not necessarily achieved, however, because silica hydrogel is typically around three times as expensive as bleaching clay.

Because of the high adsorbency of soap onto silica hydrogels, a modification has been proposed to the traditional refining and water washing process [22]. In the modified process, residual soap in oil leaving the primary separator is adsorbed onto silica hydrogel instead of being washed out by hot water. The main advantage of this process is the significant reduction in the quantity of water requiring treatment in the effluent plant. It is not known to what extent this process has been adopted in commercial practice.

Somewhat surprisingly it has been found that rather than soap competing with phosphatides for active sites on the silica hydrogel, soap actually enhances the adsorptive capacity for phosphatides. Combined soap and phosphatide capacities in excess of 100% on a weight basis can be achieved [25].

A comparable process using silica hydrogel has been developed for use in physical refining [26]. This involves adding a very small quantity of caustic soda solution with the silica hydrogel (e.g., 0.5% of 18°Baumé/3.6 Normal caustic) to produce some soap to enhance phospholipid removal prior to addition of bleaching clay as normal (see Section III.F.1).

A silica-based process has also been developed for acid refining high phosphorus oils such as canola and soybean oils. The process [27,28] involves adding a degumming acid such as citric acid. This increases the phosphatide adsorption capacity of silica hydrogel to about 65% by weight (dry basis) at a residual phosphorus content of 5 mg/kg in the oil [29]. This capacity is used to calculate the amount of silica hydrogel required to reduce the phosphorus content of an oil of known phosphorus content to below 5 mg P/kg oil. For oils containing more than 200 mg/kg of phosphorus, the quantity of silica hydrogel required is impractically high. An improved process [30] involves mixing citric acid (as a 50% solution) into heated oil, then, after a suitable holding period, adding sufficient sodium silicate (10% solution) to neutralize about 70% of the citric acid introduced originally. This limits the level of soap produced while ensuring good hydration of the phosphatides. Silica hydrogel is then added at approximately 80°C (100°C max) to adsorb the gums and the soaps. Thereafter, the oil may, if desired, be bleached with clay prior to filtration to remove the silica and the bleaching clay. Laboratory work has shown that this process can be satisfactorily applied to oils containing levels of phosphatides up to 400 mg P/kg oil. As with the other silica refining processes reported in the literature, it is not known whether, or to what extent, this process is being used in commercial practice.

H. Deodorization

Almost invariably, deodorization is the final processing stage in the refining sequence. After deodorization the oil is fully fit for human consumption and is ready for shipment in bulk to food manufacturers or for conversion into a range of oil and fat products.

The purpose of deodorization is to remove from the oil those trace components that give the oil its characteristic, and generally unpleasant, taste and odor. Because of the process conditions that have to be applied to achieve this objective, deodorization also effects a reduction in free fatty acid and decolorization of some pigments (mostly carotenoids) by thermal degradation; removes portions of the natural tocopherol and tocotrienol antioxidants and other unsaponifiable components such as sterols, hydrocarbons, and alcohols; and almost completely removes any pesticide residues, polycyclic aromatic hydrocarbons (only those containing four or less benzene rings in their structure), final traces of any extraction solvent, and any volatile sulfur compounds [31,32].

Deodorization is, in effect, a steam distillation process that is carried out at high temperature (200–260°C) and low pressure (2–7 torr; 2.5–9.2 mbar). Theoretical treatments of batch deodorization have been developed by Norris [33] (subsequently modified to include an activity coefficient A [34] to reflect the deviation of fatty acid solutions from "ideal"):

$$S = \frac{PO}{EP_v A} \ln\left(\frac{V_1}{V_2}\right) \tag{1}$$

or

$$\ln\left(\frac{V_1}{V_2}\right) = \frac{SEP_v A}{PO} = K\left(\frac{SP_v}{PO}\right) \tag{2}$$

where S = moles of stripping steam, O = moles of oil, P = total system pressure, P_v = vapor pressure of free fatty acid, V_1 = initial number of moles of free fatty acid in the oil, V_2 = the final number of moles of free fatty acid in the oil, E = vaporization efficiency, A = an activity coefficient, and K = EA, an experimental constant.

Gavin [35] developed a similar equation for continuous deodorization:

$$\frac{V_1}{V_2} = 1 + K\frac{P_v S}{PO} \tag{3}$$

where V_1, V_2 = moles of free fatty acid in the oil entering and leaving the deodorizer, respectively, S = moles of stripping steam per hour, O = moles of oil per hour, and K, P, and P_v are as specified above for batch deodorization.

These equations show that the main process variables influencing free fatty acid removal are the absolute pressure of the system, the amount of stripping steam used in relation to the amount of oil, and the vapor pressures of the free fatty acids and other volatile components. These are heavily influenced by the oil temperature.

The activity coefficient is determined experimentally. It is influenced by free fatty acid type and concentration. Alternatively, it can be calculated from an equation developed by Wilson [36]. Stage et al. [37] used this equation to produce a curve for mixtures containing up to 5% stearic acid in tristearin and showed that the activity coefficient A lay between 0.63 and 0.65.

The vaporization efficiency, E, reflects the effectiveness of steam–oil mixing in transferring free fatty acid and other volatile components into the steam. For a modern continuous or semicontinuous plant, E is approximately 1.0. For batch deodorizers the vaporization efficiency is much less, probably closer to 0.5.

1. Influence of Process Variables

From Eq. (1), it is clear that the quantity of stripping steam required to remove a specified quantity of volatile components (e.g., free fatty acid) is directly proportional to the total system pressure P. Thus the lower the system pressure, the smaller the quantity of stripping steam required. There are, however, limits to the vacuum pressure that can be achieved economically. Thus commercial deodorizers usually operate in the range 2–5 torr (2.5–6.5 mbar) to avoid excessive steam consumption in ejector systems.

It is the volume of the stripping steam rather than the mass that is important in stripping out volatile components. A combination of high temperature with a low system pressure ensures expansion of the stripping steam supplied once it enters the deodorizer itself.

Temperature is crucially important. The higher the temperature, the higher the vapor pressures of the volatile components that are to be removed. There are practical limits to the temperatures used in deodorizers. Production of high temperatures is expensive, requiring the use of high pressure steam or thermal fluids. Very high temperatures [in excess of 250°C (485°F)] are undesirable because they increase the conversion of cis to trans isomers in highly unsaturated liquid oils [38–41]. For this reason, deodorization temperatures for liquid oils are being limited to 245°C maximum.

Deodorization effectiveness is also time-dependent, although this is not obvious from the above equations. This is principally because it takes time to expose each "element" of the oil to the required quantity of stripping steam for the prevailing oil temperature and system pressure. The move from batch to semicontinuous or fully continuous deodorization has resulted in a reduction in deodorization times. This has been due, in part, to improved systems for contacting the oil with the stripping steam that, among other features, have permitted the use of higher steam velocities without problems resulting from excessive frothing or unacceptably high oil losses because of oil droplet entrainment into vapor streams.

2. *The Deodorization Process*

A commercial semicontinuous or fully continuous deodorization process may be viewed as a sequence of several steps: deaeration, preheating, high temperature heating, steam stripping (deodorization), heat recovery, chelation, final cooling, and polish filtration. For a batch deodorizer, the preheating and heat recovery steps do not apply. Also, all the steps, except polish filtration, are carried out within the deodorizer vessel itself. With semicontinuous and continuous deodorizers, several steps may be carried out externally to the deodorizer.

Deaeration. Because deodorization is carried out at relatively very high temperature, it is essential that the oil to be deodorized be almost completely free of dissolved oxygen. Failure to remove dissolved oxygen will lead to oxidation under the high temperature conditions in the process.

In batch deodorizers, where the oil is normally charged to the vessel at a moderate temperature, say 40–80°C, deaeration takes place naturally as the oil is heated by steam coils within the vessel because the oil is held under a low absolute pressure. Heating rates in batch deodorizers are relatively low because there are practical limitations to the surface areas of the heating coils through which high pressure steam is passed. Thus batch deodorized oil is likely to be completely deaerated under vacuum before the oil temperature exceeds 100–120°C (210–250°F).

With semicontinuous or fully continuous deodorizers, deaeration has to be provided as a discrete process step. Undeodorized oil is heated to approximately 100°C and then passed into a deaeration vessel operating at a moderate vacuum pressure. The deaerator may draw its vacuum from the main deodorizer plant or, more likely, from an independent vacuum source such as a liquid ring pump.

Preheating. To maximize heat recovery, deaerated oil is further heated by hot deodorized oil either directly or indirectly and either inside or external to the deodorizer itself. Effective use of heat interchange significantly reduces the heat that must be added via high temperature heating.

High Temperature Heating. As mentioned already, high pressure steam (70–80 bar) or a thermal heating fluid (such as Thermex, Dowtherm A, Santotherm, Therminol, Syntrell) is used to raise the undeodorized oil to the desired temperature.

In recent years the European refiners trade association (FEDIOL) has implemented a voluntary policy for members that all new installations will use only high pressure steam and that the use of thermal oils in older installations will be phased out. These decisions were based on concerns that thermal heating oil degraded by continuous use may pass undetected into the food product stream. Although such leaks are unlikely and the most commonly used diphenyl–diphenyl oxide eutectic mixtures are easily detected in taste testing of deodorized oils, a possibility of food contamination nevertheless exists.

Rossell [42,43] reviewed the safety of a range of thermal heating fluids and concluded that although steam remains the preferred high temperature heating medium, the best alternatives are Syltherm 800, which is a dimethylsiloxane polymer, and Dowtherm A and Thermex, which are diphenyl–diphenyl oxide eutectic mixtures. Based on toxicity, ease of detection, and chemical composition after extended use, mineral oils, synthetic organic fluids, and hydrogenated terphenyls are not recommended.

The attraction of thermal fluid systems from an engineering viewpoint is that they operate at low pressures relative to the high steam pressures required for high temperature heating. Because of thermal degradation, diphenyl–diphenyl oxide mixtures should be regularly returned to the supplier for reprocessing.

Whichever heat transfer medium is used, high temperature heating may be carried out inside or external to the deodorizer itself. If it is carried out within the deodorizer, it is important that the oil be sparged with stripping steam to avoid local overheating. External heating is usually carried out in heat exchangers, where the oil is in motion and localized overheating is avoided.

In batch deodorizers, it is usually impractical to achieve any heat interchange between deodorized and undeodorized oil. All the heating and cooling is carried out within the deodorizer, using high pressure (25–30 bar) steam and then cold water.

Heat Recovery. The heat recovery step balances the preheating. Heat is interchanged with deaerated but undeodorized oil either directly or indirectly. It is desirable to cool the deodorized oil as rapidly as practical to maintain the highest oil quality. If heat interchange takes place within the deodorization vessel, the oil should be sparged with steam to provide agitation and to maintain the oil's freshness.

In modern semicontinuous and continuous deodorizers, heat recovery can save between 50 and 85% of the heating costs. Interchange can be achieved in the following ways.

1. Inside the deodorizer
   a. Direct interchange between undeodorized and deodorized oils
   b. Indirect interchange using a heat exchange medium in a closed circuit, probably operating on the thermosiphon principle
2. External to the deodorizer
   a. Direct interchange between undeodorized and deodorized oil in an interchanger, e.g., of the spiral design
3. Inside/external
   a. Indirect interchange between deodorized oil in the deodorizer and undeodorized oil in a vessel that is external to the deodorizer

Chelation. Once the oil is cooled below 120°C, it is almost universal practice to dose it with a low level of citric acid (usually 50–100 mg/kg). The acid is preferably added as a 30–50% solution in water to minimize the risk of blocking transfer lines and nozzles with crystalline citric acid.

The function of the citric acid is to chelate trace metals present in the oil so that the chelates may be removed by subsequent polish filtration. Even at 50 mg/kg, citric acid is present in excess. Thus a low level of citric acid remains dispersed in the deodorized oil. Experience has shown that this improves the oil's resistance to oxidation, particularly if this is measured by an accelerated method such as the one employing the Rancimat instrument.

Cooling. Deodorized oil is finally cooled to an appropriate temperature for storage, shipment to customers, or blending. Cold water is usually used to achieve the desired temperatures. It is common practice to cool liquid oils to 30–35°C and hard fats or hydrogenated oils to approximately 10°C above their melting points. If liquid oils are to be used for bottling it will be necessary to cool them further to below 25°C. This may be done after deodorization or prior to bottling. Depending on the cold water temperature and the heat exchanger design, it may be necessary to use chilled water when ambient temperatures are high (above 30°C).

Polish Filtration. Metal chelates, particles of polymerized oil, and/or traces of bleaching clay should be removed from the cooled, deodorized oil prior to storage and shipment or further processing. This is typically achieved by the use of bag or cartridge filters. In a majority of plants, duplex filter units are employed so that filter elements can be changed without interruption to the oil flow.

The grade of filter element varies between refineries but is generally in the range 10–15 μm. Indication that a filter element needs to be changed may be by differential pressure measurement, reduction in oil throughput rate, or operator judgment.

Many deodorization plant laboratories run standardized filter tests to check the effectiveness of polish filtration. Typically these involve passing a fixed quantity (say, 1 lb) of oil at a specified fixed temperature (say, 50°C) through a specific grade of filter medium (e.g., 5–10 μm nylon sediment test disks) in a laboratory vacuum filter unit operating at a fixed vacuum pressure. When all the oil has been filtered, the filter paper is compared with preprepared standard papers to indicate whether the filter grade is acceptable, borderline, or unacceptable.

Similar polish filtration, with or without filter grade assessment, is usually repeated on the oil or blend immediately prior to tank car or railcar filling and shipment from the refinery. This represents an additional safeguard to ensure that only clear bright oil is sent to customers or for further processing.

3. *Deodorization Equipment*

Semicontinuous deodorizers are probably the most widely used types at present. Fully continuous deodorizers are better suited to refineries processing a single type of oil or a small number of similar or compatible oils. Increasingly, batch deodorizers are being reserved for small capacity plants or for low throughput oils, which are often of high relative value and incompatible with any high throughput oils processed at the same facility.

The term "semicontinuous" indicates that the oil progresses through the deodorization process sequence, outlined above, as a series of small discrete batches by automatic sequencing. Thus oil flow into and out of the plant is discontinuous. "Continuous" or "fully continuous" indicates that oil flows at a steady rate through each of the steps in the process sequence and discharges at the same steady rate from the plant.

A minority of deodorizer designs combine semicontinuous and continuous features. Oil may flow continuously through the deaerating, preheating, and final heating stages to be treated as a batch in the steam stripping or deodorization stage and then be returned to continuous flow through heat recovery, chelation, final cooling, and polish filtration. Alternatively, the above may be reversed such that only the steam stripping stage is fully continuous.

Semicontinuous Deodorizers. Semicontinuous deodorizers normally consist of a series of trays arranged vertically in a tall cylindrical vacuum vessel. Each tray represents a process stage in the sequence, although it is common practice to use two or three trays for the steam stripping or deodorization stage.

Figure 6 shows a process diagram for a typical semicontinuous deodorizer. Undeodorized oil is passed through one of two bag filters to a measuring and deaeration tank mounted above the deodorizer vessel. This tank is used to control the quantity of oil that is passed as a small discrete batch through the process. As the automatic sequencing valve opens, the measured quantity is drawn into the top tray within the deodorizer shell. Here it is preheated indirectly by hot deodorized oil lower down the deodorizer, by means of a thermosiphon arrangement.

After a predetermined interval, the heated oil passes into the second tray, where it is heated to deodorization temperature by high pressure steam or a thermal fluid. The next two trays down the column shown are deodorization trays, in which the oil is sparged with steam using mammoth pumps or similar arrangements to increase stripping efficiency.

Following the deodorization tray, the hot deodorized oil first gives up heat, via the thermosiphon, to incoming undeodorized oil and then to high pressure demineralized water, which is subsequently flashed to produce stripping steam. Citric acid is dosed into oil in the final tray.

Cool oil leaving the bottom of the deodorizer column is water-cooled and then polish-filtered prior to storage. It is becoming more common, as shown in the figure, to use nitrogen injection to return the oil to atmospheric pressure, in preference to allowing it to dissolve air.

The measuring tank, every tray in the deodorizer, and the pool of oil at the base of the deodorizer are all continuously sparged with steam to provide both agitation and some steam stripping.

Fully Continuous Deodorizers. Figure 7 shows one design for a fully continuous deodorizer. In this example, cool undeodorized oil is pumped through a bag filter to a deaeration tank mounted vertically above the main deodorizer column. The oil is sprayed into the tank, which is held under vacuum. The oil is sparged with steam and is preheated by steam coils within the tank. Heated, deaerated oil passes to the top tray of the deodorizer, where it is further heated by a thermosiphon arrangement from the penultimate tray lower down the deodorizer column. Note that all the trays within the deodorizer and the pool of oil at the base of the column are sparged with stripping steam.

The second tray from the top of the column is the main heating tray, where the oil temperature is raised to deodorization temperature by high pressure steam generated locally to the deodorizer in a specialist boiler. Below the main heating tray are two steam stripping or deodorization trays. The design of the steam–oil contact arrangements in these trays varies from one equipment supplier to another, each having its preferred means of enhancing stripping efficiency.

Hot deodorized oil interchanges heat with the incoming deaerated oil and is partially cooled by this process. Further cooling is effected in the bottom tray, in which high pressure water flowing through the cooling coils is partially converted to steam, which is subsequently used as sparge or stripping steam. Citric acid solution is dosed into the oil in this bottom tray, water being flashed off in the low vacuum pressure environment.

Deodorized, precooled oil is pumped through a cold water cooled heat exchanger and then via polish filters to storage. Pressure is brought back to atmospheric by introducing high purity nitrogen into the oil stream prior to polish filtration. This has the important added benefit of

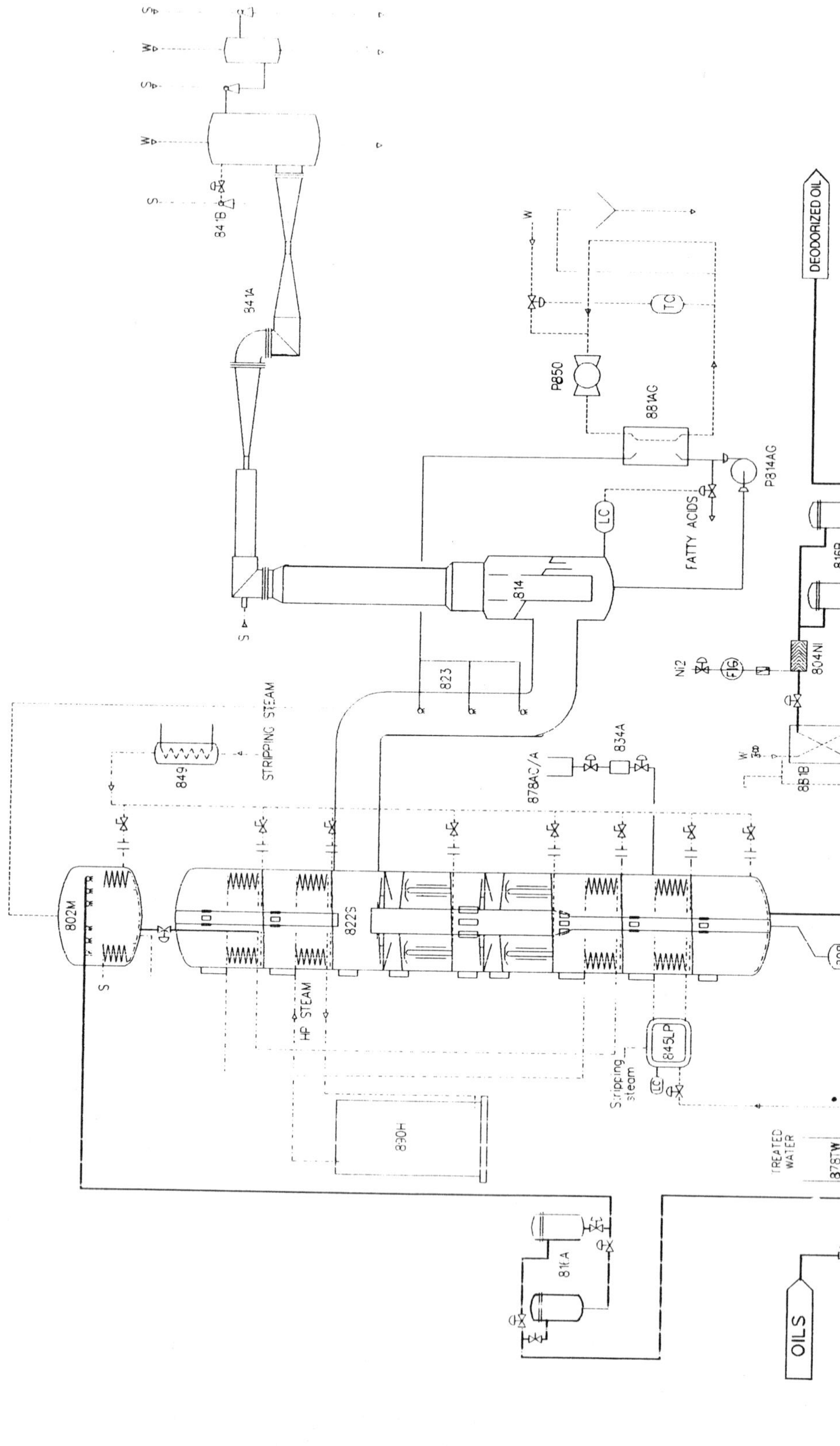

Fig. 6 Flow diagram of a semicontinuous deodorizer. (Courtesy of Extraction de Smet NV/SA.)

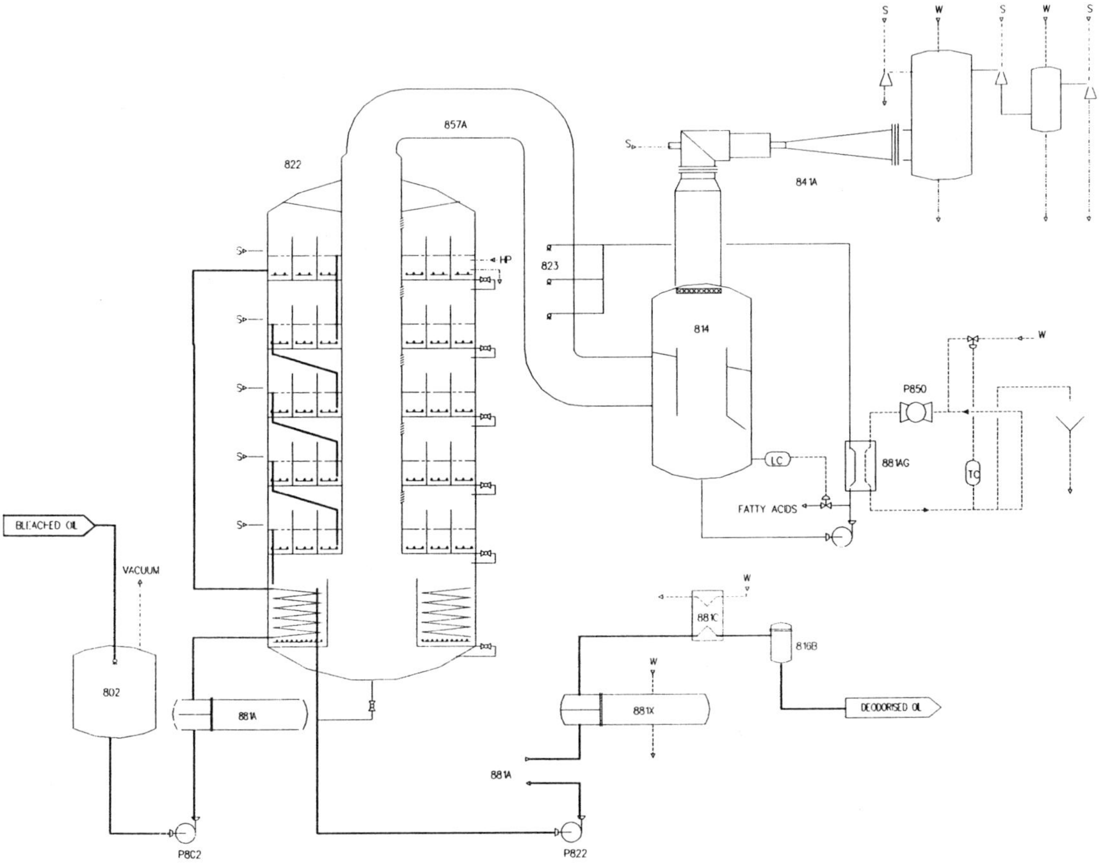

Fig. 7 Flow diagram of a fully continuous deodorizer. (Courtesy of Extraction de Smet NV/SA.)

ensuring that the oil dissolves nitrogen and not air prior to storage. If the oil is stored in nitrogen-blanketed tanks, then any oxidation during storage should be negligible.

Deodorizer Designs. As indicated, the vertically stacked tray design is the most common for both semicontinuous and fully continuous deodorizers. There are a number of different tray designs, providing alternative means of achieving good stripping steam efficiencies. There are also various designs for stacked trays. Again their objectives are to enhance stripping efficiency by optimizing factors such as pressure drop, oil depth, and stripping steam rate. This latter factor is important in that it strongly influences the size of the vacuum system required to sustain the low pressures that are essential for good deodorization. With ever more stringent limits being set for the quality of water discharged by refineries and with higher costs for fresh water, the amount of motive steam required for ejectors and augmentors is becoming critically important.

Several deodorizer designs set out to achieve a minimum oil film thickness so as to eliminate pressure drop and maximize stripping steam–oil contact. Such designs include falling film deodorizers [44], packed column deodorizers [45], and climbing film deodorizers [46].

Batch Deodorization. Batch deodorizers are often used for processing small quantities of oil that are not consistent with the relatively high throughput rates of modern semicontinuous or continuous deodorizers. They can also be used for oils whose values and characteristics are significantly different from those of the major oils being processed, because there is very little chance of accidental commingling of oils in a batch deodorizer.

Modern batch deodorizers are likely to be fabricated from stainless steel, whereas older units are probably of carbon steel. Heating is by high pressure steam or thermal fluid applied to coils within the vessel itself. Capacities are in the range 5–30 t, most commonly 15–25 t.

One problem with batch deodorizers has been the relatively low vaporization efficiencies achievable because of the large depth of oil present compared with the depths encountered in semicontinuous deodorizers, for example. These efficiencies have been improved by fitting devices known as mammoth pumps that greatly increase the intimacy of mixing between steam and oil.

The tops of batch deodorizers are insulated in the newer installations. This assists in ensuring that the top of the vessel is kept hot, at almost deodorization temperature, thereby avoiding the problem of refluxing, in which free fatty acid vapors condense on cool surfaces and drop back into the deodorizing oil.

Higher sparge steam rates are required to achieve satisfactory oil quality by batch deodorization. This can lead to significant oil losses by entrainment of oil droplets into the vapors being drawn into the vacuum system. This effect is reduced by allowing a significant "disengagement" space above the level of the oil in the deodorizer and by fitting a demister either across the deodorizer vessel or, more usually, in the vapor exit pipe.

In batch deodorizers, 2–5 h is the typical deodorization time. Because all the process stages are carried out in the deodorizer vessel (see Section H.2), cycle times for batches are likely to fall in the range of 6–10 h, giving an effective production rate of only approximately 3 t/h.

IV. PHYSICAL REFINING

A. Intensive Degumming

As mentioned in Section II, physical refining is an increasingly attractive process compared with chemical refining. It is suitable only for low phosphatide oils. In order to make more oils suitable for satisfactory physical refining, processes have been developed that remove much greater proportions of phosphatides from liquid seed oils, in particular, to produce feedstocks for physical refining. These contain less than 30 mg/kg of phosphorus and often less than 10 mg/kg. Best known of these processes are the Superdegumming process, patented by Unilever [47] and the TOP process patented by Vandemoortele [48]. The Superdegumming process has been extended and improved subsequent to the initial patent to be capable of producing oils containing less than 5 mg/kg phosphorus, thus matching the capability of the TOP process. The extended Superdegumming process is often known as Unidegumming [49].

A summary of the total degumming process (TOP) is given in Fig. 8. The essential elements of the process [50] are (1) virtually complete decomposition of metal–phosphatide complexes by mixing with a dilute acid that is capable of binding metal ions very finely into the oil being degummed, (2) complete removal of reaction products by partial neutralization of the degumming acid, while avoiding soap formation, (3) removal of residual gums and recycling of oil-rich gums, (4) removal of inorganic compounds and gum residues by water washing, and (5) final removal of water by vacuum drying. The original process was equipment-intensive, requiring four centrifuges. Further operating experience at Vandemoortele has led to the development of the TOP Dry process (Fig. 9), which now operates satisfactorily

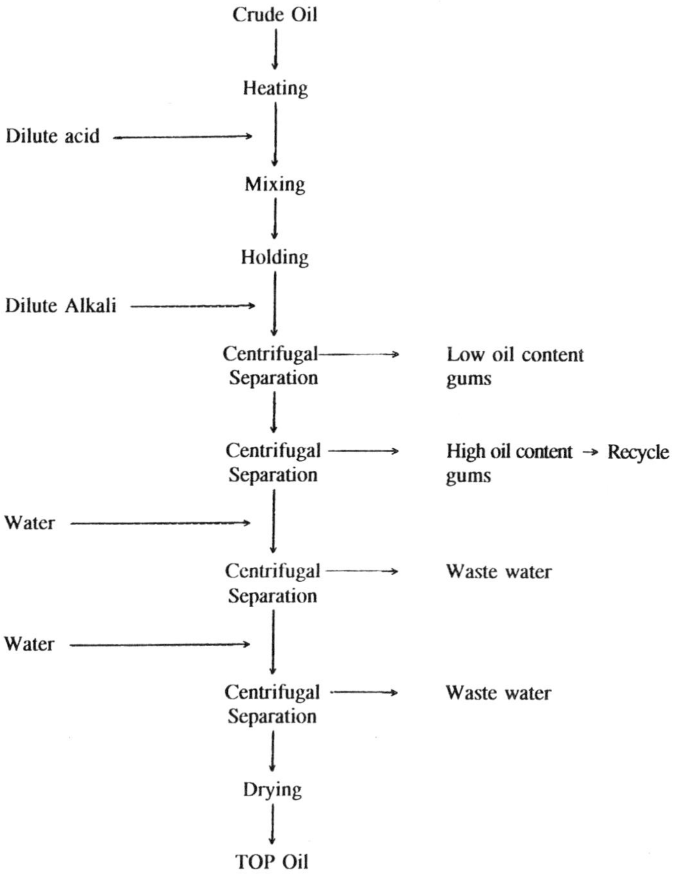

Fig. 8 Total degumming process (TOP).

with two centrifuges [51]. TOP oil typically contains less than 5–10 mg/kg phosphorus and less than 0.2 mg/kg iron.

The Superdegumming process sequence [52,53] is shown in Fig. 10. The process appears rather complex because of the number of stages. It offers the possibility, however, of increasing the amount of easily hydrated phospholipids, which facilitates the removal of phospholipids containing more phosphatidic acid [53]. Citric acid added to warm oil (~70°C) is dispersed under high shear mixing. Once phospholipid hydrate particles are formed, only special low shear mixing is used so as to prevent damaging these particles.

Unidegumming involves further processing of oil after Superdegumming. Superdegummed oil is cooled slowly to 25°C, at which temperature an alkali solution is added to promote the formation of larger phospholipid particles. Further residence time at 25°C allows the desired particle agglomeration to take place. The mix is then rapidly heated to 60–65°C immediately prior to centrifugal separation. The separator used is of the clarification type, capable of removing 1–2% solids from the oil. Unidegummed oil contains less than 8 mg/kg phosphorus. For some oils, levels as low as 1 mg/kg have been achieved [49].

Superdegumming, Unidegumming, and TOP degumming processes all produce oils that are suitable for physical refining, i.e., they contain less than 30 mg/kg phosphorus and typically less than 15 mg/kg when the latter two processes are employed.

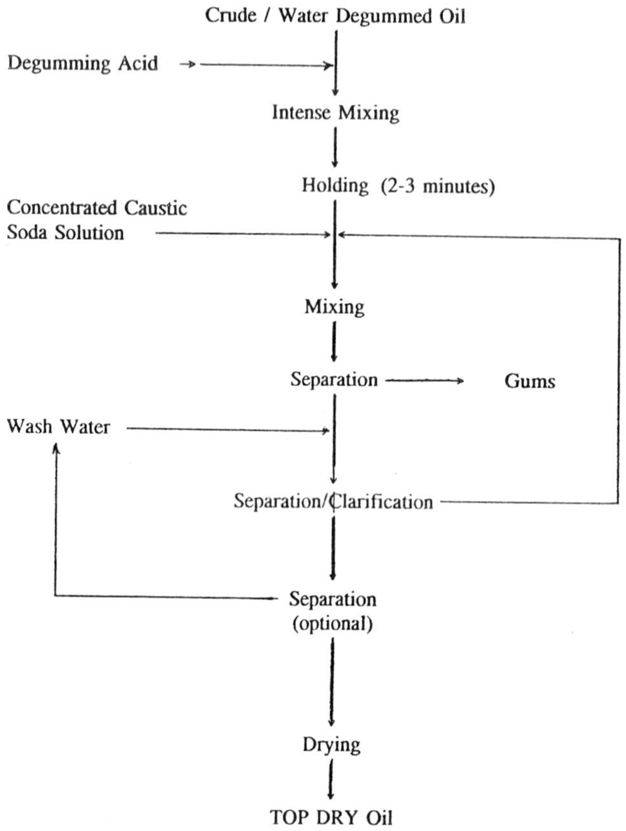

Fig. 9 Total degumming process (TOP Dry).

Because Superdegumming and Unidegumming are carried out at low temperatures (25°C), some waxes will be removed together with the phosphatides when sunflower oil is being processed. This significantly reduces the load on subsequent dewaxing operations.

The TOP process focuses on iron removal as well as on phosphatide removal. This is claimed to give a superior oil because of the prooxidant effect of iron.

Successful adoption of these processes on a commercial scale is determined by several factors, including the additional capital and operating costs of the processes themselves and the cost savings from reduced effluent treatment or disposal. As effluent disposal costs are expected to continue to rise, at least in western countries, over the next few years as environmental restrictions are tightened further, it is anticipated that intensive degumming processes for liquid seed oils will become more attractive. This will be particularly true if these processes are designed into new facilities, as opposed to being retrofitted into existing refineries.

Intensively degummed oils can be caustic refined, if desired. In this case neutral oil losses in the soapstock may be slightly reduced because of the virtually complete absence of phosphatides in the oil and thus a much reduced tendency to emulsion formation. An overall refining loss of 0.5% from crude to fully refined oil has been noted for caustic refining of TOP degummed oil [50].

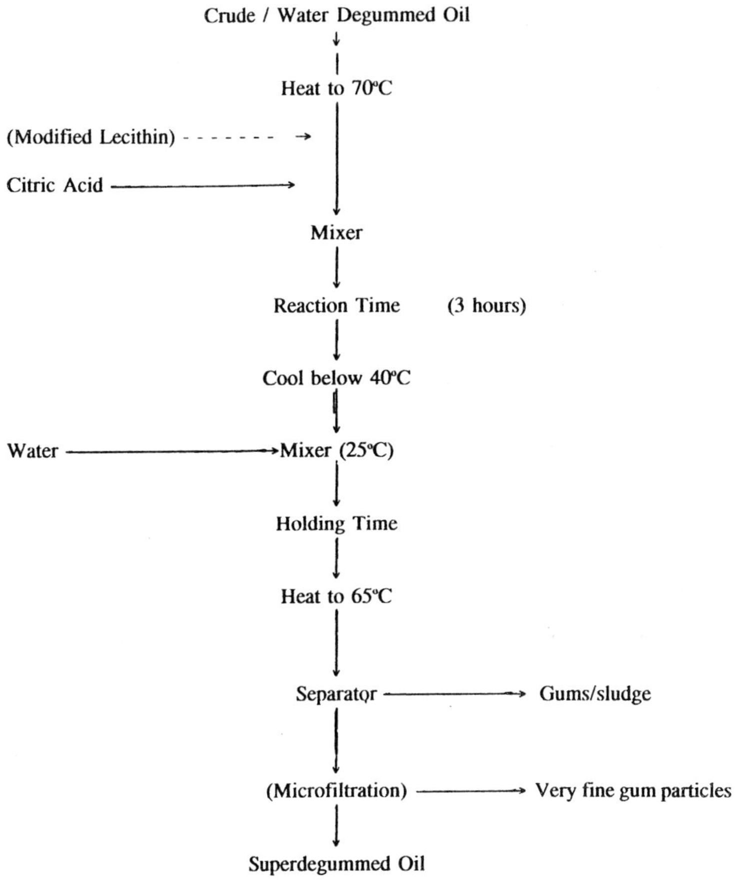

Fig. 10 Superdegumming process.

B. Pretreatment

The objectives of pretreatment are similar to those of intensive degumming. Pretreatment is applied to all oils, whether or not they require intensive degumming to be made suitable for physical refining.

The main functions of the pretreatment process are to reduce the Totox value (Anisidine Value + 2x Peroxide Value), the trace metal content, and the phosphorus content of the oil. This is usually achieved by intimately mixing the oil with a low level (0.1–0.2%) of concentrated phosphoric acid before adding a relatively high level of bleaching clay to the oil (usually at least 1.5 times the level used for bleaching refined or neutralized oils). The bleaching clay is used to adsorb trace metals and oxidation products and to adsorb hydrolyzed phosphatides.

C. Physical or Steam Refining

As already indicated, the principal difference between physical or steam refining and deodorization is the level of free fatty acids present in the feed material. Oils that have been neutralized and bleached, possibly followed by modification through hydrogenation, interesterification, or fractionation, will have free fatty acid levels below about 0.3% and probably below

0.1%. Oils that have been pretreated only in preparation for steam or physical refining will have much higher levels of free fatty acids in the range of 1–5%, depending on oil type and quality. As a result, physical refining plants have to be designed to be able to strip these high levels of free fatty acid from the oil. In comparison with deodorizers, steam or physical refining deodorizers have larger vacuum systems to handle the additional free fatty acid vapor and to maintain a lower absolute pressure than is normally required for deodorization of neutralized oils. They also achieve a longer residence time for the oil at deodorization temperature. They may, in addition, heat the oil to slightly higher temperatures and use greater quantities of stripping steam. Current concerns about isomerization of cis bonds to the trans configuration during deodorization is limiting the temperatures used for physical refining in the same way that deodorization temperatures are being reduced by 10–15°C.

The equipment used for physical refining is very similar to that used for deodorization and may be either semicontinuous or fully continuous. Designs are virtually identical to those used for deodorization, with detailed differences only as discussed above. The whole range of deodorizer designs are used for physical refining.

Palm oil is the oil that is most commonly physically refined. Care must be taken in selecting palm oil for physical refining. Oil that has been oxidized should not be physically refined. Satisfactory refined oil will be achieved only by chemically refining very oxidized palm oil. Totox value is taken as the principal means of assessing whether or not a particular palm oil consignment is suitable for physical refining. Crude palm oils with totox values in excess of 10 are not generally regarded as appropriate for physical refining. Crude palm oils with totox values less than 10 should be suitable for physical refining provided that trace metal and phosphorus contents are low—iron below 3 mg/kg, copper below 0.2 mg/kg, and phosphorus below 20 mg/kg. Ideally the content of carotene should be high (over 550 mg/kg) because this is widely regarded as a further indicator of overall good palm oil quality.

Because of its high carotene content, palm oil has a characteristic deep orange-red color. This is little affected by pretreatment with phosphoric acid and bleaching clay. Once in the high temperature environment of the physical refiner, the carotene is subjected to thermal decomposition. Loncin et al. [54] showed that at 240°C the carotene content of palm oil is reduced to only approximately 1% of its original level after a period of 35–40 min. Thus dark-colored palm oil will be "heat bleached" during the physical refining process, usually yielding a fully refined product that is lighter than 3 Red (Lovibond 5.25 in. cell) in color.

Coconut and palmkernel oils are also commonly physically refined because they are less prone to oxidative deterioration in their crude states and because they are naturally low in phosphorus content. Good quality, high grade tallows are also well suited to physical refining.

Batch deodorizers are used only infrequently for physical refining. This reflects the practical difficulties of achieving sufficiently high temperatures and of economically providing large enough vacuum systems to cope with the relatively high sparge steam rates. Physical refining in batch vessels is normally limited to oils that have low free fatty acids levels, approximately 1–2%. Because heat interchange is much more difficult to achieve in batch processes than in either semicontinuous or continuous processes, the energy costs of batch deodorization and especially batch steam refining are much higher. This further reduces the incentive to attempt steam refining as a batch process.

V. FUTURE TRENDS

Concerns to minimize the environmental impact of oil refining processes will become more significant in the future. It is certain that legal requirements will impose even more stringent limits on discharges to the atmosphere and to water. This pressure, coupled with an ongoing objective of reducing processing costs and/or improving oil quality, is likely to encourage equipment manufacturers and process material suppliers to further improve their products.

Within the limitations of trying to forecast the future, the following approaches seem likely to be significant in contributing to the achievement of the above objectives.

Increased use of physical refining. Physical refining does not involve the production of soapstock with the consequent need to acidulate it to produce acid oil. This conversion leads to the production of relatively large quantities of (usually) sodium sulfate solution of variable pH and varying levels of oil contamination, requiring extensive treatment prior to discharge into public sewer systems. In addition to significantly reducing the quantity of water requiring treatment, physical refining leads to higher yields, because of lower losses of neutral oil, and produces a higher concentration of free fatty acid in the distillate by-product. It is likely that more widespread use will be made of current intensive degumming processes (Section IV.A) or other innovative processes will be developed to make possible satisfactory physical refining of a wider range of crude oil types and qualities.

Use of lower adsorbent levels for bleaching. The introduction of silica hydrogels has focused attention on the relative abilities of different adsorbents to remove the range of unwanted components from oils and fats. Since the initial approach that saw silica hydrogel as an alternative or substitute for bleaching clay, it has been recognized that each medium has its own strengths and weaknesses. From this has come a recognition that the use of different adsorbents either together or in sequence can yield improved oil qualities and lower the total level of adsorbent usage. From this have also come renewed efforts on the part of bleaching clay manufacturers to further enhance the abilities of bleaching clay to remove unwanted components at lower levels of clay in oil. Bleaching clay and other solid adsorbents are expected to pose more demanding disposal problems in the future, with high associated costs. Attempts to improve current adsorbents and the introduction of other, better adsorbents will reduce solid waste disposal quantities. Allied with this will probably be a reduction in the quantity of oil disposed in the spent clay, with a resultant modest improvement in refinery yield.

Use of regenerable bleaching media. Some work has already been reported [13,14,55] on the possibility of economically regenerating bleaching clays for reuse through several cycles, possibly with augmentation by fresh bleaching clay. Clearly, if this approach can be developed into a commercial process, the quantity of absorbent to be disposed of will be very significantly reduced compared with the current situation.

Improved deodorizer steam efficiencies. Further improvements in steam stripping efficiencies will result in reduced steam consumption in stripping demand and vacuum generation by steam ejectors. Falling film deodorizers, for example, require less stripping steam than tray designs [1]. Mention of vacuum systems leads to consideration of odor emissions from deodorizer distillate and/or hot wells. The present trend toward indirect condenser systems and recycling of greasy water is expected to continue and to accelerate, giving further improvements to water management and odor elimination.

Integrated facilities. Where refining is carried out on the same site as crushing or extraction, there is greater flexibility to optimize processing to maximize yield or to minimize energy costs. There are opportunities also to blend certain of the waste products from refinery processes into the meal product from the crushing plant. Examples are gums, soapstock, and spent bleaching clay. It is important to recognize, however, that not all these refinery wastes can be blended into the meal simultaneously because the recipient meal can be put outside specified limits for oil content, for example, by indiscriminate blending. In the future, it is possible that the practice of mixing spent bleaching clay into oilseed meals may be restricted or even prohibited because of the presence of antinutritive components.

A further benefit from operating a large integrated crushing and refining facility may be the economic justification for a combined heat and power (CHP) plant, which permits steam raising and power generation to be optimized so as to reduce the overall energy consumption of the facility and reduce, or even remove, its dependence on externally supplied electricity.

In addition to the specific points noted above, it is anticipated that through ongoing development work, equipment and process material suppliers will continue to make small but important improvements to their products to permit achievement of high yields, improved oil quality, greater throughputs, or less plant downtime.

VI. GENERAL

The processes described here are used successfully to produce high quality refined oils for a wide range of applications. These may be supplied as individual oils or as blends of oils chosen to meet particular specifications.

In addition to the refining processes described here, a refiner may need to be able to use modification processes (described in Chapters 8–10) in order to provide the full range of oil products required by customers. It may also be necessary to include additional processes such as dewaxing or winterizing that are necessary to ensure that certain oils remain completely clear even at low temperatures. This is important for bottled oils supplied for consumer use and oils used in mayonnaise production. Where oils are being produced for bottling or for mayonnaise production, it is important that there be no crystallization of stearins or waxes at low temperatures. For bottling, this is an aesthetic requirement. For mayonnaise production it is essential in order to avoid instability in the emulsion when the product is refrigerated.

Dewaxing may be carried out in a variety of ways. The simplest process involves cooling partially refined or fully refined oil to approximately 4–5°C, holding the oil at that temperature under very gentle agitation for a long enough period for the high melting waxes (fatty alcohols) to crystallize, and then heating the mixture to approximately 10°C to reduce the viscosity prior to filtration through a filter that has been precoated with a filter aid and/or using a filter aid as a body feed with the oil.

More sophisticated processes have been developed in attempts to reduce the yield losses associated with the basic process. It is quite common to take oil from the continuous neutralization plant after the primary separator, at which stage the oil may contain about 1000 mg/kg of soap, or to dose the oil with a low level of caustic soda prior to the secondary separator. Rapid chilling of this oil will crystallize the waxes, with the soap particles acting as nuclei. After storage at low temperature for extended periods (12–24 h) under low shear agitation, the mixture is heated in-line to approximately 10–15°C and then water washed as normal. This process sequence claims to keep yield losses to a minimum. In most cases, however, dewaxed oils produced this way do not meet the stringent chill test requirements and a secondary chilling and filtration process is also required. Here the loading of waxes on the filter is very low and yield losses are modest.

The term "winterizing" is often used as an alternative to dewaxing to describe the removal of high melting point waxes. More correctly, winterizing is a very similar process to dewaxing but its purpose is to remove traces of high melting stearins from, for example, lightly hydrogenated oils or oils such as peanut oil that naturally contain low levels of stearins. In winterizing plants, the filtration step is carried out in cold rooms because the stearins are much more likely to melt if exposed to warm temperatures than are waxes, whose melting points are often in excess of 65–70°C.

Oils that are to be hydrogenated are usually partially refined and occasionally fully refined prior to hydrogenation. This is because hydrogenation catalysts are relatively easily poisoned by some of the non-triacylglycerols found in crude oils, especially sulfur compounds. Catalyst poisoning, or partial deactivation, leads to extended reaction times and, more important, changes in relative reaction rates, affecting selectivity and trans isomerization. Where fish oils and canola oils are being partially refined prior to hydrogenation, it is very important to ensure that the bleaching process has effectively removed almost all of the sulfur compounds that may have been present in the crude oil. For canola oil it can occasionally be crucially important to reduce the chlorophyll content to low levels (0.1 mg/kg or less) prior to hydrogenation in order to avoid formation of a green coloration in the hydrogenated oil, which is almost impossible to remove completely.

Overall, modern refineries have a wide choice of processing alternatives, which should enable them to produce high quality oils and fats from a range of types and qualities of crude oils. Indications are that the range of alternatives available will increase even further in the future.

REFERENCES

1. K. F. Carlson and J. D. Scott, Recent developments and trends: processing of oilseeds, fats and oils, *INFORM 2*: 1034 (1991).
2. A. Hvolby, Removal of non-hydratable phospholipids from soybean oil, *J. Am. Oil Chem. Soc. 48*: 503 (1971).
3. H. Pardun, Deacidification of vegetable oils with ammonia—an environmentally safe refining method, *Fette Seifen Anstrichm. 75*: 297 (1979).
4. F. V. K. Young, Caustic refining. Trace impurities, refining methods and edible oil quality, Presented at Second ASA Symp. on Soybean Processing, Antwerp, The Netherlands, June 1981.
5. R. Fahn, Influence of the structure and morphology of bleaching earths on their bleaching action on oils and fats, *Fette Seifen Anstrichm. 75*: 77 (1973).
6. T. K. Mag, Bleaching—theory and practice, in *Edible Fats and Oils Processing: Basic Principles and Modern Practices* (D. R. Erickson, Ed.), AOCS Press, Champaign, IL, 1990, p. 107.
7. F. V. K. Young, Processing of fats and oils: production of edible oils, in *The Lipid Handbook* (F. D. Gunstone, J. L. Harwood, and F. B. Padley, Eds.), Chapman and Hall, New York, 1986, p. 189.
8. A. N. Sagredos, D. Sinha-Ray, and A. Thomas, Determination, occurrence and composition of polycyclic aromatic hydrocarbons (PAH) in oils and fats, *Fat Sci. Technol. 90*: 76 (1988).
9. F. A. Norris, Refining and bleaching, in *Bailey's Industrial Oil and Fat Products*, 4th ed., Vol. 2 (D. Swern, Ed.), Wiley, New York, 1982, p. 292.
10. U. I. Brimberg, Kinetics of bleaching vegetable oils, *J. Am. Oil Chem. Soc. 59*: 74 (1982).
11. J. H. Henderson, A laboratory study of the press effect in adsorptive bleaching, Presented at 83rd AOCS Annual Meeting, Toronto, Canada, May 1992.
12. R. Fahn and W. Zschau, Possibilities for the use of spent bleaching earth, Presented at 75th AOCS Annual Meeting, Dallas, TX, 1984.
13. R. Nebergall, D. Taylor, and B. Jaynes, Bleaching: solving the problem of disposal, Presented at 84th AOCS Annual Meeting, Anaheim, CA, May 1993.
14. G. Toeneboehn and W. A. Welsh, Regenerable adsorbent process for refining, Presented at 84th AOCS Annual Meeting, Anaheim, CA, May 1993.
15. W. A. Welsh and Y. O. Parent, U.S. Patent 4,629,588 (W. R. Grace & Co.) (1986).
16. P. M. Parker and W. A. Welsh, U.S. Patent 4,734,226 (W. R. Grace & Co.) (1988).

17. J. N. Pryor, J. M. Bogdanor, and W. A. Welsh, U.S. Patent 4,781,864 (W. R. Grace & Co.) (1988).
18. J. N. Pryor, J. M. Bogdanor, and W. A. Welsh, U.S. Patent 487765 (1989) (W. R. Grace & Co.)
19. W. A. Welsh, U.S. Patent 4880574 (1989) (W. R. Grace & Co.)
20. A. Nock, Factors affecting the performance of silica hydrogel in edible oil refining, Presented at 86th AOCS Annual Meeting, San Antonio, TX, May 1995.
21. W.-L. Siew, Y.-A. Tan, and T.-S. Tang, Silica refining of palm oil, *J. Am. Oil Chem. Soc. 71*: 1013 (1994).
22. J. M. Bogdanor, Modified caustic refining, Presented at the International Conference on Fats for the Future, sponsored by the Royal Society of New Zealand, Auckland, February 1989.
23. J. M. Bogdanor, Modified bleaching, Presented at 80th AOCS Annual Meeting, Cincinnati, OH, May 1989.
24. J. M. Bogdanor and G. J. Toeneboehn, Silica refining of oils with little or no chlorophyll, Presented at 80th AOCS Annual Meeting, Cincinnati, OH, May 1989.
25. W. A. Welsh, J. M. Bogdanor, and G. J. Toeneboehn, Silica refining of oils and fats, in *Edible Fats and Oils Processing: Basic Principles and Modern Practices* (D. R. Erickson, Ed.), AOCS Press, Champaign, IL, 1990, p. 189.
26. J. M. Bogdanor, Modified physical refining—commercial considerations, Presented at 84th AOCS Annual Meeting, Anaheim, CA, April 1993.
27. L. O. F. Schumutzler, U.S. Patent 5,248,799 (1993).
28. A. Nock, A novel acid refining process using silica adsorbent, Presented at 84th AOCS Annual Meeting, Anaheim, CA, April 1993.
29. W. A. Welsh and P. M. Parker, Eur. Patent 0 234 221 (W. R. Grace) (1991).
30. A. Nock, Acid refining using silica—a process for refining high phosphorus containing oils, Presented at 85th AOCS Annual Meeting, Atlanta, GA, May 1994.
31. A. Thomas, Desired quality attributes in winter and summer rapeseed, *J. Am. Oil Chem. Soc. 49*: 1 (1982).
32. A. Thomas, The removal of trace contaminants from oils and fats by deodorisation, *Fette Seifen Anstrichm. 84*: 133 (1982).
33. F. A. Norris, Deodorization, in *Bailey's Industrial Oil and Fat Products, Vol. 3* (T. H. Applewhite, Ed.), Wiley, New York, 1985, p. 127.
34. D. S. Szarbo Sakardi, Laboratory deodoriser with a vaporization efficiency of unity, *J. Am. Oil Chem. Soc. 35*: 472 (1958).
35. A. M. Gavin, Edible oil deodorization, *J. Am. Oil Chem. Soc. 55*: 783 (1978).
36. G. M. Wilson, Vapour liquid equilibrium. Part XI. A new expression for the excess free energy of mixing, *J. Am. Oil Chem. Soc. 86*: 127 (1964).
37. H. Stage, W. Kuhns, and H. Hammer, U.K. Patent 2 046 606 (1979).
38. R. L. Wolff, Trans polyunsaturated fatty acids in French edible rapeseed and soybean oils, *J. Am. Oil Chem. Soc 69*: 106 (1992).
39. L. H. Wiederman and D. R. Erickson, Soyabean oil: modern processing and utilisation, *INFORM 2*: 200 (1991).
40. R. G. Ackman, S. N. Hooper, and D. L. Hooper, Linolenic acid artifacts from the deodorization of oils, *J. Am. Oil Chem. Soc. 51*: 42 (1974).
41. G. Devinat, L. Scamaroni, and M. Naudet, Isomerization of linolenic acid during deodorization of rapeseed and soya oils, *Rev. Fr. Corps Gras 27*: 283 (1980).
42. J. B. Rossell, Heat transfer fluids in the oils and fats industry, 1. Specification, types and toxicity, *Lipid Technol. 5*: 110 (1993).
43. J. B. Rossell, Heat transfer fluids in the oils and fats industry, 2. Acceptable media, characteristics and analysis, *Lipid Technol. 5*: 134 (1993).
44. D. R. H. Stage, Countercurrent falling film steam distillation using externally applied temperature field for efficient deodorization of fatty acids and edible oils, *Fette Seifen Anstrichm. 78*: 457 (1976).
45. W. B. Hendrix and P. Sjoberg, Evaluation of the effects of packed column deodorization on seed oil, Presented at 86th AOCS Annual Meeting, San Antonio, TX, May 1995.

46. R. S. Jickling, Deodorization in climbing thin films, Presented at 68th AOCS Annual Meeting, New York, 1977.
47. H. J. Ringers and J. C. Segers, U.S. Patent 4,049,686 (to Unilever) (1977).
48. A. J. Dijkstra and M. Van Opstal, U.S. Patent 4,698,185 (to Vandemoortele) (1987).
49. J. C. Segers and R. L. K. M. Van de Sande, Degumming of rapeseed oil, Presented at 83rd AOCS Annual Meeting, Toronto, Canada, May 1992.
50. A. J. Dijkstra and M. Van Opstal, The total degumming process, in *Edible Fats and Oils Processing: Basic Principles and Modern Practices* (D. R. Erickson, Ed.), AOCS Press, Champaign, IL, 1990, p. 176.
51. A. J. Dijkstra, Degumming, refining, washing and drying fats and oils, Presented at World Conf. on Oilseed Technology and Utilisation, Budapest, Hungary, September 1992.
52. J. C. Segers, Possibilities and limitations for decreasing the environmental load from the refining of edible oils, *Fette Seifen Anstrichm. 87*: 541 (1985).
53. J. C. Segers and R. L. K. M. Van de Sande, Degumming—theory and practice, in *Edible Fats and Oils Processing: Basic Principles and Modern Practices* (D. R. Erickson, Ed.), AOCS Press, Champaign, IL, 1990, p. 88.
54. M. Loncin, B. Jacobsberg, and G. Evrard, Palm oil, *Rev. Fr. Corps Gras 18*: 69 (1971).
55. T. Jacobson and R. Nebergall, Spent bleaching earth regeneration—commercial scale results, Presented at 86th AOCS Annual Meeting, San Antonio, TX, May 1995.

7

Oil Storage, Transport, and Handling

Martin H. Hilder
Unilever Research Laboratory, Vlaardingen, The Netherlands

I. INTRODUCTION

Oil storage and handling are no more than a necessary evil. They cost money and result in deterioration of oil quality. At best, if the deterioration in oil quality is not kept under control, extra costs are incurred; in extreme cases the oil can become unprocessable or unsalable.

Considering the number of times an oil may be handled or stored during processing, storage and handling must be the most widely used processes in the oil industry (Fig. 1). Even this is a simplified picture, as oil will be stored by the supplier and often transported to and stored at the port of loading, and after arrival sea cargoes often undergo further storage and transport before they reach their destination. At each storage or transport step there is a risk of degradation; each transfer into storage or transport involves an oil handling operation that is equally as critical, if not more so.

The objectives of good oil storage and handling are to minimize costs and minimize oil deterioration. Sometimes these are mutually consistent objectives, but in the end the art is to strike an optimal balance. This requires an understanding of the source and magnitude of the costs of oil storage and handling and the factors involved in oil degradation, which also provide a sound basis for the proper design, operation, and management of oil storage and handling facilities.

The following sections contain a summary of the principles involved, covering

The relative costs of capital, operation, and degradation and the factors that determine them
The principles of oil hydrolysis and oxidation, which are the main causes of degradation
The application of these principles to the practical requirements of oil storage, transport, and handling

For more detailed information several publications are worth referring to. Basil Patterson's book [1], which also covers oilseeds and meal, is the most comprehensive. The chapter by List and Erickson in the *Handbook of Soy Oil Processing and Utilisation* [2] provides a useful

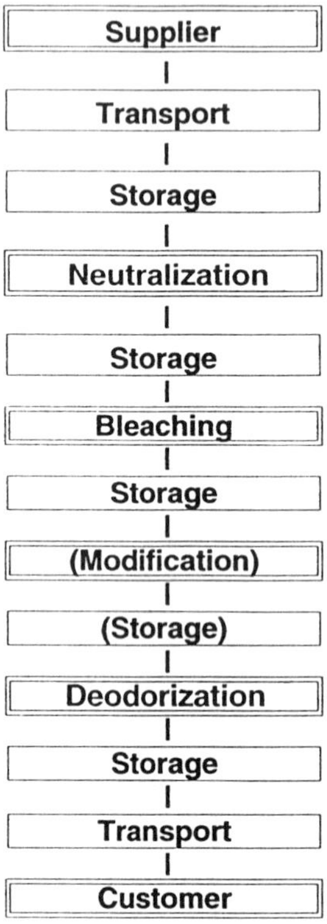

Fig. 1 Storage and handling in oil processing.

review of the literature and state of the art in the U.S. soybean industry up to 1980. This review also formed the basis for List's chapter in Volume 3 of the latest edition of Bailey [3]; previous editions of Bailey (e.g., Ref. 4) are not so informative.

A final point regarding definitions. In the United States the term "refined" oil is often used to describe what in Europe is known as neutralized oil. In this chapter the following European definitions are used:

Neutralized oil is referred to as "neutral" or "neutralized" oil.
"Fully refined" or "refined" oil always refers to the final deodorized oil.

II. COSTS

The costs associated with oil storage and handling can be divided into three parts:

1. Capital cost: both investment and working capital
2. Operating costs: the costs of labor, heating, pumping, cleaning, etc.
3. Oil degradation costs: any extra oil loss or processing or reprocessing costs and the loss in final oil quality

For the present purpose it is more important to understand the relative contributions rather than to know the actual costs; any cost data in real money are soon out of date, and every company will have its own unit costs and financial criteria. Costs are therefore discussed here in relative terms based on the "triglyceride dollar (TG$)," with an average price of oil of 1000 TG$/ton or 1 TG$/kg.

A. Capital Costs

The capital costs of oil storage and handling are primarily determined by the quantity to be stored. One week's oil (buffer) stock for a 100,000 tons-per-annum (tpa) plant represents 2000 metric tons (t) of oil. On average, tanks are only half-full, and therefore the required tank capacity is double the average stock requirement—in this example, 4000 t tank capacity per week oil stock requirement. The actual stock requirement will depend on company policy and supply-and-demand logistics.

Crude oil storage capacity is normally larger than refined oil storage capacity because

1. Crude oil is usually delivered in larger quantities and is more stable.
2. Refined oil is usually less stable and is usually transported in smaller amounts, and once refined, quality (freshness) is paramount.

Ideally, the buffer requirement should be minimized to what is essential for operational purposes. Any extra quantities should be justified on their own merits.

1. Investment

The capital invested in storage facilities will obviously depend on the oil storage requirements and the scale of operation (plant size) but will ultimately be determined by the number and size of the tanks. The type of operation, and in particular the number of crude oils and refined products to be handled, will therefore also play a major role. One to two tanks are required for each crude or refined oil. The choice will depend on how much segregation is required to separate oil qualities, ensure first-in, first-out operation, and other factors.

Assuming for the sake of simplicity that 1 week's crude oil and 1 week's refined oil buffer are required, there is a significant difference between a 100,000 tpa plant for

1. One crude + one refined oil: say, 4×2000 t tanks
2. Five crude + 20 refined oils: say, 32 × (on average) 250 t tanks

General cost engineering data for the United Kingdom and The Netherlands show that tankyard costs increase with the 0.5–0.7 power of the tank size. For our present purposes, we will assume total investment based on Table 1.

Table 1 Effect of Tank Size on Tankyard Costs

Tank size (t)	Investment per tank (1000 TG$)
100	200
250	300
500	450
1000	600
2500	1000

The crude and refined oil tankyard investment for our two plants is therefore about

3.5 million TG$ for 4×2000 t tanks
10 million TG$ for 32×250 t tanks

Every company will have its own criteria, but a target pretax return on investment plus depreciation of 20–25% per year is not unusual; say at least 5% depreciation and 15% return on investment (ROI).

2. *Working Capital*

The cost of the oil held as a buffer stock also has to be financed. If, for example, interest rates of 10% have to be paid, the cost of financing our 2000 t of crude oil and 2000 t of refined oil buffer stocks will be about 0.4 million (4×10^5) TG$ per year.

B. Operating Costs

Labor costs will depend on the organization rather than on technology, so we will assume a nominal cost.

Heating costs based on a steam requirement of 600 kg/yr per square meter of tank surface area to maintain the oil at 50°C in tanks with 50 mm insulation would be about 15,000 TG$/yr for a 2000 t tank (1000 m^2) and 3500 TG$ for a 250 t tank (250 m^2), although the single oil plant is likely to be dealing with only liquid oils, which would not require heating.

The pumping costs based on 0.25 kWh/t for a 4 bar pressure drop amount to about 0.025 TG$/t oil, or, for pumping crude oil in and pumping refined oil out, about 0.05 TG$/t.

The agitation costs for crude oil based on an installed power of 3–5 W/t tank capacity, depending on the tank size, will amount to 0.1–0.15 TG$/t oil processed.

The nitrogen consumption for deodorized oil sparging (0.5 m^3/t), refined oil storage blanketing (max 1 m^3/m^3 oil), and line blowing, etc., would amount to about 2 m^3/t oil, giving a cost of about 0.5 TG$/t oil.

The repair and maintenance (R&M) costs can be estimated as, say, 5% per year on investment.

Cleaning costs and disposal of waste will depend on the local situation and possibilities.

C. Degradation Costs

The cost of any hydrolysis in crude oil storage is relatively simple to estimate. Each 0.1% free fatty acid (FFA) increase plus the neutral oil it takes with it during processing represents a loss of, say 1.5 kg/t, but the recovered (acid) oil can be sold at about half the price of the original oil. With liquid oils at ambient temperature, hydrolysis should be negligible; but with high FFA crude that has to be stored at higher temperatures (e.g., crude palm oil), some hydrolysis is virtually unavoidable. With poor practice, examples where the FFA increase was 0.5% or more are not unknown.

The cost of oxidation is more difficult to quantify. Depending on where it occurs during processing, it can result in extra processing costs (e.g., extra bleaching earth) or total reprocessing, with all the costs and loss in effective capacity that this involves.

Based on the example given by Strecker and his colleagues [5], which is summarized in Section V.A.2, we will assume that with good oil storage and handling practice (time and conditions), oxidative degradation can be reduced to a minimum level, which can be compensated for by an additional 0.1% bleaching earth during normal processing. The costs associated with this (earth, oil loss, etc.) amount to, say, 1.5 TG$/t oil more than if the crude oil were processed directly on arrival.

As Strecker demonstrated, long storage times and poor storage conditions can result in much greater extra earth requirements; with poor practice, 0.5% or more extra earth can be required. We can add to this the oil loss associated with any "tank-bottoms," the sludge which separates out during storage. Crude oils should not normally contain more than about 0.05% dirt. If this is allowed to settle in storage it can entrain many times its own weight of oil; as a paste or slurry, tank-bottoms will incur oil losses of at least 1–1.5 kg/kg dirt. This oil loss can be avoided to a great extent if tanks are agitated.

Finally, we should not forget the loss in refined oil quality due to oxidative degradation either during or after processing. This is even more difficult to quantify, but it can be expected to affect oil stability or shelf life and therefore may show itself in the form of reduced market share or increased trade returns. For the present purpose, we will assume a cost equal to the cost of oxidation during processing.

D. Total Costs

Adding up the individual items for the two 100,000 tpa examples, assuming good practice, we arrive at the totals in the first column of Table 2. The column headed "Poor practice" has been added to show the possible effect of long storage times and poor storage and handling conditions. These figures are simply intended to show the relative effects of individual items and practices and do not necessarily follow conventional accounting rules.

The first conclusion that can be drawn is that items related to capital, including R&M, make up a large part of the total, even if we are satisfied with a lower return on investment.

Table 2 Hypothetical Costs (TG$/t oil) of Crude and Refined Oil Storage

Item	Good practice	Poor practice
Average stock (weeks)	2	4
Capital (TG$/tpa)	35–100	70–200
Capital costs		
Depreciation	2–5	3.5–10
ROI	5–15	10–30
Working capital	4	8
Operating costs		
Labor	3	4.5
Heating	0–1	0–2
Pumping	0.05	0.05
Agitation	0.1–0.15	—
Nitrogen	0.5	—
R&M	2–5	3.5–10
Tank cleaning	—[a]	—[b]
Degradation costs		
Hydrolysis	0–0.75	2.5
Oxidation	1.5	5
Tank bottoms	—[a]	1
Final oil quality	1.5	5
Total (TG$/t)	20–37	43–78

[a]minimal cost.
[b]significant extra cost.

The capital cost of storage is high, and this is the first good reason for minimizing storage capacity and stocks. The other good reason is the inevitable degradation associated with any storage. Once the capital has been invested, the capital costs are more or less fixed, but at least half of the remaining (controllable) costs are associated with oil degradation. With good practice, the cost of degradation can be reduced to a tolerable minimum, but with long storage and poor conditions this cost can increase more than proportionally.

Up to now we have considered only crude and refined oil storage, but the same arguments apply to intermediate storage. For the same reasons, any intermediate storage should be reduced to an absolute minimum. A minimal process buffer between processing steps is the ideal situation, so that intermediate storage as such can be eliminated.

III. HYDROLYSIS OF OILS

The reaction between water and glycerides (hydrolysis) results in the formation of free fatty acid (FFA) and a diglyceride, monoglyceride, or glycerol. All these degradation products can cause a financial or quality loss:

1. Any increase in FFA and, to a lesser extent, di- and monoglycerides usually results in a direct financial (oil) loss.
2. In some products, the presence of di- or monoglycerides has an adverse effect on product properties.
3. Glycerol in effluent will contribute to any COD or BOD (chemical or biological oxygen demand) costs.

Under storage conditions the FFA will normally be less than 10% and the predominant reaction is [6]

$$\text{Trigyceride} + H_2O \rightarrow \text{diglyceride} + \text{FFA}$$

Enzymatic hydrolysis will not be considered; with normal levels of impurities in crude oil and the proper (heat) pretreatment at an earlier stage, this can be neglected.

Fatty acids of chain length C_{16} and above are essentially flavorless and do not affect the taste of oils and fats. However, short-chain fatty acids can cause "hydrolytic rancidity" in butter (rancid flavor) and in lauric oils and fats (soapy flavor); the flavor thresholds of lauric and capric fatty acids are about 0.07% and 0.02%, respectively [1, p. 84; 3, p. 274].

A. Analyses

The determination of FFA and moisture are standard analyses in the oils and fats industry. For accurate measurement of moisture contents, the Karl Fischer method is normally used. With good laboratory practice, levels down to 0.01% H_2O can be detected, and with more care an accuracy of $\pm 0.001\%$ H_2O is achievable [7], but with poor laboratory practice it can be difficult to obtain reliable results below 0.1% H_2O.

B. Solubility and Diffusivity

Hydrolysis takes place only with water dissolved in the oil. The solubility of water in oil and the rate of diffusion of "free" water (liquid or vapor) into the oil may therefore be relevant in understanding the rate and extent of hydrolysis.

Values of the approximate solubility and diffusivity of water in edible oil deduced from published data are given in Table 3. The solubility of water in oil has been measured in the

Table 3 Solubility and Diffusivity of Water in Oil

Temp. (°C)	Water saturated (%)	Water vapor at 100 mbar (%)	Viscosity (cP)	Diffusivity ($m^2/s \times 10^9$)
0	0.1	—	—	—
25	0.15	—	62	0.25
50	0.2	0.15	24	0.7
75	0.3	0.065	12.5	1.4
100	0.45	0.035	7.5	2.6
125	0.65	0.02	5.1	4.0
150	~1	0.015	3.6	6.1

temperature range 0–265°C. The saturation solubility increases with temperature, but the solubility of water vapor decreases with temperature [7,8].

The mole fraction solubilities are the same for all oils. The weight percent solubilities in Table 3 (for an average oil of MW 850) are therefore dependent on oil type; for example, in coconut oil the percent solubility is 25–30% higher. The solubility of water vapor obeys Henry's law, and the (mole fraction) solubility is proportional to the partial pressure of water vapor. The solubility at 50°C and 10 mbar is therefore about 0.015%, and at 100°C and 200 mbar, about 0.07%.

The diffusivity of water in oil was measured at 24°C [9]. The other values in Table 3 have been estimated based on the Stokes–Einstein relation; i.e., the product of the diffusivity and viscosity divided by the absolute temperature is roughly constant.

C. Factors Affecting Hydrolysis

The rate and extent of hydrolysis under storage conditions was extensively studied by Loncin [6,7]. He and his colleagues also made a significant contribution to the study of the solubility of water in oil, especially on the effect of FFA [6,7,11]. The rate of hydrolysis is dependent upon the FFA level, the concentration of (dissolved) water, the temperature, and the type of oil.

1. Free Fatty Acid

Hydrolysis is an autocatalytic reaction; the rate is proportional to the level of FFA. If the hydrolysis of an oil with 1% FFA results in a final FFA of 1.5%, an oil with 5% FFA will hydrolyze to about 7.5% FFA under the same conditions. Loncin [6,7] provides many examples to illustrate this effect.

2. Moisture

It is reasonable to expect that the rate of hydrolysis is proportional to the concentration of (dissolved) water, but direct experimental evidence for this is scarce. Sarkadi [12] showed that the increase in FFA during deodorization is proportional to the water vapor pressure. Based on the solubility relations, this suggests that the rate of hydrolysis is basically dependent on the water concentration. The situation is made more complicated by the fact that the solubility of water in fatty acids is greater than in the corresponding triglycerides, and the solubility of water in oil also increases with increasing FFA.

Table 4 Effect of FFA on the Saturation Solubility of Water in Palm Oil

	Temperature (°C)		
	40	60	80
FFA (%)	Saturation solubility (%)		
0.2	0.18	0.23	0.32
2.5	0.21	0.27	0.36
6.5	0.25	0.32	0.42
15	—	0.42	0.52

The effect of FFA on the saturation solubility of water in palm oil [6,7,11] is shown in Table 4. However, it is important to note that Loncin [6,7] concluded that the autocatalytic effect is not due to the enhanced solubility of water with increasing FFA.

Moisture is also consumed in the hydrolysis reaction. Complete conversion of 0.1% H_2O will result in the formation of about 1.2–1.6% FFA, depending on the molecular weight of the triglycerides. If no more moisture is supplied, the dissolved water content will decrease and the reaction will slow down. With an excess of ("free") water, more water can dissolve. In this case the rate of hydrolysis may be dependent on the rate of transfer or diffusion of the water into the oil.

All these effects illustrate the importance of minimizing the moisture content during transport and storage:

1. At moisture contents at or below the saturation solubility, the hydrolysis reaction will tend to be self-limiting.
2. With even only 1% of "free" water, the rate and extent of hydrolysis can be many times greater.

3. *Temperature*

As with any chemical reaction, the effect of temperature on the rate of hydrolysis is significant; the rate approximately doubles with each 10°C rise in temperature [6,7]. From the above description of the effects of FFA and moisture, the rate of hydrolysis can be formulated as

$$\frac{d\,[\mathrm{FFA}]}{dt} = k\,[\mathrm{H_2O}][\mathrm{FFA}]$$

Loncin and his colleagues worked mainly with *saturated* oil–FFA mixtures in the presence of excess ("free") water; this is not always clear from the published results [6,7] but is a reasonable assumption. They evaluated the effect of temperature based on the equation

$$\frac{d\,[\mathrm{FFA}]}{dt_{10}} = k'\,[\mathrm{FFA}]$$

with time (t_{10}) in units of 10 days; their results are summarized in Table 5. The rate constant k' is therefore the product $k\,[H_2O]_{sat}$ and is not a pure reaction rate constant; it includes the effect of temperature (and FFA) on the saturation solubility of water. k' therefore indicates the maximum rate of hydrolysis of an oil–FFA mixture under storage conditions.

Table 5 Effect of Temperature on the Rate of Hydrolysis of Palm Oil

Temperature (°C)	Rate constant k' [(10 days)$^{-1}$]
37	0.025
50	0.05
60	0.125
70	0.29
80	0.51
100	1.5

4. *Oil Type*

Loncin [6,7,11] also investigated the effect of the type of oil and the type of fatty acid on the rate of hydrolysis. The results show that the rate of hydrolysis increases in the order

Groundnut oil < palm oil < coconut oil

Palm acid < coconut acid < butter acid

This is also the order of decreasing molecular weight and increasing saturation solubility of water. The influence of oil or fatty acid type is therefore at least qualitatively due to its effect on the level of dissolved water.

The pH or addition of anionic or cationic surface-active compounds to the water phase, or the presence of protein, pectin, carotene, cellulose, phosphatides, monoglycerides, or metal soaps have no effect on the rate of hydrolysis [6,7].

5. *Overall Rate of Hydrolysis*

The effect of FFA, moisture, and temperature on the rate of hydrolysis can therefore be described by

$$\frac{d\,[\mathrm{FFA}]}{dt} = k' \frac{[\mathrm{H_2O}]}{[\mathrm{H_2O}]_{\mathrm{sat}}}\,[\mathrm{FFA}]$$

Where the time t and rate constant k' are expressed in the same units. The maximum hydrolysis occurs when the oil–FFA mixture is saturated with water, and can be expressed as

$$\ln \frac{[\mathrm{FFA}]}{[\mathrm{FFA}]_0} = \ln \left(1 + \frac{[\mathrm{FFA}]_{\mathrm{inc}}}{[\mathrm{FFA}]_0}\right) = k't$$

where $[\mathrm{FFA}]_0$ is the initial FFA concentration and $[\mathrm{FFA}]_{\mathrm{inc}}$ is the increase in FFA concentration. The maximum increase in FFA over initial FFA can therefore be estimated. The effect of storage time and temperature is illustrated in Fig. 2.

The exponential increase with time shown in Fig. 2 assumes that the oil–FFA mixture is always saturated, i.e., that additional "free" water dissolves to make up for the increase in saturation solubility with increasing FFA and for the moisture consumed by the reaction. In practice this will not always happen, and the reaction will slow down as a result, giving the approximately linear increase in FFA often observed in practical situations. Even under satu-

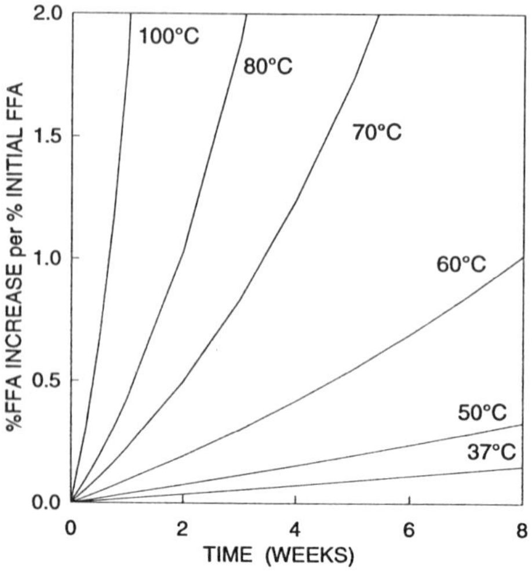

Fig. 2 The hydrolysis of palm oil saturated with water.

rated conditions the increase may appear to be linear over short periods of time within the limits of experimental accuracy.

The rate and extent of hydrolysis at lower moisture levels or with other types of oil can be estimated by adjusting the "rate constant" (k') for the actual moisture content and/or saturation solubility. For example, after 4 weeks storage or transport of crude palm oil with 5% FFA at 50–60°C and 0.1–0.2% H_2O (50% saturated), one would expect an FFA increase of 0.37–0.96%. This is probably representative for the storage and transport conditions in 1953 when Loncin [6] quoted an increase of about 0.1–0.25% FFA per week. For neutralized or "rbd" (refined–bleached–deodorized) oils and with modern transport and storage practices, the increase should be (and generally is [13]) much lower.

IV. OXIDATION OF OILS

The reaction between oxygen and glycerides is far more complex than hydrolysis, but in general it will also lead to an increase in processing costs, oil losses, and a loss in quality. In extreme cases it can result in an oil becoming impossible to process or impossible to sell.

The oxidation of edible oils proceeds via three steps:

1. Absorption of O_2
2. Formation of peroxides
3. Degradation of peroxides to secondary oxidation products:

$$O_2\ (g) \rightleftharpoons O_2\ (l) \rightarrow \text{peroxides} \rightarrow \text{secondary products}$$

Much has been written about the mechanisms of peroxide formation and degradation. Far less attention has been paid to the absorption step, which is probably the most important in industrial practice.

Oxidation can take place only if there is dissolved oxygen in the oil. If the absorption of oxygen is prevented or minimized, oxidation cannot take place or is limited.

A. Analyses

Many methods for analyzing the extent of oxidation of edible oils have been proposed and tested [15,16]. The two most popular and useful are those determining the peroxide value (PV) and the anisidine value (AV).

1. *Peroxide and Anisidine Values*

The peroxide value is the result of a chemical analysis of the hydroperoxide content of the oil. One millimole of hydroperoxide per kilogram of oil will give PV = 2; a PV of 1 therefore represents the conversion of 16 mg O_2/kg oil into hydroperoxides. An alternative measure is the Lea value (LV), which is equal to the number of millimoles of hydroperoxide per kilogram of oil and is therefore related to PV by PV = 2 LV.

The anisidine value (previously called the benzidine value) is a physical measure of the aldehyde content based on the extinction at 350 nm. Aldehydes are the most important degradation products of hydroperoxides with respect to final taste and odor. The AV does not measure all the secondary products but is accepted throughout the industry as the best standard analysis.

In practice, often only the PV is measured, but this can be misleading. Referring to the reaction scheme above, it can be seen that (1) a low PV may be the result of fast breakdown of hydroperoxides rather than slow formation and (2) a high PV during processing is never good, but the effect on final taste and stability may be limited by selective breakdown to secondary products with less taste impact.

Another problem with the PV is the need for careful sampling and sample handling. It is difficult to avoid contact with or absorption of air, but it should be remembered that every 8 mg dissolved or absorbed oxygen per kilogram of oil may give a false PV = 1 due to the so-called oxygen error [16, p. 220]. However, dissolved oxygen is less reactive than the "active oxygen" in hydroperoxides, and the "oxygen error" should be less than 0.5 PV unit if the analysis is correctly executed.

2. *Totox Value*

A measure often used to express the total extent of oxidation of an oil is the Totox value, which by definition is

$$\text{Totox} = 2\ \text{PV} + \text{AV}$$

Although this empirical combination of chemical and physical analyses may seem strange, in practice it has been found to be very useful for evaluating the effects of processing (including storage and handling) on oxidation.

For pure thermal degradation of PV or physical separation of Totox, I have found that a "Totox balance" usually works.

As an example of the use of the Totox value, Fig. 3 includes the Totox values corresponding to the PV and AV figures already reported by Mag [17, p. 112] to illustrate the effect of clay (earth) dosage on the oxidation values of soybean oil. The Totox value provides a better physical appreciation of Mag's conclusion that "adsorption of secondary products is relatively inefficient"; the adsorption capacity in this case was about 3.5 Totox per percent of clay. We shall see further examples of the use of the Totox value later.

Strecker and his colleagues [5, p. 318] recognize the importance of PV and AV in corn (maize) oil processing but consider that "the significance of the AV [in the Totox value] must be enhanced to properly express the oxidative state of an oil throughout the processing

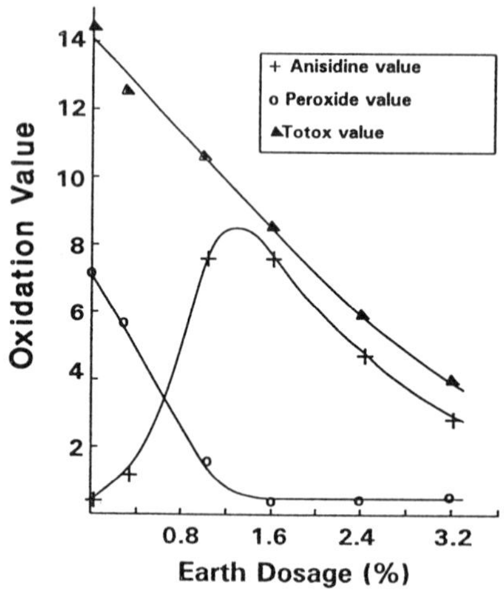

Fig. 3 Oxidation values of bleached soybean oil.

scheme." They proposed an alternative index, but their reasons for this conclusion are not clear to me.

3. *Oxygen Content*

The dissolved oxygen content of an oil may eventually cause oxidation and therefore contributes to the potential oxidation of the oil. It can also be used to measure the extent of oxygen absorption during processing. The oxygen content is not easy to measure, but various methods have been developed in the past [15, p. 223]. For example, dissolved oxygen analysis has been used to follow the rate of oxidation and the corresponding formation of volatile components [18]; Fedeli and Brillo [19] used an oxygen analyzer to measure the oxygen profile during absorption in a column of olive oil.

Portable oxygen-in-oil analyzers are also available [e.g., 20] that have been successfully used to compare and evaluate alternative oil storage and handling procedures [13,21]. These analyzers usually measure a percentage saturation or partial pressure of dissolved oxygen, which, if the solubility of oxygen is known, can be converted to oxygen content.

B. Solubility and Diffusivity

Oxidation takes place only with dissolved or absorbed oxygen. The solubility and rate of diffusion of oxygen in the oil are therefore relevant in understanding the rate and extent of oxidation. The solubility of nitrogen is also relevant because nitrogen is often used to protect oils from oxidation.

A summary of approximate solubilities and the diffusivity of oxygen deduced from published data is given in Table 6.

The solubility of nitrogen in various oils has been determined between 25 and 150°C [22,23]; there is no systematic effect of oil type, and other authors report similar results. There is far more variation in the published data on the solubility of oxygen in oil, especially with unsaturated oils and at temperatures above 60°C, where oxidation could affect the results. The figures in Table 6 are realistic average values based on measurements with hardened lard and olive oil in the range 25–85°C [22,23].

Table 6 Solubility of Oxygen and Nitrogen and the Diffusivity of Oxygen in Oil

Temp. (°C)	Oxygen at 1 bar (mg/kg)	Nitrogen at 1 bar (mg/kg)	Viscosity (cP)	Diffusivity of oxygen ($m^2/s \times 10^9$)
0	170	80	—	—
25	180	85	62	0.8
50	185	90	24	1.3
75	190	95	12.5	1.9
100	200	105	7.5	2.5
125	—	110	—	—
150	—	115	—	—

The solubility of gases usually obeys Henry's law: the solubility is proportional to the partial pressure of the gas. The solubility of oxygen from the air (21 vol% oxygen) at 25°C is therefore about $0.21 \times 180 = 38$ mg/kg oil. Berger [13,, p. 439] quotes 36 mg/kg for palm olein at 60°C, and Fedeli and Brillo [19] imply a value of 39 mg/kg for olive oil at 20°C. These solubility data will be seen to be useful when discussing the rate or extent of degradation.

Data on the diffusivity of oxygen in edible oils are surprisingly scarce. Crowe and coworkers [24] (working in the Water and Effluents Laboratory of a civil engineering department!) reported diffusivities in olive oil of 0.87×10^{-9} m^2/s at 25°C and 1.06×10^{-9} m^2/s at 37°C. They also measured the diffusion resistance of triolein and tributyrin. Schrader et al. [25] determined the solubility and diffusivity in vegetable oil (probably soybean and/or sunflower) to be 0.6×10^{-9} m^2/s at 20°C. They also measured the solubility and diffusivity in butter, mayonnaise, and two well-known brands of German margarine.

The specific diffusion resistance [= 1/(diffusivity × solubility)] in olive oil and triolein is about the same, as would be expected, but in tributyrin it is about 50% lower, which suggests that the solubility and/or diffusivity of oxygen in oils is dependent on the chain length of the triglyceride, in a yet to be quantified manner.

The temperature dependence of the diffusivity of oxygen in oil also still needs be clarified. Based on measurements of bimolecular collision frequencies of oxygen in olive oil in the range of about 5–60°C, Subczynski and Hyde [27] concluded that the diffusivity should depend approximately on the square root of the value of the absolute temperature divided by the viscosity. In the absence of any better evidence this relation has been used to estimate the values in Table 6.

C. Absorption of Oxygen

Any oxygen dissolved or absorbed into the oil is a potential risk; it is the starting point for any oxidation.

1. *Dissolved Oxygen Content*

The dissolved gas content of oil saturated with air (25°C, 1 bar), calculated from the component solubilities, is shown in Table 7. The dissolved oxygen is sufficient to produce a potential PV rise of about 2.5, and therefore the first precaution is to avoid contact with air wherever possible. For example, according to Pardun [15, p. 223], the oxygen content of freshly filled refined oil is indeed 36–38 mg/kg. Measurements of the dissolved oxygen content during

Table 7 Gas Content of Oil Saturated with Air

Gas	Solubility [mg/(kg.bar)]	Air (vol %)	Dissolved in oil	
			mg/kg	Nl/kg
Oxygen	180	21	38	0.026
Nitrogen	85	78	66	0.053
Argon	270[a]	1	3	0.002
Total O_2, N_2, Ar		100	107	0.081

[a]From Ref. 23.

refining [13] show that similar values are approached during processing, in particular in storage; see Table 14 in Section V.

2. *Headspace Volume*

The air headspace above oils, for example in closed tankcars or sample bottles, is potentially even more damaging. The potential PV rise due to the oxygen in air in an enclosed headspace depends on the degree of filling, as illustrated in Table 8.

3. *Rate of Absorption*

The rate of absorption is more difficult to assess. In general, the increase in average oxygen content C in a volume V of oil will depend on (1) an absorption or mass transfer coefficient, K, (2) the contact area between gas and oil, A, and (3) the concentration difference between oxygen in the gas, expressed in terms of the equivalent solubility in oil (C_o), and the oxygen content in the bulk of the oil.

$$V\frac{dC}{dt}=KA\ (C_o-C)$$

The factors that would be expected to affect the rate of absorption, apart from the dissolved oxygen content, are therefore (1) the surface area-to-volume ratio of the oil, (2) the oxygen concentration (in the gas/headspace) above the oil, and (3) the effect of agitation (natural or mechanical), especially near the oil surface, on the absorption coefficient.

4. *Surface Area-to-Volume Ratio*

If the rate of absorption is the determining factor, we would expect the increase of the oxygen content of the oil to depend on the area-to-volume ratio (A/V). The area-to-volume ratio for

Table 8 Potential PV Rise Due to Air in an Enclosed Headspace

Container filling (vol % oil)	Headspace (m^3 air/m^3 oil)	Potential PV rise (meq/kg)
99	0.01	0.2
90	0.1	2
50	1	20

oil in any normal container, from sample bottle to storage tank, is also related to the filling height H.

$$\frac{A}{V}=\frac{1}{H}$$

The effect and importance of the area-to-volume ratio was quite clearly demonstrated by Going [28], who measured the PV rise in neutral, bleached soybean oil after 5 weeks of storage at 49°C in containers of various sizes. Similar results have been obtained in laboratory tests [29] with fully refined soybean oil stored for 2 weeks at 23°C and 50% RH. Both sets of results, with approximate contact area and filling height values, are summarized in Table 9. Similar effects have repeatedly been observed during thermogravimetric studies of oil oxidation [30,31].

The oxidation rate is not directly proportional to contact area. Factors other than absorption play a role, and, as Going [28] pointed out, it is inadvisable to extrapolate these figures. However, it is clear that contact area or filling height are very important. For example, oxidation rates in a sample bottle (filling height 10 cm) may be similar to those in 10 cm of oil in a storage tank, which we would obviously not consider ideal storage conditions!

5. *Oxygen Concentration in Gas/Headspace*

The use of nitrogen to decrease the oxygen content in the "air" above oil will reduce both the rate of absorption and the rate of the subsequent oxidation reaction. Going [28] clearly showed the positive effect of nitrogen in storage. Neutralized bleached soybean oil (PV = 1) was stored for 5 months, at 24 ± 8°C as 340 t lots in half-filled 750 m^3 tanks. The PV of the oil stored under air (21 vol % O_2) increased by 4 units; the PV of the other half, stored under nitrogen (1.5 ± 0.5 vol % O_2), increased by only 0.5 unit. Both oils were subsequently lab-deodorized, hydrogenated and lab-deodorized, and plant processed to give a hydrogenated shortening; in all cases the flavor stability of the products made from the oil stored under nitrogen was significantly better.

Similar results were obtained during extensive long-term storage trials with commercially produced and packed cottonseed and soybean salad oils sponsored by the U.S. Defense Department [32]. Eleven different oils in bottles or cans filled under air or nitrogen were evaluated during 52 wk of storage in the dark at 26 and 38°C. The flavor of all the oils kept under

Table 9 Influence of Contact Area on the Rate of Oxidation

Container	Contact area (m^2/m^3)	Filling height (mm)	Final PV (meq/kg)	
			Ref. 28	Ref. 29
Drum (200 L)	1.2	850	28	—
Can (25 L)	3	350	39	—
Cylinder	8	120	—	11
Can (1 L)	13	75	55	—
Bottle	35	30	—	16
Petri dish	95	10	—	19
	400	2.5	—	22
	4000	0.25	—	44

nitrogen (zero O_2) was still good after 52 wk, and better than that of the oils kept under air (21% O_2) or nitrogen (with 2% O_2) for only about 5 or 15–30 wk, respectively.

The two examples given in detail show that

1. The PV of the oil kept under air increased to 6 after about 24 wk and then stayed constant; the flavor development showed a similar trend.
2. There was no increase in PV in the oil kept under nitrogen (zero O_2), and the flavor decreased only very gradually.

D. Hydroperoxide Formation

Hydroperoxide formation is a complex subject, and I do not intend to go into detail. Lundberg edited a two-volume review in 1961 [33], and he presented a useful summary of the mechanisms [34] at about the same time.

Hydroperoxides are formed by an autocatalytic reaction between oxygen and unsaturated oils involving free radicals. The reaction starts with the formation of free radicals (initiation). Each free radical can combine with oxygen and an unsaturated group to produce a hydroperoxide and another free radical, which starts a chain reaction (propagation). This continues until the free radicals are destroyed, usually by combining to form nonreactive molecules (termination). In practice this results in (1) a slow initial rate of reaction during the so-called induction period, often followed suddenly by (2) a much faster rate of hydroperoxide formation. The length of the induction period is often used as a measure of the basic stability of the oil.

Many factors that affect the reaction have a similar effect on both the induction period and the subsequent, faster rate of reaction, but this is not always true.

1. *Type of Oil*

The type of oil, in particular the degree of polyunsaturation, has an important effect. The relative oxidation rates of fatty acids (Table 10) based on results at 20–100°C [4, p. 74] also illustrate the effect of conjugation and isomeric form. Conjugated double bonds, which can be detected by extinction measurements at 232 nm (dienes) and 268 nm (trienes), are particularly susceptible to oxidation, and to make matters worse, they also tend to be formed as a result of oxidation.

The induction periods Active Oxygen Method (AOM hours) of laboratory-extracted and fully refined safflower and sunflower seed oils from plant breeding trials, with widely differing 18:1 and 18:2 contents [35], provide a good example. Purdy noted the dependence on 18:2

Table 10 Relative Oxidation Rates of Fatty Acids AOM- "Active Oxygen Method"; office method of the AOCS

Fatty acid	Isomer	Conjugated	Rate
18:0 (stearic)	—	—	1
18:1 (oleic)	1 × *cis*	—	10
18:2 (linoleic)	2 × *cis*	No	100
18:3			
Linolenic	3 × *cis*	No	200
Elaidolinolenic	3 × *trans*	No	150
α-Eleostearic	2 × *trans*	Yes	1000
β-Eleostearic	3 × *trans*	Yes	400
20:4 (arachidonic)	4 × *cis*	No	400

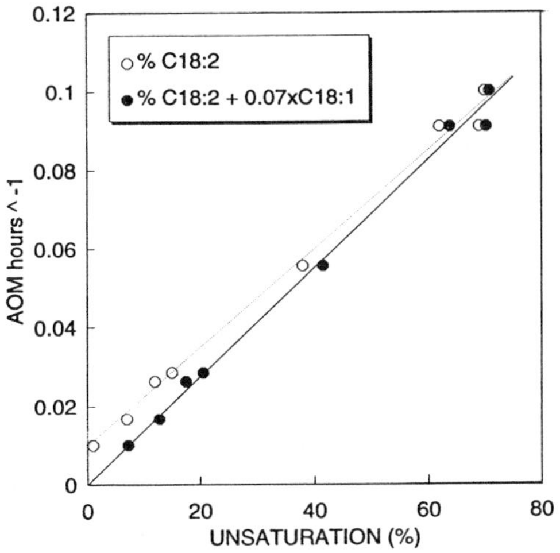

Fig. 4 Effect of unsaturation on the rate of deterioration.

content, but if the reciprocal of the AOM hours is taken as a measure of the rate of deterioration, the physical relationship is more obvious (Fig. 4). When the contribution of the 18:1 content is included as 7% of the rate of oxidation of 18:2 rather than the 10% indicated in Table 10, the relationship appears to account for the total rate of oxidation.

The state of an oil also affects the potential for oxidation, as was illustrated by Going [28] for soybean oil. After 7 wk storage of the oils at 38°C in 1 L cans, he noted a PV of about 50 for neutral bleached, 30 for neutral bleached hydrogenated (IV 25), and only 5 for crude soybean oil.

2. *Temperature*

In practical terms the effect of temperature is even more important than that of the type of oil because we can influence it. As with any chemical reaction, temperature has a very strong effect on the rate of oxidation. Published data [4, p. 75; 36] show a doubling of the rate of oxidation for every 10–15°C rise in temperature. This is well illustrated by figures quoted in Ref. 37 for the influence of storage temperature on the time taken to develop off-flavor in a fully refined soybean oil (Table 11).

Table 11 Effect of Temperature on Off-Flavor Development in Fully Refined Soybean Oil

Storage temp. (°C)	Time to off-flavor (h)
0	500
20	180
40	42
60	16
80	4

3. *Oxygen Concentration*

The concentration of oxygen in the oil also has a direct effect on the rate of hydroperoxide formation and is again an effect we can influence. The overall, practical effect has already been mentioned in Section IV.C.5.

In model studies the rate of oxidation was shown to be more or less proportional to the pressure of oxygen at pressures below about 200 mbar [34], which is approximately the partial pressure of oxygen in air. Another example was provided by Min and Wen [18], who measured the decrease in dissolved oxygen in sealed oil samples. Analysis of their data shows that the rate of decrease of dissolved oxygen is directly proportional to the dissolved oxygen concentration.

4. *Light*

Radiation in most forms, including sunlight and artificial light, accelerates oxidation. A whole chapter could be written on this subject alone, but I will simply note that

1. Light promotes the formation of free radicals and therefore influences the initiation of autoxidation.
2. Photo-oxidation occurs by a different mechanism involving singlet oxygen at rates 1000 or more times as fast as autoxidation; it is promoted by "sensitizers" such as chlorophyll and inhibited by singlet oxygen "quenchers" such as carotene.
3. It is therefore advisable to store oils and fats in the dark.
4. Lea's study on the effect of (artificial) light on the oxidation of animal fat [38] is of interest not only for the elegant experimentation but also because it laid the foundation for the PV analysis.

5. *Metals*

Some metals are pro-oxidants even at very low levels; the levels in Table 12 were found to halve the shelf life of lard at 98°C [39]. The simplest explanation for the effect of these metals is that they catalyze the formation of free radicals, for example via hydroperoxide breakdown [34]. It is clear from Table 12 why copper in any form should be avoided and iron and nickel should be removed during refining. Much lower levels are usually recommended to avoid pro-oxidant effects; for refined oils and fats, maximum levels of 0.01 ppm Cu, 0.1 ppm Fe, and

Table 12 Trace Metal Levels at Which a Significant Effect on Oxidation Occurs

Metal	Level (ppm)
Copper	0.05
Manganese	0.6
Iron	0.6
Chromium	1.2
Nickel	2.2
Vanadium	3.0
Zinc	20
Aluminum	50

0.1 ppm Ni are typical requirements. This also explains the antioxidant effect of processing aids such as citric and phosphoric acid, which chelate with or otherwise aid in the removal of metals.

6. *Antioxidants*

The study of antioxidants is also a major subject in its own right, which is made even more complicated by the pro- and antioxidant effects and the synergistic effects of some of the compounds involved. The natural antioxidants such as the tocopherols, carotene, lecithin, and gossypol show that nature usually ensures a suitable level of antioxidant in the crude oil.

The simplest explanation for the effect of antioxidants is that they suppress or reduce the formation of free radicals by promoting termination reactions. Most natural antioxidants are also oxidized themselves; this loss of natural protection is one reason for the decrease in refined oil stability often observed as a result of (feedstock) oil oxidation. Oxidation of the pigments such as carotene and gossypol also leads to "color fixation," which is very difficult to remove in subsequent processing.

The best course of action is to try (where possible) to maintain the natural level of antioxidants by good oil processing practice. Many synthetic antioxidants have been proposed, produced, and tested, and some have been accepted for use with edible oils [1, p. 61], for example, for the protection of palm oil during transport. But this is time-consuming and expensive. In addition to confirming their efficacy as antioxidants, they must be thoroughly tested for product safety.

7. *Moisture*

The effect of moisture has always caused confusion. The consensus of opinion seems to be that water activities of about 0.5 ± 0.2 (i.e., water contents of about 50% of the saturation solubility in Table 3) are the best. At this level water tends to stabilize hydroperoxides and facilitate free radical termination reactions. At lower levels this "hydration protection" effect is negligible, leading to higher oxidation rates. At higher levels there is a risk of formation of free fatty acids (which oxidize faster than glycerides), hydrolytic rancidity, enzymatic attack, and leaching of iron. Whatever the cause, oxidation rates tend to be higher at moisture contents approaching or in excess of saturation.

E. Hydroperoxide Breakdown

The breakdown of hydroperoxides is, if anything, more complex than their formation. Depending on the conditions, decomposition of hydroperoxides can produce a variety of compounds, including aldehydes, ketones, esters, alcohols, acids, hydrocarbons, and polymers.

Initial decomposition often involves fission at the hydroperoxide position, giving a low molecular weight (relatively volatile) aldehyde and a high molecular weight (nonvolatile) aldehydic triglyceride residue. The anisidine value includes the contributions of both volatile and nonvolatile aldehydes.

A useful summary of the main volatile compounds produced during oxidation of eight types of oil was provided by Snijder et al. [40].

The hydroperoxides themselves are tasteless and odorless, but the short-chain aldehydes are particularly obnoxious even at low concentrations and are organoleptically detectable at levels on the order of 1 ppm in oil and 0.01 ppm in water. The formation of aldehydes initially follows roughly the same trend as the formation of hydroperoxides, and they can reach levels that are organoleptically detectable by the end of the induction period. Even

Table 13 Analyses of Oxidized Soybean Oil Before and After Refining

	Before refining			After refining		
Sample	PV	AV	Totox	PV	AV	Totox
1	0	7.5	7.5	0.1	8.1	8.3
2	5.0	9.5	19.5	0.1	8.9	9.1
3	10.7	10.4	31.8	0.1	13.8	14.0
4	24.6	9.2	58.4	0.1	29.9	30.1
5	50	9.2	109.2	1.1	49.2	51.4
6	105	12.2	222.2	1.0	104.2	106.2

without further formation, the decomposition of hydroperoxides will continue, and it is particularly rapid at high temperature.

The aldehydes are an important cause of oxidative rancidity in oils, which explains the interest in the edible oils industry for (1) peroxide and anisidine values, (2) induction periods, (3) taste and flavor stability, and (4) the whole subject of oil oxidation.

1. *Effect of Processing*

It is not easy to generalize on the trend of PV, AV, and Totox during processing, especially when storage is also involved. A decrease will occur during hydrogenation (PV and AV) and interesterification (PV) owing to the specific reactions involved. The combined effect of other processing steps often results in a net increase in Totox before deodorization. During deodorization, all the hydroperoxides decompose (PV = 0) and the volatile AV components are stripped out. The fully refined oil should contain only relatively stable, nonvolatile oxidation products.

Any deterioration of the fully refined oil will therefore depend on renewed absorption of oxygen, hydroperoxide formation, and decomposition. But the rate of deterioration is also influenced by the oxidation history and the processing of the oil, and experience has shown that the stability/keepability can be optimized by optimal processing (not overprocessing) but that in the end the greater the extent of oxidation of the oil (as crude or during processing), the worse the refined oil stability will be.

Pardun [15, p. 233], a chief chemist with a lifelong experience of oils and fats, stresses the importance of Totox and the relation between the increase in AV during refining and the decrease of the original PV in assessing the keepability of oils and fats. He illustrates this with examples of the PV and AV figures of soybean oil (to which I have added Totox) given in Table 13. These figures also illustrate the general experience that deodorization reduces Totox by about 50%; the experience with corn oil, for which a 45–50% reduction in AV during deodorization is claimed [5, p. 139], seems to be different, unless the PV is already low.

V. TRANSPORT AND STORAGE REQUIREMENTS

Transport and storage of oils are essential for the operations of any oil user whether he is adjacent to a plantation, oil mill, extraction plant, or refinery or relies on these as his supplier of crude or refined oils. Both supplier and customer require efficient transport and storage at an acceptable quality. Apart from normal economic considerations, the main concerns are therefore related to contamination and degradation. By contamination I mean any form of admixing with a foreign substance or another oil or fat during transport or storage or during the associated handling before or after.

Transport is really no more than mobile storage. The risk of degradation is the same or perhaps even greater due to the inherent movement and agitation. But there are two factors that increase the chance of contamination during the transport stages. The first is that transport is not usually under the direct control of either supplier or receiver. The second is that it cannot be assumed that transport, especially by ship, is used for the transportation of edible oils alone. This means that strict procedures and inspections are required at all stages to avoid or identify any contamination.

As regards degradation, the basic principles of hydrolysis and oxidation provide an excellent guideline as to what and what not to do, both when deciding on the requirements and when designing the facilities for oil storage, transport, and handling. The same principles apply to the (crude or refined) oil before it comes under your control; how have your suppliers handled the oil?

The type of oil will determine the degree of unsaturation or polyunsaturation and the normal level of natural antioxidants, and these factors should be borne in mind when deciding on the requirements for transport and storage. Most other factors affecting degradation are more or less controllable:

1. Minimize contact with or absorption of air.
2. Avoid excessive moisture.
3. Avoid contact with and minimize content of pro-oxidants.
4. Keep oils in the dark.
5. Keep time and temperature to a minimum.
6. Avoid unnecessary agitation.

However, the precautions taken must also be economically justifiable, and they are usually related to the commercial risk involved in *not* taking them.

The usual ranking of the oxidative stability of oils [1, p. 268] is

Crude > deodorized > neutral > bleached

but the ranking with respect to commercial risk closely follows the reverse order of processing:

Deodorized > bleached > neutral > crude

This is mainly because the further one gets, the less chance there is to rectify any degradation. For example, the use of nitrogen to protect deodorized oil is quite common, but although the positive effect of nitrogen in crude [5] and bleached [28] oil storage has been clearly demonstrated, it is rarely used.

Another basic factor that cannot be stressed enough is that every stage of transport or storage involves an oil-handling step, or two oil-handling steps if the ideal alternative is degradation-free transfer between processing stages. These oil-handling steps probably account for a major part of the potential oxidative degradation. This is illustrated by the dissolved oxygen contents during palm oleine processing reported by Berger [13], which are summarized in Table 14. These figures are the average (and minimum-maximum) measurements on 12 separate days, or only 4 days for deodorized storage. As Berger commented, "Filling [of the bleached oil buffer and deodorized oil storage tank] involved splashing the oil from the top of the tank and resulted in a substantial increase in oxygen level"; crude oil storage was the next most critical stage.

Table 14 Dissolved Oxygen During Processing

Process step	Dissolved O_2 (mg/kg)[a]
Crude storage	15 (12–20)
Outlet bleacher	8 (5–18)
Bleached buffer	26 (18–33)
Outlet deaerator	12 (0.3–31)
Outlet deodorizer	8 (4–13)
Deodorized storage	31 (30–32)

[a]Range given in parentheses.

The requirements for transport, storage, and handling of crude and fully refined oils should therefore be carefully considered.

In the following subsections on crude and fully refined oil, I have tried to avoid repetition, and it should therefore be read as a whole. For example, the comments made about crude oil transport and refined oil storage are also relevant to refined oil transport but are not repeated.

A. Crude Oil

Crude oil is relatively stable and is transported and stored in relatively large lots. The negative effects of any poor practice will therefore tend to be diluted by the bulk of oil or take longer to show themselves. However, all the evidence and indications we have show that any degradation of the crude oil results in higher processing costs and a poorer final product quality. It is therefore worthwhile specifying crude oil transport and storage requirements properly, rather than just waiting to see what happens.

1. Transport

Crude oil is transported by ship, barge, rail, and road. Large quantities are transported by sea. Lauric oils and palm oil products, for example, are typically transported in lots of anywhere between 250 and 10,000 t in ships carrying a total of 25,000–30,000 t [41]. The journey from Malaysia to northern Europe takes about a month.

We will therefore take marine transport as an example; the same principles apply to other forms of transport.

The supplier or receiver (seller or buyer) will book space with a shipowner, and both will appoint their own surveyor or superintendent to supervise the loading and/or unloading by the crew and terminal personnel. With so many independent players, the rules of the game must be clearly defined and agreed on to avoid mistakes being made, especially with regard to contamination.

There are many national sets of rules, but most of the international trade in oils and fats occurs under FOSFA (Federation of Oil Seeds and Fats Associations) contracts.

In the 1980s, requirements were reevaluated and updated. Berger [13,14] summarized some examples of deterioration and presented the recommended practice for storage and transport of edible oils and fats proposed by the Palm Oil Research Institute of Malaysia (PORIM). This document influenced FOSFA's update of their codes of practice, which also suggested stricter rules on adulteration and contamination [42]. Ludwiczak [41] probed the relation between shipper and shipowner and gave an excellent step-by-step account of procedures from the port of loading to the port of discharge.

It is impossible to go into details here, and I restrict myself to some examples illustrating the principles behind good practice. A useful summary of the details is given in Farrer's *The Shipment of Edible Oils* [43].

Contamination. Contamination from previous cargoes can be disastrous. Information on at least the three previous cargoes should be checked. All three should preferably be on the FOSFA "approved" list, and none should be on the FOSFA "banned" list. In intermediate situations, the cargo should be segregated and analyzed. The minimum requirement is that the last cargo is not on the "banned" list. If the second or third previous cargoes are on the "banned" list, the cargo must be segregated and analyzed by a competent laboratory before processing.

Contamination can also occur in transit or during handling (loading, unloading, etc.). Berger [13] refers to a study by M. S. A. Kheiri of samples taken during unloading of fully refined palm oil. During the first 25 s (about 1 t) he found about 0.015% impurities, 0.9% moisture, and Lovibond red (5.25 in.) = 20–25, compared with the bulk of oil, which contained about 0.003% impurities, <0.04% moisture, and Lovibond red (5.25 in.) = 2.5.

Air Contact. Partially filled compartments should be avoided. The oil should be loaded through a "drop line" and not allowed to free fall from deck level; the figures in Table 14 for crude storage, bleached buffer and deodorized storage illustrate the effect of free fall into a tank.

Temperature and Heating. The international standard is the IASC (International Association of Seed Crushers) handbook, which includes the PORIM recommendations [14], equivalent to

Unloading temperature: 5–15°C above the melting point or, for liquid vegetable oils, 15–25°C to ensure good pumpability and drainage

Storage and shipping temperature: 0–5°C above the melting point or, for liquid vegetable oils, ambient temperature

This is a realistic compromise between ensuring that oils are pumpable and homogeneous for unloading and minimizing storage temperatures and heating during transport.

The recommended maximum heating rate of 5°C per day for the last few days of the journey is sufficient to reach unloading temperatures. However, Berger [13] suggests that even then local overheating may cause deterioration; he gives an example of the PV of a shipment of crude palm oil that increased by less than 1.5 units during the first 18 days at about 40°C but then increased by more than 4 units while heating at 2°C per day for the last 6 days. A maximum heating coil temperature would probably be a better criterion, and hot water heating is definitely preferable to steam heating.

Metal Contact. Any copper-containing metal is totally unacceptable. Berger [13] quotes the study by M. Pike of crude palm oil samples taken using a brass sampling device and then stored at 50°C in glass bottles. An instantaneous sample (0.1 ppm Cu) had an AV of 6 and had lost less than 10% of its carotene and tocopherol after 5 days. Samples left in the sampling device for 5–20 min at 50°C showed a roughly exponential increase in copper content to 0.2–1.4 ppm Cu; after 5 days these samples had an AV of 11–20 and had lost 75–90% and 94–100% of their carotene and tocopherol, respectively.

Ships' tanks should preferably be of stainless steel or coated mild steel with stainless steel heating coils [14,41]. Any coating should be food grade, and for effective coating the metal surface must be properly prepared.

Postscript. The above are just a few examples of transport requirements and the reasoning behind them. A complete definition of good working practice inevitably contains a lot of detailed instructions and requirements. For these details you should consult the cited literature and the relevant codes of practice.

2. *Storage*

Crude oil is relatively stable and is stored in relatively large tanks. Both requirements and good construction practice will therefore differ from those for final product tanks. Nevertheless, it still makes good economic sense to bear the principles of oil degradation in mind.

Strecker et al. [5] provide a good example of the combined effect of storage time, temperature, and air or nitrogen blanketing on the cost of degradation. They stored two different crude corn oils under various conditions and determined the extra earth (clay) dosage required to achieve a final color of 2.0 red (1 in. cell) compared with the fresh oils, all on a laboratory scale. The results are reproduced in Table 15 because they suggest some interesting relationships.

Apart from the usual increase in oxidation values with time, the results suggest that

1. The required earth (clay) dosage depends roughly linearly on the Totox of the crude oil.
2. The earth (clay) requirement for physical refining is more sensitive to oxidation of the crude oil than with alkali refining.
3. The decrease in tocopherol content (TOCO) is directly related to the increase in Totox value.

We also see once again that storage under nitrogen or at low temperatures significantly reduces the rate of deterioration and extra earth (clay) required.

Whatever the actual relationships, it is clear that oxidative degradation in storage results in increased processing costs, and a loss of natural antioxidant that will probably be reflected in the stability of the final product. These results suggest an extra earth (clay) requirement of 0.1–0.2% after 2 weeks of storage and a corresponding PV increase of about 0.5 unit. Although these were laboratory-scale results, the assumption of 0.1% extra bleaching earth consumption to cover the cost of oxidative degradation in storage and during processing is not unrealistic.

Some aspects of good crude oil storage practice have already been covered directly or indirectly in the section on transport. Some of the following points on storage are also relevant to transport.

Table 15 Effect of Crude Oil Storage Conditions on Refinability

	Physical refining			Alkali refining			
Storage conditions							
Time, mo	0	2	2	0	3	3	6
Temp., °C	—	30	30	—	10	50	50
Atm.	—	N_2	Air	—	Air	Air	Air
Quality							
PV	0.5	1.0	2.7	0.4	—	3.7	18.6
AV	2.7	3.0	3.7	3.5	—	3.6	4.4
Totox	3.7	5.0	9.1	4.3	—	11.0	41.6
Toco (ppm)	1690	1500	1200	2000	—	1460	1160
Clay (%)	1.0	1.2	2.0	0.5	0.5	1.1	1.9

Contamination and Segregation. Crude oil tanks should be easily and completely drainable, not only for the removal of tank bottoms but also to minimize intermixing of different lots of oil. Intermixing of old and new deliveries of crude oil or different types of oil is seldom a good idea; the oils lose their identity and purity. Apart from the contamination issue, it makes it difficult to keep track of the quality history of the oil, which can help in predicting required or optimal processing conditions.

Where necessary, oils should be segregated according to type: vegetable, animal, lauric, etc.

Air Contact. The size of a tank should be chosen in relation to the requirements. A tank that is never more than 50% full obviously has an unnecessarily large surface/volume ratio and headspace.

The method of filling will have a strong influence on the absorption and initial level of oxygen in oil. A drop line or dip pipe to the bottom of the tank (with a siphon breaker if necessary) can also be used for storage tanks. The use of bottom filling or even a floating inlet that discharges just below the oil surface can also be considered [13,14].

Moisture and Impurities. The accepted levels of moisture and dirt for trading and good merchantable quality are often very lenient. Crude oil with higher levels should be rejected. Moisture levels lower than the saturation solubility at storage temperature should be preferred.

Any undissolved water, waxes, lecithin, and impurities will inevitably tend to settle and form tank bottoms. Tank bottoms, especially when they are near heating coils, provide an ideal environment for various degradation reactions and should therefore be regularly removed. These effects must also be allowed for in the design and operation of crude oil storage tanks.

The classical method for removing tank bottoms involves "sweeping" the bottom of the tank at regular intervals. This involves labor, creates a waste disposal problem, and results in an oil loss and a temporarily "unproductive" tank. Modern processing machines can cope with a reasonable (preferably constant) level of moisture and impurities. Agitation of crude oil tanks, sufficient to maintain any potential sediment in suspension but not to promote oxidation, is therefore a viable alternative. This will minimize the tank bottoms problem and has the added advantage of creating a constant quality feedstock.

Temperature and Heating. Crude oils that are regularly used should be stored at about 10°C above their melting point. Liquid vegetable oils can usually be stored at ambient temperature in uninsulated tanks.

Heated tanks should always be insulated. This is not only economically sound but also prevents the extra degradation due to excessive heating.

Hot water heating is definitely preferred for a number of reasons:

It is sufficient for the hardest fats (in insulated tanks).
It is easier to control than steam heating.
It prevents excessive coil surface temperatures.
It reduces heat losses from steam connections and appendages.

Even old steam-heated tanks can usually be easily modified for hot water heating if they are insulated.

The whole tank should be insulated, including the roof. Uninsulated roofs will result in condensation, corrosion, and the risk of additional iron (rust) and moisture in the oil. Even insulation underneath the tank can be considered if feedstock fractionation is a problem.

Oils not immediately required for processing are sometimes stored without heating and allowed to "set up" (solidify). The pros and cons of this practice should be carefully evaluated. The heat required for remelting and the extra degradation caused by the extreme heating conditions required will only outweigh the savings of normal heating and degradation if the oil is to be stored cold for a long time.

Metal Contact. It is still common practice to use mild steel tanks and mild steel heating coils. With liquid vegetable oils, a polymerized oil "coating" will gradually build up on the tank surfaces, which will protect both the metal and the oil. It would be even better to properly prepare and coat all the surfaces with vegetable oil when the tank is installed.

For oils and fats stored at higher temperatures the choice is more difficult. The higher the temperature, the greater the chance of corrosion and iron pick-up, especially at high moisture levels and with high FFA and high lauric oils. Under these conditions stainless steel heating coils would seem to be a sensible choice. With palm oil deliveries, quality differences can be detected that may be partly due to the tank material used by the supplier.

3. *Handling*

Some of the comments on transport and storage are related to handling; the following are equally relevant.

Any form of air ingress via pumps, pipelines, and fittings or vortices in tanks should be avoided.

Pipelines should be designed to be emptied and to allow gravity to help rather than hinder drainage, whatever system of line clearing is used. Pigging systems are often recommended [13,14], but with proper design other systems employing line blowing, drain pumps, and gas lift-lines can be equally effective.

Steam blowing of lines can never be recommended because it only introduces unwanted moisture. Liquid oils certainly do not require it. Transfer lines for higher melting point oils and fats should be properly trace-heated; heated lines should always be insulated. With these precautions and proper pipeline design, controlled air (or nitrogen) blowing combined where necessary with gas lift-lines should be quite adequate. Trace heating should be turned on only when it is needed, to prevent both unnecessary energy consumption and degradation of any oil (film) left in the lines.

B. Fully Refined Oil

The requirements for the transport, storage, and handling of refined oils are of necessity more stringent than those for crude oils. For the refiner, fully refined oil is the final product. There is no further processing to "put things right," but there is plenty of chance for things to go wrong.

1. *Storage*

With few exceptions, the principles and comments applying to crude oil also apply to refined oil. Moisture and impurities should not be a problem.

Contamination and Segregation. Complete emptying and draining of refined oil storage tanks is essential. The tanks should therefore be constructed with (insulated) conical or dished bottoms. In general, mixing of old and new oil should be avoided, but there is an economic risk vs. cost trade-off. For regular, fast-moving oils, topping up can be acceptable.

Segregation of refined oil types should be considered. Dedicated tanks may be required for vegetable, animal, lauric, or hardened products. For example, contamination of dewaxed sunflower oil, intended for bottling, with any hardened oil would be disastrous.

Air Contact. The results of improper handling of deodorized oil were given in Table 14. A deodorizer is a very efficient degasser, and the oil leaving it should be oxygen-free. If the oil contains oxygen, it has probably come from either air leaks in the discharge system or contact with air during sampling.

Deodorized oil is like a sponge; it will "suck up" any air it comes in contact with. If the degassed oil is top-filled into a storage tank containing air, the oil will be more or less saturated with air, 80–85% saturated according to Table 14.

The preferred procedure is to sparge the oil with nitrogen before or during discharge and to bottom-fill the storage tank. The nitrogen, typically 0.5–0.9 Nm^3/t oil (although suppliers may recommend more than 1.0 Nm^3/t [2, p. 312]), is sparged into the oil under pressure. According to Table 6, only 85 mg/kg (about 0.07 Nm^3/t) will remain dissolved at atmospheric pressure; the rest will come out of solution and help to flush the air out of the headspace of the storage tank.

Considering the solubility of nitrogen, it is questionable whether such large volumes are required for sparging. The nitrogen purity should also be carefully specified. If 0.5 Nm^3/t nitrogen of 99% purity (max. 1.0% oxygen) is sparged into oil at 4 bar and 50°C, only about 0.3 Nm^3/ton (and 5 mg/kg oxygen) will dissolve. On discharge to 1 bar, the excess nitrogen will flash off and strip out part of the oxygen, leaving, say, 2.5 mg/kg oxygen (and 0.07 Nm^3/ton nitrogen) dissolved. This is equivalent to a potential PV increase of about 0.15 and is only slightly more than the equilibrium content of 1.85 mg/kg with 99% nitrogen at 1 bar and 50°C. For line blowing and blanketing, which affects only a small proportion of the oil, nitrogen of 99% purity is certainly quite sufficient.

It is important to remember that sparging the oil with nitrogen does not protect it forever. If the oil comes into contact with air again, oxygen will still be absorbed, giving a renewed risk of oxidation.

Agitation of refined oil tanks should not be necessary and is not recommended.

Temperature and Heating. The same rules apply as for crude oil, except that a refined oil should never be allowed to "set up" or be remelted. It is essential that the deodorized oil be cooled to storage temperature before storage; otherwise the oil can start to deteriorate before it has cooled down. This is particularly important for liquid vegetable oils, which should never be stored at more than 30°C.

Metal Contact. The simplest and safest solution is to choose stainless steel or coated steel tanks. However, economics still have to be considered. The minimum acceptable choice is

Mild steel for oils at $<45°C$
Stainless or coated tanks for oils and fats at $>45°C$

Example. The following example [44] illustrates the combined effect of some of these parameters on the stability of deodorized sunflower oil.

Sunflower oil, deodorized under two different conditions A and B, was stored in open 1 L cans. The results therefore represent an "accelerated oxidation test" rather than the actual storage or shelf life of the oil.

In the first trial, the oil leaving the deodorizer was not sparged with nitrogen and was stored at 40°C. In the second trial, the oil leaving the deodorizer was sparged with nitrogen and was stored at 40°C and 27°C. The PV of the samples during 15 days storage is shown in Fig. 5. The taste results were similar.

The effect of the processing conditions is less significant than the effect of the storage and handling conditions.

1. Under the more stringent deodorization conditions, the "induction period" was about 1 day longer.
2. The influence of nitrogen sparging was negligible; this is not surprising, as the subsequent storage was in contact with air.
3. The effect of storage temperature was the most significant. At 27°C the PV increase and taste decrease were negligible; at 40°C they were unacceptable.

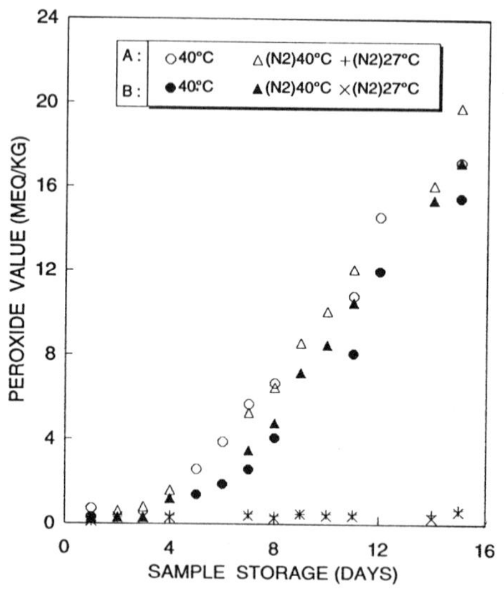

Fig. 5 Effect of processing and storage conditions on the stability of refined sunflower oil.

2. *Transport*

Transport of fully refined oil by ship, barge, road, or rail are all feasible. Road and rail transport over up to 1000–2000 km lasting up to 2 weeks (with the necessary precautions!) has been common practice in the United States for many years. The same principles apply as for crude oil transport, but the actual procedures and precautions will be more stringent and will depend on the exact application. The remarks on refined oil storage are also relevant, since transport is also a form of storage. It is also worth remembering that problems are often the result of a seemingly trivial oversight or mistake.

Since large amounts of refined oil travel by road, we will take road transport as an example, but the same principles apply to other forms of transport.

Contamination and Segregation. Dedicated transport is preferable. Otherwise, the tanker should be cleaned and dried before filling each load.

Air Contact. Here again, partially filled tankers must be avoided (minimum headspace), and bottom-filling either via a dip pipe or via the bottom outlet is essential.

Sparging the oil with nitrogen during filling is a sensible precaution, especially where long transport distances or times or sensitive oils are involved. Nitrogen blanketing should then not be necessary, except in extreme circumstances.

Temperature and Heating. The same temperatures apply as for storage. For short distances or times, insulated and unheated tankers are adequate. If heating is required, external coils are preferred; there is less risk of local overheating or contamination with the heating medium, and it is easier to keep the tanker clean.

Metal Contact. Stainless steel should be considered mandatory for all forms of refined oil transport except in exceptional cases. The requirements concerning material of construction are in general more stringent for transport than for storage. This is understandable, considering that storage is normally "dedicated" whereas transport tanks usually have to be cleaned before each load. The combination of mild steel and cleaning increases the risk of oil degradation. However, if liquid vegetable oils, for example, are to be transported in dedicated ships, the "coating" of polymerized oil may provide sufficient protection.

3. Handling

Most aspects of handling relevant to fully refined oil have already been mentioned. As regards line blowing, however, the clear choice is nitrogen, not air—and definitely not steam.

C. Intermediate Oil Storage

By now it will be obvious that (1) intermediate storage should be reduced to an absolute minimum and (2) a minimum buffer between processing stages is the ideal situation. If intermediate storage is still required, the principles of oil degradation should be considered. The storage (or transport) requirements will lie somewhere between those for crude oil and fully refined oil. Although in-process buffer tanks are less critical owing to the limited residence time of the oil, air contact, heating, etc. should still be carefully considered.

ACKNOWLEDGMENTS

My thanks go to Ir J. C. Segers, Ir A. Leniger, B. P. Backlog, Rosita van der Burg, Irma Knieriem, Mirto Oduber, Hans Houtman, and Anne Hilder, for their help.

REFERENCES

1. H. B. W. Patterson, *Handling and Storage of Oilseeds, Oils*, Fats and Meal, Elsevier Science, London, 1989.
2. G. R. List and D. R. Erickson, Storage, handling and stabiliziation, in *Handbook of Soy Oil Processing and Utilisation* (D. R. Erickson, E. H. Pryde, O. L. Brekke, T. L. Mounts, and R. A. Falb, Eds.), American Soybean Association, St. Louis, MS, and AOCS, Champaign, IL, 1980, pp. 267–353.
3. G. R. List, Storage, handling and stabilization of edible fats and oils, in *Bailey's* Industrial Oil and Fat Products, Vol. 3, 4th ed. (T. H. Applewhite, Ed.), Wiley, New York, 1985, pp. 273–309.
4. D. Swern (Ed.), *Bailey's* Industrial Oil and Fat Products, 3rd ed., Interscience, New York, 1964.
5. L. Strecker, A. Maza, and G. Winnie, in *World Conference Proceedings: Edible Fats and Oils Processing* (D. R. Erickson, Ed.), Am. Oil Chem. Soc., Champaign, IL, 1990, pp. 309–323.
6. M. Loncin, Die spontane autokatalytische Spaltung der Triglyceride, *Fette Seifen 55*(1): 7 (1953).
7. M. Loncin, l'Hydrolyse spontanée autocatalytique des triglycérides, *Oleagineux 7*(12): 695 (1952).
8. M. H. Hilder, The solubility of water in edible oils and fats, *J. Am. Oil Chem. Soc. 45*: 703 (1968).
9. M. H. Hilder, The solubility of water in edible oils and fats above 100°C, *J. Am. Oil Chem. Soc. 48*: 296 (1971).
10. M. H. Hilder and M. van den Tempel, Diffusivity of water in groundnut oil and paraffin oil, *J. Appl. Chem. Biotechnol. 21*: 176 (1971).
11. M. Loncin, Die Löslichkeit des Wassers in Fetten und Ölen und der Dampfdruck des gelösten Wassers, *Fette, Seifen, Anstrichm. 57*(6): 113 (1955).
12. D. S. Sarkadi, Hydrolysis during deodorisation of fatty oils; catalytic action of fatty acids, *J. Am. Oil Chem. Soc. 36*: 143 (1953).
13. K. G. Berger, Quality control in storage and transport of edible oils, *J. Am. Oil Chem Soc. 62*: 438 (1985).
14. K. G. Berger, *Recommended Practices for Storage and Transport of Edible Oils and Fats*, Palm Oil Res. Inst. of Malaysia, Malaysia, 1985.

15. H. Pardun, *Analyse der Nahrungsfette*, Verlag Paul Parey, Berlin, 1976.
16. V. C. Mehlenberger, *The Analysis of Fats and Oils*, Gerrard, Champaign, IL, 1960.
17. T. K. Mag, Bleaching—theory and practice, in *World Conference Proceedings: Edible Fats and Oils Processing* (D. R. Erickson, Ed.), AOCS, Champaign, IL, 1990, pp. 107–116.
18. D. B. Min and J. Wen, Effects of dissolved free oxygen on the volatile compounds of oil, *J. Food Sci. 48*: 1429 (1983).
19. E. Fedeli and A. Brillo, Meccanismo dei fenomeni di autossidazione delle sostanze grasse, *Riv. Ital. Sostanze Grasse 52*(3): 88 (1975).
20. *Dissolved Oxygen in Oils and Fats*, Brochure, Orbisphere Laboratories, A. Woerden, Holland.
21. B. A. Bol, Loders Croklaan BV, Internal report, 1992.
22. L. B. Parsons, Private communication, 1938, cited in *Bailey's* Industrial Oil and Fat Products, 3rd ed. (D. Swern, Ed.), Interscience, New York, 1964, p. 127.
23. R. Battino, F. D. Evans, and W. F. Danforth, The solubilities of seven gases in olive oil, *J. Am. Oil Chem. Soc. 45*: 830 (1968).
24. M. Crowe, P. E. O'Connor, and B. Masterson, Oxygen diffusion in natural oils and pure triglycerides, *Biochem. Soc. Trans. 15*(2): 260 (1987).
25. U. Schrader, K. Becker, and R. Heiss, Der Einfluss von Diffusion und Löslichkeit auf die Reaktion von Sauerstoff mit kompakten Lebensmittein, *Verpack.-Rundsch. 31*(6): 33 (1980).
26. R. Heiss, U. Schrader, and G. R. Reinelt, The influence of diffusion and solubility on the reaction of oxygen in compact food, in *Food Process Engineering*, Vol. 1 (P. Linko, Y. Mälkki, and J. Olkku, Eds.), Applied Science, London, 1980, pp. 364–370.
27. W. K. Subczynski and J. S. Hyde, Diffusion of oxygen in water and hydrocarbons using an electron spin resonance spin-label technique, *Biophys. J. 45*: 743 (1984).
28. L. H. Going, Oxidative deterioration of partially processed soyabean oil, *J. Am. Oil Chem. Soc. 45*: 632 (1968).
29. J. Koch and J. Bittorf, Unilever Research Laboratory, Vlaardingen, Internal report, 1983.
30. H. S. Olcott and E. Einset, A weighing method for measuring the induction period of marine and other oils, *J. Am. Oil Chem. Soc. 35*: 161 (1958).
31. H. Mikula and A. Khayat, Reaction conditions for measuring oxidative stability of oils by thermogravimetric analysis, *J. Am. Oil Chem. Soc. 62*: 1694 (1985).
32. C. D. Evans, G. R. List, H. A. Moser, and J. C. Cowan, Long term storage of soyabean and cottonseed salad oils, *J. Am. Oil Chem. Soc. 50*: 218 (1973).
33. W. O. Lundberg (Ed.), *Autooxidation and Antioxidants*, Interscience, New York, 1961.
34. W. O. Lundberg, in *Symposium on Foods: Lipids and Their Oxidation* (H. W. Schultz, Ed.), Avi, Westport, CT, 1962, pp. 31–47.
35. R. H. Purdy, Oxidative stability of high oleic sunflower and safflower oils, *J. Am. Oil Chem. Soc. 62*: 523 (1985).
36. S. P. Kochhar, *Influence of Temperature on Accelerated Stability Testing of Oils and Fats: A Literature Survey*. Sci. Tech. Surveys No. 161, BFMIRA, Leatherhead, UK, 1988.
37. *Speciality Fat Technology: Quality I*, Loders Croklaan BV, Wormerveer, The Netherlands.
38. C. H. Lea, The effect of light on the oxidation of fats, *Proc. Roy. Soc. Lond. 108B*: 175 (1931).
39. M. H. Gordon, The mechanism of antioxidant action in vitro, in *Food Antioxidants* (B. J. F. Hudson, Ed.), Elsevier, London, 1990, pp. 1–18.
40. J. M. Snijder, E. N. Frankel, and E. Selke, Capillary gas chromatographic analyses of headspace volatiles from vegetable oils, *J. Am. Oil Chem. Soc. 62*: 1675 (1985).
41. J. E. Ludwiczak, Marine transportation of edible oils and fats, in *World Conference Proceedings: Edible Fats and Oils Processing* (D. R. Erickson, Ed.), AOCS, Champaign, IL, 1990, pp. 25–27.
42. B. P. Backlog, Storage, handling and shipping practices: buyer/recipient viewpoint, in *World Conference Proceedings: Edible Fats and Oils Processing* (D. R. Erickson, Ed.), AOCS, Champaign, IL, 1990, pp. 28–30.
43. K. T. H. Farrer, *The Shipment of Edible Oils*, IBC Tech. Services Ltd., London, 1990.
44. P. J. Curtis, Van Den Berghs & Jurgens Ltd., Internal report, 1991.

8

Fractionation

Ralph E. Timms
The Cottages, Swinderby, Lincoln, England

I. INTRODUCTION

In the industrial processing of oils and fats, the term fractionation without further qualification usually refers to the process of fractional crystallization. In this process, the oil is first crystallized and then the crystals are separated from the remaining liquid by a variety of means. Fractionation separates the various triglycerides into one or more fractions by using the different solubilities of triglycerides, which depend on their molecular weights and degree of unsaturation.

Fractionation may also be effected by other methods such as molecular distillation, solvent extraction, or cyclodextrin complexing. However, none of these alternative processes is economical for large-scale processing of oils and fats; nor do these processes offer the same separation efficiency as fractionation crystallization. In recent years, much interest has also been shown in the use of supercritical or liquid carbon dioxide for the fractionation of oils and fats [1]. Although some interesting results have been obtained, it is unlikely that this process could replace fractionation crystallization because of its higher costs and no real advantage in separation efficiency. Where these alternative processes can have an advantage is in the selective removal of minor, nontriglyceride components, e.g., cholesterol [2], β-carotene [3], or free fatty acids [4], or the enrichment of very unsaturated triglycerides in liquid oils where crystallization would require very low temperatures [5].

Fractionation of fats is carried out for four reasons:

1. The removal of minor components detrimental to the application of the oil, e.g., the dewaxing of sunflower oil
2. The enrichment of a desirable triglyceride, e.g., POP in palm oil
3. The separation into two or more fractions of wider application and hence greater value than the parent fat, e.g., fractionation of beef tallow, palmkernel oil, or palm oil to liquid (olein) and solid (stearin) fractions

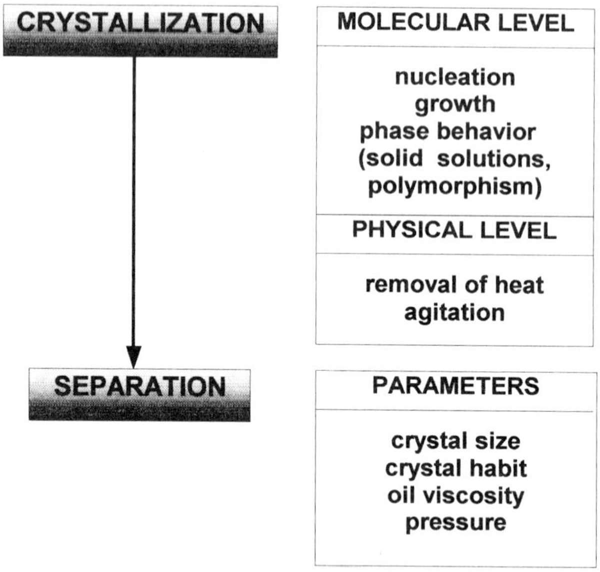

Fig. 1 Schematic diagram of fractionation process indicating factors that are important at each stage.

4. As an alternative to hydrogenation, e.g., in the fractionation of palm oil to yield fractions that can replace hydrogenated soybean or rapeseed oils in margarine hard-stocks

Several of these applications are considered in detail later.

The process of fractionation consists of two steps:

1. Crystallization to produce solid crystals in a liquid matrix
2. Separation of the crystals from the liquid matrix

Different factors are important at each stage as shown in Fig. 1, and these are considered in the next section. At this point it is useful to note that the quality of the liquid fraction depends only on the crystallization step, whereas the quality of the hard fraction depends on both the crystallization and separation steps. By "quality" is meant the degree of concentration of the desired triglycerides in the separated fraction. Quality is usually assessed by physical criteria such as cloud point or solid fat content.

The crystallization step is usually crystallization of the melted fat (crystallization from the melt), but crystallization of the fat dissolved in a solvent is also practiced industrially. These processes are commonly referred to as dry and wet fractionation.

The economics of fractionation are not considered in this chapter. Some useful information is given in a recent review by Hamm [6].

II. PRINCIPLES

A. Phase Behavior and Solubility

When a liquid fat is cooled, a solid phase separates whose composition and amount depend principally on the temperature. The situation is illustrated in Fig. 2, which shows the phase diagram of a binary mixture of triglycerides A and B that form a continuous solid solution,

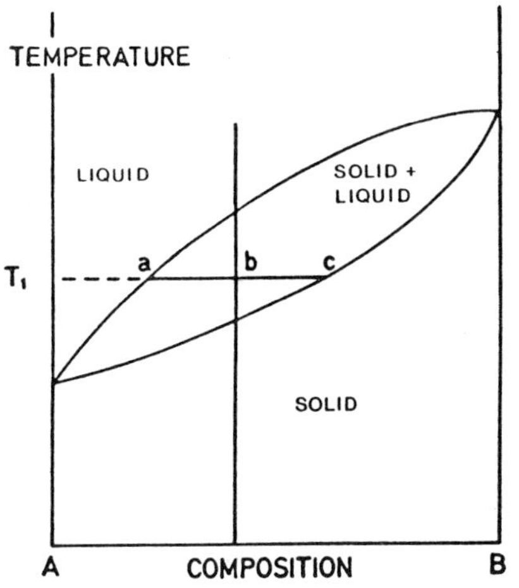

Fig. 2 Schematic phase diagram of a binary mixture of triglycerides A and B showing a continuous solid solution.

i.e., they are completely miscible in the solid state. Holding the mixture at temperature T_1 results in the formation of a solid phase (crystals) of composition c in a liquid of composition a. The fraction of solid phase is ab/ac).

The phase behavior of fats and the principles of phase diagrams have been reviewed [7]. Where solid solutions form, as is always the case in real fats, both the composition and the amount of crystals vary with the temperature of crystallization. This is a quite separate effect from polymorphism, which is considered later.

Another way of looking at the same situation is to plot a solubility curve as shown in Fig. 3. Conventionally, temperature now varies along the x axis, and the concentration of the higher melting component or solute along the y axis. A solution whose concentration at a given temperature is exactly that given by the solubility curve is said to be a saturated solution.

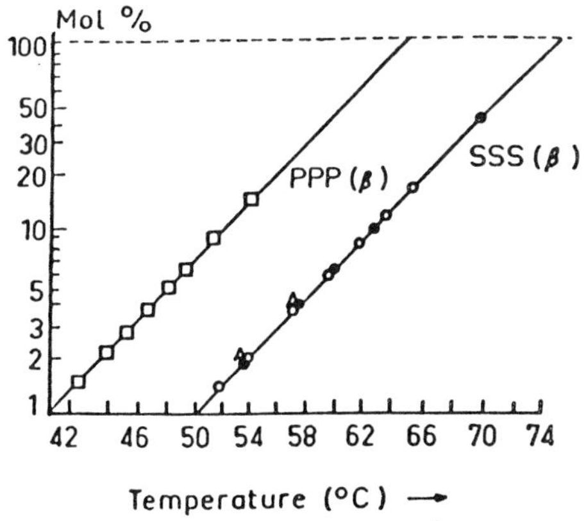

Fig. 3 Solubility of tristearin (SSS) and tripalmitin (PPP) in triolein, trilinolein, and paraffin oil. (Redrawn from Ref. 37.) (○) SSS in triolein; (●) SSS in trilinolein; (△) SSS in paraffin oil; (□) PPP in triolein.

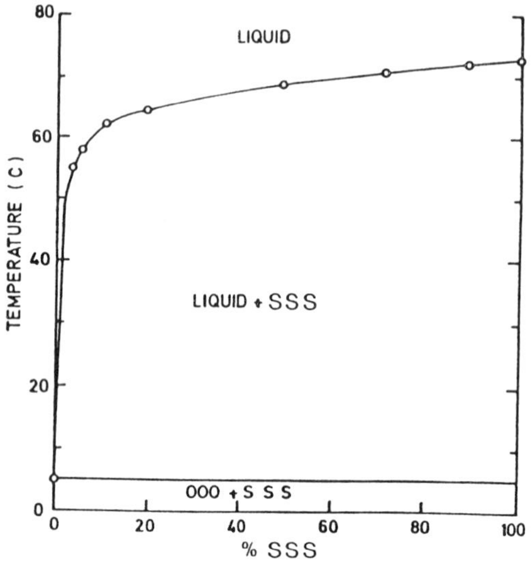

Fig. 4 Phase diagram of mixtures of tristearin (SSS) in triolein (OOO). (Redrawn from Lutton as given by Rossell [38] with solidus line added.)

Consider the solubility of tripalmitin and tristearin as shown in Fig. 3. It can be seen that the solubility is independent of the liquid oil or solvent used. The solubility depends only on the solute, in this case, tripalmitin or tristearin. This is an example of ideal solubility, and such solutions are called ideal solutions. Ideal solubility is observed when solute and solvent do not interact, i.e., no solid solutions are formed. The phase diagram then looks different from Fig. 2, as shown in Fig. 4.

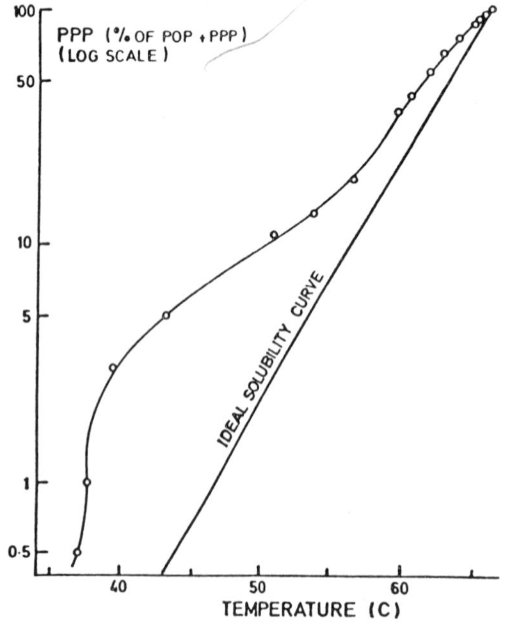

Fig. 5 Solubility of tripalmitin (PPP) in 2-oleodipalmitin (POP).

Ideal solubility can be described and defined by the equation

$$\ln x = \Delta H_{\mathrm{m}} \frac{1/T_{\mathrm{m}} - 1/T}{R}$$

where ΔH_{m} = molar heat of melting/crystallization, T_{m} = melting point of solute, and x = solubility at temperature T expressed as the mole fraction. This equation was used to draw the lines in Fig. 3, and it can be seen that it describes the experimental results closely.

In practice, the triglycerides in real fats do interact to form solid solutions, and this affects the solubility curve. Tripalmitin (PPP) and 2-oleo-dipalmitin (POP), two of the major triglycerides in palm oil, do form a solid solution, and the effect on the solubility curve is shown in Fig. 5. It can be seen that at all temperatures the observed solubility is higher than the ideal solubility predicted by the equation. The solid phase that crystallizes out is now not pure tripalmitin but a mixture of PPP and POP. This effect of increased solubility is what is meant when one fat or triglyceride is described as "solubilizing" another.

So far, we have considered a triglyceride as having a single melting point and a single solubility. However, fats and triglycerides exhibit polymorphism and occur in any one of three basic polymorphs designated α (alpha), β' (beta prime), and β (beta). α is the least stable, most loosely packed, lowest melting, and most soluble; β is the most stable, most densely packed, highest melting, and least soluble. Fats and triglycerides commonly crystallize in the α polymorph and then quickly transform to a β' or β polymorph.

B. Supersaturation

To obtain crystallization it is necessary to increase the concentration of the triglycerides to be crystallized above the saturated solution concentration at a given temperature. In practice this is not sufficient to cause crystallization, and solutions can exist indefinitely with concentrations above the saturation level without forming any crystals. Such solutions are said to be supersaturated.

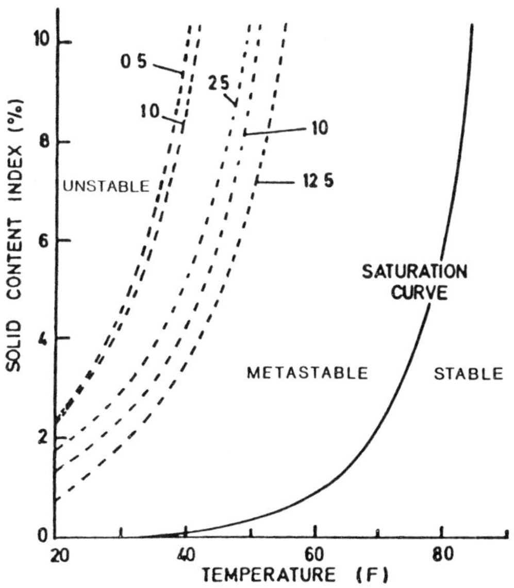

Fig. 6 Saturation-supersaturation diagram for crystallization of partially hardened soybean oil. Effect of cooling rate on metastable/unstable boundary (dashed line). Numbers are rates of cooling (°F/min) at an agitation speed of 120 rpm. (Adapted from Ref. 33.)

For any system we can draw a saturation/supersaturation diagram as shown schematically in Fig. 6 for the crystallization of partially hardened soybean oil. The continuous line is the normal solubility or saturation curve. Below this line, crystallization is impossible because the solution is not saturated, and the situation is stable indefinitely. The dashed lines divide a metastable zone from an unstable or crystallization zone. In the metastable zone, crystallization is possible but will not occur spontaneously or immediately without assistance such as stirring or seeding. In the unstable zone, crystallization will occur spontaneously and immediately. It can be seen that the position of the dashed line boundary between the metastable and unstable zones is variable and depends on process variables such as cooling rate and agitation. The position of the continuous line boundary between the stable and metastable zones depends on thermodynamic factors, whereas the position of the dashed line boundary depends on kinetic factors.

The reason for the existence of the metastable zone can be understood if crystallization is treated as a two-step process: nucleation followed by crystal growth.

C. Nucleation

A crystal *nucleus* is the smallest crystal that can exist in a solution of a certain concentration and temperature. Aggregates of molecules smaller than a nucleus are called *embryos* and will redissolve if formed [8].

When molecules come together to form a crystal there are two opposing forces. First, energy is evolved due to the heat of crystallization, which favors the process. Second, the surface of the crystal increases as the molecules aggregate. Just as when a balloon is blown up, increasing the surface requires energy to overcome the surface tension or pressure, a stable crystal will form only when the energy due to the heat of crystallization exceeds that required to overcome the surface energy. Since the surface energy is proportional to the surface area and hence to size to the second power (size^2) and the heat of crystallization is proportional to the volume and hence to size to the third power (size^3), it is clear that the solubility must depend on the size of the crystal. (Size is the linear dimension of the crystal, e.g., the diameter for spherical crystals.) Using the equation derived by van den Tempel [9], data for a typical triglyceride are shown in Table 1. *Supercooling* is the decrease in temperature below the solubility temperature that is required to get small crystals (in a supersaturated solution) to crystallize. The *critical size* is the minimum size of a crystal that is stable at the prevailing temperature.

In practice, spontaneous or homogeneous nucleation rarely occurs in fats [8,9]. Instead, heterogeneous nucleation takes place on solid particles such as dust, walls of the container,

Table 1 Variation of Solubility and Supercooling with Radius of Crystals of a Triglyceride

Radius of crystal			
μM	Å	Supercooling (°C)	Increase in solubility
10	100000	0.004	1.001
1	10000	0.036	1.007
0.1	1000	0.36	1.1
0.01	100	3.6	2.1
0.001	10	7.2	1380

Source: Ref. 10.

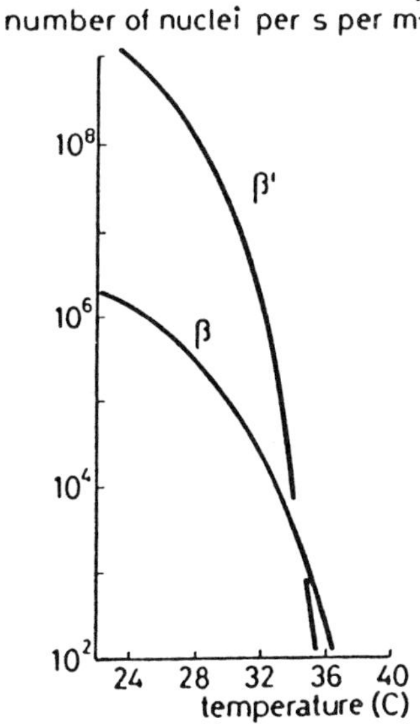

Fig. 7 Primary nucleation rate of triunsaturated triglyceride β' and β crystals in refined palm oil as a function of temperature. (From Ref. 11.)

or foreign molecules. Crystallization at a wall can lower the surface energy barrier to crystallization. The smaller the contact angle between the crystallizing fat and the surface on which nucleation is to take place, the easier nucleation becomes.

Once crystals have formed due to primary nucleation, secondary nucleation can occur. Secondary nuclei form whenever small pieces of crystal are removed from the growing crystal surface. If the pieces are smaller than the critical size, they redissolve; if larger, they act as nuclei and grow to become crystals. Secondary nucleation is undesirable in fractionation. Stirring is the primary cause, and stirring is usually kept to the minimum required to facilitate heat transfer.

As we have seen, a fat may crystallize in different polymorphs that have different stabilities and solubilities. The least stable α polymorph has the lowest surface tension as well as the lowest heat of crystallization. A small difference in surface tension can result in a large difference in nucleation rate in favor of the unstable phase [8], despite the fact that the stable phase is less soluble and hence more supersaturated. The result is that the nucleation rate of fat crystals is in the order $\alpha > \beta' > \beta$. In Fig. 7 a large difference in nucleation rates is observed for the crystallization of the trisaturated triglycerides from palm oil in both the β' and β polymorphs [11]. The difference in nucleation rates between the two polymorphs indicates why rapid cooling of a fat leads to unstable α or β' crystals, even though the more stable β polymorph is favored thermodynamically because of its lower solubility.

D. Growth

Once a crystal nucleus has formed it will start growing by the incorporation of other molecules. These molecules are taken from the adjacent liquid layer, which is replenished continuously from the surrounding, supersaturated, liquid by diffusion. The rate-determining step is incorporation of the new molecule in the correct configuration at the correct place on the growing crystal surface. The growth rate is proportional to the amount of supercooling and

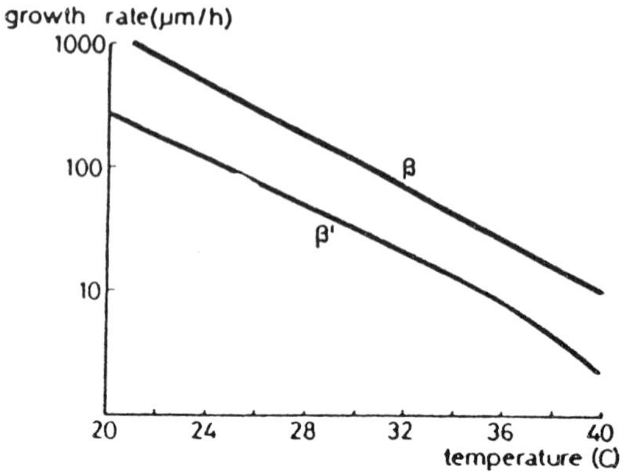

Fig. 8 Growth rate of triunsaturated triglyceride β' and β crystals in refined palm oil as a function of temperature. (From Ref. 11.)

inversely proportional to the viscosity, which affects the rate of diffusion. In crystallization from the melt, viscosity increases as the temperature falls and the growth rate can go through a maximum and then decrease with increasing supersaturation.

If a small degree of supercooling is applied, incorporation will take place only if the new molecule is in exactly the right configuration, because the molecule will have time to orient itself correctly. Thus lower supercooling tends to produce crystals that are more perfect. At higher degrees of supercooling, molecules are attached to the crystal face at a faster rate. A new molecule may become attached when its neighbor is not in a perfect position, and dislocation to the regular crystal packing may occur. As a result, perfect crystals are difficult to obtain at high degrees of supercooling, and different triglycerides can cocrystallize if they are comparable in chain length and melting point. Mixed crystals and unstable solid solutions are formed. This conclusion was confirmed by Dafler [12], who used X-ray diffraction to show that when tristearin was crystallized from the melt the crystals formed at 37°C were less perfect than those crystallized at 52°C under similar conditions.

When growth is underway, there is a substantial evolution of heat. In particular, in the absence of stirring, local temperature rises can be substantial. As a result, the volume adjacent to a growing crystal surface may cease to be saturated and/or existing nuclei may redissolve, because the critical size has increased with the rise in temperature. The nucleation and growth sequence may become erratic, leading to imperfect crystals of variable size. In the industrial fractionation of fats it is an important aim of the process to limit this exotherm by varying the rate of heat removal and agitation with temperature.

As with nucleation, growth rates depend on the particular polymorph crystallized. Since growth rate is proportional to supersaturation and the more stable forms are always the more supersaturated at any given temperature, it follows that the more stable form has the higher growth rate. This has been observed in practice for the crystallization of the trisaturated triglyceride fraction from palm oil as shown in Fig. 8.

III. PRACTICE

In practice, nucleation and growth do not occur sequentially and separately. Growth of the first nuclei will start while the temperature is still falling, thus inducing further nucleation and growth of these new nuclei. Nevertheless, the aim should be to minimize further nucleation and instead grow the existing crystals to a large uniform size, enabling easy separation from the liquid.

A. Crystallization

The fat to be crystallized is first heated above its melting point to destroy all crystal embryos. Usually, a temperature 20°C above the melting point is satisfactory. The melted fat is then cooled until crystallization starts. Until that point the cooling medium removes only sensible heat from the thermal mass. After crystallization starts, the cooling medium must also remove the latent heat of crystallization, which is much greater than the sensible heat.

The rate of cooling and crystallization are determined by the commercial need to expedite the process and the thermodynamic need to keep in step with the rate of nucleation and growth. Because the rate of nucleation increases so much more rapidly than the rate of growth (Figs. 7 and 8), rapid cooling leads to large numbers of nuclei and hence a lot of small, imperfect crystals. These crystals are then difficult to separate from the liquid, leading to a hard fraction of overall poor quality.

Crystallizers are therefore designed to provide for slow rates of cooling during crystallization; 1–3°C/h is typical. For small-scale crystallizers, a simple cooling jacket surrounding the crystallizer can provide adequate cooling capacity. The problem in the design of larger scale industrial crystallizers (5–50 t) is the need to ensure adequate cooling surface and heat transfer without excessive agitation, which would induce secondary nucleation. Clearly, the volume of a crystallizer vessel increases as the cube of its linear size, whereas the surface area of the vessel increases only as the square of the linear size. There comes a point in increasing the size of crystallizers when additional cooling surface has to be provided [13]. This additional surface can take the form of internal coils, plates, or a cooled stirrer as shown in Fig. 9.

An alternative to increasing the cooling surface is to lower the temperature of the cooling medium, thus increasing the differential between the temperature of the oil and that of the cooling medium. The disadvantage of this is that crystals formed at the cold surface are far from equilibrium with the bulk oil. They will contain too high a proportion of lower melting triglycerides. For a good quality fractionation they must eventually equilibrate with the bulk oil. Since the attainment of solid–liquid equilibrium is slow, it is unlikely to be achieved in practice. Additionally, the crystals formed will tend to stick to the cooling surface, thus further impeding heat transfer.

A practical crystallizer of optimum design will have the following features:

- Adequate cooling surface (at least 2 m^2/m^3 vessel; 3–4 m^2/m^3 is typical).
- Small temperature differential between cooling medium and oil, e.g., maximum 3°C, preferably 1°C, although this can be higher in the initial cooling period before crystallization starts, and it is certainly useful to be able to vary the temperature differential systematically.
- Gentle but effective agitation to assist heat transfer and maintenance of a uniform temperature while avoiding damage to the crystals.
- Slow cooling to ensure that the crystallization takes place under conditions as close as possible to equilibrium. Crystallization times can be 10–30 h in a typical fractionation process.

The overall aim of the crystallization step is to produce large, uniform, perfect crystals in equilibrium with the liquid. If adequate time is allowed for the crystals to grow, not only will

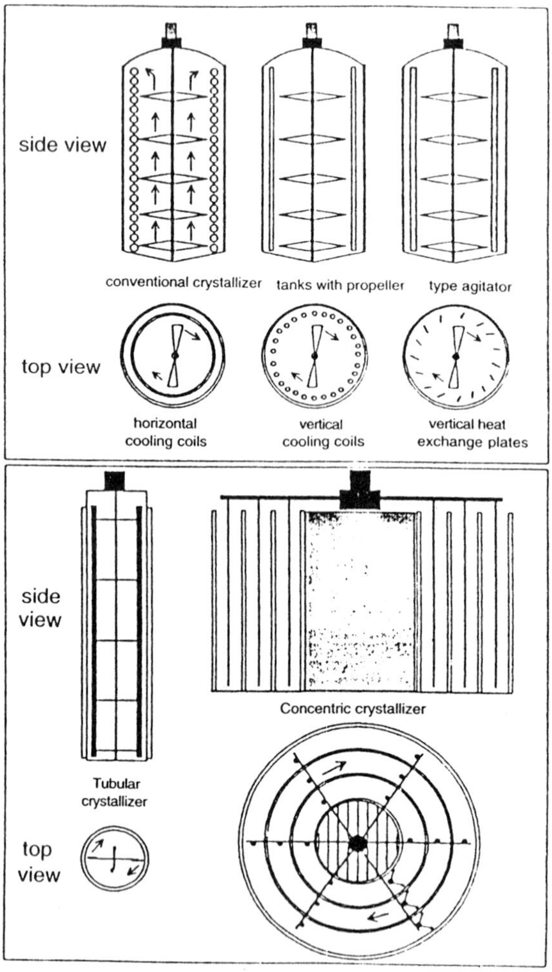

Fig. 9 Schematic representation of the most common types of crystallizer vessels used in dry fractionation. (From Ref. 13.)

the molecular level factors of crystallization (Fig. 1) have been optimized, but so also will the crystal parameters of size and habit, which have a major effect on the efficiency of the separation process.

B. Separation

To complete the fractionation process, the desired solid triglycerides in the crystals need to be separated from the triglycerides that are liquid at the temperature of crystallization. These liquid triglycerides are distributed in three locations: (1) in the uncrystallized bulk oil, (2) in the uncrystallized oil that is physically trapped or entrained in the crystals, and (3) in solid solution with the solid triglycerides.

The extent and type of solid solutions formed depend primarily on the fundamental phase behavior of the fat being crystallized. High degrees of supercooling tend to increase the extent of formation of solid solutions. (At the necessarily lower temperature a greater proportion of liquid triglycerides may take part in the solid solution.) As we saw earlier, increased solid solution formation leads to increased solubility, so in practice a higher yield of lower quality

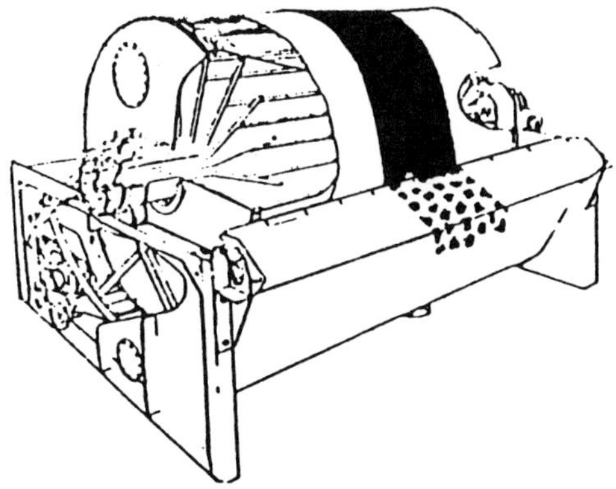

(a)

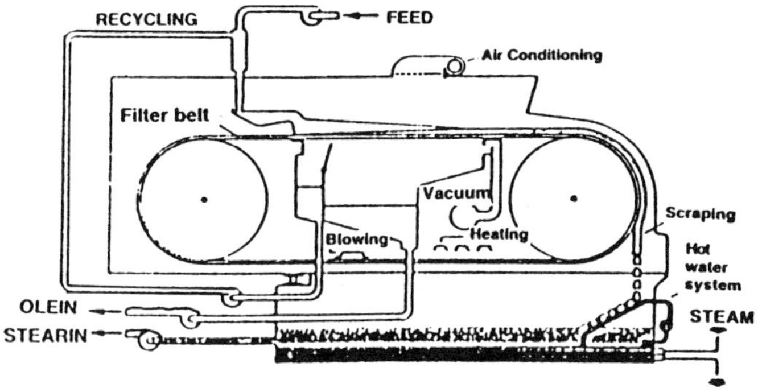

(b)

Fig. 10 Schematic representation of the two most common types of vacuum filters used in dry fractionation. (a) Rotary drum filter; (b) vacuum belt filter. (From Ref. 13.)

crystals is produced. However achieved, once the solid solution has formed in the crystals, the separation step can do nothing to change the composition of the solid phase of the crystals.

The uncrystallized bulk oil is relatively easily removed from the crystals. The entrained oil is more difficult to remove. The separation step must address the problem of reducing the level of entrained oil in the final solid fraction. Three methods have been developed to achieve this.

1. Vacuum Filtration

Two types of vacuum filters are in use—the rotary drum and the belt filter—as shown in Fig. 10. Such filters operate in two stages. In the first stage the crystallized bulk oil is removed and the crystal cake builds up. In the second stage the residual liquid is sucked out and air or nitrogen is sucked through the cake. It is in this second stage that the entrained oil is reduced. Although more complex and expensive, the belt filter has the advantage that more time can be given to this second stage and the differential pressure can vary along the belt. As a result, the belt filter yields a solid cake fraction with a lower level of entrainment than the drum filter. Comparative results are given in Table 2.

2. Centrifugal Separation

Because of the viscosity of the uncrystallized oil at the crystallization/separation temperature and the large amount of solids to be separated, industrial centrifugal separators cannot be used to efficiently separate crystals from the bulk oil. A solution to this problem is to use a surface-

Table 2 Estimates of Entrained Oil in Stearin from Different Fractionation Processes

Company	Process	Yield (%)	Entrained oil (% of stearin)
Alfa Laval	Detergent (Lipofrac)	17–23	35–52
Tirtiaux	Dry + belt filter	28–33	60–67
De Smet	Dry + drum filter	37–40	70–73
Bernadini	Hexane	37–40	70–73
De Smet	Dry + membrane filter	20–21	47–50

Source: Ref. 14.

active agent (detergent) to "wet" the crystals so they are transferred into an aqueous phase. Separation of a light oil phase and a heavy aqueous phase containing the crystals is then easy. The concept was patented in 1905 by Fratelli Lanza and then developed as a proprietary process using centrifugal separation (the Lipofrac Process) by the Alfa-Laval Co. The Lanza, Lipofrac, or detergent process enables more entrained oil to be removed than can be achieved using vacuum filtration (Table 2). Until the development of membrane pressing, the Lipofrac process was a popular alternative to vacuum filtration. The higher costs of the process were outweighed by the higher yield of liquid fraction and the better quality of the solid fraction. The process is now less favored on environmental grounds related to the disposal of the spent detergent solution containing surface-active agent and lost oil.

3. *Pressing*

Vacuum filters are necessarily limited to a pressure on the crystals of less than 1 bar. If the pressure is increased, it is possible to squeeze out more of the entrained oil. As noted in the Introduction, improved separation improves the yield but not the quality of the liquid fraction, whereas for the solid fraction the separation step has a decisive effect on quality.

The mechanical or hydraulic pressing process was therefore developed particularly for the production of palmkernel stearin, where the quality of the stearin determines its utility and value as a replacement for cocoa butter. Conventional vertical hydraulic presses capable of 100 bar pressure have been used for many years.* The process is applied in practice by crystallizing the fat statically in blocks, e.g. in trays, in a cold room with cold air circulating around the blocks to cool the fat. The crystallized blocks of fat are then wrapped in filter cloths of a mesh size able to retain the crystals while allowing the entrained oil to flow out, and hydraulic or mechanical pressing is applied. The practical limit on pressure is determined by the point at which no further oil can be squeezed out and/or the solid crystals begin to extrude through the cloth. Such a process is labor-intensive and, because of the need to surround the crystals completely with the cloth, it is feasible only for solid crystalline masses with at least ~25% solid fat content.

If a crystallized slurry of the consistency produced in a stirred crystallizer is to be pressed, a totally different design of press is required. Such presses, membrane filter presses, were developed in the early 1980s for the fractionation of palm oil. Their main features are that:

*Palmkernel can be used as one, or two separate words in the literature. It appears here as one term, and should not be confused with palm oil.

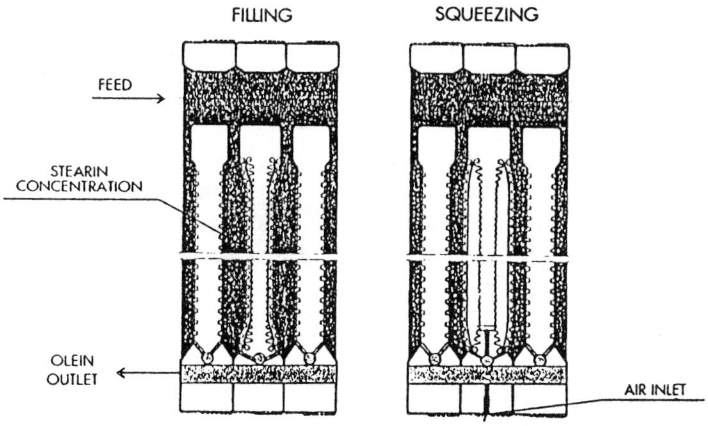

Fig. 11 Operation of filtration sequence in membrane press. (From Ref. 13.)

1. A crystal slurry may be pumped into the press chambers.
2. Pressure may be applied to the cake in the press chambers.
3. Squeeze pressures of 6 bar (and more recently 15 bar) are easily achieved.
4. The filling, pressing, and emptying of the press are automated, making the process relatively cheap to operate.

The design and operation of a membrane filter press are shown schematically in Fig. 11.

As a result of the development of the membrane filter press, entrainment levels as low as in the Lipofrac process can be achieved at lower capital, running, and environmental costs. This separation method is now the overwhelmingly preferred choice for new fractionation plants.

More recent developments reported by Willner and Weber [15] have shown the dramatic effect on entrainment levels and hence on the quality of palm midfractions that can be achieved by increasing the pressure up to 50 bar (Fig. 12). However, it does not look as if

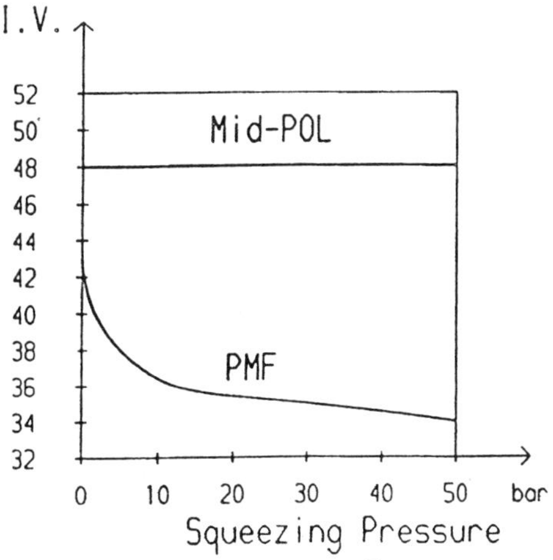

Fig. 12 Pressing of soft palm midfraction (mid-POL) to produce palm midfraction (PMF) suitable for CBE production. Effect of pressure on iodine value (iv) of PMF. (From Ref. 15.)

pressures of 100 bar as achieved in vertical hydraulic presses will prove feasible due to the limitations of the flexible pressure membrane and filter materials.

C. Multistage Dry Fractionation

A single fractionation consists of crystallization and separation steps. In many cases more than one fractionation is carried out to give a multistage fractionation process. Multistage fractionation processes are used where the desired fraction is a middle-melting fraction or where improved purity of the fraction is required—greater than can be obtained in a single-stage fractionation. Such processes are illustrated in Section IV.

D. Solvent Fractionation

If the fat to be crystallized is dissolved in an organic solvent, the fractionation process can be substantially improved. There are several benefits of using a solvent.

1. Nucleation and growth are faster, so higher degrees of supercooling and faster rates of cooling can be used. This favors short crystallization times (less than 1 h) and continuous crystallizers.
2. Lower viscosity of the liquid leads to easier filtration.
3. Dilution of the fat makes heat transfer easier and the amount of oil in the entrained liquid smaller.
4. The ability to wash the cake with fresh solvent leads to very low levels (less than 10%) of entrained oil.

Unfortunately, the extra costs of solvent fractionation compared to dry fractionation have made the process uneconomical except for the production of specialized fractions to replace cocoa butter [16].

The choice of solvent has been discussed by Hamm [16] and by Timms [17]. The selectivity of triglyceride separation in the crystallization step is little affected by the choice of solvent, and indeed solvent fractionation has no advantage over dry fractionation here. The choice of solvent does affect the selectivity of separation between lipid classes, particularly the separation of triglycerides from diglycerides, free fatty acids, and other lipids more polar than triglycerides. Of the two commonly used solvents, acetone and industrial hexane, the more polar acetone effects a better separation of triglycerides from diglycerides and free fatty acids than does the nonpolar hexane.

E. Countercurrent Fractionation

Although countercurrent crystallization is commonly applied in the chemical industry, it has rarely been applied to the fractionation of oils and fats. This is attributable to the long crystallization times required and to the relatively inefficient solid–liquid separation methods available. With the introduction of membrane filter presses, countercurrent fractionation becomes feasible; the possibilities have been discussed by van den Kommer and Keulemans [18].

In Fig. 13 is shown a hypothetical phase diagram for the first four fractionations of a two-stage countercurrent process. Feed of composition F1 is fractionated at temperature T_1 to give an olein O1 and stearin S1. Note that O1 and S1 are different from O_1 and S_1, the equilibrium compositions, due to the system not having reached equilibrium (for the olein) and the contamination of the stearin with entrained olein.

Olein O1 then becomes feed F2 and is refractionated at temperature T_2 to give olein O2 and stearin S2. Stearin S2 is then returned to the previous stage and mixed with F1 to give a

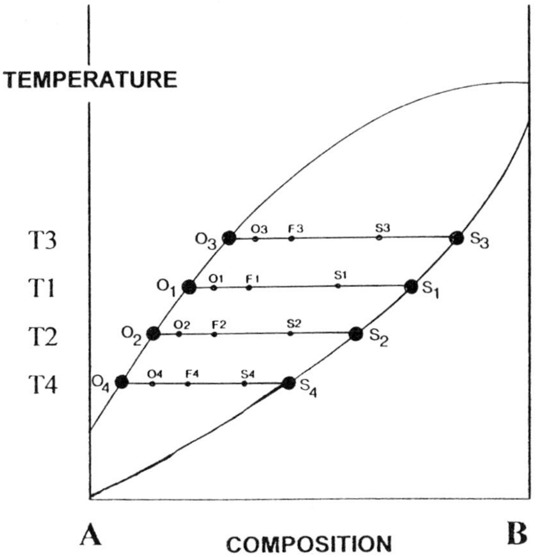

Fig. 13 Schematic hypothetical phase diagram showing the first four fractionations of a countercurrent process. (Adapted from Ref. 18.)

new feed F3, which is fractionated at T_3. The resulting stearin S3 is enriched in component B (compared with S1). Olein O3 is returned to the previous stage and mixed with feed F2 to give a new feed F4, which is fractionated at T_4 to give an olein O4 enriched in component A.

At least three cycles are required to achieve a steady state, and the process can be time-consuming to establish, so in practice only a two-stage or perhaps three-stage countercurrent process could be commercially practicable.

The advantage of a two-stage countercurrent process over a two-stage consecutive fractionation process is shown in Fig. 14 for the fractionation of palm oil. The two stearins from the

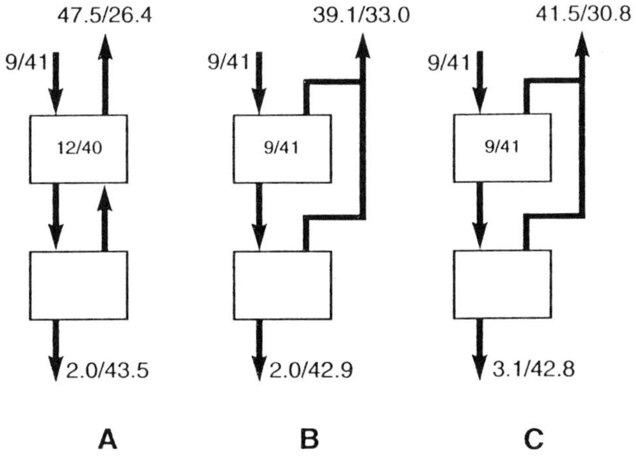

Fig. 14 Two-stage countercurrent fractionation (A) compared to a two-step process with the same olein composition (B) or same stearin yield (C), respectively. The numbers refer to SSS/SSO concentrations. (From Ref. 18.)

consecutive fractionations are combined so that both processes result in only two output streams, one liquid and one solid. There is little difference in the compositions of the oleins but a large difference in the compositions of the stearins.

F. Process Economics

The commercial viability of a fractionation process depends on

The value of the raw material feedstock
The value and yield of all products
The cost of each fractionation stage
The number of fractionation stages

For simplicity assume a main product A and a by-product B, which comprises all the by-product fractions; the cost of each fractionation stage is the same. Then,

Raw material→product A + by-product(s) B

For commercial viability,

$$x_A V_A + x_B V_B > V_{RM} + nF$$

where x_A,x_B are fractions of A and B, V_A,V_B are values of A and B, V_{RM} is the cost of raw material, n is the number of fractionations, and F is the cost of each fractionation.

Rearranging this equation we find that for break-even,

$$\text{Break-even } V_A = V_B + \frac{nF}{x_A} + \frac{V_{RM} - V_B}{x_A}$$

In Table 3, values are given for V_A for different values of n, x_A, and V_B. It can be seen that although the break-even value increases with the number of fractionations and decreasing yield, an even more important parameter is the value of the by-product(s) in relation to the value of the raw material. The advantage of starting with a relatively low cost raw material such as palm oil is clear.

Table 3 Calculation of Break-Even Value of Product A (V_A) According to Equation Given in the Text[a]

Number of fractionations n	Fractional yield of main fraction (x_A)	Break-even value of A depending on value of by-products (V_B) in relation to value of raw material (V_{RM})		
		$V_B = 1.1\ V_{RM}$	$V_B = V_{RM}$	$V_B = 0.9\ V_{RM}$
1	0.5	1000	1100	1300
1	0.3	933	1167	1500
2	0.5	1100	1200	1400
2	0.3	1100	1333	1667
3	0.5	1200	1300	1500
3	0.3	1267	1500	1700

[a] Assuming value of raw material, V_{RM} = \$1000/t; cost of each fractionation stage = \$50.

IV. APPLICATIONS

A. Palm Oil

Palm oil is a semisolid fat with about 20% solid fat crystals at 20°C. Fractionation to separate these crystals as a stearin leaving an olein with a melting point of 20–24°C is carried out on a large scale throughout the world, but mainly in Malaysia and Indonesia. In 1994, world production of palm oil was almost 14 megatonnes, and production of palm olein is estimated at 5.1 Mt. In Malaysia alone, 3.4 Mt of palm olein is estimated to have been produced, of which 3.1 was exported.

Because palm oil is ideal for fractionation and because of the massive growth in palm oil production in the last 30 years, modern fractionation processes have been developed and optimized mainly for palm oil. Palm oil consists essentially of three groups of triglycerides:

About 8% trisaturated triglycerides, principally PPP
About 45% disaturated triglycerides, principally POP and PLinP
About 40% monosaturated triglycerides, principally POO and PLinO

The main fractionation process is carried out to produce an olein that is liquid below 24°C, making it an ideal cooking oil in tropical countries. Essentially the process is designed to separate PPP from the palm oil. Because of solid solution formation between PPP and POP

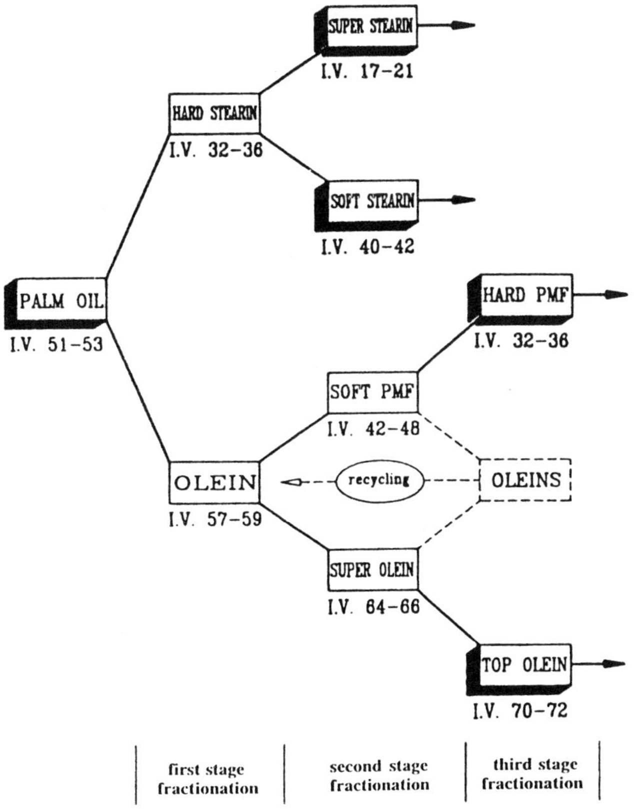

Fig. 15 Dry multiple fractionation of palm oil. (From Ref. 21.)

Table 4 Properties of Palm Oleins Produced by Dry Fractionation

Olein type	Iodine value	Cloud point (°C)	Percent soybean oil required to meet cold test at 0°C
"Traditional"	57	8.0	100
Superolein 1	63	4.0	90
Superolein 2	65	1.3	70
"Top" olein	71	−2.0	0

Source: Ref. 20.

and because of entrainment, the palm stearin produced contains substantial quantities of disaturated and monosaturated triglycerides. Dry fractionation processes are used, with the membrane filter press now the preferred method of separation. Using this process, an olein of IV 56–58 with PPP content 1% maximum can be produced in a yield of 75–80%. Deffense [20] compared different fractionation processes for the production of this palm olein.

Palm olein from a single-stage fractionation has now become a commodity product and one of the world's major oils in its own right. To develop products with more added value, multistage fractionation has been applied to produce more-liquid palm oleins as shown in Fig. 15 and Table 4.

Multistage fractionation necessarily produces middle-melting fractions, usually called midfractions. Palm midfractions (PMFs) concentrate the disaturated triglycerides. A major use of PMF has been in the production of cocoa butter equivalent fats (CBEs) to replace cocoa butter in chocolate. Essentially, POP is required, and solvent fractionation using acetone, hexane, or 2-nitropropane has been used to get PMF of the required quality. (The use of 2-nitropropane has been discontinued for toxicological reasons.)

With the improved crystallization and separation methods now applied to palm oil, it is possible to produce a PMF by dry fractionation that is equivalent to that produced by solvent fractionation, albeit in somewhat lower yield. If the belt vacuum or low pressure membrane filter is used, then three fractionation steps are required. This process is shown schematically in Fig. 15 to produce a "hard PMF." Alternatively, if a high pressure (25–50 bar) membrane filter press is used for separation, then only two fractionation stages are required [15]. In Table 5 the various sorts of PMF now commonly available are compared with a solvent-fractionated PMF. Iodine value is used as a measure of the total unsaturation, SFC as

Table 5 Properties of Palm Midfractions Produced by Dry and Solvent Fractionation

PMF type	Iodine value	SFC at 30°C	Triglyceride with Carbon No. 52 (%)
Dry "soft"	50	1	50
Dry "hard"	40	18	64
Dry "super"	34	45	74
Solvent	34	46	75

a measure of the steepness of melting (SFC at 35°C is less than 5%), and the triglyceride carbon no. at 52 as a measure of the POP and PLinP content. It can be seen that the "super PMF" obtained from the high pressure membrane press is similar to the solvent-fractionated PMF.

B. Palmkernel and Coconut Oils

Palmkernel (PK) oil is fractionated in one step to concentrate the triglycerides containing medium-chain fatty acids (lauric and myristic) into a stearin fraction. Coconut oil can also be fractionated similarly, but the tonnages are very small in comparison with PK oil. The fractionation of lauric oils has been reviewed in detail by Rossell [22].

Palmkernel stearin is widely used as a cocoa butter substitute to replace cocoa butter in compound chocolate. Palmkernel oil of IV 18 is fractionated to produce a stearin of IV 7–8 [23]. Because the stearin is the main product and olein the by-product, separation of the olein from the stearin crystals is critical to the production of a good quality product. A dry fractionation process with drum or belt vacuum filtration cannot be used because the resulting stearin contains too much entrained olein. Instead, solvent fractionation, dry fractionation with detergent (Lipofrac process), or dry fractionation with high pressure pressing are used as discussed in Section III. In the detergent process, it is necessary to recycle about 30% of the olein to the feedstock in order to lower the solid fat (crystal) content of the crystallized slurry to about 20%. If this is not done, it is necessary to accept a low yield of stearin in order to be able to pump the slurry and not overload the centrifugal separator.

A comparison of these three fractionation processes is given in Table 6. In Fig. 16 the melting curves of palmkernel oil, olein, and stearin are compared. The steep melting curve of the palmkernel stearin, similar to the melting curve of cocoa butter, is clearly shown. Recently, the use of a high pressure membrane filter press has been described. These presses seem to offer the advantages of the traditional hydraulic presses without their high labor costs and difficulty of operating a clean process [24].

Table 6 Comparison of Fractionation Processes for Palmkernel Oil

Process	Capital cost	Operating costs	Stearin yield (%)	Olein value	Remarks
Solvent fractionation	High	High	~50	Lowest	Uses nitropropane, acetone, or hexane as solvent.
Crystallization + pressing	Low	Medium to high	~40	Intermediate	A labor-intensive and "messy" process as usually practiced. Operating costs depend much on labor costs.
Detergent fractionation	Medium	Low to medium	~30	Highest	Continuous process, usually well automated.

Source: Ref. 22.

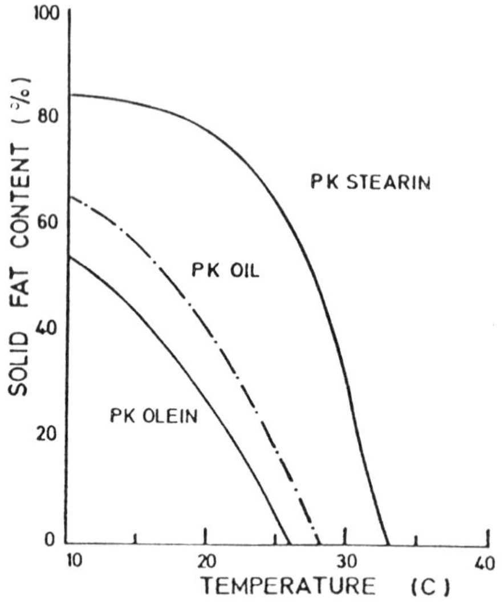

Fig. 16 Solid fat contents (%) of palmkernel (PK) oil, olein, and stearin. (From Ref. 23.)

C. Milk Fat

Milk fat is a semisolid fat like palm oil and can be easily fractionated into a similar range of products. The range of fatty acids and triglycerides present is much greater than in palm oil, so the fractions produced are less distinct than palm fractions. Relatively little milk fat is fractionated, perhaps about 250 kt/yr, due to its high price in relation to the value of the by-product fractions as discussed in the previous section. Most milk fat fractionation is carried out not where milk fat is cheapest, e.g., New Zealand, but in Europe, where the raw material is expensive even with subsidies. This anomaly exists as a result of the European Common Agricultural Policy price support scheme, whereby the value of the by-products is maintained artificially high in relation to the raw material cost.

Although solvent and detergent fractionation processes have been investigated [25,26] and are perfectly feasible, only dry fractionation with vacuum or membrane filtration is practiced

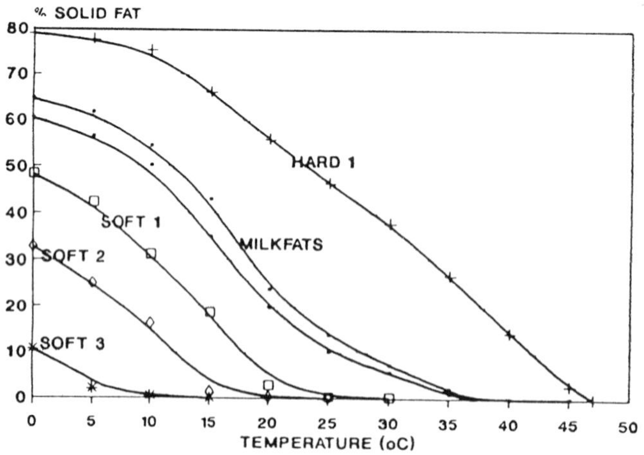

Fig. 17 Melting properties of milk fat fractions from three-step fractionation. (From Ref. 28.)

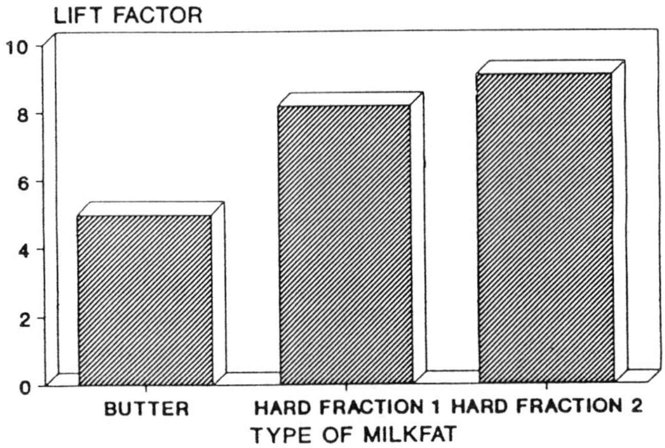

Fig. 18 Lift of puff pastry (48 layers) using butter and plasticized milk fat fractions. (From Ref. 28.)

commercially [27]. Some of the wide range of milk fat fractions that can be produced by three-stage fractionation are shown in Fig. 17. By blending the soft (olein) and hard (stearin) fractions, a "butter" that is spreadable at refrigeration temperatures can be produced; this butter has been marketed by the New Zealand Dairy Board [27,28]. Milk fat stearins are used to make pastry butter [29] with substantial improvement in performance as shown in Fig. 18.

D. Beef Tallow

Beef tallow can be fractionated into an olein and a stearin in a similar manner to the fractionation of palm oil. Beef tallow has mostly been fractionated in Australia using the Lipofrac detergent process. Some properties of commercial Australian beef tallow olein and stearin in comparison with palm olein and stearin are shown in Table 7. Tallow stearin is used to formulate pastry shortenings. Tallow olein is used as a frying fat or to formulate margarines.

The solvent fractionation of beef tallow has also been proposed as a method to produce tallow midfractions for use in CBEs and other speciality fats [30,31]. Some useful fractions can be obtained, but the process has never become commercial because of the general decline in the use of animal fats for food, especially to replace vegetable fats, and the ready availability of palm oil, which is better suited to the production of a CBE component.

Table 7 Properties of Australian Beef Tallow Stearin and Olein Compared with Palm Oil Stearin and Olein

	Beef tallow		Palm oil	
Property	Stearin	Olein	Stearin	Olein
Iodine value	35	48	36	57
Melting pt (°C)	52	37	52	21
SFC at 10°C	73	43	74	40
SFC at 20°C	66	28	62	6
SFC at 30°C	49	12	41	0
SFC at 40°C	30	3	28	0
SFC at 50°C	9	0	12	0

E. Other Fats

Many other fats are also fractionated. Oils such as cottonseed oil and hydrogenated soybean oil are fractionated to remove a small amount of solid so that the oils can be used as salad oils that remain clear at refrigerator temperatures. The process is often called winterization, because crystallization occurs naturally with the onset of winter, providing the ideal slow crystallization conditions. Winterization is a basic dry fractionation with a relatively simple vacuum filtration. In the winterization of cottonseed oil the stearin removed consists mainly of PLinP, whereas for soybean oil the stearin removed contains mainly di- or triunsaturated triglycerides containing trans acids. The process has been reviewed by List and Mounts [32]. Unlike the fractionation of natural oils, in the fractionation of hydrogenated soybean oil the properties of the feedstock itself can be varied by the processor according to the degree of hydrogenation. Clearly the greater the extent of hydrogenation (and the lower the IV), the less olein of a particular IV can be produced. However, Singh [33] studied the crystallization of hydrogenated soybean oil in detail and concluded, rather surprisingly, that there is an optimum concentration of solid fat in the feedstock for the production of olein with the best cold test.

In a process similar to the winterization of cottonseed and soybean oils, sunflower and ricebran oils are fractionated to give oils that will not cloud at refrigerator temperatures. Unlike all the other fractionation processes we have considered, the solid phase removed is not a triglyceride but a wax, and the process is usually called dewaxing.

Sal fat, a fat extracted from the seed of Indian forest trees, is fractionated to produce sal stearin, which is rich in SOS triglycerides for use as a component of CBEs. Although solvent fractionation is practical and is used in Europe, sal fat is now more commonly dry fractionated in India. The fat is crystallized in trays and pressed in a hydraulic press in a way similar to the process described earlier for palm kernel oil. The yield of sal stearin by the dry fractionation process is typically 55–60%.

Shea fat, a fat extracted from the nut of trees in the dry tropical areas of West Africa, is also fractionated to produce a shea stearin rich in SOS triglycerides. Compared with sal fat, the content of stearin/SOS triglycerides is lower, and a solvent process is usually preferred to give a yield of stearin of 35–40%. If acetone is used, an undesirable isoprenoid gum may also be removed in an initial "degumming" step. Only where the product, in this case shea stearin, has a high value can the cost of solvent fractionation be justified. With cocoa butter at about US$4000/t, CBEs can be sold at up to US$3000/t, several times higher than the price of most other edible oils and fats.

V. PROSPECTS

After hydrogenation, fractionation is the second most important process for modifying oils and fats. Hydrogenation is used to modify a wide range of liquid oils, principally soybean, rapeseed, sunflower, and fish oils, to produce semisolid plastic fats for use in margarine and shortenings. Fractionation is used overwhelmingly to modify a single oil—palm oil—to produce liquid fractions for use as frying oils and middle and higher melting fractions for use in a wide variety of fat-based food products as well as oleochemicals. The use of fractionation is likely to grow relative to hydrogenation for the foreseeable future because

1. Palm oil production is growing quickly, and palm oil and its associated oil, palm kernel oil, are forecast to be the oils produced in the greatest volume in the world by the year 2002, overtaking soybean oil [34].

2. There are nutritional and health concerns about the level of trans fatty acids in the diet [35]. Some of these acids come from animal fats, but the major contribution in the diet is from hydrogenated oils.
3. Fractionation is a mild process, whereas hydrogenation is a chemical process conducted at high temperatures using nickel and hydrogen gas. Fractionation is thus more environmentally benign and is likely to be increasingly favored by today's "greener" consumers.

REFERENCES

1. W. Hamm, Trends in fractionation practice for edible oils, Presented at Soc. Chem. Ind. Symp. Fractional Crystallization of Fats, London, 9 March 1994, SCI Lecture Paper No. 0037.
2. E. Schlimme, Removal of cholesterol from milk fat, *Eur. Dairy Mag. 4*: 12, 13, 16–21 (1990).
3. C. K. Ooi, Y. M. Choo, S. C. Yap, Y. Basiron, and A. S. H. Ong, Recovery of carotenoids from palm oil, *J. Am. Oil Chem. Soc. 71*: 423–426 (1994).
4. W. Hamm, Liquid–liquid extraction in food processing, in *Science and Practice of Liquid Extraction* (J. D. Thornton, Ed.), Clarendon Press, Oxford, 1992.
5. Unilever Ltd., Br. Patent 1 444 551 (1976).
6. W. Hamm, Trends in edible oil fractionation, *Trends Food Sci. Technol. 6*: 121–126 (1995).
7. R. E. Timms, Phase behaviour of fats and their mixtures, *Prog. Lipid Res. 23*: 1–38 (1984).
8. R. Boistelle, Fundamentals of nucleation and crystal growth, in *Crystallization and Polymorphism of Fats and Fatty Acids* (N. Garti and K. Sato, Eds.), Marcel Dekker, New York, 1988, pp. 189–226.
9. M. van den Tempel, Effects of emulsifiers on the crystallization of triglycerides, in *Surface-Active Lipids in Foods*, SCI Monograph No. 32, Soc. Chem. Ind., London, 1968, pp. 22–36.
10. R. E. Timms, Crystallization of fats, *Chem. Ind.*, 20 May, 342–345 (1991).
11. K. P. A. M. van Putte and B. H. Bakker, Crystallization kinetics of palm oil, *J. Am. Oil Chem. Soc. 64*: 1138–1143 (1987).
12. J. R. Dafler, Polymorphism behaviour in fully hydrogenated mono acid triglycerides, *J. Am. Oil Chem. Soc. 54*: 249–254 (1977).
13. M. Kellens, Developments in fractionation technology, Presented at Soc. Chem. Ind. Symp. Fractional Crystallization of Fats, London, 9 March 1994, SCI Lecture Paper No. 0042.
14. R. E. Timms, Principles of fractionation, Presented at Soc. Chem. Ind. Symp. Fractional Crystallization of Fats, London, 9 March 1994, SCI Lecture Paper No. 0039.
15. T. Willner and K. Weber, High-pressure dry fractionation for confectionery fat production, *Lipid Technol. May/June*: 57–60 (1994).
16. W. Hamm, Fractionation—with or without solvent?, *Fette Seifen Anstrichm. 88*: 533–537 (1986).
17. R. E. Timms, Choice of solvent for fractionational crystallization of palm oil, in *Palm Oil Product Technologies in the Eighties* (Pushparajah and Rajadurai, Eds.), Kuala Lumpur, 1983.
18. M. van den Kommer and C. N. M. Keulemans, Developments in dry fractionation, Presented at Soc. Chem. Ind. Symp. Fractional Crystallization of Fats, London, 9 March 1994, SCI Lecture Paper No. 0040.
19. E. Deffense, Fractionation of palm oil, *J. Am. Oil Chem. Soc. 62*: 376–385 (1985).
20. E. Deffense, Superolein production and analysis of fractions, Presented at Soc. Chem. Ind. Symp. Fractional Crystallization of Fats, London, 9 March 1994, SCI Lecture Paper No. 0036.
21. E. Deffense, Dry multiple fractionation: trends in products and applications, *Lipid Technol. March*: 34–38 (1995).
22. J. B. Rossell, Fractionation of lauric oils, *J. Am. Oil Chem. Soc. 62*: 385–390 (1985).
23. R. E. Timms, Processing of palm kernel oil, *Fette Seifen Anstrichm. 88*: 294–300 (1986).
24. T. Willner, High-pressure dry fractionation, Presented at Soc. Chem. Ind. Symp. Fractional Crystallization of Fats, London, 9 March 1994, SCI Lecture Paper No. 0038.

25. P. C. Chen and J. M. de Man, Composition of milk fat fractions obtained by fractionation crystallization from acetone, *J. Dairy Sci. 49*: 612 (1966).
26. R. Norris, I. K. Gray, A. K. R. McDowell, and R. M. Dolby, The chemical composition and physical properties of fractions of milk fat obtained by a commercial fractionation process, *J. Dairy Res. 38*: 179 (1971).
27. E. Deffense, Milk fat fractionation today: a review, *J. Am. Oil Chem. Soc. 70*: 1193–1199 (1993).
28. C. Versteeg, L. N. Thomas, Y. L. Yep, M. Papalois, and P. S. Dimick, New fractionated milkfat products, *Aust. J. Dairy Technol. 49*: 57–61 (1994).
29. W. Seibel, Use of fractionated butterfat (butter pastry fat), *Zucker Süsswarenwirtsch. 40*: 122–126 (1987).
30. F. E. Luddy, J. W. Hampson, S. F. Herb, and H. L. Rothbart, Development of edible tallow fractions for specialty fat uses, *J. Am. Oil Chem. Soc. 50*: 240–244 (1973).
31. H. H. Taylor, F. E. Luddy, J. W. Hampson, and H. L. Rothbart, Substitutability of fractionated beef tallow fractions for other fats and oils in the food and confectionery industries: an economic evaluation, *J. Am. Oil Chem. Soc. 53*: 491–495 (1976).
32. G. R. List and T. L. Mounts, Partially hydrogenated-winterized soybean oil, in *Handbook of Soy Oil Processing and Utilization* (D. R. Erickson, E. H. Pryde, O. L. Brekke, T. L. Mounts, and R. A. Falb, Eds.), Am. Soyabean Assoc., St. Louis, MS, and Am. Oil Chem. Soc., Champaign, IL, 1987, pp. 193–216.
33. G. Singh, Analysis of crystallization systems with applications to continuous fractional crystallization of fatty acid triglycerides, *AIChE Symp. Ser. 72*: 100–109 (1974).
34. S. Mielke, Trends in supply, consumption and prices, in *Oils and Fats in the Nineties* (V. K. S. Shukla and F. D. Gunstone, Eds.), Int. Food Sci. Centre, Lystrup, Denmark, 1992, pp. 10–22.
35. British Nutrition Foundation, *Trans* Fatty Acids, London, July 1995.
36. J. Hannewijk, A. J. Haighton and P. W. Hendrikse, Dilatometry of fats, in *The Analysis and Characterization of Oils, Fats and Fat Products*, Vol. 1 (H. A. Boekenoogen, Ed.), Interscience, London, 1964, pp. 119–182.
37. J. B. Rossell, Phase diagrams of triglyceride systems, *Adv. Lipid Res. 5*: 353–408 (1967).

9

Interesterification of Oils and Fats

A. Rozendaal
Unilever Research Laboratory, Vlaardingen, The Netherlands

A. R. Macrae
Unilever Research Laboratory, Sharnbrook, Bedfordshire, England

I. INTRODUCTION

Interesterification is a catalytic process involving the exchange of fatty acids between existing esters to form new esters. When applied to a mixture of triacylglycerols (TAGs), the available fatty acids are redistributed over all the possible triacylglycerol types. Due to this rearrangement the physicochemical properties of the mixture, in particular the melting and crystallization or recrystallization properties, can be changed and steered into a desired direction. Interesterification, next to hydrogenation and fractionation, is therefore an important tool in modifying the physical and functional properties of oil and fat mixtures. Unlike hydrogenation, it leaves the overall fatty acid composition of the mixture unchanged. The wide range of triacylglycerols present in interesterified fats usually leads to a more regular crystallization during the manufacture of spreads and shortenings. Moreover, the rate of recrystallization during storage of such products may be retarded, resulting in stabilization of the generally preferred β' modification and a reduced tendency to develop product defects. Interesterification or rearrangement processes therefore find ever wider application for converting oils and fats into blend components that are optimally suited for incorporation into various types of edible fat-based consumer products.

Among the catalysts nowadays available for rearrangement are alkali metal derivatives (such as sodium methoxide and other alcoholates) and enzymes (lipases). The latter are effective under mild reaction conditions and in many cases are specific in exchanging only the fatty acids from the 1,3-positions of the triacylglycerols, thereby offering new possibilities for the manufacture of tailor-made TAGs (structured lipids).

Interesterification is usually carried out in a single phase (liquid). With a nonselective catalyst, rearrangement of the fatty acids will then ultimately result in their statistical or random distribution over all possible positions. However, it is also possible to carry out the rearrangement under such conditions that part of the reaction products is continually removed

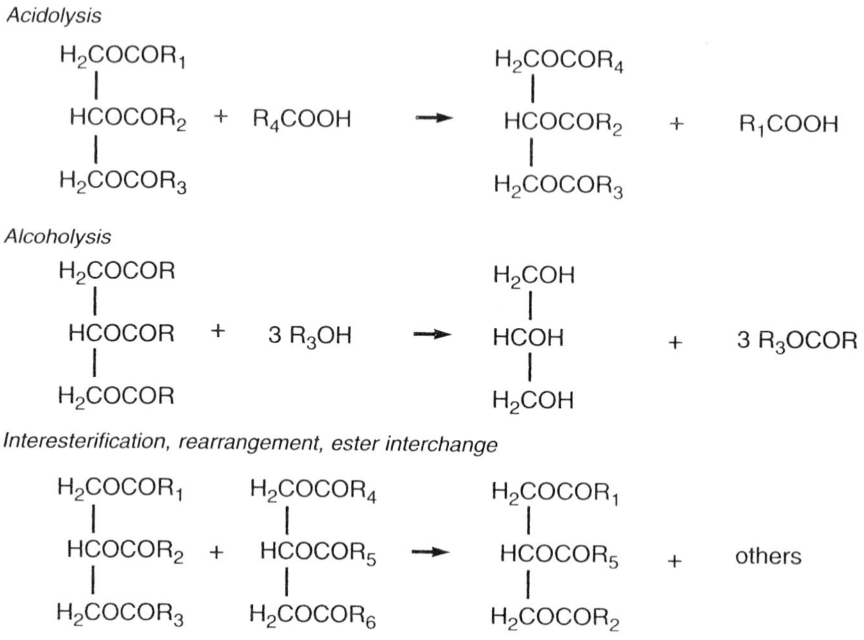

Fig. 1 Types of interesterification reactions.

from the reacting mixture, usually by crystallization (directed interesterification). This process is essentially the reverse of randomization as it tends to convert mixed triacylglycerols into high melting, crystallizable, fully saturated TAGs on the one hand and triunsaturated TAGs on the other hand:

$$SU_2 + S_2U \rightleftharpoons S_3 \downarrow + U_3$$

The term "interesterification" involves various types of reactions in which an oil, fat, or other material composed of fatty acid esters is caused to react with fatty acids, alcohols, or esters. It therefore includes (see Fig. 1)

1. The reaction of an ester with an acid, more specifically known as acidolysis
2. The reaction of an ester with an alcohol, known as alcoholysis (e.g., glycerolysis or methanolysis)
3. The reaction of an ester with another ester, referred to as interesterification, ester interchange, rearrangement, or sometimes transesterification.

Several excellent reviews on interesterification, covering both processing aspects and product applications, are available [1–11].

II. CHEMICAL BACKGROUND AND PRODUCT COMPOSITION

As already indicated, single-phase interesterification with a nonspecific catalyst or lipase will eventually lead to the so-called random distribution in which the fatty acids are statistically distributed over all possible positions (hydroxyl groups) in the system. The background to this is that changes in free energy due to fatty acid exchange between esters are negligibly small. The statistical nature of the fatty acid distribution implies that the equilibrium triacylglycerol composition and thereby the physicochemical properties of interesterified fat blends are highly predictable from the overall fatty acid composition of the mixture.

In naturally occurring oils and fats, on the other hand, the constituting fatty acids are distributed according to specific patterns. For example, in vegetable oils, saturated and long-chain monounsaturated fatty acids (such as 22:1 in the former high-erucic rapeseed oils) preferentially occupy the external positions of the triacylglycerol molecules while the 2-position is almost exclusively occupied by unsaturated fatty acids, especially linoleic but also oleic and linolenic acid. Detailed stereospecific analyses [12,13] have demonstrated that the 1- and 3-positions are not identical and may show significantly different ratios of the fatty acids attached to them. On first approximation, however, the fatty acid distribution in vegetable oils obeys the so-called 1,3-random, 2-random distribution [14,15].

In animal fats the differences between the fatty acid compositions at the 1- and 3-positions, are generally larger than in vegetable oils. Moreover, the 2-position may be enriched in saturated fatty acids, notably palmitic acid. The presence of a considerable amount of StPU triacylglycerols in lard and, as a consequence of this, a rather irregular crystallization behavior, has historically been one of the main driving forces for applying interesterification in the United States (preferably directed interesterification in this case [16]).

Returning to vegetable oils, up to a saturated fatty acid content of about 30–35%, the proportion of palmitic (P) and stearic (St) acids at the external positions can be approximated by

$$P_{1,3} = 1.47\ P_{ov}, \qquad St_{1,3} = 1.47\ St_{ov}$$

in which ov refers to the overall fatty acid composition. Saturated fatty acids will thus be present predominantly as asymmetric SUU and symmetric SUS type triglycerides, whereas fully saturated triacylglycerols will be virtually absent. Only in the case of a high overall content of saturated fatty acids such as in palm oil will part of the saturated fatty acids occupy the 2-position. As a consequence, trisaturated glycerides will be present (7–10%, mainly tripalmitin) and the SUS/SSU ratio will typically be in the range of 4–8.

Applying hydrogenation and fractionation to oils and fats will disturb the 1,3-random, 2-random distribution. Fractionation will remove the high melting triacylglycerols; e.g., the tripalmitin content will be reduced to below 1% upon fractionation of palm oil at temperatures of about 27°C or lower. Hydrogenation is unselective with respect to the position of the fatty acids in the triacylglycerol molecule; the unsaturated fatty acids will thus be attacked in proportion to their relative concentrations at the 1,3- and 2-positions. In any case, we may conclude from the above that natural oils and fats as well as their hydrogenated or fractionated derivatives have a nonrandom distribution pattern, and the same is, of course, true for their physical mixtures. Random distributions will be obtained only by interesterification (rearrangement) at temperatures above the melting point of the triacylglycerol mixture. Due to the randomization and thereby the formation of increased amounts of fully saturated TAGs, the melting point of several oils, including soybean, cottonseed, palm, and cocoa butter, is raised. On the other hand, mixtures comprising high melting components (e.g., hydrogenated oils and fats, stearin fractions, and animal fats such as lard and tallow) show a reduction in melting point and solid-phase content when interesterified with soft oils. Although randomization often leads to fats with flat melting curves and a long plastic range, a steepening of the solid-phase content versus temperature profile may occur at a sufficiently wide chain length distribution of the fatty acids present in the mixture. Tables 1 and 2 show typical examples of changes in TAG composition as a result of interesterification.

If two monoacid triacylglycerols, AAA and BBB, are interesterified, the redistribution of the fatty acids will result in an equilibrium mixture of six possible triacylglycerols (Fig. 2). At equimolar amounts of A and B, the final mixture will consist of

Table 1 TAG Composition (mol %) of Palm Oil Before and After Interesterification

		After random esterification	
Triacylglycerol	Initial	Actual	Calc.
SSS	8	13	14
SSU	9	13	13
SUS	35	25	26
SUU	36	23	24
USU	9	14	12
UUU	7	12	11

S = saturated, U = unsaturated fatty acid.
Source: Ref. 17.

12.5% of each of the parent triacylglycerols AAA and BBB
37.5% of each of the intermediate triacylglycerols A_2B and AB_2, of which two-thirds will be asymmetric and one-third symmetric

If three different fatty acids A, B, C are present in a mixture at molar fractions a, b, and c, respectively, in principle 27 different TAGs will be formed during interesterification. Disregarding stereoisomers, this number falls to 18 (Table 3). The monoacid glycerides AAA, BBB, and CCC will be present in molar amounts of a^3, b^3, and c^3. The molar fraction of each of the triacid glycerides ABC, BAC, and ACB will correspond to $2abc$. Diacid glycerides such as BBC and BCB will be present in amounts of $2b^2c$ and b^2c, respectively. When positional isomers are taken together, the amounts of each of the diacid glycerides will equal $3a^2b$, $3a^2c$, $3b^2c$, etc., and the triacid glyceride ABC will account for a mole fraction of $6abc$.

Further expressions for the theoretical composition of randomized mixtures [6] containing n different fatty acid types are given in Table 4.

The above predictions with regard to the composition of randomly interesterified fats and oils have been confirmed extensively. Classical studies include those of Norris and Mattil [18] and Naudet and Desnuelle [19], who found excellent agreement between calculated and experimental compositions for mixtures of tristearin and triolein interesterified with sodium methoxide as catalyst (Fig. 3). Other studies involving real oils and fats have led to a similar

Table 2 TAG Composition (mol %) of Lard after Random and Directed Interesterification

		Mole % TAG	
		Interesterification	
Triacylglycerol	Initial	Random	Directed
SSS	2	5	14–17
SSU + SUS	26	25	15–12
SUU + USU	54	44	32–37
UUU	18	26	39–34

Source: Refs. 4 and 17.

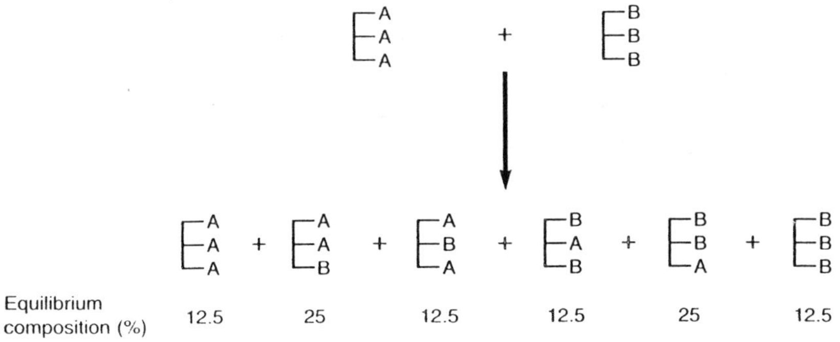

Fig. 2 Interesterification of a binary mixture (50/50) of triacylglycerols.

conclusion [20,21]. List et al. [22] recently applied interesterification between various liquid oils (cottonseed, peanut, soybean, corn, and canola) and fully hydrogenated soybean or cottonseed oil hardstocks for the manufacture of zero-trans margarine fats. Generally speaking, excellent agreement was found between the experimentally determined TAG compositions and those calculated for random distribution, from which it was therefore concluded that under the reaction conditions employed, complete randomization of the triacylglycerol structure had occurred in most cases.

Even if there are very large differences in the chain lengths of the constituting fatty acids, randomness of the triacylglycerol composition after interesterification appears to remain preserved. This follows from studies by Klemann et al. [23] on Salatrim fats, which are composed of very short chain fatty acids (C_2–C_4) and long-chain saturated fatty acids. These fats are claimed to be low in caloric value, due to poor adsorption of long-chain fatty acids on the one hand and low combustion energy of short-chain fatty acids on the other hand. Klemann et al. applied sodium methoxide catalyzed interesterification to mixtures of C_2–C_4 triacylglycerols and completely hydrogenated canola or cottonseed oils and subsequently removed unreacted short-chain triacylglycerols from the reaction mixture by steam distillation. The recovered compositions were examined by several analytical techniques and found to be in good agreement with values calculated by the statistical model for random distribution [23].

For specific applications one may aim at a maximum amount of a certain triacylglycerol or

Table 3 Random Triacylglycerol Composition of an Oil Mixture Containing Fatty Acids A, B, and C in Molar Fractions a, b, and c, respectively

Triacylglycerol	Molar fraction	Triacylglycerol	Molar fraction
AAA	a^3	CAC	ac^2
BBB	b^3	ACC	$2ac^2$
CCC	c^3	BCB	b^2c
ABA	a^2b	BBC	$2b^2c$
AAB	$2a^2b$	CBC	bc^2
ACA	a^2c	BCC	$2bc^2$
AAC	$2a^2c$	ABC	$2abc$
BAB	ab^2	BAC	$2abc$
ABB	$2ab^2$	ACB	$2abc$

Table 4 Random Triacylglycerol Composition of an Oil Mixture Composed of n Fatty Acids A, B, C, D, . . . in Molar Amounts a, b, c, d, . . .

Triacylglycerol	Number with positional isomers	Number without positional isomers	Amount without positional isomers
Monoacid (AAA, BBB, . . .)	n	n	a^3, b^3, . . .
Diacid (AAB/ABA, AAC/ACA, . . .)	$2n(n-1)$	$n(n-1)$	$3a^2b$, $3a^2c$, . . .
Triacid (ABC, ACD, . . .)	$(1/2)n(n-1)(n-2)$	$(1/6)n(n-1)(n-2)$	$6abc$, $6abd$, . . .
Total number	$(n^3+n^2)/2$	$(n^3+3n^2+2n)/6$	
$n=2$	6	4	
$n=3$	18	10	
$n=4$	40	20	
$n=5$	75	35	
$n=6$	126	56	

class of triacylglycerols after completed interesterification. From Fig. 3 it can be seen that for a pseudobinary system the monoacid triacylglycerols show a continuous decrease or increase, whereas the diacid triacylglycerols show a maximum value at fatty acid molar fractions of one-third and two-thirds, respectively. The maximum amount of these intermediate triacylglycerols equals

$$3 \times 2/3 \times 2/3 \times 1/3 = 4/9$$

or 44 mol %. For a ternary system one finds a maximum for only the triacylglycerol ABC, which is obtained at fatty acid mole fractions $a=b=c=1/3$.

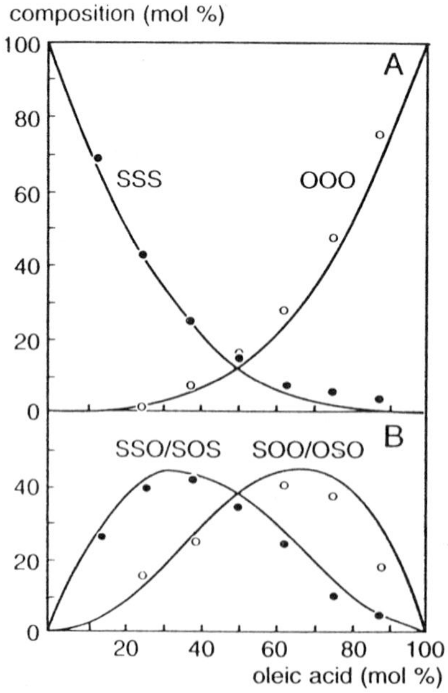

Fig. 3 Theoretical and experimental composition of interesterified triacylglycerols containing stearic and oleic acids in different proportions. (From Ref. 19.)

III. REARRANGEMENT CATALYSTS

Since the finding of interesterification of oils and fats, a very large number of potentially useful catalysts have been investigated as a means to enhance the reaction rate. Without added catalyst, some degree of rearrangement can be obtained, but only under extreme conditions of temperature and time, leading to undesirable effects such as isomerization, polymerization, and decomposition. Apart from the now generally preferred basic catalysts (e.g., sodium methoxide and other alcoholates), the following substances have been claimed to catalyze interesterification or related reactions [2,4,8,24]:

Acids, including mineral acids, sulfonic acids, cation exchangers, and zeolites
Metals and their oxides, hydroxides, and salts, e.g., the oxides of zinc, magnesium, and calcium, the hydroxides and chlorides of tin and zinc, the acetates and stearates of calcium and magnesium, aluminates of alkali metals and anion exchangers
Organometallic compounds, including acetylacetonates, organotin and titanium compounds, metal alkyl aluminates, etc.

The above list is by no means complete. Further details can be found in the cited review papers [1–9]. Generally speaking, none of these materials is sufficiently active or otherwise attractive for the interesterification of edible oils and fats. In most cases reaction temperatures of at least 200°C were required to achieve substantially complete interesterification, which would imply a risk of undesirable side reactions and traces of catalyst being left in the oil. On the other hand, the alkali metal alcoholates generally applied nowadays are sufficiently active at much lower temperatures and can also easily be washed out after completion of the reaction. Formo [25] estimated that alcoholysis catalyzed with sodium alkoxide proceeds at least 4000 times as fast as with an equivalent amount of hydrochloric acid.

As so many substances show at least some catalytic effect at elevated temperatures, their presence should be minimized in cases where even a slight degree of interesterification is undesirable, e.g., during physical refining of olive oil and palm oil for the production of palm midfraction. In both cases partial inter- or intraesterification will result in an increased amount of saturated fatty acids at the 2-position and a shift in the ratio of symmetric to asymmetric TAGs. Willems and Padley [26] studied the effects of temperature and time on the SOS/SSO ratio of palm oil; their results are reproduced in Fig. 4. A distinct drop in the above ratio from about 8 to 6 was observed after holding times of a few hours at 240°C. Such changes, although significant when palm oil is to be used as a feedstock for sophisticated fractionation products, appear to be irrelevant for oils and fats to be applied in spreads and shortenings.

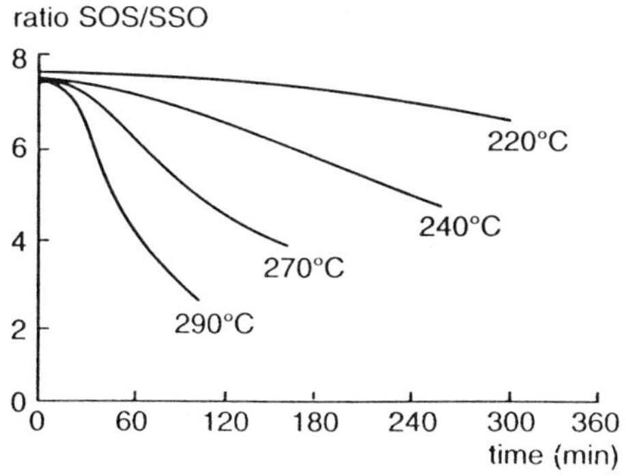

Fig. 4 Effect of heating time and temperature on the SOS/SSO ratio of palm oil. (From Ref. 26.)

Palm oil, because of its relatively high content of diacylglycerols, is probably also more sensitive to partial interesterification than most other oils and fats.

As indicated above, alkaline catalysts are now generally preferred for interesterification of oils and fats. The following compounds have been proposed and/or actually used:

Alkali metals	Na, K, Na-K alloys
Alkoxides (alcoholates)	$NaOCH_3$, $NaOC_2H_5$
Hydroxides	NaOH, KOH
Glycerolates	From NaOH + glycerol
Others	E.g., NaH, $NaNH_2$, NaR (R = alkyl or phenyl)

Most of these allow a fast rearrangement at temperatures <100°C and—provided the feedstock is sufficiently low in moisture, free fatty acids, and oxidized compounds—at very low catalyst doses, e.g., down to 0.1% [2,8] or even 0.05% [11] of sodium methoxide or ethoxide.

Among the alkali metals proposed for interesterification, metallic sodium for random rearrangement [27] and Na-K alloy for directed interesterification [16,28] are best known. Sodium-potassium alloys are liquid at room temperature (the system is eutectic with a melting point of −12°C at the molar composition NaK_2) and can therefore easily be dispersed at the low reaction temperatures required for directed interesterification with crystallization as the driving force. Metallic sodium has a melting point of 98°C and has been dosed by pressing the soft metal through orifices into a stream of preheated oil in which the sodium wires are melted off, followed by further reduction of droplet size in a high shear mixer to ensure

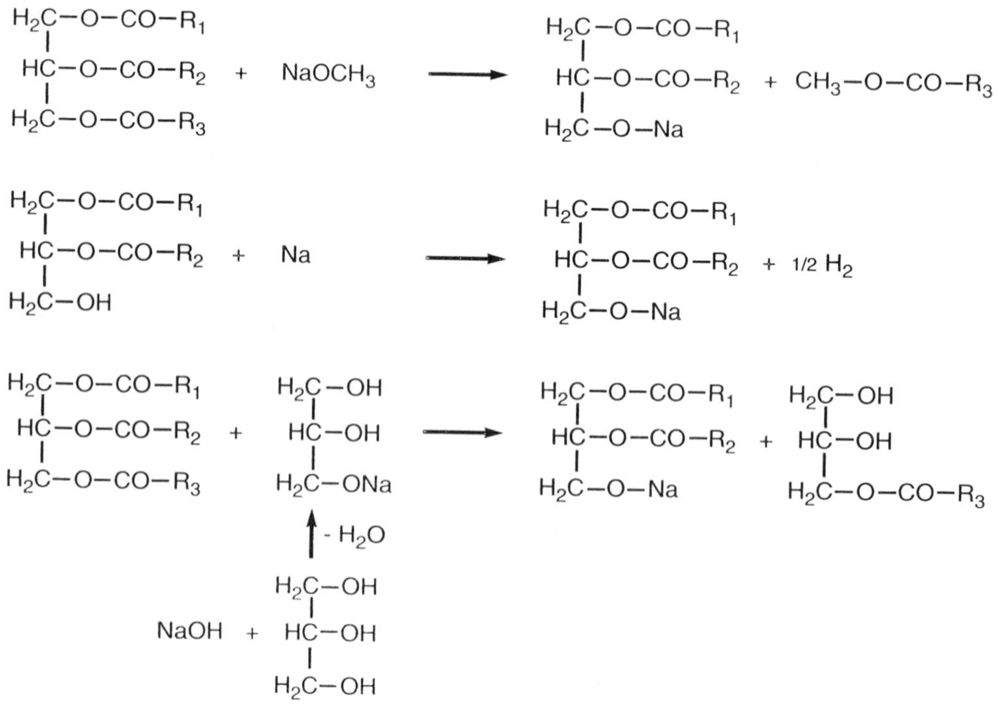

Fig. 5 Formation of sodium diacylglycerolate from various precatalysts.

effective reaction. Application of metallic catalysts also has a number of disadvantages, i.e., high reactivity, the need for extensive safety precautions during storage and in use, generation of some hydrogen during catalyst activation (see Section IV. A and Fig. 5), and risk of ignition of unreacted metal during catalyst inactivation with water after completion of the reaction.

Alcoholates (sodium methylate, ethylate, or glycerolate) are therefore most widely applied in current rearrangement processes. Sodium or potassium glycerolate is preferably formed in situ from an aqueous solution of the corresponding hydroxide and glycerol, either in a two-stage batch process with catalyst addition and drying at low temperature followed by interesterification at higher temperature [29] or as a one-stage continuous process in which catalyst addition, activation, and rearrangement are all carried out at the same temperature [30]. Sodium and potassium hydroxide, without addition of glycerol, are distinctly less active, and in general temperatures of at least 160–180°C appear to be necessary.

Sodium ethylate or methylate can in principle be dosed in several ways, e.g., as a powder, dispersed in inert solvent or fat, or dissolved in the corresponding alcohol (e.g., as 25% solution). For industrial use, dosing as dry powder is the most practical method. Application of alcoholic solutions will lead to very high oil losses, because most, if not all, of the added solvent is converted into ethyl or methyl esters. Predispersion in liquid oils should be avoided or handled with care, as high catalyst concentrations may cause heating up and side reactions.

IV. MECHANISM OF ALKALI-CATALYZED INTERESTERIFICATION

A. Overall Description

Several stages can be distinguished in the total interesterification process:

1. Formation of the genuine interesterification catalyst by gradual dissolution of the added catalytic material into the oil (induction period of the reaction)
2. Interesterification by exchange of fatty acids between triacylglycerols and the genuine catalyst (maximum reaction rate period)
3. Declining rate period due to approach of the equilibrium (random) distribution.

Much evidence is now available to prove that the genuine catalyst responsible for fatty acid exchange in current interesterification processes of oils and fats consists of the sodium or potassium compound of a diacylglycerol molecule, i.e.,

H_2CONa

|

HCOCOR

|

H_2COCOR

This species is structurally similar to simple alcoholates (sodium methoxide or ethoxide) but soluble in oil. Interesterification is thus a homogeneously catalyzed reaction, with the active catalyst being formed in the initial stages of the process by reaction of the added (pre-)catalyst with

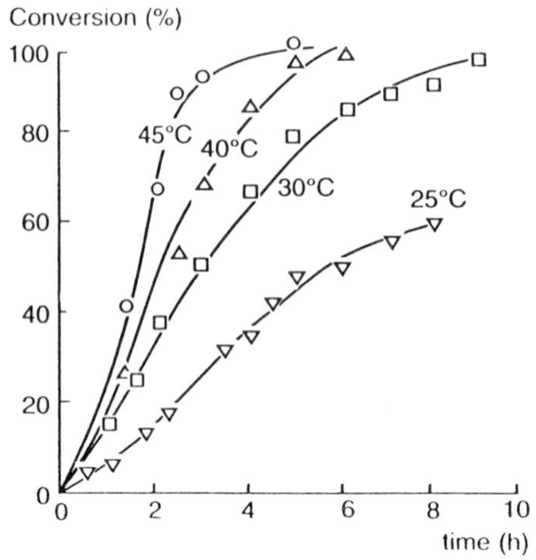

Fig. 6 Kinetics of interesterification of methyl myristate and butyl palmitate. (From Ref. 31.)

tri- and/or diacylglycerols from the oil. The genuine catalyst is identical for all types of precatalysts; only the way it is formed during the catalyst activation period is different (Fig. 5).

Several kinetic studies support the above view of catalyst activation as a necessary first step in interesterification reactions. To minimize analytical complexity, most of these early kinetic studies were carried out in model systems. Jordan de Urries and Utrilla [31], for example, studied the interesterification of methyl myristate and butyl palmitate using sodium metal as catalyst in the 25–45°C temperature range. Their results (Fig. 6) indicate that the degree of conversion (relative to the equilibrium composition) goes up rapidly with increasing reaction temperature, and, especially at the lower temperatures, there is evidence that there is an induction period. In a similar study by Coenen [7] into the interesterification of glycol esters of short-chain fatty acids, very pronounced induction periods were observed; the actual redistribution reaction appeared to be fast even at temperatures as low as 30–40°C (Fig. 7). A third example refers to palm oil where the solids content at 40°C was taken as a measure of the degree of interesterification (Fig. 8). Again it can be seen that the reaction speeds up at higher temperatures and that at lower temperature there is an induction period.

All these data strongly suggest that the added catalysts were indeed only precursors for the genuine interesterification catalyst, which is slowly formed during the initial stages of the reaction. This was confirmed by the fact that no induction period is observed if the catalyst is

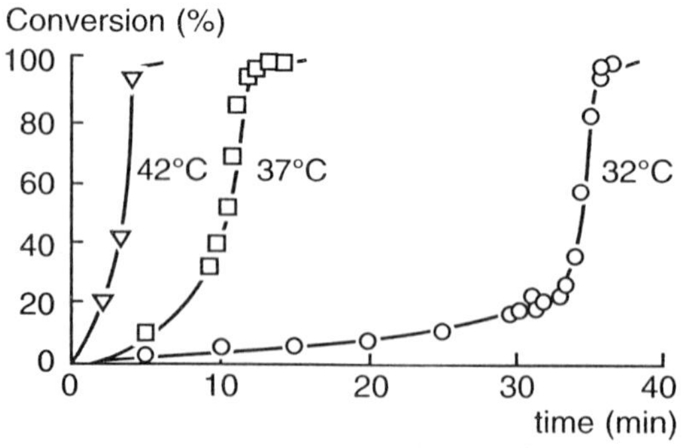

Fig. 7 Kinetics of interesterification of glycol esters of C_6 and C_8 fatty acids. (From Ref. 7.)

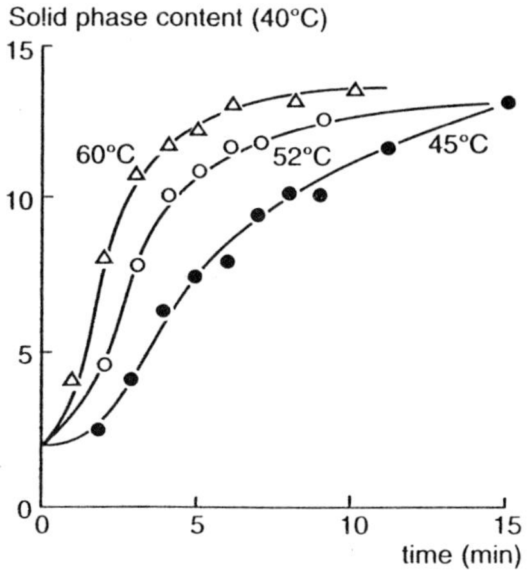

Fig. 8 Kinetics of interesterification of palm oil at these different reaction temperatures (60°C, 52°C, and 45°C). (From Ref. 7.)

predissolved in part of the oil or dosed as a molecular solution of, e.g., sodium methylate in methanol [7].

Due to the heterogeneous nature of the catalyst activation step, the length of the induction period will depend on numerous factors, including

1. Type of catalyst and its physical state (liquid, solid)
2. Particle size and size distribution
3. Input of mechanical energy into the system
4. Oil impurities that affect the surface condition of the catalyst particles

In the actual practice of oil and fat interesterification, the induction phenomenon may be virtually absent, due to effective dispersion of catalyst, well refined and dried oil quality, and small catalyst particle size. Temperature is a factor also, because the activation energy for the formation of the genuine interesterification catalyst is higher than that of the interesterification reaction itself.

Coenen [7] established that the interesterifications of esters of mono-, di-, and trihydric alcohols as depicted in Figs. 6, 7, and 8, respectively, can all be described (after the induction period) by first-order kinetics and activation energies of 13–15 kcal/(mol · °C).

Weiss et al. [32], in their study of sodium methoxide catalyzed rearrangement of lard, in fact distinguished three reaction phases:

1. Active catalyst formation
2. Crystal modification (associated with a reduction of S_2U content)
3. Interesterification (approach to random distribution)

The activation energy for the first phase, i.e., formation of the active catalyst, was estimated at 26 kcal/(mol · °C) against 14–17 kcal/(mol · °C) for the other two phases, which is in good agreement with Coenen's data.

B. Detailed Mechanism

Several investigators [2,6–8,32–36] have attempted to understand the mechanisms involved in catalytic interesterification, and it appears now to be almost generally accepted [7,11,34,35]

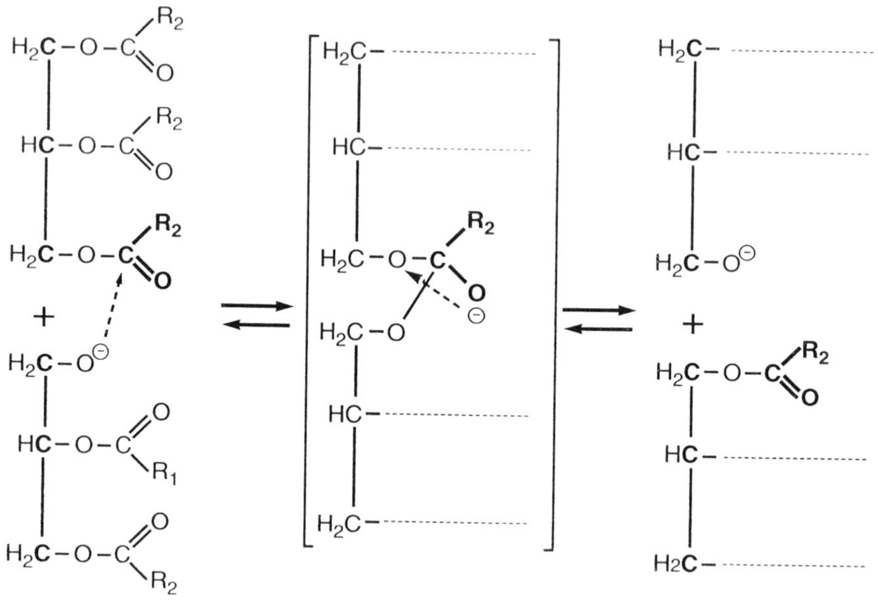

Fig. 9 Mechanism of interesterification. (From Ref. 7, 11, 34, 35.)

that the fatty acid exchange occurs by interaction of triacylglycerols with the genuine catalytic species (sodium or potassium compound of diacylglycerols) according to the mechanism schematically shown in Fig. 9. This is a so-called bimolecular, nucleophilic substitution reaction, in which the slightly negatively charged diacylglycerol anion with fatty acids R_1 is supposed to attack on the slightly positive carbonyl carbon atom of one of the fatty acids R_2 of the triacylglycerol molecule. In this way a transition complex is formed that can subsequently decompose to the original compounds or form a new triacylglycerol molecule ($R_2R_1R_1$) and a new active catalyst ion ($R_2R_2O^-$). This anion is then available for reaction with other triacylglycerols, until eventually the reaction reaches an equilibrium in which fatty acid exchanges still continue but do not lead to further net changes in composition.

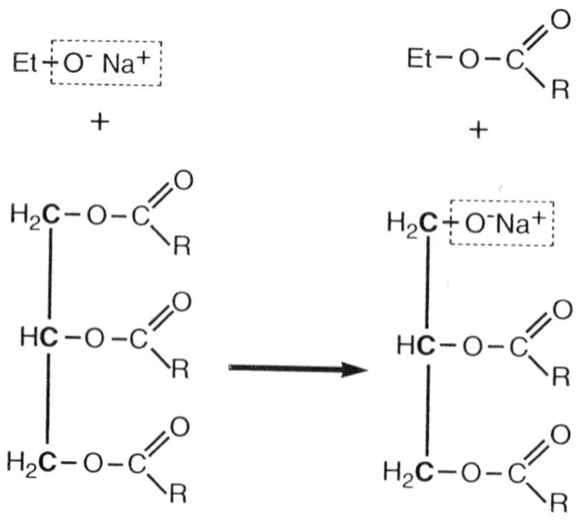

Fig. 10 Mechanism of catalyst activation. (From Ref. 11, 33, 34.)

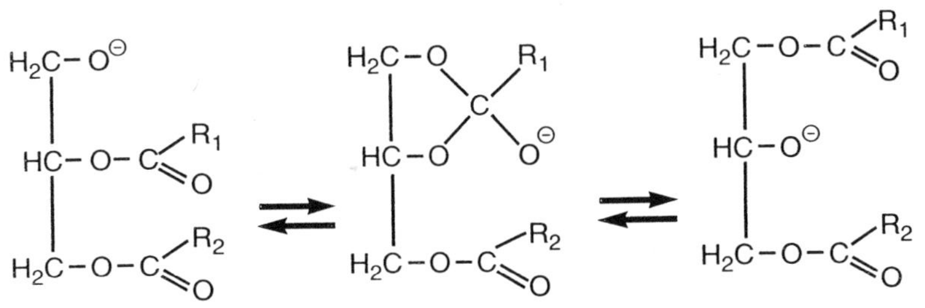

Fig. 11 Mechanism of intramolecular interesterification. (From Ref. 33.)

The mechanism of catalyst activation can be described in a very similar way. For example, with sodium ethoxide as catalyst, the ethoxide anion will attack on one of the carbonyl carbon atoms of a triacylglycerol molecule, thus forming the catalytically active diacylglycerol anion and an equivalent amount of fatty acid ethyl ester (Fig. 10).

Intramolecular interesterification, i.e., fatty acid exchange within the same molecule, is assumed to proceed more quickly [37] than the intermolecular reaction, but again the mechanism appears to be quite similar (Fig. 11).

Weiss et al. [32] proposed a slightly different mechanism, the first step of which is the formation of an enolate ion by hydrogen abstraction from the CH_2 group adjacent to the carbonyl group (Fig. 12). This enolate ion would then interact with the slightly positive carbonyl carbon atom to produce a β-ketoester intermediate and, through a number of other steps, an inter- or intramolecularly rearranged triacylglycerol (Fig. 13). Infrared spectral data were in agreement with the proposed mechanism, although it was shown later [8] that the measured IR adsorptions could also be attributed to other compounds such as soaps. Further evidence against this mechanism was obtained independently by van der Plank [36] and Hei-

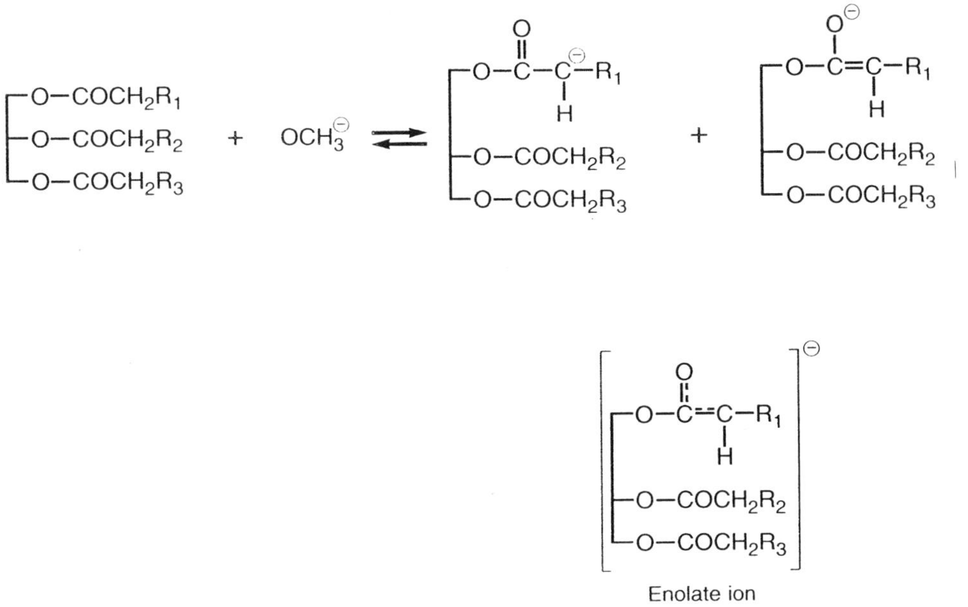

Fig. 12 Enolate ion formation as proposed by Weiss et al. [32].

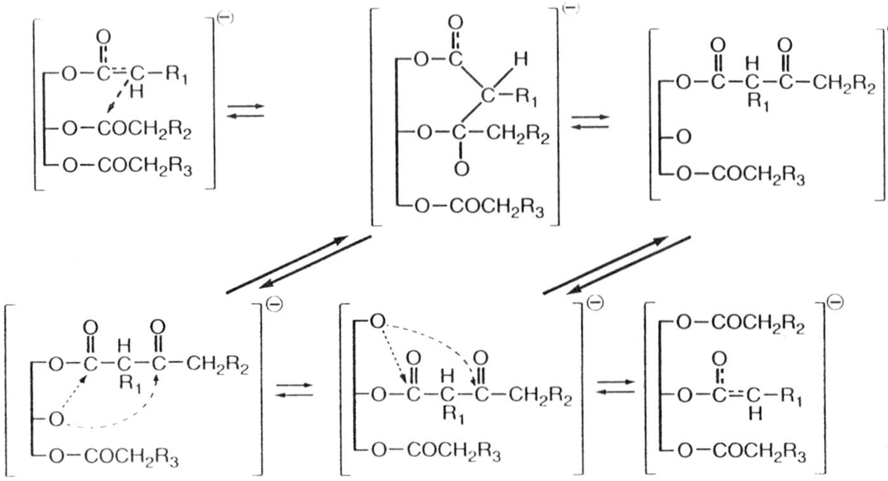

Fig. 13 Mechanism of (intra-)molecular interesterification as proposed by Weiss et al. [32].

dal and Mørk [35], who showed that the β-ketoester anion is so strongly resonance-stabilized that it is unable to catalyze interesterification reactions.

Strong support for the mechanism involving sodium diacylglycerolate as active species has been provided by Steenhoek (cited in Refs. 7 and 11), who made a detailed study into the kinetics of interesterification of the glycol esters of saturated C_8 and unsaturated C_{10} fatty acids with predissolved catalyst. The system was described by a set of six equations between all possible triacylglycerols and all possible diacylglycerol anions (Fig. 14). Each equation was considered as an equilibrium reaction with in principle the same rate constant $3K$. Corrections, however, were applied to account for the fact that only part of the exchange reactions result in a net change of composition. Therefore, effectively the rate constants of some of the

$$S_3 + U_2ONa \underset{K}{\overset{3K}{\rightleftharpoons}} SU_2 + S_2ONa$$

$$U_3 + S_2ONa \underset{K}{\overset{3K}{\rightleftharpoons}} S_2U + U_2ONa$$

$$SU_2 + U_2ONa \underset{3K}{\overset{2K}{\rightleftharpoons}} U_3 + SUONa$$

$$S_2U + S_2ONa \underset{3K}{\overset{2K}{\rightleftharpoons}} S_3 + SUONa$$

$$S_2U + U_2ONa \underset{K}{\overset{2K}{\rightleftharpoons}} SU_2 + SUONa$$

$$SU_2 + S_2ONa \underset{K}{\overset{2K}{\rightleftharpoons}} S_2U + SUONa$$

Fig. 14 Kinetics of interesterification via diacylglycerolate as catalytic species. (From Ref. 7, 11.)

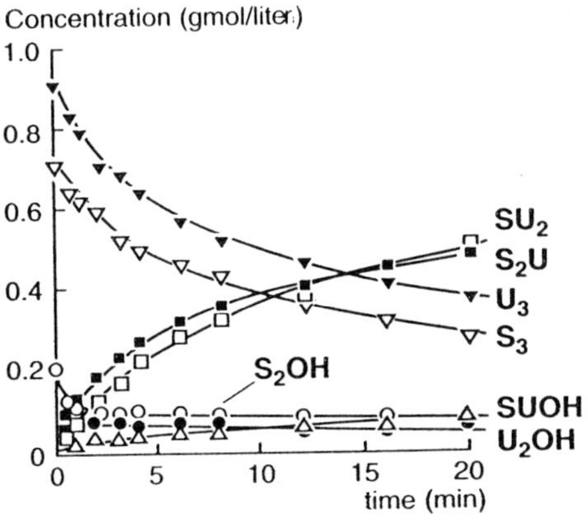

Fig. 15 Calculated and experimental triacyl and diacylglycerol compositions for reaction of S_3 and U_3 with predissolved catalyst. (From Ref 7, 11.)

reactions are equal to *K* or 2*K*. Results of this simulation study are given in Fig. 15, which shows excellent agreement between calculated and experimental data not only for the triacylglycerols but also for the diacylglycerol anion composition, thus clearly supporting the proposed mechanism.

Two concluding remarks need to be made in this section on the kinetics and mechanism of interesterification. The first refers to the characteristic brown color that usually develops in the early stages of the reaction and is generally seen as an indication that the reaction is likely to proceed to completion. The nature of the coloring compound, however, does not seem to have been unambiguously identified. The second comment refers to a certain regioselectivity effect as demonstrated by Konishi et al. [39], who found that under specific reaction conditions the ester interchange rate at the 1,3-positions may be up to a factor of 2 faster than at the 2-position. The system investigated was the sodium methoxide catalyzed interesterification of soybean oil with methyl stearate in hexane solvent. Regioselectivity was observed at 30–40°C but disappeared at 60°C. A kinetic factor, i.e., steric hindrance of the fatty acids attached to the central position of the glycerol molecule, is likely to be responsible for the observed regioselectivity.

V. DIRECTED INTERESTERIFICATION

Directed interesterification, as first described in some detail by Eckey [40], can be a valuable extension to random rearrangement. The directed process can, for example, be applied for

1. Extending the plastic range of fats and improving their consistency properties
2. Preparing oleins from oils like palm oil for use as salad or frying oil
3. In situ preparation of hardstock components from liquid oils such as sunflower and cottonseed oils
4. Producing low SAFA oils by directed interesterification and subsequent fractionation

The directing effect is always due to the presence of a second phase (gas or solid), into which reaction products are transported and thus withdrawn from the liquid phase in which fatty acid rearrangement proceeds to continually restore the disturbed equilibrium. One example of the removal of reaction products via the gas phase is described by Schmulinzon et al.

[41], who interesterified hydrogenated coconut oil with longer chain fatty acid methyl esters in vacuum at 140°C. The low boiling esters of caprylic and capric acid were thus continuously distilled off during the reaction. Another example is given by Barsky [42], who, by simply heating, replaced short-chain fatty acids from a fat with free fatty acids having at least two carbon atoms more than the fatty acids to be displaced. However, the liquid–solid system in which high melting triacylglycerols are forced to crystallize by application of sufficiently low temperatures is by far the most important system for practical directed interesterification.

Directed interesterification is a relatively slow process because of (1) the decreased rate of reaction as a result of the low temperatures required for crystallization and (2) the small driving force for crystal growth during the process. Depending on the fatty acid composition of the system and the applied conditions of temperature and catalyst concentration, either the interesterification or the crystallization will be the principal rate-determining step. For instance, an oil containing 85% unsaturated and 15% saturated fatty acids will have the following TAG composition at random distribution:

$$U_3 = 61\%, \quad SU_2 = 33\%, \quad S_2U = 6\%, \quad S_3 = 0.3\%$$

The actual concentration of S_3 and, depending on the degree of cooling, that of S_2U will exceed the solubility of these triacylglycerols when the randomized mixture is cooled down to the directed interesterification temperature, resulting in crystallization of S_3 and S_2U, or their mixed crystals. The reduced concentration of these TAGs will then be compensated for by the reactions

$$SU_2 + SUNa \rightleftharpoons S_2U \downarrow + U_2Na$$

$$S_2U + U_2Na \rightleftharpoons U_3 + S_2Na$$

$$SU_2 + S_2Na \rightleftharpoons S_3 \downarrow + U_2Na$$

In this example, temperatures as low as 0–10°C will be required to effect crystallization. With feedstocks like lard and palm oil, the S_3 content after randomization will be on the order of 5–14%, and considerably higher crystallization/reaction temperatures can be applied. In that case directed interesterification will result in both SU_2 and S_2U being reduced in concentration [16,43].

To compensate as much as possible for the low reaction temperatures, highly active catalyst types such as sodium-potassium alloy [16,28] or sodium ethylate or methylate must be applied for directed interesterification. The catalyst activation step is normally carried out at somewhat elevated temperatures, typically 40–60°C, to avoid lengthy induction periods (see Section IV.A).

Adding small amounts of glycerol to oils and fats appears to be beneficial for speeding up directed interesterification [44], presumably due to the fact that saturated mono- and diacylglycerols are less soluble in oil than the corresponding triacylglycerols. Various other methods to enhance the rate or degree of completion of the process have been used also, e.g., a gradual

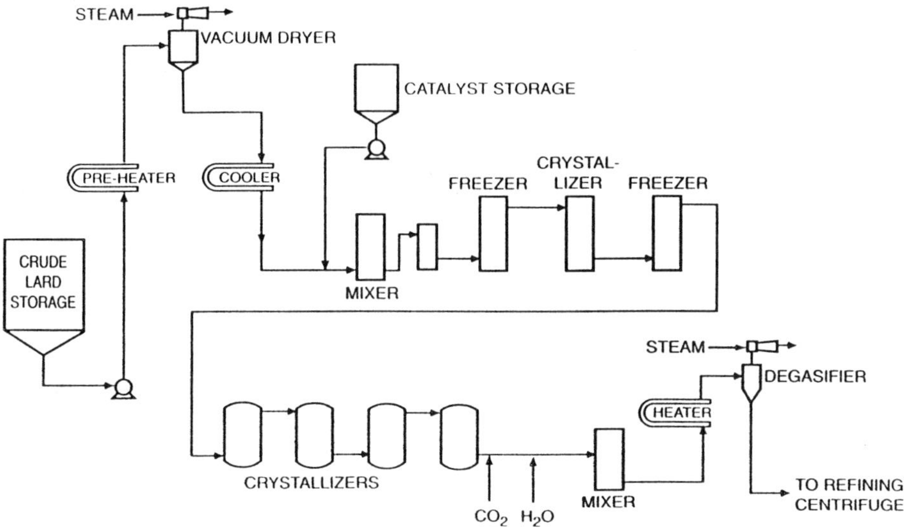

Fig. 16 Flow chart of directed interesterification process of lard. (From Ref. 16.)

reduction of temperature during the directed interesterification [40] and a rapid chilling from the catalyst activation temperature to the directed interesterification temperature [16,45] to create a large number of crystal nuclei on which further crystal growth is facilitated. Periodic variation of temperature (cycling) between an upper and a lower temperature has also been claimed [45,47] to enhance the overall rate of the coupled interesterification–crystallization processes.

Determinations of triacylglycerol composition after directed interesterification of lard [48] have shown that formation of fully saturated TAGs is favored at relatively high temperatures (20–38°C), whereas at lower temperatures (0–10°C) S_2U formation is more important. In this fraction the asymmetric triacylglycerols also dominate the symmetric ones. The rate-enhancing effect of a stepwise reduction of temperature was confirmed in this study.

A flow diagram for continuous directed interesterification of lard [4,16] is reproduced in Fig. 16. Liquid Na-K alloy catalyst is metered into a stream of dried oil and dispersed to a droplet size below 50μm in a high shear mixer. The oil–catalyst mixture is then rapidly cooled in a scraped surface heat exchanger to induce nucleation and crystallization in a first crystallizer. Owing to the liberation of crystallization heat, a second cooling step is necessary, after which the oil flows through a series of crystallizers to bring the process to the required degree of completion. The design of the crystallizers must ensure a gentle movement throughout the crystal dispersion, preventing any stagnant zones.

VI. RANDOM INTERESTERIFICATION—PROCESSING ASPECTS

A. Feedstock Requirements

Prior to interesterification, oils should be thoroughly pretreated and dried to remove any catalyst-poisoning substances such as free fatty acids, moisture, and oxidized compounds (peroxides) as far as reasonably possible. The reactions of FFA and moisture with strongly alkaline catalysts such as sodium methylate, ethylate, or diacylglycerolate (the genuine active catalyst, see Section IV.B) occur quantitatively:

$$\text{Nadi} + H_20 \rightarrow \text{NaOH} + \text{diacylglycerol}$$

$$CH_3ONa + H_2O \rightarrow NaOH + CH_3OH$$

$$CH_3ONa + RCOOH \rightarrow RCOONa + CH_3OH$$

Consequently, each 0.01% moisture in the feedstock will inactivate about 0.03% (by weight on oil) sodium methylate. Similarly, 0.1 wt % free fatty acids is equivalent to an additional catalyst requirement of 0.02 wt %. For peroxides also an equimolar reaction has been proposed [6,8], implying an additional catalyst requirement of about 0.03 wt % for each 5 POV units, but this may be somewhat overestimated.

Based on the above, the following feedstock specifications are recommended:

Free fatty acids	<0.1%, preferably <0.05%
Moisture	<0.01%
POV	<10, preferably <5 mgaeq/kg

The required FFA level can easily be obtained by conventional neutralization processes. Bleaching is not generally required except when the peroxide value is too high and color specifications of the final product are very tight.

Predrying is most effectively carried out by (1) spray drying in a vacuum chamber (continuous processes) or (2) batch drying with oil recirculation and spray drying into the evacuated headspace. Some guidance for the required (minimum) drying conditions can be derived from Fig. 17, which shows equilibrium moisture contents as a function of oil temperature and pressure. Under optimal conditions, 0.05–0.06 wt % sodium ethylate or methylate, i.e., 0.5–0.6 kg catalyst per tonne of oil, is sufficient to achieve complete interesterification.

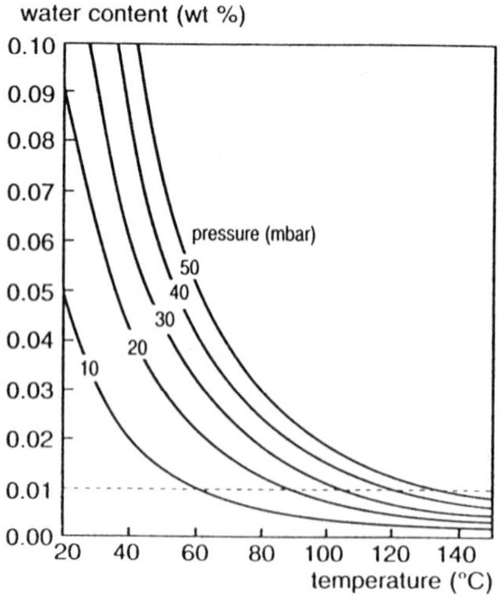

Fig. 17 Equilibrium moisture contents of oils and fats as a function of temperature and pressure.

B. Process Options

In the literature, many types of random interesterification processes have been described:

1. Batch, with sodium alcoholates (CH_3ONa, C_2H_5ONa) or glycerolates (in situ from NaOH or KOH and glycerol) as catalyst
2. Continuous, with sodium [27], predispersed sodium methylate [4], aqueous sodium hydroxide [4,49], and aqueous sodium hydroxide plus glycerol [30] as catalyst
3. Semicontinuous, again catalyzed by the sodium hydroxide and glycerol combination [50]

1. Batch Processes

The sodium or potassium hydroxide–glycerol catalyzed batch process [29,51] consists of essentially two stages that are both carried out in vacuum. The first stage is carried out at about 60–80°C and comprises the steps of catalyst addition, catalyst dispersion, and removal of most of the water added with the catalyst solution. Vigorous boiling and use of a sufficiently dilute catalyst solution (e.g., weight ratio NaOH:glycerol:water = 1:2:7) are important factors for obtaining good catalyst dispersion. In the second stage of the process, the oil–catalyst mixture is heated up to first form sodium glycerolate, then sodium diacylglycerolate (Fig. 5), the active species that catalyzes the rearrangement reaction.

Sodium ethoxide or methoxide catalyzed batch interesterification is applied more widely. The process can be carried out in conventional neutralizing-bleaching equipment [8,9] provided with adequate facilities for catalyst dosing and oil drying. As explained in Section VI.A, the neutral oil must first be dried to a moisture content below 0.01 wt %. Care should be taken to prevent any subsequent ingress of water; i.e., valves and coils should be absolutely leak-free. Then the catalyst is sucked in, preferably from a nitrogen-flushed dosing funnel, through a dip pipe that ends below the oil surface near the tip of the stirrer to ensure optimum catalyst dispersion and fast formation of the genuine active catalyst. The generally preferred reaction and catalyst addition temperature is between 80 and 100°C, and the allowed reaction period is normally between 10 and 30 min, although in many cases the actual rearrangement may be largely completed in a few minutes.

Sodium alcoholates are toxic and highly reactive materials and should therefore be handled with care. Sodium methylate normally has a higher bulk density, a longer shelf life, and a somewhat coarser, more uniform particle size distribution than sodium ethylate. It causes less dust and is easier to handle. For both catalysts it is essential to prevent deterioration of activity by avoiding contact with moisture and air. The catalysts should therefore be stored under cool and dry conditions in closed bags and containers until the moment of use. The contents of the catalyst bags should be such that an integral number of them are used per interesterification batch; the presence of half-empty bags in the refinery should be avoided. When emptying catalyst bags, safety goggles, powder masks, and gloves should be worn. Fires should never be extinguished with water, but with sand and chemical foam.

2. Continuous Processes

The plant designs that have been proposed or actually used for continuous interesterification are in principle all rather similar:

1. A suitable amount of catalyst is added to a purified and preheated oil stream.
2. Measures are taken to obtain a fine dispersion of catalyst particles.

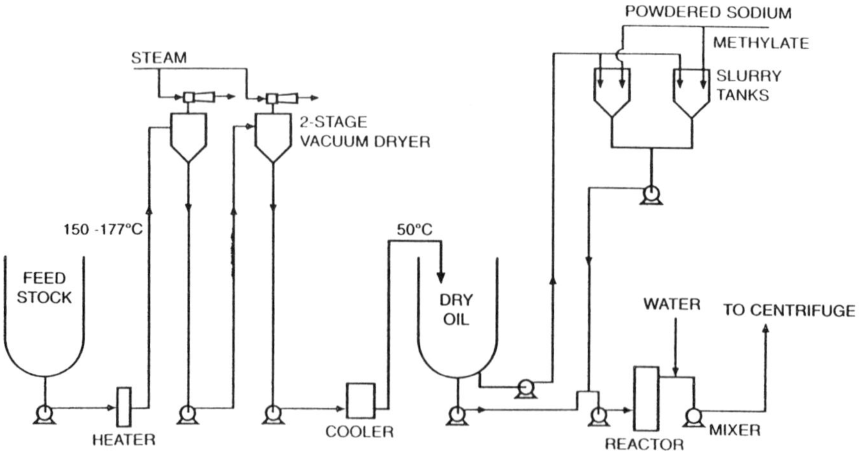

Fig. 18 Flow chart of continuous interesterification process with sodium methylate catalyst. (From Ref. 5.)

3. Sufficient residence time is provided to allow for catalyst activation and fatty acid rearrangement.
4. The catalyst is inactivated (below 100°C) by addition of water followed by separation of the oil and aqueous phases.

The various processes differ mainly with respect to the applied catalyst and the associated reaction temperature and regarding the relative importance of thorough drying before or after catalyst addition. Two examples are shown in Figs. 18 and 19.

The required reaction temperatures are lowest for alkali metals [27] and predissolved metal alcoholates [4], but, as indicated in Section III, these highly active catalysts also have some disadvantages in terms of safety of handling and stability in use. Sodium hydroxide, generally

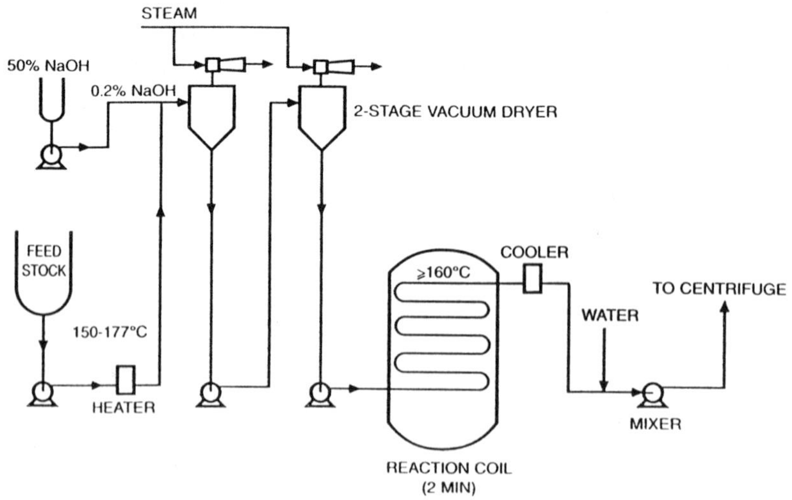

Fig. 19 Flow chart of continuous interesterification process with aqueous sodium hydroxide as catalyst. (From Ref. 5.)

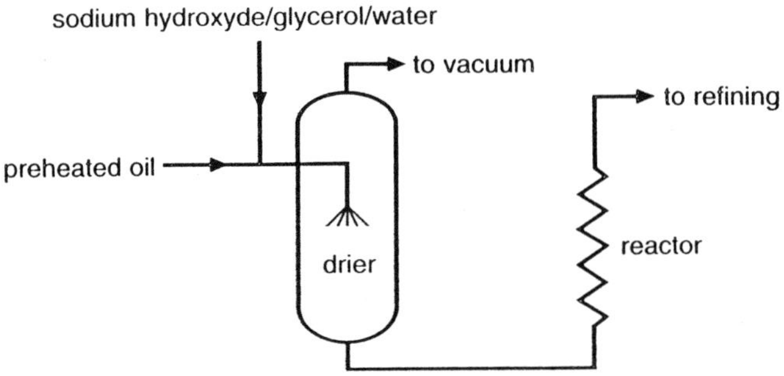

Fig. 20 Principle of continuous sodium hydroxide–glycerol interesterification process.

applied as a 50% aqueous solution, has a rather low activity, and reaction temperatures in the range of 160–180°C therefore appear to be necessary. The combination of sodium hydroxide and glycerol is intermediate in activity, and the preferred reaction temperature is therefore about 125–140°C [30].

It is essential that immediately after these aqueous catalyst solutions have been dosed, the mixture of oil and catalyst be thoroughly dried to prevent loss of catalytic activity and excessive oil losses due to saponification. This is most conveniently achieved by spray drying in a vacuum chamber (Fig. 20). In this unit also the genuine interesterification catalyst is formed. The actual redistribution reaction is very fast and can be completed in a few minutes. It is preferably carried out in a coil reactor with a narrow residence time distribution to minimize backmixing effects. A specific feature of this process is that the feedstock need not be carefully neutralized. Up to a certain level of free fatty acids, neutralization and interesterification can be combined in a single operation just by increasing the amount of sodium hydroxide solution added to the oil.

Glycerol in the catalytic solution has been found [11] to suppress the saponification reaction that is always in competition with the formation of the genuine active catalyst. This is due to the fact that the equilibrium reaction

$$\text{glOH} + \text{OH}^- \rightleftharpoons \text{Glo}^- + \text{H}_2\text{o}$$

lies far to the right-hand side, as a result of which the OH^- concentration is lower than in systems without glycerol.

C. Posttreatment and Process Yield

All interesterification processes with alkaline catalysts have in common that on completed reaction the catalyst is inactivated by addition of water:

$$\text{Na diacylglycerolate} + H_2O \rightarrow \text{NaOH} + \text{diacylglycerol}$$

The sodium hydroxide thus formed may cause oil losses due to saponification, especially when the separation of oil and aqueous phase is delayed, as may be the case in batch processing. Hawley and Holman [16] showed that coaddition of some acid (citric, phosphoric) or carbon dioxide minimizes such saponification losses. Further refining steps include conventional washing, bleaching, and deodorization.

As a first approximation one may assume that all added catalyst is ultimately converted into soap. At a catalyst dose of, say, 0.06 wt % sodium methoxide, the theoretical oil loss as soaps is therefore as low as 0.3%. In practice somewhat higher losses will be observed owing to saponification and entrainment of neutral oil with the water phase (more significant in batch settling processes than in the case of continuous centrifugal separation). The loss as ethyl or methyl esters is also equivalent to the amount of catalyst used, again typically 0.3% at a catalyst dose of 0.06 wt % on oil. Naturally, the actual process step in which the latter loss occurs is the deodorization, not the washing stage. Sodium hydroxide from NaOH-glycerol catalyst will also more or less quantitatively be converted to soaps (hence 0.1% NaOH will yield about 0.7% soaps); the glycerol will be recovered in the rearranged oil as diacylglycerols.

D. Interesterification Control Methods

A large number of analytical methods can be used [6,8,9,52] to establish whether an interesterification reaction has proceeded to the required degree of completion. The degree of interesterification is defined as

$$\alpha = \frac{P - P_0}{P_e - P_0}$$

in which P_0 and P_e represent the value of a characteristic property before and after interesterification (equilibrium situation) and P the actual value. The property P can be either a chemical or a physical parameter. It should be realized that not all parameters vary in the same way during an interesterification process, hence the value of α will to some extent depend on the parameter that is chosen for evaluation. However, these mutual differences become less important when the degree of interesterification approaches 1 (virtually complete randomization, the normally desired situation).

Chemical methods include

1. Determination of the fatty acid composition at the 2-position (pancreatic lipase method). On full randomization the 2-position composition will be the same as the overall composition.
2. Determination of the triacylglycerol composition by chromatographic methods (TLC, HPLC, capillary gas chromatography, or combinations thereof). A characteristic triacylglycerol or a certain class of related triacylglycerols must be selected for quantification of α.
3. Determination of the free sterol content, for example, by TLC. This is a semiquantitative method based on the observation that sterols take part in the fatty acid exchange process at a rate comparable to that of triglycerides. Virtual disappearance of free sterols is thus an indication that TAG interesterification is complete. (Note: Tocopherols also participate in the fatty acid exchange process but at a considerably lower rate. Hence, interesterification will reduce the free tocopherol content, which may have an effect on the oxidative stability but not on the vitamin E activity.)

Physical methods for estimating the degree of interesterification are generally based on changes in melting properties or crystallization behavior and therefore include determination of

1. Melting point or drop point.
2. Solid-phase content as a function of temperature (NMR).
3. Differential scanning calorimetry.
4. Turbidity. In this case the fat is first melted to destroy all crystal nuclei, then cooled

under standardized conditions to a fixed, optimum temperature (dependent on the nature of the mixture). Crystallization will start after a certain period, and the development of turbidity is determined photometrically.
5. Cooling curve (similar principle as for turbidity, but assessment now based on deflection of cooling curve due to liberation of heat of crystallization).

For all these physical methods it is essential that the measured parameter value be compared to those of reference samples of identical composition before and after complete interesterification (e.g., prepared on laboratory scale with excess catalyst).

Several of the above methods, although suitable for a final assessment, are too complex or time-consuming for process control and application on the factory floor. For the latter purpose the following strategy is recommended:

1. Ensure that all critical feedstock and process parameters are clearly specified and adhered to (e.g., temperature, time, and vacuum conditions for oil drying; catalyst quality and dosing rate; etc.)
2. Control end point of reaction by using a quick and simple method (e.g., melting point or crystallization-turbidity test method).

VII. ENZYMATIC INTERESTERIFICATION

A. Introduction

During the past 15 years there has been much interest in the development of enzyme-catalyzed processes for the interesterification of oils and fats. The selectivity of enzymatic catalysis can be exploited to generate tailored TAGs that are difficult to obtain by conventional chemical or physical methods, and the mild conditions used for enzymatic interesterification enable reactions with unstable materials, such as polyunsaturated fatty acids, to be run without the generation of by-products associated with the use of more severe chemical procedures.

Extracellular microbial lipases from bacteria, yeasts, or molds are used as catalysts for the interesterification of oils and fats. The natural function of these enzymes is to catalyze the hydrolysis of acylglycerols and other fatty acid esters to provide nutrients for the microorganisms, but these reactions are easily reversed, and consequently lipases are also effective catalysts for various esterification reactions. The properties and industrial applications of lipases have been described in some recent articles [53–57]. Because of the reversibility of the lipase reaction, hydrolysis and resynthesis of ester bonds occurs when lipases are incubated with TAGs, TAGs plus free fatty acids (FFAs), or TAGs plus alcohols, giving the interesterification, acidolysis, and alcoholysis reactions shown in Fig. 1.

Microbial lipases are produced industrially by growing yeasts, molds, or bacteria in large fermenters. The enzymes are excreted by the microorganisms in their growth media, from which, after removal of the cells, they can be recovered by conventional enzyme processing techniques. In the past, because of low fermentation yields and small markets for the enzymes, lipases were expensive in comparison to other more widely used extracellular enzymes such as proteases and carbohydrases. However, in the last few years successful application of recombinant DNA techniques for the production of lipases and an increasing market for the enzymes, particularly as additives to laundry products, has enabled enzyme manufacturers to produce lipases more efficiently. The availability of some cheap lipases is widening their potential for application as catalysts for interesterification of oils and fats and a variety of other industrial processes and products.

B. Lipase Catalysts

Enzymatic interesterification reaction systems consist of the lipase catalyst together with a very small amount of water dispersed in a continuous organic phase comprising the reactants (usually mixtures of TAGs or TAGs plus FFA) and if necessary a solvent such as hexane. The water in these microaqueous reaction systems is dissolved in the organic phase and absorbed onto the enzyme catalyst particles, and there is no distinct aqueous phase. Lipases catalyze reactions at interfaces, and to obtain high interesterification rates a large interfacial area between the reactants and the more hydrophilic enzyme phase is required. This can be achieved either by production of a fine dispersion of lipase in the organic phase, e.g., by use of a surfactant, or by immobilizing the enzyme on the internal surfaces of macroporous support particles. Immobilized lipases are usually preferred for interesterification processes, because immobilization generally improves lipase stability. Immobilized enzyme catalysts can also be used in packed bed reactors and easily recovered from stirred tank reactors for reuse. To produce effective interesterification catalysts, appropriate lipases and support materials must be selected. The main criteria for selection of lipases for use as interesterification catalysts are specificity, activity, and stability in the microaqueous reaction system and acceptability for use in food processing.

Microbial lipases can show specificity with respect to both the fatty acid and alcohol parts of their substrates [53], and for interesterification processes, lipases with appropriate specificity on reaction with acylglycerols must be selected. A large group of microbial lipases including those produced by *Rhizopus*, *Humicola*, *Rhizomucor*, and *Aspergillus* species catalyze reactions regiospecifically at the outer 1- and 3-positions of glycerol, the 2-position being unaffected (Fig. 21). This type of specificity can be exploited to generate TAGs that are difficult to obtain by conventional chemical and physical methods. For example, high value TAGs used in the formulation of confectionery fats can be produced from cheaper oils such as palm midfraction or high oleate sunflower oil by an interesterification reaction with stearic acid using a 1,3-regiospecific lipase [58–60] (Fig. 22).

A second group of lipases show no significant regioselectivity and catalyze reactions at all three glycerol positions (Fig. 21). The lipases produced by *Candida rugosa*, *C. antarctica*, and *Geotrichum candidum* are examples from this group of enzymes [53,62,63]. When these lipases are used as interesterification catalysts, the TAG products have a random distribution of fatty acyl groups and are similar to those produced by chemical interesterification (see Section II) [64]. Many lipases, such as those from several *Pseudomonas* species, show regioselectivity and catalyze reactions faster at the 1,3-positions than at the 2-position of glycerol [58]. When these lipases are used as interesterification catalysts, prolonged reaction can give TAGs with a random distribution of fatty acyl groups, but in practice products intermediate between those formed by 1,3-regiospecific and nonselective lipases are obtained. Recent work has shown that some microbial lipases can display stereoselectivity with TAGs, i.e., the reaction rate at the 1-hydroxy group is significantly different from the rate at the 3-hydroxyl group [65]. This type of selectivity can be exploited in interesterification reactions to produce chiral TAGs (Fig. 23) [66]. Chiral TAGs do not occur to any significant extent in most common oils and fats and are very difficult to produce by conventional physical or chemical techniques.

In general, extracellular microbial lipases show little selectivity as regards the type of fatty acid released from common oils and fats. Oils containing very long chain polyunsaturated fatty acids (PUFAs) are exceptional, because these fatty acids are only slowly released from the oils by some lipases. This selectivity has been used to enrich acylglycerol fractions in these PUFAs [67]. A lipase excreted by the mold *Geotrichum candidum* is unusual, because it shows extreme fatty acid specificity, preferentially releasing unsaturated fatty acids with a

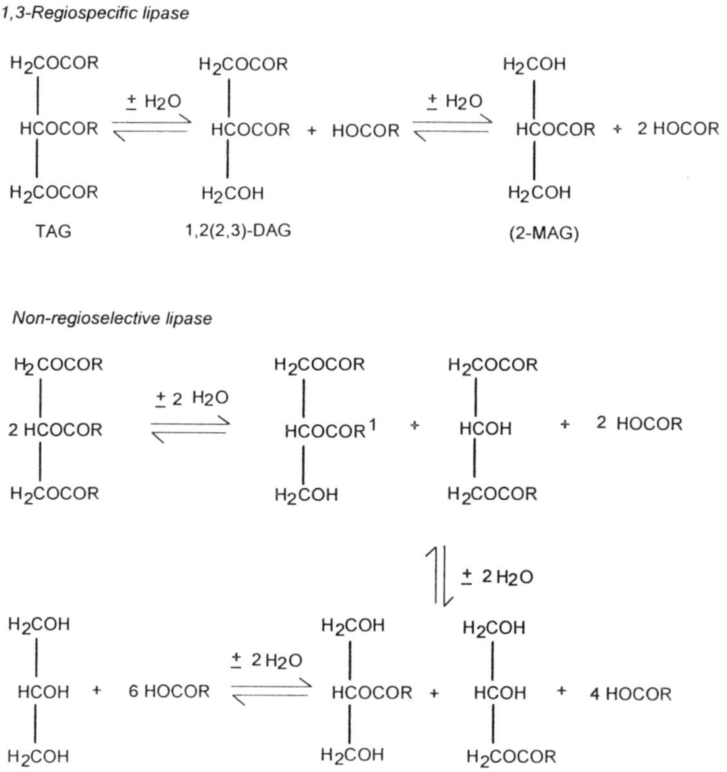

Fig. 21 Hydrolysis reactions catalyzed by lipases.

double bond in the Δ9 position from TAGs. Saturated fatty acids and unsaturated fatty acids without a double bond in the Δ9 position are only slowly released [63]. When *G. candidum* lipase is used as an interesterification catalyst, only Δ9 fatty acyl groups are exchanged in mixtures of TAGs or TAGs plus free fatty acid, and the enzyme has been used to enrich olive oil in linoleoyl groups at the expense of oleoyl groups, the saturated fatty acyl content of the oil remaining unchanged [64].

Microbial lipases vary greatly in the activity expressed in interesterification reaction systems. The reactions are run under conditions of low water activity (A_w), and one reason for the variation in interesterification activity lies in the different responses of the enzymes to low A_w. Some immobilized lipases have good catalytic activity at low A_w (<0.2), whereas others are essentially inactive under these conditions and express high catalytic activity only when A_w is raised to >0.5 [68,69]. For interesterification catalysts, enzymes in the former category, e.g., *Rhizomucor miehei* lipase, are preferred. A second reason for variation in interesterification activity probably lies in differences between lipases in the kinetic parameters of the various component reactions of the overall interesterification process (Fig. 24). Subtle variations in protein structure can alter the relative rates of hydrolysis and interesterification reactions.

The recovery and reuse of lipase catalysts from interesterification reactors is required for cost reasons. Consequently lipase stability in the reactors is an important parameter. Loss of activity during interesterification is caused by two factors: heat inactivation of the lipase and poisoning by minor components in the reactants [70,71]. Poisoning can be largely prevented

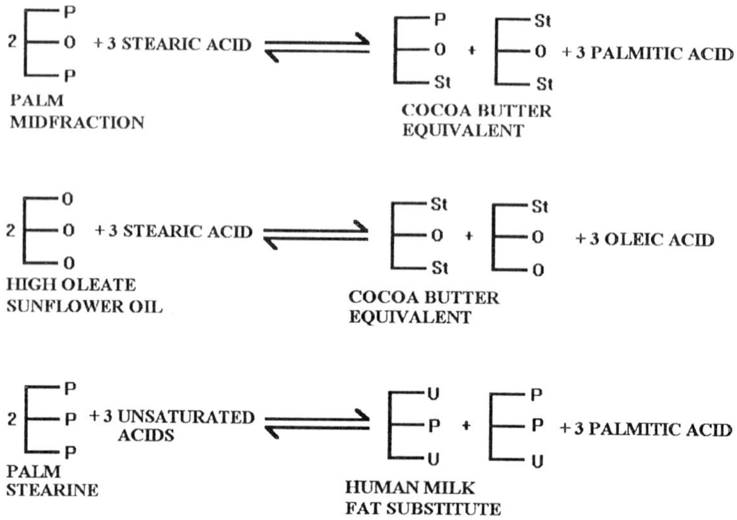

Fig. 22 Interesterification reactions catalyzed by 1,3-regioselective lipases.

by careful refining of the reactants, but a lipase must be used that is resistant to heat inactivation at the reaction temperature. Operation without a solvent is advantageous, and a reaction temperature of 70°C is often necessary to ensure that the reactants and products are fully liquid. Fortunately, the immobilized 1,3-regiospecific *Rhizomucor miehei* and the nonselective

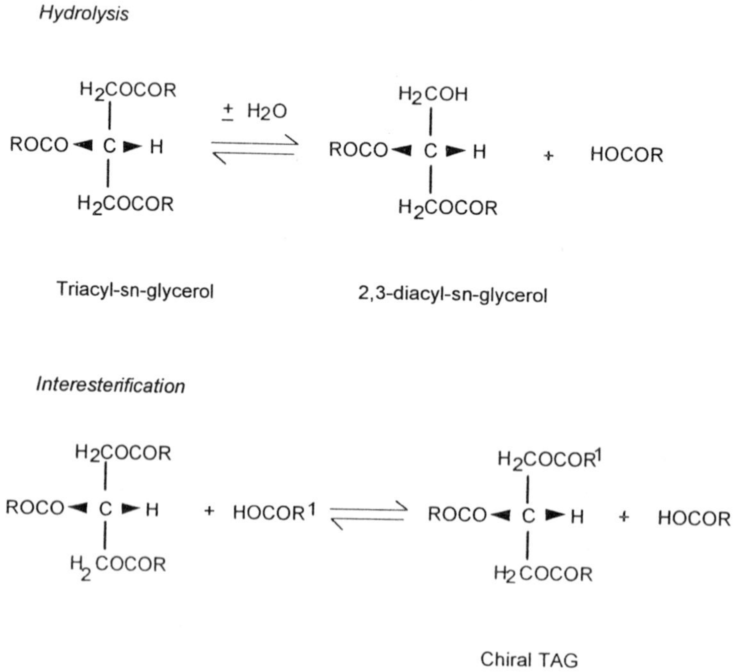

Fig. 23 Reactions catalyzed by a stereospecific lipase.

Candida antarctica lipases are stable at 70°C and can be used as catalysts in solvent-free processes [62,72]. Most other immobilized lipases are less stable, and the use of a lower reaction temperature, often with addition of a solvent, is required to maintain catalyst stability.

The most effective catalysts for use in enzymatic interesterification processes are prepared by adsorption of enzyme onto macroporous particles by ionic or hydrophobic interactions. Because protein is insoluble in interesterification reaction mixtures, tight irreversible association of lipase with the support materials is not required, and immobilization by covalent bonding is unnecessary. The physical characteristics of the support particles have an important influence on their effectiveness, and some important parameters have been investigated by

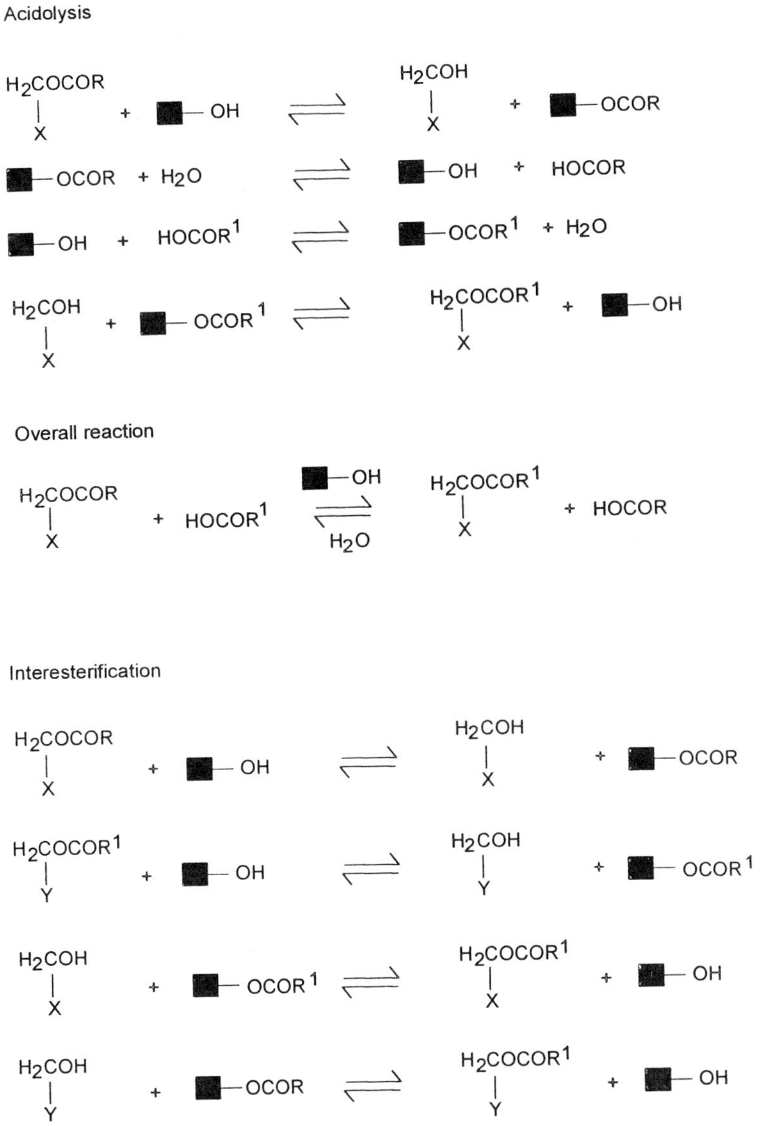

Fig. 24 Mechanisms of lipase-catalyzed interesterification reactions.

Overall

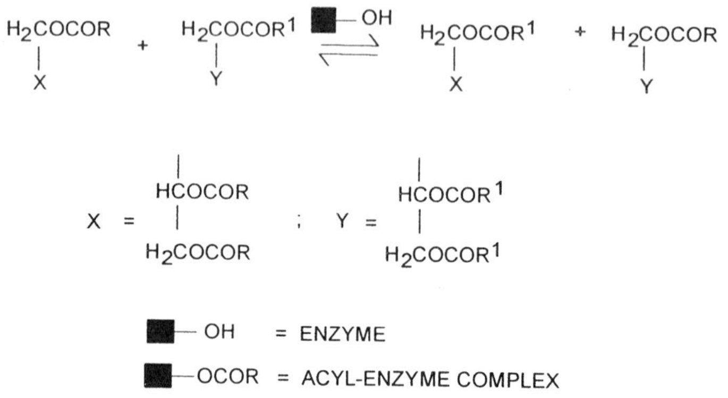

Fig. 24 Continued

Bosley and Clayton [73]. The macroporous particles should have sufficient internal surface area to allow adsorption of a substantial amount of lipase, and materials with a surface area in the range of 10–100 m^2/g are normally used. The pores of the particles must be large enough to allow enzyme molecules access to the internal surfaces of the particles. Support materials with a mean pore diameter of >100 nM are preferred. The chemical nature of the particle surface is also important. With ionic materials, weak anion exchangers are generally more effective than cation or strong anion exchangers, while for hydrophobic particles saturated surfaces are better than aromatic or unsaturated surfaces. Co-immobilization of lipase with an inert protein such as sodium caseinate on ion-exchange or hydrophobic particles can give more effective interesterification catalysts [74]. The presence of another protein probably reduces the unfolding of the lipase into an inactive form during immobilization.

The mechanical properties and size of the support particles have to be taken into account when preparing enzymatic interesterification catalysts. In large stirred tank reactors, fragile brittle materials can be damaged by attrition, and consequently catalysts used in those reactors must be prepared from support materials that are strong and preferably have some elasticity. For packed bed reactors, rigid, essentially incompressible support materials are required. Because of mass transfer effects the efficiency of interesterification catalysts decreases when the particle size is increased above a certain diameter. With large particles the reaction rate is often limited by the rates of diffusion of the reactants and products within the pores [75]. In stirred tank reactors, to avoid diffusion effects, small particles (0.1–0.2 mm) can be used, but in packed bed reactors these give unacceptably high pressure drops across the beds. For packed reactors a particle size of ~0.5 mm is recommended [76]. With particles of this size, acceptable pressure drops are obtained and any diffusional limitations are small.

Novo-Nordisk produces a highly active interesterification catalyst consisting of *Rhizomucor miehei* lipase immobilized on a phenol-formalydehyde weak anion-exchange resin. This catalyst is particularly suited for application in packed bed reactors [72]. Very effective interesterification catalysts suitable for use in both stirred tank and packed bed reactors can also be prepared by immobilization of lipase onto hydrophobic macroporous polypropylene particles [77].

Because of the high cost of macroporous materials, catalysts prepared by immobilization of lipases on ion-exchange or hydrophobic particles are expensive. Cheaper catalysts are obtained by deposition of lipase onto flux-calcined diatomaceous earth particles. Deposition can be achieved by solvent precipitation of enzyme in the presence of the particles or by mixing lipase solution and diatomaceous earth to form a paste that is then vacuum dried at low temperature [78,79]. Although they are cheaper to prepare, catalysts based on diatomaceous earth are less active than those based on macroporous ion-exchange or hydrophobic particles. The diatomaceous earth catalysts are also less suited to application in large-scale reactors because of their susceptibility to attrition in stirred tanks and excessive pressure drops in packed beds.

As mentioned previously, high interesterification reaction rates can be achieved by using surfactants to disperse lipases in reactant mixtures. For one type of system, lipase solution is dispersed in the continuous organic phase by formation of thermodynamically stable microemulsion droplets using surfactants such as AOT [sodium bis-(2-ethylhexyl) sulfosuccinate] or lecithin [80,81]. In another method, complexes of lipase and a nonionic surfactant such as sorbitan-monostearate are prepared. The lipase-surfactant complexes are then finely dispersed or dissolved in an interesterification reaction mixture [82,83]. With both types of reaction systems, rapid interesterification rates are achieved, but so far neither has been adopted for large-scale processes. Expensive membrane reactors are required for recovery of the lipase catalysts for reuse, and contamination of the reaction products by small amounts of surfactant is a problem.

C. Reaction Systems

The reactants for enzymatic interesterification processes are usually mixtures of TAGs or TAGs plus FFAs. Fatty acid methyl or ethyl esters can also be used instead of FFAs [60,84–86]. The advantages of using esters rather than the free acids are that with some lipases faster reaction rates are obtained and, because the esters have lower melting points, lower reaction temperatures are possible. The main problems with using esters are their higher cost and the inevitable occurrence of some hydrolysis during enzymatic interesterification processes. This gives small amounts of FFA in addition to esters, TAGs, and DAGs as reaction products and complicates product recovery and by-product recycle procedures.

Enzymatic interesterification without use of a solvent is preferable for cost reasons, but in some cases a solvent must be added to ensure that the reactants and products are completely liquid at the required reaction temperature. Lipase catalysts are active in the presence of a wide range of water-immiscible solvents, but for enzymatic interesterification processes hexane should be selected because this solvent is already used industrially for the extraction and processing of oils and fats. The presence of hydrocarbon solvents in enzymatic interesterification reaction mixtures has little effect on the reaction rate or product yield [85]. The application of supercritical carbon dioxide as a solvent in enzymatic interesterification reactions has been investigated by several groups [87,88]. Good reaction rates were observed, although the solubility of the reactants in the fluid is rather low. In the future, supercritical carbon dioxide could be used as an environmentally friendly nontoxic solvent in interesterification processes, particularly if its unusual properties can be exploited to improve product yields [87].

Enzymatic interesterification reaction systems must contain a small amount of water. This water has two functions. First, it is required by the enzyme catalyst for the maintenance of an active hydrated state. As mentioned previously, the amount and thermodynamic activity of the water needed for this purpose can be very low. Second, water is a reactant that generates partial glycerides and FFAs from TAGs by hydrolysis. The hydrolysis reactions reduce the final yield of interesterified TAGs but accelerate the interesterification reactions by raising the concentration of DAGs, which are essential intermediates (Fig. 24). In operating enzymatic interesterification processes, the water content of the reaction mixture must be carefully controlled to achieve a balance between acceptable reaction rates and product yields.

Enzymatic interesterification reactions can be run using both stirred tank and packed bed reactors. For large-scale processes, continuous packed bed reactors are preferred because they give higher product yields and enable more oil to be processed with a given quantity of lipase catalyst. In stirred tank reactors, catalyst particles can be damaged by attrition and not fully recovered during recycling procedures, but with packed bed reactors these problems are avoided and greater catalyst productivity is achieved. When regio- or stereospecific lipases are used as interesterification catalysts, side reactions resulting in a decrease in the yield of valuable TAGs occur if long reaction times are used. These side reactions are caused by migration of fatty acyl groups between the 1,3- and 2-positions of partial glycerides. The acyl migration reactions in combination with reesterification reactions give additional DAG and unwanted TAGs (Fig. 25). Because the rate-limiting steps in the side reactions are chemical rather than enzyme-catalyzed isomerizations of monoglycerides, increasing the lipase catalyst concentration and reducing the reaction time essentially eliminate the unwanted side reactions [86,90]. High catalyst concentrations and short contact times between the catalyst and re-

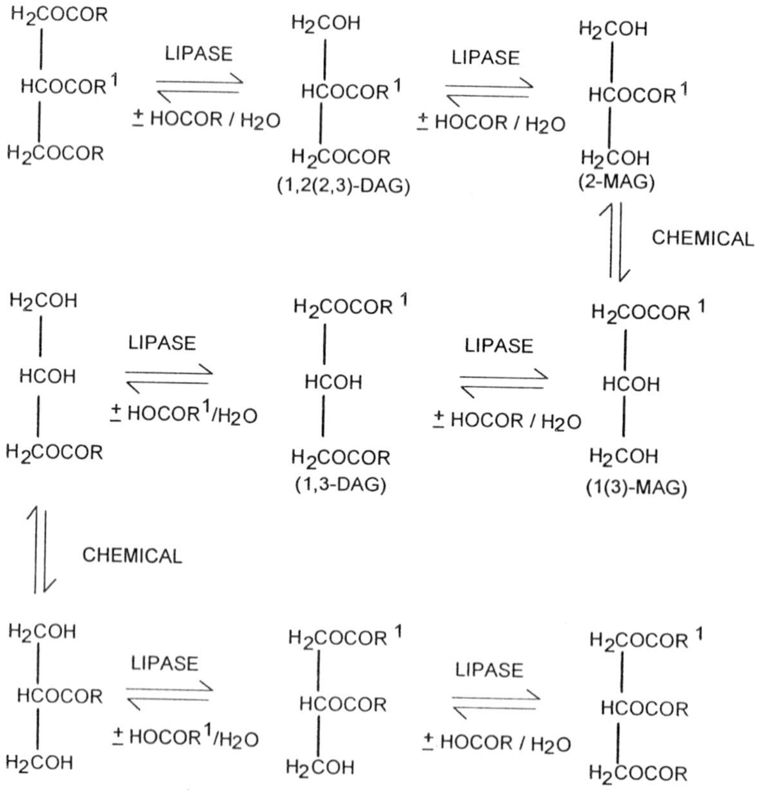

Fig. 25 Acyl migrations occurring during interesterification catalyzed by 1,3-regiospecific lipases.

actants are characteristic of packed bed reactors, whereas lower catalyst concentrations and longer reaction times are used in stirred tank reactors.

For operation of packed bed reactors, the reactants are treated to remove lipase catalyst inhibitors and poisons, and, if required, they are dissolved in hexane. A small amount of water is dissolved in the reactant mixture, which is then pumped continuously through a bed of catalyst particles. Operation of a typical reactor is illustrated by interesterification of a mixture of palm oil midfraction (1 part) and stearic acid (0.4 part) dissolved in hexane. A small amount of water (0.06%) was dissolved in the reactant mixture, which was then pumped through a bed of immobilized *Rhizomucor miehei* lipase to give a mean residence time of approximately 10 min. The temperature of the reactor was maintained at 40°C. High interesterification activity as measured by an increase in the stearoyl content of the TAGs was obtained through 13 days of continuous reactor operation (Fig. 26). Analysis of the product showed that, as expected for a 1,3-regiospecific lipase, stearoyl groups were incorporated exclusively into the 1- and 3-positions of the TAGs with the formation of the valuable TAGs, 1,3-distearoyl-2-oleoyl glycerol (StOSt) and 1(3)-stearoyl-3(1)-palmitoyl-2-oleoylglycerol (POSt) [59] (Table 5). After an initial equilibration period, the DAG and FFA contents of the product stream remained constant, and 5% additional DAG and 2% additional FFA were formed as by-products in the reactor (Fig. 26). During the equilibration period, which lasted for a few hours, water was removed from the hydrated lipase catalyst by the formation of higher concentrations of DAG and FFA until steady-state conditions were attained. In the steady state some of the water in the feedstream was consumed to give the additional DAG and FFA, setting up an equilibrium between TAG, water, DAG, and FFA. Therefore, there was variation in the water content and thermodynamic water activity (A_w) of the reactant and catalyst phases through the bed (Fig. 27). On entry to the reactor, the feedstream was almost saturated with water, and at the top of the catalyst bed the A_w of both phases was ~0.9. Limited but rapid hydrolysis occurred in the topmost part of the bed, leading to a reduction in the water content and A_w of both phases. Further down the reactor no further hydrolysis occurred (i.e., equilibrium was reached) and the A_w of both phases was <0.5. Interesterifica-

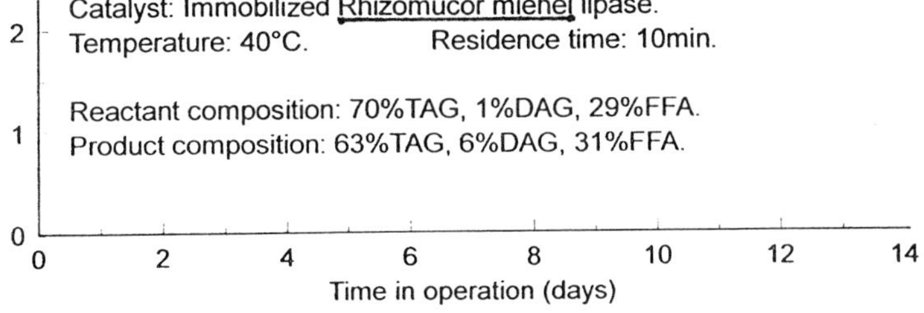

Fig. 26 Interesterification of a mixture of palm midfraction and stearic acid in a packed bed reactor.

Table 5 TAGs Formed by Reaction of a Mixture of Palm Oil Midfraction (1 part) and Stearic Acid (0.4 part) Using Immobilized *Rhizomucor miehei* Lipase as Catalyst

	Amount in palm oil midfraction (%)			Amount in interesterified TAG (%)		
Fatty acyl group	Total TAG	1,3-Position	2-Position	Total TAG	1,3-Position	2-Position
16:0	57.2	81.1	9.5	39.4	54.4	9.4
18:0	6.0	8.6	0.7	24.4	36.3	0.7
18:1	32.3	7.8	81.4	31.2	5.9	81.8
18:2	3.2	0.8	7.9	3.1	0.7	7.8
Others	1.3	1.7	0.5	1.9	2.7	0.3
TAG[a]						
SSS	2			7		
POP	65			28		
POSt	15			37		
StOSt	2			15		
SSO	6			2		
Others	10			1		

[a] S = saturated fatty acyl group; P = palmitoyl; St = stearoyl; O = oleoyl.

tion occurred throughout the whole of the catalyst bed, and therefore the catalyst was active under the low A_w conditions present in all but the top part of the reactor.

The effect of varying the water content of the feedstream to packed bed interesterification reactors has been studied [85,91]. Typical results obtained using a mixture of olive oil and methyl palmitate dissolved in petroleum ether as reactants are given in Table 6. As expected, both the interesterification reaction rate and by-product (DAG) formation decreased as the feedstream water content was lowered. With a dry reactant mixture the rate of reaction was very low and no DAG by-products were formed. For operation of commercial reactors, the

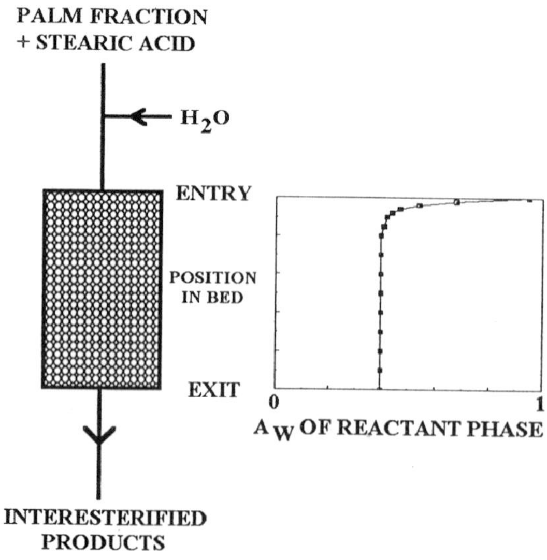

Fig. 27 A_w profile in a packed bed reactor.

Table 6 Effect of Feedstream Water Content on Interesterification and By-Product Formation in a Packed Bed Reactor[a]

Water content of feedstream (% saturation)	Interesterification activity of catalyst (g TAG/h/g catalyst)	DAG in product
100	3.4	3.7
52	1.4	2.0
31	1.0	1.4
10	0.5	<1.0
0	0.1	<1.0

[a] Various concentrations of water were dissolved in a mixture of olive oil (1 part), methyl palmitate (0.96 part) and 100–120°C petroleum ether (0.85 part). The resulting solutions were pumped through beds of immobilized *Rhizomucor miehei* lipase to give a mean residence time of ~1 h.

feedstream water content is selected to give the minimum level of by-product formation while maintaining an acceptable reaction rate.

Immobilized lipase catalysts can be very stable under the conditions found in packed bed reactors, and operation of reactors at temperatures as high as 70°C is possible when immobilized heat-stable lipases such as those produced by *Rhizomucor miehei* or *Candida antarctica* are used [62,72]. An important cause of catalyst inactivation can be the presence of deleterious minor components in the oils such as phospholipids and fat oxidation products [70,71]. These components inhibit or inactivate lipase catalysts, and to obtain good catalyst stability and productivity, effective refining of the reactants is essential. With well-refined oils and fats and heat-stable lipase catalysts, reactors can be run continuously at temperatures up to 70°C for many weeks. During operation of packed bed reactors, a constant product composition is maintained by varying the flow rate to take account of decreasing catalyst activity. When the flow rate becomes unacceptably low because of catalyst inactivation, the reactor is stopped and the inactivated catalyst is replaced with fresh immobilized lipase.

The progress of interesterification reactions between TAGs and FFA or fatty acid esters can be followed by the change in the fatty acyl compositions of the various species. The composition of the interesterified TAGs can also be examined by HPLC or capillary gas chromatography. For many lipases a very simple pseudo-first-order kinetic model based on FAME analysis of the TAGs or FFA can be used to calculate interesterification reaction rates and catalyst activity (Fig. 28). This simple model works well for 1,3-regiospecific or completely nonselective lipases but cannot be used for fatty acid or regio- or stereoselective lipases. With these types of catalysts, more complex kinetic models are required. When mixtures of TAGs are used as reactants, the progress of the interesterification reaction can be monitored by analysis of TAGs by HPLC and/or capillary gas chromatography. With most mixtures of oils and fats the complexity of the TAG composition is such that simple kinetic models for calculation of reaction rates cannot be used.

Kinetic models of varying complexity and sophistication have been developed by several groups to describe enzyme interesterification processes [92–95]. These models can be applied to reactions catalyzed by different types of lipase and can be used to simulate and predict the effects of parameters such as water and DAG concentrations, reactant concentrations, mass transfer limitations, and catalyst activity on reaction rate and product yield. These models will be useful in the future for designing and predicting the behavior of large-scale reactors under a variety of conditions.

Stirred tank reactor

$$k = \frac{1}{t} \ln \left(\frac{A - B}{A - X} \right)$$

$$\text{catalyst activity} = \frac{kV}{W} = \frac{V}{Wr} \ln \left(\frac{A - B}{A - X} \right)$$

Packed bed reactor (PBR)

$$k = \frac{1}{r} \ln \left(\frac{A - B}{A - X} \right)$$

$$\text{catalyst activity} = \frac{kV}{W} = \frac{V}{Wr} \ln \left(\frac{A - B}{A - X} \right) = \frac{F}{W} \ln \left(\frac{A - B}{A - X} \right)$$

k = reaction rate
t = reaction time
r = residence time in PBR
A = Calculated mole % of reactant fatty acid in TAG at completion of reaction (t or r = ∞). For non-selective lipases calculated assuming a fully random distribution of fatty acyl groups amongst all three positions of TAG and FFA. For 1,3-regiospecific lipases calculated assuming a random distribution of fatty acyl groups amongst FFA and 1,3-positions of TAG
B = mole % of reactant fatty acid in TAG at start of reaction (t or r = 0)
X = mole % of reactant fatty acid in TAG at time = t or residence time = r
V = volume of reaction mixture in reactor
W = weight of catalyst in reactor
F = flow rate through PBR

Fig. 28 Simple kinetic model for calculation of interesterification reaction rate and catalyst activity.

D. Applications

To date, enzymatic interesterification has been applied for the production of TAGs used for confectionery fat formulations and nutritional applications. A process for the production of confectionery fats is outlined in Fig. 29. In this process a small amount of water is dissolved in a 1:1 mixture of high oleate sunflower oil and stearic acid and the mixture is pumped continuously through a bed of immobilized *Rhizomucor miehei* lipase at 70°C. On passage through the catalyst bed, interesterification of the reactant mixture occurs and stearate residues are incorporated into TAG largely at the expense of oleate residues, with the formation of 1,3-distearoyl-2-monolein (StOSt), 1(3)-monostearoyl-3(1)2-diolein (StOO), and oleic acid (Fig. 22 and Table 7). The reaction products are then passed through a distillation unit to remove FFAs, and the resulting acylglycerols are fractionated by crystallization from acetone to give a stearine containing a high concentration of the valuable TAG, StOSt. This stearine can be used as a substitute for expensive fats such as shea stearine and illipé butter in the formulation of cocoa butter equivalents. The FFA by-product of the reaction, which consists mostly of stearic and oleic acids, can be hydrogenated and then recycled for reaction with fresh high oleate sunflower oil, so there is no net consumption of stearic acid in the overall process. The oleine by-product, which contains StOO, can be reacted with additional stearic

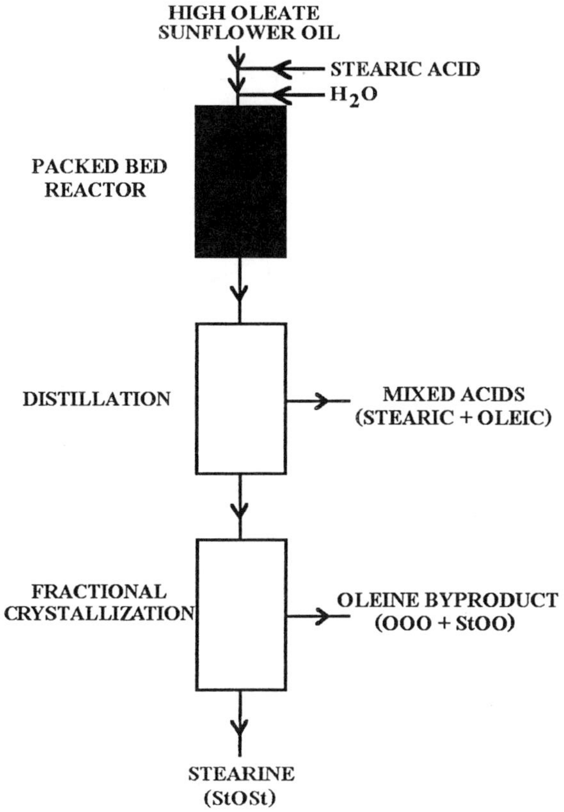

Fig. 29 Process for the production of confectionery fats.

acid to generate more StOSt and increase the overall yield of stearine from the original high oleate sunflower oil.

Other reactants can be used for the production of TAGs useful as components of confectionery fats. In particular, there are many reports on the enzymatic interesterification of mixtures of palm oil fractions and stearic acid or stearic acid esters to produce fats containing high concentrations of POSt and StOSt. These TAGs are the main components of cocoa butter, and enzymatic interesterification processes can produce fats with compositions and physical properties very similar to cocoa butter (Table 8) [59]. Olive, safflower, and high oleate rapeseed oils have also been interesterified with stearic acid or methyl stearate to give TAGs enriched in POSt and StOSt [96–98]. Similar products were obtained by reaction of shea and sal fats with stearic and palmitic acids or their methyl esters [84,97]. Enzymatic interesterification of mixtures of high oleate sunflower oil and ethyl esters obtained from hydrogenated high-erucate rapeseed oil gave fats enriched in 1,3-dibehenoyl-2-monolein and related very long chain fatty acid ($>C_{20}$) TAGs. These fats are useful as cocoa butter substitutes, particularly because it is claimed that they inhibit bloom formation in chocolate products [99].

In the area of nutritional fats, enzymatic interesterification is used to produce a human milk fat substitute for use in infant formula [76]. Interesterification of a mixture of tripalmitin and unsaturated fatty acids using a 1,3-regiospecific lipase as catalyst gives TAGs derived entirely

Table 7 TAGs Formed by Interesterification of a Mixture of High Oleate Sunflower Oil and Stearic Acid

	Amount in TAG fraction (%)		
Component[a]	High oleate sunflower oil	Reaction product before fractionation	Stearine
SSS	—	1	2
SOS	2	36	79
SSO	—	1	3
SLnS	—	3	5
SOO	26	41	10
OOO + others	72	18	1

[a] S = Saturated fatty acyl group; O = oleoyl; Ln = linoleoyl.

from vegetable oils that are rich in 2-position palmitate with unsaturated fatty acyl groups in the 1- and 3-positions (Fig. 22). These TAGs closely mimic the fatty acid distribution found in human milk fat, and when they are used in infant formula instead of conventional fats the presence of palmitate in the 2-position of the TAGs has been shown to improve digestibility of the fat and absorption of other important nutrients such as calcium.

Enzymatic interesterification can also be used for production of oils and fats containing nutritionally important PUFAs, such as eicosapentaenoic acid (EPA) and docosahexaenoic acid (DHA). For example, various vegetable and fish oils have been enriched in EPA and DHA by enzymatic interesterification reactions [100,101]. Use of the technique to produce structured lipids with medium-chain fatty acids (C_8 and C_{10}) and PUFAs located specifically in either the 2- or 1,3-positions of the TAGs has been described [89,102,103]. Lipids of this type have been reported to have beneficial effects on a range of metabolic parameters, including immune function, nitrogen balance, and improved lipid clearance from the bloodstream [76]. Enzyme processes are particularly suitable for the production and modification of lipids containing PUFAs, because these unstable fatty acids are susceptible to damage under the more severe conditions used for chemical processing.

Table 8 Compositions of Cocoa Butter and a Confectionery Fat Produced by Enzymatic Interesterification[a]

TAG	Amount in cocoa butter (%)	Amount in enzymatically produced confectionery fat (%)
POP	16	16
POSt	41	39
StOSt	27	28
SLnS	7	8
SOO	6	4
Others	3	5

[a] The confectionery fat was produced by interesterification of a mixture of palm oil midfraction (1 part) and stearic acid (0.5 part) using immobilized *Rhizomucor miehei* as catalyst. After removal of the FFAs by distillation, the interesterified glycerides were fractionated from acetone to give the confectionery fat as a midfraction.

The possible application of enzymatic interesterification for the production of lower value nonspecialty oils and fats such as margarine hardstocks and cooking oils has been investigated by several groups. When nonspecific lipases such as *Candida rugosa* and *C. antarctica* are used as catalysts for interesterification of oil blends, the TAGs produced are very similar to those obtained by chemical interesterification [64,104]. Therefore, replacement of chemical interesterification by an enzyme process giving similar products is technically feasible, although it has not been adopted on a commercial scale to date largely because of the comparatively high process and catalyst costs.

If regio- or stereospecific lipases are used to interesterify oil blends, the products formed are different from those obtained by chemical interesterification, and they can have better functional properties. For example, interesterification of blends of canola and palm oils using the 1,3-regiospecific *Rhizopus delemar* lipase as catalyst gave oils with improved fluidity compared with the original blends or chemically interesterified products. Because of their higher fluidity the enzymatically interesterified oils were most effective as frying oils [105].

Interesterification of blends of palm and hyrogenated canola oils and cottonseed and hydrogenated soybean oils using 1,3-regiospecific lipases as catalysts gave fats with a low *trans* acid content that were effective as margarine hardstocks [106,107]. Reaction of mixtures of palm stearine and lauric fats using immobilized *Rhizomucor miehei* as catalyst also produced fats that were functional as margarine hardstocks [72,108]. With these enzymatically interesterified fats, margarine could be formulated without using hydrogenated fats.

Because of high catalyst and processing costs, enzymatic interesterification has not been used to date for the large-scale commercial production of margarine hardstocks, cooking oils, or other nonspecialty oils and fats. The technique will be adopted for the manufacture of these types of product only if the comparatively high processing costs can be offset by addition of extra value to the oils and fats. Higher value could come from improved functionality as indicated above or issues such as naturalness. In the future development of novel improved processing techniques and more effective and cheaper lipase catalysts could increase the opportunities for application of enzymatic interesterification in the manufacture of edible oils and fats.

REFERENCES

1. N. O. V. Sonntag, Fat splitting, esterification and interesterification, in *Bailey's Industrial Oil and Fat Products,* 4th ed., Vol. 2, Wiley-Interscience, New York, 1979, pp. 97–173.
2. K. S. Markley, Esters and esterification, in *Fatty Acids: Their Chemistry, Properties, Production and Uses*, 2nd ed., Interscience, New York, Vol 2, 1961, pp. 757–984.
3. K. S. Markley, Esters and esterification, in *Fatty Acids: Their Chemistry, Properties, Production and Uses*, 2nd ed., Interscience, New York, Vol. 5, 1968, pp. 3459–3560.
4. R. O. Feuge, Derivatives of fats for use as foods, *J. Am. Oil. Chem. Soc. 39*: 521 (1962).
5. L. H. Going, Interesterification products and processes, *J. Am. Oil Chem. Soc. 44*: 414A (1967).
6. DGF (German Society for Fat Research), Gemeinschaftsarbeiten der DGF, 57. Mitteilung, Die Umesterung von Speisefetten I, II, III, *Fette, Seifen, Anstrichm. 75*: 467, 587, 663 (1973).
7. J. W. E. Coenen, Fractionement et interesterification des corps gras dans la perspective du marché mondial des matières premières et des produits finis. II. Interesterification, *Rev. Fr. Corps Gras 21*: 403 (1974).
8. B. Sreenivasan, Interesterification of fats, *J. Am. Oil Chem. Soc. 55*: 796 (1978).
9. S. J. Laning, Chemical interesterification of palm, palmkernel and coconut oils, *J. Am. Oil Chem. Soc. 62*: 400 (1985).
10. F. V. K. Young, Interchangeability of fats and oils, *J. Am. Oil Chem. Soc. 62*: 372 (1985).
11. A. Rozendaal, Interesterification of oils and fats, Proc. AOCS World Conf. on Edible Fats and Oils Processing, Champaign, IL, 1990, pp. 152–157.

12. H. Brockerhoff, A stereospecific analysis of triglycerides, *J. Lipid Res. 6*: 10 (1965).
13. H. Brockerhoff, Stereospecific analysis of triglycerides; an alternative method, *J. Lipid Res. 8*: 167 (1967).
14. R. J. van der Wal, Calculation of the distribution of the saturated and unsaturated acyl groups in fats from pancreatic lipase hydrolysis data, *J. Am. Oil Chem. Soc. 37*: 18 (1960).
15. M. H. Coleman and W. C. Fulton, *Enzymes of Lipid Metabolism*, Pergamon, New York, 1961, p. 127.
16. H. K. Hawley and G. W. Holman, Directed interesterification as a new processing tool for lard, *J. Am. Oil Chem. Soc. 33*: 29 (1956).
17. G. Hoffmann, *The Chemistry and Technology of Edible Oils and Fats and Their High Fat Products*, Academic Press, New York, 1989, pp. 264–277.
18. F. A. Norris and K. F. Mattil, Interesterification reactions of triglycerides, *Oil Soap 23*: 289 (1946).
19. M. Naudet and P. Desnuelle, Phenomena of transesterification between triglycerides. III. Detailed study of the equilibria of transesterification, *Bull. Soc. Chim. Fr. 14*: 323 (1947).
20. A. Kuksis, M. J. McCarthy, and J. M. R. Beveridge, Triglyceride composition of native and rearranged butter and coconut oils, *J. Am. Oil Chem. Soc. 41*: 201 (1964).
21. D. Chobanov and R. Chobanova, Alterations in glyceride composition during interesterification of mixtures of sunflower oil with lard and tallow, *J. Am. Oil Chem. Soc. 54*: 47 (1977).
22. G. R. List, T. L. Mounts, F. Orthoefer, and W. E. Neff, Preparation and properties of zero trans soyabean oil margarines, *J. Am. Oil Chem. Soc 72*: 379, 383 (1995).
23. L. P. Klemann, K. Aji, M. M. Chaysam, R. P. D'Amelia, J. M. Henderson, A. S. Huang, M. S. Otterburn, and R. G. Yarger, Random nature of triacylglycerols produced by the catalyzed interesterification of short- and long-chain fatty acid triglycerides, *J. Agric. Food Chem. 42*: 442 (1994).
24. H. P. Kaufmann, F. Grandel, and B. Grothues, Interesterification in fats. I. Theoretical background and literature. Interesterification by hydrogenation, *Fette, Seifen, Anstrichm. 60*: 919 (1958).
25. M. W. Formo, Ester reactions of fatty materials, *J. Am. Oil Chem. Soc. 31*: 548 (1954).
26. M. G. A. Willems and F. B. Padley, Palm oil: quality requirements from a customer's point of view, *J. Am. Oil Chem. Soc. 62*: 454 (1985).
27. W. Herman de Groot and M. H. Hilder (Unilever Ltd), Preparation and use of alkali metal dispersions, Br. Patent 1,236,233 (1968).
28. C. W. Placek and G. W. Holman, Directed interesterification of lard, *Ind. Eng. Chem., Ind. Ed. 49*: 162 (1957).
29. J. Burgers, C. W. Molt, and P. Seiden (Procter and Gamble Co.), Interesterification process, U.S. Patent 3,170,798 (1965).
30. C. N. M. Keulemans and G. Smits (Lever Brothers Co), Interesterification process and apparatus, U.S. Patent 4,585,593 (1986).
31. M. P. Jordan de Urries and R. M. Utrilla, Studies on transesterification. I. Kinetics and mechanism of the reaction between fatty acid monoesters catalysed by alkali metals, *Grasas Aceites 21*: 281 (1970).
32. T. J. Weiss, G. A. Jacobson, and L. H. Wiedermann, Reaction mechanisms of sodium methoxide treatment of lard, *J. Am. Oil Chem. Soc. 38*: 396 (1961).
33. M. Naudet, Modifications de structure des corps gras, B. Modifications affectant leur structure glycéridique, *Rev. Fr. Corps Gras 21*: 35 (1974).
34. M. Naudet, Interesterification et esterification. Mécanismes réactionnels et conséquences sur la structure, *Rev. Fr. Corps Gras 23*: 387 (1976).
35. J. A. Heidal and P. C. Møork, On the mechanism of the base catalysed interesterification of fats, Proc. 11th Scand. Symp. on Lipids, 1982, pp. 147–152.
36. P. van der Plank, unpublished results.
37. I. P. Freeman, Interesterification. I. Change of triglyceride composition during the course of interesterification, *J. Am. Oil Chem. Soc. 45*: 456 (1968).

38. A. Steenhoek, unpublished data cited in Refs. 7 and 11.
39. H. Konishi, W. E. Neff, and T. L. Mounts, Chemical interesterification with regioselectivity for edible oils, *J. Am. Oil Chem. Soc. 70*: 411 (1993).
40. E. W. Eckey, Directed interesterification in glycerides, *Ind. Eng. Chem. Ind. Ed. 40*: 1183 (1948).
41. C. Schmulinzon, A. Yaron, and A. Letan, Directed transesterification of hardened coconut oil with several saturated fatty acid methyl esters, *Riv. Ital. Sostanze Grasse 48*: 440 (1971).
42. G. Barsky, Restructuring glyceride oils, U.S. Patent 2,182,332 (1940).
43. R. C. A. Lago and L. Hartman, Directed interesterification of a Brazilian palm oil and analysis of the original and interesterified oil and its fractions, *J. Sci. Food Agric. 37*: 689 (1986).
44. E. W. Eckey and M. W. Formo, Directed interesterification in glycerides. Reaction in presence of added glycerol, *J. Am. Oil Chem. Soc. 26*: 207 (1949).
45. R. de Lathauwer, M. van Opstal, and A. J. Dijkstra (Safinco BE), Process for the directed interesterification of a triglyceride oil or oil mixture, U.S. Patents 4,284,578 (1981) and 4,419,291 (1983).
46. H. R. Kattenberg, Beschleunigung der gelenkten Umesterung mittels periodischer, variabeler Temperaturfuehrung, *Fette, Seifen, Anstrichm. 76*: 79 (1974).
47. A. J. Haighton and H. R. Kattenberg (Lever Brothers Co), Interesterification process for glycerides, U.S. Patent 3,855,254 (1975).
48. D. G. Chobanov and M. R. Topoalova, Alterations in glyceride composition during directed interesterification of lard, *J. Am. Oil Chem. Soc. 56*: 581 (1979).
49. G. W. Holman, T. B. Laycock, and R. L. Wille (Procter and Gamble Co), Catalytic interesterification, U.S. Patent 2,738,278 (1956).
50. C. N. M. Keulemans and A. Rozendaal (Lever Brothers Co), Process and apparatus for the interesterification of a triglyceride oil and products therefrom, U.S. Patent 4,790,962 (1988).
51. W. E. Dominick and D. Nelson (Swift and Co), Lard crystal modification process, U.S. Patent 2,625,485 (1953).
52. R. Duterte, Méthodes analytiques applicables au controle de l'estérification et de l'interestérification, *Rev. Fr. Corps Gras 23*: 547 (1976).
53. A. R. Macrae and R. C. Hammond, Present and future applications of lipases, *Biotechnol. Gen. Eng. Rev. 3*: 193–217 (1985).
54. P. Wooley and S. F. Petersen, *Lipases—Their Structure, Biochemistry and Application*, Cambridge Univ. Press, 1994.
55. S. E. Godtfredsen, Microbial lipases, in *Microbial Enzymes and Biotechnology*, 2nd ed., W. M. Fogarty and C. T. Kelly, eds., Elsevier Applied Science, London, 1990, pp. 255–274.
56. M.-P. Egloff, S. Ransac, F. Marguet, E. Rogalska, H. van Tilbeurgh, G. Buono, C. Cambillau, and R. Verger, Enzymes lipolytiques et lipolyse, *Oleagineux Corps Gras Lipides, 2*: 52 (1995).
57. P. Adlercreutz, Enzyme-catalysed lipid modification, *Biotechnol. Gen. Eng. Rev. 12:* 231–254 (1994) (see ref. 53).
58. M. Matori, T. Asahara, and Y. Ota, Positional specificity of microbial lipases, *J. Ferment. Bioeng. 72*: 397 (1991).
59. A. R. Macrae, Interesterification of fats and oils, in *Biocatalysis in Organic Syntheses*, J. Tramper, H. C. van der Plas and P. Linko, eds., Elsevier, Amsterdam, 1985, pp. 195–208.
60. S. Bloomer, P. Adlercreutz, and B. Mattiasson, Triglyceride interesterification by lipases. 1. Cocoa butter equivalents from a fraction of palm oil, *J. Am. Oil Chem. Soc. 67*: 519 (1990).
61. A. R. Macrae, Modifying oils—enzymic methods, in Oils and Fats in the Nineties, V. K. S. Shukla and F. D. Gunstone, eds., Int. Food Sci. Centre, Lystrup, Denmark, 1992, pp. 199–208.
62. H. P. Heldt-Hansen, M. Ishii, S. A. Patkar, T. T. Hansen, and P. Eigtved, A new immobilised positional non-specific lipase for fat modification and ester synthesis, in *Biocatalysis in Agricultural Biotechnology*, J. R. Whitaker and P. E. Sonnet, eds., American Chemical Society, Washington, DC, 1989, pp. 158–172.
63. E. Charton and A. R. Macrae, Substrate specificities of lipases A and B from *Geotrichum candidum* CMICC 335426, *Biochim. Biophys. Acta 1123*: 59 (1992).

64. A. R. Macrae, Lipases catalysed interesterification of oils and fats, *J. Am. Oil Chem. Soc. 60*: 243A (1983).
65. E. Rogalska, C. Cudrey, F. Ferrato, and R. Verger, Stereoselective hydrolysis of triglycerides by animal and microbial lipases, *Chirality 5*: 24 (1993).
66. I. C. Chandler and P. T. Quinlan (Loders Croklaan BV), Enzymic triglyceride conversion, PCT Patent, WO 9410326 (1994).
67. K. D. Mukherjee, Fractionation of fatty acids and other lipids via lipase-catalysed reactions, *Oleag. Corps Gras Lipides 2*: 365 (1995).
68. R. H. Valivety, P. J. Halling, A. D. Peilow, and A. R. Macrae, Lipases from different sources vary widely in dependence of catalytic activity on water activity, *Biochim. Biophys. Acta 1122*: 143 (1992).
69. R. H. Valivety, P. J. Halling, and A. R. Macrae, *Rhizomucor miehei* lipase remains highly active at water activity below 0.0001, *FEBS Lett. 301*: 258 (1992).
70. Y. Wang and M. H. Gordon, Effect of phospholipids on enzyme-catalysed transesterification of oils, *J. Am Oil Chem. Soc. 68*: 588 (1991).
71. Y. Ohta, T. Yamane, and S. Shimizu, Inhibition and inactivation of lipase by fat peroxide in the course of batch and continuous glycerolysis of fat by lipase, *Agric. Biol. Chem. 53*: 1885 (1989).
72. L. H. Posorske, G. K. LeFebvre, C. A. Miller, T. T. Hansen, and B. L. Glenvig, Process considerations of continuous fat modification with an immobilised lipase, *J. Am. Oil Chem. Soc. 65*: 922 (1988).
73. J. A. Bosley and J. C. Clayton, Blueprint for lipase support: use of hydrophobic controlled-pore glasses as a model, *Biotech. Bioeng. 43*: 934 (1994).
74. J. A. Bosley (Unichema Chemie BV), Supported enzyme, Eur. Patent 0424130 (1991).
75. H. J. Jung and W. Bauer, Determination of process parameters and modelling of lipase-catalysed transesterification in a fixed bed, *Chem. Eng. Technol. 15*: 341 (1992).
76. P. Quinlan and S. Moore, Modification of triglycerides by lipases: process technology and its application to the production of nutritionally improved fats, *INFORM 4*: 580 (1993).
77. A. R. Macrae, J. A. Bosley, and A. D. Peilow (Unilever NV), Process for the preparation of fatty acid esters, Eur. Patent 0322213 (1989).
78. R. A. Wisdom, P. Dunnill, and M. D. Lilly, Enzymic interesterification of fats: the effect of non-lipase material on immobilised enzyme activity, *Enzyme Microb. Technol. 7*: 567 (1985).
79. T. Matsuo, N. Sawamura, Y. Hashimoto, and W. Hashida (Fuji Oil Co. Ltd.), Method for enzymic interesterification of lipid and enzyme used therein, Eur. Patent 0035883 (1984).
80. K. Holmberg and E. Osterberg, Enzymatic transesterification of a triglyceride in microemulsions, *Prog. Colloid Polym. Sci. 74*: 98 (1987).
81. A. G. Marangoni, R. D. McCurdy, and E. D. Brown, Enzymatic interesterification of triolein with tripalmitin in canola lecithin-hexane reverse micelles, *J. Am. Oil Chem. Soc. 70*: 737 (1993).
82. S. Basheer, K.-I. Mogi, and M. Nakajima, Surfactant-modified lipase for the catalysis of the interesterification of triglycerides and fatty acids, *Biotech. Bioeng. 45*: 187 (1995).
83. M. Goto, M. Goto, N. Kamiya, and F. Nakashio, Enzymatic interesterification of triglyceride with surfactant-coated lipase in organic media. *Biotech. Bioeng. 45*: 27 (1995).
84. T. Matsuo, N. Sawamura, Y. Hashimoto, and W. Hashida (Fuji Oil Co. Ltd.), Producing a cacao butter substitute by transesterification of fats and oils, U.K. Patent 2035359 (1980).
85. A. R. Macrae, Tailored triacylglycerols and esters, *Biochem. Soc. Trans. 17*: 1146 (1989).
86. S. Bloomer, P. Aldercreutz, and B. Mattiasson, Triglyceride interesterification by lipases: 2. Reaction parameters for the reduction of trisaturated impurities and diglycerides in batch reactions, *Biocatalysis 5*: 145 (1991).
87. K. Nakamura, Y. M. Chi, Y. Yamada, and T. Yano, Lipase activity and stability in supercritical carbon dioxide, *Chem. Eng. Commun. 45*: 207 (1986).
88. D. A. Miller, H. W. Blanch, and Y. M. Prausnitz, Enzymatic interesterification of triglycerides in supercritical carbon dioxide, *Ann. N.Y. Acad. Sci. 613*: 534 (1990).
89. A. Shishikura, K. Fujimoto, T. Suzuki, and K. Arai, Improved lipase-catalysed incorporation

of long-chain fatty acids into medium-chain triglycerides assisted by supercritical carbon dioxide extraction, *J. Am. Oil Chem. Soc. 71*: 961 (1994).
90. A. R. Macrae and P. How (Unilever N.V.), Interesterification with a lipase enzyme as an interesterification catalyst, Eur. Patent 0093602 (1983).
91. P. Forssell, P. Parovouri, P. Linko, and K. Poutanen, Enzymatic transesterification of rapeseed oil and lauric acid in a continuous reactor, *J. Am. Oil Chem. Soc. 70*: 1105 (1993).
92. H. J. Jung and W. Bauer, Determination of process parameters and modelling of lipase-catalysed transesterification in a fixed bed, *Chem. Eng. Technol. 15*: 341 (1992).
93. S. Basheer, K. Mogi, and M. Nakajima, Interesterification kinetics of triglycerides and fatty acids with modified lipase in *n*-hexane, *J. Am. Oil Chem. Soc. 72*: 511 (1995).
94. H. R. Reyes and C. G. Hill, Kinetic modelling of interesterification reactions catalysed by immobilised lipase, *Biotech. Bioeng. 43*: 171 (1994).
95. S. R. Moore and J. Davies, The kinetics of lipase catalysed interesterification of fats and oils, *Inst. Chem. Eng. Res. Event 1*: 526 (1994).
96. T. Tanaka, E. Ono, and K. Takinami (Ajinimoto Co. Ltd.), Method of producing improved glyceride by lipase, U.S. Patent 4,275,011 (1981).
97. T. Matsuo, N. Sawamura, Y. Hashimoto, and W. Hashida (Fuji Oil Co. Ltd.), Method for modification of fats and oils, U.S. Patent 4,420,560 (1983).
98. T. Gitlesen, I. Svensson, P. Adlercreutz, B. Mattiasson, and J. Nilsson, High oleic acid rapseed oil as starting material for the production of confectionery fats via lipase-catalysed transesterification, *Ind. Crops Prod. 4*: 167 (1995).
99. N. Sagi, T. Izumi, H. Kida, and H. Maeda (Fuji Oil Co. Ltd.), Hard butter composition, Eur. Patent 0227364 (1985).
100. G. G. Haraldsson and P. A. Hostuldsson, The preparation of triglycerides highly enriched with ω-3 polyunsaturated fatty acids via lipase catalysed interesterification, *Tetrahedron Lett. 30*: 1671 (1989).
101. K. H. Huang and C. C. Akoh, Lipase-catalysed incorporation of *n*-3 polyunsaturated fatty acids into vegetable oils, *J. Am. Oil Chem. Soc. 71*: 1277 (1994).
102. F. Bjorkling, S. E. Gotfredsen, and O. Kirk, The future impact of industrial lipases, *Trends Biotechnol. 9*: 360 (1991).
103. C. C. Akoh, Structured lipids—enzymatic approach, *INFORM 6*: 1055 (1995).
104. P. Kalo, P. Parviiainen, K. Vaara, S. Ali-Yrrko, and A. Antila, Changes in the triglyceride composition of butter fat induced by lipase and sodium methoxide catalysed interesterification reactions, *Milchwissenschaft 41*: 82 (1986).
105. J. Kurashige, N. Matsuzaki, and Takahashi, Enzymatic modification of canola/palm oil mixtures: effects on the fluidity of the mixture, *J. Am. Oil Chem. Soc. 70*: 849 (1993).
106. C. France and J. M. deMan, Physical properties and composition of low trans canola/palm blends modified by continuous enzymic interesterification, *Elaeis 6*: 39 (1994).
107. H. M. A. Mohamed and K. Larsson, Modification of fats by lipase interesterification. 2. Effect on crystallisation behaviour and functional properties, *Fat Sci. Technol. 96*: 56 (1994).
108. J. Muderhwa, M. Pina, and J. Graille, Interestérification catalysée par les lipase régiosélectives 1,3 en milieu fondu: applications à l'huile de palme et à sa fraction concrète, *Rev. Fr. Corps Gras 36*: 11 (1989).

10

Hydrogenation of Edible Oils

Technology and Applications

Wicher T. Koetsier
Unichema International, Emmerich, Germany

I. INTRODUCTION

The hydrogenation of triglyceride oils is a heterogeneously catalyzed reaction that is induced commercially in three-phase reactors. The three phases are oil, hydrogen gas, and solid, porous, Ni catalyst particles. In principle, the following reactor configurations can be used: tank reactors (well-stirred autoclaves or loop reactors), bubble columns, or fixed bed reactors. Patterson [1] gives an extensive overview of early and current reactor designs.

Autoclaves operated batchwise are the preferred type of reactor for edible oil hydrogenation and are most likely used in more than 95% of all edible oil hydrogenation processes. In some countries, for example in the former USSR, continuous fixed bed trickle flow reactors or a continuous process in a sequence of three autoclaves in series are used. The trickle flow process requires the use of dedicated catalyst particles, otherwise the quality of the partially hydrogenated product will be unacceptable.

In the hydrogenation of triglyceride oils using nickel catalysts various reactions occur on the nickel surface. These include consecutive hydrogenation reactions as depicted by the simple reaction equation

$$A \xrightarrow{+\ H_2} B \xrightarrow{+\ H_2} C \xrightarrow{+\ H_2} D$$

In edible oil hydrogenation it is important that a high quality product be obtained within a reasonable time. In practice this means an acceptable rate of the hydrogenation reaction, appropriate selectivity of the chemical reactions, and adequate ease of removal of the catalyst from the oil by filtration. To obtain an acceptable hydrogenation rate one has the option of controlling the concentration of the reacting components, the temperature, and the amount and quality of the catalyst. The concentration of the oil molecules is a given fact, but the mean concentration of hydrogen in the bulk of the oil phase is determined by several factors that need to be well understood.

A main objective of this chapter is to discuss two correlations: the hydrogenation rate and the hydrogenation time required to drop the initial iodine value, IV_o, to a specific end iodine value, IV_{end}.

The overall rate of the hydrogenation process [which is usually expressed as ΔIV/min] is reflected by the rate at which the hydrogen is absorbed by the oil. It can be derived from the definition of the iodine value, that

$$1 \text{ IV} = 36 \text{ mol H}_2/\text{m}^3_{\text{oil}}$$

and

$$1 \text{ IV/min} = 0.6 \text{ mol H}_2/(\text{m}^3_{\text{oil}} \cdot \text{s})$$

$$\text{which generates} \cong 17 \text{ kcal}/(\text{m}^3_{\text{oil}} \cdot \text{s}) = 71 \text{ kJ}/(\text{m}^3_{\text{oil}} \cdot \text{s}) \text{ or kW/m}^3_{\text{oil}} \text{ energy}$$

This overall rate is determined by a number of physical steps, such as transfer of hydrogen from the gas phase to the oil phase and diffusion of hydrogen through the intricate texture of the catalyst particles, and the reaction rate on the nickel surface.

These aspects have been discussed in many papers and are at least qualitatively well understood by the hydrogenators. However, in many papers the hydrogenation rate results, depicted as IV versus hardening time, for example, are presented in graphs containing the data not only of relevant parameters such as temperature, hydrogen pressure, and amount of catalyst, but also of parameters that cannot be easily interpreted for use in other systems, for example stirring speed. This latter parameter does indeed greatly affect the hydrogenation rate, but the stirring speed is not a meaningful parameter because its effect is also determined by the size, type, and position of the impeller in the vessel.

It is the intention of this chapter to review the following aspects:

- In the first part the technology of hydrogenation is discussed. For example, it is explained which parameter should be used as a meaningful indication of the effect of the hydrogen transfer rate for which we use the concept of minimum hydrogenation time.
- In the second part of a number of aspects are reviewed that determine the properties of the hydrogenated product.

The filtration rate is determined by particle size distribution, catalyst particle morphology, and the presence of filter aid and of components in the oil that easily absorb onto the catalyst particles and affect the characteristics of the filter cake, such as soaps and phospholipids. These aspects are not reviewed in this chapter.

II. TECHNOLOGY OF HYDROGENATION

In the following paragraphs a simple general equation for the overall rate of the hydrogenation process is derived that we have used successfully to study catalyst activity in a number of different types of reactors. Moreover, we discuss whether the results obtained in small-scale reactors can be applied to large-scale reactors.

A. Transfer of H_2 from the Gas to the Catalyst Phase

In edible oil hydrogenation the solubility (maximum concentration at a given pressure and temperature) of hydrogen in the oil is much lower than the concentration of the double bonds. From Ostwald solubility coefficients given in Bailey [2] or from the data published by Andersson et al. [3], it can be derived that at 1 bar the following hydrogen concentrations exist at equilibrium in an edible oil.

Temp.	[2] (mol/m^3)	[3] (mol/m^3)
100°C	2.60	2.76
120°C	2.76	2.91
140°C	2.92	3.05
160°C	3.07	3.2
180°C	3.20	3.3
200°C	3.36	3.4

Due to the low solubility and the low pressure that are used in these hydrogenation processes, the concentration of hydrogen is always some orders of magnitude lower than the concentration of the molecules to be hydrogenated. For example, in the hydrogenation of a fish oil with an IV_0 of 180 the overall concentration of the $-C{=}C-$ bonds is on the order of 7000 $mol/m^3{}_{oil}$, while at 5 bar and 180°C the hydrogen solubility is only 16 $mol/m^3{}_{oil}$.

Such substantial differences in concentration between the two components have two important consequences. First, hydrogen has to be continuously supplied to the liquid phase by effectively dispersing a large volume of hydrogen bubbles through the liquid. Second, the hydrogen concentration in the bulk liquid phase, C_{bulk}, will always be lower than its solubility, C_{max}, because this bulk concentration, as explained in the next sections, is determined by the equilibrium of two processes, namely, (1) the rate at which hydrogen is transferred into the oil phase and (2) the reaction and diffusion rate within the catalyst particles—the rate at which hydrogen is removed again from the oil phase to the catalyst phase.

The bulk liquid-phase concentration is one of the important factors determining the hydrogenation rate as well as the selectivity. Because there are no analytical techniques available to measure this hydrogen concentration in the bulk liquid phase, it cannot be continuously monitored. It is possible to calculate this concentration throughout the hydrogenation reaction, but because certain assumptions have to be made there will always remain uncertainty as to its correct value. This hydrogen concentration is not necessarily constant during hydrogenation. For example, during isothermal hydrogenation in which the hydrogen gas is continuously recirculated by using a compressor, the bulk concentration will continuously increase because the reaction rate decreases as hydrogenation proceeds.

These aspects are discussed first before coming to the general formulas expressing the overall hydrogenation rate.

B. Rate of H_2 Transfer from the Gas to the Liquid Phase

The rate J [4], expressed in moles per cubic meter of oil per second, at which the hydrogen is transferred from the dispersed hydrogen bubbles to the bulk oil phase is directly proportional to the driving force $(C_{max} - C_{bulk})$:

$$J = k_L a(C_{max} - C_{bulk}) \tag{1}$$

In this equation C_{max} is the maximum concentration at the temperature and hydrogen pressure used in the reactor. Its value can be calculated from the pressure and the solubility of hydrogen by using the Ostwald solubility coefficients. In fact, this concentration exists at the gas/

liquid interface, but within a thin layer it drops to the hydrogen concentration C_{bulk}, which exists in the bulk of the liquid. The product k_La (1/s) is the characteristic rate constant. In chemical engineering practice this parameter is often referred to as the volumetric liquid-side mass transfer coefficient. It is the product of the liquid-phase mass transfer coefficient k_L (m/s) and the specific interfacial area $a(m^2/m^3_{oil})$ of the hydrogen bubbles in the oil phase. For various reasons it is important to understand which factors affect this characteristic rate constant. As well as for commercial reactor design, upscaling, and the choice of the lab reactor size, for example, used to determine the catalyst performance, it is important to properly understand what determines this characteristic rate constant. For example, important questions are whether k_La remains constant during the entire hydrogenation process and whether it is a function of the location within the reactor. In many reactor types this is the case. But before going on to answer these questions, let us first review the parameters determining this characteristic rate constant.

One last remark: This equation is the equation to use if one wants to calculate the hydrogen concentration within the bulk liquid phase at any time t. However, it requires that the volumetric liquid-side mass transfer coefficient, k_La, be determined first. Then

$$C_{bulk}(t) = C_{max} - J(t)/k_La \tag{2}$$

1. *The Liquid-Phase Mass Transfer Coefficient k_L*

The liquid-phase mass transfer coefficient, k_L, is usually expressed as the dimensionless Sherwood number $Sh = k_Ld_b/D_{H_2}$ and is determined from:

1. The diffusion coefficient of hydrogen in oil, D_{H_2}, several authors have published data on the value of this coefficient as a function of temperature [5]. It appears that its value is on the order of 1×10^{-8}–3×10^{-8} m²/s, within a temperature range of 100–200°C.
2. The hydrodynamics of the liquid flowing around a free rising bubble having a diameter d_b and a velocity v_b in a highly turbulent liquid; these hydrodynamics are determined in particular by the conditions at the gas/liquid interface and the effect of the turbulence.

The hydrodynamics are in most cases not well understood, but in general two extreme cases exist for free rising bubbles, one for small (d_b =<3 mm) bubbles and one for bubbles larger than 3 mm. The two cases are the result of the two different hydrodynamic situations at the gas/liquid interface, which have been observed for free rising single bubbles and are determined by surface phenomena occurring at the gas/liquid interface. Small bubbles behave like rigid spheres, while bubbles larger then 3–5 mm do not. This effect is caused by small differences in the surface tension of liquid elements just arriving at the interface and those that have already been present at the interface for some time. These differences in surface tension cause a liquid flow along the surface from low to high surface tension, hence in the direction of the newly arriving liquid elements, which is the direction opposite that of free rise. This surface flow causes a flow resistance of the free rising bubble that is equal to that observed for a solid sphere.

For free rising single bubbles that behave like rigid spheres the well-known Frössling [6] equation is applied to calculate the Sherwood numbers:

$$Sh = 2 + 0.6\ Re^{0.5}\ Sc^{0.33} \tag{3}$$

It can be theoretically derived that the Sherwood number is 2 if there is no flow around the bubble. This condition exists for very small (d_b =<0.1 mm) bubbles, because then the effect of the free rise velocity can be neglected. If the effect of the free rise velocity is taken into account, Sh will be larger by the factor given by the second term, which is a function of the Reynolds ($Re = \rho_{oil} v_b d_b / \eta_{oil}$) and Schmidt ($Sc = \eta_{oil} / \rho_{oil} D_{H2}$) numbers.

The free rise velocity, v_b, can be calculated from the equation

$$\text{Buoyancy force} = \text{drag force}$$

$$(1/6)\pi d_b^3 \Delta\rho g = C_w (1/2) \rho_{oil} v_b^2 (1/4) \pi d_b^2 \tag{4}$$

or

$$v_b = [(4/3) g d_b / C_w]^{1/2} \tag{5}$$

in which C_w is the drag coefficient, which is a function of the Reynolds number Re [7] and $\Delta\rho = \rho_{oil} - \rho_{H_2} = \rho_{oil}$.

For bubbles larger than 3 mm, the above-mentioned flow along the surface cannot, as explained above, exhibit the effect that is observed for small bubbles. In this case the flow pattern around the bubbles becomes quite different. No friction exists at the bubble surface, and the liquid flows almost frictionless along the interface. As a result, liquid elements flow from the top of the bubble to the bottom and disappear again in the bulk of the oil. The contact time of these liquid elements is the ratio of the bubble diameter to the free rise velocity. This free rise velocity as calculated from Eq. (4) can be higher by a factor of 2, because the drag coefficient C_w is much smaller. The liquid-phase mass transfer coefficient k_L should now be calculated from penetration theory [4]. In this theory it has been shown that

$$k_L = 2(D_{H_2}/\pi\tau)^{1/2} \tag{6}$$

where D_{H_2} is the liquid-phase diffusion coefficient for hydrogen and τ is the mean contact time of the liquid elements flowing along the hydrogen/oil interface. This mean contact time is approximately equal to

$$\tau = d_b / v_b \tag{7}$$

Rearranging the two equations into a Sherwood correlation gives

$$Sh = k_L d_b / D_{H_2} = 1.13\ Re^{0,5}\ Sc^{0,5} \tag{8}$$

From a comparison of the Frössling equation with the Sherwood equation, which stems from penetration theory, it can be concluded that Eq. (8) results in a Sherwood number that is a factor of 1.9 $Sc^{0.17}$ larger than calculated from Eq. (3). Schmidt numbers are on the order of 1500. Thus this calculation results in Sherwood numbers that are $1.9 \times 1500^{0.17} = 6.6$ times larger than calculated from the Frössling equation. It reflects the change in the flow pattern around a free rising single bubble. Usually this change occurs for bubbles that are larger than 3 mm.

Table 1 summarizes these two calculations of k_L for six bubble sizes in an oil with a temperature of 150°C, a dynamic viscosity of $1.8 \times 10^{-2}\ N \cdot s/m^2$, a liquid density of 830 kg/m^3 (then the kinematic viscosity $\nu = 2.2 \times 10^{-5}\ m^2/s$), and a hydrogen diffusivity of 15×10^{-9} m^2/s. Hence the Schmidt number is 1445, $Sc^{0.5} = 38$, and $Sc^{0.333} = 11.3$

The result of these calculations shows that for a wide range of bubble sizes the liquid-phase mass transfer coefficient at 150°C and for free rising single bubbles ranges around 2×10^{-4} m/s. It is more a function of liquid temperature than of the size of the bubbles. Only if bubbles become smaller than 0.1 mm or larger than 3 mm does the value of k_L start to increase significantly. As the majority of the bubble sizes present in the hydrogenation reac-

Table 1 Liquid-Phase Mass Transfer Coefficient for Various Bubble Sizes

d_b (mm)	C_W	V_b (mm/s)	Sh	$k_L \times 10^4$ (m/s)	τ (s)	$k_L \times 10^4$ (m/s)
0.1	20,870	0.25	2.3	4.1	0.2	na
0.2	2,680	1	2.7	2.4	0.1	na
0.5	167	6	3.8	1.4	0.042	
1.0	22	24	10	1.8	0.021	na
2.0	3	90	25	2.3	0.011	na
4.0	1	160	41	1.8	0.025	12

na = not applicable.

tors are within the range for which k_L has the above-mentioned constant value, it will be understood that an improvement in k_L cannot be easily achieved. The above estimate holds for a free rising single bubble. For a swarm of bubbles it can be expected to give a slightly lower k_L value, because the free rise velocity will be smaller.

The above calculation does not take into account the effect of turbulence, presence of catalyst particles, or the frequent collisions occurring in such dispersions.

The effect of turbulence can be estimated from the Brian–Hales [8] or Sänger–Decker [9] theory. Brian and Hales have shown that the Sherwood number $k_L d_p / D_{H_2}$ is a function of the parameter $\epsilon_T d_{3.2}{}^4/\nu^3$, which is a Reynolds number using the characteristic parameters describing homogeneous turbulence and in which $d_{3.2}$ is the Sauter [10] mean bubble diameter, ϵ_T is the mean energy dissipation per unit mass, and ν is the kinematic viscosity. For example, Sänger and Decker derived the following Sherwood correlation:

$$\mathrm{Sh} = 2 + 2.309(\epsilon_T d_{3.2}{}^4/\nu^3)^{0.221}\mathrm{Sc}^{0.178} \tag{9}$$

From this equation the effect of the turbulence on the liquid phase mass transfer coefficient can be estimated. In a well-stirred tank reactor the local energy dissipation is a strong function of the position within the vessel. In the impeller region it can be higher than the mean (over the whole vessel) value by a factor of 50, whereas in other parts it can be lower by a factor of 10. Usually the mean energy dissipation is 2 kW/m^3_{oil} or 2.4 kW/kg. From Eq. (9) it can be derived that $k_L \sim \epsilon_T{}^{0.221}$, a proportionality that is far from linear. In the impeller region the effect on k_L for 1.1 mm bubbles will be higher by a factor of 2.4 versus the value for a free rising single bubble, and in the other regions it will be higher by a factor of only 1.7. All in all, if all the above-mentioned effects are taken into account, the liquid-phase mass transfer coefficient in a well-stirred reactor in which the mean energy dissipation is 2.4 kW/kg is expected to vary between 4×10^{-4} and 6×10^{-4} m/s.

The zone within which the hydrogen concentration drops from its maximum value to its value in the bulk is very thin. It can be estimated from the penetration depth as given by the equation

$$\delta = 4(D_{H_2}\tau)^{1/2} \tag{10}$$

This depth of penetration is about 10^{-5} m, thus extremely small compared to the size of the bubbles, but of a size equal to the catalyst particle size. So if a catalyst particle is within this small zone, it will recognize the hydrogen concentration as higher than in the bulk of the oil phase. The mixing outside the bubbles and within the bulk of the liquid is so good that it will have no effect on the overall rate of the hydrogenation process.

2. *The Specific Interfacial Area a*

The specific interfacial area of the hydrogen bubbles in the oil, a, is given by the equation

$$a = 6\epsilon_{gas}/d_{3.2} \quad (11)$$

where ϵ_{gas} is the gas hold-up and $d_{3.2}$ is the Sauter mean diameter (volume/surface ratio) of the gas bubbles.

Van Dierendonck et al. [15] empirically determined scale-up correlations for gas hold-up in well-stirred tank reactors having diameters from 0.3 to 2.6 m and a $H/T = 1$. The equations have been experimentally determined for variations within the following properties:

Viscosity:

$$5 \times 10^{-4} < \eta_{oil} < 5 \times 10^{-3}\ \mathrm{N \cdot s/m^2}$$

Density:

$$800 < \rho_{oil} < 1300\ \mathrm{kg/m^3{}_{oil}}$$

Surface tension:

$$20 \times 10^{-3} < \sigma < 75 \times 10^{-3}\ \mathrm{N/m}$$

The correlation for the gas hold-up at which he arrived, if Rushton-type turbine impellers are used, is given by the equation

$$\epsilon_{gas} = 0.31\left(\frac{V_s}{(\sigma g/\rho_{oil})^{0.25}}\right)^{0.67} + \frac{0.45V_L}{(gT)^{1/2}} \quad (12)$$

where V_s is the linear gas velocity, T the tank diameter, D the impeller diameter, H the liquid height, h_R the liquid height above the impeller, and σ the surface tension at the interface H_2/oil [12] ($\sigma \approx 25 \times 10^{-3}$ N/m), and

$$V_L = (N - N_O)D^2/T \quad (13)$$

in which N_O is the minimum stirrer speed that is required to draw in bubbles from the liquid level in the reactor,

$$N_O = 2(h_R/T)^{0.5}(\sigma g/\rho_{oil})^{0.25}(T/D^2) \quad (14)$$

At 150°C,

$$\sigma g/\rho_{oil} = (25 \times 10^{-3})(10)/830 = 3.6 \times 10^{-4} \quad (15a)$$

so

$$(\sigma g/\rho_{oil})^{0.25} = 0.132 \quad (15b)$$

For example, for a commercial reactor that uses one Rushton-type six-blade impeller and has a tank diameter $T = 2.5$ m, an impeller diameter $D = 0.9$ m, a liquid height $H = T$, $h_R = 1.7$ m, and no recirculation of hydrogen, then $N_O = 0.7\ \mathrm{s^{-1}}$, and if the power $P(= 6\ \rho_{oil}N^3D^5)$ is 2 kW/$\mathrm{m^3_{oil}}$, then $N = 2.8\ \mathrm{s^{-1}}$. Hence, $V_L = 0.68$ m/s and ϵ_{gas} ranges between 0.09 and 0.06 if the linear gas velocity V_s drops from 0.004 to 0 m/s, which occurs as a result of the decrease in the hydrogenation rate. The value of 0.004 m/s is calculated from an initial hydrogenation rate of 1 ΔIV/min.

Van Dierendonck gives the following correlation for the mean bubble diameter:

$$d_{3.2} = (0.41\sigma/\rho_{oil}g)^{0.5} \pm 25\% \quad (16)$$

A gas hold-up of 9% ($\epsilon_{gas} = 0.09$) and a Sauter diameter of 1.1 mm are the values calculated from the above equations. The specific interfacial area a will then be 490 m^2/m^3_{oil}.

3. *The Maximum Achievable Rate Constant k_La*

The maximum obtainable value of the characteristic rate constant k_La depends very much on the type and size of the reactor. In small-scale reactors it is possible to use very high stirring speeds such that the mean energy dissipation is on the order of 25 kW/kg. Under these conditions one can obtain a gas dispersion of up to 30 vol %. The specific interfacial area can then be as high as 1500 m^2/m^3_{oil} and k_L, 8×10^{-4} m/s. On the basis of these predictions a maximum value for k_La of 1.2 s^{-1} would seem to be achievable. It has been demonstrated that in our 400 mL open lab reactors with stirring speeds of 3000 rpm and continuous introduction of hydrogen gas below the impeller, a k_La of 1 s^{-1} can indeed be achieved.

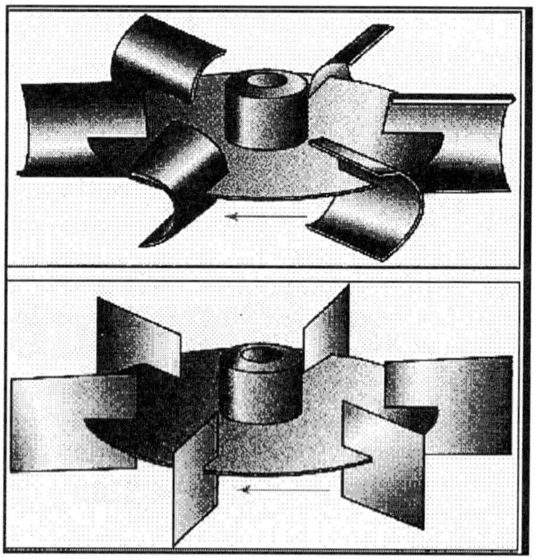

(a)

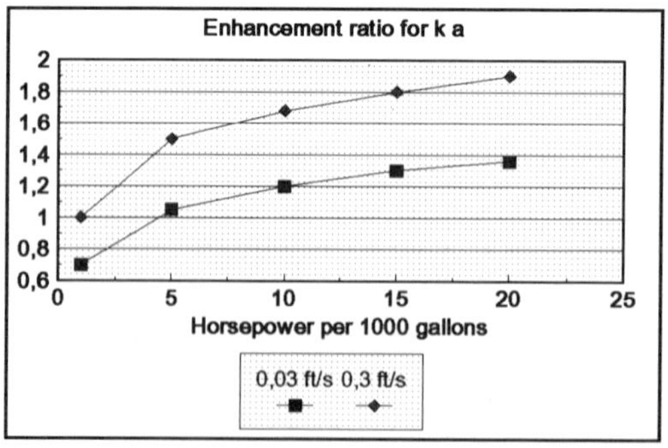

(b)

Fig. 1 Top: (a) The CD-6 and (b) Rushton impeller designs. Bottom: The k_La enhancement factor for the CD6 versus the Rushton impeller type as a function of the power input. (Reprinted with permission of Chemineer Inc.)

In the example of a commercial reactor discussed above, the volumetric liquid-phase mass transfer coefficient k_La would be between 0.16 and 0.25 s^{-1}. This estimate results in a value that is slightly higher than what is commonly seen in commercially used reactors, namely k_La ranging from 0.10 to 0.15 s^{-1}, if a stirring power of 2 $kW/m^3{}_{oil}$ is applied. The maximum achievable value for k_La depends on the maximum power input that can be used in such reactors and on a large number of factors that are discussed below.

4. *Effect of Reactor Design on* k_La

Reactor design has always been focused on getting an excellent hydrogen transfer rate, for which it is important for the characteristic rate constant k_La to be as high as possible. This can be achieved by ensuring

1. The presence of adequate baffles, proper position of the impellers, number, type and size of impellers, liquid height, presence and position of heating coils, hydrogen introduction through spargers, etc....
2. The quality of the oil; for example, one usually obtains higher values of k_La in fish oils than in vegetable oils.

In commercial practice there are two reactor designs in use. These operate on different principles. One is the well-stirred tank reactor with one or two Rushton-type or Lightnin [13] impellers and optionally a second smaller impeller at the height of the gas/liquid interface to suck in gas bubbles, four baffles, hydrogen sparger below the bottom impeller, and sometimes with continuous gas recirculation. Bakker et al. [14] of Chemineer, Inc., discussed a new impeller design that should show better mass transfer performance than the conventional Rushton-type design. The two designs are illustrated in Fig. 1. It is claimed that the k_La factor could be higher by a factor 1.3–2 depending on the superficial gas velocity. The other is the so-called Buss loop [15] reactor in which the liquid is recirculated and reintroduced into the reactor through a Venturi tube into which the hydrogen is also fed into the system. In this Venturi tube the hydrogen gets extremely well dispersed into the liquid. In this type of reactor it seems to be possible to achieve higher values for k_La.

Even today examples have come to our attention where the hydrogen absorption rate is the rate-determining step because the k_La value is far too low.

C. Rate of H_2 Transfer from the Liquid Phase to the Catalyst Particles

The rate J at which the hydrogen is transferred from the bulk oil phase to the catalyst particles is directly proportional to the driving force $C_{bulk} - C_{cat}$:

$$J = k_s a_{cat}(C_{bulk} - C_{cat}) \tag{17}$$

In this equation the product $k_s a_{cat}$ (s^{-1}) is again the characteristic rate constant. It is the product of the solid-phase mass transfer coefficient k_s (m/s) and the specific contact area between the catalyst particles and the oil a_{cat} ($m^2{}_{cat}/m^3{}_{oil}$). The H_2 concentration C_{cat} is that existing at the outer surface of the catalyst particles.

1. *The Characteristic Rate Constant* $k_s a_{cat}$

The solid-phase mass transfer coefficient k_s is a property that, in well-stirred solid–liquid systems, can be calculated after the Brian–Hales or Sänger–Decker theory [8,9]. For catalyst particles that have a mean Sauter diameter of 5 μm, the corresponding value of the solid-phase mass transfer coefficient is 0.9×10^{-2} m/s. It is, in fact, mainly determined by the 2 in Eq. (9) and not by the turbulence of the liquid elements flowing around these particles. The contribution of the second term to the Sherwood number is only 0.23 if $\epsilon_T = 2.4$ kW/kg and 0.45 if $\epsilon_T = 50$ kW/kg. The explanation is that the size of the smallest eddies, that is, the eddies in which the energy is lost due to friction, is several times larger than the size of the particles. Kolmogorov [16] gives the following expressions for the size η_e and velocity v_e of these small eddies:

$$\eta_e = (\nu^3/\epsilon_T)^{0.25} \tag{18}$$

$$v_e = (\nu\epsilon_T)^{0.25} \tag{19}$$

For an energy dissipation of 2.4 W/kg, the Kolmogorov scale is 0.26 mm and the velocity is 85 mm/s. This scale is 50 times larger than the mean size of the particles. Therefore, the turbulence will not have a great effect on the mass transfer rate to these particles.

The specific interfacial area a_s between the catalyst particles and the oil is given by the equation

$$a_s = 6\epsilon_{cat}/d_{3.2,cat} \tag{20}$$

in which ϵ_{cat} is given by the equation

$$\epsilon_{cat} = \frac{\rho_{oil}}{\rho_{part}}\left(\frac{\text{mg Ni in oil}}{(\text{\% Ni in particles})(10^4)(\text{kg oil})}\right) = \frac{(\text{wppm Ni})(\rho_{oil}/\rho_{part})}{\text{\% Ni in particles} \times 10^4} \tag{21}$$

If 100 wppm Ni is used, then ϵ_{cat} is 1.4×10^{-4} $m^3{}_{cat}/m^3{}_{oil}$ and a_s is 170 $m^2{}_{cat}/m^3{}_{oil}$.

Hence, the characteristic rate constant $k_s a_s$ is 1.5 s^{-1}. It is one order of magnitude larger than the characteristic rate constant $k_L a$ for the absorption rate of hydrogen into the oil phase. The effect of mass transfer rate from the bulk to the catalyst particles can therefore be neglected!

D. Reaction Rate Within the Catalyst Particles

Several reactions occur on the surface of the Ni crystallites that are randomly distributed throughout the catalyst particles. Hydrogenation and positional and configurational isomerization of double bonds are the main reactions. As a result of the isomerization reactions, a variety of isomers are formed as hydrogenation proceeds.

Rate equations are in general of a very complex nature. For practical purposes it is usually sufficient to have a simple equation that describes the rate as a function of the concentration of the reacting components and the temperature. Hydrogenation of edible oils and fatty acids is characterized by the following facts.

1. The concentration of hydrogen, as already stated, is considerably lower than the concentration of the double bonds. It is our experience that the hydrogenation rate can be approximated by a process first order in hydrogen if the hydrogen concentration remains below 10 mol/m^3_{oil}.
2. The oil contains the double bonds in different types of molecules, and during hydrogenation the number of different types of molecules increases even further due to positional and configurational isomerization.
3. The hydrogenation rate is of zero order with respect to each type of component containing a double bond and has a zero-order hydrogenation rate constant k_r that is differ-

ent for each type of unsaturated component. Some are hydrogenated fast, while others are hydrogenated much more slowly. Such similar systems often occur in practice. Examples are the hydrogenation and desulfurization of a number of petrochemical feedstocks [17]. The desulfurization rate is first-order in a pure component, but for a mixture of many different components the "apparent" order in "sulfur" is second-order if the reaction is carried out in a batch process and of 1.5 order if the reaction is carried out in a continuous, well-stirred, tank reactor.

4. A similar situation exists for the hardening rate in our set of reactions. All these different unsaturated oil components are each hydrogenated at their specific rate and initially as a reaction of zero order in oil. One could describe such a system by using a distribution of rate constants if all these rate constants were known.

A rigorous analysis of the reaction rate constants for each of these components requires an understanding of the hydrogenation rate of each individual component. Such an analysis is too time-consuming because it demands a detailed component analysis, and it is questionable whether the resulting information is of much use. However, many results have indicated that the hydrogenation of edible oils is of first order in the iodine value over a wide range of IV. In this case the IV reflects the concentration of all components containing an unsaturated bond. It is also an "apparent" order. It is an order of reaction that is determined by both the type of process and the difference in reaction rates of the various components. Some of the components react very fast because their reaction rate constants are high. The reaction rate constants of other components may be much smaller, and as a consequence the hydrogenation rate is much lower. And in this case also the apparent order, if the rate is expressed in terms of IV, can be considered a first-order reaction. However, it should be understood that this conclusion is based on the assumption that during the hydrogenation reaction the hydrogen concentration C_{bulk} remains constant. As already mentioned, this is usually not the case. Even under isothermal conditions the hydrogen concentration in the bulk liquid phase will continuously change. Examples are given in Section II.H. From our results we have come to the conclusion that the correlation between the hydrogenation rate *at constant hydrogen concentration* and the iodine value have to be expressed by two terms, one term reflecting the rate for the iodine values above IV = 80 and one for the regime below 80.

All in all, we have concluded that the hydrogenation rate per unit volume of catalyst can be approximated by the following equations.

$$R(t) = R_{\text{poly}} = k_{\text{r,p}}[\text{IV}(t) - 75]C_{\text{bulk}}(t)\eta_{\text{p}}(t), \qquad \text{for IV} > 80 \tag{22}$$

$$R(t) = R_{\text{mono}} = k_{\text{r,m}}\text{IV}(t)C_{\text{bulk}}(t)\eta_{\text{m}}(t), \qquad \text{for IV} < 80 \tag{23}$$

and

$$R(t) = J(t)/\epsilon_{\text{cat}} \qquad (\text{mol H}_2/(\text{m}^3{}_{\text{cat}} \cdot \text{s})) \tag{24}$$

in which $C_{\text{bulk}}(t)$ is also the hydrogen concentration at the liquid/catalyst interface, because we demonstrated that $k_s a_s >> k_L a$, $\eta(t)$ is the effectiveness factor, which is a function of the Thiele modules [18], and ϵ_{cat} is the volume fraction of catalyst particles.

Equations (22) and (23) reflect the correlation between the reaction rate R and the iodine value. An example of these two correlations is shown in Fig. 2, in which the hydrogenation rate is expressed as J $(=\epsilon_{\text{cat}}R/0.6)$ ΔIV/min, in which the factor 0.6 is the proportionality constant between IV/min and mol $H_2/(\text{m}^3{}_{\text{oil}} \cdot \text{s})$, has been plotted as a function of the iodine value of a fish oil hydrogenation. However, Fig. 2 does not give any information about the mechanism. And one should remember that the reaction rate for two different oils of equal iodine values will be different because IV can be the result of different fatty acid composi-

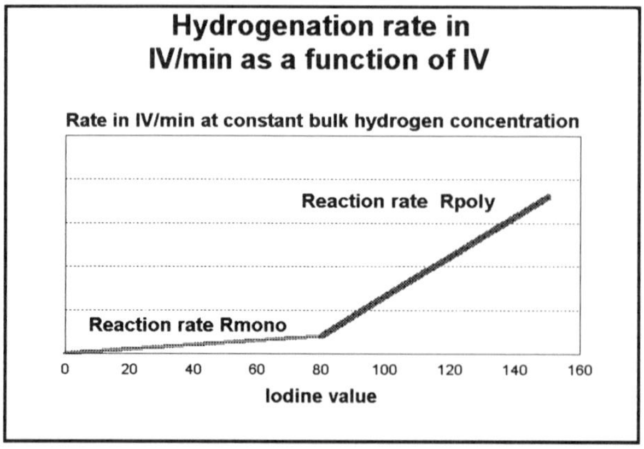

Fig. 2 Hydrogenation rate in IV/min as a function of IV.

tions. Finally, it should be borne in mind that once the concentration of these unsaturated molecules has become significantly lower, the hydrogenation rate will become first-order instead of zero-order in each individual component. This change will cause the "apparent" order of the reaction to become second-order.

E. Factors Affecting the Overall Rate of the Hydrogenation Process

In Sections II.C and II.D, it has been explained that the overall rate of the hydrogenation process can be described by the rate equations (1) and (22,23). Moreover, it has been estimated that the rate of hydrogen transfer from the bulk of the oil to the surface of the catalyst particles is not expected to affect the overall rate of this hydrogenation process. There are, however, three aspects that have not yet been discussed. First a general equation is derived to describe the overall rate of the hydrogenation process.

1. Initial Saturation Rate of Oil

In edible oil hydrogenation processes the feedstock is heated under vacuum to a specific temperature at which the hydrogen is introduced into the reactor. From time zero onward, the bulk hydrogen concentration is determined by the mass balance

$$\begin{aligned} V_{\mathrm{oil}}\frac{dC_{\mathrm{bulk}}(t)}{dt} &= k_{\mathrm{L}}a[C_{\mathrm{max}} - C_{\mathrm{bulk}}(t)]V_{\mathrm{oil}} - R(t)V_{\mathrm{cat}} \\ &= k_{\mathrm{L}}a[C_{\mathrm{max}} - C_{\mathrm{bulk}}(t)]V_{\mathrm{oil}} - k_{\mathrm{r,p}}[\mathrm{IV}(t) - 75]\eta_{\mathrm{p}}(t)C_{\mathrm{bulk}}(t)V_{\mathrm{cat}} \end{aligned} \tag{25}$$

At $t = 0$ the bulk hydrogen concentration is 0.

Upon integration the following equation is obtained:

$$\frac{C_{\mathrm{bulk}}(t)}{C_{\mathrm{max}}} = \frac{K}{K+L}\{1 - \exp[-(K+L)t]\} \tag{26}$$

in which

$$K = k_{\mathrm{L}}a \tag{27}$$

and

$$L = k_{r,p}[\mathrm{IV}(t) - 75]\eta_p(t)\epsilon_{cat} \qquad (28)$$

In Fig. 3, examples of three different solutions to Eq. (26) are given:

$$K = 0.1 \text{ and } L = 0 \qquad (29)$$

$$K = 0.1 \text{ and } L = 0.1 \qquad (30)$$

$$K = 0.1 \text{ and } L = 0.1\,[1 - \exp(-kt)] \qquad (31)$$

In which the rate constant k reflects the rate at which the catalyst is reactivated to its maximum activity. The third solution [Eq. (31)] has to be obtained numerically because an analytical solution cannot be derived.

Once the hydrogen concentration has reached its equilibrium value, which will occur after a time $t = 4/k_L a$, the derivative becomes almost zero. This derivation neglects the fact that in the initial phase the specific gas/liquid interface a is also a function of time. It may be expected that this time will be much shorter than the time constant $1/k_L a$ for the absorption process.

2. *Reactivation of Catalyst Activity*

The second process is the reactivation of the nickel that has been passivated as a result of unavoidable contact with oxygen. Nickel is dispersed into hardened fat at the end of the manufacturing process to prevent reoxidation of the reduced nickel surface. However, the pastilles occasionally come into contact with air, and then oxygen diffuses into the hardened fat and reoxidizes a small fraction of the reduced nickel. This reoxidation is extremely slow, because the permeability of oxygen through hardened fat is low. The oxygen that permeates will always start to passivate the most active part of the nickel first, which consists of the nickel sites at the outer surface of the micrometer size particles. These nickel sites are the most active ones, because there is no diffusional limitation for the oil molecules to reach them.

3. *Poisoning of Edible Oil Hydrogenation Catalysts*

Nickel catalysts can be poisoned as a result of various reactions.

1. Absorption of components within the porous structure of the catalyst particle. For example, components such as phospholipids, in particular those containing the basic

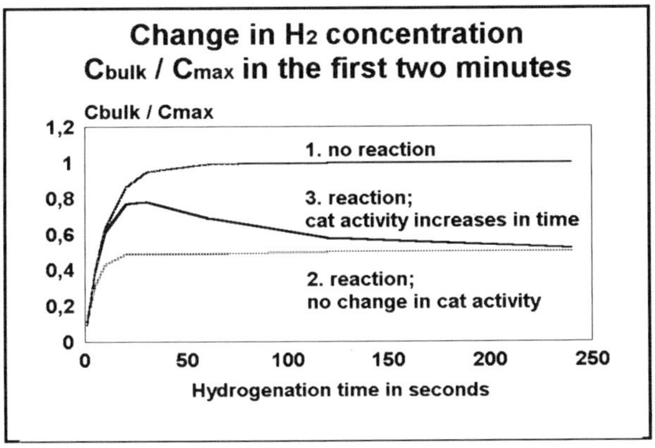

Fig. 3 Change in H_2 concentration C_{bulk}/C_{max} in the first 2 min.

NH groups, will be adsorbed onto the acid sites of the carrier. This will affect the accessibility of the nickel sites.

2. Dissolution of the active nickel surface due to the presence of fatty acids. Free fatty acids react with nickel sites according to the reaction

$$Ni + 2\ FFA \leftrightharpoons Ni(FFA)_2 + H_2 \tag{32}$$

This is an equilibrium reaction, and therefore the higher the hydrogen pressure the less nickel will dissolve. In edible oil hydrogenation it will result in only 1–2 wppm Ni if 0.2 wt % FFA is present. In fatty acid hydrogenation it is an important contribution to the deactivation of the catalysts during hydrogenation.

3. Poisoning of the nickel surface by reaction with sulfur components. In many oils minor amounts of sulfur are still present in the oil after refining. These components are desulfurized, and the nickel surface is sulfided to the extent that one S atom is present per two atoms of surface Ni atoms [19]. The reaction with sulfur components is a chemical reaction the rate of which depends on the type of component and the temperature of the reaction. If the conditions chosen are such that the desulfurization rate is high, the nickel present within the region at the outer side of the particles will be sulfided first, then the sulfur will penetrate further to the center of the particle. This mechanism is responsible for the fact that small particles are poisoned much faster than larger particles. Since small particles are the most active particles this leads to a serious loss of activity. One would prefer all the nickel sites, irrespective of where they are located within the particles, to be poisoned at the same rate.

The reaction rate of the sulfided nickel is much lower than the rate observed for pure nickel. For a sulfided nickel surface, Eqs. (22) and (23) become

$$R(t) = R_{s,poly} = k_{rs,p}[IV(t) - 75]C_{bulk}(t)\eta_s(t), \qquad \text{for IV} > 80 \tag{33}$$

and

$$R(t) = R_{s,mono} = k_{rs,m}\ IV(t)C_{bulk}(t)\eta_s(t), \qquad \text{for IV} < 80 \tag{34}$$

Equations (22) and (33) can be combined as

$$R_{ov}(t) = \left[x + (1-x)\left(\frac{k_{rs,p}}{k_{r,p}}\right)\left(\frac{\eta_s(t)}{\eta_p(t)}\right)\right]\{k_{r,p}[IV(t) - 75]\eta_p(t)\}C_{bulk}(t) \tag{35}$$

in which the x is the fraction of the nickel surface that has not yet been sulfided, or

$$R_{ov}(t) = X\{k_{r,p}[IV(t) - 75]\eta_p(t)\}C_{bulk}(t) \tag{36}$$

in which X has been substituted for $[x + (1-x)(k_{rs,p}/k_{r,p})\eta_s(t)/\eta_p(t)]$.

As already said, the sulfiding (and any other poisoning) of the nickel surface area is not expected to occur homogeneously over all the catalyst particles or within a catalyst particle. Sulfiding of the whole nickel surface area would be homogeneous only if the reaction were slow and if all the particles were of the same size. However, the sulfiding is a fast reaction. First all the nickel at the outer surface will be sulfided and subsequently the nickel that is located deeper in the particle. In the case of a particle size distribution the rate of sulfiding of the nickel will also be a function of the particle size. The sulfiding of the smallest particles is completed much faster than that of the larger particles. To reflect these effects in the rate equation, the factor X has to have the exponent n. Equation (36) becomes

$$R_{ov}(t) = X^n\{k_{r,p}[IV(t) - 75]\eta_p(t)\}C_{bulk}(t) \tag{37}$$

In one of our fish oil experiments (see Section II.H.3) it is demonstrated that the data could be adequately fitted by using a value of 2 for n.

F. The Overall Rate Equation of the Hydrogenation Process

Once the catalyst has reached its maximum activity after the initial induction period, the hydrogenation rate has also reached its maximum value. Consequently, the hydrogenation rate will slowly drop as time passes because the rate is first order in IV. As the rate drops over time, the hydrogen concentration in the oil phase will slowly increase over time. The rate of this increase is rather slow, and for a short time the overall hydrogenation rate can be considered almost constant.

If we now eliminate the unknown concentration $C_{\mathrm{bulk}}(t)$ from the equations (1), (24), and (37), we arrive at the following equation for the hydrogenation rate $J(t)$ for IV>80:

$$J(t) = C_{\max}\left[\frac{1}{k_{\mathrm{L}}a} + \frac{1}{X^{n}\{k_{\mathrm{r,p}}[\mathrm{IV}(t) - 75]\eta_{\mathrm{p}}(t)\}\epsilon_{\mathrm{cat}}}\right]^{-1} \tag{38}$$

This equation expresses the absorption rate as a function of the driving force, which is the maximum hydrogen concentration $C_{\max}$, and the term in brackets, which expresses the total resistance.

The first term in the large brackets in Eq. (38) represents the resistance of the physical transfer rate of hydrogen from the gas phase to the oil phase. This term can be experimentally determined if the hydrogenation rate is measured by using an excess of catalyst so that the second term is no longer relevant.

The second term in the large brackets reflects the effect of the amount of catalyst and the parameters describing the quality in terms of the chemical reaction rate. In other words this term depends on the intrinsic rate, the extent of poisoning, the iodine value, and the extent of diffusional limitation within the catalyst particles, but not on the hydrogen concentration, if the above assumption of first-order kinetics in hydrogen is correct. The effect of this second term is therefore complex. All three important parameters, namely k_{r}, IV, and η, change in time as the hydrogenation proceeds. The reaction rate constant k_{r}, decreases as a result of poisoning of the nickel surface area, IV drops continuously, and $\eta_{\mathrm{p}}(t)$ increases to a maximum value of one as the rate decreases.

Equation (38) can be rearranged as follows:

$$\frac{C_{\max}}{J(t)} = \frac{1}{k_{\mathrm{L}}a} + \frac{1}{X^{n}\{k_{\mathrm{r,p}}[\mathrm{IV}(t) - 75]\eta_{\mathrm{p}}(t)\}\epsilon_{\mathrm{cat}}} \tag{39}$$

The left-hand term is the quotient of the maximum hydrogen concentration, which is determined by the temperature and the pressure applied in the reactor and the absorption rate $J(t)$, a rate that can be continuously monitored.

Equation (39) is a very useful equation to describe the initial absorption rate, $J_{\max}$, as a function of the most relevant parameters:

$$\frac{C_{\max}}{J_{\max}} = \frac{1}{k_{\mathrm{L}}a} + \frac{1}{X^{n}\{k_{\mathrm{r,p}}[\mathrm{IV}(t=0) - 75]\eta_{\mathrm{p}}(t)\}_{\mathrm{t}=0}}\left(\frac{1}{\epsilon_{\mathrm{cat}}}\right) \tag{40}$$

or the mean absorption rate, J_{mean}, where the initial iodine value, $\mathrm{IV_O}$, drops to the final value, $\mathrm{IV_{end}}$.

$$\frac{C_{\max}}{J_{\mathrm{mean}}} = \frac{1}{k_{\mathrm{L}}a} + \frac{1}{X^{n}\{k_{\mathrm{r,p}}[\mathrm{IV}(t) - 75]\eta_{\mathrm{p}}(t)\}_{\mathrm{mean}}}\left(\frac{1}{\epsilon_{\mathrm{cat}}}\right) \tag{41}$$

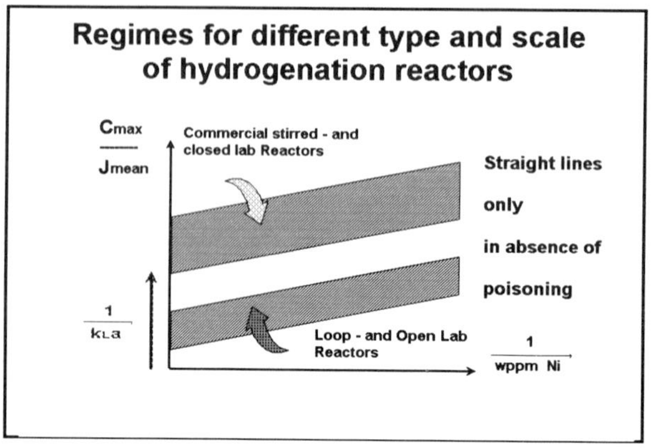

Fig. 4 Regimes for different types and scales of hydrogenation reactors.

If the results of C_{max}/J_{mean} of any hydrogenation are plotted as a function of 1000/wppm Ni ($\approx 1/\epsilon_{cat}$), then a line will be obtained as illustrated in Figs. 4–6.

In Fig. 4 the difference between reactors operating under conditions of excellent hydrogen transfer rate (low value of $1/k_La$) and of moderate mass transfer rate (higher values for $1/k_La$) are reflected. The intersection with the C_{max}/J_{mean} axis indicates the value of the parameter $1/k_La$. On the basis of $1/k_La$ it is possible to distinguish two regimes, one in which $1/k_La$ varies between 1 and 3 and one in which it varies between 5 and 50. This latter value is obtained in closed reactors in which baffles are not used.

It is our experience that two different types of lines are obtained:

1. Straight lines if oils are hydrogenated that do not contain contaminants that poison the nickel catalyst particles
2. Lines that bend upward as the parameter 1000/wppm Ni increases (which occurs if poisons are present)

These two different lines are illustrated in Fig. 5. The curved lines are typically obtained for the results of fish oil and rapeseed oil hydrogenations. The less catalyst is used, the higher the value 1000/wppm Ni and the greater the effect of poisoning will be.

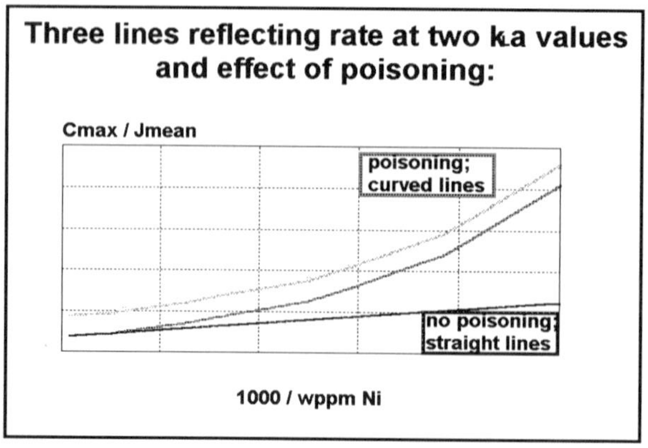

Fig. 5 Typical lines obtained for C_{max}/J_{mean} vs 1000/wppm Ni when hydrogenated oils are (top two lines) and are not (bottom line) contaminated by poisons.

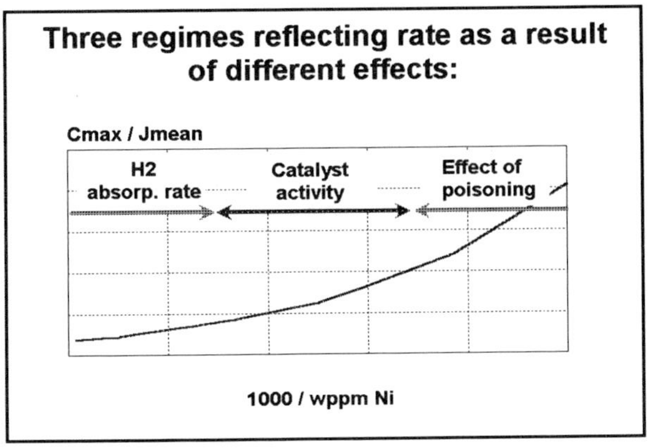

Fig. 6 Three regimes reflecting changes in hydrogenation rate as a result of effect of hydrogenation rate, catalyst activity, or poisoning.

Another important observation is that if two experiments are performed at the same pressure but under lower or higher values of $k_L a$, the lines move upward or downward parallel to each other. The difference between the two experiments is that in the case of higher values of $k_L a$, the actual hydrogen concentration in the liquid is higher and the reaction proceeds faster. The fact that the lines move parallel to each other implies that the parameter determining the slope remains constant. This is the parameter $X^n\{k_{r,p}[\mathrm{IV}(t)-75]\eta_p(t)\}_{\mathrm{mean}}$, the parameter that describes the reaction rate. Moreover, the reaction can indeed be considered first-order in hydrogen.

However, if the pressure is changed considerably then these lines tend to bend upward more sharply if the pressure is higher. Apparently, if the bulk hydrogen concentration between two experiments is different because higher pressures are used, then the slope is more significant at the higher pressures. Usually, higher pressures will result in a significant increase in the hydrogen concentration. The stronger curvature of the slope implies that the poisoning occurs faster as a result of which the rate becomes slower.

In Fig. 6 three regimes are shown where the effect of hydrogen absorption rate, catalyst activity, or poisoning dominates the process. At low values of 1000/wppm Ni the predominant effect is the hydrogen absorption rate, and at high values of this parameter the predominant effect is poisoning.

G. General Equation for the Hydrogenation Time of the Hydrogenation Process

In most cases the hydrogenators are interested in the time required to get the iodine value reduced from the initial value $\mathrm{IV_o}$ to a particular $\mathrm{IV_{end}}$. A graph in which this hydrogenation time is plotted on the vertical axis against the amount of catalyst, preferably expressed as 1000/wppm Ni, on the horizontal axis can be very helpful. For that reason we use the concept of hydrogenation time. The equation expressing the hydrogenation time, HT, is

$$\mathrm{HT} = 0.6(\mathrm{IV_o} - \mathrm{IV_{end}})/J_{\mathrm{mean}} \qquad (\mathrm{min}) \tag{42}$$

The proportionality constant of 0.6 is required if HT is in minutes and J_{mean} is in moles per cubic meter of oil per second.

By substitution of Eq. (41) for J_{mean}, the following equation is obtained:

$$\text{HT} = \left(\frac{1}{k_{\text{L}}a} + \frac{1}{\epsilon_{\text{cat}} X^{n}\{k_{\text{r,p}}[\text{IV}(t) - 75]\eta_{\text{p}}(t)\}_{\text{mean}}}\right)\left(\frac{0.6(\text{IV}_{\text{o}} - \text{IV}_{\text{end}})}{C_{\text{max}}}\right) \tag{43}$$

or

$$\text{HT} = \text{HT}_{\text{min}} + \frac{1000}{\text{wppm Ni}} B \tag{44}$$

in which

$$\text{HT}_{\text{min}} = 0.6(\text{IV}_{\text{o}} - \text{IV}_{\text{end}})/(k_{\text{L}}a\, C_{\text{max}}) \qquad (\text{min}) \tag{45}$$

and

$$\frac{1000\ \text{B}}{\text{wppm Ni}} = 0.6(\text{IV}_{\text{o}} - \text{IV}_{\text{end}})\left\{\left[\frac{\rho_{\text{oil}}/\rho_{\text{part}}}{\%\ \text{Ni in particle} \times 10^{4}}\right][X^{n}\{k_{\text{r,p}}[\text{IV}(t) - 75]\eta(t)\}_{\text{mean}}\, C_{\text{max}}]\right\}^{-1} \tag{46}$$

The first term on the right-hand side of Eq. (44) is the minimum hydrogenation time HT_{min}. It is determined by process parameters only and not by the type or quality of the catalyst. HT_{min} decreases linearly with increases in C_{max} and $k_{\text{L}}a$.

In a commercial soybean oil hydrogenation process with $k_{\text{L}}a = 0.15$, a pressure of 2 bar (=200 kPa), and ΔIV (=$\text{IV}_{\text{o}} - \text{IV}_{\text{end}}$) =50, the minimum hydrogenation time is already 32 min, and in fish oil hydrogenation with $k_{\text{L}}a$ =0.15, a pressure of 4 bar (=400 kPa), and $\Delta\text{IV} = 80$, the minimum hardening time will be 25 min. If in the latter case a very high stirring speed were used, then $k_{\text{L}}a$ would be 1 and the minimum hydrogenation time only 4 min.

It was explained in Section II.B.4 that the $k_{\text{L}}a$ value for a specific power input can be increased by a factor of 2 if an optimally designed impeller is used. Then the minimum hydrogenation time is lower by a factor of 2. Moreover, if a loop-type reactor is used, the minimum hydrogenation time can probably be reduced further.

The second term on the right in Eq. (44) is determined by $\text{IV}_{\text{o}} - \text{IV}_{\text{end}}$; the type, quality, extent, and rate of poisoning; and the amount of catalyst and hydrogen pressure used in the system, but not by other process parameters such as the stirring speed.

All in all, the minimum hydrogenation time is a meaningful parameter reflecting all factors that affect the hydrogenation rate and are not determined by the quality or amount of the catalyst. The performance of two reactors of different sizes can be considered almost equal if they are operated under the same temperature and pressure and the minimum hydrogenation times are equal. The only difference between two reactors of different size and type originates from the difference in the distribution of the local energy dissipation, which can cause minor differences in the concentration of hydrogen in the liquid phase. It is not expected that these differences will have a major impact on the product qualities obtained after hydrogenation.

The following table summarizes the drop in hydrogenation time for three cases.

Reactor type	Impeller type	$k_{\text{L}}a(\text{s}^{-1})$	HT_{min} (min)	HT (min)
Tank	Rushton	0.15	32	64
Tank	CD-6	0.22	22	54
Loop		0.30	16	48

In cases where the mean value of the volumetric liquid-phase mass transfer coefficient for a loop reactor is twice as high as for a tank reactor using a conventional impeller, then the overall hydrogenation time will be lower by only 16 min. To get the same product quality it may be necessary to operate the loop reactor under a slightly lower pressure such that the hydrogenation times are equal again. An advantage of the loop reactor remains that it can be operated at a lower pressure to get the same product quality in the same hydrogenation time.

H. Examples Obtained from Laboratory-Scale Reactors

In this section we review a number of examples of the hydrogenation of two soybean oils, one canola oil, and three fish oils. Let us first discuss the main features of the equipment used for our catalyst performance tests. An adequate catalyst performance test consists of two important elements: reliable equipment and appropriate test procedures.

We selected Medimex autoclaves, which have been operating to our full satisfaction since 1990. The rate of hydrogenation is monitored and controlled by a press flow control unit as depicted in Fig. 7. These units were originally manufactured by Peteric, but have been manufactured for the past few years by Büchi. This system is fully automated and computer-controlled.

The hydrogen supply unit controls the hydrogen supply rate by keeping the hardening reactor at a fixed pressure by continuously injecting a known small amount of hydrogen. A computer monitors the amount of hydrogen. In this way the hydrogen absorption rate can be monitored. Consequently, the iodine value and the rate of change in it can be calculated as a function of hydrogenation time.

Our test reactor is operated as a dead end reactor. This has the advantage that in combination with the computer control, the oil can be hydrogenated to a fixed end IV because all the hydrogen fed into the reactor is used for hydrogenation.

Once the test is completed, the hydrogenated product can be subjected to several analyses. The proper functioning of the equipment is controlled by measuring the change in the refractive index of the oil. A second advantage of the computer control is that not only the pressure, but also the temperature and stirring speed, are continuously monitored and adjusted. The temperature can be kept isothermic or raised adiabatically.

1. Hydrogenation of a Soybean Oil

Figure 8 depicts the results of hardening "soybean oil 1" at a temperature that was increased stepwise from 150°C to 180°C at atmospheric pressure and two different stirring speeds. The volumetric liquid-phase mass transfer coefficient, $k_L a$ was 1 and 0.3 s^{-1}, respectively. This difference will cause a significant difference in the bulk hydrogen concentration.

The amount of catalyst was varied between 30 and 500 wppm Ni. The two lines are almost

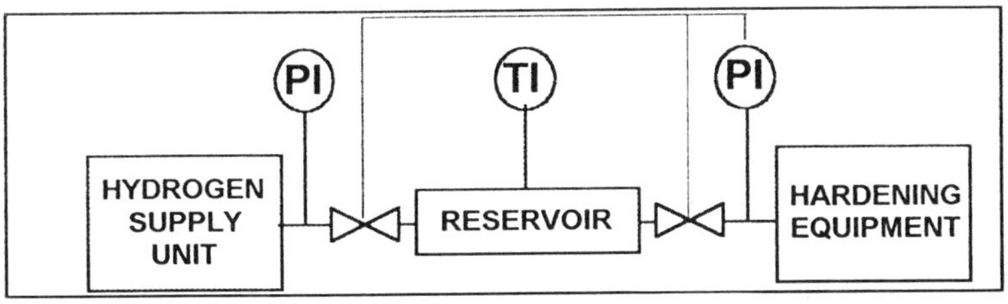

Fig. 7 Control scheme to monitor the rate of hydrogen supply to the Medimex autoclave.
PI = pressure indicator.
TI = temperature indicator.

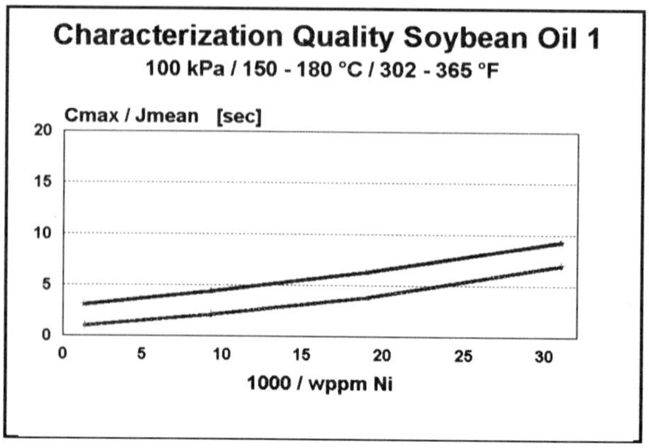

Fig. 8 Characterization of soybean oil 1 at 100 kPa and 150–180°C (302–365°F).

straight and exactly parallel. There was no effect of poisoning, and it can also be concluded that the activity of the catalyst is the same at the two different hydrogen concentration levels.

In Fig. 9 the results of an isothermal experiment of hardening of "soybean oil 2" at 180°C and 150 kPa are depicted. In this case also the data show an almost straight line. In this case the $k_{L}a$ value is much smaller, 0.14 s^{-1}.

2. *Hydrogenation of a Canola Oil*

Figure 10 depicts the results of hardening a canola oil at a constant temperature and three different pressures, with a volumetric liquid-phase mass transfer coefficient $k_{L}a$ of 0.13 s^{-1}. The amount of catalyst was varied between 100 and 6000 wppm Ni. In this case straight lines are not obtained. The effect of poisoning causes the lines to bend upward. The higher the pressure, the more the curve bends. Hence at higher pressures poisoning seems to have a stronger effect.

In Fig. 11 the results for C_{max}/J_{mean} for two different catalysts, PRICAT 9920 and PRICAT 9910 (see Section III.B), are depicted. The higher the nickel content (the lower the value 1000/wppm Ni), the less difference there is between the two catalysts. Only at low wppm Ni can a

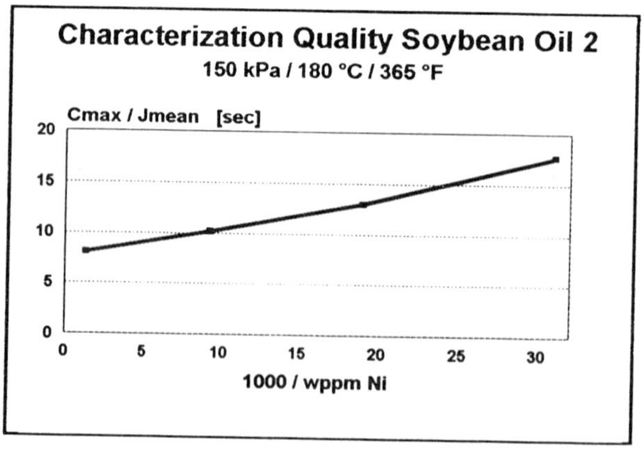

Fig. 9 Characterization of soybean oil 2 at 150 kPa and 180°C (365°F).

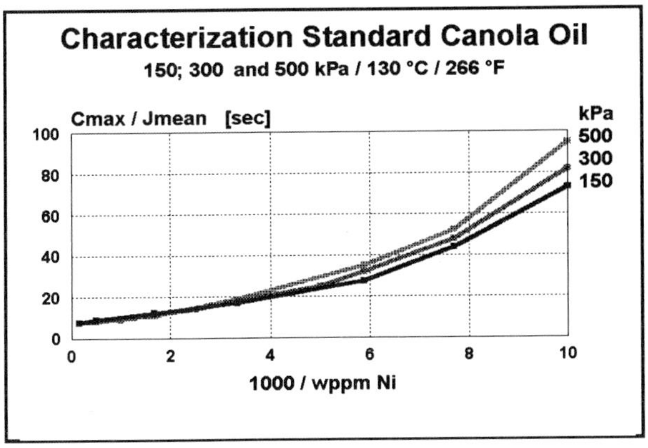

Fig. 10 Characterization of standard canola oil hardened at 150, 300, and 500 kPa and 130°C (266°F).

difference be observed for the parameter C_{max}/J_{mean}. For example, at 100 wppm Ni this difference is 12 s versus 16 s. It can also be concluded from Figs. 10 and 11 that at equal stirring speed the $k_L a$ value increases if the temperature rises from 130 to 180°C, namely from 0.13 s^{-1} to 0.19 s^{-1}.

3. *Hydrogenation of Three Fish Oils*

Figures 12–17 depict the results for "fish oil 1." The amounts of catalyst used in these experiments were 1760 wppm Ni (line 1), 880 wppm Ni (line 2), 660 wppm Ni (line 3), and 440 wppm Ni (line 4). The operating pressure was 900 kPa, and the temperature was adiabatically increased from 160°C to 190°C and thereafter kept constant.

In Figs. 12 and 13, IV is plotted as a function of time, in Fig. 12 on a linear scale and in Fig. 13 on a logarithmic scale. These results are in line with findings that have been reported by others [20]. The plot on the logarithmic scale shows straight lines. It is wrong, however, to conclude that this observation proves that the hydrogenation reaction is of first order in the iodine value. Such a conclusion can be drawn only if it is certain that the bulk hydrogen concentration had a constant value during the entire reaction time.

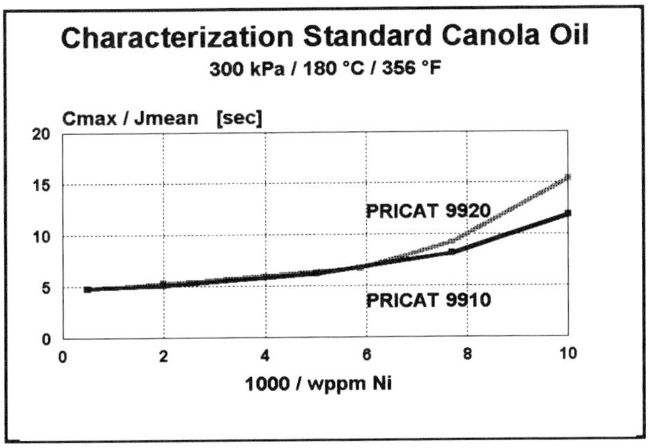

Fig. 11 Characterization of standard canola oil hardened at 300 kPa and 180°C (365°F).

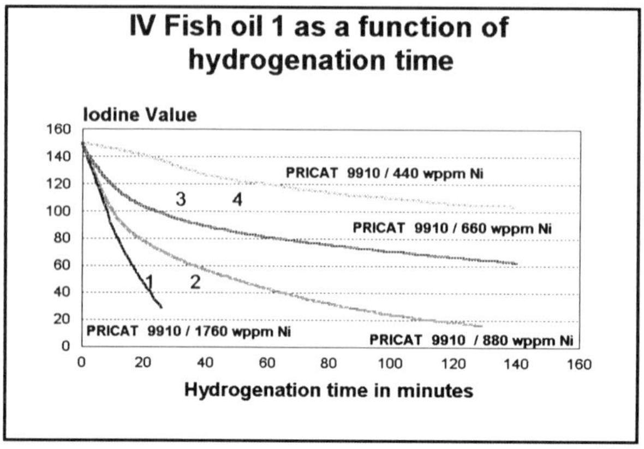

Fig. 12 Iodine value of fish oil 1 as a function of hydrogenation time with PRICAT 9910 and various amounts of Ni as catalysts.

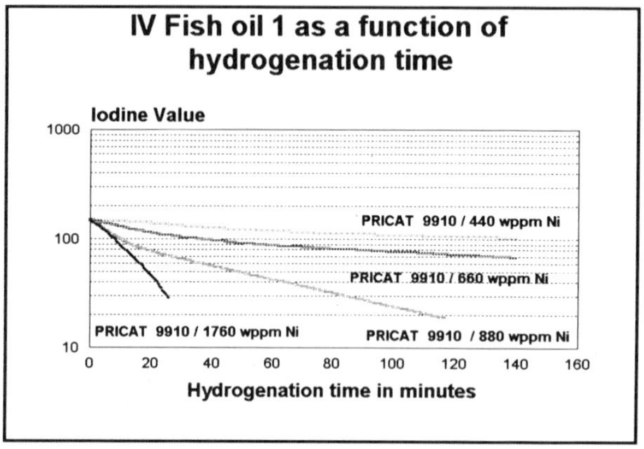

Fig. 13 Iodine value of fish oil 1 plotted logarithmically as a function of hydrogenation time with PRICAT 9910 and various amounts of Ni as catalysts.

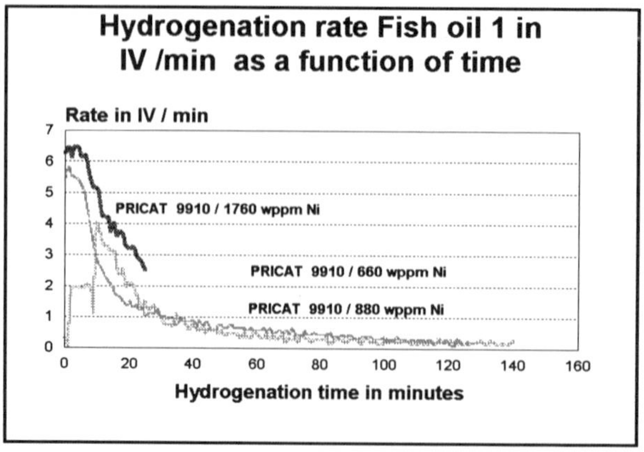

Fig. 14 Hydrogenation rate of fish oil 1 as a function of hydrogenation time.

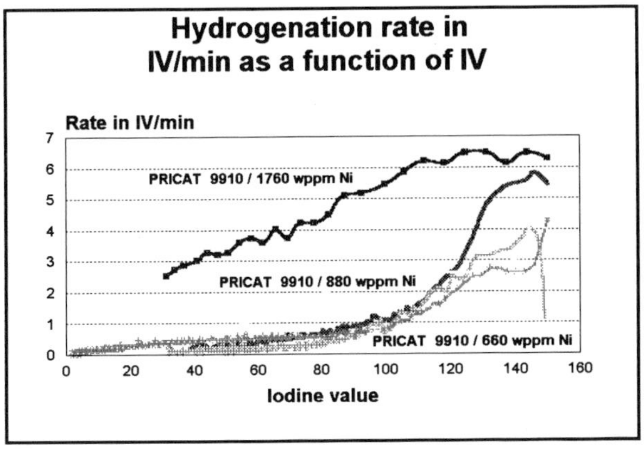

Fig. 15 Hydrogenation rate as a function of iodine value.

In Figs. 14 and 15, hydrogenation rate is plotted as function of the hydrogenation time (Fig. 14) and as function of the iodine value (Fig. 15). Figure 15 shows that the hydrogenation rate increases if IV increases, but the increase is not in linear proportion to IV.

It still cannot be concluded that this simple linear correlation exists over the whole IV range, because during this hydrogenation the bulk hydrogen concentration changes also. We therefore now first calculate the bulk hydrogen concentration for the four cases. These results are plotted in Fig. 16.

The hydrogen concentration in the bulk of the liquid increases as hydrogenation proceeds. In the first 30 min the increase is rapid, but it then slows down. The change in the bulk hydrogen concentration is, as expected, different for all three cases. The higher the amount of nickel used, the lower the hydrogen concentration at the start of hydrogenation. In the case of the lowest nickel content the maximum hydrogen concentration is obtained almost directly after the start of the hydrogenation process. In the other two cases the bulk hydrogen concentration changes from a low value to the maximum value. In the case of the highest nickel content this maximum has not yet been achieved when hydrogenation is complete at an IV of 30 (see Fig. 16).

We can now calculate the hydrogenation rate assuming that the hydrogen concentration in

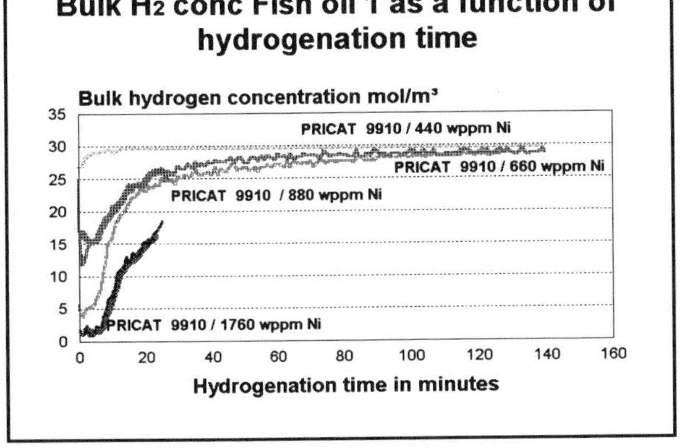

Fig. 16 Bulk hydrogen concentration in fish oil 1 as a function of hydrogenation time.

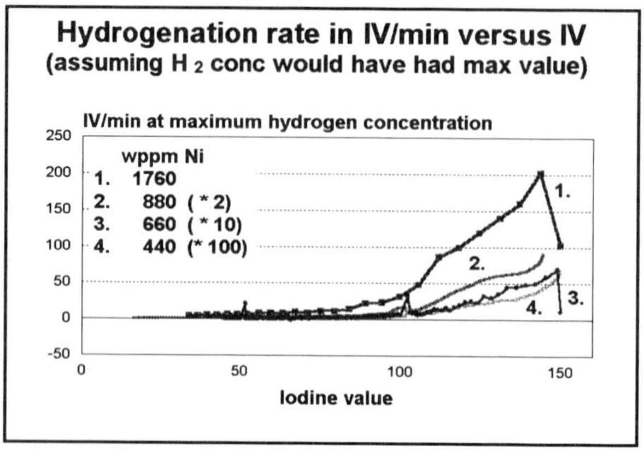

Fig. 17 Bulk hydrogen concentration in fish oil 1 in IV/min vs IV (assuming that the H_2 concentration would have had a maximum value).

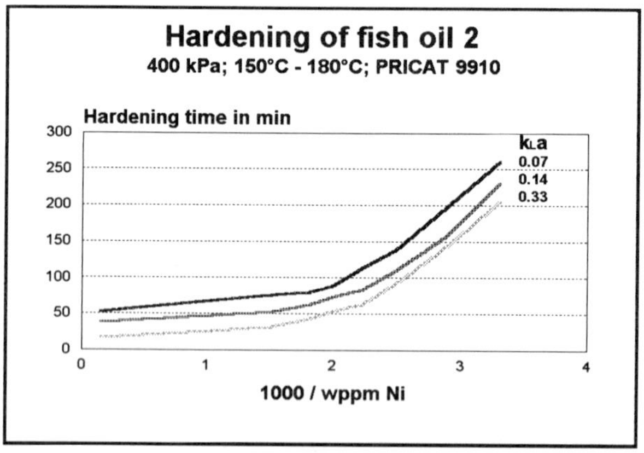

Fig. 18 Hardening time for fish oil 2 vs concentration of Ni catalyst for PRICAT 9910 at 400 kPa and 150–180°C for three values of $k_L a$.

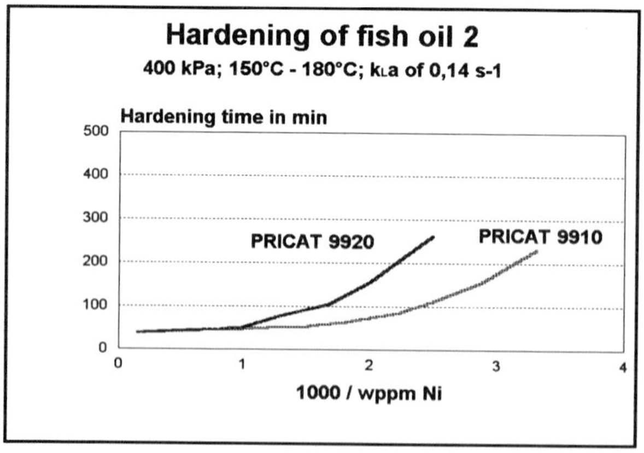

Fig. 19 Hardening time for fish oil 2 vs Ni concentration with PRICAT 9910 and 9920 at 400 kPa, 150–180°C, $k_L a = 0.14 s^{-1}$.

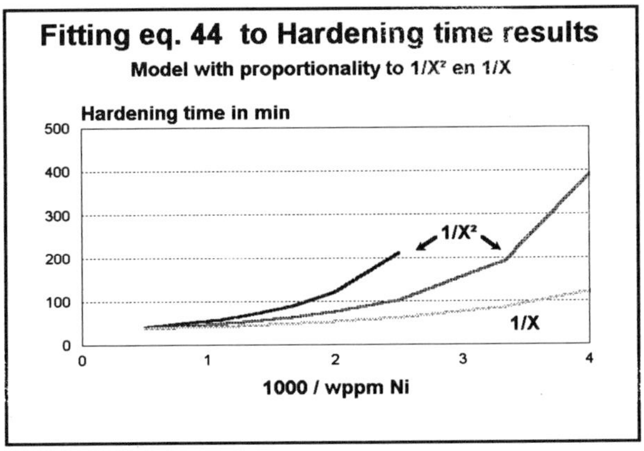

Fig. 20 Fitting eq. 44 to hardening time results. Model with proportionality to $1/X^2$ en $1/X$.

the bulk of the liquid reached its maximum value right at the start of the hydrogenation process. This calculation is, however, based on the assumption that the hydrogenation rate is indeed first order in the hydrogen concentration. These results are plotted in Fig. 17.

From the results depicted in Fig. 17 it can be concluded that for this fish oil the hydrogenation rate is certainly not first order in the iodine value. In all three cases there are two distinct regimes. For an IV of 0 to 80, the increase in the hydrogenation rate is indeed linear, but above IV-80 the rate increases linearly with the quantity IV-75. This is the regime where most commercial edible oil hydrogenations are carried out. In order to reflect all these data in one picture, the data for the lower wppm Ni had to multiplied by the factor given in parenthesis.

The next two figures depicted the results of the hardening time required to drop the IV of fish oil 2 from 165 to 82 as a function of 1000/wppm Ni. The conditions applied were 400 kPa, 150°–180°C, with the switch point at the iodine value calculated from $\Delta IV = 0.002(IV_o)^2$, and three different stirring speeds resulting in $k_L a$ values of 0.07, 0.14, and 0.33 s^{-1} (Fig. 18). All three lines are parallel, indicating that the assumption of a reaction being first order in hydrogen is correct for this fish oil hardening reaction as well.

In Fig. 19 the results of a similar test at $k_L a = 0.14$ are plotted for the two catalysts PRI-

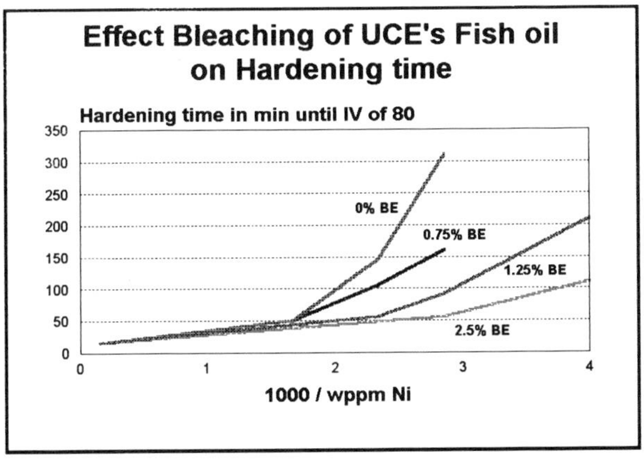

Fig. 21 Effect bleaching of UCE's Fish oil on hardening time.

CAT 9920 and PRICAT 9910. It is clearly demonstrated that PRICAT 9910 is much more poison-resistant than PRICAT 9920. This feature is ascribable to the lower specific nickel surface area of this catalyst.

The next question to be answered is whether the lines in Fig. 19 can be described by Eq. (37). This analysis requires an assumption for $(k_s/k_r)(\eta_s/\eta_p)$ as given in Eq. (37). A value of 0.1 has been used, because the reaction rate constant for sulfided nickel is about one-tenth that for pure nickel, and η_s/η_p will be about 1. The value of x has been calculated assuming that the surface of 200 wppm Ni of catalyst A and 275 wppm Ni of catalyst B has been poisoned by the sulfur content of the oil. The result is depicted in Fig. 20.

In Fig. 20 the lower line reflects the correlation for $n=1$ in X^n, and the two upper lines (black and gray) are for $n=2$. There is surprisingly good agreement with the experimental results if $n=2$, for both PRICAT 9920 (black line) and PRICAT 9910 (gray line).

Figure 21 depicts the correlation between the hardening time and the reciprocal amount of nickel for four different cases: one for fish oil 2 and three for nb fish oil 2, fish oil 2 that was bleached using 0.75, 1.25, and 2.5 wt % bleaching earth. This figure clearly demonstrates the reduction of catalyst poisoning if the fish oils are bleached before they are hardened.

III. PRODUCT PROPERTIES AFTER HYDROGENATION

A. Modification of Oils

The use of liquid oils for a variety of fat-based products requires modifications of two properties: the oxidative stability of the oil and its melting characteristics. Three different modification processes can be used for this purpose. Among these three basic fat modification processes—hydrogenation, fractionation, and interesterification—hydrogenation is, even today, more than 90 years after its commercial introduction, the most widely used. It effectively improves the oxidative stability of the oil, promotes the interchangeability of raw materials, and creates great diversity in the melting characteristics of the resulting fats. However, due to several factors, such as the increased awareness of possible negative effects of trans isomers on human health, the continuous change in the availability of the raw material mix, an increased preference for soft margarines, the recognition of environmental issues, and the use of technically improved esterification processes, the picture of hydrogenation is changing.

The question of how oxidative stability and melting characteristics can be changed such that optimal product properties will be obtained has been studied over the past 70 years by many scientists, for example, in the early 1950s by Waterman and coworkers [21] at Delft Institute of Technology, in the 1960s and 1970s by Coenen and colleagues [22] at the Unilever Research Laboratory at Vlaardingen, by Albright at Purdue University, and by others. These studies led to the conclusion that when porous Ni catalysts are used, the selectivity of the edible oil hydrogenation reactions is the key factor determining melting characteristics.

Different selectivities have been defined to characterize the complexity of this hydrogenation reaction. To explain the effects of selectivity, it must be understood which types of reactions occur during hydrogenation. It is well known that around 80% of all fatty acids in edible oils have a carbon chain length of 18 atoms. Therefore, the consecutive hydrogenation reactions are usually depicted by the simple reaction equation

$$18{:}3 \xrightarrow{+H_2} 18{:}2 \xrightarrow{+H_2} 18{:}1 \xrightarrow{+H_2} 18{:}0$$

For such a set of consecutive reactions it has been observed [23] for many years that the selectivity toward the intermediate products is strongly determined by the diffusional limitation of the large triglyceride molecules and by the hydrogen concentration within the pores.

Usually it has been assumed that there is no significant diffusional limitation of hydrogen within the pores. This assumption is correct for low temperature hydrogenation, but not for temperatures as high as 180°C, which will be discussed further in this section.

Four different selectivities have been defined:

- First, the linoleic and oleic acid selectivities, which express the preference for linoleic acid over oleic acid and for oleic over stearic acid hydrogenation, respectively. If these preferences are high, stearic acid is not formed until almost all polyene has been hydrogenated. The result is a product high in linoleic and oleic acid and low in stearic acid.
- Second, triglyceride selectivity is a measure of the extent to which the three fatty acid chains in one fat molecule behave independently. This can be illustrated by the following three hypothetical cases.

  1. The case of extreme triglyceride selectivity is given by the presence of only one saturated fatty acid in each triglyceride molecule. This case will result in a very low melting point.
  2. In a fully random situation the same amount of saturated fatty acids will not have been distributed equally over all the triglyceride molecules, but it will have been distributed in accordance with a Poisson distribution. There will be a small fraction of fat molecules containing two and three saturated fatty acids, and there will also be a small fraction containing no saturated fatty acid molecules. This situation yields a melting point about 10°C higher.
  3. If all the saturated fatty acid molecules are concentrated in only trisaturated molecules mixed with unhardened oil, the result will be a mixture of high melting fat crystals and liquid oil, usually a highly undesirable situation.

  In practice, the highest possible triglyceride selectivity yields a fully random distribution. If one-third of the fatty acids is saturated (S) and two-thirds is unsaturated (U), the fraction of SSS will be almost 4%, and the fractions of UUU, SUU, and SSU will be 30%, 44%, and 22%, respectively.
- Finally, the trans selectivity describes the tendency for the formation of trans isomers, defined as percent trans formed per unit IV reduction. Trans fatty acids greatly raise the solid fat content below 30°C, resulting in steep melting products.

All types of selectivity, and hence melting behavior, are determined by three parameters: temperature, bulk hydrogen concentration in the oil, and type of catalyst.

Temperature. The effect of temperature is that the rate by which each hydrogenation reaction is increased is different, because the activation energies for each hydrogenation reaction are not expected to be the same. Moreover, the effect of a distribution of hydrogen concentration within the catalyst particles will start to affect selectivity, because as temperature increases, the diffusional limitation of hydrogen will start to exert its effect on selectivity. These concentration distributions will be the result of diffusional limitation of hydrogen within the catalyst particles. The lower the concentration, the higher the selectivity will be.

Bulk Hydrogen Concentration in the Oil. This parameter is not an independent parameter, but it has been explained in the previous sections that the bulk hydrogen concentration is determined by the hydrogen pressure, the volumetric liquid-phase mass transfer coeficient $k_L a$, and the amount and type of catalyst used in the hydrogenation. The higher the pressure, the higher the hydrogen solubility in the oil. It will increase in proportion to the pressure in the pressure range used for hydrogenation. The higher the $k_L a$ value, the faster the hydrogen is transferred from the gas to the oil phase, which will result in a higher bulk hydrogen concentration. It has been explained in Section II.B that higher k_La values can be obtained by a number of options, one of them being higher stirring speeds.

The more catalyst is used, the faster the hydrogen reacts away from the bulk, and the lower its concentration will be. Equation 2 expresses the latter two effects quantitatively.

It has been experimentally observed for several systems of consecutive hydrogenation reactions that the selectivity for the intermediate product is better if the hydrogen concentration is low [24]. A proper explanation for this observation has not yet been found.

Type of Ni Catalyst. The main features of edible oil hydrogenation catalysts are (1) the mean particle size and the particle size distribution, (2) the mean pore size and the pore size distribution, and (3) the specific nickel surface area. The main reason for using small Ni catalyst particles with a mean size of 4–10 μm is to minimize the phenomenon of diffusional limitation. The molecules that are hydrogenated on the active Ni sites, which are located throughout the whole particle volume, diffuse from the outside of the particle through the intricate texture within the particle toward the center. At the same time there is a chance of them reacting while passing an active site. As a result of these two simultaneously occurring processes, a gradient will be obtained in the concentration of these two components.

The rate of diffusion of the triglyceride (TG) and hydrogen molecules through the outer surface of the (spherical) particles (at $r=R$) is given by Fick's first law of diffusion:

$$J_{TG} = -\epsilon_{pore} D_{TG,eff} \left. \frac{d\,TG}{dr} \right|_{r=R} \tag{47}$$

$$J_{H_2} = -\epsilon_{pore} D_{H_2,eff} \left. \frac{dC_{H_2}}{dr} \right|_{r=R} \tag{48}$$

where ϵ_{pore} is the porosity of the catalyst particle, a property that also reflects the fraction of the outer surface through which the molecules diffuse, $D_{TG,eff}$ is an effective diffusion coefficient of the oil molecules diffusing through the intricate structure of the catalyst particle, and $D_{H_2,eff}$ is the same property for hydrogen. The concept of effective diffusion coefficient has been defined by van Krevelen [25]. The molecules diffusing from the outer surface of the particle in the direction of the pores have to follow a zigzag path and pass several narrow passages, and they can be adsorbed as well. The zigzag factor τ reflects the decrease in the diffusion coefficient due to these three obstacles.

$$D_{H_2,eff} = D_{H_2}/\tau \tag{49}$$

The value of τ has been measured for several different types of extrudates and pellets 0.8–3 mm in size. Usually these values are on the order of 5. For highly porous catalysts one might expect the value of τ to approach 1, because it becomes easier and easier for the molecules to diffuse through the intricate texture of the catalyst particle.

The ratio of the rate of diffusion to the rate of reaction is decisive for the steepness of this concentration gradient. If the diffusion rate cannot keep up with the reaction rate, then there will be a steep concentration gradient. At the outer surface the concentration will be high, but toward the center it can become very low, causing a decrease in the reaction rate. Such a situation is usually described by an effectiveness factor η, defined by the ratio

$$\eta = \frac{\text{actual conversion rate within the catalyst particles}}{\text{conversion rate in the absence of diffusional limitation}} \tag{50}$$

In edible oil hydrogenation three regimes are observed:

1. No diffusional limitation. There is no diffusional limitation of oil molecules or hydrogen at very low temperatures, for example at 80°C. An excessive amount of catalyst has to be used to obtain a rate of hydrogenation that is sufficiently high, and it is therefore not of commercial interest. In lab experiments it can be demonstrated that under these conditions the quality of the product is not determined by the type of Ni catalyst as long as the catalyst particles are in the micrometer size range. For example, the properties of the product obtained by using colloidal unsupported Ni will be the same as if it were obtained by using Ni supported by an inert carrier.

2. Diffusional limitation of oil molecules only. Once the temperature increases to above 100°C, the reaction rate increases to such an extent that the diffusion rate of the oil molecules cannot keep up with the reaction rate. The quality of the hydrogenated product is now determined by the type of catalyst, namely by the accessibility of the Ni sites for the large triglyceride molecules, the size of which is 1.5–2 nm. If the pores are long and narrow, the diffusion rate of large triglyceride molecules cannot cope with the reaction rate, resulting in steep concentration gradients along the length of the pores. Even at the beginning of the reaction with polyunsaturated fatty acids still present, the deepest pore bottom only sees fully hydrogenated products and the nickel surface area hardly contributes any more to activity. Less deep in the pore, intermediate products may accumulate, and if the supply of the original raw material is insufficient, through slow diffusion, intermediates cannot escape quickly from the catalyst particle, and highly saturated products are formed. In other words, the composition inside the pore runs ahead of that in the bulk of the liquid. Thus, narrowing of pores impairs not only activity but selectivity as well. It may be necessary to increase pore radius much further than 3 nm if high selectivity is required or if unselective conditions such as high H_2 pressure, low temperature, or hydrogenation with low amounts of catalyst are used.

3. Diffusional limitation of oil and hydrogen molecules. At still higher temperatures, for example at 180°C, the reaction is so rapid that the diffusional limitation of hydrogen also starts to affect the reaction rate. A steep concentration gradient of hydrogen is then also obtained, and the hydrogen concentration at the center of the catalyst particle will be considerably lower than the concentration at the outer surface. Despite the fact that a vast amount of literature exists on the hydrogenation of edible oils, very little quantitative data have been published on the degree of diffusional limitation of hydrogen within the nickel catalyst particles. For optimization of the size of the catalyst particles, it is of interest to understand whether any diffusional limitation of hydrogen exists.

To estimate the extent of diffusional limitation of hydrogen within the catalyst particles, one needs to know the following properties:

Liquid phase diffusion coefficient of hydrogen in oil, D_{H_2}; see Ref. 5.

The tortuosity factor, τ. One of the features of edible oil catalysts is the very high porosity. Porosities of 70–85% are common. For such highly porous catalysts the tortuosity factor is assumed to vary from 3 to 2, respectively.

The particle porosity, ϵ_{pore}, which can be calculated from the nitrogen physisorption data and the skeletal density.

The volume of catalyst particles per wppm Ni. This can be calculated from the composition and the particle porosity.

The Sauter mean diameter $d_{3.2}$, which is measured by the laser diffraction technique.

The mean hydrogen concentration within the bulk of the oil, which can be calculated using Eq. (2).

The initial hydrogen absorption rate [mol/($m^3_{oil} \cdot s$)], which is calculated from the rate at which IV changes over time.

The hydrogenation rate per unit volume of catalyst is expressed by Eq. (24), which we repeat here:

$$R(t)=J(t)/\epsilon_{cat} \tag{24}$$

The flux J_{cat} in mol/(m$^2{}_{cat}\cdot$s), diffusing into the particles at $r=R$ is

$$J_{cat}(t)=R(t)d_{3,2}/6 = \epsilon_{pore}\frac{D_{H_2}}{\tau}\frac{\Delta C_{H_2}}{\Delta x}\bigg|_{\text{at } r=R} \tag{51}$$

or

$$\Delta x=\frac{\epsilon_{pore}}{\tau}D_{H_2}\frac{\Delta C_{H_2}}{J_{cat}(t)} \tag{52}$$

And the smallest value of Δx, Δx_{min}, is obtained for the highest value of the absorption rate. Namely, for J_{cat} $(t=0)$,

$$\Delta x_{min}=\frac{\epsilon_{pore}}{\tau}D_{H_2}\frac{\Delta C_{H_2}}{J_{cat}(t=0)}$$

The physical meaning of the value of Δx is the depth of hydrogen penetration if the decrease in concentration had been linear. Of course, this decrease will not be linear, but in any case this distance Δx can be considered a characteristic depth.

In Table 2 the value of Δx_{min} has been calculated for two different catalyst samples A and B for the initial hydrogenation rate of a soybean oil. The data calculated for Δx_{min} show hardly any diffusional limitation for hydrogen within particles of a size of 5 μm at a temperature of 120°C, but at 180°C a severe diffusional limitation does indeed exist. Figure 22 depicts the hydrogen concentration gradient at $x=R$ at 120° and 180°C, as can be estimated from Δx_{min} for catalyst A.

At 120°C a gradient at $x=R$ can be observed, but this gradient will not result in severe hydrogen depletion at $x=0$. The gradient at 180°C is very steep, and this gradient will result in severe hydrogen depletion at $x=R$. Moreover, it has been clearly demonstrated that under these conditions hydrogenation occurs mainly within a thin layer at the outer surface of the catalyst particle. Of course, as hydrogenation proceeds and the rate drops in proportion to IV − 75, Δx_{min} will increase.

Table 2 The Depth of H_2 Penetration for Two Catalysts and Two Hydrogenation Temperatures

Temp. (°C)	Bulk H_2 concn. (mol/m³)	$J(t=0)$ [mol/(m$^3{}_{oil}$/s)]	a_S (m$^2{}_{cat}$/m$^3{}_{oil}$)	J_{cat} $(t=0)$ [mol/(m$^2{}_{cal}$/s)]	D_{H_2} (m²/s)	ΔX_{min} (μm)
Catalyst A, $\epsilon_{pore}/\tau=0.26$						
120	1.5	0.44	400	1.1×10^{-3}	1.3×10^{-8}	4.6
180	1.3	0.71	50	14.2×10^{-3}	2.7×10^{-8}	0.64
Catalyst B, $\epsilon_{pore}/\tau=0.42$						
120	0.9	0.52	608	1.9×10^{-3}	1.3×10^{-8}	5.4
180	1.4	0.68	76	24×10^{-3}	2.7×10^{-8}	0.67

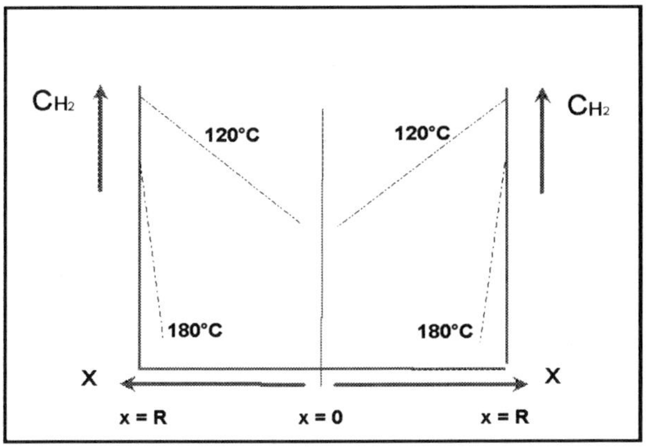

Fig. 22 The slope of the H_2 concentration gradient at r = R if T = 120°d and 180°d (both of cat A).

B. Current Types of Ni Catalysts for Edible Oil Hydrogenation

Over the past 10 years the development of new or improved versions of Ni catalyst systems by most catalyst companies has been the result of a better understanding of the factors affecting activity, selectivity, catalyst poisoning, and filtration rate in combination with an understanding of customer needs. Activity and poison resistance have in many cases been improved to the extent that these catalysts can be used at starting temperatures as low as 120°C. New catalysts have been introduced onto the market that exhibit considerably better selectivity and are very useful in vegetable oil hydrogenation.

At Unichema Int. this development has led to the introduction of PRICAT 9910 and PRICAT 9920. At Engelhard it has led to the introduction of Nysosel 222 and 545 and Nysosel 325 and 645; at UCI, of G135 and G95D; at Mallinckrodt, of Calsicat 428 and Calsicat 472; and at Hoechst, of 882 OF.

PRICAT 9910 is a multipurpose, sulfur-resistant, medium-pore catalyst with a high nickel surface area suitable for all types of oils under quite selective conditions. PRICAT 9920 is a very selective, wide-pore catalyst that has been particularly developed for vegetable oil hydrogenation. The better poison resistance of PRICAT 9910 has been shown by the results depicted in Figs. 11 and 19. To illustrate the difference between these two catalysts with respect to their selectivity, their performance will be compared in a number of hydrogenation processes.

1. Examples of Catalyst Performance in the Modification of Oils

The first example below reviews the difference between the performances of the two catalysts PRICAT 9910 and PRICAT 9920 in hydrogenating an nb-soybean oil. These data should not be considered as characteristic of an nb-soybean oil, because the quality of an nb-soybean oil is determined by several factors, such as phospholipid content. With other oil qualities, melting points 1–2°C higher or lower may be obtained.

The data give the results of two sets of experiments, one at 150 kPa and the other at 400 kPa. In both cases the experiments were performed adiabatically at 120 and 180°C. The volumetric liquid-phase mass transfer coefficient, k_La, was 0.13 and 0.20 s^{-1}, respectively.

Table 3 Characteristic Data for IV = 80 of a Hydrogenated Soybean Oil with IV_o = 132.8 and a Feed Composition of 4% C18:0, 23.1% C18:1, 53.1% C18:2 and 7.4% C18:3

	120°C; 150 kPa		180°C; 150 kPA	
	PRICAT 9910	PRICAT 9920	PRICAT 9910	PRICAT 9920
Hydrog. time, min	75	85	74	83
Melting point, °C	34.0	29.5	31.5	30.5
Trans content, wt %	29.0	30.0	40.0	40.0
Solid fat content by NMR				
N_{35}	1.5	0.2	0.2	0.1
GC analysis of fatty acids				
18:0, %	9.5	7.5	6.8	7.0
18:1, %	64.0	70.5	71.5	71.8
18:2, %	13.5	10.0	9.2	9.1
18:3, %	0.2	0.0	0.0	0.0

In Tables 3 and 4, the data for the slip melting point, N_{35}, and GC analysis of the C_{18} fatty acid composition, clearly indicate the higher selectivity of PRICAT 9920 versus PRICAT 9910. These differences are particularly significant when the hydrogenation is performed under nonselective conditions.

In Figs. 23 and 24 the slip melting points of the hydrogenated oils are plotted as a function of the iodine value for temperatures of 120 and 180°C respectively. In Fig. 23 it is seen that by using the wide pore catalyst PRICAT 9920 the lowest slip melting points are obtained; whether the pressure is 150 kPa or 400 kPa; the slip melting points at IVs ranging from 60 to 105 are not a function of pressure. This is not observed when PRICAT 9910 is used. Then the slip melting points at a particular IV are several degrees higher. The highest values are obtained at the highest pressure. Under the most selective conditions, namely 180°C and 150 kPa, the differences between the slip melting points at a given IV are only on the order of 1°C.

The same conclusions with respect to the difference in selectivity between these two catalysts can be made if one considers the other characteristic data, which are summarized in Tables 3 and 4 for an IV of 80. For example, the solid fat content at 35°C, N_{35}, has its

Table 4 Characteristic Data for IV = 80 of a Hydrogenated Soybean Oil with IV_o = 132.8 and a Feed Composition of 4% C18:0, 23.1% C18:1, 53.1% C18:2 and 7.4% C18:3

	120°C; 400 kPa		180°C; 400 kPa	
	PRICAT 9910	PRICAT 9920	PRICAT 9910	PRICAT 9920
Hydrog. time, min	30	29	30	32
Melting point °C	37.0	29.0	35.0	31
Trans content, wt %	28.0	27.0	39.0	40.0
Solid fat content by NMR				
N_{35}	3.0	0.2	1.0	0.15
GC analysis of fatty acids				
18:0, %	12.0	9.0	9.5	8.0
18:1, %	63.0	69.0	67.5	69.0
18:2, %	13.0	10.5	10.5	9.5
18:3, %	0.1	0.0	0.3	0.1

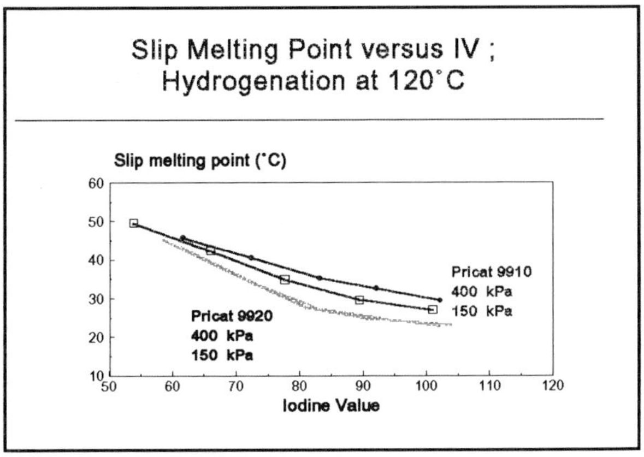

Fig. 23 Slip melting point versus IV; hydrogenation at 120°C.

highest value if PRICAT 9910 is used at 120°C and a pressure of 400 kPa, and the lowest values are obtained for PRICAT 9920. Similar conclusions can be drawn from the GC analyses of the four C_{18} fatty acids. A minimal difference is obtained if the reaction is performed under the most selective conditions, namely at 180°C and 150 kPa, but a considerable difference in selectivity is seen at 120°C and 400 kPa. Then a much higher selectivity is obtained if the wide-pore catalyst is used. All in all, these results convincingly show the higher selectivity of the wide-pore catalyst PRICAT 9920.

Finally, the solid fat content curves at IV = 80 are depicted in Figs. 25 and 26. For the experiments at 120°C it can be concluded that, using PRICAT 9920, the oil can be hydrogenated to a lower IV to get the same curve as obtained by using PRICAT 9910. The advantage can be that a lower 18:3 content is obtained—in other words, an oil with higher oxidative stability.

Figure 26 depicts the difference in steepness of the solid fat content curve at IV = 80 for the two cases where (1) the most selective catalyst is used under the most selective conditions and (2) the medium-pore catalyst is used under the least selective conditions.

The advantage of a low 18:3 content while the solid fat content at 10°C, N_{10}, is still low is depicted by the results given in Fig. 27. In liquid table oils a low N_{10} in combination with

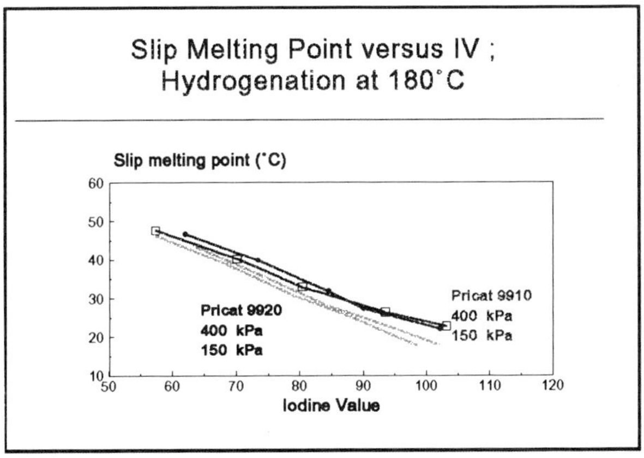

Fig. 24 Slip melting point versus IV; hydrogenation at 180°C.

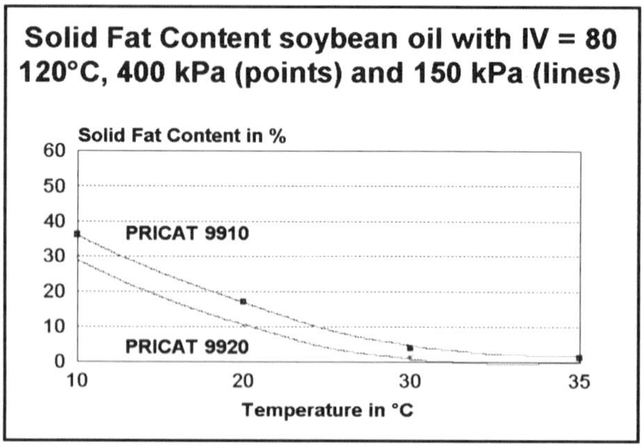

Fig. 25 Solid fat content soybean oil with IV = 80; 120°C, 400kPa (points) and 150 kPa (lines).

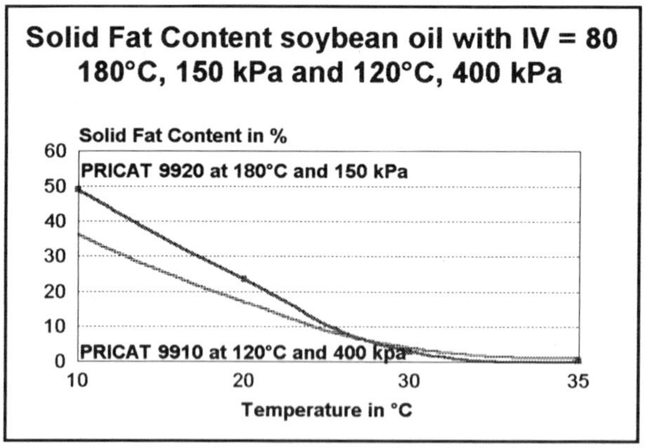

Fig. 26 Solid fat content soybean oil with IV = 80; 180°C, kPa and 120 C, 400 kPa.

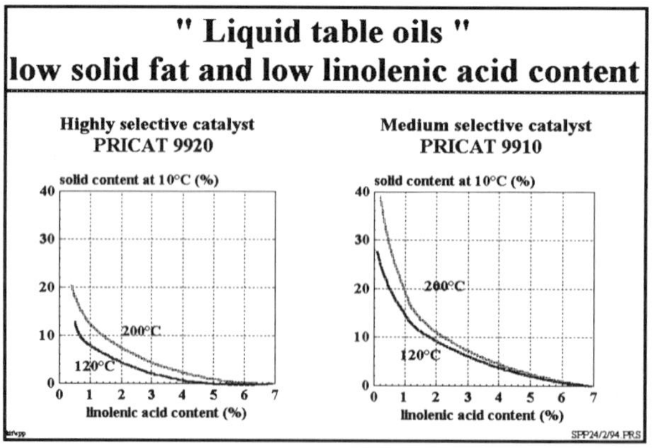

Fig. 27 Liquid table oils: low solid fat and low linolenic acid content.

Table 5 Characteristic Data for IV = 80 of a Hydrogenated Rapeseed Oil with IV_o = 113.7 and a Feed Composition of 1.7% C18:0, 60.6% C18:1, 20.7% C18:2 and 8.8% C18:3

	120°C; 400 kPa; 440 wppm Ni		180°C; 400 kPa; 110 wppm Ni	
	PRICAT 9910	PRICAT 9920	PRICAT 9910	PRICAT 9920
Hydrog. time, min	42	45	37	45
Melting point °C	37.0	28.0	38.5	30.5
Trans content, wt %	22.0	22.0	36.0	43.0
Solid fat content by NMR				
N_{35}	1.5	0.1	4.4	2.5
GC analysis of fatty acids				
18:0, 1.7	12.0	11.0	12.0	10.0
18:1, 60.6	72.0	73.0	73.2	76.0
18:2, 20.7	8.0	8.0	6.6	6.0
18:3, 8.8	0.5	0.5	0.2	0.1

a low 18:3 content is an important quality criterion. This criterion can be better achieved if PRICAT 9920 is used.

Table 5 summarizes the same characteristic data for the product properties of a rapeseed oil, hydrogenated to IV = 80. Again it is demonstrated that the highest selectivity is obtained by using the wide-pore catalyst PRICAT 9920. However, in this hydrogenation the presence of 3 wppm S in the canola oil causes, at 180°C, a significantly larger increase in the trans content due to the fact that on the basis of the same Ni content a larger fraction of the active nickel surface of PRICAT 9920 has been sulfided.

2. *Effect of Phospholipids on Product Properties*

The effect of phospholipids, added as lecithin, on the performance of these two types of catalysts is demonstrated by the data given in Table 6. Soybean oil was hydrogenated

Table 6 Data of *a*, to an IV of 80 Hydrogenated Soybean Oil with IV_0 = 132.8 and a Feed Composition of 4% C18:0, 23.1% C18:1, 53.1% C18:2 and 7.4% C18:3

	180°C; 300 kPa; 100 wppm Ni			
	PRICAT 9910	PRICAT 9920	PRICAT 9910	PRICAT 9920
wppm P	0	0	10	10
Melting point, °C	31.5	29.5	42.5	27.5
Trans content, wt %	39.5	40.0	32.0	34.5
Solid fat content by NMR				
N_{10}	45	38	48	32
N_{20}	22	17	31	13
N_{30}	6.5	4	14.5	2
N_{35}	1.5	0.0	7.5	0.0
GC analysis of fatty acids				
18:0, 4.0	4.0	7.5	11.5	7.5
18:1, 23.1	69.5	70.0	62.5	70
18:2, 53.1	10.5	11	13	10.5
18:3, 7.4	0.0	0.0	1.5	0.0

at 180°C and 300 kPa to an IV of 80, in one set of experiments without the addition of lecithin and in one set in which lecithin was added in an amount such that 10 wppm P was present.

Lecithin has a dramatic effect on the hydrogenated product properties if PRICAT 9910 is used but hardly any when PRICAT 9920 is used. The following effects are observed for the two catalysts PRICAT 9910 and PRICAT 9920 if lecithin is added in amounts of 10 wppm P:

1. The slip melting point increases by 11°C and drops by 2°C, respectively.
2. Trans content decreases by 7.5% and 6%, respectively.
3. Flatter dilatation curves are obtained.
4. Considerably less selective hardening is seen for PRICAT 9910.

In other words, if PRICAT 9920 is used in such a case, then the steepness of the dilatation curves at a particular IV will be less, but only because of the lower trans content. Hardly any difference can be observed in the distribution of the C_{18} fatty acids between the two cases. Hence, if one wants to hydrogenate a soybean oil containing only a few parts per million P by weight, then a catalyst such as PRICAT 9920 is the preferred type.

Two other aspects should also be noted. One is the reduction in the hydrogenation rate, and the other is the extent to which lecithin is adsorbed onto the catalyst. The hydrogenation times (HT) have been increased by the addition of 10 wppm P from 29 min to 44 min for PRICAT 9910 and from 32 min to 55 min for PRICAT 9920. PRICAT 9920 usually quantitatively adsorbs all the lecithin from the oil.

C. Reduction of Trans Content

Today an important issue in hydrogenating edible oils is to reduce the trans content as much as possible. Low temperature and high pressure are the preferred conditions to get a low trans content. This effect is reflected in Fig. 28 for a fish oil hydrogenation. In this figure the trans content is plotted as a function of the slip melting point. At a slip melting point of 35°C it is possible to get a reduction of 30% in the trans content. However, it is possible to achieve a much lower IV if PRICAT 9920 is used. This aspect is depicted by Fig. 29, in which the slip melting point is plotted as a function of the iodine value. The same melting point is achieved at iodine values of 72 and 83 by using PRICAT 9920 and PRICAT 9910, respectively. The melting curve of the oil that was hydrogenated to an IV = 72 by using PRICAT 9920 is much

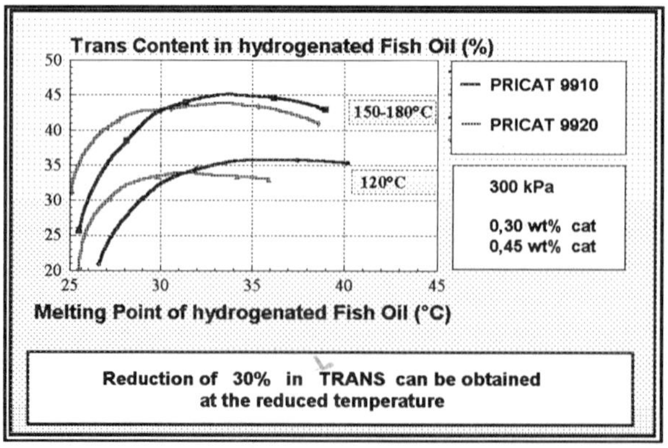

Fig. 28 Reduction of 30% in TRANS can be obtained at the reduced temperature.

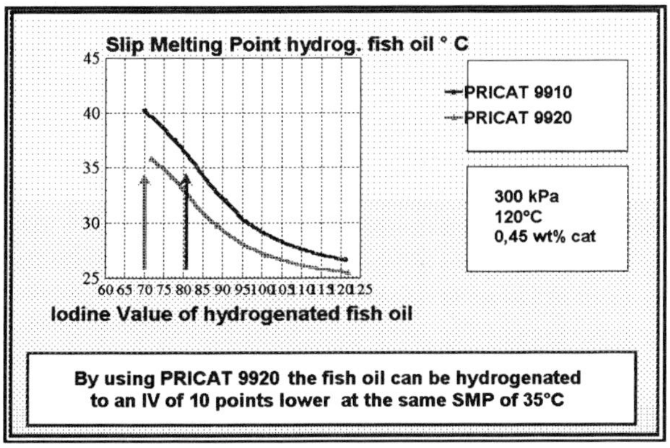

Fig. 29 By using PRICAT 9920 the fish oil can be hydrogenated to an IV of 10 points lower at the same SMP of 35 C.

steeper than the curve of the oil that was hydrogenated to IV = 83 by using PRICAT 9910. PRICAT 9920 produces at low temperature an oil with the same melting curve steepness as is obtained by using PRICAT 9910 at 180°C, but with 30% less trans isomers.

IV. DISCUSSION AND CONCLUSION

The rate of hydrogenation, expressed as the hydrogenation time (Eq. (44)) required to drop the initial IV value, IV_o, to an end IV value, IV_{end}, can be expressed by a simple equation consisting of two terms—the minimum hydrogenation time and a term reflecting all effects that determine the actual hydrogenation rate within the particles. The minimum hydrogenation time reflects all factors that affect the rate at which the hydrogen is transferred from the gas to the liquid phase, such as stirring speed and hydrogen pressure, and the drop in IV. It is a meaningful parameter, and it provides more information than the stirring speed alone, the parameter that has been used by many authors. To give the stirring speed as a relevant parameter is misleading, because it does not give any valuable information. Size, type, position of the impeller, and presence of baffles and coils are other factors affecting the minimum hydrogenation time.

As has already been said, the second term in the hydrogenation time equation reflects all the factors affecting the reaction rate within the catalyst particles. It has been demonstrated that for feedstocks that do not poison the catalyst this correlation yields a straight line if the hydrogenation time is plotted versus the reciprocal amount of catalyst. This observation was made by Coenen [26] in 1969. If there are any components in the oil that will poison the Ni catalyst, this correlation will no longer result in a straight line but will bend upward. The extent to which this line deviates from a straight line is a measure of the quality of the feedstock with respect to the amount of Ni catalyst poisons present. For example, it has been demonstrated that for nb-soybean oils this correlation results in an almost straight line, but for nb-canola oils it is no longer straight. The most severe poisoning occurs when fish oils are used. However, in this case also the extent to which bleaching has a favorable effect on the quality of the feedstock has been demonstrated.

Once the correlation between the hydrogenation time and 1000/wppm Ni has been determined for one particular minimum hydrogenation time, it has been demonstrated in a number of cases that this line can be moved upward/downward if the process is operated at other minimum hydrogenation times. This is a very important observation, because it can be used to predict the hydrogenation time in different types of large-scale reactors if all processing conditions are the same. The only prerequisite is the knowledge of the minimum hydrogenation time.

The quality of the hydrogenation catalysts has been improved over the past decade. In addition to medium-pore catalysts, wide-pore catalysts are now commercially available. By using these wide-pore catalysts in vegetable oil hydrogenation, considerably better selectivity can be obtained. All in all, we have come to the conclusion that the technique described in this chapter is a valuable method of correlating hydrogenation time with process parameters.

ACKNOWLEDGMENTS

I would like to express my appreciation for the valuable comments made by Mr. Senno van der Velde, Unichema's Business Unit Manager Catalyst; Dr. Peter Scherm, Catalyst Development Scientist; Mr. Wolfgang Geuking, Market Application Manager, and Mr. Robert Rütjes, Catalyst Characterization Engineer. Finally I thank Mrs. Mary Stephens, Deutsche Unilever GmbH, Hamburg, for help in correcting my English.

REFERENCES

1. H. B. W. Patterson, *Hydrogenation of Fats and Oils: Theory and Practice*, AOCS Press, Champaign, IL, 1994, pp. 52–68.
2. D. Swern (Ed.), *Bailey's Industrial Oil and Fat Products*, 4th ed., Wiley, New York, 1979, Vol. 1, Chap. 3.
3. K. Andersson, M. Hell, L. Lowendahl, and N. H. Schöön. Diffusivities of hydrogen and glyceryl trioleate in cottonseed oil, *J. Am. Oil Chem. Soc. 51*: 171–77 (1974).
4. J. M. Coulson and J. F. Richardson, *Chemical Engineering*, 3rd ed., Pergamon, New York, 1977, Vol. 1, Chap. 8.
5. K. L. Ganguli. Measurements of H_2/edible oil interfacial area in an agitated hydrogenator using a Ziegler–Natta catalyst, Ph.D. Thesis, TH-Delft, 1975; see also Ref. 3.
6. N. Frössling, Über die Verdunstung fallender Tropfen, *Beitr. Geophys. 52*: 170 (1938).
7. J. M. Coulson and J. F. Richardson, *Chemical Engineering*, 3rd ed., Pergamon, New York, 1977, Vol. 2, Chap. 3.
8. P. L. T. Brian and H. B. Hales, *A. I. Ch. E. J. 15*: 419 (1969).
9. P. Sänger and W. -D. Decker, Liquid–solid mass transfer in aerated suspensions, *Chem. Eng. J. 22*: 179 (1981).
10. M. Alderliesten, A nomenclature for mean particle diameters, *Anal. Proc. 21*: 167–172 (1984).
11. L. L. van Dierendonck, The specific contact area in gas–liquid reactors, Proc. 4th Eur. Symp. Reaction Eng., Brussels, 1971, p. 205.
12. *Bailey's Industrial Oil and Fat Products*, 4th ed. (D. Swern, Ed.), Wiley, New York, 1979, Vol. 1, Chap. 3.
13. P. Varey, Towards the flexible impeller, *Chem. Eng.*, Sept. 24: 577–33 (1992).
14. A. Bakker, J. M. Smith, and K. J. Myers, How to disperse gases in liquids, *Chem. Eng. 101*(12): 98–104 (1994).
15. D. Urosevic, Buss technology for oils and fats and fatty acid derivatives, presented at the ISF-JOCS World Congr. 1988, Tokyo.
16. A. N. Kolmogorov, The local structure of turbulence in incompressible viscous fluid of very large Reynolds numbers, *C. R. Acad. Sci. URSS 30*: 301.
17. K. R. Westerterp, *Chemical Reader Design and Operation*, Wiley, New York, 1984, p. 146.
18. E. W. Thiele, Relation between catalytic activity and size of particle, *Ind. Eng. Chem. 31*: 916 (1939).

19. M. Perdereau, Structure, mécanisme de formation et stabilité de la ouche d'adsorption du soufre sur le nickel, *Surf. Sci. 20*: 80 (1970).
20. L. F. Albright, in *Hydrogenation*, Proceedings of an AOCS Colloquium, Ed (R. Hastert, Ed.), AOCS Press, Champaign, IL, 1987.
21. H. I. Waterman, *Hydrogenation of Fatty Oils*, Elsevier, New York, 1951.
22. J. W. E. Coenen and B. G. Linssen, in *Physical and Chemical Aspects of Adsorbents and Catalysts* (B. G. Linssen, Ed.), Academic, New York, 1970.
23. L. F. Albright, Quantitative measure of selectivity of hydrogenation of triglycerides, *JAOCS 42*: 350 (1965).
24. C. Niklasson, B. Andersson, and N. Schöön, Influence of hydrogen pressure on selectivity in consecutive hydrogenation reactions, *Ind. Eng. Chem. Res. 26*: 1459–1463 (1987).
25. D. W. van Krevelen, Stoftransport en kinetiek bij contact katalyse, *Chem. Weekbl. 48*: 427 (1951).
26. J. W. E. Coenen; Hydrogenation of oils and fats, in *Margarine Today*, Proc. of a seminar held at Dijon Univ., March 1969, pp. 62–91.

11

Butter, Margarine, Spreads, and Baking Fats

Eric Flack
Danisco Ingredients (UK) Limited, Bury St. Edmunds, Suffolk, England

I. BUTTER

A. History

The processing of milk to produce butter has been known for thousands of years. A Sumerian frieze from about 4000 BC shows milk being processed and, according to archaeologists, the milk was being whipped to form butter [1]. It is also recorded in the Hindu Veda, written over 3500 years ago, that Hindus valued their cows in relation to the amount of butter that could be obtained from their milk [2]. Of course, the principle involved in these ancient processes, i.e., phase inversion, still applies today although the milk used for making butter has been substituted by more concentrated milk fat in the form of cream.

B. Early Developments

Initially, butter was produced manually by agitating and beating the milk in simple wooden churns. Wood was long considered ideal for the construction of churns as, apart from being readily available and easily fashioned, it was also one of the natural materials to which butter would not adhere during production, that would not impart off-flavors to the product, and that would not be adversely affected by the milk or the salt used in processing. Unfortunately, wood is difficult to sterilize, which is a severe disadvantage, and wears out with use. Thus, the introduction of all-metal equipment to overcome these difficulties was welcomed. Other developments in the late nineteenth century also led to the production of microbiologically safer butter on a large scale.

The mechanical separation of cream, the pasteurization of milk products, and the introduction of butter starter cultures in the 1880s enabled the establishment of dairies that could produce butter in much greater quantities and thus heralded the beginning of modern buttermaking.

C. The Process

The process of manufacturing butter basically transforms cream into butter grains and buttermilk by agitation and by beating air into the cream [3]. The individual stages of butter grain formation, i.e., phase reversal from oil in water to water in oil, are explained as

1. Building up foam from the skim milk
2. Adsorption of fat globules at the foam lamellae
3. Agglomeration of fat globules
4. Destabilization of the foam and mechanical clumping of the agglomerates to form butter grains

The microstructure of the butter grains as well as the finished butter is characterized as a dispersion of water, air, fat crystals, and fat globules in oil. The destabilization of the cream and the subsequent phase separation involve a series of overlapping dynamic processes.

During the early stages of churning, air is whipped into the cream and the proteins (lactoglobulin, lactalbumin, and casein) form an interfacial film around the air bubbles [4]. As a result, an unstable foam is formed, and the mechanical stress during churning causes the fat globules to lose their membrane so that the globule surface becomes hydrophobic [5]. Due to their lower surface tension compared to milk proteins, they accumulate at the air/serum interfaces of the air bubbles. Through the action of shear forces on the semisolid fat phases, the globules become deformed and release liquid fat, which promotes agglomeration of the fat globules.

Although large-scale batch churning is still in operation, it has been replaced to a very large extent by the continuous buttermaking process introduced in the 1950s by Fritz.

D. Cream Preparation

Cream is prepared by the centrifugal separation of whole milk at 40–50°C to a fat content of 36–42% for continuous buttermaking and 25–35% for batch production. It is pasteurized at 85–110°C for 1–30 s to destroy microbial contamination and to inactivate enzymes. The cream is then left to rest at varying temperatures for several hours while it undergoes physical ripening. If cultured or lactic butter is desired (as against sweet cream butter), starter culture is added at this stage to initiate bacteriological ripening, during which lactose is partially converted to lactic acid, thus reducing the pH from, say, 6.5 to 4.6–4.8. Simultaneously, the culture will develop the typical lactic butter flavor through the generation of diacetyl.

E. Buttermaking

The process of buttermaking may be illustrated in Fig. 1 [6].

The main elements of the process may thus be summarized as follows:

1. Cream preparation. See Section D.

2. Churning. During the churning phase, inversion from o/w to w/o takes place and buttermilk is separated off. Air is whipped into the cream, creating a foam at the air/serum interface where the fat globules are concentrated. Due to mechanical forces, some of the globules become damaged, releasing liquid fat, which promotes agglomeration of the remaining fat globules and the fat crystals from the damaged globules. This process continues until pea-sized butter granules are formed and the buttermilk is separated.

3. Draining and Washing. The separated buttermilk is drained off. However, where draining off is insufficient, the butter granules may be washed with cool, clean water to

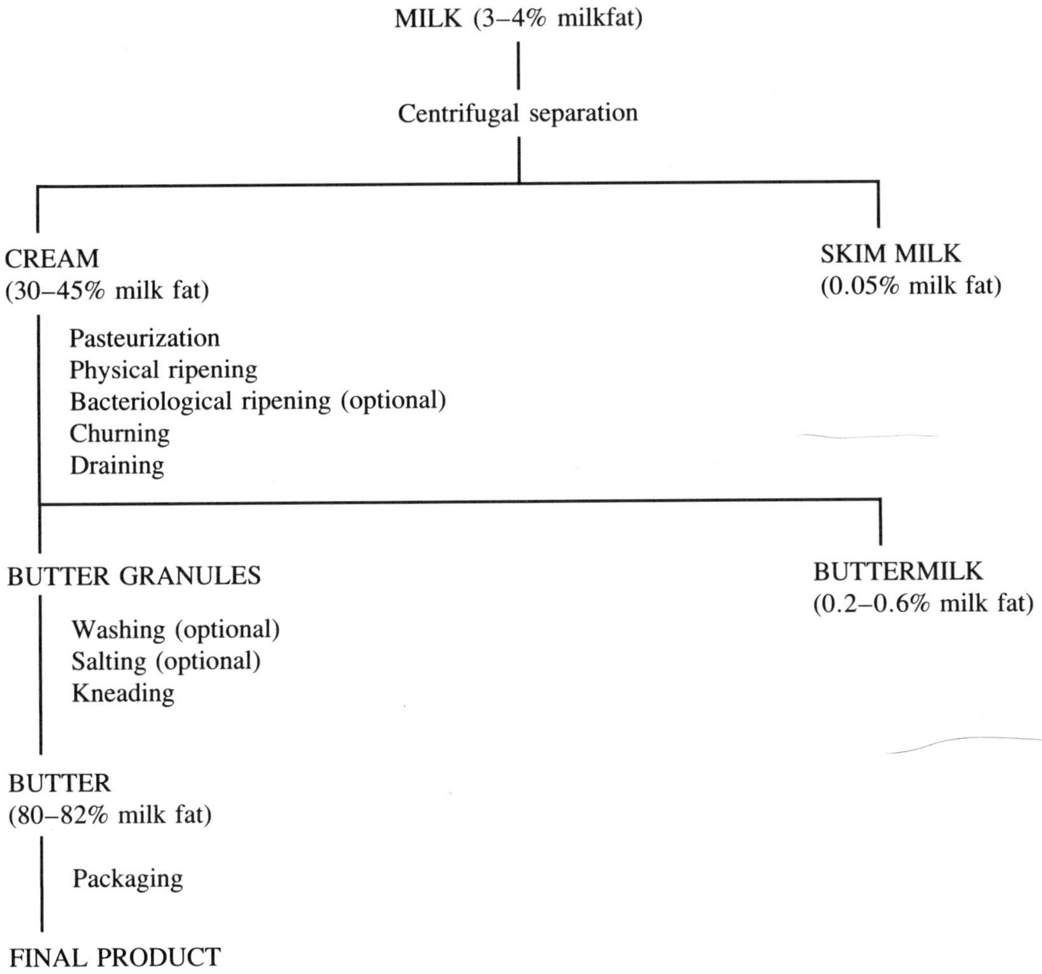

Fig. 1 The process of buttermaking.

reduce the buttermilk solids content. Removal of buttermilk and wash water is controlled to allow for the convenient adjustment of the final moisture content of the butter.

4. Salting. Invariably, butter, whether sweet cream or lactic, will be salted at levels from 0.1% to 2% by the addition of brine prior to kneading.

5. Kneading. The butter granules, together with brine and additional water when required, are kneaded either by continued rotation of the churn in batch processing or by means of worm-screw conveyors in continuous processing. This ensures a fine water distribution and a smooth texture in the finished butter. In some processes, kneading is carried out in two stages and also includes vacuum treatment to remove occluded air.

F. Final Composition

The legal requirements for butter and variations containing lower fat contents are listed later in Table 2.

II. MARGARINE

A. Introduction

Margarine is a fatty food resembling butter in appearance, character, and composition that is used as a substitute for or alternative to butter. Statutory definitions may vary in different countries (see later), but usually the term "margarine" is applied to fatty foods similar to butter except that the fat is not, or only to a minor extent, derived from milk fat.

B. History and Early Developments

The invention of margarine in the 1860s is attributed to the French chemist Hippolyte Mège Mouriés, in response to a competitive challenge initiated by the French government under Napoleon III for a less expensive and less perishable substitute for butter. Mège had already undertaken research in this field at the imperial farms, where he had observed that although underfed or hungry cows lost weight, they still produced milk, though of lower than normal yield, and the milk contained fat. Thus, he deduced that milk fat was derived from normal body fat, i.e., tallow. However, as tallow itself does not possess the melting property of milk fat, he construed that through some metabolic process a fractionation of the fat occurred, the lower melting fractions of which were transported to the mammary glands of the udder whereby, through the enzymatic action of pepsin, it became transformed to butterfat and dispersed as an emulsion in the milk plasma, while the harder fractions were used by the animal as a source of energy [7].

Having arrived at these conclusions, Mège set out to imitate the natural process by carefully rendering fresh tallow at body heat (about 45°C) using artificial gastric juices to facilitate separation of tissue from pure fat and then, following crystallization at 25–30°C, to extract under pressure about 60% of a soft semifluid fraction—oleomargarine—and about 40% of a hard white fat—oleostearine. The softer fraction had a melting point similar to butterfat, a pleasant flavor not unlike melted butterfat, and a pale yellow color. It could easily be plasticized and, although different in fatty acid composition from butterfat, it offered a useful basis for the production of a substitute. He assumed, therefore, that this new butterfat substitute consisted of glycerides of margaric and oleic acids (the former then considered to be a C_{17} homologue of the saturated fatty acids although now known to be a eutectic mixture of palmitic and stearic acids) that solidified in crystals having a pearly luster. The name margarine was derived from the Greek word *margarites* meaning pearl.

Mège's process was to take a proportion of the soft fat with appropriate quantities of milk and water, into which a small amount of udder extract was stirred. Following agitation, a stable emulsion similar to thick butter cream formed that, on further churning, thickened to resemble butter. Although this process may now sound rather rudimentary, it nonetheless comprised all the elements of margarine production as we know them today.

Mège took out patents in France and England, but the next major development took place in Holland. In 1871, in the absence of patent law in Holland at that time, he sold his invention to the butter merchants Johannes and Anton Jurgens in the village of Oss. Also in the same locality were another family of butter merchants—Simon and Henry Van den Bergh—who commenced production shortly afterwards, later joined forces with the Jurgens, and eventually become the largest margarine manufacturers in the world [8].

Production followed in most countries in Europe—Austria (1873/4), Italy (1874), Germany (1874), Norway (1876) and in Denmark (1883) which was later to become the largest per capita consumer of margarine. Mège was granted a U.S. patent in 1874.

Many variations were developed in the following years, and patents on new formulations and processes taken out. At the same time there was strong opposition to the introduction of margarine to the market from the farming community, and in some places anti-margarine legislation was introduced. Opposition from such quarters continued well into the 20th century and is not entirely unknown today.

C. Definitions and Descriptions

The basic composition of margarine, or oleomargarine, has been more or less established since Mège's original formulation—that is, it should be similar to butter. However, with the new developments during more recent years involving different components and varying combinations of vegetable oils, animal fats, and milk fat, a range of new descriptions have come into being such as low-fat spreads, low-calorie spreads, and yellow-fat spreads. In 1993, Moran [9] listed the variety then available as shown in Table 1.

In Europe, the adoption of Council Regulation (EC) No. 2991/94 in December 1994 brought all products, including butter, under the one description "spreadable fats" and described them as "products in the form of a solid, malleable emulsion—with a fat content of at least 10% but less than 90% by weight. The fat content, but excluding salt, must be at least two-thirds of the dry matter." These terms are further qualified as "products which will remain solid at a temperature of 20°C and which are suitable for use as spreads." The definitions and sales descriptions are detailed in Table 2.

Clause 2 of Article 3 of this regulation further states that the sales descriptions "minarine" and "halverine" may be used in place of "half-fat margarine." Fat spreads falling outside these standards, for instance those with fat contents below 10% and concentrated products with fat contents of 90% or more, may not be classed as spreadable fats.

European legislation does not specify vitamin fortification, which is left for decision at the national level. However, in the United Kingdom the requirement will remain as previously specified in the Margarine Regulations (SI 1867) 1967. That is, each 100 g of margarine shall contain 800–1000 μg vitamin A and 7.05–8.82 μg vitamin D. This does not apply to other

Table 1 Some Current Spreads

Spread	Approx fat content (%)
High fat spreads	
Butter	80
Margarine, packet	80
soft	80
Polyunsaturated fatty acids (PUFA) spreads	80
Vegetable fat/butterfat blended spreads	80
Low fat spreads	
Vegetable fat spreads	70, 60, 40
Vegetable fat/butterfat blended spreads	70, 40
Butterfat spreads	40
Very low fat spreads	20–30
Water continuous spreads	15, 9, 5

Source: Ref. 9.

Table 2 EC Definitions and Descriptions for Spreadable Fats

Fat Group Definitions	Sales description	Product categories[a]
A. Milk fats. Products derived exclusively from milk and/or certain milk products for which the fat is the essential constituent of value.	1. Butter	Milk fat content not less than 80% but less than 90%. Max. water 16%; max dry nonfat milk materials 2%.
	2. Three-quarter fat butter	Milk fat content not less than 60% but not more than 62%.
	3. Half fat butter	Milk fat content not less than 39% but not more than 41%.
	4. Dairy spread *x*%	Products with milk fat contents less than 39%; more than 41%, less than 60%; more than 62%, less than 80%.
B. Fats. Products derived from solid and/or liquid vegetable and/or animal fats suitable for human consumption, with a milk fat content of not more than 3% of the fat content.	1. Margarine	Vegetable and/or animal fats not less than 80%; less than 90%.
	2. Three-quarter fat margarine.	Not less than 60%; less than 62%.
	3. Half fat margarine	Not less than 39%; more than 41%.
	4. Fat spreads *x*%	Products with fat contents: less than 39%; more than 41%, less than 60%; more than 62%, less than 80%.
C. Fats composed of plant and/or animal products. Products derived from solid and/or liquid vegetable and/or animal fats suitable for human consumption, with a milk fat content of between 10% and 80% of the fat content.	1. Blend	Product from a mixture of vegetable and/or animal fats with a fat content of not less than 80% but not more than 90%.
	2. Three-quarter fat blend	Not less than 60%; not more than 62%.
	3. Half fat blend	Not less than 30%; more than 41%.
	4. Blended spread *x*%	Products with fat contents: less than 39%; more than 41%, less than 60%; more than 62%, less than 80%.

Source: *Official Journal of the European Communities* L316, Vol. 37, Dec. 9, 1994.
[a] Additional description with indication of % fat content by weight.

spreads, so vitamin fortification in these cases are matters for individual decision by the manufacturers and/or retailers.

Although legislation in the United States and Canada is currently less detailed than in Europe, both also require a minimum of 80% fat. In the United States FDA 166.110 describes margarine (or oleomargarine) as "food in plastic form or liquid emulsion" containing not less than 80% fat and describes a method for fat determination. It specifies vitamin A fortification to be not less than 15,000 I.U. per pound but leaves vitamin D as an optional ingredient not less than 1500 I.U./lb. Although spreads are not specified in the legislation, low fat spreads constituted the fastest growing category in 1990, with spreads containing 20–72% fat on the market [10].

In the United States, the National Association of Margarine Manufacturers describes spreads as having fat contents ranging from 18% to 79%.

Canadian Standard B.09.016 states that margarine shall be "a plastic or fluid emulsion of water in fat, oil or fat and oil that are not derived from milk," and shall contain not less than 80% fat and not less than 3300 I.U. vitamin A and 530 I.U. of vitamin D. Calorie-reduced margarine is specified in Standard B.09.017 as containing not less than 40% fat and having 50% of the calories normally present in margarine.

The conversion of international units of vitamins A and D to microgram equivalents is complicated in the case of vitamin A as it depends on the origin of the fat. Thus, to convert vitamin A from I.U. to μg retinol in foods of animal origin, the I.U. of retinol should be multiplied by 0.3. However, for foods of plant origin, the I.U. of β-carotene should be divided by 10. This arises because 1 I.U. of vitamin A is equivalent to 0.3 μg retinol or 0.6 μg β-carotene. The conversion of the units for vitamin D is straightforward: 1 I.U. vitamin D=0.025 μg.

D. Structure and Raw Materials

Margarine is a water-in-oil (w/o) emulsion; that is to say, the water (the disperse phase) is distributed as droplets within the oil (the continuous phase). In margarine, the levels of each largely simulate that found in butter and are regulated in most places by legislation: minimum 80% fat, maximum 16% water, the remainder consisting of salts, proteins, emulsifiers, vitamins, colors, and flavors.

Whereas production initially was based upon the use of fractionated animal fats such as tallow and lard, these were superseded later by vegetable oils and, for a period, marine oils, both of which offer a much greater flexibility in physical characteristics and economy.

Being the main and most expensive ingredient, the fat blend is also the most important factor in the formulation of margarine. As oil prices fluctuate, it is essential to have a fat blend providing the desired quality at minimum cost. Thus, it is vital to be able to utilize alternative formulations based on prevailing price conditions while still maintaining quality. Therefore, a full understanding of the physical characteristics of the individual oils composing the blend is critical. In this respect, factors of importance for the fat phase include the solid/liquid fat ratio, crystallization rate, and melting properties.

Linear programming has been used for many years to formulate to optimum cost, performance, and specification. Hydrogenation has allowed greater flexibility in fat blend formulation. The negative aspect of *trans* acids as perceived by the consumer has, however, resulted in manufacturers turning to natural sources of hardstocks (solid phase) to achieve the desired texture and meltdown of products. The EU margarine manufacturers, for example, plan to reduce *trans* acids to below 5% in margarine by the end of 1996.

In the early days of margarine production, large quantities of milk and water in almost equal proportions were used for emulsification with the fat, with much of the excess water being pressed out during kneading to arrive at a final moisture content of 16% or lower. Nowadays, the aqueous phase is added at optimum levels and consists usually of a simple solution of skim milk powder or whey powder in water, with the addition of salt, and the pH is adjusted to 5.5–6.0 by using, for instance, citric acid.

E. Fat Crystallization

The structure and physical properties of lipids have been covered in earlier chapters, but it is worth reviewing the important issue of fat crystallization. Variations in the crystallization properties of fats and differences between batches of a similar origin may cause problems in production even though processing techniques are fully controlled and highly automated [11]. The crystallization rates of some individual and blended fats are shown in Table 3.

Oils and fats are polymorphic and can crystallize in more than one form, which can differ in terms of melting point, density, heat of fusion, and rate of crystallization. The three crystal forms are commonly known as alpha (α), beta-prime (β'), and beta (β), with the α form having the lowest values, β' intermediate, and the most stable β form the highest values. Transition from one form to another is $\alpha \rightarrow \beta' \rightarrow \beta$ and is not reversible without remelting the fat.

The polymorphism of crystalline fats may cause problems with the consistency of margarine and spreads. During production, the fats initially crystallize in the α form and normally will rapidly transform to the β' form. This is the desirable form for spread production as the small needle-shaped β' crystals (about 1 μm long) impart good plasticity. Should the crystals transform from the β' form to the much larger β form (>20 μm) they will give the spread a grainy consistency known as sandiness, as can be seen in Fig. 2 [12].

Sandiness may also be caused by the physical interaction of triglycerides to produce higher melting systems (compound formation). This can occur, for example, with triglycerides of the forms POP/OPP (P = palmitic, O = oleic) when lard and palm oil are blended. Some care must be taken, therefore, when formulating hardstocks from such blends [13]. At the same time, the specific area of the crystal surface will reduce, allowing the liquid oil to penetrate to the surface of the margarine, which may then lead to oiling out, especially if the margarine comes under pressure. Some vegetable oils such as partially hardened sunflower or low-erucic rapeseed are particularly prone to form β crystals and thus can cause sandiness in spreads. In this case, sorbitan tristearate at 0.3% of the fat has been found to inhibit the transition from β' to β form.

The crystal-modifying effect of a range of emulsifiers—sorbitan esters, ethoxylated sorbitan esters, ethoxylated fatty alcohols, citric acid esters of monoglycerides (Citrem), diacetyl tar-

Table 3 Crystallization Rates of Fats

Fat	Time (min.)	Temp. (°C)
Coconut	3	20
Coconut/palm 1:1	4	15
Coconut/palm, interesterified	5	18
Palm hardened	5	17
Palm hardened/palm 2:1	8	13
Lard	14	10
Lard/palm 1:1	15	10
Palm	27	10
Shea fat	45	10

Source: Ref. 11.

taric acid esters of monoglycerides (Datem), sucrose monostearate, sodium stearoyl lactylate, and polyglycerol esters—on the polymorphism of tristearin (glyceryl tristearate) has been investigated [14]. In these trials it was found that sorbitan monostearate and Citrem (C_{16}–C_{18} fatty acids) were the most effective in preventing the recrystallization (from α to β form) of tristearin.

It should be noted that when used in emulsions, surface-active materials will be absorbed at the oil/water interface and that only lipophilic emulsifiers with high solubility in the oil phase can perform as crystal inhibitors in emulsions.

Varying the level of saturated fat affects the degree of crystallinity of the oil phase [10]. Spreads with 40% fat were made using oil phases of varying crystallinity by blending either liquid soybean oil or stick margarine oil with soft margarine oil. The textures of the spreads made with these blends were visibly very different, increasing in softness in the same direction as the oil itself. As a consequence of increased firmness, the stability of the resulting spread was expected to increase with increasing crystallinity in the oil phase. A centrifugation procedure was used to measure the stability of the emulsions (Fig. 3). The higher amount of water and/or oil (especially water) released, the less stable the emulsion.

In summary, three main types of margarine can be distinguished according to the different criteria for which they are formulated: table margarine (which includes the distinct categories of tub and packet margarine), industrial or bakery (cake) margarine, and pastry margarine. Apart from specific compositional criteria such as "all vegetable," the main product characteristics are those of performance. This, in the case of table margarines, includes consistency at the temperature of use, and for bakery and pastry margarines, creaming and plasticity, respectively [15]. The performance of the margarine in each case will relate closely to the solids content (dilatation) of the fats used, as illustrated in Fig. 4.

F. Emulsifiers

The main function of emulsifiers in processed foods is to reduce the interfacial tension between the phases of an emulsion, usually oil and water. In such two-phase systems, one phase is dispersed as large droplets within the other. They are called oil-in-water (o/w) emulsions, when the continuous phase is water (such as in milk or ice cream) and water-in-oil (w/o) emulsions when the continuous phase is oil (as in butter and margarine).

The stability of two-liquid phase emulsions is kinetic stability, i.e., the system is not thermodynamically stable [16]. Thermodynamically stable emulsions would spontaneously reform following separation by, for instance, centrifugation, whereas experience shows that an emulsion that has separated remains so unless mixed by some external action. In reality, the two separated phases are in the most stable state to which all emulsions will tend. Thus, a stable emulsion is one in which this inevitable trend has been retarded so that it is not noticeable during the normal life of the product, be it several years, even.

Margarine is a w/o emulsion in principle only because in reality it is a dispersion of water droplets in a semisolid fat phase containing fat crystals and liquid oil [17]. Preparation of the emulsion requires considerable energy to reduce the droplet size of the disperse phase, thereby creating an increase in the surface area between the two immiscible phases.

The initial w/o emulsion is prepared in mixing tanks with vertical or horizontal stirrers to ensure emulsification but without incorporating too much air. It is rather coarse in water droplet size and is fairly unstable if not kept agitated.

In modern continuous production, the emulsion exists only for a brief period before passing into the chilling unit where final emulsification and crystallization of the fat phase take place. Thus, the emulsion need not be very stable against coalescence, as the water droplets become

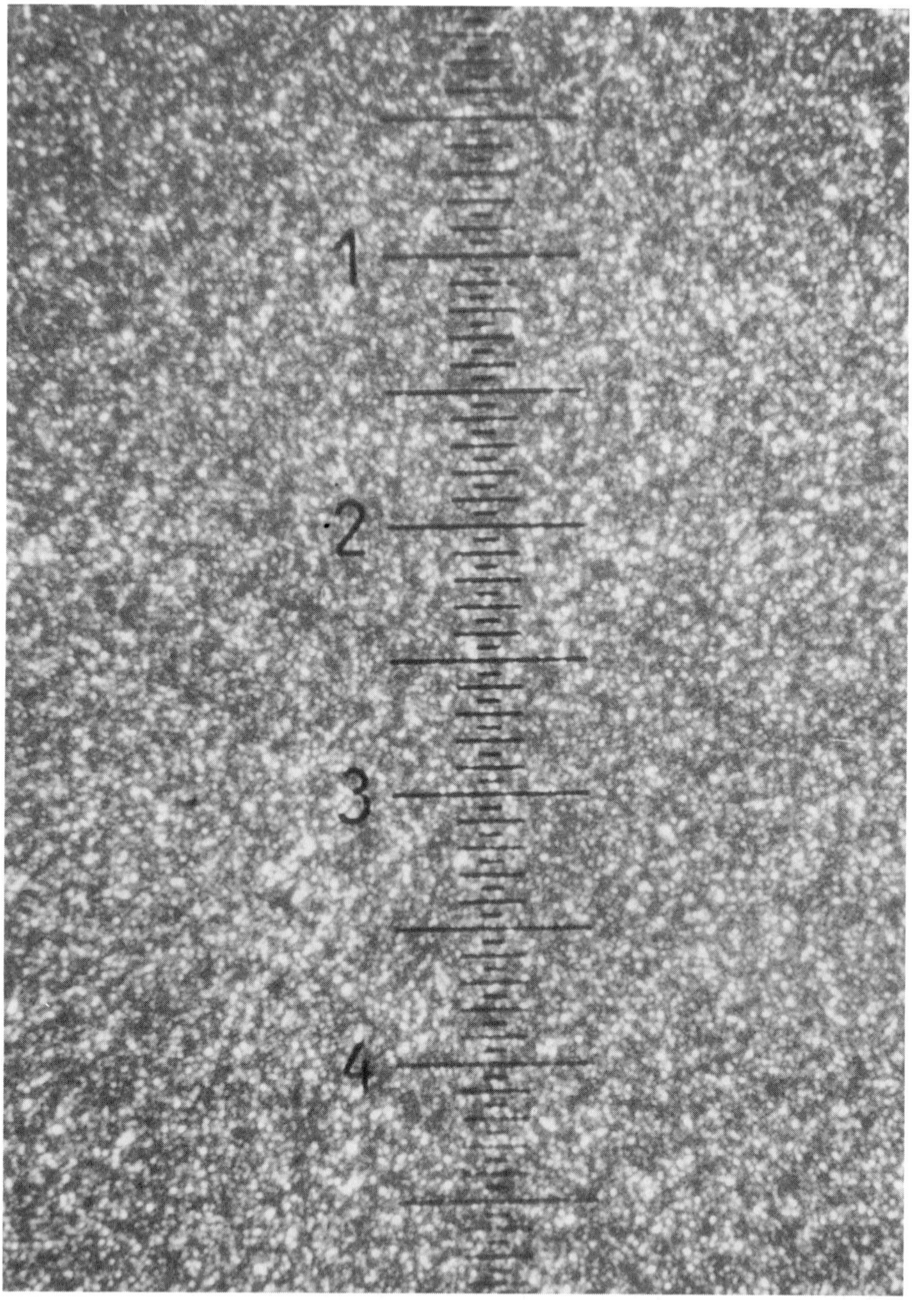

(a)

(b)

Fig. 2 Fat crystals in (a) normal margarine with good consistency and (b) a margarine with large β crystals causing a "sandy" texture (magnification 200X). (Courtesy of Danisco Ingredients, Denmark.)

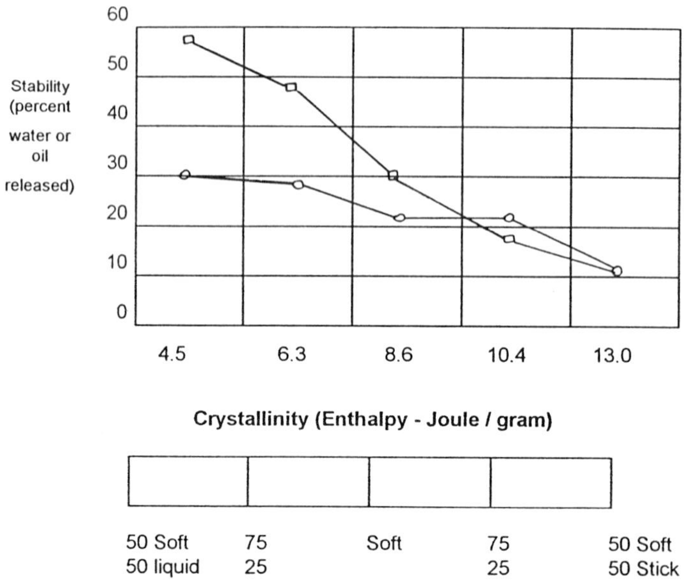

Fig. 3 Effect of the degree of crystallinity on the emulsion stability of 40% fat spreads as measured by ultracentrifugation. (□) Oil; (○) water. (From Ref. 10.)

fixed within the semisolid phase. However, the droplet size is important when considering flavor release and microbiological spoilage. Thus, emulsifiers are used to lower the interfacial tension between the oil and water phases, which will generally result in a smaller water droplet size. Distribution of the water droplets in the range 2–4 μm will give better stability against deterioration such as mold growth. However, it is still desirable for some of the water droplets to be larger—10–20 μm—to give a better flavor release in the mouth. For these purposes, therefore, lipophilic emulsifiers such as monodiglycerides of long-chain fatty acids (C_{16}–C_{18}) are used at levels of 0.1–0.3%, often in combination with 0.05–0.1% refined soy lecithin.

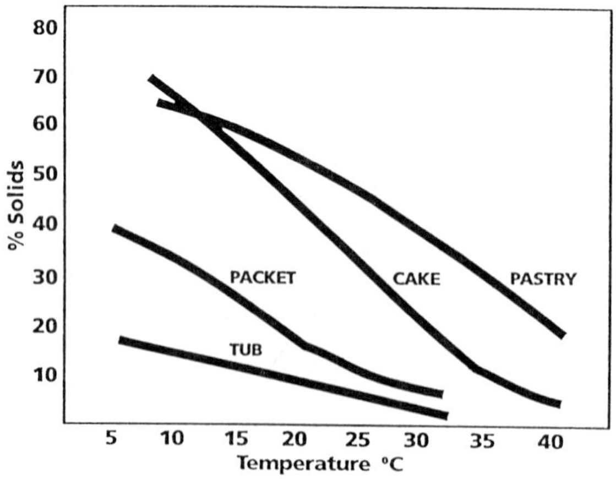

Fig. 4 Typical solids contents of margarine fats. (From Ref. 15.)

Comparisons of the water droplet distribution in margarine emulsion and in finished margarine are shown in Fig. 5. The difference in water droplet size shows that the droplet size in the emulsion is further reduced during the cooling and kneading processes in the tube chiller.

The water droplets in finished margarine are stabilized by adsorbed fat crystals as seen in Fig. 6, which shows the microstructure of margarine by freeze-fracture transmission electron microscopy. It is clearly seen that the water droplets are covered with fat crystals, oriented flatly along their surface [18].

Margarine is frequently used for frying when it is especially important that it does not spatter. Spattering is caused when the margarine melts in the frying pan, the emulsion breaks, and the coalesced water droplets will, due to gravity, form a film of water covered with molten fat on the base of the frying pan. When the temperature reaches the boiling point of water, the increase in vapor pressure will cause the water phase to spatter—sometimes explosively. It is desirable, therefore, that the margarine allow gradual evaporation of the water from the small water droplets and the formation of a fine, golden brown sediment that does not adhere to the frying pan. Both formulation and processing play important roles in reducing the tendency to spattering. The presence of salt and milk are desirable, as is a high pH, up to say 6, whereas sugars and starches will increase the tendency to spatter. The selection and addition of the emulsifier is of considerable importance in producing a margarine with good frying properties.

A combination of monodiglycerides and lecithin will have only a limited effect in reducing spattering in low salt margarine, but there are other emulsifiers that, either alone or with lecithin, will help prevent coalescence of water droplets during frying. In this respect, citric acid esters of mono- and diglycerides (Citrem), polyglycerol esters, and thermally oxidized soybean oil interacted with mono- and diglycerides used at 0.3–0.4% together with soy lecithin can be very effective.

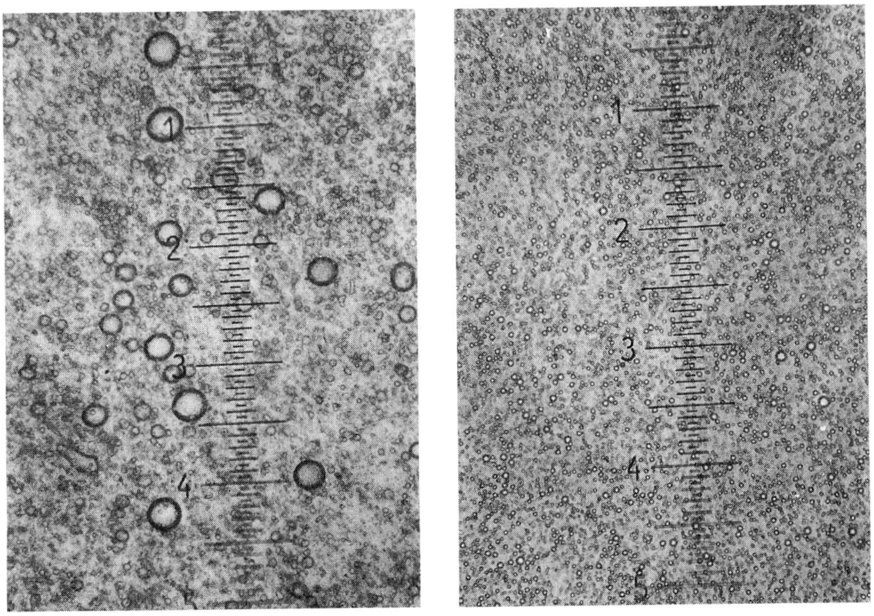

Fig. 5 Water droplet size distribution in (a) margarine emulsion and (b) the finished margarine (magnification 200X). (Courtesy of Danisco Ingredients, Denmark.)

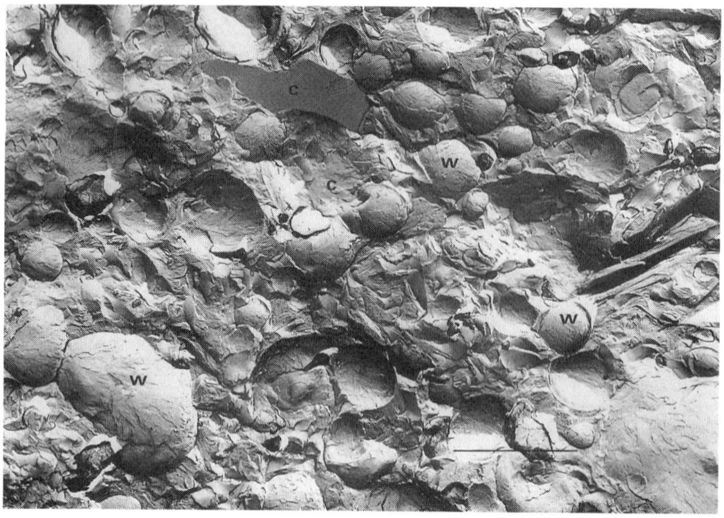

Fig. 6 Freeze-fracture electron micrograph showing the microstructure of margarine. Water droplets (w) are covered with fat crystals. Bar: 1 μm. (Courtesy of Dr. W. Buchheim, Keil.)

G. Processing

Modern processing methods involve weighing the major components by means of computer-controlled load cells, which ensure accuracy, fast throughput, and constant composition. The essential elements include

Aqueous phase preparation tanks
Fat blend preparation vats
Emulsion make-up tanks
Intermediate holding tanks
Cooling and kneading equipment
Worker units and resting tubes

The typical plant layout shown in Fig. 7 would be suitable for most types of both table and industrial margarines, low fat spreads, and shortenings. The entire plant should be free of copper and copper alloys to avoid the high risk of oxidation they would present.

The aqueous phase components may include salt, milk components such as whey powder or skimmed milk powder, thickeners such as gelatin or sodium alginate, water-phase flavors where appropriate, and preservatives such as potassium sorbate for low fat products. These can be incorporated in the aqueous phase preparation or make-up tanks or via the emulsion make-up tanks. The final oil blend will also contain the oil-soluble components such as emulsifiers, colors, flavors, and vitamins.

The aqueous phase should be pasteurized before production of the emulsion, which requires strong agitation, avoiding aeration, to ensure a good dispersion of the water droplets, as discussed earlier.

Cooling and kneading can be carried out by either of two methods: (1) tube chiller or (2) chilling drum-complector. In the former, as shown in Fig. 7, cooling and kneading take place in a closed system and in a single process. In the latter, the cooling and kneading are carried out separately—cooling on the chilling drum and kneading in the complector. The advantage

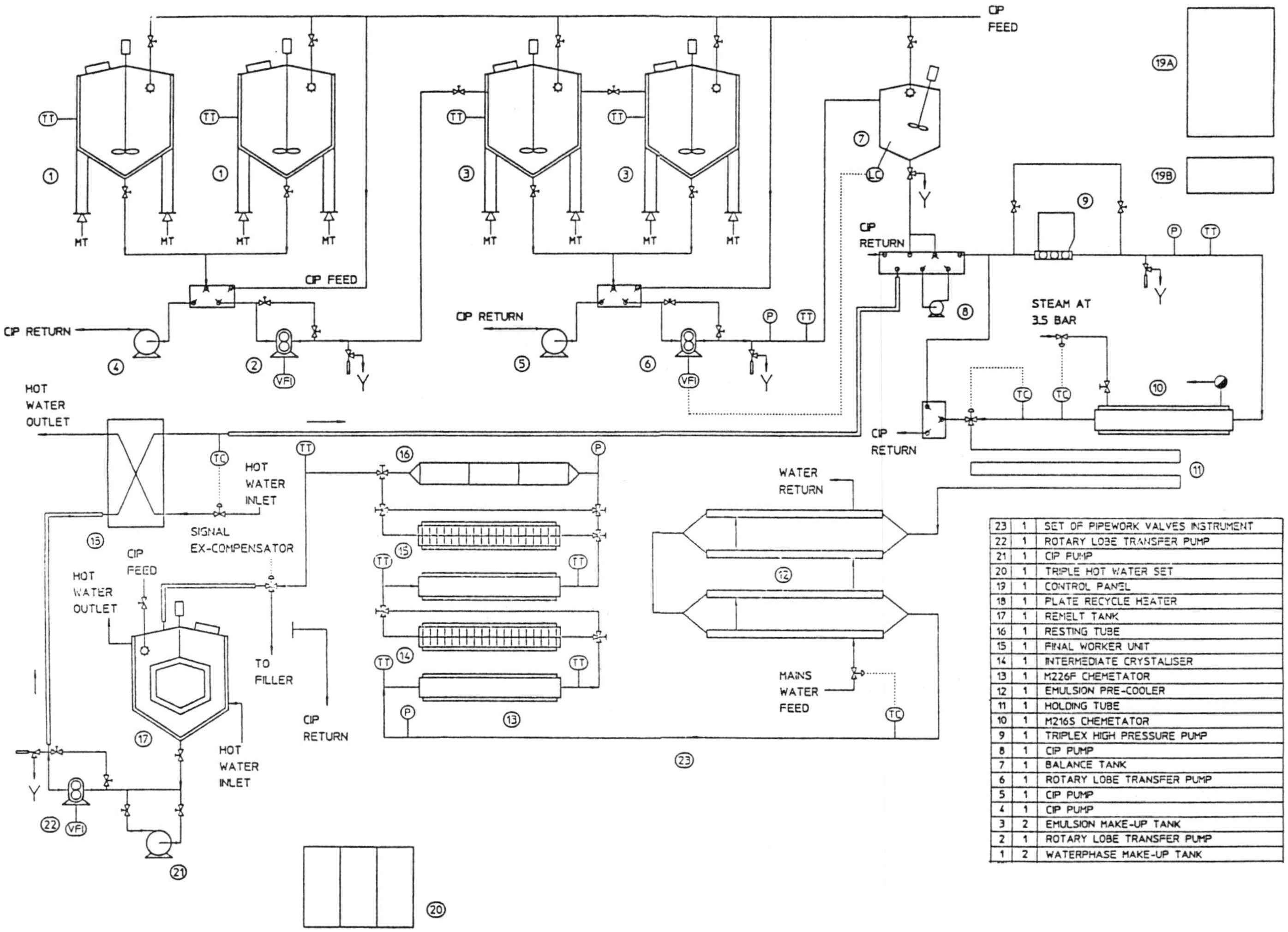

23	1	SET OF PIPEWORK VALVES INSTRUMENT
22	1	ROTARY LOBE TRANSFER PUMP
21	1	CIP PUMP
20	1	TRIPLE HOT WATER SET
19	1	CONTROL PANEL
18	1	PLATE RECYCLE HEATER
17	1	REMELT TANK
16	1	RESTING TUBE
15	1	FINAL WORKER UNIT
14	1	INTERMEDIATE CRYSTALISER
13	1	M226F CHEMETATOR
12	1	EMULSION PRE-COOLER
11	1	HOLDING TUBE
10	1	M216S CHEMETATOR
9	1	TRIPLEX HIGH PRESSURE PUMP
8	1	CIP PUMP
7	1	BALANCE TANK
6	1	ROTARY LOBE TRANSFER PUMP
5	1	CIP PUMP
4	1	CIP PUMP
3	2	EMULSION MAKE-UP TANK
2	1	ROTARY LOBE TRANSFER PUMP
1	2	WATERPHASE MAKE-UP TANK

Fig. 7 Flow diagram of margarine production line. (Courtesy of Crown Chemtech Limited, Reading, UK.)

Table 4 Comparison of Processing Methods

Type	Tube chiller	Chilling drum/ complector
Block/stick margarine	XXX	XXX
Soft table margarine	XXX	X
Low fat spread	XXX	X
Cake and creaming margarine	XXX	XXX
Puff pastry margarine	XX	XXX
Shortening	XXX	X
W/o emulsion with high water content	XXX	X

X = Poor; XX = acceptable; XXX = excellent.
Source: Ref. 19.

of the chilling drum-complector method is that it allows the product to rest between cooling and kneading, which is important with formulas based on slower crystallizing fats such as puff pastry margarine. By comparison, the tube chiller is relatively more compact considering throughput, is easy to operate, and reduces the possibility of spoilage.

Most types of margarine can be made very satisfactorily using the tube chiller method, although in some cases extra working units (pinning machines) may prove advantageous. However, the chilling drum-complector was the preferred method for puff pastry margarine and is perfectly satisfactory for normal table margarine and for cake margarine. The suitability of both methods is summarized in Table 4. After cooling and kneading, industrial margarines may be filled directly. Table margarines may require additional resting before going to the filling unit.

In the case of cake margarine, the most important points are that the fat blend must be soft and easy to incorporate into the batter and have good creaming properties. This therefore suggests the use of lauric fats (coconut, palm kernel), which crystallize quickly and thereby facilitate creaming. However, due to cost, lauric fats are mostly used only as components at 10–20% of blends, which may include hardened and liquid oils such as palm, palm fractions, and soybean to provide the particular requirements of the end product.

III. REDUCED AND LOW FAT SPREADS

As has been previously highlighted, low fat spreads have been the only growth area in a gradually declining market. They can have fat contents of 10–79% and, except for very low fat spreads, are w/o emulsions. However, with low fat products, it is necessary to strike a balance between stability and mouthfeel, which can be affected by both the composition and the method of processing.

Margarine is essentially a stable product, and the appropriate fat blend and the combination of milk proteins, lecithins, and/or saturated mono- and diglycerides used in production ensure both stability and the desired eating characteristics [20]. However, in reduced and low fat spreads in which the disperse phase is in much greater proportion to the continuous phase, problems can arise with stability, meltdown, and flavor release [9].

Although milk proteins are used in high fat products to improve mouthfeel and flavor, they also act as o/w emulsion stabilizers, and thus their use in low fat products, together with the considerable energy input needed, could result in phase inversion.

These problems can be overcome by a combination of formulation and processing. Important factors in the formulation include the melting characteristics of the fat blend, the type and level of emulsifier, and the addition of thickeners such as gelatin, sodium alginate, pectin, and carrageenan to the aqueous phase. Low levels of whey protein may be used to improve flavor release, with the further advantage that the pH of the aqueous phase can be lowered to improve keeping properties because, unlike casein, whey proteins do not precipitate at low pH. The speed of processing, i.e., the rate of throughput, and the emulsion temperature are also important factors in the stability of the spread.

Examples of the effect of various types of emulsifiers on the stability of low fat spreads with fat contents of 40% and 20% are shown in Tables 5 and 6, respectively. The trial batches in both cases were made on a three-tube Perfector pilot plant using a fat blend composed of 60 parts soy oil, 30 parts soy 41, and 10 parts coconut oil. Both examples indicate the potential instability of these emulsions and the variation in the stabilizing effects of different emulsifier blends even when used at low dosage levels.

More recently, the concept of "mixed gels" has been shown to be of interest for the stabilization of reduced and low fat spreads. This concept suggests that two or more gelling components that are not compatible normally may, under certain conditions, interact and thereby increase their individual gelling capacities, thus reducing the amounts of the components in combination compared to their use individually [21]. In particular, interactions between polysaccharides (hydrocolloids) and proteins are of special interest.

Mixed gels are obtained when the two components, e.g., starch hydrolysates and gelatin, of a single-phase system form continuous networks separately and can thus be a case of homogeneous polymeric networks [22]. In this respect, one component of the mixed gel will form a thermosetting gel while the other will be set by the action of ions. The behavior of individual components and their use in combinations in mixed gel systems composed of gelatin and maltodextrins have been fully investigated [23] and have also formed the basis for many patents relating to the stabilization of water-in-oil emulsion spreads.

Table 5 Effect of Types and Blends of Emulsifiers on the Stability of a 40% Fat Spread

		Water separation (%) after			
	Emulsifier	5 min	10 min	15 min	20 min
0.6%	Distilled monoglyceride from vegetable oil, IV ~ 80.	1.6	2.6	9.5	15.8
0.6%	Distilled monoglyceride from vegetable oil, IV ~ 80	0	1.0	2.6	6.3
plus					
0.2%	Polyglycerol ester of fatty acids, IV ~ 80.				
0.4%	Distilled monoglyceride from vegetable oil, IV ~ 80	0	0	0	0
plus					
0.2%	Polyglycerol ester of interesterified ricinoleic acid, IV ~ 85.				

IV = Iodine value.
Formulation of spread:
Water phase (pH 4.5) 56.4% water, 0.5% whey powder, 1.5% gelatin, 1.5% salt, 0.1% potassium sorbate
Fat phase 39.2–39.4% fat blend, 0.6–0.8% emulsifier (as above), 4 ppm β-carotene

Table 6 Effects of Type and Blend of Emulsifier on the Stability of a 20% Fat Spread

Levels and types of emulsifiers	Results
0.8% distilled monoglyceride from sunflower oil, IV ~ 105.	Emulsion split into separate phases in tube chiller.
0.5% distilled monoglyceride from sunflower oil, IV ~ 105 plus 0.5% polyglycerol ester of interesterified ricinoleic acid, IV ~ 85.	Produced a fine, stable spread with good spreadability and mouthfeel.

Formulation of spread:
Water phase (pH 6.8) 69.9% water, 4.0% skim milk powder, 3.0% gelatin, 1.5% sodium alginate, 1.5% salt, 0.1% potassium sorbate
Fat phase 19.0–19.2% fat blend, 0.8–1.0% emulsifier (as above), 8 ppm β-carotene

IV. DAIRY SPREADS AND BLENDS

The fat component of butter-containing spreads may be composed solely of milk fat as listed in part A of Table 2 or may consist of blends of milk fat and other fats, primarily liquid vegetable oils, as in the case of the products listed in part C of Table 2. In the case of blends, the spread can be produced by normal buttermaking methods by adding liquid oil to the cream prior to churning or through the normal margarine production method.

It is rather more difficult to produce low fat butter spread from butter oil due to the relative hardness of the fat at low temperatures. Nonetheless, spread with a butterfat content of about 40% can be produced using 5% sodium caseinate and 2% sodium alginate in the water phase and 0.5% distilled monoglyceride, IV approximately 55, in the fat phase. However, in this case it is not possible to lower the pH of the aqueous phase without precipitating the caseinate, which will reduce its emulsifying properties and thereby limit the keeping properties of the spread. However, a satisfactory low fat butter spread can be produced from dairy cream, using a distilled monoglyceride with high iodine value (80–105) in the cream and a thickener such as sodium alginate. In this case, phase inversion from o/w to w/o can be achieved in the tube chiller, using normal or slightly reduced cooling and operating at 40–50% of the usual capacity. To obtain a satisfactory working effect, a high rotor speed in the tube chiller cooling cylinder would be preferable [24].

Factors that may affect the efficiency of phase inversion in low fat emulsions are listed in Table 7. Most of these factors concern the increase in rate of collision of oil droplets upon which the rate of coalescence is dependent [3].

Table 7 Factors Promoting the Inversion of Oil-in-Water Emulsions

1. Increased rotor speed
2. Controlled temperature (maintenance of optimum fat solids level)
3. Use of low HLB emulsifiers
4. Increased oil droplet size entering unit

Source: Ref. 9.

V. OIL-IN-WATER SPREADS

According to Moran [9], the advantages of o/w spreads over the more conventional w/o spreads may be attributed to the following factors.

1. The product structure is not dependent on the type of fat used.
2. Any level of fat can be used, from 1 to 50% or more.
3. High levels of protein are possible.
4. Processing is easier and cheaper.
5. Flavor release is quicker on the palate.

Against this, however, one major drawback is that unless products are prepared at comparatively low pH, then ultrahigh temperature processing and possibly an aseptic filling procedure must be followed if shelf lives comparable to those of conventional spreads are required.

Oil-in-water spreads remain a relatively unexplored area of the spreadable fats market, possibly due to the problem of microbiological deterioration. They may offer a potential for future expansion in markets where adequate levels of preservatives are permitted.

VI. LIQUID MARGARINE

Liquid margarine is used primarily for frying and is available particularly in markets such as the United States, Germany, and Sweden, where butter or margarine are normally used for frying. In markets such as the United Kingdom that have traditionally used solid fats such as lard or cooking fat or, more recently, liquid vegetable oils, there is less interest in using liquid margarine.

A typical liquid margarine composition has a fat phase of about 82% based on soy or sunflower oil, in which emulsifiers perform two functions. First, an emulsifier such as citric acid ester of monoglyceride (Citrem) at, say, 0.4% plus 0.2% soy lecithin produces a stable water dispersion with limited spattering during frying; second a selected mono- and triglyceride blend prevents oiling out on storage and thus gives a homogeneous product with low viscosity. If the margarine is not intended for frying, the Citrem can be replaced with 0.2% distilled monoglyceride. The water phase may contain proportions of skimmed milk powder and salt as well as a preservative such as potassium sorbate. Flavoring may be added to both phases.

The aqueous phase should be adjusted to pH 4.5–6.5 and pasteurized at 80°C. The fat phase should be tempered to about 60°C and the emulsifiers melted into a small amount of the liquid oil to a temperature of 65–70°C, then stirred into the main part of the fat phase. The water phase is added to the fat phase under continuous agitation before cooling through the tube chiller with an outlet temperature of approximately 5–7°C.

The emulsion should be rested for 2 h and then stirred vigorously for 15 min before tapping off for packaging.

VII. BAKING FATS

Apart from margarines produced specifically for cake and pastry production, baking fats can also take the form of shortenings or compounds.

Fat fulfills a number of roles when used in baked products. At its simplest, fat provides lubricity in the mouth and allows products with a "short" texture to be produced. This "crumbly" texture also associated with easy breakdown and dispersion in the mouth arises because the fat interrupts the gluten network of the dough, thus preventing a hard cohesive texture from developing. Hodge [25] listed the range of functions fat plays during the bakery process and in the final product:

Better aeration
Emulsifying properties (greatly enhanced by emulsification)
Provision of an impervious layer (puff pastry)
Improvement in antistaling
Enhancement of flavors (particularly butter)
Texture modifier—shortening power and lubricity

Some particular applications are described below.

A. Shortenings in Cakes

The term "shortening" arises from the use of this type of fat to impart "shortness" in the preparation of flour confectionery products such as shortbread, short pastry, and cakes [26]. Shortenings contain no moisture and consist of plastified fat or a blend of fats, often together with emulsifiers suitable for the intended end products. The consistency and plasticity of shortenings are therefore important to their effectiveness. For instance, high ratio shortenings are used in cake formulas to enable a higher proportion of sugar to fat by ensuring a greater dough strength arising from the presence of higher levels of emulsifiers, for instance 3–6% distilled monoglycerides. In this case, it is essential that the fat blend be of a consistency that enables good distribution of the fat and excellent creaming properties in the batter.

An important development in the use of plastified fats such as lard was the production of pumpable shortenings to enable delivery to the factory in bulk. From bulk containers, where the shortening is held at temperatures of 23–25°C, it can then be pumped to any part of the production plant.

The first stage in cake making is the batter preparation. The objective is to obtain a fine dispersion of air in the batter. This is achieved by mixing the ingredients sequentially, e.g., initial creaming of sugar and fat or fat and flour, or by the all-in mixing method. The first approach has been practiced over many years whereas the all-in method is comparatively recent and is now widely used both domestically and by large commercial bakeries.

In the initial batter phase the air cells are stabilized largely by fat crystals, although some air will undoubtedly be in the viscous aqueous phase. At this stage the emulsion is complex, being a mixture of domains, some water-continuous and others oil-continuous.

During the baking process the fat melts as the temperature rises and will be completely molten by about 37°C depending on the fat. At this point the water-in-oil emulsion inverts and the air remains trapped in the aqueous phase. With increasing temperature the starch is increasingly hydrated and gelatinized, egg protein starts to coagulate, and the air cells form a nucleus for further expansion by steam and carbon dioxide (e.g., from baking powder). The baking process has been described by Shepherd and Yoell [27].

Butter and margarine are often preferred in cake making because they are able to convey additional flavor characteristics not commonly found in shortenings (100% fat). The prime function of fat in cake making is therefore to assist in the aeration of the batter and modify the texture of the final product.

B. Short Pastry

As indicated earlier, fat is able to interrupt the gluten network of the flour, greatly improving the texture of the final product. This is particularly true for short pastry where aeration is secondary.

The fat needs to be of a reasonably firm consistency such that it distributes throughout the dough as a thin film, creating many "faults" in the dough structure. A variety of fats are used for short pastry manufacture including lard, beef tallow olein, hardened fish oils, and hardened vegetable oils. Margarines and butter are also used. They are generally less effective weight for weight because they contain less fat but will generally impart an improved flavor.

C. Puff Pastry

These products rely on the barrier properties of fat, separating the individual dough layers from one another. The expansion of gas or steam during the bakery process creates a layer structure with a unique texture.

The key to puff pastry making is in having conditions such that the consistency of the butter or margarine is sufficiently plastic to permit it to move coherently with the dough layers. As these are rolled out, relayered, and rolled again to produce the multilayered dough, the fat has to "stretch" with the dough layers, ensuring that they will remain apart during baking. For this reason puff pastry margarine fats are normally higher melting (slip melting point 42°C) than conventional margarine fats and have a higher solid fat content. Plasticity is created in the margarine by postworking (work softening) of the margarine after chilling and crystallizing the molten premix through tubular heat exchangers. Alternatively, chilled drum followed by complector (mixing and work softening) are used [19].

In the case of butter, which has been widely used, additional functionality has been achieved by using butter oil stearins.

D. Bakery Compounds

Bakery compounds or bread fats are water-in-oil emulsions with high water contents used in yeast-raised white bread and producing a softening effect comparable to fat. Their composition is relatively simple (see Table 8.).

Emulsifiers are used to reduce the interfacial tension between the fat and water phases, to stabilize the liquid emulsion, and to impart a fine, stable water dispersion to the bakery compound.

Production is carried out through emulsification of the two phases, held separately at 40–50°C, by adding the water phase slowly to the fat phase, preferably in a process involving a high capacity circulation pump. It is essential that a water-in-oil emulsion be maintained throughout the process. The emulsion is then cooled and kneaded via the tube chiller. Such compounds are used at levels of 2–5% of the bread dough.

E. General

A number of more detailed reviews on the baking process and the function of fats and emulsifiers and lipids in baking have been published. For details on the use of butter and butter fractions, the reader is referred to Kaylegian and Lindsay [28] and Mann [29].

Table 8 Composition of Bakery Compound

Water Phase	
Water	20–60%
Sucrose or glucose syrup	0–10%
Fat blend	
Soft fat (MP 30–35 °C)	40–80%
Emulsifier	1.2%

Baking science and technology is described by Kamel and Stauffer [30] and Pyler [31]. Fats in baking systems are reviewed by La Guardia [32] and Podmore [33]. Lipid interactions in breadmaking are discussed in detail by Carr et al. [34], Eliasson and Larsson [35], and Barnes [36].

REFERENCES

1. E. Huber, Cloister dairies in the fourth pre-Christian milleniumn (in German), *Deut. Molkerie-Z. 45*: 2610 (1931).
2. L. M. Lamport, *Modern Dairy Products*, Chemical Pub. Company, New York, 1965, p. 282.
3. E. Frede and W. Buchheim, Buttermaking and the churning of blended fat emulsions, *J. Soc. Dairy Technol. 47*(1): (1994). p. 17.
4. M. Anderson and B. E. Brooker, Dairy foams, in *Advances in Food Emulsions and Foams* (E. Dickinson and G. Stainsby, Eds.), Elsevier, London, 1988, p. 221.
5. W. Buchheim, Electron microscopic investigation of the structure of whipped cream, Proceedings 19th Int. Dairy Congr., New Delhi, 1974.
6. C. Poot and E. Biernoth, Margarine and butter production, in *The Lipid Handbook*, 2nd ed., (F. D. Gunstone, J. L. Harwood, and F. B. Padley, Eds.), Chapman and Hall, London, 1994, p. 293.
7. A. J. C. Andersen and P. N. Williams, *Margarine*, 2nd ed., Pergamon, Oxford, 1954, p. 2.
8. M. K. Schwitzer, *Margarine and Other Food Fats*, Leonard Hill, London, 1956, p. 60.
9. D. P. J. Moran, Reduced calorie spreads, *Porim Technol.* No. 15 Palm Oil Research Institute of Malaysia, February 1993.
10. R. P. Borwanker and G. S. Buliga, Emulsion properties of margarines and low fat spread, Proc. Food Emulsions and Foams—Theory and Practise, San Francisco, 1989, pp. 44–52.
11. J. Madsen, Product formulation and processing of margarine and yellow fat spreads, Proc. Margarine and Yellow Fat Seminar, Coventry, 1983.
12. J. Madsen and G. Als, Sandiness in table margarine and the influence of various blends of triglycerides and emulsifiers thereon, Proc. IXth Int. Soc. Fat Res. Congr., Rotterdam, 1968.
13. R. E. Timms, Phase behaviour of fats and their mixtures, *Prog. Lipid Res. 23*: 1–38 (1984).
14. N. Garti, E. Wellner, and S. Sarig, *J. Am. Oil Chem. Soc. 59*: (1981).
15. K. Berger, Palm oil products—why and how to use them, *Food Technol.*, September 1986, p. 72.
16. S. E. Friberg, R. F. Goubron, and I. H. Kayali, Emulsion stability, in *Food Emulsions* (K. Larrson and S. E. Friberg, Eds.), Marcel Dekker, New York, 1990.
17. N. Krog, T. Riisom, and K. Larsson, Applications in the food industry, in *Encyclopedia of Emulsion Technology*, Vol. 2 (K. Larrson and S. E. Friburg, Eds.), Marcel Dekker, New York, 1988, p. 321.
18. N. Krog, Interactions of emulsifiers with other components in foods, in *Ingredient Interactions—Effects on Food Quality* (A. G. Gaonkar, Ed.), Marcel Dekker, New York, 1995, p. 380.
19. J. Madsen, Puff pastry margarine: a comparison of the chilling drum and tube chiller methods of production, TP 101, Danisco Ingredients, Denmark.
20. E. A. Flack, The role of emulsifiers in reduced fat and fat-free foods, in *Food Technology International Europe* (A. Turner, Ed.), Sterling, London, 1992, pp. 179–181.
21. A. H. Clark, R. K. Richardson, S. B. Ross-Murphy, and J. M. Stubbs, Structural and mechanical properties of agar/gelatin co-gels. Small deformation studies, *Macro Molecules 16*: 1367 (1983).
22. V. B. Tolstoguzov, in *Functional Properties of Food Macromolecules* (J. R. Mitchell and D. A. Ledward, Eds.), Elsevier, London, 1985, p. 385.
23. S. Kasapis, E. R. Morris, I. A. Norton, and R. T. Brown, Phase equilibria and gelation in gelatin/maltodextrin systems. Parts I–IV, *Carbohydrate Polym. 21*: 243 (1993).
24. J. Madsen, Low calorie spread and melange production in Europe, Proc. World Conf. on Edible Oils and Fats Processing, Maastricht, 1989.
25. D. G. Hodge, Fats in bakery products, *Br. Nutr. Found. Nutr. Bull. 11*(3): 153 (1986).

26. M. Gordon, Fats and fatty foods, in *Food Industries Manual*, 23rd ed. (M. D. Ranken and R. C. Kill, Eds.), Blackie, London, 1993, p. 322.
27. I. S. Shepherd and R. W. Cake Emulsions Yoell, in *Food Emulsions* (S. Friberg, Ed.), Marcel Dekker, New York, pp. 217–275. 1976.
28. K. E. Kaylegian and R. C. Lindsay, *Handbook of Milk Fat Fractionation and Technology*, AOCS Press, Champaign, IL, 1995.
29. E. Mann, Modified butters and fat spreads, *Dairy Ind. Int. 60*: 21 (1995).
30. B. S. Kamel and C. E. Stauffer, *Advances in Baking Technology*, Blackie, London, 1993.
31. E. J. Pyler, Emulsions in *Baking Science and Technology*, Vol. I, Sosland,Chicago 1988.
32. M. K. La Guardia, High performance fat systems in baked products, *Cereal Foods World 39*: 147 (1994).
33. J. Podmore, Fats in bakery and kitchen products, in *Fats in Food Products* (D. P. J. Moran and K. K. Rajah, Eds.), Blackie, London, 1995, p. 220.
34. N. O. Carr, N. W. R. Daniels, and P. J. Frazier, Lipid interactions in breadmaking, *Rev. Sci. Nutr. 31*(3): 237 (1992).
35. A. C. Eliasson and K. Larsson, Intereactions between components, in *Cereals in Breadmaking: A Molecular Colloidal Approach* (A. C. Eliasson and K. Larsson, Eds.), Marcel Dekker, New York, 1993, pp. 161–210.
36. P. J. Barnes, *Lipids in Cereal Technology*, Academic, London, 1983.

12

Ice Cream

H. Douglas Goff
University of Guelph, Guelph, Ontario, Canada

I. INTRODUCTION

The importance of ice cream and related products as a food commodity is very significant, with perhaps an annual production of 136 million hectoliters worldwide, increasing at 4–5% per annum. Annual production figures from some of the major world producers are approximately 54 million hL in the United States, 5.5 million hL in Germany, 5.2 million hL in the United Kingdom, 4.0 million hL in Australia, 3.9 million hL in Canada, and 2.3 million hL in Sweden. Annual per capita consumption figures are approximately 18 L in Australia, New Zealand, and the United States; 14 L in Canada and Sweden; 9 L in the United Kingdom; 7 L in Germany; and 6 L in Norway and Finland [1]. Given also that, on average, 5–12% of the weight of ice cream is fat [2], it becomes readily apparent that the ice cream industry is a large consumer of fat, worldwide, approximately 800,000 tonnes per annum. Fat not only provides flavor to ice cream and related products but is also vital in creating the smooth-eating texture that is both desired by and expected from the consumer.

The term "ice cream" is often taken to mean a family of whipped dairy products that are manufactured by freezing and are consumed in the frozen state, including ice cream that consists of either dairy or non-dairy fats; premium, higher fat versions; "light," lower fat versions; ice milk; sherbet; frozen yogurt; etc. The manufacturing processes for most of these products are similar and involve the preparation of a liquid mix; concomitantly whipping and freezing this mix dynamically under high shear to a soft, semifrozen slurry; incorporation of flavoring ingredients to the partially frozen mix; packaging the product; and further quiescent freezing (hardening) of the product in blast air. Many comprehensive reviews of ice cream manufacture are available [2–7]. This chapter briefly summarizes the manufacture and composition of ice cream and related products and focuses on the use and role of fats in these products.

II. FORMULATIONS

Table 1 presents the range of compositional variables found in most ice cream mix formulations. The composition, minimum fat content, total solids content, and weight per volume of frozen dairy products in most countries are standardized and regulated. Ice cream and ice milk mix formulations specify the content of fat, milk solids-not-fat, sweeteners, stabilizers, emulsifiers, and moisture that are desired. Usually one mix is used for the production of a variety of flavors. Dairy and other ingredients used to supply these components are chosen on the basis of availability, cost, legislation, and desired quality. Ingredients available for each of the major components are listed in Table 2. It must be remembered, however, that in addition to supplying the major component, most of the dairy ingredients also supply minor amounts of the other components. Thus a detailed calculation is necessary to convert the desired formula into a recipe, based on the ingredients chosen. Algebraic methods for the

Table 1 A Typical Compositional Range for the Components Used in Ice Cream Mix Formulations

Component	Range (%)
Milkfat	10–16
Milk solids-not-fat	9–12
Sucrose	9–12
Glucose syrup solids	4–6
Stabilizers/emulsifiers	0–0.5
Total solids	36–45
Water	55–64

Table 2 Components of an Ice Cream Mix and the Ingredients That Supply Them

Component	Ingredients[a]
Milkfat	Cream (msnf, water)
	Butter (msnf, water)
Milk solids-not-fat	Skim milk powder (water)
	Condensed skim milk (water)
	Condensed milk (water, fat)
	Sweetened condensed skim milk (water, sugar)
	Sweetened condensed milk (water, sugar, fat)
	Whey powder (water)
Water	Skim milk (msnf)
	Milk (fat, msnf)
	Water
Sweetener	Sucrose
	Glucose syrup solids
	Liquid sugars (water)
Stabilizers/emulsifiers	

msnf = milk solids-not-fat.
[a]Minor suppliers of the component listed in parentheses.

solution of these problems are available [2,5], and computer programs to solve mix calculation problems, based on the solution of simultaneous equations, are in widespread use in many commercial operations.

III. INGREDIENTS

A. Fat

The fat component of the mix increases the richness of flavor in ice cream, produces a characteristic smooth texture by lubricating the palate, helps to give body, and aids in producing desirable melting properties [2,4]. The fat content of a mix also aids in lubricating the freezer barrel while the ice cream is being manufactured. Limitations on excessive use of fat in a mix include cost, a hindered whipping ability, decreased consumption due to excessive richness, and high caloric value. Fat contributes 9 kcal/g to the diet, regardless of its source. During freezing of ice cream, the fat emulsion that exists in the mix will partially coalesce (destabilize) as a result of emulsifier action, air incorporation, ice crystallization, and high shear forces of the blades [4,8,9]. This partial coalescence is necessary to set up the structure and texture in ice cream, which is similar to the structure in whipped cream [10,11]. This process is discussed in Section V.A. The fat content is an indicator of the perceived quality and/or value of the ice cream. Ice cream must have a minimum fat content of 8–10% in most legal jurisdictions. Premium ice creams may have fat contents as high as 14–18%. It has also become desirable, however, to create light ice creams, <5% fat, with similar perceived quality. In addition to structure formation, fat contributes a considerable amount of flavor to ice cream, which is difficult to reproduce in lowfat ice creams. Fat content must be altered by at least 2–3% before any noticeable difference appears in the taste or texture [2].

1. Dairy Fat

Milk fat as a fat source for ice cream formulations is in widespread use in North America, Australia, New Zealand, and parts of Europe. The triglycerides in milk fat have a wide melting range, +40° to −40°C. The crystallization patterns of milk fat are also very complex, due in part to the large variation in fatty acids and large numbers of different triglycerides present [12–17]. Consequently, there is always a combination of liquid and crystalline fat at chilled and subzero temperatures. Alteration of this solid/liquid ratio at freezer barrel temperatures, through natural variation or fat fractionation, may affect the ice cream structure formed [9,18]. The best source of butterfat in ice cream for high quality flavor is fresh sweet cream from fresh sweet milk [2]. Other sources of butterfat include anhydrous milk fat, sweet (unsalted) butter, frozen cream, and condensed milk blends. Whey creams have also been used but may lead to flavor or texture problems.

2. Non-Dairy

Vegetable fats are used extensively as fat sources in ice cream in the United Kingdom, parts of Europe, the Far East, and Latin America but only to a very limited extent in North America. Five factors of great interest in selection of fat source are the crystal structure of the fat, the rate at which the fat crystallizes during dynamic temperature conditions, the temperature-dependent melting profile of the fat, especially at chilled and freezer temperatures, the content of high melting triglycerides (which can produce a waxy, greasy mouthfeel), and the flavor and purity of the oil [4]. As will be discussed in Section V.A, it is important that the fat droplet contain an intermediate liquid/solid fat ratio at the time of freezing. It is difficult to

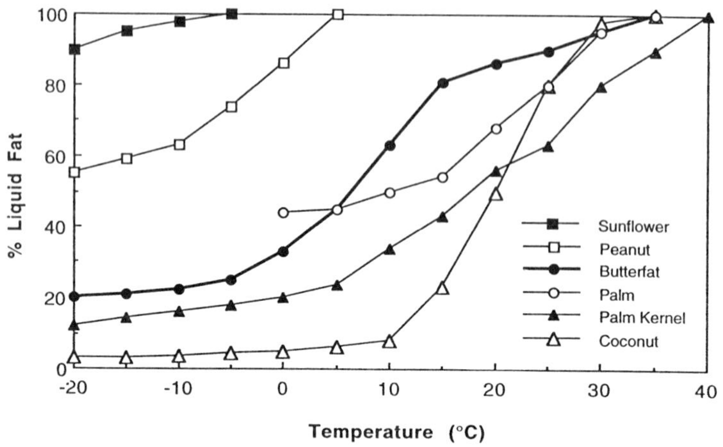

Fig. 1 The variation of liquid fat content with temperature of fats suitable for use in ice cream. (Courtesy of D. Finney, Unilever Research Laboratory, Colworth House, United Kingdom.)

quantify this ratio as it is dependent on a number of composition and manufacturing factors; however, one-half to two-thirds crystalline fat at 4–5°C is a good working rule [4]. Crystallization of fat occurs in three steps: undercooling to induce nucleation, heterogeneous or homogeneous nucleation (or both), and crystal propagation. In bulk fat, nucleation is predominantly heterogeneous, with crystals themselves acting as nucleating agents for further crystallization, and undercooling is usually minimal. However, in an emulsion each droplet must crystallize independently of the next. For heterogeneous nucleation to predominate, there must be a nucleating agent available in every droplet, which is often not the case. Thus in emulsions, homogeneous nucleation and extensive undercooling may be common [4,13].

Figure 1 illustrates the melting profiles for a range of fats that are used in ice cream formulations. Blends of oils are often used in ice cream manufacture, selected to take into account physical characteristics, flavor, availability, stability during storage, and cost. Hydrogenation is often necessary to achieve the appropriate melting characteristics. Palmkernel oil, coconut oil, palm oil, and fractions thereof, plus their hydrogenated counterparts, can all be used. Tong et al. [19] substituted up to 40% milk fat in ice cream with safflower oil, a highly unsaturated oil, in an attempt to lower the saturated fatty acid content of the final product. They reported that increasing the concentration of safflower oil decreased overrun (the incorporation of air into the ice cream) but had little effect on the extent of fat destabilization. However, Berger [4] reported that ice cream made entirely with liquid vegetable oil as the fat source was unacceptable. The oil spread at the air surface, leading to collapse and undesirable texture. A partly crystalline fat globular structure is thus necessary for structure development.

3. *Fat Substitutes*

There has been great interest in the marketplace for the development of lower fat alternatives to traditional ice cream products. As a result, a large amount of product development time has been used in searching for a combination of ingredients that will replace the textural and flavor characteristics of fat in ice cream. These often involve the use of fat substitutes. Such products may be formulated with starch or other polysaccharides, proteins, or lipids, but their main requirement is to provide fewer calories to the product than traditional fat sources in the diet. A great deal of technical literature exists on the various properties of the products being marketed by a number of commercial firms. Schmidt et al. [20] studied the rheological, freezing, and melting properties of ice milks manufactured with protein-based or maltodextrin-based fat alternatives. They concluded that the carbohydrate-based alternatives resulted in greater effects on mix rheology while the protein-based alternatives were more similar to the control, due in part to the functional contributions of proteins to food systems, especially in the area of emulsification and air incorporation. Ice cream products are very complex systems, both in structure and in flavor. In creating products that are meant to deliver to the consumer the same attributes but with less fat or calories, it is imperative that the structural element be considered to the same extent as flavor in order to deliver high quality products and develop market share for these products.

B. Nonfat Milk Ingredients

The milk solids-not-fat (msnf) or serum solids improve the texture of ice cream, aid in giving body and chew resistance to the finished product, are capable of allowing a higher overrun without the characteristic snowy or flaky textures associated with high overruns, and may be a cheap source of total solids [21]. The msnf contain the lactose, caseins, whey proteins, minerals (ash), vitamins, acids, enzymes, and gases of the milk or milk products from which they were derived. The content of serum solids used in a mix can vary from 10% to 14% or more. Limitations on their use include off-flavors from certain products and an excess of lactose, which may lead to numerous problems as discussed below [22]. Traditionally, the best sources of serum solids for high quality products are fresh concentrated skimmed milk or spray-dried low-heat skim milk powder. Others include sweetened condensed whole or skimmed milk, superheated condensed skimmed milk, buttermilk powder or condensed buttermilk, and dried or condensed whey [21]. Buttermilk powder is a by-product of the butter industry. It contains a higher concentration of fat globule membrane phospholipids than skim milk. Thus it can be used for its emulsifying properties to reduce the need for emulsifiers [2]. Whey is a by-product of the cheese industry and contains fat, lactose, whey proteins, and water but no casein. Whereas skim milk powder contains 54.5% lactose and 36% protein, whey powder contains 72–73% lactose and only about 10–12% protein [12]. Thus it can promote some of the problems associated with high lactose as described below as well as flavor problems associated with the high salt content. However, there are an increasing number of whey products available that have higher protein and lower lactose and mineral contents. Many of these can provide much higher quality than the traditional whey ingredients [23]. It is also possible to produce concentrated protein products from the casein portion of milk proteins, which have potential for use in ice cream [23,24]. These caseinates are formulated to give different functional properties, such as degrees of emulsification, foaming, gelation, or water holding [25].

Proteins contribute much to the development of structure in ice cream, including emulsification, whipping, and water-holding capacity [25–27]. Emulsification properties of proteins in the mix arise from their adsorption to fat globules at the time of homogenization [23]. Whipping properties of proteins in ice cream contribute to the formation of the initial air bubbles in the mix [27,28]. The water-holding capacity of proteins leads to enhanced viscosity in the mix [25], which imparts a beneficial body to the ice cream, increases the meltdown time of ice cream, and contributes to reduced iciness [23]. The protein content of a mix is usually about 4%. When replacing traditional msnf sources in ice cream with newer sources with the ratio of proteins altered from the native state, it is important to ensure that these functional roles are being met.

Lactose is a disaccharide of glucose and galactose. Lactose does not contribute much to sweetness; it is only one-fifth to one-sixth as sweet as sucrose [12]. Lactose content of ice cream mix is about 6% if no whey powder has been used in the formulation. Excessive levels of lactose in ice cream mix leads to reduced freezing point, causing a softening of the ice cream and the potential for development of iciness and a greater potential for lactose crystallization [29]. Lactose is relatively insoluble, and lactose crystals, particularly the α form, which takes on a characteristic tomahawk shape [22], can produce the defect known as sandiness when they are allowed to grow to a large size. The lactose solubility in water at room temperature is about 11% [12]. During freezing, this concentration is exceeded as a result of freeze concentration (water removal in the form of ice). When 75% of the water is frozen in a mix consisting originally of 11% msnf (6% lactose), the lactose content in the unfrozen phase corresponds to 39.3%. Probably much of the lactose in ice cream exists in a supersaturated, amorphous (noncrystalline) state, however, due to extreme viscosity [30]. Stabilizers help to hold lactose in a supersaturated state because of viscosity enhancement.

C. Sweeteners

Sweet ice cream is usually desired by the consumer. As a result, sweetening agents are added to ice cream mix at a rate of usually 12–17% by weight. Sweeteners improve the texture and palatability of the ice cream, enhance flavors, and are usually the cheapest source of total solids [2]. Their ability to lower the freezing point of a solution imparts a measure of control over the temperature–hardness relationship. In determining the proper blend of sweeteners for an ice cream mix, the total solids required from the sweeteners, the sweetness factor of each sugar, and the combined freezing point depression of all sugars in solution must be calculated to achieve the proper solids content, the appropriate sweetness level, and a satisfactory degree of hardness [3,4,31]. The most common sweetening agent used is sucrose, alone or in combination with other sugars. Sucrose and lactose are most commonly present in ice cream in the supersaturated or glassy state, with few crystals being present [4,30]. It has become common practice in the industry to substitute sweeteners derived from starch hydrolysate syrup (glucose syrup) for all or a portion of the sucrose [2,31]. A typical sweetener blend for an ice cream mix usually includes 10–12% sucrose and 4–5% glucose syrup solids [2,31].

The use of primarily corn, but also wheat, potato, or tapioca starch hydrolysis products in ice cream is generally perceived to provide greater smoothness by contributing to a firmer and more chewy body, to provide better meltdown characteristics, to bring out and accentuate fruit flavors, to reduce heat shock potential, which improves the shelf life of the finished product, and to provide an economical source of solids [31,32]. The higher molecular weight saccharides (dextrins) are effective stabilizers and provide maximum prevention against the formation of coarse ice crystals, which is reflected in improved meltdown and heat shock resistance. They also improve cohesive and adhesive textural properties. Smaller sugars provide smoothness, sweetness, and flavor enhancement. Dextrose offers sweetness synergism with sucrose but has the greatest freezing point depression of all the sugars, being a monosaccharide. The ratio of higher to lower molecular weight fractions can be estimated from the dextrose equivalent (DE) of the syrup. Ice cream manufacturers usually use a 28–42DE syrup, either liquid or dry [31]. When dextrose is converted to fructose, the resultant syrup is much sweeter than sucrose, although it has half the molecular weight and thus contributes more to freezing point depression than sucrose [32].

D. Stabilizers

Ice cream stabilizers are a group of ingredients (usually polysaccharides) commonly used in ice cream formulations. The primary purposes for using stabilizers in ice cream are to produce smoothness in body and texture; retard or reduce ice and lactose crystal growth during storage (or mask the effects of crystal growth), especially during periods of temperature fluctuation, known as heat shock [33]; and to provide uniformity to the product and resistance to melting [2,3]. They also increase mix viscosity, stabilize the mix to prevent separation of a clear serum on meltdown, known as wheying off (e.g., carrageenan), aid in suspension of flavoring particles, produce a stable foam with easy cut-off and stiffness at the barrel freezer for packaging, slow down moisture migration from the product to the package or the air, and help to prevent shrinkage of the product volume during storage [34]. Stabilizers must also have a clean, neutral flavor, not bind to other ice cream flavors, contribute to acceptable meltdown of the ice cream, and provide desirable texture and mouthfeel upon consumption [34]. Limitations on their use include production of undesirable melting characteristics, excessive mix viscosity, and contribution to a heavy, soggy body. Although stabilizers increase mix viscosity, they have little or no impact on freezing point depression.

Gelatin, a protein of animal origin, was once used almost exclusively in the ice cream industry as a stabilizer but has gradually been replaced with polysaccharides of plant origin due to their greater effectiveness and lower cost [2,5]. Stabilizers currently in use include (1) carboxymethyl cellulose, derived from the bulky components or soluble fiber of plant material; (2) locust bean gum (carob bean gum), which is derived from the beans of the tree *Ceratonia siliqua* grown mostly in the Mediterranean; (3) guar gum, from the guar bush, *Cyamoposis tetragonolba*, a member of the legume family grown in India and Pakistan for centuries and now grown to a limited extent in Texas; (4) carrageenan, an extract of *Chondus crispis*, Irish moss, a red algae, originally harvested from the coast of Ireland, near the village of Carragheen; and (5) sodium alginate, an extract of another seaweed, kelp, a brown algae. Each stabilizer has its own characteristics, and often two or more of these stabilizers are used in combination to lend synergistic properties to each other and improve their overall effectiveness. Guar, for example, is more soluble than locust bean gum at cold temperatures; thus it finds more application in HTST pasteurization systems. Carrageenan is a secondary colloid used to prevent the wheying off of mix, which is usually promoted by one of the other stabilizers [2–5].

The mechanism of action of stabilizers in enhancing frozen stability is related primarily to their effect on the ice phase. Stabilized ice creams have been observed with electron microscopy techniques to have smaller ice crystals than unstabilized ice cream both before and after storage at fluctuating temperatures [33,35,36]. The control of iciness by stabilizers has been examined by several researchers [36–44]. It was originally thought that stabilizers bound water, rendering it unfreezable and thereby exerting their stabilizing mechanism by reducing the quantity of ice present [2], but recent research has failed to demonstrate this effect [39–44]. The stabilizing effect of the polysaccharides presently appears to result from an alteration of the diffusion properties of water and solutes within the unfrozen phase following freeze-concentration [3,36,45–47].

E. Emulsifiers

Emulsifiers have been used in ice cream mix manufacture for many years [48]. They are sometimes integrated with the stabilizers in proprietary blends, but their function and action are very different from those of the stabilizers. They are used for improvement of the whipping quality of the mix; production of a drier ice cream to facilitate molding, fancy extrusion, and sandwich manufacture; smoother body and texture in the finished product; superior drawing qualities at the freezer to produce a product with good stand-up properties and melt resistance; and more exact control of the product during freezing and packaging operations [2,49]. Their mechanism of action can be summarized as follows: They lower the fat/water interfacial tension in the mix, resulting in protein displacement from the fat globule surface, which in turn reduces the stability of the fat globule to the partial coalescence that occurs during the whipping and freezing process, leading to the formation of a fat structure in the frozen product that contributes greatly to texture and meltdown properties [8]. The extent of protein displacement from the membrane, and hence the extent of dryness achieved, is a function of the emulsifier concentration [4,50]. Their role in structure formation is described in detail in Section V.A.

Egg yolk was formerly commonly used as an ice cream emulsifier. Emulsifiers used in ice cream manufacture today are of two main types: the mono- and diglycerides and the sorbitan esters. Mono- and diglycerides are derived from the partial hydrolysis of fats of animal or vegetable (important for certain religious or ethnic considerations) origin. The sorbitan esters are similar to the monoglycerides in that the sorbitan esters have a fatty acid molecule such as stearate or oleate attached to a sorbitol molecule whereas the monoglycerides have a fatty acid molecule attached to a glycerol molecule. To make the sorbitan esters water-soluble, polyoxyethylene groups are attached to the glucose molecule. Thus Polysorbate 80, the most common of these sorbitan esters, is actually polyoxyethylene sorbitan monooleate [48]. Polysorbate 80 is a very active drying agent in ice cream [8] and is used in many commercial stabilizer/emulsifier blends.

IV. PROCESSING

A. Mix Manufacture

1. Blending and Pasteurization

Ice cream processing operations can be divided into two distinct stages: mix manufacture and freezing operations. Ice cream mix manufacture consists of the following unit operations: combination and blending of ingredients, batch or continuous pasteurization, homogenization, and mix aging [6]. Ingredients are usually preblended prior to pasteurization, regardless of the type of pasteurization system used. Blending of ingredients is relatively simple if all ingredients are in the liquid form, as automated metering pumps or tanks on load cells can be used. When dry ingredients are used, powders are added either through a pumping system under high velocity or through a liquefier, a large centrifugal pump with rotating knife blades that chops all ingredients as they are mixed with the liquid [6].

Pasteurization is the biological control point in the system, designed for the destruction of pathogenic bacteria. In addition, it serves a useful role in reducing the total bacterial load and in solubilization of some of the components (proteins and stabilizers). Both batch and continuous (high temperature, short time, or HTST) systems are in common use [6]. In a batch pasteurization system, proper ingredient amounts are blended in large jacketed vats equipped with some means of heating, usually saturated steam or hot water. The product is then heated in the vat to at least 69°C (155°F) and held for 30 min to satisfy legal requirements for pasteurization, necessary for the destruction of pathogenic bacteria [51]. Various time–temperature combinations can be used, depending on the legal jurisdiction. The heat treatment must be severe enough to ensure destruction of pathogens and to reduce the bacterial count to a particular maximum (e.g., 100,000 per gram) depending on the legal jurisdiction [51]. Following pasteurization, the mix is homogenized using high pressures and then is passed across some type of heat exchanger (plate heat exchanger or double- or triple-tube heat exchanger) for the purpose of cooling the mix to refrigerated temperatures (4°C).

Continuous pasteurization is usually performed in an HTST heat exchanger following the blending of ingredients in a large insulated feed tank. Some preheating, to 30–40°C, may be necessary for solubilization of the components. The HTST system is equipped with a heating section, a cooling section, and a regeneration section (Fig. 2). Mix first enters the raw regeneration section, where cold or preheated mix is heated to as warm as possible on one side of a plate heat exchanger while the pasteurized hot mix is cooled to as low as possible running countercurrent on the opposite sides of the plates. Following the raw regeneration section, mix may enter the homogenizer, depending on the particular configuration. The location of the homogenizer at this point may be preferred, as it may act as the timing pump, it makes use of the heating effect of homogenization (2–3°C), and it ensures that any potential contamination is maintained on the raw side. Mix then flows to the heating section, where a minimum temperature of 80°C is obtained [51]. Mix is held at this temperature for at least 25 s [51] by flowing either through a series of holding tubes or through an additional set of plates in the HTST unit. Holding times much longer than the minimum can be accomplished with longer holding tubes. Holding times of 2–3 min may produce superior mixes that retain many of the advantages of batch pasteurization but may also leave the mix prone to cooked flavors [4,5]. After leaving the holding tube, mix flows back through the pasteurized side of the regeneration section and enters the cooling plates, where a chilled brine or glycol solution or chilled water brings the mix down to around 4°C.

2. *Homogenization*

Homogenization is responsible for the formation of the fat emulsion by forcing the hot mix through a small orifice under pressures of 15.5–18.9 MPa (2000–3000 psig), depending on the mix composition. Table 3 show the effects of homogenization on the fat emulsion [12,52]. The actual mechanism of fat disruption within the homogenizer is thought to result from turbulence, cavitation, and velocity gradients (energy density) within the valve body [53].

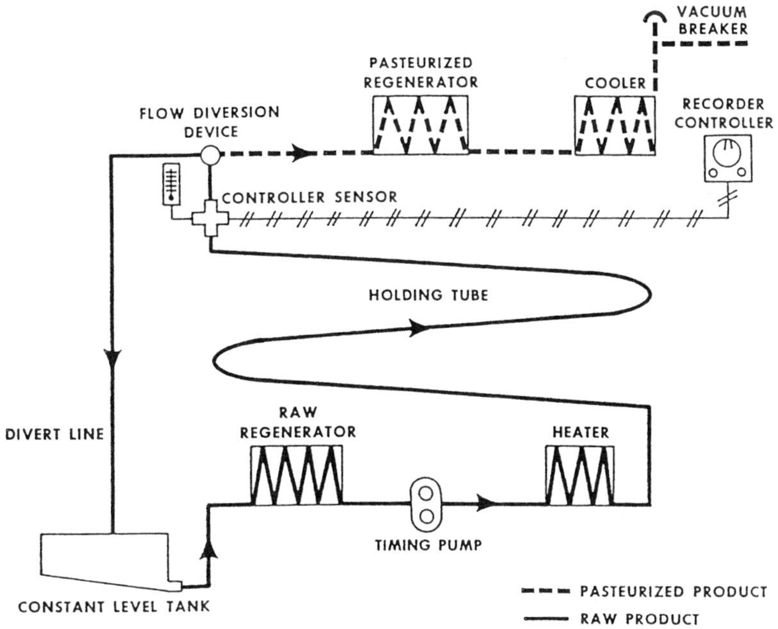

Fig. 2 A schematic illustration of the equipment used for HTST pasteurization of milk, frozen dairy dessert mixes, and other fluid dairy products. (From Ref. 51.)

Table 3 Effect of Homogenization of Milk on the Size Distribution of Fat Globules

	No homogenization	15 MPa (2500 psig)
Average fat globule diameter (μm)	3.3	0.4
Maximum diameter (μm)	10.0	2.0
Surface area (m^2/mL at 3.5% fat)	0.08	0.75
Number of globules (μm^{-3})	0.015	12

Source: Refs. 12 and 52.

The 8–10-fold increase in the surface area of the fat globules is responsible in part for the formation of the fat globule membrane, composed of adsorbing materials attempting to lower the interfacial free energy of the fat globules [54,55]. With single-stage homogenizers, fat globules tend to cluster as bare fat surfaces come together or adsorbed molecules are shared. Therefore, a second homogenizing valve is frequently placed immediately after the first with applied backpressures of 3.4 MPa (500 psig) [6], allowing more time for surface adsorption to occur. The net effects of homogenization are the production of smaller, more uniform fat droplet size resulting in a greater stability of fat droplets during aging, better whipping ability, and a smoother, more uniform final product with a greater apparent richness [2]. Homogenization also decreases the danger of churning the fat in the freezer and makes it possible to use products that could not otherwise be used, such as butter and frozen cream.

3. *Aging*

An aging time of 4 h or greater at 2–4°C is recommended following mix processing prior to freezing. This allows for hydration of milk proteins and stabilizers (some viscosity increase occurs during the aging period), crystallization of the fat globules, and a membrane rearrangement, to produce a smoother texture and better quality product [56]. Non-aged mix exhibits a loose stand-up, is very wet at extrusion, and exhibits variable whipping abilities [2,56]. The appropriate solid/liquid fat ratio must be attained at this stage, a function of temperature and the triglyceride composition of the fat used, as a partially crystalline emulsion is needed for partial coalescence in the whipping and freezing step [57], as discussed in Section V.A. Emulsifiers generally displace milk proteins from the fat surface during the aging period [8,23,58], and this is also discussed in detail in Section V.A. The whipping qualities of the mix are usually improved with aging. Aging is performed in insulated or refrigerated storage tanks, silos, etc., or in single-walled tanks in chilled rooms, where valves and pipework can also be kept cold. Mix temperature should be maintained as low as possible without freezing.

B. Ice Cream Freezing

Ice cream freezing also consists of two distinct stages: (1) passing mix through a swept-surface heat exchanger under high shear conditions to promote extensive ice crystal nucleation and air incorporation and (2) freezing the packaged ice cream under conditions that promote rapid freezing and small ice crystals [3,6]. The freezing and whipping process is one of the most important unit operations for the development of quality, palatability, and yield of finished product, due to the incorporation of air creating the foam, the formation of the ice phase, and the partial destabilization of the fat emulsion. The objective of ice cream manufacturers is to produce ice crystals that are below, or at least not significantly above, the threshold of sensory detection at the time of consumption. This threshold has been suggested by Arbuckle [2] to be between 40 and 50 μm. Consequently the freezing steps of the manufacturing process and the temperature profile throughout the distribution system are critical factors in meeting this objective. Flavoring and coloring can be added as desired to the mix prior to passing it through the barrel freezer, and particulate flavoring ingredients such as nuts, fruits, candy pieces, or ripple sauces can be added to the semifrozen product at the exit from the barrel freezer prior to packaging and hardening.

1. Dynamic Freezing and Aeration

Continuous freezers dominate the ice cream industry [6]. In this type of process, mix is drawn from the flavoring tank into a scraped surface heat exchanger, which is jacketed with a liquid, boiling refrigerant [usually ammonia; a chlorinated fluorinated hydrocarbon (CFC) such as R-12, R-22, or R-502; or one of the newly developed CFC substitutes]. Incorporation of air into ice cream, termed the overrun, is necessary to produce desirable body and texture. Overrun is calculated as [2]

$$\%\ \text{Overrun} = \frac{\text{volume of ice cream produced} - \text{volume of mix used}}{\text{volume of mix used}} \times 100\%$$

As a guide, maximum overrun should be 2.5–3 times the total solids content to avoid possible defects in the finished ice cream [2].

There are three types of air incorporation systems used in continuous freezers. In older systems, the pump configuration resulted in a vacuum either at the pump itself or on the mix line entering the pump. Air was then incorporated through a spring-loaded, controllable needle valve. Newer types of freezers use compressed air, which is injected into the mix. This type of air handling system allows for ultrafiltration (0.65 μm micropore filter) of the air prior to its entering the mix [6]. Additionally, recent experimentation with preaeration systems, in which the mix is whipped independently prior to entering the barrel freezer, has demonstrated that much smaller air bubble sizes and improved body and texture can be attained [59]. Following aeration, the water in the mix is partially frozen as the mix–air combination passes through the barrel of the heat exchanger. Rotating knife blades and dashers keep the product agitated and prevent freezing on the side of the barrel. Residence time for mix through the annulus of the freezer varies from 0.4 to 2 min, although some product may remain for much longer times, especially with open-cage dashers; freezing rates can vary from 5 to 27°C per min, and draw temperatures of −6°C can easily be achieved [4].

Batch freezing processes differ slightly from the continuous systems just described. The barrel of a batch swept-surface heat exchanger is jacketed with refrigerant and contains a set of dashers and scraper blades inside the barrel. It is filled to about one-half volume with the liquid mix. Barrel volumes usually range from 2 to 12 L. The freezing unit and agitators are then activated, and the product remains in the barrel for sufficient time to achieve the desired degree of overrun and stiffness. Whipping increases with time and cannot exceed the amount that will fill the barrel with product (i.e., 100% overrun when starting half-full). Batch freezers are used in smaller operations where it is desirable to run individual flavored mixes on a small scale or to retain an element of the "homemade"-style manufacturing process. They are also operated in a semicontinuous mode for the production of soft-serve desserts. A hopper containing the mix feeds the barrel as product is removed.

The formation of the ice phase during this stage is responsible in part for the structure and texture of the final product [3]. The crystallization of water to ice involves two major steps: nucleation and crystal growth [60–62]. Rapid heat removal, which results from low temperatures (large ΔT) in the freezing medium, and rapid agitation, which are both present in the barrel freezer, promote extensive nucleation and create numerous tiny ice crystals. Further temperature reduction during hardening accounts for continued growth of the preformed crystals [3]. As water freezes out of solution in a relatively pure form, the formation of ice acts to concentrate the dissolved sugars, lactose, milk proteins, salts, and hydrocolloids in an ever-decreasing amount of water. Water and its dissolved components are referred to as the serum or matrix of the mix [2]. Because the freezing point of the serum is a function of the concentration of dissolved solids, the formation of more ice concentrates the serum and results in an ever-decreasing freezing temperature for the remaining serum. Thus at temperatures several degrees below the initial freezing temperature, there is always an unfrozen phase present.

Bradley [63,64] presented a method of calculating initial freezing points and equilibrium freezing curves for sweetened dairy mixes based on their composition. An equilibrium freezing curve for an ice cream mix, which plots the amount of water frozen as a function of temperature, is shown in Fig. 3. This represents the maximum amount of ice that can form at any given temperature, based on freezing point considerations. Ice cream hardness is a function of temperature due to its effect on this conversion from unfrozen water to ice and further concentration of the serum phase surrounding the ice crystals, which helps to give ice cream its ability to be scooped and chewed at freezer temperatures.

2. *Static Hardening*

Ice cream following dynamic freezing, ingredient addition, and packaging is immediately transferred to a hardening chamber (−30°C or colder, either forced convection or plate-type conduction freezers), where the majority of the remaining water freezes [6]. As mentioned earlier, rapid initial freezing is essential to set up as many crystal nuclei as possible so that during the maturation or growth stage their size stays small. In the same context, rapid hardening is also necessary to keep ice crystals small. When hardening is slow, there is too great an opportunity for water still remaining in the ice cream to migrate to crystal centers already formed, resulting in large ice crystals [3]. Many factors need to be considered during the hardening process. The main factors affecting heat transfer are the temperature difference between the product and the freezing medium, the area of product being exposed to the freezing medium, and the heat transfer coefficient for the particular operation [6]. Following

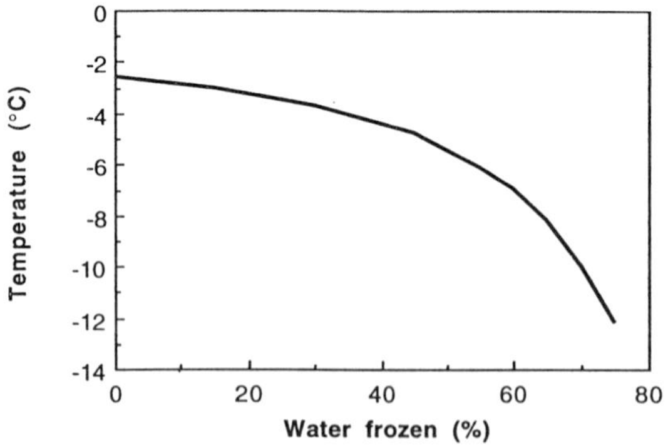

Fig. 3. An equilibrium freezing curve for a typical ice cream mix as calculated by the method of Bradley [63,64] illustrating the relationship between temperature and unfrozen water as determined by freezing point depression characteristics. This relationship illustrates the maximum amount of ice that may form at any temperature, but due to kinetic considerations the amount of ice may be considerably less as a result of freezing rate considerations.

rapid hardening, ice cream storage should occur at low, constant temperatures, usually −25°C.

V. ROLE OF FAT IN PHYSICAL STRUCTURE AND FLAVOR

A. Structure Formation

The development of structure and texture in ice cream is sequential, basically following the manufacturing steps, and results from the interactions of several components. Ice, air, and fat each occupy distinct but interrelated phases (Fig. 4), all surrounded by a temperature-dependent unfrozen phase [30,65]. Ice crystals and air bubbles are usually in the range of 20–50 μm and 50–100 μm, respectively [4].

To properly describe the role of fat, it is necessary to begin with the formation of the emulsion at the time of homogenization and the role of the ingredients present at that time, with particular reference to the fat, proteins, and emulsifiers. Following homogenization, the emulsion is further affected by changes occurring during the aging step, that is, crystallization of the fat and rearrangement of the fat globule membrane to the lowest free energy state. This emulsion then undergoes both whipping and ice crystallization during the dynamic freezing process, which contributes to the development of the three main structural components of the frozen product: a discontinuous foam, a network of partially coalesced fat surrounding the air bubbles, and ice crystals.

Homogenization begins the process of fat structure formation. It has been shown that a finely divided fat emulsion with little fat globule clustering after homogenization is essential for the manufacture of a smooth, dry ice cream [66,67]. After preheating or pasteurization, the mix is at a temperature sufficient to have melted all the fat and emulsifier present, and the fat passes through one or two homogenizing valves. Immediately following homogenization, the newly formed fat globule is practically devoid of any membranous material and readily adsorbs amphiphilic molecules from solution [68–70]. The immediate environment supplies the surfactant molecules, which include caseins, undenatured whey proteins, phospholipids, lipoprotein molecules, components of the original milk fat globule membrane, and any added

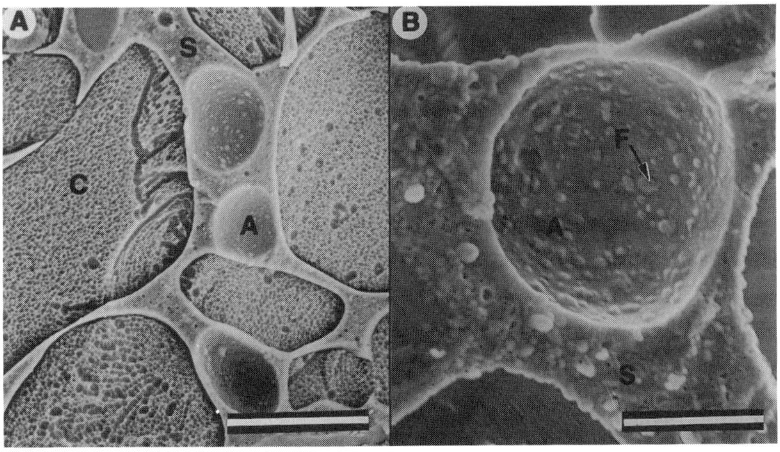

Fig. 4 The structure of frozen, aerated dairy dessert products showing ice crystals (C), air bubbles (A), fat globules (F), and the freeze-concentrated unfrozen serum phase (S), as evidenced from cryo-scanning electron microscopy. (From Ref. 30.) (a) Bar = 30 μm; (b) bar = 7.5μm.

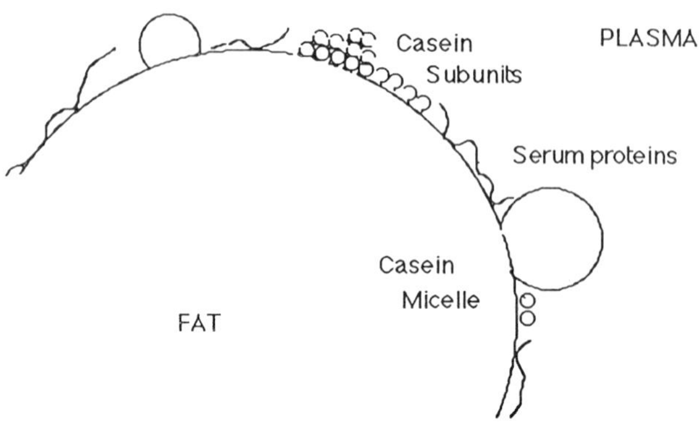

Fig. 5 A schematic representation of the adsorption of casein and whey proteins to the fat globule following homogenization. Because of the differences in conformational structure, the resulting surface excess is much lower with the whey proteins than with the caseins.

chemical surfactants [55,68]. These all compete for space at the fat surface, shown schematically in Fig. 5 [55,70].

By selecting and/or limiting the adsorbing material present at the time of homogenization, it may be possible to predetermine the adsorbing substances and thus create a membrane with more functional attributes, using natural proteins rather than relying on the chemical surfactants [23,71]. The membrane formed during homogenization continues to develop during the aging step, and rearrangement occurs until the lowest possible energy state is reached [58,68]. The transit time through a homogenization valve is on the order of 10^{-5}–10^{-6} s, and velocities on the order of 12,000 cm/s are reached [53]. Protein adsorption or unfolding at the interface may take minutes or even hours to be complete [52]. It is clear, therefore, that the immediate membrane formed upon homogenization is a function of the microenvironment at the time of its creation and that the recombined membrane of the fat globule in the aged mix is not fully developed until well into the aging process [68].

Emulsifiers are not needed in an ice cream mix to stabilize the fat emulsion (Table 4), due to an excess of protein and other amphiphilic molecules in solution [49,72]. If a mix is homogenized without any emulsifier, both the whey proteins and the caseins will form this new fat globule membrane, with the caseins contributing much more to the bulk of the adsorbed protein (Fig. 6a). However, if surfactants such as the added emulsifiers are present,

Table 4 Comparison of the Creaming Rates of Ice Cream Mix Emulsions Either With No Surfactant or With 0.08% Polysorbate 80 When Both Have Sat Quiescently for 10 Days[a]

	No emulsifier	Polysorbate 80
Butterfat (%) in mix	10.00	10.00
Butterfat (%), top of 100 mL cylinder, 10 days	11.25 ± 0.43	11.28 ± 0.43
Butterfat (%), bottom of 100 mL cylinder, 10 days	9.47 ± 0.11	9.46 ± 0.11

Butterfat determined by Mojonnier ether extraction.
[a] $n = 4$.
Source: Ref. 75.

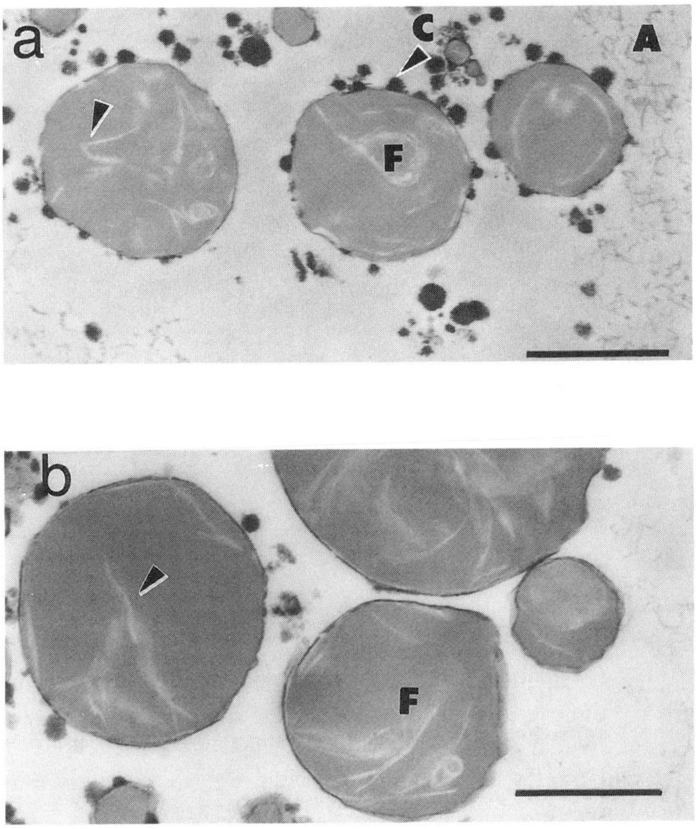

Fig. 6 Transmission electron micrograph of the ice cream mix emulsion in the absence (a) and presence (b) of Polysorbate 80. (From Ref. 75.) F = fat globule, C = casein micelle, A = agarose matrix; arrow points to fat crystal within the globule. Bar = 1 μm.

they have the ability to lower the interfacial tension between the fat and water phases more than the proteins (Table 5). Thus they become preferentially adsorbed to the surface of the fat [8,58,73–75]. As the interfacial tension is lowered and proteins are eliminated from the surface of the fat, the surface excess (quantity of adsorbed material, mg/m^2) is reduced (Table 6) and the actual membrane becomes weaker to subsequent destabilization [75,76]. This is due to the fact that the protein molecules, particularly the caseins, are considerably larger than the emulsifier molecules so that a membrane made up entirely of emulsifier is very thin (Fig. 6b), i.e., there is lower surface excess, although the emulsion is thermodynamically favored due to the lowering of the interfacial tension and net free energy of the system [75]. However, Dickinson et al. [77] point out that subsequent instability to shear is not induced by the protein displacement per se but by some change in the adsorbed layer properties caused by the replacement of a pure protein monolayer by a mixed protein–surfactant monolayer.

Crystallization of fat also occurs during aging, creating a highly intricate structure of needle-like crystals within the globule (Fig. 7). The high melting point triglycerides crystallize first and continue to be surrounded by liquid oil of the lower melting point triglycerides. The process of fat crystallization during aging of an ice cream mix was investigated by Iversen and Pedersen [56]. Examining the crystallization of pure butterfat and butterfat in an emulsion

Table 5 Interfacial Tension Values (dyn/cm) Between 11% Milk Solids-Not-Fat Solutions and Anhydrous Butter Oil in the Presence of Various Emulsifiers[a]

Emulsifier	Interfacial tension[b] (dyn/cm)
No emulsifier	6.16
GMS	5.52 a
GMO	5.09 b
Span 60	5.64 a
Span 80	5.02 b
Tween 60	2.42
Tween 80	2.24

[a]$n=4$. Emulsifiers were dispersed in their appropriate soluble phase (0.08%) at 70 °C. Interfacial tension measured by a duNuoy ring.
[b]Values followed by the same letter are not significant at $p < 0.05$.
Source: Ref 8.

by pulsed NMR techniques, they reported that fat crystallization had completed within 1.5 h. A partially crystalline fat is necessary for clumping to occur [52,78]. Van Boekel and Walstra [79] found emulsion stability of a paraffin oil in water emulsion to be reduced by six orders of magnitude when crystals were present in the dispersed phase. This has been attributed [73,75,79] to the protrusion of crystals into the aqueous phase causing a surface distortion of the globule (Fig. 8). The crystal protrusions can then pierce the film between two globules upon close approach. As the crystals are preferentially wetted by the lipid phase, clumping is thus inevitable [78]. This phenomenon may account for partial clumping of globules under a shear force (Fig. 9). The clusters thus formed actually hold the ice cream serum in their interstices, resulting in the observed dryness. These fat globule chains may also envelop the air cells, thus improving overrun and adding stability to the foam structure [30,78]; however, fat crystals have also been shown to impair overrun development in whipped cream [80]. It is possible that in addition to crystallization, polymorphic transitions of the fat crystals are occurring. Fat crystals exist in three polymorphic forms, α, β', and β, in order of stability [52]. The rotation of the triglycerides from one form to another is important in many food fats, especially those of a relatively few fatty acids, and takes time to occur [81].

Table 6 Differences in the Amount of Protein Adsorbed[a] to the Fat Globules in the Same Mixes as in Table 4

	No emulsifier	Polysorbate 80 (0.08%)
Mix protein (%)	4.08±0.01	4.08±0.01
Protein in serum (% of mix)	3.43±0.07	3.76±0.17
Adsorbed protein (% of total protein)	15.9	7.84

[a]$n=4$. Determined by centrifugal separation and Kjeldahl analyses.
Source: Ref. 75.

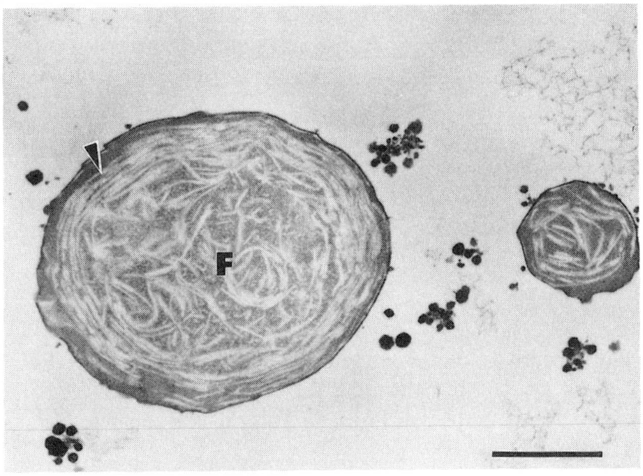

Fig. 7 Milk fat globule from an ice cream mix showing the intricate structure of needle-like crystals present in the cross section of the globule, preserved by fixation at 4°C. (From Ref. 75.) F = fat globule; arrow points to fat crystal. Bar = 1 μm.

The next stage of structure development occurs during the concomitant whipping and freezing step. Air is incorporated either through a lengthy whipping process (batch freezers) or by being drawn into the mix by vacuum or injected under pressure (continuous freezers) [2]. This process causes the emulsion to undergo partial coalescence or fat destabilization, during which clumps and clusters of the fat globules form and build an internal fat structure or network into the frozen product [2,9,78,82] in a manner very analogous to the whipping of heavy cream [11,52,73,83]. In studying the whipping of heavy cream, Schmidt and vanHooydonk [10] stated that the proteinaceous membrane that envelops the air bubble is penetrated by fat globules as the process proceeds, and this fat penetration offers foam stability to the

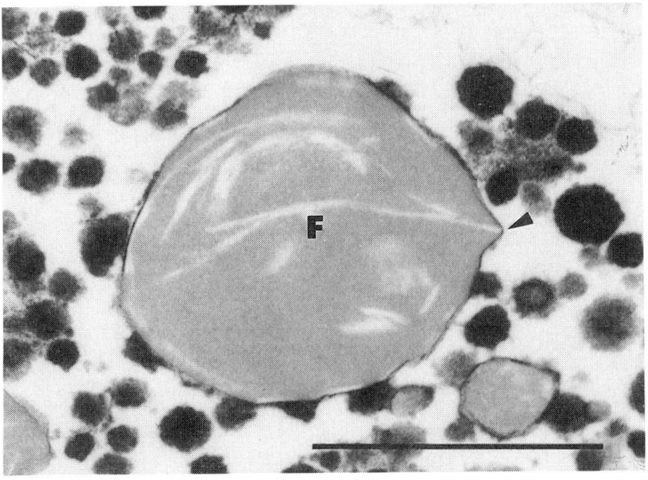

Fig. 8 Milk fat globule from an ice cream mix. Arrow points to surface distortion of the globule caused by the presence of the internal crystalline structure. (From Ref. 75.) Bar = 1 μm.

Fig. 9 A schematic representation of the network of partially coalesced fat globules formed during whipping, illustrating the important role of fat crystals in cementing the coalescing globules together (visualized as the straight lines within the globule). (Courtesy of Prof. P. Walstra, Wageningen, Netherlands.)

whipped product. Brooker et al. [11] offered a more detailed explanation of the whipping and foam stabilization process in heavy cream. During the initial stages of whipping, air bubbles were stabilized primarily by β-casein and whey proteins with little involvement of fat. Adsorption of fat to air bubbles occurred when the fat globule membrane coalesced with the air/water interface. Only rarely did fat spread at the air/water interface. The final cream was stabilized by a cross-linking of fat globules surrounding each air cell to adjacent air cells, thus building an infrastructure in the foam. In skim milk foams, the initial air/water interface is also formed by the serum proteins and soluble β-casein with little involvement of micellar casein. Micelles became attached as a discontinuous layer but were not deformed or spread [28]. It can be postulated that air cell incorporation into ice cream mix follows a similar mechanism. Cross-linking of fat globules from one air cell to the next, thus forming an infrastructure, is less likely due to the reduction in dispersed phase volume from the heavy cream system to the ice cream mix system; however, it must also be remembered that the air bubbles, fat globules, and aqueous phase are being freeze-concentrated at the same time.

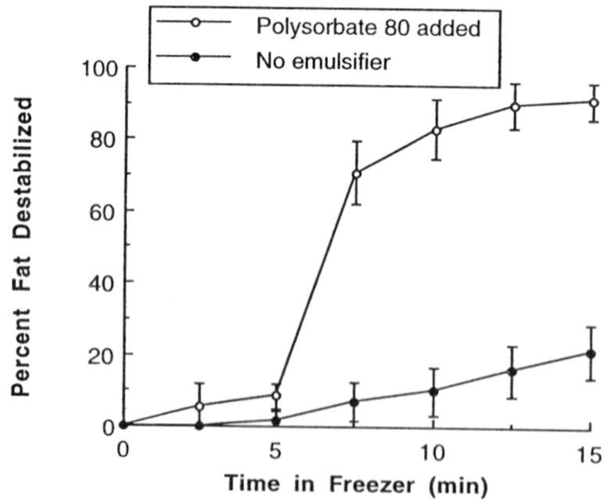

Fig. 10 The effect of emulsifier addition on the percent of fat destabilized (determined by turbidity of a 1:500 dilution at 5°C) during the period of time the ice cream mix was held in the barrel of the batch freezer. Ice crystallization began after about 5 min. (From Ref. 8.)

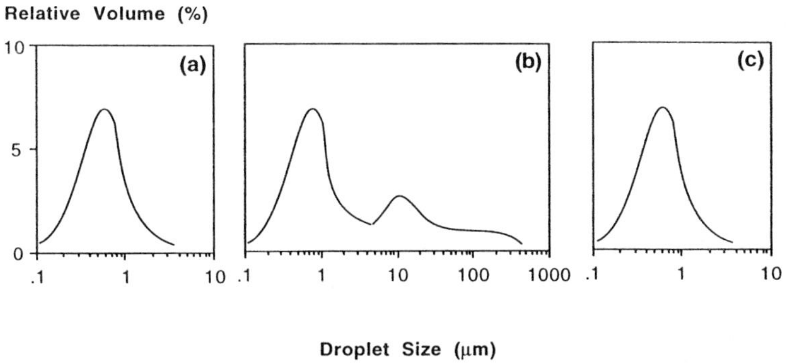

Fig. 11 Droplet size distribution in ice cream with butteroil as a fat source measured at room temperature as a function of the stage of the process. (a) For an aqueous solution of homogenized, aged mix. (b) For an aqueous solution of hardened, thawed ice cream. (c) For hardened, thawed ice cream measured in 1 wt % SDS, used as a dissociating medium. (From Ref. 87.)

The fat globule clusters formed during the process of partial coalescence are responsible for surrounding and stabilizing the air cells and creating a semicontinuous network or matrix of fat throughout the product, resulting in the beneficial properties of dryness upon extrusion during the manufacturing stages (aids in packaging and novelty molding, for example), a smooth-eating texture in the frozen dessert, and resistance to meltdown or good stand-up properties (necessary for soft serve operations) [4,9,84,85]. Kloser and Keeney [86] presented the first evidence that fat destabilization was enhanced by some of the emulsifiers in common use, and this has been confirmed by numerous researchers (Fig. 10) [4,8,72,78,84]. When the emulsion is subjected to the tremendous shear forces in the barrel freezer, the thin membrane created by the addition of surfactant is not sufficient to prevent the fat globules from colliding and coalescing, thus setting up the internal fat matrix. If an ice cream mix is subjected to excessive shearing action or contains too much emulsifier or not enough protein,

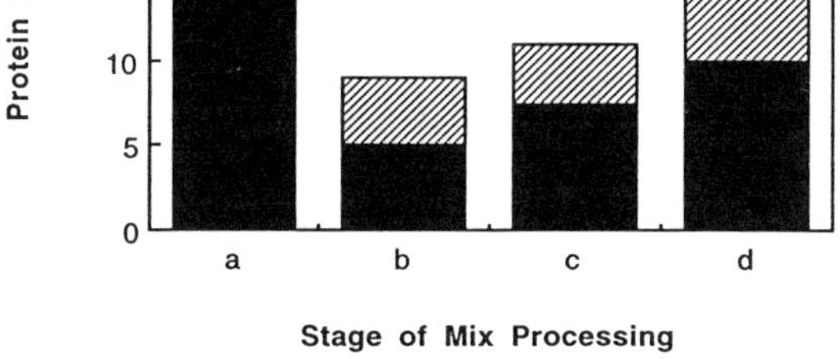

Fig. 12 Changes in protein composition of aqueous phase as a function of the stage of the process in the manufacture of ice cream with butteroil as a fat source. (a) Initial aqueous phase; (b) homogenized mix; (c) aged mix; (d) thawed ice cream. (From Ref. 87.)

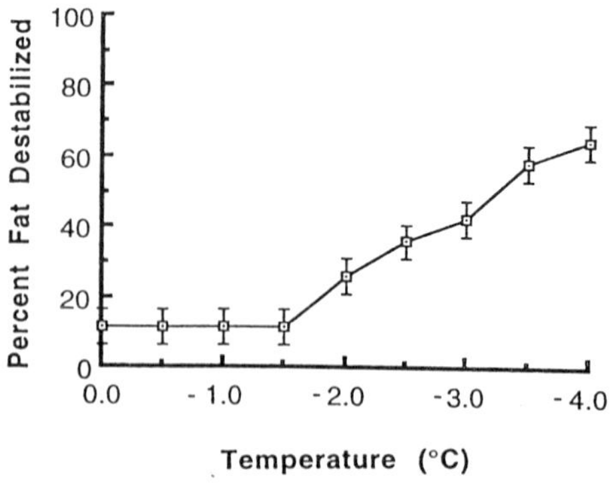

Fig. 13 The effect of ice crystallization as demonstrated by a reduction in draw temperature in a continuous ice cream freezer on the fat destabilization process in ice cream. (From Ref. 8.)

the formation of objectionable butter particles can occur as the emulsion is churned beyond the optimum level. It should be possible to predict the destabilizing power of an emulsifier in the barrel freezer by measuring the interfacial tension between the fat and water phases in the presence of the emulsifier to determine the extent of protein coverage that will result and by considering the molecular size (and surface excess) of the emulsifier (Table 5) [8]. Polysorbate 80, having a small molecular weight and producing the lowest interfacial tension, displaces more protein, resulting in a very thin membrane, and thus produces the maximum amount of fat destabilization [8,75]. This can lead to the development of desired dryness or to excessive fat clumping beyond the optimum level if not properly controlled.

Changes to the fat emulsion have recently been studied by Gelin et al. [87]. They demonstrated through light scattering measurements of fat globule size distribution and aggregation (see Fig. 11) that the freezing step is responsible for considerable fat aggregation (Fig. 11b) but that this aggregation is reversible through dissociation with SDS (Fig. 11c). They also showed the changes occurring in the protein distribution between the aqueous and adsorbed states (Fig. 12). It is obvious from this study that the homogenization step accounts for a large amount of adsorbed protein and that casein is preferentially adsorbed over whey protein. The aging and freezing-hardening-thawing steps each accounted for subsequent protein desorption, again mostly of the caseins. The emulsifier used in these mixes contained saturated mono- and diglycerides but no Polysorbate 80. According to the results of Goff and Jordan [8], Polysorbate 80 may have led to greater amounts of protein desorption during the aging period. The sequential process of partial coalescence during ice cream freezing has also been examined [8]. The incorporation of air alone, or the shearing action alone, independent of freezing, was not sufficient to cause the same degree of fat destabilization as when ice crystallization occurs at the same time (Fig. 13). The freezing process causes an increase in concentration of the mix components in the unfrozen water phase. It is believed that the ice crystals contribute to the shearing action on the fat globules, owing to their physical shape, and that the concentration of components also leads to enhanced destabilization. However, to create the desired fat destabilization, whipping and freezing must occur simultaneously [8,49].

The static freezing step is responsible for the conversion of most of the remaining unfrozen water into ice and for the continued crystallization of the fat [3,4]. A concentrated, unfrozen solution of sugars and salts remains, although the sugars are probably supersaturated or in the glassy (amorphous solid) state [3,4,30,46,47]. It is likely that a glass transition temperature is reached at temperatures less than −30 to −35°C (composition-dependent) where the remaining solution solidifies but does not crystallize [36,45]. Below this glass transition temperature, it is unlikely that any further physical or chemical phenomena occur.

B. Flavor and Structure Defects

There can be numerous flavor and textural defects associated with ice cream. This chapter discusses only those associated with the fat content. However, an excellent review on ice cream defects has been published by Bodyfelt et al. [88]. Defects in ice cream flavor associated with the fat phase are usually related to either autoxidation of the fat, resulting in oxidized flavors, or, especially in the case of milk fat, lipolysis of free fatty acids from triglycerides by the action of lipases (known in the dairy industry as rancidity). These defects tend to be present in the raw ingredients used in ice cream manufacture rather than promoted by the manufacturing process itself. The fat phase can also account for textural defects associated with content (too high or too low) or degree of fat destabilization (affecting melting properties).

Oxidation of fat proceeds by the well-known autoxidation reaction [89] in three stages: initiation, propagation, and termination. Free radicals formed during the initiation step, especially from unsaturated fatty acids or phospholipids, are then able to initiate oxidation of triglycerides, especially in the presence of copper and proteins. During propagation, antioxidant compounds such as tocopherols and ascorbic acid are depleted while peroxide derivatives of fatty acids accumulate. Peroxides, which have little flavor, undergo further reactions to form a variety of carbonyls, some of which are potent flavor compounds, especially some ketones and aldehydes. Milk may oxidize as a result of factors either extrinsic or intrinsic to the milk [12,90–92]. Important extrinsic factors include contamination with metals, temperature of storage, oxygen tension, heat treatment, agitation, light, and acidity. Both copper and iron may catalyze lipid oxidation, but probably only copper is significant in milk. Added copper is much more potent than natural copper because a significant portion of added copper goes directly to the fat globule [12]. Significant intrinsic factors affecting milk fat oxidation include metalloproteins such as milk peroxidase and xanthine oxidase, endogenous ascorbic acid, which acts as a cocatalyst with copper to promote oxidation, endogenous copper content, and endogenous antioxidants, mainly tocopherols [93]. Fresh forage is well known to control spontaneous oxidation, as indicated by obvious seasonal effects on the incidence of oxidized flavor. This effect is probably due to increased levels of endogenous antioxidants.

Hydrolysis of fatty acid esters by the action of lipases results in the common flavor defect known as lipolytic or hydrolytic rancidity and is distinct from oxidative rancidity [90]. Lipolysis in dairy fats can be extremely detrimental due to the number of highly volatile short-chain fatty acids present, especially butyric acid. A comprehensive review of flavor impairment of milk and milk products due to lipolysis was published by the International Dairy Federation [94]. Lipases are unique among enzymes in that they are active at the lipid/serum interface. In milk, lipases are ineffective unless the fat globule membrane is damaged or weakened in some way. Lipolysis may be caused by the lipoprotein lipase (LPL) that is endogenous to milk or by bacterial lipases. The properties of the fat globule membrane are most important to lipolysis. The observed lactation effects may be due to reduced content of phospholipids in the fat globule membrane in late lactation [95]. Mastitis, which alters milk composition, also increases sensitivity of the fat globule to lipolysis [96]. Other factors that destabilize the fat globule membrane, especially agitation and/or foaming, also promote lipolysis. Churned fat does not appear to be a good substrate for LPL [95], but lipolysis is accelerated by the replacement of the native membrane with surface-active material (mainly casein micelles and whey proteins) from the plasma [97]. This effect is at least partly due to redistribution of LPL from the plasma to the fat globule membrane and accounts for greatly increased lipolysis after homogenization. In the milk from some animals, lipolysis may proceed without subsequent thermal or mechanical activation. This effect, frequently referred to as spontaneous lipolysis, is unlikely to occur in herd milks or in pooled milks because it is prevented by mixing affected milk with three to five times its volume of normal milk. The major conditions that influence spontaneous lipolysis are late lactation, insufficient fresh forage, and low-yielding cows.

Textural defects attributed to the fat component of the ice cream are associated with either the fat content or its degree of destabilization [88]. Fat contributes smoothness to the finished product. Lowfat mixes must therefore compensate for this lack of inherent smoothness by altering the ratio of other components, particularly the protein, polysaccharide stabilizer, and emulsifier components. On the other hand, mixes high in fat such as the premium products typically have a heavier, dense texture related to both high fat and lower than normal air content. The fat destabilization process also affects the structure and hence the texture and meltdown of the ice cream. If too much destabilization has occurred, a defect known as "does not melt" may occur. This results from a network of fat that gives sufficient structure to the product to hold its shape in the absence of the ice phase after warming to temperatures sufficient to melt the ice [2]. Some slowness of melt is desirable and can be achieved by fat destabilization, but it is also considered desirable for the product to melt down or collapse to a consistency not unlike the unfrozen mix after the ice phase has melted.

VI. CONCLUSIONS

In summary, the fat content of ice cream and related products is vital both for its flavor and for structural formation and resulting texture. While it may be possible to replace the former attribute in creating lower fat products, it is much more difficult to replace the latter attribute. The partial coalescence process is a complicated one, with important roles being served by both the proteins and the emulsifiers, in addition to the fat, and occurring sequentially during several steps: homogenization, aging, and whipping/freezing. Consumers have come to expect a smooth-eating, creamy product with good meltdown characteristics. There is no doubt that control of the size distribution of the ice phase is critical in producing a smooth texture [3], but so too is the fat content. With consumer interest in lower fat products increasing, especially in North America, the challenges for the ice cream manufacturer lie in developing products that meet consumer expectations for lower fat while not compromising textural quality.

REFERENCES

1. International Dairy Federation, *Consumption Statistics for Milk and Milk Products*, Int. Dairy Fed. Bull. 295, Brussels, 1994, p. 1.
2. W. S. Arbuckle, *Ice Cream*, 4th ed., AVI, Westport, CT, 1986, p. 1.
3. H. D. Goff and M. E. Sahagian, Freezing of dairy products, in *Freezing Effects on Food Quality* (L. E. Jeremiah, Ed.), Marcel Dekker, New York, 1996, 299–335.
4. K. G. Berger, Ice cream, in *Food Emulsions*, 2nd ed. (K. Larsson and S. Friberg, Eds.), Marcel Dekker, New York, 1990, p. 367.
5. R. Jimenez-Flores, N. J. Klipfel, and J. Tobias, Ice cream and frozen desserts, in *Dairy Science and Technology Handbook*, Vol. 2, *Product Manufacturing* (Y. H. Hui, Ed.), VCH, New York, 1993, p. 57.
6. H. L. Mitten and J. M. Neirinckx, Developments in frozen products manufacture, in *Modern Dairy Technology*, Vol. 2, *Advances in Milk Products* (R. K. Robinson, Ed.), Elsevier, New York, 1986, p. 215.

7. P. G. Keeney and M. Kroger, Frozen dairy products, in *Fundamentals of Dairy Chemistry*, 2nd ed. (B. H. Webb, A. H. Johnson, and J. A. Alford, Eds.), AVI, Westport, CT, 1974, p. 873.
8. H. D. Goff and W. K. Jordan, The action of emulsifiers in promoting fat destabilization during the manufacture of ice cream, *J. Dairy Sci. 72*: 18 (1989).
9. P. G. Keeney, Development of frozen emulsions, *Food Technol. 36*(11): 65 (1982).
10. D. G. Schmidt and A. C. M. vanHooydonk, A scanning electron microscopical investigation of the whipping of cream, *Scanning Electron Microsc. III*: 653 (1980).
11. B. E. Brooker, M. Anderson, and A. T. Andrews, The development of structure in whipped cream, *Food Microstruct. 5*: 277 (1986).
12. P. Walstra and R. Jenness, *Dairy Chemistry and Physics*, Wiley, New York, 1984.
13. P. Walstra, Physical chemistry of milkfat globules, in *Developments in Dairy Chemistry*, Vol. 2, *Lipids* (P. F. Fox, Ed.), Applied Science, New York, 1983, p. 119.
14. J. M. deMan, Physical properties of milkfat. II. Some factors influencing crystallization, *J. Dairy Res.* 28: 117 (1961).
15. J. Vasic and J. M. deMan, Effect of temperature history on the solid fat content of milkfat, *J. Dairy Sci. 48*: 1277 (1965).
16. J. E. Schaap and G. A. M. Rutten, Effect of technological factors on the crystallization of bulk milk fat, *Neth. Milk Dairy J. 30*: 197 (1976).
17. I. Soderberg, L. Hernqvist, and W. Buchheim, Milk fat crystallization in natural milk fat globules, *Milchwissenschaft 44*: 403 (1989).
18. E. L. Thomas, Structure and properties of ice cream emulsions, *Food Technol. 35*: 41 (1981).
19. P. S. Tong, W. K. Jordan, and G. Houghton, Response surface methodology to study fat destabilization and development of overrun in ice creams produced with polyunsaturated safflower oil and milkfat blends, *J. Dairy Sci. 67*: 779 (1984).
20. K. Schmidt, A. Lundy, J. Reynolds, and L. N. Yee, Carbohydrate or protein based fat mimicker effects on ice milk properties, *J. Food Sci. 58*: 761 (1993).
21. H. D. Goff, Examining the milk solids-not-fat in frozen dairy desserts, *Mod. Dairy 71*(3): 16 (1992).
22. P. A. Morrissey, Lactose: chemical and physicochemical properties, in *Developments in Dairy Chemistry*, Vol. 3, *Lactose and Minor Constituents* (P. F. Fox, Ed.), Elsevier, New York, 1985, pp. 1–34.
23. H. D. Goff, J. E. Kinsella, and W. K. Jordan, The influence of various milk protein isolates on ice cream emulsion stability, *J. Dairy Sci. 72*: 385 (1989).
24. J. G. Parsons, S. T. Dybing, D. S. Coder, K. R. Spurgeon, and S. W. Seas, Acceptability of ice cream made with processed whey and sodium caseinate, *J. Dairy Sci. 68*: 2880 (1985).
25. J. E. Kinsella, Milk proteins: physicochemical and functional properties, *CRC Crit. Rev. Food Sci. Nutr. 21*(3): 197 (1984).
26. M. C. Phillips, Protein conformation at liquid interfaces and its role in stabilizing emulsions and foams, *Food Technol. 35*(1): 50 (1981).
27. P. J. Halling, Protein stabilized foams and emulsions, *CRC Crit. Rev. Food Sci. Nutr. 15*: 155 (1981).
28. B. E. Brooker, Observations on the air serum interface of milk foams, *Food Microstruct. 4*: 289 (1985).
29. D. E. Smith, A. S. Bakshi, and C. J. Lomauro, Changes in freezing point and rheological properties of ice cream mix as a function of sweetener system and whey substitution, *Milchwissenschaft 39*: 455 (1984).
30. K. B. Caldwell, H. D. Goff, and D. W. Stanley, A low-temperature SEM study of ice cream. I. Techniques and general microstructure, *Food Struct. 11*: 1 (1992).
31. H. D. Goff, R. D. McCurdy, and G. N. Fulford, Advances in corn sweeteners for ice cream, *Mod. Dairy 69*(3): 17 (1990).
32. H. D. Goff, R. D. McCurdy, and E. A. Gullett, Replacement of carbon-refined corn syrups with ion-exchanged corn syrups in ice cream formulations, *J. Food Sci. 55*: 827 (1990).
33. H. D. Goff, Heat shock revisited, *Mod. Dairy 72*(3): 24 (1993).

34. H. D. Goff and K. B. Caldwell, Stabilizers in ice cream. How do they work?, *Mod. Dairy 70*(3): 14 (1991).
35. K. B. Caldwell, H. D. Goff, and D. W. Stanley, A low-temperature SEM study of ice cream. II. Influence of ingredients and processes, *Food Struct. 11*: 11 (1992).
36. H. D. Goff, K. B. Caldwell, D. W. Stanley, and T. J. Maurice, The influence of polysaccharides on the glass transition in frozen sucrose solutions and ice cream, *J. Dairy Sci. 76*: 1268 (1993).
37. E. K. Harper and C. F. Shoemaker, Effect of locust bean gum and selected sweetening agents on ice recrystallization rates, *J. Food Sci. 48*: 1801 (1983).
38. Y. Shirai, K. Nakanishi, R. Matsuno, and T. Kamikubo, Effects of polymers on secondary nucleation of ice crystals, *J. Food Sci. 50*: 401 (1985).
39. A. H. Muhr and J. M. V. Blanshard, The effect of polysaccharide stabilizers on ice crystal formation, in *Gums and Stabilisers for the Food Industry*, Vol. 2, *Applications of Hydrocolloids* (G. O. Phillips, D. J. Wedlock, and P. A. Williams, Eds.), Pergamon, New York, 1983, p. 321.
40. A. H. Muhr, J. M. V. Blanshard, and S. J. Sheard, Effects of polysaccharide stabilizers on the nucleation of ice, *J. Food Technol. 21*: 587 (1986).
41. A. H. Muhr and J. M. V. Blanshard, Effect of polysaccharide stabilizers on the rate of growth of ice, *J. Food Technol. 21*: 683 (1986).
42. E. R. Budiaman and O. Fennema, Linear rate of water crystallization as influenced by temperature of hydrocolloid suspensions, *J. Dairy Sci. 70*: 534 (1987).
43. E. R. Budiaman and O. Fennema, Linear rate of water crystallization as influenced by viscosity of hydrocolloid suspensions, *J. Dairy Sci. 70*: 547 (1987).
44. N. Buyong and O. Fennema, Amount and size of ice crystals in frozen samples as influenced by hydrocolloids, *J. Dairy Sci. 71*: 2630 (1988).
45. M. E. Sahagian and H. D. Goff, Thermal, mechanical and molecular relaxation properties of stabilized sucrose solutions at sub-zero temperatures, *Food Res. Int. 28*: 1 (1995).
46. H. Levine and L. Slade, Principles of cryo-stabilization technology from structure/property relationships of carbohydrate/water systems. A review, *Cryo-Letters 9*: 21 (1988).
47. H. Levine and L. Slade, Cryostabilization technology: thermoanalytical evaluation of food ingredients and systems, in *Thermal Analysis of Foods*, (C. Y. Ma and V. R. Harwalker, Eds.), Elsevier, London, 1988, p. 221.
48. W. H. Knightly, The role of the liquid emulsifier in relation to recent research on ice cream emulsification, *Ice Cream Trade J. 55(6)*: 24 (1959).
49. H. D. Goff, Emulsifiers in ice cream: how do they work?, *Mod. Dairy 67*(3): 15 (1988).
50. A. Tomas, J. L. Courthadon, D. Paquet, and D. Lorient, Effect of surfactant on some physicochemical properties of dairy oil-in-water emulsions, *Food Hydrocoll. 8:* 543 (1994).
51. Food and Drug Administration, *Milk Pasteurization Controls and Tests*, U.S. Dept. of Health and Human Services, Cincinnati, OH, 1981, p. 25.
52. H. Mulder and P. Walstra, *The Milk Fat Globule*, Commonwealth Agricultural Bureaux, Farnham Royal, Bucks., England, 1974.
53. W. D. Pandolfe, Development of the new Gaulin micro-gap homogenizing valve, *J. Dairy Sci. 65*: 2035 (1982).
54. H. Oortwijn, P. Walstra, and H. Mulder, The membranes of recombined fat globules. 1. Electron microscopy, *Neth. Milk Dairy J. 31*: 134 (1977).
55. H. Oortwijn and P. Walstra, The membranes of recombined fat globules. 2. Composition, *Neth. Milk Dairy J. 33*: 134 (1979).
56. E. K. Iversen and K. S. Pedersen, Ageing of Ice Cream, Grinsted Technical Paper TP 33-1e, 1982, p. 1.
57. N. M. Barfod and N. Krog, Destabilization and fat crystallization of whippable emulsions (toppings) studied by pulsed NMR, *J. Am. Oil Chem. Soc. 64*: 112 (1987).
58. N. M. Barfod, N. Krog, G. Larsen, and W. Buchheim, Effect of emulsifiers on protein–fat interaction in ice cream mix during ageing I. Quantitative Analyses, *Fat Sci. Technol. 93*: 24 (1991).
59. B. W. Tharp, T. V. Gottemoller, and A. Kilara, The Role of Processing in Achieving Desirable Properties in Health-Responsive Frozen Desserts, Germantown Manufacturing Co. Tech. Paper 100, Broomall, PA, 1992.

60. F. Franks, Complex aqueous systems at subzero temperatures, in *Properties of Water in Foods* (D. Simatos and J. L. Multon, Eds.), Martinus Nijhoff, Dordrecht, Netherlands, 1985, p. 497.
61. F. Franks, *Biophysics and Biochemistry at Low Temperatures*, Cambridge Univ. Press, New York, 1985, pp. 21, 37, 167.
62. J. M. V. Blanshard and F. Franks, Ice crystallization and its control in frozen food systems, in *Food Structure and Behavior* (J. M. V. Blanshard and P. Lillford, Eds.), Academic, New York, 1987, p. 51.
63. R. Bradley and K. Smith, Finding the freezing point of frozen desserts, *Dairy Rec. 84*(6): 114 (1983).
64. R. Bradley, Plotting freezing curves for frozen desserts, *Dairy Rec. 85*(7): 86 (1984).
65. K. G. Berger, B. K. Bullimore, G. W. White, and W. B. Wright, The structure of ice cream, *Dairy Ind. 37*: 419, 493 (1972).
66. L. M. Klotzek and J. G. Leeder, Relationship of emulsifier and homogenization pressures to fat destabilization of soft serve ice cream, *J. Dairy Sci. 46*: 592 (1963).
67. L. M. Klotzek and J. G. Leeder, Effect of high second stage homogenization and added salts on churning of soft ice cream, *Ice Cream Rev. 50*(12): 26 (1966).
68. K. G. Berger and G. W. White, The fat globule membrane in ice cream, *Dairy Ind. Int. 41*: 199 (1976).
69. D. F. Darling and D. W. Butcher, Milkfat globule membrane in homogenized cream, *J. Dairy Res. 45*: 197 (1978).
70. P. Walstra and H. Oortwijn, The membranes of recombined fat globules. 3. Mode of formation, *Neth. Milk Dairy J. 36*: 103 (1982).
71. P. R. Musselwhite and D. A. Walker, The effect of the colloidal state of the emulsion on ice cream structure, *J. Text. Stud. 2*: 110 (1971).
72. R. Govin and J. G. Leeder, Action of emulsifiers in ice cream utilizing the HLB concept, *J. Food Sci. 36*: 718 (1971).
73. D. F. Darling, Recent advances in the destabilization of dairy emulsions, *J. Dairy Res. 49*: 695 (1982).
74. J. Chen, E. Dickinson, and G. Iveson, Interfacial interactions, competitive adsorption and emulsion stability, *Food Struct. 12*: 135 (1993).
75. H. D. Goff, M. Liboff, W. K. Jordan, and J. E. Kinsella, The effects of Polysorbate 80 on the fat emulsion in ice cream mix: evidence from transmission electron microscopy studies, *Food Microstruct. 6*: 193 (1987).
76. H. Oortwijn and P. Walstra, The membranes of recombined fat globules. 4. Effects on properties of the recombined milks, *Neth. Milk Dairy J. 36*: 279 (1982).
77. E. Dickinson, R. K. Owusu, and A. Williams, Orthokinetic destabilization of a protein-stabilized emulsion by a water-soluble surfactant, *J. Chem Soc. Faraday Trans. 89*: 865 (1993).
78. K. G. Berger and G. W. White, An electron microscopical investigation of fat destabilization in ice cream, *J. Food Technol. 6*: 285 (1971).
79. M. A. J. S. van Boekel and P. Walstra, Stability of oil-in-water emulsions with crystals in the disperse phase, *Colloids Surf. 3*: 109 (1981).
80. B. E. Brooker, The adsorption of crystalline fat to the air-water interface of whipped cream, *Food Struct. 9*: 223 (1990).
81. S. Lee and J. M. deMan, Effect of surfactants on the polymorphic behavior of hydrogenated canola oil, *Fette, Seifen, Anstrichm. 86*: 460 (1984).
82. P. G. Keeney and D. V. Josephson, A measure of fat stability in ice cream, *Ice Cream Trade J. 54*(5): 32 (1958).
83. M. Kalab, Microstructure of dairy foods. 2. Milk products based on fat, *J. Dairy Sci. 68*: 3234 (1985).
84. P. M. Lin and J. G. Leeder, Mechanism of emulsifier action in an ice cream system, *J. Food Sci. 39*: 108 (1974).
85. W. Buchheim, N. M. Barfod, and N. Krog, Relation between microstructure, destabilization phenomena and rheological properties of whippable emulsions, *Food Microstruct. 4*: 221 (1985).

86. J. J. Kloser and P. G. Keeney, A study of some variables that affect fat stability and dryness in ice cream, *Ice Cream Trade J. 55*(5): 26 (1959).
87. J. L. Gelin, L. Poyen, J. L. Courthadon, M. Le Meste, and D. Lorient, Structural changes in oil-in-water emulsions during the manufacture of ice cream, *Food Hydrocolloids 8*: 299 (1994).
88. F. W. Bodyfelt, J. Tobias, and G. M. Trout, *The Sensory Evaluation of Dairy Products*, Van Nostrand Reinhold, New York, 1988, p. 166.
89. W. N. Wassef, Lipids, in *Food Chemistry* (O. R. Fennema, Ed.), Marcel Dekker, New York, 1985, pp. 176–205.
90. H. D. Goff and A. R. Hill, Dairy chemistry and physics, in *Dairy Science and Technology Handbook*, Vol. 1, *Principles and Properties* (Y. H. Hui, Ed.), VCH, New York, 1993, p. 1.
91. J. C. Bruhn, A. A. Franke, and G. S. Goble, Factors relating to development of spontaneous oxidized flavor in raw milk, *J. Dairy Sci. 59*: 828 (1976).
92. W. F. Shipe, R. Basette, D. D. Deane, W. L. Dunkley, E. G. Hammond, W. J. Harper, D. H. Kleyn, M. E. Morgan, J. H. Nelson, and R. A. Scanlan, Off flavors of milk: nomenclature, standards and bibliography, *J. Dairy Sci. 61*: 855 (1978).
93. R. L. King, Oxidation of milk fat globule membrane material, *J. Dairy Sci. 45*: 1165 (1962).
94. W. K. Downey, Flavour Impairment of Milk and Milk Products Due to Lipolysis, Int. Dairy Fed. Document 118, Brussels, 1980.
95. H. C. Deeth and C. H. Fitzgerald, Lipolytic enzymes and hydrolytic rancidity in milk and milk products, in *Developments in Dairy Chemistry*, Vol. 2, *Lipids* (P. F. Fox, Ed.) Applied Science, London, 1983, pp. 195–239.
96. S. C. Murphy, K. Cranker, G. F. Senyk, D. M. Barbano, A. I. Saeman, and D. M. Galton, Influence of bovine mastitis on lipolysis and proteolysis in milk, *J. Dairy Sci. 72*: 620 (1989).
97. M. Anderson, Milk lipase and off-flavour development, *J. Soc. Dairy Technol. 36*(1): 3 (1983).

13

Cream Alternatives

Iain J. Campbell and Malcolm G. Jones
Unilever Research Laboratory, Sharnbrook, Bedfordshire, England

I. INTRODUCTION

Cream alternatives first came onto the market in Europe as short shelf life products for the bakery and catering sectors. They were formulated such that they could be used in both pouring and whipping applications, with the emphasis being on physical rather than organoleptic quality. The whippable products generally had high overruns and had very good shape retention, although the flavor and oral melt properties were poor. More recently products have been developed for the retail market to compete directly with dairy cream, and in the United Kingdom this growth has been particularly significant as exemplified by the Van den Berghs Foods product Elmlea, which has achieved almost 20% of the total cream market by volume.

The dairy cream market is segmented in a number of ways. The main differentiations made are between fresh (pasteurized), long-life (sterilized at ultrahigh temperature, UHT), and aerosols. It can also be divided in terms of fat level, and in the United Kingdom, single cream for pouring contains not less than 18% fat, whipping cream not less than 35% fat, and double cream not less than 48% fat. In many other European countries whipping cream contains not less than 30% fat, and double and single creams are less common. Cream alternatives, or non-dairy creams as they are sometimes called, have been developed to compete in all these categories, and many are close in flavor and textural quality to the dairy equivalents. Nearly all are UHT products, hence giving the combined benefits of product performance and extended shelf life.

Outside of Europe the market for cream alternatives to compete directly with dairy cream is less well developed, although there is some growth of whipping creams in Japan, while in the United States the interest is mainly for coffee creamers.

Because cream alternatives are fabricated products it is possible to tailor the physical and organoleptic properties of the emulsion by manipulation of the fat phase, the aqueous phase, and the emulsifier system. This chapter describes the role of the various components, how they interact with each other and with the processing conditions, and how this knowledge can be used to develop new product options.

II. COMPOSITION AND PROPERTIES

The development of cream alternatives with the desired storage and subsequent in-use properties presents many challenges to the product developer. There is a requirement for an extensive knowledge of the role of the individual ingredients that make up the complete formulation, and since all cream alternatives are oil-in-water (o/w) emulsions, there is also the requirement for a complete understanding of the factors that affect the stability of such emulsions. A complicating factor in formulating creams is that emulsion stability is always required during the storage shelf life of the products. However, for the cream to function during use, for example whipping, the emulsion must destabilize in a controlled manner. The emulsion scientist is thus faced with the problem of controlling the instability of the cream. This section discusses some of the important issues that need to be considered in formulating a cream alternative.

A. Liquid Creams

Cream alternatives consist of a dispersion of small (normally < 1 μm) oil droplets in an aqueous phase containing milk proteins. A typical formulation (Table 1) contains a blend of vegetable oils (typically palmkernel and coconut oils), a milk protein source, one or more lipid emulsifiers, and one or more polysaccharide stabilizers. The milk proteins impart both stability against recoalescence of the oil droplets and a desirable creamy flavor to the product. Common sources of milk proteins are skim milk powder and buttermilk powder, although in principle any milk source can be used. Homogenization of a preemulsion with coarse (> 20 μm) oil droplets results in small droplets coated in milk protein. The thickness of the protein coat controls the stability of the droplets to recoalescence by the formation of a steric barrier preventing film drainage and film rupture between adjacent oil droplets. When two oil droplets approach each other to close distances, the protein molecules become entangled, resulting in an osmotic pressure that forces the droplets apart. Thus, the protein coat ensures emulsion stability against coalescence during storage.

Oil droplets that have densities less than that of the aqueous phase are prone to creaming during storage. Stokes' law predicts the creaming rate of particles in a dilute system, and is given by

$$\text{Creaming rate } U = (\rho_c - \rho_d)R^2/\eta_c$$

where ρ_c and ρ_d are the continuous phase and disperse phase densities, respectively, R is the oil droplet radius, and η_c is the continuous phase viscosity. Although this equation is inadequate in predicting the creaming rate in concentrated emulsions it nevertheless shows the

Table 1 Typical Cream Alternative Formulations

Ingredient	Whipping cream	Pouring cream
Buttermilk powder (%)	7.00	7.00
Guar gum (%)	0.10	0.15
Hardened palmkernel oil (PK38) (%)	17.50	9.00
Coconut oil (CN) (%)	17.5	9.00
Tween 60 (%)	0.35	0.30
β-Carotene (%)	0.25	0.25
Water (%)	57.3	74.3

parameters that affect creaming. It should be appreciated that most cream alternatives at typical storage temperatures (approximately 5°C) consist of partially crystallized droplets, which means that the density difference between the oil and aqueous phases is less than might be expected. It is known that very small (0.5 μm) oil droplets with a high protein surface coverage can be more dense than the aqueous phase and will therefore sediment. In reality most creams are prone to creaming, especially if the storage temperature is high, and for this reason a polysaccharide "stabilizer" is usually added to the cream to raise the aqueous phase viscosity and so reduce the creaming rate as predicted by Stokes' law. A secondary function of the polysaccharide is to impart a more viscous or creamy mouthfeel to the product. Typical polysaccharides used in cream alternatives are locust bean gum, guar gum, alginates, and carrageenan. Care has to be exercised in the level of usage of many polysaccharides in protein-containing emulsions as protein and polysaccharide compete for available water, resulting in a type of depletion flocculation [1]. While depletion flocculation leads only to weakly attracted oil droplets, flocculated droplets in dilute emulsions (<20% fat) will cream much faster than single droplets. At higher phase volumes of fat, a percolated network of flocculated oil droplets can form a weak gel that prevents creaming. It is often noticed in cream alternatives that the emulsion becomes less viscous on stirring, particularly after a period of storage. This is due to the action of stirring being sufficient to break the weak gel formed by oil droplet flocculation.

A much stronger form of flocculation that can occur in cream alternatives is via a reduction in the electrostatic charge on the interfacially adsorbed protein. Above the isoelectric point (pH 4.7 for casein), protein carries a net negative charge. This negative charge imparts stability against flocculation by creating an electrostatic repulsion between adjacent oil droplets. If the charge is reduced by lowering the pH, or screened by the addition of calcium ions, the electrostatic repulsion is reduced or eliminated (at the isoelectric point) and rapid strong flocculation occurs. This effect is always witnessed when either fresh dairy cream or a cream alternative is added to acidic fruit juice. Flocculation due to a reduction in the electrostatic charge is the basis for the formation of a yogurtlike texture in soured cream. In soured cream the pH is reduced by the formation of lactic acid, which is normally produced microbiologically. Flocculation in cream alternatives can be a major problem in creams designed to be added to coffee. Any flocculation already present in the cream combined with the heat from the coffee can cause a "feathering" appearance. Sodium citrate is often added to creams, especially to so-called coffee creamers, to help sequester calcium ions and reduce flocculation during storage.

The factors affecting the stability of cream alternatives can conveniently be shown by a simplistic energy diagram (Fig. 1) similar to those given by DLVO theory (after Deryaguin, Landau, Verwey, and Overbeek [2,3]).

A special type of oil droplet aggregation that can be used to produce viscous or slightly gelled creams termed "clustering," is employed to produce cream sold as extra thick or spoonable. Clustering is produced by high pressure homogenization, particularly in systems containing milk protein. Early work by Darling and coworkers [4,5] showed that clustering is produced when the interfacial area of the emulsion is greater than the surface-covering capacity of the protein, leading to protein molecules sharing two adjacent droplets. In this case the clustering will be eliminated by increasing the concentration of protein in the emulsion. We have found that this type of clustering is prevalent in lower fat emulsions, while the opposite effect is found in higher fat (>30%) emulsions, where increasing the protein level can create more oil droplet aggregation. Figure 2 shows a plot of emulsion viscosity versus buttermilk powder concentration in creams containing 45% fat (50% coconut oil, 50% hardened palmkernel oil) and either 1% or 2% sodium caseinate. In both cases it can be seen that increasing

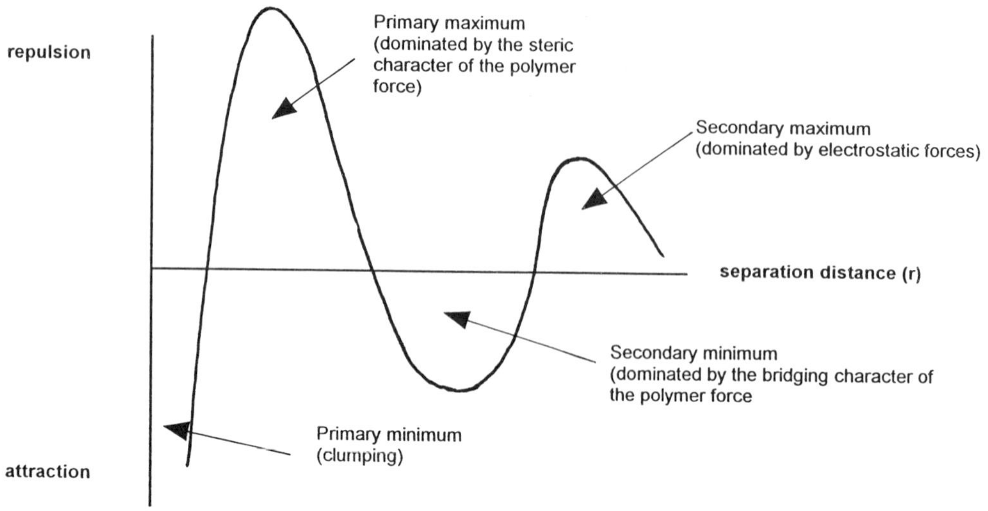

Fig. 1 Conceptual energy diagram showing aggregation states possible in cream alternatives.

levels of buttermilk powder lead to more oil droplet aggregation and hence higher viscosities. The higher level of sodium caseinate led to less aggregation. Electron microscopy of the emulsions showed that this form of aggregation was due to casein micelles adsorbing to two adjacent oil droplets as also shown by Darling and Butcher [5] and Ogden [6]. Increasing numbers of casein micelles led to more bonds between aggregated droplets, resulting in higher viscosities. It is clear that the mechanism for this type of aggregation is different from that described by Darling and Ogden and also by Mulder and Walstra [7], although the microstructure of the emulsion may be similar. We propose a mechanism in which oil droplets are disrupted by the shear in the homogenizer valve, and during the recoalescence process (the oil droplet size produced by homogenization is a result of rapid droplet breakup and recoalescence) casein micelles prevent film drainage by adsorbing almost simultaneously to two adjacent oil droplets. One could describe this type of aggregation, which we refer to as high fat/high protein clustering, as a form of resticted coalescence in which the film thickness is determined by the thickness of the adsorbed casein micelles. We prefer to describe the aggregation due to insufficient protein as bridging flocculation.

Clustering such as we have described is a very strong form of aggregation and is not broken down by stirring. High power ultrasonics are, however, effective at declustering and can reduce an almost solid cream to a low viscosity liquid. It is interesting to note that clusters will not re-form after disruption because there is no mechanism for casein micelles to readsorb to two oil droplets simultaneously. We have shown in other work that a wide range of smaller molecule emulsifiers such as monoglycerides, polysorbates, lecithin, and sucrose esters, when used at a low level (0.3%), are very effective in preventing clustering. It is believed that the mechanism is similar to that by which sodium caseinate reduced clustering in the examples shown in Fig. 2. During oil droplet breakup in the homogenizer there is a large increase in interfacial area, and the rate of recoalecence of the new oil droplets will depend on the rate of adsorption of surface-active materials from the aqueous phase. If adsorption is slow, much oil droplet coalescence will occur, whereas if casein micelles are present between the thinning films, clustering will occur. Small surface-active molecules will adsorb more rapidly than large protein molecules and will therefore prevent recoalescence and clustering by covering

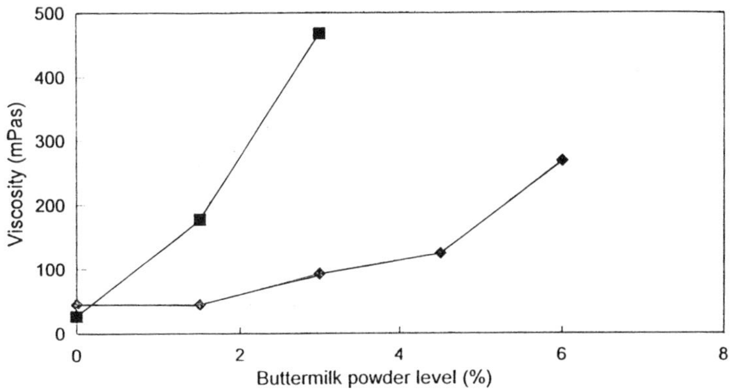

Fig. 2 Effect of buttermilk powder level on the viscosity of creams containing 45% fat and either 1% or 2% sodium caseinate.

the interface. This use of emulsifiers has an important yet often poorly understood role in the formulation of most cream alternatives.

B. Whipping Creams

A large proportion of creams are sold for use as whipping creams. It is the development of these creams that presents the emulsion scientist with the greatest challenge. If one considers the energy diagram shown in Fig. 1, it is desirable to avoid all forms of instability during storage, yet on the application of shear and the incorporation of air the oil droplets must overcome the steric barrier to form an oil droplet network. The oil droplet network shown surrounding air cells in Fig. 3 is responsible for imparting a desirable firm and creamy texture to the whipped cream. Anyone who has attempted to whip cream will bear witness to the

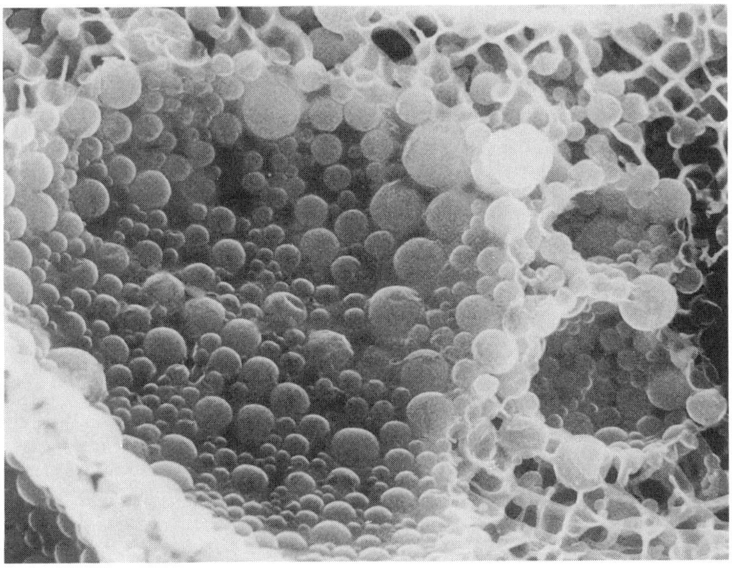

Fig. 3 Electron micrograph of freeze-fractured sample of whipped cream alternative.

difficulty of obtaining exactly the required texture, not too soft yet not overwhipped. Many cooks will have experienced the indignation of producing butter rather than a perfect whipped cream. It is for this reason (and, of course, convenience) that the cream alternative market has seen a huge increase in the use of aerosol creams. The following section explores the factors affecting the stability and "whippability" of creams.

The difficulty in controlling the whipping process is well illustrated by a whipping curve generated using a Mohr-type apparatus [8]. The cream is stirred at a fixed temperature by a constant-speed motor. Use of a torque meter on the motor or a simpler measurement of the motor load yields a whipping curve of cream viscosity versus time. A typical curve is shown in Fig. 4. As air is whipped into the cream a gradual rise in cream viscosity is witnessed until a maximum is obtained in the curve. The optimum whipped structure is normally considered to be close to this viscosity maximum. On further whipping, a decrease in viscosity occurs, eventually leading to a further viscosity increase and the phenomenon of buttering. This type of whipping curve is often used to examine the influence of processing or formulation variables on the time to achieve an optimum whipped cream structure. Some of the important factors in this process are discussed below.

Before one can fully appreciate the factors that influence both the whip time and the texture of a whipped cream, it is important to understand the mechanisms involved in the formation of a whipped cream structure. If a solution is stirred vigorously with a whipper beater, air is rapidly incorporated and is disrupted into smaller air cells by the action of the beater. The creation of small stable air cells is aided by a lowering of the air/water interfacial tension by surface-active materials in the aqueous phase. The surface-active ingredients then play a vital role in preventing film drainage and consequent film rupture. The ability of egg white to make a stable foam structure is shown in Fig. 5. It can be seen that as the time of whipping increases, both the phase volume and the firmness of the foam increase. A similar plot for a whipping cream containing 45% fat (Fig. 6) shows many significant differences compared to the whipping behavior of egg white. In the case of a cream, a maximim overrun is achieved and further whipping only serves to exclude air from the whipped structure, whereas for egg white no decrease in overrun was observed. A second difference can be seen when overrun is plotted against firmness (Fig. 7), showing a direct correlation between these two parameters

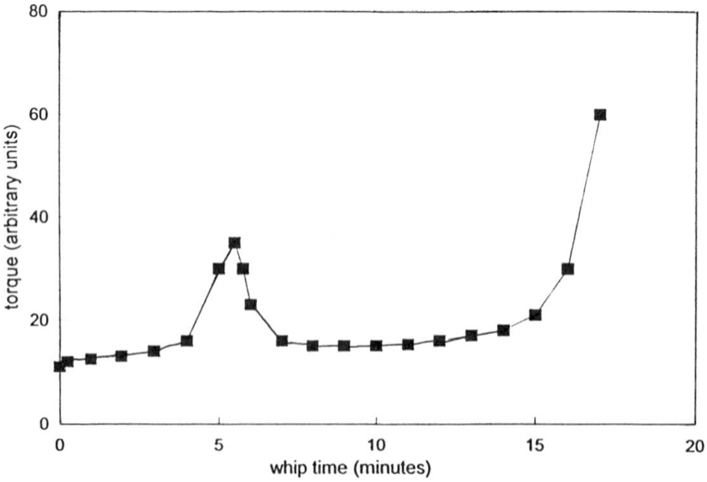

Fig. 4 A typical Mohr-type curve of torque versus whip time.

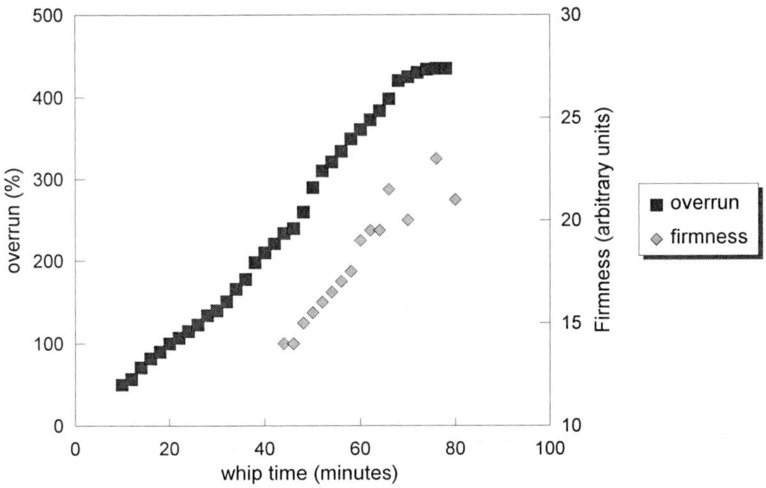

Fig. 5 (■) Overrun and (◆) firmness versus whip time for an egg white solution.

for egg white but not for cream. The firmness of cream continues to increase even after the maximum overrun is achieved. The conclusion from these results is that the microstructures of an egg white foam and a whipped cream are very different.

A whipped cream relies for its rigidity and also its stability against air cell coalescence on the presence of a fat droplet network, as previously shown in Fig. 3. The network is formed by shear-induced aggregation of oil droplets resulting from rupture of the interfacial film between adjacent droplets during the whipping process. It is essential that the oil droplets contain fat crystals at this stage to give them mechanical rigidity such that full droplet coalescence is prevented. For this reason it is always necessary to cool the cream in a refrigerator before whipping to ensure that sufficient solid fat is present. Because the whipped cream structure depends on the formation of a fat droplet network, and because the strength of the

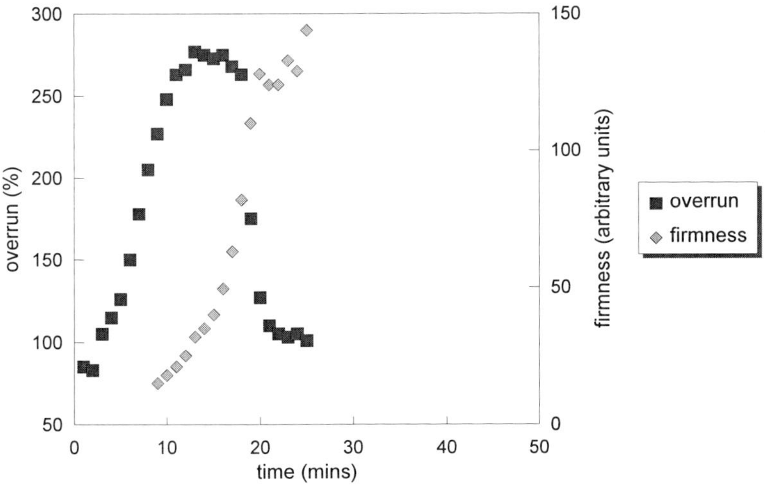

Fig. 6 (■) Overrun and (◆) firmness versus whip time for a 45% fat cream alternative.

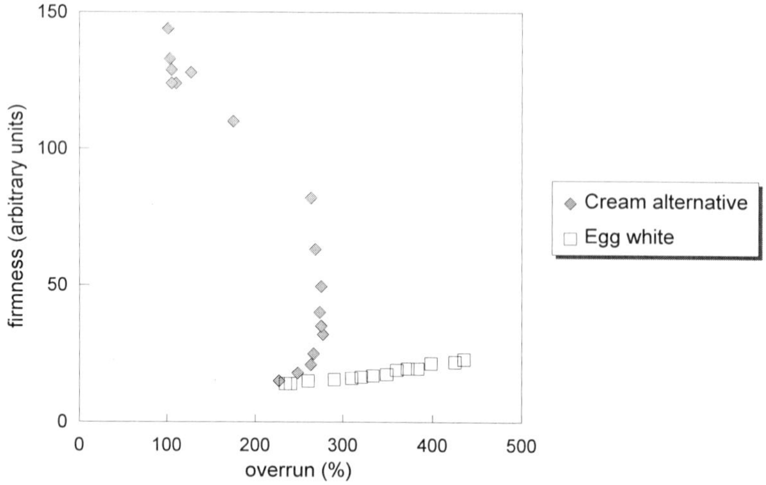

Fig. 7 Overrun versus firmness plots for (□) egg white solution and (◆) a cream alternative.

network will ultimately depend on the strength of the oil neck between two partially coalesced droplets, one would expect the nature of the fat blend to affect the whipped firmness. This is confirmed by the plots of overrun versus firmness, shown in Fig. 8, for a series of 45% fat creams with fat blends with different ratios of hardened palmkernel oil (PK38) and rapeseed oil (RP). The figure shows that the maximum obtainable overrun and the firmness are directly related to the amount of PK38 in the blend.

The importance of the fat blend composition in determining whipped cream properties makes it imperative that full crystallization of the blend occur within the oil droplets. It is important to appreciate that the crystallization behavior of fat in bulk and in emulsion droplets can be very different. Figures 9 and 10 show isothermal crystallization curves at a temperature of 5°C for PK38 and coconut oil (CN), respectively, in three different systems: bulk oil, an emulsion with droplets similar in size to dairy cream ($D_{3,2} = \sim 3 \mu m$), and an emulsion with very small droplets ($D_{3,2} = \sim 1 \mu m$) produced by high pressure homogenization. These data

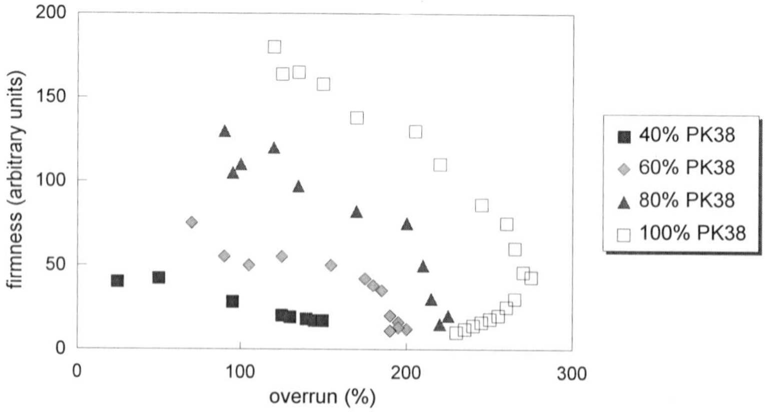

Fig. 8 Overrun versus firmness for cream alternatives containing different levels of PK38. (■) 40% PK38; (◆) 60%; (▲) 80%; (□) 100%.

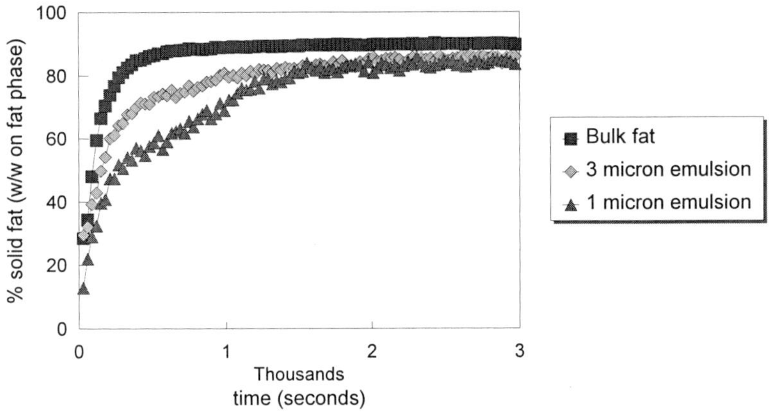

Fig. 9 Isothermal crystallization curves for PK38 at 5°C. (■) Bulk fat; (▲) 1 μm emulsion; (◆) 3μm emulsion.

show that the crystallization rate of PK38 decreases with decreasing oil droplet size as shown earlier by Walstra and Van Beresteyn [9], although an equilibrium level of solids is achieved in each case. With coconut oil the case is quite different in that equilibrium solids are not achieved in either of the emulsions. In fact, no crystallization occurred in the smaller oil droplets even after several weeks of storage. The explanation for the different crystallization behavior between bulk fat and emulsions lies in the manner in which crystallization is nucleated [9]. In bulk fat and large oil droplets, nucleation is heterogeneous, which means that a single nucleus is sufficient to promote total crystallization. In very small oil droplets the probability of heterogeneous nucleation occurring is very low owing to the lack of nuclei within the droplets. Nucleation therefore has to be homogeneous, which requires a significantly lower temperature. For this reason coconut oil droplets with a diameter less than 1 μm can remain supercooled for very long times. The equilibrium level of fat solids achieved in the emulsions shown in Figs. 9 and 10 is controlled by the number of droplets that were homogeneously nucleated, although it should be noted that once a droplet is nucleated, full

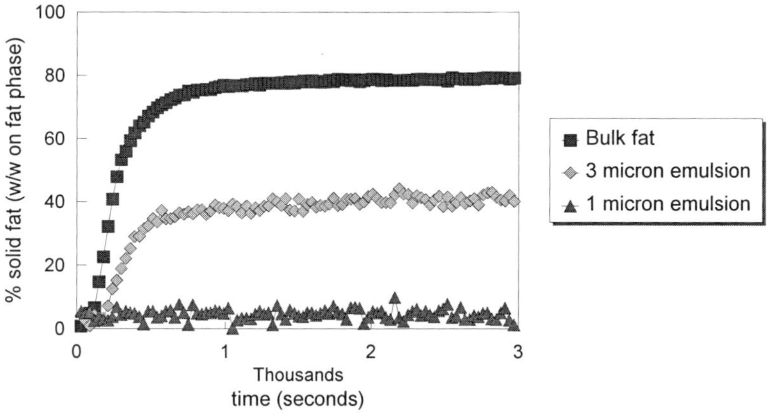

Fig. 10 Isothermal crystallization curves for CN at 5°C. (■) Bulk fat; (▲) 1 μm emulsion; (◆) 3 μm emulsion.

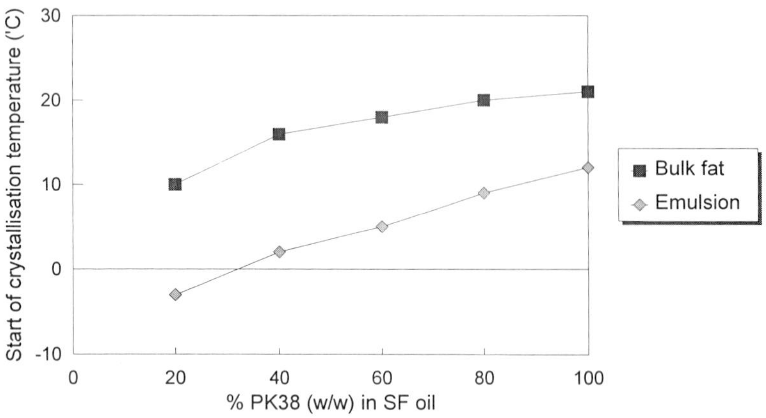

Fig. 11 Crystallization temperature for fat blends with different levels of PK38 in SF for bulk fat and (◆) an emulsion.

equilibrium solids for a given temperature will be achieved within that droplet. Figure 11 shows a plot of crystallization temperature (measured by differential scanning calorimetry, scan rate −5°C/min) as a function of PK38 level in sunflower oil (SF) in the fat blend for bulk fat and a homogenized emulsion. The emulsion had an approximate mean oil droplet size ($D_{3,2}$) of 1 μm. It can be seen that the emulsion requires a temperature approximately 10–12°C lower to initiate crystallization than that required for the bulk fat. This factor can present a major problem to the product developer, particularly with healthy fat blends containing less saturated fat, since oil droplet crystallization may not occur at normal processing or storage temperatures. One method to overcome this problem is to "seed" the oil droplets with a minor (<2%) fraction of a high melting triglyceride (e.g., fully hardened rapeseed oil). This fraction will be homogeneously nucleated at normal processing temperatures and will then heterogeneously nucleate the remainder of the fat blend to ensure that full crystallization takes place.

III. PROCESSING

Although cream alternatives have been manufactured in the past using HTST (high temperature ~85°C, short time ~20 s) methodology, it is only sterilized (UHT) products that are of significant commercial interest today.

There are five main steps in the UHT processing of creams, each of which can have a significant effect on the properties of the final product. The steps are premixing, sterilization, homogenization, cooling, and packing, and a typical flowchart is shown in Fig. 12.

Premixing is a critical part of cream production and needs to be carried out with care. The aqueous phase must be sufficiently agitated to avoid clumping of gums and sufficiently heated to ensure complete hydration without going to temperatures that may lead to alteration of protein functionality. After the initial mixing it is normal to allow sufficient time (at least 1 h) for all components to fully hydrate. The oil phase is prepared in liquid form before mixing with the aqueous phase. Most lipid emulsifiers, e.g., monoglycerides, Tweens, lecithins, and polyglycerol esters, should be melted into the oil before addition to the premix. The oil phase is added to the aqueous phase, normally at 55–65°C, and agitated sufficiently to maintain a homogeneous mixture until homogenization. It is essential that the elapsed time between addi-

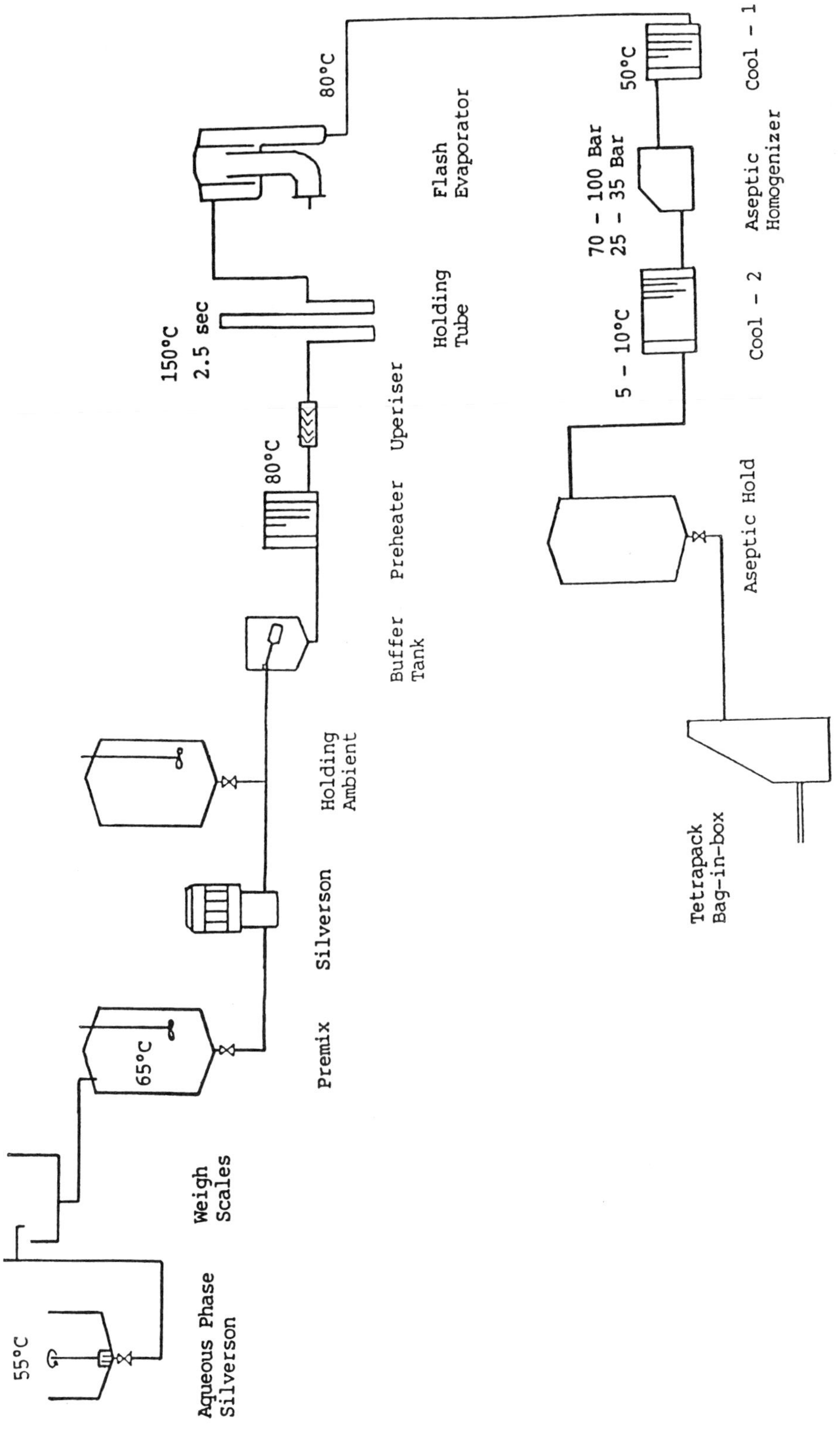

Fig. 12 Typical flow diagram for a cream alternative UHT process.

tion of powders and completion of processing of the premix not exceed 3 h, as growth of thermophilic bacteria could occur, which would result in spoilage of the premix.

Sterilization of the premix normally involves a preheating stage to approximately 80°C to bring the mix temperature to a point where the sterilizer can cope with the ΔT required to ensure complete sterilization. The temperature–time regime required for sterilization is defined in terms of the conditions required to ensure that *Clostridium botulinum* spores cannot be present in the final product. The process equipment must therefore be heated with steam to a sufficiently high temperature for a sufficiently long time to achieve a "*botulinum* cook" at the coldest part of the line. This normally requires time–temperature combinations such as 130°C for 1 h depending on the plant design. The hydrated raw materials are given a full "*botulinum* cook" in-line, and this is normally by direct steam injection to a temperature of 140–150°C for 2–5 s. The in-line sterilization can also be carried out by indirect heating via tubular heaters or plate heat exchangers. This results in very different amounts of heat being received by the products. In the case of the direct steam process, the product temperature rises from 80 to 150°C and back to 80°C equally quickly. In the indirect process, more time is required to raise the product to sterilization temperature and more time is required to cool it down again as shown in Fig. 13. It is generally believed that direct steam injection results in better flavored products than indirect heating, and certainly there can be differences in other product properties linked to emulsion stability.

Homogenization is normally carried out with an aseptic high pressure valve homogenizer. The temperature of homogenization is normally in the range 50–90°C, where the fat phase is liquid, and the pressure is selected to give the desired droplet size. In some instances a two-stage homogenization is given in which the second stage is used to reduce the size of any large droplets that escaped the first stage, hence giving a narrower size distribution. A second homogenizer valve can sometimes be used downstream of the main homogenizer to apply a small pressure (5–10 bar) for declustering of fat droplets that have become clustered during the initial high pressure treatment.

Cooling of the homogenized cream emulsion is an area that is not well defined and one that is often not well controlled in factory production, particularly if aseptic holding tanks are used prior to packing. The primary function of cooling is to ensure that crystallization of the fat phase takes place, because fat solids are required if the cream is to be whippable. Also, if

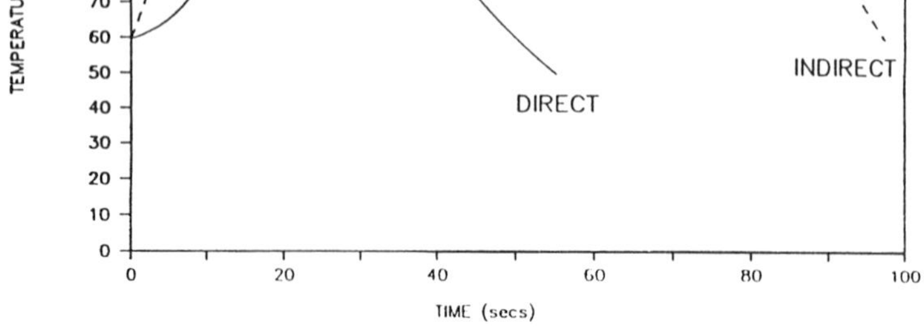

Fig. 13 Temperature–time profiles for direct and indirect UHT process regimes.

a cream is collected at too warm a temperature, the resultant viscosity can lead to creaming upwards of the fat droplets. The final cooling temperature must be low enough to ensure that fat crystal nucleation occurs in all droplets, and the subsequent storage temperature must be low enough to ensure that crystal growth will proceed.

In most industrial processes aseptic holding tanks are used as buffers between the processing lines and packing machines. They are frequently only partially jacketed so that cooling can be applied but not always uniformly, and they are rarely stirred. During storage of cream in the tank after cooling, crystallization proceeds with concomitant production of heat. Lack of stirring means that the emulsion next to the cooled tank wall will experience a different environment from the emulsion in the center of the tank. It is also known that temperature-induced aggregation of oil droplets can occur in the aseptic tank, hence cream coming from the tank may often have a shorter whip time than cream that has gone directly to the packing machine. This is a result of the tempering that it has received in the tank. A temperature rise in-pack of as much as 10°C can be observed when a holding tank is used, whereas the in-pack temperature is normally within one degree of the final cooled temperature when product is packed directly. A variety of packaging regimes are used for UHT products, but for cream alternatives most products are filled into 1 L Tetrapaks, 200 mL (or 250 mL) *Tetrabriks*, or 142 and 284 mL Cuplets. The packaging materials are normally sterile when made and are given a relatively mild heat treatment combined with chemicals (usually hydrogen peroxide) to provide an adequate level of safety assurance. Shelf life of products is limited by the physical or taste stability of the products, not by microbiological considerations. Many products sold in the retail market are placed in the chill cabinet for image reasons, but also because a constant storage temperature helps remove the uncertainty associated with large temperature fluctuations and the resultant effects on product stability.

IV. CURRENT MARKET

The growth of the cream alternatives market has taken place to a greater extent in the United Kingdom than in the other dairy cream consuming countries of the world. Some European countries have followed the U.K. trend (e.g., Sweden and Italy), but countries where consumption of dairy cream is large (e.g., Germany) have not generally followed suit. The reason for this may be that fresh dairy cream is considered a commodity rather than a luxury product, hence the attraction of a cheaper alternative is not so great. Alternatively, a positive product attribute perhaps related to health may be required in order to expand the market in the same way as happened with low fat spreads. This will certainly be the case for the United States, where dairy products have suffered severely from a negative health image for some time now. The overall dairy market in the United States is forecast to fall in value terms by about 30% by the year 2000, with the greatest losses being in those segments where the health issue is important.

About two-thirds of the U.K. dairy cream market is accounted for by fresh cream, although its share has remained fairly static since 1989. Sales of UHT cream have expanded at a slightly faster rate largely due to the success of aerosols, which have been the main area of growth within the cream market and now command around 40% of the UHT sector.

After an initial impact on the market, the cream alternative sector appears to have stabilized at around 11%. The market is dominated in the United Kingdom by Van den Bergh Foods, which markets a full fat Elmlea range, and a half-fat Delight range. Elmlea was launched in 1984 and has grown to almost 20% of the total cream market by volume, while Delight is a much smaller brand with around 2% of the market.

Aerosols are the main growth area within the cream market, with sales having grown by around 20% since 1991. Anchor claims to have an 80% share of the aerosol UHT creams market by value and is the fastest growing brand with its full fat and half-fat products, both of which consist of real cream. Branded cream alternative aerosols are also on the market, but they account for only a small proportion of the total. In the United States, aerosol creams are the most widely used form of whipped cream, but this is small in comparison to the usage of ice cream as a desert topping.

V. FUTURE TRENDS

Future products in the cream alternatives category are likely to be marketed on the basis of added value, in order to attract higher profit margins. The added value will most probably be in terms of health issues, such as low SAFA or high PUFA, or even fat-free. These are areas where it is difficult for the dairies to compete, particularly with products for whipping. Technically, low SAFA/high PUFA and low fat/fat-free are difficult to achieve for a whipping cream.

Cream alternatives tailored for culinary use form another area where there is scope for technical developement of products with added value. Control of oil droplet flocculation during cooking or in acidic products where "feathering" can give an unacceptable appearance is an opportunity for technical development.

Ready whipped cream for convenience has been achieved in the past with frozen dairy cream, but for the chill cabinet or even ambient storage the target is much more difficult. Stabilization of air cells against disproportionation and shrinkage may require something other than the generation of the aggregated fat particle network that is currently present in whipped cream structures.

REFERENCES

1. E. Dickinson, *An Introduction to Food Colloids*, Oxford Univ. Press, New York, 1992, pp. 101–107.
2. E. J. W. Verwey and J. Th. G. Overbeek, *Theory and Stability of Lyophobic Colloids*, Elsevier, Amsterdam, 1948
3. J. Th. G. Overbeek, The interaction between colloidal particles, in *Colloid Science* (H. R. Kruyt Ed.), Elsevier, New York, 1952, pp. 271–277.
4. D. F. Darling and R. J. Birkett, Food colloids in practice, in *Food Emulsions and Foams* (E. Dickinson, Ed.), Royal Society of Chemistry, London, 1987, p. 31.
5. D. F. Darling and D. W. Butcher, Milk fat globule membrane in homogenised cream, *J. Dairy Res. 45*: 197–208 (1978).
6. L. Ogden, Homogenisation induced clustering of fat globules in cream and model systems, Ph.D. Thesis, University of Minnesota, 1973.
7. H. Mulder and P. Walstra, The Milk Fat Globule, Centre for Agric. Pub. Doc., Wageningen, The Netherlands, 1974.
8. W. Mohr and E. Mohr, *Molk. Kaseneiz 6*: 34 (1955).
9. P. Walstra and E. C. H. van Beresteyn, *Neth. Milk Dairy J. 29*: 35–65 (1975).

14

Ghee, Vanaspati, and Special Fats in India

K. T. Achaya
CSIR Emeritus Scientist (Retired), Bangalore, India

I. USAGE OF FATS IN INDIA

A. Historical

The Indus Valley civilization of north India, spanning the millenium 2500–1500 BC, has yielded carbonized remains of three oilseeds—sesame, mustard, and linseed—and clay models of the coconut. Beautiful steatite seals depict fine breeds of dairy cattle that still flourish in that region and imply a knowledge of milk fat. Numerous animal and fish bones would point to familiarity with animal body fats. Copper frying pans of thoroughly modern design testify to the prevalence of frying, and tandoor clay ovens, to the prevalence of baking. Use of edible fats in India has thus a hoary lineage [1].

B. Current Patterns of Fat Use

National usage today is made up of ghee 3%, hydrogenated fat or vanaspati 18%, and vegetable oil 79% [2]. In recent years, the average per capita daily intake has been about 13 g, with a statewise range of 3–20 g depending on regional culinary practices, income, and rural or metropolitan residence. Ghee is consumed in only 7.6% of Indian households, and two-thirds of these are confined to two large northern states [3]. Household dietary surveys reveal a per capita daily intake of ghee that averages only 0.36 g [2], but estimates based on the total milk that is converted into ghee give the far higher figure of 2.5 g [3], implying considerable indirect consumption of ghee as products. Vanaspati availability calculated from production data and population figures yields a figure of about 2.0 g per capita per day [2]. By far the largest use is of edible vegetable oil [2], whose proportion in the total fat mix varies according to the location of states: southern 97%, eastern 86%, western 83%, and northern, where both vanaspati and ghee have larger shares, 56%. Most traditional oils are consumed in unrefined form; these include, for example, groundnut, rape-mustard, safflower, sesame, coconut, and niger. Barely 15–20% is in the form of refined oils; this includes all newer entrants such as cottonseed, soybean, sunflower, ricebran, and palm, while some traditional oils are also being refined [4].

Refining practices in India are those employed everywhere and are covered elsewhere in this volume. For larger scale manufacture of ghee and vanaspati, distinctive technologies have been developed on the Indian subcontinent. To some extent this is also true of bakery fats, margarines, and fat spreads, which are also briefly noted in this chapter.

II. GHEE

A. Traditional Preparation

The oldest Sanskrit text extant, the *Rigveda* of about 1500 BC, refers to a cake of cereal flour fried in *ghrta* (ghee), and indeed the Aryan culture would countenance no other cooking fat. To start with, its origin from the sacred cow gave it special sanctity; further, being fire-derived from butter, golden, shining ghee was deemed to be pure in itself and to confer ritual purity on dishes cooked with it [1]. To make ghee, boiled and cooled milk from the cow or the buffalo was set overnight using a pat of mixed starter culture from a previous run. The firm curd was diluted and churned in various domestic devices [5, pp. 137–138] to yield buttermilk and butter granules, which were patted together into lumps. Very little butter was used in the home; material pooled for a few days was boiled down over a fire while stirring with a ladle by hand to yield ghee. It was estimated in 1951 that about 42% of all the milk produced in India was converted into ghee [5, pp. 299–307]. At present the figure is about 28%, which would yield annually an estimated 850,000–900,000 tonnes of ghee [3].

B. Nature of Ghee

1. *Ghee Attributes*

The shelf life of ghee is 6–8 months even at high ambient tropical temperatures, with cow ghee having an edge over buffalo ghee. This high stability is attributed to the low moisture content (0.2%) and the high content of phospholipids (~40 mg/100 g) and perhaps of free amino acids, which are liberated from the phospholipid–protein complex into the fat phase [6]. Flavor precursors in ghee develop in part during the souring of milk to form curd (if omitted, the flavor is flat and oily), but in the main during the rendering step by interaction between a protein, which may be a casein degradation product, the reducing sugar lactose, and minerals. A range of compounds have been identified, including ketones (propanone and butanone-2), alcohols (ethanol, *n*-butanol), 2-ketohydrocarbons of 3–9 carbon chain length, monocarbonyls, ketoglycerides, oct- and dec-2-enols, and δ- and γ-lactones of various chain lengths [6].

Aroma is highly prized, with preferences in various parts of the country for undercooked or overcooked flavors. Also valued is a granular appearance, which generally takes the form of solid grains floating in a liquid mass. Buffalo ghee, which is more saturated, yields larger crystals in the form of irregular clusters, and cow ghee forms smaller spheroids [7]. After initial seeding, slow undisturbed cooling at a temperature 1°C above the melting point, about 29–31°C, yields larger crystals than if the cooling is too rapid.

2. *Ghee Composition*

Milk fat makes up 99.0–99.5% of ghee. The fatty acids range in carbon chain length from 4 to 22, and buffalo ghee carries the following types as weight percentages: saturated 46, *cis*-monoene 29, *trans*-monoene 7, diene 13, and polyene 5 [8]. Cow ghee would have about 5% less saturates and correspondingly more *cis*-monoenes. In consequence, the following glyceride types are present in cow and buffalo ghee, respectively: SSS 42, 49; SSU 42, 39; SUU 14, 11; and UUU 2, 1 [7]. Thin-layer chromatography yielded three distinct molecular weight groups in which various fatty acids were preferentially associated. Thus in the high molecular weight triglycerides, hardly any very short chain fatty acids were associated with the stearic and oleic acids. In the low molecular weight group, the SSS component consisted of two short- or medium-chain fatty acids and one long-chain fatty acid, and in the diene and polyene triglycerides one short- or medium-chain fatty acid was associated with two long-chain unsaturated fatty acids [8]. Ghee carries about 4% diglycerides and 0.6% monoglycerides, probably developed during the several steps of its isolation from milk.

The unsaponifiable matter of ghee contains, among other things, 300 mg of cholesterol, 9 mg of lanosterol, and 4, 60, 9, 28, and 6 μg of lutein, squalene, vitamin A, vitamin E, and ubiquinones, respectively, per 100 g. Cow ghee has a distinct yellow color and carries 72 μg of carotenoids per 100 g; the latter are totally absent in buffalo ghee, which contains biliverdin and bilirubin and therefore has a greenish cast [6].

3. *Standards for Ghee*

Two broad categories of ghee are specified under the AGMARK rules of the Agricultural Produce (Grading and Marking) Act of 1937. These permit maximum free fatty acid levels of 1.4% and 2.5% (as oleic acid) and a maximum moisture level for both of 0.3%. The marked effects of feeding cottonseed to cattle, in hardening milk fat and depressing lower fatty acid levels, especially in summer, are reflected in the following AGMARK specifications for ghee:

		Regional (cottonseed tract)	
	All-India	Winter	Summer
Butyrorefractometer reading at 40°C	40.0–43.0	41.5–44.0	42.5–45.0
Reichert–Meissl value, min	28	23	21
Polenske value	1.0–2.0	0.5–1.2	0.5–1.0

Under the Prevention of Food Adulteration Act and Rules of 1976, addition of antioxidants to ghee is not permitted.

C. Preparation of Ghee

1. *Principles of Ghee-Making*

Three distinct stages are recognized in the conversion of butter or cream into ghee by the application of heat [9]. In the first stage, the temperature of the product must be raised rather gradually to the boiling point of water while ensuring adequate stirring to control frothing. In the second stage of heating, free water evaporates as steam with a crackling sound. This requires a considerable amount of heat, so the rate of heat application can be increased while ensuring sufficient agitation to prevent local overheating and sputtering. As most of the water evaporates, a fine stream of bubbles rises from the cast-out solids-not-fat (SNF) at the base. The rate of heating should be reduced to prevent their charring, since that would impart bitter flavors and a brown color to the ghee. Overheating could also drive off desirable volatile flavor materials and also appears to alter the subsequent ability of the fat to form suitable grains on cooling. A temperature of about 103°C should be maintained throughout this stage. In the third stage, water bound to the non-fat solids must be removed without heat damage either to the fat or the SNF; this is done by raising the temperature to between 105 and 118°C with constant agitation. A light amber color in the precipitated SNF and a pleasant aroma are traditional criteria to signal the end of the operation. Regional preferences range from a buttery, uncooked flavor in the south to a slightly burned, overcooked flavor in the west. Cream rendering takes much longer than when butter is used. Sometimes betel or curry leaves, which contain antioxidant principles, are added at the end of the boiling down.

The yield of ghee in domestic or village preparation is about 82% of the fat in milk; about 13% is lost in the buttermilk produced during curd churning, about 2% is held in the ghee residue, and handling losses are about 2.5% [10]. Dairy butter yields about 92% ghee, with a handling loss of 4.5%, and when cream is boiled down, ghee recovery is 88%, with a 9% loss of fat in the large ghee residue.

2. *Larger Scale Traditional Production*

Ghee was sold in bulk in the past by wholesale merchants in towns and cities [11]. Partly clarified ghee was assembled from small producers for final desiccation in large iron vats kept over wood fires and stirred with iron ladles. A new development at this level of operation is the proliferation in Indian cities and towns of minidairies owning cream separators, in which cream of 55–60% fat content is produced from milk [12]. Such cream is ripened for a day in vats of 50–100 kg capacity, after which it is diluted and skillfully churned using both hands to yield butter. This in turn is melted into ghee in pans that have a simple tilting device or a hollow central tube to facilitate ghee removal into 17 kg tins. These are then delivered to bulk users such as traditional sweetmeat makers, hostels, hotels, hospitals, and the armed forces.

As much as 186,000 t of ghee, comprising 22% of all the ghee made in India, is produced by the unorganized sector [3].

3. *Ghee Production in the Organized Sector*

Batch Production Commonly, a steam-jacketed stainless steel vessel of 500–1000 kg capacity is employed. This is charged with butter, and the temperature is raised to 90°C with agitation to remove water. When almost all of the water has gone, the temperature will tend to rise sharply, and careful control is needed. The final temperature lies between 105 and 120°C depending on the flavor desired. To remove sediment, the ghee is either centrifuged or filtered through fine muslin or even through filter paper. The use of glass bottles and tin cans of various sizes to pack ghee is yielding to the use of plastic jars and flexible film laminates, which are easier to handle and comparatively impermeable to oxygen and water [6,13].

An alternative starting material for ghee manufacture is cream of 50% fat content obtained from a standard cream separator. This contains about 5% solids-not-fat and foams copiously when heat is applied. Washing it with an equal volume of water will reduce the SNF content to 1–2%, not only reducing foaming but also raising the yield of ghee by reducing the quantity of ghee residue and hence the amount of fat held up in it. After washing, the cream can be raised to an 80% fat level using a plastic cream separator. To improve the flavor of the ghee made, the cream can be ripened using a lactic starter to a lactic acid level of 0.2% or simply acidified with citric acid. Rendering cream is a slower operation than when butter is used, and precautions are necessary to achieve both an acceptable color and flavor [9,13].

A procedure involving prestratification can effect considerable economy in steam consumption [14,15]. Butter is held in a ghee kettle with a bottom valve at 80°C for 30 min, when three layers are formed. The top layer is a scum of curd particles, the middle bulky layer is essentially melted fat, and the bottom layer is serum or buttermilk, which will contain 80% of the original water and 60% of the SNF. The bottom layer is run off, and the other two are boiled down together to yield ghee of good flavor thanks to the scum.

Continuous Ghee Making. Many advantages are claimed for continuous preparation methods that employ scraped-surface heat exchangers [16]. These are a low steam consumption, 0.35 kg/kg of ghee prepared compared with 0.48 kg in a conventional batch unit; low electricity consumption, 0.01 kWh/kg; and a high thermal efficiency, 71% versus 37% for batch operation. The very high heat transfer coefficients also permit a compact design.

The equipment is shown in Fig. 1. The butter is melted in a receiving and heating vat; thereafter a gravity pump moves part of it to a gravity separator and part to a balance tank [13]. In the gravity separator the butter is separated into two portions, rich in fat and serum, respectively. The fat-rich fluid containing a little buttermilk goes back to the same balancing tank. If cream is the starting material, the gravity separator is bypassed. The molten butter or cream then passes to the first scraped-surface heat exchanger. Here it is heated with steam, after which the superheated liquid (butter or cream) is flashed to the vapor separator, where water is separated. The fat then passes for further moisture removal to the second scraped-surface heat exchanger, where the ghee flavor develops. If cream is the raw material, passage to a third heat exchanger may be necessary. After centrifuging or filtering, the clear ghee is packed.

Removal of water from cream or butter using heat is expensive and not very efficient. Mechanical means can be used to break the emulsion and free the bulk of the water, leaving heating entirely for flavor development. In a recent development, shown in Fig. 2 [13], the fat-in-water emulsion is broken down using centrifugal forces, after which a scraped-surface heat exchanger and vacuum vessel serve to generate flavor. The plant is claimed to be less energy-intensive, to result in lower losses of both fat and SNF, and to cause less environmental thermal pollution. Moreover, much larger volumes can be handled.

Some 110,000 t of ghee, or 13% of total production, derives annually from the organized sector, under a score of brand names. Five brands are each produced in quantities of 5000–13,000 t [3]. This development is just a decade old.

III. VANASPATI

A. Growth of the Industry

1. Historical

Edible hydrogenated fats came into India as imports from the Netherlands after World War I, for supply initially to bulk users of ghee like traditional sweetmeat makers and restaurants [17]. The origin of the term "vanaspati," from *pati* (lord) and *vana* (forest), has been lost, but the name was almost certainly intended to emphasize its vegetable origin in contrast to the animal product ghee. Even the later official designation for vanaspati was Vegetable Oil Product, or V.O.P.

Between 1927 and 1932, some 20,000 t annually was imported. Production in India commenced in 1930, and within a decade imports ceased completely as the domestic annual production climbed to 66,709 t, produced in nine factories. World War II gave a tremendous boost to production, following the decision to include vanaspati as an approved cooking fat for the armed forces to supplement the use of ghee, which was in short supply. This led in October 1943 to a control on price with a formula based on ruling vegetable oil prices and conversion costs. A year later a comprehensive V.O.P. control order was promulgated under the Defence of India Rules to (1) promote the manufacture of vanaspati on a priority basis, (2) regulate its procurement and distribution, (3) control its price, and (4) lay down minimum quality standards [18]. In course of time it became the practice, as part of price regulation, for the government to import edible oils

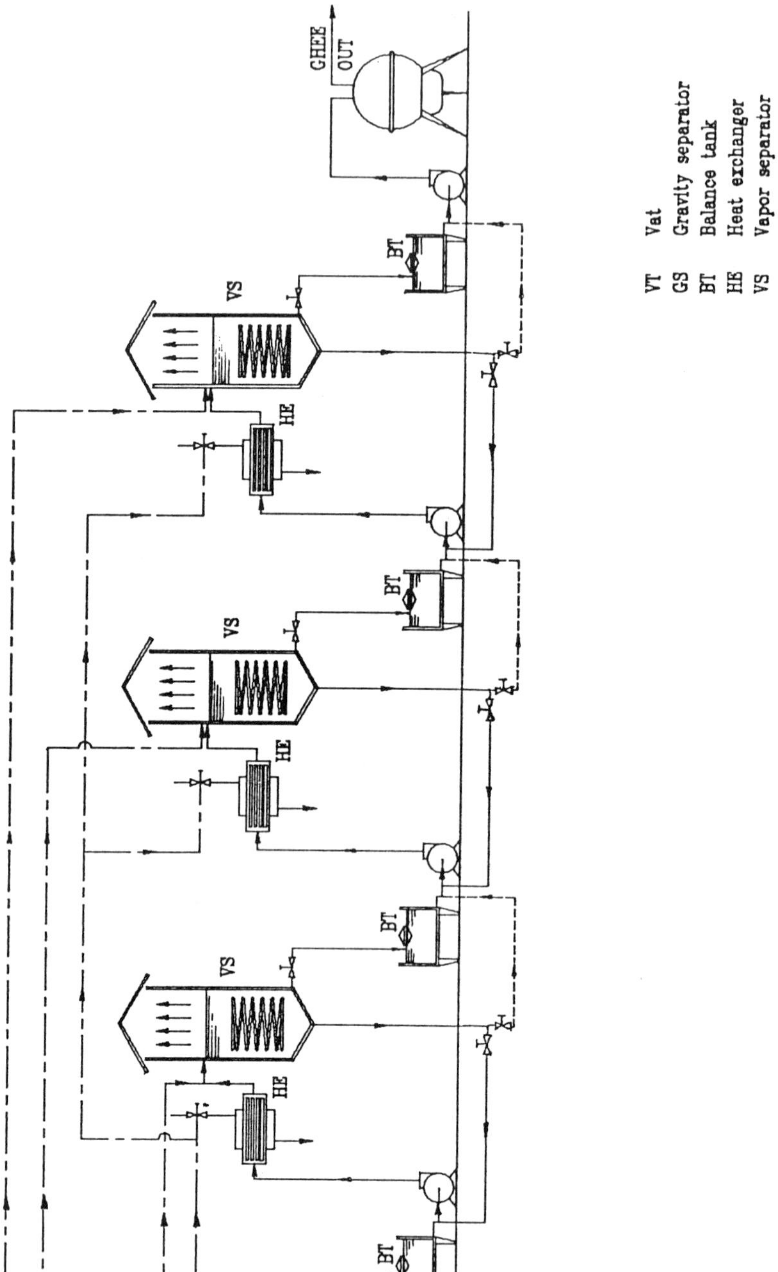

Fig. 1 Continuous ghee-making plant. (From Ref. 13.)

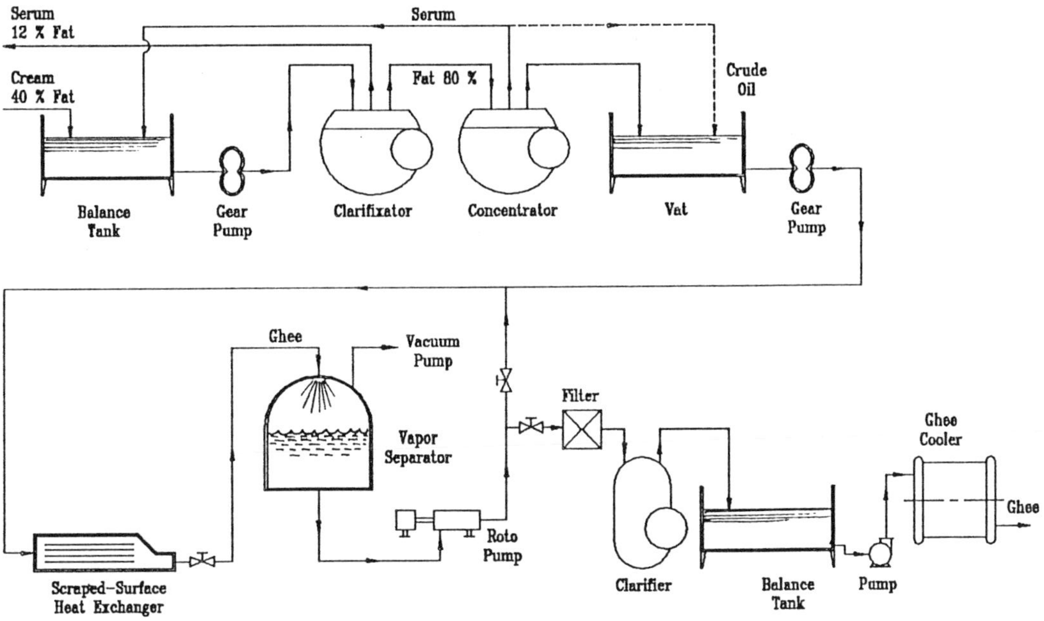

Fig. 2 Continuous ghee-making process using cream. (From Ref. 13.)

for supply to vanaspati manufacturers at prices lower than those ruling in the Indian market. This practice came to an end in December 1988; with it the control on the price of vanaspati also ceased, and the product was obliged to compete for its place in the market.

2. *Usage of Base Oils*

At the start, the only oil used as a base for hydrogenation was screw-pressed groundnut oil, following its refining, bleaching, and deodorization in the vanaspati factory itself. In consequence, groundnut oil production rose dramatically from a figure of 1.5×10^6 t in 1930 to 4×10^6 t in 1947, and it took its place as India's leading edible oil. Addition of 5% sesame oil as a latent marker was enforced in 1947. The decades that followed witnessed three developments. Production of vegetable oils in India grew far too slowly to keep up with population increase and the market demand for edible oils and vanaspati. This meant in turn that vegetable oils had to be imported from world supplies, and the least expensive of these, e.g., soybean, rapeseed, and palm, were naturally favored. Third, a number of new oil sources came to be developed and exploited in the country, such as cottonseed, ricebran, and soybeans. Marketing these unfamiliar oils in India posed problems, and vanaspati manufacture offered a ready outlet in a known and standard product.

Oils permitted for vanaspati manufacture are under constant review [4]. Even rebates and incentives have been offered from time to time to encourage the use of particular oils in vanaspati, such as cottonseed oil in the 1960s and ricebran oil in the 1980s. Since screw-pressed groundnut oil is greatly in demand for direct edible use, its use for vanaspati manufacture is not now permitted. After much hesitation, the use of three oils that are solvent-extracted from expeller oilcakes, namely groundnut, rape-mustard, and sesame, is now allowed. Thirteen other oils are also permitted. Five of these are obtained by direct solvent extraction, namely ricebran, soybean, sunflower, maize, and sal (use of which should not exceed 10%). The other eight permitted oils are cottonseed, mahua, nigerseed, palm, palmolein, sunflower, safflower, and watermelon seed.

Figures of actual usage reflect these shifts [4]. When oil imports were extensive, as in 1986–1987, the major oils used in vanaspati manufacture were soybean (22%) and palm (17%), and, a year later, palm (31%) and rapeseed (9%), besides, of course, indigenous oils. Once indigenous oil production began to increase and imports became notional, the order of usage of oils in making vanaspati in 1992–1993 was soybean 26%, cottonseed 23%, ricebran 16%, rape-mustard 12%, sunflower 8%, and mandatory sesame 5%, for a total of 90%. The remaining oils included mahua, nigerseed, maize, and others [4,19].

3. *Regulations and Standards*

Over the years, the original Vegetable Oil Products Control Order 1942 has been subject to frequent revision [4,19]. As it now stands, vanaspati should conform to the following specifications: moisture content 0.25% (max.); capillary slip melting point 31–41°C (from an earlier upper limit of 37°C, which is nevertheless largely observed even now); butyrorefractometer reading at 40°C, 48.0 (max.); free fatty acids as oleic 0.25% (max.); unsaponifiable matter 2.0% (max.); and nickel content 1.5 ppm (max.). No coloring or flavoring matter, nor any antioxidant, synergist, emulsifier, or similar substances, nor any matter injurious to health can be added to the product. Addition of refined sesame oil is mandatory, the quantity being such that the vanaspati should give a Baudouin color of at least 2 red units in a 1 in. Lovibond cell when mixed in a proportion of 20:80 with refined groundnut oil and examined under specified conditions. Also compulsory is the addition of at least 25 IU of synthetic vitamin A; addition of vitamin D is optional, but label declaration is necessary. The product has to be manufactured in premises maintained under hygienic conditions, and no refined oils other than those permitted should be stocked on the premises.

There are two provisos. Vanaspati for use by the defense forces may contain up to 4 ppm of diacetyl. If more than 30% ricebran oil is used in manufacture, the permitted level of unsaponifiable matter is 2.5% (max.).

4. *Vanaspati Production Data*

During the last decade, vanaspati production has varied between 816 and 985, average 906, thousand tonnes annually [4,19]. In 1993, there were 133 factories for vanaspati manufacture, with an installed capacity of 2041 thousand tonnes, hence capacity utilization is just 44%. The location of factories is strongly regional, so that local demands, as shown below, can be met without excessive transport costs.

	Percentage of total vanaspati production			
	North	West	East	South
1966	57	19	16	8
1993	57	27	9	7

The trend is toward less consumption in eastern India and greater usage in the traditionally ghee-consuming western region.

B. Hydrogenation Process Parameters

Chapter 9 deals with the hydrogenation of oils. Accordingly, only outlines of Indian practices relating to vanaspati production are set down here.

1. *Equipment and Operation*

The batch hydrogenator is essentially a dish-ended autoclave of mild steel varying in capacity from 5 to 15 t [20,21]. It is fitted with closed and open steam coils and variable turbine or baffle stirrers. At the top is a vacuum connection and a sight glass. Oil from a storage tank and catalyst from a doser are drawn in from the top. Compressed hydrogen gas from a holder is admitted at the bottom through a perforated ring or a fritted ceramic disk. Temperature indicators are provided. It is usual to have a number of hardening autoclaves so that various oils can be separately hydrogenated.

Alkali-refined, bleached, filtered, and deodorized oil is drawn into the hydrogenator to about two-thirds its capacity, either cold or preheated in a heat exchanger. A measured quantity of the nickel catalyst in the form of pellets or flakes is introduced from the catalyst doser. Stirring is commenced, and steam is slowly introduced through the closed coils. As the temperature rises to within 10°C of the hydrogenation temperature, vacuum is applied to draw off dissolved moisture, which could otherwise retard the hydrogenation reaction. When the oil is deemed dry, compressed hydrogen at a pressure of 1.1–1.3 kg/cm^2 is let into the autoclave. The hydrogenation reaction being exothermic, the temperature rises, and usually no further heat need be supplied; in fact, if too vigorous, cold water may even be passed through the coils. The degree of hardening is followed by noting the consumption of hydrogen with time based on previous experience with the same oil in the same unit. When the process is deemed to be nearly complete, a sample of the hardened fat is drawn out using vacuum and its melting point is determined [22–25].

Cold water is now drawn through the coils until the temperature drops to about 95°C, after which the oil passes for further cooling to a heat exchanger to warm up the next batch of oil. This is followed by filtration in a plate-and-frame or leaf filter press at 50–60°C to remove the suspended catalyst.

Postrefining operations include neutralization, bleaching, and deodorization. During hydrogenation, the level of free fatty acids will show an increase, and these are neutralized using dilute lye of 5° Beaumé. Bleaching with 0.25–0.5% bleaching earth serves to absorb both coloring matters and any nickel that may be present in the form of soap. Characteristic odors develop during hydrogenation, which have been attributed primarily to short-chain aldehydes, ketones, alcohols, and lactones. Deodorization is effected by intimate contact between the hydrogenated fat held in thin layers and dry superheated steam at a high vacuum and temperature. In batch units this can take the form of shallow trays or pans standing one below the other, with the oil moving downward by gravity. In continuous units the stripping steam itself spreads the flowing oil in very fine layers on tubular surfaces, yielding turbulence and high interfacial contact for efficient and rapid deodorization. After deodorization the vanaspati is sometimes filtered again.

Hydrogenated products made from various permitted oil components are then blended in a stirred tank, into which refined and deodorized sesame oil, appropriate quantities of vitamin A, and vitamin D, where this is added, are also introduced. The finished vanaspati then passes at 50–60°C to storage tanks. After being filled into containers, the latter are allowed to stand undisturbed while cooling gradually to near ambient temperature. Fresh air is now blown into the room to hasten the rate of cooling. When an incipient cloudiness is noticed, refrigerated air is blown in to hasten cooling and induce a uniform and fairly firm consistency [20].

2. *Types of Operation*

Two systems are in vogue for hydrogenation [21,22–25]. In the recirculation system, hydrogen gas that passes unused through the oil and collects at the top is returned to the gas holder after passing through a cooler, a scrubber to absorb moisture, a degreaser to trap entrained fat, and a compressor. In the dead-end system, the gas that thus accumulates, along with any volatile matter, is simply vented away; this constitutes both a waste and a fire risk but eliminates the need for costly gas purification equipment. Dead-end operation is also well-suited to monitoring the hydrogen uptake.

Individual hydrogenator units in India are frequently of 10 t capacity. Average factory output is close to 50 tpd [4]; about half the units turn out less, and the other half more, than this figure (Fig. 4, pp. 385).

3. *Consumption of Hydrogen*

For a unit drop in iodine value, 0.9–1.1 m^3 of hydrogen is consumed by 1 t of oil. Thus the quantities of hydrogen needed to bring about the drop in iodine value required to meet vanaspati specifications for any batch size of oil can be calculated. Both theoretical calculations and practical experience show that to convert 10 t of oil to vanaspati, the requirement of hydrogen in cubic meters at NTP for various oils is as follows: soybean 645, cottonseed 555, ricebran 530, rapeseed 500, and palm 220. For groundnut oil, which had a long history of use for making vanaspati in India, though at present it is not permitted, the figure is 280 m^3 of hydrogen [22].

4. *Monitoring of Hydrogenation*

Hydrogen uptake during the hardening of any specific oil is the main criterion by which the endpoint is judged, and this is confirmed by determination toward the end of the run of the capillary slip point on a sample that is withdrawn from the reactor and chilled in ice [22–25]. The filtration rate of the product after the hydrogenation run is over is also a pointer to any major deviation from normal in the reactor. During a run, the iodine value, refractive index, and solid fat index are not generally employed for monitoring but are determined on the finished vanaspati product.

5. *Catalysts*

Up to a decade ago it was usual to generate the nickel catalyst from its salts in the vanaspati factory [21, Chap. 10]. Dry-reduced catalysts were prepared by first precipitating nickel hydroxide or carbonate on diatomaceous earth or other refractory support material and then reducing the ground, dried material to the metal in a current of hydrogen. To make wet-reduced catalysts, nickel was first converted into its formate; this was then easily reduced in a vegetable oil medium at a relatively low temperature to give carbon dioxide, hydrogen, water, and the metal, which again was supported on siliceous material. At present, commercial catalysts take the form of pastilles or free-flowing flakes that contain 18–26% reduced nickel metal precipitated on a siliceous or silica-alumina support and suspended in almost fully hydrogenated soybean oil. In general, oil hydrogenation nickel catalysts are designed to exhibit a high degree of selectivity toward linoleate over oleate during hydrogenation and a high capacity to generate *trans* fatty acids. Use of a high catalyst concentration also favors selectivity as well as a high temperature and low hydrogen pressure. For 1 t of oil, 0.5–1.0 kg of catalyst pellets are commonly employed.

Oils that are solvent-extracted from screw press oilcakes after storage, in particular from rape-mustard oilcakes, are known to carry catalyst poisons. Special catalysts have been developed for such oils that have sufficient active sites still left for the absorption and hydrogenation of unsaturated fatty acids even after a proportion of them have been nullified by catalyst poisons. A higher concentration of 2 kg of catalyst per tonne of oil is also generally employed.

Nickel catalyst is expensive, and nickel resources in India are scarce. It is therefore customary to use, along with fresh catalyst, 10–15% of once-used catalyst. Too high a proportion of spent catalyst will produce less saturation and more isomerization and result in products that, even at the specified melting point, will show considerable separation of liquid glycerides. In practice, the catalyst from, say, two plates of a 20 plate filterpress can be set aside and the rest used again with fresh make-up catalyst [23].

Hydrogen uptake with time during a hydrogenation run under set conditions is employed as a sensitive measure not only of the activity of the catalyst being used but also of the proportion of fresh catalyst that will be needed for the next run [22]. Yet another indicator of catalyst activity during a hydrogenation run is the filtration rate of the hardened product in the filter press. Too high a rate implies an excessively liquid oil, and too low a rate, oversaturation. This must be corrected when various individually hydrogenated components are finally blended to vanaspati [22–29].

6. *Oils Employed for Hydrogenation*

Groundnut oil was for decades the single base stock for manufacture of vanaspati in India. Though over a dozen vegetable oils are now permitted for conversion to vanaspati, as noted earlier, the belief in the industry is that no single oil other than groundnut can yield a satisfactory vanaspati product.

Each oil presents special problems and needs an appropriate approach [22–25].

When soybean oil is to be hardened, the level of gums present through insufficient degumming in the refinery is important, because they are powerful catalyst poisons. Degumming may therefore have to be carried out using dilute acids in the vanaspati factory before the hydrogenation of soybean oil. During the hardening operation, it is often the practice to start the hydrogenation at a lower temperature, 140°C, so as to selectively reduce the linolenic acid component, which averages about 5% in India-grown soybean oils, before raising the temperature to the usual 180°C. Postoperation steps of alkali neutralization, bleaching, and deodorization of hardened soybean oil also call for stronger measures than for groundnut oil.

Cottonseed oil is offered in India both as a fully refined and bleached product and as a "washed" oil. In the latter, most of the free fatty acids present in the crude oil, but only part of the coloring matter, have been removed by washing with dilute lyes at slightly elevated temperatures. Frequently, washed cottonseed oils are color-fixed, and even after alkali-refining, bleaching, and deodorization in the vanaspati factory, the product obtained after hardening will tend to be darker in color. Use of higher than usual levels of bleaching earths and active carbons at all stages is fairly routine.

Ricebran oil also tends to have fixed red colors. If crude oils are procured, the free fatty acid level is frequently of the order of 4%, and such oils need degumming, chilling, alkali-refining, and bleaching before hardening. Even refined ricebran oils when procured tend to carry gums and waxes and need to be thoroughly degummed at the vanaspati factory. Some factories treat the oil with spent nickel catalysts to remove catalyst poisons prior to hardening. Others recommend postbleaching of the hydrogenated product with, say, 1% of bleaching earth and 0.3% of active carbon to reduce the red color component.

Rape-mustard oil that has been solvent-extracted from screw press oilcakes, frequently after some delay, is known to contain catalyst poisons. Some factories treat the oil with spent catalyst prior to hardening, and others deodorize it at 140°C to reduce the content of volatile sulfur components. Hydrogenation of solvent-extracted rape-mustard oil is recognized as being sluggish, with little generation of exothermic heat during hardening. Not more than 50% of hardened rapeseed oil is employed as a vanaspati component.

Palm oil and palmolein require only slight hardening. This is performed at low temperatures, using mainly spent catalyst to slow down the hydrogenation and avoid excessive saturation.

Different oils are never hydrogenated together. To have better control of the process, each is hydrogenated separately to a vanaspati specification end product, followed by blending. Only if an oil component is very minor will it be incorporated in a major component before the latter is hardened.

7. *Generation of Hydrogen*

Earlier it was common practice to generate hydrogen in vanaspati factories by the steam iron process [21, Chap. 10], but this has largely been given up. One large vanaspati plant in India uses hydrogen from caustic chlorine cells, but a majority use hydrogen from alkaline electrolytic cells [30, 31]. These can be worked at capacities varying from 10 to 100% merely by varying the voltage, which is very convenient. Two types of cells are in use. In the unipolar or tank type, each cell is independent and self-contained; electrodes are suspended in the electrolyte in an iron tank, and alternate electrodes are connected in parallel. It requires large currents at low voltages, in addition to massive transformers, rectifiers, and copper bus bars. In the bipolar or filter press type, a series of flat, platelike cells are stacked together and connected in series, with a voltage supply at either end. Operating pressures vary from 5 to 35 kg/cm^3, so gas compressors can be dispensed with. The electrolyte solution in both types of cells is 25–30% aqueous sodium hydroxide; this requires no replacement, and only the concentration of water needs to be automatically made up as it is consumed. The purity of electrolytic hydrogen gas is over 99.5%; the hydrogen is scrubbed to remove droplets of alkali and compressed for storage in high pressure cylinders. The by-product oxygen, produced in half the volume of the hydrogen, is compressed into cylinders and finds a ready outlet in welding operations.

At 2500 A and with the electrodes at 2 V, such a cell can produce 1.142 m^3 of hydrogen and 0.571 m^3 of oxygen per hour [30].

8. *Bleaching Earths and Active Carbons*

Certain natural silicoaluminates, when suitably treated with acid, heat, and steam, become both porous and charged and are capable of adsorbing the coloring matters present in fatty materials [21, Chap. 9]. Special types are made in India for specific oils (depending on the nature of the pigments present) and for vanaspati. High capacity for adsorption of pigments, low fat absorption, easy filterability, and economy in use are the qualities looked for.

Active carbons are adsorbents prepared in India from a wide range of carbonaceous materials by treatment at elevated temperatures with air, steam, or metal salts. They are particularly effective in removing the red color component of fats as well as picking up both traces of soap and the earthy or musty flavors imparted to the oil by bleaching earth. Frequently they are mixed in about 10% concentration with the earth, as they cause a loss of far more oil by adsorption than do the latter [21, Chap. 9].

Postbleaching of vanaspati is carried out in cylindrical mild steel vessels. Steam coils raise the temperature of the fat from 50 to 100°C in about 30 min in the presence of about 0.5% of the earth–carbon mixture, with stirring at 80–120 rpm under 700 mm pressure. Filtration thereafter is in plate-and-frame or metal leaf filters using heavy duty filter cloths, polyester weaves, or stainless steel wire gauze [32].

9. *By-Products of Vanaspati Manufacture*

A small quantity of soapstock is produced during the postneutralization of vanaspati. This is acidified to yield acid oil, which finds a ready outlet as a hard fat component in soapmaking. Spent nickel catalyst finally discarded from the filter presses carries between 4 and 14% nickel held in a mass of fat (45–85%) and siliceous matter. Batches of discarded catalyst are pooled and transported to catalyst recovery units in the country where the fat is extracted away and the nickel is leached out with acid. Salt solutions are then reduced to the metal in a current of hydrogen for deposition on siliceous supports suspended in a hardened fat. The fatty matter carried over in vanaspati deodorizer distillates may be recovered and added to soapstocks. By-product oxygen is compressed and sold.

10. *Energy Conservation*

Fuel and electricity consumption in Indian vanaspati units is far higher than in modern installations elsewhere [33], and installation of heat exchangers at several points in the total system is on the increase. Thus, hot hardened oil from the autoclave can be dropped into a heat exchanger to carry the next batch of oil from a temperature of 50–60°C to even 190°C, thus eliminating the need for heat prior to hardening [34].

There is a trend for new vanaspati units to use the Buss loop reactor [34]. This employs two loops in which the processing occurs. In the first, the oil and suspended catalyst are put through a heat exchanger, and in the second loop, hydrogen in very fine form is distributed through the suspension. Reaction times are short (80–120 min, compared to up to 12 h in batch units), and as many as eight batches can be run in a day. Moreover, post-hardening operations are conducted separately and do not affect plant capacity. No heat energy has to be supplied, and hardly any cooling water, because of excellent heat efficiency. The process is fully automated and even generates some low pressure steam for various uses [34].

11. *Effluent Treatment*

Administration of the Environment (Pollution) Act 1986 in India is the responsibility of central or state pollution boards [35]. Specifications laid down by the Bureau of Indian Standards are followed, the chief requirements of which are shown in Table 1 along with data for vanaspati plant effluents.

Vanaspati factory effluents are treated by employing physical, chemical, and biological methods [36]. Physical methods include vibratory and rotary screening, sedimentation of solids that are capable of settling or floating, flotation using air bubbles, and filtration by gravity or under

Table 1 Specifications for Wash Waters and for Effluents from Vanaspati Units

	Wash water for discharge into			Effluent from vanaspati factories
	Inland water	Land for irrigation	Public sewers	
pH value	5.5–9.0	5.5–9.0	5.5–9.0	2–3
Suspended solids, mg/L	100 max	200 max	600 max	5000–25,000
Oil and grease, mg/L	10 max	10 max	20 max	500–10,000
Biochemical oxygen demand (5 days at 20 °C), mg/L	30 max	100 max	350 max	2000–15,000
Chemical oxygen demand, mg/L	250	—	—	5000–20,000

pressure. Chemicals used are lime, alum, ferric chloride, ferrous sulfate, and anionic or cationic polymers. Biological treatment involves the use of living bacteria in activated sludges.

In practice, the effluent is mixed with dosed quantities of chemicals in a flash mixer, with continuous removal of the dregs that are formed. The liquid passes into collecting tanks and thence for dewatering to drying beds or filter presses. The clear filtrate that overflows the weir is treated with the chemicals earlier noted, which greatly reduce its total solids content as well as its biological oxygen demand (BOD) and chemical oxygen demand. For a further 90% reduction in BOD, the liquid is brought into contact in a mechanically aerated reactor containing nutrients with the aerobic bacterial culture. The treated effluent is fit for discharge, while the sludge can be dewatered in drying beds, filter presses, or centrifuges for use as a landfill [36].

Few vanaspati units now employ the full complement of treatment, which requires considerable capital outlay and physical space, but will eventually have to fall in line.

C. Product Characteristics and Composition

1. Selectivity Concepts During Hydrogenation

Selectivity concepts have been dealt with elsewhere in this volume and are therefore only outlined here in relation to vanaspati production.

When a *cis* double bond in an oil is adsorbed during the hydrogenation process at an active center on the catalyst surface and encounters a hydrogen atom also dissociated at an active catalyst center, it can become saturated and then desorbed. However, if the hydrogen pressure is low, the reaction temperature high, and the quantity of catalyst also high, it is more likely to be desorbed with the double bond still intact but largely twisted into the *trans* configuration. Preferential adsorption on the catalyst of linoleate, with two *cis* double bonds, and partial hydrogenation leads to both *trans*- and *cis*-monoenes in which the double bond has shifted along the chain as a result of the formation of conjugated diene intermediates. Thus selective hydrogenation of an oil containing linoleate will lead to an accumulation of *trans*-monoenes with double bonds scattered from carbons 6 to 16. These have melting points of around 44°C, and in effect the oil has been partially saturated and hardened. Some dienes may also be desorbed without saturation, and these could also carry *trans* double bonds. Fortuitously, a selectively hydrogenated fat high in *trans*-monoenes can be made to physically resemble ghee, the prized culinary fat of India, and early promotion of vanaspati was based on this circumstance.

2. Crystallization Behavior and Granularity

The ghee-like grain structure desired by consumers is achieved by careful cooling of vanaspati [20]. As the hardened product cools from, say, 50°C, the *trans*-glycerides in the liquid phase become supersaturated and nucleation commences; there is considerable evolution of heat, which can even cause remelting if external cooling is insufficient. The large saturated glyceride crystals give the product its granular texture, and the large number of *trans* nuclei, its firmness. Photomicrographs of vanaspati [37] show large granules consisting of radially arranged needle-like crystals. Such a crystallization network has enough surface area to firmly entrap the liquid phase and prevent it from separating out at ambient temperatures as an oily layer on top.

3. Nature and Properties of Vanaspati

Detailed studies on 20 samples of Indian vanaspati were carried out at the Palm Oil Research Institute of Malaysia [37]. The results, shown in Table 2, reflect the nature of the products that were being made in the early 1980s. The vanaspati being made at present uses much the

Table 2 Characteristics of an Average Indian Vanaspati

Parameter	Average	Standard dev
Capillary slip melting point, °C	35.6	1.78
Iodine value	76.5	4.31
Yield value, kg/cm^2		
15°C	7.4	1.5
12°C	3.9	0.98
Solid fat content, %		
10°C	75.1	7.63
15°C	64.6	6.86
20°C	52.6	7.51
25°C	37.1	6.98
30°C	20.7	5.80
35°C	6.8	3.79
Fatty acids, wt %		
16:0	14.8	2.62
18:0	8.1	2.27
18:1 *t*	43.2	3.89
18:1 *c*	18.8	2.56
18:2 *t,t*	5.0	1.80
18:2 *c,t*	3.5	1.25
18:2 *t,c*+20.0	1.3	1.40
18:2 *c,c*	3.4	1.16
Total *trans* acids	53.7	4.29
Total saturated acids	23.8	3.00
Total unsaturated acids	76.2	2.92

Source: Based on analysis of 20 vanaspati products, early 1980s [37].

same bulk raw materials (soybean, cottonseed, and rape-mustard oils; only ricebran oil is used in larger proportions now) and should not be vastly different.

The capillary slip melting point averaged 35.6°C with a small standard deviation of 1.78°C and a coefficient of variation of only 5%. The specification of a maximum of 37°C prevalent at that time was thus being strictly observed. So firmly is this limit ingrained that it still tends to be followed in vanaspati manufacture though the legal upper limit has been raised to 41°C to permit manufacture of harder products that will not be totally liquid during the hot summer months. The best correlation of slip melting point was with total saturated acids (0.71). The iodine value averaged 76.5, again with only a 5.6% coefficient of variation.

Indian vanaspati exhibited a steep solid fat content profile, as shown in Fig. 3 [37]. This is a well-known consequence of a high content of *trans* fatty acids (here averaging 53.7%) coupled with low saturated and *cis*-unsaturated fatty acid levels [37]. Any increase in saturated acids would tend to push the melting point of the product outside the permitted range, so that linoleate selectivity is imperative during hardening. This is in contrast to vanaspati products made in neighboring Pakistan [37], which showed higher melting points (mean 37.2°) and a softer consistency as a result of higher contents of saturated and *cis*-unsaturated acids and considerably lower levels of *trans* acids (mean 27.4%).

The solidification curve at 20°C determined by wide-line NMR (Fig. 4) shows that Indian vanaspati crystallizes rapidly and attains a high final solids content of about 35% [37]. This would make for reduced oiling out at higher ambient temperatures.

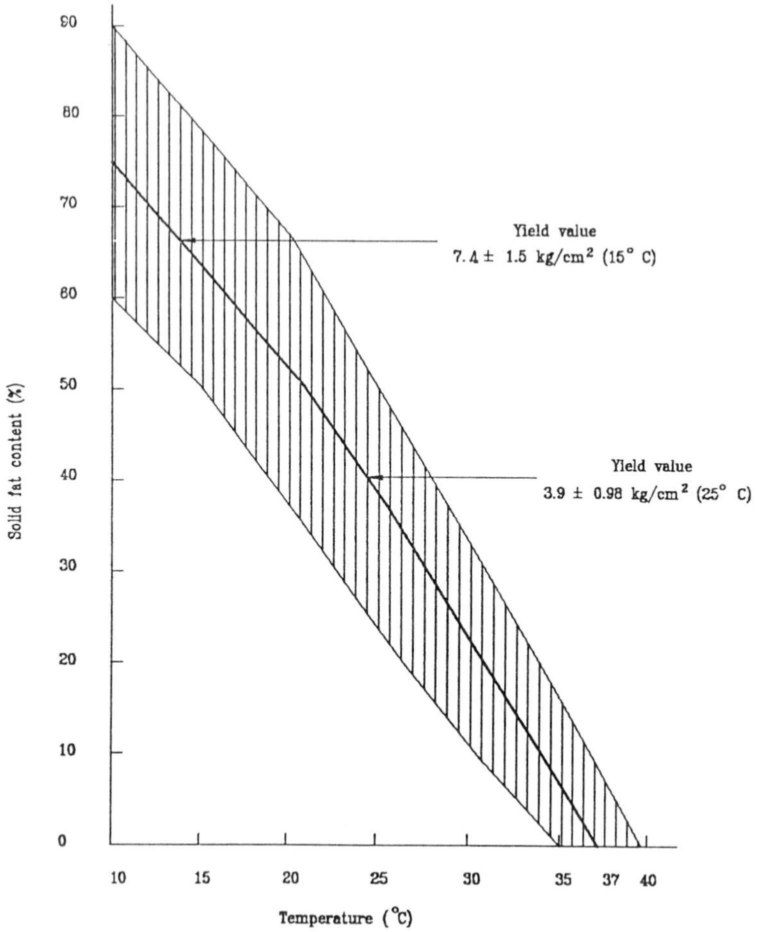

Fig. 3 Solid fat content profile (95% confidence level) of Indian vanaspati products. (From Ref. 37.)

4. *Sesame Oil as a Latent Color*

To deter use of vanaspati as an adulterant of ghee, early legislation specified that not less than 5% of sesame oil, whether raw or refined, be finally blended into vanaspati, with the further proviso of a minimum color of 2 red units in the Baudouin test carried out under specified conditions. Since raw sesame oil would give a markedly higher response than the refined product, much smaller quantities would serve to answer the color test. At present, only the addition of refined sesame oil is permitted; the quantity to be used is not specified, and only a color response of 2 red units is laid down in the Baudouin test. It is usual to use a small overage of sesame oil.

Considerable scientific effort has been expended to find other possible markers for vanaspati [5, pp. 299–307]. These include (1) various natural pigments like ratanjot, curcumin, kamala dye, chlorophyll, copper pheophytin, alkanet, annatto, several carotenoids, and lactoflavin; (2) synthetic and coal tar dyestuffs (yellow, orange, red, blue); and (3) latent colors such as fluorescent dyes, tyrosine ethyl ester, quinine, vitamin K, alkylated dihydroxydiphenylamine, acetylated methylene quinine, and butylated hydroxyanisole [5]. None of these has been adopted or stood the test of time.

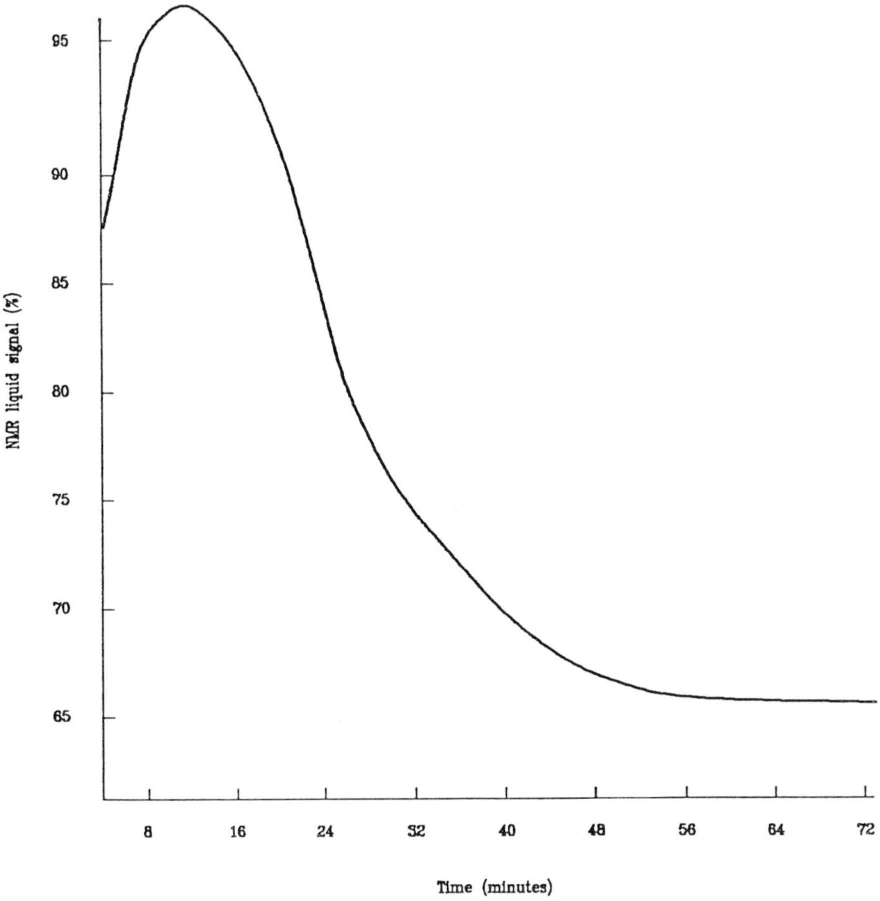

Fig. 4 NMR solidification curve of an Indian vanaspati. (From Ref. 37.)

5. *Addition of Vitamin A*

The compulsory addition of 25 IU of synthetic vitamin A to vanaspati in May 1951 was intended to equate the product in this respect with ghee. Observations show that half or more of the vitamin is lost on storage of vanaspati for 6 months whatever the container or temperature [38]. Further, vanaspati is used mostly as a frying medium at high temperature, which would doubtless destroy any surviving vitamin A.

6. *Nickel Contents*

Though once thought to be innocuous, nickel is now regarded as an essential nutrient for humans [39]. Little of it is absorbed from the diet, and impossibly high levels of 100 mg/100 g bodyweight, or, say, 5 g of nickel a day, are needed to produce deleterious effects. The daily intake of nickel from customary Indian diets has been estimated at 300–600 μg [39].

Commercial vanaspati samples varied in nickel content from 0.30 to 4.54, average 1.55 ppm [39]. Since the higher contents could reflect poor manufacturing practice, the legal specification was recently fixed at 1.55 ppm (max) [4].

IV. BAKERY FATS

A. Definition

Even the legal definition of a bakery shortening, in the Prevention of Food Adulteration Rules 1955 as it currently stands, states that it "means vanaspati conforming to the standards prescribed," with certain qualifications. These are (1) a capillary slip melting point of 41°C (max), (2) a content of 12% (max) by volume of nitrogen, air, or any inert gas where aeration has been employed, and (3) optional use of mono- and diglycerides as emulsifying agents [4].

B. Production Parameters

1. Earlier Technologies

The aim at first was simply to convert a hardened vanaspati product from the regular production unit into a material that was smooth, fine, homogeneous, and nongrainy. In the early stages, this was achieved by chilling the hardened fat very rapidly as it emerged from the deodorizer. When the product had become semisolid, but still pourable and pumpable, it was filled into tins, and no further cooling was applied [20]. In a later development, the vanaspati was cooled continuously on the surface of slowly revolving (10 rpm) hollow rolls chilled internally with brine or ammonia. Fat was scraped from these rolls to fall into an open trough in which a conveyor or paddle provided with blades beat air into the product as it passed in semisolid form into the filling machine [20]. In 1968, some 38,000 t of such all-purpose bakery fat was produced in India [21, Chap. 10].

2. Current Processes

Basis. The present practice in India is to manufacture three types of hardened products and then blend and sometimes aerate them to yield bakery fats suited to specific end uses. The three hydrogenated fats have melting points of 28, 36, and 42°C, and in blending and plasticizing them the critical factors in achieving the desired product characteristics are (1) the components and composition of the blend, (2) the chilling conditions, (3) whether aeration is employed, and (4) the filling temperature [26]. It is usual in Europe to adopt a common temperature of 20°C for the operations of filling, transport, and storage, but in the warmer Indian ambience this common temperature is 30°C, which would in turn influence product characteristics.

Hardening. Although all the edible oils used in the manufacture of vanaspati can be employed to make the components of bakery shortenings, experience has led to certain preferences for each of the three hardened products. For the product of melting point 28°C, soybean oil is the choice; for that of melting point 36°C, cottonseed, sunflower, and soybean oils; and for the product of melting point 42°C, ricebran oil [23]. In preparing the product melting at 28°C, a high proportion of used catalyst and a low hydrogenation temperature of 160–170°C appear to be preferred [23].

Blending and Plasticization. Blending the three hardened fats in different proportions yields various types of bakery fats. Thus the preferred solids content index at 20°C for table margarines in India lies between 4 and 20%; for cake shortenings, 20–40%; and for puff pastry, 60%; so that an ever greater proportion of the hardest component will be required for each product type [40].

The right amounts of fats and the permitted additives are mixed in a heated and stirred vessel. The next steps of converting this mixture into plastic bakery fats is described in Chapter 11. In brief this involves its passage through various arrangements of scraped and chilled open tubes to achieve very fine crystallization. The chilled end products are variously packaged in India: in tins of 0.5, 1, 2, and 15 kg; in polybags of 1 and 5 kg; in polyjars of 1, 5, 10, and 15 kg; and in bags-in-boxes of 15 kg [26].

Products Manufactured. Nonaerated, high melting, but plastic products go into the manufacture of puff pastry and patties and also that of a popular, highly layered, crisp and eggless salted snack called Bombay khari. Aerated products of high creaming power are chosen for making crisp cookies, wafers, and cream biscuits. Margarines themselves are employed in considerable quantity in India to make cakes and cream fillings [26].

The total production of bakery fats in India is on the order of 20,000 t annually, and the quantity is rising fast as commercial baking expands.

Related fat-based products made for the baking industry are emulsifying agents like glycerol monostearate, mixtures of mono- and diglycerides, and pan-releasing agents employed in baking bread [26].

V. MARGARINES AND FAT SPREADS

A. Background

Butter was too perishable a product to gain widespread traditional use in India, and it was mostly rendered into ghee. Packaged table butter is now made in organized dairies only to the extent of about 30,000 t annually. Fat spreads are in concept and practice substitutes for table butter; they have only recently made a beginning in India, with production in 1992–1993 of about 2000 t of margarine and 5000 t of table spreads [4,27]. The distinction is one of water content; the maximum permissible in a margarine is 16%, and if greater than this the product qualifies as a table fat spread [36].

B. Specifications

Three types of fat spreads are recognized by the food laws of India [4]. A milk fat spread should have only that commodity as the fat component. In a mixed fat spread, the fat should consist of a mixture of milk fat with one or more hydrogenated or unhydrogenated refined edible vegetable oils or interesterified oils. A vegetable fat spread will carry only the latter. The fat content has to be declared on the label, and in a mixed fat spread the content of milk fat must also be declared. The word "butter" should not be associated with the labeling. The fat content should lie between 80 and 40% by weight, and the moisture content between 16 and 56%. The fat extracted from a vegetable fat spread should have a melting point of 7°C (max), an acid value of 1.5% (max), and a content of unsaponifiable matter of 1.5% (max); the latter value is 1.0% (max) for fat extracted from a milk fat spread or a mixed fat spread. The vegetable fat spread should contain 25 IU per gram of synthetic vitamin A when packed and show a positive response to the Carr–Price antimony trichloride test. A vegetable fat spread should contain sufficient raw or refined sesame oil to answer the same Baudouin color test that is prescribed for vanaspati. The package size of any fat spread should not exceed 500 g, and loose sale is prohibited.

Certain additives are permitted but not mandatory. These include (1) common salt, 2% by weight (max); (2) the solids-not-fat of milk; (3) starch at levels of 100–150 ppm; (4) diacetyl at a 4 ppm (max) level; (5) BHA and TBHQ as antioxidants at a level of 0.02% (max) on the fat content; (6) sorbic acid (as its sodium, potassium, and calcium salts) and benzoic acid (as its sodium and potassium salts), singly or in combination, at a 1000 ppm (max) level; and (7) annatto and/or carotene as coloring agents.

One commercial milk fat spread contains 60% fat, of which one-sixth is butter. It is designed to be spreadable when taken out of the refrigerator and has a solids content index of 30 at 10°C, 18 at 20°C, 2.5 at 30°C, and nil at 40°C, which are all higher than for a similar product in colder countries [27]. Two commercial table margarines are in production, and a low calorie diet margarine is stated to be in development [29].

C. Manufacture

1. *Fat Component*

The fat component of a margarine or a vegetable fat spread is a blend of refined and hardened edible vegetable oils [4,27–29]. A mixed fat spread will also contain milk fat. The refined oils that can be used cover a wide range, and in India they chiefly include sunflower, soybean, cottonseed, and sesame oils. The hydrogenated oil component could include those made from soybean, cottonseed, ricebran, and palmolein. Soybean oil that has been hydrogenated to two different degrees of hardness could be employed. To achieve the desired smooth texture and physical stability, one component of the hydrogenated oil mix should have a B′ crystal habit, with considerable palmitic acid in the 2-position; these include in India cottonseed, palm, and rape-mustard oils [27] (see Chapter 11).

Selective hydrogenation conditions are employed in hardening the vegetable oils. However, the pressure and temperature used may be slightly lower than for a vanaspati-type hydrogenation, and the catalyst employed could contain a higher proportion of used catalyst.

A table margarine in India could well be a blend of three parts of liquid oil and two parts of the solid component [40].

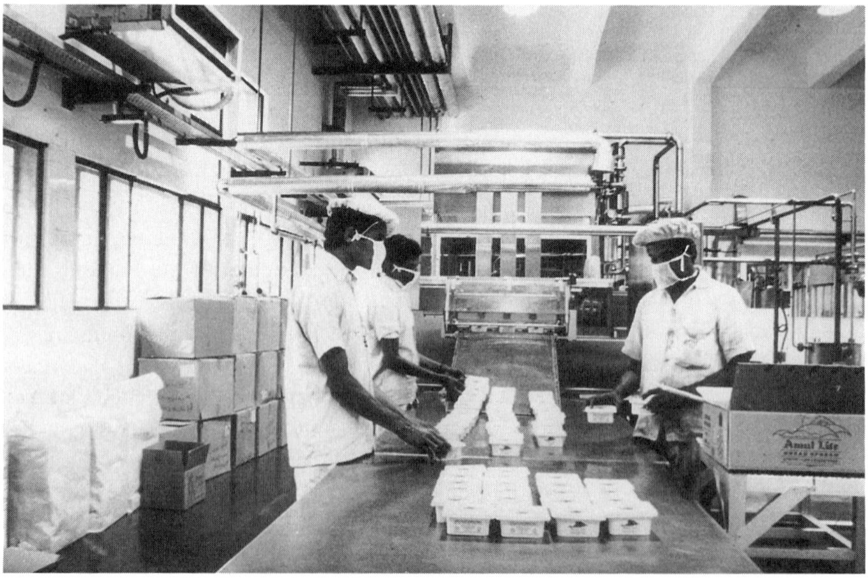

Fig. 5 Packing section of an Indian mixed fat spread unit. (Courtesy of National Dairy Development Board, Anand.)

2. *Emulsification*

The emulsion is prepared in two stages [40]. A coarse water-in-oil emulsion is first formed by adding pasteurized salted water or milk containing the water-soluble ingredients to the fat blend carrying the emulsifiers and fat-soluble ingredients under strong agitation at 45°C. This is cooled and then homogenized by passing the mass under high pressure through fine orifices to yield a fine, uniform emulsion. A part of the fatty matter crystallizes around the aqueous droplets in the form of a fine network and keeps them from coalescing.

The next step of chilling and working the emulsion is carried out in the annular space between a shaft and a chilled insulated outer jacket at very high speed to achieve nucleation (see Chapter 11). In the second half of the unit, the mass is worked vigorously to complete crystal formation, followed by a resting phase to achieve phase stabilization and the stiff consistency needed for packaging. Such units are manufactured in India [28]. The extruded margarine or fat spread is packed in plastic tubs (Fig. 5), frequently of 100 g size for convenience of domestic use.

REFERENCES

1. K. T. Achaya, *Oilseeds and Oilmilling in India: A Cultural and Historical Survey*, Oxford and IBH Publishing, New Delhi, 1960, p. 157.
2. S. L. Rao (Ed.), *Consumer Market Demographics in India*, rev. ed., National Council of Applied Economic Research, New Delhi, 1994, Chap. 23, pp. 268–272.
3. Anon., *Ghee in India: Production, Statewise Consumption and Product Profile*, Gujarat Cooperative Milk Marketing Federation, Anand, 1993.
4. M. K. Kundu, Orders and production data, Directorate of Vanaspati, Vegetable Oils and Fats, Ministry of Civil Supplies, Consumer Affairs and Public Distribution, Government of India, New Delhi, Personal communication, February 1995.
5. K. S. Rangappa and K. T. Achaya, *Indian Dairy Products*, Asia Publ. House, Bombay, 1974.
6. R. P. Aneja, Dairy process and product development under Operation Flood, *Indian J. Dairy Sci. 42*: 53 (1991).
7. C. Arumughan and K. M. Narayanan, Grain formation in ghee (butterfat) as related to structure of triglycerides, *J. Food Sci. Technol. 16*: 242 (1979).
8. C. Arumughan and K. M. Narayanan, Triacylglycerol composition of buffalo milk fat, *J. Dairy Res. 49*: 81 (1982).
9. J. N. Warner, *Principles of Dairy Processing*, Wiley Eastern, New Delhi, 1976, Chap. 13, pp. 253–259.
10. R. C. Misra and N. S. Kushwaha, Studies in methods of preparation of ghee, *Indian J. Dairy Sci. 23*: 115 (1970).
11. S. De, *Outlines of Dairy Technology*, Oxford Univ. Press, Delhi, 1980, pp. 433–466.
12. K. L. Gaba, Recent trends in rural marketing, packing and trading of ghee, *Indian Dairyman 32*: 783 (1980).
13. J. S. Punjrath, Indigenous milk products of India, *Indian Dairyman 42*: 75 (1991).
14. B. Bahadur, M. R. Srinivasan, and S. C. Ray, Utilisation of induced stratification of butter melt in ghee making, *Indian J. Dairy Sci. 3*: 94 (1950).
15. S. C. Ray and M. R. Srinivasan, *Pre-Stratification Method of Ghee Making*, Indian Council of Agricultural Research, New Delhi, Res. Ser. No. 8, 1976.
16. H. Abhichandani, S. P. Agrawala, R. D. Verma, and B. S. Sector, Advances in continuous ghee making, *Indian Dairyman 30*: 769 (1978).
17. K. T. Achaya, *The Food Industries of British India*, Oxford Univ. Press, New Delhi, 1994, Chap. 12, pp. 211–218.
18. A. C. Chhatrapati, *The Vanaspati Industry: A Historical Review*, Popular Prakashan, Bombay, 1985, pp. 9–21.

19. Vanaspati Manufacturers Association of India, New Delhi, Statistical data, 1994.
20. G. S. Hattiangdi, *The Vanaspati Industry*, Indian Central Oilseeds Committee, Hyderabad, 1958, Chap. 3, pp. 18–32.
21. K. S. Murti and K. T. Achaya, *Cottonseed Chemistry and Technology*, Publications and Information Directorate, Council of Scientific and Industrial Research, New Delhi, 1975, repr. 1984.
22. T. N. Murthi, National Dairy Development Board, Anand, Personal communication, 1994.
23. K. Chandrasekaran, Brooke Bond Lipton India Ltd., Bangalore, Personal communication, 1994.
24. R. S. Vaidyanathan, Rasoi Ltd., Calcutta, Personal communication, 1994.
25. K. S. Holla, Tata Oil Mills Co. Ltd., Bombay, Personal communication, 1994.
26. S. Gupta and R. Somnath, Masterline Bakery Service, Brooke Bond Lipton India Ltd., Bangalore, Personal communication, 1994.
27. A. P. Mahadevan, National Dairy Development Board, Anand, Personal communication, 1994.
28. J. K. Khaitan, Amrit Banaspati Co. Ltd., Rajpura, Personal communication, 1994.
29. M. M. Chakrabarty, Calcutta, Personal communication, 1994.
30. S. C. Jain and O. P. Narula, Hydrogen through water electrolysis, in *Treatise on Fats, Fatty Acids and Oleochemicals* (O. P. Narula, Gen. Ed.), Industrial Consultants (India), New Delhi, 1994, Chap. I-1.
31. A. K. Mehra and R. Viswanath, Hydrogen production processes—a review, in *Treatise on Fats, Fatty Acids and Oleochemicals* (O. P. Narula, Gen. Ed.), Industrial Consultants (India), New Delhi, 1994, Chap. I-3.
32. O. P. Narula, Economics and practice of oil bleaching, in *Treatise on Fats, Fatty Acids and Oleochemicals* (O. P. Narula, Gen. Ed.), Industrial Consultants (India), New Delhi, 1994, Chap. F-3.
33. R. K. Khullar, Energy conservation in vanaspati industry, in *Treatise on Fats, Fatty Acids and Oleochemicals* (O. P. Narula, Gen. Ed.), Industrial Consultants (India), New Delhi, 1994, Chap. S-1.
34. O. P. Narula and R. K. Chopra, Heat recovery and energy savings in oils and fats plants, in *Treatise on Fats, Fatty Acids and Oleochemicals* (O. P. Narula, Gen. Ed.), Industrial Consultants (India), New Delhi, 1994, Chap. S-2.
35. R. K. Suri and A. Mehrotra, Environmental protection and pollution control in oil processing industry, in *Treatise on Fats, Fatty Acids and Oleochemicals* (O. P. Narula, Gen. Ed.), Industrial Consultants (India), New Delhi, 1994, Chap. T-3.
36. K. C. Mathur and Surinder Kumar, Effluent treatment in oil processing industry, in *Treatise on Fats, Fatty Acids and Oleochemicals* (O. P. Narula, Gen. Ed.), Industrial Consultants (India), New Delhi, 1994, Chap. T-4.
37. M. S. A. Kheiri, *A Survey of Indian and Pakistani Vanaspati Products*, Palm Oil Res. Inst. of Malaysia, Kuala Lumpur, August 1982.
38. N. Nasir and K. V. Nagaraja, Loss of vitamin A in vanaspati under different storage conditions, *J. Oil Technol. Assoc. India 19*:44 (1987).
39. I. S. Shenolikar, Nickel content of vanaspati, *J. Oil Technol. Assoc. India 23*:48 (1991).
40. T. K. Roy and S. K. Banerjee, Margarine manufacture—technology and plant design, in *Treatise on Fats, Fatty Acids and Oleochemicals* (O. P. Narula, Gen. Ed.), Industrial Consultants (India), New Delhi, 1994, Chap. N-1.

15

Chocolate and Confectionery Fats

Fred B. Padley
Bedford, Bedfordshire, England, Consultant, Loders Croklaan bv., Division of Quest International, Wormerveer, The Netherlands

I. INTRODUCTION

Fats are widely used in the confectionery industry mainly to convey specific textural attributes and as a vehicle for fat-soluble flavors. The application that has received most scientific attention has been in the area of chocolate and chocolate substitutes for ambient confectionery. The major part of this chapter therefore focuses on the use of fats in chocolate, although mention is made of other relevant applications including cream fillings, toffee, and ice cream couvertures.

It is interesting to note that of all the edible oils and fats, cocoa butter has probably received the most detailed examination, no doubt because it combines scientific interests in relating physical behavior with chemical composition. For this reason cocoa butter, cocoa butter equivalents, and substitute fats are discussed in some detail in this chapter, but first a short description of the major confectionery product, chocolate.

II. CHOCOLATE

A. Chocolate Composition and Manufacture

The main component of chocolate is derived from the cocoa bean, whose continuing attraction is due to its flavor and melting characteristics. The Aztecs of South America created a spiced drink from the cocoa bean called *chocolatl* and believed it to have aphrodisiac properties, an image continued today with the romantic associations created by advertisers of chocolate.

The extraction of cocoa butter from the cocoa bean and the formulation of chocolate was first carried out in Holland by Van Houten in 1828. Milk chocolate was developed later by Peters of Vevey in Switzerland in 1876. These innovations subsequently led to the development of a major industry throughout the world including the grinding of 2.5×10^6 tonnes of cocoa beans and the production of approximately 4.5×10^6 tonnes of chocolate in 1995 [1].

Chocolate, as defined within the EU [2] is "the product obtained from cocoa nib, mass, powder and sucrose with or without added cocoa butter, having a minimum dry cocoa solids content of 35% and at least 14% of dry non-fat cocoa solids and 18% of cocoa butter." Other types of chocolate, including plain chocolate and milk chocolate, are also defined. For example, milk chocolate, among other requirements, must contain

A minimum of 25% total dry cocoa solids including at least 2.5% of dry non-fat cocoa solids
At least 14% of milk solids obtained by evaporation, including at least 3.5% of butterfat
Not more than 55% sucrose
At least 25% fat

This is, however, a simplification of what chocolate can contain, and the reader is recommended to read the appropriate legislation. Chocolate is essentially a mixture of finely ground cocoa beans and sugar plus milk powder in the case of milk chocolate. However, the use of only the cocoa beans with sugar creates a dry, powdery product because there is insufficient fat in the mixture to create a truly fat-continuous product. To create a product with the right texture and mouthfeel it is essential to add fat. This additional fat is obtained from additional cocoa beans by either pressing (expelling) or extraction. The addition of extra cocoa butter at elevated temperatures creates a fluid mass that can be used for molding or enrobing, the cocoa butter being essentially solid at ambient temperature.

Some typical chocolate compositions are given in Table 1. The residual cocoa powder left after fat extraction is used to formulate chocolate or cocoa drinks much lower in fat than those prepared by the Aztecs. Cocoa powder is widely used in formulating chocolate substitutes in which the expensive cocoa butter is replaced by cheaper sharp melting fats (see later).

Chocolate technology has been described in a number of textbooks, notably those of Mini-

Table 1 Some Chocolate Formulations

	Milk chocolate				
	High (%)	Average (%)	Low (%)	milk crumb (%)	Dark chocolate (%)
Cocoa mass	14.00	11.78	10	12.00	39.62
Whole milk powder[a]	23.60	19.08	18		
Whole milk (roller)					
Whole milk (spray)					
Enzyme-treated (WMP)					
Sugar	46.00	48.73	55		48.08
Added cocoa butter	16.00	19.98	16.50	18.00	11.75
Lecithin	0.40	0.35	0.38		0.35
Milk crumb[b]				70.00	
Salt					0.06
Vanillin		0.08	0.06		0.14
Total	100.00	100.00	100.00	100.00	100.00
Total fat	30.00	31.50	27.00	31.32	31.1

[a]Whole milk powder is sometimes substituted with skimmed milk powder and butter oil.
[b]Milk crumb FCM (Full Cream Milk) solids 37%, sugar 54%, cocoa mass solids 8%, water 1%, total 100%, total fat 13.32%.
Source: Ref. 139.

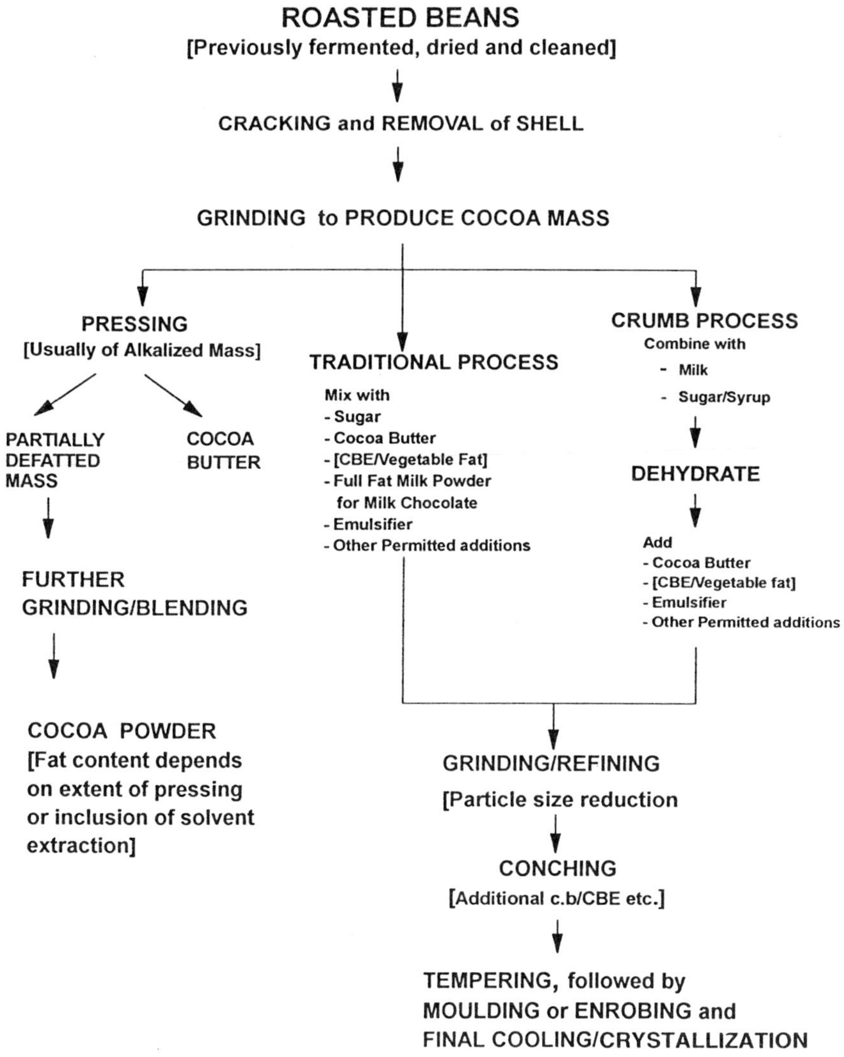

Fig. 1 Traditional chocolate processing. (Adapted from Beckett [4].)

fie [3] and Beckett [4]. An overview of the process is given in Fig. 1. The key features of traditional chocolate making are

Flavor development of beans by fermentation and roasting
Separation of cocoa shell from the bean to give cocoa nib
Grinding cocoa nib to give cocoa mass
Pressing cocoa mass to obtain cocoa butter
Mixing cocoa mass, sugar, cocoa butter (and milk powder for milk chocolate), and emulsifier
Grinding and conching to reduce particle size to less than approximately 25 μm
Conching to remove acidic volatile flavors and to modify flavor

Tempering to achieve the appropriate crystal nuclei
Molding or enrobing
Packaging and long-term storage

There are, of course, variants of this, in particular the so-called crumb process, which was developed in the United Kingdom and is now used in many other countries. In this case whole milk is dehydrated under vacuum at an elevated temperature in the presence of either sugar to give white crumb or sugar and cocoa mass to give chocolate crumb. Processing in this way gives a unique partially caramelized, chocolate flavored product. The crumb is then refined down to a given particle size. Each stage is critical in providing the customer with an acceptable product. The fat component plays a greater or lesser part at different process stages. The key aspects of chocolate technology where fat plays a major role are in influencing the viscosity of molten and tempered chocolate, the nucleation and crystallization behavior during tempering, molding, enrobing, storage, and the dependence on fat crystal stability for adequate shelf life. Finally the melting behavior of the stabilized fat phase and the overall flavor perception depend to a large extent on the blends of fats used. The wider aspects of chocolate technology are therefore not covered here, and the reader is referred to more comprehensive texts as mentioned above. The reader is also referred to specialist papers on chocolate flavor [5], quality control of chocolate [6], cocoa beans [7], chocolate manufacture [8], and cocoa powder [9]. The areas where fat has a major role to play are described below.

B. Chocolate Rheology

Chocolate is a dispersion of finely divided sugar, protein, and cocoa particles in a continuous fat phase. The particle size distribution is normally such that <20% of the particles are greater than 20 μm to ensure that the product texture is acceptable and does not taste grainy [10,11]. Some typical ranges of particle size found in some commercial chocolates are given in Table 2. Chocolate rheology is normally characterized using a range of instruments including viscometers chosen from various types, e.g., falling ball, cone-and-plate, and rotating cylinder. Tensile strength is also measured using appropriate equipment, e.g., Instron or Bohlin. Molten chocolate is a non-Newtonian fluid, that is, a finite force (yield value) is required before the chocolate flows, and there is not a linear relationship between shear rate and shear stress. The interpretation of shear rate–shear stress curves has been studied by a number of workers. Steiner [12] and NCA [13] proposed that an approach devised by Casson for paint and printing ink was also very useful for chocolate, and this has been widely adopted by the industry.

Table 2 Particle Size Distribution of a Range of Milk chocolate masse (lower milk content)

Size of sieve (μm)	Average residues (% by weight)		
	USA (ref.)	UK	Mainland Europe
>800	0.003	0.002	
>500	0.025	0.015	0.010
>400	0.050	0.030	0.020
>300	0.200	0.090	0.070
>200	0.500	0.450	0.400
>100	10.000	9.000	10.000

Source: Ref. 11.

The Casson equation relates the square roots of shear rate (D) and shear stress (τ) in the following way:

$$\sqrt{\tau}=K_0+K_I\sqrt{D}$$

where τ is the shear stress, D the shear rate, and K_0^2 the yield value.

In the case of rotational viscometers, corrections are made depending on the geometry of the cylinders used [12,13], giving the equation

$$(1+a)\sqrt{D_N}=\frac{1}{K_1}[(1+a)\sqrt{\tau}-2K_0]$$

where

$$D_N=\frac{2WR^2}{R^2-r^2}=\frac{2W}{1-a^2}$$

$$a=\frac{\text{radius of inner cylinder } (r)}{\text{radius of outer cylinder } (R)}$$

W = angular velocity

N = revolutions per minute

From this relationship $(1/K_1)^2$ = plastic viscosity and K_0^2 = yield value (Casson yield value).

The rheology of molten chocolate is dependent on the relative proportions of fat and dispersed solid phase (Fig. 2) [14]. It is also very significantly influenced by the addition of certain emulsifiers or, more correctly in this case, surface-active agents such as lecithin and polyglycerol esters. Both influence viscosity; polyglycerol esters in particular greatly reduce the yield value. Their effect is explained on the basis that the emulsifier coats the sugar

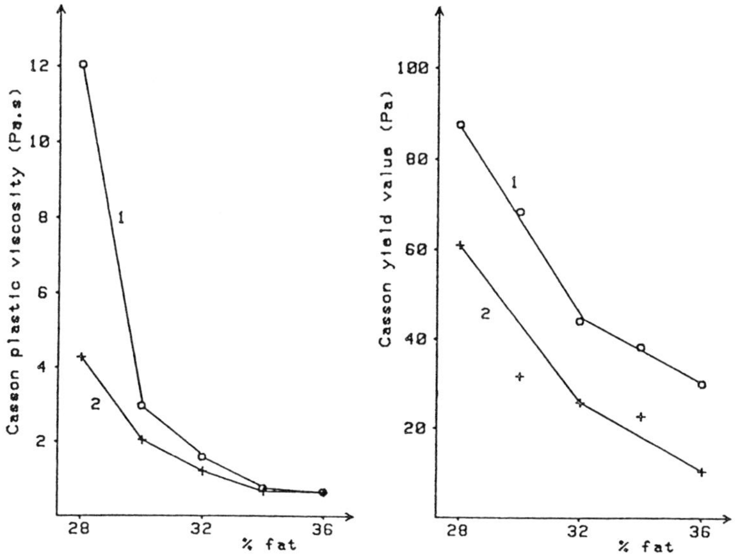

Fig. 2 The effect of fat content on chocolate rheology. Graphs depict the Casson parameters of two milk chocolates with 0.25% lecithin. 1, Fine chocolate with 5.7% particles >20 μm; 2, moderately coarse chocolate with 16% particles >20 μm. (According to Chevalley [14].)

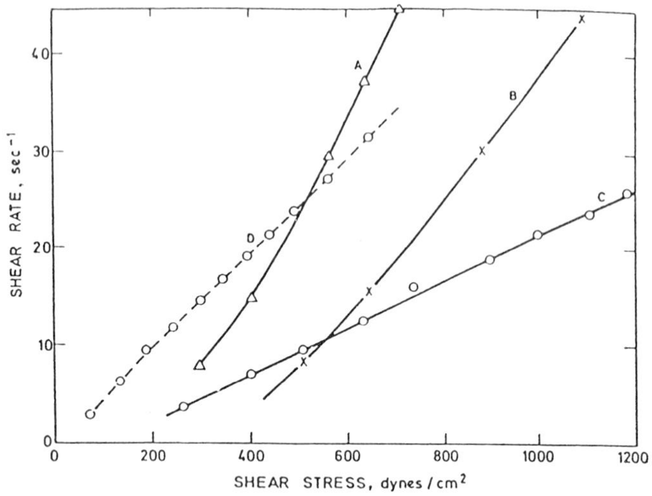

Fig. 3 The effect of surface-active lipids on chocolate depicting plastic flow and yield value at 50°C. A (△), with 0.3% soy lecithin; B (x), with 0.3% sucrose dipalmitate; C (o), with 0.3% polyglycerol polyricinoleate; D (o), with 0.8% polyglycerol polyricinoleate. (From Ref. 15.)

particles, preventing flocculation and thus minimizing particle–particle interaction. The influence of emulsifiers on chocolate rheology is demonstrated in Fig. 3 [15,16]. The use of emulsifiers therefore provides the manufacturer with a method of influencing chocolate viscosity in an advantageous way that also allows a reduced fat level to be used to achieve a given rheology. The reduction in yield value, especially by polyglycerol polyricinoleate, is important in aiding the removal of air during the molding process and in assisting some enrobing processes. Note that viscosity is recorded in pascal-seconds (1 Pȧs = 10 poise) and yield value in pascals (1 Pa = 10 dyn/cm^2).

Some typical viscosities of molten and tempered chocolate are given in Table 3.

C. Chocolate Tempering

The need for a chocolate tempering stage arises from the complex polymorphic behavior of cocoa butter. Six polymorphs numbered I to VI have been described by Wille and Lutton [17]. The most stable, forms V and VI, are β polymorphs, also designated β_2 and β_1, respectively [18–20].

Table 3 Some Typical Ranges of Viscosity[a] and Yield Value of Chocolate

Application	Molten (40°C) plastic viscosity (Pa's)	Yield value (Pa)
Molding	3–12	3–12
Enrobing	1.5–2	10–30

[a]Viscosity at temper will increase by a factor of up to 2 depending on the fat content and degree of temper.
Source: Ref. 21.

Of the various confectionery fats used, only cocoa butter and cocoa butter equivalents exhibit this complex polymorphic behavior and therefore require tempering. Other fats used in substitute chocolate crystallize into an essentially stable β' form and do not need tempering. Products that have a stable melting behavior and retain good texture and surface gloss are produced when cocoa butter crystallizes in the β_2 form or form V (see Section V). The tempering process is used to generate a sufficient number of stable fat crystal nuclei onto which the bulk of the fat phase of the chocolate can crystallize in a form that remains stable over long periods of time. This is achieved by the tempering process in which the chocolate is partially crystallized initially in an unstable form. The temperature is then raised sufficiently to melt out all the unstable nuclei yet retain a small but sufficient level of stable form V, β_2 seed crystals. These promote the formation of form V crystals during the bulk crystallization of the fat phase when molded or enrobed chocolate is allowed to cool, normally with the assistance of a cooling tunnel. On the production scale, tempering is normally carried out using a multistage continuous tempering unit with a minimum of three stages—one for cooling, the second to generate fat crystals, and the third to raise the temperature of the chocolate, melting out all crystals that are not form V nucleators. Batch tempering units are also used for both production and laboratory-scale work. A small batch tempering kettle designed by the Leatherhead Food Research Association, England, can be used for experimental work. The various types of tempering machines are described in detail by Nelson [21].

The amount and type of fat in crystalline form at temper is open to some debate. Some authors [21] quote 2–4%; however, others suggest significantly less. In a detailed article Reade [22] discusses the solidification process of chocolate, relating the number of crystal nuclei to crystal growth rate and solidification. As little as 0.1–0.2% fat as seed nuclei would appear to be adequate for good temper. These apparently conflicting statements explain to a large extent why the viscosity increase at temper is also reported to be either very small or as much as twice that of molten chocolate.

The state of chocolate temper is normally determined using a tempermeter [6,23,24], which essentially determines the temperature of a standard weight of chocolate as it crystallizes when cooled in a controlled way. The Greer tempermeter is one such piece of apparatus. The interaction of the temperature on the different stages of the tempering unit with the flow rate of chocolate and chocolate temper has been studied in detail by Cebula et al. [25] using a laboratory-scale three-stage continuous temper machine. The interrelationship between the temperature of the first and third stages of a temper unit with the degree of chocolate temper is shown in Fig. 4. Essentially this demonstrates that a standard degree of chocolate temper can be achieved under a very wide variety of conditions. For a range of flow rates there is a range of conditions defined in terms of the temperatures of the different stages. However, it is only at the more elevated temperature of the final stage that the chocolate will remain in a relatively stable state of temper, with little change in viscosity, for some time. Tempered chocolate emerging from the unit at the lower temperatures will continue to crystallize. Information of this type can be used to relate results obtained by simple batch-scale laboratory experiments to larger scale continuous procedures.

D. Crystallization During Enrobing and Molding of Chocolate

An understanding of the role of crystal nuclei and subsequent crystal growth is important in the ultimate optimization of the tempering and heat transfer/cooling tunnel processes to achieve stable products with good gloss. This is still an area that has been relatively under-researched. For further discussions on nucleation, see publications by Larsson [19] and Cebula et al. [26].

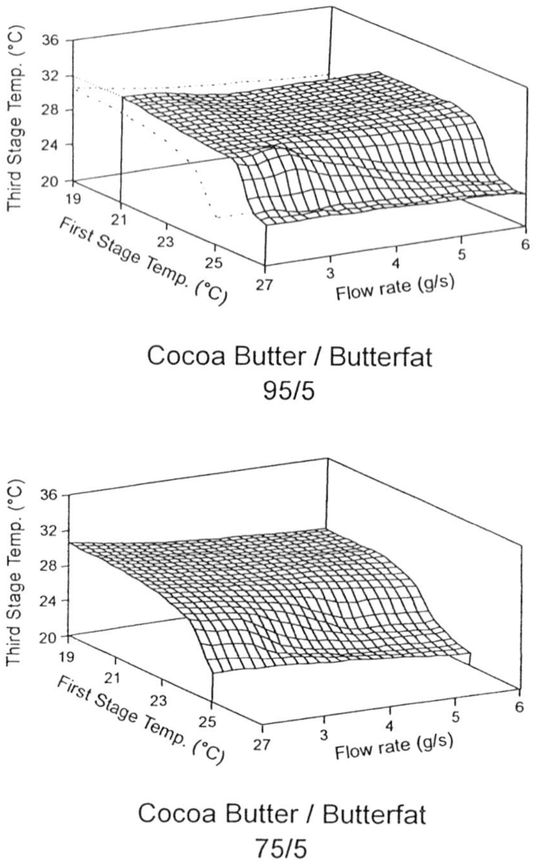

Fig. 4 The effect of fat phase on the continuous tempering performance of chocolate. Results obtained on a small-scale continuous three-stage unit. For simplicity only the temperatures of the first and third (last) stage are depicted. (According to Cebula et al. [25].) The "surface" indicates those conditions giving equivalent chocolate temper.

The role of the tempering process as indicated above is to generate seed crystals that on subsequent cooling of the semimolten mass promote the crystallization of the "stable" form V polymorph. The fundamental questions are, How many seed crystals? and What rate of subsequent cooling can be permitted? To a very large extent the answers are still arrived at empirically by each manufacturer by controlling temperatures and flow rates, etc., such that a satisfactory product emerges from the cooling tunnel. For example, very rapid undercooling of tempered chocolate may result in an initially very glossy product, but this will be relatively unstable, leading to loss of gloss and bloom.

A mathematical treatment [22] shows that the final crystal size is approximately 8 to 19 times the size of the original seed crystal assuming that the seed crystal occupies 0.2 to 0.01%, respectively, of the final volume of the solidified phase. To take this analysis further ideally requires information on the number and size of crystal nuclei and the linear growth rate of the crystals. However, in one example Reade [22] assumes that 0.01% of the fat is present as seed crystals with the dimensions in 10^{-3} μm of $10 \times 1 \times 1$. These would grow by approximately 19-fold to achieve complete solidification, producing final crystal sizes well below those that would interfere with gloss and dispersion of light (wavelength of light 500–700 μms $\times 10^{-3}$). Again, this analysis has largely ignored the role of other particulate material in chocolate and its possible involvement with crystallization.

Numerous types of molding and enrobing equipment are available [3,21,27]. The way in which chocolate is cooled to crystallize the bulk of the fat will depend on the type of chocolate being used. In general, tempered chocolate is not cooled too quickly during the first stage of the cooling tunnel. In the case of some of the nontempered substitute chocolates, the cooling tunnel conditions have to be tailored somewhat differently; e.g., for palmkernel stearines, relatively cool inlet temperatures of 12°C are used. Products emerging from the tunnel should be at a high enough temperature to prevent moisture condensation, which in extreme cases would promote sugar bloom. Residence times can be on the order of 10–30 min.

Chocolate used in molding processes needs to exhibit good contraction prior to demolding. This has been studied by a number of workers but is a difficult property to fully characterize [3]. In modern processing plants, the product is wrapped immediately after it emerges from the cooling tunnel. This demands that the chocolate be essentially solid and resistant to mechanical damage. Solid fat content is normally associated with a degree of hardness or firmness of a product. For fully stabilized products formulated using similar glyceride types this almost certainly occurs. However, during the solidification of chocolate this was observed not to be the case [28]. Simultaneous measurements of solid fat content by NMR and of hardness using a penetrometer were made on chocolate as it crystallized from the tempered state. This showed that the hardness and mechanical strength of the chocolate developed at a significantly later stage after most of the fat had been crystallized (Fig. 5). This is an intriguing observation when one considers that in addition to the crystallized cocoa butter there is already >60% dispersed solids in the chocolate. One explanation might be that large random agglomerates of solid sugar or cocoa particles and fat are initially formed that then are "cemented" together by the sintering of fat crystals. Further care must be taken during the enrobing process. The component to be enrobed should be at a sufficiently high temperature to prevent further supercooling of the chocolate and the crystallization of less stable polymorphic forms.

The only case where this is not adhered to is in ice cream technology where the chocolate is invariably nontempered. In this case the fat phase crystallizes in one of the less stable (e.g., form IV) states, which has a lower melting point and more acceptable mouthfeel when eaten as a frozen product. An alternative to conventional tempering and cooling to achieve the stable crystal form of cocoa butter is to use high pressure [29]. However, even in this work,

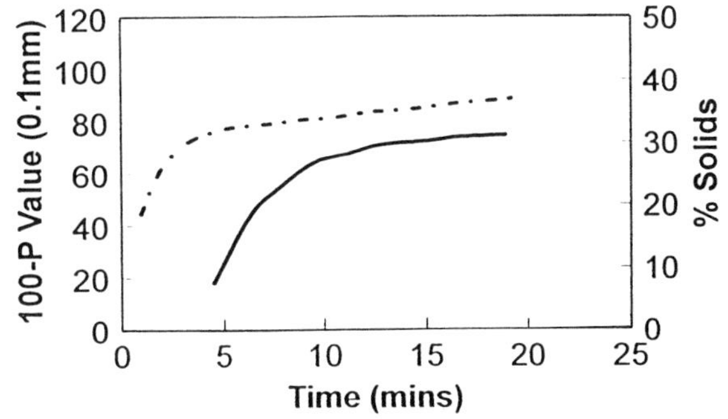

Fig. 5 The relationship between solid fat content and hardness of chocolate during crystallization in a cooling tunnel.(——)Hardness measured by penetrometry; (–·–·–) solid fat content measured by indirect NMR. (According to Cebula and Dilley, private communication, with permission of Loders Croklaan.)

some stable nuclei were added to the crystallizing fat. High pressure extrusion of "solid" chocolate has also been described and patented [30].

III. CONFECTIONERY FATS

A wide range of confectionery fats are used in the chocolate confectionery industry. However, for the uninitiated not all that appears to be chocolate is chocolate. For a product to be called chocolate it has to adhere to the definition as given under the legislation of the particular country concerned. Other products that do not conform to the legislation will not (should not) be labeled chocolate. These latter products are subsitute chocolates in which the main fat phase is based on palmkernel, coconut, or partially hydrogenated fats. The flavor and color is obtained normally from defatted cocoa powder, but some products, e.g., "supercoatings" based on cocoa butter equivalents (CBE) can be formulated using some cocoa mass.

As will be seen from the detail on individual fats, cocoa butter and CBE exhibit a more complex polymorphic behavior than palmkernel, coconut, or partially hydrogenated fats. For this reason, genuine chocolate requires a more complex method of crystallization, i.e., a tempering stage, than the β' stable chocolate substitute fats, which can be crystallized directly from the melt. These latter fats therefore provide the advantage of simpler processing and can also be tailored to have higher melting points more appropriate for warm climates despite the fact that the product cannot be called chocolate.

Whereas in the past there was a strong commercial interest to find alternatives to cocoa butter, there is now also, in Europe at least, a strong interest to define milk fat alternatives in those countries where vegetable fats are legally permitted at a defined low level, normally 5% in chocolate [2]. The main groups of fats used in chocolate confectionery are summarized in Table 4. This grouping is made mainly on the basis of the chemistry of the different fats that are used. The simplest in chemical terms are cocoa butter and cocoa butter equivalents; the most complex is probably butterfat. Perhaps surprisingly, the difference in chemical composition is also reflected in an inverse way in physical behavior; that is to say, the simpler the chemical composition the more complex the physical behavior. Each of these main groups is described in separate sections below.

Table 4 Chocolate Fat Types

Symmetrical SOS type	Lauric	Trans hardened oils	Butterfat
Cocoa butter	Palmkernel (PK)	Midfractions and nonfractionated *trans* hardened liquid oils	and its fractions
Cocoa butter fractions	Stearins		
Cocoa butter equivalents	Hydrogenated PK stearins		
Illipe	Hydrogenated PKs		
Sal			
Kokum	Coconut oil and stearins		
Palm fractions			
Shea and enzyme equivalents			
β-Stable	β'-Stable	β'-Stable	β'-Stable
Require tempering	Non-temper	Non-temper	

Note: Liquid oils are also used in confectionery and behave physically as a solvent for higher melting fats and also as a carrier for flavor, e.g., hazelnut oil.

Table 5 Fatty Acids (wt %) of Cocoa Butter

Origin		16:0	18:0	18:1	18:2	Quotient (% 16:0)/(% 18:0)
Brazil	8	25.1	33.9	36.1	3.3	0.75
		(24.6–26.0)	(30.2–35.5)	(35.2–38.6)	(2.4–4.8)	(0.70–0.81)
Ecuador	6	26.3	35.4	33.9	2.7	0.75
		(26.0–27.4)	(34.8–35.8)	(33.5–34.3)	(2.3–3.0)	(0.73–0.79)
Venezuela	3	26.8	34.5	33.9	2.6	0.78
		(25.5–30.5)	(32.2–36.5)	(33.2–34.8)	(2.6–2.6)	(0.70–0.95)
Trinidad, Grenada, Jamaica	5	26.8	34.9	34.0	3.1	0.77
		(25.3–27.3)	(33.2–35.1)	(33.7–35.9)	(2.7–4.0)	(0.72–0.80)
Ghana	6	25.2	34.3	35.1	3.5	0.73
		(23.6–25.6)	(33.9–35.4)	(34.8–37.6)	(3.3–3.6)	(0.70–0.75)
Nigeria	4	25.8	35.4	33.9	3.3	0.74
		(25.5–26.2)	(33.9–35.7)	(33.7–35.1)	(2.9–3.8)	(0.71–0.76)
Ivory Coast	4	25.9	34.6	34.6	3.2	0.75
		(25.4–26.2)	(34.2–35.2)	(34.2–35.1)	(3.0–3.4)	(0.72–0.76)
Cameroon	3	25.9	34.3	34.8	3.4	0.75
		(24.0–26.3)	(34.2–34.6)	(34.8–36.0)	(3.2–4.1)	(0.69–0.76)
Togo	3	24.9	34.3	35.2	3.3	0.72
		(24.6–26.0)	(33.9–35.1)	(35.1–35.3)	(3.2–3.6)	(0.72–0.77)
Congo	1	25.0	34.2	35.6	3.4	0.73
St Thomas	1	25.3	34.2	35.2	3.4	0.74
Madagascar	1	25.7	36.1	34.0	2.4	0.71
New Guinea, Borneo	9	26.1	35.4	34.2	2.6	0.74
		(24.6–28.2)	(33.8–36.4)	(33.3–34.4)	(2.2–2.9)	(0.69–0.83)
Samoa, Java	3	26.1	34.6	34.6	3.0	0.75
		(25.8–27.8)	(33.2–34.6)	(34.4–34.7)	(2.7–3.2)	(0.75–0.84)

Note: Range in parentheses.
Source: Ref. 145.

In understanding the behavior of confectionery fats it is important to be able to relate physical properties to chemical composition. The chemistry of cocoa butter and cocoa butter equivalents (CBE) is relatively simple compared to that of other oils and fats. Much of what is described about cocoa butter also relates directly to CBE. Because cocoa butter is a natural product there is a natural variation in its composition depending on a range of factors including climate.

A. Cocoa Butter[1]

The chemical composition of cocoa butter can be described in a number of ways based either on individual glycerides, glycerides grouped according to degree of unsaturation characterized by silver-phase chromatography, or glycerides grouped according to carbon number as determined by high temperature gas-liquid chromatography [31–37]. Some typical data are given in Tables 5–9.

The chirality of POSt has been determined and is seen to be essentially racemic [38]. The major glycerides in cocoa butter are 1,3-distearoyl-2-oleoyl glycerol (StOSt), *r*-1-palmitoyl-2-

[1]See also Chapter 2.

Table 6 Fatty Acids (wt %) of Some Cocoa Butter Equivalent Components

Origin	16:0	18:0	18:1	18:2	Others
Palm fraction	56	6	32	4	2
Shea stearin	5	57	33	3	2
Illipé butter	18	46	35	1	tr
CBE	28	34	33	3	2

tr = trace.

Table 7 Triacylglycerol Composition of Cocoa Butters from Different Countries of Origin

Country of origin	n^a	Triacylglycerol (%)								
		PLP	POP	PLSt	POP	StOO	StLSt	POSt	StOSt	StOA
Bolivia	1	1.1	3.3	3.5	22.6	4.0	2.1	40.4	22.8	0.5
Brazil	6	0.9	3.9	3.7	17.9	6.7	3.2	37.1	26.0	0.04
Colombia	2	1.1	3.3	3.6	20.4	4.4	2.3	39.4	25.0	0.6
Ecuador	3	1.2	3.0	3.2	19.2	5.4	2.3	38.4	26.9	0.4
Peru	1	1.5	4.3	3.9	18.3	7.4	3.7	35.8	24.6	0.4
Venezuela	1	0.9	1.0	3.1	20.4	2.8	1.9	40.4	28.8	0.8
Costa Rica	1	1.0	2.6	3.5	17.8	5.5	3.0	38.7	27.4	0.4
Dominican Republic	4	1.1	3.3	3.2	18.4	6.1	2.7	38.2	26.5	0.6
Guatemala	1	1.0	2.3	3.4	19.3	4.9	2.2	39.0	27.5	0.4
Mexico	1	1.1	2.4	3.5	19.1	4.1	3.0	38.8	27.8	0.6
Panama	1	1.0	1.5	3.0	19.1	3.1	2.7	41.4	27.3	0.8
Cameroon	2	1.0	3.0	3.4	17.9	5.8	2.5	38.3	27.7	0.5
Gabon	1	0.9	3.7	3.5	17.5	7.3	3.0	37.1	26.5	0.4
Ghana	3	1.2	2.2	3.4	17.8	4.9	2.2	39.0	27.5	0.4
Ivory Coast	9	1.0	1.9	3.0	19.0	3.9	2.5	39.6	28.5	0.6
Nigeria	2	1.0	2.3	3.6	17.9	5.2	3.0	38.8	27.8	0.5
Indonesia	2	1.1	1.6	3.0	19.9	3.6	1.7	40.6	28.1	0.5
Malaysia	20	0.7	1.2	2.8	18.4	2.9	2.2	40.0	31.1	0.8
Solomon Islands	1	1.0	0.9	3.0	19.3	2.8	2.0	40.7	29.5	0.7

[a] n = number of samples.
P, palmitic (16:0); St, stearic (18:0); A, arachidic (20:0); O, oleic (18:1); L, linoleic (18:2). S, saturated.
Source: From Ref. 146.

Table 8 Triacylglycerol Composition of CBE and Components

	S_3	SOS	SSO	SLS	SOO	Others
Palm fraction	3	80	6	6	3	2
Shea stearin	2	83	—	7	5	3
Illipé	1	94	—	1	3	1
CBE	2	83	3	7	4	1
Brazil cocoa butter	1	83	—	5	9	2

Table 9 Triglyceride Composition of Cocoa Butter 1976–1979

	Triglyceride carbon number[a]					
Origin	C_{48}	C_{50}	C_{52}	C_{54}	C_{56}	C_{58}
Ghana	0.3	18.1	46.5	33.4	1.6	0.1
	0.4	18.9	46.6	32.4	1.5	0.2
Ivory Coast	0.4	16.9	46.3	35.0	1.6	0.2
	0.3	18.1	46.4	33.1	1.8	0.2
	0.2	17.9	46.3	34.1	1.6	0.1
Nigeria	0.4	18	46.6	33.4	1.4	0.1
	0.2	17.8	46.2	34.1	1.5	0.2
Cameroon	0.3	17.9	46.1	33.6	1.8	0.2
Togo	0.2	18.3	46.3	33.4	1.7	0.1
Bahia	0.3	17.5	45.7	34.5	1.7	0.2
	0.3	15.2	44.6	38.0	1.7	0.2
	0.3	17.4	46.2	34.7	1.6	0.1
Ecuador	0.4	19.4	46.2	31.5	2.1	0.3
	0.3	19.8	46.7	31.4	1.8	0.1
Venezuela	0.3	22.7	47.3	27.1	2.2	0.4
	0.4	19.4	46.9	31.2	1.7	0.2
	0.5	19.6	46.4	31.9	1.5	0.2
Trinidad	0.3	21.2	47.0	30.0	1.3	0.2
Grenada	0.4	18.4	46.4	32.5	1.7	0.2
Dominican Republic	0.2	17.7	46.0	34.5	1.5	0.1
Papua New Guinea	0.4	18.9	46.5	32.3	1.6	0.2
	0.2	18.0	46.3	33.8	1.6	0.1
	0.3	16.7	45.9	35.5	1.6	0.2
Malaysia	0.3	18.1	45.8	33.7	1.8	0.2
	0.3	17.0	45.8	34.8	1.9	0.2
	0.2	17.6	46.0	34.5	1.6	0.1
Sri Lanka	0.2	17.9	46.3	34.0	1.5	0.1
Indonesia	0.2	18.2	46.2	33.6	1.6	0.2
	0.3	17.8	46.0	34.1	1.6	0.2
Samoa	0.4	17.2	46.1	34.6	1.7	0.2

[a]Carbon number is the number of carbon atoms in the fatty acid chains of the triglyceride, e.g., C_{48} is all triglycerides in which the fatty acid carbon atoms add up to 48. A number of samples have been analyzed from some sources.
Source: Ref. 129.

oleoyl-3-stearoyl glycerol (POSt), and 1,3-dipalmitoyl-2-oleoyl glycerol (POP). These glycerides largely dictate the physical behavior of cocoa butter, but other minor glycerides, e.g., diglyceride and trisaturated glycerides, can influence melting and crystallization behavior [25,39]. The polymorphism of the major cocoa butter triglycerides and of cocoa butter has been described by Wille and Lutton [17] and more recently by Yano et al. [41]. In a recent review, Sato [20] brought all the relevant information together. Melting point and X-ray diffraction data of the triglyceride polymorphs are given in Table 10. The effect of liquid oil on the crystallization behavior of cocoa butter has also been described [42].

The chocolate processing stages of both tempering and cooling tunnel crystallization aim to produce chocolate crystallized in form V (β_2). This is stable for most practical purposes and ensures that the chocolate retains its gloss. Under conditions of high ambient temperature

Table 10 Polymorphic Forms and Melting Points of StOSt, POSt, and POP

	StOSt						
Property	α	γ	β'	β_2	β_1		
LS (nm)	4.83	7.05	7.00	6.50	6.50		
T_m (°C)	23.5	35.4	36.5	41.0	43.0		
	POSt						
	α	δ	β'	β			
LS (nm)	4.76	7.06	6.79	6.40			
T_m (°C)	19.5	28.3	31.6	35.5			
	POP						
	α	γ	δ	β'_2	β'_1	β_2	β_1
LS (nm)	4.65	6.54	6.25	4.24	4.24	6.10	6.10
T_m (°C)	15.2	27.0	29.2	30.3	33.5	35.1	36.7
Filer et al.	—	—	—	β'	β'	β	β
Malkin and Wilson	α	β''	—	β'	β	β	
Lutton and Jackson	α-2	—	subβ-2	β'-2	β'-2	β-3	β-3
Lavery	α-2	β''	—	β'-2	β'-2	β-3	β-3
Lovegren et al.	5	4.3	—	2	2	1	1
Gibon et al.	α-2	L_2L_3	subβ-2	pseudo-β'-2		β-3	β-3

For original references refer to Sato [20].
Source: Ref. 20.

(>25°C, say), the metastable form V transforms in a matter of weeks to form VI (β_1). This transformation is associated with the onset of bloom. This aspect of chocolate behavior is discussed later.

The stabilized solid content for the ternary mixture POP–POSt–StOSt is depicted in an isosolids diagram [40] (Fig. 6). This illustrates that the melting behavior of cocoa butter will be reduced with increasing POP contents and raised with increasing StOSt.

Recent work by Savage and Dimick [43] shows that phospholipids influence the nucleation and crystallization behavior of cocoa butter. In particular, an increase in phosphatidylcholine content increased nucleation and crystal growth. The converse was reported for lysophosphatidylcholine and phosphatidylinositol. In almost simultaneous publications from the same group [44,45], however, the triglyceride composition is observed to have an overriding effect, fast crystallizing cocoa butters being characterized as having high 2-oleodistearin (StOSt) contents. This is in line with the generally acknowledged relationship between composition and crystallization behavior [33,46]; i.e., increased StOSt content raises the melting point and the crystallization rate at a given temperature as a consequence of the higher degree of supersaturation. Taking all these data into account, they suggest that phospholipids play a secondary role in influencing crystallization behavior, the major influences being triglyceride and partial glyceride composition. In relation to phospholipids one should also note that the Dimick and coworker studies were on cocoa butter alone. The real-life situation is concerned with chocolate, which will contain added lecithin, a significant part of which will almost certainly have adsorbed onto the sugar particles.

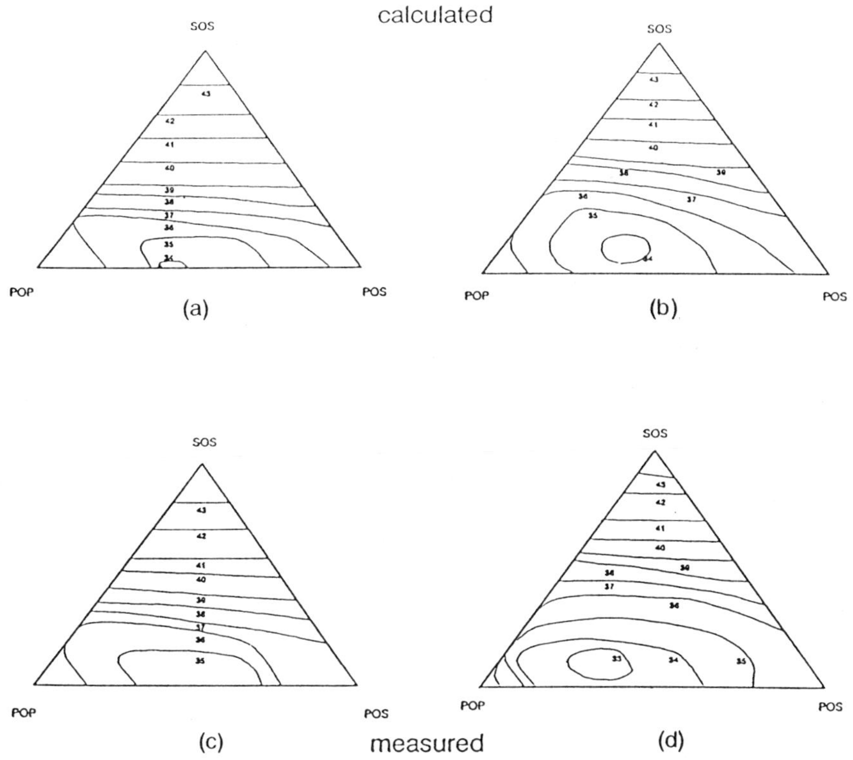

Fig. 6 Ternary phase diagram for a ternary mixture of POP, POSt, and StOSt. Calculated and measured ternary isosolids diagrams of SOS–POS–POP. (a, c) Clear point diagrams; (b, d) isotherms with 25% solid fat. The PPP–POP–POO and PPP–PPO–POO ternaries also show good agreement between measurements and calculations. The agreement near the PPP corner at the PPP–POO side of the diagrams is somewhat less good. (From Ref. 40.) (Measured data K. Smith).

Perhaps the main point to note here is that cocoa butter is a natural product and exhibits a significant natural variability. The sources of cocoa and cocoa butter are now extremely numerous, and as a consequence of different climatic and growing conditions the glyceride content and hence the melting behavior of cocoa butter can vary quite widely [46] (Table 11). This was demonstrated quite elegantly by Lehrian and Keeney [47]. (See also Section VII.) In addition, the various grades of cocoa butter also depend on the quality of the beans (FFA, etc.) and on

Table 11 Solid Fat Content (%) by NMR of Three Cocoa Butters and Fractions[a]

	Brazilian	Malaysian	West African	West African stearin	West African olein
N20	66	81	76	91	48
N25	60	76	70	95	36
N30	37	55	45	73	0.0
N35	0.0	0.0	0.0	16	0.0

[a]Samples stabilized at 26°C then 0°C. N20 = NMR solids at 20°C.

how the cocoa butter has been extracted from the bean—by hydraulic pressing or via solvent extraction. There is also the potential contamination of the pure cocoa butter with cocoa shell fat [6,48,49]. Shell and germ fats are normally considered waste fats by responsible manufacturers and are not permitted in pressed cocoa butter as legally defined. Shell fat is more unsaturated than cocoa butter, containing 10–12% linoleic acid (18:2) and a number of unique compounds such as alkyl tryptamides, 5-hydroxytryptamides, and fluorescent sterol derivatives [49].

Fractions of cocoa butter (Table 11) have also been developed and offer the chocolate manufacturer another degree of freedom in controlling chocolate properties and in optimizing formulations [50]. The properties of blends of cocoa butter stearins and oleins with unfractionated cocoa butter and milk fat are described. The stearins can be used to raise the melting behavior of cocoa butter, creating more heat-resistant products. Both stearin and olein can be used to provide greater flexibility in using milk fat [50].

B. Cocoa Butter Equivalents

The relatively high price of cocoa butter stimulated the invention of cocoa butter equivalents [51]. These products were developed at a time when the chemistry of fats was still poorly understood and without the advantage of chromatographic techniques. Nevertheless, it proved possible to mimic the chemical composition and physical properties of cocoa butter extremely well by blending together fractions of the required triglycerides from different oils and fats. This invention by Crossley and coworkers transformed the world of chocolate confectionery in a number of countries and acted as a catalyst for later technical achievements in the area (Fig. 7).

Palm oil and shea nut oil are fractionated from solvent, usually acetone, to obtain fractions with the necessary purity and with the right melting and crystallization behavior. In the case of palm oil fractions, increasing interest is being shown in the use of palm fractions obtained by dry fractionation, especially for the formulation of milk fat replacers in chocolate. In all these fractionation processes, significant amounts of by-product fractions, i.e., high melting fraction, mainly tripalmitin and a liquid olein fraction, are produced. The viability of this technology therefore depends as much on their economic disposal as on the value of the CBE. The various aspects of fat fractionation are covered in Chapter 8.

The control of palm oil quality as a raw material feedstock is very important in CBE applications [52]. The fractionation conditions and choice of solvent are also important in determining the levels of diglyceride and tripalmitin in the midfraction rich in 2-oleodipalmitin (POP) [33]. Alternative sources of cocoa butter triglycerides are given in Table 12. Kokum, illipé, and to some extent sal can be used without resorting to fractionation. All the fats apart

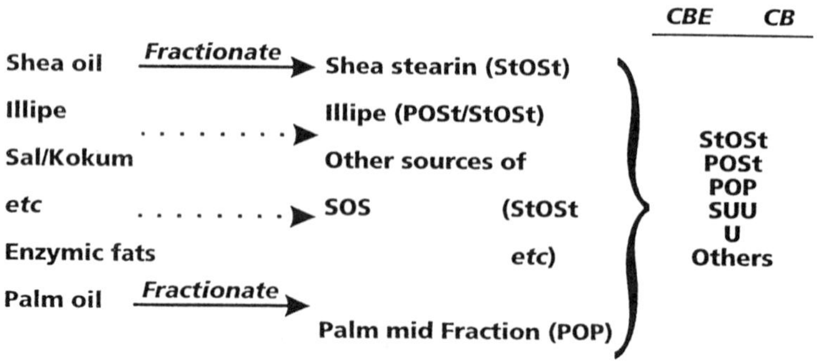

Fig. 7 Simplified scheme illustrating CBE formulation from oils and fats and their fractions.

Table 12 The 2-Position and Overall Fatty Acid Composition of Some SUS-Rich Fats

	FAC: 2-position						FAC: Overall					
	16:0	18:0	18:1	18:2	18:3	20:0	16:0	18:0	18:1	18:2	18:3	20:0
Cocoa butter	2	2	85	11	—	0	27	33	35	3	—	2
Aceituno acid	2	4	89	4	—	0	10	28	57	3	—	1
Sal fat	0.6	4.4	94.7	—	—	0.3	7.1	44.8	42.5	—	—	5.6
Mowrah tallow	3.2	3.4	57.0	35.0	1.4	0	23.1	22.0	38.5	15.8	tr	—
Malabar tallow	1.1	3.9	90.7	3.9	0.4	0	9.0	46.9	41.4	1.3	tr	—
Kokum tallow	—	4.4	94.4	1.3	—	0	1.7	61.0	36.8	0.5	—	1.4
Palm oil	11	2	65	22	—	—	45	6	40	9	—	—
Shea butter	—	3.5	81.4	15.1	—	0	3.4	41.1	47.1	6.6	—	—
Allanblackia (Dunkwa)	2	2	96	—	—	—	2.0	53	45	—	—	—
Chinese vegetable tallow	11	0.5	84.5	4.0	0	0	65.0	2.0	32.0	1.0	tr	—
Illipe butter	2.3	2.1	94	1.6	—	0	17.0	46.0	35	1.0	—	2

tr = trace.
Source: Ref. 76.

from palm oil (source of POP) are uncultivated and have to be harvested by local people retrieving nuts from either native forest trees or natural groves. Illipé, shea, and sal are the major sources of POSt and StOSt for the confectionery industry. Because of the diverse conditions associated with the harvesting and storage of the nuts, the quality and availability of these oils is very variable [53,54]. Free fatty acid content can be very high, even up to 30%. Diglycerides, an initial by-product of the hydrolyses, can be present at 10% or more in the neutral fat. On the other hand, if good husbandry is followed from collection through intermediate storage, shipment, and subsequent storage prior to extraction, then oils with low free fatty acids and diglycerides can be obtained. Diglycerides have particular relevance to CBE manufacture because their presence lowers melting point and retards crystallization [25,39].

In the case of sal, further complications can arise through microbial attack on the seed oil during harvesting and storage. This introduces a range of modified glycerides in which the oleic acid has been converted first to an epoxide, via ring opening to a dihydroxy fatty acid and by partial esterification of the hydroxy acid to an estolide containing four fatty acids. The triglycerides containing epoxy, dihydroxy, and estolide components are designated P, Q, and R, respectively [55,56].

Chinese vegetable tallow (*Sapium sebiferum*), the outer coating of the Chinese vegetable tallow seed, has been proposed as a rich source of pure POP. Examination of the tallow and the seed has, however, demonstrated that there is a small but significant level of stillingia oil contamination. This arises from the migration of stillingia oil from the center kernel to the outer coating [57]. The presence of toxic phorbol esters in the seed has also been described [58].

The shortage of natural sources of StOSt and POSt stimulated research to seek alternatives. In the 1960s and 1970s synthetic routes were explored that enabled the individual contributions of the major glycerides to be defined [46,59]. No commercial exploitation of this technology occurred. Fermentation has also been explored in detail but again without commercial success [60].

More recently a completely new technology has been developed, enzymatic interesterification [61–65]. This uses 1,3-specific lipases to catalyze the interesterification of, for example, high oleic sunflower oil with stearic acid to generate 2-oleodistearin (StOSt):

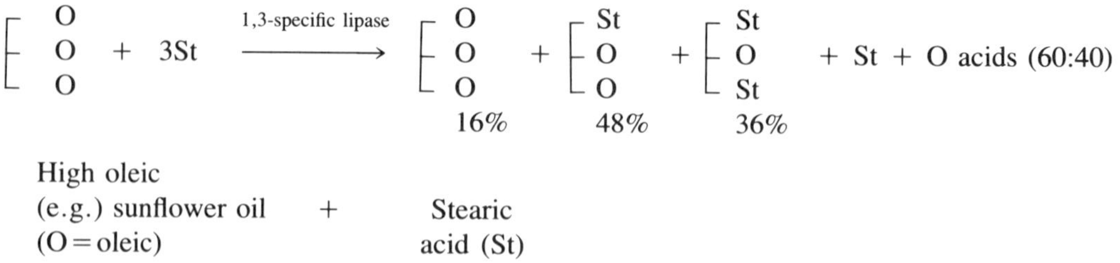

The advantage of such a process is that it utilizes readily available raw materials and very mild interesterification conditions that are lipase-catalyzed. Commercial implementation relies on a number of problems being resolved, notably the cost of the catalyst, including catalyst life and the use of by-products. The by-products, some not shown in the above simplified equations, include diglycerides generated by lipolysis. This is an undesirable but inevitable by-product that arises because the lipase requires a finite amount of water for it to exhibit catalytic activity. Diglyceride levels can be controlled, however, by the appropriate choice of process and process conditions. From this new technology have emerged a range of other applications that are being exploited commercially [64]. That technology is covered in detail in Chapter 9.

Cocoa butter equivalents can be tailored to meet a variety of requirements. Those containing higher amounts (≥40%) of StOSt and POSt will be more expensive but much closer in physical behavior to cocoa butter than those rich in palm midfraction (POP). Indeed, those rich in StOSt and POSt (>60–70%) will raise the melting range, thus improving the melting and crystallization behavior of softer, inferior cocoa butters when used at the 15–20% level of the fat phase. The use of high palm fraction CBE will, for example, lower the tempering temperature and extend crystallization times to some degree. CBE formulated to closely match cocoa butter have virtually no effect on the melting properties when blended with cocoa butter chocolate, and tempering will also be essentially the same. The chemical and physical properties of some typical CBE and of cocoa butter are given in Tables 5–10. The phase behavior of cocoa butter–CBE mixtures has been described [66]. The isosolids ternary diagram of StOSt, POSt, and POP given in Fig. 6 provides some guide to the effects of these major glycerides on the physical behavior of cocoa butter–CBE mixtures.

The similar chemistry of cocoa butter and CBE has led to the wider use of CBE in so-called supercoatings. This takes advantage of the total compatibility of the two fats and permits the formulation of coatings containing high levels of cocoa mass, thus ensuring that the flavor and meltdown of the product is extremely close to that of real chocolate. Some typical formulations are given in Table 13 [67]. CBE and vegetable fats are legally allowed at up to the 5% level in chocolate in the United Kingdom, Ireland, and Denmark and a number of other European and non-European countries. The EU debates this issue periodically, and a wider acceptance may occur eventually.

One of the issues that has arisen from time to time in this debate has been the inability to determine the level of vegetable fat addition, particularly CBE, because of their close similarity to cocoa butter. This problem is discussed in Section VII.

C. Milk Fat[2]

The fatty acid composition of milk fat is given in Chapter 2. From the confectionery viewpoint we are primarily concerned with the physical behavior of milk fat, its fractions, and blends

[2] See also Chapter 2.

of milk fat with cocoa butter [68,69]. Some typical solid fat content–temperature relationships are shown in Table 14 and Fig. 8. When determining the solid fat content of cocoa butter–milk fat blends, care should be taken to adopt appropriate stabilization procedures [70]. The important features to note are that butterfat is mainly liquid at temperatures between 20 and 35°C. The liquid oil component therefore acts as a solvent, leading to a significant melting point depression when mixed with cocoa butter simply due to cocoa butter dissolving in the liquid component of milk fat. The solid glyceride components, particularly those isolated in the midfraction, also depress the melting range of cocoa butter and form a eutectic mixture [69,71].

Milk chocolate therefore has a significantly softer texture than plain chocolate depending on the level of milk fat added. This influences not only the texture but also the processing behavior of the chocolate. Tempering performance is significantly affected as illustrated in Fig. 4, which depicts the interrelationship between the temperatures of the first and third stages of a three-stage pilot-scale temper unit while still retaining chocolate in temper as it emerges from the third stage. Twenty percent milk fat in the fat phase lowers the tempering temperature by approximately 2°C [25].

It has also been observed over many years that the addition of milk fat to chocolate inhibits bloom formation by delaying the form V to form VI transition (see later). Milk fat is added to chocolate in a variety of ways, directly as milk fat or full cream milk powder or even as whole milk, which is dehydrated during the crumb process. Depending on the method of addition and the degree to which the chocolate is refined, all or only part of the milk fat will be released from the milk fat globule membrane and will form part of the continuous fat phase [72]. This is relevant when trying to optimize fat content, the most expensive major ingredient, and viscosity.

Table 13 Composition (%) of Coatings Using CBE and Cocoa Butter Substitutes

Coating	Cocoa mass (10–12%)	Cocoa powder	Full cream milk powder	Skim milk powder	Coating fat	Sugar	% Fat overall
Lauric[a]							
Dark	—	14		7	31	48	32
Milk A	—	5	10	10	29	46	32
Milk B	—	5	0	17.5	31.5	46	32
Non-lauric[b]							
Dark	—	20	—	—	33	47	35
Milk A	10	—	6	12	28	44	35
Milk B	—	5	—	17	34	44	34.5
Supercoating[c]							
Dark A	40	—	—	—	10	50	32
Dark B	20	10	—	—	20	50	32
Milk A	10	—	20	—	22	48	32.5
Milk B	10	—	—	15	27	48	32

[a]Palmkernel stearin based (non-temper). Cocoa powder should contain as low a level as possible of cocoa butter.
[b]Trans hardened fats and fractions (non-temper).
[c]Cocoa butter equivalent (temper).
Note: Overall fat content will vary depending on the particular application in terms of viscosity/weight control.
Source: Ref. 139.

Table 14 Solid Fat Content (%) of Cocoa Butter, Milk Fat, and High, Middle, and Low Melting Fractions of Milk Fat[a]

Temp. (°C)	Cocoa butter	Milk fat	HMF	MMF	LMF
0	86.0	56.8	99.7	92.1	26.1
10	81.9	42.0	99.5	90.5	11.6
15	79.1	30.2		88.6	8.4
20	76.0	19.5	99.0	86.2	6.1
25	68.9	13.1		80.2	0
30	31.6	8.2	98.6	51.2	
32.5	8.3	5.6		26.4	
35	0	3.5		16.4	
40		0	93.0	0	
45			78.9		
50			38.1		
55			0		

[a]For stabilization procedures used, refer to original paper.
HMF = High melting fraction; MMF = midfraction; LMF = low melting fraction.
Source: Ref. 69.

In the past the relative cost of cocoa butter and milk fat has induced some manufacturers to develop milk chocolate containing high levels of milk fat (>20% on fat phase). In recent years this has become less commercially attractive. In those countries where 5% vegetable fats are legally permitted in chocolate, there has been increasing interest in partially replacing the milk fat rather than the cocoa butter. A new group of products are therefore emerging aimed at retaining the texture and mouthfeel of the original formulation but replacing part of the milk fat with an appropriately tailored vegetable fat blend.

An alternative approach has been to consider using milk fat fractions. The oleins will behave as a liquid phase, softening chocolate in a similar way to milk fat itself. The stearin will lower the melting range, moving toward a eutectic at higher levels of addition (~50%). Because of the

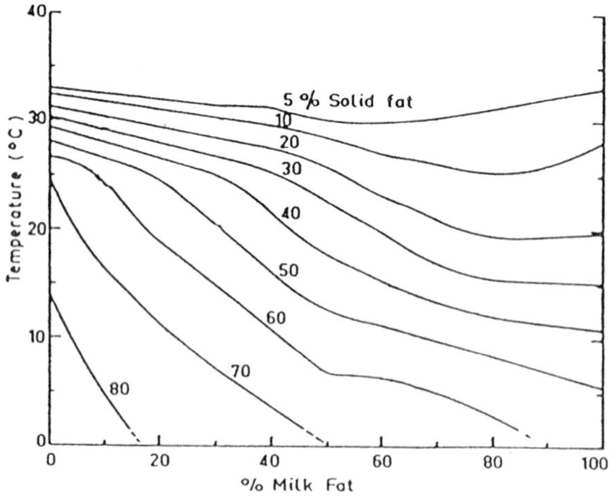

Fig. 8 Isosolids diagram for cocoa butter–milk fat mixtures after 4 weeks of storage at 13°C. (From Ref. 69.)

presence of high melting glycerides there will also be a tendency to increase the viscosity of tempered chocolate. In most European countries milk fat fractions are now regarded as being legally equivalent to milk fat. Milk fat stearin has been used, in some cases, as a bloom inhibitor. The addition of high melting stearin will tend to increase the viscosity of the tempered chocolate. Other derivatives of milk such as hydrogenated milk fat have also been suggested as antibloom agents but with little practical success to our knowledge [73].

D. Palmkernel Stearins and Other Lauric-Rich Confectionery Fats

Palmkernel oil can be easily fractionated using either solvent-based or non-solvent-based processes. Today most processes are non-solvent and rely on processes that produce a partially crystallized fat that is amenable to separation into solid and liquid phases using hydraulic or membrane presses [74,75]. Coconut oil can be processed in a similar way. For most confectionery applications palmkernel stearins, either hydrogenated or nonhydrogenated, are preferred because of their sharp melting behavior. The slightly lower melting coconut stearins find application in low melting fillings. Coconut oil is also widely used in formulating couvertures for ice cream. Some typical compositions and the melting behaviors of lauric stearins are given in Tables 15 and 16. Babassu and tucum oils are also used as alternative sources of lauric fats, but they are in limited supply. Readers are also referred to Soon [76].

Both palmkernel and coconut oils and their stearins are complex mixtures of triglycerides. The high level of lauric acid (~55 wt %) means that most of the glycerides will contain at least one lauric acid group, and approximately one-third will contain two lauric acids.

The complexity of the triglyceride mixture results in a fat that is β'-stable. The significance of this is that the fat can be crystallized directly from the melt into its β' form. The triglycerides of palmkernel stearin are chemically very different from those in cocoa butter. The phase behavior of palmkernel stearin–cocoa butter mixtures [66] therefore exhibit eutectic behavior with partial miscibility in the solid phase (Fig. 9). In practical terms this means that only small amounts of palmkernel stearin (<2) can be mixed with cocoa butter before the melting

Table 15 Typical Lauric Fat Compositions

			Fatty acids (wt %)					
Fat	IV	C_6–C_{10}	12:0	14:0	16:0	18:0	18:1	18:2
Palmkernel oil	17.5	7	48	16	9	2	15	2
Palmkernel stearins	9	6	53	21	9	2	8	1
	7	4	55	22	9	2	7	1
	2	3	50	30	12	3	2	0
	4	4	55	22	9	6	5	0
Palmkernel stearin (IV = 7) hydrogenated to lower IV	0.4	4	55	22	9	9	0.5	0
Palmkernel stearin (IV = 7) refractionated to lower IV	3	3	56	26	9	2	3	0.5
Coconut oil	8.5	15	48	18	8	3	6	2
Coconut stearin (25% yield)	4	9	55	22	9	2	3	0.5
Hydrogenated coconut stearin	1.5	9	55	22	9	5	2	0
Palmkernel olein	21	9	45	13	9	3	19	2
Coconut olein	10	17	46	17	7	3	7	3

Source: Ref. 75.

Table 16 Typical Lauric Fat Properties

Fat	IV	Solid fat content (% by pulse NMR)					Slip MP (°C)
		N20	N25	N30	N35	N40	
Palmkernel oil	17.5	44	20	0	—	—	28
Palmkernel stearins	9	70	48	8	0	—	30
	7	82	68	29	0	—	32
	2	94	84	37	3	0	34
	4	87	70	24	1	—	31
Palmkernel stearin (IV = 7) hydrogenated to lower IV	0.4	95	90	50	5	1	35
Palmkernel stearin (IV = 7) refractionated to lower IV	3	94	90	60	1	0	33
Coconut oil	8.5	36	0	0			24
Coconut stearin	4	84	53	2	0	—	30
Hydrogenated coconut stearin	1.5	92	57	8	2	—	32
Hydrogenated palmkernel stearin	1.0	73	—	27	14	6	41

Source: Ref. 75.

point is noticeably depressed. Over a period of time phase separation will also be observed. Conversely, only low amounts of cocoa butter (<5%) can be used in chocolate substitutes in which palmkernel stearin is the main fat phase. In both cases chocolate bloom will be observed if excess amounts of the minor component are used in the blend, i.e., palmkernel bloom on cocoa butter chocolate, cocoa butter bloom on palmkernel chocolate 15 ~20°C or lower (see Section V). It is the presence of these low amounts of cocoa butter in palmkernel stearin substitute chocolate that no doubt accelerate the loss of gloss. However, even in the absence of cocoa butter, palmkernel-based products still lose their gloss more rapidly than cocoa butter chocolate. At elevated temperatures, however, the bloom on palmkernel chocolate is lauric-rich. Minor amounts of cocoa butter remain dissolved, but growth of lauric-rich crystals is encouraged by holding at elevated temperatures. The interference by cocoa butter in palmkernel stearin–based chocolate substitute has been largely removed with the develop-

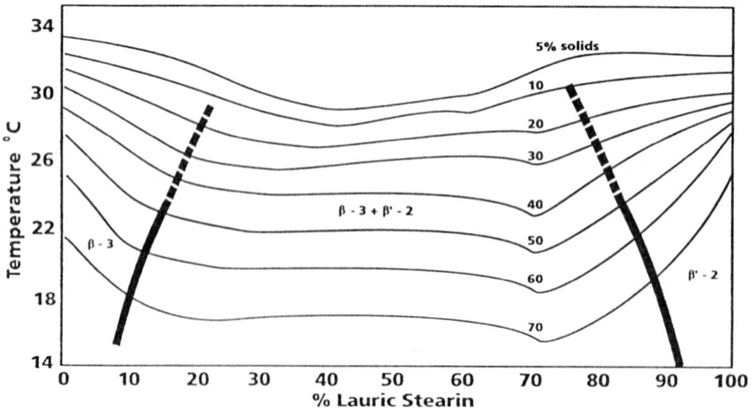

Fig. 9 Phase diagram of cocoa butter–palmkernel stearin mixtures. A similar phase diagram is obtained with partially hydrogenated, high *trans* fats but with greater miscibility, i.e., the phase boundaries move slightly toward the center. (From Ref. 66.)

ment of fat-free cocoa powders [9]. Bloom in these products can also be alleviated to some extent by the addition of surface-active agents such as sorbitan tristearate.

The processing of palmkernel stearin chocolate is greatly simplified by the elimination of the tempering stage. Molten chocolate at, say, 45°C can be used directly for enrobing or molding. The chocolate is then solidified by cooling rapidly at approximately 12°C in a cooling tunnel, avoiding condensation as the product emerges by raising the temperature of the final section.

The use of palmkernel stearins has also been associated with the development of soapy rancidity in some confectionery products processed under nonhygienic conditions. This aspect is discussed under rancidity, but suffice to say the wide application of these fats demonstrates that by taking reasonable precautions this problem is easily overcome.

E. Fractions of Partially Hydrogenated Oils and Fats

The ready availability of liquid oils has stimulated confectionery fat manufacturers to use hydrogenation, with or without the additional step of fractionation, to generate suitable products for the confectionery industry.

With the appropriate choice of catalyst and under conditions of limited hydrogen availability it is possible to minimize the level of fully saturated acids and to maximize the level of *trans* fatty acids in the product (see chapter on hydrogenation). *Trans* fatty acids and their triglycerides have an intermediate melting point between saturated and natural *cis*-unsaturated acids. The generation of *trans*-rich fats therefore offers an alternative approach to obtaining sharp melting fractions suitable for confectionery purposes. The melting behavior of unfractionated, *trans*-hardened fats is not very sharp, and the high melting components give the product a waxy mouthfeel. For some confectionery applications, however, this is acceptable. Significantly improved products are produced by fractionating the hydrogenated fat from solvent and isolating a midfraction, i.e., removing high and low melting components. Some typical compositions and melting profiles are given in Tables 17 and 18 [77].

Chemically these fats are very dissimilar to palmkernel stearins. The hardened fats are based on C_{16} and C_{18} fatty acids. Their glyceride composition is much more complex than that of cocoa butter, and they can crystallize from the melt directly into the B′ form, which is also the stable polymorph for these fats. These fats therefore do not need tempering and are widely used in the manufacture of substitute chocolate. Generally they display good gloss retention, although fats based mainly on C_{18} fatty acids have a tendency to lose gloss more rapidly. This can be counteracted to some degree by the addition of sorbitan esters of stearic acid.

The phase behavior of *trans*-hardened fats with cocoa butter has been studied [66]. The phase diagram is similar to that shown for palmkernel stearin–cocoa butter mixtures, but there is a greater degree of miscibility in the solid phase. These fats are therefore more compatible with cocoa butter than palmkernel stearins. Eutectic behavior and melting point depression does occur, however. Another aspect relevant to these fats in particular is that of postcrystallization. For a reason that is not fully understood, the solid fat content of many of these fats

Table 17 Typical Fatty Acid Compositions of *Trans* Hardened Fats

Fat	16:0	18:0	18:1*t*	18:1*c*	Σ Others <1[a]
A	28	6	44	21	1
B	12	11	59	15	3
C	19	11	34	33	3
D	23	13	39	22	3

[a]Sum of all others occuring at <1% individually.

Table 18 Typical Solid Fat Content of *Trans*-Hardened Fats[a]

Fat	% Solid Fat Content				
	N20	N25	N30	N32.5	N35
A	86	78	41	21	3
B	82	77	62	34	4
C	70	59	38	24	12
D	62	50	28	18	6

[a] N20 indicates solid fat content measured by NMR at 20°C, etc. Measurement made on a Brucker Minispec p20i on unstabilized samples.

increases significantly during storage at ambient or slightly elevated temperatures. This results in a deterioration in melting properties. This effect is, however, tolerable, possibly beneficial, in warmer climates for those products for which a reasonable degree of heat stability is required and for those enrobed products where a slight elevation in melting point can be beneficial to the overall mouthfeel of the product. Also see discussion by Soon [76] and a patent describing these types of fats and nonpostcrystallizing alternatives [78]. Because of consumer pressure there has been a move to seek alternatives to partially hydrogenated fats, especially as filling fats. Nutritionally this is a difficult area because replacing *trans* fatty acids will in most cases result in increasing saturated acids.

F. Liquid Oils

Although not regarded strictly as confectionery fats, the liquid oils are widely used in chocolate confectionery. Typical examples are peanut and hazelnut oils present in nut-paste fillings. The addition of up to 5% finely ground hazelnuts, peanuts, etc. is legally allowed within the European Union without the need for declaration. A knowledge of the influence of liquid oils on physical behavior is therefore important. The effect of liquid oil on the solid fat content of cocoa butter is illustrated in Fig. 10.

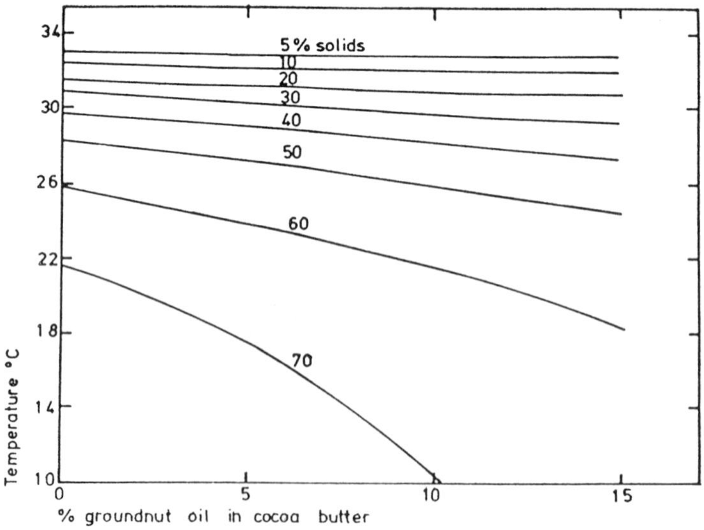

Fig. 10 The effect of liquid oil on the solid fat content of cocoa butter.

The addition of liquid oil reduces the overall solid fat content, which can be advantageous when trying to achieve softer textures. There is, however, a significantly greater risk of bloom formation (see later). The effect of liquid oils on the polymorphic behavior of cocoa butter has been studied [42].

IV. METHODS OF CHARACTERIZING MELTING AND CRYSTALLIZATION BEHAVIOR

The complexity of fat crystallization and melting behavior arises because fats are polymorphic. The most widely applied technique for determining the polymorphs present in a crystallized fat is X-ray diffraction. Various books and papers [19,20,79,80] describe the technique and its application to fats, including the more recent application of a position-sensitive detector (PSD) for more dynamic measurements [20].

Four main methods are used with differing degrees of success to characterize the melting and crystallization behavior of fats. These are dilatometry, wide-line nuclear magnetic resonance (NMR), particularly pulse NMR, differential scanning calorimetry (DSC), and empirical cooling curves. Dilatometry, prior to the advent of NMR, was the preferred method for determining melting behavior [53] and relies on measuring the volumetric change arising from the change from solid to liquid. The method has been very useful but requires trained operators to obtain reproducible results. In addition, the volumetric change in going from solid to liquid is different for fats widely different in fatty acid type. For example, lauric fats rich in C_{12} exhibit a smaller volume change than more conventional fats based on C_{16} and C_{18} fatty acids for the same amount of solid. A 100% solid fat in palmkernel stearin will give a dilation as measured by the IUPAC method of ~1800 whereas cocoa butter will give ~2500 (mol/25 g fat). Solid fat index is approximately obtained by dividing the IUPAC dilation by 25.

"Pulsed NMR" [81–83] is now the preferred method, and results are given in percent solid fat. Again, care needs to be taken in making the measurement because the signals from different fats decay at different rates, thus influencing the way in which the instrument has to be set up.

In the case of both dilatometry and NMR, standard methods of precrystallizing and tempering the fat (stabilization) have to be observed to obtain reproducible results. These stabilization conditions also depend on whether the fats are β'- or β-stable, different procedures being used for palmkernel stearins (β'-stable) than for cocoa butter and cocoa butter equivalents (β-stable) [53]. Care is also needed in defining appropriate stabilization regimes for cocoa butter–milk fat mixtures [70].

Differential scanning calorimetry (DSC) and differential thermal analysis (DTA) have been widely used to determine the phase behavior of fats. These techniques have not met with wide acceptance as methods of measuring solid fat content. This is because both polymorphic changes and melting can occur at the same time, inducing thermal effects in different directions. Even more complex is the influence of heating and cooling rates on melting and crystallization behavior. These effects have been amply demonstrated by Cebula and Smith [84,85], who studied the effect of heating and cooling rates on the DSC curves for the triglycerides POP, POSt, and StOSt. A detailed study of the effect of composition on the crystallization and melting behavior of a cocoa butter equivalent also demonstrates the need to understand the chemistry of the fat in order to make rational interpretations of the data [85]. A study by Ali and Dimick [86] on palm midfraction, cocoa butter, and milk fat also highlights some of the difficulties associated with obtaining fully stabilized systems to obtain meaningful phase diagrams. The technique potentially offers a useful in-house method for quality control of routinely produced products. Careful standardization would be required for more general quality control purposes across laboratories. The reader should also refer to other detailed publications [71,87–89].

The empirical crystallization cooling curves are typified by the Jensen and Pichard cooling curves. Standard procedures have been described for both methods [3,51]. The Jensen cooling curve is normally applied to cocoa butter and CBE. It is obtained by recording the temperature of the fat as it crystallizes from the molten state under periodic agitation. The aim of the test is to nucleate and crystallize the form V polymorph of cocoa butter or CBE. After supercooling, the temperature of the fat rises to a maximum as the fat crystallizes. The characteristics of the curve are used by some manufacturers to anticipate the performance of the fat during tempering and crystallization in the cooling tunnel. The Pichard cooling curve has been used for a similar purpose, but because it is obtained without agitation it is more suited to fats that crystallize and are stable in the β' form.

V. CHOCOLATE BLOOM

There are two basic forms of bloom that appear on chocolate, sugar bloom and fat bloom. Sugar bloom will arise on the surface of products produced under conditions that promote moisture condensation on the surface of the product as it emerges from the cooling tunnel. Alternatively, it will arise during storage in conditions of high humidity and fluctuating temperature. The main purpose of this section, however, is to discuss fat bloom.

A. Bloom on Chocolate Substitutes

There have been numerous studies on the nature of fat bloom. We encounter bloom with those products that include in their formulations a blend of cocoa butter with fats having a very dissimilar chemical composition. This invariably arises because the two fat phases separate [66]. This has been briefly discussed above under palmkernel stearins (see Fig. 8), and it is in this area that phase-separated bloom is most likely to occur. The boundaries between the single solid solutions of either cocoa butter in β' palmkernel stearin or palmkernel stearin dissolved in β-3 stable cocoa butter and the area of formulation containing both solid phases defines the extremes of the safe limits for blends of both fats to ensure that bloom does not occur. The problem will be exacerbated by temperature cycling and, perhaps surprisingly, by storing products at low temperatures. So-called ghost bloom arises at low temperatures, for example, in chocolate containing low but significant levels (3–5%) of palmkernel oil or stearin. In this case temperatures of 10–15°C are sufficient to encourage phase separation and the emergence of bloom rich in lauric triglycerides.

The main application of palmkernel stearins and trans hardened fat fractions is, however, as the major fat phase in substitute chocolate. In this case the level of cocoa butter must be carefully controlled, particularly for palmkernel stearins.

Cocoa powder is usually the only source of cocoa butter in chocolate substitute formulations. For this reason both the fat level in the cocoa powder and the absolute level of cocoa powder used must be controlled in palmkernel stearin–based "chocolate" such that the level of cocoa butter in the fat phase is less than 3%. This will ensure the avoidance of rapid bloom formation at ambient temperatures of ~20°C or less arising from the separation of a cocoa butter–rich phase. At higher storage temperatures the loss of gloss and development of bloom on palmkernel-based products is almost invariably caused by the recrystallization of lauric triglycerides. Despite taking all reasonable precautions it has been widely observed that palmkernel-based products and some *trans* fats lose their gloss more rapidly than conventional chocolate. As mentioned earlier, the development of fat-free cocoa powders does offer a solution to the problem of cocoa butter interference in these products. Bloom can also be alleviated to some extent by the addition of some surface-active components such as sorbitan esters.

B. Bloom Arising from Cocoa Butter

The problem of chocolate bloom for systems based primarily on cocoa butter has been widely studied [17,73,90–94]. The starting point has been the phase diagram of the stable forms of the major cocoa butter triglycerides StOSt, POSt, and POP (Fig. 6). Various hypotheses have been made, including the presence of a eutectic that encourages phase separation on prolonged storage or storage at elevated temperatures [95]. There is, however, a general consensus arising from the work of Wille and Lutton [17] that loss of gloss and bloom on correctly tempered chocolate (form V, β_2 polymorph) is directly related to a polymorphic change from form β_2 to the most stable form β_1 (form VI).

The structure of β_2 and β_1 crystals (most stable form) of the pure symmetrical triglycerides and of cocoa butter has been studied recently [20,41,96–99]. In the Hagemann and Rothfus paper [96], computer modeling is used to calculate the most stable (β_1) configuration of the triglyceride molecule StOSt. This structure shows the familiar alignment of the saturated acid chains alongside each other. The oleic chains pack in a separate layer in a bent configuration (Fig. 11). Sato and coworkers [20,41,97,98] have studied the β_2 and β_1 polymorphs of the pure symmetrical triglycerides using a range of spectroscopic techniques. Interestingly, a β_2

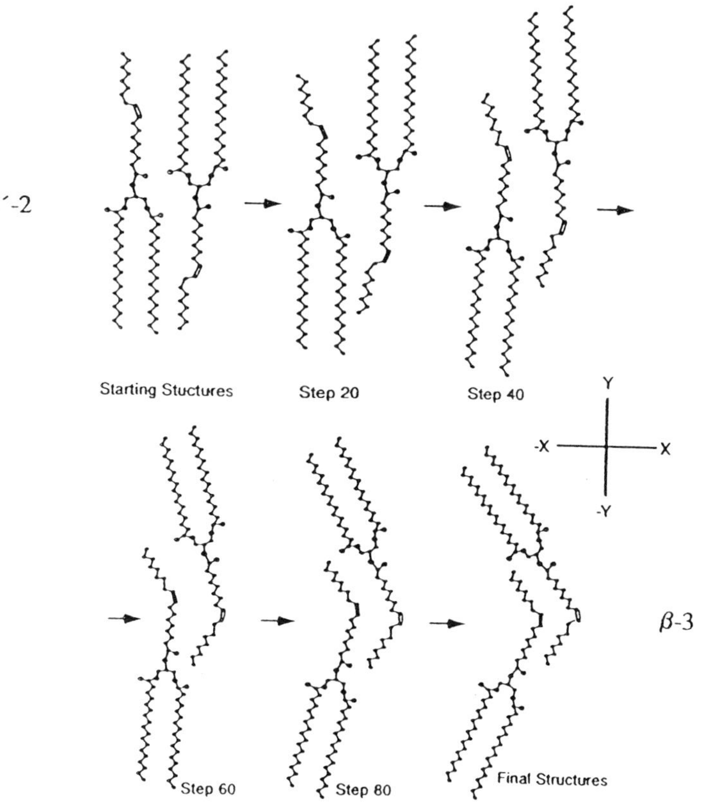

Fig. 11 Speculative scheme achieved via computer modeling for the double chain-to-triple chain transition of StOSt. (From Ref. 96.)

form of POSt was not observed [98]. The confirmation of the oleic chain in the β_2 and β_1 polymorphs is of specific interest. The oleic acid moiety is invariably depicted as having a "bent configuration" around the cis double bond in the β-stable crystal lattice. However, a number of studies have shown that a straight configuration closer to that observed for saturated hydrocarbons can also exist. The differences in free energy for the bent and straight configurations is sufficiently small as to make a definitive choice difficult [100] (Fig. 12). In fact, the oleoyl chain of crystalline cholesteryl oleate is known to be in the straight configuration. In their model study of the β_1 form of the symmetrical triglyceride SOS, de Jong et al. [100] concluded that the oleic acid was in the bent configuration.

The "straight" configuration of the oleoyl chain is proposed both by Sato et al. [99] and by Hagemann and Rothfus [96] for those crystal states where SOS triglycerides have adopted the double chain length packing arrangement as found in certain γ and β' states. The transformation of SOS from β' to β has been modeled by computer-aided techniques [96], and the realignment of the straight β' oleoyl chain to the bent β form is clearly depicted. Could it be that the oleoyl chain in β_2 (form V) is in a straight configuration, moving to the bent configuration in the most stable β_1 form (form VI)? Recent work by Sato's group [41] indicates that the difference between the β_2 and β_1 forms of SOS does in fact lie in the conformation of the oleoyl chain. However, their evidence suggests that the oleic acid is "bent" in both β_2 and β_1 forms but transforms from an orthorhombic to a triclinic form of packing. Any attempt to delay cocoa butter bloom must therefore address the problem of finding a method of delaying the reorientation of the oleoyl chain, thus preventing the β_2-to-β_1 transition. The polymorphism and structure of the major triglycerides in cocoa butter has been recently reviewed by Sato [20].

The precise conformation of the oleoyl chain in cocoa butter crystals and the mechanism of the β_2-to-β_1 transformation may still be open to debate. However, based on our current knowledge it is possible to propose various mechanisms that would lead to a delay in the onset of

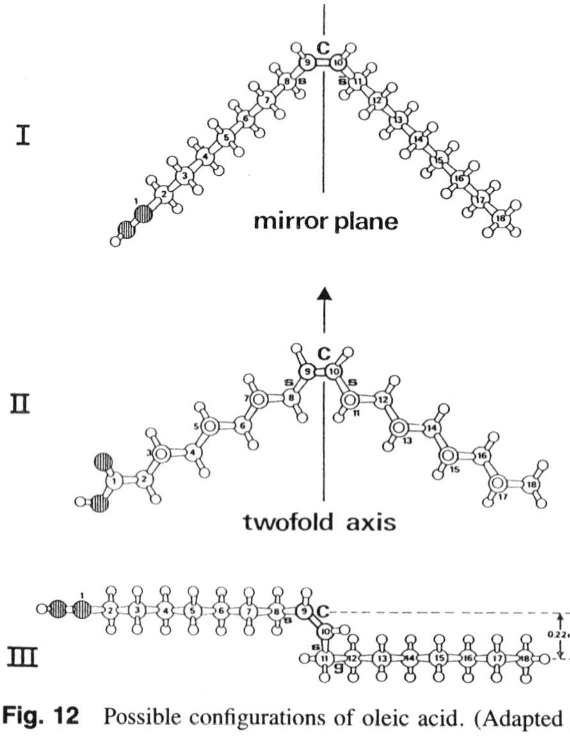

Fig. 12 Possible configurations of oleic acid. (Adapted from Ref. 100.)

bloom. The development of chocolate bloom is initiated by the β_2-to-β_1 transition. This can occur via either solid–solid or solid–liquid–solid transformations. In the case of solid–solid rearrangements it is highly probable that the initial changes will occur either at the surface or at dislocations within the crystal where there is greater room for molecular mobility. The remainder of the crystal can then be reordered by a concerted action somewhat like a "Mexican wave" passing through the crystal. Bloom would be delayed if an appropriate foreign molecule interfered either with the initial surface transformations or with the bulk transformation. This could occur, for example, if the foreign molecule interferes specifically with the oleic acid realignment, i.e., if a molecular "splint" is inserted to hold the oleic acid in the β_2 shape.

In the case of bloom being formed by solid–liquid–solid transformations, two approaches can be envisaged. The first is to add stable β_2 seed crystals. This would encourage β_2 growth to the detriment of β_1. The second approach is to include a foreign molecule that either prevents β_1 nucleation or significantly delays β_1 crystal growth, i.e., a nucleation or crystal growth inhibitor specific for β_1 (form VI). The approach of ensuring that there are always sufficient β_2 nuclei and seed crystals available, even at elevated temperature, to direct any recrystallization of fat that occurs into the β_2 form has met with some success. The use of the high melting symmetrical triglyceride 1,3-dibehenoyl-2-oleoyl glycerol (BOB) employs this effect [101–103]. Using the β_2 form of BOB added in a finely powdered form to nontempered chocolate at 33°C at levels of 1–5% inhibited bloom on chocolate cycled between 20 and 32°C. Interestingly, the most stable β form of StOSt (form VI) and cocoa butter was also effective, which to some extent negates the thoughts about β_2 seeds. See also Giddey and Clerc [92] for studies on seeding effects. This suggests that the mechanism is not simply one of ensuring that only β_2 nuclei are present. The role of tempering in delaying bloom [91] relates to the need to ensure the presence of the right type and number of crystal nuclei for β_2 growth. The success of extended tempering regimes such as those proposed by Kleinert is almost certainly influencing this aspect of fat crystallization.

In the second approach the success of certain other fats in delaying bloom is explained by the mechanisms that rely either on a "blocking" mechanism to prevent β_2-to-β_1 transitions or on inhibiting β_1 nucleation and growth. Milk fat, for example, has been widely used in chocolate formulations for many years both as a component of milk chocolate and as a bloom inhibitor. It almost certainly operates as a crystallization inhibitor or blocking agent [3,73,94,104].

A group of triglycerides based on two long-chain saturated fatty acids (C_{16}, C_{18}) and one short-chain saturated acid have proven to be eminently successful in inhibiting bloom [105,106]. These triglycerides have been tailored to act either as crystal inhibitors for β_1 or as retainers of β_2 configuration. The intriguing observation is that these materials are effective not only by direct addition to chocolate but also via fat migration, e.g., from a center filling. This suggests that they are operating either by adsorption on the crystal surface, thus preventing the β_2–β_1 transition, or as crystal growth or nucleation poisons for β_1.

With the continuing development of techniques such as atomic force microscopy [107] for characterizing crystal structure, we can anticipate further advances in this area. Retention of good gloss is clearly very important. Of equal importance are the retention of good flavor and the prevention of rancidity.

VI. RANCIDITY IN CONFECTIONERY

A. Oxidative Rancidity

There are essentially three types of rancidity that can be encountered arising from fats in confectionery products [108]. These are oxidative, lipolytic, and ketonic. All confectionery fats have potential for oxidative rancidity. Lipolytic rancidity is limited to the lauric fats and butterfat, and ketonic rancidity is limited primarily to palmkernel and coconut oils but could also be anticipated in butterfat-containing products. For a comprehensive view of rancidity in foods the reader is referred to Allen and Hamilton [109]. The general topic of oxidative rancidity has been reviewed by a number of workers, including Labuza [110] and Frankel [111]. Factors influencing the shelf life of cocoa products have been discussed by Kattenberg and de Muijnck [112].

Confectionery fats are relatively stable to oxidative rancidity provided that sufficient antioxidant is present. The majority of confectionery fats contain only low levels of the oxidatively unstable polyunsaturated acids such as linoleic or linolenic acid. The lauric fats are essentially saturated and contain only low levels of unsaturated acids. Despite the fact that there is a low level of risk, it is nevertheless necessary to take reasonable and sensible precautions to minimize autoxidation, which can occur under extreme conditions of low antioxidant, elevated temperatures, and ready access to oxygen.

Cocoa butter is primarily based on saturated and oleic acids. It is also protected, however, by a range of naturally occurring antioxidants, tocopherols, and flavonoids. These two factors combine to give a fat that is extremely stable to oxidative rancidity. Related products such as cocoa butter equivalents may need additional protection (100–200 ppm tocopherol) as these are normally midfraction or stearins and may be depleted in natural antioxidants. This comment applies generally to all fractionated products apart from the olein fraction which will be enriched in the antioxidants of the parent oil. Additional protection can also be obtained by adding synthetic antioxidants such as butylated hydroxyanisole (BHA), *tert*- butylhydroquinone (TBHQ, in the United States), or butylated hydroxytoluene (BHT), customer and legislation permitting. In the case of commercial CBE, these have virtually identical induction periods as measured using the Rancimat accelerated oxidative stability test [113] to cocoa butter.

The relatively saturated lauric fats should also be handled with some care. These saturated fats contain only low levels of natural antioxidant. Typical levels of tocopherol in palmkernel and coconut oils are <20 ppm. Storing these fats and their fractions at elevated temperatures

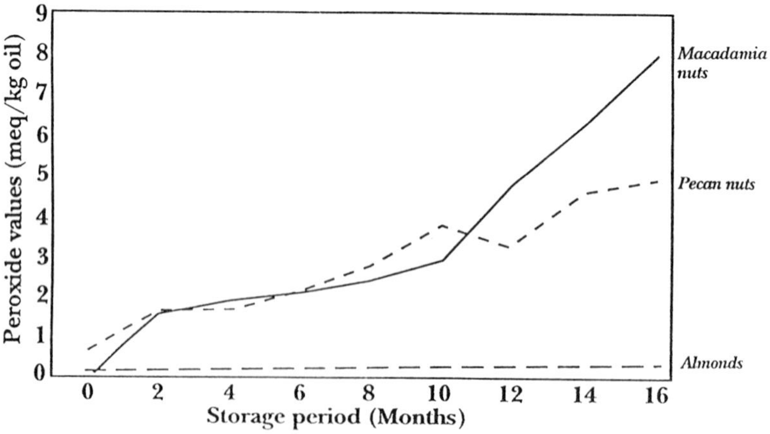

Fig. 13 Changes in peroxide value of oil in stored nuts. Mean values of 10 determinations on nuts stored at 30°C and 55% RH. (From Ref. 114.)

in the presence of air and in the absence of any supplemented antioxidant will lead to fat autoxidation. This will influence oil quality in terms of both flavor and physical performance, depending on the degree of oxidation.

Lipid autoxidation also arises in a number of confectionery components, particularly nuts. These may, of course, be a primary source of confectionery fat such as shea, illipé, or sal or used as a key component, unmodified, of a confectionery product, e.g., as whole, small pieces or ground nutmeats. Oxidation in nuts can occur via at least two pathways. The first is by simple autoxidation arising from poor storage of the nuts over a period of time, i.e., ambient and higher temperatures in the presence of oxygen (Fig. 13) [114,115]. The relative stability of almonds compared to pecans can be explained in part by the approximately five times higher level of tocopherol in almonds compared to pecans.

Oxidation via the action of lipoxygenase is the second route most common in nuts and seeds. The lipoxygenase can either be indigenous to the nuts as in the case of sunflower and soy [116] or introduced by microbial contamination as has been observed in the case of sal seeds [53,55,56,117]. The products of oxidation under the influence of lipoxygenase are usually epoxides, which can be further modified to dihydroxy derivatives of the peanut fatty acid. In the case of sal fat, the combined action of lipoxygenase and general hydrolytic degradation of the oil adversely affects the physical properties of the fat, and in extreme cases the fat becomes unsuitable for confectionery applications.

B. Lipolytic and Ketonic Rancidity

Lipolytic and ketonic rancidity are grouped together in this discussion because they both primarily affect the lauric oils. Short-chain fatty acids have low flavor thresholds (Table 19). Only a low level of hydrolysis, catalyzed by microbial lipases, of either palmkernel or coconut oil is sufficient to release an organoleptically detectable level of free fatty acid. This type of rancidity has been most prevalent in chocolate substitutes formulated with poor quality, microbially contaminated cocoa powder and lauric stearins. It does not occur in products manufactured under good standards of hygiene as demonstrated by the wide use of lauric stearins in substitute chocolate. Ketonic rancidity arises through microbial contamination and has been studied by Kinderlerer and coworkers [118,119]. The microorganisms identified as being responsible are xerophilic fungi of the type *Eurotium amstelodami*, *E. chevalieri*, *E. herbariorium*, and *P. citrinum*. These fungi reduce even-chain-length fatty acids to a homologous series of odd-chain-length aliphatic methyl ketones and secondary alcohols (C_5–C_{11}). Some of the volatile off-flavor compounds that have been identified are listed in Table 20.

Table 19 Flavor Characteristics and Flavor Threshold Values of Short-Chain Free Fatty Acids

Fatty acid	Flavor charactistic (in milk)	Flavor threshold (ppm)		
		Milk	Oil	Water
Butyric	Butyric	25.0	0.6	6.8
Caproic	Cowy (goaty)	14.0	2.5	5.4
Caprylic	Cowy (goaty)	—	350	5.8
Capric	Rancid, unclean, bitter, soapy	7.0	200	3.5
Lauric	Rancid, unclean, bitter, soapy	8.5	700	—

Source: Ref. 147.

Table 20 Volatile Off-Flavor Compounds in Rancid Coconut Oil

Off-flavor compounds	Consumer return (mg/g oil)	Rancid coconut (mg/g) S.D.	Flavor note	Intensity 1:4
Pentan-2-one	Trace	Trace	Pear drops	3
Pentan-2-ol	N.D.	0.58 ± 0.21	Ethereal	2
Hexan-2-one	9.36 ± 5.01	Trace	Ethereal	4
Hexan-2-ol	4.88 ± 5.45	N.D.	Turpentine	3
Heptan-2-one	10.72 ± 3.55	5.05 ± 0.68	Rancid almonds	4
Heptan-2-ol	9.29 ± 1.11	7.05 ± 0.18	Rancid coconut	4
Octan-2-one	2.57 ± 0.44	N.D.	Weakly ethereal	2
Nonan-2-one	19.63 ± 0.88	2.97 ± 0.07	Weakly turpentine	1
Nonan-2-ol	2.91 ± 0.88	0.35 ± 0.15	Musty, stale	2
Undec-2-one	6.77 ± 0.20	1.29 ± 0.12	Weakly turpentine	1
Total	66.13	17.29		

N.D. = not detected; S.D. = standard deviation.
Source: Ref. 119.

VII. COCOA BUTTER ADULTERATION AND THE DETERMINATION OF VEGETABLE FAT IN CHOCOLATE

The adulteration of food and food commodities has occurred since time immemorial. This is very true of fats, particularly cocoa butter, over the last 100 years or more. There is no comprehensive publication on fat adulteration, but oil and fat analysis has been well documented in standard methods of analysis agreed on by international bodies such as British Standard Methods, IUPAC, AOAC, and ISO. These have also been described in various textbooks [53,120]. Accepted ranges of fatty acid compositions of oils and fats have been published by FAO/WHO [121].

A recent and notorious case of adulteration occurred in Spain when olive oil was adulterated with refined aniline-denatured rapeseed oil [122,123]. Olive oil is a typical case where there is a strong commercial advantage to diluting an expensive oil with cheaper commodity oils. Palm oil, palm fractions, and palmkernel oil have also been the subject of fraud investigations. At certain times, for example, there has been a commercial advantage to adulterating palm oil with palm fractions.

In the confectionery area the adulteration of cocoa butter has been widespread. This has become much less prevalent since the development of new analytical techniques and the internal controls exercised by large responsible companies. The use of fatty acid compositions to detect adulteration is of very limited value because of the natural variation encountered with most oils and fats. Analyses for triglyceride and minor components, including sterols, offer the more satisfactory approaches. Sterol analysis is useful for detecting adulteration but not for quantitative estimation. The reason for this is that there is a wide variation in the sterol content of individual oils and fats. This arises in part by natural variation and in part through their partial removal during refining. Where the sterol content is widely different from that of cocoa butter, the level of detection can be quite low, as in the case of illipé and shea, for example [124]. The sterols and minor components of a wide range of oils and fats have been determined [53,120].

The analysis of triglycerides, on the other hand, offers a good approach for quantifying the composition of many fat mixtures. Triglycerides represent the bulk of the oil, and their overall level is virtually constant, but their actual composition is subject to natural variations. Triglyceride analysis has been particularly useful in characterizing pure and adulterated cocoa butter [31,32,34,125–129]. The method is based on analyzing triglycerides by high temperature gas-liquid chromatography. This technique separates triglycerides on the basis of carbon number, i.e., the total number of carbon atoms in the fatty acid chains (POP ≡ 50; POSt ≡ 52; StOSt ≡ 54). When the data for a wide range of genuine cocoa butters are plotted with P50 versus P54 (C_{50}, C_{52}, C_{54} normalized to 100), then a straight line is observed [124] (Fig. 14). This is, in fact, a small part of the overall curve representing the C_{50}/C_{54} relationship for all naturally occurring vegetable fats [130,131].

The straight-line relationship for cocoa butter can be used as a vehicle to detect adulteration by a large range of fats, in particular cocoa butter equivalents. A standard method was developed for the European Union for the determination of cocoa butter equivalents in chocolate [2]. Triglyceride analysis has also been used successfully to quantify milk fat in cocoa butter and chocolate [132,133].

The determination of vegetable fats in chocolate presents additional problems. The total fat has to be first determined using methods defined by the OICC, for example. A combination of chromatographic methods including fatty acid, triglyceride, and sterol analysis will have to be used to characterize "unknown" samples [134,135]. The determination of 2-oleo-disaturated (SOS) glycerides also provides a way of characterizing the type of fats present [38,136].

The CBE method of analysis can also be used to make reasonable estimates of the level of other fats, especially those based mainly on c_{18} fatty acids, or their origin can be determined by other means. Vegetable oils including soybean, corn, rapeseed, sunflower, many ground-

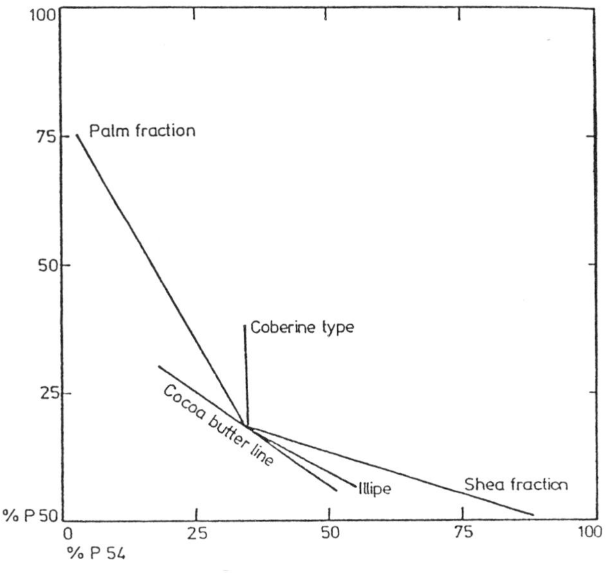

Fig. 14 The determination of cocoa butter equivalents (CBE) in cocoa butter. The triglycerides with carbon numbers 50, 52, and 54 (total number of carbon atoms in fatty acid chains) are normalized to 100%, giving P_{50}, P_{52}, P_{54} values. (From Ref. 125.) Coberine is a commercial CBE.

groundnuts, and hazelnut oils fall into this category. As the C_{50} content of the oil increases, the quantitative aspect of the analysis diminishes. The CBE method in this case would determine the total level of added fat but would not say specifically what was added.

Certain palm fractions can be detected in mixtures with cocoa butter using the techniques described above. An unscrupulous operator could choose judicious blends of palm fractions with or without other fats to try to hide the adulteration. Certain blends will be more difficult to analyze than others. However, the combined use of silver-phase HPLC chromatography and gas-liquid chromatographic techniques should provide a solution—at least for the more sophisticated oil and fat laboratories.

The potential development of genetically engineered seed oils and/or enzymatically modified fats with essentially the same fatty acid and triglyceride compositions as cocoa butter would present a challenge to the analyst. These materials are not available today. Methods of detection would rely on minor component analysis, such as sterols, hydrocarbons, and waxes.

Cottonseed oil and other oils and fat blends that also have triglyceride carbon numbers similar to cocoa butter would also require special treatment. In these cases the use of high resolution capillary gas chromatography triglyceride analysis and silver ion chromatography [77] should provide a solution. An example of high resolution capillary GC is given in Fig. 15 [35].

Many confectionery products are made up of a number of components—for example, chocolate with inclusions and other "countlines" in which a variety of centers are enrobed in chocolate. Many of these centers contain significant levels of other vegetable fats that rapidly migrate into the chocolate coating [137]. Any interpretation of data must therefore attempt to take this into account.

If a semiquantitative determination of chocolate adulteration by excessive amounts of vegetable fats is required, then a combination of modern analytical methods should provide the answer in simple cases. The unscrupulous blending of particular fats and their fractions will always present problems, as it no doubt does today, and the interpretation of analytical data will depend on the analyst and the degree of complexity of the blend.

The rigorous characterization of any chocolate, e.g., cocoa solids, milk solids, and vegetable fat content, can be carried out only with the aid of factory inspection, now a legal obligation within the EU. In the case of factory inspection, the actual components used to fabricate

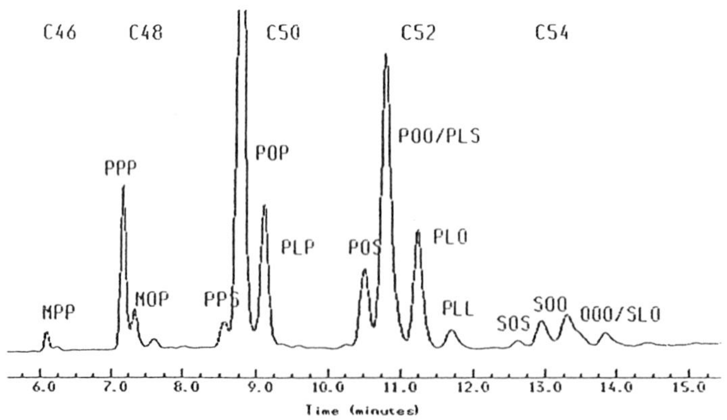

Fig. 15 The determination of triacylglycerols by high resolution wide-bore capillary gas chromatography. (From Ref. 35.)

products will be available and the composition of the final product can be verified precisely using existing analytical methods.

VIII. ADDITIONAL ASPECTS OF CONFECTIONERY FAT APPLICATION

This chapter has been largely restricted to chocolate fat applications. Some other applications of fats in confectionery are much less demanding as regards the physical behavior of the fat; other applications have different physical requirements altogether. Some brief examples are given below. The reader is referred to a number of specialized textbooks, notably those of Manley [140] and Jackson [148].

The specialized area of low calorie fats is covered in Chapter 19.

A. Toffee Fats

One of the main attributes required from the fat in recipes such as for caramel toffee is good flavor, in particular butterlike flavor. In other formulations, as in caramel, because aqueous sugar solution or sugar glass is the continuous phase, the physical properties of the fat are not critical provided that the melting behavior is not detrimental to mouthfeel, i.e., significant solid fat at about 37°C. For this reason butterfat might be regarded as the preferred fat from the flavor point of view but certainly not from the economic aspect. A wide range of fats are therefore used depending on the application and processing/marketing requirements. Typically, partially hardened fats with slip melting points in the range 27–40°C are used in many products, some with added flavors or flavor precursors to provide a buttery flavor note.

B. Biscuit Cream Fillings

Biscuit cream fillings are typically flavored mixtures of aerated fat and sugar. A major requirement is good mouthfeel, ideally cool melting, coupled with good aeration properties and ability to adhere to the biscuit. Fats fulfilling these requirements range from palmkernel and partially hardened palmkernel oils to trans-hardened oils and fats with appropriate slip melting points, e.g., 27–29°C, depending on the application.

C. Ice Cream Couvertures

The most widely used couvertures for ice cream are based either on chocolate, i.e., cocoa butter based, or on substitute chocolate in which the fat phase is almost universally coconut oil. Some other fat blends such as mixtures of palm oil, palmkernel oil, and partially hardened fats with slip melting points of approximately 27°C etc. are also used, however.

Whether the ice cream product is enrobed by dipping or in a conventional enrober, the important requirement is to be able to control the weight of chocolate used for enrobing at a time when the chocolate is crystallizing rapidly. This is done by a combination of viscosity control (fat, emulsifier level) and process manipulation, e.g., air jets to remove excess chocolate from the products. For example, in those operations where ice cream is enrobed by dipping, the fat content of the couverture will be between 45 and 60%.

In the case of cocoa butter it is interesting to note that for enrobed ice cream products the chocolate is not usually tempered and crystallizes in the less stable form IV (which is stable at ice cream temperatures). Thus the chocolate melts in the mouth at a lower and more acceptable temperature.

D. Center Filling Fats

Pralines are combinations of sugar, cocoa, nut paste, and confectionery fat. In this case the selection of the confectionery fat is dependent on the product application and the mouthfeel required. In some cases the "confectionery fat" will in fact be a liquid oil, as in hazelnut or peanut paste. Where firmer textures are required, the melting profile of the fat can be raised progressively, ultimately to that of cocoa butter. The fats used are therefore very diverse and range from butterfat to fractions of coconut and palmkernel, fractions of palm oil, hardened fats and their fractions, etc. Apart from mouthfeel and its retention during the life of the product, another preferred requirement is that these fats can be crystallized without the need to temper the fat as in the case of cocoa butter.

An additional consideration with many of these products in which a fat-continuous center is enrobed in chocolate is the problem of fat migration [137]. This occurs primarily from the softer center to the chocolate coating, although the migration occurs in both directions via the liquid oil phase. This migration causes a significant softening of the chocolate coating and in many cases increases the susceptibility to bloom (see Section V). The rate of fat migration is clearly dependent on the product, solid fat content, and temperature. However, migration will be virtually complete in most systems held at 28°C for 14 days.

To minimize any textural changes or reduction in melting point, then, wherever feasible a center filling fat that is compatible, or most compatible, with cocoa butter should be chosen (assuming cocoa butter is the fat phase of the coating). For example, center filling fats based on lauric oils will, on migration, significantly reduce the hardness of the chocolate compared to a similar fat based at least in part on SOS-type triglycerides.

The adverse publicity given to a range of fats on health grounds has encouraged the industry to develop unhardened, non-lauric alternatives based on palm fractions. The postcrystallization and subsequent change in mouthfeel for many of these products has recently been overcome by the judicious addition of unsymmetrical glycerides to the POP-rich fraction [138].

E. Fats as Moisture Barriers

The subject of fats as moisture barriers is covered to some extent in Chapter 17. However, one major aspect is mentioned here. For those confectionery products that rely on the appeal of texture contrast created by the use of crisp wafers or similar material together with chocolate and fat fillings, it is important to prevent, or at least minimize, the uptake of water by the wafer. This will retain product quality and extend the acceptable shelf life. This is not always achieved, and the final result strongly depends on product design. Invariably, fat-based coatings, particularly chocolate or chocolate couverture (substitute chocolate), are used as a moisture barrier in confectionery products. Moisture transfer is clearly dependent on the permeability of the coating used. This depends primarily on the solid fat content; the higher the percent solid fat, the lower the permeability. A severe limitation to this is, however, the potential for the coating to crack, thus allowing rapid ingress of moisture. A compromise must be reached between overall product appeal, moisture barrier thickness, and melting profile and cost. For further background reading there are numerous papers and reviews [141–144].

As a final comment the reader will have noted that no reference has been made to the use of animal fats, notably beef tallow. The presence of unsymmetrical triglycerides, i.e., of the SSO type, means that these fats form eutectics with cocoa butter. In addition there is a virtual embargo, based on consumer response, to the use of these fats in high quality confectionery.

Those interested in gaining further insight into the interaction of glycerides, particularly binary interactions of pure triglycerides and the phase behavior of fats, are directed to reviews by Rossell [149] and Timms [150].

IX. CONCLUSION

The use of fats in confectionery products in becoming increasingly sophisticated with our improved understanding of the factors that influence physical properties and the increased demand from the marketplace. New developments will continue, driven by new technologies and changing attitudes to nutrition.

ACKNOWLEDGMENT

I am very grateful for the support and encouragement given over many years by Loders Croklaan.

REFERENCES

1. J. J. Scheu, The changing patterns of cocoa bean supply and demand, *Manuf. Confect.* p. 41 (1995). Also Cocoa report, *Manuf. Confect. 75*: 74 (September 1995).
2. EU Council Directive 73/241/EEC of 24 July 1973 on The Approximation of the Laws of the Member States Relating to Cocoa and Chocolate Products Intended for Human Consumption and EU Working Paper, 1984. Also Commission of the European Communities; Directorate General for Internal Market and Industrial Affairs. Working Group on Methods of Analysis for Foodstuffs (Cocoa and Chocolate). Proposal by UK Delegation III/A/3.
3. B. W. Minifie, *Chocolate, Cocoa and Confectionery: Science and Technology*, 2nd ed., AVI, Westport, CT, 1980.
4. S. T. Beckett, *Industrial Chocolate Manufacture and Use*, Blackie, London, 1994.
5. J. J. Urbanski, Chocolate flavour—origins and descriptions, *Manuf. Confect. 72*: 69 (November 1992).
6. J. Kleinert, Quality control on chocolate, *Zucker Süss. Wirtsch. 44*: 52 (1991).
7. W. J. Shaughnessy, Cocoa beans: planting, fermentation and effect on flavour, *Manuf. Confect. 72*: 51 (November 1992).
8. E. Minson, Chocolate manufacture—beans through liquor production, *Manuf. Confect. 72*: 61 (Nov. 1992).
9. H. R. Kattenberg, The application of cocoa powders in chocolate confectionery, *Manuf. Confect. 75*: 73 (May 1995).
10. W. Rostagno, Particle size of non-fat solids in chocolate, *Manuf. Confect. 49*: 81–85 (May 1969).
11. E. A. Niediek, Particle size reduction, in *Industrial Chocolate Manufacture and Use* (S. T. Beckett, Ed.), Blackie, London, 1994, pp. 83–89.
12. E. H. Steiner, A new rheological relationship to express the flow properties of melted chocolate, *Rev. Int. Choc. 13*: 290 (1958).
13. NCA (National Confectioners Association), Method for determination of Casson yield value and plastic viscosity of chocolate, *Manuf. Confect. 68:* 72–75, 76–81 (January 1988).
14. J. Chevalley, Chocolate flow properties, in *Industrial Chocolate Manufacture and Use* (S. T. Beckett, Ed.), Blackie, London, 1994, p. 139.
15. T. L. Harris, *Surface Active Lipids in Chocolate*, SCI Monograph 32, 1968, pp. 109–122.
16. H. F. Banford, K. G. Gardiner, G. R. Howat, and A. F. Thomson, The use of polyglycerol polyricinoleate in chocolate, *Confect. Prod.* 359–365 (1970).
17. R. L. Wille and E. S. Lutton, Polymorphism of cocoa butter, *J. Am. Oil Chem. Soc. 43*: 491 (1966).
18. K. Larsson, Molecular arrangement in glycerides, *Fette Seifen Ansstrichm. 74*: 136 (1972).

19. K. Larsson, *Lipids—Molecular Organisation, Physical Functions and Technical Applications*, Oily Press, Dundee, Scotland. 1994.
20. K. Sato, Polymorphism of pure triacylglycerols and natural fats, in *Advances in Applied Lipid Research*, Vol. II (F. B. Padley, Ed.), JAI Press, New York, in press, 1996, pp. 213–268.
21. R. B. Nelson, Pumps and tempering, in *Industrial Chocolate Manufacture and Use* (S. T. Beckett, Ed.), Blackie, London, 1994, p. 211.
22. M. G. Reade, The natural rate of solidification of chocolate, *Manuf. Conf. 65*: 59 (January 1985).
23. P. G. Allen, Chocolate temper measurement, *Manuf. Confect. 85*: 91 (1995).
24. M. G. Reade, Temper and temper testing. Parts I and II, *Manuf. Confect. 60:* January and February (1980), pp. 29–32 and 51–54, respectively.
25. D. J. Cebula, K. M. Dilley, and K. W. Smith, Continuous tempering studies on model confectionery systems, *Manuf. Confect.* 131 (1991).
26. D. J. Cebula, D. J. M. Clements, M. J. W. Povey, and P. R. Smith, Neutron diffraction studies of liquid and crystalline trilaurin, *J. Am. Oil Chem. Soc. 69*: 130 (1992).
27. R. A. Martin, Chocolate, *Adv. Food Res. 31*: 211 (1987).
28. D. J. Cebula and K. M. Dilley, Personal communication; published by agreement with Loders Croklaan b.v., Holland.
29. A. Yasuda and K. Mochizuki, The behaviour of triglycerides under high pressure, in *High Pressure and Biotechnology* (C. Balny, R. Hayashi, and K. Heremans, Eds.), Colloque INSERM, John Libby Eurotext Ltd. *224*, 1992, pp. 255–259.
30. M. R. Mackley, Extrusion process, Br. Patent GB 2270828A (September 1992).
31. F. B. Padley and R. E. Timms, The determination of cocoa butter equivalents in chocolate, *Chem. Ind. (Lond.) 178*: 918.
32. A. Fincke, Triglyceride distribution by carbon number in cocoa butter substitutes, *Deut. Lebens. Rundsch. 76*: 187–192 (1980).
33. H. R. Kattenberg, The quality of cocoa butter, *Manuf. Confect. 61*: 32 (January 1991).
34. E. Geeraert and P. Sandra, Capillary GC of triglycerides in fats and oils, Part II. The analysis of chocolate fats, *J. Am. Oil Chem. Soc. 64*: 100–105 (1987).
35. G. J. Sassano and B. S. J. Jeffrey, Gas chromatography of triacyl glycerols in palm oil fractions with medium polarity wide bore columns, *J. Am. Oil Chem. Soc. 70*: 111 (1993).
36. S. Chaiseri and P. S. Dimick, Cocoa butter—its composition and properties, *Manuf. Confect. 67:* p. 115 (1987).
37. B. S. J. Jeffrey, Silver-complexation liquid chromatography for fast high resolution separations of triacylglycerols, *J. Am. Oil Chem. Soc. 68*: 289–293 (1991).
38. T. Tagaki and Y. Ando, Stereospecific analysis of mono-unsaturated triacylglycerols of cocoa butter, *J. Am. Oil Chem. Soc. 72*: 1203 (1995).
39. T. Okawachi and N. Sagi, Confectionery fats from palm oil, *J. Am. Oil Chem. Soc. 62*: 421 (1985).
40. L. Wesdorp, Liquid-multiple solid phase behaviour in fats, Ph.D. Thesis, Technical Univ., Delft, Holland, 1990.
41. J. Yano, S. Ueno, and K. Sato, FTIR study of polymorphic transformations in SOS, POP, POS, *J. Phys. Chem. 97*: 12967–12973 (1993).
42. N. V. Lovegren, M. S. Gray, and R. O. Feuge, The effect of liquid fat on melting point and polymorphic behaviour of cocoa butter and cocoa butter fractions, *J. Am. Oil Chem. Soc. 52*: 108 (1976).
43. C. M. Savage and P. S. Dimick, Influence of phospholipids on the crystallisation of hard and soft cocoa butters, *Manuf. Confect. 75*: 127 (May 1995).
44. S. Chaiseri and P. S. Dimick, Dynamic crystallisation of cocoa butter. I. Characterisation of simple lipids in rapid and slow nucleating cocoa butters and their seed crystals, *J. Am. Oil Chem. Soc. 72*: 1491 (1995).
45. S. Chaiseri and P. S. Dimick, Dynamic crystallisation of cocoa butter. II. Morphological, thermal and chemical characteristics during crystal growth, *J. Am. Oil Chem. Soc. 72*: 1497 (1995).
46. F. B. Padley, C. N. Paulussen, C. J. Soeters, and D. Tresser, The improvement of chocolate using mono-unsaturated triglycerides SOS and POS, *Rev. Int. Choc. 27*: 226 (1972).

47. D. W. Lehrian and P. G. Keeney, Triglyceride characteristics of cocoa butter from cacao fruit matured in a microclimate of elevated temperature, *J. Am. Oil Chem. Soc. 57*: 66 (1980).
48. J. Kleinert, Pressed and extracted cocoa butter, *Rev. Int. Choc. 19*: 142 (1964).
49. U. Bracco, R. Horman, and R. Viani, Isolation and identification of new compounds in cocoa shell fat, *Riv. Ital. Sost. Grasse 53*: 167 (1976).
50. M. Weyland, Cocoa butter fractions: a novel way of optimising chocolate performance, *Manuf. Confect. 72:* p. 53 (1992).
51. R. L. Best, A. Crossley, S. Paul, H. Pardun, and C. J. Soeters, Improvements in or relating to cocoa butter substitutes, Br. Patent 827,172 (1956).
52. M. G. A. Willems and F. B. Padley, Palm oil: quality requirements from a customers point of view, *J. Am. Oil Chem. Soc. 62*: 454 (1985).
53. F. D. Gunstone, J. L. Harwood, and F. B. Padley, *Lipid Handbook*, 2nd ed., Chapman and Hall, London, 1994, p. 94.
54. S. J. Kershaw and E. Hardwick, Heterogenicity within commercial contract analysis of samples of shea nut kernels, *J. Am. Oil Chem. Soc. 58*: 706 (1981).
55. N. V. Bringi, *Non Traditional Oilseeds and Oils of India*, Oxford Univ. Press and IBH, New Delhi, 1987.
56. N. V. Bringi, R. E. Timms, and F. B. Padley, Fatty acids and glyceride composition of sal fat, *Chem Ind.* (*Lond.*) *1982*: 805 (1982).
57. B. S. J. Jeffrey and F. B. Padley, Chinese vegetable tallow and its contamination with stillingia oil, *J. Am. Oil Chem. Soc. 68*: 123 (1991).
58. G. Brooks, N. A. Morrice, C. Ellis, A. Aitken, A. T. Evans, and F. J. Evans, Toxic phorbol esters from chinese tallow, *Toxicon 25*(11): 1229–1233 (1987).
59. C. J. Soeters, C. N. Paulussen, F. B. Padley, and D. Tresser, Br. Patent 1,390,936 (April 1971).
60. R. J. Davies and J. E. Holdsworth, Synthesis of lipids in yeast: biochemistry, physiology, production, in *Advances in Applied Lipid Research*, Vol. I (F. B. Padley, Ed.), JAI Press, New York, 1992, pp. 65–194.
61. A. R. Macrae, Fat process and composition, Br. Patent 1577933 (1976).
62. A. R. Macrae, Lipase catalysed interesterification of fats, *J. Am. Oil Chem. Soc. 60*: 291 (1983).
63. P. Eigtved, Enzymes and lipid modification, in *Advances in Applied Lipid Research*, Vol. I (F. B. Padley, Ed.), JAI Press, New York, 1992, pp. 1–64.
64. S. R. Moore and P. T. Quinlan, Modification of triglycerides by lipases, *INFORM 4*: 580 (1993).
65. Y. Hashimoto, Production of cocoa butter like fats by enzymic transesterification *Bioprocess Technol. 16*: 337–352 (1993).
66. M. H. Gordon, R. E. Timms, and F. B. Padley, Factors influencing the use of vegetable fats in chocolate, *Fette Seifen Anstrichm. 81*: 116 (1979).
67. G. Talbot, Vegetable fats, in *Industrial Chocolate Manufacture and Use*, 2nd ed. (S. T. Beckett, Ed.), Blackie, London, 1994, pp. 242–257.
68. G. M. Chapman, E. E. Akehurst, and W. B. Wright, Cocoa butter and confectionery fats. Studies using programmed temperature, X-ray diffraction and differential scanning calorimetry, *J. Am. Oil Chem. Soc. 48*: 824 (1971).
69. R. E. Timms, The phase behaviour of mixtures of cocoa butter and milkfat, *Lebensm. Wiss. Technol. 13*: 61 (1980).
70. B. Peterssen, Pulsed NMR method for determining solid fat content, Part II. Cocoa butters and equivalents in blends with milkfat, *Fette Seifen Anstrichm. 88*: 128 (1986).
71. J. Schlichter-Aronhime, J. S. Sarig, and N. J. Garti, Reconsideration of polymorphic transformations in cocoa butter using DSC, *J. Am. Oil Chem. Soc. 65*: 1140 (1988).
72. J. G. P. Verhey, Physical properties of dried milk in relation to chocolate manufacture, *Neth. Milk Dairy J. 40*: 261–268 (1986).
73. L. B. Campbell, D. A. Anderson, and P. G. Keeney, Hydrogenated milk fat as an inhibitor of fat bloom, *J. Dairy Sci. 52*: 976 (1969).
74. R. E. Timms, Processing of palm kernel oil, *Fette Seifen Anstrichm. 88*: 294 (1986).
75. J. B. Rossell, Fractionation of lauric oil, *J. Am. Oil Chem. Soc. 62*: 385 (1985).

76. W. Soon, *Speciality Fats versus Cocoa Butter*, Privately published, Malaysia, 1991. Fax: 60 373 47162.
77. K. W. Smith, J. M. Perkins, B. S. J. Jeffrey, and D. L. Phillips, Separation of cis and trans triacylglycerols in trans hardened confectionery fats by silver ion high performance chromatrography, *J. Am. Oil Chem. Soc. 71*: 1219 (1994).
78. A. D. Hughes and K. W. Smith, Non-post hardening fats, WO Patent 9424882 (1994).
79. D. Chapman, *The Structure of Lipids*, Methuen, London, 1965.
80. M. Garti and K. Sato, *Crystallisation and Polymorphism of Fats and Fatty Acids*, Marcel Dekker, New York, 1988.
81. D. Waddington, *Fats and Oils: Chemistry and Technology*, Applied Science, London, 1981, p. 25.
82. J. C. van der Enden, A. Haighton, K. van Putte, L. F. Vermaas, and D. Waddington, *Fette Seifen Anstrichm. 80*: 180 (1978).
83. J. C. van der Enden, J. B. Rossell, L. F. Vermaas, and D. J. Waddington, Determination of the solid fat content of hard confectionery butters, *J. Am. Oil Chem. Soc. 59*: 433 (1982).
84. D. J. Cebula and K. W. Smith, DSC of confectionery fat triglycerides, *J. Am. Oil Chem. Soc. 68*: 591 (1992).
85. D. J. Cebula and K. W. Smith, DSC of confectionery fats. Part II. Effect of blends and minor components, *J. Am. Oil Chem. Soc. 69*: 992 (1992).
86. A. R. M. Ali and P. S. Dimick, Thermal analysis of palm mid fraction, cocoa butter and milk fat blends by differential scanning calorimetry, *J. Am. Oil Chem. Soc. 71*: 299 (1994).
87. D. G. Manning and P. S. Dimmick, Interpreting the thermal characteristics of cocoa butter using differential scanning calorimetry, *Manuf. Confect. 63*(8): 73 (1983).
88. V. P. Gibon, P. Blainpain, F. Durant, and C. Deroanne, *Belg. J. Food Chem. Biotech. 40*(5): 119 (1985).
89. E. Kaisersberger, The application of heat flux DSC for the characterisation of edible fats, *Anal. Proc. 27*: 64 (1990).
90. J. Kleinert, Bloom on chocolate, *Rev. Int. Choc. 8*: 72 (1953).
91. J. Kleinert, Studies on the formation of bloom and methods of delaying it, Paper presented at the 15th Annual Production Conf. of the Pennsylvania Manufacturing Confectioners Assoc., Lancaster, PA, April 1961.
92. C. Giddy and E. T. Clerc, Polymorphism of cocoa butter and its importance to the chocolate industry, *Int. Choc. Rev. 16*: 548 (1961).
93. J. S. Aronhime and N. Garti, Solidification and polymorphism in cocoa butter and bloom problems, in *Crystallisation and Polymorphism of Fats and Fatty Acids* (N. Garti and K. Sato, Eds.), Marcel Dekker, New York, 1988, p. 363.
94. L. B. Campbell and P. G. Keeney, Developments in fat bloom research on dark chocolate coatings, *Manuf. Confect. 48*: 77 (1968).
95. W. Andersson, Fat bloom and phase changes, *Int. Choc. Rev. 18*: 92 (1963).
96. J. W. Hagemann and J. A. Rothfus, Computer modelling of packing arrangements and transitions in saturated–cis-unsaturated mixed triglycerides, *J. Am. Oil Chem. Soc. 69*: 429 (1992).
97. T. Arishima and K. Sato, Polymorphism of POP and SOS. Part III. Solvent crystallisation of β_2 and β_1 polymorphs, *J. Am. Oil Chem. Soc. 66*: 1014 (1989).
98. T. Arishama, N. Sagi, H. Mori, and K. Sato, Polymorphism of POS. I. Occurrence and polymorphic transformation, *J. Am. Oil Chem. Soc. 68*: 710 (1991).
99. K. Sato, T. Arishima, Z. H. Wang, K. Ojima, N. Sagi, and H. Mori, Polymorphism of POP, SOS. Part I. Occurrence and polymorphism transformation, *J. Am. Oil Chem. Soc. 66*: 664 (1989).
100. S. de Jong, T. C. van Soest, and M. A. van Schaick, Crystal structure and melting points of unsaturated triacylglycerols in the β-phase, *J. Am. Oil Chem. Soc. 68*: 371 (1991).
101. T. Koyano, I. Hachiya, and K. Sato, Fat polymorphism and crystal seeding effects on fat bloom stability of dark chocolate, *Food Struct. 9*: 231 (1990).
102. I. Hachiya, T. Koyano, and K. Sato, Seeding effects on the behaviour of cocoa butter and dark chocolate solidification. Part I. Kinetics of solidification, *J. Am. Oil Chem. Soc. 66*: 1757 (1989).

103. I. Hachiya, T. Koyano, and K. Sato, Part II. Physical properties of dark chocolate, *J. Am. Oil Chem. Soc. 66*: 1763 (1989).
104. D. J. Cebula and G. Ziegleder, Studies of bloom formation using X-ray diffraction on chocolates after long-term storage, *Fat Sci. Technol. 95*: 340 (1993).
105. G. Talbot, F. Cain, and A. Hughes, Chocolate anti-bloom fat, Eur. Patents EP 530864 and EP 521549 (1991).
106. G. Talbot, Chocolate bloom, the cause and the cure, *Int. Food Ing.* 40–45 (1995).
107. P. J. M. W. L. Birker and J. C. G. Blonk, Alkyl chain packing in the β' triacyl glycerol measured by atomic force microscopy, *J. Am. Oil. Chem. Soc. 70*: 319 (1993).
108. F. B. Padley, Rancidity in confectionery fats, in *Rancidity in Foods*, 3rd ed. (D. Allen and R. J. Hamilton, Eds.), Blackie, London, 1994, pp. 230–255.
109. D. Allen and R. J. Hamilton, *Rancidity in Foods*, 3rd ed., Blackie, London, 1994.
110. T. P. Labuza, Kinetics of lipid oxidation in foods, *CRC Crit. Rev. Food Technol. 2*: 355 (1971).
111. E. N. Frankel, Recent advances in lipid oxidation, *J. Sci. Food Agric. 54*: 495 (1991).
112. H. R. Kattenberg and L. de Muijnck, The shelf life of cocoa products as ingredients for the food industry, *Dev. Food Sci. 33:* pp.311–340. (1993).
113. J. B. Rossell, Measuring resistance to oxidative rancidity, *Lipid Technol. 4*: 39 (1992).
114. P. C. Fourie and D. S. Basson, Predicting the occurrence of rancidity in stored nuts by means of chemical analysis, *Lebensm. Wiss. Technol. 22*: 251 (1989).
115. T. Keme, M. Messerli, J. Sheijbal, and F. Vitali, Storage of hazelnuts at room temperature under nitrogen, *Rev. Choc. Conf. Baker 8*(2): 15 (1983).
116. T. Meshedani, J. Pokorny, and J. Panek, Interaction of oxidised lipids with protein, *Nahrung 34*: 915 (1990).
117. S. Y. Reddy and J. V. Prabhakar, Effect of triglycerides containing 9,10-dihydroxystearic acid on the solidification properties of sal fat (*Shorea robusta*), *J. Am. Oil Chem. Soc. 62*: 1126 (1985).
118. J. L. Kinderlerer and S. Johnson, Rancidity in hazelnuts due to volatile aliphatic aldehydes, *J. Sci. Food Agric. 58*: 89 (1992).
119. B. Kellard, D. M. Busfield, and J. L. Kinderlerer, Volatile off-flavour components in desiccated coconut, *J. Sci. Food Agric. 36*: 415 (1985).
120. J. B. Rossell and J. L. R. Pritchard, *Analysis of Oilseeds, Fats and Fatty Foods*, Elsevier, London, 1991.
121. Codex Alimentarius Commission (1993), Report of the 14th Session of the Codex Committee on Fats and Oils, September 1993, London. Alinorm 95/17. Issued by the Joint FAO/WHO Food Standards Programme, Rome.
122. A. Pastana and E. Munoz, *Nature 298*: 608 (1982).
123. Toxic oil syndrome references archive, Fondo de Investigaciones Sanitarias, Ministero de Sanidady-consumo, Madrid, 1990.
124. D. Gegiou and K. Staphylakis, Detection of cocoa butter equivalents in chocolate, *J. Am. Oil Chem. Soc. 62*: 1047 (1985).
125. F. B. Padley and R. E. Timms, The determination of cocoa butter equivalents in chocolate, *J. Am. Oil Chem. Soc. 57*: 286–293 (1980).
126. A. Fincke, Possibilities and limits of gas chromatography of triglycerides for detecting foreign fats in cocoa butter and chocolate fat, *Deut. Lebensm. Rundsch. 78*: 389–396 (1982).
127. A. Fincke, Studies on the diglyceride content of cocoa butter, *Gordian 82*: 246 (1982).
128. C. C. Young, New interpretation of GLC triglyceride data for the determination of cocoa butter equivalents in chocolate, *J. Am. Oil Chem. Soc. 61*: 576 (1984).
129. A. Fincke, Triglyceride analysis of cocoa butter and chocolate fats according to carbon number, *Deut. Lebens. Rundsch. 76*: 162–167 (1980).
130. F. D. Gunstone, The distribution of fatty acids in natural glycerides of vegetable origin, *Chem. Ind. (Lond.) 1962*: 1214 (1962).
131. A. Fincke and F. B. Padley, Theoretical aspects of triglyceride composition of natural fats expressed in terms of carbon number, *Fette Seifen Ansstrichm. 83*: 461–465 (1981).

132. D. O. Biltcliffe and R. Wood, The determination of milkfat in admixture with cocoa butter, *J. Assoc. Pub. Anal. 20*: 69 (1980).
133. D. von. Precht, Quantitative determination of milk fat in chocolate, *Fat Sci. Technol. 92*: 275 (1990).
134. K. Grob, A. M. Giuffré, U. Leuzzi, and B. Mincione, Recognition of adulterated oils by direct analysis of the minor components, *Fat Sci. Technol. 96*: 286–290 (1994).
135. K. Grob, H. P. Neukom, and R. Battaglia, Triglyceride analysis with glass capillary gas chromatography *J. Am. Oil Chem. Soc. 57*: 282–286 (1980).
136. H. Chaveron, M. Adenier, F. B. Padley, M. S. J. Dallas, and G. A. Jongenotter, Method of determining mono-unsaturated symmetrical and unsymmetrical triglycerides, *Ann. Falsif. Exp. Chim. Toxicol. 74*: 455–464 (1981).
137. M. Wooton, Fat migration in chocolate enrobed biscuits, *Gordian 72*: 95 (1972).
138. F. W. Cain, A. D. Hughes, and J. H. Pierce, Br. Patent Appl. 94 12045 (1994).
139. K. Jackson, Recipes, in *Industrial Chocolate Manufacture and Use* (S. T. Beckett, Ed.), Blackie, London, 1994, pp. 258–280.
140. D. Manley, *Technology of Biscuits, Crackers and Cookies*, 2nd ed., Ellis Horwood, New York, 1991.
141. T. H. McHugh and J. M. Krochta, Water permeability properties of edible whey protein–lipid emulsions, *J. Am. Oil Chem. Soc. 71*: 307 (1994).
142. E. Baldwin, J. M. Krochta, and M. O. Nisperos-Carriedo, *Edible Film and Coatings to Improve Food Quality*, Technomic, Lancaster, PA, 1994.
143. I. Greener Donhowe and O. Fennema, The effect of relative humidity gradient on water vapour permeance of lipid and lipid–hydrocolloid bilayer films, *J. Am. Oil Chem. Soc. 69*: 1081 (1992).
144. I. Greener Donhowe and O. Fennema, Water vapour and oxygen permeability of wax films, *J. Am. Oil Chem. Soc. 70*: 867 (1993).
145. A. Fincke, Fatty acid composition of cocoa butter, *Gordian 76*: 325–327 (1976).
146. S. Chaiseri and P. S. Dimick, Lipid and hardness characteristics of cocoa butters from different geographical regions, *J. Am. Oil Chem. Soc. 66*: 1772 (1989).
147. J. E. Kinsella, The flavour chemistry of milk lipids, *Chem Ind.* (*Lond.*) *1969*: 36 (1969).
148. E. B. Jackson, *Sugar Confectionery Manufacture*, Blackie, London, 1990.
149. J. B. Rossell, The phase behaviour of glycerides, *Adv. Lipid Res. 5*: 353–408 (1967).
150. R. E. Timms, Phase behaviour of fats and their mixtures, *Prog. Lipid Res. 23*: 1 (1984).

16

Frying Oils and Salad Oils

Timothy L. Mounts
National Center for Agricultural Utilization Research, Agricultural Research Service, U.S. Department of Agriculture, Peoria, Illinois

I. INTRODUCTION

The preparation of fat-containing food products uses two classes of lipid material based on their physical state at about 25°C (74°F); (1) liquid oils, which include soybean, canola, rapeseed, corn, cottonseed, sunflower, safflower, peanut, and olive oils; and (2) semisolids and solids, including lard, palm oil, coconut, and palmkernel oils.

Frying and salad oils are produced in many countries as indicated in Table 1. Soybean oil is used in the greatest quantity in the world fats and oils industry.

II. DETERIORATION OF OILS AND FATS

All fats and oils are subject to deterioration. Hydrolysis, the reaction of fats and oils with water, and oxidation, the chemical reaction in which oxygen combines with another substance with the liberation of heat, are the two basic processes that result in the deterioration of oils and fats. Oxidation is responsible for much more of the deterioration of fats and oils than hydrolysis.

During the use of fats and oils for the frying of foods, food moisture promotes the splitting of triglycerides to form free fatty acids (Fig. 1). Essentially, hydrolysis is the reverse of making a fat molecule.

The deterioration of a fat or oil in the presence of oxygen is termed oxidative rancidity. The initial step in the oxidation of an oil or fat is the addition of oxygen at or near the double bond of a fatty acid chain to form unstable compounds generally designated as peroxides (Fig. 2) [1]. When the oxidation of an oil is followed experimentally, either by measuring the amount of oxygen absorbed or by determining the peroxide value, the course of oxidation is defined by the oxygen absorbed in the oil as shown in Fig. 3 [2]. During the initial or induction phase, oxidation proceeds at a relatively slow and more or less uniform rate. Peroxides are formed during this period much faster than they are destroyed so their content increases

Table 1 Production of Salad/Frying Oils in Major Countries

Country	1989	1990	1991	1992	1993
Germany	474	524	714	710	726
Austria	65	75	75	81	84
Czechoslovakia	97	99	84	86	90
Hungary	66	74	65	95	81
Poland	112	77	70	94	110
Romania	224	222	217	220	220
Former Yugoslavia	253	208	205	190	185
East Europe	752	680	642	685	685
Egypt	362	370	380	410	430
S.Africa Rep.	162	194	220	231	229
Canada	230	241	275	293	338
United States	2777	2785	2862	2944	2829
Israel	66	69	73	75	76
Japan	165	181	174	165	170

Source: Oil World Statistics Update, Sept. 2, 1994.

more or less in parallel with the oxygen absorption. Thus, after a certain critical amount of oxidation has occurred, the reaction enters a second phase characterized by a rapidly accelerating rate of oxidation. The point at which the sample begins to smell and taste rancid coincides with the beginning or early part of the second phase. As oil oxidation continues, with time, the peroxides that are formed decompose to generate volatile and nonvolatile compounds that contribute to flavor and odor deterioration of oils and fats. The extreme stages of oxidation, polymerization, and degradation are accompanied by rapid increase in the viscosity of the oil.

There is considerable variation among fats in the manner in which their oxidation and accompanying flavor deterioration proceeds. The amount of oxygen that must be absorbed to produce rancidity will vary according to (1) the composition of the oil, (2) the presence or absence of antioxidants and prooxidants, and (3) the conditions of oxidation. Generally, oxygen absorbed will amount to about 15–150% of the oil by volume or 0.02–0.20% by weight. Fats high in oleic acid and low in linoleic acid will become rancid after the absorption of less oxygen than fats in which the ratio is reversed.

The relative oxidation rates of pure fatty acid esters are shown in Table 2 [3]. This table is based on a common factor of 1 that was arbitrarily assigned to the oxidation rate of oleic

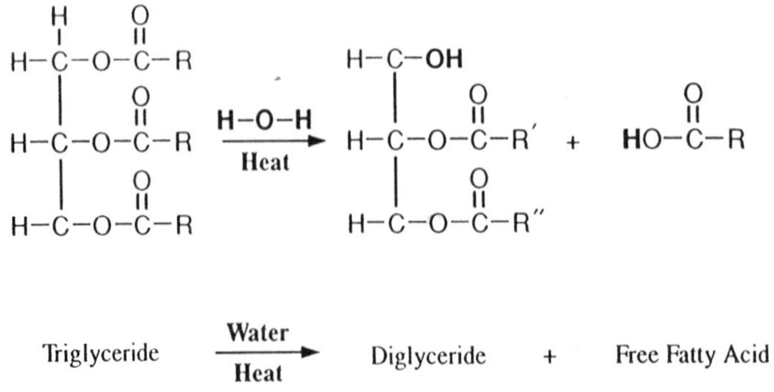

Fig. 1 Hydrolysis of fats and oils.

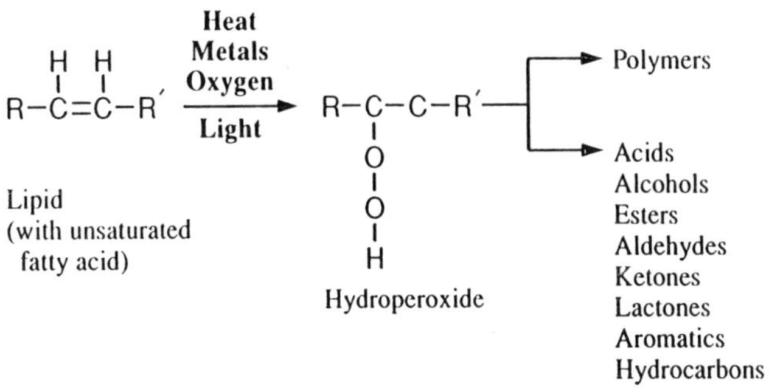

Fig. 2 Oxidation of fats and oils.

acid. The rate at which oxygen is absorbed is markedly accelerated by heat and also by exposure to light, particularly in the ultraviolet and near-ultraviolet range [4–6].

III. FLAVOR, ODOR, AND STABILITY

Many fats and oils develop off-flavors during storage. This has been called "reversion," because the oil flavor is reverting to the original flavor of the crude oil. However, as discussed by Smouse [7], it is more accurately called "retrogression" because the flavor retrogrades from a better to a worse condition. Also, aged oils do not develop off-flavors identical with those of the crude oils, but for any given type of oil the off-flavor is always specific and characteristic. Although all oils develop off-flavors, some do so at a faster rate than others. Soybean and marine oils develop "fishy" flavors, lard and tallow develop "animal" flavors, rapeseed and linseed oils develop "grassy/painty" flavors, and palm oil develops "painty/rancid" flavors. All of the efforts to prevent the development of off-flavors in oils are aimed at slowing down the rate at which they develop. Multivalent metals, such as iron or copper, act as prooxidants when present with triglycerides. Early research [8] showed that metal inac-

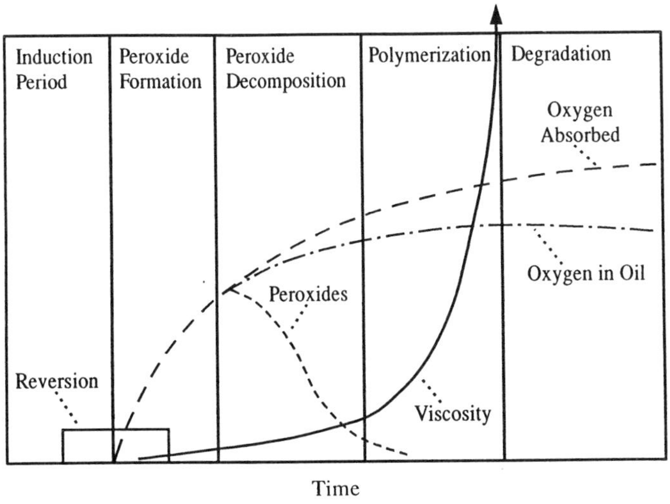

Fig. 3 Stages of progressive deterioration of fats through oxidation.

Table 2 Relative Reaction Rates of Fatty Acids with Oxygen

Fatty acid	Approximate relative oxidation rate
Oleic acid (18:1)	1
Linoleic acid (18:2)	10
Linolenic acid (18:3)	25

Source: Ref. 3.

tivators, which chelate or sequester these metals, are very effective at reducing the rate of reversion in soybean oil. This was a major improvement in processing for soybean oil stability and is practiced universally by the industry. Research [9] has indicated that the addition of citric acid to canola oil is important for flavor stability in the oil.

A good discussion of the theories of soybean oil reversion is presented by Smouse [7], who includes 121 citations to pertinent research. Each theory is discussed in detail. These theories concern linolenic acid, isolinoleic acid, oxidative polymers, phospholipids, nonsaponifiables, multivalent metals, and singlet oxygen. Each of the theories plays a role in soybean oil flavor deterioration.

IV. ADDITIVES AND STABILITY

A. Metal Chelating Agents

Perhaps the most important additive used with edible oils to enhance stability during storage and use is citric acid or a similar metal chelating agent [10]. Traces of metals in a reactive form can act as catalysts that promote the autoxidation of fats and oils. Concentrations of iron as low as 0.1 ppm or of copper as low as 0.01 ppm can accelerate the deterioration of soybean oil [11]. If an oil has higher concentrations of metals, it is necessary to first lower the content by treating the oil with a vigorous acid wash, e.g., with very dilute phosphoric acid, or by using activated bleaching earth to absorb metals and metal complexes from the oil. A small amount of citric acid (i.e., 0.05% w/w citric acid/oil) can be added to enhance the effectiveness of a bleaching earth. Other acids that have been used to chelate metals include ascorbic, phosphoric, tartaric, and ethylenediamine tetraacetic acid. It should be noted that citric acid does not improve the stability of olive oil [12].

B. Antioxidants

Antioxidants effectively improve the stability of fats and oils with lower levels of natural antioxidants such as tocopherols and with lower levels of polyunsaturation. With oils that have a high content of polyunsaturated fatty acids with significant levels of tocopherols, such as soybean oil, antioxidants minimize the accumulation of peroxides but do not improve the flavor stability [13]. The qualities desired in an antioxidant used to stabilize edible fats and oils ensure that (1) its use in a product is safe; (2) it contributes no odor, flavor, or color to the product in which it is used; (3) it is effective in the product at low concentrations; (4) it is easily incorporated into the product; (5) it is retained after cooking processes such as baking and frying; and (6) it is available at low cost for the application [14]. Lauric, olive, and palm oils contain only small amounts of tocopherols compared to soybean, corn, and cottonseed oils. Synthetic antioxidants were shown to be effective in the stabilization of virgin olive oil [12]. Butylated hydroxyanisole (BHA), butylated hydroxytoluene (BHT), and *tert*-butylhydroquinone (TBHQ) all improved the oxidative stability of olive oil during storage at 50°C (120°F) in the dark but were not effective in preventing photooxidation of olive oil. It should be noted that addition of antioxidants to oils that are already deteriorated will not improve the quality or stability of the oil.

C. Nitrogen Treatment

As mentioned above, oxidation has the most detrimental effect on the quality of finished products held in storage. Thus, exclusion of oxygen during storage is highly desirable and practical for preventing quality deterioration [15–18]. Sparging (Fig. 4) is a practical method for protecting finished oils from oxidative deterioration. The principle involved is saturation of the oil with nitrogen while it is completely free of air and oxygen, i.e., after deodorization. A sparger introduces tiny bubbles of nitrogen into the oil stream; as the gas-saturated oil falls into the package, the effusing gas sweeps the headspace and thus removes most of the air and oxygen from the vessel [15]. Another benefit of nitrogen sparging is the reduction of oxygen picked up from natural leakage of air into the package. The gradual flushing action of the sparge gas reduces the reabsorption rate of oxygen because of the pressure differential between the liquid and the headspace above it [17]. A nitrogen gas sparger is illustrated in Fig. 4.

V. OBJECTIVE METHODS FOR EVALUATING OXIDATIVE AND FLAVOR STABILITY

A. Shelf Storage Test

The most realistic estimate of the storage life of fats and oils may be obtained by simply storing the test oil under conditions identical to or resembling normal or anticipated storage and handling conditions. Periodic evaluation by the available methods for detecting and measuring oxidation serves to pinpoint the storage time at which the oil is harmed or rendered unacceptable by oxidation. The principal objection to this type of testing is the extensive amount of time required to develop information.

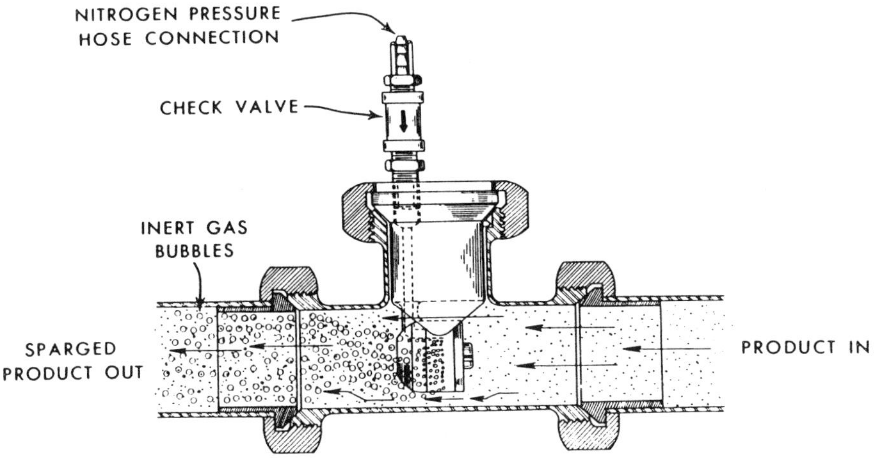

Fig. 4 An in-line nitrogen gas sparger.

B. Schaal Oven Test

The accelerating effect of heat on fat and oil oxidation is a commonly used factor in speeding up stability tests of edible oils. This is based on the estimation that for each 10°C rise in temperature in the range of 97–110°C, the rate of peroxide formation approximately doubles [19]. The test is usually conducted in electrically heated convection ovens at about 60–63°C. Such oven testing permits estimation of fat and oil stability and antioxidant effectiveness in days or hours rather than the weeks or even month required for normal shelf storage. Adaptations of the Schaal oven test have been used for the accelerated aging of edible oils prior to sensory evaluation to assess the relative flavor stability of the oil [20].

C. Active Oxygen Method

The active oxygen (AOM) has been widely used in the fats and oils industry for measuring the oxidative stability of fats and oils. The method measures the time (in hours) required for a sample of fat or oil to attain a predetermined peroxide value under the specific conditions of the test. The length of this period of time is assumed to be an index of resistance to rancidity. The AOM is highly empirical, and close attention to details is required if reproducible results are to be expected. The details of the test are included in the AOCS Official Method Cd 12-57 [21]; however, the method has been declared surplus, effective 1996, and is no longer recommended.

D. Oil Stability Index

The Oil Stability Index test is recommended to replace the AOM. All oils and fats have a resistance to oxidation that depends on the degree of saturation, natural or added antioxidants, prooxidants, and prior abuse. Oxidation is slow until this resistance is overcome, at which point oxidation accelerates and becomes very rapid. The length of time before this rapid acceleration of oxidation is the measure of the resistance to oxidation and is commonly referred to as the induction period. In the oil stability index method for determining the induction period, a stream of purified air is passed through a sample of oil or fat that is held in a thermostated bath. The effluent air from the oil or fat sample is then bubbled through a vessel containing deionized water. The conductivity of the water is continually monitored. The effluent air contains volatile organic acids, swept from the oxidizing oil, that increase the conductivity of the water as oxidation proceeds. Formic acid is the predominant organic acid formed [22]. The conductivity of the water is monitored by a computer or stripchart recorder. The oil stability index is defined as the point of maximum change of the rate of oxidation, and this method was adopted in 1993 as AOCS Official Method Cd 12b-92 [21].

E. Volatiles Analyses

Numerous publications have shown that volatiles analysis can correlate with flavor panel evaluations [23–26]. AOCS Recommended Practices Cg 1-83 and Cg 2-83 (21) have been adopted for application of these techniques. These data have identified total volatiles and/or individual volatile compounds such as pentane, pentanal, hexanal, and 2,4-decadienal that may serve as markers for oil stability. These volatiles are not individually responsible for the total flavor of the oil as in a cause-and-effect relationship; however, they do increase rapidly with increasing oil storage time and are indicative of oxidation levels. Volatile analysis by gas chromatography is described in AOCS Recommended Practice Cg 4-94 [21]. The three basic gas chromatographic procedures described in the recommended practice are direct injection, static headspace, and dynamic (purge and trap) headspace analysis. The choice of method will depend on available equipment and/or on the type of volatile compounds of interest. The two headspace procedures described in the recommended practice require an additional piece of instrumentation attached to the gas chromatograph to collect volatile compounds.

The static headspace method is selective toward the low molecular weight compounds such as pentane, whereas dynamic headspace is selective toward medium and higher molecular weight compounds such as 2,4-decadienal. The entire range of low to high molecular weight compounds are detected by direct injection [27].

Flavor problems from nonoxidative sources are not detected by volatiles analysis. Data on volatiles found in the literature are difficult to extrapolate to other chromatographic conditions to determine oil stability, because the data are usually expressed in integrator counts for peaks rather than in parts per million (ppm). Integrator counts can vary widely depending on the chromatographic conditions used.

VI. SENSORY EVALUATION METHODS

A. Salad Oils

Salad-type oils that are liquid at room temperature are evaluated for odor and flavor to determine their initial quality and/or flavor stability after storage. The American Oil Chemists' Society (AOCS) has methods and recommended practices (Cd 8-53, Cg 1-83, and Cg 2-83) [21], as does the American Society for Testing and Materials (ASTM) [28] for preparation and presentation of oils for sensory testing. These practices include basic information such as serving containers, oil temperature, and heating procedures for the oils.

The AOCS recommended practice also includes scoring scales based on a 1–10 system for rating the quality or intensity of liquid vegetable oils. Two scales are given, one for intensity and the other for quality. The intensity scale is based on a 1 = strong, 10 = bland (no odor/flavor) scale. This is recommended for rating oils that have blandness as their ultimate goal. Oils such as soy, sunflower, cottonseed, palm, low erucic acid rapeseed (canola), and safflower that have little or no flavor immediately after a proper deodorization are rated on this scale. On the other hand, a quality scale could be used for oils such as corn, peanut, and olive that have natural distinctive, although desirable, flavors. Rating intensity may be less subjective than rating an oil for quality because more exact reference standards for intensity can be used to train judges. Flavorful oils can be rated for intensity of individual flavors; however, overall intensity scores for these oils are not indicative of quality as the natural flavors in addition to any off-flavors would affect the score.

The descriptors represent flavor and odors that are present from various sources ranging from crude or partially processed oil to freshly processed oil to one that has deteriorated. For example, crude soybean oil has beany, grassy, and hay flavors. After sufficient processing and deodorization, soy oil has little or no odor or flavor and is usually described as bland or as slightly nutty or buttery. If the processing or deodorizing is not adequate, then the soybean oil may have a weak grassy or beany or hay flavor. As the oil oxidizes, odors and flavors become detectable. In the early stages of oxidation, soybean oil may have weak to moderate buttery odors and flavors. As oxidation continues, odors and flavors develop that are typical of a crude or partially processed oil. Sometimes this is referred to as a reverted flavor or retrograded flavor [7]. For soybean oils, these odors and flavors are grassy, hay, or beany.

Flavors detected in later stages of oxidation of soybean oil and canola oil include both rancid and painty, because both linoleic and linolenic acid are present in these oils. Oils such as peanut oil that contain no linolenic acid do not develop a painty flavor, as this is a flavor characteristic of linolenate-containing oils such as soybean and canola. Oils such as cottonseed with high linoleic acid levels will develop a predominant rancid flavor in highly oxidized samples. Frankel [29,30] presents reviews of fat and oil oxidation that explain both the mechanism of formation and the flavor significance of oxidation products.

B. Frying Oils

Methods to evaluate the room odor characteristics of heated and frying oils were developed in the 1970s by Evans and coworkers [31,32]. The initial method included simply heating a pan of oil in an empty laboratory. This method was upgraded by constructing small rooms ($5 \times 8 \times 10$ ft) with controlled air flow and temperature [20,33]. Judges were instructed to rate the overall intensity of room odor and the intensities of the individual odors detected. The intensity scoring scale is based on a 0–10 scale (no odor to strong odor). High oleic acid sunflower or low linolenic acid soybean oil (nonhydrogenated) heated to 190°C usually have the weakest overall odor [34]. Low erucic acid rapeseed (canola) oil and olive oil receive ratings of 7.0 or 8.0. Predominant odors for canola oil are fishy, acrid, and burnt. Hydrogenated oils have a characteristic "hydrogenation" odor. Descriptions such as fruity, flowery, and waxy are typical of hydrogenated oils.

VII. SALAD OILS, MAYONNAISE, AND SALAD DRESSINGS

Salad oil is generally a clear, light-colored vegetable oil that remains liquid at refrigerator temperatures (40°F or 4.4°C). Most oils are refined, bleached, and deodorized to remove flavor. Olive oil is generally not deodorized because it contributes a special flavor to salads and foods that is considered to be desirable. Cottonseed oils are generally winterized so that they remain liquid at refrigerator temperatures. Crystal inhibitors provide further protection against graining at low temperatures. Oil crystallization can accelerate mayonnaise or salad dressing separation at storage temperatures.

An excellent discussion of the preparation and use of mayonnaise and salad dressings has recently been published [35]. Mayonnaise and mayonnaise-type dressings are emulsified semisolid foods containing a minimum of 65% vegetable oil (one or a blend of two or more); acidifying agents (vinegar mixed with less than 25% by weight of citric acid); lemon juice, lime juice, or both; egg yolks; and, as optional ingredients, salt, sugar, corn syrup, honey or other syrups, dextrose, and other seasonings such as mustard, paprika, other spices or spice oils; and may contain ethylenediaminetetraacetate (EDTA) and antioxidants.

Stable mayonnaise emulsions can be prepared by both batch and continuous processes, and the time and degree of agitation exert a profound influence on the stability and taste of the final emulsion. The viscosity and stability of the product is influenced by the temperature of the oil and other ingredients during the mixing. Because emulsion fails at temperatures above 24°C (75°F), the initial temperatures should be 10–15°C (50–60°F) and the final product temperature should be 15–21°C (60–70°F).

Salad dressings are also emulsified semisolid foods but differ from the mayonnaise type by using a lower level of vegetable oil (30–40%) and by employing cooked starch materials, emulsifiers, and gums to provide stability and thickness. These food products are sometimes confused with mayonnaise as they are similar in taste and appearance, but they are generally cheaper.

French dressings are temporary emulsions of oil, vinegar and/or lemon juice, and seasonings. Shortly after mixing, these temporary emulsions separate, and they require shaking before each use. Egg yolk and/or other emulsifying agents are used to keep the oil in suspension to make a nonseparating French dressing.

Finished product quality is first evaluated by its appearance. The whiteness of the mayonnaise or dressing depends on the color of the eggs and oils used and the presence of some of the optional ingredients. Injected nitrogen gas, used to control the final viscosity of the product, can help to improve the whiteness of the product. The second consideration is the flavor and odor of the finished product, which are usually evaluated by a trained panel. Standard microbiological procedures test the product for safety as well as quality assurance.

VIII. FRYING OILS

A. Partial Hydrogenation

Partial hydrogenation of edible oils is practiced to increase stability by the selective reduction of linolenic acid. Commercially, a dual-purpose salad/cooking oil is prepared by hydrogenation of soybean oil with nickel catalysts under selective conditions as described in Chapter 10. The oil is hydrogenated to an iodine value of 110–115 and must be winterized to meet the requirements of the standard American Oil Chemists' Society cold test, Official Method Cc 11-53 [21]. This test calls for the oil to remain clear for a minimum of 5.5 h at 0°C. Stearine, high melting glycerides, and palmitic and stearic acid fractions are removed by a winterization process [36]. Oil is chilled slowly to about 6°C during a 24 h period; at this point cooling is stopped and the oil–crystal mixture is allowed to stand for 6–8 h. The yield of liquid oil is approximately 75–85%. By-product stearine is generally used for shortening manufacture. Lowering of the linolenic acid content does provide increased stability during the use of soybean oil as a cooking oil. It is still, however, only a partial solution to the problem, and thermal oxidation will eventually produce objectionable odors and flavors.

B. *Trans* Fatty Acids

Trans isomers of naturally occurring *cis*-unsaturated fatty acids are produced when liquid vegetable or marine oils are chemically hydrogenated to produce frying fats or other hardened fat products. An intense discussion has centered on the trans isomers formed during hydrogenation and their potential role in coronary heart disease. The health debate has centered most recently on work of Dutch researchers that showed that the *trans* fatty acids increase blood cholesterol in humans to the same extent as saturated fatty acids [37]. In addition, researchers at Harvard Medical School published an epidemiological study in 1993 claiming that *trans* fatty acid intake was directly related to the risk of coronary heart disease (CHD) [38]. The conclusions of the Harvard Medical School studies were criticized for biased estimates of intake, and the scientific community was warned to be cautious about these conclusions [39]. Two subsequent studies that based their assessment of *trans* fatty acid intake on analysis of the fatty acid composition of the adipose tissue of the subjects concluded that there was no convincing evidence that *trans* fatty acids are important causes of CHD [40] and that there were no correlations found between adipose tissue *trans* fatty acid content in men who had died suddenly from CHD compared to living controls [41]. While there continue to be some serious questions concerning conclusions that link *trans* fatty acids with CHD, a task group of the Danish Nutrition Council recommended that the consumption of *trans* fatty acids be reduced as much as possible [42]. Perhaps the most reasoned approach in this area is the recent emphasis that the priority in dietary management of CHD should remain the reduction of saturated fatty acids, which are the major cause of elevated cholesterol and coronary heart disease and not be diverted by the *trans* fatty acid factor, which fails to match the risk for CHD that saturates pose [43].

C. Deep Frying

Deep frying is a most useful method of cooking food because it is a fast method of cooking and most foods can be fried in under 5 min. Because of the short cooking period, it is possible to prepare food as it is required. In this way, waste is kept to a minimum. In deep frying the food is immersed in a very hot material that sears and seals the food, keeping in more flavor than many other methods of cooking. The layer of fat or oil deposited on the food during frying improves the eating quality of the food. Frying also imparts a characteristic flavor that is different from that produced by other methods of cooking.

Multiple changes in an oil occur during its use at frying temperatures. Colors are formed according to the type of food fried and any breading used. Accelerated oxidative deterioration of the oil occurs, with development of peroxides and polymers. Excessive foaming and an increase in oil viscosity with formation of gummy deposits are results of oil polymerization. Differences in oil stabilities are more apparent at high temperatures. As mentioned earlier, hydrolysis occurs in the presence of moisture-containing foods, and this results in an increase in free fatty acids.

Polymerization can result in foaming during the frying of foods. With the development of very high molecular weight polymers, the frying oil contains fatty acids of considerably different chain lengths. This difference in chain lengths results in foaming. Foaming is the formation of small bubbles that creep slowly up the sides of the fryer. If the fat foams excessively, it should be discarded, because foaming fat can pose a serious safety and fire hazard.

The effect of the amount of food on the foaming of a frying fat is illustrated in Fig. 5 [44]. The "zero rate" (no food frying, heating only) results in a rather high rate of oxidation, polymerization, and foaming. Under these conditions the frying oil is unprotected by the steam generated from moisture-containing foods or by turnover from fat absorption. More deterioration occurs at the low rate of frying because not enough fat adsorption occurs. At the high rate of frying, the more food fried, the more fat adsorption there is and the more rapid the replacement of used fat with fresh fat.

As shown in Fig. 6, the amount of free fatty acid formed during frying is directly proportional to the amount of steam released by the food into the fat. Differences in free fatty acid levels in frying fats do not necessarily constitute a differentiation between good and poor

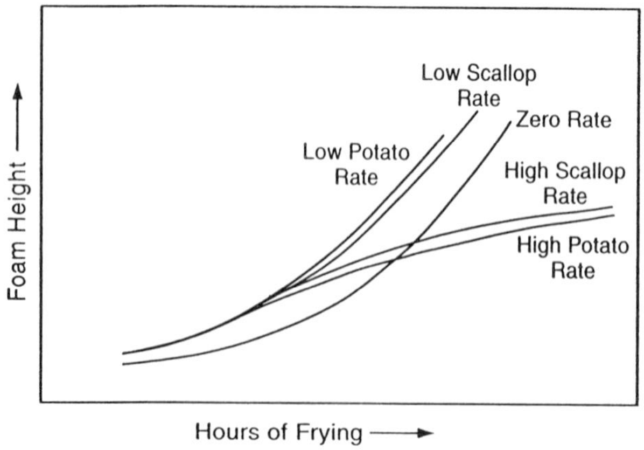

Fig. 5 Foam development during deep fat frying. (Reprinted with permission from Ref. 44.)

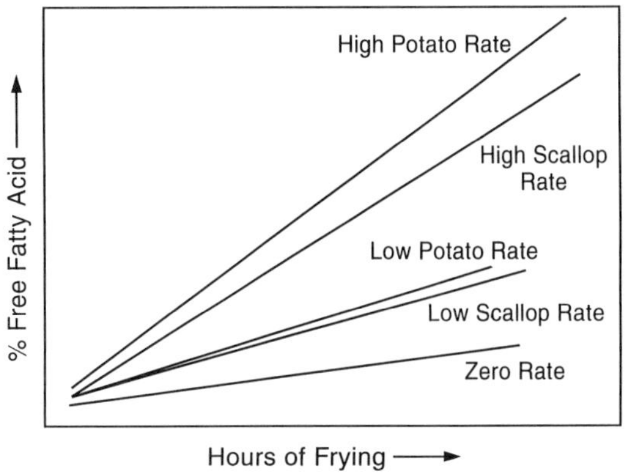

Fig. 6 Free fatty acid development during deep fat drying. (Reprinted with permission from Ref. 44.)

quality fats in the frying system. Free fatty acid level does not correlate well with fried food quality. Furthermore, free fatty acids are somewhat volatile at frying temperatures. Although frying large quantities of high moisture foods tends to increase the rate of free fatty acid development, it has a very positive effect on the life of the frying fat, because the rate at which fresh fat is added is increased.

Methyl silicone at 0.5–2.0 ppm effectively retards oxidation and polymerization, thus reducing foaming. Frying stability increases of three- to tenfold with the antifoamer have been confirmed in controlled laboratory frying evaluations. The degree of frying stability increase depends on the original stability of the frying oil before the antifoamer is added; the more stable original products show a greater increase in stability with the additive [45,46].

The mechanism by which silicone oil on the oil/air interface suppresses the thermal deterioration of frying oil was clarified in recent experiments [47]. The temperature of the air/oil interface was measured by a surface thermometer while heating the frying oil and was usually lower than that of the bulk oil in the fryer. The difference in temperatures of the surface and bulk was greater in oil with silicone oil than in that without it. It thus appears that silicone oil effectively slows down the convection current of frying oil.

Oil quality during frying and the quality of food produced in it are intimately related. The degradation process has been described by Blumenthal [48] as consisting of five oil phases (Fig. 7). The initial phase, break-in oil, is characterized by no cooked odors, no crisping of the surface, little oil pickup by the food, and generally a white product with raw, ungelatinized starch at the center of the fry. In the subsequent phase, fresh oil, a slight browning at the edges of the fry is observed as well as partially cooked (gelatinized) centers, crisping of the surface, and slightly more oil absorption. The next phase is optimum oil, in which is obtained the golden-brown color; crisp, rigid surfaces; delicious potato and oil odors; fully cooked centers; and optimal oil absorption. On the down side of the curve is the degrading oil phase with darkened and/or spotty surfaces, excess oil pickup, limp product, and case-hardened surfaces. The final phase is runaway oil, recognized by dark, case-hardened surfaces, excessively oily products, surfaces collapsing inward, centers not fully cooked, and off-odors and flavor (burned).

Liquid shortenings containing the methyl silicone antifoamer were introduced with claims of the convenience of a pourable product and longer frying stabilities than the specially hydrogenated solid frying shortenings. Frying tests confirmed these claims until the antifoamer was also incorporated into the specially hydrogenated solid frying shortenings. However, the

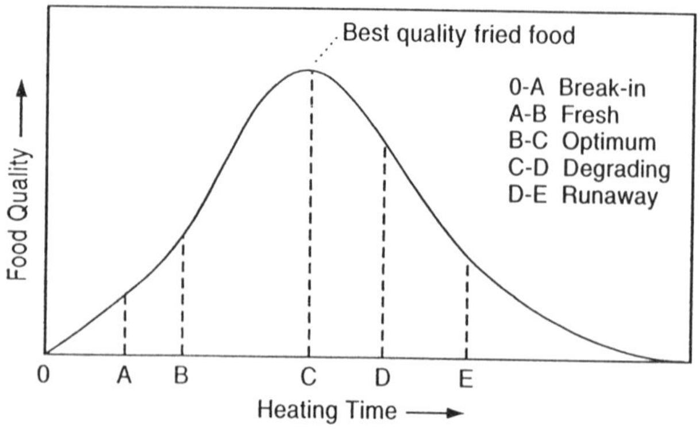

Fig. 7 Frying oil quality curve showing the five phases that a frying oil passes through during the degradation process. (Reprinted with permission from Libra Laboratories.)

convenience of the liquid shortening, the long frying stability, and the slightly oilier food appearance have made the liquid frying shortening type a major food service ingredient.

Before the addition of antifoamers to frying oils, foaming was the indicator used by food service operators to decide when to replace their frying oil. Now, a frying fat that reaches the persistent foaming stage may have gone beyond the discard point. Currently, food service operators use several frying oil discard indicators, which include the following [49]:

Some operators rely on color, discarding the oil when it reaches a certain darkness or when visibility is impaired at a definite distance under the oil surface.

Smoke is caused by hydrolysis and oxidation, which lower the smoke point, so some operators replace their frying oil when smoking reaches a certain intensity.

Some operators replace their frying oil after a prescribed time has elapsed or a specified quantity of food has been fried.

A number of test kits have been introduced to measure color, conductivity, free fatty acid, and other factors.

Foaming is still used as an indicator by some operators and is acceptable if the amount of foaming is not excessive.

A measure of the heat abuse taking place in an oil and an understanding of the products formed during deep fat frying are of great interest and importance to the consumer and to the food industry. Therefore, a review of the methods available for studying changes that take place during deep fat frying is in order.

D. Standard Methods

Several methods to establish the quality of used frying fats have been widely used and subsequently adopted as standard procedures.

1. Polar Components

Total polar materials are determined by dissolving a weighed amount of fat (2.5 g) in light petroleum ether–diethyl ether (87:13) and running it through a silica gel column that absorbs the polar compounds [50,51]. After evaporation of the eluted solvent, the nonpolar fat is weighed and the total polar material is estimated by difference. If desired, the polar material remaining on the column can be eluted with diethyl ether. A level of 27% polar components has been suggested as the limit beyond which a restaurant should discard its frying oil [52].

2. *Conjugated Dienoic Acids*

When polyunsaturated fatty acids are oxidized, a shift in one of the double bonds occurs, producing a conjugated diene that can be measured by ultraviolet absorption at 232 nm [50]. The absorbance increases initially and then plateaus during frying. This has been related to the establishment of an equilibrium between the rate of formation of conjugated dienes and the rate of formation of polymers formed by a Diels–Alder reaction involving conjugated dienes. This test is most useful in measuring heat abuse of polyunsaturated oils but is less applicable to fats containing few unsaturates [53].

3. *Fatty Analyses and 18:2/16:0 Ratio*

Thompson and Aust [54] found that the total linoleic and linolenic acid content of a lightly hydrogenated frying oil, after 100 h of frying, dropped by 50%. The relative amount of saturated fatty acids thus increased. Augustin et al. [55] found good correlations between the 18:2/16:0 ratio in used frying oil and results from other methods.

E. Quick Tests

Several commercially available procedures have been developed to provide a rapid means of testing for frying oil abuse. The tests are based on one chemical change taking place in the oil and are not always accurate. However, a speedy test for estimating oil abuse may be all that is necessary.

1. *Dielectric Constant*

A quick method that measures the dielectric constant has been used to estimate frying oil degradation [56,57]. As the oil degrades, there is an increase in the dielectric constant. This procedure is useful in some settings, but it would be difficult to adapt it to real situations because the reading is influenced by many outside factors such as water or fat extracted from the fried food. Nonetheless, the method is useful in defined circumstances. A portable electronic instrument, the food oil sensor (FOS, Northern States Instrument Corp., Lino Lakes, MN), measures the dielectric constant in frying fat relative to fresh oil [58]. It is suggested that an FOS reading of 4.0 should be the limit beyond which a used fat should be discarded [59].

2. *Oxifrit-Test*

The Oxifrit-Test (previously the RAU test, E. Merck, Darmstadt, Germany) [60] is a colorimetric test kit that contains redox indicators that react with the total amount of oxidized compounds.

3. *Fritest*

The Fritest is a colorimetric test kit that is sensitive to carbonyl compounds (E. Merck, Darmstadt, Germany) [61]. The mixture of sample and reagent is compared to three diagnostic colors, with 1 representing good, 2 intermediate, and 3 bad.

4. *Spot Test*

The spot test is a colorimetric procedure [62]. To test the oil, a drop is placed on a silica gel–covered slide that has bromocresol green incorporated into the gel as a pH indicator. The test monitors the free fatty acid content as an indicator of hydrolytic rancidity. The diagnostic colors of the pH indicator are blue, green, and yellow.

5. *Alkaline Contaminant Materials, Quick Test*

A colorimetric quick test produced by Libra Labs (Piscataway, NJ) shows the presence and semiquantitative concentration of alkaline contaminant materials (ACMs), such as soaps that accumulate in used frying fats from the interaction of food materials with oil degradation products [63]. A test kit solution containing bromophenol blue as a pH indicator, sodium hydroxide, and hydrochloric acid is added to the oil, and the oil and solution are shaken in a test tube. The oil quickly forms a lower layer, and the transparent upper layer assumes a color directly proportional to the amount of soaps or ACMs extracted from the oil. Studies [64] indicated that at 43 ppm ACM, frying oil is ready to be discarded.

In a study of nine analytical methods [65] the polymer/FOS ratio not only proved to be adequate for monitoring the quality of the used oil but was also affected minimally by replenishment.

White [66] summarized recent research on comparing procedures for evaluating frying oil abuse, i.e., polar components, conjugated dienoic acids, fatty acid analyses and 18:2/16:0 ratio, and several quick tests: dielectric constant, RAU test, Fritest, spot test, and alkaline contaminant materials. In most studies the determination of polar components was proven to be accurate, simple, and reproducible. White's conclusion concerning the choosing of a method was that the choice will depend on the accuracy of measurement desired, the money available for establishing and running the tests, and the time allowed. Perhaps more important, the choice will depend on the purpose for measuring the frying oil.

The Swedish National Food Administration (NFA) prepared a document in 1989 presenting advice and guidelines on handling deep frying fats. An English edition was issued in June 1990 [67].

1. All the fat in the deep fat fryer must be changed before it starts smoking or foaming. Use an indicator such as a food oil sensor or Oxifrit test to indicate when it is time to change.
2. Strain the fat and clean the fryer once a day. Rinse carefully after cleaning. Solid material in the fat and detergent residues accelerate breakdown of the fat. Store strained fats at room temperature or at lower temperatures in a covered stainless steel vessel. If iron pots are used, they should be rinsed only with hot water. Detergents remove the protective film of polymerized fat that builds up during use.
3. The frying temperature should be 160–180°C (320–356°F). At lower temperatures, the product absorbs more fat. At higher temperatures the fat deteriorates more quickly.
4. Use fat that is specially intended for frying.
5. Avoid salting or seasoning the fried food over the fryer. Salt or seasoning can accelerate breakdown of the fat.
6. Lower the temperature when not frying, and protect the fat from light.
7. The fryer should have no iron, copper, or brass parts that come into contact with the heated fat.
8. Keep a constant level of fat in the fryer. Fry a little at a time to keep the temperature as even as possible. Prefry when large amounts are to be prepared.

9. Use a separate fryer, if possible, for frying potatoes. The fat deteriorates more rapidly when meat or fish is fried than when only potatoes are fried.

Caution: Do not overheat. If the fat temperature rises above 300°C (572°F), the fat may start to burn.

The purpose of the guidelines was to encourage employees in food establishments to prepare high quality fried foods. In Sweden, antifoam agents, such as silicones, are not permitted in frying oils because they mask the natural foaming in deteriorated oils [68].

IX. SAFETY ASPECTS OF FRYING FATS AND OILS

Clark and Serbia [69] present an excellent review of the scientific literature concerning the safety of heated fats and oils. It is well established that the heating of fats can result in the formation of compounds with antinutritional properties. Also, studies involving exaggerated stress of fat or oil, well in excess of the conditions actually encountered in home or commercial frying, have produced toxic reaction products. Cooking at such conditions yields totally unpalatable foods. The conclusion was that although excessive cooking conditions can produce toxic compounds in fats and oils, the fats and oils become unacceptable from a sensory standpoint much earlier.

Recent research has indicated mutagenic activity of polar fractions isolated from repeatedly used deep frying fats [70]. In these studies, samples were collected from restaurants and snack bars after sensory indication of abuse. Of the 20 deep frying fat samples obtained, levels of diglycerides and polymeric triglycerides in 13 of the samples exceeded the threshold level of 10% above which fats are rejected for use. The significance of these findings remains to be determined.

X. NEW OILSEED VARIETIES

Soybean and canola varieties that yield oils with low linolenate content have been produced by hybridization. At cooking and frying temperatures, these modified oils lack the objectionable fishy odor generated by normal soybean oil and the hydrogenation odor of partially hydrogenated cooking oil.

The development of these genetically modified varieties has generated much interest in the food oil industry. Currently, soybean and canola oils must be chemically altered by catalytic hydrogenation for improved stability and extended use at high cooking and frying temperatures in restaurants, institutions, and elsewhere in the food industry. Theoretically, soybean and canola oils that are naturally low in linolenate should not require hydrogenation for most applications. In addition, foods produced from such oils would not contain the positional and geometrical isomers of poly- and monounsaturated fatty acids produced by catalytic hydrogenation. Studies [71–75] have reported the oxidative stability and flavor quality of modified soybean and canola oils having less than 3% linolenate for use as cooking and frying oils.

Other efforts have focused on raising the oleate content. High oleic sunflower production has experienced rapid growth since 1988, with high oleic sunflower seed currently accounting for 10–15% of the total U.S. oil seed crop. Industry researchers believe that the potential for specially tailored sunflower oil, particularly high oleic oil, may overshadow that of conventional sunflower oil [76].

Recent research [77] evaluated blends of commercial cottonseed oil and high oleic sunflower oil to achieve ranges of linoleic acid of 9–54% and of oleic acid, 20–80%. Results indicated that potato chips and fried potatoes prepared in 100% cottonseed oil had the highest intensities of fried food flavor; however, this positive flavor characteristic decreased with decreasing levels of cottonseed oil in the blends. This work suggested that there may be a limitation on the level of the oleic acid content of sunflower oil consistent with obtaining flavor quality of the foods prepared therein.

REFERENCES

1. D. R. Erickson and G. R. List, Storage, handling and stabilization of edible fats and oils, in *Bailey's Industrial Oil and Fat Products*, Vol. 3 (T. H. Applewhite, Ed.), Wiley, New York, 1985, p. 275.
2. E. G. Perkins, Formation of non-volatile decomposition products in heated fats and oils, *Food Technol. 21*: 611 (1967).
3. F. D. Gunstone and T. P. Hilditch, The union of gaseous oxygen with methyl oleate, linoleate and linolenate, *J. Chem. Soc. 1945*: 836.
4. G. R. Greenbank and G. E. Holm, Photochemical oxidation of cottonseed oil, *Ind. Eng. Chem. 25*: 167 (1933).
5. G. R. Greenbank and G. E. Holm, Promoting the oxidation of fats and oils. Relative effectiveness of different bands of the visible spectrum, *Ind. Eng. Chem. 33*: 1058 (1941).
6. S. W. Thompson, Proceedings of the Committee on Food Research, Conference on Deterioration of Fats and Oils, Quartermaster Corps Manual 17-7, 1945.
7. T. H. Smouse, Flavor reversion of soybean oil, in *Flavor Chemistry and Fats and Oils* (D. B. Min and T. H. Smouse, Eds.), Am. Oil Chem. Soc. Champaign, IL, 1985, p. 85.
8. H. J. Dutton, A. W. Schwab, H. A. Moser, and J. C. Cowan, The flavor problem of soybean oil. IV. Structure of compounds counteracting the effect of prooxidant metals, *J. Am. Oil Chem. Soc. 25*: 385 (1948).
9. K. Warner, E. N. Frankel, and T. L. Mounts, Flavor and oxidative stability of soybean, sunflower and low erucic acid rapeseed oils, *J. Am. Oil Chem. Soc. 66*: 558–564 (1989).
10. K. S. Law, A review on the use of citric acid in the processing of oils and fats, *Oleagineaux 39*: 89 (1984).
11. J. C. Cowan, Advances in research on the flavor stability of edible soybean oil—a review, *Food Technol. 19*: 1413 (1965).
12. A. K. Kiritsakis, C. M. Stine, and L. R. Dugan, Effect of selected antioxidants on the stability of virgin olive oil, *J. Am. Oil Chem. Soc. 60*: 1286 (1983).
13. T. L. Mounts, K. A. Warner, G. R. List, J. P. Friedrich, and S. Koritala, Flavor and oxidative stability of hydrogenated and unhydrogenated soybean oils: effects of antioxidants, *J. Am. Oil Chem. Soc. 55*: 345 (1978).
14. P. P. Coppen, in *Rancidity in Foods* (J. C. Allen and R. J. Hamilton, Eds.), Elsevier, London, 1983, p. 67.
15. L. M. Wright, Finished oil handling and storage, *J. Am. Oil Chem. Soc. 53*: 408 (1976).
16. T. W. Eselgroth, Inert gas: safeguard of quality, *Food Eng. 23*: 72 (1951).
17. Anonymous, *Treatment and Protection of Edible Oils to Preserve Quality and Increase Storage Life*, Tech. Data Proc. Bull., Air Reduction Sales Co., Madison, WI, 1961.
18. L. H. Going, Oxidative deterioration of partially processed soybean oil, *J. Am. Oil Chem. Soc. 45*: 632 (1968).
19. V. C. Mellenbacher, *The Analysis of Fats and Oils*, Garrard, Champaign, IL, 1960.
20. T. L. Mounts and K. Warner, Evaluation of finished oil quality, in *Handbook of Soy Oil Processing and Utilization* (D. R. Erickson, E. H. Pryde, O. L. Brekke, T. L. Mounts, and R. A. Falb, Eds.), Am. Soybean Assoc. and Am. Oil Chem. Soc., Champaign, IL, 1980, p. 262.
21. *Official Methods and Recommended Practices of the American Oil Chemists' Society*, 4th ed., AOCS, Champaign, IL, 1990.
22. J. M. deMan, F. Tie, and L. deMan, Formation of short chain volatile organic acids in the automated AOM method, *J. Am. Oil Chem. Soc. 64*: 993 (1987).

23. C. D. Evans, G. R. List, R. L. Hoffman, and H. Moser, Edible oil quality as measured by thermal release of pentane, *J. Am. Oil Chem. Soc. 46*: 501 (1969).
24. K. Warner, C. D. Evans, G. R. List, H. P. Dupuy, J. I. Wadsworth, and G. E. Goheen, Flavor score correlation with pentanal and hexanal contents of vegetable oil, *J. Am. Oil Chem. Soc. 55*: 252 (1978).
25. H. W. Jackson, G. K. Mirmira, and W. M. Wagner, Oxidative oil stability using volatiles methodology, *J. Am. Oil Chem. Soc. 63*: 117 (1986).
26. K. Warner, E. N. Frankel, and K. J. Moulton, Flavor evaluation of crude oil to predict the quality of soybean oil, *J. Am. Oil Chem. Soc. 65*: 386 (1988).
27. J. Snyder, E. N. Frankel, E. Selke, and K. Warner, Comparison of gas chromatographic methods for volatile lipid oxidation compounds in soybean oil, *J. Am. Oil Chem. Soc. 65*: 1617 (1988).
28. American Society for Testing and Materials, STP 434, *Manual on Sensory Testing Methods*, ASTM, Philadelphia, PA, 1968.
29. E. N. Frankel, Chemistry of autoxidation: mechanism, products and flavor significance, in *Flavor Chemistry of Fats and Oils* (D. Min and T. H. Smouse, Eds.), Am. Oil Chem. Soc., Champaign, IL, 1985, p. 1.
30. E. N. Frankel, Volatile lipid oxidation products, *Prog. Lipid Res. 22*: 1 (1982).
31. C. D. Evans, H. A. Moser, G. R. List, H. J. Dutton, and J. C. Cowan, Edible oil evaluation by room odor tests, *J. Am. Oil Chem. Soc. 48*: 711 (1971).
32. C. D. Evans, K. Warner, G. R. List, and J. C. Cowan, Room odor evaluation of oils and cooking fats, *J. Am. Oil Chem. Soc. 49*: 578 (1972).
33. T. L. Mounts, Odor considerations in the use of frying oils, *J. Am. Oil Chem. Soc. 56*: 659 (1979).
34. K. Warner, Sensory evaluation of flavor quality of oils, in *Flavor Chemistry of Fats and Oils* (D. Min and T. H. Smouse, Eds.), Am. Oil Chem. Soc. Champaign, IL, 1985, p. 207.
35. A. Moustafa, Salad oil, mayonnaise, and salad dressing, in *Practical Handbook of Soybean Processing and Utilization* (D. R. Erickson, Ed.), Am. Oil Chem. Soc., Champaign, IL, 1995, p. 314.
36. G. M. Numenz, Old and new in winterizing. *J. Am. Oil Chem. Soc. 55*: 396A (1978).
37. R. P. Mensink and M. B. Katan, Effect of dietary *trans* fatty acids on levels of high-density and low-density lipoprotein cholesterol in healthy subjects, *N. Engl. J. Med. 323*: 439 (1990).
38. W. C. Willett, M. J. Stampfer, J. E. Manson, G. A. Colditz, F. E. Speizer, B. A. Rosner, L. A. Sampson, and C. H. Hennekens, Intake of *trans* fatty acids and risk of coronary heart disease among women, *Lancet 341*: 581 (1993).
39. P. McKeigue, *Trans* fatty acids and coronary heart disease: weighing the evidence against hardened fat, *Lancet 345*: 269 (1995).
40. A. Aro, A. F. M. Kardinaal, I. Salminen, J. D. Kark, R. A. Riemersma, M. Delgado-Rodriguez, J. Gomez-Aracena, J. K. Huttunen, L. Kohlmeier, B. C. Martin, J. M. Martin-Moreno, V. P. Mazaev, J. Ringstad, M. Thamm, P. van't Veer, and F. J. Kok, Adipose tissue isomeric *trans* fatty acids and risk of myocardial infarction in 9 countries: the EURAMIC study, *Lancet 345*: 273 (1995).
41. T. L. Roberts, D. A. Wood, R. A. Riemersma, P. J. Gallagher, and F. C. Lampe, *Trans* isomers of oleic and linoleic acid in adipose tissue and sudden cardiac death, *Lancet 345*: 278 (1995).
42. Transfedtsyrers betydning for sudhenen—Report of a task force of the Danish Nutrition Council, Copenhagen, Denmark.
43. P. J. Nestel, Comment on *trans* fatty acids and coronary heart disease risk, *Am. J. Clin. Nutr. 62*: 522 (1995).
44. H. Lawson, *Food Oils and Fats—Technology, Utilization, and Nutrition*, Chapman & Hall, New York, 1994, p. 66.
45. I. P. Freeman, F. B. Padley, and W. L. Sheppard, Use of silicones in frying oils, *J. Am. Oil Chem. Soc. 50*: 101 (1973).
46. S. Ohta, Effect of silicone oil on the thermal degradation of frying oil, *Yushi 42*: 79 (1989).

47. S. Ohta, Y. Shimizu, M. Kinugawa, K. Tosaki, and Y. Tottori. Functions of silicone oil in frying oil. X. Reconsideration of previous assumptions, *Yukagaku 39*: 23 (1990).
48. M. M. Blumenthal, A new look at the chemistry and physics of deep-fat frying, *Food Technol. 45*: 68 (1991).
49. R. D. O'Brien, Food service use of fats and oils, *INFORM 4*: 913 (1993).
50. Anonymous, Evidence of purity and determination from ultraviolet spectrophotometry, Method 2.505 and Determination of polar compounds in frying fats, Method 2.507, in *Standard Methods for Analysis of Oils, Fats and Derivatives*, 7th ed. (C. Paquot and A. Hautfenne, Eds.), Blackwell, Oxford, England, 1987.
51. Anonymous, Polar components of frying fats, Method 18.074, in *Official Methods of Analysis*, 14th ed., AOAC, Washington, DC, 1984.
52. A. J. Paradis and W. W. Nawar, Evaluation of new methods for the assessment of used frying oils, *J. Food Sci. 46*: 449 (1981).
53. M. Peled, T. Gutfinger, and A. Letan, Effect of water and BHT on stability of cottonseed oil during frying, *J. Sci. Food Agric. 26*: 1655 (1975).
54. L. U. Thompson and R. Aust, Lipid changes in french fries and heated oils during commercial deep frying and their nutritional and toxicological implications, *Can. Inst. Food Sci. Technol. J. 16*: 246 (1983).
55. M. A. Augustin, A. Telingai, and L. K. Heng, Relationships between measurements of fat deterioration during heating and frying in RBD olein, *J. Am. Oil Chem. Soc. 64*: 1670 (1987).
56. V. J. Graziano, Portable instrument rapidly measures quality of frying fat in food service operations, *Food Technol. 33*: 50 (1979).
57. C. W. Fritsch, D. C. Egberg, and J. S. Mangnuson, Changes in dielectric constant as a measure of frying oil deterioration, *J. Am. Oil Chem. Soc. 56*: 746 (1979).
58. C. W. Fritsch, Measurements of frying fat deterioration: a brief review, *J. Am. Oil Chem. Soc. 58*: 272 (1981).
59. L. M. Smith, A. J. Clifford, C. L. Hamblin, and R. K. Creveling, Changes in physical and chemical properties of shortenings used for commercial deep-fat frying, *J. Am. Oil Chem. Soc. 63*: 1017 (1986).
60. L. B. Croon, A. Rogstad, T. Leath, and T. Kiutamo, A comparative study of analytical methods for quality evaluation of frying fat, *Fette Seifen Anstrichm. 88*: 87 (1986).
61. H. V. Von Zeddelmann, Probleme bei der Beurteilung von Fritierfetten, *Lebensm.-Rundsch. 69*: 81 (1973).
62. H. Robern and L. Gray, A colorimetric spot test for heated oils, *Can. Inst. Food Sci. Technol. J. 14*: 150 (1981).
63. M. M. Blumenthal, J. R. Stockler, and P. J. Summers, Alkaline contaminant materials in used frying oils: a new quick test, *J. Am. Oil Chem. Soc. 62*: 1373 (1985).
64. M. M. Blumenthal and J. R. Stockler, Isolation and detection of alkaline contaminant materials (ACM) in used frying oils, *J. Am. Oil Chem. Soc. 62*: 1373 (1985).
65. P.-F. Wu and W. W. Nawar, A technique for monitoring the quality of used frying oils, *J. Am. Oil Chem. Soc. 63*: 1363 (1986).
66. P. J. White, Methods for measuring changes in deep-fat frying oils, *Food Technol. 45*: 75 (1991).
67. Anon., General advice on handling frying fats (SLV FS 1990:2), National Food Administration, Uppsala, Sweden, 1990.
68. D. Firestone, Worldwide regulation of frying fats and oils, *INFORM 4*: 1366 (1993).
69. W. L. Clark and G. W. Serbia, Safety aspects of frying fats and oils, *Food Technol. 45*: 84 (1991).
70. G. Hageman, R. Kikken, F. Ten Hoor, and J. Kleinjans, Assessment of mutagenic activity of repeatedly used deep-frying fats, *Mutagen Res. 204*: 593 (1988).
71. T. L. Mounts, K. Warner, G. R. List, R. Kleiman, W. R. Fehr, E. G. Hammond, and J. R. Wilcox, *J. Am. Oil Chem. Soc. 65*: 624 (1988).
72. H. Liu and P. J. White, High-temperature stability of soybean oils with altered fatty acid composition, *J. Am. Oil Chem. Soc. 69*: 533–538 (1992).

73. K. Warner and T. L. Mounts, Effect on frying stability of canola oils with modified fatty acid compositions, *J. Am. Oil Chem. Soc. 70*: 983 (1993).
74. T. L. Mounts, K. Warner, and G. R. List, End-use performance evaluation of genetically modified soybean oils, *J. Am. Oil Chem. Soc. 71*: 157 (1994).
75. T. L. Mounts, K. Warner, G. R. List, W. E. Neff, and R. W. Wilson, Low-linolenic acid soybean oils—alternatives to hydrogenated cooking oils?, *J. Am. Oil Chem. Soc. 71*: 495 (1994).
76. B. Fitch Haumann, Modified oil may be key to sunflower's future, *INFORM 5*: 1198 (1994).
77. K. Warner, P. Orr, and M. Glynn, Effect of fatty acid composition on fried-food flavor quality and stability, *INFORM 6*: 507 (1995).

17

Edible Coatings and Film Barriers

Thomas H. Shellhammer and John M. Krochta
University of California–Davis, Davis, California

I. INTRODUCTION

An edible coating can be defined as a thin film of edible material that is formed on a food surface. With some materials, a stand-alone film can be formed for placement between food components. The purpose of an edible coating or film is to control mass transfer, improve mechanical handling, and/or modify external appearance. Controlling mass transfer involves inhibiting the movement of moisture, permanent gases, aromas, and lipids among heterogeneous food components or between a food and its external environment. Release of food ingredients such as flavors, antioxidants, or antimicrobial agents is a possibility from within the film or coating itself. Edible coating and film materials comprise three general categories of materials: polysaccharides, proteins, and lipids. The first two categories are hydrophilic and are therefore effective at minimizing the transport of gases (at low relative humidity) but are poor moisture barriers. Lipids, on the other hand, are very effective moisture barriers and somewhat ineffective gas barriers.

Lipid coatings are used primarily to inhibit moisture loss from foods and also to improve consumer appeal by adding a glossy finish to the coated product. The application of lipid film coatings to retard moisture loss, induce flavor changes, or alter product appearance is not a new practice. As early as the twelfth or thirteenth centuries, the Chinese used wax to ferment oranges and lemons [1]. Enrobing food in fat, a process called larding, was used in England during the sixteenth century to minimize moisture loss [2]. For many years, waxes have been used as protection against contamination and moisture loss [3]. More current roles for lipid coatings include carrying ingredients to a food surface, protecting encapsulated materials from moisture uptake, and reducing gas transport for the purpose of decreasing respiration in fresh produce. Lipids are used as coating agents for a broad and diverse set of products. They are used on fruits and vegetables, confectioneries, nuts, dried fruit, refrigerated and frozen meat, starch-based baked products, and moisture-sensitive encapsulated ingredients. While conventional packaging can offer protection from mass transfer for bulk materials and individually packaged products, edible lipid coatings offer continued protection for the product after the conventional package has been opened. Yet the incomparable advantage of lipid coatings and films comes from their ability to retard moisture transport within heterogeneous food materials. It is not feasible to include synthetic, inedible packaging materials within food products. However, edible lipid coatings and films can conveniently perform a moisture barrier function within a heterogeneous food without otherwise affecting food quality.

Many material-specific properties determine the appropriateness of a particular lipid for use as a coating or film. These include blocking (self-adhering), color, compatibility with other ingredients, crystalline structure, ductility, edibility, emulsifiability, flavor, flexibility, gas permeability (oxygen, carbon dioxide, nitrogen, and ethylene), hardness, melting point, odor, plasticity, resistance to bacteria and fungi, sheen and reflectivity, stability to light and oxidation, surface tension, and water vapor permeability [4]. Additionally, other non-material-specific factors such as availability, price, ease of handling (i.e., bulk, slab, flakes, or powder), and legal restrictions (i.e., health, safety, or patents) must be considered.

II. FACTORS AFFECTING MASS TRANSPORT IN LIPID FILMS

Barrier properties of film and coating materials are generally described through a permeability coefficient. The permeability of a penetrant, such as water or oxygen, through a film is a function of three factors: (1) the nature of the film material, (2) the nature of the penetrant, and (3) interactions between the penetrant and the film [5]. The transport process is a combination of sorption of the penetrant on the high concentration side of the film, diffusion of the penetrant through the film, and desorption of the penetrant from the film on the low concentration side (Fig. 1).

Sorption is a general term used to describe dispersal of a penetrant into a matrix to form a mixture [6]. The equilibrium condition existing between the concentration of the penetrant outside a film matrix and its concentration in the matrix (assuming a dilute concentration) can be described through Henry's law [7]:

$$c_A = S\, p_A \tag{1}$$

The molecular concentration outside the film material is expressed in terms of its partial pressure, p_A (e.g., kilopascals), and the concentration of the penetrant has units of moles per unit volume (mol/cm^3). The constant describing the equilibrium relationship is a solubility coefficient, or Henry's constant. Figure 1 displays this relationship graphically. The concen-

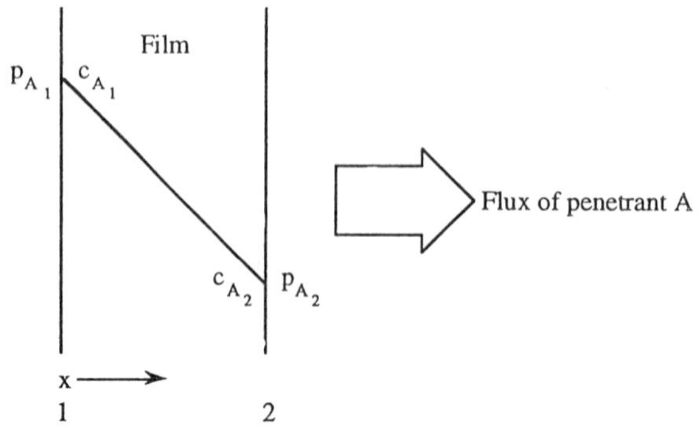

Fig. 1 The transport of a penetrant through a film. P_A = partial pressure of A; c_A = molar concentration of A.

tration in the film at point 1 or point 2 is merely the partial pressure of A outside the film divided by the material-specific solubility coefficient.

Diffusion is the process by which a penetrant molecule travels from one position in a matrix to another position as a result of random molecular motions and because of an imposed concentration gradient. This process can be described as a Fickian diffusion [8], which can be represented by the following equation in one dimension:

$$J = -D\frac{\partial c_A}{\partial x} \tag{2}$$

The flux J of species A is proportional to the concentration driving force ($\partial c_A/\partial x$) via the diffusion coefficient or diffusivity D. Assuming both D and S are constant, Eqs. (1) and (2) can be combined and integrated to yield a third equation that describes the permeability of the penetrant through the film solely in terms of the partial pressure driving force,

$$J = P\frac{\Delta p_A}{\Delta x} \tag{3}$$

where P = permeability [g·mm/(m^2·day·kPa)], Δp_A = partial pressure driving force (kPa), Δx = film thickness (mm), and J = flux of molecule A [g/(m^2·day)].

The permeability coefficient is the product of the diffusivity and solubility coefficients ($P = DS$). Thus, the transport through the film can be characterized by controlling only the concentration of the penetrant outside the film rather than having to determine its concentration within the film. Determining the concentration of the penetrant in the film matrix is achieved through sorption and desorption isotherms. The process of measuring the isotherms is time-consuming and tedious. Therefore, researchers often focus their efforts on measuring the permeability of the penetrant through the film material, since the partial pressures of the components outside the film are somewhat easier to control.

Because the transport of a molecule through a film depends on both its solubility in and diffusivity through the film, factors that affect either of these two properties influence the barrier properties of the film. With regard to solubility, the state of the film material structure influences the penetrant's compatibility in the matrix. Relative polarities of the penetrant and the matrix material strongly influence solubility, as the general principle of "like dissolves like" is applicable [9]. Nonpolar species such as permanent gases and organic vapors are readily soluble in hydrophobic films, whereas water, which is strongly polar, is sparingly soluble. Consequently, hydrophobic films are permeable to hydrophobic penetrants but are effective barriers to moisture. Such is the case with lipid films. Of course, the converse is true, as in the case of hydrophilic films such as ethyl vinyl alcohol and edible protein films. In these films, water is very soluble, whereas hydrophobic species are not [10,11].

Anything that affects the "tightness" of the molecular structure of the film material network will have a significant effect on the diffusivity, and therefore the permeability, of the penetrant molecule. Factors affecting the permeability, especially the diffusivity, are the film material density, its free volume, coefficient of expansion, cohesive energy density, and crystallinity (morphology and orientation) [12]. Increasing temperature increases molecular diffusivity [13]. The increased thermal motion of lipid molecules in their liquid state, along with changes in their solubility characteristics, increases the movement of the penetrant through the matrix [5]. The barrier properties of lipid films are highly temperature-dependent for these reasons. Many lipids crystallize at slightly elevated temperatures and can be a mixture of crystalline and liquid phases at room temperature. Thermal motion of the lipid molecules is much higher in the liquid phase than in the crystalline phase; hence the transport of penetrants in these mixtures is much higher than in completely crystalline systems.

III. LIPID MATERIALS AVAILABLE FOR EDIBLE COATINGS AND FILMS

Many types of lipids have been examined for use as coating or film-forming materials. These include waxes, fats, oils, and lipid-containing materials such as chocolate liquor and chocolate. Table 1 lists the lipid materials approved by the U.S. Food and Drug Administration (FDA) for use as edible coatings, their applications, and their federally regulated limits as listed in the *Code of Federal Regulations* (CFR). This table was produced from lists of food ingredients in CFR title 21, chapter I, subchapter B, parts 172 and 184 [14,15]. CFR 21.172 lists food additives that are permitted for direct addition to food for human consumption, while CFR 21.184 refers to food substances that have been reaffirmed as generally recognized as safe (GRAS). Table 1 is a representative list of lipid coating materials rather than a comprehensive list. Lipid materials, such as hydrogenated soybean oil, that are not listed in Table 1 can be used as coatings but are not explicitly stated as coating materials in CFR 21.172 or 21.184. Table 2 lists the physical properties of common lipid coating materials.

Waxes are naturally occurring esters of long-chain carboxylic acids (C_{16} or greater) [16]. Waxes can be classified on the basis of their source into five general groups: mineral (fossil and petroleum), vegetable, animal, synthetic, or compound [4,17]. The mineral wax category is dominated by paraffin wax, to a lesser degree by microcrystalline wax, but also comprises petroleum, montan, ozocerite, and peat waxes. Of these materials, only paraffin and petroleum wax are approved for use as a food coatings, although microcrystalline and montan wax are approved for use on food contact surfaces as in packaging materials [18]. Paraffin wax is a purified mixture of hydrocarbons obtained from petroleum that has a crystalline structure and no color, odor, or taste [4]. It is sold in various grades, with the main difference being melting point. Refiners produce paraffin waxes covering a wide range of melting points from 27 to 69°C. Higher melting waxes, which are sold as microcrystalline, amorphous, and petroleum waxes, are derived from heavier lubricating oil fractions of petroleum distillates.

Of the vegetable waxes, candelilla, carnauba, and ricebran wax are approved for food use at a level defined as "good manufacturing practices." Cotton wax, esparto (or reed) wax, fir wax, flax wax, Japan wax, palm waxes, and sugar cane wax can be classified as vegetable waxes but do not have U.S. federal approval for use in or on foods. Candelilla wax is produced by reedlike plants (*Euphorbia antisiphilitica*, *Euphorbia cerifera*, and *Pedilanthus pavonis*) that grow wild in northwestern Mexico and southern Texas [4]. It is composed in large part of hydrocarbons with some wax acid esters and alcohols. Carnauba wax is a plant exudate from the Brazilian "tree of life" (*Copernica cerifera*). It is the hardest, highest melting, natural commercial wax [4] and is composed almost entirely of wax acid esters of C_{24} and C_{28} carboxylic acids and the C_{32} and C_{34} straight-chain primary alcohols [16]. Rice bran wax is refined wax from rice bran.

The animal waxes comprise beeswax, insect (or Chinese) wax, shellac wax, spermaceti wax, and wool wax. Beeswax is produced by honeybees for building honeycomb cells. It is a mixture of wax esters, wax acids, and hydrocarbons. Hydrolysis of these esters yields mainly C_{26} and C_{28} carboxylic acids and the C_{30} and C_{32} straight-chain primary alcohols [16].

Although not considered a lipid material, shellac is often used as a protective and decorative coating for foods because it forms a barrier to moisture and gas and because of its physical toughness. "Shellac" is the word used today referring to all forms of lac. Lac is the hardened resinous secretion of the scale insect, Laccifer lacca, and is the only commercial resin of animal origin [19]. It is composed of more than 90% resin and less than 5% wax [20]. Shellac has a molecular weight of approximately 1000 and forms films because of extensive hydrogen bonding [21]. Shellac was reaffirmed as GRAS for use as food coatings in 1990 by the U.S. FDA [15].

Synthetic waxes are waxes only in the sense that they possess physical properties similar to those of natural waxes. They are made of long-chain polymers of ethylene or ethylene oxide, halogenated hydrocarbons, montan wax derivatives, hydrogenated waxes, and many reaction products involving fatty acids [17]. Some of these materials are accepted for use in or on food. The primary material used from this category is oxidized polyethylene or polyethylene wax. Oxidized polyethylene is the basic resin produced by the mild air oxidation of polyethylene. The polyethylene used in the oxidation process must conform to the density, maximum *n*-hexane-extractable fraction, and maximum xylene-soluble fraction specifications prescribed by section 177.1520(c) in title 21 of the CFR in order to be permitted on selected foods. The oxidized polyethylene has a minimum number average molecular weight of 1200, as determined by high temperature vapor pressure osmometry, contains a maximum 5% w/w total oxygen, and has an acid value of 9–19 (CFR 21.172.260).

Wax materials containing a combination of various waxes or containing different substances are considered "compounded" waxes [22].

The other major group of lipid coatings other than waxes consists of fatty acids and their glyceride and sucrose esters. Of particular interest as coatings are the acetylated glyceride monoesters of long-chain fatty acids, otherwise known as acetylated monoglycerides or commercially as Myvacet. Distilled acetylated monoglycerides form flexible films at room temperature and retain their flexibility at low temperatures. They are fairly effective barriers to moisture and moderate barriers to permanent gases. The oxidative stability of some of these products has seldom, if ever, been approached by any other fat [4]. Also contained in this general category of lipids are triglycerides and mixtures thereof. These are generally termed "fats" and "oils," with the only difference being their melting point [23]. Fats are solid at room temperature; oils are not. Full hydrogenation of cottonseed, coconut, peanut, soybean, or other vegetable oils or animal oils yields a waxlike substance [4]. Increasing the melting point of these materials through hydrogenation produces a lipid that has less liquid-like and more viscoelastic mechanical properties, greater chemical stability, and increased moisture barrier properties.

IV. BARRIER PROPERTIES OF LIPID FILMS

The primary use for lipid coatings is due to their moisture barrier properties. Features necessary for a lipid film to have a low water permeability include a regular chemical structure, with a high degree of saturation such that crystallization or close packing is encouraged, and the absence of highly polar groups [24]. Purified triglycerides and waxes have low water vapor barrier properties for these very reasons [25]. Waxes, in general, are the best moisture barriers because of their high melting point and high degree of hydrophobicity. Compared to fatty acids, fatty alcohols, and hydrogenated or crystalline triglycerides, waxes are superior moisture barriers [26–28]. The best barrier to moisture of all naturally occurring lipids is candelilla wax [29,30]. Because of its hydrophobic nature and highly crystalline character, the moisture barrier properties of candelilla wax are better than those of low density polyethylene, the industry standard for efficient, inexpensive moisture barrier polymers (Table 3). However, with regard to oxygen permeability, waxes and high melting point lipids have barrier properties 100–1000 times poorer than polyvinylidene chloride (Saran) or ethyl vinyl alcohol [29,31] (Table 4). These latter two polymers represent the forefront of gas barriers for synthetic polymers. This relationship between moisture and gas barrier properties indicates that lipid materials are well suited as coating materials for fresh fruit and vegetables. Shellac,

Table 1 U.S. Food and Drug Administration Approvals of Lipid Materials Used for Edible Coatings

Material (CFR citation)	Uses and limitations
Acetylated monoglyerides (21.172.828)	The food additive is used at a level not in excess of the amount reasonably required to produce its intended effect in food, or in food processing, food packing, or food storage equipment.
Beeswax (21.184.1973)	GRAS Current good manufacturing practice results in a maximum level, as served, of 0.065%—chewing gum 0.005%—confections and frostings 0.04%—hard candy 0.1%—soft candy 0.002% max.—all other food categories.
Carnauba wax (21.184.1978)	GRAS No limitations other than good manufacturing practice.
Candelilla wax (21.184.1976)	GRAS No limitations other than good manufacturing practice.
Castor oil (21.172.876)	As an antisticking agent in hard candy production at levels not to exceed 500 ppm. As a component of protective coating for vitamin and mineral tablets.
Cocoa butter substitute from coconut oil, palmkernel oil, or both oils (21.172.861)	GRAS As coating material for sugar, table salt, vitamins, citric acid, succinic acid, and spices. In compound coatings, cocoa creams, cocoa-based sweets, toffees, caramel masses, and chewing sweets.
Fatty acids, including salts of fatty acids (21.172.860)	As a component in the manufacture of food grade additive at levels not to exceed good manufacturing practice.
Methyl esters of fatty acids produced from edible fats and oils (21.172.225)	The additive is used or intended for use at a level not to exceed 3% by weight in an aqueous emulsion in dehydrating grapes to produce raisins, whereby the residue of the additive on the raisins does not exceed 200 ppm.
Mineral oil (21.172.878)	On raw fruits and vegetables at levels not to exceed good manufacturing practice. In frozen meat, as a component of hot-melt coating at levels not to exceed 0.095% of meat. As release agent and as sealing and polishing agent in the manufacture of confectionery at levels not to exceed 0.2% of confectionery.
Mono- and diglycerides (21.184.1505)	GRAS
Morpholine (21.172.235)	Used as the salt(s) of one or more of fatty acids, as a component of protective coat-

Table 1 Continued

Material (CFR citation)	Uses and limitations
	ings applied to fresh fruits and vegetables at a level not in excess of that reasonably required to produce its intended effect.
Odorless light petroleum hydrocarbons (21.172.884)	As a coating on shell eggs at levels not exceeding good manufacturing practice.
Paraffin (21.172.274)	Used or intended for use as a protective coating or component of protective coatings for fresh grapefruit, lemons, limes, muskmelons, oranges, sweet potatoes, and tangerines.
Petrolatum (21.172.880)	In confectionery; as release agent and as sealing and polishing agent at levels not to exceed 0.2% of confectionery.
Petroleum wax (21.172.886)	On cheese and raw fruits and vegetables as a protective coating and as a component of microcapsules for spice-flavoring substances at levels not to exceed good manufacturing practice.
Petroleum wax, synthetic (21.172.888)	On cheese and raw fruits and vegetables as a protective coating at levels not to exceed good manufacturing practice.
Polyethylene, oxidized or polyethylene wax (21.172.260)	Used or intended for use as a protective coating or component of protective coatings for fresh avocados, bananas, beets, coconuts, eggplant, garlic, grapefruit, lemons, limes, mango, muskmelons, onions, oranges, papaya, peas (in pods), pineapple, plantain, pumpkin, rutabaga, squash (acorn), sweet potatoes, tangerines, turnips, watermelon, Brazil nuts, chestnuts, filberts, hazelnuts, pecans, and walnuts (all nuts in shells).
Ricebran wax (21.172.890)	Candy, 50 ppm Fresh fruits and vegetables, 50 ppm
Sperm oil (21.172.210)	As an adjuvant for citrus fruit coatings.
Stearic acid (21.184.1090)	GRAS
Sucrose fatty acid esters (21.172.859)	As components of protective coatings applied to fresh apples, avocados, bananas, banana plantains, limes, melons (honeydew and cantaloupe), papaya, peaches, pears, pineapples, and plums to retard ripening and spoiling. This material is to be used in accordance with current good manufacturing practice and in an amount not to exceed that reasonably required to accomplish the intended effect.
Synthetic isoparaffinic petroleum hydrocarbons (21.172.882)	As a component of coatings on fruits and vegetables and as a coating on shell eggs at levels not to exceed good manufacturing practice.

Table 2 Physical Properties of Common Lipid Coating Materials

Material	Melting point (°C)	Specific gravity (g/mL)	Refractive index	Iodine number
Cocoa butter	30–35	0.858–0.864	1.454–1.458	35–40
Acetylated monoglycerides	32–43	0.96–1.00	1.447–1.442	1–14
Paraffin	50–51	0.88–0.915	1.410–1.450	0
Stearyl alcohol	54–57	0.85	—	<2
Beeswax	61–65	0.950–0.960	1.434–1.445	8–11
Candelilla wax	66–71	0.982–0.9856	1.456	19–44
Stearic acid	69	0.847	1.430	0
Carnauba wax	82–86	0.996–0.998	1.463	7–14
Polyethylene wax	124–127	0.96	—	—

Source: Ref. 4.

which has an oxygen permeability similar to the high barrier waxes, delivers a high gloss but can easily cause coated fruit to respire anaerobically [32]. Shellac coatings tend to block pores in the citrus peel where gas exchange occurs [33]. The result is off-flavor development rather than an extension of storage life. Due to its low moisture permeability and very high gas permeability, oxidized polyethylene appears to be the best material for coating high respiration rate produce [33–36].

With regard to lipids other than wax, many factors determine their ability to retard the transport of moisture, including crystal morphology, polarity, and carbon chain length. In general, the higher the melting point of the lipid, the greater is its barrier to moisture transport [26]. Tempering lipid films can decrease oxygen and moisture permeabilities due to healing of crystal imperfections and the development of a more extensively linked arrangement of crystalline platelets [37]. Lipid crystal polymorphology and, in turn, molecular density might affect moisture transport in lipid films as it does in polymeric films [24,38]. The α form is the least stable polymorph and has hydrocarbon chain packing characterized by a hexagonal subcell [39]. A more stable form is the β' polymorph, followed by the β polymorph. The β' has a more dense hydrocarbon packing characterized by an orthorhombic subcell. Thus, the order of thermodynamic stability is $\beta > \beta' > \alpha$, which is also the order of decreasing molecular density [23]. Although one might expect the β form to have a lower moisture permeability because of its denser, more restricted structure, Kester and Fennema [40] conversely found that the moisture permeability increased following transition from α to β' in a blend of fully hydrogenated soybean and rapeseed oil. They hypothesized that the β' polymorph may have had a higher hydration capacity (moisture sorption) and hence the higher permeability. The oxygen permeability, however, was lowest for the β form.

Of the non-wax lipids, acetylated monoglycerides have been used in a broader range of applications than any other lipid material. Lovegren and Feuge [41,42] were the first to report the oxygen and moisture barrier properties of acetylated monoglycerides as well as their potential use as protective coatings. Compared to other lipid materials, particularly waxes, it is clear that acetylated monoglycerides are rather poor barriers to both moisture and permanent gases (see Tables 3 and 4). Nevertheless, many foods have been coated with them with positive results. One possible reason for these results may be the mechanical properties of this material. Acetylated monoglycerides produce flexible films even at low temperature, whereas wax films crack under similar conditions.

Table 3 Water Vapor Permeabilities of Lipid Coating Materials

Material	Thickness (mm)	WVP [g · mm/(m^2 · day · kPa)]	RH difference	Temp. (°C)
Lipid films				
Candelilla wax [30]	0.14	0.0120	0–100%	24.9
Paraffin [42]	0.66	0.0194	0–100%	25
Beeswax [30]	0.14	0.0888	0–100%	25.9
Carnauba wax [25]	0.130	0.098	0–100%	27.5
Tripalmitin [25]	0.130	0.194	0–100%	27.5
OPE–oleic acid emulsion [36]	0.0127	0.61	0–75%	30
Anhydrous milk fat fraction [25]	0.130	0.886	0–100%	24.9
Chocolate [135]	0.593	2.313	0–80%	26
Hydrogenated peanut oil [42]	3.39	3.33	0–100%	25
Acetylated monoglyceride [42]	1.31–2.73	1.91–12.8	0–100%	25
Synthetic polymers				
PVDC [25]	0.019	0.024	0–100%	27.60
LDPE [25]	0.010	0.031	0–100%	27.6
Polyester [25]	0.025	0.168	0–100%	25
PVC [25]	0.012	0.617	0–100%	27.6
Multicomponent emulsion films containing lipids				
45% SA–55% HPMC [50]	0.019	0.0267	0–85%	27
BW on gluten–DATEM [192]	0.090	0.0360	0–32%	30
SA, PA, HPMC, PEG [26]	0.041	0.0480	0–85%	25
BW on SA/PA/MC/HPMC/PEG [45]	0.056	0.0572	0–85%	25
BW on MC/PEG [44]	0.051	0.0953	0–100%	25
44% Paraffin–44% MC–12% PEG [52]	0.087	0.2646	22–84%	25
SA–PA–HPMC–PEG [43]	0.041	0.3429	0–97%	25
40% PA–60% corn zein [58]	0.081	1.64	50–100%	25
37.5% BW–62.5% NaCas [54]	0.075	3.60	0–97%	25
2% AM–68% gluten–30% Gly [193]	0.066	4.84	0–11%	23
28% BW–56% WPI–16% Sorb [56]	0.140	5.28	0–97%	23
40% BW–56.3% WPI–3.7% Gly [30]	0.140	10.82	0–98%	25
Chitosan–lauric acid [61]	0.0254–0.0380	154.28	0–100%	25

AM = Acetylated monoglyceride; BW = beeswax; DATEM = diacetyl tartaric acid esters of monoglycerides; Gly = glycerol; HPMC = hydroxypropyl methylcellulose; MC = methylcellulose; NaCas = sodium caseinate; OPE = oxidized polyethylene; PA = palmitic acid; PEG = polyethylene glycol; SA = stearic acid; Sorb = sorbitol; WPI = whey protein isolate.

V. MULTICOMPONENT FILMS CONTAINING LIPIDS

Since lipid materials generally have poor mechanical properties and are often brittle, incorporation of other film-forming materials can be used to aid the integrity and performance of the film. Many studies have examined the barrier properties of multicomponent films composed of chemically modified cellulose and lipids [26,43–52] or protein and lipids [30,53–58]. These efforts have been focused more on improving the moisture barrier properties of the various hydrophilic film-forming materials than on improving the mechanical integrity of the lipid films. From the standpoint of the lipid film, unless the multicomponent film is produced as a bilayer, added hydrophilic materials act as pinholes in the lipid substrate, thereby reducing its moisture barrier properties [51,52].

Table 4 Oxygen Permeabilities of Lipid Coating Materials

Material	Thickness (mm)	O_2 perm. [$cm^3 \cdot \mu m/(m^2 \cdot day \cdot kPa)$]	Relative humidity	Temp. (°C)
Lipid films				
Carnauba wax [194]	0.04–0.05	157.2	0%	25
Candelilla wax [194]	0.04–0.05	175.4	0%	25
Beeswax [194]	0.04–0.05	931.4	0%	25
Acetylated monoglyceride [42]	1.74	1360.9	0%	26
Microcrystalline wax [194]	0.04–0.05	1536.2	0%	25
OPE:oleic acid emulsion [36]	0.0127	8523.1	50%	30
Synthetic Polymers				
EVOH (70% VOH) [10]		0.1	0%	23
EVOH (70% VOH) [10]		12.0	50%	23
Polyvinylidene chloride (Saran) [195]		3.2	0%	25
Polyester [195]		19.7	0%	25
High density polyethylene [195]		493.5	0%	25
Low density polyethylene [195]		2960.8	0%	25
Multicomponent emulsion films containing lipids				
Chitosan-lauric acid [61]	0.025–0.038	6.28–9.40	0%	25
29% BW:55% WPI:16% Sorb [11]	0.110	11.6	50%	23

Delamination of the lipid layer from the hydrophilic support can be a problem because of the high surface energy existing between these two phases. Alternatively, the lipid component can be dispersed throughout the hydrophilic phase in the form of an emulsion. Proteins, especially casein and whey proteins, are effective emulsifiers because of their amphiphilic nature and therefore lend themselves to forming protein–lipid emulsions. When cellulose ethers are used, a separate emulsifier must be incorporated to stabilize a lipid emulsion. The stability of film-forming emulsions plays an important role in their final moisture barrier properties. Unstable emulsions lead to phase separation during drying of the film. Consequently, the orientation of the film during moisture permeability testing results in differences in the measured values [54–56].

Lipid content in multicomponent films plays a significant role in determining the final water vapor permeability (WVP) of the film [26,30,59]. Lipid type also has a significant effect on the WVP of emulsion films. Film-forming solutions that contain wax generally offer the best resistance to moisture transport [54]. However, in some cases medium-chain fatty acids (myristic, palmitic, or stearic) or triglycerides perform better than waxes in composite systems [30,47,55].

Both the barrier properties and mechanical properties of the dispersed phase contribute significantly to the barrier properties of an emulsion film. For whey protein emulsion films made of four separate lipids (candelilla wax, carnauba wax, beeswax, and a high melting industrial milk fat fraction) and with a wide range of lipid concentrations (0–80%), the trend in moisture barrier properties of the emulsion films was contrary to those of the bulk lipid material [30]. The low moisture-transmitting lipids (i.e., candelilla wax and carnauba wax) yielded emulsion films with the largest moisture permeabilities, while the higher moisture-transmitting lipids (beeswax and the milk fat fraction) provided emulsion films that were better barriers. These results correlated well with the degree of viscoelasticity of the bulk lipids. The beeswax and milk fat fractions were more viscous, less elastic, and more easily deformable lipids that either the candelilla or carnauba wax [60]. As the emulsions dried, the softer lipid particles may have come together and yielded to the internal forces of the shrinking, collapsing protein structure. As a result, these films may have contained an intact internal network of lipid that provided the better moisture barrier. The significance of an internal lipid network within a hydrophilic support with respect to the moisture barrier properties of edible films has also been suggested by Koelsch and Labuza [47].

Chitosan, like many carbohydrate polymer films, has poor water vapor barrier properties. The moisture permeability of chitosan–lipid films was most effectively reduced by forming an emulsion with lauric acid. Other fatty acids and esters did not have the same effect of significantly reducing moisture permeability [61]. These differences may have been due more to morphological differences in the film microstructure of the various chitosan–lipid films than to the barrier properties of the emulsified lipids.

VI. APPLICATIONS OF LIPID FILMS AND COATINGS

A. Fruit Coatings

Waxing fruit is one of the largest uses of lipid coatings. The goal of using these coatings is not only to reduce moisture loss but also to extend storage life by reducing the fruits' respiration and, in some cases, by carrying fungicides to the fruits' surface. Additionally, the appearance of the fruit is improved in the eyes of the consumer by the high gloss of some lipid coatings. The waxing of citrus fruit is so widespread that it is difficult for American consumers to find oranges and lemons that have not been coated. In some instances, the goal of using these coatings may only be to impart a glossy shine to the fruit [62]. Although a broad range of lipid-coated fruits are reviewed here, we go into depth in discussing citrus coating because it deals with all of the issues found in waxing nearly any fruit.

Waxing of citrus fruits is not a process new to the twentieth century. As early as the twelfth or thirteenth century, oranges and lemons were coated with a thick layer of molten wax to ferment the fruit [1]. In the 1920s and 1930s, citrus growers began waxing fruits and vegetables to reduce moisture loss during storage [63–65]. More recently, the emphasis in waxing fresh produce has been less on extending shelf life and more on aesthetics. Consumers desire produce that has a bright shine. Kaplan [66] summarizes four reasons why fruit is waxed:

1. To reduce shrinkage due to water loss
2. To provide a barrier to free gas exchange
3. To improve appearance and marketability by application of a shiny, sweat-resistant film
4. To provide a carrier for decay control fungicides, growth regulators, and coloring agents

From the viewpoint of extending the product's shelf life, waxing citrus fruits is performed, in part, because the natural waxes are removed during washing. The fruits' natural wax, a mixture of fatty alcohols and fatty acids [67], picks up dust, dirt, fungal and bacterial spores, and preharvest sprays before and during harvesting. These contaminants, along with the wax, are removed by washing; after a brief drying step, the fruit is waxed with either a water-based or solvent-based wax to replace the original waxy coating.

Wax coatings are applied to citrus fruit by using either a solvent wax system or, more commonly now, a water wax emulsion. A solvent wax system consists of the coating material dissolved in a highly volatile petroleum solvent typically composed of 70–80% aliphatic hydrocarbons with up to 25% aromatic hydrocarbons and possibly including solvents such as acetone or ethyl acetate [68]. A water wax system is an oil-in-water emulsion of the hydrophobic coating material. An additional difference between these two systems is that the solvent wax requires that the fruit be dried prior to coating but may not require drying after coating because of the volatility of the solvent. This is not the case with the water wax emulsion, since the fruit can be wet prior to coating but must be dried following application of the emulsion [66]. However, total drying times using water waxes may not necessarily be longer than with solvent waxes, since the fruit in the latter case does not need to be dried prior to coating with the emulsion [69]. Lipid and hydrophobic materials used for both waxing systems on citrus fruit are listed in Table 5. Section 21.172.210 of the U.S. CFR lists all coatings (including those not specifically lipid-based) that may be applied to fresh citrus fruit.

Citrus fruit storage life is extended because of a reduction in moisture loss and a decrease in respiration rate [70–72]. Lipid-coated citrus fruit displays an increase in carbon dioxide and a decrease in oxygen compared to noncoated fruit [33,73–76]. One of the problems associated with waxing citrus fruits is the application of too thick a coating. Providing too effective a gas barrier will certainly reduce moisture loss, but it can cause internal oxygen concentrations to drop below a critical level such that ethanol and off-flavors are produced [32,35,75,77,78]. There is some evidence that lipid coatings may enhance the flavor of oranges by causing a buildup of components important to orange flavor (acetaldehyde, ethyl acetate, ethyl butyrate, and methyl butyrate) [76]. However, the same study indicated that the oranges coated with partially hydrogenated vegetable oil or propylene glycol esters of fats and fatty acids displayed a significant loss in limonene, a twofold increase in methanol, and a sixfold increase in ethanol. A minimal coating sufficient to provide a glossy appearance and reduce weight loss has been suggested, with a maximum level of 0.2–0.3 mg/cm^2 [35].

Table 5 U.S. Food and Drug Administration Approvals of Principal Film Formers Used in Citrus Fruit-Coating Formulations

Material	21 CFR citation[a]
Beeswax	184.1973, 184.1975
Carnauba wax[b]	184.1978
Candelilla wax	184.1976
Cottonseed oil	172.894
Fatty acids	172.860
Mineral oil[b]	172.878
Montan wax	178.3770
Oleic acid	172.862
Paraffin[c]	172.275
Petrolatum[c]	172.880
Petroleum wax[c]	172.886, 172.888
Polyethylene, oxidized[c,d]	172.260
Rice bran wax	172.890
Sperm oil	172.210
PVDC	172.210

[a]Code of Federal Regulations, Title 21.
[b]Not permitted on citrus exported to the United Kingdom.
[c]Not permitted on citrus exported to Japan.
[d]Not permitted for use in Italy.

Waxing may partially or completely plug stomatal pores in citrus fruit, restricting the transport of ethylene, oxygen, and carbon dioxide while not restricting the movement of moisture [79]. Kalmar [80] suggests that the wax coating be matched to the respiratory quotient (ratio of carbon dioxide production to oxygen consumption) of the particular fruit or vegetable. Those products having a low respiratory quotient can be more completely "sealed" than those that have a high quotient. By spraying a solvent wax on citrus fruit, a glossy appearance is achieved, but the film coating is discontinuous and porous, thereby allowing sufficient gas exchange to prevent the fruit from respiring anaerobically. Commercial fruit waxing materials have a wide range of gas and moisture permeabilities [75]. It is important that a producer of waxed fruit understand the differences in these materials so that the desired effect is achieved (gloss and extended storage life) while the undesirable effects of anaerobic respiration are prevented.

Compared to wax coatings, polyethylene-based coatings have high gas permeabilities and therefore do not cause a buildup of ethanol and off-flavors in the coated citrus fruit, but they still manage to impart a glossy appearance and reduce water loss in these products [33–36]. This makes sense, since polyethylene is a very good moisture barrier but a rather ineffective gas barrier [36].

Lipid coatings have also been used in conjunction with fungicides to extend the storage life even further than with lipid coatings alone. A wide range of fungicides have been tested in wax emulsions on citrus products, including 2-aminopyridine [81], thiabendazole [82], benomyl [83], sodium orthophenylphenate, and hexamine [69,84,85]. Treating oranges with either of the two hormones 2,4-D (2,4-dichlorophenoxyacetic acid) or 2,4,5-T (2,4,5-trichlorophenoxyacetic acid) at levels of 1000–2000 ppm followed by waxing with a paraffin–carnauba wax emulsion [86] resulted in a delay of ripening and a reduction in fungal attack, which increased the storage life by 900% over a control [87,88].

An additional issue that must be considered when selecting wax materials for coating is the degree of gloss. Natural waxes alone are not necessarily effective at producing a high gloss finish [66,68]. Hydrocarbon wax provides low gloss, and polyethylene wax provides intermediate gloss, while shellac and resin esters can give a high gloss appearance on citrus fruit [32]. The glossy finish of some lipid coatings can be affected by exposure to moisture. When fruit has been removed from cold storage, moisture will condense on the exterior of the fruit; this "sweating" may irreversibly dull the finish of some coatings. For example, carnauba wax and polyethylene wax coatings containing palmitic or stearic acids produced a whitish blush upon wetting with water, while the same waxes emulsified using oleic acid did not produce the whitish blush [74]. Further issues include requirements for buffing of the coating to achieve a glossy appearance and the temperature of the fruit during coating. Brushing of coated produce can result in damage to the tissue [89]. Furthermore, problems occur when trying to coat refrigerated fruit, because moisture condensation on the exterior of the coating surface can stall or reverse the drying process. A patented process claims to overcome this problem [90]. The surface of refrigerated fruit is preheated with hot (100–160°F) water, the fruit is coated and then dried, and the core temperature of the fruit does not rise above 15°F.

B. Deciduous Fruit Coatings

Apple coating represents another large area of lipid-coating use. The minimization of moisture loss in apples has been achieved using various lipid coatings. Several factors affect the effectiveness of the coating in extending the storage life of apples: (1) maturity of the fruit (the greater the maturity of the fruit, the greater the gas resistance of the skin); (2) type of coating (waxes are better than oils); (3) coating thickness; (4) temperature of storage (affecting the apples' rate of respiration); and (5) variety of apple [91]. The order of decreasing moisture loss prevention was found to be paraffin wax > carnauba wax > beeswax; peanut oil, corn oil, and light mineral oils enhanced moisture loss [91]. Commercial coatings are often not a single lipid system. A typical apple wax may contain any of the following ingredients: water (solvent), isopropanol (cosolvent), shellac, carnauba wax, propylene glycol (plasticizer), morpholine or ammonium oleates, casein, soy protein (binders), and methyl siloxanes (antifoam agents) [92].

Red Delicious apples coated with wax and stored at room temperature have displayed reduced respiration rate, increased sugar and soluble solids, and slower development of soluble pectic substances over noncoated controls [93]. As with citrus fruit, too thick a wax coating on apples and pears reduces respiration to such an extent that alcoholic flavors develop [94]. Methyl esters of fatty acids, fatty acids, and triglyceride-type oils applied to Jonathan apples after harvest reduced levels of hexanal in the fruit and the incidence of "soft scald" that developed during cold storage [95]. However, the use of pure tripalmitin had little effect on the occurrence of soft scald. This may have been due to tripalmitin's limited solubility in ethanol and hence an incomplete coverage with a solvent system, or it may have been due to crack development in this very brittle substance during storage. Waxing Golden Delicious apples increases their color and reduces their weight loss during storage, yet firmness was not affected. Increasing the drying temperature to 60°C improved firmness and color but increased weight loss [96]. Consumer acceptability scores were linearly related to the amount of wax applied to apples, with good scores for appearance directly related to high level of wax application [97].

Bananas ripen quickly at high temperature and therefore have a short shelf life. Lipid coatings have been used on bananas chiefly as a means of controlling gas transport and to a lesser degree for preventing moisture transport. Coating bananas with sucrose esters of fatty acids, glycerides, or paraffin–sugarcane wax emulsions had a significant effect on reducing moisture loss and delaying the ripening process [98–100], while polyethylene wax coatings had no significant effect on speed of ripening [100].

In addition to citrus fruit, apples, and bananas, many different types of fruit have responded positively to lipid coatings. Emulsion coatings of carnauba wax, paraffin wax, polyethylene wax, and mixtures thereof had a significant effect on reducing moisture loss, altering metabolic activities, extending postharvest storage life, and improving the appearance of apples [97], apricots [101], avocados [102], bananas [103], grapes [101], peaches [101], pears [104], pineapples [105], plums [101], mangoes [106–108], and nectarines [101]. Sucrose fatty acid esters have been used effectively on many of the previously mentioned fruit plus durian fruit [109], guavas [110], and melons (honeydew and cantaloupe). The addition of fungicides to the wax coatings has significantly increased storage life by reducing fungus-associated decay in addition to moisture loss in apples [71,111,112], bananas [113], guavas [114,115], mangoes [116], melons [117], pears [118], and rambutan [119]. Baldwin and coworkers [92,120] have reviewed the application of coatings to a wide range of fresh and lightly processed fruits and vegetables with many references to commercial coatings. A similar review exists on coatings for processed foods [121].

C. Vegetable Coatings

Vegetable coating, like fruit coating, is directed toward reducing moisture loss, reducing the respiration rate (in some cases), and improving appearance. Vegetables that have low respiration rates are better candidates for lipid coatings than those with high respiration rates. Although the coating may have a significant effect on the moisture loss, the gas barrier properties of the coating may have varying effects on the respiration rate. Coating a highly respiring vegetable with a lipid film will tend to make it respire anaerobically, and this leads to off-flavor development. Commonly coated vegetables are cucumbers, rutabagas, and mature green tomatoes [92]. Bell peppers, cucumbers, eggplant, squash, and tomatoes are typically coated with a mineral oil–based coating, while paraffin wax emulsions are frequently used to coat root crops such as rutabaga, turnip, and yucca [122]. Other crops that are occasionally given wax coatings include asparagus [123], avocados [75], beets, carrots, celery [123], eggplant [75], green beans [124], kohlrabi, okra, parsnips, potatoes, radishes, summer squash, turnips, and winter squash [123]. Waxing leafy vegetables does not extend their storage life [125].

The application of surfactants to potato tubers is associated with a resistance to carbon dioxide transport in the surfactant films, increased carbon dioxide levels in the tuber, and prevention of greening during cold storage [126]. Addition of plant acids, such as tartaric, ascorbic, or citic, reduced the carbon dioxide permeability of the surfactant layer, thereby allowing a thinner coating to achieve the same antigreening results [127].

The whitish blush that develops on the surface of peeled carrot pieces (manufactured "baby carrots") is a defect noticeable by the consumer. The white appearance comes from a combination of irreversible dehydration and lignification as a result of wounding [128]. An optimized sodium caseinate–stearic acid emulsion eliminated the formation of the white blush [129]. The optimal formulas consisted of 91% sodium caseinate and 9% stearic acid (dry basis). Although the emulsions contained stearic acid, the authors hypothesized that the hydrophilic protein phase contributed to moisturizing the carrot surface, thereby reducing surface whiteness.

D. Candy and Confectionery Coatings

The largest use for lipid coatings in the candy and confectionery industry is for polishing panned products. A panned confection is produced by repeated application and drying of a sugar syrup or chocolate coating on candy pieces. These pieces are contained in a rotating drum, called a pan, such that they tumble against each other, which ensures an even distribution of the coating material. Application of a wax coating in the final stage of panning candies provides numerous benefits such as preventing them from sticking to surfaces and other candies; improving their appearance by providing a glossy finish; preventing cracking, splitting, or crumbling; and improving shelf life through the reduction in moisture loss or pickup [130,131]. For the polishing stage of panning, the pan is first coated with a layer of molten beeswax; depending on the coating thickness and the volume of product being polished, this coating will last from 2 weeks to 6 months [132]. When hard sugar-coated candies, such as Jordan almonds, are to be polished, they are placed in the coated polishing pan and tumbled with added carnauba wax or beeswax [132,133]. During the polishing stage it is important to maintain a steady flow of cold, dry air (less than 60°F and 50% RH) over the candy pieces in the pan [134]. Wax usage is low, approximately 0.5 oz per 300 lb of candies.

Since the polishing process can take some considerable time, a faster technique involves melting beeswax and carnauba wax and then adding them to odorless volatile mineral spirits [130]. The polishing liquid is added to the finished pieces in the polishing pans, and the pieces are tumbled until they are uniformly coated. The evaporated solvent is exhausted, and the pieces are then polished.

Soft panned candies, such as jelly beans or marshmallow eggs, must be first coated with a sugar syrup that is allowed to dry. This final sugar coating provides a thin shell over the candy on which the polishing material can deposit [132]. It also prevents dusting of the soft finish underneath. The candies are then polished as if they were hard candies.

Wax is not the exclusive material for polishing candies. Shellac, dissolved in ethanol, is another common polishing agent, which is referred to as confectioners glaze. Relative to shellac, wax polishes impart a less glossy matte finish [133]. A combination of shellac and wax may be used to improve the glossy appearance. Other waxes sometimes used for polishing are spermaceti, ouricouri, and candelilla wax [133]. Mineral oil can be used to prevent sticking in hard gum candies made from gelatin or starch, such as Ju Ju Bees or Gummi candies. The drawback from using mineral oil is that it is easily transferred from one surface to another; thus the inner surface of the package containing these coated candies will end up covered with mineral oil [130].

Chocolate coatings, although not solely lipid, are used throughout the candy and confectionery industry to provide a moisture barrier on nuts and to minimize the transport of moisture among candy components of differing water activity. The WVP of chocolate lies between that of hydrogenated oils and those of high melting waxes and is a function of the amounts of lipid and carbohydrates present in the mixture [135]. Incorporation of alkali derivatives of soy protein [136], calcium stearate [137], or hydroxy lecithin [138] can reduce the moisture transmission rate in chocolate by 30–60%. Additionally, the type of tempering and, in turn, the nature of the fat crystals influence the moisture barrier properties of chocolate coatings [139]. Rapidly cooled cocoa butter has a moisture permeability higher than that of the tempered fat [140]. Unlike the other coatings used on candies, chocolate imparts it own flavor or additional flavors such as milk or mint.

In addition to chocolate, similar materials called compound coatings can be used. Compound coatings are typically white and made from refined vegetable butters that are often based on imported lauric oils such as palmkernel and coconut oil [130]. There are a number of drawbacks to the use of lauric-based fats. Hydrolysis into lauric acid can produce an undesirable "soapy" flavor, and fats containing laurin do not provide a good moisture barrier [130]. Furthermore, most fats used for compound coatings are not compatible with cocoa butter in that they produce rapid bloom formation in the mixture. This can be alleviated by using an anhydrous milk fat fraction [133] or modified fats [141]. Use of water-in-oil emulsions for chocolate and other fat-containing confectionery coatings has been patented as a means of incorporating water-soluble flavoring and coloring in the fat-based coating [142].

A coating of corn zein and acetoglycerides is able to resist moisture transfer and oil migration in candies [130]. This coating can provide sufficient gloss that it can be used as a replacement for confectioner's glaze or mineral oil coatings. The main drawback to using corn zein is that the coating dissolves slowly and unevenly during the eating of hard candies [130].

E. Nut and Raisin Coatings

The two main defects in stored nuts are oxidative rancidity and moisture uptake. Lipid coatings have been effective at minimizing or delaying the onset of these defects. Mixtures of various antioxidants (TBHQ, BHA, BHT) and citric acid in hydrogenated vegetable oil or acetylated monoglycerides have been effective at minimizing oxidative rancidity in peanuts [143]. When the nuts are granulated for use in food items, the increased surface-to-volume ratio decreases the time it takes for the product to become unacceptable because of oxidation and moisture uptake. Coating granulated peanuts just after roasting has proven to delay these deteriorative processes [144]. Coating nuts with acetylated monoglycerides helped them retain crispness when used in ice cream and also reduced rancidity [145]. A similar study examining the effects of coating pecans with acetylated monoglycerides also showed that the coating reduced oxidative rancidity and extended the shelf life of the product, with or without antioxidants [146]. Film-forming solutions of corn zein and acetoglycerides can retard moisture sorption and rancidity in nuts and other items [130,147,148].

Mineral oil has traditionally been used to prevent the clumping of raisins as well as to improve their appearance [149]. However, the major deteriorative process for raisins is excessive moisture loss. Raisins can be stabilized against moisture loss by coating with beeswax. Mixtures of beeswax and cottonseed oil or other vegetable oils have increased water vapor permeability and are not as effective as beeswax coating alone [150,151]. Precoating raisins with polysaccharides that have a molecular weight above 5000 followed by wax coating improved the integrity of coating and reduced the moisture loss of the product [150]. Coating dehydrated fruit pieces with acetylated monoglycerides protected them from moisture loss or uptake in cake mixes [152]. Foods containing dehydrated fruits or other intermediate moisture products have been coated with a patented film-forming dispersion of protein, polysaccharides, a plasticizer, and an immiscible liquid such as an edible oil [153,154].

F. Meat, Fish, and Cheese Coatings

Postmortem changes in the meat of warm-blooded animals results in considerable dehydration of meat protein. Rapid freezing to −40°F and low temperature storage can extend the storability of the product. However, a considerable amount of moisture is lost during frozen storage; this is known as freezer burn. Coating frozen meat with an oil-in-water emulsion composed of plant oils, waxes, and methyl cellulose retained moisture during frozen storage [155,156]. Ethyl cellulose and oil [157], mono- and triglycerides [158], fatty acids and cetyl alcohol in a glycerol–water base [159], and acetylated monoglycerides [145] have been patented as coating materials for the prevention of moisture loss in frozen or refrigerated meat.

Dehydration and oxidation are the primary sources of degradation in frozen fish. Ice glazing can slow both of these processes [160], but the loss of an ice barrier due to sublimation or cracking and chipping requires that the fish be periodically reglazed [161]. Whey protein isolate and acetylated monoglycerides, alone and in combination, have been examined as a means of reducing these types of degradation [162]. Spraying the frozen fish with a low melting point (−13°C) acetylated monoglyceride had a significant effect on reducing moisture loss and a noticeable, though not significant, effect in reducing peroxide values of the frozen stored fish.

Coating meats with lipids does not come without its own set of problems. Acetylated monoglycerides have been tested on beef and lamb cuts [163,164]. When used on meats that were vacuum-packaged, the coated cuts purged less moisture and had lower microbial counts; however, the authors of the study felt that the uncharacteristic aroma and oily appearance might detract from the benefits of improved shelf life [165]. Addition of chlortetracycline to the coating further decreased microbial loads and improved appearance during storage [166]. Yet the consumption of antibiotics with coated product remains a controversial issue. Coating sub-scalded chicken broiler parts with acetylated monoglycerides prevented dehydration and skin darkening in storage at 40°F [167]. Yet chicken parts coated with the same material had unacceptable flavor scores when the coated meat was stored in the presence of other foods [168]. The addition of cellulose acetate butyrates to acetylated monoglyceride coatings for meat and fish improved the performance of the coating by making it a more mechanically resilient package. This film must be peeled from the meat prior to cooking and consumption [169].

Cheese has historically been coated with wax for the purpose of minimizing moisture loss and protecting the product from contamination. Single or multiple dips in paraffin wax or mixtures containing paraffin and petrolatum have been used on cheddar and gouda. The use of microcrystalline wax helps to improve the moisture barrier properties of the wax coating, while the incorporation of elastomers (such as polystyrene or polyethylene) in the wax mixture aids in peeling the coating from the wax [17]. The addition of antimicrobial agents, such as sorbic acid, to edible hydrophilic films has proven to reduce the surface concentration of micoorganisms on a cheese analog [170]. The permeability of composite films of cellulose esters and various lipids to sorbic acid and potassium sorbate has been studied with a similar intention [49,171].

G. Coating Starch-Based Products

Moisture uptake in dry cereal-based foods is the leading mode of deterioration for these products. Above a water activity of 0.40–0.45, enough water is present to plasticize dry cereal-based snacks to the extent that they lose their crispness [172]. Minimizing moisture uptake through the application of lipid coatings has proven effective in a number of instances. Applying edible coconut oils to cereal flakes resulted in a number of advantages including improved storage stability, enhanced sweetness when sugar-coated, and improved texture and flavor [173]. Yeast-raised doughnuts suffer such rapid moisture loss and starch retrogradation that their shelf life is only a matter of hours. Coating this type of doughnut with an oil-in-water emulsion of acetylated monoglycerides, triglyceride shortening, and ricebran wax minimized moisture loss and extended its shelf life [173]. Coating the interior surface of ice cream cones with two layers of a lipid mixture containing high melting triglyceride shortening, tristearin, and cocoa improved the moisture repellency and extended the crispness of these products [174]. In practice the interior of cones is often sprayed with chocolate or couverture to provide an effective moisture barrier. In addition to ice cream cones, ice cream sandwich wafers, pie shells, and some types of candy bars have been successfully coated with acetylated monoglycerides [145].

H. Fat Encapsulation of Moisture-Sensitive Ingredients

Moisture-sensitive food ingredients can be protected by encapsulation in a lipid material. Coating food grade acids with hydrogenated vegetable oils prevents the acid from accelerating food component deterioration, such as starch hydrolysis, color degradation, and flavor loss, until the food is heated. The coating helps reduce the acid's hygroscopicity, reduces its dusting, and improves its flowability by minimizing clumping [175]. Sodium acid pyrophosphate, glucono-δ-lactone, and lactic acid have been encapsulated with lipids to improve their use in foods such as restructured pork meat [176,177]. Ascorbic acid has been protected from water and oxygen by lipid encapsulation for use as a dough conditioner [178].

Sweeteners represent another large group of lipid-encapsulated ingredients. Similar to coating acids, encapsulating sweeteners reduces their hygroscopicity, improves their flowability, and prolongs their perceived sweetness. Encapsulated sugar can be used in chewing gum to extend the perception of sweetness. The coated sugar requires a higher temperature and shearing to release its sweetness than the uncoated sugar [179]. Several techniques for encapsulating aspartame (L-aspartyl-L-phenylalanine) to improve its stability against moisture and heat in chewing gum have been patented [180–183].

In addition to acids and sweeteners, a wide range of flavorings and vitamins [184,185] have been encapsulated in fat for protection. Nucleoside-5′-phosphate is used in flavoring a wide range of foods and beverages, but it is susceptible to decomposition by phosphatase. In cases where the enzyme phosphatase has not been inhibited and the flavoring agent is desired (such as with miso), or when it is not feasible to add the flavoring agent after heat (ham, sausage), the flavoring agent must be protected. Coating the flavoring agent with a high melting fat protects it until phosphatase has been inactivated or until the food has been heated and is ready to be consumed (instant soups) [186].

Salt in some meat products can act a prooxidant and result in rancidity development during storage. Such is the case with pork sausage. This problem is confounded during frozen storage by the concentration of unfrozen salt-containing liquid. The potential for rancidity can be reduced by coating the salt with a lipid substance that itself has excellent flavor stability. When the meat is cooked later, the fat melts and releases the salt to flavor the product. The lipid source can be hydrogenated triglyceride oils or mono- and diglycerides [187]. It must have a sufficiently high melting point to protect the salt during storage, but not so high a melting temperature that some of it remains crystalline even during cooking. Approximately 20–60% by weight of fat is used in practice. A similar objective was achieved for use with salt or sugar using a multicomponent emulsifier mixture consisting of polyoxyethylene (20) sorbitan monostearate (Polysorbate 60), polyoxyethylene (20) sorbitan monooleate (Polysorbate 80), and a blend of mono- and diglycerides of fat-forming fatty acids [188].

I. Coating Fat with Fat

In some cases it is desirable to coat bars of cooking fat so that they may be stored in contact with each other without sticking inseparably. The patent describing this procedure is sufficiently vague to state that any suitable fat material may be used provided it has been "hardened" to a melting point greater than 43°C. Only 0.25–2 mm thickness is required, and the patent suggests coextruding the two fats using an attachment on the end of a Votator scraped surface heat exchanger [189].

VII. CONCLUSION

The development and use of edible coatings is motivated by quality, aesthetic, and economic factors [3]. Minimizing moisture loss, enhancing appearance, and reducing the overall cost and complexity of packaging systems are goals that can be achieved by using edible lipid coatings. Lipid coatings have proven themselves as moisture barriers for encapsulating moisture-sensitive ingredients and as a means of extending the postharvest life of fresh produce. The use of edible lipid coatings on and in foods has, without question, improved the salability of many products by extending their shelf life and, in some cases, improving their appearance. Yet consumers remain somewhat unaware of the extent to which these materials are used. The FDA's ruling in January 1993 on postharvest coatings used for fresh produce requires shippers to list the origin of the coating, rather than its subingredients, on the shipping container and on prominently and conspicuously displayed signs at retail outlets [190]. These regulations became effective on May 8, 1994. The consequences of these regulations are far-reaching, since most consumers have been unaware of the materials used for coating fresh produce. Should consumers come to believe that their food is coated with unsafe "chemicals," any benefit these coatings may have provided will be of little consequence because the coated product will not be purchased. Currently there are certain countries that will not accept fruit coated with specific waxes. Norway will not import fruit coated with any wax [191]. In this light, the future for coated fruit is somewhat uncertain.

The future of edible lipids will be affected by their use in multicomponent edible coatings and films. This area has not yet been fully explored. Considerable research is being focused on improving the moisture barrier properties of polysaccharide and protein edible films by incorporating lipid materials. With the proper materials and film-forming techniques, an edible film will be produced that can control gas, aroma, and lipid transport but also remain an effective moisture barrier.

REFERENCES

1. Anonymous, Invisible package—long life to fruits and vegetables, *Mod. Pack. 18*: 75–80 (1944).
2. T. P. Labuza and R. Contreras-Medellin, Prediction of moisture protection requirements from foods, *Cereal Foods World 26*: 335 (1981).
3. S. Guilbert, Technology and application of edible protective films, in *Food Packaging and Preservation: Theory and Practice* (M. Mathlouthi, Ed.), Elsevier Applied Science, London, 1986.
4. H. Bennett, *Industrial Waxes*, Chemical Publishing, New York, 1975.
5. C. E. Rogers, J. A. Meyer, V. Stannett, and M. Szwarc, Some factors affecting the permeability constant, Part II, in *Permeability of Plastic Films and Coated Paper to Gases and Vapors* (TAPPI Monograph Ser. No. 23) (V. Stannett, M. Szwarc, R. L. Bhargava, J. A. Meyer, A. W. Myers, and C. E. Rogers, Eds.), Technical Association of the Pulp and Paper Industry, New York, 1962.
6. C. E. Rogers, Permeation of gases and vapours in polymers, in *Polymer Permeability* (J. Comyn, Ed.), Elsevier Applied Science, London, 1985, Chap. 2.
7. W. J. Moore, *Physical Chemistry*, Prentice-Hall, Englewood Cliffs, NJ, 1972.
8. J. Crank, *The Mathematics of Diffusion*, Oxford Univ. Press, New York, 1975.
9. C. E. Rogers, Solubility and diffusivity, in *Physics and Chemistry of the Organic Solid State*, Vol. II (D. Fox, M. M. Labes, and A. Weissberger, Eds.), Interscience, New York, 1965. pp. 510–635.
10. M. Salame, Barrier polymers, in *The Wiley Encyclopedia of Packaging Technology* (M. Bakker, Ed.), Wiley, New York, 1986. pp. 48–54.
11. T. H. McHugh and J. M. Krochta, Sorbitol- versus glycerol-plasticized whey protein edible films: integrated oxygen permeability and tensile property evaluation, *J. Agric. Food Chem. 42*: 841–845 (1994).
12. V. T. Stannett, The permeability of plastic films and coated papers to gases and vapors, *Convert. Pack.* 22–26 (1985).
13. R. B. Bird, W. E. Stewart, and E. N. Lightfoot, *Transport Phenomena*, Wiley, New York, 1960.
14. *Code of Federal Regulations*, Title 21, Part 172, 1977 ed., amended 1989.
15. *Code of Federal Regulations*, Title 21, Part 184, 1977 ed., amended 1983.
16. A. Streitwieser and C. H. Heathcock, *Introduction to Organic Chemistry*, Macmillan, New York, 1985.
17. A. H. Warth, *The Chemistry and Technology of Waxes*, Reinhold, New York, 1956.
18. *Code of Federal Regulations*, Title 21, Part 175, 1977 ed.
19. J. W. Martin, Shellac, in *Treatise on Coatings*, Vol. 1, Part 3, *Film-Forming Compositions* (R. R. Myers and J. S. Long, Eds.), Marcel Dekker, New York, 1972. pp. 442–478.
20. P. K. Bose, Y. Sankaranarayanan, and S. C. S. Gupta, *Chemistry of Lac*, Indian Lac Research Institute, Namkum, India, 1963.
21. M. Weiss and S. Beinfest, Systematic analysis of shellac, *Soap and Chemical Specialities* 38: 135, 137, 139, 141, 149 (1962).
22. H. Bennett, *Industrial Waxes*, Chemical Publishing, New York, 1975.
23. A. E. Bailey, *Melting and Solidification of Fats*, Interscience, New York, 1950.
24. J. A. Barrie, Water in polymers, in *Diffusion in Polymers* (J. Crank and G. S. Park, Eds.), Academic, New York, 1968. pp. 259–314.
25. T. H. Shellhammer and J. M. Krochta, Water vapor barrier and rheological properties of simulated and industrial milkfat fractions, *Trans. ASAE* (In press) (1997).
26. S. L. Kamper and O. Fennema, Water vapor permeability of edible bilayer films, *J. Food Sci. 49*: 1478–1481, 1485 (1984).
27. J. J. Kester and O. Fennema, Resistance of lipid films to water vapor transmission, *J. Am. Oil Chem. Soc. 66*: 1139–1146 (1989).

28. O. Fennema and J. J. Kester, Resistance of lipid films to transmission of water vapor and oxygen, in *Water Relationships in Food* (H. Levine and L. Slade, Eds.), Plenum, New York, 1991. pp. 703–719.
29. I. G. Donhowe, *Water Vapor and Oxygen Permeability of Wax Films*, The American Chemical Society Edible Film Symposium, St. Louis, MO. August, 1992.
30. T. H. Shellhammer and J. M. Krochta, The performance of whey protein emulsion films as affected by the viscoelasticity and water vapor permeability of the dispersed phase, *J. Food Eng.*: (1996).
31. J. J. Kester and O. Fennema, Resistance of lipid films to oxygen transmission, *J. Am. Oil Chem. Soc. 66*: 1129–1138 (1989).
32. R. D. Hagenmeier and R. A. Baker, Layered coatings to control weight loss and preserve gloss of citrus, *HortScience 30*: 296–298 (1995).
33. R. D. Hagenmeier and R. A. Baker, Reduction in gas exchange of citrus fruit by wax coatings, *J. Agric. Food Chem. 41*: 283–287 (1993).
34. S. Ben-Yehoshua, Some physiological effects of various skin coatings on range fruit, *Israel J. Agric. Res. 17*: 17–27 (1967).
35. P. L. Davis and R. C. Hofman, Effects of coatings on weight loss and ethanol buildup in juice of oranges, *J. Agric. Food Chem. 21*: 455–458 (1973).
36. R. D. Hagenmeier and P. E. Shaw, Permeability of coatings made with emulsified polyethylene wax, *J. Agric. Food Chem. 39*: 1705–1708 (1991).
37. J. J. Kester and O. Fennema, Tempering influence on oxygen and water vapor transmission through a stearyl alcohol film, *J. Am. Oil Chem. Soc. 66*: 1154–1157 (1989).
38. B. Pascat, Study of some factors affecting permeability, in *Food Packaging and Preservation* (M. Mathlouthi, Ed.), Elsevier Applied Science, London, 1986. pp. 7–24.
39. K. Larsson and P. Dejmek, Crystal and liquid structures of lipids, in *Food Emulsions* (K. Larsson and S. E. Friberg, Eds.), Marcel Dekker, New York, 1990. pp. 97–125.
40. J. J. Kester and O. Fennema, The influence of polymorphic form on oxygen and water vapor transmission through lipid films, *J. Am. Oil Chem. Soc. 66*: 1147–1153 (1989).
41. N. V. Lovegren and R. O. Feuge, Permeability of acetostearin products to carbon dioxide, oxygen, and nitrogen, *J. Agric. Food Chem. 4*: 634–638 (1956).
42. N. V. Lovegren and R. O. Feuge, Permeability of acetostearin products to water vapor, *J. Agric. Food Chem. 2*: 558–563 (1954).
43. S. L. Kamper and O. Fennema, Water vapor permeability of an edible, fatty acid, bilayer film, *J. Food Sci. 49*: 1482–1485 (1984).
44. I. K. Greener and O. Fennema, Barrier properties and surface characteristics of edible, bilayer films, *J. Food Sci. 54*: 1393–1399 (1989).
45. J. J. Kester and O. Fennema, An edible film of lipids and cellulose ethers: barrier properties to moisture vapor transmission and structural evaluation, *J. Food Sci. 54*: 1383–1392 (1989).
46. V. Sapru and T. P. Labuza, Dispersed phase concentration effect on water vapor permeability in composite methyl cellulose–stearic acid edible films, *J. Food Process. Preserv. 18*: 359–368 (1994).
47. C. M. Koelsch and T. P. Labuza, Functional, physical and morphological properties of methyl cellulose and fatty acid-based edible films, *Lebensm. Wiss. Technol. 25*: 404–411 (1992).
48. D. C. Rico-Peña and J. A. Torres, Oxygen transmission rate of an edible methylcellulose–palmitic acid film, *J. Food Process. Preserv. 13*: 125–133 (1990).
49. D. C. Rico-Peña and J. A. Torres, Sorbic acid and potassium sorbate permeability of an edible methyl cellulose–palmitic acid film: water activity and pH effects, *J. Food Sci. 56*: 497–499 (1991).
50. R. D. Hagenmeier and P. E. Shaw, Moisture permeability of edible films made with fatty acids and (hydroxypropyl) methylcellulose, *J. Agric. Food Chem. 38*: 1799–1803 (1990).
51. M. Martin-Polo, C. Mauguin, and A. Voilley, Hydrophobic films and their efficiency against moisture transfer. 1. Influence of the film preparation technique, *J. Agric. Food Chem. 40*: 407–412 (1992).

52. F. Debeaufort, M. Martin-Polo, and A. Voilley, Polarity homogeneity and structure affect water vapor permeability of model edible films, *J. Food Sci. 58*: 426 (1993).
53. J. M. Krochta, A. E. Pavlath, and N. Goodman, Edible films from casein–lipid emulsions for lightly-processed fruit and vegetables, in *Preservation Processes and Related Techniques* (W. Speiss and H. Schubert, Eds.), Elsevier Applied Science, London, 1990. pp. 329–340.
54. R. J. Avena-Bustillos and J. M. Krochta, Water vapor permeability of caseinate-based edible films as affected by pH, calcium cross-linking and lipid content, *J. Food Sci. 58*: 904–907 (1993).
55. T. H. McHugh and J. M. Krochta, Water vapor permeability properties of edible whey protein–lipid emulsion films, *J. Am. Oil Chem. Soc. 71*: 307–312 (1994).
56. T. H. McHugh and J. M. Krochta, Dispersed phase particle size effects on water vapor permeability of edible emulsion films, *J. Food Process. Preserv. 18*: 173–188 (1994).
57. N. Gontard, C. Duchez, J. L. Cuq, and S. Guilbert, Edible composite films of wheat gluten and lipids: water vapour permeability and other physical properties, *Int. J. Food Sci. Technol. 29*: 39–50 (1994).
58. J. W. Park, R. F. Testin, H. J. Park, P. J. Vergano, and C. L. Weller, Fatty acid concentration effect on tensile strength, elongation, and water vapor permeability of laminated edible films, *J. Food Sci. 59*: 916–919 (1994).
59. C. M. Koelsch, The functional, physical and morphological properties of edible fatty acid and methyl cellulose based barriers, Ph.D. Thesis, Univ. Minnesota, 1992.
60. T. H. Shellhammer, T. R. Rumsey, and J. M. Krochta, The viscoelastic properties of lipid film-forming materials, *J. Food Eng.* (in press).
61. D. W. S. Wong, F. A. Gastineau, K. S. Gregorski, S. J. Tillin, and A. E. Pavlath, Chitosan-lipid films: microstructure and surface energy, *J. Agric. Food Chem. 40*: 540–544 (1992).
62. C. Lecos, How to shine an apple, *FDA Conserv. 16*: 8–11 (1982).
63. F. Ahulberg, Preserving oranges and other fruit by coating with scale wax, U.S. Patent 1,618,159 (1927).
64. E. M. Brogden, Preserving fresh fruit, U.S. Patent 1,585,370 (1926).
65. E. M. Brogden, Preserving fresh fruit, U.S. Patent 1,585,371 (1926).
66. H. J. Kaplan, Washing, waxing, and color-adding, in *Fresh Citrus Fruits* (W. F. Wardowski, S. Nagy, and W. Grierson, Eds.), AVI, Westport, CT, 1986. pp. 379–398.
67. T. Robinson, *The Organic Constituents of Higher Plants*, Cordus, Amherst, MA, 1975.
68. D. J. Hall, Innovations in citrus waxing—an overview, *Proc. Fla. State Hort. Soc. 94*: 258–263 (1981).
69. W. F. Newhall and W. Grierson, A low-cost, self-polishing, fungicidal water wax for citrus fruits, *Proc. Am. Soc. Hort. Sci. 66*: 146–154 (1955).
70. W. A. Farooqi, M. S. Ahmad, and Zain-ul-Adbin, Effect of wax coatings on the physiological and bio-chemical aspects of Kinow fruit, *Pak. J. Sci. Ind. Res. 31*: 142–145 (1988).
71. M. Meheriuk and S. W. Porrit, Effects of waxing on respiration, ethylene production, and other physical and chemical changes in selected apple cultivars, *Can. J. Plant. Sci. 52*: 257–259 (1972).
72. O. Paredes-López, E. Camargo-Rubio, and Y. Gallardo-Navarro, Use of coatings of candelilla wax for the preservation of limes, *J. Sci. Food Agric. 25*: 1207–1210 (1974).
73. I. L. Eaks and W. A. Ludi, Effects of temperature, washing, and waxing on the composition of the internal atmosphere of orange fruits, *Am. Soc. Hort. Sci. 76*: 220–228 (1960).
74. R. D. Hagenmaier and R. A. Baker, Wax microemulsions and emulsions as citrus coatings, *J. Agric. Food Chem. 42*: 899–902 (1994).
75. R. D. Hagenmeier and P. E. Shaw, Gas permeability of fruit coating waxes, *J. Am. Soc. Hort. Sci. 117*: 105–109 (1992).
76. M. Nisperos-Carriedo, P. E. Shaw, and E. A. Baldwin, Changes in volatile flavor components of pineapple orange juice as influenced by the application of lipid and composite films, *J. Agric. Food Chem. 38*: 1382–1387 (1990).
77. P. L. Davis and J. J. Smoot, Effect of polyethylene and wax coatings, with and without fungicides, on rind breakdown and decay of citrus, *Citrus Ind. 41*: 6–7,9 (1960).

78. S. M. Norman, The role of volatiles in storage of citrus fruits, *Proc. Int. Soc. Citriculture 1*: 238–242 (1977).
79. S. Ben-Yehoshua, S. P. Burg, and R. Young, Resistance of citrus fruit to mass transport of water vapor and other gases, *Plant Physiol. 79*: 1048–1053 (1985).
80. A. F. Kalmar, Wax coating preserves citrus for the market in field-fresh condition, *Citrus Grower 316:* 7–11 (1960).
81. J. R. Winston, 2-Aminopyridine, a promising inhibitor of decay in oranges, *Citrus Ind. 29*: 5–7,16–17 (1948).
82. P. T. Pesler, Decay control in Washington Navel and Valencia oranges by application of TBZ of benomyl suspended in water or incorporated into waxes, *Citrus. Subtrop. Fruit J. 463*: 12–13 (1972).
83. D. J. Hall, Comparative fungicidal activity of benomyl and its breakdown product 2-benzimidazole carbamate (MBC) on citrus, *Proc. Fla. State Hort. Soc. 94*: 341–344 (1980).
84. E. F. Hopkins and K. W. Loucks, Prevention of the phytotoxic action of sodium orthophenylphenate on citrus fruits by hexamine, *Science 112*: 720–721 (1950).
85. V. B. Dalal, S. D'Souza, H. Subramanyam, and H. C. Srivastava, Wax emulsions for extending the storage life of Sathgudi oranges, *Food Sci. 11*: 232–235 (1962).
86. S. A. Trout, The preparation of emulsions for coating fruits and vegetables., *J. Council Sci. Ind. Res. 15*: 26–32 (1942).
87. W. S. Stewart, Effect of 2,4-dichlorophenoxy acetic acid and 2,4,5-trichlorophenoxy acetic acid on citrus fruit in storage, *Proc. Am. Soc. Hort. Sci. 54*: (1949).
88. S. B. Lodh, S. De, S. K. Mukherjee, and A. N. Bose, Storage of mandarin oranges, II. Effects of hormones and wax coatings, *J. Food Sci. 28*: 519–524 (1963).
89. K. W. Loucks and E. F. Hopkins, Commercial packinghouse operations in relation to decay in citrus fruits, *Citrus Ind. 30*: 5–7, 9 (1949).
90. G. U. Glascow and A. J. Wright, U.S. Patent 3,488,200, cited in Gloss coating for fruits, *Edible Coatings and Soluble Packaging* (R. Daniel), Noyes Data Corporation, Park Ridge, NJ, 1973, pp. 85–92.
91. S. A. Trout, E. G. Hall, and S. M. Sykes, Effects of skin coatings on the behavior of apples in storage. I. Physiological and general investigations, *Aust. J. Agric. Res. 4*: 57–81 (1952).
92. E. A. Baldwin, Edible coatings for fresh fruits and vegetables: past, present and future, in *Edible Coatings and Films to Improve Food Quality* (J. M. Krochta, E. A. Baldwin, and M. O. Nisperos-Carriedo, Eds.), Technomic, Lancaster, PA, 1994. pp. 25–64.
93. J. D. Tewari, N. Pandey, R. M. Rai, and C. B. Ram, Effect of wax emulsion on physiological factors influencing the storage behaviour of Red Delicious apples, *Ind. J. Plant Phys. 23*: 257–265 (1980).
94. R. M. Smock, Certain effects of wax treatments on various varieties of apples and pears, *Proc. Am. Soc. Hort. Sci. 33*: 284–289 (1936).
95. R. B. H. Wills, K. J. Scott, and G. Hopkirk, Reduction of soft scald in apples with methyl linoleate, *HortSci. 12*: 72–73 (1977).
96. S. R. Drake and J. W. Nelson, Storage quality of waxed and nonwaxed 'Delicious' and 'Golden Delicious' apples, *J. Food Qual. 13*: 331–341 (1990).
97. J. K. Fellman, S. R. Drake, and J. M. Mattheis, Shellac and carnauba waxing of Delicious apples held in refrigerated storage studied, *Good Fruit Grower 43*: 16–18,20–21 (1992).
98. S. Krishnamurthy and C. G. Kushalappa, Studies on the shelf life and quality of Robusta bananas as affected by post-harvest treatments, *J. Hort. Sci. 60*: 549–556 (1985).
99. S. Goburdhun, Chemical ripening of Dwarf Cavendish bananas (cv. Naine). I. Effects of ethrel and ethylene on ripening. II. Extension of shelf-life of ripened fruits., *Rev. Agric. Sucr.re-de-l'Ile-Maurice 73*: 36–43 (1994).
100. V. B. Dalal, P. Thomas, N. Nagaraja, G. R. Shah, and B. L. Amla, Effect of wax coating on bananas of varying maturity, *Ind. Food Pack. 24*: 36–39 (1970).
101. L. L. Claypool, The waxing of deciduous fruit, *Proc. Am. Soc. Hort. Sci. 37*: 443–447 (1939).

102. Anonymous, Facelift for avocados, *Food Ind. So. Africa*, March, 1975, pp. 31, 33, 47.
103. S. Muthuswamy, R. Sadasivam, J. S. Sundararaj, and V. Vasudevan, Storage studies on Dwarf Cavendish banana, *Ind. J. Agric. Sci. 41*: 479–484 (1971).
104. G. Suemnue and L. Bayindirli, Effects of Semperfresh® and Johnfresh® fruit coating on post-storage quality of Ankara pears, *J. Food Process. Preserv. 18*: 189–199 (1994).
105. R. E. Paull and K. G. Rohrbach, Juice characteristics and internal atmosphere of waxed 'Smooth Cayenne' pineapple fruit, *J. Am. Soc. Hort. Sci. 107*: 448–452 (1982).
106. R. B. Patel and N. P. Goswami, Post harvest diseases of mango fruits (*Mangifera indica*) and their control, *Ind. J. Micro. 24*: 140–141 (1984).
107. M. Ullah, M. A. Haq, and M. S. Bhatti, A new technique to boost export of grafted mango fruit by expanding its marketing season from June to November by the use of water soluble wax and bioregulants at room temperature, *J. Agric. Res. Pakistan 15*: 311–314 (1977).
108. L. Gomez-Brito and M. Peleg, The effect of coating with TAG on postharvest ripening of mangoes, *Arch. Latin Nut 24*: 513–525 (1974).
109. S. Sriyook, S. Siriatiwat, and J. Siriphanich, Durina fruit dehiscence—water status and ethylene, *HortScience 29*: 1195–1198 (1994).
110. R. G. McGuire and G. J. Hallman, Coating guavas with cellulose- or carnauba-based emulsions interferes with postharvest ripening, *HortScience 30*: 294–295 (1995).
111. R. B. Patel, Post harvest diseases of apple (*Pyrus malus*) fruits and their control, *Ind. J. Micro. 24*: 142–143 (1984).
112. G. Bombiex and F. Morgat, Anti-scald fungicidal waxes and storage of apples, *Fruits 32*: 189–196 (1977).
113. C. A. Shillingford, Effect of plastic coating and polyethylene wraps on banana fruit rot and quality, *Tropic. Agric. 55*: 173–179 (1978).
114. R. V. Singh, M. C. Joshi, H. B. Ram, and N. S. Bisht, Effect of wax coating and prepackaging on the storage behaviour of guava cv. Allahabad Safeda, *Ind. Food Pack. 38*: 80–85 (1984).
115. R. N. Agnihotri and H. B. Ram, Storage behaviour of guava variety Allahabad Safeda (*Psidium guajava* L.) treated with fungicidal wax-emulsion, *Food Ind. J. 5*: 13–16 (1972).
116. H. B. Ram and W. B. Date, Studies on the storage behaviour of two late varieties of mangoes treated with fungicidal wax emulsion at room temperature, *Ind. J. Hort. 28*: 206–209 (1971).
117. J. P. Singh and D. K. Bhatnagar, Controlling post-harvest decay of muskmelon fruits by coating wax-emulsion and fungicides, *Haryana Agric. Univ. J. Res. 12*: 295–297 (1982).
118. M. Meheriuk and O. L. Lau, Effect of two polymeric coatings on fruit quality of 'Bartlett' and 'd'Anjou' pears, *J. Am. Soc. Hort. Sci. 113*: 222–226 (1988).
119. D. B. Mendoza, E. B. Pantastico, and F. B. Javier, Storage and handling of rambutan (*Nephelium lappaceum* L.), *Phil. Agric. 55*: 322–332 (1972).
120. E. A. Baldwin, M. O. Nisperos-Carriedo, and R. A. Baker, Edible coatings for lightly processed fruits and vegetables, *HortScience 30*: 35–38 (1995).
121. R. A. Baker, E. A. Baldwin, and M. O. Nisperos-Carriedo, Edible coatings and films for processed foods, in *Edible Coatings and Films to Improve Food Quality* (J. M. Krochta, E. A. Baldwin, and M. O. Nisperos-Carriedo, Eds.), Technomic, Lancaster, PA, 1994. pp. 65–88.
122. L. A. Grant and J. Burns, Application of coatings, in *Edible Coatings and Films to Improve Food Quality* (J. M. Krochta, E. A. Baldwin, and M. O. Nisperos-Carriedo, Eds.), Technomic, Lancaster, PA, 1994. pp. 189–200.
123. R. E. Hardenburg, *Wax and Related Coatings for Horticultural Products—A Bibliography*, U.S. Dept. of Agric., Agric. Res. Service, Washington, DC, 1967.
124. R. P. Singh, F. H. Buelow, and D. B. Lund, Storage behaviour of artificially waxed green snap beans, *J. Food Sci. 38*: 542–543 (1973).
125. H. Platenius, *Wax Emulsions for Vegetables*, Cornell Agricultural Experiment Station Bulletin No. 723, 43 pgs. (1939).
126. P. A. Poapst and F. R. Forsyth, The role of internally produced carbon dioxide in the prevention of greening in potato tubers, *Acta Hort. 38*: 900–903 (1974).
127. P. A. Poapst, I. Price, and F. R. Forsyth, Prevention of post storage greening in table stock potato tubers by application of surfactants and adjuvants, *J. Food Sci. 43*: 900–903 (1978).

128. L. Cisneros-Zevallos, M. E. Saltveit, and J. M. Krochta, Mechanism of surface white discoloration of peeled (minimally processed) carrots during storage, *J. Food Sci. 60*: 320–323, 333 (1995).
129. R. J. Avena-Bustillos, J. M. Krochta, and M. E. Saltveit, Optimization of edible coating formulations on zucchini to reduce water loss, *J. Food Eng. 21*: 197–214 (1994).
130. J. J. Alikonas, *Candy Technology*, AVI, Westport, CT, 1979.
131. D. H. Wing, Enrobing of bakery products, *Proc. Am. Soc. Bakery Eng. 51:* 136–141 (1975).
132. C. D. Barnett, *Candy Making—As a Science and Art*, Gussow, New York, 1960.
133. B. W. Minifie, *Chocolate, Cocoa, and Confectionery Science and Technology*, AVI, Westport, CT, 1980.
134. M. J. Lynch, Panning: an overview, *Manuf. Confect. 72:* 59–64 (1992).
135. B. Biquet and T. P. Labuza, Evaluation of the moisture permeability characteristics of chocolate films as an edible moisture barrier, *J. Food Sci. 35*: 989–997 (1988).
136. F. P. Colten and M. Kaplan, U.S. Patent 3,784,713, cited in Chocolate-coated confections with improved moisture barrier, *Candies and Other Confections* (M. T. Gillies), Noyes Data Corp, Park Ridge, NJ, 1974, pp. 2–4.
137. J. L. McReynolds and H. H. Topalain, Chocolate-coated confections with improved moisture barrier, U.S. Patent 3,784,714, cited in *Candies and Other Confections* (M. T. Gillies), Noyes Data Corp, Park Ridge, NJ, 1974.
138. F. P. Colten and M. Kaplan, U.S. Patent 3,895,104, cited in Chocolate-coated confections with improved moisture barrier, *Candies and Other Confections* (M. T. Gillies), Noyes Data Corp, Park Ridge, NJ, 1979, pp. 4.
139. N. W. Kempf, The conging process—tests on the moisture barrier properties of coatings, *Manufac. Confect.* 38–40 (1967).
140. W. Landman, N. V. Lovegren, and R. Feuge, Permeability of some fat products to moisture, *J. Am. Oil Chem. Soc. 37*: 1–4 (1960).
141. S. A. Matz, *Cookie and Cracker Technology*, AVI, Westport, CT, 1992.
142. H. A. Pinkalla, G. L. Neuser, L. R. Cook, and R. F. Korfhage, U.S. Patent 3,223,532, cited in Coating emulsions for confectionery, *Edible Coatings and Soluble Packaging* (R. Daniel), Noyes Data Corporation, Park Ridge, NJ, 1973. 140–145.
143. R. Shea, New edible protective coating keeps nut fresh, *Food Eng. 26*: 148 (1965).
144. M. W. Hoover and P. J. Nathan, Influence of tertiary butylhydroquinone and certain other surface coatings on the formation of carbonyl compounds in granulated roasted peanuts, *J. Food Sci. 47*: 246–248 (1981).
145. G. T. Luce, Acetylated monoglycerides as coatings for selected foods, *Food Technol. 21*: 1462–1468 (1967).
146. S. D. Senter and W. R. Forbus, Effects of acetylated monoglyceride coatings on pecan kernel shelf-life, *J. Food Sci. 44*: 1752–1755 (1979).
147. H. B. Cosler, Stroud Jordan medal winner extends nut candy shelf life with a coating, *Candy Ind. Confect. J. 3*: 17,18,22,35,36 (1958).
148. H. B. Cosler, Prevention of staleness, rancidity in nut meats and peanuts, *Peanut J. Nut World 37:* 10,11,15 (1958).
149. S. P. Kochar and J. B. Rossell, A vegetable oiling agent for dried fruits, *J. Food Technol. 17*: 661 (1982).
150. G. G. Watters and J. E. Brekke, U.S. Patent 2,909,435, cited in Raisins for cereal products, *Edible Coatings and Soluble Packaging* (R. Daniel), Noyes Data Corporation, Park Ridge, NJ, 1973, pp. 109–112.
151. G. G. Watters and J. E. Brekke, Stabilized raisins for dry cereal products, *Food Technol. 15*: 236–238 (1961).
152. R. A. Shea, U.S. Patent 3,516,836, cited in Fat coated fruits for cake mixes, *Edible Coatings and Soluble Packaging* (R. Daniel), Noyes Data Corporation, Park Ridge, NJ, 1973, pp. 127–133.
153. J. R. Durst, U.S. Patent 3,323,922, cited in Film-forming dispersions, *Edible Coatings and Soluble Packaging* (R. Daniel), Noyes Data Corporation, Park Ridge, NJ, 1973, pp. 250–255.

154. J. R. Durst, U.S. Patent 3,434,843, cited in Coating of food bars, *Edible Coatings and Soluble Packaging* (R. Daniel), Noyes Data Corporation, Park Ridge, NJ, 1973, pp. 255–259.
155. C. D. Bauer, G. L. Neuser, and H. A. Pinkalla, U.S. Patent 3,406,081, cited in Simple application and solidification of water-in-oil emulsion, *Edible Coatings and Soluble Packaging* (R. Daniel), Noyes Data Corporation, Park Ridge, NJ, 1973, pp. 196–200.
156. C. D. Bauer, G. L. Neuser, and H. A. Pinkalla, U.S. Patent 3,483,004, cited in Simple application and solidification of water-in-oil emulsion, *Edible Coatings and Soluble Packaging* (R. Daniel), Noyes Data Corporation, Park Ridge, NJ, 1973, pp. 200–204.
157. L. I. Berkowitz, U.S. Patent 2,957,772, cited in Simple application and solidification of water-in-oil emulsion containing ethycellulose, *Edible Coatings and Soluble Packaging* (R. Daniel), Noyes Data Corporation, Park Ridge, NJ, 1973, pp. 212–214.
158. J. Scheide, U.S. Patent 3,667,970, cited in Meat coating containing mono and triglycerides, *Edible Coatings and Soluble Packaging* (R. Daniel), Noyes Data Corporation, Park Ridge, NJ, 1973, pp. 208–212.
159. T. R. Anderson, U.S. Patent 2,989,402, cited in Meat coating containing fatty acids and cetyl alcohol, *Edible Coatings and Soluble Packaging* (R. Daniel), Noyes Data Corporation, Park Ridge, NJ, 1973, pp. 204–206.
160. A. Khayat and D. Schwall, Lipid oxidation in seafood, *Food Technol. 37*: 130–140 (1983).
161. F. W. Wheaton and T. B. Lawson, *Processing Aquatic Food Products*, Wiley, New York, 1985.
162. Y. M. Stuchell and J. M. Krochta, Edible coatings on frozen king salmon: effect of whey protein isolate and acetylated monoglycerides on moisture loss and lipid oxidation, *J. Food Sci. 60*: 28–31 (1995).
163. M. Stemmler and H. Stemmler, Composition for the preparation of coatings on meats and sausage goods, U.S. Patent 3,936,312 (1976).
164. M. Stemmler and H. Stemmler, Method for preserving freshly slaughtered meat, U.S. Patent 3,851,077 (1974).
165. J. T. Keeton, R. Leu, C. Vanderzant, J. J. Bohac, D. B. Giffin, J. W. Savell, and H. R. Cross, Evaluation of fresh vacuum-packaged beef steaks coated with an acetylated monoglyceride, *J. Food Sci. 53*: 701–704 (1988).
166. J. C. Ayres, Use of coating materials of film impregnated with chlortetracycline to enhance color and storage life of fresh beef, *Food Technol. 13*: 512–515 (1959).
167. C. W. Woodmansee and O. J. Abbott, Coating sub-scalded broiler parts in order to afford protection against dehydration and skin darkening in fresh storage, *Poultry Sci. 37*: 1367–1373 (1958).
168. M. E. Zabik and L. E. Dawson, The acceptability of cooked poultry protected by an edible acetylated monoglyceride coating during fresh and frozen storage, *Food Technol. 23*: 87–91 (1963).
169. W. L. Clark and R. J. Shirk, A hot-melt transparent peelable coating for food, *Food Technol. 19*: 1561–1567 (1965).
170. A. J. Torres and M. Karel, Microbial stabilization of intermediate moisture food surfaces. III. Effects of surface preservative concentration and surface pH control on microbial stability of an intermediate moisture cheese analog, *J. Food Process. Preserv. 9*: 107–119 (1985).
171. F. Vojdani and J. A. Torres, Potassium sorbate permeability of methylcellulose and hydroxypropyl methylcellulose multi-layer films, *J. Food. Process. Preserv. 13*: 417–430 (1989).
172. E. E. Katz and T. P. Labuza, Effect of water activity on the sensory crispness and mechanical deformation of snack food products, *J. Food Sci. 46*: 403 (1981).
173. R. K. Scharschmidt and L. Murphy, Flake cereal coated with edible oil, U.S. Patent 4,211,800, cited in *Edible Oils and Fats, Development Since 1978* (S. Torrey), Noyes Data Corp, Park Ridge, NJ, 1983, 364–366.
174. S. Werbin, I. H. Rubenstein, and D. Weinstein, U.S. Patent 3,526,515, cited in Fat coatings for ice cream cones, *Edible Coatings and Soluble Packaging* (R. Daniel), Noyes Data Corporation, Park Ridge, NJ, 1970. pp. 280–282.
175. L. E. Werner, Encapsulated food acids, *Cereal Foods World 25*: 102 (1980).
176. J. C. Cordray and D. L. Huffman, Restructured pork from hot processed sow meat: effect of encapsulated food acids, *J. Food Prot. 48*: 965 (1985).

177. P. J. Percel and D. W. Perkins, Process of preparing a particulate food acidulant, U.S. Patent 4,537,784, August 27, 1985.
178. Anonymous, Specially coated natural ingredient improves bread, *Food Process. 34*: 10 (1973).
179. F. Shahidi and X.-Q. Han, Encapsulation of food engredients, *CRC Rev. Food Sci. Nutr. 33*: 501–547 (1993).
180. T. Cea, J. D. Posta, and M. Glass, Encapsulated APM and method of preparation, U.S. Patent 4,384,004, May 17, 1983.
181. A. M. Schoebel and R. K. Yang, Encapsulated sweeteners composition for use with chewing gum and edible products, U.S. Patent 4,824,681, April 25, 1989.
182. R. K. Yang and A. M. Schoebel, Chewing gum composition with encapsulated sweetener having extended flavor release, U.S. Patent 4,911,934, March 27, 1990.
183. S. R. Cherukuri and G. Mansukhani, Sweetener delivery systems containing polyvinyl acetate, U.S. Patent 4,816,265, March 28, 1989.
184. A. Koff, U.S. Patent 3,080,293, cited in Encapsulating with fatty acids, *Edible Coatings and Soluble Packaging* (R. Daniel), Noyes Data Corporation, Park Ridge, NJ, 1973. pp. 329–331.
185. Anonymous, Fat matrix encapsulation controls ingredients release—reactions are temperature-specific, *Food Process. 37*: 72 (1976).
186. E. Fujita, K. Ishii, H. Nakatani, Y. Nakai, and K. Katagiri, U.S. Patent 3,389,000, cited in Fat encapsulation, *Edible Coatings and Soluble Packaging* (R. Daniel), Noyes Data Corporation, Park Ridge, NJ, 1973. pp. 306–308.
187. S. S. Chang and P. X. Hoynak, U.S. Patent 3,261,692, cited in Encapsulating fat, *Edible Coatings and Soluble Packaging* (R. Daniel), Noyes Data Corporation, Park Ridge, NJ, 1973, pp. 314–319.
188. L. J. Henthorn, U.S. Patent 3,507,665, cited in Encapsulating emulsion coating, *Edible Coatings and Soluble Packaging* (R. Daniel), Noyes Data Corporation, Park Ridge, NJ, 1973, pp. 326–327.
189. H. W. L. Westenberg, U.S. Patent 1,238,336, cited in Hardened fat coating, *Edible Coatings and Soluble Packaging* (R. Daniel), Noyes Data Corporation, Park Ridge, NJ, 1973, pp. 344–345.
190. *Federal Register*, Vol. 58, No. 3, January 6, 1993.
191. Anonymous, General news: waxed fruit banned in Norway, *Postharvest News Info. 1*: 433 (1990).
192. N. Gontard, S. Marchesseau, J. L. Cuq, and S. Guilbert, Water vapor permeability of edible bilayer films of wheat gluten and lipids, *Int. J. Food Sci. Technol. 30*: 49–56 (1995).
193. A. Gennadios, C. L. Weller, and R. F. Testin, Property modification of edible wheat, gluten-based films, *Trans. ASAE 36*: 465–470 (1993).
194. I. K. Greener, Physical properties of edible films and their components, Ph.D. Thesis, Univ. of Wisconsin—Madison, 1992.
195. M. Karel, Protective packaging of foods, in *Principles of Food Science*, Part II, *Physical Principles of Food Preservation* (O. R. Fennema, Ed.), Marcel Dekker, New York, 1975. pp. 399–466.

18

Spray Processing of Fat-Containing Foodstuffs

Spray Drying and Cooling

Keith Masters
NIRO A/S, Soeborg, Denmark

I. INTRODUCTION

Spray drying is one of the few continuous industrial drying techniques that converts heat-sensitive liquid feedstocks into powders without the product quality degradation often associated with dryers that use hot air to promote moisture evaporation. This important characteristic of spray drying is further enhanced by the ability of a spray dryer not only to produce powders, but to produce powders having specific desired properties defined by their particulate structure (powders, agglomerates, granules), particle size distribution, bulk density, moisture content, and other properties that define overall powder quality.

It is no surprise, therefore, that the food industry has long recognized the potential of applying spray drying whenever liquid food concentrates are required in powder form. The ability to handle production capacities of any size in one dryer and to adopt the latest process automation ideas to continuous operation while achieving reproducibility of powder quality during production are also factors that have played a part in firmly establishing spray drying in food processing.

Spray drying [1] involves the atomization or spraying of a pumpable fluid food formulation into hot air entering a drying chamber through an air disperser. Various configurations of air disperser and drying chamber designs can be used, but whatever the design it is essential that a controlled air flow is created in the drying chamber and that the droplets in the spray created by the atomization stage are contained in this air flow. In this way, the best conditions are created for rapid completion of moisture evaporation from each droplet and the formation of a stable, dried particulate form.

The process description of spray drying indicates a sense of simplicity of process, but in reality spray drying is a highly complex operation. Spray formation involves the creation and handling of billions and billions of droplets, all of different sizes and moving in the confined volume of a drying chamber. The movement of these droplets must then be so controlled that all are subjected to similar heat and humidity environments. This will give the best possible

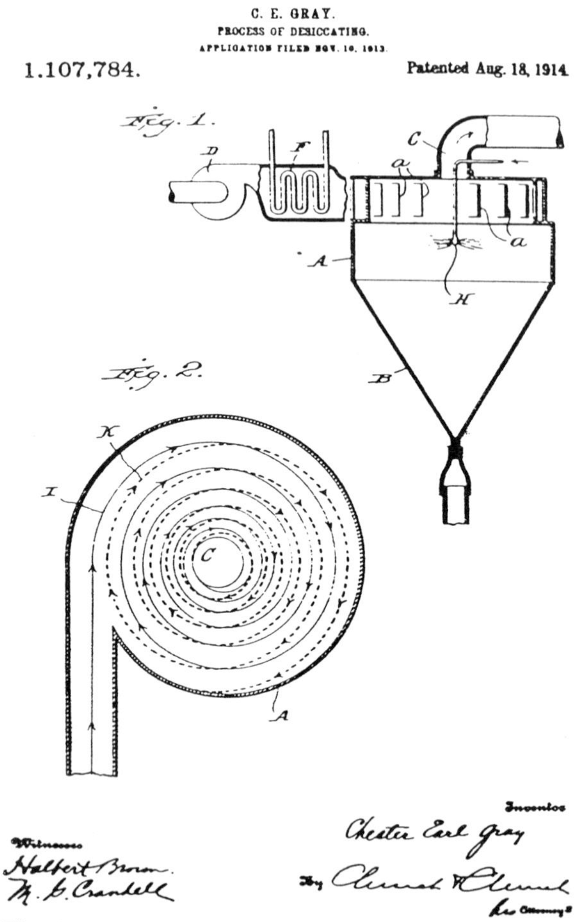

Fig. 1 The first commercially successful spray dryer for food formulations: The Gray Jensen design, patented 1914.

circumstances to obtain uniformity of the physical properties and overall quality of the powder.

Every spray dryer is characterized by the flow of droplets, particles, and air within the dryer layout. To achieve a continuous operation, extreme care has to be taken in selecting the design of the air disperser, atomizer, and drying chamber and in specifying the operating conditions in relation to the product being dried.

Different products have different drying histories, but any product that is hygroscopic or has a sticky phase during its drying history is particularly problematic if it has to be spray dried. Full control of the air flow pattern in the spray dryer is therefore essential to prevent droplets from reaching the walls of the drying chamber too quickly. There must be sufficient airborne drying time available to carry out the physical conversion of wet droplets into dried particles before the resulting particulates reach the vicinity of the wall. If the surfaces of the particulates, on contact with the wall, are too soft due to excessive moisture or heat, then wall deposits will occur. This is an unwanted phenomenon of spray drying, as deposit formation leads to product quality degradation (misshapen, discolored, hard particles) and poor dryer performance (too frequent plant shutdown for wet or dry cleaning). Furthermore, a safety hazard can be created as overheated deposits can be a potential ignition source for a powder explosion or fire. All powdered foodstuffs, by their very organic chemical structure, and especially those containing fats, exhibit explosive and inflammable properties.

Many foodstuffs or food ingredients required in powder form contain constituents that give rise to stickiness, and into this category falls the range of fat-containing products. The drying of such products provides some of the greatest challenges to the spray dryer designer and plant operator. Successful spray drying is not as easy as it looks.

Spray drying was first used in industrial production at the beginning of the twentieth century, and the earliest products handled included milk and eggs. Although the spray drying concept was first patented in the 1860s, the early designs failed due to the lack of appreciation of the importance of how droplets are contacted with hot air and the requirements of atomization and air distribution to achieve this. In fact, it took nearly 50 years to establish the technology as industrially feasible. The Grey Jensen design (Fig. 1) of 1914 was the first to show the way and started a development that has accelerated up to the present day. With the range of designs now available,[1] it is possible to spray dry the most difficult products, and these include products having a high fat content.

II. PRINCIPLES OF SPRAY DRYING

Spray drying involves the process stages of atomization and hot air introduction, which combine to give the conditions of spray–air contact necessary to accomplish droplet drying and particle shape formation during the residence time made available in the drying chamber. Atomization can be conducted by a rotating wheel (rotary) atomizer or nozzles.

A. Atomizers

For drying high fat products, pressure nozzle atomizers (Fig. 2a) are widely used. In such a nozzle atomizer, the liquid feed is fed to the nozzle under pressure (50–300 bar depending on the feed properties, nozzle capacity, and droplet size distribution required). Pressure energy, imparted to the feed by the high pressure pump, is converted to kinetic energy in the form of a high speed thin conical film of liquid issuing from the nozzle orifice. A swirl insert within the nozzle body creates the necessary liquid rotation to form this film. The film is very unstable and disintegrates into droplets under the influence of internal forces and external turbulence effects. However, the atomized liquid has a distinctive spray (cone) angle on leaving the nozzle, and this angle varies according to the product application, the required nozzle conditions of operation, and the sizes of orifice and swirl inserts selected. Spray angles are normally in the range of 50–110°. The size of the droplets in the spray varies inversely with the pressure (exponential 0.6) and directly with feed rate and viscosity (exponential 0.2). However, these exponentials are valid only over the somewhat limited capacity range of the nozzle. As the turndown ratios of these nozzles are low, multinozzle assemblies must be used in spray dryers designed for high feed capacities. However, within the capacity range of a given nozzle, the feed rate varies with the square root of pressure.

Rotary (wheel) atomizers (Fig. 2b), on the other hand, utilize centrifugal energy. The liquid feed is fed onto the center of the wheel, which is rotating with a high peripheral speed (60–160 m/s). The liquid moves to the edge of the wheel under the influence of centrifugal force, passing along vanes that prevent liquid slippage over the wheel surface. The liquid thus achieves the peripheral velocity of the wheel and is able to use the available centrifugal energy

[1]Major companies supplying spray dryers to the food industry include APV Anhydro A/S Copenhagen, Denmark; NIRO A/S Copenhagen, Denmark; and Stork Friesland BV, Gorredijk, The Netherlands.

(a)

(b)

Fig. 2 (a) Pressure nozzle in operation, showing its distinctive spray angle. (b) Rotary atomizer in operation, showing the horizontal trajectory of spray droplets from the wheel. (Courtesy of the NIRO Group.)

to the full. The liquid is ejected from the wheel edge as a thin film that readily disintegrates into droplets involving a mechanism similar to film break-up from nozzle atomizers. Air turbulence caused by the wheel rotation is also a major factor in the atomization mechanism (film break-up). The size of the droplets created depends on the wheel speed (peripheral velocity). The higher the speed, the smaller the droplet size (exponential 0.8). Rotary atomizers have infinite turndown potential, and there is very little effect of feed rate on droplet size within the feed rate variations occurring in a given spray drying operation. Viscosity becomes an influential factor only at high viscosity values.

B. Spray Dryer Chamber Designs

The design of the drying chamber in which the atomizer and air disperser are located depends on

1. The droplet size distribution required from the atomizer that will form the basis for the end product powder specification (size and structure) (Fig. 3).
2. The time droplets and particles must remain in the chamber (residence time) to complete drying down to the specified value of residual moisture content in the end powder.
3. The maximum temperature to which dried and semidried particles can be subjected and reach before heat degradation occurs.
4. The air flow pattern required in the drying chamber to minimize wall deposit formation.

For the drying of fat-containing products, two categories of design are applicable:

1. Cocurrent, in which droplets, particles, and air pass through the drying chamber in the same direction.
2. Mixed flow, in which droplets and particles pass downward through the drying chamber. The air initially flows downward only for the flow to be reversed so as to enable a top exhaust.

Drying chambers that conform to these categories are shown in Fig. 4.

Figure 4a. The standard cocurrent, conical base chamber operates with a two-point discharge system. Powder is discharged from the chamber base, whereas air and entrained fines (particles from the lower end of the size distribution) are exhausted from the side of the cone into cyclones for fines collection and reprocessing. The air disperser located centrally in the roof of the drying chamber creates a cyclonic air flow around the atomizer and down into the cone of the drying chamber. Both nozzle and rotary atomizers can be utilized. Pneumatic powder-conveying systems can be installed at the base of the chamber to give the so-called one-stage drying layout, i.e., powder leaves the drying chamber at the residual moisture content required in the end product. However, vibrating fluidized beds can also be installed at the base of the chamber for achieving two-stage drying. This lowers the product temperature during drying by operating the spray dryer at a lower exhaust air temperature. This in turn results in powder leaving the drying chamber at a higher residual moisture content than that specified for the end product. Therefore, drying is completed in the second-stage vibrating fluidized bed, which also includes powder cooling prior to discharge. Two-stage drying produces agglomerated powders of improved flowability and wettability and creates a spray drying operation of higher thermal efficiency. Drying chambers of this design are suited for only low fat content products and products having virtually no sticky and hygroscopic characteristics.

The air temperature profile in this design is given in Fig. 5a. The temperature of powders

Fig. 3 Spray-dried, fat-containing food ingredients. (a) Typical agglomerated free-flowing powders. (b) High fat powder ingredients: Butter (sample shown), fish oil, lard, lecithin, tallow, vegetable oil. (c) Dairy foods: Baby formula, butter, buttermilk, casein, caseinate, cheese, cream, lactose hydrolyzed whey (sample shown), permeate, skim milk, sweetened condensed milk, whey, whole milk, yogurt. (Courtesy of the NIRO Group.)

discharged from the chamber is approximately 20–25°C lower than the exhaust air temperature required to achieve the required residual moisture content in the powder.

Figure 4b. The tower (tall-form) cocurrent, conical base chambers are used exclusively with nozzle atomizers. The chamber can be supplied with an oversize lower cone section (bustle) to maximize powder discharge and minimize powder carryover to the cyclones. The air disperser creates a streamlined air flow with a minimum of turbulence down the chamber. This maintains both droplets and particles airborne, keeping them away from the surrounding walls until well into the lower cone section. A typical air temperature distribution through the chamber is shown in Fig. 5b.

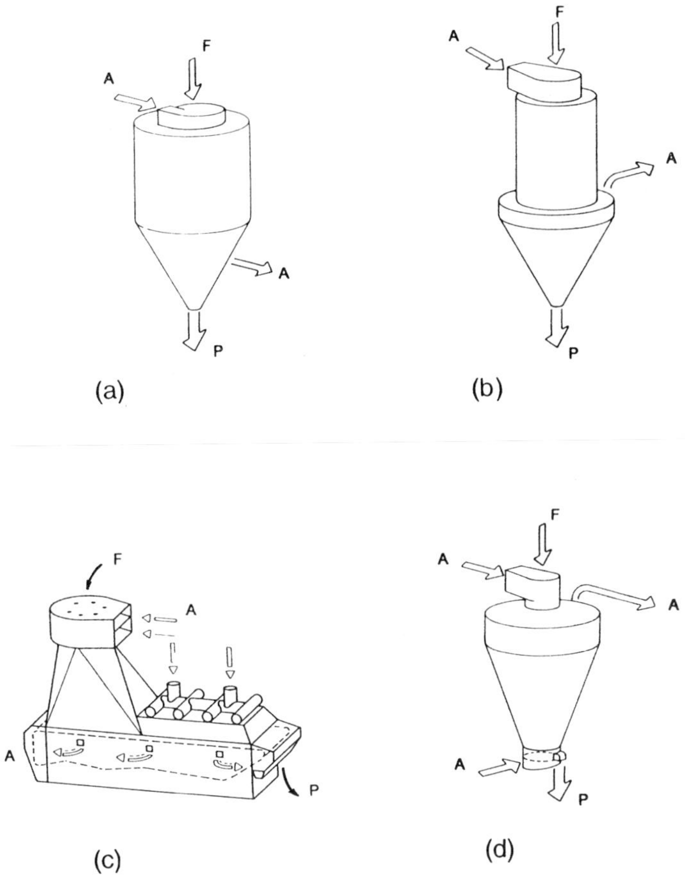

Fig. 4 Chamber designs used in the spray drying of foodstuffs. (a) Cocurrent, conical base, wide body; (b) cocurrent, conical base, tower body; (c) cocurrent, integrated belt, Filtermat body; (d) mixed flow, integrated fluidized bed, MSD body. A = air, F = feed, P = dried product.

This design is used for low to medium fat content products, baby foods, and milk and coffee powders, especially when large individual particles are required to maximize bulk density and flowability. The drying chamber design can be operated with a vibrating fluidized bed attachment acting as powder cooler/conditioner for fat-containing powders and as an afterdryer agglomerator/cooler for instant, agglomerated powders.

Figure 4c. Integrated belt, cocurrent chambers feature nozzles spraying feed down into hot air and onto a moving belt. The first section of the drying chamber resembles a short nozzle tower with an air disperser creating a streamline, nonturbulent air flow pattern. Semi-dried powder accumulates on the belt and moves with the belt into further sections of the drying chamber where completion of drying and cooling is achieved with secondary warm and cool air flows. All incoming air passes through the powder layer and belt fabric prior to exhausting. In this way, the powder acts as an effective particulate filter and hence negligible amounts of fines are collected in the cyclone. In many cases, further air cleaning equipment such as scrubbers or bag filters is not required after the cyclone, as environmental emission standards can be met with just a cyclone.

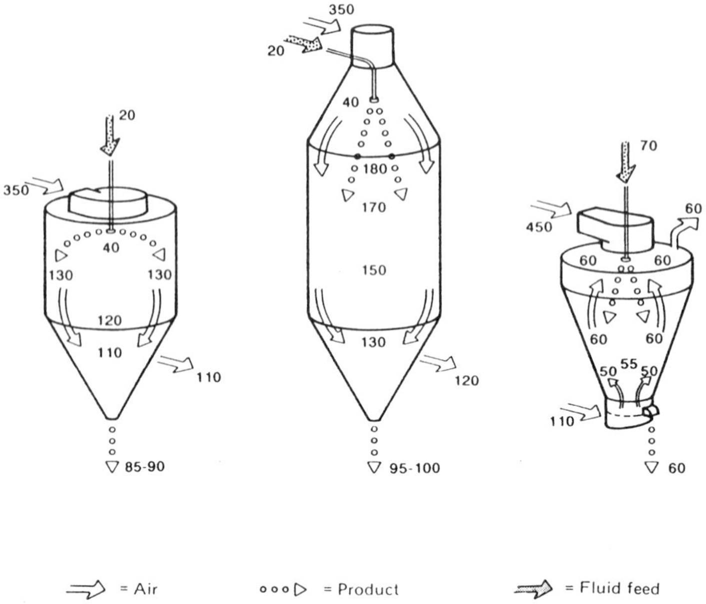

Fig. 5 Air temperature profiles in spray dryer chambers. Numbers are temperatures in degrees Celsius. *Note*: Profiles are for operating conditions with inlet hot air temperatures much higher than those used in spray drying fat-containing formulations. This is to illustrate that even with these higher inlet levels, air temperatures in the drying chamber are maintained low and the corresponding product temperatures even lower.

With the previously described cocurrent designs, air and powder residence times in the drying chambers are comparable. In the integrated belt design, the residence time available for the powder as it moves slowly on the belt through the drying chamber is much longer than the time required for the air to flow from the air disperser to the cyclone. This makes this design very suitable for high fat content products and products that slowly crystallize, i.e., sugar-containing foodstuffs.

The fact that all particles during their most soft-surface (stickiest) state are lying on the belt and not swirling in air flows makes the design ideal for products that cause the greatest problems of wall deposits in more conventional chamber designs. Powder adhesion on the belt is prevented by the selection of the belt material. A further important feature of operation is the very low exhaust drying air temperature required to achieve completion of drying. Powder temperatures are maintained very low, which is very important for fat-containing powders. Powders produced on this type of design have an agglomerate form.

Figure 4d. The mixed flow integrated fluidized bed drying chambers represent a design frequently used to handle the more difficult drying operations, i.e., with fat-containing products or foodstuffs exhibiting hygroscopic and sticky characteristics. However, for the highest of fat contents and for sugar-containing products, the integrated belt design is preferred.

Nozzle atomizers are normally used, although in large-diameter drying chambers, rotary atomizers are equally successful. The air disperser creates a strong downward flow of air so that the spray is projected deep down into the lower cone of the chamber toward the fluidized bed that is installed in the chamber base. Operation of a drying chamber with an integrated fluidized bed enables lower outlet temperatures to be used than with the standard cocurrent and tall-form tower designs and thus leads to lower powder temperatures and higher thermal efficiencies. A typical temperature distribution through the chamber is shown in Fig. 5c. The drying air from the roof-mounted air disperser together with the air supplied to the fluidized bed becomes mixed in the central part of the drying chamber and then flows back up the chamber for exhaust via ducts mounted at the top of the chamber walls or through the roof, away, however, from the air disperser. The fluidized bed operates as a totally mixed fluidizing system as this allows powder that cannot be directly fluidized at the moisture content it possesses on entering the fluidized layer to be successfully and continuously handled at the base of the chamber, thereby achieving completion of drying and agglomeration. This type of drying chamber is used for heat-sensitive, sticky products where agglomerated free-flowing powders are required.

All four designs described above have full cleaning-in-place (CIP) capability.

C. Drying Mechanisms

Irrespective of the design of the drying chamber, the mechanism of droplet drying follows the same principles.

The moisture in a spray droplet can be present in two forms: bound and unbound. The nature of the solid and accompanying moisture determines the drying characteristics. The bound moisture in the solid exerts an equilibrium vapor pressure lower than that of the pure water at the same temperature. Water retained in small capillaries in the solid absorbed at solid surfaces as solutions in cells or fibers falls into the category of bound moisture. The unbound moisture in a hygroscopic material is that moisture in excess of the bound moisture. All water in a nonhygroscopic material is unbound water that exerts an equilibrium vapor pressure equal to that of pure water at the same temperature. The equilibrium moisture is the moisture content of a product when it is at equilibrium with the partial pressure of the water vapor of the surroundings. The free moisture is the moisture in excess of the equilibrium moisture and consists of unbound and some bound moisture. Free moisture is evaporated during spray drying.

The mechanism of moisture flow through a droplet during spray drying is diffusion supplemented by capillary flow. The drying characteristics of the droplet depend upon whether bound or unbound moisture is evaporated, as each has distinct features. As long as unbound moisture exists, drying proceeds at a constant rate and will continue while the rate of moisture diffusion within the spray droplet is fast enough to maintain saturated surface conditions. However, for food-based products, constant-rate periods are very short. When diffusional and capillary flows can no longer maintain these conditions, the critical point is reached, and the drying rate will decline until an equilibrium moisture content condition is attained. However, in practice, the residence time for particles in the spray dryer is too short to allow equilibrium conditions to be established with the exhaust air.

The residual moisture content for a given dryer operation is thus related to the temperature and humidity levels of the exhaust air. In the majority of industrial spray dryer operations, saturation conditions at the powder discharge area are never approached, and thus maintaining the exhaust air temperature at a set point is control enough for the residual powder moisture content. This outlet temperature is maintained constant either by (1) adjustment to the feed rate to the atomizer at fixed heat input to the air heater supplying air to the air disperser or (2) adjustment of the heat input at the air heater at constant feed rate to the atomizer. To compensate for humidity changes (from day to night, or from winter to summer), the outlet temperature set point may have to be readjusted up and down as required to maintain a constant residual moisture content value.

Control systems that account for those temperature and humidity variations are available if such a degree of control is required for continuity of operation and reproducibility of powder quality. Systems for continuous measurement of moisture content are also available, but it is seldom that these measurements are used to control the spray dryer. It is the outlet temperature measurement that is the important control variable in all spray dryers for foodstuffs.

As the resistance to moisture transfer from within each spray droplet to the surface differs from product to product, the shape of the particle also differs, even under the same atomization and air dispersion conditions. Furthermore, droplets of the same product are always subjected to different local temperature, humidity, and velocity environments, and therefore spray-dried particles are never homogeneous in shape and size. Powder viewed under a microscope displays a wide range of structures including individual and agglomerated particles and misshapen and broken particulates. The inability to produce homogeneous, uniform particles in the chamber is no longer an important criterion, because with today's development of spray drying systems, the original atomized droplet and particle size is used as the starting material for aftertreatment techniques in integrated fluidized beds and belts or externally mounted vibrating fluidized beds to create the uniformity of the final agglomerate size and structure required by today's quality standards.

III. TYPICAL HIGH FAT SPRAY-DRIED POWDERS

One of the most important groups of fat-containing foodstuffs is based on milk powders produced by spray drying. Such products are intended for both human consumption and animal feeding and often have a fat content of more than 35%. Products for human consumption can either serve the needs in direct household use or act as raw materials for further industrial processing. Few high fat milk powders employ only milk fat as the fat constituent. To this group belong high fat, full cream powders, dried cream, and dried butter powder. A common feature of all other products is that the cream has been removed and the milk fat has been replaced by vegetable or animal fat. Regardless of the origin or kind of fat used, the composition of these products can vary greatly.

Historically, the production of fat-filled milk powders started in the 1960s. These were used as milk replacers for feeding calves. The original milk replacers were produced with a fat content of 18–22%. Such levels of fat enabled conventional one-stage cocurrent spray dryers with pneumatic transport systems to be used (see later, Fig. 6). Typical installations had evaporative capacities of 500 kg/h and used moderate inlet drying air temperatures. It was soon recognized that the installed spray dryer capacity could be better utilized by processing skim milk in the peak season and filled milk powder of high fat content during the off-peak season. This production trend created the need to better understand the drying mechanisms of high fat milk formulations and stimulated development in design to achieve improved continuous and high quality production. This involved dry mixing and recombination techniques. High fat powder was blended with skim milk powder to obtain a product of the same composition as that produced by drying the original raw material directly. High fat content concentrates prepared for spray drying contained initially about 35% fat, a percentage that has increased over the years with the development of the technology to handle higher fat levels. Today, fat levels of 60% are by no means uncommon, and the spray dryers suited for handling such formulations incorporate fluidization or belt drying technology for the necessary operational stages of drying, cooling, and powder handling (Figs. 4c and 4d).

Milk replacers for calves can be considered the pioneer product that opened the way to more sophisticated fat-containing products for both household and industrial use. Many branches of the food industry have always used dairy-based ingredients in their classical form, i.e., skim milk, full cream milk, cream, butter, and whey, as components in their recipes. However, automation of these industries has demanded modified raw materials that are better suited for dosing and transportation and have a longer shelf life and improved functional properties.

The driving force of the milk powder industry to develop new products was the demand and pressure from food manufacturers—i.e., bakeries and manufacturers of soups and sauces in liquid and dried form, ice cream, confectioneries, sweetened condensed milk, cream-based liquors, chocolate, and meat products—in fact, all industries who were seeking new, innovative, and functional food ingredients that could assist in extending their range of products of improved quality, all of course at reduced manufacturing costs.

IV. SPRAY DRYING HIGH FAT FORMULATIONS

Spray-dried powders of milk-based formulations containing fat include

Full cream milk powder with fat content greater than 35%
Full milk powder of high free fat content
Fat-filled milk powder with fat content of 40–60%
Fat-filled milk powder with glycerine monostearate content of 30%
Fat-filled whey powder with fat content of 40–60%
Butter powder with fat content of up to 80%
Soft-serve ice cream mix powders
Coffee whiteners
Cream liquor powders

Each of these products appears in numerous compositional variations with functional properties tailormade to the needs of a particular customer. The raw materials available for use include

Caseinates	Whey (sweet/acid)	Glucose
Cream	Whey protein concentrate	Lard
Emulsifiers	Whole milk	Sequestering agents
Free-flowing agents	Animal fat	Soybeans
Gelatin	Antioxidants	Sugar
Skim milk	Coconut	Sunflower seed
Stabilizers	Corn	Tallow
Starches	Cottonseed	Vegetable oil
		Vitamins and minerals

A. Fat-Filled Powders

Irrespective of the formulation, the mix preparation is very important to ensure a trouble-free drying operation and production of a quality powder. Most formulations contain skim milk or whey solids as the non-fat component. A typical processing line based on skim milk is as follows.

Skim milk is concentrated in a multiple-effect, falling film evaporator to achieve 50–55% total solids level after mixing in the fat. The fat components, together with emulsifiers and possibly vitamins, are stored at 50–55°C, but not for longer than 24 h. The fats and concentrated skim milk are then mixed in the ratio necessary to achieve the desired final fat content. The resulting mix is stored in agitated tanks prior to homogenizing and spray drying.

Homogenization is conducted in two stages. For mixes containing approximately 50% fat in the total solids, homogenizing pressures in the first stage are in the range of 70–80 bar and in the second, 20–30 bar. At higher fat contents, pressures should be proportionally lowered.

Following homogenization, the mix concentrate is spray dried. Moderate inlet drying air temperature levels of 180°C are used for formulations having fat contents of less than 50%. Should the fat contents be higher, lower inlet drying air temperatures are recommended. Skim milk powder containing glycerine monostearate as the fat component is processed in a similar way.

For the processing of formulations using whey powder as the non-fat constituent, similar procedures are used, but it is advantageous to precrystallize the whey concentrate and reheat to ~50°C before mixing with the fat. The process has been subjected to much fundamental study including the stability of emulsion when fat-filled whey powders of about 50% are applied for producing milk replacers. Two problems have been highlighted during reconstitution: foaming and excessive creaming of the emulsion. The explanation of these phenomena relates to cluster formation during homogenization when using high pressures and at low protein/fat ratios. This problem is reduced with the addition of a calcium sequestering agent prior to homogenization.

B. Butter Powder

With its 70–80% milk fat (approximately the same level as in normal table butter), butter powder is most difficult to dry and handle. Mix preparation involves the addition of sodium caseinate, non-fat milk solids, emulsifiers, and stabilizing salts. Spray dryers with belt or fluidized bed coolers are essential, as is also a final moisture content of the powder, which must be below 1%. If the powder is to be used in further processing, a free-flowing agent is added. The product is used exclusively in bakeries as a source of milk fat in dry form for making croissants.

C. Whole Milk Powder with High Free Fat Content

High free fat content powder having a total fat content of 28% does not by definition belong to the category of high fat milk powders. However, the characteristics of this product and the difficulties experienced in drying it justify its mention in the coverage of high fat powder production.

The free fat content of such powder should be more than 90% of the total fat, and this is achieved by precrystallizing the lactose in the whole milk concentrate. Lactose in conventional whole milk powder is present in its amorphous form and provides the continuous phase of the milk powder particle encapsulating the fat globule. Higher degrees of crystallization and even higher free fat content are achievable in the spray dryer designs that feature an integrated fluidized bed or belt where the extended residence time of the moist powder in the fluidized bed and on the belt enables continuing aftercrystallization.

D. Coffee Whiteners and Coffee Creamers

Coffee whiteners and coffee creamers are available to the consumer with varying recipes, but most of them do not contain any milk solids. These products were developed to avoid the flocculation that appears when whole milk powder is used in hot coffee or tea. This flocculation was attributed to the whey protein. The pH of coffee and tea is quite low, sometimes well below 5, which together with the temperature of the beverage creates favorable conditions for denaturing of whey proteins on the surface of particles before acceptable dispersion and dissolving could take place. Typical constituents that make up coffee whiteners include glucose (corn syrup solids), sodium caseinate, vegetable fats, emulsifiers, stabilizers, stabilizing salts, and possibly sucrose and some non-fat milk solids. The formulation is spray dried in dryers with or without integrated fluidized beds to produce a fairly agglomerated product that will give good functional properties to the powder. Compared with high fat formulations, coffee whiteners and creamers are classified as among the easier products to spray dry.

E. Ice Cream Powder

An ice cream powder can be considered a whole milk powder with a fat content of up to 45%. However, part of this fat can be vegetable fat. Besides milk solids, powders also consist of sucrose and emulsifiers. Ice cream mix powders can be flavored and colored prior to the preparation of the ice cream or left neutral. Owing to the fat and sugar content, ice cream powder is produced on spray dryers featuring integrated fluidized beds or belts.

F. Cream Liquor Powder

Cream liquor powder is produced by spray drying a feed formulation consisting of sodium caseinate, cream (48% fat), sugar, starch, glucose, coloring and flavoring ingredients, emulsifiers, and milk or vegetable fat, forming a mixture containing approximately 60% fat on a dry weight basis. This is homogenized prior to spray drying to produce a stable emulsion. The high fat content requires drying in spray dryers with integrated fluidized beds or belts.

V. SUITABILITY OF SPRAY DRYER DESIGN: GUIDE TO SELECTION

The suitability of a given spray dryer design depends on the feed formulation and its total fat content. The higher the fat content, the more difficult the spray-drying operation becomes, and the more sophisticated the design of the plant required. For fat contents up to 35%, conventional layouts suffice (Fig. 6), but for higher fat contents, fluidized bed cooling systems become essential (Figs. 7 and 8), and the problem of handling powder fines in the cyclones must be addressed. Cyclones have to be sized for low pressure drop operation so as to prevent excessive deposit formation and bridging of the powder at the outlet. All mechanical handling of such fine particles

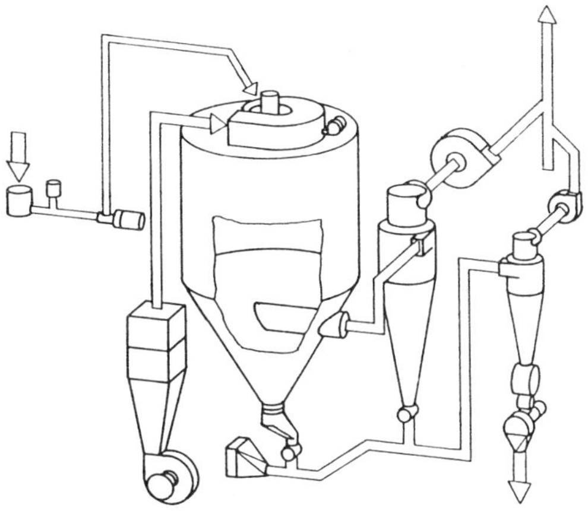

Fig. 6 Spray dryer layout with pneumatic conveying of powder from drying chamber and cyclone.

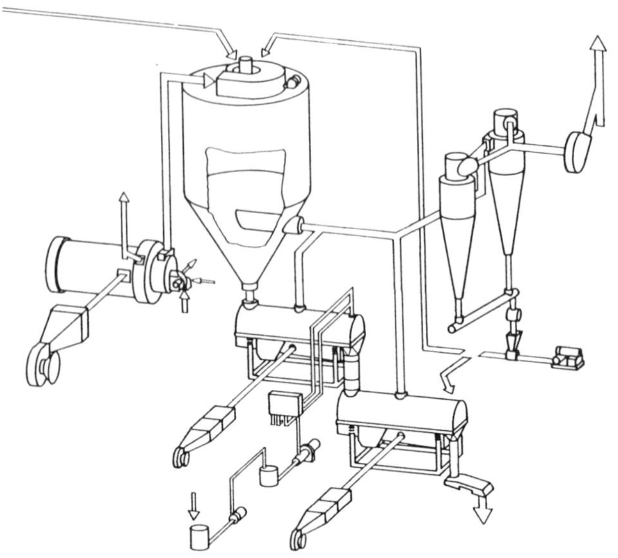

Fig. 7 Spray dryer layout featuring external vibrating fluidized beds as powder cooler/afterdryer-cooler.

must be substantially reduced or eliminated, and therefore the plant layout must allow for gravity flow of these fine particles from the cyclones, etc., to the cooling bed. Formulations with up to 60% fat can be handled on spray dryers with rotary atomizers; however, there is the added risk of increased wall deposits at increasing fat content. This is because in spray dryers that use rotary atomizers and air dispersers creating a cyclonic air flow in the drying chamber, particles cannot be prevented from moving radially toward the drying wall under the influence of the cyclonic

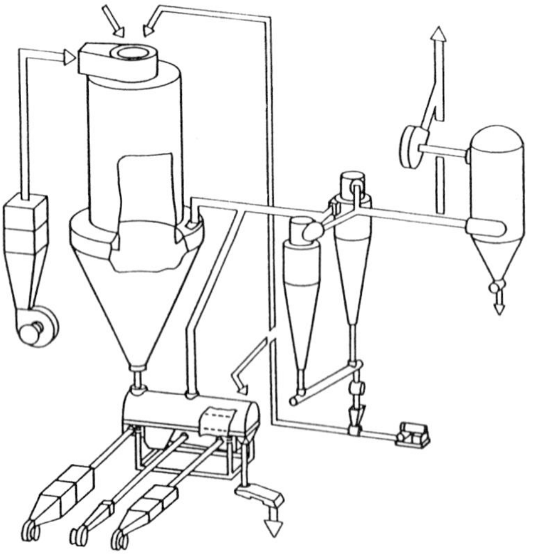

Fig. 8 Spray dryer layout featuring tower (tall-form) drying chamber 1, external vibrating fluidized bed, and powder fines recycle from cyclones. Final exhaust air cleaning in bag filter optional.

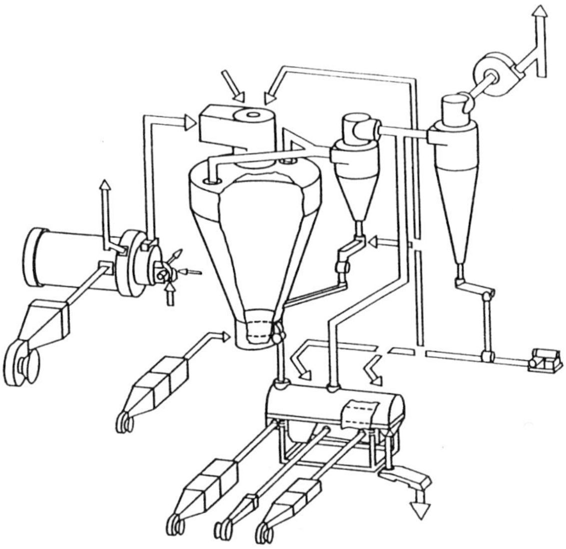

Fig. 9 Spray dryer layout (MSD) featuring drying chamber with integrated fluidized bed, external vibrating fluidized bed, and powder fines recycle from cyclones.

turbulent air flow. If the particles move to the wall too fast and contact is made while the particles are soft and in a sticky state, then excessive deposits will occur. For this very reason, when powders of fat contents higher than 60% are required, dryers with alternative spray and air flow patterns are necessary, i.e., nozzles must be used and a more streamlined air flow established down the center of the dryer as in the tower-type design or in the spray dryer with an integrated fluidized bed or belt (see Figs. 8–10).

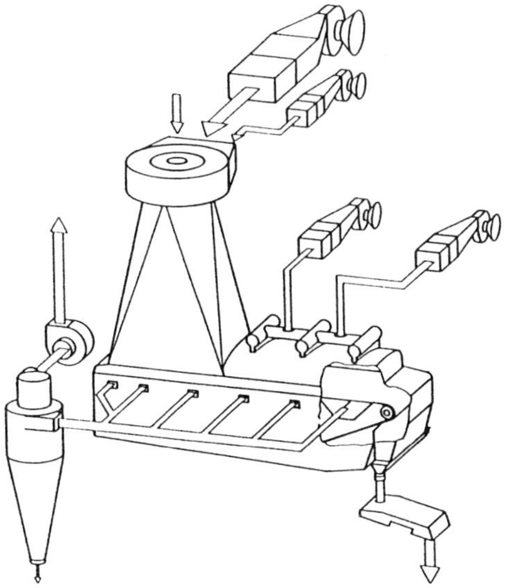

Fig. 10 Spray dryer layout (Filtermat) featuring drying chamber with integrated belt and powder fines collection in cyclone. No fines recycle required.

Table 1 Suitability of Design

Product	Fat content (%)	Spray dryer design[a] A (Fig. 6)	B (Fig. 7)	C (Fig. 8)	D (Fig. 9)	E (Fig. 10)
Milk replacer	~35	1	2	3	3	4
	~40–50	0	2	3	3	4
	~50–60	0	1	3	3	4
Fat-filled whey	~35	1	2	3	3	4
	~40–50	0	2	3	3	4
	~50–60	0	1	3	3	4
Coffee whitener	~40–45	1	3	4	4	4
Butter	>70	0	0	1	2	4

[a]Classification: 0, Not recommended. 1, Possible, especially as a multipurpose plant with non-fat or low fat powders as main products. 2, Suited. 3, More suited. 4, Most suited.

As a general rule, the higher the fat content, the more difficult the drying operation due to the increased tendency for wall deposits to build up. This tendency can be reduced by using lower inlet drying air temperatures, lower homogenizing pressures in the feed system, and higher feed concentrations and cooling the chamber walls. Furthermore, powder flowability can be improved by incorporating the recycling of the very fine particles collecting in the cyclone back to the spray drying chamber to coat the spray droplets with powder as they leave the atomizer. The fines return system can also be used to add other solid ingredients

Fig. 11 Spray dryer with installed safety suppressant system. (Courtesy of the NIRO Group.)

such as skim milk powder in a planned powdering technique. The use of such techniques means that the feed must have a higher fat content than is required in the final product. The suitability of spray dryer designs available to the food processing industry for the handling of fat-containing feedstocks is summarized in Table 1. The spray dryer with integrated belt is considered the most suitable for fat-containing products. An industrial-size dryer is shown in Fig. 11.

VI. SAFETY ASPECTS

Spray dryers operating on fat-containing products have a fine safety record, but there are examples where explosion and fire have caused structural damage. Fat-containing powders can form explosive mixtures in air, and it is impossible to prevent the likelihood of dangerous powder–air concentrations existing in some part of the spray dryer chamber and auxiliary powder handling equipment. The fat content itself presents a fire hazard under any circumstances. Therefore, during the spray drying of fat-containing formulations every precaution must be taken to eliminate ignition sources. Experience has shown that the usual source is overheated deposits falling through the swirling powder cloud in the spray drying chamber and into the fluidized bed integrated within or attached to the chamber. Therefore close attention must be given to the prevention of deposits on the roof air disperser and atomizer assembly, as these are located in the hottest zones of the dryer. Furthermore, the use of systems

Fig. 12 Spray dryer chamber with removable air insulation panel. (Courtesy of the NIRO Group.)

Fig. 13 Filtermat spray dryer designed for food ingredients including fat-based formulations. (Courtesy of the NIRO Group.)

that return fines via the atomizer must not permit interference of the overall air flow pattern in the drying chamber and create a turbulence that can direct the fines, which consist of powder of small particles, back into the air disperser and hot air zone. Deposits on the lower wall areas are a lesser danger as they are at much lower temperatures.

All suppliers of spray dryers for high fat formulations have to address the safety issue by selecting and providing the most suitable design. For these types of products, designs have to be quite sophisticated; therefore they represent a fairly expensive investment. As a minimum the design must be equipped with an automatic fire extinguishing system for protection against the spreading of fire, and overpressure venting for protection against a sudden pressure rise in an explosion situation. Installation of overpressure vents inevitably influences plant performance by increasing the possibilities of air leakage and liquid seepage during CIP cleaning. There is always an advantage, therefore, to minimizing the venting areas, and this is done without compromising safety by strengthening the chamber structure to a higher pressure shock–resistant rating so that a smaller vent area can be used. An alternative to pressure shock–resistant designs is the use a suppressant system, but this approach is only beginning to be considered as a practical alternative within food drying. Pressure systems involve canisters of a fire extinguishing chemical that is dispersed throughout the spray dryer as soon as an explosion hazard situation arises. Figure 12 shows a spray dryer equipped with such a system. The canisters are on the cone and can be clearly seen. However, venting and strengthened structures still remain the most widely used safety features on spray dryers handling fat-containing products. Whenever venting is used, it is most important that all venting areas be ducted to safe areas, preferably outside the building, so as to prevent the possibility of a secondary explosion and fire occurring just outside the spray dryer structure. Ducting also protects the operator in the working areas and protects the building against partial or total collapse.

Aspects of safety always focus on degrees of deposit formation that can be expected with a given fat-containing formulation, since the nature of the products gives a high possibility of deposits forming at the walls. Deposit formation is usually minimized through the correct adjustment of the air disperser and location of the atomizer assembly, but the quality of the wall insulation and effective overall cooling also has an effect on the positioning of deposits. In recent years, the whole situation regarding deposit formation has been improved by the use of air insulation panels that are also removable. This gives not only the advantage of reducing deposit tendencies but also the ability to inspect the wall easily for cracks or pinholes in the stainless steel plate that can cause safety problems associated with bacterial contamination. A dryer chamber equipped with such removable insulation panels is shown in Fig. 13.

ACKNOWLEDGMENTS

I thank the management of the NIRO Group of Companies for permission to use NIRO photographs and flow diagrams in illustrating the chapter. I am also grateful to Dr. J. Pisecky for his advice and technical input.

REFERENCE

1. K. Masters, *Spray Drying Handbook*, 5th ed., Longmans, London, 1991.

BIBLIOGRAPHY

Bucheim, W., Electron microscopic localization of solvent-extractable fat in agglomerated spray-dried whole milk powder particles, *Food Microstruct. 1*: 233–238 (1982).

Buma, T. J., Free fat in spray dried whole milk. I. General introduction and brief review of literature, *Neth. Milk Dairy J. 25*: 33–41 (1971).

Buma, T. J., Free fat in spray dried whole milk. II. An evaluation of methods for the determination of free fat content, *Neth. Milk Dairy J. 25*: 42–52 (1971).

Buma, T. J., Free fat in spray dried whole milk. III. Particle size. Its estimation, influence of processing parameters and its relation to free fat content, *Neth. Milk Dairy J. 25*: 53–72 (1971).

Buma, T. J., Free fat in spray dried whole milk. IV. Significance of free fat for other properties of practical importance, *Neth. Milk Dairy J. 25*: 88–106 (1971).

Buma, T. J., Free fat in spray dried whole milk. V. Cohesion; determination; influence of particle size, moisture content and free fat content, *Neth. Milk Dairy J. 25*: 107–122 (1971).

Buma, T. J., Free fat in spray dried whole milk. VIII. The relation between free fat content and particle porosity of spray dried whole milk, *Neth. Milk Dairy J. 25*: 123–140 (1971).

Buma, T. J., Free fat in spray dried whole milk. IX. The size distribution of fat globules in concentrated milk and spray dried milk, *Neth. Milk Dairy J. 25*: 151–158 (1971).

Buma, T. J., Free fat in spray dried whole milk. X. A final report with a physical model for free fat in spray dried milk, *Neth. Milk Dairy J. 25*: 159–174 (1971).

Faeldt, P., and B. Bergenstahl, Fat encapsulation in spray-dried food powders, *J. Am. Oil Chem. Soc. 72*: 171–176 (1995).

Faeldt, P., B. Bergenstahl, and G. Carlsson, The surface coverage of fat on food powders analyzed by ESCA (electron spectroscopy for chemical analysis), *Food Struct. 12*: 225–234 (1993).

Hansen, P. S., Production of agglomerated fat-filled milk powder, *J. Soc. Dairy Technol. 33*: 19–23 (1980).
Hilaire, J. R., Spray-drying: principle and recent developments, *Via-Lactea 8*: 13–15,17–20 (1976).
Horn, J. D., Spray dried fats, *Int. Food Flavor. Food Addit. 7*: 64–65, 68 (1976).
Hussmann, P., Process for producing emulsified, edible fat preparations of good keeping quality, German Patent Appl. 1 417 445 (1970).
Kautz, K., Method for manufacture of free flowing powders from fat and flour, German Patent Appl. 1 417 553 (1968).
Kharitonov, V. D., Distribution of free fat on the surfaces of dried milk particles, *Molochn.-Prom. 7*: 12–15 (1975).
Kubicki, M., and S. Misicki, Spray drying of emulsion in powdered fat manufacture, *Tluszcze-Jadalne 18*: 225–238 (1974).
Lipatov, N. N., G. B. Dvoretskii, L. A. Kabanov, and N. N. Mizeretsikii, Method for production of dried rapidly soluble fat-containing milk products, USSR Patent 428 734 (1974).
Loo, C. C., Process for the production of rapidly dispersible dried milk, or of a milk based fat containing dried product, German Patent Appl. 1 492 773 (1969).
McMahon, T., Powdered lipid composition and its production and use, UK Patent Appli. 2 031 937A (1980).
Mol, J. J., The milk fat globule membrane and the solubility of whole milk powder, *Neth. Milk Dairy J. 29*: 212–224 (1975).
Nava, L. J., J. T. Hutton, J. Shields, and C. Kempf, Process for manufacturing instant fat containing milk, German Patent Appl. 1 492 784 (1969).
Onwulata, C., P. W. Smith, J. C. Craig, Jr., and V. H. Holsinger, Physical properties of encapsulated spray-dried milk fat, *J. Food Sci. 59*: 316–320 (1994).
Onwulata, C. I., and V. H. Holsinger, Thermal properties and moisture sorption isotherms of spray-dried encapsulated milk fat, *J. Food Process. Preserv. 19*: 33–51 (1995).
Pisecky, J., The new generation of spray-driers for milk products, *Deut. Molkerei Z. 104*: 886, 888–891 (1983).
Pyle, J. R., Instant milk powders and the principles of high-fat powder manufacture, in *Winter School on Spray Drying*, see FSTA (1976) 8 2P275, 1975.
Rayner, P. B., Spray dried fats—their nature and applications, *Flavour Ind. 4*: 379–380 (1973).
Sadini, V., Modification of organoleptic properties and presentation of butter, *Int. Dairy Congr. E*: 891 (1978).
Tamsma, A., F. E. Kurtz, A. Kontson, and C. Sutton, Reconstitutability of spray dried milk products containing 26 or 13% fat, *J. Dairy Sci. 58*: 1433–1436 (1975).
Tamsma, A., and A. Kontson, Preparation of a foam spray dried whole milk type product with good sinkability, *J. Dairy Sci. 57*: 1149–1151 (1974).

19

Low Calorie Fats

John W. Finley,* A. McDonald,† and L. P. Klemann
Nabisco Foods, East Hanover, New Jersey

I. INDUSTRIAL SIGNIFICANCE

The development of food products with flavor and quality attributes similar to their full fat and calorie counterparts but reduced in both fat and calories has presented a major technological challenge for food manufacturers. Reducing fat content requires replacement of part or all of the fat by an ingredient or system of ingredients that can effectively duplicate the diverse functional roles of the fat it replaces in a wide variety of product applications. Reducing fat content is the most efficient means for reducing caloric content because, with a caloric contribution of more than twice that of carbohydrate or protein, less than half as much fat has to be removed to achieve an equivalent caloric reduction. Consumer demand for high quality products low in both fat and calories has stimulated the development of a number of new technologies and innovative ingredients aimed at replacing fat in food products without compromising its functional contributions to flavor delivery, textural characteristics, and heat stability.

II. APPROACHES TO REDUCTION OF CALORIES FROM FAT

Caloric contribution from fat can be reduced by one or more approaches including fat reduction by emulsion technology, fat mimetics, low calorie fats, and fat substitutes. Emulsifiers are also lipid-based, but some may include carbohydrate moieties as well. Fat mimetics do not have any lipid components, being derived entirely from either protein or carbohydrate sources. Low calorie fats are fats that exhibit only partial absorption. Fat substitutes are fatlike materials that are not digested but function like fat in food applications. In addition to use as food ingredients, some fat replacers may also function by forming films that serve as barriers

---

**Current affiliation*: Monsanto Company, Chesterfield, Missouri.
†*Current affiliation*: Consultant, Chicago, Illinois.

to fat absorption. These types of fat replacers are useful in frying applications to reduce the amount of fat retained by batter-coated or breaded products.

Fat replacement and caloric reduction are readily achieved by fat mimetics because they are not lipid compounds. Because they are not fats, their applications are sometimes limited by their lack of heat stability and true fat functionality. Emulsifiers generally serve as fat extenders, enabling reduced levels of fat to perform as effectively as larger amounts by maximizing functionality. Most emulsifiers have the caloric value of fat on a weight basis but contribute fewer calories to finished food because only small amounts are used. Low calorie fats and fat substitutes, on the other hand, must accomplish fat reduction through structural modifications that reduce digestibility. If less fat is available for absorption, its metabolic caloric value is reduced, as is the possibility of other metabolic effects typically associated with high fat intakes.

A. Low Calorie Fats and Fat Substitutes

Low calorie fats are genuine triacylglycerol mixtures that have been chemically and physically modified to provide fewer than the usual 9 cal/g normally contributed by lipids. This reduction in caloric value is accomplished through designating a specific fatty acid compositional mix and the positional arrangement of these fatty acids on the triacylglycerol backbone to reduce availability. In contrast, fat substitutes are engineered lipids that bear only a resemblance to the natural triacylglycerols of food lipids, although edible fats and oils are frequently used as starting materials in the synthesis of these compounds. Fat substitutes are actually structural analogues of natural lipids that either completely or almost completely resist hydrolysis by digestive enzymes and thus contribute minimal or no caloric value. Although fat substitutes can technically be considered low calorie fats because they have lower caloric values than regular fats, they are distinguished from low calorie fats in this chapter on the basis of their origins. As such, use of the term "fat" is applied only to those compounds that are true fats in the sense of their structural similarity to the triacylglycerol moieties of edible fats and oils.

B. Emulsifiers

Emulsifiers have been in wide use in fat replacement systems for some time. Incorporating emulsifiers such as mono- and diacylglycerols into the formulation of fat-reduced products is appealing because these compounds are familiar ingredients with an established record of safety. Emulsifiers perform as fat replacers most optimally when used in combination with other emulsifiers and as part of a fat replacement system with other ingredients such as water or flavorings. Classification of the structural, functional, and metabolic properties of emulsifiers place these compounds somewhere between carbohydrates and lipids. The extent to which they exhibit properties of lipids is a function of the degree of esterification present [1].

Emulsifiers may be used as fat extenders in partial replacement of fat or to completely replace fat in a product. Emulsifiers most commonly used in fat replacement systems at the present time are those surface-active molecules that have had a previous history of successful use as stabilizers in a variety of food products. Examples of such emulsion stabilizers are polyglycerol esters, polysorbate, and propylene glycol mono- and diesters [2]. When emulsifiers are used in fat replacement, properties not considered of consequence in conventional applications may become increasingly more significant as the amount of fat is reduced. Adjustments to formulations may need to be made when emulsifying agents are added to replace fat or extend fat functionality to accommodate the emergence of properties not normally of concern in full fat products. These properties include interactions with proteins, complexing of starch, and modification of the crystallization characteristics of other fats in the system [2].

C. Fat Mimetics

Fat mimetics function in fat replacement by simulating the physical attributes of fat. These types of fat replacers are structurally engineered to form spheroidal particles that will slide over one another upon hydration. The desirable diameter for reproducing the mouthfeel of fat is between 0.1 and 3.0 μm. Larger particles are too powdery or grainy, and smaller particles are imperceivable on the tongue [3]. Fat mimetics are effective at calorie reduction because their protein or carbohydrate natures render them intrinsically lower in calories on a dry weight basis than the fat they replace. The caloric value of these compounds is reduced further by dilution when reconstituted with the water required for functional activity. The hydrated state of fat mimetics lowers the caloric value by one-fourth to one-third of the value of the dry powder, to 1–1.33 cal/g. Carbohydrate-based mimetics may be used to replace 50–100% of fat in a product, while protein-based mimetics may replace 75–100%, but neither of these substitutions can replace fat in a gram-equivalent exchange.

The protein-based fat mimetics currently available have been derived from egg, skim milk, whey concentrate, or grain proteins. These proteins are subjected to microparticulation or denaturation processes to shape spherical particles similar in diameter to natural lipid particles. The prolific carbohydrate-based fat mimetics include hydrocolloid compounds such as starches, dextrins, maltodextrins, fibers, and gums. These compounds may also be processed to further enhance their intrinsic water-holding capacity. It is the hydrating property of these hydrocolloid particles that slows their clearance from the mouth, allowing the lubricant and flow characteristics of fat to be effectively duplicated.

Like the emulsifiers in current use, fat mimetics are widely used as fat replacers because they too are derived from substances with prior histories of safe use in food systems for purposes other than fat replacement. The carbohydrate-based fat mimetics are the most frequently used fat replacers presently on the market. The utility of protein-based fat mimetics for fat replacement is more limited by the instability of these compounds to high temperatures and the potential of the protein core components to provoke allergic responses. As with emulsifiers, fat mimetics must frequently be used in a fat replacer system with other ingredients before all of the functions of fat can be effectively replaced. Because large amounts of water are integral to the functional properties of fat mimetics, flavor release may be altered when these compounds are used, making it also necessary to adjust or add new flavor delivery systems. The effectiveness of fat mimetics in achieving functional properties of fat may vary with the particular application, making it unworkable to use the same fat mimetic satisfactorily in every category of fat-reduced products.

III. STRUCTURAL AND TECHNICAL ASPECTS OF LOW CALORIE FATS AND FAT SUBSTITUTES

The structural modifications of lipid-based compounds enabling them to function as fat replacers with lower bioavailability and caloric value than natural lipid compounds take one of two general approaches. In the first approach, the basic triacylglycerol structure is preserved but the composition and positional sequence of the fatty acid esters is dictated by requirements for decreased bioavailability or low metabolic energy value. Alternatively, the basic structures are synthesized from chemicals unrelated to triacylglycerols but possess the physical and chemical attributes of natural lipids that allow them to function as lipids in food systems while resisting the hydrolytic digestive enzymes that make them available metabolically. With either approach, the end products are lipid-based compounds that possess all of the organoleptic properties of conventional lipids but deliver fewer calories per gram. Most low calorie fats and fat substitutes are more accurately described as "families" of compounds related by their core structural elements. Minor adjustments in the physical or chemical natures of substituents on this core can expand the number of applications for a particular low calorie fat or fat substitute by creating different physical forms with a range of temperature or other physicochemical characteristics.

Low calorie fats are manufactured using the first approach, which modifies the identity and position of fatty acid substituents on the glycerol backbone of triacylglycerols obtained from edible fats and oils. Also referred to as structured triacylglycerols, these compounds retain the conventional properties of natural food lipids but with a bioavailability reduced by about 45%. During the synthesis of low calorie fats, selected fatty acids are esterified in a positional sequence on the glycerol backbone of the triacylglycerol defined according to random interesterification theory and designed to create a lower caloric profile [4]. These fatty acids may be of short- or medium-chain lengths (C_8–C_{10}), which are handled metabolically in a manner more similar to glucose than to fatty acids, thus yielding a lower fuel value and a decreased propensity for storage as body fat. Saturated fatty acids of longer chain lengths ($\geq C_{16}$) may also be used in the synthesis of low calorie fats. These long-chain saturated fatty acids, which are relatively more poorly absorbed in their free forms, are esterified preferentially at the first and third hydroxyl groups on the glycerol backbone, where they will be more readily hydrolyzed by digestive enzymes to their poorly absorbed free forms.

Fat substitutes are synthesized using the second approach, where structures that are unlike natural triacylglycerols are forged from a variety of chemical compounds and fatty acids, which are often obtained from edible fats and oils [5]. These synthetic structural analogues of triacylglycerols may include, but are not exclusive to, compounds where the glycerol backbone has been replaced with alternative alcohols such as neopentyl alcohol or sucrose polyol to sterically protect the ester bonds from enzymatic hydrolysis [6,7]; compounds such as branched triacylglycerols where the fatty acids have been replaced with alternative acids that also sterically protect the ester bonds [8]; compounds such as retrofats where the ester linkage has been reversed so that a polyfunctional acid serves as the backbone esterified with long-chain alcohols [9]; or compounds such as glyceryl ethers where the ester linkage has been reduced to an ether linkage [10,11]. Although they are more accurately labeled lipid analogues, fat substitutes may be loosely considered low calorie fats because they are lipid-based structures that deliver fewer calories per gram than conventional lipids. Because of their similarities to low calorie fats, fat substitutes are included here in the discussion of this type of fat replacer.

The advantages that the use of low calorie fats has over the use of other fat replacers stem from the authentic triacylglycerol nature of these compounds. Fat substitutes offer the same technical advantages as low calorie fats, but the synthetic basis of these lipid analogues raises questions about safety. Both low calorie fats and fat substitutes can perform identically to conventional fats in food systems and thus are better able to preserve the textural and flavor attributes imparted by fat. Both compounds also provide thermal stability at the high temperatures needed for baking and frying. Low calorie fats and fat substitutes are well suited for fat replacement in low moisture baking applications such as cookies and crackers. Either type of lipid-based fat replacer may also be used with equally good results in high moisture food systems such as ice cream and cheese. Fat replacement in low moisture systems has been among the most troublesome technical hurdles for food manufacturers to overcome because the common fat replacers presently used depend on hydration for functional effectiveness.

The fact that low calorie fats are partially digested and absorbed does not permit the degree of reduction in fat calories that fat substitutes may achieve by their indigestibility and that nonlipid-based fat replacers may achieve by completely replacing fat calories with calories from protein or carbohydrate. Fat substitutes are unique among fat replacers in their ability to completely reduce total fat calories in food systems on a weight-equivalent basis with little or no additional caloric contributions associated with their use. By being virtually noncaloric, fat substitutes are able to exceed the fat calorie reduction utility of low calorie fats, emulsifiers, and fat mimetics. As lipids, they can substitute for equivalent amounts of fat without requiring adjustments in formulations. The unique ability of fat substitutes to completely reduce the total amount of fat and its corresponding caloric contribution in a product without compromising product attributes is functionally important, but it raises a number of safety issues. The metabolic impact from ingesting the lipid analogue structures of fat substitutes is less predictable and less well understood than that of low calorie fats and other fat replacers that consist of or have been derived from conventional food lipids or common food ingredients with a prior history of safe use.

A. Low Calorie Fats

1. *Caprenin*

Developed jointly by the Procter and Gamble Company (Cincinnati, OH) and Grinstead Products, Inc. (Kansas City, MO) as a substitute for cocoa butter, caprenin (caprocaprylobehenin) is a structured triacylglycerol formed by esterification of glycerol with the medium-chain saturated fatty acids caprylic acid (C_8) and capric acid (C_{10}) and the very long chain saturated fatty acid behenic acid (C_{22}). All of these fatty acids are derived from natural food sources. Caprylic and capric acids are obtained by fractionation of palmkernel and coconut oils. Behenic acid is produced from complete hydrogenation of (high erucic) rapeseed oil and is also found in peanuts and marine oils. Behenic acid is only partially absorbed and metabolized, whereas caprylic and capric acids are metabolized rapidly for energy, yielding a caloric value similar to that of carbohydrate. The total caloric value of caprenin is 56% of the caloric value of conventional fats at approximately 5 cal/g.

Caprenin is limited by its particular fatty acid compositional mix to confectionery applications as a replacement for cocoa butter. There have been few early applications for this fat replacer because of high initial production costs. Caprenin has been used in combination with the fat mimetic polydextrose in several commercially available reduced fat chocolate candy bars. A GRAS self-affirmation petition has been submitted for caprenin for use as a fat replacer in soft candy and confectionery coatings [12]. Further information is available in Refs. 13–16.

2. *Medium-Chain Triacylglycerols*

Medium-chain triacylglycerols (MCTs) were originally developed for therapeutic purposes to provide a source of energy for individuals with compromised gastrointestinal systems. MCTs have been used extensively in parenteral nutrition formulations. MCT are structured triacylglycerols composed of fatty acids with chain lengths between 8 and 10 carbons obtained from splitting, distilling, and fractionating coconut oil. Caprylic and capric acids comprise more than 96% of the fatty acids in MCT. The presence of these medium-chain saturated fatty acid substituents create a triacylglycerol with a low melting point that is liquid at room temperature and stable to oxidative rancidity. MCTs are readily hydrolyzed by digestive enzymes, and the fatty acid end products are rapidly absorbed directly into the bloodstream following a postprandial pattern similar to that observed for glucose [17]. Medium-chain fatty acids are used as immediate sources of energy by the liver, yielding fewer calories per gram than longer chain fatty acids. The caloric value of MCT preparations is 7 kcal/g, or 22% lower than that of the long-chain triacylglycerols present in vegetable oils [18].

3. *Salatrim*

Salatrim comprises a family of structured triacylglycerols consisting of short- and long-chain saturated fatty acids randomly distributed on the glycerol backbone. Salatrim is synthesized using a sodium methoxide-catalyzed interesterification process developed by Nabisco Brands, Inc. (East Hanover, NJ) that reacts triacetin, tributyrin, and tripropionin with triacylglycerols of long-chain fatty acids from hydrogenated canola, soybean, or cottonseed oils [19,20]. Unreacted short-chain fatty acid triacylglycerols are removed from the reaction mix by vacuum steam deodorization [19]. Each newly formed triacylglycerol contains one short-chain and one long-chain fatty acid and either a short- or long-chain fatty acid at the third site, in proportions that can be theoretically predicted from the compositional profile of the starting materials [21]. Nuclear magnetic resonance imaging is consistent with the chromatographic characterizations of the compounds, confirming that the interesterification reaction is random and that fatty acids are distributed during the process at attachment sites on the glycerol backbone predictable by random reaction statistics [19].

A maximum of five variants of triacylglycerols can be produced from this interesterification reaction encompassing all possible combinations of short- and long-chain fatty acid arrangements to be derived from the starting materials [4,21]. Short-chain fatty acids typically make up 33–66 mol % of the mix of triacylglycerol end products, with the balance made up of long-chain saturated fatty acids [22]. Stearic and palmitic acids are the dominant long-chain saturated fatty acids in the mix. The ratio of stearate to palmitate in the final product mix reflects the proportions of these fatty acids in the original source of hydrogenated oil. Canola oil has the highest ratio of stearate to palmitate at 17.2 followed by soybean at 5.5 and cottonseed at 2.8 [22]. Stearic acid is absorbed much less efficiently than palmitic acid, and thus a higher proportion of stearate would contribute to a lower caloric value for Salatrim.

The ratio of stearic acid to palmitic acid is one of two determinants of the metabolic caloric value assessed for Salatrim. The other determinant is the intrinsically lower metabolic fuel value of the short-chain fatty acids, which, at 4 kcal/g, is more consistent with values for carbohydrate than with the 9 kcal/g value typical for fat. The absorption efficiency for Salatrim is a function of the molar ratio of short- to long-chain fatty acids, which may from vary from 0.51 to 1.99. As values approach the upper end of the range, absorptive efficiency increases from 15% to 70% and the caloric value derived per gram increases correspondingly from 2.56 to 6.39 kcal. According to both human clinical studies and rat bioassays, the metabolic caloric value observed for Salatrim ranges from 4.5 to 5.5 kcal/g. Based on these data, an average caloric value of 5 kcal/g has been assigned to Salatrim for labeling purposes [23].

The potential for preparing various combinations of fatty acid compositional mixes and positional arrangements in Salatrim increases the versatility of this fat replacer, expanding its uses to a wide range of applications including confectionery coatings, bakery cream fillings, and milk fat and shortening replacements. The ability to modify the ratio of short-chain to long-chain fatty acids in the Salatrim mix permits the physical properties of this fat replacer to be tailored to match those of a number of different conventional food fats [20]. Because Salatrim is completely saturated, it is highly stable to the effects of oxidation. The initial focus for Salatrim has been the development of a preparation that closely matches the melting profile of cocoa butter to create the cooling mouthfeel and smooth melt associated with chocolate. The preparation currently available will facilitate a 45% reduction in calories from fat and a 25% reduction in calories in chocolate products [24]. Pfizer Food Science Group (New York, NY) has licensed from Nabisco the exclusive rights for the manufacture and commercialization of Salatrim internationally. Nabisco has retained certain rights in North America associated with their core businesses.

B. Fat Substitutes

1. Carboxyl/Carboxylate Esters

Hydroxycarboxylic acids such as malic or tartaric acids form the backbone of the carboxyl/carboxylate ester group of fat substitutes. Carboxy, methyl carboxy, carboxylate, or methyl carboxylate groups extend from the carbon core. Fatty acid esters are formed with the acid groups, and fatty alcohol esters with the hydroxyl groups. These fat replacers are partially digested, yielding about 4% of the calories potentially available depending on the particular fatty alcohol and fatty acid substituents. Physical forms may also vary from oil-like to hard fats. This compound may be particularly useful as a coconut oil mimetic. Possible applications for carboxyl/carboxylate esters include butter cream, ice cream, vanilla wafers, and crackers. The process patent for producing these compounds belongs to Nabisco Brands, Inc. (East Hanover, NJ) [25,26].

2. Dialkyl Dihexadecylmalonate

Dialkyl dihexadecylmalonate (DDM) is a fatty alcohol ester of malonic and alkylmalonic acids patented by Frito-Lay, Inc. for use as a fat substitute in high temperature frying media [27]. This compound is a dimer consisting of a dicarboxylic acid carbon backbone substituted with hydrogen or C_1–C_{20} alkyls and esterified with alkyl, alkenyl, or dienyl groups [28]. Lower molecular weight forms may be synthesized by reacting a malonyl dihalide with a fatty alcohol such as oleyl. To produce higher molecular weight forms, an alkyl halide is added in basic solution. Impurities are removed by vacuum distillation or silica gel chromatography. The unique structure of DDM is almost completely resistant to digestive enzymes as a result of steric protection provided for the ester linkages. About 0.1% of the compound consumed is absorbed. A blend of DDM and soybean oil was used to prepare potato chips and tortilla chips to be tested by panelists. The products were found to be as crisp as those fried in conventional vegetable oils but without being as oily. DDM itself contributes no calorie value, but in a soybean oil blend it yields a fried product reduced in fat by 60% and reduced in calories by 33% [29].

3. Esterified Propoxylated Glycerol

The propoxylated glycerols are analogues of triacylglycerols in which a propoxyl group has been inserted between the glycerol backbone and the fatty acid side chain to replace the ester linkage with an ether linkage. Glycerol is reacted with propylene oxide to form a polyether glycol and then esterified with fatty acids obtained from any edible fat or oil. Lard, tallow, soybean, cottonseed, corn, and canola oils have all been used in preparing esterified propoxylated glycerol (EPG). Digestibility of the molecule decreases with the number of ether linkages formed. One to three polyoxypropylene groups of various lengths may be added to the glycerine backbone prior to esterification. Variations in the length of the propoxyl units, in types of fatty acids, and in the vegetable oil blends in which it may be contained contribute to the adaptability of this family of compounds to a wide range of applications. EPG is thermally stable and suitable for use in baking and frying applications as well as in formulated products. It is currently under development as a noncaloric fat and oil substitute for general use in all applications in a joint effort by ARCO Chemical Company and CPC International/Best Foods [30].

4. *Polyorganosiloxanes*

Polyorganosiloxanes are chemically inert, linear polymeric structures of silica. They are stable compounds that maintain their viscosity over a wide range of temperatures. They are also resistant to oxidation, hydrolysis, and degradation. Polysiloxanes exhibit solubility characteristics similar to those of nonpolar lipids. These compounds are not absorbed and thus do not contribute any caloric value. Phenyl and methyl substitution of the polysiloxane backbone forms an oil having a viscosity similar to soybean oil (Dow Corning 550 Fluid, Contour Chemical, North Reading, MA). The preferred structure for this molecule takes the general formula $(CH_3)_3SiO[(CH_3)_2SiO]_xSi(CH_3)_3$, where x has an average value between 25 and 500 [31]. Phenylmethylpolysiloxane has been tested as a noncaloric fat substitute in animals [32]. Although safety and toxicity have been established by Dow Corning Corporation (Midland, MI), anal leakage appears to be a problem when the product is fed to rats at level of 6.5% by weight [33]. This problem may be corrected by use of antianal leakage agents such as long-chain saturated fatty acids or cellulose [34].

5. *Sorbestrin (Sorbitol Polyester)*

Sorbestrin is one of the family of carbohydrate-based fatty acid polyesters that also includes sucrose polyesters [35]. Sorbitol and sorbitol anhydrides serve as the backbone of this compound, which is esterified with fatty acids of varying chain length and degree of saturation. Sorbitol hexaoleate can be synthesized by heating sorbitol at 180°C with a fivefold excess of ethyl oleate in dimethylacetamide for 5 h with sodium methoxide as a catalyst [7]. The final structure must have more than four fatty acid esters to provide the steric hindrance needed to inhibit digestion. Thus polyesters with less than this number of sorbitol units are not useful fat replacers but function well as emulsifiers. Sorbitol polyesters with less than four fatty acid esters can be synthesized by enzymatic as well as nonenzymatic methods, but the higher number of substitutions that have utility as fat substitutes cannot. The functional uses for sorbitol polyesters as fat substitutes are determined by the composition of the fatty acid esters on the sugar–alcohol backbone. Sorbestrin is a thermally stable liquid fat substitute suitable for use in all vegetable oil applications including salad dressing, mayonnaise, baked goods, and fried foods. It is currently under investigation by Pfizer Food Science Group (New York, NY).

6. *Sucrose Polyester (Olestra)*

The process for chemically linking lipophilic fatty acids to hydrophilic sugars was originally patented in 1952 for preparation of a nonionic surfactant to function as a detergent [35]. The development of sucrose polyesters currently being tested for use as fat substitutes has evolved from this technology. These sucrose polyesters may consist of glycosides with between 4 and 14 acetate or hydroxyl groups available for substitution with fatty acids. Olestra is the generic name assigned to sucrose polyesters with six to eight fatty acid esters, reflecting the identity of the fatty acid methyl ester of oleic acid used in its synthesis. Soybean oil is currently the most common source of fatty acids used in preparation of olestra.

The resistance of sucrose polyesters to lipolytic enzymes is increased by the steric hindrance resulting when the degree of substitution, or the number of fatty acids esterified at the acetal and hydroxyl sites on the glycoside unit, exceeds three. The types of fatty acids esterified may also add to an even greater steric hindrance and contribute further to decreased digestibility of the molecule. The extent to which lipophilic attributes are expressed by sucrose polyesters is also a function of the degree of substitution or esterification of the molecule. These functional properties may be more finely tuned by altering the chain length and degree of saturation of the fatty acid substituents. Olestra is a completely indigestible, noncaloric fat substitute that exhibits thermal stability at the high temperatures used in baking and frying. Its physical properties are parallel to those of the partially hydrogenated soybean oil used in its preparation. Liquids are generated from lightly hydrogenated soy oil, and solids are generated from the more heavily hydrogenated forms of this oil [18].

A number of methods for preparing sucrose polyesters have been reported [35]. Solvent-free interesterification processes involve admixing long-chain fatty acid methyl esters synthesized from edible oils with molten sucrose at temperatures between 110 and 180°C using 1–2% sodium metal as a catalyst [36,37]. This method eliminates the need for using toxic solvents such as dimethylacetamide, dimethylformamide, and dimethylsulfoxide polyesters that promote formation of the homogeneous melts that increase product yields but that also make this process unsuitable for preparing food grade polyesters [38].

Olestra has been under investigation by the Procter and Gamble Company (Cincinnati, OH) for use as a fat substitute since the original patent was filed in 1971 [7]. The first food additive clearance petition was filed with the FDA in April 1987 for replacement of 35% of fat in shortening and oils for home use and up to 75% of fats used industrially for deep fat frying and manufacture of snack foods [39]. The petition has since been modified a number of times. In August 1990, the focus of the petition was narrowed to restrict the intended use of olestra to savory snacks for the purpose of accelerating the approval process. In late 1995, an advisory panel to the FDA recommended approval for the petition, and an official ruling by the agency is expected in early 1996.

7. *Trialkoxytricarballyate*

Trialkoxytricarballyate (TATCA) is a retrofat or a structural analogue of a triacylglycerol in which the ester linkages have been reversed. In these molecules, the glycerol moiety has been replaced by thermally stable polycarboxylic acids such as tricarballylic, malonic, or citric acids, which are esterified with saturated or unsaturated long-chain alcohols instead of fatty acids [9]. Trialkoxycitrate (TAC) displays viscosity and surface tension properties similar to those of corn oil [40]. Polymorphic behavior was displayed upon melting for both compounds. Both TATCA and TAC are virtually indigestible oil-like compounds and thus contribute no caloric value. The potential for thermal decomposition of citrate esters in TAC prevent it from being used in frying oils. The primary interest in TATCA has been used as a fat replacer for vegetable oils in formulating fat-free and fat-reduced margarine, mayonnaise, and similar products. TATCA has been successfully used in the formulation of mayonnaise-type products and of a fat-reduced margarine that was softer and melted more quickly than regular margarine. The patent for this group of compounds is held by CPC International, Division of Best Foods (Englewood Cliffs, NJ) [9].

IV. UTILITY OF LOW CALORIE FATS IN FAT-REDUCED FOOD PRODUCTS

The ideal fat replacer has been described as one that is safe, retains all of the organoleptic and functional properties of fat, substitutes for fat on a one-to-one basis, and contributes significantly fewer calories than the fat it replaces [2]. It should also be readily available at a reasonable cost. The quality, and thus consumer acceptance, of food products reduced in fat is dependent on the degree to which each of the many functions of fat is successfully duplicated by the specific fat replacer used. Among the contributions fats make to foods are a pleasing mouthfeel, flavor and aroma, moisture and lubricity, smoothness and spreadability, elasticity, thickness, body, heat and structural stability, emulsification, dispersion, aeration, processing ease, color, and gloss [41,42]. The difficulty of accomplishing all of these tasks without fat is enormous.

Low calorie fats and fat substitutes are the simplest fat replacement systems because they allow for substitution of fat in weight-equivalent amounts. Both compounds exhibit the chemical and physical properties of edible fats and oils and express the same functional properties of conventional lipids in food systems. When low calorie fats and fat substitutes are used for fat replacement, product formulations do not need to be adjusted to account for the loss of fat functionality. Fat replacement systems must be used with emulsifiers and fat mimetics because these nonlipid compounds cannot replace all of the functions of fat if used alone. To satisfy the need for a desired set of fat functional attributes, fat mimetics and emulsifiers must frequently be accompanied by additional ingredients such as water, additional emulsifiers, bulking agents, and flavor enhancers in amounts and proportions that depend on the particular application. Consequently, weight-equivalent replacement of fat with fat mimetics or emulsifiers is not possible.

Use of fat mimetics provides a direct approach to reducing fat and calories because it involves substitution of fat calories with nonlipid components intrinsically lower in calories. Structural manipulations of proteins and carbohydrates enable the modified forms of these compounds to simulate many of the properties of fat but with fewer calories. Mouthfeel can be reproduced by dispersing spherical particles with diameters of 0.1–3.0 μm in water as achieved during microparticulation of proteins [3]. This process creates the sensory perception of an oil-in-water suspension. Fat mouthfeel may also be duplicated by the films formed when interfacial tension between water and other phases is relieved by compounds such as methylcellulose [2]. Gelatin-based stabilizers or other thickeners can provide the viscous stimuli perceived as the creamy or smooth mouthfeel that is usually associated with fat [41]. The effectiveness of protein- and carbohydrate-based mimetics as fat replacers depends on individual structural features that determine how each will perform in a particular application. Mouthfeel and texture have been successfully duplicated by most of the currently used emulsifiers and fat mimetics in many applications, but flavor delivery, heat stability, and shelf life are often compromised [1].

Preserving the flavor attributes of products reduced in fat is paramount to ensuring a high degree of quality and acceptability of these products. The large amount of water integral to the structures of fat mimetics needed to create the mouthfeel of fat can upset the flavor balance, potency, and release in fat-reduced products [43]. The presence of fat in a food system solubilizes and lowers the vapor pressures of flavor components. Consequently, when the amount of fat in the system is reduced, the lack of a solvent medium makes the lipid-soluble flavors stand out in the profile. The increased vapor pressures that also result intensify the sensory perceptions of these flavors relative to water-soluble flavor components and change the order in which the flavors are perceived. Flavor imbalances create the off-flavors frequently experienced with fat-reduced products compared to their full fat counterparts.

The extent to which the functionality of fat is successfully imitated by a particular fat replacer may vary considerably from product to product, even among those in the same category, depending on the set of attributes preferred for the product and the degree of fat replacement desired. As with the conventional fats they replace, some fat replacers are preferred over others in a particular application on the basis of a particular property desired such as melting point, smoke point, or viscosity. Minor alterations in the core molecule or the substitution of fatty acids with different chain lengths and degrees of saturation can be employed to prepare a family of compounds for low calorie fats and fat substitutes. Oils, semisolids, and hard fats can then be made available for a particular low calorie fat or fat substitute to support different fat preferences, thus covering a broader range of applications. With fat mimetics and emulsifiers, entirely different compounds must often be chosen for different applications, and the adjustments in formulations needed to accommodate each one will have to vary accordingly.

Low calorie fats and fat substitutes may be used for partial or total fat replacement in all applications, giving them the maximum utility for fat replacement. Of value technically is the ability of these compounds to be used in low moisture baking systems, which has presented the most difficult challenge in fat replacement for the food industry. Fat mimetics and emulsifiers are prevented from being used in low moisture systems by the hydration required to attain their functional properties. Of value nutritionally is the ability of low calorie fats and fat substitutes to replace all of the fat in cooking oils and much of the fat in salad oils, which account for a large proportion of total fat in the American diet. Use of fat replacers in cooking oils in particular could contribute considerably to the extent of fat reduction possible in a number of applications. The flavor attributes and instability at high temperatures exhibited by many fat mimetics makes these fat replacers inappropriate for either partial or total fat replacement in cooking oils.

V. NUTRITIONAL IMPLICATIONS OF LOW CALORIE FATS

Fat replacers can make a significant contribution to improving the healthfulness of the food supply if their use increases consumer acceptance of fat-reduced products. A number of health policy groups have recommended that dietary fat intake be decreased to reduce risk for coronary heart disease, adult-onset diabetes, and breast, colon, and prostate cancers [44,45]. Moreover, a high fat intake has also been linked to weight gain and obesity, and a reduction in fat intake has been suggested as a means for facilitating weight control [46,47]. The number of adults who are considered to be overweight has been increasing over the past 20 years, and the rise in obesity among children over this same time period has also been considerable [48,49].

The wider the range of applications that can utilize a particular fat replacer, and the more closely it is able to duplicate the functional and organoleptic properties of fats, the more successful this compound will be in helping to improve nutritional health. Low calorie fats and fat substitutes have more versatility in fat replacement applications than other fat replacers because they are heat-stable and can be tailored molecularly to yield different physical forms offering different sets of defined functional characteristics appropriate for a wide range of applications.

A. Blood Cholesterol Levels

One nutritional goal that most fat replacers may achieve equally well is the contribution to a reduction in total dietary fat intake in amounts that could significantly influence blood cholesterol levels and reduce the associated risk of coronary heart disease [50–53]. One dietary assessment study has estimated that substitution of fat with a fat replacer in frozen desserts, processed cheese, and commercial baked goods could reduce total fat intake by more than 8% for people who regularly consumed these products [52]. When dairy foods and salad dressings incorporating the same fat replacer were also included in the assessment, the potential reduction in total fat intake for these people was doubled to approximately 17%. Either of these magnitudes of reduction in total fat intake could have a considerable impact on closing the gap between current fat intake and the recommendations made by most health policy groups for a total fat intake of no more 30% of total calories [44].

Displacement of saturated fatty acids contributed by conventional fats and oils with fatty acids that do not adversely affect cholesterol levels is one means by which low calorie fats may exert a positive influence on blood cholesterol levels. In this regard, the impact of low calorie fats on blood cholesterol would be greatest when used to replace the more highly saturated fats in food products. The short- and medium-chain fatty acids typically contained in low calorie fats have not been found to elevate blood cholesterol levels [54], and the dominant long-chain saturated fatty acid in Salatrim, stearic acid, is considered to be neutral in its effects on blood cholesterol levels [55]. The poor absorptive efficiency for stearic acid also lowers its bioavailability to participate in this or any potentially adverse metabolic effect.

Fat substitutes may contribute to reducing dietary fat intake not only through displacement of fat calories but also through reducing amounts of fat absorbed. Poorly digestible lipid compounds cause a lipid phase to persist in the small intestines, providing a solvent medium for lipophiles that transports these compounds through the gastrointestinal tract unabsorbed [56]. Solubilization of endogenously secreted bile acids does not occur to the extent that bile acid pool size or bile composition, saturation, and reabsorption are adversely affected [57]. However, the amount of dietary cholesterol absorbed intestinally and the amount of endogenously secreted cholesterol reabsorbed enterohepatically in bile are reduced when olestra is consumed [57]. Ingestion of olestra in amounts consistent with its use as a fat replacer was associated with lower levels of triacylglycerols and lower levels of both total and LDL cholesterol in human subjects than what would have been expected from caloric reduction alone [58].

B. Weight Control

Regardless of whether low calorie fats and fat substitutes are able to achieve a physiologically significant reduction in saturated fat to affect blood cholesterol levels, these compounds could alternatively produce favorable blood lipid profiles by assisting in weight loss and facilitating weight control. Caloric reduction is a primary objective in the development of fat-reduced products and is accomplished by low calorie fats and fat substitutes through decreased or lack of digestibility. Use of these fat replacers will often achieve a greater reduction in calories from fat than in total calories per se depending on the contribution originally made by fat to the total calories in the product. For example, incorporation of Salatrim into a chocolate coating formulation reduces total calories from fat in the finished product by 45% while reducing total calories by just 25% [24]. Nevertheless, a growing body of evidence suggests that fat calories may make a greater contribution to body fat weight gain than calories from other sources [46].

The success with which weight can be controlled by replacing full fat products with their fat-reduced counterparts will depend on whether compensation for caloric dilution will occur as a consequence of ingesting these products. It will also depend on whether the use of fat replacers in fat-reduced products results in a caloric dilution of a significant enough magnitude to trigger compensatory responses. If a reduction in total energy intake does occur when foods reduced in fat are ingested, individuals may compensate for this energy loss to different degrees [59–62]. Ingestion of foods containing fat replacers by healthy adults over a 12 week period produced a weight loss similar to that observed with ingestion of foods naturally low in fat [63]. The fact that the control group in this study gained a small but not statistically significant amount of weight suggests that caloric compensation did not occur among individuals who reduced their fat intake by either method. Only minimal caloric compensation was observed among adults consuming fat-reduced products containing olestra over a 20 day period [64]. Inconsistencies found in these studies between the weight loss observed and that predicted on the basis of the expected caloric reduction may have resulted as much from the inaccuracies of measuring caloric intake using current dietary methods as from caloric compensation and thus should not be interpreted to suggest unequivocally that caloric compensation had occurred [47].

The probability of compensation for caloric dilution from ingesting foods containing fat replacers may depend on age, sex, degree of obesity, and motivational factors. Children and lean adult men may be more likely to compensate for calories displaced by fat replacers than obese adult men or women [65,66]. When compensation for energy dilution occurs, it appears to be specific for energy and is not selective for a particular macronutrient such as fat [67,68]. Energy is the primary factor driving appetite among children to order to ensure an adequate caloric intake to support growth [65]. This essential drive for caloric density may be manifested by children as a preference for fat [66]. For some adults, psychological factors may also be a powerful motivation to compensate for caloric dilution [52,67,69]. Studies examining the effects of fat replacers on caloric intake have not satisfactorily taken into account whether the knowledge that fat-reduced foods were being consumed may have influenced the decision to eat more of these foods or may have provided the discipline needed to consume less.

C. Other Nutrient Intakes

Consumption of fat-reduced products will assist consumers in attaining a healthier dietary intake only if the use of these products promotes a lower fat intake, and the corresponding decrease in calories, without compromising nutritional health in other ways. Fat is important nutritionally as a source of linoleic acid and other long-chain unsaturated fatty acids. It also serves as a solvent medium that assists in the conveyance of fat-soluble vitamins to intestinal absorptive surfaces and facilitates their uptake by mucosal cells. Long-chain unsaturated fatty acids have important physiological roles as structural constituents of cell membranes and as precursors of functional compounds regulating cellular reactions.

At the present time, the common uses of fat replacers result most frequently in displacement of vegetable oils [70]. These oils are the primary sources of unsaturated fatty acids in the diet, including both the essential ω-6 fatty acid linoleic acid and the important ω-3 fatty acid linolenic acid. The requirement for linoleic acid can be met with intakes as low as 1–3% of total calories (2–7 g/day) [71]. This level of linoleic acid can usually be ingested if vegetable oils account for a minimum of 10% of total caloric intake. The applications most frequently using low calorie fats and fat substitutes will involve food categories such as baked goods, confectioneries, and salty snacks, which are not considered to be important contributors of essential amounts of fat to the diet. Use of fat substitutes and low calorie fats to replace all or most of the fat in salad and cooking oils could make it more difficult to achieve the required level of linoleic acid intake. Replacement of soybean oil in particular would remove a principal source of linolenic acid from the diet.

Fat-soluble vitamins depend on lipids in foods to assist in their transport and absorption. The presence of fat also stimulates the secretion of bile acids required for the uptake of these vitamins by mucosal cells. Displacement of fat by low calorie fats is unlikely to interfere with this absorptive process because only small amounts of fat are required to satisfy this need. A reduction in total fat intake to below the critical amount of 10% of calories is not likely to occur with consumption of low calorie fats in all their potential applications. As partially absorbed lipids, low calorie fats should function in the transport and absorption of fat-soluble vitamins in the same manner as conventional food lipids. Feeding Salatrim to minipigs for 4 weeks at levels equivalent to 3, 6, and 10% of the diet did not significantly affect serum levels of vitamin A or vitamin E compared with controls fed 10% corn oil [72]. Prothrombin time was also unaffected in these animals, indicating that vitamin K status was unimpaired. In addition, concentrations of calcium and phosphorus in bone were similar between Salatrim-fed minipigs and controls, indicating that vitamin D status was also unimpaired.

Use of fat substitutes that are virtually indigestible raises other issues regarding the absorption of fat-soluble vitamins. Theoretically, poorly absorbed or completely unabsorbed fat substitutes could solubilize fat-soluble vitamins, keeping some fraction of these vitamins in solution where they would be eliminated in the feces along with the undigested fat substitute [57]. Ingestion of olestra by human subjects in amounts of 17–27 g/day over a 16 week period was not found to adversely affect serum levels of vitamins A or D or prothrombin time, but serum vitamin E levels were notably decreased [58]. Concern about the possible depletion of vitamin E reserves over time has generated a recommendation that olestra preparations be supplemented with this vitamin. Because vegetable oils are the primary contributor of vitamin E in the diet, providing approximately one-third of the total intake of this vitamin, use of olestra as a fat replacer in salad oils may not be appropriate [73].

D. Adjustment to Low Fat Diets

One of the most significant contributions that low calorie fats and fat substitutes may make toward improving health is in assisting consumers to become accustomed to low fat diets. The preference for fat is a powerful response regulated by both sensory and postingestive factors [70,74]. It may be possible to dissociate the sensory aspects controlling fat intake from the metabolic properties and to develop a preference for low fat foods if the organoleptic qualities of fats are successfully duplicated by fat replacers [69,74]. Low calorie fats and fat substitutes may make an additional contribution to helping consumers achieve better adherence to low fat diets over the long term by enabling them to maintain their usual food selection patterns [70,75]. Products containing fat replacers were consumed by adult women participating in the Women's Health Trial who were able to maintain a 20% fat intake over a period of several years [76]. Much more remains to be learned about the relative contributions made by fat taste and fat–energy interactions in acquiring a preference for fat before the contribution that fat replacers may make in helping consumers maintain a low fat intake can be fully understood [69].

VI. REGULATORY ASPECTS OF LOW CALORIE FATS

As food ingredients, low calorie fats and fat substitutes are classified as food additives and therefore must receive regulatory approval before they may be used in fat-reduced foods. The approach to obtaining this approval will vary depending upon whether the fat replacer was developed as a result of creating a new molecule, applying new technology to make an old molecule serve new purposes, or expanding old technologies to apply to old molecules for the first time [77]. Fat replacers derived from traditional food sources are also handled differently from fat replacers produced through chemical synthesis or derived from novel food sources such as jojoba oil or the membrane lipid phytanyl phosphatidylglycerol phosphate produced by *Halibacterium halobium* [31,70,78].

A. Safety

The creation of a new molecule by chemical synthesis was the method used to develop olestra and the other fat substitutes. Regulatory approval for these new molecules requires the submission of a food additive clearance petition using guidelines for safety testing as outlined in the FDA monograph commonly referred to as the "Red Book" [79]. Fat replacers developed by either old or new technologies involving traditional molecules with prior histories of safe use, or by chemical synthesis from these traditional molecules, need only be shown to be safe in their new forms and/or applications. These methods are generally used to develop low calorie fats as well as emulsifiers and fat mimetics. For these types of fat replacers, a GRAS (generally regarded as safe) affirmation petition must be submitted for approval. Whether regulatory approval is sought through a food additive clearance petition or a GRAS affirmation petition, the burden of proof of safety is on the manufacturer [77,78].

The procedures for safety testing outlined in the first version of the "Red Book" did not address the additional considerations raised in testing additives such as fat replacers that are added to foods in substantially larger amounts than has been typical for food ingredients in the past. The standards established for determining the amount of preclinical and clinical testing needed for microadditives such as preservatives, colorants, and flavorings upon which the original guidelines were based cannot be interpreted in a practical manner for macroadditive fat replacers. Consequently, traditional safety concerns and standard toxic endpoints are difficult to apply [78].

The amount of clinical testing required to assess food additive safety is based on guidelines found in the most recent version of the Red Book in the section "Concern Levels as Related to Chemical Structure and Exposure" [80]. If these numbers are used, macroadditives would be placed at the highest concern level for safety at almost any appreciable exposure, which would almost always require extensive human testing to establish the safety of fat replacers [81]. The extensive testing triggered by the relatively high average exposure levels typically found for these ingredients compared with microadditives unfairly imposes a greater burden on manufacturers of macroadditives [81]. Although traditional evaluation of food additive safety does not allow for risk/benefit assessment, this approach may be more appropriate for macroadditives [78].

B. Labeling

1. Ingredients Listing and Nutrition Label Information

Food products formulated with low calorie fats or fat substitutes must carry information on ingredients and nutritional content on the package label in compliance with current labeling regulations [82]. Ingredients must appear in the list of ingredients according to their contributions by weight to the final product. If the fat replacer also contributes caloric value, it must be included in the calculations of total fat, fat calories, and total calories used to provide nutrient information for the label using values supported by experimental data provided by the manufacturer. The preferred method for estimating caloric value of a new ingredient is to establish an Atwater factor using data on weight gain obtained from animal bioassays.

2. Nutrient Content Claims and Health Claims

The presence of low calorie fats and fat substitutes in a packaged food product may qualify the product for a nutrient content claim. Many of these products may qualify for a health claim as well [78]. Nutrient content claims allow use of descriptors such as "fat-free," "low fat," "reduced fat," and "light" or "lite" to appear on the label, but the product must meet a specified level of fat reduction per reference amount and label serving size relative to the fat content of the same type of product or that of a standard value for the product. The "reduced fat" descriptor allows claims on products with at least a 25% reduction in fat, which can be readily achieved with low calorie fats and fat substitutes. To qualify for "light" claims, the food must be reduced in fat by 50% if the comparative product is high in fat (more than 50% of calories) or it must be reduced in calories by 33% or in fat by at least 50% if the comparative product is not high in fat.

The "low fat" definition can be met if the product contains no more than 3 g of fat per reference amount. "Low saturated fat" requires that no more than 1 g of saturated fat be present in the reference amount and no more than 15% total calories come from saturated fat. The rigorous definition of "fat free," i.e., trivial or negligible amounts of fat and no added ingredient that is fat, may allow only foods containing nonlipid fat mimetics to qualify. Health claims regarding cancer and coronary heart disease may also be made for products formulated with low calorie fats and fat substitutes as long as these products meet the definition for low fat. Additional qualifiers for saturated fat content or fiber content may also be needed depending upon the particular health claim being made.

VII. FUTURE TRENDS IN DEVELOPMENT OF LOW CALORIE FATS

Continued investment in the development of new low calorie fats and fat substitutes in the future will depend on whether these new ingredients are embraced or dismissed by consumers and health policymakers. In spite of how these ingredients may be received by consumers and health policymakers, incentives for developing new fat replacers will still be limited by the willingness of the regulators to modify traditional food ingredient safety testing to accommodate macroadditives. The basis of consumer acceptance will evolve from hedonic appeal of fat-reduced products and attainment of concrete health benefits such as controlling weight by consuming these products. Acceptance by health policymakers will be forthcoming if results of studies convincingly and consistently demonstrate the effectiveness of these products in promoting specific health outcomes without compromising nutritional health. Changes in regulatory barriers are the most difficult to predict, but proactive pressure applied to regulators by consumer groups and health policymakers can assist manufacturers in promoting a more favorable regulatory climate for obtaining approval for macroadditives.

Future innovations used in the synthesis of low calorie fats and fat substitutes must be managed in ways that will minimize toxicological concerns. Fat replacers derived from novel food sources may receive more attention in the future. Biotechnology may also play an increasingly important role in generating low calorie fats and fat substitutes from natural sources. For example, genetic engineering may allow bacterial synthesis of sucrose polyesters with degrees of substitution greater than 4, which is currently impossible because steric hindrance interferes with this ability [35]. Fat substitutes produced through biotechnology would not present the toxicological concerns associated with the presence of chemical impurities that are frequently introduced during industrial synthesis.

The demand for fat-reduced products should continue to grow well into the next decade as the aging population begins to increase disproportionately in numbers. Health concerns will remain or increase in importance as a dominant motivating factor in selection of foods. Weight control will in all likelihood be the overriding health concern. The availability of foods reduced in fat and calories but hedonically comparable to full fat foods will allow the need for indulgence to be satisfied while fulfilling the drive toward a more healthful diet. The use of low calorie fats and fat substitutes that successfully duplicate the organoleptic and functional attributes of fat with fewer calories should give fat-reduced food products an even stronger competitive edge in the marketplace.

REFERENCES

1. J. D. Gussow, Are we really fixing up the food supply?, *J. Am. Diet. Assoc. 93*: 1300–1304 (1993).
2. Calorie Control Council, *Fat Reduction in Foods*, 1994.
3. G. E. Gaull, Role of microparticulated protein fat substitutes in food and nutrition, *Ann. N.Y. Acad. Sci. 623*: 350–355 (1991).
4. L. P. Klemann, J. W. Finley, and G. A. Leveille, Estimation of the absorption coefficient of stearic acid in SALATRIM fats, *J. Agric. Food Chem. 42*: 484–488 (1994).
5. D. J. Hamm, Preparation and evaluation of trialkoxytricarballylate, trialkoxycitrate, trialkoxyglyceryl ether, jojoba oil, and sucrose polyester as low calories replacements of edible fats and oils, *J. Food Sci. 49*: 419–428 (1984).
6. A. Minich, Dietetic compositions, U.S. Patent 2,962,419 (1960).
7. F. H. Mattson, M. Healthy, and R. A. Volpenhein, Low calorie fat-containing compositions, U.S. Patent 3,600,186 (1971).
8. D. D. Whyte, Triglyceride esters of branched carboxylic acids, U.S. Patent 3,579,548 (1971).
9. D. J. Hamm, Low calorie edible oil substitutes, U.S. Patent 4,508,746 (1985).
10. V. W. Trost, Low calorie fat substitutes, Can. Patent 1,106,681 (1981).
11. H. K. Mangold, Biological effects and biomedical applications of alkoxylipids, in *Ether Lipids, Chemistry, and Biology*, 1972, p. 154.
12. A. M. Thayer, Food additives, *Chem. Eng. News 70*: 26 (1992).
13. The Proctor and Gamble Co., Cincinnati, Ohio, GRAS Affirmation Petition: Caprenin, to Foods and Drug Administration, Washington, DC (4.1.1991).
14. K. D. Fisher (Life Science Research Office, Federation of American Societies for Experimental Biology, Bethesda, Maryland): Evaluation of the Health Aspects of Caprenin; prepared for P&G (Jan. 1991).
15. D. R. Webb and R. A. Sanders. Caprenin I. Digestion and rearrangement in thoracic diet cannulated rats, *J. Am. Coll. Toxicol. 10*: 325 (1991).
16. D. R. Webb, J. C. Peters, R. J. Jandacek, and N. E. Fortier. Caprenin 2. Short-term safety and metabolism in rats and hamsters, *J. Am. Coll. Toxicol. 10*: 341 (1991).
17. V. K. Babayan, Modification of food to control fat intake, *J. Am. Oil Chem. Soc. 51*: 260–264 (1974).
18. C. E. Stauffer, Fats that cut calories, *Baking Snack*, May 1995, pp. 44–49.
19. L. P. Klemann, K. Aji, M. M. Chrysam, R. P. D'Amelia, J. M. Henderson, A. S. Huang, M. S. Otterburn, R. G. Yarger, Random nature of triacylglycerols produced by the catalyzed interesterification of short- and long-chain fatty acid triglycerides, *J. Agric. Food Chem. 42*: 442–446 (1994).
20. R. E. Smith, J. W. Finley, G. Leveille, Overview of SALATRIM, a family of low-calorie fats, *J. Agric. Food Chem 42*: 432–434 (1994).
21. B. J. Softly, A. S. Huang, J. W. Finley, M., Petersheim, R. G. Yarger, M. M. Chrysam, R. L. Wieczorek, M. S. Otterburn, A. Manz, G. J. Templeman, *J. Agric. Food Chem. 42*: 461–467 (1994).
22. J. M. Talbot, S. A. Anderson, and K. D. Fisher, The evaluation of the health aspects of using certain triacylglycerols as food ingredients, Life Sciences Research Office, Federation of American Societies for Experimental Biology, Rockville, MD, 1993.
23. J. W. Finley, L. P. Klemann, M. S. Oterburn, and C. G. Walchak, Caloric availability of SALATRIM in rats and humans, *J. Agric. Food Chem. 42*: 495–499 (1994).
24. J. Stowell, Salatrim—a versatile family of reduced-calorie fats, Paper prepared for *Die Ernahrungs Industrie*, 1995.
25. L. P. Klemann and J. W. Finley, Low calorie fat mimetics comprising carboxy/carboxylate esters, Eur. Patent 303,503 (1989).
26. L. P. Klemann and J. W. Finley, Low calorie fat mimetics comprising carboxy/carboxylate esters, Int. Patent 89/01293 (1989).

27. M. E. Spearman and J. G. Fulcher, Malonate Esters: New Synthetic Fat Substitutes for Food Use. Annual Meeting of the Oil Chemists Society, Cincinnati, Ohio, May 3–7, (1989).
28. J. Fulcher, Synthetic cooking oils containing dicarboxylic acid esters, U.S. Patent 4,582,927 (1986).
29. J. D. Dziezak, Fats, oils, and fat substitutes, *Food Technol. 43*(7): 66–74 (1989).
30. Charles F. Cooper, Reduced calorie fat substitute, U.S. Patent 528888 4A.
31. W. E. Artz, and S. L. Hansen, Other fat substitutes, in *Carbohydrate Polyesters as Fat Substitutes* (C. C. Akoh and B. G. Swanson, Eds.), Marcel Dekker, New York, 1994, pp. 197–236.
32. E. F. Bracco, N. Baba, and S. A. Hashim, Polysiloxane: potential noncaloric fat substitute; effects on body composition of obese Zucker rats, *Am. J. Clin. Nutr. 46*: 784–789 (1987).
33. J. W. Ryan, Prevention of anal leakage of polyorganosiloxane fluids used as fat substitutes in foods, Eur. Patent 368,534 (1989).
34. J. W. Ryan, Food composition containing a siloxane polymer and a particulate silica, U.S. Patent 4,925,692 (1990).
35. C. C. Akoh, Synthesis of carbohydrate fatty acid polyesters, in *Carbohydrate Polyesters as Fat Substitutes* (C. C. Akoh and B. G. Swanson, Eds.), Marcel Dekker, New York, 1994, pp. 9–35.
36. G. P. Rizzi and H. M. Taylor, A solvent-free synthesis of sucrose polyesters, *J. Am. Oil Chem. Soc. 55*: 398–401 (1978).
37. C. C. Akoh and B. G. Swanson, Optimization of sucrose polyester synthesis: comparison of sucrose polyesters, raffinanose polyesters and salad oils, *J. Food Sci. 55*: 236–243 (1990).
38. R. O. Fuege, H. J. Zeringue, Jr., T. J. Weiss, and M. Brown, Preparation of sucrose esters by interesterification, *J. Am. Oil Chem. Soc. 47*: 56–60 (1970).
39. Procter and Gamble, Olestra as a calorie free substitute for fats and oils, Petition submitted to the FDA, 1987.
40. S. D. Huhn, P. S. Given, Jr., and L. P. Klemann, Ether bridged polyesters and food compositions containing ether bridged polyesters, U.S. Patent 4,888,195 (1989).
41. A. Drewnowski, Sensory properties of fats and fat replacements, *Nutr. Rev. 50*: 17–20 (1992).
42. L. Schinhofen, Fat substitutes, *Fat and Cholesterol Issues Monitor*, March 1995, pp. 16–26.
43. J. P. Schirle-Keller, G. A. Reineccius, and L. C. Hatchwell, Flavor interactions with fat replacers: effect of oil level, *J. Food Sci. 59*: 813–875 (1994).
44. National Research Committee (U.S.), Committee on Diet and Health, Fats and other lipids, in *Diet and Health*: Implications for Reducing Chronic Disease Risk/Committee on Diet and Health, Food and Nutrition Board, Commission on Life Sciences, National Research Council, National Academy Press, Washington, DC, 1989, pp. 205–231.
45. J. A. Marshall, R. F. Hamman, and J. Baxter, High-fat, low-carbohydrate diet and the etiology of non-insulin-dependent diabetes mellitus, *Am. J. Epidemiol. 134*: 590–603 (1991).
46. J. P. Flatt, Use and storage of carbohydrate and fat, *Am. J. Clin. Nutr. 51*: 952S–959S (1995).
47. A. Kendall, D. A. Levitsky, B. J. Strupp, and L. Lissner. Weight loss on a low-fat diet: consequence of the imprecision of the control of food intake in humans, *Am. J. Clin. Nutr. 53*: 1124–1129 (1991).
48. S. L. Gortmaker, W. H. Dietz, A. M. Sobol, and C. A. Wehler, Increasing pediatric obesity in the United States, *Am. J. Dis. Child. 141*: 535–540 (1987).
49. National Institutes of Health, Consensus Development Panel on the Implications of Obesity, Health implications of obesity, *Ann. Intern. Med. 103*: 1073–1077 (1985).
50. D. M. Hegsted, R. B. McGandy, M. L. Meyers, and F. J. Stare, Quantitative effects of dietary fat on serum cholesterol in man, *Am J. Clin. Nutr. 17*: 281–295 (1965).
51. A. Keyes, J. T. Anderson, and F. Grande, Serum cholesterol response to changes in the diet. IV. Particular saturated fatty acids in the diet, *Metabolism 14*: 776–787 (1965).
52. B. J. Lyle, K. E. McMahon, and P. A. Kruetler, Assessing the potential dietary impact of replacing dietary fat with other macronutrients, *J. Nutr. 122*: 211–216 (1992).
53. National Heart, Lung, and Blood Institute, Consensus Development Conference, Lowering blood cholesterol to prevent heart disease, *JAMA 253*: 2080–2086 (1985).

54. S. M. Grundy and M. A. Denke, Dietary influences on serum lipids and lipoproteins, *J. Lipid Res. 31*: 1149–1172 (1990).
55. S. M. Grundy, Influence of stearic acid on cholesterol metabolism relative to other long-chain fatty acids, *Am. J. Clin. Nutr. 60* (Suppl.): 986S–990S (1994).
56. C. M. Bergholz, Olestra and the potential role of a nonabsorbable lipid in the diets of children, *Ann. N.Y. Acad. Sci. 623*: 356–368 (1991).
57. R. J. Jandacek, The effect of nonabsorbable lipids on the intestinal absorption of lipophiles, *Drug Metab. Rev. 13*: 695–714 (1982).
58. M. J. Mellies, C. Vitale, R. J. Jandacek, G. E. Lampkin, and C. J. Glueck, The substitution of sucrose polyester for dietary fat in obese, hypercholesterolemic outpatients, *Am. J. Clin. Nutr. 41*: 1–12 (1985).
59. L. Lissner, D. A. Levitsky, B. J. Strupp, H. J. Kalkwarf, and D. A. Roe, Dietary fat and the regulation of energy intake in human subjects, *Am. J. Clin. Nutr. 46*: 886–892 (1987).
60. K. P. Porikos, G. Booth, and T. B. Van Italie, Effect of covert nutritive dilution on the spontaneous food intake of obese individuals: a pilot study, *Am. J. Clin. Nutr. 30*: 1638–1644 (1977).
61. K. P. Porikos, M. F. Hesser, and T. B. Van Italie, Caloric regulation in normal-weight men maintained on a palatable diet of conventional foods, *Physiol. Behav. 29*: 293–300 (1982).
62. R. D. Mattes, C. B. Pierce, and M. I. Friedman, Daily caloric intake of normal-weight adults: response to changes in energy density of a luncheon meal, *Am. J. Clin. Nutr. 48*: 214–219 (1988).
63. R. D. Mattes, Fat preference and adherence to a reduced-fat diet, *Am. J. Clin. Nutr. 57*: 373–381 (1993).
64. C. J. Glueck, M. M. Hastings, C. Allen, E. Hogg, L. Baehler, P. S. Gartside, D. Phillips, M. Jones, E. J. Hollenbach, A. S. Braun, and J. V. Anastasia, Sucrose polyester and overt caloric dilution, *Am. J. Clin. Nutr. 35*: 1352–1359 (1982).
65. L. L. Birch, S. L. Johnson, M. B. Jones, and J. C. Peters, Effects of a nonenergy fat substitute on children's energy and macronutrient intake, *Am. J. Clin. Nutr. 58*: 326–333 (1993).
66. L. L. Birch, Children's preferences for high fat foods, *Nutr. Rev. 50*: 283–290 (1992).
66. R. G. Campbell, S. A. Hashim, and T. B. Van Italie, Responses to variation in nutritive density in lean and obese subjects, *N. Engl. J. Med. 285*: 1402–1407 (1971).
67. R. W. Foltin, M. W. Fischman, T. H. Moran, B. J. Rolls, and T. H. Kelly, Caloric compensation for lunches varying in fat and carbohydrate content by humans in a residential laboratory, *Am J. Clin. Nutr. 52*: 969–980 (1990).
68. B. Rolls, P. A. Pirraglia, M. B. Jones, and J. C. Peters, Effects of olestra, a noncaloric fat substitute, on daily energy and fat intake in lean men, *Am. J. Clin. Nutr. 56*: 84–92 (1992).
69. B. Rolls, Changing the preference for fat in foods, *Nutr. Rev. 52*: 21–23 (1993).
70. D. J. Mela, Nutritional implications of fat substitutes, *J. Am. Diet. Assoc. 92*: 472–476 (1992).
71. Committee on Dietary Allowances, Food and Nutrition Board, National Research Council, *Recommended Dietary Allowances*, 10th ed., National Academy Press, Washington, DC, 1989, pp. 44–51.
72. J. R. Hayes, N. H. Wilson, M. C. Roblin, P. C. Mann, and A. L. Kiorpes, 28-day continuous dosing study in minipigs with a SALATRIM structured triacylglycerol composed of stearate, acetate, and propionate, *J. Agric. Food Chem. 42*: 563–571 (1994).
73. S. P. Murphy, A. F. Subar, and G. Block, Vitamin E intakes and sources in the United States. *Am. J. Clin. Nutr. 52*: 361–367 (1990).
74. B. J. Rolls, and D. J. Shide, The influence of dietary fat on food intake and body weight, *Nutr. Rev. 50*: 283–290 (1992).
75. American Dietetic Association, Position of the American Dietetic Association: fat replacements, *J. Am. Diet. Assoc. 91*: 1285–1288 (1991).
76. A. R. Kristal, E. White, A. L. Shattuck, S. Curry, G. Anderson, A. Fowler, and N. Urban, Long-term maintenance of a low-fat diet: durability of fat-related dietary habits in the Women's Health Trial, *J. Am. Diet. Assoc. 92*: 553–559 (1992).
77. C. J. Glueck, P. A. Streicher, E. K. Illig, K. D. Weber, Dietary fat substitutes, *Nutr. Res. 14*: 1605–1619 (1994).

78. W. H. Glinsmann and J. E. Vanderveen, Fat substitutes: a regulatory perspective, *Annu. Rev. Nutr. 12*: 473–487 (1992).
79. Food and Drug Administration, Toxicological principles for the safety assessment of direct food additives and color additives used in foods, Bureau of Foods, Washington, DC, 1982.
80. Food and Drug Administration, *Toxicological Principles for the Safety Assessment of Direct Food Additives and Color Additives Used in Foods*, 2nd ed., Bureau of Foods, Washington, DC, 1995.
81. W. H. Glinsmann, Commentary—focus on substitutes that alter nutrition and gastrointestinal physiology, ILSI Workshop on Safety and Regulatory Aspects of Macronutrient Substitutes, 1994.
82. Food and Drug Administration, Nutrition labeling of food, Code of Federal Regulations 21.101.9, Office of the Federal Register, National Archives and Records Administration, Washington, DC, 1990.

20

Food Emulsifiers

Niels Krog
Danisco Ingredients, Brabrand, Denmark

I. INTRODUCTION

Modern industrial food production requires surface-active lipids (emulsifiers, surfactants) as processing aids to facilitate uniform quality and ensure long shelf life of the finished products. Foods are very complex colloidal systems that may undergo changes during storage, resulting in deterioration of appearance and texture and loss of flavor characteristics. Food formulations and composition in terms of calories, fat content, vitamins, or minerals change all the time due to new insights into the effect of eating patterns or health aspects. This leads to a constant demand for optimization of product formulations, including optimal selection of raw materials. A range of food grade emulsifiers based on edible fats and oils are available. They possess many specific applications in foods.

Food emulsifiers are polar lipids needed to increase colloidal stability and provide interfacial interactions between food components such as lipids, proteins, and carbohydrates. Such interactions are important factors in obtaining emulsion stability, foam formation of whipped products, and increased shelf life in many foods.

Food emulsifiers are esters of edible fatty acids and polyvalent alcohols such as glycerol, propylene glycol, sorbitol (sorbitan), and sucrose. These esters may be further esterified with organic acids such as acetic acid, lactic acid, tartaric acid, succinic acid, or citric acid to modify their hydrophilic properties and functional effects.

The chemistry of food emulsifiers is described in detail by Lauridsen [1], Schuster [2], Benson [3], and Krog [4].

The total world production of food emulsifiers is not known exactly due to lack of statistical information from many countries. However, it is estimated that the amount of food grade emulsifiers produced is about 250,000 metric tons. Mono- and diglycerides, including distilled monoglycerides and their organic acid esters, amount to approximately 75% of the total production [5], making monoglycerides the most important product from both an economic and a functional point of view.

All food emulsifiers are regulated by the health authorities in each country. Emulsifiers for foods are evaluated by the Joint Expert Committee on Food Additives (JECFA) of the Food and Agriculture Organization/World Health Organization (FAO/WHO). In Europe the legislation is administered by the Commission of the European Communities' Scientific Committee for Food, and in the United States by a department under the Food and Drug Administration (FDA). Data from toxicological studies are used by FAO/WHO's Codex Alimentarius Commission to issue recommendations to national health authorities concerning the use of emulsifiers in foods. A compendium of food additive specifications was issued by JECFA in 1992 [6], and it is currently being updated.

II. MONO-DIGLYCERIDES, EU 471, US/FDA/CFR §184.1505

Industrial production of monoglycerides began in the 1930s, when their use in margarine production started [7]. Mono-diglycerides is a common term describing a mixture of mono- and diglycerides, which is produced by glycerolysis [8] of fats as shown schematically in Fig. 1 or by esterification of glycerol with fatty acids [9]. Depending on the glycerol/fat ratio in the reaction blend, the monoglyceride content in the equilibrium mixture obtained after glycerolysis may vary from 10% to 60%. Commercial mono-diglycerides usually contain 45–55% monoglyceride, 38–45% diglyceride, and 8–12% triglyceride.

The main applications of mono-diglycerides in foods are typically in fat-based products, such as margarine, spreads and bakery fats (shortenings), and cake mixes. Mono-diglycerides are added by way of the fat phase, often in combination with other more hydrophilic emulsifiers. In dairy emulsions, mono-diglycerides are used in ice cream and recombined milk in combination with hydrocolloids (see Chapters 11–13).

A. Distilled Monoglycerides

Monoglycerides are separated from the di- and triglycerides by a process referred to as molecular distillation, a thin-film, high vacuum technique [1]. The total monoglyceride content of

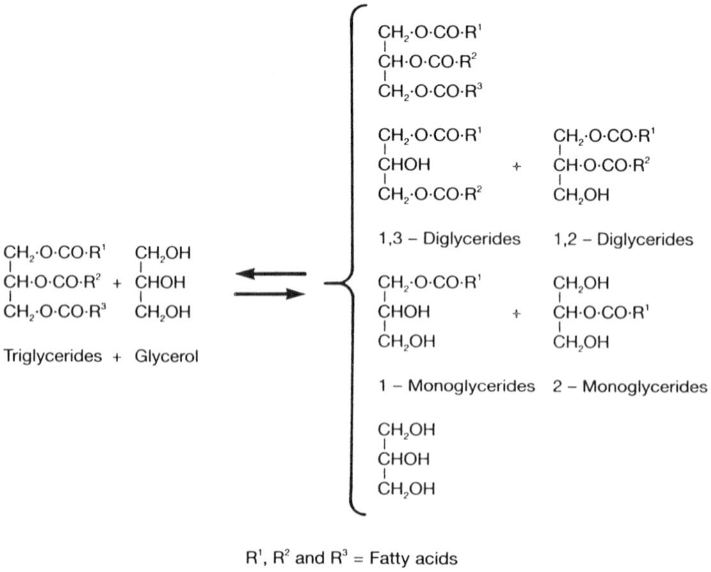

Fig. 1 Reaction scheme for production of mono-diglycerides by glycerolosis.

distilled products is 93–97%, with the 1-monoester content being at least 90%. An alternative monoglyceride production method is to esterify fatty acids with glycerol, followed by molecular distillation. High purity distilled monoglycerides with a single fatty acid composition, for example glycerol monopalmitate or glycerol monooleate, can be manufactured using this process.

B. Physical Properties

Monoglycerides always have higher melting points than their corresponding triglycerides [10]. With distilled monoglycerides the increase in melting point can be as much as 10°C compared to the triglycerides used in the glycerolysis process.

Monoglycerides are polymorphic like triglycerides [11] and solidify from melt in a metastable α crystal form, which transforms to a sub-α form on further cooling. Both α forms are metastable and transform into a stable high melting β crystal form when stored. Typical melting points of some distilled monoglycerides with different fatty acid compositions are shown in Table 1.

Distilled monoglycerides have better dispersibility in water than mono-diglycerides, because distilled monoglycerides form liquid crystalline mesomorphic phases [4], whereas mono-diglycerides in most cases form emulsions due to a relatively high content of triglycerides. The mesomorphic behavior of distilled monoglycerides in water is caused by their purity and well-defined molecular structure. This allows distilled monoglycerides to form ordered structures in water with bilayers of the fatty acid chains separated by water layers associated with the polar groups [4]. A schematic structure of the lamellar phase is shown in Fig. 2. The thickness of the lipid bilayers (d_a) and that of the water layers (d_w), as well as the specific surface area of the polar group (S) can be measured by X-ray diffraction analysis, and the water layers can be up to several hundred angstroms in thickness [4].

Table 1 Melting and Crystallization Points and $\alpha \rightarrow$ sub-α Transition Temperature of Distilled Monoglycerides

Monoglyceride	Melting point, β form (°C)	Crystallization point, α form (°C)	Transition point, $\alpha \rightarrow$ sub-α form (°C)
Monocaprin (90% C_{10} fatty acid)	52	16	—
Monolaurin (92% C_{12} fatty acid)	58	38	19
Monomyristin (92% C_{14} fatty acid)	64	50	26
Monopalmitin (92% C_{16} fatty acid)	68	62	34
Monostearin (85% C_{18} fatty acid)	74	68	42
Monoolein (80% 18:1 fatty acid)	35	2	—
Monolinolein (65% 18:2 fatty acid)	12	−6	—
Monobehenin (90% C_{22} fatty acid)	84	78	54

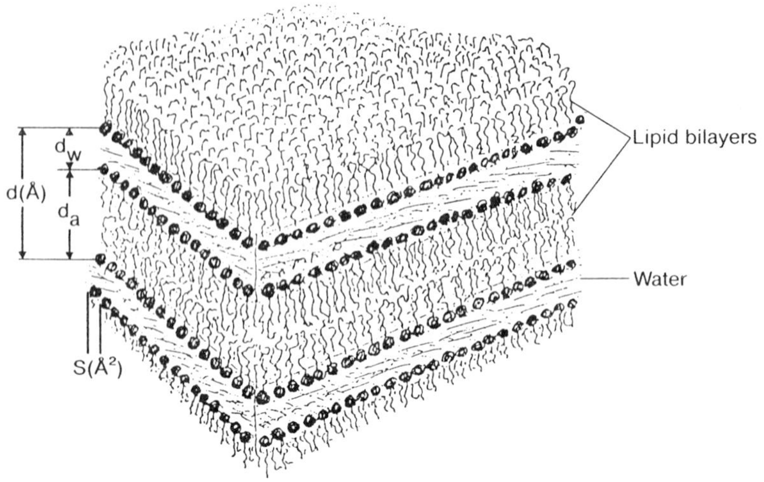

Fig. 2 Schematic structure of the lamellar mesophase. d, d_w, d_a, and S are structural parameters measured by low-angle X-ray diffraction analysis. (Courtesy of Danisco Ingredients, Denmark.)

Such ordered structures have a long-range order but no short-range order similar to the crystalline state, and such structures are referred to as liquid crystals or mesomorphic phases. The fatty acid chains are in a liquid-like state with rotating hydrocarbon chains. The association of distilled monoglycerides with water results in formation of various mesomorphic phases with lamellar, cubic, or reversed hexagonal structures depending on the fatty acid composition of the monoglyceride, temperature, and concentration [4]. A detailed description of lyotropic, liquid crystalline phases of monoglycerides, phospholipids, and glycolipids was recently given by Larsson [12].

Formation of lamellar phases of saturated C_{16}–C_{18} distilled monoglycerides in water takes place at temperatures above the Krafft point (T_c) as shown in Fig. 3a. This figure shows a binary phase diagram of saturated, distilled monoglycerides in water. The Krafft point for a monoglyceride is 20–25°C lower than its bulk melting point. At high water concentrations, "dispersions" of lamellar units are formed. These lamellar units have unique properties in low fat foods, where monoglyceride dispersions can be added by way of the water phase. This is used to facilitate interactions with starch components in cereal-based products or processed potato products, etc. The use of mesomorphic phases of monoglycerides to improve the texture of fat-free foods such as spreads, dressings, and bakery products was recently described in the patent literature [13].

When cooling a mesophase below the Krafft temperature, an α gel phase is formed due to crystallization of the fatty acid chains. The water layers remain between the polar head groups as in the lamellar phase, so the only difference is in the crystallinity of the fatty acids.

The gel phase is an important physical state for the function of monoglycerides or other emulsifiers used in whippable emulsions [14] or as texturizing agents in fat-free foods. The gel phase has water-binding capacity, enhances destabilization of whippable emulsions, or increases the relative viscosity of low fat products.

The use of monoglyceride "hydrates" as antistaling agents or crumb-softening agents in wheat bread has been known for decades [15], and this application accounts for a large proportion of the world production of distilled monoglycerides. The so-called hydrates are produced by hydrating 20–45% distilled monoglycerides in water above the Krafft point where

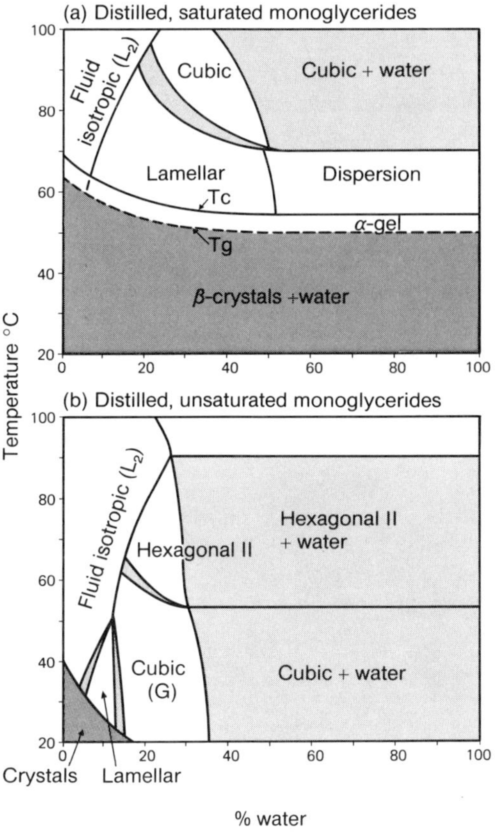

Fig. 3 Binary phase diagrams of (a) saturated, distilled monoglycerides (DIMODAN P, Danisco Ingredients, Denmark) and (b) unsaturated, distilled monoglycerides (DIMODAN LS, Danisco Ingredients, Denmark) in water. T_c = Krafft temperature; T_g = temperature where the α gel is formed. (Courtesy of Danisco Ingredients, Denmark).

they form a lamellar phase. This is cooled under agitation to a temperature at which the α gel phase is formed. The mix is then acidified to a pH of about 3.5 under continuous agitation, resulting in a transformation from the gel phase to a suspension of β crystals in water. The final product is a white, pastelike product that is easy to incorporate into bread dough. The advantage of hydrating the monoglycerides before use is that the specific surface area of the β crystals formed by recrystallization in water is much greater than what can be made by mechanical processes (spray-cooling or milling). The hydrated aqueous preparations of monoglycerides are more effective as starch complexing agents than any powdered products [16].

Unsaturated distilled monoglycerides (mainly monooleates or monolinoleates) form cubic mesophases in water at ambient temperature as shown in Fig. 3b. The interaction of unsaturated monoglycerides with water has a specific function, particularly in whippable emulsions, where the affinity of unsaturated monoglycerides to water provides partial destabilization of the fat globules [17]. This is desired to improve the aeration properties and foam stability of whipped cream or ice cream products. The mesomorphic behavior of polar lipids in water can be an important factor in many different applications of emulsifiers in foods (e.g., bakery products).

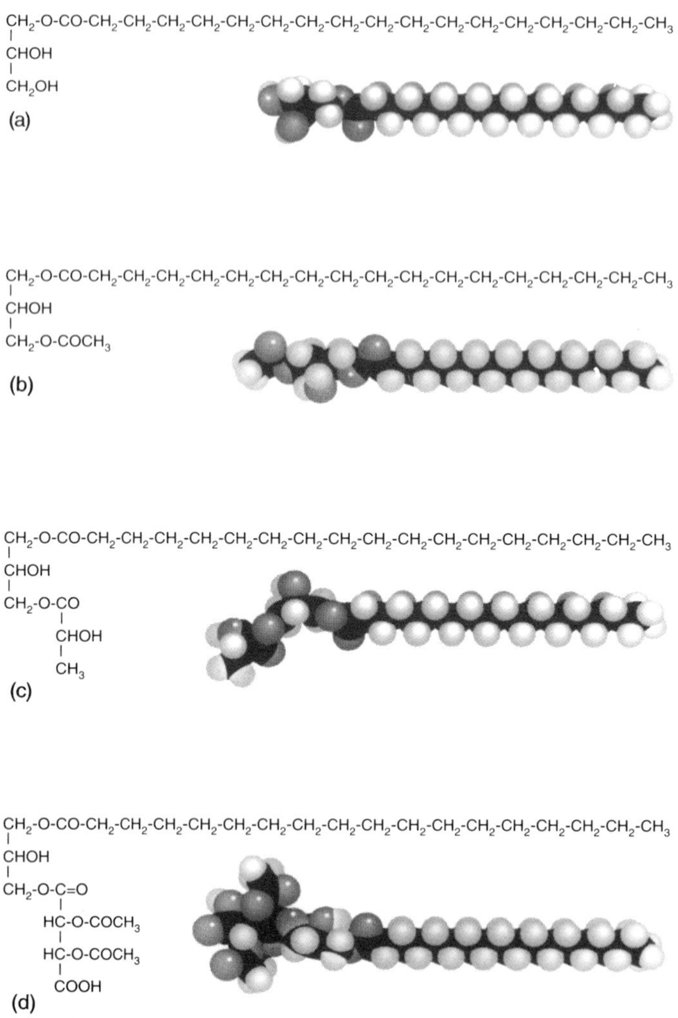

Fig. 4 Chemical structure of organic acid esters of monoglycerides. (a) Glycerol 1-monostearate; (b) acetic acid ester of glycerol monostearate; (c) lactic acid ester of glycerol monostearate; (d) diacetyltartaric acid ester of glycerol monostearate. (From Ref. 21.)

The uses of distilled monoglycerides in foods are numerous, and replacement of mono-diglycerides with distilled monoglycerides has occurred in many products. This is the case in dairy products, where distilled monoglycerides are used instead of mono-diglycerides in ice cream, coffee whiteners, and whippable creams. Combinations with lecithin or other more polar emulsifiers are used in many emulsions.

III. ORGANIC ACID ESTERS OF MONOGLYCERIDES

Derivatives of monoglycerides containing organic acids such as acetic acid, lactic acid, diacetyltartaric acid, or citric acid exhibit properties very different from those of monoglycerides in terms of crystalline behavior and surface activity (polarity). Emulsifiers with different functional properties in foods can therefore be made by reacting monoglycerides with selected organic acids.

Organic acid esters are in principle made by reacting mono-diglycerides or distilled monoglycerides with anhydrides of the organic acid in question on a mole/mole basis, yielding a mixture of components where the organic acid is esterified to one of the free hydroxyl groups on the monoglyceride molecule [4]. Chemical structures and molecular models of organic acid esters compared to distilled monoglycerides are shown in Fig. 4.

A. Acetic Acid Esters of Monoglycerides—EU 472a, US/FDA/CFR § 172.828

Acetylated monoglycerides (ACETEMs) are based on distilled monoglycerides, and products with 50, 70, and 90% acetylation of free hydroxyl groups are available. Acetylation of monoglycerides lowers their melting point to about 25–30°C below that of the corresponding monoglycerides. Furthermore, ACETEMs are stable in the α crystal form and show no polymorphism. The chemical structure of an ACETEM is shown in Fig. 4b.

Fully acetylated ACETEMs are not surface-active but behave like plastic fats. Such products are used as low calorie fats in emulsions and some bakery products (see Chapter 19). Fully acetylated ACETEMs form flexible films with a low degree of permeability of oxygen and water vapor [18], and they are therefore used as coating agents for frozen meat or fruit.

Partially acetylated ACETEMs have low surface activity depending on their degree of acetylation, which is usually between 50% and 70%. Such products can form α gel structures with water, and this is the basis for their application in specialty fats for toppings, whippable emulsions, or cake mixes.

B. Lactic Acid Esters of Monoglycerides—EU 472b, US/FDA/CFR § 172.852

Mono-diglycerides or distilled monoglycerides can be esterified with 15–35% lactic acid, forming lactic acid esters (LACTEMs). The chemical structures vary, but a main component of a LACTEM is shown in Fig. 4c. Together with ACETEMs, LACTEMs are often referred to as "α-tending" emulsifiers due to their monomorphic properties and stability in the α crystal form. The melting point of LACTEMs is ~45°C, and they form α gel structures with water [19]. LACTEMs are mainly used in specialty fats for dessert products (non-dairy creams, toppings, cake mixes).

C. Diacetyltartaric Acid Esters of Monoglycerides—EU 472e, US/FDA/CFR § 182.4101

Diacetyltartaric acid esters of monoglycerides (DATEMs) are made by reacting mono-diglycerides or distilled monoglycerides with diacetyltartaric acid anhydride. Many variations in composition exist, depending on the ratio of monoglyceride to tartaric acid anhydride used during esterification, and the final products may contain more or less nonesterified monoglycerides. A main component of DATEMs is shown in Fig. 4d. The physical form may vary from a viscous liquid to a solid, depending on the fatty acid composition of the monoglycerides used. DATEMs based on C_{16}–C_{18} monoglycerides melt at ~45°C and crystallize from melt to a stable α crystal form. DATEMs are anionic-active, very hydrophilic emulsifiers and are dispersible in water. A DATEM has a free carboxyl group, and its acidity in water lowers the pH to below 2. If the pH is adjusted to above 4.5, the DATEM is partially neutralized, and under such conditions it forms stable lamellar mesophases in water. DATEMs have multiple applications in foods, mainly as dough strengtheners in yeast-raised bakery products (bread, rolls, etc.), but they are also effective emulsifiers in many emulsions.

D. Citric Acid Esters of Monoglycerides—EU 472c, § US/FDA/CFR 172.832

Citric acid esters of monoglycerides (CITREMs) are produced by reacting citric acid in an amount of 12–20% by weight of the final product with mono-diglycerides or distilled monoglycerides. CITREMs are anionic-active, hydrophilic emulsifiers forming a milky dispersion in water and are only slightly soluble in oils and fats.

The typical use of CITREMs is in emulsions. They are used in frying margarine as antispattering agents and in meat emulsions to prevent fat separation.

IV. FATTY ACID ESTERS

Several emulsifiers are produced by esterification of polyvalent alcohols with fatty acids. The main fatty acids are palmitic, stearic, and oleic acids, but products with other fatty acids are available.

A. Polyglycerol Esters of Fatty Acids—EU 475, US/FDA/CFR § 172.854

Esterification of condensed glycerol, mainly tri- and tetraglycerol, with fatty acids, usually palmitic, stearic, or oleic acids, yields products with 20–40% mono-fatty acid esters. The degree of polymerization of the glycerol and the degree of esterification can vary considerably depending on application and legal aspects.

Polyglycerol esters of fatty acids (PGEs) are stable in the α crystal form, are water-dispersible, and form reversed hexagonal mesomorphic phases above their melting point (55–60°C). PGEs are slightly more polar than monoglycerides, and they are mostly used in emulsions, often in bakery products such as cakes.

B. Propylene Glycol Fatty Acid Esters—EU 477, US/FDA/CFR § 172.856

Propylene glycol can be esterified with palmitic and stearic acids, forming a blend containing 55% monoester and 45% diesters. This can be purified by molecular distillation, yielding a product containing a minimum of 90% monoesters. Propylene glycol monostearates (PGMSs) have a melting point of about 40°C. They are capable of forming α gel phases in water below the melting point, making them effective emulsifiers in whippable emulsions (toppings) [20], bakery shortenings, cake mixes, etc. A PGMS is often used in combination with monoglycerides to make stable α crystalline blends of emulsifiers. Such combinations are mostly used in non-fat cake systems.

C. Sorbitan Monostearate—EU 491, US/FDA/CFR § 172.842

Sorbitan is derived from sorbitol by dehydration and by esterification with palmitic and stearic acid blends, yielding products with different hydrophilic and lipophilic properties depending on the degree of esterification. Sorbitan mono-fatty acid esters, usually referred to as sorbitan monostearates (SMSs), are hydrophilic, water-dispersible emulsifiers. SMSs can be reacted with ethylene oxide to form polyoxyethylene sorbitan esters (POEMS), which are among the most hydrophilic, water-soluble emulsifiers available. SMSs are used in many types of emulsions (dressings, sauces, etc.) often in combination with POEMS.

Blends of monoglycerides and POEMS are used in dairy products (ice cream), but the use of POEMS in foods is limited due to restrictive legislation regarding these products in many countries.

D. Sorbitan Tristearate (STS)—EU 492

By using a high proportion of fatty acids relative to the amount of sorbitan during esterification, sorbitan tristearate (STS) is produced. STS is a fat-soluble emulsifier with a molecular structure close to that of triglycerides [21], as shown in Fig. 5. It is stable in the α crystal form and can cocrystallize with fats and form solid solutions. This makes STS an effective crystal modifier in fat products. STS is specifically used as a crystal modifier in margarine blends, stabilizing the β' crystal form, or as a bloom inhibitor in chocolate and confectionery fats. STS has no surface-active properties and is not used as an emulsifier for stabilizing emulsions.

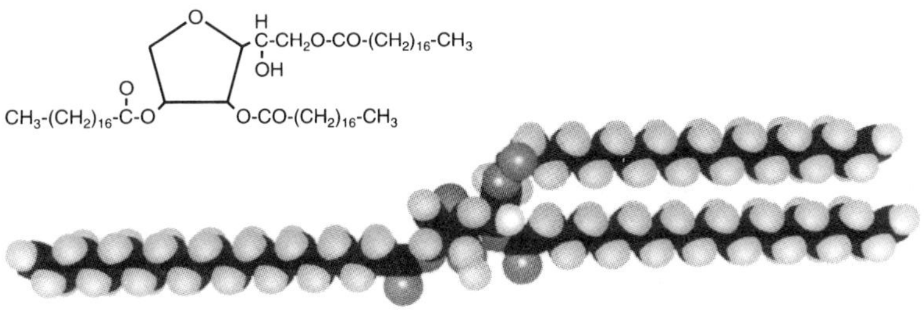

Fig. 5 Chemical structure of sorbitan tristearate. (From Ref. 21.)

E. Sucrose Esters of Fatty Acids—EU 473, US/FDA/CFR § 172.859

Sucrose esters are a new generation of food emulsifiers made by esterification of sucrose with fatty acids. By varying the ratio between fatty acids and sucrose, a number of esters with different hydrophilic/lipophilic properties can be made. Sucrose esters containing up to 70% monoesters are water-dispersible, and sucrose esters with a low monoester content (10–30%) are oil-soluble.

Sucrose esters can be used in many types of foods from emulsions to starch- or fat-based products. However, sucrose esters are relatively new to the food industry. The widest scope of application is found in Japan, where these products are generally permitted for use in foods. In Europe and the United States, the use of sucrose esters is still somewhat limited.

A mixture of sucrose esters and mono-diglycerides, so-called sucroglycerides (EU 474), made by transinteresterification of triglycerides and sucrose, is also available but not widely used.

F. Sodium Stearoyl-Lactylate—EU 482, US/FDA/CFR § 172.844

Reaction of stearic acid with lactic acid in the presence of sodium hydroxide yields sodium stearoyl-lactylates (SSL), which is a water-dispersible, anionic active emulsifier with a melting point of ~45°C. SSL is a very versatile emulsifier with many different applications. Due to its anionic properties it is an effective emulsifier in various types of emulsions, and it also works as a dough strengthener and starch-complexing agent in bakery products.

V. SPECIFIC EMULSIFIER INTERACTIONS IN FOODS

The function of emulsifiers in foods is very complex and can be only briefly described here. In principle, functions of emulsifiers can be divided into interactions taking place during food processing (e.g., emulsification, aeration, dough strengthening) and interactions with food components leading to extended shelf life and improved texture of finished products (e.g., crumb softening of bread, texture of ice cream, emulsion stability). Very often it is not possible to correlate physical properties of emulsifiers with their functions in foods. It is also difficult to make reliable predictions about the functionality of a specific type of emulsifier in a given type of food. Although more and more basic knowledge about functions and interactions of emulsifiers is available [22], selection and optimization of an emulsifier system are still based on experimental tests.

A. Emulsion Stability and Destabilization

The effect of emulsifiers on emulsion stability is related to their ability to form viscoelastic interfacial films, preventing droplet coalescence [23]. Most food emulsions contain proteins that are responsible for the emulsion's stability in, for example, milk or cream, due to adsorption of proteins to fat globule surfaces. In combination with emulsifiers, mixed interfacial films are formed [24], thus increasing the stability.

When anionic emulsifiers (DATEM, SSL) are used, their affinity to proteins may further bind proteins to the interface [25] and consequently provide increased emulsion stability.

In whippable emulsions (ice cream, cream), partial desorption of interfacially bound protein is desired to enhance aeration properties. This is provided by monoglycerides, which have higher surface activity than proteins at low temperatures and therefore replace the proteins at the surface of the fat globules [26]. The competition between emulsifiers and proteins for adsorption at the surface of fat globules can be studied by interfacial tension measurements as demonstrated in Fig. 6. From this it appears that the interfacial tension, in the presence of monoglycerides, is reduced when the temperature is lowered, whereas a system in which only

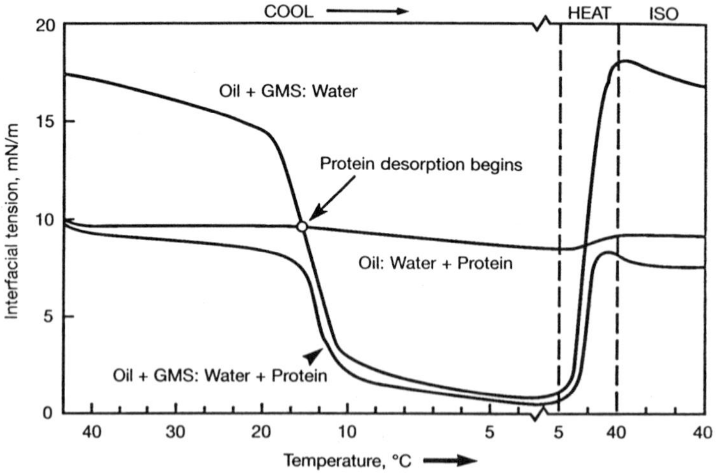

Fig. 6 Measurements of interfacial tension versus temperature of purified sunflower oil–water systems containing (a) oil with 0.2% monoglycerides (GMS) against pure water, (b) pure oil against water with 0.01% milk proteins, and (c) oil with GMS against water with protein. Temperature program: Cool from 40°C to 5°C at 0.3°C/min, hold 60 min at 5°C, heat to 40°C at 2°C/min, and hold for 90 min. (Courtesy of Danisco Ingredients, Denmark.)

proteins are present is not affected by a change in temperature. This means that the interfacial concentration of the monoglycerides is increased at low temperatures. Consequently, when both monoglycerides and proteins are present, the monoglycerides will displace the proteins from the interface. When the protein concentration at the surface is reduced, the fat globules tend to flocculate and form clusters, and this enhances the whipping properties and improves the texture and stability of the finished whipped cream. It is well known that agglomerated fat particles (clusters) stabilize the air cells in whipped emulsions such as ice cream or dairy/non-dairy creams [27,28]. (See Chapters 12 and 13.)

B. Interactions with Starch Components

Distilled monoglycerides have specific functions in starch-based foods that cannot be obtained with mono-diglycerides. Due to their water dispersibility, distilled monoglycerides are more effective in interactions with water-soluble food ingredients. Consequently, they work better than mono-diglycerides as starch-complexing agents [29] in potato products or bakery products.

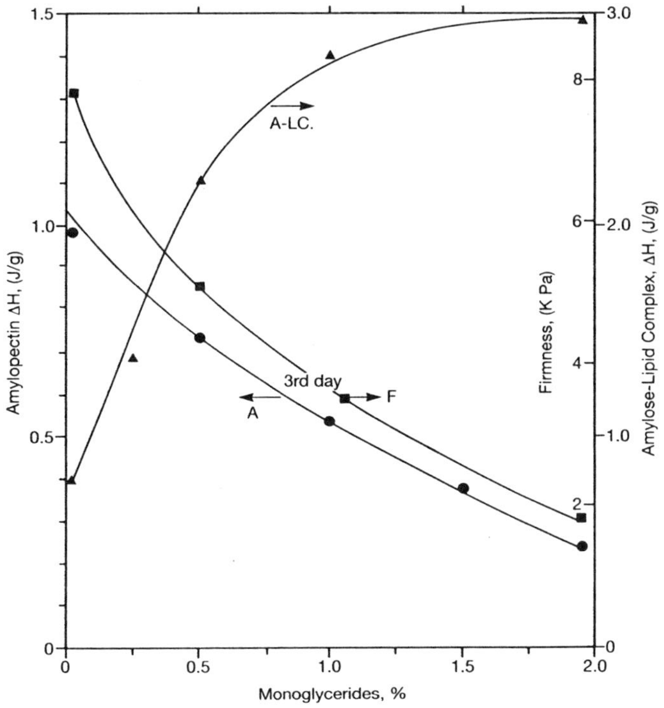

Fig. 7 Effects of monoglycerides on the amylose–lipid complex formation (A-LC), the retrogradation of amylopectin (A), and the crumb firmness (F) of bread after 3 days of storage at 22°C. (From Ref. 22.)

The shelf life of wheat bread (toasting bread, high fiber bread, etc.) is very important for the baking industry and has great economic aspects. The complex formation between the water-soluble amylose and saturated monoglycerides delays the staling process and increases the crumb softness of bread [30]. The interaction of monoglycerides with the starch components amylose and amylopectin is a function of the concentration of the monoglycerides added to the flour part of the dough. Their influence on the crumb firmness of the bread is shown in Fig. 7. It appears that at low concentrations the monoglycerides interact primarily with the amylose fraction, and this complex formation decreases the crumb firmness in fresh bread. At higher monoglyceride concentrations, complex formation with the amylopectin fraction takes place, thus reducing the rate of staling during storage. Consequently, the optimal concentration of monoglycerides is between 1% and 2% based on flour weight. The complex formation with starch components also improves the texture of extruded cereals, processed potato products, pasta, etc.

C. Dough-Strengthening Effects

Anionic-active emulsifiers, such as DATEM or SSL, are used as processing aids in dough to increase the strength of the gas cell membranes formed during fermentation and to provide improved dough elasticity to prevent the dough structure from collapsing due to mechanical abuse during production in automated plants. The resistance of dough to mechanical shock is important when producing bread or rolls with uniform volume and texture.

The relative baking effect of flour is related to the content of the polar lipids in it [31]. The polar lipids in flour consist of phospholipids and glycolipids, and it has been demonstrated that emulsifiers such as lecithin and DATEM have the same effect in bread dough as the polar flour lipids [32]. This can be seen from Fig. 8, which shows the relative effect of flour lipids, lecithin, and DATEM on loaf volume of bread made from defatted wheat flour. The function of both the polar flour lipids and emulsifiers such as DATEM, SSL, or lecithin in dough is related to the formation of lamellar, liquid crystalline phases that are part of the gas cell membrane [33].

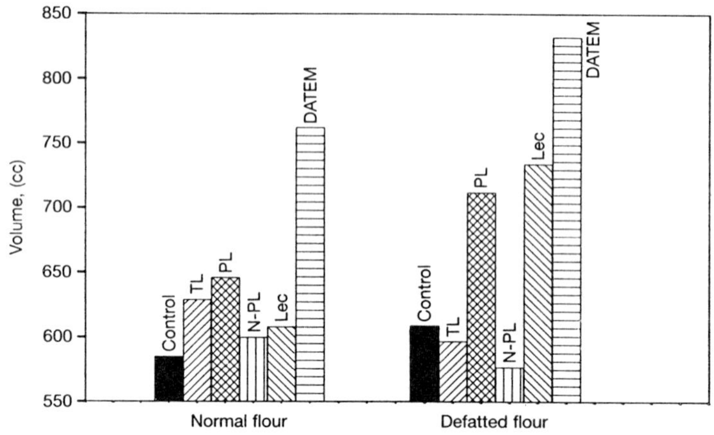

Fig. 8 Influence of flour lipids and emulsifiers (lecithin, DATEM) on bread loaf volume. The following amounts of lipids were added based on flour weight: 0.7% total wheat lipid (TL), 0.15% nonpolar wheat lipid (N-PL), 0.3% lecithin (Lec), and 0.3% DATEM. The wheat flour (12.3% protein) was used before and after partial extraction of lipids. No emulsifiers were added to the control test. (From Ref. 22.)

IV. CONCLUDING REMARKS

The variety of emulsifiers and their properties provides selective choices for various food products, with applications ranging from film coatings that protect foods from oxidation or loss of humidity, through stabilizing or destabilizing emulsions, dough strengthening and crumb softening of bakery products, texturizing of starch-based foods, and modification of fat crystallization. Progress in research on the interfacial and functional properties of emulsifiers has greatly increased the understanding of how emulsifiers work in various food systems. The need for emulsifiers is evident in most industrially manufactured foods, particularly to ensure uniform quality and prolonged shelf life during storage. Low fat products certainly need emulsifiers as well as other texturizing ingredients to provide textural characteristics similar to those of full fat products.

The development of new chemical types of emulsifiers is not likely to happen due to the high cost of toxicological tests necessary to ensure their safety for use in foods. Instead, developments can be expected to deal with improving the purity of the existing range of emulsifiers and making better defined products in terms of chemical composition.

New developments in biochemical synthesis of emulsifiers, such as monoglycerides, using lipolytic enzymes are in progress. Different enzymatic methods for glycerolysis, selective hydrolysis of fats and oils, or esterification of fatty acids with glycerol or other polyols are possible [34].

ABBREVIATIONS

ACETEM	Acetic acid ester of monoglyceride
LACTEM	Lactic acid ester of monoglyceride
DATEM	Diacetyltartaric acid ester of monoglyceride
CITREM	Citric acid ester of monoglyceride
PGE	Polyglycerol ester of fatty acid
PGMS	Propylene glycol monostearate
POEMS	Polyoxyethylene sorbitan monostearate
SMS	Sorbitan monostearate
STS	Sorbitan tristearate
SSL	Sodium stearoyl-lactylate

REFERENCES

1. J. B. Lauridsen, Food emulsifiers: surface activity, edibility, manufacture, composition and application, *J. Am. Oil Chem. Soc. 53*: 400 (1976).
2. G. Schuster (Ed.), *Emulgatoren für Lebensmittel*, Springer Verlag, Berlin, 1985.
3. F. R. Benson, in *Nonionic Surfactants* (M. J. Schick, Ed.), Marcel Dekker, New York, 1987, p. 247.
4. N. J. Krog, Food emulsifiers and their chemical and physical properties, in *Food Emulsions* (K. Larsson and S. Friberg, Eds.), Marcel Dekker, New York, 1990, p. 127.
5. G. Als and N. Krog, Emulsifiers as processing aids, in *Proceedings. World Conference on Oleochemicals into the 21st Century* (T. H. Applewhite, Ed.), Am. Oil Chem. Soc. Press, Champaign, IL, 1990, p. 67.
6. *Compendium of Food Additive Specifications*, FAO Food and Nutrition Paper 52, Food and Agriculture Organisation of the United Nations, Rome, 1992.
7. W. Clayton, Margarine, in *The Theory of Emulsions and Their Technical Treatment* (W. Clayton, Ed.), J. & A. Churchill, London, 1943, p. 313.
8. R. O. Feuge and A. E. Bailey, Modification of vegetable oils. VI. The practical preparation of mono- and diglycerides, *Oil Soap 23*: 259 (1946).
9. N. O. V. Sonntag, Fat splitting, esterification, and interesterification, in *Bailey's Industrial Oil and Fat Products*, Vol. 2 (D. Swern, Ed.), Wiley, New York, 1982, p. 97.
10. M. W. Formo, Physical properties of fats and fatty acids, in *Bailey's Industrial Oil and Fat Products*, Vol. 1 (D. Swern, Ed.), Wiley, New York, 1979, p. 192.

11. K. Larsson and P. J. Quin, Physical properties: structural characteristics, in *The Lipid Handbook*, 2nd ed. (F. D. Gunstone, J. L. Harwood, and F. B. Padley, Eds.), Chapman & Hall, London, 1994, p. 401.
12. K. Larsson, Liquid-crystalline lipid-water phases, in *Lipids—Molecular Organization, Physical Functions and Technical Applications* (K. Larsson, Ed.), The Oily Press, Dundee, Scotland, 1994, p. 47.
13. I. Heertje, Eur. Patent Specification No. 0 558 523 B1 (1994), to Unilever N.V., Rodderdam (NL).
14. N. Krog, The role of Low-polar emulsifiers in protein-stabilized food emulsions, in *Emulsions—A Fundamental and Practical Approach* (J. Sjöblom, Ed.), Kluwer, Boston, 1992, p. 61.
15. G. H. Doerfert, Improved hydrates of hard monoglycerides, *Bakers Dig. 42*: 56 (1968).
16. N. Krog and B. Nybo Jensen, Interaction of monoglycerides in different physical states with amylose and their anti-firming effects in bread, *J. Food Technol. 5*: 77 (1970).
17. N. M. Barfod, N. Krog, G. Larsen, and W. Buchheim, Effects of emulsifiers on protein–fat interaction in ice cream mix during ageing. I: Quantitative analysis, *Fat Sci. Technol. 93*: 24 (1991).
18. O. Fennema and J. J. Kester, Resistance of lipid films to transmission of water and oxygen, in *Water Relationships in Food* (H. Levine and L. Slade, Eds.), Plenum, New York, 1991, p. 703.
19. J. M. M. Westerbeek and A. Prins, Function of alpha-tending emulsifiers and proteins in whippable emulsions, in *Microemulsions and Emulsions in Foods* (M. El-Nokaly and D. Cornell, Eds.), Am. Chem. Soc., Washington, DC, 1991, p. 146.
20. W. Buchheim, N. M. Barfod, and N. Krog, Relation between microstructure, destabilization phenomena and rheological properties of whippable emulsions, *Food Microstruct. 4*: 221 (1985).
21. N. Krog and D. Last, Palm-oil based food emulsifiers, *Malaysian Oil Sci. Technol. 4*: 93 (1995).
22. N. Krog, Interactions of emulsifiers with other components in foods, in *Ingredient Interactions: Effects on Food Quality* (A. G. Gaonkar, Ed.), Marcel Dekker, New York, 1995, p. 377.
23. E. H. Lucassen-Reynders, Interfacial viscoelasticity in emulsions and foams, *Food Struct. 12*: 1 (1993).
24. G. Doxastakis and P. Sherman, The interaction of sodium caseinate with monoglyceride and diglyceride at the oil-water interface and its effect on interfacial rheological properties, *Colloid Polym. Sci. 264*: 254 (1986).
25. E. Dickinson and S.-T. Hong, Surface coverage of beta-lactoglobulin at the oil-water interface: influence of protein treatment and various emulsifiers, *J. Agric. Food Chem. 42*: 1602 (1994).
26. N. Krog, Thermodynamics of interfacial films in food emulsions, in *Microemulsions and Emulsions in Foods* (M. El-Nokaly and D. Cornell, Eds.), Am. Chem. Soc., Washington, DC, 1991, p. 138.
27. W. Buchheim and P. Dejmek, Milk and dairy-type emulsions, in *Food Emulsions*, 2nd. ed. (K. Larsson and S. Friberg, Eds.), Marcel Dekker, New York, 1990, p. 203.
28. B. E. Brooker, M. Anderson, and A. T. Andrews, The development of structure in whipped cream, *Food Microstruct. 5*: 277 (1986).
29. N. Krog, Amylose complexing effect of food grade emulsifiers, *Stärke/Starch 23*: 206 (1971).
30. N. Krog, S. K. Olesen, H. Törnæs, and T. Jönsson, Retrogradation of the starch fraction in wheat bread, *Cereal Foods World 34*: 281 (1989).
31. F. MacRitchie, Flour lipids and their effects in baking, *J. Sci. Food Agric. 28*: 52 (1977).
32. Y. Pomeranz, El-Baya, W. Seibel, and H. Stephan, Toast bread from defatted wheat flour, *Cereal Chem. 61*: 136 (1984).
33. A.-C. Eliasson and K. Larsson, Physicochemical behaviour of components, in *Cereals in Breadmaking*, (A.-C. Eliasson and K. Larsson, eds.), Marcel Dekker, New York, 1993, p. 42.
34. U. Bornscheuer, Lipase-katalysierte Synthese von Monoglyceriden, *Fat Sci. Technol. 97*: 241 (1995).

21

Lipid Emulsions for Intravenous Nutrition and Drug Delivery

Robert T. Lyons and Eric G. Carter
Pharmacia & Upjohn Inc., Clayton, North Carolina

I. INTRODUCTION: THE STRUCTURE AND PROPERTIES OF OIL-IN-WATER EMULSIONS

A liquid emulsion is a dispersion of one immiscible phase within another, with a surfactant used to stabilize the dispersed droplets. Water-in-oil emulsions for vaccine delivery are given by intramuscular injection and are designed for sustained release. In contrast, fat emulsions for parenteral nutrition need to be compatible with blood and therefore are of the oil-in-water type [1]. Those currently in clinical use are composed of either soybean or safflower oil, and some products contain oil blends with medium-chain triglycerides. The latter are composed mainly of saturated C_8 and C_{10} fatty acids. Suitable vegetable oils are emulsified with either egg yolk or soybean phospholipids. The emulsified lipid droplets typically range in size from about 0.2 to 0.6 μm and are metabolized in a manner similar to that of natural chylomicrons [1,2]. Commercial examples in the United States include Intralipid (Pharmacia) and Liposyn (Abbott Laboratories). In Europe, several others are also available, including Lipofundin (B. Braun), and in Japan, Soyacal (Green Cross). These products are described in more detail in Section III. Their relatively low toxicity and excellent safety record, exemplified by extensive use in admixtures for total parenteral nutrition, have made this type of fat emulsion an attractive choice as a carrier system or vehicle for lipophilic drugs and vitamins [3–7]. These applications are described in Section II.

Purified egg yolk phospholipids, the most commonly used emulsifier, stabilize the dispersed submicron oil droplets by two important mechanisms, both of which act at the oil/water interface to inhibit aggregation and coalescence. Trace quantities of ionized molecular species, such as nonesterified (free) fatty acids and phosphatidic acid, impart a net negative surface charge (zeta potential) that stabilizes the dispersion by charge–charge repulsion, even at elevated temperatures. Simultaneously, zwitterionic species such as phosphatidylcholine and phosphatidylethanolamine and neutral lipids such as cholesterol establish a monomolecular interfacial film that binds water molecules at the surface and serves as a physical barrier

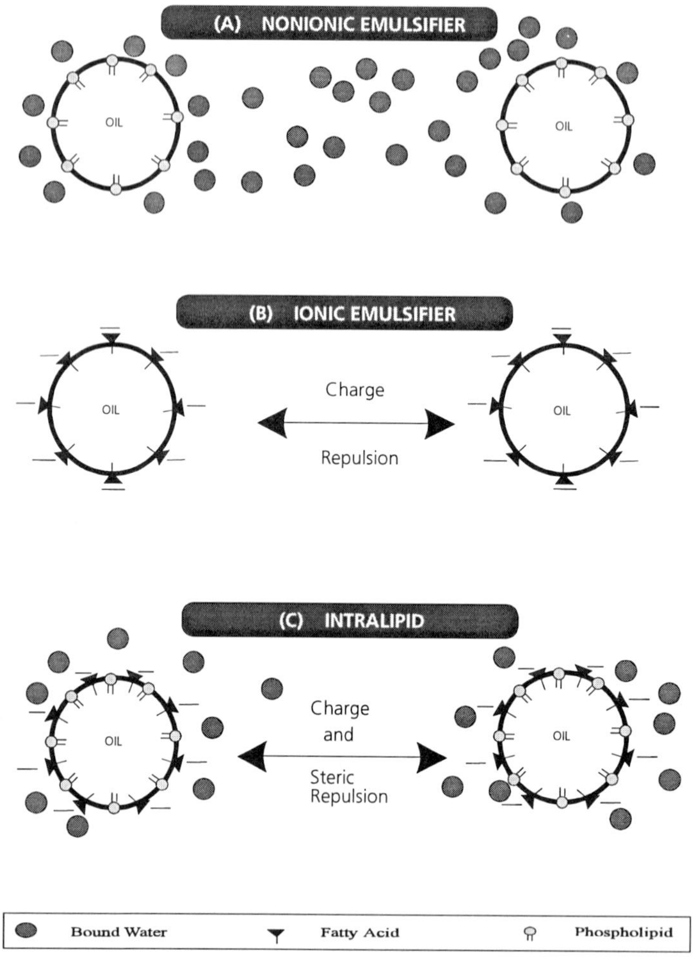

Fig. 1 Stabilization of oil-in-water emulsions. (A) Nonionic emulsifiers at the oil/water interface stabilize dispersed droplets by forming a physical barrier consisting of a monomolecular film and bound water molecules. (B) Ionic emulsifiers provide charge–charge repulsive forces. (C) Intralipid is made with egg yolk phospholipids, which combine both nonionic and ionic stabilizing principles.

against coalescence [8–10]. These stabilizing principles are summarized in Fig. 1. Fat emulsions designed for parenteral nutrition are chemically and physically stable enough to be terminally heat sterilized in steam autoclaves at 121°C.

During prolonged storage, e.g., over 18 months, certain physical changes may be observed. These events, including flocculation, coalescence, and creaming [6,8–10], are summarized in Fig. 2. Creaming is the result of lower density lipid droplets floating upward in the aqueous phase and is readily reversible by gentle shaking. Flocculation is the formation of droplet clusters or aggregates. These are not readily redispersed by agitation. Coalescence is also irreversible, and the resulting large oil droplets can make an emulsion product unsuitable for injection [11].

Possible chemical changes that may occur during long-term storage of these emulsions include hydrolysis and/or peroxidation of the unsaturated fatty acid side chains on the phospholipid emulsifier as well as in the triglyceride oil phase [12,13]. However, these reactions

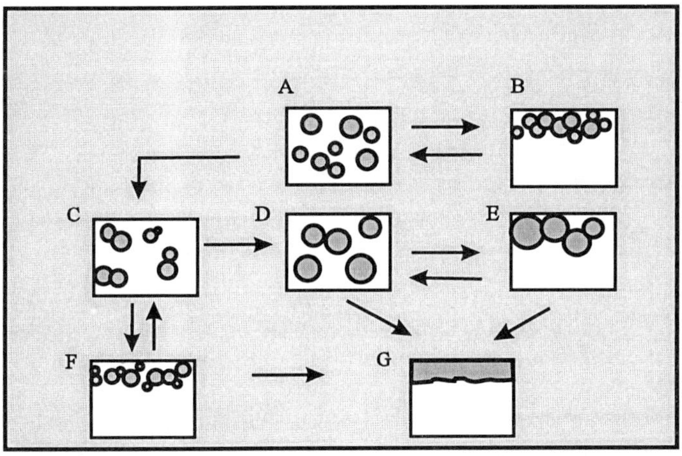

Fig. 2 Physical changes possible in a lipid emulsion. (A) Freshly prepared emulsion. (B) Creaming is easily reversible. (C) Flocculation and (D) coalescence are irreversible. (E) Rapid creaming after coalescence. (F) Rapid creaming after flocculation. (G) Total separation of oil and water phases ("broken" emulsion) is irreversible.

may be minimized by careful pH adjustment during homogenization and by stringent protection from oxygen. For example, continuous nitrogen blanketing during all production steps is standard practice to minimize oxidation of unsaturated lipids and of the drug component, when present.

II. INTRAVENOUS LIPIDS: AN ALTERNATIVE DELIVERY SYSTEM FOR DRUGS AND VITAMINS

A. Potential Advantages of Using Lipid Emulsion Vehicles for Lipophilic Drugs

The most obvious reason to consider using a lipid emulsion as a vehicle is to solubilize lipophilic drugs. Although chemists can usually derivatize any compound to enhance water solubility, lipophilic forms are often more effectively taken up by target cells. Amphipathic molecules, which contain both strongly nonpolar and strongly polar groups, are only partially lipid-soluble. Yet amphipathic compounds may also be attractive candidates because they tend to localize at the oil/water interface and intercalate in the phospholipid monolayers [5,14].

In contrast, total immersion in an oil phase may stabilize certain otherwise labile drugs against hydrolysis and/or oxidation during processing, sterilization, and storage [7]. In certain cases, inclusion of pH buffering agents such as amino acids, antioxidants such as α-tocopherol, and metal chelators such as EDTA should be helpful. Although most nutritional fat emulsions are stable for up to 18 months at room temperature, lipid–drug emulsions may require controlled storage at lower temperatures, often 4–8°C. In practice, the shelf life of a lipid–drug emulsion is more often limited by degradation of the incorporated drug than by deterioration of the emulsion vehicle. Loss of drug potency is usually monitored by HPLC-based analysis of degradants; hydrolytic and/or oxidative reactions are often involved. Therefore, a perennial challenge for formulation chemists and process engineers is to optimize combinations of excipients, pH, drug concentration, and process conditions in order to mini-

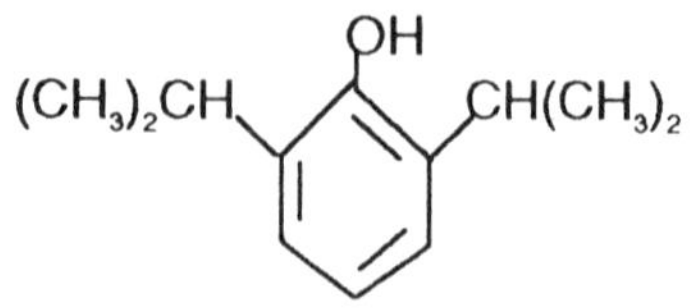

Fig. 3 Propofol (2, 6-diisopropylphenol). This intravenous anesthetic agent is highly lipophilic. It has been formulated in a 10% soybean oil emulsion that is terminally heat sterilized and is successfully marketed as Diprivan.

mize rates of degradation while retaining those properties that keep the emulsion physically stable.

Reduction of drug toxicity by means of association with lipids has been well documented; examples include the antifungal agent amphotericin B [15,16] and the antineoplastic drug penclomedine [17]. Utilization of a lipid emulsion vehicle also eliminates the need for organic solvents such as propylene glycol, *tert*-butanol, or dimethyformamide. These solvents are often associated with pain on injection, phlebitis, and other acute toxicities. Propofol (Fig. 3), an injectable anesthetic formerly in a solvent vehicle, was reformulated in a 10% soybean oil emulsion about 15 years ago to prevent vein irritation [18,19]. Now being marketed as Diprivan, this product represents a classical example of the benefits of lipid-based drug delivery.

Another potential advantage of using lipid emulsions as a vehicle for certain lipophilic agents is to reduce adsorption onto plastic infusion sets [7]. Documented examples of such losses include Perilla ketone, an investigational cytotoxic drug [20], and the fat-soluble vitamins A, D, E, and K_1 (see Table 2).

Possibilities for improved pharmacokinetics, such as some form of sustained release and/or directed drug delivery to various organs, is a complex and still evolving field. Controlled release systems require the preparation of emulsified lipid droplets with long circulation times in the blood and slow drug transport rates from the dispersed to the continuous phases. In general, the clearance of lipid droplets from circulation depends on size (smaller particles clear more slowly), surface charge (highly charged particles clear more rapidly), and surfactant composition [1]. For example, Jeppsson and Rossner [21] observed a fractional clearance rate of about 7%/min following intravenous administration of an egg yolk phospholipid-stabilized emulsion in rabbits. The addition of an amphipathic polymeric surfactant, specifically poloxamer 338 (Pluronic F108), resulted in a sixfold prolongation of circulation time.

B. Potential Disadvantages of Using Lipid Emulsion Vehicles for Lipophilic Drugs

Potential problems with the use of lipid drug emulsions include the risk of embolism. Particles greater than about 7 μm in diameter may lodge in capillary beds, especially those in the lungs [22]. Therefore, manufacturing processes for such parenteral products must be under stringent quality control. Intravenous fat emulsions stabilized by egg phospholipids have an excellent safety record and are an integral part of total nutrient admixtures, along with dextrose and amino acids [23]. However, inclusion of diverse surfactants or the use of alternative oils may increase vehicle toxicity. For example, although medium-chain triglycerides are an excellent solvent for many lipophilic drugs, too rapid infusion of these emulsions causes metabolic acidosis and central nervous system toxicity in dogs [24].

Another important concern is the interaction of emulsified oil droplets with plasma proteins. For example, a strong association with fibronectin could lead to opsinization and rapid uptakc by phagocytic cells in the reticuloendothelial system (RES) [25]. Alternatively, a rapid dissociation of surfactant from the surface of these droplets could induce flocculation in vivo, which in turn would increase the risks of RES involvement or even of lung embolism. Potential biological effects such as these must be monitored carefully during controlled animal studies in at least two species before human trials can begin.

Yet another consideration is possible drug crystallization during prolonged storage and shipping. Emulsions with crystals in the oil droplets tend to be destabilized if polar domains pierce and weaken the interfacial film, leading to unsuitable flocculation or coalescence [26]. Also, crystal formation in any parenteral formulation will result in unacceptable particulate counts [27].

C. The Design, Testing, and Manufacture of Lipid Emulsions for Drug Delivery

There are basically two ways to prepare a lipid drug emulsion. One is referred to as the extraneous[1] addition of a lipophilic compound to a preformed, sterile fat emulsion, usually with the assistance of an organic solvent such as ethanol or dimethyl sulfoxide. This method is not recommended, for several reasons. First, even small amounts of solvent can damage emulsions and produce large, poorly emulsified oil droplets. Second, undissolved crystals of drug are not easily detected in an opaque white matrix. Third, it is difficult to ensure sterility after entering a previously sterilized container. The preferred method of preparation is an ab initio[2] emulsification of a drug-containing oil phase, either aseptically or preferably in conjunction with terminal heat sterilization. In this manner, an emulsion system is custom-designed to accommodate the unique requirements of a particular therapeutic agent.

The major steps required for the production of a drug-containing fat emulsion are summarized in Fig. 4. All manipulations are performed under a protective blanket of nitrogen gas. Typically, the oil phase is first heated together with drug, surfactant, and cosurfactant (if any). Then the aqueous phase, containing a tonicity agent such as glycerin, is added slowly while running a high shear overhead mixer. Next, this preemulsion is transferred to a homogenizer in which the liquid is pumped through a spring-loaded valve system under very high pressure. The resulting shear and cavitation break large oil drops into submicron droplets. Multipassing through these valves creates a more uniform and less polydisperse size distribution. For every formulation, we find that there is a set of optimum process conditions that includes pH, temperature, pressure, and even the number of homogenizer passes. Finished emulsion is then filtered, vials are filled under nitrogen, and rubber stoppers with overseals are applied. Finally, vials are sterilized in a steam autoclave [28,29].

A sequence of characterizations is normally performed on each batch of lipid drug emulsion to assess stability and biocompatibility in the case of new formulations under development and to maintain quality and ensure safety for products already approved for marketing. Some examples of analytical testing commonly performed on lipid drug emulsions are summarized in Table 1 [29].

D. First Generation Injectable Lipid Drug Emulsions

"First generation" emulsions are those already approved and being marketed in Europe and/or the United States. As shown in Table 2, only five lipid drug emulsions are in this category.

[1] "Extraneous" here means "coming from the outside."
[2] "Ab initio" here means "from the beginning."

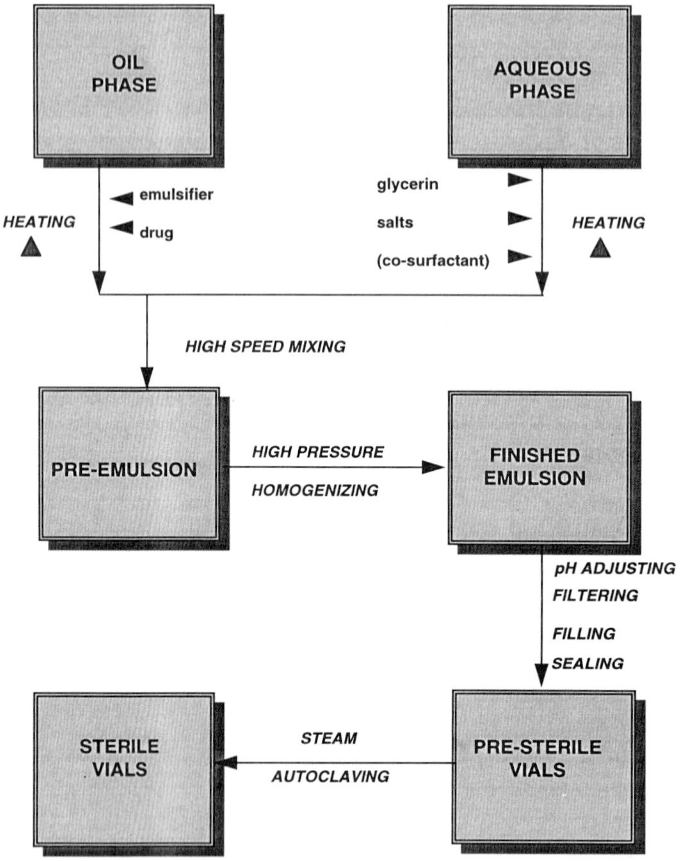

Fig. 4 Major production steps for a lipid drug emulsion. The preparation of a commercial lipid drug emulsion is technically demanding because of the requirement that droplet mean diameter remain in the submicrometer range. Repeated passing through a high pressure piston homogenizer reduces polydispersity in the drop size distribution. At all steps, processing is performed under a nitrogen blanket to minimize oxidative degradation of unsaturated components.

Alphabetically, the first is Diazemuls, which contains the sedative diazepam (Fig. 5) at 5 mg/mL in a soybean oil base. Drug solubility is enhanced with diacetylated monoglycerides, and the emulsifier consists of egg yolk phospholipids. There are no quantitative or qualitative differences in the pharmacological properties of diazepam, whether administered in an ethanol-propylene glycol solution or in this emulsion. However, clinical studies demonstrate that the lipid formulation reduces the incidence of local pain and thrombophlebitis seen with the solution. Diazemuls is used preoperatively to relieve anxiety and provide sedation and light anesthesia. It is also used as an anticonvulsant [30,31].

The second is Diprivan, which contains the anesthetic propofol (Fig. 3) at 10 mg/mL, using Intralipid 10% as vehicle. Following injection, propofol exhibits a high metabolic clearance rate and a large steady-state distribution volume, suggesting that the drug rapidly dissociates from the oil droplets in vivo. Diprivan is a short-acting sedative hypnotic agent, suitable for induction and maintenance of anesthesia or sedation. There is swift recovery with little postoperative nausea and vomiting [18,19].

Table 1 Common Analytical Procedures for Lipid Drug Emulsions

Analytical procedure	Purpose of evaluation	Possible defects
Visual appearance	Uniformity	Presence of nonemulsified surface oil
Phase-contrast microscopy	Uniformity	Large droplets or aggregates; drug crystals
Laser light scattering	Drop size distribution	Large droplets or aggregates
Single-particle optical counting	Concentration of particles > 1 μm	High counts due to crystals, oil droplets
Hydrogen ion electrode	Aqueous phase pH	Out of specified physiological range
Osmolality	Compatibility with blood	Out of specified physiological range
HPLC	Drug potency and impurities	High levels of degradation products
Free fatty acids	Hydrolysis of oil and phospholipid	High levels cause blood cell hemolysis
Peroxides	Oxidation of oil and phospholipid	High levels toxic and may degrade drug
Zeta potential	Electrostatic surface charge	Low charge predicts poor stability
Limulus amebocyte lysate	Endotoxin concentration	High levels pyrogenic (cause fever)

Table 2 First Generation Injectable Lipid Drug Emulsions

Trade name	Manufacturer	Marketer	Drug	Activity	Comments
Diazemuls (Dizac in U.S.)	Pharmacia & Upjohn	Dumex in Europe Ohmeda in U.S.	Diazepam	Sedative	Reduced venous irritation
Diprivan	Pharmacia & Upjohn	Zeneca in Europe; Stuart in U.S.	Propofol	Anesthetic	Induction and maintenance
Fluosol-DA	Green Cross	Green Cross in Japan	Perfluorodecalin	Oxygen delivery	O_2 delivery during balloon angioplasty
Vitalipid	Pharmacia & Upjohn	Pharmacia & Upjohn in Europe	Vitamins A, D, E, K_1	Parenteral nutrition	Infused as admixture
Lipo-PGE,	Green Cross	Green Cross in Japan	Lipo-PGE, Prostaglandin E_1	Vasodilator, platelet inhibitor	Stable to heat sterilization

Fluosol-DA is the most extensively tested example of a perfluorocarbon emulsion. This product is composed of perfluorodecalin (14% w/v) and perfluorotripropylamine (6% w/v), emulsified with a mixture of Pluronic F-68, egg yolk phospholipids, and potassium oleate. The Fluosol emulsion is supplied in a 500 mL flexible plastic container but is unstable and must be stored frozen and then thawed and mixed with electrolytes just before use. It is approved only for intracoronary administration, to provide oxygen to heart muscle during balloon angioplasty [32].

Vitalipid is also a 10% soybean oil emulsion, essentially Intralipid with added fat-soluble vitamins A, D, E, and K_1. This emulsion is first used to reconstitute a second vial of lyophilized water-soluble vitamins. The mixture is then diluted in regular Intralipid prior to infusion

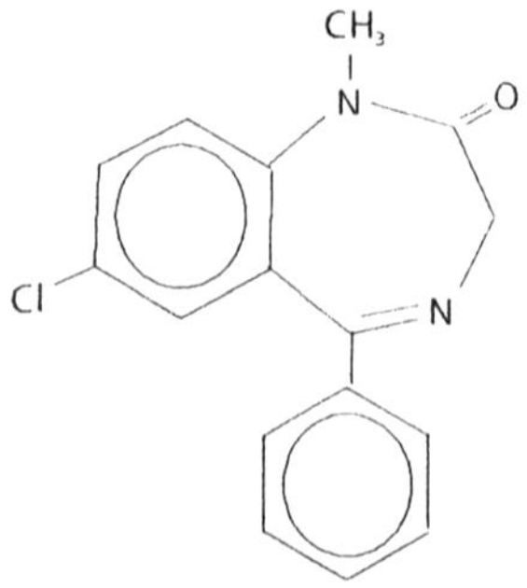

Fig. 5 Diazepam. This sedative drug of the benzodiazepine class occurs as a white crystalline powder and is only slightly soluble in water (3 mg/mL).

for adult and pediatric patients. Vitalipid is terminally heat sterilized and has a 2 year shelf life, yet it contains no added solubilizers, antioxidants, or antimicrobial preservatives [33].

Prostaglandin E_1 (PGE_1) is a potent vasodilator as well as an inhibitor of platelet aggregation. Incorporation of PGE_1 into a 10% soybean oil emulsion is reported to increase biological activity and reduce undesirable side effects [34].

E. Second Generation Injectable Lipid Drug Emulsions

It should be apparent by now that first generation products are, with the exception of Fluosol-DA, essentially modified versions of Intralipid. However, these products have important market niches, and they have validated the concept of lipid-based drug delivery. In contrast, "second generation" projects demonstrate increasing divergence from the original Intralipid concept. By "second generation" we mean formulations that are currently being evaluated for tolerance and efficacy in human clinical trials. Some examples are summarized in Table 3. Trade names should be considered tentative and for identification only, as these are not approved products.

Eltanolone (Fig. 6) is a steroid with hypnotic properties intended for induction and maintenance of intravenous general anesthesia. This endogenous metabolite of progesterone has no hormonal action. The emulsion product contains 4 mg/mL pregnanolone in a 20% soybean oil base with diacetylated monoglycerides as cosolubilizer and egg yolk phospholipids as emulsifier. The clinical hypothesis being tested is that potential advantages over existing prod-

Table 3 Second Generation Injectable Lipid Drug Emulsions

Trade name	Company	Drug	Activity	Comments
Eltanolone	Pharmacia & Upjohn	Pregnanolone	Anesthetic	Similar to Diprivan
Imagent	Alliance Pharm.	Perflubron	Imaging agent	Liver and spleen
Intraiodol	Pharmacia & Upjohn	Iodinated oil	Imaging agent	Liver and spleen
Kynacyte	Sphinx/Eli Lilly	Safingol	PKC inhibitor	Chemopotentiator
Oncosol	HemaGen/PFC	Perfluorocarbon	Oxygen delivery	Radiation sensitizer
Supercytes	HemaGen/PFC	Perfluorocarbon	Oxygen delivery	Heart surgery
None	NCI	Penclomedine	Antineoplastic	Cytotoxic drug
None	PDT Pharmaceuticals	Porphyrin derivative	Antineoplastic	Photodynamic therapy
Freedox	Pharmacia & Upjohn	Tirilazad	Neuroprotective	Stroke, CNS trauma

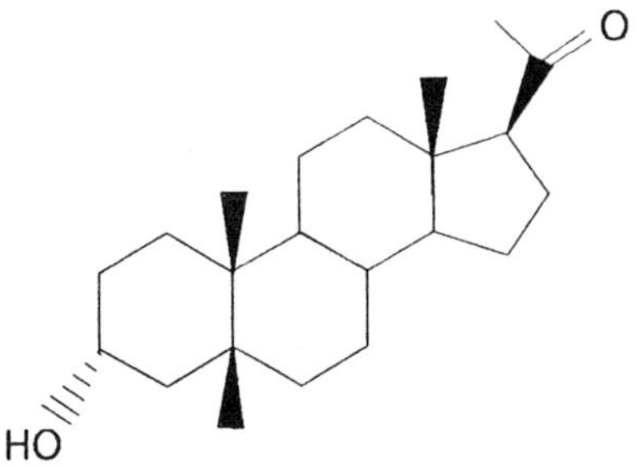

Fig. 6 Eltanolone. Also known as pregnanolone, this lipophilic anesthetic agent is intended for intravenous administration. Eltanolone is a nonhormonal, water-insoluble steroid derivative.

ucts will include absence of pain on injection, wide therapeutic margin, favorable cardiovascular profile, minimal respiratory depression, and rapid recovery [35].

Imagent is an emulsion containing perflubron (perfluorooctylbromide), a perfluorocarbon-type contrast agent developed for CT imaging of either lymph nodes (LN version) or blood pool (BP version). The latter is designed to enhance the detection of small (<1 cm) tumors in the liver and spleen [36]. Intraiodol is another emulsified contrast agent for liver imaging that is now in clinical trials. It is composed of iodinated ethyl esters of fatty acids derived from poppyseed oil, emulsified with egg yolk phospholipids plus 0.1% phenylalanine for improved stability [37,38].

Kynacyte emulsion contains as its active ingredient safingol, a stereoisomer of dihydrosphingosine. This sphingolipid is an inhibitor of protein kinase C and a potential adjunctive treatment for cancer. Preclinical studies have demonstrated that Kynacyte can enhance the effectiveness of approved anticancer drugs such as doxorubicin or cisplatin without increasing their toxicity [39].

Oncosol and Supercytes are emulsions that exploit the ability of perfluorocarbons to dissolve 20 times more oxygen than water. Malignant tumors often have a subpopulation of hypoxic cells that are radiation-resistant. By delivering oxygen to these tissues, Oncosol should enhance the therapeutic effect of ionizing radiation. Supercytes will be used to promote gas exchange during open heart surgery.

Penclomedine, a cytotoxic drug being developed by the National Cancer Institute, shows significant activity against leukemia and mammary tumors in preclinical studies. Penclomedine is very lipophilic but is also heat-sensitive. It has been formulated into an Intralipid-like emulsion that is prepared aseptically [7,17].

Another example of a novel emulsion for oncology is being developed by PDT Inc. A proprietary lipophilic porphyrin derivative serves as a photosensitizer for photodynamic therapy and has been formulated in a lipid emulsion. Following intravenous administration, this green porphyrin drug is selectively retained by tumor tissue. Irradiation by red laser light then induces the formation of intracellular oxygen radicals, which destroy the host cells [40].

Pharmacia & Upjohn Company is developing a second generation emulsion delivery system for tirilazad. This 21-aminosteroid antioxidant is being evaluated as an intravenous injectable drug for the treatment of central nervous system trauma and ischemia [41,42].

F. Conclusions

Although the concept of lipid-based drug delivery has been discussed and evaluated for over 20 years, relatively few such products have actually been approved and marketed. However, product development pipelines currently contain a large assortment of novel lipid emulsions for drug administration as well as for the delivery of vaccines and diagnostic agents. It is now apparent that the unique requirements of each drug and therapeutic regimen will require customized emulsion technology if these new lipid-based carrier systems are to achieve their true potential: clinically significant improvements in drug efficacy, preferably accompanied by reductions in toxicity.

III. LIPID EMULSIONS FOR INTRAVENOUS NUTRITION

A. Historical Overview

Intravenous or parenteral nutrition (PN), the provision of a sterile solution of nutrients by intravenous administration, has evolved in step with the acquisition of knowledge of chemistry, physiology, and microbiology. Although the injection of substances into the bloodstream dates back to the seventeenth century, it was not until the middle of the nineteenth century that physicians recognized the therapeutic value of infusing large volumes of saline to counter fluid losses incurred from cholera infections [43]. Subsequently, attempts to supplement nutrients intravenously were unsuccessful until the concepts of microbial infection became widely accepted. Early this century, as principles of nutrition and metabolism became recognized, attempts were made to replenish protein deficits parenterally. Initially this was hampered by allergic complications, but in one of the first successful experiments of parenteral nutrition, in 1915, Henriques and Anderson [44] maintained nitrogen equilibrium in goats for over 2 weeks by the infusion of a mixture of hydrolyzed protein together with glucose and salts. Protein hydrolysates were subsequently shown to be a safe and effective source of nitrogen for human beings.

In due course, numerous complicating and limiting metabolic and technical issues became apparent when clinicians sought improvements in the parenteral delivery of nutrients. The provision of basal energy and nitrogen was not always sufficient to address catabolic stress. On the other hand, excess volume would result in pulmonary edema, and concentrating the glucose solution for caloric support was not the answer, as this resulted in venous thrombophlebitis.

The infusion of lipids was an obvious answer to the question of how to effectively meet energy needs. The high caloric density and low thrombogenic potential of lipid emulsions made these substrates obvious candidates for PN. Murlin and Riche infused a lipid emulsion into animals in 1915, and in 1920 Yamakawa was able to provide a lecithin-stabilized lipid emulsion for human parenteral use [43]. In 1945, McKibbin et al. [45] advocated the provision of calories from lipid emulsions, and shortly thereafter the first commercial preparations based on castor and cottonseed oils became available.

The tendency of these early emulsions to form fat globules and the high incidence of fevers, back pain, and liver dysfunction led to their disuse and to their eventual withdrawal in the United States in 1964 [46]. Meanwhile, in Europe, Wretlind and colleagues were developing an emulsion based on soybean oil stabilized with egg yolk phospholipids that was shown in clinical trials to be safe as well as effective as a source of calories and essential fatty acids. Supporting clinical results from the United States [47] led to FDA approval of Intralipid and Liposyn in 1981. To date in excess of 100 million units of lipid emulsion have been administered to patients as an integral part of PN.

B. Uses in Parenteral Nutrition

1. Composition

Lipid emulsions have three main components:

1. Triacylglycerols (TAG), which make up the bulk of the emulsion particle, provide the substrate for energy and the fatty acids that are essential for health. TAG are characterized by their chain length:
   a. Long-chain TAG (LCT) are provided by soybean or safflower seed oils, which are rich in C_{16}–C_{18} polyunsaturated fatty acids of the *n*-6 (or ω-6) series.
   b. LCT may also be of the *n*-3 (or ω-3) series, derived from fish oil.
   c. Medium-chain TAG (MCT) are provided by coconut or palmkernel oil and consist of fatty acids with chain length C_8–C_{10}.
2. Phospholipids (PL) derived from egg yolk act as the emulsifying agents.
3. Glycerol renders the solution isotonic.

Triacylglycerol

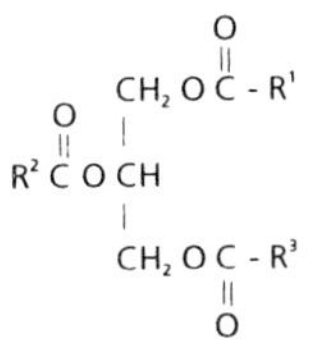

R^1, R^2 and R^3 are saturated and unsaturated fatty acid residues

Fatty Acid Linoleic Acid ($C_{18}H_{32}O_2$ 18:2n-6)

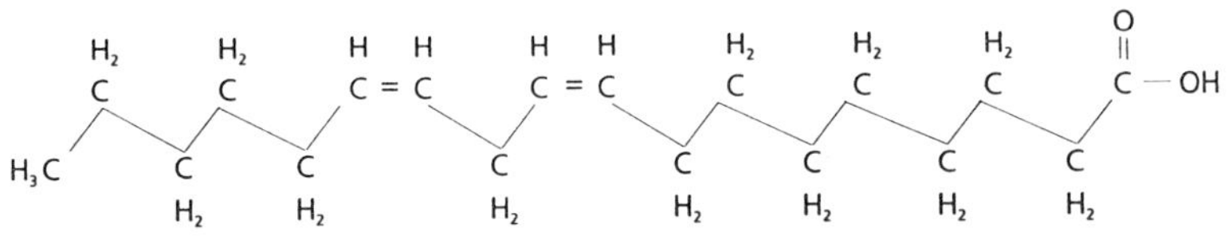

Phospholipid

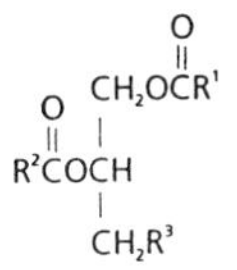

R^1 and R^2 are saturated fatty acid residues

R^3 is primarily either choline or the ethanolamine ester of phosphatidic acid

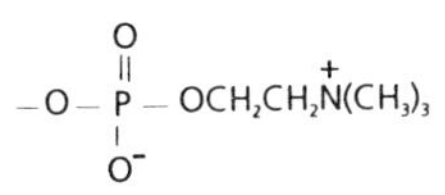

Phosphatidylcholine

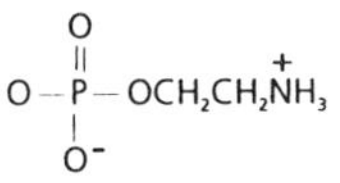

Phosphatidylethanolamine

Fig. 7 Chemical and structural formulas of the main components of lipid emulsions.

Table 4 Composition of Selected Commercial Lipid Emulsions

Name (Manufacturer)	Oil (%)			Phospholipid (%)	Glycerin (%)	Fatty acid (%)						
	Soybean	Safflower	Coconut/palm			Linoleic (18:2, *n*-6)	Oleic (18:1, *n*-9)	Palmitic (16:0)	Linolenic (18:3, *n*-3)	Stearic (18:0)	Caprylic (8:0)	Capric (10:0)
Intralipid 10% (Pharmacia & Upjohn)	10	—	—	1.2	2.25	50	26	10	9	3.5	—	—
Intralipid 20%	20	—	—	1.2	2.25	50	26	10	9	3.5	—	—
Intralipid 30%	30	—	—	1.2	2.25	50	26	10	9	3.5	—	—
Liposyn II 10% (Abbott)	5	5	—	1.2	2.5	65.8	17.7	8.8	4.2	3.4	—	—
Liposyn II 20%	10	10	—	1.2	2.5	65.8	17.7	8.8	4.2	3.4	—	—
Liposyn III 10%	10	—	—	1.2	2.5	54.5	22.4	10.5	8.3	4.2	—	—
Liposyn III 20%	20	—	—	1.2	2.5	54.5	22.4	10.5	8.3	4.2	—	—
Nutrilipid 20% (Kendall McGaw)	20	—	—	1.2	2.21	49-60	21-26	9-13	6-9	3-5	—	—
Lipofundin 20% (Braun)	10	—	10	1.2	2.5	~25	~12	~5	~4	~3	~35	~15

The chemical and structural formulas of the main components of lipid emulsions are shown in Fig. 7.

Commercially available lipid emulsions vary in their composition depending primarily on the source and concentration of the TAG and PL. Likewise, the exact composition of the PL will depend on the origin of the egg yolk source. The compositions of selected commercial lipid emulsions are shown on Table 4.

Lipid emulsions were designed such that the particles resembled chylomicrons, which could then be transported in the circulation and provide TAG fatty acids to different tissues. Chylomicrons belong to a family of lipoprotein particles that are involved in the transport of lipid in the circulation. The relative concentrations of constituents of a typical lipid emulsion particle compared with that of native chylomicrons and other lipoproteins are shown in Fig. 8 [48].

The emulsion particle, like the chylomicron, has a diameter ranging from about 0.2 to 0.6 μm and consists of a TAG core stabilized by a PL surface layer. In contrast, there are no cholesterol esters or apoproteins on the surface of lipid emulsion particles, and the overall concentration of PL is typically greater. Differences in surface properties may reflect subtle changes in the distribution, uptake, and metabolism of emulsion particles and native chylomicrons [49].

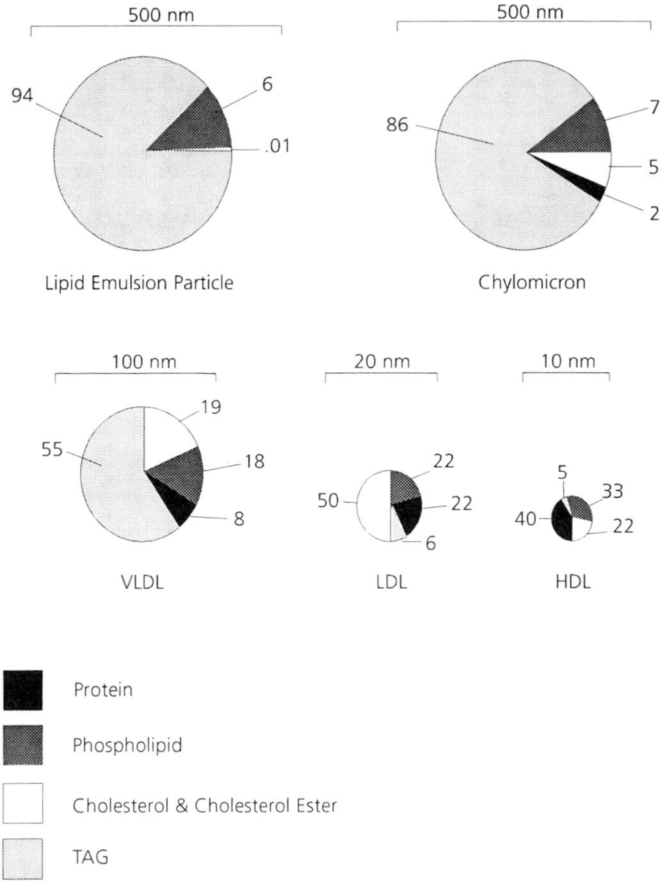

Fig. 8 Relative concentrations of constituents of a typical lipid emulsion particle compared with that of native chylomicrons and other lipoproteins.

Due to the relatively elevated PL concentration, lipid emulsions contain two particle populations that can be separated by centrifugation. The low density TAG-rich population resembles chylomicrons. A higher density liposome-like population is rich in PL. The proportion of TAG-rich particles to liposome-like particles depends on the proportion of TAG to PL in the emulsion [49].

2. *Lipid Metabolism*

Metabolism of Lipid Emulsion Particles. The traditional view is that, owing to their structural similarity, lipid emulsions are metabolized like chylomicrons [49]. Acquisition of surface C and E apoproteins allows the particle to interact with and activate lipoprotein lipase at the capillary endothelial surface. In turn, TAG hydrolysis is initiated with the release of fatty acids (FA) at the 1 and 3 positions and subsequent uptake of the resultant monoglyceride and some of the FA into the adjacent tissues. The remaining FA remain in the circulation bound to albumin.

As more TAG is hydrolyzed from the particle by peripheral lipoprotein lipase, the surface apoproteins change, and as the particle shrinks in size it becomes enriched by cholesterol esters. The remnant, now TAG-poor but cholesterol-rich, is further hydrolyzed by hepatic lipase, preferentially taken up by hepatocytes and degraded. Other types of cells, notably tissue macrophages, are capable of taking up emulsion particles, particularly when in the remnant form.

The composition of the emulsion particle will influence the rate of hydrolysis and uptake, and this is determined not only by the FA chain length but also by its relative position on the glycerol backbone, with FA at position 2 having the most influence. LCT emulsions are cleared at the same rate as chylomicrons [50, 51], whereas the clearance of MCT emulsions is more rapid [52].

Recent animal research with labeled emulsion particles suggests that (1) the traditional view may be incomplete in that, in contrast to chylomicrons, emulsion particles may be taken up before substantial lipolysis has taken place and (2) a large proportion, more than 50%, of this uptake occurs in extrahepatic tissues [53].

The amount of PL relative to TAG can influence the hydrolysis of lipid emulsions. Thus 20% emulsions, or 10% emulsions with a lower PL content than regular 10% emulsions, are able to clear TAG more quickly [54].

Phospholipid-rich particles may interfere with the binding of TAG-rich particles to lipoprotein lipase, thereby decreasing the rate of TAG hydrolysis as well as the binding of other lipoprotein species to their receptors [54,55]. In this way and by their affinity to cholesterol, these particles may influence intravascular lipid metabolism in ways that remain to be fully elucidated.

Phosphatidylcholine, a component of PL in emulsion particles, has recently been promoted as a source of choline. Choline is an essential component of membrane PL. It is a precursor of the neurotransmitter acetylcholine and a source of methyl groups. Choline deficiency leads to fatty liver and, in some animal models, to cancer [56].

Oxidation of Fatty Acids. Free fatty acids are transported to tissues in the circulation bound to albumin. After cellular uptake, the long-chain FA are taken up by mitochondria in a carnitine-dependent manner and undergo β oxidation to acetyl CoA and hence to carbon dioxide and water, via the citric acid cycle with the concomitant formation of high energy ATP. Acetyl CoA formed in excess will be converted to ketone bodies (acetoacetate, β-hydroxybutyrate), primarily in the liver and these may be exported to other organs to be used as an alternative source of fuel. Medium-chain FA are metabolized in a different manner. MCT are cleared faster than LCT from plasma. Medium-chain FA have a lower carnitine requirement than long-chain FA and so are oxidized more rapidly with increased ketone body formation. Reesterification of medium-chain FA is negligible, and as a result, administration of MCT is less likely than LCT to result in tissue deposition of lipid [57]. High levels of plasma medium-chain FA have been associated with neurological toxicity, and this occurs in a dose-dependent manner [58]. In part to minimize this risk, MCT has been combined with LCT in certain lipid emulsions (see Section III.C).

3. Intravenous Nutrition

Lipid emulsions form an integral part of intravenous nutrition as a source of energy and fatty acids essential to life. Energy from lipids is delivered to tissues in the form of free FA, which are released from adipose tissue by lipolysis and distributed in the circulation bound to albumin [59]. Fatty acids are the primary fuel source for resting skeletal muscle, heart, and liver. In contrast, brain, skin, and elements of the blood are dependent on carbohydrates for energy as they cannot readily oxidize FA.

Certain FA are termed "essential" in that they cannot be synthesized endogenously to meet demand. Linoleic and linolenic acids are classic examples because a well-characterized essential fatty acid deficiency (EFAD) syndrome results when they are omitted from the diet. This occurs primarily as a result of changes in the structure of biological membranes of which essential FA are important constituents. Eicosanoids are active metabolites of membrane-bound EFA, and they play a significant role in orchestrating inflammatory responses. The minimum requirement of EFA is 1–3% of dietary caloric intake [60].

Principles of Administration. PATIENT SELECTION. Malnutrition leads to wasting and functional impairment, and increasing severity of malnutrition is associated with increasing morbidity and mortality. Adequate nutritional status can be achieved only by timely and appropriate nutritional support. Parenteral nutritional support is required to maintain nutritional status or to correct status deficiencies when there is preexisting or anticipated malnutrition in the setting of gut failure and/or the inability to receive adequate nutrients orally or enterally. Typical conditions include short bowel syndrome, prolonged postoperative ileus, acute inflammatory bowel disease, fistula, and chronic wasting diarrhea of AIDS.

PARENTERAL FORMULATION. Parenteral preparations consist of a mixture of solutions or emulsions of amino acids, glucose, lipids, electrolytes, and other minerals, vitamins, trace elements, and water. Certain drugs such as insulin or H2 blockers, compatible with the nutrition admixture, may also be added. These nutritional admixtures are tailored and formulated to meet the precise nutritional needs of individual patients as well as some or all of their fluid and electrolyte requirements.

Protein needs are determined on the basis of body weight and degree of metabolic stress experienced by a patient. This can be further refined by ascertaining nitrogen balance, the difference between intake and losses and assessing the needs accordingly. Fluid and electrolyte needs are likewise determined on the basis of close monitoring of the clinical status. Caloric requirements and protein needs during disease and catabolic stress are interrelated. Inadequate caloric support from glucose and/or fat will result in the utilization of the amino acids as a source of energy. Caloric needs are based on calculated or measured estimates of caloric expenditure. In critical illness the aim is usually to provide sufficient calories to meet expenditures. During recovery, extra calories may be administered to correct remaining deficits.

Calories are typically provided from carbohydrate and lipid sources. This mixture is analogous to a mixed diet and is felt to be more physiological than the provision of a large excess of one of the components. Dextrose solutions are available in a wide range of concentrations and provide about 3.5 kcal/g dextrose. Lipid is available as 10, 20, and 30% emulsions providing 1.1–3.0 kcal/mL. There is no clear recommendation as to the proportion of calories that comes from carbohydrate and from lipid. Typically, 20–50% of the non-protein calories are provided as lipid. (It is traditional to exclude parenteral protein from caloric content calculations because protein is supplemented to meet protein rather than caloric needs. Ultimately, though, the breakdown of exogenous protein will contribute to the energy pool.) Approximately 100 g of glucose is required to meet the needs of the glucose-dependent central nervous system, so some carbohydrate has to be provided. Excess glucose administration, on the other hand, results in hyperglycemia, which is associated with osmotic diuresis, protein glycosylation leading to immune dysfunction, and other impairments. Fatty deposition, particularly in the liver, and excessive generation of carbon dioxide requiring greater respiratory effort will also occur.

Parenteral nutrition admixtures are typically administered via a canula inserted into a central vein. The high blood flow and quick dispersion allow a concentrated admixture to be delivered over 12–24 h each day for prolonged periods of time, enabling nutritional goals to be met with a low risk of fluid overload.

ADMIXTURES. Total parenteral nutrition (TPN) is the provision of a complete nutritional admixture adequate to support life. This may be formulated and infused as one admixture, which typically contains protein in the form of amino acids, carbohydrate in the form of glucose, and lipid in the form of soybean or safflower oil emulsion as well as micronutrients (electrolytes and minerals, vitamins, and trace elements), in which case it is referred to as a 3-in-1 admixture. The stability of these complex admixtures is governed by the physicochemical interactions of the constituents. The presence of a lipid emulsion can either stabilize or destabilize the admixture, depending on a number of factors. Because of this, some practitioners prefer to administer the lipid independently of the rest of the admixture constituents. The lipid emulsion is then either piggybacked onto the main TPN line via a Y-connector at the time of infusion or administered independently via a peripheral vein. This is referred to as a 2-in-1 admixture.

Admixtures have to be prepared aseptically using compatible types and amounts of ingredients, and strict mixing rules must be followed to ensure that the mixture remains stable and free of large particulate matter. Automatic compounders are available that can be preprogrammed to deliver the correct proportion of ingredients in an appropriate sequence into a plastic bag to ensure that a stable and sterile admixture is prepared. In the United States, the compounding of admixtures is typically done immediately prior to their administration to avoid changes in stability that may occur with storage. European regulations allow for a storage period prior to administration. Increasingly, an in-line filter is used in the infusion of parenteral solutions as added safety against particulate matter that may have formed during the preparation stage as a result of incorrect mixing or storage or incompatible ingredients.

PERIPHERAL ADMINISTRATION. Parenteral nutrition may be delivered by infusion via a peripheral vein. Peripheral parenteral nutrition (PPN) is indicated for patients with a nonfunctioning gut who are not in catabolic stress and who require nutritional support for less than 2 weeks. The composition of the PPN admixture differs from that of TPN administered though a central line. The main complication of PPN is phlebitis, which can be minimized by infusing a solution with a low osmolarity. To achieve this, the proportion of lipid emulsion is increased to approximately 70% of non-protein calories, the relative proportions of glucose and amino acids are decreased, and the mixture is further diluted with water for injection.

Complications Associated with Lipid Emulsion Administration. Intravenous administration of lipid emulsions is well tolerated, and the incidence of adverse events is less than 0.1%.

Most complications arise when the rate of administration exceeds the rate of TAG clearance and hypertriglyceridemia ensues. For this reason lipid emulsions are contraindicated in the rare individuals with genetic or acquired conditions that predispose them to hyperlipemia.

Abnormality in liver function is one of the more common adverse events reported following TPN administration, and lipid emulsions have been implicated as part of this process, particularly in patients requiring prolonged support. Typically, fatty infiltration of the liver occurs when lipids are absent or are provided in excess of need [61]. Deficiencies of essential fatty acids [62] and of choline [56] as well as an excess of phospholipid-rich particles have been associated with liver disease.

Respiratory difficulties have been reported in some patients infused with lipid emulsions. This has been blamed on eicosanoids and other metabolites of FA acting via a vasoactive effect on pulmonary vasculature [63]. Lipid accumulation in lung tissue macrophages and changes in oxygen-diffusing capacity of red blood cells have also been implicated in the process [64]. Whether or not lipids or other factors are involved in pulmonary compromise remains controversial. Patients with pulmonary insufficiency may actually benefit from lipids as a source of energy because fat oxidation is associated with a lower rate of carbon dioxide production (and so a decreased need for elimination) than carbohydrate oxidation. If there is a negative effect of lipid administration on the lungs, it most likely only occurs with excessive dosing [65].

By accumulating in macrophages and other cells involved in the immune response and by modulating the inflammatory process through the elaboration of active metabolites such as eicosanoids, lipid emulsions consisting predominantly of LCT have also been associated with impairment of the immune system. These conclusions were drawn primarily from animal and early human studies. A review of more recent human trials proposes that other factors occurring concomitantly could play a more important role in immune dysfunction in patients sick enough to require TPN and that there is actually no evidence of a detrimental effect of lipids on the immune system [66].

Conflicting results have been published on the effect of lipid emulsions on the ability of blood to clot. However, it appears that in the absence of significant hyperlipemia there is no effect on platelet function or on the clotting cascade [67].

Polyunsaturated fatty acids (PUFA) are susceptible to damage by reactive oxygen species and may participate in a chain reaction process of lipid peroxidation [68]. Very long chain FA are particularly susceptible. When this process occurs in cell membrane phospholipids, cellular damage may result. To minimize this, micronutrients (vitamins A, C, and E and selenium) and enzymes (catalase, superoxide dismutase) act as antioxidants. Whether or not the administration of lipid emulsions containing PUFA contributes to the overall oxidative stress of sick patients with additional detrimental health effects is not known.

C. Future Trends

For over a decade, attempts have been made to manipulate the composition of lipid emulsions to improve disease outcomes without compromising the ability of the emulsions to provide a concentrated source of energy and adequate essential fatty acids to meet needs.

An early approach was to combine medium-chain and long-chain triacylglycerols. The aim (as detailed above) was to benefit from the specific properties of MCT, such as faster clearance from the bloodstream, more rapid oxidation, a lesser tendency to tissue incorporation, and improved protein sparing compared to LCT provided isocalorically [69,70]. The outcome was an emulsion composed of a 1:1 physical mixture of MCT and LCT (Table 4). This product has been marketed in Europe but is not currently available in the United States. A second mixture with a higher concentration of MCT than LCT (3:1) has been submitted for

U.S. approval. It is claimed that the higher proportion of MCT makes the emulsion better suited for critically ill patients with an enhanced need for more rapid lipid fuel utilization [71]. Whether or not this is the case has yet to be confirmed. Nevertheless, it has been demonstrated not only that the rate of release of total free FA by emulsion hydrolysis is greater for MCT than for LCT but that it also depends on the amount of MCT in the mixture [72].

Structured triglycerides are an alternative to the mixture of MCT and LCT. These triacylglycerols are the products of MCT and LCT oils that have been hydrolyzed and reesterified with random redistribution such that any one glycerol backbone contains a mixture of both medium-chain and long-chain FA [73]. Studies in animals and human beings have indicated that structured triglycerides may have clinical advantages over the MCT–LCT mixes [74]. The rate of hydrolysis of structured triglycerides is moderated with a less pronounced rise in plasma medium-chain FA and therefore in the subsequent risk of ketoacidosis and adverse neurological effects [24]. Overall, it would appear, for reasons that are not completely understood, that the administration of structured triglycerides, compared with that of LCT or a mixture of LCT and MCT, is associated with an improved protein metabolic profile [75,76]. It must be noted, however, that large, well-controlled comparative trials have yet to be published, so definitive conclusions on the relative merit of one type of emulsion versus another cannot be drawn at this time.

Attempts at defining the optimal proportion of essential FA to nonessential FA, of monounsaturated FA to polyunsaturated FA, and of *n*-6 series FA to *n*-3 series FA is ongoing, as is the role of long-chain ($>C_{20}$) PUFA and the benefits of supplementation of antioxidant relative to the content of PUFA. The physiological rationales for these approaches vary. The most notable is the attempt to modify the eicosanoid profile by changing the ratio of *n*-3 FA to *n*-6 FA in the lipid emulsion. Eicosanoids are important in moderating communication between cells and modulating the inflammatory response that coexists with disease. Metabolites of the *n*-6 FA series have been more closely associated with inflammation and immunosuppression than those of *n*-3 series FA, and this has been the rationale for inclusion of fish oil rich in *n*-3 series FA into lipid emulsions [77,78]. Unfortunately, convincing clinical trials have yet to confirm the early promising animal and human studies [79].

For a new emulsion to become widely accepted into medical practice, it will have to demonstrate, in a cost-effective manner, that its use results in a positive clinical outcome. Adequate and well-controlled trials will have to be designed and carried out in patients with a broad spectrum of diseases sharing a need for lipids. This is a formidable task given the heterogeneity of the patient population, the large sample size needed, the difficulties in defining accepted clinically relevant and economically appropriate endpoints, and the wide variety of possible types and concentrations of FA and TAG that are available to be tested.

ABBREVIATIONS

CNS	central nervous system
CT	computed tomography
EDTA	ethylenediaminetetraacetic acid
EFA	essential fatty acid
EFAD	essential fatty acid deficiency
FA	fatty acid
LCT	long-chain triacylglycerols
MCT	medium-chain triacylglycerols
PKC	protein kinase C

PL	phospholipids
PN	parenteral nutrition
PPN	peripheral parenteral nutrition
PUFA	polyunsaturated fatty acid
RES	reticuloendothelial system
TPN	total parenteral nutrition

REFERENCES

1. S. S. Davis, J. Hadgraft, and K. Palin, Medical and pharmaceutical applications of emulsions, in *Encyclopedia of Emulsion Technology* (P. Becher, Ed.), Marcel Dekker, New York, 1985, Vol. 2, pp. 159–238.
2. R. L. Juliano, Factors affecting the clearance kinetics and tissue distribution of liposomes, microspheres, and emulsions, *Adv. Drug Deliv. Rev. 2*: 31–54 (1988).
3. P. K. Hansrani, S. S. Davis, and M. J. Groves, The preparation and properties of sterile intravenous emulsions, *J. Parent Sci. Technol. 37*: 145–150 (1983).
4. M. Singh and L. J. Ravin, Parenteral emulsions as drug carrier systems, *J. Parenter. Sci. Technol. 40*: 34–41 (1986).
5. S. S. Davis, C. Washington, P. West, L. Illum, G. Liversidge, L. Sternson, and R. Kirsh, Lipid emulsions as drug delivery systems, *Ann. N.Y. Acad. Sci. 507*: 75–88 (1987).
6. L. C. Gold, R. T. Lyons, and L. C. Bartholow, Parenteral emulsions for drug delivery, *Adv. Drug Deliv. Rev. 5*: 189–208 (1990).
7. R. J. Prankerd and V. J. Stella, The use of oil-in-water emulsions for parenteral drug administration, *J. Parenteral Sci. Technol. 44*: 139–149 (1990).
8. J. T. Rubino, The influence of charged lipids on the flocculation and coalescence of oil-in-water emulsions, *J. Parenter. Sci. Technol. 44*: 210–215, 247–252 (1990).
9. G. Zografi, H. Schott, and J. Swarbrick, Disperse systems, in *Remington's Pharmaceutical Sciences* (A. Gennaro, Ed.), Mack, Easton, PA, 1990, Chap. 19, pp. 298–309.
10. B. J. Carroll, The stability of emulsions and mechanisms of emulsion breakdown, in *Surface and Colloid Science* (E. Matijevic, Ed.), Wiley, New York, 1976, pp. 1–67.
11. S. S. Davis, L. Illum, P. West, and M. Galloway, Studies on the fate of fat emulsions following intravenous administration to rabbits and the effect of added electrolyte, *Clin. Nutri. 6*: 13–19 (1987).
12. I. Hakansson, Physico-chemical changes in artificial fat emulsions during storage, *Acta Chem. Scand. 20*: 2267–2281 (1966).
13. I. M. A. Kemps and D. J. A. Crommelin, Chemical stability of phospholipids in pharmaceutical preparations. I. Hydrolysis of phospholipids in aqueous solutions; II. Peroxidation of phospholipids in aqueous solutions, *Pharm. Weekbl. 123*: 355–363, 457–469 (1988).
14. S. S. Davis, Eur. Patent EP 456106A (1991).
15. W. R. Perkins, S. R. Minchey, L. T. Boni, C. E. Swenson, M. C. Popescu, R. F. Pasternack, and A. S. Janoff, Amphotericin B–phospholipid interactions responsible for reduced mammalian cell toxicity, *Biochim. Biophys. Acta 1107*: 271–282 (1992).
16. M. R. Lance, C. Washington, and S. S. Davis, Structure and toxicity of amphotericin B/triglyceride emulsion formulations, *J. Antimicrob. Chemother. 36*: 119–128 (1995).
17. R. J. Prankerd, S. G. Frank, and V. J. Stella, Preliminary development and evaluation of a parenteral emulsion formulation of penclomedine: A novel, practically water insoluble cytotoxic agent, *J. Parenter. Sci. Technol. 42*: 76–81 (1988).
18. N. Mackenzie and I. S. Grant, Comparison of the new emulsion formulation of propofol with methohexitone and thiopentone for induction of anaesthesia in day cases, *Br. J. Anaesth. 57*: 725–731 (1985).
19. J. Dundee and R. Clarke, Propofol, *Eur. J. Anaesth. 6*: 5–22 (1989).
20. M. Paborji, C. Riley, and V. Stella, A novel use of intralipid for the parenteral delivery of perilla

ketone, an investigational cytotoxic drug with a high affinity for plastic, *Int. J. Pharm. 42*: 243–249 (1988).
21. R. Jeppsson and S. Rossner, The influence of emulsifying agents and of lipid soluble drugs on the fractional removal rate of lipid emulsions from the blood stream of the rabbit, *Acta Pharmacol. Toxicol. 37*: 134–144 (1975).
22. W. R. Burnham, P. K. Hansrani, C. E. Knott, J. A. Cook, and S. S. Davis, Stability of a fat emulsion-based intravenous feeding mixture, *Int. J. Pharm. 13*: 9 (1983).
23. K. Y. Warshawsky, Intravenous fat emulsions in clinical practice, *Nutr. Clin. Pract. 7*: 187–196 (1992).
24. J. M. Miles, M. Cattalini, F. W. Sharbrough, L. E. Wold, et al., Metabolic and neurologic effects of an intravenous medium-chain triglyceride emulsion, *J. Parenter. Enteral Nutr. 15*: 37–41 (1991).
25. D. L. Seidner, E. A. Mascioli, N. W. Istfan, et al., Effects of long-chain triglyceride emulsions on reticuloendothelial system function in humans, *J. Parenter. Enteral Nutr. 13*: 614–619 (1989).
26. D. Johansson, B. Bergenstahl, and E. Lundgren, Water-in-triglyceride emulsions. Effect of fat crystals on stability, *J. Am. Oil Chem. Soc. 72*: 939–949 (1995).
27. M. Groves, *Parenteral Technology Manual*, 2nd ed., Interpharm, Prairie View, IL, 1988, pp. 159–179.
28. E. Dickinson, Emulsions and droplet size control, in *Controlled Particle, Droplet, and Bubble Formation* (D. J. Wedlock, Ed.), Butterworth-Heinemann, Oxford, UK, 1994, pp. 190–216.
29. S. Benita and M. Y. Levy, Submicron emulsions as colloidal drug carriers for intravenous administration: comprehensive physicochemical characterization, *J. Pharm. Sci. 82*: 1069–1079 (1993).
30. R. Jeppsson and S. Ljungberg, Anticonvulsant activity in mice of diazepam in an emulsion formulation for intravenous administration, *Acta Pharmacol. Toxicol. 36*: 312–320 (1975).
31. R. Jeppsson, Plasma levels of diazepam in the dog and the rabbit after two different injection formulations, emulsion and solution, *J. Clin. Pharm. 1*: 181–187 (1976).
32. K. Tremper, *Perfluorochemical Oxygen Transport*, International Anesthesiology Clinics, Vol 23, Little, Brown, Boston, MA, 1985.
33. H. H. Wu, G. W. Dombi, and R. L. Poland, Parenteral delivery of vitamins with lipid emulsion, *Pediatr. Res. 27* (Part 2): 293A (1990).
34. Y. Mizushimna, A. Yanagawa, and K. Hoshi, Prostaglandin E_1 is more effective when incorporated in lipid microspheres for treatment of peripheral vascular diseases in man, *J. Pharm. Pharmacol. 35*: 666–667 (1983).
35. S. Hogskilde, J. W. Nielsen, P. Carl, and M. B. Sorensen, Pregnanolone emulsion, *Anaesthesia 42*: 586–590, 1045–1050 (1987).
36. Alliance Pharmaceutical Corp., San Diego, CA, *Annual Report*, 1993.
37. K. Ivancev, A. Lunderquist, R. McCuskey, P. McCuskey, and A. Wretland, Experimental investigation of a new iodinated lipid emulsion for computed tomography of the liver, *Acta Radiol. 30*: 1–7 (1989).
38. K. Ivancev, A. Lunderquist, R. McCuskey, P. McCuskey, and A. Wretland, Effect of intravenously injected iodinated lipid emulsion on the liver, *Acta Radiol. 30*: 291–298 (1989).
39. Sphinx Pharmaceuticals Corp., Durham, NC, *Annual Report*, 1992.
40. PDT Inc., Santa Barbara, CA, *Photodynamic Therapy Report*, 1993.
41. J. C. Fleishaker, G. R. Peters, K. S. Cathcart, and R. C. Steenwyk, Evaluation of the pharmacokinetics and tolerability of tirilazad mesylate, a 21-aminosteroid free radical scavenger, *J. Clin. Pharm. 33*: 182–190 (1993).
42. E. D. Hall, The role of oxygen radicals in traumatic injury: clinical implications, *J. Emerg. Med. 11*: 31–36 (1993).
43. A. W. Wilkinson, Historical background of intravenous feeding, *Nutr. Dieta 5*: 295–297 (1963).
44. V. Henriques and A. C. Andersen, Über parenterale Ernährung durch intravenöse Injektion, *Hoppe-Seylers Z. Physiol. Chem. 88*: 357–369 (1913).
45. J. M. McKibbin, A. Pope, S. Thayer, R. M. Ferry, Jr., and F. J. Stare, Parenteral nutrition: studies on fat emulsions for intravenous alimentation, *J. Lab. Clin. Med. 30*: 488–497 (1945).

46. H. L. Lehr, O. Rosenthal, H. M. Rawnsley, J. E. Rhoads, and M. B. Sen, Clinical experience with intravenous fat emulsions, *Metabolism 6*: 666–672 (1957).
47. L. M. Hansen, W. R. Hardie, and J. Hidalgo, Fat emulsion for intravenous administration: clinical experience with intralipid 10%, *Ann. Surg. 184*: 80–88 (1976).
48. E. B. Feldman, Nutrition and diet in the management of hyperlipidemia and atherosclerosis, in *Modern Nutrition in Health and Disease* (M. E. Shils, J. A. Olson, and M. Shike, Eds.), Lea and Febiger, (Philadelphia, 1994).
49. Y. A. Carpentier, Intravascular metabolism of fat emulsions: the Arvid Wretlind Lecture, *ESPEN, Clin. Nutr. 8*: 115–125 (1988).
50. D. Hallberg, Elimination of exogenous lipids from the blood stream. An experimental methodological and clinical study in dog and man, *Acta Physiol. Scand. 65* (Suppl. 259): 1 (1965).
51. D. Hallberg and J. Wersall, Electron-microscopic evaluation of chylomicrons and fat emulsions for intravenous use, *Acta Chir. Scand. 325* (Suppl): 23 (1964).
52. M. J. Ball and K. White, Metabolic effects of intravenous medium- and long-chain triacylglycerols in critically ill patients, *Clin. Sci. 76*: 165–170 (1989).
53. M. Hultin, C. Carneheim, K. Rosenqvist, and T. Olivecrona, Intravenous lipid emulsions, removal mechanisms as compared to chylomicrons, *Clin. Nutr. 14*: 2 (1995).
54. Y. A. Carpentier, M. Richelle, and J. Bury, Phospholipid excess of fat emulsion slows triglyceride removal and increases lipoprotein remodelling, *Arteriosclerosis 7*: 541a (1987).
55. C. Rojas, T. Olivecrona, and G. Bengtsson-Olivecrona, Comparison of the action of lipoprotein lipase on triacylglycerols and phospholipids when presented in mixed liposomes or in emulsion droplets, *Eur. J. Biochem. 197*: 315–321 (1991).
56. S. H. Zeisel, Phosphatidylcholine: the unappreciated nutrient in total parenteral nutrition, in *Organ Metabolism and Nutrition* (J. M. Kinney and H. N. Tucker, Eds.), Raven Press, New York City, 1994.
57. N. Sato, R. J. Deckelbaum, G. Neeser, Y. A. Carpentier, and J. M. Kinney, Hydrolysis of mixed lipid emulsions containing medium-chain and long-chain triacylglycerol with lipoprotein lipase in plasma-like medium, *JPEN 18*: 112–118 (1994).
58. J. M. Miles, M. Cattalini, F. W. Sharbrough, L. E. Wold, R. E. Wharen, J. E. Gerich, and M. W. Haymond, Metabolic and neurologic effects of an intravenous medium-chain triglyceride emulsion, *JPEN 15*: 37–41 (1991).
59. S. W. Coppack, M. D. Jensen, and J. M. Miles, The in vivo regulation of lipolysis in humans, *J. Lipid Res. 35*: 177–193 (1994).
60. R. T. Holman and S. B. Johnson, Essential fatty acid deficiency in man, in *Dietary Fats and Health* (E. G. Perkins and W. J. Visek, Eds.), Am. Oil Chem. Soc., Champaign, IL, 1983.
61. I. Tulikoura and K. Huikuri, Morphological fatty acid changes and function of the liver, serum free fatty acids and triglycerides during parenteral nutrition, *Scand. J. Gastroenterol. 17*: 177–185 (1982).
62. N. L. Keim and J. A. Mares-Perlman, Development of hepatic steatosis and essential fatty acid deficiency in rats with hypercaloric, fat free parenteral nutrition, *J. Nutr. 114*: 1807–1815 (1984).
63. T.-L. Hwang, S.-L. Huanng, and M.-F. Chen, Effects of intravenous fat emulsion on respiratory failure, *Chest 97*: 935–938 (1990).
64. D. B. Allardyce, The postmortem interval as a factor in fat embolism, *Arch. Pathol. 92*: 248–253 (1971).
65. D. Haumont, C. Rössle, and A. Cercx, Modifications of surfactant phospholipid pattern in premature infants treated with Curosurf: clinical and nutritional correlations, *Biol. Neonate 61* (Suppl.): 37–43 (1992).
66. J. Palmblad, Intravenous lipid emulsions and host defense—a critical review, *Clin. Nutr. 10*: 303–308 (1991).
67. V. C. Herson, C. Block, and L. Eisenfeld, Effect of intravenous fat emulsion on neonatal neutrophil and platelet function, *JPEN 13*: 620–622 (1989).
68. M. Logani and R. Davies, Lipid oxidation. Biologic effects and antioxidants: a review, *Lipids 15*: 485–495 (1980).

69. A. C. Bach, D. Storck, and Z. Meraihi, Medium-chain triglyceride based fat emulsions: an alternative energy supply in stress and sepsis, *JPEN 12*: 835–845 (1988).
70. Z. Jiang, S. Zhang, and X. Wang, A comparison of medium-chain and long-chain triglycerides in surgical patients, *Ann. Surg. 217*: 175–184 (1993).
71. M. Jeevanandam, N. J. Holaday, T. Voss, R. Buier, and S. R. Petersen, Efficacy of a mixture of medium-chain triglyceride (75%) and long-chain triglyceride (25%) fat emulsion in the nutritional management of multiple-trauma patients, *Nutrition 11*: 275–284 (1995).
72. N. Sato, R. J. Deckelbaum, G. Neeser, Y. A. Carpentier, and J. M. Kinney, Hydrolysis of mixed lipid emulsions containing medium-chain and long-chain triacylglycerol with lipoprotein lipase in plasma-like medium, *JPEN 18*: 112–118 (1994).
73. R. Sandstrom, A. Hyltander, U. Korner, and K. Lundholm, Structured triglycerides to postoperative patients: a safety and tolerance study, *JPEN 17*: 153–157 (1993).
74. A. Hyltander, R. Sandstrom, and K. Lundholm, Metabolic effects of structured triglycerides, *Nutrition in Clinical Practice 10*: 91–97 (1995).
75. K. Magnusson, G. Sandberg, and H. Ronnholm, Effect of structured triglycerides on protein synthesis in liver and muscle tissue during TPN in an injured animal model, *Clin. Nutr. 5* (Suppl.): 141 (1986).
76. K. T. Mok, A. Maiz, and K. Yamazaki, Structured medium-chain and long-chain triglyceride emulsions are superior to physical mixtures in sparing body protein in the burned rat, *Metabolism 33*: 910–915 (1984).
77. M. M. Gottschlich, Selection of optimal lipid sources in enteral and parenteral nutrition, *Nutrition in Clinical Practice 7*: 152–165 (1992).
78. C. H. Simoens, M. Richelle, C. Rossle, M. Derluyn, R. J. Deckelbaum, and Y. A. Carpentier, Manipulation of tissue fatty acid profile by intravenous lipids in dogs, *Clin. Nutr. 14*: 177–185 (1995).
79. M. Katan, Fish and heart disease (editorial), *N. Engl. J. Med. 332*: 1024–1025 (1995).

22

The Role of Lipids in Animal Feeds

Julian Wiseman and Phil C. Garnsworthy
University of Nottingham, Sutton Bonington Campus, Loughborough, England

I. INTRODUCTION

Modern high performing genotypes of farm livestock (particularly pigs, poultry, and dairy cattle) have greater nutrient and dietary energy requirements than former strains. Fats and oils have approximately twice the dietary energy-yielding capacity of carbohydrates. They may contain essential fatty acids and fat-soluble vitamins, their physical texture reduces dust in feed mills, and they promote palatability of diets (which is of particular importance in the pet food industry, where diet acceptance is of major importance in marketing and the value of fats and oils as energy-yielding ingredients is perhaps of less concern). These materials, accordingly, have assumed considerable importance as ingredients in compound animal feeds in recent years although it is perhaps pertinent to mention that, with the exception of essential fatty acids, there is no obligatory requirement to include fats and oils in diets. Rather, it is their ability to provide dietary energy that dictates whether they are employed.

It is the purpose of this review to consider the role of fats and oils in the context of their nutritional value and their influence on the quality of animal products (for example, carcass and milk quality). Decisions on whether fats and oils will be incorporated into animal diets will also be based on other factors including their cost and the presence of appropriate milling technology, which often limits the actual amounts that may be used. Thus fats and oils employed in animal feeding are frequently blends of a number of individual ingredients providing a final mixture with a melting point in the region of 40–50°C. Storage of blends is usually within this temperature range, requiring specialized facilities (which should be stainless steel or polymer-based) and they are usually added in the liquid state to other dietary ingredients. Most compound feeds are pelleted subsequent to mixing, and this process is not effective if added fat levels are excessive (beyond approximately 40 g fat/kg diet). Addition of further amounts of fat, which is frequently cost-effective, is through sophisticated liquid fat spraying equipment after pelleting.

To overcome these technological problems, so-called dry fat products are becoming increasingly available; these include fat and oil blends absorbed onto a solid carrier, spray-dried products, and calcium soaps of fatty acids (of value only to ruminants). Finally, the use of oilseeds prior to oil extraction (e.g., soybeans) is a well-established means of adding oil to animal diets, a process that has the additional advantage of simultaneously providing other nutrients (e.g., amino acids).

The nutritional value of any dietary raw material is governed primarily by its chemical composition (i.e., its ability to provide energy-yielding compounds and specific nutrients) and the degree to which it is digested by the animal. In terms of the latter, nonruminants and ruminants are very different and accordingly are considered separately in terms of both nutritional value and product quality.

II. UTILIZATION OF FATS AND OILS BY NONRUMINANTS

The detailed chemical structure of fats and oils together with their major sources have been considered earlier in this book (Chapters 1 and 2). In addition, the various processes to which fats and oils are subjected have been described (Chapters 4–9). The relevance of this to the nutrition of nonruminants is that fats and oils available for incorporation into animal diets are of diverse origin and chemical composition—the animal feed industry is frequently the recipient of by-products from other processes. A wide range of commodities are thus available, including crude vegetable oils, soapstocks, hydrogenated materials, rendered animal tallows, recovered vegetable oils from human food production, and even products from the refining of oils such as bleaching earths, which are sometimes available as dry fat products.

Although all these materials (with the exception of the last category) probably have similar total fat contents approaching 1000 g/kg, there is considerable variability in chemical composition, which has a pronounced influence on their digestibility. This topic has been the subject of a number of previous reviews, which have considered both the physiological basis for the digestion and absorption of fats and the factors that are responsible for the large differences in their subsequent dietary energy values [e.g., 1–4].

A. Influence of Chemical Structure of Fats and Oils on Digestibility and Dietary Energy Value

It is customary in most countries to express dietary energy values of compound diets or raw materials for poultry and pigs in terms of their apparent metabolizable energy (AME) and digestible energy (DE) value, respectively. The SI units for AME and DE are MJ/kg, although the dietary calorie (kcal or Cal) is still found, as is the pound (lb) as the unit of mass. Descriptions of the detailed methodologies employed in deriving AME and DE values are outside the scope of this chapter but are considered elsewhere [5].

There is considerable confusion relating to the AME and DE values of fats, primarily because descriptions of the products evaluated are confined to names and origins with no accompanying chemical characterization of a more precise nature. Systematic studies on the influence of chemical composition of fats and oils on dietary energy values for nonruminants, in terms of the quantitative contribution of the variables involved, are limited. Thus tables of nutritional value of fats and oils are limited to descriptions such as "beef tallow" and "vegetable oil." The need for greater precision in defining source materials in a chemical rather than descriptive manner is of particular importance to animal feeding as fats and oils are rarely fed as a single commodity but rather as blends of a variety of materials. The two variables of most importance are the degree of saturation and content of free fatty acids, with chain length of the constituent fatty acids being of secondary concern.

1. *Degree of Saturation and Free Fatty Acid Content*

Physiologically, the mechanisms of fat digestion and absorption in nonruminants are well documented [2]. The major site of fat digestion in pigs and poultry is the duodenum, and basically it consists of emulsification of dietary fat by conjugated bile salts, followed by hydrolysis of triglycerides by pancreatic lipase into mixtures consisting essentially of 2-monoglycerides and free fatty acids. The subsequent absorbability of these products is dependent on their solubility in bile salt micelles. Polar solutes are more readily incorporated into micelles, and this explains the relatively higher absorbabilities of unsaturated fatty acids compared to saturated fatty acids and the well-established observation that unsaturated fatty acids are more digestible than those that are saturated [e.g., 6]. Accordingly, oils have a higher dietary energy value than the more saturated fats; this also explains why hydrogenation of oils (even partial) is associated with a reduction in dietary energy value.

However, this is an overly qualitative description, and to be of any value in assigning dietary energy values to the wide range of commodities available, a quantitative measurement of the degree of saturation is essential. Although reference has been made to the digestion of individual fatty acids, some measure of the overall fat or oil is more important. What has been used [5,7] is the ratio of unsaturated (U) to saturated (S) fatty acids (giving U/S). Increasing the degree of unsaturation of a fat by mixing a saturated fat with an unsaturated oil, which is associated with higher U/S ratios, appears to be associated with a nonlinear improvement in dietary energy value [7].

The relative superiority of an intact triglyceride compared to hydrolyzed fat in terms of dietary energy value is also well known [e.g., 8,9]. Increasing the proportion of FFA would appear to be associated with a linear reduction in fat digestibility [1]. The FFA content is expressed in terms of grams per kilogram of fat.

Systematic studies on the influence of U/S ratio and FFA content of fats and oils on AME and DE values have recently been concluded [10–14]. The approach adopted was to select raw materials of known U/S and FFA content that covered a range sufficiently wide that they represented commodities that would be employed in the feeding of nonruminants, and all blends were evaluated for AME and DE content. Chain length of constituent fatty acids was predominantly in the range of 16–20 carbon atoms. Data generated confirmed that the response of AME and DE to the U/S ratio was curvilinear (Fig. 1), with the greatest improve-

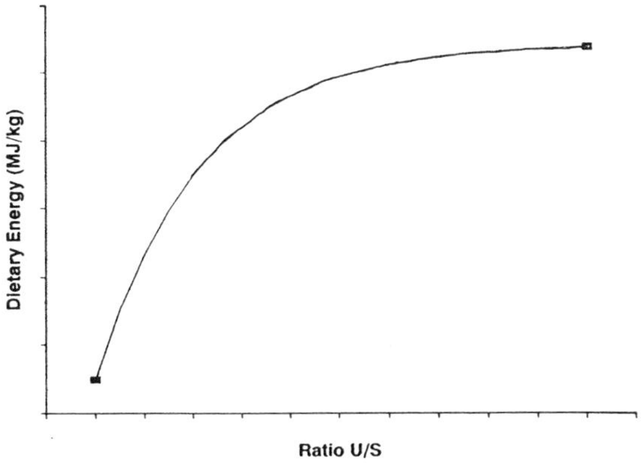

Fig. 1 Response of dietary energy of fats and oils to increasing degree of unsaturation, as measured by the ratio of unsaturated (U) to saturated (S) fatty acids (U/S).

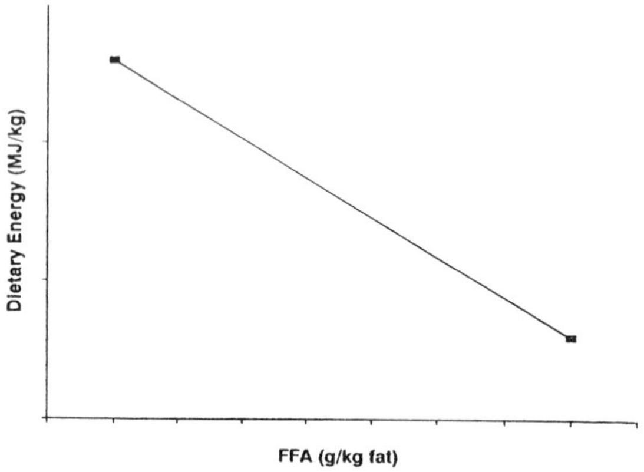

Fig. 2 Response of dietary energy of fats and oils to increasing free fatty acid content.

ment in dietary energy value occurring over the lower range of increase in U/S. The response of dietary energy to FFA was linear (Fig. 2).

Nutritional evaluation through biological experimentation is, of necessity, a lengthy and costly procedure. Accordingly there have been many studies attempting to predict dietary energy value from chemical composition. Having identified the two major chemical components that have nutritional relevance and generated AME and DE values, regression analysis

Table 1 Prediction Equations Relating the Apparent Metabolizable Energy (AME, MJ/kg—Poultry) and Digestible Energy (DE, MJ/kg—Pigs) to Ratio of Unsaturated to Saturated Fatty Acids (U/S) and Free Fatty Acid Content (FFA, g/kg Fat) of Fats and Oils[a,b]

AME or DE (MJ/kg fat) $= A + B \times \text{FFA} + C \times e^{(D \times \text{U/S})}$

a. Poultry (AME)

Constant	Age 1 (PV = 0.816)	Age 2 (PV = 0.925)
A	38.112 ± 1.418	39.025 ± 0.557
B	−0.009 ± 0.002	−0.006 ± 0.001
C	−15.337 ± 2.636	−8.505 ± 0.746
D	−0.506 ± 0.186	−0.403 ± 0.088

b. Pigs (DE)

Constant	Age 1 (PV = 0.802)	Age 2 (PV = 0.768)
A	37.890 ± 1.690	36.898 ± 0.501
B	−0.005 ± 0.002	−0.005 ± 0.001
C	−8.200 ± 1.750	−7.330 ± 2.700
D	−0.515 ± 0.376	−0.906 ± 0.452

[a] Age 1 and 2 in poultry refer to 1.5 and 7.5 weeks of age, respectively, and to 10–20 kg and 35–85 kg live weight in pigs, respectively.
[b] PV = proportion in variance of dependent variables accounted for by function.

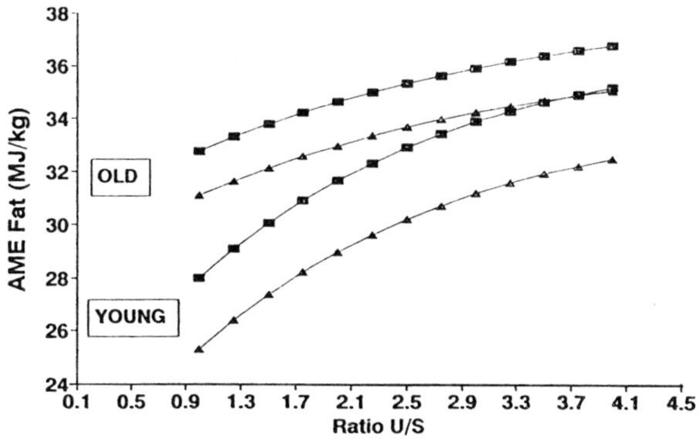

Fig. 3 Prediction of the apparent metabolizable energy (AME) of fats for poultry as influenced by age, ratio of unsaturated (U) to saturated (S) fatty acids (U/S), and FFA content (employing two levels—100 and 400 g/kg fat). Young and old birds aged 1.5 and 7.5 weeks.

Table 2 Mean Chemical Analyses of Coconut–Palmkernel Oil Blend (CP) and Its Acid Oil (CPAO)—All Data Expressed as g/kg Oil

Fatty acid profile[a]	CP	CPAO
8:0	95	42
10:0	71	41
12:0	494	462
14:0	171	170
16:0	78	101
18:0	24	26
16:1	0	0
18:1	53	129
18:2	12	21
18:3	0	0
20+	1	7
U/S[b]	0.07[c]	0.19
	2.66[d]	2.36
	8.78[e]	6.86
Free fatty acids (g/kg fat)	13.8	839.0

[a]Notation indicates length of carbon chain followed by number of double bonds.
[b]Ratio of unsaturated (U) to saturated (S) fatty acids. Calculation of U/S ratio:
[c]All saturated fatty acids included in "saturated" fraction.
[d]Only 14:0, 16:0, and 18:0 appearing in "saturated" fraction.
[e]Only 16:0 and 18:0 appearing in "saturated" fraction.

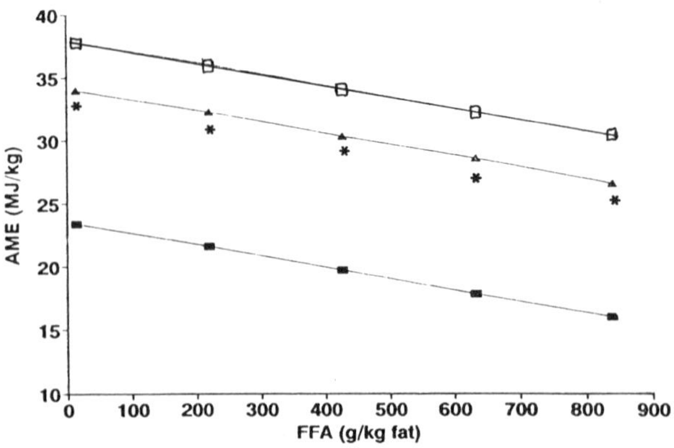

Fig. 4 Prediction of the apparent metabolizable energy (AME) of fats for poultry as influenced by ratio of unsaturated (U) to saturated (S) fatty acids (U/S) using the linear responses ratios described in Table 2 and FFA content compared with data (as indicated by the *) generated for mixtures of coconut or palmkernel oil and the respective acid oil.

was undertaken to relate dietary energy value (dependent variable) to both U/S and FFA content (independent variables).

Those functions combining both independent variables are presented in Table 1. Separate functions were derived for both "young" and "old" animals, it having been established that the ability to digest fats improves with age [15–17]. As an example of the influence of both U/S and FFA together with age, solutions to functions, for poultry, are presented in Fig. 3. These functions do represent a considerable improvement, in terms of accuracy of prediction of AME and DE, over those based on rather more empirical approaches. An example of the latter would be prediction equations employing iodine value as the independent variable. Thus the iodine values for soybean oil and rapeseed oil, for example, would be different (the former would be higher), whereas in terms of dietary energy value the two commodities would be very similar.

The functions described in Table 1 are, however, not without their problems. Thus it may not be appropriate to consider applying the functions derived to those fats containing saturated fatty acids of chain length shorter than those of the current study (i.e., below C_{16}), as these may be associated with improved digestibility [6]. This problem has been studied [18] where a combination of coconut and palmkernel oils (together with the respective acid oil and with mixtures of the two to give blends of intermediary free fatty acid content) was evaluated. Both commodities consist predominantly of saturated fatty acids, but of chain lengths shorter than C_{14} (Table 2).

The AME data generated are presented in Fig. 4, which also illustrates the responses as predicted employing the functions in Table 1 for the three U/S ratios. It is evident that content of saturated fatty acids used to calculate the U/S ratio should be based on the sum of myristic (14:0), palmitic (16:0), and stearic (18:0) acids, but that lauric acid (12:0) behaves like an unsaturated fatty acid.

2. *Heat-Damaged Fats and Oils*

As described above, fats and oils destined for use in animal feeding are usually by-products of other industrial processes. As such it is likely that they have undergone some form of heat treatment, often in the presence of oxygen. Fats and oils are generally relatively unstable (the more so the greater the degree of unsaturation) and are therefore prone to some form of degradation under a variety of conditions (which explains why protection of fats and oils through the use of antioxidants, added as early as possible in the manufacturing process because oxidative degeneration is irreversible, is crucial for animal feeding). This has prompted numerous studies on the chemical commodities produced during heating and the nutritional implications (see, e.g., the reviews in Refs. 19–21).

A large number of modifications to the chemical structure of fats and oils following heating have been identified, ranging from simple oxidation products through to dimerization and polymerization (linear and cyclic) of both fatty acids and triglycerides depending on the substrate in question and the conditions operating.

The biological effects of feeding these modified structures are also extremely varied in terms of both the actual response in the animal and its severity. It should be noted that even minor adverse biological consequences would have serious repercussions for output from intensive livestock enterprises. Initially, digestibility and hence dietary energy value would be reduced. The consequences would be that the animals would not perform to expectations and also that an increasing amount of dietary fat would pass through the gastrointestinal tract and be excreted. This may have serious implications for the conditions under which animals and birds are maintained. Thus, for example, the litter in housed broiler chickens, if allowed to deteriorate through becoming "greasy," will give rise to a poorer environment and problems of both bird welfare and product quality. It is therefore possible that the presence of modified fat structures within the gastrointestinal tract may interfere with the overall digestive process such that general nutrient uptake is impaired. Thus there may be overall nutrient deficiency. Furthermore, an actively oxidizing fat or oil will destroy other nutrients present, including some vitamins.

Perhaps more concern has been expressed over whether any toxic products are generated following fat and oil heating or oxidation. It does appear, however, that the majority of these products are only sparingly absorbed and would thus not present a threat to the metabolic process. However, in the case of oxidized fats and oils, defense mechanisms in the gut mucosae to prevent absorption may be stretched such that overall nutrient absorption might be reduced. It is certainly the case that death in laboratory animals fed heated fats and oils has been recorded [22], but these are extreme cases.

Because of the potential adverse effects of feeding heat-damaged/oxidized fats and oils, there has been considerable interest in developing chemical methods for the detection of such damage. It is important to note that any method adopted has to be one that measures all products collectively if it is to have any practical application. Peroxide value (PV) has been employed widely for this purpose, but it is an unsound method. In tracing the change in PV over time [23], an increase followed by a reduction was observed. Thus a low PV value may indicate on the one hand a commodity that had not undergone any degradation or, on the other hand, one that had been seriously denatured. Measurement of "oxidized fat" has been employed, but the evaluation is solvent-dependent and would not measure complexes that are insoluble in polar solvents.

Free fatty acid content has been employed frequently to assess damage to oils used in the human food industry but is inappropriate for fats and oils for animal feeding. This is because soapstocks, for example, are perfectly acceptable ingredients for blends (although of lower energy value than the original triglyceride) even though the free fatty acid content is invariably high. This also explains why assessments of molecular weights or sizes are inappropriate. Thus a fatty acid trimer (of no dietary energy-yielding value) would generate data similar to those of a triglyceride (of high value).

A technique that has found favor is one based on estimating the total nonelutable material (NEM) of a fat or oil through quantitative gas-liquid chromatography [21,24]. Although this method merely measures collectively most degraded structures within fat or oil, it does at

Table 3 Mean Chemical Analyses of Refined (RS) and Sunflower Acid Oil (SAO)—All Data Expressed as g/kg Oil

Fatty acid profile[a]	RS	SAO
14:0	0.7	1.4
16:0	69.3	82.4
17:0	2.0	2.7
18:0	44.7	47.9
16:1	1.1	2.0
18:1	196.1	190.5
18:2	651.1	619.6
18:3	1.3	2.0
20+	16.4	25.3
Other	17.3	26.2
U/S[b]	7.32	6.25
Nonelutable material	41.0	136.6
Free fatty acids	0.0	388.0
Water	2.0	8.2
Unsaponifiable material	0.6	82.0
Oxidized fatty acids	0.4	14.8
Impurities	0.01	0.1

[a]Notation indicates length of carbon chain followed by number of double bonds.
[b]Ratio of unsaturated (16:1 + 18:1 + 18:2 + 18:3) to saturated (16:0 + 18:0) fatty acids.

least provide guidance as to whether the commodity has been damaged and has proved useful in identifying those commodities (e.g., some recovered vegetable oils from frying operations) that are liable to have been excessively heated.

In a trial designed to examine the reduction in dietary energy value likely to result from damage [25], a refined sunflower oil was extensively heat damaged (Table 3). Chemical analysis revealed an increase in FFA and NEM content following such treatment but little difference in the proportion of individual fatty acids. The two commodities (together with mixtures of the two) were evaluated for AME, and data generated were compared with predicted values based on the regression equations described above and in Table 1, which employed FFA and U/S. From analysis of the data it was evident that the dietary energy value of the NEM fraction, in this material, was on the order of zero. This indicates the problems identified with heat-damaged fats, although no account was taken of associated issues of the presence of the NEM fraction (e.g., reduction in general nutrient uptake).

3. *Contaminants Within Fats and Oils and Naturally Occurring Fatty Acids with Nonnutritional Activity*

In addition to the products arising from heat damage and oxidation, fats and oils may contain contaminants that are fat-soluble. Perhaps the most important group of contaminants are pesticides. Not only are they potentially damaging to the animal itself, residues within the product are also of concern to the human consumer. Furthermore, many of these commodities are rendered even more dangerous following the action of heat [e.g., 26]. Levels of many such contaminants within commodities destined for use in animal feeds are strictly controlled. Other contaminants found in fats and oils include unsaponifiable matter (plant waxes) that acts as a diluent, water (also acting as a diluent but also having a role in the oxidation process), and polythene (which would compromise fat-spraying equipment). Effective quality control procedures should evaluate fats and oils for these products. Finally it is of interest to note that, in any process designed to purify a fat or oil, the materials employed in such purification will themselves become contaminated—this is of particular relevance to the products employed in the animal feed industry.

There are numerous examples of naturally occurring fatty acids that have been associated with antinutritional activity. Erucic acid (22:1) is an important component of some varieties of rapeseed, and it is alleged that it might be responsible for fatty acid infiltration of heart muscle [e.g., 27]. While this is a contentious issue (a fatty acid imbalance in rapeseed oil rather than erucic acid per se was thought responsible [27], it has resulted in a major plant breeding program designed to reduce the levels of erucic acid in rapeseed cultivars.

Finally the cyclopropenoid fatty acids found in, for example, cottonseed oil have been implicated in the inhibition of fatty acid metabolism [28], which led, ultimately, to discoloration of egg yolks [29].

B. Effect of Dietary Fats and Oils on Meat Quality

An important feature of modern livestock production is increasing interest in quality of product, and this section proposes to consider the role of dietary lipids in the meat quality of nonruminants. Carcass fatty acids in pigs and poultry can arise from two discrete sources—de novo synthesis or direct deposition from dietary sources. The latter is the route by which polyunsaturated fatty acids (principally linoleic acid, which is nutritionally essential for nonruminants) appear in the carcass.

The degree to which dietary factors influence the fatty acid profile of carcass fat is controlled by a number of factors, including the actual fatty acid profile of the dietary fat and the energy status of the diet (direct deposition of dietary fat is more likely the higher the overall dietary energy level). The basic effects of dietary fat have been recognized for some considerable time, and it was demonstrated [30,31] that diets based on corn and soybean oil have a greater softening effect than peanut and rice oils. Carcass linoleic acid levels, specifically, have been identified as perhaps the major determinant of the degree of softness of carcass fat. Investigations into the role of linoleic acid have led to recommendations that concentrations above 150 g/kg fatty acid in body fat are to be avoided if excessive softness of backfat is to be prevented [32].

1. Effect on Physical Texture and on Keeping and Eating Quality

The increased use of diets with high AME and DE levels (particularly as it has been associated with including significant amounts of relatively unsaturated dietary fats) has been implicated in promoting softer carcass fat. There is an additional problem with pigs because, with the progressive move to leaner carcasses, what adipose tissue is left has a higher concentration of unsaturated fatty acids. The consequences of this are significant for a number of reasons. Meat from pig carcasses with softer fat does not "set" (it is floppy), and the tissues separate easily, making slicing and packaging for retail sales difficult [33].

In addition to softness, unsaturated carcass fat in both pig and poultry meat is accompanied by a risk of a negative effect on the keeping quality of the carcass because of the risk of oxidative breakdown of unsaturated fatty acids resulting in development of peroxides and rancidity [34]. This is because oxidation products, being volatile, give rise to off-odors that will reduce the shelf life of meat. However, the evidence that carcass fatty acid profiles influence the eating quality of meat is less conclusive. In a trial conducted at the University of Nottingham (unpublished), a range of 67.8–203.6 g linoleic acid per kilogram of carcass fatty acids in the backfat (subcutaneous fat) was produced, with the highest level therefore extending well beyond that at which it is suggested that problems of meat quality would become apparent. However, influences on the subsequent eating quality of the meat were not evident although tests were conducted on fresh samples, and it is possible that storage characteristics might alter depending on linoleic acid levels in backfat. A subsequent trial (unpublished) examining linolenic acid levels in backfat did establish a negative correlation between levels in the adipose tissue of pigs and eating quality. It would appear that the more unstable the fatty acid, the greater the risks. Certainly, feeding oils with high levels of long-chain polyunsaturated fatty acids (e.g., fish oil) is not advisable.

Although current recommendations appear to suggest that the polyunsaturated fatty acid content of pig and poultry adipose tissue should be limited (on both textural and eating quality grounds), an opposing viewpoint is occasionally put forward by human nutritionists. Thus a British government report [35], in recommending that total dietary fat should represent no more than 33% of total energy intake of the national diet, also proposed that the proportion of this total fat that was saturated should be no more than one-third. Recommendations were also established in connection with unsaturated fatty acid intake, which represents an increasing awareness of the importance of different classes of unsaturated fatty acids in human nutrition. The proposed cis-monounsaturated/cis-polyunsaturated fatty acid ratio was 2:1, and furthermore, the ratio of *cis*-α-linoleic acid (of the n-6 or ω-6 essential fatty acid family) to *cis*-α-linolenic acid (n-3 or ω-3 essential fatty acid family) was 5:1.

These recommendations apply to the total diet and should not necessarily be used to define optimum fatty acid ratios for individual food items in the human diet. Nevertheless, there is growing interest in modifying the fatty acid profiles of animal products, which is comparatively easy in pigs and poultry, because of the somewhat negative reputation that these products have of being high in saturated fatty acids, which has been implicated in consumer avoidance of these food items. It is possible that production of animal commodities that are aligned better to the "optimum" would have a significant impact on consumer acceptance of them. Nevertheless, such a development should not proceed unless there is clear evidence that such modifications to carcass adipose tissue levels are not accompanied by deterioration in meat quality as defined by technological and organoleptic criteria. In this context the protective function of dietary α-tocopherol acetate has recently been studied [36], and it was considered that although it would not improve performance, it had an important effect on meat stability. Such investigations are proceeding at the University of Nottingham.

3. *Rate of Change of Carcass Adipose Tissue Fatty Acids*

Problems of unsaturated fatty acids within adipose tissue need to be placed alongside the higher dietary energy values associated with more unsaturated fat blends. Such seemingly mutually exclusive objectives might be more easily reconciled if the speed with which carcass fatty acid profiles respond to changes in dietary levels could be established. The more rapid these changes, the later the introduction of the more saturated lower dietary energy fat. The responsiveness of carcass fat to dietary fat has been reported [37], where, feeding pigs a diet based on 100 g tallow/kg following 5 weeks on a diet containing 100 g safflower oil/kg, it was possible to restore the fatty acid pattern of safflower oil–fed pigs to that of the control animals. Data indicated that the major changes in fatty acid composition due to diet occurred

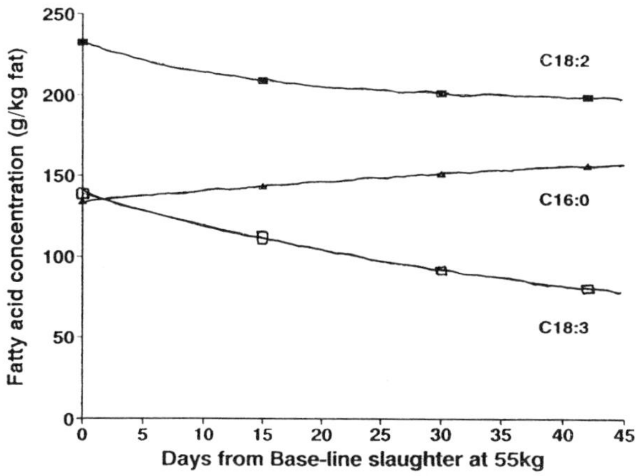

Fig. 5 Rate of change of pig adipose tissue fatty acids (for linoleic, C18:2; palmitic, C16:0; and linolenic, C18:3 acids) following a dietary change from rapeseed oil to tallow at 55 kg live weight.

within 4–5 weeks. It seems reasonable, therefore, that carcass quality of pigs can be altered through dietary fat change during the growing and finishing period, although the response remains to be adequately quantified.

A further report [38] presented data from a trial where pigs had been fed diets based on soybean oil (S, 60 g/kg), rapeseed oil (R, 60 g/kg), or tallow (T, 65.5 g/kg) while their weight increased from 25 to 55 kg. Thereafter the animals were either maintained on the same diet or transferred to either of the other two diets, giving nine treatment combinations in total. The pattern of change of fatty acid concentration in adipose tissue for one of the experimental treatments is given in Fig. 5, from which it is evident that the bulk of the change in carcass fatty acid profile takes place within approximately 25 days of a dietary change. Similar studies with poultry (University of Nottingham, unpublished) employing diets based on tallow and safflower oil indicate responses comparable to those of pigs, although somewhat more rapid (Fig. 6).

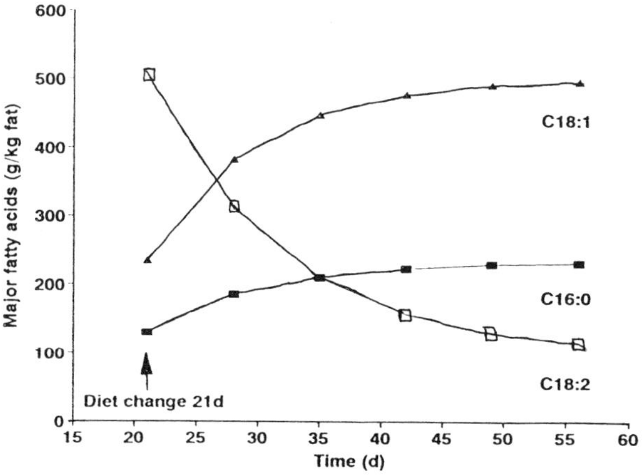

Fig. 6 Rate of change of poultry abdominal fat pad tissue fatty acids (for oleic, C18:1; palmitic, C16:0; and linoleic, C18:2 acids) following a dietary change from safflower oil to tallow at 21 days of age.

III. UTILIZATION OF FATS AND OILS BY RUMINANTS

Ruminants have evolved an extremely efficient digestive system, principally for using the poor quality forages found in their natural diet. The rumen is a large fermentation chamber containing billions of microorganisms that break down plant material into volatile fatty acids (principally acetate, propionate, and butyrate) that can be used by the host animal as energy sources. During this process, microorganisms grow and multiply, converting nitrogenous substances into high quality microbial protein. This protein is made available to the animal when the microorganisms pass out of the rumen for digestion in the small intestine. Although moderate levels of performance can be achieved from diets containing only grass or forage, high producing animals require additional energy and protein from supplementary sources. This is particularly true for dairy cows in early lactation, whose appetite is usually reduced to such an extent that the cow cannot eat enough to satisfy her nutrient requirements for milk production. Traditional sources of supplementary energy include starchy materials such as cereals, or fibrous materials such as sugar beet pulp, and other food processing by-products. The proportion of starchy materials in the diet must be limited because they are digested by different sorts of bacteria than forages. High starch diets can result in rapid fermentation in the rumen, and the low pH induced may inhibit forage-digesting bacteria, which are less tolerant of low pH conditions than amylolytic microorganisms. This results in lower feed intake and reductions in the butterfat content of the milk. For diets with very high energy concentrations it is necessary to use fats.

A. Rumen Fermentation, Digestion, and Absorption of Fats

Fats have a gross energy content about twice that of grass and cereals. In ruminants they are absorbed from the intestine and used very efficiently (Table 4), and their digestibility is little affected by chain length and degree of unsaturation.

The fat content of natural diets for ruminants is less than 50 g/kg. However, if added fat increases the total fat content of the diet to more than about 100 g/kg, digestive problems can occur. Rumen microorganisms cannot use large quantities of fats, and long-chain unsaturated fatty acids have a detergent effect on bacterial cell walls [43]. Under normal circumstances, the ester linkages of triacylglycerides are rapidly hydrolyzed by bacterial lipases in the rumen. Once released from ester combination, unsaturated fatty acids are subsequently hydrogenated to detoxify them [44]. Fats also have a physical effect on fiber in the rumen; fiber particles become coated with fat, rendering them inaccessible to microbial attack [45]. The magnitude of these effects depends on level, source, and type of fat; dietary carbohydrate source; and feed intake. A further problem is that long-chain free fatty acids can form soaps with calcium

Table 4 Gross Energy and Metabolizable Energy of Selected Feeds for Ruminants (MJ/Kg Dm)

Feed	Gross energy	Metabolizable energy	Net energy for milk production
Grass	18.7	11.2	6.9
Grass silage	20.9	11.0	6.8
Barley grain	18.5	13.3	8.1
Fat prills	39.0	33.0	26.4
Tallow	39.3	30.5	24.4

Sources: Refs. 39–42.

and magnesium in the rumen. This will detoxify the fatty acids, but it can also reduce the availability of the minerals.

Unlike the situation for nonruminants, it is not usually possible to alter the fatty acid composition of carcass fats by altering the fatty acid profile of the diet. This is because hydrogenation in the rumen results in the predominance of stearic acid. Rumen microorganisms do synthesize shorter chain fatty acids, so there is some scope for altering carcass or milk fatty acids by dietary manipulation, but for unsaturated fatty acids to be directly incorporated into milk and meat it is necessary to protect them against rumen modification.

1. Protected Fats

Various products have been developed that are referred to as "protected fats." This is really a misnomer, as the main requirement is to protect the rumen from the detrimental effects of fat, rather than to protect the fat from rumen degradation. However, protected fats do offer the potential for increasing the absorption of unsaturated fatty acids from the small intestine.

The three methods of protection that have received the most attention are formaldehyde protection, fat prills, and calcium soaps. Formaldehyde protection involves spray drying an emulsion of fat and casein that has been treated with formaldehyde to render it undegradable in the rumen. Fat prills rely on selected fatty acids that have a high melting point, thus remaining solid at rumen temperatures, and small particle size, thus passing through the rumen quickly. Calcium soaps are produced by attaching calcium to the carboxyl groups of fatty acids, rendering them inert in the rumen; the calcium is detached when the fat reaches the low pH environment of the abomasum. The cost of formaldehyde-treated casein makes the first method of protection very expensive. The majority of protected fats used are in the form of calcium soaps or prills.

In addition to manufactured protected fats, there are certain naturally occurring feed materials that have a high oil content but do not cause digestive problems, even when the quantity of added fat is up to 0.5 kg/day for dairy cows. These feeds are oilseeds, such as soybeans, rapeseed, sunflower seed, and cottonseed. They must be fed in the full fat form, although they can be heat-treated and rolled or ground. Extracted oil from these products can be used in only small quantities before rumen disturbances occur. In the unextracted form, rumen bacteria digest the fiber and protein of the oilseeds slowly and therefore release the oil at a rate that is compatible with normal rumen function. Cottonseed is often fed with the lint still attached to the seed. This provides plenty of attachment points for rumen bacteria and increases digestibility. Soybeans contain trypsin inhibitors that may interfere with intestinal digestion of protein. Therefore, if large quantities of whole soybeans are being fed, it is better if they are heat-treated to denature the trypsin inhibitors. Soybeans should not be used in conjunction with urea, because the urease in whole soybeans leads to rapid release of ammonia from the urea, which can cause urea toxicity and feed refusal.

A study was conducted on the effects of free fatty acids (as palm acid oil) and protected fats (in the form of a calcium soap or fat prills) on the digestibility of fiber and fat in sheep, where fats replaced barley and tapioca to provide 15% of digestible energy intake [46] (Table 5). At the maintenance level of feeding, fiber digestibility was not affected by the inclusion of fat in the diet. Fiber digestibility was lower at the twice-maintenance level of feeding and was significantly higher for the diet containing calcium soap than for the control diet and diets containing free fatty acids or fat prills. The digestibility of added fat was also higher at the maintenance level of feeding. At the twice-maintenance level, the fat added as a calcium soap was more digestible than free fatty acids or fat prills. In a further study comparing fat prills and calcium soaps [47], it was found that with pelleted diets the fat digestibility of the two

Table 5 Apparent Digestibilities of Fiber and Added Fat in Sheep Fed at Two Levels of Intake

	Maintenance			Twice maintenance			
	Control	Free fatty acids	Calcium soap	Control	Free fatty acids	Calcium soap	Fat prills
Fibre	0.71	0.71	0.71	0.55	0.55	0.61	0.53
Added fat		0.82	0.76		0.56	0.63	0.48

Source: Ref. 42.

forms was similar (prills 0.902, soap 0.965), but when diets were not pelleted the digestibility of fat in the form of prills was considerably lower than that of calcium soaps (prills 0.506, soap 0.932). It is possible that heat generated during the pelleting process altered the nature of the fat prills.

B. Use of Fats in Commercial Situations

1. *Sheep*

The use of fat supplements for sheep under commercial conditions is limited because the animal production level is much lower than that of dairy cows. If supplementary feeding is required to complement grazing, the most common form is barley with a protein balancer. There is some scope for the use of fats with ewes in late pregnancy, when food intake is limited. It is important that high intakes of energy be maintained during the last month of pregnancy, particularly with ewes in poor body condition, but caution is needed because it is undesirable to make the ewes too fat at lambing. Under intensive conditions there may be a case for supplementing housed ewes, or those on restricted pasture, in early lactation. It has been shown [48] that a protected fat supplement in the form of a calcium soap fed at levels of 75, 150, 200, and 250 g/day significantly increased milk yield and milk fat concentration in Finn Dorset ewes, leading to increased growth rates of lambs, which were on average 1.0 kg heavier at 5 weeks of age.

2. *Beef Cattle*

Fats are not used extensively in diets for beef cattle in the United Kingdom. The normal objective is to achieve high carcass weights with a minimum amount of fat cover [49]. Although the efficiency of utilization of metabolizable energy for live-weight gain increases with the energy content of the diet, the rate of carcass fat deposition also increases. Extensive use of added dietary fat would result in overfat carcasses or carcasses with the desired fat content but at a lighter weight. In systems where heavier and fatter carcasses are preferred, e.g., in northern France, supplementary fats are used to achieve faster live-weight gains and heavier weights at slaughter.

Fats are used for preruminant calves in milk replacers, which are based on skim milk or artificial milk with added fat in the form of vegetable oil or tallow to increase the fat content to that of whole milk [50]. This is essential because the calf's capacity for carbohydrate digestion is limited [51]. It is important that the fat be broken down into small particles (3–4 μm in diameter) by homogenization to ensure efficient reconstitution of the milk and utilization by the calf. The fat source influences calf performance through a combination of factors [52]. Digestibility decreases with increasing chain length and increases with the degree of unsaturation at any given chain length. Short-chain fatty acids are absorbed faster than medium- or long-chain fatty acids. Palmitic acid is used more efficiently when present in the middle position of the triglyceride molecule, as in lard, than when occupying positions 1 or 3. With fats of low digestibility, there is an increase in digestibility with age.

3. Dairy Cows

By far the greatest value of fats for ruminants is in the feeding of dairy cows. Estimates of the upper limit to the absorption of fatty acids have ranged from 900 to 2180 g/day [53]. It would be very difficult to achieve these large quantities without the use of protected fats.

The response of dairy cows to supplementary fat is complex and not always predictable. Responses that have been reported include increased milk yields, increased or decreased milk fat content, decreased milk protein content, increased live-weight gain, and decreased live-weight loss. The observed response will depend on the quantity of fat, its fatty acid profile and degree of protection, the other components of the diet, and the overall feeding level, stage of lactation, and genetic merit of the cow.

Milk yield responses can normally be explained by the increase in total energy intake when fats are given and the increased efficiency of utilization of energy from fats. The response will not be as great as expected if rumen fermentation is disrupted, because digestion of the non-fat components of the diet, and subsequent energy release, will be reduced. Stage of lactation and genetic merit affect both the milk yield response and the live-weight response. Dairy cows in early lactation and those of higher genetic merit partition energy toward milk production at the expense of body fat reserves. Cows normally lose 0.5–1.0 kg of body weight each day for the first 8 weeks of lactation, and this is mostly body fat. Therefore increased energy intake at this stage of lactation could result in further increases in milk yield if the cow's genetic potential has not been reached and/or a reduction in the daily amount of body fat mobilized [54,55]. In later lactation, when appetite is greater and milk yield less,

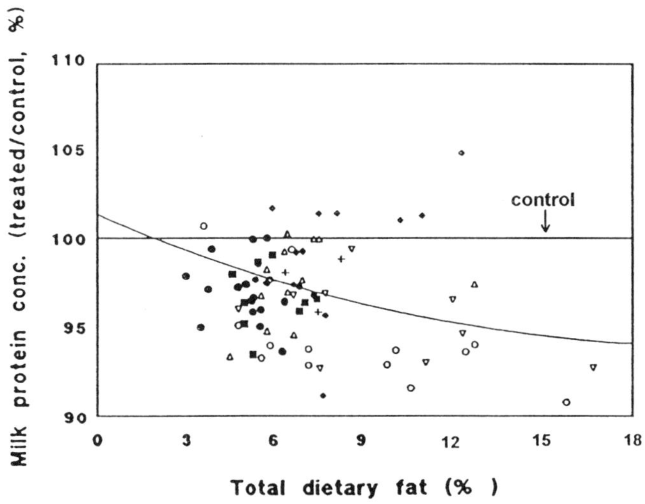

Fig. 7 Milk protein concentrations of cows fed added fat relative to controls (control = 100). $Y = 101.1 - 0.6381X + 0.0141X^2$, $r^2 = 0.24$. Added fat sources: animal–vegetable fat blend (+); calcium salts of fatty acids (△); prilled fat (◆); protected tallow (○); tallow (▽); oilseeds (●); yellow grease (■). (From Ref. 51.)

partition of energy switches more toward body reserves, so the increase in the energy supply results in greater fat deposition in body reserves. Cows of low genetic merit have a greater propensity for fat deposition and will partition a greater proportion of surplus energy in this direction.

Milk fat is derived from intramammary synthesis using acetate and β-hydroxybutyrate and from triglycerides or fatty acids absorbed directly from the bloodstream. Short-chain fatty acids, up to 10 carbon atoms in length, are virtually all synthesized. Acids with 18 carbon atoms are absorbed, and intermediate chain length acids originate from both sources. If added dietary fats interfere with normal fiber digestion in the rumen, acetate production will be reduced and the shortage of precursors in the mammary gland may lead to reduced de novo synthesis of milk fat. On the other hand, added fat may increase the quantity of fatty acids available for absorption and secretion in milk. In both cases, the proportion of C_{4-16} fatty acids in milk fat will decrease and the proportion of longer chain acids will increase. Total milk fat output will depend on the relative magnitude of the effects on synthesis and absorption, which are affected by protection and degree of unsaturation.

The biggest problem in using supplementary fats is the depression in milk protein concentration. In a review of 47 published experiments [56] involving 1396 cows, it was found that milk protein concentration decreased by 1–5% of control values when dietary fat concentration was increased within the range 3–15% (Fig. 7). Factors affecting milk protein secretion have received a lot of attention in recent years because this component has a greater effect on milk price than milk fat. The prediction of response is complicated by the fact that dietary protein supply, except in deficiency cases, has less effect on milk protein secretion than dietary energy supply or, more precisely, glucose and its precursors. The system is illustrated in Fig. 8, which explains the priorities for nutrient use in the dairy cow [57]. The three main nutrients of interest are amino acids, glucose, and energy substrates including acetate, butyrate, and long-chain fatty acids. The first priority for amino acids is to satisfy the nitrogen requirements for maintenance of the animal. The next priority is to provide glucose if it is in short supply, followed by the requirements of the growing fetus. Milk protein comes fairly low on the list of priorities for amino acids and also requires energy-yielding substrates to support protein synthesis. Alterations in the diet that affect glucose supply (which is mostly from propionate absorbed across the rumen wall) or requirements will alter the extent of utilization of amino acids for gluconeogenesis. There are two possible scenarios in which dietary fat can reduce glucose supply. First, direct effects on rumen fermentation may reduce propionate production. This is not likely to be a major effect, as fats tend to reduce the amount of fiber digested which would be expected to cause a reduction in the production of acetate. Propionate is produced mainly by the digestion of starch in the rumen, which is not greatly affected by fats. The second possibility is that since fats normally replace starch in the ration, the total quantity of glucose precursors may be reduced (both propionate in the rumen and starch digested in the small intestine). Where fat supplements have replaced grains on an isoenergetic basis [58], milk yield was similar for both treatments but milk protein content was reduced. However, it has been reported [59] that abomasal infusion of glucose increased the milk yield of cows but reduced the milk protein content. Another possible mechanism that has been proposed to explain the depression of milk protein content with high fat diets is insulin resistance. It has been noted that plasma insulin is elevated in high fat diets and that the rate of glucose removal is negatively related to insulin concentration [60]. Therefore, high fat diets may cause insulin resistance, which would retard the uptake of amino acids by mammary tissue for protein synthesis [56]; however, changes in plasma insulin concentrations in response to fat supplementation have been inconsistent [61–63].

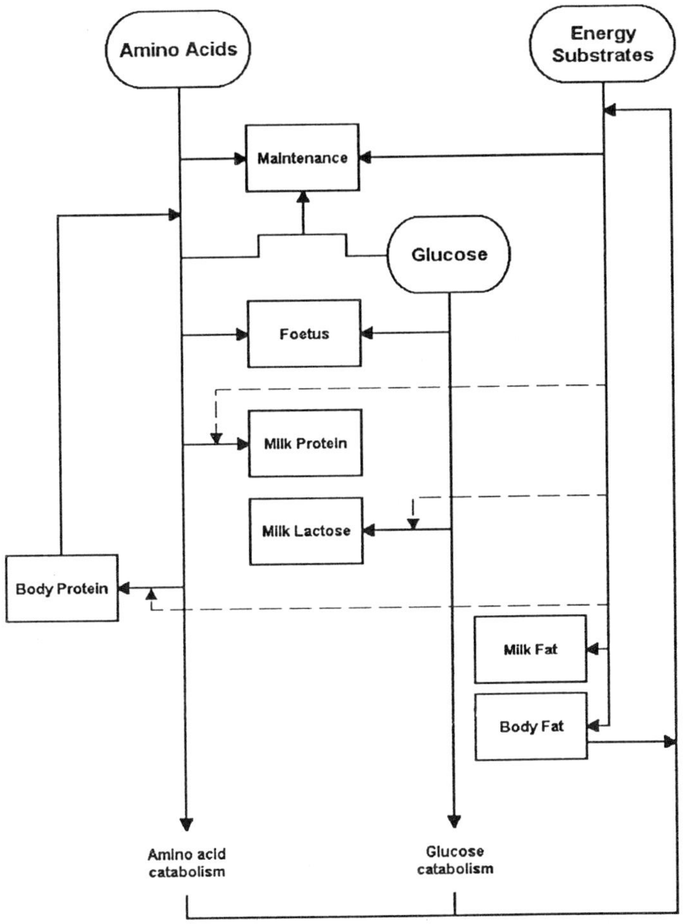

Fig. 8 Order of priority for use of major classes of nutrients by dairy cows. Glucose includes absorbed glucose and absorbed propionate. Energy substrates include absorbed acetic and propionic acids plus absorbed long-chain fatty acids. Solid lines indicate pathways for use of substances in meeting defined animal functions. Dashed lines indicate use of resources in support of synthetic processes. (From Ref. 52.)

There is one further way in which dietary fat may affect milk protein: through miscalculations of protein supply during ration formulation. The Metabolizable Protein system for protein evaluation recently proposed by AFRC [39] calculates the supply of metabolizable protein to the animal from digestible undegraded dietary protein (DUP) and microbial protein produced in the rumen. Microbial protein production, where it is limited by energy intake, is calculated from fermentable metabolizable energy. Although dietary fats contribute to the metabolizable energy supply of the animal, they are not fermented in the rumen, so they do not provide energy for microbial growth. Diets with added fat often need extra DUP or fermentable carbohydrates to avoid induced metabolizable protein deficiencies. In an experiment conducted at Nottingham University [64], it was found that increasing the fat content of the diet from 25 to 76 g/kg DM using a calcium soap increased milk yield and protein yield but

Table 6 Milk Yield and Composition of Four Groups of Cows Over Weeks 4–15 of Lactation

	Treatment group				
	Control	Protected fat	Lactose	Protected fat and lactose	SED
Milk yield, kg/day	24.6[a]	27.7[b]	25.6[a,b]	26.5[a,b]	1.20
Milk fat, g/kg	46.3	46.9	46.8	48.0	2.99
kg/day	1.13	1.30	1.20	1.27	0.09
Milk protein, g/kg	32.3[a,b]	30.7[a]	32.7[b]	31.9[a,b]	0.79
kg/day	0.79	0.85	0.83	0.85	0.05

SED = Standard error of the difference.
[a,b]Means within a row with different superscripts are significantly ($P<0.05$) different.
Source: Ref. 60.

decreased milk protein concentration. When lactose, a readily fermentable carbohydrate, was included in the diet, the milk protein concentration was the same as control values (Table 6). In a further experiment [65] it was found that protected protein also partially alleviated the negative effect of dietary fat on milk protein concentration. However, when a combination of protected fat, protected protein, and lactose was used in the diet, the results were complementary. A significant increase in milk yield was found without any depression in milk protein concentration (Table 7).

Table 7 Milk Production, Milk Composition, and Dry Matter Intake of Dairy Cows

	Treatment group				
	Control	Protected fat & protein	Lactose	Protected fat & protein + lactose	SED
Milk yield, kg/day	21.3[a]	23.4[a,b]	22.7[a,b]	27.8[b]	2.18
Milk fat, g/kg	42.6	41.7	42.5	41.8	2.68
Milk protein, g/kg	33.7	33.6	33.7	33.8	1.38
Dry matter intake, kg/day	17.3	17.6	18.9	17.9	1.3

SED = Standard error of the difference.
[a,b]Means with different superscripts are significantly ($P<0.05$) different.
Source: Ref. 61.

REFERENCES

1. C. P. Freeman, Digestion and absorption of fat, in *Digestion and Absorption in the Fowl* (K. N. Boorman and B. M. Freeman, Eds.), British Poultry Science, Edinburgh, UK, 1976, p. 117.
2. C. P. Freeman, The digestion, absorption and transport of fat—non-ruminants, in *Fats in Animal Nutrition* (J. Wiseman, Ed.), Butterworths, London, 1984, p. 105.
3. J. Wiseman, Assessment of the digestible and metabolisable energy content of fats for non-ruminants, in *Fats in Animal Nutrition* (J. Wiseman, Ed.), Butterworths, London, 1984, p. 277.
4. A. Krogdahl, Digestion and absorption of lipids in poultry, *J. Nutr. 115*: 675 (1985).

5. J. Wiseman, Variability in the nutritive value of fats for non-ruminants, in *Feedstuff Evaluation* (J. Wiseman and D. J. A. Cole, Eds.), Butterworths, London, 1990, p. 215.
6. R. Renner and F. W. Hill, Utilisation of fatty acids by the chicken, *J. Nutr. 74*: 259 (1961).
7. T. S. Stahly, Use of fats in diets for growing pigs, in *Fats in Animal Nutrition* (J. Wiseman, Ed.), Butterworths, London, 1984, p. 313.
8. R. J. Young, The energy value of fats and fatty acids for chicks. 1. Metabolisable energy, *Poult. Sci. 40*: 1225 (1961).
9. D. Sklan, Digestion and absorption of lipids in chicks fed triglycerides or free fatty acids. Synthesis of monoglycerides in the intestine, *Poult. Sci. 58*: 885 (1979).
10. J. Wiseman and F. Salvador, The influence of free fatty acid content and degree of saturation on the apparent metabolisable energy value of fats fed to broilers, *Poult. Sci. 70*: 573 (1991).
11. J. Wiseman, F. Salvador, and J. Craigon, Prediction of the apparent metabolisable energy content of fats fed to broiler chickens, *Poult. Sci. 70*: 573 (1991).
12. J. Powles, J. Wiseman, D. J. A. Cole, and B. Hardy, Effect of chemical structure of fats upon their apparent digestible energy value when fed to growing/finishing pigs, *Anim. Prod. 57*: 137 (1993).
13. J. Powles, J. Wiseman, D. J. A. Cole, and B. Hardy, Effect of chemical structure of fats upon their apparent digestible energy value when given to young pigs, *Anim. Prod. 58*: 411 (1993).
14. J. Powles, J. Wiseman, D. J. A. Cole, and S. Jagger, Prediction of the apparent digestible energy value of fats given to pigs, *Anim. Sci. 61*: 149 (1995).
15. L. B. Carew, R. H. Machemer, R. W. Sharp, and D. C. Foss, Fat absorption by the very young chick, *Poult. Sci. 52*: 738 (1972).
16. M. R. Fedde, P. E. Waibel, and R. E. Burger, Factors affecting the absorbability of certain dietary fats in the chick, *J. Nutr. 70*: 447 (1960).
17. J. Wiseman and F. Salvador, The influence of age, chemical composition and rate of inclusion on the apparent metabolisable energy of fats fed to broiler chicks, *Br. Poult. Sci. 30*: 653 (1989).
18. J. Wiseman and A. Blanch, The effect of free fatty acid content on the apparent metabolisable energy content of coconut/palm kernel oil for broiler chickens aged 12 and 52 days, *Anim. Feed Sci. Technol. 47*: 225 (1994).
19. N. R. Artman, The chemical and biological properties of heated and oxidised fats, *Adv. Lipid Res. 7*: 245 (1969).
20. J. Wiseman, Antinutritional factors associated with dietary fats and oils, in *Recent Advances in Animal Nutrition—1986* (W. Haresign and D. J. A. Cole, Eds.), Butterworths, London, 1986, p. 47.
21. B. K. Edmunds, Chemical analysis of lipid fractions, in *Feedstuff Evaluation* (J. Wiseman and D. J. A. Cole, Eds.), Nottingham Univ. Press, Nottingham, UK, 1990, p. 197.
22. J. S. Andrews, J. F. Mead, and W. H. Griffith, Toxicity of lipid peroxides in the rat, *J. Nutr. 70*: 199 (1960).
23. C. E. Poling, W. D. Warner, P. E. Mone, and E. E. Rice, The influence of temperature, heating time and aeration upon the nutritive value of fats, *J. Am. Oil Chem. Soc. 39*: 315 (1962).
24. A. E. Waltking, W. E. Seery, and G. W. Bleffert, Chemical analysis of polymerisation products in abused fats and oils, *J. Am. Oil Chem. Soc. 52*: 96 (1975).
25. J. Wiseman, B. K. Edmunds, and N. Shepperson, The apparent metabolisable energy of sunflower oil and sunflower acid oil for broiler chicks, *Anim. Feed Sci. Technol. 36*: 41 (1992).
26. L. D. Metcalfe, Proposed source of chick edema factor, *J. Assoc. Off. Agric. Chem. 55*: 542 (1972).
27. J. K. G. Kramer, H. W. Hulan, H. L. Trenholm, and A. H. Corner, Growth, lipid metabolism and pathology of two strains of rats fed high fat diets, *J. Nutr. 109*: 202 (1979).
28. R. Jeffcoat and M. R. Pollard, Studies on the inhibition of the desaturases by cyclopropenoid fatty acids, *Lipids 12*: 480 (1977).
29. F. S. Shenstone, J. R. Vickery, and A. R. Johnson, Studies on the chemistry and biological effects of cyclopropenoid compounds, *J. Agric. Food Chem. 13*: 410 (1965).

30. N. R. Ellis and H. S. Isbeli, Soft pork studies. II. The influence of the character of the ration upon the body composition of body fat of hogs, *J. Biol. Chem. 69*: 219 (1926).
31. N. R. Ellis and H. S. Isbeli, Soft pork studies, III. The effect of food fat upon body fat as shown by the separation of the individual fatty acids of the body fat, *J. Biol. Chem. 69*: 239 (1926).
32. F. M. Whittington, N. J. Prescott, J. D. Wood, and M. Enser, The effect of dietary linoleic acid on the firmness of backfat in pigs of 85kg live weight, *J. Sci. Food Agric. 37*: 753 (1986).
33. A. J. Kempster, A. W. Dilworth, D. G. Evans, and K. D. Fisher, The effect of fat thickness and sex on pig meat quality with special reference to the problems associated with overleanness, *Anim. Prod. 43*: 517 (1986).
34. M. Enser, in *Proceedings of Meat Research Institute Symposium No. 3* (C. L. Cutting, Ed.), Meat Research Institute, Bristol, 1974.
35. COMA, Dietary reference values for food and nutrients in the United Kingdom, Department of Health and Social Services, Report on Health and Social Subjects 41, HMSO, London, 1991.
36. I. Bartov and M. Frigg, Effect of high concentrations of dietary vitamin E during various age periods on performance, plasma vitamin E and meat stability of broiler chicks at 7 weeks of age, *Br. Poult. Sci. 33*: 393 (1992).
37. D. E. Koch, A. M. Pearson, N. T. Magee, J. A. Hoefer, and B. S. Schweigert, Effect of diet on the fatty acid composition of pork fat, *J. Anim. Sci. 27*: 360 (1968).
38. J. Wiseman, J. Agunbiade, and D. J. A. Cole, The influence of changes in dietary oil content on the fatty acid profile of backfat in pigs, *Anim. Prod. 54*: 497 (1992).
39. AFRC, *Energy and Protein Requirements of Ruminants*, An advisory manual prepared by the AFRC Technical Committee on Responses to Nutrients, CAB International, Wallingford, 1993.
40. MAFF, *UK Tables of Nutritive Value and Chemical Composition of Feedingstuffs*, Rowett Research Services, Aberdeen, Scotland, 1990.
41. P. McDonald, R. A. Edwards, J. F. D. Greenhalgh, and C. A. Morgan, *Animal Nutrition*, Longman, Harlow, UK, 1995.
42. NRC, *Nutrient Requirements of Dairy Cattle*, National Academy Press, Washington, DC, 1989.
43. H. Galbraith and T. B. Miller, Effect of metal cations and pH on the antibacterial activity and uptake of long chain fatty acids, *J. Appl. Bacteriol. 36*: 635 (1973).
44. C. G. Harfoot, Lipid metabolism in the rumen, in *Lipid Metabolism in Ruminant Animals* (W. W. Christie, Ed.), Pergamon, Oxford, 1981, pp. 21–55.
45. C. Devendra and D. Lewis, The interaction between dietary lipids and fibre in sheep. 2. Digestibility studies, *Anim. Prod. 19*: 67.
46. H. Ainsworth and E. L. Miller, The effects of protected and unprotected fats in rations fed to sheep at two intakes on the apparent digestibilities of fat and fibre, *Anim. Prod. 40*: 534 (1985).
47. J. C. Mathers, N. Sadler, and E. L. Miller, Some effects of dietary fats on rumen metabolism and digestion in sheep, *Proc. Nutr. Soc. 42*: A64 (1983).
48. M. Perez Hernandez, J. J. Robinson, R. P. Aitken, and C. Fraser, The effect of dietary supplements of protected fat on the yield and fat concentration of ewe's milk and on lamb growth rate, *Anim. Prod. 42*: 455 (1986).
49. D. Allen, *Planned Beef Production and Marketing*, BSP Professional Books, Oxford, UK, 1990.
50. W. Thickett, D. Mitchell, and B. Hollows, *Calf Rearing*, Farming Press, Ipswich, UK, 1986.
51. J. B. Owen, *Cattle Feeding*, Farming Press, Ipswich, UK, 1991.
52. J. H. B. Roy, *The Calf*, Butterworths, London, 1980.
53. J. E. Storry, The effect of dietary fat on milk composition, in *Recent Advances in Animal Nutrition—1981* (W. Haresign, Ed.), Butterworths, London, 1981, pp. 3–33.
54. G. P. Jones and P. C. Garnsworthy, The effects of dietary energy content on the response by dairy cows to body condition at calving, *Anim. Prod. 49*: 183 (1989).
55. P. C. Garnsworthy and C. D. Huggett, The influence of the fat concentration of the diet on the response by dairy cows to body condition at calving, *Anim. Prod. 54*: 7 (1992).
56. Z. Wu and J. T. Huber, Relationship between dietary fat supplementation and milk protein concentration in lactating cows: a review, *Livestock Prod. Sci. 39*: 141 (1994).

57. J. D. Oldham and G. C. Emmans, Prediction of responses to protein and energy yielding nutrients, in *Nutrition and Lactation in the Dairy Cow* (P. C. Garnsworthy, Ed.), Butterworths, London, 1988, pp. 76–96.
58. N. E. Smith, W. L. Dunkley, and A. A. Franke, Effects of feeding protected tallow to dairy cows in early lactation, *J. Dairy Sci. 61*: 747 (1978).
59. R. A. Frobish and C. L. Davis, Effects of abomasal infusions of glucose and propionate on milk yield and composition, *J. Dairy Sci. 60*: 204 (1977).
60. D. L. Palmquist and E. A. Moser, Dietary fat effects on blood insulin, glucose utilisation and milk protein content of lactating cows, *J. Dairy Sci. 64*: 1664 (1981).
61. K. A. Cummins and J. L. Sartin, Response of insulin, glucagon and growth hormone to intravenous glucose challenge in cows fed high fat diets, *J. Dairy Sci. 70*: 277 (1987).
62. A. D. Kennedy, F. R. Tekpetey, J. R. Ingals, and W. M. Palmer, Effect of stage of lactation and diet on serum insulin level and mononuclear leukocyte insulin receptor characteristics in dairy cows, *Can. J. Anim. Sci. 67*: 721 (1987).
63. G. Gagliostro, Y. Chilliard, and M. J. Davicco, Duodenal rapeseed oil infusion in early and mid lactation cows. 3. Plasma hormones and mammary apparent uptake of metabolites, *J. Dairy Sci. 74*: 1893 (1991).
64. P. C. Garnsworthy, The effects on milk yield and composition of incorporating lactose into the diet of dairy cows given protected fat, *Anim. Sci. 62*: 1 (1996).
65. P. C. Garnsworthy, Effects of feeding lactose with protected fat and protein on milk production of cows in mid lactation, *Anim. Sci. 63*: in press (1996).

23

Anionic Detergents

Maurice R. Porter
Maurice R. Porter and Associates, Sully, Vale of Glamorgan, South Wales

I. INTRODUCTION

Anionic surfactants are manufactured and used in greater volume than all other types of surfactants. Soap, the alkali metal salt of a carboxylic acid, derived from natural oils and fats, has been used as a detergent for thousands of years. It is produced by reacting an alkali with an animal fat or vegetable oil that contains the glyceryl esters of the long-chain fatty acids, the process known as saponification. Soap was used by the Phoenicians and the Romans, and a soap-making process was recorded by Plinius in A.D. 70.

The very early soaps were made from animal fats and wood ash. Saponification is shown in Fig. 1. Improved products were introduced in the Middle Ages by the Arabs, who used vegetable oils. By the twelfth century, soap was being made commercially in Bristol, England.

Saponification requires alkali. All the early processes used wood ash and needed considerable quantities of timber to produce the ash. It was not until Leblanc invented his process for the manufacture of caustic soda in 1778 that large-scale soap manufacture became possible. Crosfields was founded in 1814, Pears in 1879, and Lever in 1884, so that the latter half of the nineteenth century was the main period when the production of soap from lipids was carried out on an industrial scale.

Up until the nineteenth century, soapmaking was an empirical science, but by the 1800s inorganic chemistry was well developed so that the industrial production of alkalis and acids was on a scientific basis. The organic chemistry of lipids was understood, the variations of lipid raw materials were being investigated, and the sulfonation of castor oil and olive oil was being carried out industrially in the period 1850–1870. The cheaper sulfonated castor oil was employed by Crum to assist in dyeing with alizarin dyestuffs, the so-called Turkey red style. Since that time sulfonated castor oil has been known as Turkey red oil. Later (about 1900), the sulfonation of oleic acid was commercialized and was used in fat splitting during soap manufacture.

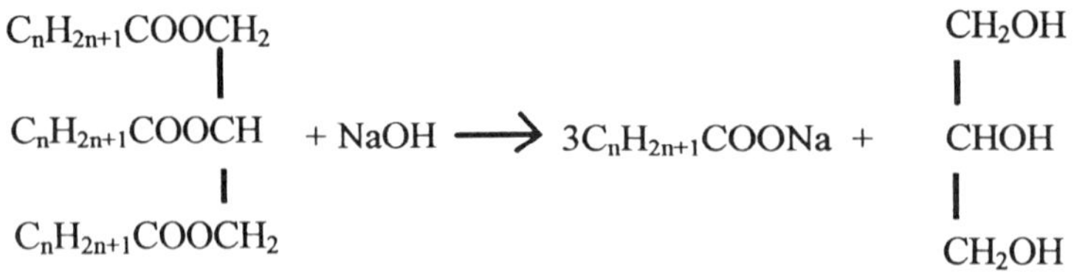

Fig. 1 Saponification of fats with caustic soda.

Soap is an excellent general household detergent and was and still is widely used for personal washing (toilet soaps), clothes washing, and wall and floor washing, but only in soft water. In hard water, soap is not soluble; it precipitates and thereby loses its detergent properties. Soap will also precipitate in the presence of acids. At the beginning of the twentieth century the reason for this was found to be the length of the hydrocarbon chain of the lipid that was used in the manufacture of the soap. The calcium or magnesium salts are insoluble in water if n in Fig. 1 is greater than 16. As animal fats, the cheapest lipids used in soap manufacture have principally alkyl chains of 16 and greater, soaps made from tallow give precipitates in hard water. If $n = 10–14$, the solubility improves considerably. However, at the turn of the century the availability of lipids containing alkyl chains of 10–14 was limited. Such lipids are the vegetable oils such as coconut and palmkernel oil. Soap made with alkyl chains of 10–14 gave superior hard water detergency but at a considerably higher price and could only be used for toilet soaps. Even so, fatty acids of alkyl chain length 10–14 are still insoluble in water and therefore can only be used in alkaline conditions. It was found that the sulfonated vegetable oils (e.g., Turkey red oil) were stable to hard water and limited acidic conditions. The availability of suitable vegetable oils was limited for the mass market to laundry soaps, which had been developed in the early 1900s. It was the discovery of lime soap dispersing agents, builders, and bleaches that could be added to soap to overcome the hard water problem that resulted in the widespread use of soap for laundry. These additives, generally inorganic, in some way overcame the solubility properties of sodium and potassium salts of long-chain fatty acids. The discoveries came about entirely by trial and error, as the reason for the action of surfactants lies in their physical chemistry, which has only recently begun to be understood.

By the 1930s, laundry products for domestic use were formulated mixtures, with soap (generally from tallow) as the major surfactant in the formulation. Research still continued on chemical methods of modification to overcome the hard water problem, and a very large number of new synthetic chemicals were synthesized in this period. By 1940 probably all the presently known surfactants had been made on a laboratory scale. As an example, nonionic surfactants based on ethylene oxide were found to give good detergency and excellent solubility in both hard and soft water over a wide pH range, but ethylene oxide was not available on a large scale.

It was the development of the petrochemical industry in the period 1940–1970 that brought about a revolution in the type of surfactants available on the market. The reason was twofold. First, the hydrophobic group (usually a saturated paraffinic chain C_{12-18} chain length) was now available in large quantities at much cheaper prices than the equivalent hydrophobic group from lipids (animal fats and vegetable oils). Second, hydrophobic groups containing the benzene ring were now available in large quantities that allowed far more variety in attaching hydrophilic groups to the hydrophobic group, e.g., sulfonation using sulfur trioxide. As a result, a large number of "synthetic" surfactants appeared on the market from 1940

Table 1 Soap Production in Western Europe

Year	Soap production (1000 t)
1972	866
1982	776
1993	580

Source: Ref. 1.

(depending on the country) onward, and the products that were successful were those whose properties were superior to those of soap in hard water. Soap was replaced in most household and domestic detergents with the exception of toilet soap, and this decline in soap production still continues in Western Europe (see Table 1).

The principal anionic surfactants that have replaced soap in detergents are listed in Table 2. (Note the different alkyl chain lengths that are necessary to obtain a detergency equivalent to that of soap.)

It must be emphasized that the replacement of soap by a synthetic surfactant is not a simple matter of comparing the relative prices and surfactant properties of the two surfactants but concerns the price/performance ratio of a fully formulated detergent. A detergent is the product that does the washing. The surfactant is the component that has the major influence on the properties of an aqueous solution in relation to wetting, foaming, dispersing solids, emulsifying oils, and removing dirt from a fabric. A modern heavy duty laundry detergent for horizontal drum machines will have at least two surfactants, a builder, a bleaching system, an enzyme, antiredeposition agents, foam stabilizer and control additive, fluorescent whitening agents or optical brighteners, corrosion inhibitor, perfume, dyestuff, and fillers. There will be a considerable number of interactions between the various components and the surfactants, and therefore the testing and comparison of competitive surfactants is a complex procedure.

Due to the different washing habits in various countries, different formulations will gain market dominance. Therefore the major anionic products used in both domestic and industrial detergents can vary from one country to another. Nevertheless, the surfactant or the mixture of surfactants plays the major role. Linear alkylbenzene sulfonates (LABS) are undoubtedly the major anionic component used in detergents worldwide to replace soap, α-olefin sulfonates are used in the United States but not in Europe. Paraffin sulfonates originated in Europe and are only now being used in the United States. Nonionic surfactants, such as fatty alcohol

Table 2 Principal Anionic Surfactants That Have Replaced Soap in Synthetic Detergents

Chemical name	Abbreviation
Linear alkyl (C_{10-14}) benzene sulfonate, sodium salt	LABS
α-Olefin (C_{14-16}) sulfonate, sodium salt	AOS
Fatty alcohol (C_{12-14}) sulfate, sodium salt	AS
Fatty alcohol (C_{12-14}) ether sulfate, sodium salt	AES
Secondary alkane (C_{14-17}) sulfonates (paraffin sulfonates)	SAS
Alpha-sulfonated methyl esters of fatty acids (C_{16-18}), sodium salt	FES

ethoxylates, are playing a much more important role as washing temperatures decrease.

Lipids from animal fats and vegetable oils, once the dominant source of raw materials for detergents (i.e., soap) have had a declining share of the domestic and industrial detergent market. This situation could well change significantly in the future due to possible shortage in the supply of crude petroleum, the demand for "mild" detergents, the demand for "natural" products, and the environmental problems of synthetic surfactants. The overall result of these factors will probably be a slow decline in the consumption of anionic synthetic detergents based on petrochemicals, which will be replaced by new semisynthetic products based on natural products rather than soap. For instance, the major new surfactant type in the last decade is the alkyl polyglycosides, which use the polyhydroxy groups in carbohydrates as the hydrophilic entity with a long-chain fatty alcohol (from fats, oils, or petrochemicals) as the hydrophobic entity.

II. RAW MATERIAL ROUTES TO ANIONIC SURFACTANTS

Tallow soap has a hydrophobic group of $C_{17}H_{35}$ with a molecular weight of 239 and a hydrophilic group COONa with a molecular weight of 67. Sodium lauryl sulfate has a hydrophobic group of $C_{12}H_{25}$ with a molecular weight of 169 and a hydrophilic group of SO_4Na with a molecular weight of 119. These are typical surfactants used in detergents, and, as shown by the relative molecular weights of the hydrophobic and hydrophilic groups, the major portion of the surfactant molecule is the hydrophobic group. This has economic significance because the major cost component of detergents is the surfactant and the major cost component of the surfactant is the hydrophobic group; therefore the main cost component of detergents is the hydrophobic group of the surfactant used in the detergent. The cost of manufacture of the surfactant and other raw materials will rarely equal the cost of the raw materials that supply the hydrophobic group.

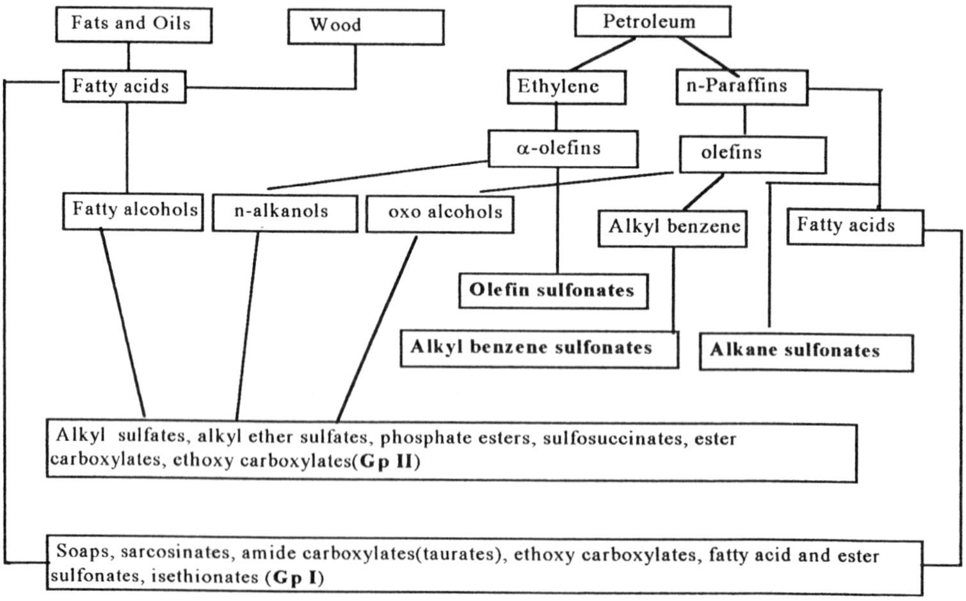

The detergent industry requires hydrophobic groups consisting of paraffinic chains, generally saturated to give chemical stability and linear to give good biodegradability. The sources of these paraffinic chains are either petroleum or animal fats [2] or vegetable oils [2]. Petroleum itself is, of course, the result of biological action on plant material. Figure 2 gives a general overview of the origin and processing of surfactant raw materials for anionic surfactants.

Figure 2 shows that there are some anionics, namely alkylbenzene sulfonates, alkane sulfonates, and olefin sulfonates, that are derived solely from petroleum. These are not discussed in this chapter, but it should be borne in mind that these three groups represent the largest production of anionic surfactants today, due mainly to their low cost and excellent detergent properties. The other two groups of anionic surfactants—group I in Fig. 2, derived from fatty acids, and group II, derived from alcohols—can be manufactured from lipids (fats and oil) or petroleum.

A. Anionic Surfactants—Fatty Acids

In practice, very little of the fatty acids made from petroleum is used in the production of surfactants and fatty acids for detergent use. In Europe, the United States, and Japan, fatty acids are made from natural fats and oils consisting of glyceryl esters of linear, saturated, or unsaturated fatty acids with an even number of carbon atoms. There are three principal groups of fats and oils:

1. Vegetable oils with a high proportion of lauric (C_{12}) and myristic (C_{14}) acids, e.g., coconut oil
2. Animal fats and oils with a high content of palmitic (C_{16}) and oleic (C_{18}) acids
3. Vegetable oils containing high quantities of mono-, di-, and triunsaturated acids [oleic (C_{18}) acid, linoleic (C_{18}) acid, linolenic (C_{18}) acid] of varying ratios.

The fats or oils will contain many other natural impurities such as sterols, seed particles, and dirt. The direct reaction of triglycerides with alkali gives a soap and glycerine; this is described more fully in Section III. The fatty acids can be separated from soap by the addition of inorganic acids, but the industrial manufacture of fatty acids is carried out by splitting the triglycerides with water. The major problem of all large-scale processes is the insolubility of water in the fat and the stepwise hydrolysis of the triglyceride to diglyceride to monoglyceride to fatty acid. At high temperatures (260°C) and pressures (60 bar), water becomes more soluble in the fat phase; therefore, fat splitting is carried out under such conditions, usually in a continuous process, for large-scale manufacture where the unsaturated content of the fatty acid is low. Less harsh conditions of hydrolysis must be used for fats and oils with polyunsaturated oils to avoid the production of polymers. Category 3 products are also obtained directly as fatty acids (known as tall oil) as a by-product in the Kraft process for paper.

Depending on the end use of the fatty acid derivative, the crude fatty acid can be used in the manufacture of a surfactant, e.g., crude tall oil soaps. The great majority of fatty acid based surfactants are made from purified fatty acids. Fatty acids are distilled to effect the removal of impurities (e.g., sterols, phosphatides), and this is usually carried out under low pressure to avoid oxidation of unsaturated acids. The fractional distillation process also removes short-chain fatty acids (e.g., C_8) and C_{10}) that can give disagreeable odors. The fatty acids can then be hydrogenated to remove any unsaturation; such a process is called hardening. For specific surfactants (e.g., sulfonated oleic acid), specific high purity fatty acids are required that cannot be obtained by fractional distillation. Fractional crystallization [2,3] with the aid of a solvent or a surfactant (known as hydrophilization [2]) processes are used in this case.

The resulting fatty acids can then be used in the manufacture of various surfactants, but fatty alcohols, many specialty surfactants, and soap are made from the methyl esters of the fatty acids. The methyl esters can be made either by esterification of the fatty acids or by transesterification of triglycerides with methanol [2]. Transesterification tends to be the dominant process as it will proceed under extremely mild conditions in the presence of an alkaline catalyst at 50–70°C and atmospheric pressure, and therefore oxidation of unsaturated groups is avoided. Higher temperature processes are often employed as these can not only transesterify but also esterify any free fatty acid mixed in with the triglyceride. The methyl esters are also easier to distill than the corresponding fatty acids. In 1982, western Europe produced 190,000 tonnes of methyl esters of fatty acids, the majority being further converted to fatty alcohols.

B. Anionic Surfactants—Fatty Alcohols

Fatty alcohols are produced from methyl esters of fatty acids by catalytic high pressure hydrogenation [2]. In principle, hydrogenation could start with the triglycerides, but the harsh conditions result in poor glycerine recovery and greater use of hydrogen and catalyst. The same objection is found with the hydrogenation of fatty acids, which requires higher temperatures and gives greater corrosion than using the methyl ester. There are various complex processes for the production of fatty alcohols from methyl esters [3].

As Fig. 2 shows, fatty alcohols are also obtained from petrochemical sources, and here there are two main routes, starting with either *n*-paraffins or ethylene.

1. Via n-*Paraffins and the Oxo Process*

By chlorination of *n*-paraffins and subsequent elimination of hydrogen chloride, a mixture of linear olefins is obtained with olefin bonds distributed over the entire chain length. The subsequent mixture of olefins is then subjected to the oxo process, also known as hydroformylation. This proceeds as shown in Fig. 3.

Figure 3 shows the addition of an aldehyde group to the double bond in a long-chain olefin (an "internal" olefin). The olefin product is usually a mixture of olefins in the range C_{10}–C_{16} and contains both odd and even-numbered carbon chains. The aldehyde group is then hydrogenated to an alcohol group. The resulting fatty alcohol is a mixture of alcohols randomly distributed along the chain. Some of this mixture will have the alcohol group at the end of the chain. Therefore the fatty alcohol produced by this method will differ significantly from the fatty alcohol produced from lipids via the methyl ester and hydrogenation in two ways:

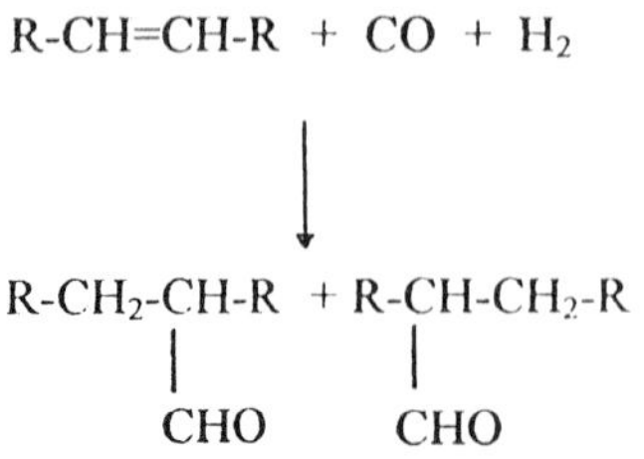

Fig. 3 The oxo process.

1. The alcohol group will be randomly distributed along the fatty alcohol chain as distinct from the fatty alcohol from lipids, which have the alcohol group at the end of the chain.
2. The paraffinic chain will have both odd and even numbers of carbon atoms, as distinct from the fatty alcohol from lipids, which has only an even number of carbon atoms.

The first difference is important to the surfactant properties, as the shape of the subsequent surfactant will differ depending on whether the hydrophilic group is at the end of the chain or within it, and this will affect surfactant properties [4, pp. 39–42]. The second difference has no significant effect on surfactant properties, but these petrochemical-derived alcohols are chemically different from "natural" products and can therefore be easily identified as such by analysis.

2. *Via Ethylene and the Alfol Process*

The differences between fatty alcohols from lipids and those from internal olefins were nearly entirely overcome by the Ziegler process discovered in 1953 where the polymerization of ethylene with triethylaluminum at high temperatures gave long paraffinic chains with even-numbered C atom chains terminating in an aluminum atom. These compounds can be oxidized and hydrolyzed (the Alfol process) to give long-chain primary alcohols very similar to those obtained from lipids. Chemically the primary alcohols from both sources are identical except for the distribution of the alkyl chain. The distribution of the fatty chain in the alcohol from lipids will depend on the original distribution in the vegetable oil(s) used as feedstock and the fractionation in the distillation stages of fatty acid, methyl ester, and fatty alcohol. In alcohols from ethylene it will depend on the distribution in the Ziegler process and any fractionation in the purification stages of the alcohol. With some degree of effort, a "synthetic" fatty alcohol from ethylene can be made with properties identical to those of a fatty alcohol derived from lipids.

The choice of which alcohol to use in the manufacture of surfactants and detergents will depend on their relative cost and availability, but the concept of "mildness," "natural materials," and "green-like" can sometimes influence the choice.

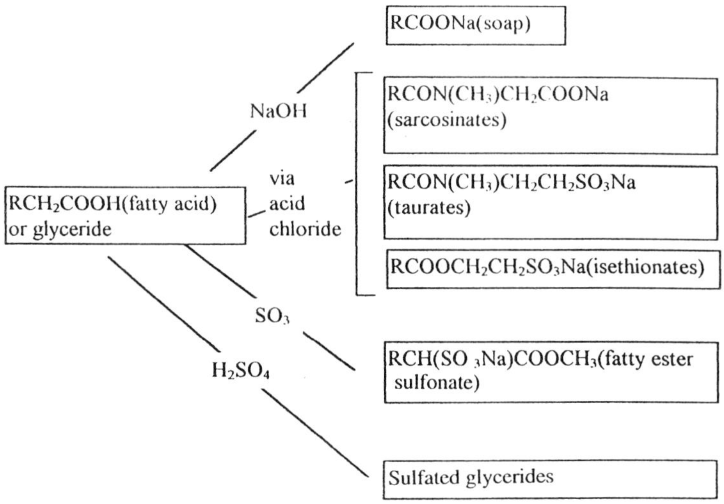

Fig. 4 Surfactants from fatty acids.

There has been no attempt here to describe the very important roles that α-olefins, ethylene oxide, and alkylbenzenes play as raw materials for the detergent industry as these are all derived from petrochemicals.

III. SURFACTANTS DERIVED FROM FATTY ACIDS AND/OR FATTY ACID ESTERS

A considerable number of surfactants can be made from fatty acids, and the main products are shown in Fig. 4. The four main groups are (1) soaps; (2) fatty ester sulfonates or alpha-sulfonated acids or esters; (3) sarcosinates, taurates, and isethionates; and (4) sulfated glycerides.

A. Soaps

Soap is defined as the product formed when glycerides (fats or oils) or fatty acids are heated with alkalis or organic amines. Personal (toilet) soaps are made by reacting animal fats and/or vegetable oils with caustic soda solutions. Glycerol is a by-product. A soap may be looked upon as a salt of a fatty acid,

RCOOM

where R is a saturated or unsaturated paraffinic chain with 8–20 carbon atoms and M is an inorganic or organic base.

The degree of ionization and hence the water solubility of the soap will depend on three factors: (1) the length of the paraffinic chain, (2) the degree of unsaturation of the paraffinic chain, and (3) the nature of the base (M).

Increasing temperature will increase the solubility of soaps in water. Table 3 shows the effect of paraffinic chain length on water solubility. It shows that the longer the paraffinic chain, the less water-soluble the soap. The incorporation of one unsaturated group (oleic) into the soap molecule increases the water solubility of the soap significantly. Substituting potassium or an organic amine (e.g., ethanolamine) for the sodium will also improve water solubility. The water solubility will determine the surfactant properties: the less soluble the soap, the better the surfactant properties such as foaming, wetting, solubilization, dispersing of solids, and detergency [5]. However, the amount of surfactant present in a solution will diminish as the solubility decreases; therefore in practice there will be an optimum solubility for a particular practical application. For a more detailed discussion, see Porter [4, pp. 26–89]. Different practical applications will need different chain lengths of fatty acids or different bases to achieve the optimum conditions. A toilet soap used in warm or even cold water will need

Table 3 The Effect of Paraffinic Chain Length and Unsaturation on the Water Solubility of Soaps

Product	Length of carbon chain	Unsaturation in carbon chain	Temperature needed to obtain a 10% saturated solution (°C)
Sodium stearate	18	None	71
Sodium palmitate	16	None	63
Sodium myristate	14	None	51
Sodium tallowate	Mixed 14/16/18	40% oleic	48
Sodium laurate	12	None	36
Sodium oleate	18	One double bond	23

shorter chain lengths or more unsaturation than soaps used at high temperatures for clothes washing.

Soap needs to be stable on storage and not affected by exposure to air (autoxidation), which could lead to rancidity or discoloration. Autoxidation is accelerated by catalysts such as copper and iron. The raw materials used in the preparation of the soap must be pretreated before saponification to remove color and insoluble impurities. The principal methods are water washing, adsorption bleaching with activated earth, and alkali refining.

The direct saponification of oils is carried out in large open pans or kettles using live steam (as a heat source) and agitation. In continuous methods the raw materials are mixed with strong caustic solution at 75°C in a sealed reactor under pressure. Saponification is complete within 2–3 min; the reaction is exothermic, and recirculation takes place in the presence of preformed soap, which acts as an autocatalyst. The soap is recovered by washing with hot brine (to remove the glycerol) and possibly other purification processes depending on the raw materials. The final product is neat soap, which is a viscous liquid at 90°C and is 30–33% water. This needs drying (usually spray drying) and appropriate additives in order to become a usable soap. The recovery and purification of the glycerol is very important for economic reasons.

The neutralization of fatty acids is a simpler process and is usually carried out by mixing with caustic soda solutions at 110–120°C in a turbodisperser with recycle. No further purification is needed.

Soap made from the methyl esters of fatty acids is made in a manner similar to the neutralization of fatty acids except that the methanol that is formed has to be separated out.

Soap manufacture was originated by empirical methods, and it was found that the solid soap produced had different properties depending on its water content and the heat treatment given to the soap. The reason for this is that soap exists in a polymorphic crystalline state at normal room temperature. At least four crystalline states have been identified by X-ray diffraction methods and designated α, β, δ, ω by Ferguson et al. [5]. Toilet soap contains predominantly the β phase and is formed in the soapmaking process by vacuum drying and rapid cooling. The β phase gives a high degree of foam and swells easily in water, whereas the other phases produce less foam and the soap tablet (bar) cracks easily.

B. Fatty Ester Sulfonates or Alpha-Sulfonated Acids and Esters

Sulfonation at the alpha position of the fatty esters does not occur readily, and sulfur trioxide must be used, leading to problems of poor color, particularly if the starting esters have unsaturated groups. The sulfonation reaction is complex (see Fig. 5), which summarizes the overall reaction and indicates the complexity of the reaction and the formation of "disulfonates." At the end of the reaction there is a mixture of alpha-sulfonated ester acid and disulfonate. Neutralization with sodium hydroxide gives the results shown in Fig. 6.

The neutralization of a sulfonated fatty acid ester will give

1. The monosodium sulfonate salt of the ester, ~80% of the active weight of constituents.
2. The disodium salt of the alpha-sulfonated carboxylate, ~20% of the active weight of constituents.
3. The sodium salt of the fatty acid produced by hydrolysis of the starting ester that was not sulfonated, that is, soap, ~3% of the active weight of constituents.

Thus commercial products will be a mixture of these three products.

Sulfonation is usually carried out in a continuous plant very similar to the one described in the discussion of the preparation of alkyl sulfates (see Section IV). Even under optimum

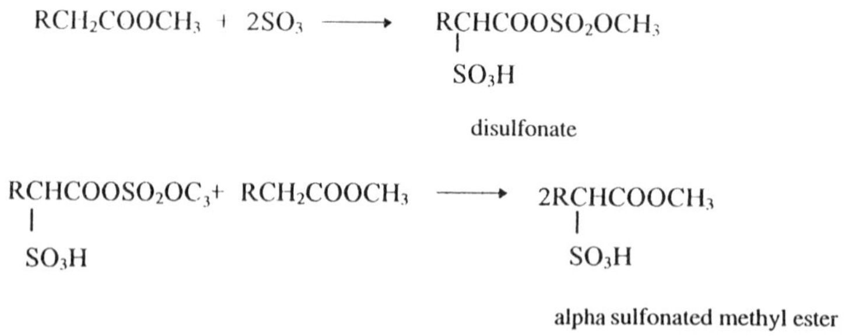

Fig. 5 Sulfonation of fatty esters.

sulfonation, dark-colored products are reported in the literature, and special bleaching of the acid and salts is often needed.

The solubility of the salts where R is a C_{12}–C_{14} alkyl chain is excellent, but the solubility of the salts where R is a C_{16}–C_{18} alkyl chain is poor compared to other detergent sulfonates. Salts of the disulfonate will tend to crystallize out of solution. The commercial products can be solubilized with hydrotropes.

The lowest surface tensions obtainable are for the C_{14}, 34 mN/m; for the C_{16}, 33 mN/m; and for the C_{18}, 33 mN/m. The foaming properties of the C_{12}–C_{14} are better than those of the C_{16}–C_{18} sulfonated fatty esters; thus alpha-sulfomethyl tallowate gives low foam. C_{16}–C_{18} based detergents showed superior detergency to LABS in the absence of polyphosphates at low concentration and at high water hardness.

C. Sarcosinates, Taurates, and Isethionates

The sarcosinates, taurates, and isethionates are produced from fatty acids by first making the acid chloride and then reacting this with the appropriate active group as shown in Fig. 7. More details of the preparation of these surfactants can be found in Porter [4, pp. 99–168]. The need to make the acid chloride intermediate has severely restricted the use of these products compared to many other surfactants, mainly on account of the extra cost. In addition, the products will contain considerable quantities of sodium chloride unless they are purified, which again adds to the cost.

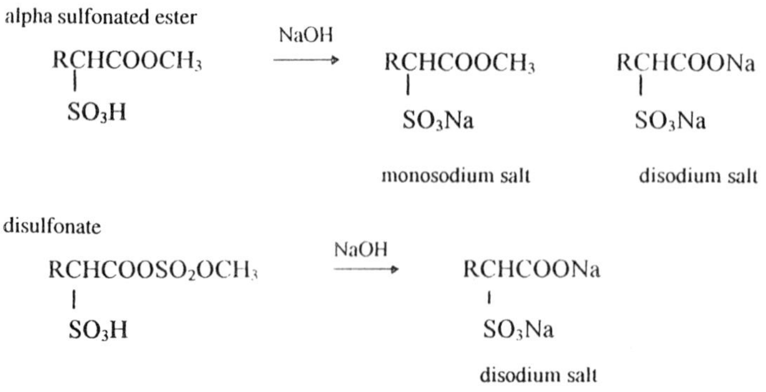

Fig. 6 Neutralization of sulfonated esters.

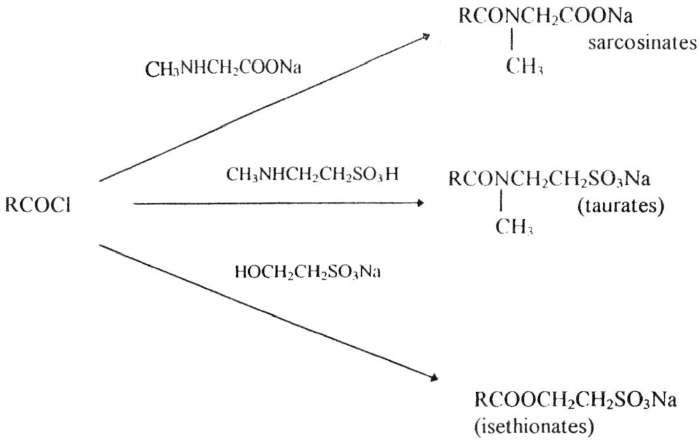

Fig. 7 Manufacture of sarcosinates, taurates, and isethionates.

The sarcosinates and taurates can be looked upon as modified soaps with the alkaline salt of the carboxy group still present but modified with an additional polar group—in the case of the sarcosinate with the amide group, and in the case of the taurates with a sulfonate group. Either of these additional groups improves water solubility and reduces the effect of hard water. In the case of the isethionate, the carboxyl group has been replaced with the sulfate group, again improving water solubility and reducing the adverse effects of hard water.

Before the introduction of synthetic surfactants from petroleum, these three products were used to overcome the disadvantages of soap in hard water. The introduction of cheaper petrochemical-based products led to the decline in their use. However, there is now considerable interest in this group of products, particularly the isethionates, due to their non-petroleum origins, and they have been found to cause little eye and skin irritation. Excellent shampoos can be formulated due to their high foaming properties.

D. Sulfated Glycerides

Reaction of sulfuric acid (usually with oleum) with the glyceride of a fatty acid, i.e., an animal fat or vegetable oil, will usually give a complex mixture of sulfates and sulfonates. The type of products formed will depend on the nature of the glycerides and the conditions of sulfation. The most common reactions are the following.

1. If a hydroxyl group (e.g., castor oil) is present, then sulfation of the hydroxyl group proceeds preferentially.

$$\mathrm{R{-}OH + H_2SO_4 \rightarrow R{-}OSO_3H}$$

2. If a carbon–carbon double bond in the hydrocarbon chain of the fatty acid (e.g., oleic acid) is present, this will sulfonate:

$$\mathrm{{-}CH{=}CH{-} + H_2SO_4 \rightarrow {-}CH(SO_3H)CH_2{-}}$$

3. If a fatty monoglyceride or diglyceride is present, then sulfation of the free —OH group proceeds as in 1, above. Hydrolysis of the ester group in triglycerides will occur in acid solution followed by sulfation of the hydroxyl group. Alternatively, transesterification of triglycerides in acid solution can take place if excess glycerol is present.

Even with one specific oil, a large number of products with different properties can be made depending on the conditions of sulfation and the addition of fatty acid, soap glycerol, glycols, or alcohols. Neutralization will give mixtures of sulfated products, soap, and free oil, very often in the form of emulsions and microemulsions.

The most common sulfated oils are Turkey red oil (ricinoleic acid triglyceride) made from castor oil; sulfated methyl and ethyl ricinoleates; sulfated methyl esters such as butyl oleate; and fish oil, lard oil, tallow, palmkernel oil, tall oil, and rapeseed oil sulfates.

Although the products are complex mixtures, they are all sulfate esters and subject to acid and alkaline hydrolysis. In addition, any ester groups from the glyceride portion of the molecule will also be susceptible to hydrolysis. The sulfated monoglycerides are excellent detergents. Partial glycerides containing unsaturated acids or hydroxy acids have more than one site available for sulfation and possess specific wetting, emulsification, and other properties.

Sulfated castor oils (e.g., Turkey red oil) will give good wetting, penetrating, and emulsifying properties if sulfated to a high degree. Sulfated methyl esters have good wetting properties but usually with low foam, which is rare in anionic surfactants.

The sulfated monoglycerides of coconut fatty acids were once made on a very large scale for use in household detergents but have been superseded by synthetic surfactants from petrochemicals. Sulfated castor oil was used as emulsifier for pine oil and creosote. The ammonium salt of cocoacid monoglyceride sulfate was used as the basis for a popular shampoo in the United States. The product is very similar to the equivalent lauryl sulfate, but due to the additional hydroxyl groups it has slightly better water solubility.

IV. SURFACTANTS DERIVED FROM FATTY ALCOHOLS

A. Fatty Alcohol Modification

The production of fatty alcohols has already been described in Section II, and many types of anionic, nonionic, and amphoteric surfactants can be made from these alcohols. However, the manufacture of anionic surfactants from fatty alcohols, e.g., by sulfation, is often preceded by modification of the fatty alcohol. Fatty alcohols will readily react with ethylene oxide, propylene oxide, butylene oxide, and mixtures of these oxides (see Fig. 8).

Figure 8 shows *n* molecules of ethylene oxide reacting with one molecule of a fatty alcohol. The overall result is the addition to the alkyl chain of the fatty alcohol (of a polyethoxy chain) of 2*n* carbon atoms joined by oxygen but with a primary alcohol group at the end of the product molecule. The resulting compound, a fatty alcohol ethoxylate, will react with chemical reagents in the same manner as the starting fatty alcohol as long as the chemical reagent does not attack the C—O—C group that is present in the polyethoxy chain. The polyethoxy chain confers increased water solubility on the fatty alcohol, and with a fatty alkyl group of C_{12-14} and $n=6-10$ the products are excellent wetting agents and detergents. By varying the amount of ethylene oxide, the water solubility of the starting fatty alcohol can be increased by increasing the value of *n* [4, p. 170]. The addition of propylene oxide or butylene oxide to fatty alcohols proceeds in exactly the same manner, i.e., the addition of a polypropoxy or polybutoxy chain to the fatty alcohol with a hydroxyl group at the end of the chain.

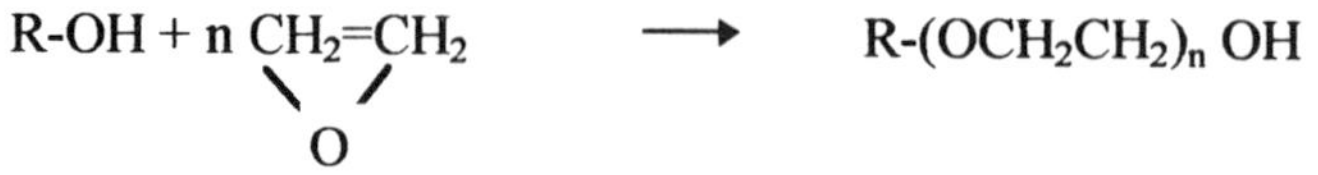

Fig. 8 Reaction of a fatty alcohol with ethylene oxide.

However, in the case of the addition of propylene oxide or butylene oxide the solubility of the fatty alcohol will decrease. The practical result of the addition of alkylene oxide is that starting with one fatty alcohol the water solubility or hydrophobic/hydrophilic properties of the raw material can be varied in a controlled manner.

B. Sulfation

In the period 1940–1960 sulfates from both lipids and petrochemicals were the largest class of surfactants, but during 1970–1990, they were replaced by the sulfonates from petrochemicals in terms of volume consumption. However, the alkyl sulfates are now being increasingly used in domestic detergents and consumption is increasing again.

Organic sulfates are the esters of sulfuric acid,

$$ROH + H_2SO_4 \rightarrow ROSO_3H + H_2O$$

The sulfur atom is joined to the carbon atom of the hydrophobic chain via an oxygen atom. The acid ester is unstable and can revert back readily to the alcohol and sulfuric acid (particularly in acid conditions), whereas the neutralized salts are stable at neutral pH. The acid hydrolysis releases sulfuric acid, which catalyzes the hydrolysis.

Acidic conditions:

$$ROSO_3H + H_2O \rightleftharpoons ROH + H_2SO_4$$

Alkaline conditions:

$$ROSO_3Na + NaOH \rightleftharpoons ROH + Na_2SO_4$$

The alcohol sulfates are therefore more stable in alkali than in acid. The stability depends on a number of factors, but for aqueous formulations, to give long-term shelf stability at room temperature, a pH range 5–9.5 is preferred. At elevated temperatures (>50°C); sulfates are unstable. It is for this reason that sulfation is generally carried out in continuous plants and the acid that is formed is immediately neutralized. Sulfation can be carried out with oleum, chlorsulfonic acid, or sulfur trioxide,

$$ROH + SO_3 \rightarrow ROSO_3H$$

The most efficient plants in terms of cost and economics use sulfur trioxide. See Fig. 9. The alcohol flows downward under gravity as a thin film on the inside walls of the tubular reactor, which is cooled. The sulfur trioxide and air pass downward cocurrently with the liquid phase. In the lower sections of the reactor, some separation of the liquid and gaseous phases takes place, but often the separation of the gases is carried out in a separate section (as shown in Fig. 9). The reaction is very fast in the tubular reactor; the air dilutes the reaction mixture and helps in cooling. The big advantage of this type of equipment is that very little degradation takes place, so very light colored materials can be produced. The limitations are in the molecular weight of the alcohol; below about a C_{10} chain length the fatty alcohol is volatile in the air stream, and at long chain lengths ($>C_{18}$) it is not liquid enough to form a thin film and flow downward. There are a number of different types of reactors, but most depend on the principles illustrated in Fig. 9.

The properties of the sulfates depend on the properties of the hydrocarbon chain and those of the sulfate group. The alkali metal salts show good solubility in water, and most surfactant sulfates show what is known as the salt effect. The addition of inorganic salts (sodium chloride, sodium sulfate) to dilute (below about 30%) solutions increases the viscosity, whereas addition to concentrated solutions decreases the viscosity. Any salts formed during the sulfation and/or neutralization can lead to viscosity changes.

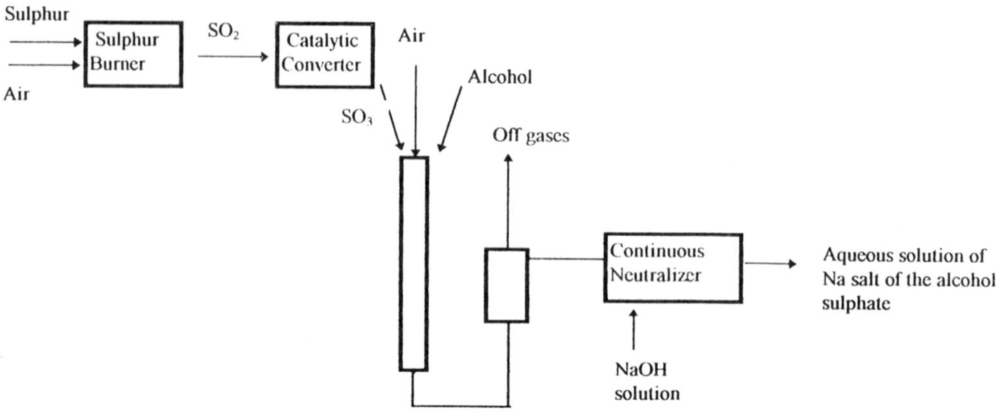

Fig. 9 Continuous sulfation plant using SO_3. (Courtesy of M. R. Porter.)

A sulfate group can replace a hydroxyl group at the end of the alkyl chain (from fatty alcohols) or the end of the polyethoxy chain (from the fatty alcohol ethoxylates). See Fig. 10. The addition of the sulfate group gives a marked increase in water solubility of the surfactant and generally gives high foam, good wetting, and excellent detergency if the alkyl chain is in the region of C_{12}–C_{16}.

The sulfated alcohols and sulfated ethoxy alcohols are by far the most important surfactants made from alcohols derived from lipids, but there are other speciality surfactants, which are summarized in Fig. 11. The R group in Fig. 11 can be either a paraffinic alkyl chain or a

$$\text{R-(OCH}_2\text{CH}_2)_n\text{OH} + \text{SO}_3 \longrightarrow \text{R-(OCH}_2\text{CH}_2)_n\text{OSO}_3\text{H}$$

Fig. 10 Sulfation of an ethoxylated alcohol.

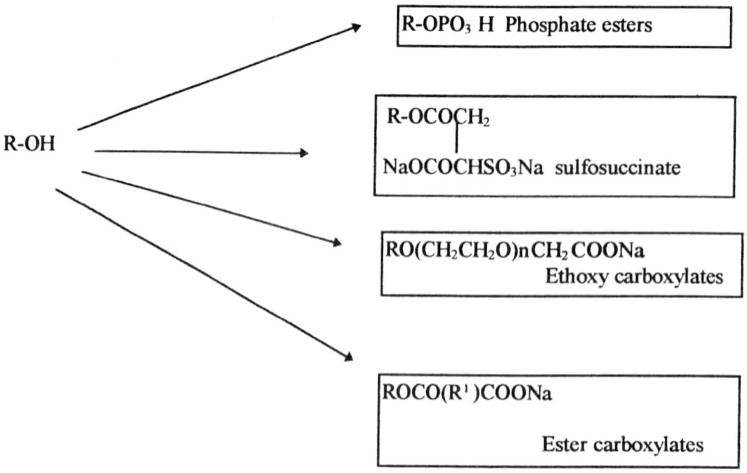

Fig. 11 Other specialty anionic surfactants.

modified chain with ethylene oxide, propylene oxide, or butylene oxide, and therefore countless variations are possible. All these specialty surfactants are manufactured and used in detergent formulations. They are very rarely used alone; they are nearly always used in conjunction with other anionic compounds as described here or with nonionic and/or amphoteric surfactants. Where mildness to skin and eyes is required, the ethoxy carboxylates have been shown to be particularly good and are also stable in electrolyte solutions [6].

V. HOUSEHOLD DETERGENTS

Approximately 50% of all synthetic detergents in western Europe are used in household products, and anionic surfactants are the major type used in detergent formulation. About 90% of soap is used in personal washing.

Within a household, many different types of detergents will be used, but there are three major groups of products, those used for

1. Personal washing, e.g., toilet soap, shampoos
2. Washing clothes
3. Washing hard surfaces, e.g., dishwashing products

Within each of these groups there are many individual products. This is in marked contrast to the situation 50 years ago, when often only one or two soap-based products would suffice. In washing clothes the development of the automatic washing machine has generated the need for quite different properties, e.g., low foam in a horizontal drum machine. Automatic dishwashers are becoming more common in the home and the modern tendency to shampoo every day has led to the need for mild shampoos. In addition, detergent manufacturers need to market different products for hard water areas than for soft water areas. Environmental demands, real or perceived, also generate different needs depending on the local legislation or public awareness. Finally, the activities of marketing departments will continually generate a need to make products "different" from those of their competitors. The result is that a typical 1990s household may have 20–30 individual detergents for household cleaning and personal use.

To satisfy all these demands, technical, economic, and environmental, requires a wide variety of surfactants possessing different properties and characteristics, but it is uneconomical to make and use a different surfactant in each different detergent. The answer has been found in formulating mixtures of chemicals that will give these different properties. The surfactant, or mixture of surfactants, is the main functional chemical in any detergent formulation. The properties of the surfactant can be modified considerably by adding other chemicals. In addition it is well known that mixtures of surfactants can often produce synergistic effects, i.e., that a mixture of surfactants can produce a particular property that cannot be obtained by any of the individual surfactants in the mixture. Thus the skill of the detergent manufacturer is not in the organic chemistry of surfactant manufacture but in the physical chemistry of mixing surfactants and other chemicals to produce the required physical properties at the appropriate cost. The formulation of detergents is still very much an empirical science, but in recent years there has developed a much better understanding of the theory of detergency, the theory of surfactant behavior, and the theory of synergism, which helps in understanding the mode of action of components in a formulation.

A. Personal Washing

Solid soap tablets (bars) are the preferred form of detergent for most personal uses with the major exception of shampoos (see Section VI). In 1990 the U.K. toilet soap market was

50,000 tonnes, and this has been slowly decreasing for some years. The reason for the reduction is the increasing popularity of non-soap-based products such as bathing and shower preparations, facial cleaners, and scrubs. Unlike the United States, there are no major brands of non-soap or "syndet" toilet bars based on synthetic detergents.

The traditional bar of soap was mainly tallow soap (80% tallow–20% coconut oil); the modern bar was originally formulated as a feminine cosmetic soap but has achieved more general use. It was originally made from a blend 50% beef tallow and 50% coconut oil with the addition of 8–9% fatty acid as the superfatting ingredient. Modern superfatted soaps are made from a blend of 60–70% tallow and 30–40% coconut oil with 5–7% coconut oil fatty acid added to the liquid soap prior to drying. Superfatted soaps give a smooth feel to the skin during and after washing due to the adsorption of the free fatty acid. They have a shorter in-use life than the traditional soap (80% tall oil–20% coconut oil) as they tend to form a mush in the presence of water. Soaps can also be superfatted with different emollient systems, e.g., light mineral oil, natural oils, fatty alcohols, polyethylene glycols. Other ingredients in soap can be dyes, perfumes, and opacifiers.

Transparent soaps are made from carefully selected blends of tallow, palm oil, coconut oil, palmkernel oil, oleic acid, and castor oil, and the glycerol is not removed but stays in the product. Translucent soaps are prepared from toilet soap formulations to which crystallization inhibitors have been added before drying and by increased mechanical working. The translucency is produced by a soap structure that has both solid and liquid phases, the two having different crystalline structures and refractive indexes. Working the soap and/or adding crystal inhibitors brings about a new equilibrium between the two phases in which the large crystals of the solid phase are reduced in size and some of the solid phase is changed into a liquid phase. Both of these increase the soap's translucency.

The most important characteristic of a toilet soap is the foam or lather. The volume and type will depend on the chain length of the fatty acids in the soap. Maximum foam production occurs with soap made from C_{12}–C_{14} saturated fatty acids, which is why coconut oil is a necessary component in toilet soaps, with the longer chain (C_{16}–C_{18}) fatty acids in the tallow soap being more responsible for the physical form of the soap. The increased level of C_{12}–C_{14} fatty acids in the superfatted soaps gives a much richer and denser foam than the old conventional soaps.

Baby soaps, those containing a minimum of nonfunctional additives, are made from the traditional (80% tallow–20% coconut) blend with the possible addition of glycerol as plasticizer. These are claimed to be "pure" soaps.

Medicated soaps are essentially toilet soaps that contain mild medicaments to assist in the treatment of minor skin problems associated with spots and pimples. Deodorant soaps are usually based on the traditional soap with a bactericide and a fragrance that is "fresh" smelling. Shaving soaps must produce a large volume of fine, stable lather. The lather must wet the beard and not dry out during use (be stable), and it must be not irritating to freshly shaven skin. Shaving soap is made from coconut oil and commercial stearin (80% stearic acid–20% palmitic acid) saponified with a mixture of sodium and potassium hydroxides with the addition of 5–6% glycerol and then adjusted to pH 7 with free fatty acid.

Liquid soaps are a small market, and most of the products found in domestic kitchens do not contain soap at all but are surfactant blends that are mild in nature (alkanolamides, ether carboxylates, betaines, etc.). True liquid soaps are made by the saponification of liquid fatty acids such as coconut or palmkernel oils and oleins with potassium hydroxide or blends of monoethanolamine and triethanolamine.

Non-soap bars are known as "syndet" (synthetic detergents) or "combo" (combinations of soap and synthetic detergents). Both types are designed to be superior to ordinary toilet soap in relation to their performance in hard water, in mildness, and in skin care (perceived or real). In Europe these products have not achieved any significant success compared to the United States, possibly due to the increased interest in specialty liquid products for bathing and showering. Most European products are based on the use of combinations of fatty alcohol sulfates (the fatty alcohol from lipids), isethionates (from fatty acids), and sulfosuccinates (from fatty alcohols, probably from lipids). Products used in the United States are based on sodium cocomonoglyceride sulfate, sodium cocoisethionate, and sodium cocoglyceryl ether sulfonate.

In recent years, showering has become the most popular form of personal hygiene. It is quick, efficient, and energy-saving and uses far less water than tub bathing. Liquid and gel shower preparations are tending to replace soap bars.

Special hand cleaners for removing oil and grease are also found in the home. These are described in Section V.D.

B. Washing Clothes

The nature of present-day textiles is extremely diverse due to the availability of various fibers, finishes, dyes, and manufacturing processes. Many fabrics cannot be easily washed in water but need to be dry-cleaned, a process that involves the use of specialty anionic surfactants (sulfosuccinates). However, the tendency is for more and more garments to be washed in automatic washing machines. Nevertheless there is still a wide variety in washable fibers and finishes. For instance, the proportion of colored materials in an average household wash is increasing. In the Federal Republic of Germany in 1970, 46% of clothes were white, but by 1980 the proportion of whites had fallen to 21%. Some textile dyes will bleed at high temperature, and this is one reason for the significant fall in average washing temperatures in the last 20 years.

Detergents for use in washing machines are known as heavy duty detergents to differentiate them from light duty detergents, which are used for hand washing temperature-sensitive garments such as woollens. Two typical formulations for a heavy duty detergent in powder form are shown in Table 4. These two formulations are typical, but actual formulations can vary widely. The chain length of the alkylbenzene, the fatty alcohol, and the soap; the amount of ethylene oxide on the fatty alcohol; and the alkali metal in the sulfonate and soap are not specified, so many variations are possible. Formulations in the United States, for instance, may well have a fatty alcohol ether sulfate instead of the fatty alcohol sulfate and probably much higher levels of the nonionic component. The two formulations shown represent mixtures containing phosphorus derivatives and those not containing phosphorus. The demand for non-phosphorus-containing formulations is entirely due to environmental needs in preventing algal growth in lakes and slow-moving rivers. The soap would be derived from lipids. The fatty alcohol for the ethoxylate and the fatty alcohol sulfate could be derived from lipids or petrochemicals; the choice would depend on their relative cost.

More recent trends have been the development of compact powders and liquid products. Liquid products have the advantage of cheaper production costs as the large spray driers used in powder manufacture are not needed and energy costs are reduced, but the packaging costs are higher for equivalent detergent performance. Liquid products are widely used in the United States but only to a small extent in Europe. The difference is due to the prevalence of a separate bleach step in the United States whereas the bleach is incorporated in the detergent in Europe. If the bleach is present in a liquid detergent it becomes unstable on storage, although there has been considerable research and development in liquid stable bleaches and enzyme systems. The major problem with liquid products is that the inorganic builders (e.g.,

Table 4 Heavy Duty Detergent Powder Formulations in Western Europe

Ingredient	Content (%) With P	Content (%) No P	Function
Linear alkylbenzene sulfonate	8	10	Main anionic surfactant; removes polar dirt
Fatty alcohol sulfate	2	—	Subsidiary anionic surfactant; synergistic effects
Fatty alcohol ethoxylate	7	5	Nonionic surfactant; removes oily dirt
Soap	3	3	Foam-controlling agent
Fatty acid monoethanolamide	1	—	Foam booster
Sodium triphosphate	30	—	Chelater and builder
Zeolite A	5	25	Ion exchange for hard water
Sodium carbonate	—	7	Alkali acts as builder
Sodium citrate	2	—	Co-builder
Sodium perborate	15	20	Bleach
Tetraacetyl ethylenediamine	1	1.5	Bleach activator
Ethylenediamine tetraacetate	0.3	0.3	Bleach stabilizer
Cellulose ether	1	1	Antiredeposition agent
Stilbene-disulfonic acid	0.2	0.2	Optical brightener
Sodium silicate	3	3	Corrosion inhibitor
Fragrance	q.s.	q.s.	Odor
Dyestuffs	q.s.	q.s.	Color
Sodium sulfate	Balance	Balance	Filler

q.s. = as necessary

phosphates) present solubility problems for the surfactants because an increase in electrolyte content will decrease the solubility of an anionic surfactant. This led to the development of liquid detergents without phosphate builders and a high concentration of surfactants. A typical formulation would have 30–40% active surfactant material of composition one-third linear alkylbenzene sulfonate (LABS), one-third soap, and one-third fatty alcohol ethoxylate; the soap would be derived from lipids, but the fatty alcohol could be derived from petrochemicals or lipids, probably the former. In addition, products based on mixtures of fatty alcohol sulfates, soap, fatty alcohol ethoxylates, and amphoteric compounds (alkyl polyaminocarboxylates) that do not contain LABS or builders have been put forward.

"Compact detergents" is the name given to heavy duty powder detergents that have a much lower filler level than the ones shown in Table 4. Such products can be based on the use of fatty alcohol sulfates to wholly or partly replace the LABS of the conventional detergents. The fatty alcohol can be derived from lipids or petrochemicals.

There are a number of specialty detergents used for washing clothes, particularly in Europe. The major ones are intended for washing wool by hand and for washing curtains. Those based on synthetic surfactants are free of anionics and are based on nonionics with added cationics to give a soft handle. There are some products based entirely on flaked toilet soaps that are used to wash very delicate fabrics; the flake form is necessary to obtain rapid solubility in warm water.

The major "workhorse" in modern heavy duty detergents has been linear alkylbenzene sulfonate (LABS), and from a cost-effective point of view it is probably still the most efficient surfactant on which a heavy duty detergent can be based. However, there have been a number of criticisms of LABS, particularly with respect to biodegradation, leading to extensive research for an environmentally friendly substitute (see Section VIII). The major candidates have been the alkyl polyglycosides (APG), the sulfonated fatty acid esters (FES), and the alkyl sulfates (AS). The APGs are apparently nonionic in constitution but behave very similarly to anionics in aqueous solution [7] and are made from glucose and fatty alcohols [10], which can be derived from lipids. The FES are derived from the glycerides via the fatty acid or ester (see Section III). The alkyl sulfates are derived from the corresponding fatty alcohols, which can be sourced from either lipids or petrochemicals (see Section IV). With alkyl sulfates, the effect of chain length and addition of nonionics is important to obtaining an efficient heavy duty detergent [8].

C. Hard Surface Cleaners

There are numerous hard surfaces inside and outside a home—walls, floors, windows, ceilings, worktops, dishes, furniture, toilets, etc. The most common are dishes, which have to be washed several times a day. The primary purpose of all dishwashing agents is the removal of soil from hard surfaces, the soil consisting of food material residues. In the cleaning of other hard surfaces, the soil composition varies much more widely. The need for different detergents for the different types of soils often means that the most efficient cleaning agents are not compatible and often cannot be combined in one universal hard surface cleaner. If the article, e.g., a dish, can be completely immersed in the detergent, then the temperature can be raised and the cleaning will be much more efficient. For example, the use of a dishwashing machine gives a degree of cleanliness not possible using manual dishwashing. However, the majority of hard surfaces in the home cannot be immersed in the detergent. The surfactants used in formulations for hard surface cleaners are basically similar to heavy duty detergents, i.e., mixtures of anionics and nonionics, but they have much lower levels of builders or none at all.

1. Dishwashing

The oldest agent for washing dishes was the running stream of water. It was found that an alkaline aqueous solution (e.g., soda ash) was much more efficient than plain water combined with agitation but had the serious shortcoming of adverse skin reactions with manual dishwashing. These disadvantages are not present in machine dishwashing, and therefore the basic detergent in machines is an alkaline solution with phosphates and silicates with a low foam nonionic wetting surfactant.

In manual dishwashing the most difficult soil to remove consists of the fats, colorants, and tannins, which are all insoluble in water. However, there is usually vigorous agitation, and it has been found that a surfactant solution that will wet the greasy plates, disperse the solid particles, and emulsify the liquid insoluble oils and greases is perfectly adequate without the need for alkali or builders. The surfactant must operate at relatively low temperatures and in hard water areas; this excludes most soaps. Surfactants with the sulfate or sulfonate group are very cost-effective dishwashing detergents; linear alkylbenzene sulfonates, α-olefin sulfonates, paraffin sulfonates, fatty alcohol sulfates, fatty alcohol ether sulfates, and fatty ester sulfonates (FES) can all be used to make efficient dishwashing detergents either alone or in mixtures. Probably the best known mixture is the 3:1 mixture of a linear sodium salt of a C_{12-15} alkylbenzene sulfonate and the sodium salt of a C_{12-14} fatty alcohol plus 3 mol of ethylene oxide sulfate. This mixture has been very widely used as it shows a synergistic detergent effect. Normally small quantities of a hydrotrope (sodium xylene sulfonate) and a foam stabilizing

Table 5 Typical Manual Washing-Up Formulation

Component	Percent (w/w)
LABS	12
C_{12-16} alcohol, $(EO)_3SO_4Na$	4
C_{12-14} diethanolamide	2
Sodium toluene sulfonate	2
Salt	As needed to thicken
Perfume/coloring	q.s.
Water	To 100

q.s. = as necessary

agent (coconut diethanolamide) are present plus perfume, coloring, and salt (which thickens anionic surfactant solutions). The surfactants are generally derived from petroleum as such surfactants are usually cheaper, but the fatty alcohol ether sulfate could originate from lipids, and FES formulations (based on fatty acids) are very efficient. A typical formulation is shown in Table 5.

In recent years there have been some changes in formulations, mainly to make milder (non-skin-irritating) products, as the surfactant mixtures described above readily defat the human skin, which can lead to skin irritation problems. The major changes are the replacement of LABS by fatty alcohol sulfates or fatty alcohol ethoxy sulfates and/or the addition of amphoterics, particularly the alkyl polyaminocarboxylates. The replacement of LABS by fatty alcohol sulfates can reduce skin irritation, but the addition of amphoterics to anionics has a profound effect on reducing skin and eye irritation [9]. There are also claims that paraffin sulfonates–amphoteric and fatty ester sulfonate–amphoteric mixtures are equally cost-effective, and it is not clear which will be the most cost effective formulations. If these new dishwashing detergents are claimed to be made from natural raw materials, then it is likely that the fatty alcohols used in the fatty alcohol sulfates will originate from lipids. Fatty acid derivatives such as the fatty ester sulfonates and the alkyl polyaminocarboxylates would also seem to have a bright future. There is also interest in formulations with products that are non-petroleum-based. A recent German patent [10] specifies alkyl polyglycosides as the nonionic component with anionic components from non-petroleum sources.

2. *General-Purpose Hard Surface Cleaners*

Hard surface cleaners for walls, floors, bathtubs, washbasins, worktops, etc. are liquids of a composition similar to that of the washing-up liquids but with the addition of a builder and a solvent for the oily material. A wide variety of surfactants can be used as long as they are soluble and the chain length is the optimum for detergency. Alcohol sulfates, fatty alcohol ether sulfates, paraffin sulfonates, and fatty ester sulfonates can all be used. A typical formulation using ether sulfates is shown in Table 6.

3. *Household Disinfectants and Detergent Sanitizers*

Household disinfectants are commonly used in many situations. A concentrated solution is poured into an excess of water to produce a cloudy appearance with the separation of insoluble germicidal products that will adhere to solid surfaces. At one time the surfactant used was a soluble soap produced by saponification of castor oil with caustic soda. The major constituent of castor oil is 12-hydroxyoleic acid, and the hydroxyl group gives increased water solubility

Table 6 General Hard Surface Cleaner

Component	Percent (w/w)
C_{12-14} fatty alcohol ether (2EO) sulfate, sodium salt	6
Tetrapotassium pyrophosphate	5
Sodium carbonate	2
Butyl cellusolve	2
Perfume/preservative	q.s.
Water	To 100

q.s. = as necessary

to the soap. The disinfectant would be a 10% active solution of the castor oil soap with 2–5% chlorinated xylenols as bactericide but would also contain a small quantity of pine oil or terpene derivatives to give a "fresh pine" smell. In recent years synthetic surfactants (ether sulfates) have replaced the castor oil soap because of the price volatility of the castor oil.

Detergent sanitizers are normal detergents that have added bactericides that will leave surfaces both clean and sterile, e.g., for toilets and bathroom tiles. The disinfectant described above will show detergent properties if the surfactant content is increased. Sodium hypochlorite is a cheap but very effective bactericide and is used in household products, but the addition of many surfactants will destabilize and deactivate the bactericide, so the choice of surfactant is very important.

VI. SHAMPOOS

A. Shampoo Formulations

Hair naturally has an oily layer that tends to collect dirt. A shampoo should remove the dirt and some or most of the oil, which has to be replaced or the hair will appear dry and be harsh to the touch. Hence the need for conditioners that will add "body" to the hair and replace the oil until the natural oil is renewed. Anionic detergents have been found to be very efficient for removing the soil and the natural oil. The most useful conditioners are the cationics, which are normally incompatible with anionics and are best applied separately. Alkanolamides, although considered nonionic surfactants, are compatible with anionic surfactants yet have enough substantivity to hair to act as conditioners. "Conditioning" is a phrase that covers prevention of dry hair static, increased body, mending of split ends, improved combability, and improved gloss. The alkanolamides also act as foam-stabilizing agents in conjunction with anionics. Many formulations therefore contain mixtures of anionics (fatty alcohol sulfates or fatty alcohol ether sulfates) and alkanolamides. A typical simple formulation for a shampoo in the form of a clear gel is given in Table 7. The addition of sodium chloride will thicken the aqueous solution to a thick gel. Most anionics, particularly fatty alcohol ether sulfates, will increase in viscosity when an electrolyte is added. This is known as the salt effect and can be very dramatic, the viscosity increase changing a low viscosity liquid into a thick gel by the addition of only 1–3% of sodium chloride. The exact amount of sodium chloride will depend on impurities in the ether sulfate, the amount and type of alkanolamide, and the amount and type of betaine present.

Shampoos are generally used straight from the bottle, and a thin liquid shampoo is not easy to use, particularly in a shower. Therefore control of the rheology of the shampoo is

Table 7 Shampoo Formulations

Component	Percent (w/w)	
	Clear gel	Frequent wash
Fatty alcohol ether (2EO) sulfate, sodium salt	15	7
Coconut diethanolamide	4	1.5
Cocoamidopropylbetaine	—	3
Sodium chloride	q.s.	q.s.
Preservative/coloring/perfume	q.s.	q.s.
Water	To 100	To 100

q.s. = as necessary

very important. Shampoos are made in the form of clear liquids, lotion, cremes, pearlized liquids, and gels. The rheology and appearance is controlled by the type of anionic detergent, the type of alkanolamide, and the addition of electrolyte or other additives. The properties of the foam generated during shampooing will depend on these same factors, and therefore the exact formulation will depend on the assessment and testing of three factors—the rheology of the shampoo, the type and stability of the foam produced during shampooing, and the cleanliness and conditioning of the hair.

The ether sulfate can be replaced by fatty alcohol sulfate, but the ammonium or triethanolamine salt is preferred to give better solubility. The fatty alcohol used in the manufacture of these surfactants can originate from either lipids or petrochemicals, which function equally efficiently, but most large shampoo manufacturers will specify a lipid origin or a composition that is indistinguishable from one derived from lipids. The optimum chain length of the fatty alcohol has been found to be in the C_{10-14} range, with C_{12} being predominant. The ether sulfates will give large copious foams. The fatty alcohol sulfates give denser, creamier foams depending on the distribution of the alkyl chain lengths.

The gel formulation shown in Table 7 is very efficient and will clean both dirt and natural oil from hair, particularly if used frequently. Reduction in the concentration of the ether sulfate and addition of a cocoamidopropylbetaine will modify the detergent properties and conditioning properties. The frequent wash formulation shown in Table 7 helps to prevent damage to the hair and improves the conditioning.

B. Frequent Wash Formulations and "Mild" Surfactants

In recent years the shampoo market has changed to meet the demands of a more discerning consumer with a busy lifestyle. There has been a trend toward milder formulations because of the tendency to wash the hair every day and to wash it while showering. Daily shampooing can cause damage to the hair (e.g., split ends) but also prevents the formation of the natural oils that lubricate and help manage the hair. In addition, shampooing during a shower can facilitate the entry of the shampoo into the eyes, and hence very low eye irritation becomes a requirement. Low irritation is generally accompanied by low skin irritation, which in itself is highly desirable. Shampoos for babies are probably those needing the mildest conditions.

There are various routes to a mild shampoo. The easiest and cheapest is to supply a much diluted normal shampoo, but these usually suffer from inadequate performance. A second

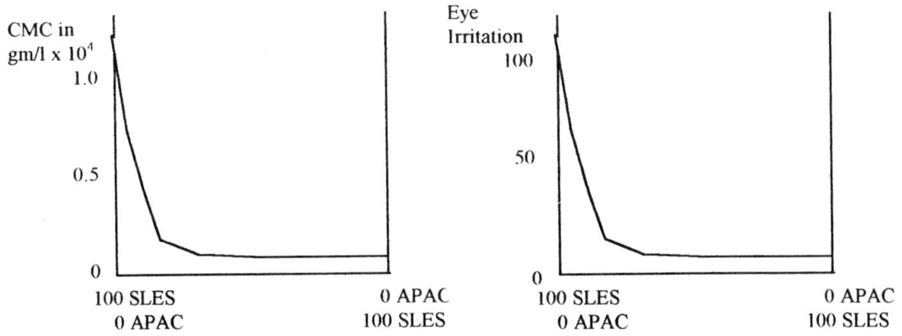

Fig. 12 Detoxification of ethyl sulfates with alkyl polyaminobetaines. (From Ref. 9.)

route is replacement of the major eye irritant in the standard shampoos. This was found to be the anionic fatty alcohol sulfate or fatty alcohol ether sulfate. Variations in the alkyl chain length, the cation (magnesium is less irritant than sodium), and the degree of ethoxylation could reduce the irritancy but not to a low enough level without adversely affecting performance. Complete replacement with imidazoline-based amphoteric surfactants, sulfosuccinates, sarcosinates, and especially ethoxycarboxylates, all of which are derived from lipids, is technically possible to give excellent shampoos but is very expensive. Nevertheless, the amphoterics have been used in the United States for many years in Johnson and Johnson baby shampoo. Finally, a mild shampoo can be achieved by partial replacement of the anionics, particularly of fatty alcohol ether sulfates by other surfactants. This was found to give a cost-effective solution, as it was found that the addition of small quantities of some surfactants to anionics reduced the eye irritation to a very large degree. Addition of betaines, sulfosuccinates, sarcosinates, ethoxycarboxylates, amidopropylbetaines, or alkyl polyaminobetaines can give this effect, known as detoxification. Figure 12 illustrates detoxification with alkyl polyaminobetaines.

SLES is the sodium salt of a C_{12-14} fatty alcohol ethoxylated with 2 or 3 mol of ethylene oxide and sulfated with sulfur trioxide. APAC is the tallow (C_{16-18}) alkyl polyaminocarboxylate. The curves show remarkable similarity and show that eye irritation is closely related to the critical micelle concentration (CMC). Lomax [9] describes this effect as due to the anionic SLES forming a mixed micelle, which gives a lower CMC. The effect of a lower CMC is that there are fewer free monomeric surfactant molecules outside the micelles (as long as the concentration is above the CMC). Eye irritation can be due to the adsorption of surfactant monomer at the eye protein/phospholipid interface, and hence the lower the CMC the lower the eye irritation. This is a simplistic view, but the experimental evidence clearly shows that eye irritation is lower at lower CMC, at least for similar chemical species. The practical effect is that anionic surfactants such as fatty alcohol ether sulfates that have excellent detergency on hair but show eye irritation may have the latter reduced by the addition of a small quantity of the tallow (C_{16-18}) alkyl polyaminocarboxylate.

C. Conditioning Shampoos

The separate application of cationic surfactants or cationic polymers is undoubtedly the most efficient way of imparting softness and conditioning to hair. Cationics are normally incompatible with anionics in solution and will precipitate out as a anionic–cationic complex. A single shampoo–conditioner mixture made with anionic or cationic surfactants is difficult to formulate. Careful choice of the cationic is possible, but a more practical solution is the discovery that some anionic surfactants are compatible with the cationic polymers used as softeners. The ethoxy carboxylates, which can be used to partly (20–40%) replace the fatty alcohol sulfates or fatty alcohol ethoxy sulfates in these formulations, will give some compatibility with cationic surfactants.

D. Bath Additives

The principal bath additive is for foam (bubble) baths. Foam bath formulations look basically very similar to shampoo formulations—there is the same need for rheology control, low skin and eye irritation, and copious foam—but there are some important differences. Bubble baths are often used with bath oils that are insoluble in water, and therefore emulsifying properties are required. Shampoos are applied directly from the bottle onto the hair, whereas the foam bath is highly diluted. This feature is important to the requirements to avoid skin and eye irritation, and therefore formulations suitable for foam baths may be quite unsuitable for mild shampoos. Toilet soap is usually used in a bath, and the foam in a bubble bath must be stable to the addition of soap. It should be stable in warm water and to both hard and soft water. However, the foam must not be too stable, as it must vanish on draining the bath and/or collapse with the addition of cold water. The typical surfactants used are the fatty alcohol ethoxy sulfates, which may be derived from either lipids or petrochemicals.

E. Shower Gels

Normal shampoos can be and often are used for washing the body, but they generally generate too much foam to be efficient. The shower gel cleanser formulations are based on the shampoo formulations but need greater control of the rheology and also less foam. The major need is to prevent an overly rapid loss of product under the shower spray. Consequently the products are usually in the form of high viscosity liquids or gels. The active concentration of surfactant is generally higher than in shampoos.

VII. INDUSTRIAL AND INSTITUTIONAL DETERGENTS

A. Definition

Industrial and institutional detergents are very often considered as one group in statistics on surfactant consumption, and there is confusion on definition of these end uses. Industrial detergents are used by industry as part of the process of manufacturing some other product, e.g., scouring raw cotton with detergent in the spinning of cotton. Such detergents have been formulated for a specific purpose, and there can be many specialized formulations depending on the particular industry. On the other hand, institutional detergents used by hotels, catering establishments, large offices, etc. are normally very similar in formulation to domestic detergents. However, very large catering establishments will have large automated equipment, e.g., dishwashing equipment that automatically dispenses the detergent, and they will need a specially formulated detergent that is not used in the home. Hospitals may insist on a higher standard of hygiene and use germicidal detergents. It is difficult to distinguish between the two market areas of institutional and industrial cleaning. Many small hotels, small restaurants, pubs, etc. use ordinary domestic detergents bought in supermarkets.

B. Disinfectants and Biocidal Cleaners

Many detergent products are disinfectant and/or biocidal cleaners. Biocides are products that either kill bacteria or control the level of bacteria and prevent them multiplying. Disinfectants usually have biocidal action but do not give any detergent action, i.e., they do not clean. They are usually used on relatively clean surfaces that have previously been cleaned with a detergent. Biocidal detergents are products that clean and have some biocidal action. Nearly all detergents will remove bacteria from surfaces, but very few detergents have a biocidal action as well. It can be very difficult to distinguish between detergent use and disinfectant use in the technical sense.

The technology ranges from the very simple to extremely complex products. In use, the concentration of the active ingredients can vary from very low (just hot, very dilute alkali solution with zero surfactant is used in some bottle-washing applications) to up to 90% in the case of dry cleaning detergents. Some hospital applications (e.g., surgical scrubs) have great demands on performance and cost is secondary. This is not general, as most users are very concerned about the cost of the cleaning operation and will assess performance more objectively than the housewife. This results in products being sold on a price/performance basis rather than by advertising.

C. Major Industrial and Institutional Areas for Detergents

1. Textile Cleaning

Natural fibers such as wool and cotton contain dirt that must be removed before processing. During processing both natural and synthetic fibers are treated with processing aids (e.g., spinning oils and lubricants) that are generally removed prior to dyeing. The process of dyeing often requires removal of excess and loose dyestuffs. Therefore many specialty detergents and surfactants are used. Foam is normally a nuisance, so formulations are seldom based on anionic surfactants and more often based on synthetic nonionics from petrochemicals.

2. Commercial Laundry

Depending on the size of the laundry, three different types of machine systems are used. The first type is simply a standard washing machine and dryer similar to the domestic versions that have been modified to take larger loads. This type of machine would use the same detergents as domestic machines. The second type is the conventional "batch" machine, and the third type is the counterflow or "continuous" machine, which has become more common in the last 20 years. The latter is much more efficient but also capital-intensive.

In hard water areas most commercial laundries have water softening equipment, and therefore soap-based products can give excellent results. The relative use of soap and synthetic surfactants does not depend only on the relative price. Soap is more difficult to rinse off clothes than linear alkylbenzene sulfonates and the fatty alcohol and fatty alcohol ethoxy sulfates. In addition, soap is an excellent detergent for cotton but not so good for synthetic textiles such as polyester. The synthetic fabrics, particularly polyester–cotton blends, are becoming a much larger fraction of commercial laundry loads, so the use of soap would be expected to decrease. Lower washing temperatures, which save energy, will also lead to a decline in the use of soap. Large commercial laundries can formulate their own products from surfactants, builders, bleach, etc. The use of soap versus synthetic detergent varies widely from country to country.

3. *Dry Cleaning*

Dry cleaning solvents contain a small quantity of water and an anionic surfactant to aid in the removal of water-soluble stains. The surfactant is usually a petrochemical-based anionic product.

4. *Hard Surface Cleaners*

A very common application of detergents in industrial and institutional settings is the cleaning of hard surfaces. Special detergents are needed when the soil is of a special nature, e.g., rust on metal, burned carbon on cooking utensils, dried blood in abattoirs, or where the soil is difficult to get at as in beer pumps and dairy equipment. With mechanical application, spray equipment, foaming must be avoided and anionics are not used; nonionics and amphoterics are more appropriate. For manual application as in offices and hospitals, formulations very similar to household products are normally used. Cream cleaners (containing anionics) are used undiluted for application on vitreous enamel, stainless steel, and ceramic surfaces.

There are many special hard surface cleaners, particularly for dairies, food processing plants, and breweries. In these applications a disinfectant is used after cleaning with the detergent, the main biocide being hypochlorite. Other disinfectants include iodophores, quaternaries (e.g., benzalkonium chloride), and chlorphenolics. Formulated biocidal detergents are blends of nonionics with quaternaries, hypochlorite, or chlorinated isocyanurates.

The choice of one type of product over another depends on a number of factors, the main ones being cost, government approval, and freedom from taste or taint. Anionics are rarely used for hard surface cleaning because of high foam.

If the soiling is of such a nature that foaming is not a problem, then standard anionics can be used. A typical powder detergent formulation is given in Table 8, and a typical liquid detergent formulation in Table 9.

Table 8 Powder Detergent for Hard Surface Cleaning

Component	Percent (w/w)
Trisodium phosphate	70
Soda ash	25
Sodium dodecylbenzene sulfonate flake	5

Table 9 Liquid Detergent for Hard Surface Cleaning

Component	Percent (w/w)
Tetrapotassium pyrophosphate	10
Sodium metasilicate	10
Sodium dodecylbenzene sulfonate	3
Phosphate ester	4
Water	To 100

5. *Metal Cleaners*

Metal cleaning in a manufacturing plant involves degreasing and derusting. This is usually done with the use of high pressure sprays, which precludes anionic surfactants except for special phosphate esters (based on fatty alcohols), which have excellent electrolyte compatibility and relatively low foam.

6. *Machine Dishwashing*

Automatic dishwashing detergents depend on the detergency and alkalinity of the detergent to remove proteinaceous soils. The hot alkali can cause saponification, which gives a foam that is stabilized by the protein present. It is essential to prevent the formation of foam, and therefore an antifoam agent (not a defoamer) must be present. In addition, the antifoam agent must be stable to alkali. EO-PO copolymers and alcohol EO-PO condensates are good antifoam agents, but if there is a high concentration of alkali, then the primary hydroxyl group must be blocked because of its instability to alkali.

7. *Food Processing Cleaners*

Of the many types of food processing equipment that require cleaning the simplest is a tank containing residues that can be filled with a detergent solution, which is heated and stirred, and then drained off. In this situation an anionic alkaline solution gives excellent service, and a typical formulation for cleaning beer tanks is shown in Table 10. However, in the majority of cases, foaming must be controlled, so anionics are seldom used.

8. *Transport Cleaners*

There are many detergent products used in transport, in car manufacture (e.g., preunderseal cleaning applied in conjunction with high pressure steam), in servicing (e.g., engine degreasing, which uses anionics), car shampoos, and traffic dirt cleaners for buses, trucks, and trains. Wherever possible the anionics (normal linear alkylbenzene sulfonates) are used, but where spray is used (e.g., for cleaning motortrucks), then amphoterics tend to be used.

D. Hand Cleaners and Gels

Hand cleaners and gels for industrial use must be capable of removing grease and oil from hands. Liquid soaps for washing hands are based on fatty alcohol sulfates or ethoxy sulfates rather than soap. A typical formulation is shown in Table 11. The fatty alcohol can be based on lipids or petrochemicals. Usually cost is the major consideration.

Industrial hand cleaners are also sold as a clear gel with a marketing gimmick of a "ringing" gel. The principal components are surfactants, a solvent (usually deodorized kerosene),

Table 10 Beer Tank Cleaner Formulation

Component	Percent (w/w)
Sodium dodecylbenzene sulfonate	7
Sodium xylene sulfonate	7
Sodium metasilicate	10
Tetrapotassium pyrophosphate	10
Water	To 100

Table 11 Liquid Soap

Component	Percent (w/w)
Monoethanolamine lauryl sulfate (33%)	50
Cocomonoethanolamide	3
Sodium chloride	0.5
Formaldehyde	0.2
Perfume and color	q.s.
Water	To 100

and an alcohol or glycol. The product is often in the form of a microemulsion. The formulation of a typical soap-based product is given in Table 12. Soap is used in this formulation, but synthetic surfactants are also used instead.

Anionics make very good detergents but with a few exceptions give high foaming products. Foam is usually a nuisance in industry, and therefore anionics tend to be used in a minor way where detergency is required.

There is a very wide and diverse use of anionics in industry in nondetergent uses involving wetting, dispersing, emulsifying, and foaming properties. The main industries are agricultural chemicals, cement manufacture, corrosion inhibitors, food, leather, paint, emulsion polymerization, paper, textiles, and polishes. Petrochemical-based anionics are mainly used. Lipid-based alcohols are perfectly satisfactory raw material bases for fatty alcohol sulfates and ethoxy sulfates, and the only determining factor is the price and availability relative to the petrochemical-based materials.

VIII. FUTURE TRENDS AND ENVIRONMENTAL PROBLEMS

In the oil crisis of 1974, the cost of crude oil escalated suddenly and increased the cost of petrochemicals relative to those of to natural fats and oils. The synthetic surfactant manufacturers suddenly found the price of their raw materials dramatically higher, and they also realized that these prices were controlled by countries who may not always be friendly to the developing countries. At the same time it was widely believed that crude oil supplies would peter out in the not-so-far future. These two factors led to increased research and the commercial development of new surfactants using raw materials not based on crude oil and with properties superior to those of soap. Some of those products have now reached large-scale use and production; an example is the sulfonated methyl esters of fatty acids (see Table 2).

Table 12 "Ringing" Hand Gel (A hand gel which "rings" when tapped or shook.)

Component	Percent (w/w)
Deodorized kerosene	20
Oleic acid	10
Lauric diethanolamide	4
Hexylene glycol	5
Triethanolamine	3.5
Potassium hydroxide	0.2
Water	To 100

Although the price of crude oil has stabilized (at the time of writing), there will undoubtedly be severe fluctuations sometime in the future, which will ensure that surfactant manufacturers will need alternative raw materials.

The early synthetic anionic surfactants based on branched long-chain paraffinic hydrophobes were nonbiodegradable in natural streams and in sewage works and produced visible stable foam in bodies of water into which they were discharged. Although this problem was soon overcome by changing to linear paraffinic chains that biodegrade easily, the biodegradation problem will not go away, as new and more stringent methods of assessing biodegradability are being examined [11]. The present linear alkylbenzene sulfonates will biodegrade satisfactorily under aerobic conditions [12] but not under anaerobic conditions. This is not necessarily a practical problem in terms of environmental acceptability [13].

Early detergents based on petrochemical-based surfactants were yellow or brown in color, had a bad odor, and were harsh to the skin. No commercial surfactant is a pure chemical compound but a mixture, and many synthetic surfactants contain by-products such as tetralins and sulfones in linear alkylbenzene sulfonates [14]. Some of these products, e.g., 1,4-dioxane in fatty alcohol ether sulfates, although present in very small amounts, are known to be toxic in the pure form. In this latter case the 1,4-dioxane occurs as a result of the sulfation process on the polyethylene oxide chain [15] and will be present whether the starting material is lipid-based or petrochemical-based. However, the surfactants now used in household products are of extremely good color, have no odor, and have been thoroughly tested for impurities, and the toxicity of any such impurities to humans or the environment is negligible.

Price, toxicity of impurities, and biodegradability can all be quantitatively measured, but there are other factors operating to replace petrochemical-based surfactants with lipid-based materials that are not easily measured. At present there is a definite market requirement for more "natural" products. This has often been interpreted as a demand for non-petroleum-based products, because even soap uses synthetic chemicals in its manufacture. There is a also a demand for "mild" detergents. Although soap can adversely affect the skin and eyes, the average consumer has grown up with soap and is well aware of the problems. Some of the early synthetic products caused more severe skin irritation than soap and, in a few cases, severe skin sensitization to a small percentage of the population. However, considerable research has now developed surfactants, particularly modified soaps and amphoterics, that irritate the skin and eyes significantly less than soap. Such products are usually manufactured from "natural" raw material, i.e., fats and oils, so they can be marketed not only as "mild" but also as "green" and "natural." Mild surfactants can be defined as those producing low toxicity, zero irritation to the eyes, zero skin irritation, and zero skin sensitization, with fast biodegradation, no toxic impurities, and no problems with the environment. Such a product probably does not exist, but some new surfactants can get nearer to meeting these requirements than soap can.

The overall results of these factors will probably be a slow decline in the consumption of anionic synthetic detergents based on petrochemicals and their replacement by semisynthetic products based on natural products rather than soap. For instance, the major new surfactant type in the last decade is the alkyl polyglycosides, which use the polyhydroxy groups in carbohydrates as the hydrophilic entity, with a long-chain fatty alcohol (from fats, oils, or petrochemicals) as the hydrophobic entity. Lipids will probably be the major source for the hydrophobic group of such surfactants.

REFERENCES

1. D. Postlethwaite, A life-cycle inventory for the production of soap in Europe, *Tenside Surfact. Deterg. 32*: 157–169 (1995).

2. D. Sudati, Overview of processing steps in the production of intermediates for detergents from natural raw materials, *Chimaoggi July–August*: 53 (1989).
3. M. Biermann, F. Lange, R. Piorr, U. Ploog, H. Rutzen. J. Schindler, and R. Schmid, Synthesis of surfactants, in *Surfactants in Consumer Products, Theory, Technology and Application* (J. Falbe, Ed.), Springer-Verlag, Berlin, 1987, pp. 33–35.
4. M. R. Porter, *Handbook of Surfactants, 2nd ed.*, Chapman and Hall, London, 1994.
5. R. H. Ferguson et al., *Ind. Eng. Chem. 35*: 1005–1012 (1943).
6. E. Stroink, Ethercarboxylates for industrial and institutional applications, in *Industrial Applications of Surfactants*, 2nd ed. (D. R. Karsa, Ed.), Royal Soc. Chemistry, London, 1990, pp. 62–75.
7. D. Balzer, Alkylpolyglycosides, their physico-chemical properties and their uses, *Tenside Surfact. Deterg. 28*: 419–427 (1991).
8. S. Tagata, A. Ishikawa, and H. Hayashi, Effect of washing temperature and added nonionic surfactants on the detergency of alkyl sulfates, *Comun. J. Com. Esp. Deterg. 25*: 251, 253–266 (1994).
9. E. Lomax, The mechanism of surfactant detoxification, *Spec. Chem.*, November 1992, pp. 37–39.
10. U. Hees, R. Jeschke and M. Weuthen, Alkyl oligoglycoside-containing liquid cleaners for hard surfaces, Ger. Patent DE 4, 311,159 (1994).
11. H. A. Painter, Testing strategy and legal requirements in *Biodegradability of Surfactants* (D. R. Karsa and M. R. Porter, Eds.), Blackie, London, 1995, p. 118.
12. H. A. Painter, The question of the anaerobic biodegradability of linear alkylbenzene sulfonates, Proc. 3rd CESIO Int. Surfactants Congress and Exhibition, Sect. E, (1–5 June 1992), pp. 34–41.
13. R. R. Birch, W. E. Gledhill, R. J. Larson, and A. M. Nielson, Role of anaerobic biodegradability in the environmental acceptability of detergent materials, Proc. 3rd CESIO Int. Surfactants Congress and Exhibition, Sect. E, (1–5 June 1992), pp. 26–33.
14. J. P. Hughes, The determination of trace components in surfactants. *Crit. Rep. Appl. Chem. 32*: 90, 99 (1991).
15. K. L. Matheson, The effect of ethylene oxide adduct distribution on the formation of 1,4-dioxane during sulfation of alcohol ethoxylates, Proc. 3rd CESIO Int. Surfactants Congress and Exhibition, Sect. C, (1–5 June 1992), pp. 211–221.

24

Cationic Surfactants

Alan D. James
Akzo Nobel Chemicals Ltd., Littleborough, Lancashire, England

I. INTRODUCTION AND MARKETS

In this review of the market, manufacturing methods, properties, and applications of cationic surfactants, only commercially important products are considered in detail. Amphoteric surfactants are excluded from the discussion.

All commercially available cationic surfactants are nitrogen derivatives, and in most cases the alkyl portion of the surfactant is derived from natural fats and oils such as tallow or rapeseed, palm, fish, tall, or coconut oils. Cationic surfactants can be classified into four product groups, the manufacture and chemistry of which is described below: fatty amines, amidoamines and imidazolines, esteramines, and etheramines. Whereas in fatty amines the hydrophobic alkyl chain is directly attached to the cationic nitrogen, in the other product groups it is separated by an amido, imidazoline, ester, or ether group, respectively (Fig. 1).

The members of each product group can be derivatized to give a wide range of cationic and amphoteric surfactants, which can be subclassified according to their head group chemistry into, for example, quaternary ammonium salts ("quats"); *N*-alkylpropylene diamines ("diamines"); *N*,*N*-dimethylalkylamines (tertiary amines), ethoxylated amines, and amine oxides. Some of these classifications may not be mutually exclusive in a strict chemical sense.

Worldwide production of cationic surfactants can be estimated at 350,000–500,000 tonnes per annum (tpa). European production of fatty amines has been estimated at 120,000–150,000 tpa [1–3], amidoamines at ~ 30,000 [2], and esteramines at 61,000 tpa [4]. Estimates of North American production of fatty amines reach 200,000–250,000 tpa; amidoamines, 50,000 tpa [2]. Worldwide production of amine ethoxylates reaches 50,000 tpa [5], amine oxides 30,000 tpa [2], and diamines 18,000 tpa [2]. But a significant portion of the fatty amines and amidoamines produced are used as intermediates in the manufacture of amphoteric surfactants, oil additives, and other industrial products in which the cationic character of the amine is lost. Table 1 lists major manufacturers of cationic surfactants. Akzo Nobel is the largest producer [2,6] and Witco the second largest.

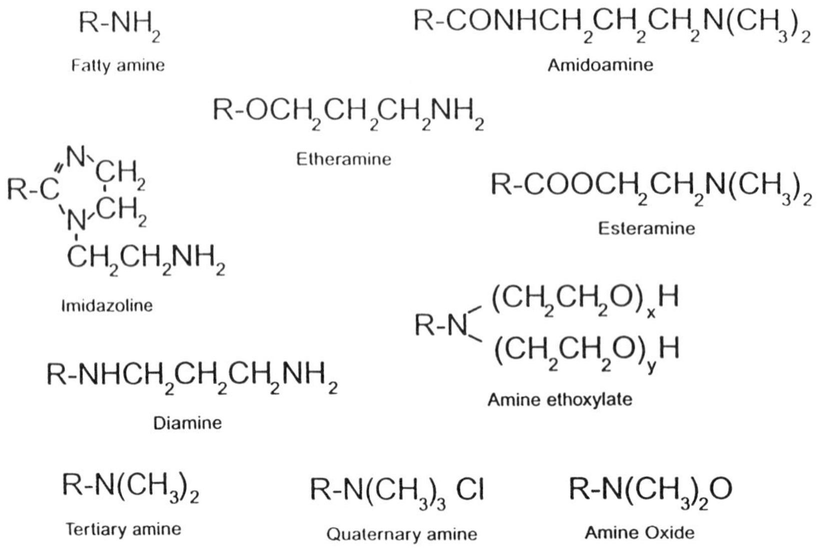

Fig. 1 Examples of cationic surfactants.

The most important market for cationic surfactants is the detergent and personal care sector, and the single most important application is as softening agents (fabric conditioners) [1] with a worldwide market of 200,000 tpa [4], 66,000 t in Europe. Industrial applications such as mineral processing, bituminous road construction, cleaning agents and biocides, agrochemicals, organoclay manufacture, and textile treatment each account for 5–10% of total cationic surfactant consumption.

Table 1 Some International Producers of Cationic Surfactants

Company [a]	Main manufacturing locations [b]	Chemical types produced [c]
Akzo Nobel	Sweden, U.K., U.S.A., Belgium, Canada, Brazil, Japan	FA, AA, EA, EST
Witco	U.S.A., Germany, Spain	FA, AA, EA, EST
Kao	Japan, Spain, Mexico, Malaysia	FA, AA, EST
Hoechst	Germany, U.K., Brazil	FA, AA, EA, EST
Elf	France	FA, AA, EST
Fina	Belgium	FA, AA, EST
Henkel	Germany, Spain	FA, EST
Stepan	U.S.A., France	EST, AA

[a] Ultimate owner.
[b] Including joint ventures.
[c] FA = fatty Amines; AA = amidoamines and imidazolines; EA = etheramines; EST = esteramines.

II. MANUFACTURE

A. Fatty Amines

1. The Nitrile Process

In the most important route to cationic surfactants, fatty acids or their methyl esters are reacted with ammonia to form nitriles, which are hydrogenated to primary and secondary amines [7]. Conversion of fatty acid to amide and then nitrile is carried out in a single batch or continuous process using alumina or zinc oxide as catalyst. Amide is not isolated.

$$RCOOH + NH_3 \rightarrow (RCONH_2) \rightarrow RCN + 2H_2O \tag{1}$$

In the batch process, nitrile is produced at 280–360°C in the liquid phase; then it is usually distilled. In the continuous process, nitrile is separated in vapor form from liquid fatty acid and amide, and further distillation is not necessary. Production of nitriles directly from fats and oils has also been investigated on a pilot scale [8].

Fatty nitriles are not cationic surfactants and have few industrial applications. They are hydrogenated to fatty amines in batch or continuous liquid-phase processes at 100–150°C using nickel or cobalt catalysts.

$$RCN + 2H_2 \rightarrow RCH_2NH_2 \tag{2}$$

Depending on the reaction conditions, dialkyl- or even trialkylamines may be formed.

$$2RCN + 4H_2 \rightarrow (RCH_2)_2NH + NH_3 \tag{3}$$

Formation of the higher alkylated products is suppressed by the addition of ammonia. Primary amines produced by the nitrile process contain some residual amide and dialkyl secondary amine (3–5%). Dialkylamines produced by this process typically contain monoalkylamine (~ 12%) and trialkylamines (~ 3%) [9].

The primary and secondary amines can be further derivatized (see below).

2. Production of Fatty Amines via Fatty Alcohols

Fatty alcohols produced by the hydrogenation of fatty acid methyl esters or fatty acids [10],

$$RCOOCH_3 + 2H_2 \rightarrow RCH_2OH + CH_3OH \tag{4}$$

or from petroleum feedstocks can be used to manufacture fatty amines.

Alcohols can react directly with an excess of ammonia and hydrogen to form dialkylamines [11], but in today's most important process, dimethyl tertiary amines are prepared from alcohol and dimethylamine.

$$ROH + HN(CH_3)_2 \rightarrow R{-}N(CH_3)_2 + H_2O \tag{5}$$

The reaction can be carried out under hydrogenation conditions. The dimethyl tertiary amines produced are important intermediates for quaternary amines, betaines, and amine oxides. A similar reaction can be used with high molecular weight dialkylamines to form trialkylamines in good yield.

In commercially less important processes, the fatty alcohol can first be converted to alkyl sulfate or alkyl halide, and this intermediate can be reacted with the secondary amine. Alkyl halides can also be used to prepare surfactants by quaternization:

$$RCl + Py^* \rightarrow R\,Py^+\,Cl^- \qquad (6)$$

B. Amidoamines and Imidazolines

The second major route from fats and oils to cationic surfactants is reaction with polyamines [12,13]. In a typical reaction, fatty acid is condensed with a polyamine such as diethylenetriamine (DETA) at a temperature of 150–170°C without catalyst in a batch process.

$$RCOOH + NH_2CH_2CH_2NHCH_2CH_2NH_2 \rightarrow RCONHCH_2CH_2NHCH_2CH_2NH_2 + H_2O \qquad (7)$$

At higher temperatures (190–250°C) and under vacuum or in the presence of solvents such as xylene or higher alkylbenzenes, which help water removal by azeotropic distillation, further condensation to imidazoline occurs.

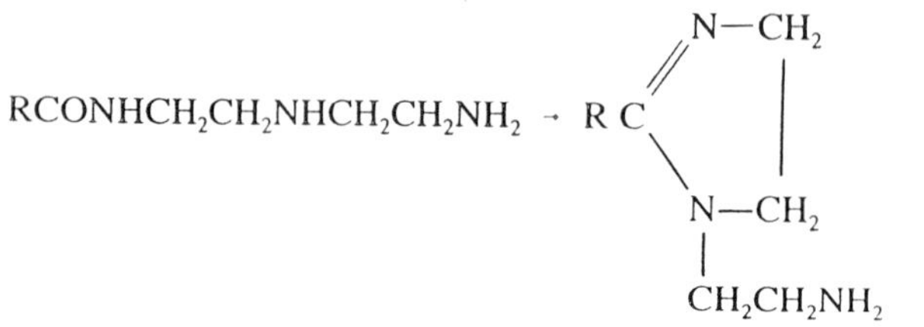

(8)

The fat itself or fatty acid methyl ester can be used instead of fatty acid, in which case the leaving group is glycerol or methanol. A wide range of polyamines is used commercially; the most important of these are diethylenetriamine, aminoethylethanolamine, aminoethylpiperazine, triethylenetetramine, tetraethylenepentamine, dimethylaminopropylamine, and ethylenediamine. Fatty acid pitches or acid oils reacted with polyamine residues such as those obtained as by-products in the manufacture of hexamethylenediamine or polyethylenepolyamines provide cationic surfactants suitable for many industrial applications.

When the polyamine contains more than one primary or secondary amine group, polyalkylated products will result. Depending on the ratio of fatty acid to polyamine, there may also be significant amounts of unreacted polyamine in the product. The relative amounts of amidoamine components and imidazolines can affect the physical properties and performance of the products and their derivatives.

C. Esteramines

Fatty acids or esters react with alkanolamines to form esteramines.

$$2RCOOH + (HOCH_2CH_2)_2NCH_3 \rightarrow (RCOOCH_2CH_2)_2NCH_3 + 2H_2O \qquad (9)$$

When the alkanolamines also contain primary or secondary amine groups, amido products are generally favored, with esteramines present only as by-products [14].

$$RCOOCH_3 + (HOCH_2CH_2)_2NH \rightarrow RCON(CH_2CH_2OH)_2 + CH_3OH \qquad (10)$$

Fatty diethanolamides are nonionic surfactants and are not discussed further in this chapter.

Reaction of fatty acids or esters at 150–250°C with triethanolamines gives a mixture of mono-, di-, and trialkylesteramines, depending on the ratio of acid and amine.

*where Py = pyridine.

Alkanolamines important commercially for the manufacture of esteramines include triethanolamine, methyldiethanolamine, diethylethanolamine, aminoethylethanolamine, and dimethyldihydroxypropylamine [15].

$$2RCOOH + HOCH_2CH(OH)CH_2N(CH_3)_2 \rightarrow RCOOCH_2CH(OOCR)CH_2N(CH_3)_2 + 2H_2O \qquad (11)$$

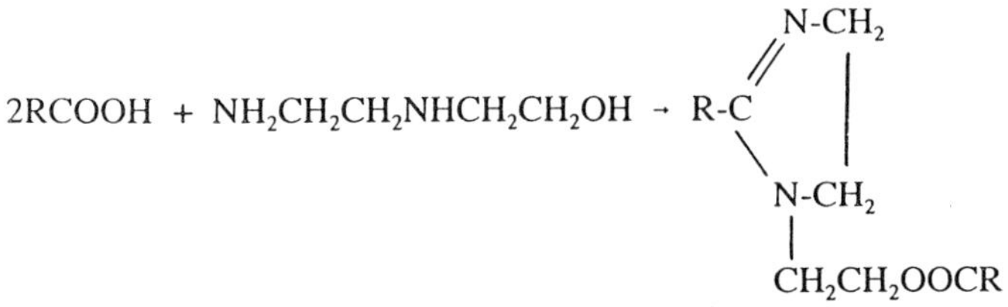

(12)

In principle, fatty alcohols can also be used to prepare esteramines by reaction with amino acids.

$$2ROH + (HOOCCH_2)_2NCH_3 \rightarrow (ROOCCH_2)_2NCH_3 + 2H_2O \qquad (13)$$

D. Etheramines

Another class of alcohol-based cationic surfactants are the so-called etheramines produced by reaction with acrylonitrile and hydrogenation:

$$ROH + CH{=}CH_2CN \rightarrow ROCH_2CH_2CN \qquad (14)$$

$$ROCH_2CH_2CN + 2H_2 \rightarrow ROCH_2CH_2CH_2NH_2 \qquad (15)$$

Hydrogenation proceeds like reaction with fatty nitriles [Eqs. (2) and (3)], and ammonia is used to suppress the formation of dialkyletheramines. A lower temperature may be used to reduce decyanoethylation, but nevertheless commercial etheramines may contain 5–15% free alcohol. Most etheramines are produced from synthetic alcohols derived from petroleum feedstocks.

E. Derivatization

1. Neutralization

Partial or full neutralization of amines with inorganic or low molecular weight organic acids gives water-soluble or dispersible products, and the amine acetates in particular are commercially supplied products. Neutralization with long-chain carboxylic acids or petroleum sulfonates gives oil-soluble products.

2. Alkoxylation

The uncatalyzed reaction of secondary or primary amines with ethylene oxide at ~100°C gives the *N*-ethanol derivatives.

(16)

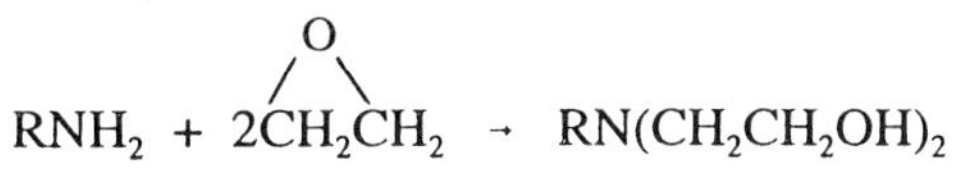

$$RNH_2 + 3CH_2CH_2O\ (\text{epoxide}) + CH_3COOH \rightarrow RN^+(CH_2CH_2OH)_3\,CH_3COO^- \quad (17)$$

Under acid conditions, quaternary amine salts are obtained [16].

Alkaline catalysis gives polyalkoxylates. To ensure complete conversion to tertiary amine, polyalkoxylates are usually prepared via the *N*-ethanol derivatives:

$$RN(CH_2CH_2OH)_2 + (x+y-2)CH_2CH_2O\ (\text{epoxide}) \rightarrow RN\begin{matrix}(CH_2CH_2O)_xH \\ (CH_2CH_2O)_yH\end{matrix} \quad (18)$$

A typical alkaline catalyst is potassium hydroxide at a reaction temperature of 120–170°C. Polyalkoxylates contain a range of components that differ in their degree of alkoxylation (Fig. 2) and often include some polyalkyleneglycol.

3. *Alkylation*

Primary and secondary amine groups can be methylated with formaldehyde under reducing conditions.

$$R_2NH + HCHO \xrightarrow{H_2} R_2NCH_3 + H_2O \quad (19)$$

The reaction is usually carried out at ~ 180°C with hydrogen over a nickel catalyst. Amines may also be alkylated with higher aldehydes under similar conditions,

$$RNH_2 + C_7H_{15}CHO \xrightarrow{H_2} RNHC_8H_{17} + H_2O \quad (20)$$

or with fatty alcohols [Eq. (5)].

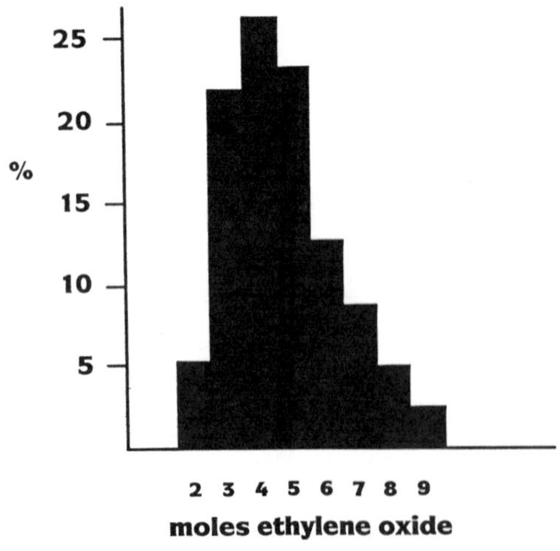

Fig. 2. Distribution of ethoxylated cocoamines in nominal 5 mol ethoxylate. (From Ref. 17.)

4. *Quaternization*

Alkylation with ethyl, benzyl, or methyl chloride, diethyl or dimethyl sulfate, or chloracetic acid leads to quaternization.

$$RN(CH_3)_2 + CH_3Cl \rightarrow RN(CH_3)_3Cl \tag{21}$$

The reactions are normally carried out at 50–120°C in water or an alcohol solvent (typically isopropanol, ethanol, or a glycol). Alkylation of primary or secondary amines leads to the formation of an acid,

$$R_2NH + 2CH_3Cl \rightarrow R_2N(CH_3)_2Cl + HCl \tag{22}$$

which must be neutralized if the reaction is to go to completion, and the resulting salts must be removed. For this reason, most quaternary amines are prepared from tertiary amines.

Quaternization of esteramines and some amidoamines and imidazolines presents particular problems because these amines may hydrolyze or react with alcoholic solvents during the quaternization process [13]. Quaternization at high activity under anhydrous conditions with careful pH control with very active quaternizing agents (e.g., dimethyl sulfate) may be favored.

Quaternization of propoxylated amines and higher ethoxylated amines may also necessitate the use of dimethyl sulfate. Quaternization with ethylene oxide [Eq. (17)] or alkyl halide [Eq. (6)] are described above. Commercially available quaternary amines often contain 1–2% of unquaternized material, inorganic salts from the neutralization of acid by-products, and traces of alcohols or ethers formed from the reaction of alkylating agent with the solvent.

5. *Cyanoethylation*

Cyanoethylation is used to add amino groups to fatty amines and etheramines.

$$RNH_2 + CH_2{=}CHCN \rightarrow RNHCH_2CH_2CN \tag{23}$$

$$RNHCH_2CH_2CN + 2H_2 \rightarrow RNHCH_2CH_2CH_2NH_2 \tag{24}$$

The addition, Eq. (23), is acid- or base-catalyzed. One or two moles of acrylonitrile can be added to a primary amine to give a diamine or a "Y" triamine product. The nitrile intermediate can be reduced as in Eq. (24), with ammonia if necessary to prevent the formation of dialkyl products. Commercially available diamines generally contain 2–10% monoamine due to decyanoethylation during reduction. The addition and reduction processes can be continued to form higher polypropylenepolyamines, and these diamines and polyamines can be derivatized by alkoxylation, neutralization, and quaternization.

6. *Oxidation*

Oxidation of tertiary amines with hydrogen peroxide gives amine oxides [18,19]:

$$RN(CH_3)_2 + H_2O_2 \rightarrow RN(CH_3)_2O + H_2O \tag{25}$$

The reaction is generally carried out at 50–90°C in aqueous or alcoholic solution to give the product activity of 30–50%. Chelating agents like EDTA may be added to prevent decomposition by metal ions [20].

III. PROPERTIES

A. Chemical

1. Thermal and Hydrolytic Stability

In the absence of oxygen, primary, secondary, and tertiary fatty amines and etheramines are generally stable to high temperatures (>180°C) even in aqueous solution, whereas quaternary amines degrade above 100°C to give tertiary amines; in alkaline conditions olefins may be formed [21]. Amine oxides can also degrade to olefins at elevated temperatures.

Amidoamines may disproportionate to dialkyl products on heating, and similar "reshuffling" of alkyl chains occurs in many amidoamine and esteramine formulations when heated.

$$2RCONHCH_2CH_2N\quad NH \rightarrow RCONHCH_2CH_2N\quad NCOR + NH_2CH_2CH_2N\quad NH \qquad (26)$$

Esteramines, amidoamines, and their quaternary amines are liable to undergo hydrolysis,

$$RCOOCH_2CH_2N(CH_3)_2 + H_2O \rightarrow RCOOH + (CH_3)_2NCH_2CH_2OH \qquad (27)$$

which is most rapid in alkaline solution. Formulations based on esteramines and ester quaternaries are generally adjusted to pH 2–4, where hydrolysis is slowest [22] (Fig. 3). Imidazoline rings open in alkaline conditions to give secondary and tertiary amidoamines. The alkaline nature of the imidazolines themselves means that they are liable to undergo hydrolysis during storage if not kept water-free, but quaternized imidazolines are stable to hydrolysis.

2. Acid–Base Behavior

With the exception of fully quaternized amines, cationic surfactants show acid–base behavior in aqueous solution.

$$R\overset{+}{N}H_3 \rightarrow RNH_2 + H^+ \qquad (28)$$

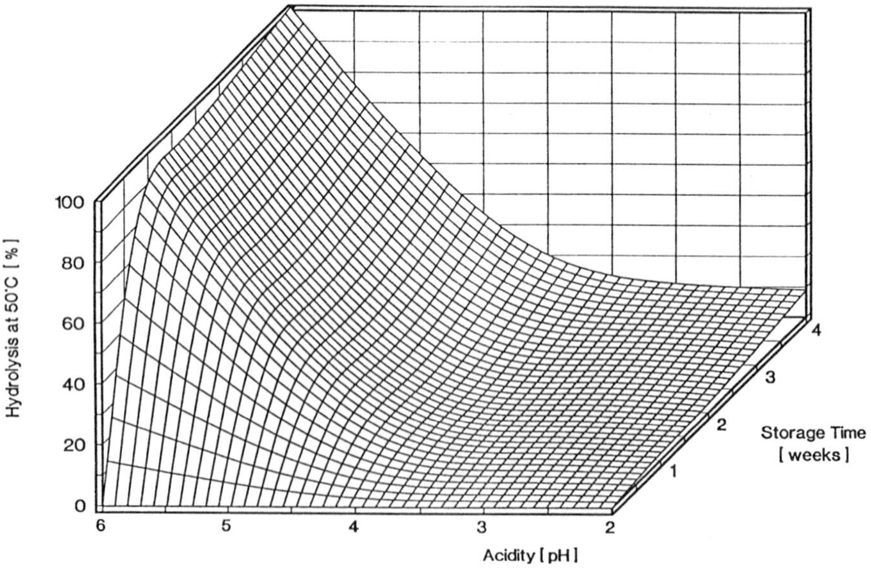

Fig. 3 Effect of storage time and pH on the hydrolysis of esterquaternary. (From Ref. 22.)

Table 2 pK_a Values of Cationic Surfactants

Base	Structure	pK_a[a]	Ref.
C_8–C_{18} alkylamines	RNH_2	10.3–10.6	17
C_5–C_{18} dialkylamines	R_2NH	11.0–11.2	23
C_5–C_8 dimethylalkylamines	$RN(CH_3)_2$	9.3–10.0	17
N-Alkylmorpholines	$RN(CH_2CH_2)_2O$	7.5–8.5	24
Dodecylamine oxide	RNO	4.8	25

[a]Of acid form.

Aqueous solutions of unquaternized cationic surfactants contain both the neutral amine form and the protonated acid form, their ratio depending on pH. The acid and base forms of the cationic surfactants have quite different surfactant properties, and it is impossible to understand the properties of cationic surfactants in aqueous solution without considering their acid–base behavior. Some typical pK_a values for the protonated amines are shown in Table 2. Most amines are strong bases with pK_a values over 9; the exceptions are amine oxides and *N*-alkyl morpholines with much lower pK_a values. The tertiary amine in substituted piperazines is also only weakly basic.

It is not possible, however, to conclude from the pK_a values that in aqueous systems the amine groups are most usually protonated, since at the surface of cationic micelles or other cationic surfactant phases or emulsion droplets, the local pH may be up to two pH units higher than the bulk pH, owing to the influence of the positive surface charge. For a similar reason the counterion at octadecylammonium phosphate monolayers was found to be HPO_4^{2-}, not $H_2PO_4^-$ [26]. In the presence of anionic surfactants, on the other hand, amine oxides are protonated well above their pK_a. Zeta potential measurements on oil droplets stabilized with cationic surfactants show that as the pH is increased, their surfaces become less charged [27–30], as expected, as the head groups become deprotonated, whereas quaternary amines are less affected by pH.

B. Physical and Surface Chemical Properties

1. Boiling and Melting Ranges

The boiling ranges of simple fatty amines lie in the range 180–300°C and can be related to the boiling points of the corresponding alkanes [31]. The melting ranges of the neutral (base) forms of cationic surfactants are related to the melting ranges of the originating fatty acid sources (Table 3). Amidoamines have melting points above those of the corresponding fatty amines [33], whereas etheramines and esteramines have lower melting points. Tertiary amines have significantly lower melting points.

The anhydrous forms of quaternary amines, amine oxides, and the salts of amines with inorganic acids have much higher melting points, in the range 100–200°C, but also often show degradation at these temperatures. Dialkyl fatty amine and esteramine derivatives in which some of the alkyl chains are branched show significantly lower melting points. Products containing 2-ethylhexyl groups are commercially available [34], and esterquats based on mixtures of tallow and 2-ethylhexanoic acid are claimed to provide liquid fabric conditioners at high activity [35].

Table 3 Melting Points of Cationic Surfactants Derived from Fats and Oils

Formula	Fatty acid source of alkyl chain[a]								
	C_{10}	C_{12}	Coco	C_{14}	C_{16}	Oleic	Tallow	HT	C_{18}
RNH_2	15[b]	26	15	38[b]	44	12	35	45	53
$RN(CH_3)_2$		−20	−10	−7	10	−10	3	20	23
$RN(CH_2CH_2OH)_2$			−10			−9	23	40	42
R_2NH	42[b]	47[b]	43	61[b]	67[b]			62	72[b]
$R_2N(CH_3)$	−5	15[c]	−2	24[c]	35[c]			33	40[c]
$RNHCH_2CH_2CH_2NH_2$			27			15	50	60	
$RCONH(CH_2)_3N(CH_3)_2$		29[d]			57[d]	5		57	63[d]
RCOOH	30	43	33[e]	54	62	6	42	59	65

[a] Carbons in fatty acid precursor.
[b] From Ref. 7.
[c] From Ref. 32.
[d] From Ref. 12.
[e] Hardened grade.
Source: Refs. 7, 12, 32 as noted. All other data from the product datasheets of Akzo Nobel (midpoint of ranges given).

2. *Phase Behavior in Aqueous Systems*

The base forms of cationic surfactants, with the exception of amine oxides and highly ethoxylated amines, are generally insoluble in water. Monoalkyl cationic surfactants formed by the acidification or quaternization of the base forms show phase behavior similar to that of anionic surfactants. There have been few studies of commercial surfactants that contain a mixture of several different alkyl chain length homologues, but studies of two-component mixtures sug-

Table 4 Critical Micelle Concentrations of Cationic Surfactants Derived from Dodecanoic Acid

Head group	CMC (mM)	Ref.
$—\overset{+}{N}H_3\overset{-}{Cl}$	15	36
$—\overset{+}{N}(CH_3)_2H\ \overset{-}{Cl}$	23	36
$—\overset{+}{N}(CH_3)_3\ \overset{-}{Cl}$	13, 21	37, 38
$—\overset{+}{N}(CH_2CH_2OH)_2CH_3\ \overset{-}{Cl}$	4.6	37
$—(CH_3)\overset{+}{N}(CH_2CH_2)_2O{\cdot}\overset{-}{Cl}$	21	38
$—\overset{+}{N}(CH_3)_2CH_2Ph\ \overset{-}{Cl}$	4.3	38
$—\overset{+}{N}(CH_2CH_2OH)_2CH_2Ph\ \overset{-}{Cl}$	1.9	37
$—\overset{+}{N}H_2CH_2CH_2CH_2\overset{+}{N}H_3\ 2\overset{-}{Cl}$	33	39
$—\overset{+}{N}(CH_3)_2\overset{-}{O}$	0.76	41
$—\overset{+}{N}(CH_3)_2OH\ \overset{-}{Cl}$	3.2	41
$—CONHCH_2CH_2CH_2\overset{+}{N}(CH_3)_2H\ \overset{-}{Cl}$	0.5	42
$—OOCCH_2\overset{+}{N}(CH_3)_3\ \overset{-}{Cl}$	5.5	40

gest relatively small influence. Molecular and micellar solutions dominate at concentrations up to 30% in water. Some representative critical micelle concentration (CMC) values are given in Table 4. Except in the case of the nonionic amine oxide, the CMC values of all the monoalkyl cationic surfactants are similar. At higher concentrations a variety of liquid crystalline phases are seen.

Monoalkyl cationic surfactants with certain counterions such as salicylate form highly viscous and viscoelastic phases that have been extensively studied [43,44]. The behavior of these surfactants is apparently due to the formation of long (1000 nm) rodlike micelles. The products have been suggested as drag-reducing agents and viscosifiers. The formation of rodlike micelles is also implicated in the use of amine oxides or ethoxylated amines as viscosifiers for acid cleaners or hypochlorite bleaches [45].

In contrast to the monoalkyl surfactants, the phase diagrams of dialkyl surfactants are dominated by lamellar phases. With the exception of products with chains shorter than 10 carbons, the surfactants form dispersions of bilayers at even low concentrations in water. This dispersion contains single- or multiwalled vesicles similar to those of phospholipids [46,47]. The vesicles can shrink or swell under the influence of osmotic forces and become leaky to solutes at clearly defined temperatures. They have been studied because of their importance in the physical properties of fabric softener formulations [48] and their potential applications in drug delivery and magnetic fluids [49].

3. *Hydrophilicity of Cationic Surfactants*

The relative importance of the hydrophobic and hydrophilic portions of a surfactant can clearly influence its properties, and there have been various attempts either to measure or to calculate an overall "HLB value" for a surfactant or to separate the contributions of the two portions. Partition measurements fall into the second approach. The contribution of the individual functional groups in a molecule to its partition between two solvents can be separated, and this is a tool used, for example, in drug design. By combining data for the head group and the methylene groups, the partition behavior of untested surfactants can be predicted. The water/heptane partition data for dodecyl derivatives calculated from measurements on shorter chain length analogues are shown in Table 5. Low values of the partition coefficient correspond to more hydrophilic surfactants. So, for example, the diethanolamine is considerably

Table 5 Partition Data for Cationic Surfactants Between Water and Heptane

Head group	Log *P* increment for head group	Calculated log *P* for dodecyl derivative
$—NH_2$	−2.9	4.8
$—N(CH_3)_2$	−1.3	6.4
$—NH_3·Cl$	−1.8	−11
$—N(CH_2CH_2OH)_2$	−4.7	3.0
$—NHCH_2CH_2CH_2NH_2$	−5.0	2.7
$—N(CH_2CH_2O)_5H_2$	−6.1	1.5
$—N(CH_2CH_2O)_{10}H_2$	−8.5	−0.8
$—N(CH_2CH_2)_2O$	−2.0	5.7
$(—CH_2—)$	(+0.64)	—
$(—CH_2CH_2O—)$	(−0.47)	—
—COOH	−3.4	3.2

Source: Ref. 50 for $C_{11}H_{23}COOH$; all others from or calculated from data in Ref. 17.

Table 6 HLB Values of Cationic Surfactants (Davies Extended Scale 0–40)

Head group	HLB increment	Calculated HLB of dodecyl derivative[a]
$—\overset{+}{N}H_3$	20.0	21.3
$—N(CH_3)_2$	8.5	9.8
$—NHCH_2CH_2CH_2NH_2$	17.4	18.7
$—\overset{+}{N}H_2CH_2CH_2CH_2NH_3$	38.6	39.9
$—\overset{+}{N}(CH_3)_3$	22.0	23.3
$—N(CH_3)_2$	22.5	19.1[b]
$—N(CH_2CH_2OH)_2$	11.3	12.6
$—\overset{+}{N}(CH_3)_2\bar{O}$	21.0	22.3[c]

[a]The HLB for the dodecyl derivative was calculated from

$$\text{HLB} = (\text{HLB increment for the head group}) - (12 \times 0.475) + 7$$

[b]Didodecyl derivative.
[c]pH 3.

more hydrophilic than the dimethylamine. The technique is not sensitive for highly hydrophilic surfactants such as the charged amine salts and quaternaries.

An electrokinetic method has been used to estimate the HLB values of the more hydrophilic surfactants, and some results are shown in Table 6 [51]. A high HLB value corresponds to a more hydrophilic material. The results show that the cationic form of the surfactant is very much more hydrophilic than the neutral base forms. Ethoxylation or additional amine groups also make a significant contribution to the hydrophilic nature.

4. *Adsorption of Cationic Surfactants at Interfaces*

The adsorption of cationic surfactants at interfaces has been reviewed [52] and is not considered in depth here. Surface and interfacial tensions of aqueous solutions containing unquaternized cationic surfactants vary with pH. Mixed films of the neutral (base) form and the cationic form give the lowest interfacial tensions at a particular concentration below the CMC.

Cationic surfactants are adsorbed on negatively charged surfaces, and this is related to their use in mineral flotation, corrosion inhibition, and textile softening. Initial adsorption is by neutralization of the negative sites. As the surfactant concentration increases, further adsorption can occur, leading to reversal of the surface charge. Adsorption generally levels out at the CMC, but further adsorption to give bi- or multilayer structures may occur in some systems, especially those containing dialkyl surfactants. Acidity and basicity can have a large effect on adsorption because in general the charge on both the surface and the surfactant will vary with (local) pH. Differences between the adsorption isotherms of different cationic surfactants in the low concentration region often disappear when plotted against reduced concentration (i.e., concentration/CMC) [17], suggesting that hydrophobic interactions are important in the adsorption process and head group structure is relatively unimportant.

IV. APPLICATIONS

A. Fabric Conditioners

Fabric conditioners (softeners) account for ~ 40% of cationic surfactant use. Adsorption of a cationic surfactant onto the surface of fabric improves handle, eases ironing, imparts antistatic

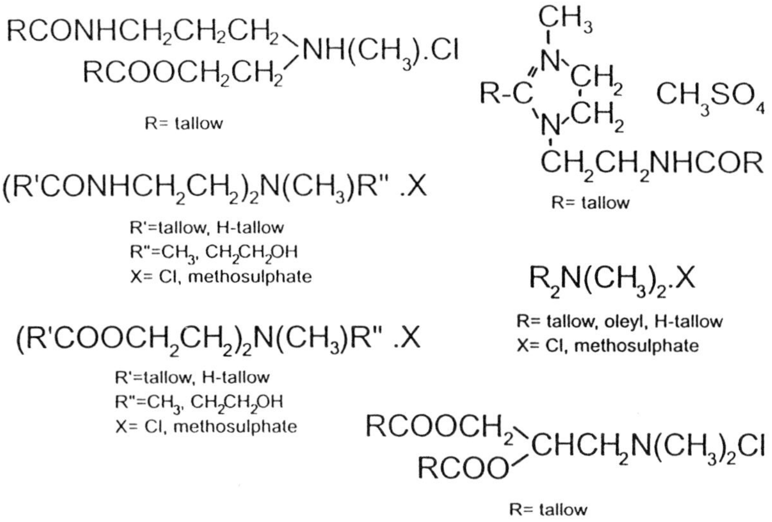

Fig. 4 Examples of fabric softener bases.

properties, and helps perfume retention [53]. Usually the softener's active ingredient is added as an aqueous dispersion to the laundry during rinsing at a level of 0.1% fabric weight. Alternatively, it may be impregnated into a fabric sheet that is added to the laundry during tumble drying, or formulated into the detergent itself and added in the wash cycle (so-called softergents).

Fabric softener bases are usually dialkylamines or quaternaries (Fig. 4). A typical rinse cycle formulation contains 5–20% active ingredients plus nonionic components, perfume, color, pH control additives, and electrolyte. The nonionic components may have a softening function or help to solubilize perfumes and control viscosity; pH control additives may help ensure the hydrolytic stability of the components; electrolytes such as sodium or calcium chloride help control the viscosity of the formulation.

There has been a trend toward high activity (>15%) products to reduce packaging requirements. The traditional softener base dihydrogenated tallow dimethylammonium chloride (DHTDMAC) is difficult to formulate into highly active products and has been partly replaced by amidoamines and imidazolines in these products [54].

Although the balance of evidence is that DHTDMAC does not seriously damage the environment, it has been largely replaced in Europe and to some extent in the United States and Japan by esterquaternary amines and amidoester products, which biodegrade faster and are less toxic to aquatic organisms [1,4,22,55]. Fabric softeners are today by far the major use for esteramine derivatives. Whereas DHTDMAC accounted for 85% of softener base in Europe as recently as 1990, it reached less than 40% in 1993 [56]. The resulting overcapacity in fatty amine manufacturing led to a major restructuring by the producers [6].

A new range of esteramine softener actives based on polyols such as glycerol and amino acids promises even lower aquatic toxicity than conventional products, especially when the aquatic toxicity of the biodegradation intermediates is considered [57].

B. Organoclays

Certain clay minerals have a layered structure in which negatively charged metal oxide platelets are neutralized by alkaline earth or alkali metal cation in between the layers. These cations can be replaced by cationic surfactants to form organoclays, which are used as thixotropic additives for oil-based drilling fluids, coatings, cosmetics, greases, and inks. Cationic surfactants are used in this application at a rate of 13,000–20,000 tpa [58]. In a typical process, a slurry of Wyoming bentonite or hectorite is centrifuged to remove impurities and optionally ion-exchanged to convert it into the sodium form. It is then mixed with quaternary amine at 60–70°C in aqueous dispersion. The relatively hydrophobic organoclay product is filtered off, dried, and milled. In an alternative process the reaction is carried out with almost dry clay. The final products may contain 25–50% organic material representing 90–120% exchange of the inorganic cations. Anionic surfactants and nonionic materials may be deliberately coadsorbed to provide special products with improved dispersibility.

The most used products in this application are dihydrogenated tallow dimethylammonium chloride, stearyldimethylbenzylammonium chloride, and dihydrogenated tallow methylbenzylammonium chloride, although other quaternary amines or fatty amine acetates are used in special products. Esteramine quaternary amines have been found useful, especially for organoclays used in oil drilling fluids based on low toxicity vegetable oils [59].

C. Mineral Flotation

Cationic surfactants are used as collectors in the separation of minerals by froth flotation [60,61], and mineral flotation accounts for perhaps 5–15% of total cationic surfactant usage [1]. The technique involves passing air through a slurry of finely ground ores that has been pretreated with typically 0.1–1 kg/t surfactant (collector). Selective adsorption of surfactant on the target minerals makes their surface hydrophobic, leading to their adhesion to the air bubbles and concentration in the froth. In direct flotation the valuable mineral is collected in the froth; in reverse or indirect flotation the impurities are collected. Selective adsorption of

Table 7 Cationic Mineral Flotation Collectors

Valuable mineral	Gangue mineral[a]	Typical collector	Ref.
Potassium chloride	Sodium chloride	Tallowamine	62, 63
Kainite	Sodium chloride	Cocoamine	
Carnallite	Sodium chloride	*N*-Cocomorpholine	63, 64
Calcite	Silicates	Tallowdiamine, dialkylquaternary	65
Magnesite	Silicates	Dialkylquaternary, tallowdiamine	66
Hematite/magnetite	Silicates	Alkyletheramine	67
	Silicates	Alkyletherdiamine	67
Phosphate rock	Silicates	Amidoamines	68
		Quaternary amine, tallowamine	69
Oxidized Zn minerals	Various	Cocoamine, tallowamine	70
Feldspar	Quartz	Tallowamine, tallowdiaminedioleate	71
Mica	Quartz	Cocoamine, alkyletheramine	72
Kaolin	Feldspar, quartz	Alkyletheramine	73
Pyrochlore	Various	Tallowdiamine	74
		Alkyl imidazoline	75

[a] Waste product.

surfactant is ensured by adjusting the conditions, especially the pH, until only the target mineral surface is negatively charged. In a typical process a slurry of ground ore is first mixed with acid or alkali to adjust pH, then mixed ("conditioned") with a solution or dispersion of cationic surfactant before air is introduced in flotation cells.

The choice of surfactant depends on the minerals to be separated. Cationic flotation is used primarily for the upgrading of industrial minerals, but the reverse flotation of iron ore and the flotation of oxidized zinc minerals are also significant. See Table 7.

Flotation performance is expressed in the grade (purity) of the mineral concentrate and its yield or recovery. As the chain length of surfactant increases, there is usually an increase in yield but a decrease in grade. Quaternary amines generally give higher grade products than primary amines, but yields may be lower.

The use of ethoxylated amines or ether amines together with quaternary amines has been claimed to improve yields [75,76], and combinations of cationic and anionic surfactants are important in specific applications, especially feldspar flotation [77,78].

D. Use in Road Construction

Bitumen (known as asphalt in the United States) is a residue from the refining of crude oil that can be used as a binder in road construction materials. Because of bitumen's high viscosity, construction materials are usually prepared and used hot. Alternative cold techniques involve the use of bitumen emulsions.

Cationic surfactants are used as adhesion agents in hot bitumen and as emulsifiers for bitumen [79], and together these applications account for ~10% of total cationic surfactant consumption [1].

Adhesion agents (also called antistripping agents or wetting agents) are added to bitumen at a level of 0.2–1%. Treated bitumen can displace water from the surface of wet aggregates (active adhesion), and road materials prepared from treated bitumen are resistant to water damage ("stripping"). Traditional adhesion agents were based on hydrogenated tallow diamine supplied in pellet form for easy addition to the bitumen, but these have been largely replaced by liquid products based on amidoamines and imidazolines. Bituminous binders are stored hot (120–200°C), and at these temperatures adhesion agents degrade either by reaction with acid components in the bitumen to form amides or through oxidation [80]. So-called heat-stable products based on less reactive tertiary amines or polyamines have been developed for situations where storage of treated binder cannot be avoided.

Bitumen emulsions typically contain 40–70% bitumen dispersed in water, as droplets 1–20 μm in diameter, stabilized with 0.15–2% surfactant. Of the approximately 7 million tonnes [81] of emulsion produced worldwide, probably more than 85% is stabilized by cationic surfactants. Contact of the cationic emulsion with the negatively charged surface of aggregates leads to its destabilization and coalescence of the bitumen droplets, a process known as breaking or setting. Emulsions are classified according to their stability in contact with aggregate into rapid-, medium-, and slow-setting grades, which contain different concentrations and types of emulsifiers and are suitable for various road construction techniques (Table 8). The actual mechanism of the setting process may involve the electrophoresis of bitumen droplets to the negatively charged aggregate surface, depletion of cationic emulsifier from the droplet surface onto the negative mineral surface leading to demulsification, and pH changes that lead to neutralization of droplet charge [82,83].

Trends in the market are toward cold emulsion processes instead of hot processes (favoring emulsifiers against adhesion agents) and toward techniques such as cold recycling, slurry surfacing, and cold mix, which use slow-setting emulsions.

Table 8 Cationic Bitumen Emulsifiers

Type of emulsion	Typical emulsifier	Use level (%)	Application
Rapid-setting	Tallowdiamine	0.2	Chip sealing
Medium-setting	Tallowdiamine	0.3–0.5	Open mixes
	Tallowquat	0.3–0.5	Tack coat
Slow-setting	Tallowdiquat	0.4–1.0	Slurry surfacing
	Amidoamines	0.8–2.0	Slurry surfacing
	Ethoxylated amines	0.8–2.0	Dense mixes
	Tallowpolypropylenepolyamines	0.6–1.0	Dense mixes

E. Biocides

Some cationic surfactants are toxic to microorganisms such as bacteria, fungi, and algae and as a result find use in disinfectants and preservatives. The multifunctionality of cationic surfactants allows biocidal properties to be combined with wetting and detergency or with corrosion inhibition and substantivity. Applications include disinfectants and sanitizing cleaners for hospitals, the food industry, animal husbandry, and around the home; antiseptic skin care preparations; control of algae and bacteria in water systems; wood preservation; and the control of bacterial corrosion in oil production and storage.

Quaternary ammonium compounds are the most used products (Fig. 5). The optimum alkyl chain length depends on the application but ranges from 12–16 carbons for monoalkyl quaternary compounds to 8–12 carbons for dialkyl products. Some data for the biostatic effect of cationic surfactants are given in Table 9. The minimum inhibitory concentration is the concentration that prevents microbial growth; higher concentrations are required to kill the organisms. A typical disinfectant may contain 1–5% active biocide and be diluted up to 100-fold before use [84]. The effectiveness of cationic disinfectants is decreased by the presence of hardness salts or soils [85]. Disinfectant and sanitizing cleaners may also contain nonionic

$R\text{-}NHCH_2CH_2CH_2NH_2.2CH_3COOH$

Fatty diamine diacetate

R= coco-

$R\text{-}N(CH_3)_3\ X$

Quaternary amine

R= coco-, X= Cl
R= hexadecyl-, X= Br
R= oleyl-, X= Cl
R= decylbenzyl-, X= Cl

$R\text{-}N(CH_3)_2(CH_2C_6H_5)\ Cl$

Benzalkonium chloride

R= coco-
R= tetradecyl-

$R\text{-}N(CH_3)_2O$

Amine Oxide

R= coco-

$R_2N(CH_3)_2.Cl$

Dialkyl quat

R=coco-
R=decyl-

Fig. 5 Examples of cationic surfactant biocides.

Table 9 Biostatic Performance of Cationic Surfactants—Minimum Inhibitory Concentrations (ppm)

Cationic surfactants—common name	Bacteria					
	Gram neg.	Gram pos.	SRB	Yeasts	Fungi	Algae
Cocoamine acetate	50–3200	10–25	400+	6–25	100–400	—
Didecyl quat	5–200	3–6	100–200	3	3–12	—
Dicoco quat	400–800	3–25	3200	400–800	6–12	—
Benzalkonium chloride	100–200	6–25	50–200	12–50	12–200	<1
Cocodiamine diacetate	25–800	12–25	50–100	6–25	50–200	—
Cocoquat	30–150	1–15	50–100	15	5	—
Cocoamine oxide	3200	50		50	50	—

SRB = Sulphate-reducing bacteria.
Source: Averaged data taken from Refs. 45, 83, 84, 86.

surfactants or amphoteric surfactants and chelating agents to improve detergency and hard water tolerance. In oil production and storage, quaternary amines or diamines can be used to control sulfate-reducing bacteria [86] that produce hazardous hydrogen sulfide gas and cause corrosion. The diamines in particular also provide direct corrosion inhibition.

In the field of wood preservation, cationic surfactants can replace the environmentally hazardous chemicals pentachlorophenol and copper chrome arsenate in some applications. Quaternary amines can be used to control sapstain (blue coloration on freshly cut wood) and provide above-ground rot protection [87].

A study of a range of quaternary amines concluded that ethoxylation had a negative effect on wood preservation [88]. Combinations of cationic surfactants and other organic or inorganic fungicides such as copper and borate may be required to totally match the performance of traditional wood preservatives [89].

F. Miscellaneous Applications

In addition to the major applications outlined above, cationic surfactants are used in a wide variety of industrial and household applications (see Table 10), which together may account for 20–30% of total consumption.

Most cationic surfactants are incompatible with anionic surfactants, and only amine oxides have widespread use in cleaning compositions such as dishwashing liquids, bath foams, and shampoos [45]. Quaternary amines and ethoxylated amines find use in consumer goods such as acid cleaners and disinfectants, hair conditioners, and vehicle cleaners. The agrochemical industry is a significant consumer of cationic surfactants, especially alkoxylates, which can improve the penetration and rainfastness of herbicides [90]. New agrochemical uses include "blossom thinning," in which an application of cationic surfactant to fruit trees in bloom stops fertilization and thus reduces the number of fruits produced. Fewer fruits means larger, more valuable fruits, and the surfactant treatment avoids hand thinning of young fruits. In another new application, "bud breaking," cationic surfactants promote early blooming in fruit trees [91].

The anticaking treatment of granular and crystalline fertilizers such as potash and ammonium nitrate typically involves applying 0.05–0.2% stearylamine to the hot fertilizer either as a solution in oil or on a talc or clay carrier [92].

Cationic surfactants such as imidazoline salts are used as emulsifiers or corrosion inhibitors in oil-based drilling fluids. Quaternary amines and diamines are used in corrosion inhibitor

Table 10 Miscellaneous Applications of Cationic Surfactants

Industry	Application	Chemical type
Cleaning and personal care	Dishwashing	Amine oxide
	Shampoo	Amine oxide
	Bath foam	Amine oxide
	Hair conditioner	Quat
	Toilet cleaners	Quat, amine oxide
	Car rinses	Dialkylquats
	Laundry	Amine oxides
Agrochemicals	Fertilizer anticaking	Stearylamine
	Herbicide adjuvant	Ethoxylated amine
	Blossom thinning	Alkoxylated amine
	Bud breaking	Alkoxylated amine
Petroleum	Oil-based drilling fluid	Alkylimidazoline
	Demulsifier for wet crude	Alkoxylated amines
	Biocides	Quats
	Corrosion inhibitors, gasoline additives	Alkylimidazolines
Textile	Viscose spin bath additives	Ethoxylated amines
	Dye leveling	Quats
	Softening/handle	Quats
Polymers	Antistatic agents	Ethoxylates, quats
	PVC adhesion promotors	Amidoamines
	PU foam stabilizer	Alkylmorpholines
	Cationic SBR latex	Ethoxylates, quats

formulations [93] and biocides for oil production. Ethoxylated amines are used in demulsifiers for crude oil [94]. Cationic surfactants are also used as diesel and gasoline additives for corrosion inhibition, and as detergents for carburetors, injectors, or valves.

In the manufacture of rayon, the cellulose xanthate solution (viscose) is spun into an acid salt bath to regenerate the cellulose in fiber form. Ethoxylated amines and amidoamines are added to both the cellulose xanthate itself and the spin bath to improve fiber quality. Other textile applications include additives to improve dyeing and softener treatments.

Ethoxylated amines and quaternary amines are used in antistatic treatments for polyethylene, propylene, styrene, and urethanes [95,96]. Ethoxylates may be incorporated in the polymer; quaternary amines are most often used in surface treatments. Amidoamines improve the adhesion of protective coatings based on PVC plastisols. Other small applications include the charge reversal of anionic rubber latex for use in the construction industry [97] and in polymer foams.

In addition to having a variety of applications in industry and in consumer products, amines of all types are important intermediates in the manufacture of other types of surfactants such as amphoterics and sulfosuccinimates as well as nonsurfactants such as epoxy resins and waxes.

G. Future Developments in Cationic Surfactants

Cationic surfactants provide the functions required for a wide variety of domestic and industrial applications. As a group their main disadvantages lie in the area of aquatic toxicity, their irritating or corrosive effects on skin and eyes, and their incompatibility with anionic surfactants and other anionic dispersants used in cleaning compositions, pigments, and coatings.

Because of their rapid degradation, esteramines provide a solution to the problem of aquatic toxicity, and with the growth in esteramine manufacturing capacity, substitution of other cationic surfactants in applications outside fabric conditioning is to be expected. Esteramine chemistry is flexible in terms of both the hydrophobic and hydrophilic parts of the surfactant molecule, which means that a wide range of product requirements can be satisfied, although their inherent hydrolytic instability may prevent their use in some cleaning and industrial applications. Cationic sugar derivatives [96] may also offer improved ecotoxicological properties, but these products are not yet commercialized.

Amphoteric surfactants can provide many of the same functions of cationic surfactants with the advantages of lower ecotoxicity, less irritant nature, and compatibility with anionic surfactants. Amphoterics may therefore also offer an alternative to cationic surfactants where the pH of the system can be adjusted to ensure expression of their cationic properties. However, regardless of any trend toward esteramines or amphoteric surfactants, the hydrophobic portion of surfactants is likely to continue to be derived mainly from natural fats and oils rather than petroleum.

REFERENCES

1. B. F. Haumann, European firms face new challenge, *INFORM 2*(5): 438–447 (1991).
2. Hewin International Inc., Industrial cationic, amphoteric and amide surfactants and related products, quoted in *INFORM 1*(11): 984 (1990).
3. Frost and Sullivan International Report, quoted in *INFORM 4*(2): 179 (1993).
4. R. B. McConnell, Worldwide trends in fabric softeners, *INFORM 5*(1): 77–85 (1994).
5. J. Baker, Surfactant prices head skywards, *European Chemical News*, Sept. 19, 1994, pp. 14–15.
6. M. Ward, Akzo Nobel marriage poised for new heights, *E.C.N.*, Sept. 19, 1994, pp. 23, 26.
7. R. A. Reck, Cationic surfactants derived from nitriles, in *Cationic Surfactants—Organic Chemistry*, Surfact. Sci. Ser. 34 (J. M. Richmond, Ed.), Marcel Dekker, New York, 1990, pp. 163–186.
8. S. Billenstein and G. Blaschke, Industrial production of fatty amines and their derivatives, *J. Am. Oil Chem. Soc. 61*(2): 353–356 (1984).
9. R. G. Laughin, *Cationic Surfactants—Physical Chemistry*, Surfact. Sci. Ser. 37 (D. D. N. Rubingh and P. M. Holland, Eds.), Marcel Dekker, New York, 1990, p. 455.
10. U. R. Kreutzer, Manufacture of fatty alcohols based on natural fats and oils, *J. Am. Oil Chem. Soc. 61*(2): 343–348 (1984).
11. Hoechst, German Patent 2255701.
12. F. E. Friedli, Amidoamine surfactants, in *Cationic Surfactants—Organic Chemistry*, Surfact. Sci. Ser. 34 (J. M. Richmond, Ed.), Marcel Dekker, New York, 1990, pp. 51–100.
13. G. W. Earl, Imidazoline surfactants, in *Cationic Surfactants—Organic Chemistry*, Surfact. Sci. Ser. 34 (J. M. Richmond, Ed.), Marcel Dekker, New York, 1990, pp. 101–144.
14. K. E. Visek, Fatty amide surfactants, in *Cationic Surfactants—Organic Chemistry*, Surfact. Sci. Ser. 34 (J. M. Richmond, Ed.), Marcel Dekker, New York, 1990, pp. 1–50.
15. R. Puchta, Cationic surfactants in laundry detergents and laundry after treatment aids, *J. Am. Oil Chem. Soc. 61*(2): 367–376 (1984).
16. Keith D. Stanley, Akzona Inc, U.S. Patent 4376736, 1983.
17. J. Wates, Solution behaviour of cationic surfactants relevant to industrial applications, Ph.D Thesis, University of Salford, UK, 1990.
18. J. D. Sauer, Amine oxides, in *Cationic Surfactants—Organic Chemistry, Surfact. Sci. Ser.* 34 (J. M. Richmond, Ed.), Marcel Dekker, New York, 1990, pp. 275–296.
19. C. J. Turney, F. E. Friedli, and P. J. Frank, Kinetics and preparation of amine oxides, *J. Am. Oil Chem. Soc. 71*(7): 793–794 (1994).

20. P. A. Vening, W. E. Adam, H. Bruil, and H. Rorig, Amines and amine derivatives, especially amine oxides, XIII Jornadas del Comite Espanol de la Detergencia tensioactives y afines Barcelona, Feb. 24–26, 1982.
21. E. H. White and D. J. Woodcock, in *The Chemistry of the Amino Group* (S. Patai, Ed.), Interscience, New York, 1968, pp. 409–427.
22. U. Sebold, Fabric softeners worldwide, 3rd World Conference and Exhibition on Detergents, Montreux, Switzerland, Sept. 26–30, 1993.
23. J. W. Smith, Basicity and complex formation, in *The Chemistry of the Amino Group*, S. Patai, Ed.), Interscience, New York, 1968.
24. A. D. James, J. M. Wates, and E. Wyn-Jones, Determination of the hydrophilicities of nitrogen-based surfactants by measurement of partition coefficients between heptane and water, *J. Colloid Interface Sci. 160*: 158–165 (1993).
25. S. Ikeda, M. A. Tsunada, and H. Maeda, The application of the Gibbs adsorption isotherm to aqueous solutions of a nonionic-cationic surfactant, *J. Colloid Interface Sci. 67*(2): 336–348 (1978).
26. D. Vollhardt, M. Wittig, G. Petrov, and G. Malewski, Infrared spectroscopic study of phosphate counter ions bonded in skimmed monolayers and transferred multilayers of octadecylamine, *J. Colloid Interface Sci. 106*(1): 28–32 (1985).
27. J. S. Laskowski and Qi Dai, Collector-extender flotation in the processing of coarse fractions of potash ores, Proc. XVIII Int. Mineral Processing Congress, Sydney, May 23–28, 1993, pp. 1045–1050.
28. A. Samadijavan, Elektrophoretische Untersuchung des Brechverhaltens von Bitumenemulsionen, *Bitumen 4*: 151–156 (1994).
29. J. M. Wates and A. D. James, Zeta potential measurements of bitumen emulsions and road aggregates, Proc. First World Congress on Emulsion, Paris, Oct. 19–22, 1993, Vol. 1, 1-40/089.
30. J. M. W. Mackensie, Electrokinetic properties of Nujol-flotation collector emulsion drops, *Trans. Soc. Mining Eng. AIME*, *244*: 393–400 (1969).
31. A. A. Woolf, Estimation of boiling points in homologous series with a new linear relationship, *Chem. Ind.*, Feb. 4, 1991, pp. 96–97.
32. K. R. Smith, Cationic surfactants from petroleum feedstocks, in *Cationic Surfactants: Organic Chemistry* (J. M. Richmond, Ed.), Marcel Dekker, New York, 1990.
33. W. M. Linfield, Fatty oxazolines and imidazolines, *J. Am. Oil Chem. Soc. 61*(2): 437–441 (1984).
34. Akzo Chemicals Inc., Industrial Surfactants Nitrogen Derivatives General Catalog, 1992.
35. Pulcra SA, German Patent 4334365-A1.
36. N. Ikeda, N. Sanefujo, K.-K. Lou, M. Aratona, and K. Moromura, Surface adsorption and micelle formation of the mixture of *N*-methylated dodecylammonium chloride and dodecylammonium chloride, *J. Colloid Interface Sci. 164*: 439–443 (1994).
37. M. te Nijenhuis, Akzo Nobel Research Centre, Deventer, Holland, unpublished work.
38. B. Rozyika-Roszak, Micellisation of some amphiphillic quaternary ammonium chlorides, *J. Colloid Interface Sci. 140*(2): 538–540 (1990).
39. H. Hagslatt and K. Fontel, Divalent surfactants—the influence of headgroup charge on the phase behaviour of alkyldiamine surfactant/water system, *J. Colloid Interface Sci. 165*: 431–444 (1994).
40. B. Rozyika-Roszak, S. Przestalskis, and S. Witek, Calorimetric studies of the micellization of some amphiphilic betaine ester derivatives, *J. Colloid Interface Sci. 125*: 80 (1988).
41. M. Tuncay and S. D. Christian, A study of the binding of dimethyldodecylamine oxide by β-cycodextrin using surface tension measurements, *J. Colloid Interface Sci. 167*: 181–185 (1994).
42. T. M. Muzyczko, S. Shore, and J. A. Loboda, Fatty Amidoamine derivatives: N, N-dimethyl-N-(3 alkylamido-propyl) amines and their salts, *J. Am. Oil Chem. Soc. 45*: 720–725 (1968).
43. N. Nemoto and M. Kuwahara, Self diffusion and viscoelasticity of elongated micelles from cetyltrimethylammonium bromide in aqueous sodium salicylate solution. II. Temperature effect, *Colloid Polym. Sci. 272*: 846–854 (1994).
44. J. Myska and P. Stern, Properties of a drag reducing micelle system, *Colloid Polym. Sci. 272*: 542–547 (1994).

45. H. Rorig and R. Stephan, Amine oxides and their applications, in *Industrial Applications of Surfactants II* (D. R. Karsa, Ed.), Spec. Publ. No. 77, Royal Society of Chemistry, London, 1990, pp. 211–234.
46. J. Matsumoto, Internal and interfacial structure of small vesicle in aqueous colloid of didodecyldimethylammonium bromide, *Colloid Polym. Sci. 270*: 492–497 (1992).
47. R. G. Laughlin, The role of swelling methods in surfactants phase science: past, present and future, *Adv. Colloid Interface Sci. 41*: 57–59 (1992).
48. A. D. James and P. H. Ogden, Effect of manufacturing variables on the viscosity of aqueous di(hydrogenated tallowalkyl)dimethyl ammonium chloride, *J. Am. Oil Chem. Soc. 56*(4): 542–547 (1979).
49. C. Monager and V. Cabuil, Synthesis of magnetic DDAB vesicles, *Colloid Polym. Sci. 272*: 1295–1299 (1994).
50. R. Smith and C. Tanford, Hydrophobicity of long chain *n*-alkylcarboxylic acids, as measured by their distribution between heptane and aqueous solutions, *Proc. Natl. Acad. Sci. U.S.A. 70*(2): 289–293 (1973).
51. O. Boen, Akzo Nobel Chemicals, Deventer, Holland, Unpublished work.
52. B. T. Ingram and R. H. Ottewill, Adsorption of cationic surfactants at interfaces, in *Cationic Surfactants—Physical Chemistry*, Surfact. Sci. Ser. Vol. 37 (D. N. Rubingh and P. M. Holland, Eds.), Marcel Dekker, New York, 1990, pp. 87–140.
53. S. D. Escher and E. Oliveros, Quantitative study of factors that influence the substantivity of fragrance chemicals on laundered and dried fabrics, *J. Am. Oil Chem. Soc. 71*(1): 31–39 (1994).
54. R. O. Keys, Formulation and production of concentrated rinse cycle fabric softeners, *Happi*, March 1995, pp. 95–97.
55. G. R. Whalley, Fabric conditioning agents, *Happi*, February 1995, pp. 55–58.
56. H. Berenbold, Aktuelle Situation der Wascheweichspuler, *SOFW-J. 120*(11): 678–684 (1993).
57. Akzo Nobel Surface Chemistry, *A New, Flexible Softener Generation*, Product data leaflet.
58. J. P. Tatum, Organophilic clays, in *Industrial Applications of Surfactants* (D. Karsa, Ed.), Royal Society of Chemistry, London, 1987, pp. 289–306.
59. O. Ponsati, A. Truis, C-P Herold, H. Müller, C. Nitsch, S. von Tapavicza, Patent applicant; Henkel Kommanditgesellschaft auf Aktien, Use of selected oleophilic compounds with quaternary nitrogen to improve the oil wettability of finely divided clay and their use as viscosity improvers, World Patent Appl. WO92/19693, 1992.
60. D. W. Fuerstenau and R. Herrera-Urbina, Adsorption of cationic surfactants and the flotation of minerals, in *Cationic Surfactants—Physical Chemistry*, Surfact. Sci. Ser., Vol. 37 (D. N. Rubingh and P. M. Holland, Eds.), Marcel Dekker, New York, 1990, pp. 407–447.
61. H. L. Shergold, Cationic surfactants in mineral processing, in *Industrial Applications of Surfactants* (D. Karsa, Ed.), Spec. Publ. No. 59, Royal Society of Chemistry, London, 1987, pp. 269–288.
62. R. J. Roman, M. C. Fuerstenau, and D. S. Seidel, Mechanisms of soluble salt flotation, Part 1, *Trans. SME*, March 1968, pp. 56–70.
63. H. Schubert, What goes on during potash flotation, *Eng. Mining J.*, March 1967, pp. 94–97.
64. D. Mörstedt, M. Winzer, R. Köhler, and F. Püschel, Verfahren zur Flotation von NaCl aus Salzgemischen, E. German Patent 64443, 1968.
65. W. J. Trahar, The concentration by flotation of cement rock from Fyansford Victoria, Report No. 612, CSIRO, Melbourne, 1961.
66. Financial Mining Industrial and Shipping Corporation, The separation of magnesite from its contaminants, U.S. Patent 3979251, 1975.
67. O. Burghardt and H. D. Wasmuth, Uber die Anwendung der indirekten kationaktiven Flotation bei der Augbereitung von Eisenerzen, *Aufbereitungs. Technik. 3*: 103–117, 1975.
68. Dow Chemical Co., *N*-Aminopiperazine condensates for the benefication of phosphate ore, U.S. Patent 4301004, 1981.
69. International Minerals and Chemical Corporation, Flotation process for the recovery of phosphate values from ore, U.S. Patent 4737273, 1988.

70. C. Caproni, R. Ciccu, M. Ghiani, and I. Trudi, The processing of oxidised lead and zinc ores in the Campo Pisano and San Giovanni plants (Sardinia), in *Processing of Oxidised and Mixed Oxide–Sulphide Lead Zinc Ores Seminar*, Proc. XIIIth Int. Mineral Processing Congress, Warsaw, 1979, (J. Laskowski, Ed.), Polish Scientific Publishers, pp. 71–91.
71. W. H. Eddy, E. W. Collins, J. S. Browning, and G. V. Sullivan, Recovery of feldspar and glass sand from South Carolina waste granite fines, Report of Investigations 7651, U.S. Dept. of the Interior, Bureau of Mines, 1972.
72. K. Hanumantha Rao, B.-M. Antti, and K. S. E. Forssberg, Flotation of mica minerals and selectivity between muscovite and biotite while using mixed anionic/cationic collectors, *Minerals Metall. Process.*, August 1990, pp. 127–132.
73. N. O. Clark, Pochin and Co. Ltd., Improvements in or relating to the treatment of clay, Br. Patent 1224877, 1967.
74. J. A. R. Faucher, Concentration of pyrochlore, *Trans. SME*, September 1964, pp. 255–258.
75. S. R. Rao, R. Espinosa-Gomez, J. A. Finch, and R. Biss, Effects of water chemistry on the flotation of pyrochlore and silicate minerals, *Minerals Eng. 1*(3): 189–202 (1988).
76. Kenobel AB, Material and method for upgrading of carbonate mineral, Swedish Patent 458747.
77. T. Jiaxing, S. Baoqi, C. Zhengbing, C. Xianguo, Z. Huiling, and Z. Shaozhong, Theoretical studies and production practice of acidless and fluoless flotation of silica sand, Proc. XVIIIth Int. Mineral Processing Congress, Sydney, May 23–28, 1993, pp. 851–855.
78. K. Hanumantha Rao and K. S. E. Forssberg, Solution chemistry of mixed cationic/anionic collectors and flotation separation of feldspar and quartz, Proc. XVIIIth Int. Mineral Processing Congress, Sydney, May 22–28, 1993, pp. 837–843.
79. A. D. James and D. Stewart, Cationic surfactants in road construction and repair, in *Industrial Applications of Surfactants*, Vol. 2 (D. Karsa, Ed.), Spec. Publ. 77, Royal Society of Chemistry, London, 1990, pp. 338–355.
80. A. D. James, P. D. McCool, D. Stewart, G. E. Varcoe, and A. R. Woodside, A gas chromatography method for the determination of antistripping agents in hot binders and the significance of this technique to surface dressing, Proc. Int. Symp., *Chemistry of Bitumens*, Rome, Italy, June 5–8, 1991, pp. 627–653.
81. Syndicat des Fabricants D'Emulsions Routieres de Bitume, A few figures about emulsions consumption around the world, Proc. 1st World Bitumen Emulsion Manufacturers Congress, Paris, Oct. 18, 1993, p. 61.
82. A. R. Lane and R. H. Ottewill, The preparation and properties of emulsions stabilised by cationic surface-active agents, *Theory and Practice of Emulsion Technology*, SCI, 1974.
83. Akzo Chemicals, Technical Data Sheet B-10, Amersfoort, Holland, 1992.
84. B. Davis and P. Jordan, An overview of the biocidal activity of cationics and ampholytes, in *Industrial Applications of Surfactants*, Vol. 2 (D. Karsa, Ed.), RSC Spec. Publ. No. 77, Royal Society of Chemistry, London, 1990, pp. 195–210.
85. R. A. Stephenson, Optimisation of the performance of quaternary ammonium compounds, in *Industrial Applications of Surfactants*, Vol. 2 (D. Karsa, Ed.), RSC Spec. Publ. No. 77, Royal Society of Chemistry, London, 1990, pp. 235–275.
86. E. Bessems and A. F. Clemmit, Quaternary ammonium compounds: evaluation and application in the control of sulphate-reducing bacteria, in *Chemicals in the Oil Industry* (P. H. Ogden, Ed.), RSC Spec. Publ. No. 45, Roy. Soc. Chem., London, 1983, pp. 159–170.
87. A. F. Preston, P. J. Walcheski, P. A. McKaig, and D. D. Nicholas, Recent research on alkylammonium compounds in the U.S., *Am. Wood Preservers Proc. 11*: 331–346 (1987).
88. H. Tang and J. N. R. Ruddick, Evaluating the potential of amine chemicals for use as wood protecting agents. Part 1. Investigation of cationic component of quaternary ammonium compounds, Proc. 25th Annu. Meeting Int. Res. Group on Wood Preservation, Bali, Indonesia, May 29–June 3, 1994.
89. R. J. Tillot and C. R. Coggins, Non-arsenical water-borne preservatives—a review of performance and properties, *Record of the 1981 Annual Convention of the British Wood Preserving Association*, pp. 32–46.

90. D. J. Turner and P. Tabbush, Studies with alternative glyphosate formulations, in *Application and Biology* (E. S. E. Southcombe, Ed.), Monograph No. 28, British Crop Protection Council, 1985, pp. 135–145.
91. The Fruits of Research, *Progress with Akzo Nobel Surface Chemistry*, Akzo Nobel Surface Chemistry AB, Stenunsund, Sweden, Autumn 1994, pp. 4–5.
92. J. R. Parks and J. Granok, Fertilizer conditioning problems solved with those versatile amines, *Farm Chem.*, October 1967, pp. 51–62.
93. A. Nestle, Corrosion inhibitors in petroleum production primary recovery, in *Corrosion Inhibitors* (C. C. Nathan, Ed.), NACE = National Association of Corrosion Engineers, Houston, 1973, pp. 61–75.
94. R. Bohm and R. Kupfer, Improved demulsifier chemistry: a novel approach in the dehydration of crude oil, Friedrich Straiss, Society of Petroleum Engineers of AIME, August 1991, pp. 334–338.
95. W. Jerzykiewicz and J. Wypych, Studies on the influence of antistatics based on tertiary and quaternary amines on the behaviour of plasticized polyvinyl chloride, *Tenside Deterg. 13*(2): 74–76 (1976).
96. B. Davis, Antistatic/conductivity additives in paint and plastics, in *Industrial Applications of Surfactants* (D. Karsa, Ed.), RSC Spec. Publ. No. 59, Roy. Soc. Chem., London, 1987, pp. 307–17.
97. Dr. Hermann Perrey, Dr. Matner Martin, and Heinz-Gerd Nonnenbruch, Bayer AG, W. German Patent Appl., DE 3628902-A1, 1986.
98. T. Kida, K. Yurugi, A. Masuyama, Y. Nakatsuji, D. Ono, and T. Takeda, Preparation and properties of new surfactants containing D-glucosamine as the building block, *J. Am. Oil Chem. Soc. 72*(7): 773–780 (1995).

25

Nonionic Surfactants

Guido Bognolo
ICI Performance Chemicals, Everberg, Belgium

I. INTRODUCTION

Nonionic surfactants are one of the most important classes of amphipathic materials, second only to the anionic surfactants in terms of volume consumed and certainly more widespread in terms of end uses and applications. Their key feature is the absence of discrete electric charges on the molecules, which confers different properties from those of the anionic, cationic, or amphoteric surfactants (Fig. 1), and this often results in unique interfacial effects.

Chemically the nonionic surfactants are esters, ethers, amides, amines, or amine oxides. Each of these classes generally includes several subclasses, depending on the nature of the hydrophilic and hydrophobic moiety. These in turn include many individual components differing in the chain length of the hydrophobe, the degree of polymerization of the hydrophile, and the relative proportion of hydrophile and hydrophobe. It follows that the number of commercial products available to the formulator is astonishingly high, probably in the region of 5000 different chemicals or more. Table 1 gives an overview of the nonionic surfactants of commercial interest.

Products of natural origin with structure and properties typical of nonionic surfactants have been used specifically or (most often) accidentally in various preparations for hundreds of years. Examples include glycerides and their hydrolysis products, which find use in cosmetics, food, and colors; and saponins from plant roots, which are used in fabric cleaning. However, the real development of nonionic surfactants as a stand-alone class of products with identified chemistry and performance properties was prompted by two factors:

On the one hand, the emergence of fat substitutes required emulsifiers that enabled the incorporation of water into oil phases, which favored the development of the chemistry of the glycerol esters and ensured a market for the new products. This paved the way for the use of polyhydric hydrophiles other than glycerol, e.g., sorbitol, which was becoming available in large quantities and at low cost as a by-product in the manufacture of mannitol, which was used in the explosives industry for the production of detonators based on mannitol hexanitrate.

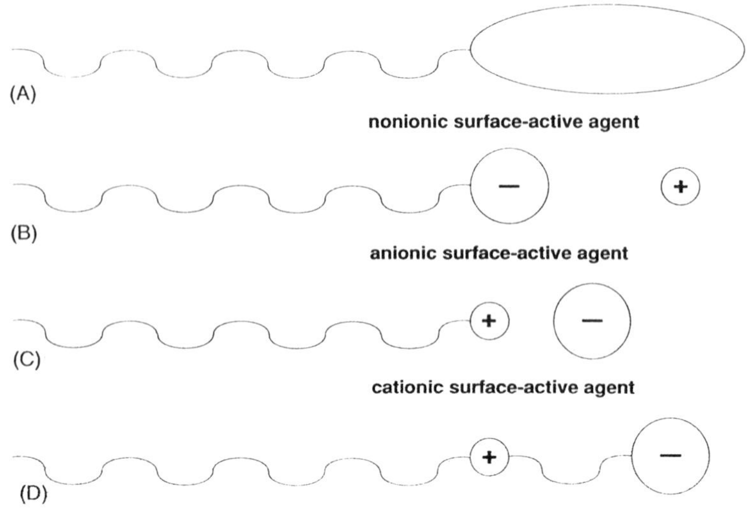

Fig. 1 Schematic representation of the four main categories of surface-active agents. (A) Nonionic surface-active agent; (B) anionic surface-active agent; (C) cationic surface-active agent; (D) amphoteric surface-active agent.

On the other hand, the industrial availability of ethylene oxide (EO) and the growing understanding of its reactivity with molecules containing active hydrogen atoms enabled the production of more water-soluble surface-active agents and emulsifiers. This broadened the range of industrial applications, and at the same time created new chemicals, providing solutions to emerging technological demands, initially from the textile and food industry, but expanding rapidly to other industrial sectors such as cosmetics, crop protection, plastics, and paints, to name just a few.

Fundamental in this respect is the work from Schoeller, Wittwer, Steindorff, Balle, Horst, and Michel, who in the early 1930s developed and patented the synthesis of ethoxylated nonionics from fatty acids, alcohols, amines, and alkylphenols [1,2]. Early in the next decade, the ethoxylation of low polarity glycerol and sorbitol esters enabled the production of more hydrophilic compounds. These could then be blended, in various proportions, with the original esters to emulsify a range of oils of different chemical structures and physical properties, both in the form of water-in-oil (i.e., with oil being the continuous external phase, abbreviated w/o) and oil-in-water (i.e., where water is the continuous external phase, abbreviated o/w).

The introduction of the hydrophilic–lipophilic balance (HLB) concept [3] and the extension of its subsequent evolutions and modifications [4–17] gave a tremendous impetus to the application technology of surfactants in general and of the nonionics in particular, creating new applications for existing products as well as meeting the demand for new chemical species capable of solving challenging wetting, dispersion, and emulsification problems. In recent years the development of new chemicals has slowed down, partly because of the exhaustion of the synthetics possibilities, partly because of the reduced funds allocated to basic research, and partly because of the ever-growing constraints from regulations and legislation. As a result, there has been a renewal of interest in chemistry and products that initially received

Table 1 Nonionic Surfactants of Commercial Interest

Chemical class	Hydrophobe	Hydrophile	Common denomination
Ester derivatives	Fatty acid	EG, POE[a]	Glycol and polyethylene glycol esters
		Gly, PGL[b]	Glycerol and polyglycerol esters
		Sor[b]	Sorbitol and sorbitan esters
		Suc[b]	Sucroesters
		Suc + Gly	Sucroglycerides
		Glu[b]	Glucoside esters
Ether derivatives	Fatty alcohol (natural or synthetic)	POE[a]	Ethoxylated alcohols
		Gly, PGL[b]	Glycidyl ethers
		Glu	Alkylpolyglucosides
	Alkyl phenol[b,c]	POE[a]	Ethoxylated alkylphenols
	Propylene oxide[d]	POE	Ethylene oxide/propylene oxide copolymers
		Glu[b]	Glucosides ethers
Amides	Fatty acid amide	POE[a]	Ethoxylated amides
	Fatty acid	Gla	Glucamides
Amines	Fatty acid amines	POE[b]	Ethoxylated amines
Amine oxides	Fatty acid amines	Oxy	Amine oxides

Note: See list of abbreviations preceding Reference at end of chapter.
[a]The hydrophobic character of the final molecule can be enhanced by adding to this moiety other hydrophobic groups (usually POP).
[b]The hydrophilic character of the final molecule can be reduced by adding to this moiety other hydrophobic groups (usually POP) or enhanced by adding other hydrophilic groups (usually POE).
[c]Includes higher aromatics, e.g., β-naphthol.
[d]Includes all block, block-etheric, and etheric POE/POP copolymers with different initiators.

only limited attention. This process is stimulated by the increasing demand for properties that were disregarded or were of low relevance in the past, prompted by changes in public opinion or by raw materials availability or by economic and ecotoxicological considerations.

For example, alkylpolyglucosides, which were used in small volumes—essentially exploiting their properties as hydrotropes in high alkalinity bottle-washing formulations—are now experiencing a rapid expansion into sectors such as household detergents and cosmetics, driven by their mildness to the skin, biodegradability, and natural origin and renewability of the constituent raw materials. Similarly, sugar esters, which were almost confined to the role of laboratory curiosities because of synthesis difficulties and the consequent high manufacturing costs, are now being paid increased attention, particularly in food and personal care applications as high HLB, ethylene oxide–free emulsifiers.

A new stimulus to the use of nonionic surfactants arose in recent years from their superior stabilization power for both emulsions and suspensions. The development of the theories of steric stabilization and the appreciation of the performance of these products as steric stabilizers enhanced that application in various fields.

II. WORLD PRODUCTION

Since the initial development of modern nonionic surfactant chemistry, this industry has experienced constant growth. It is estimated that in 1994 about 2.6 million metric tons of nonionic surfactants was produced on a worldwide basis for a value in excess of 1.8 billion pounds sterling. This represents about one-third of the total estimated surfactants market.

Because of the complex nature of the product range (often related to other surfactants or chemicals), of the routes to market, and of the industry structure and evolution, it is impossible to produce a precise list of producers of nonionic surfactants. Table 2 attempts to give a snapshot of the worldwide situation at the end of 1994.

Table 2 Major Producers of Nonionic Surfactants (1994, World Situation)

Europe	
Austria	Th. Boehme
Belgium	Oleofina
Bulgaria	Bulgarska Roza AD
	Verila
Denmark	Palsgaard Ind.
	Danisco (Grindsted)
France	Ceca (Atochem)
	Gattefosse
	ICI
	Marchon (Albright & Wilson)
	PPG Ouvrie
	Stearineries Dubois
	Seppic
	Sidobre-Sinnova (Henkel)
	Stepan
FSU (former Soviet Union)	Barva
	Biok
	Fine Organic Synthesis
	Ivhymprom
	Kaprolactam
	Kaustik
	Organic Synthesis
	Orgsteklo
	Salavatnefteorgsinthez
	Sintez
	Volzhski Kauchuck
Germany	BASF
	Bayer
	Th. Boehme
	Buna
	Chem-Y (Kao)
	Grunau (Henkel)
	Henkel
	Hoechst
	Huels
	ICI
	Peter Greven

Table 2 Continued

	Rewo (Witco)
	Stockhausen (Huels)
	Th. Goldschmidt
	Unichema
	Zschimmer & Schwarz
Holland	Dow
	Dr. Kolb
	Quest
	Servo (Huels)
	Shell
	Sybron Chemie
	Unichema
Italy	Akzo Nobel
	DAC (Condea)
	Enichem Augusta (RWE)
	Fratelli Lamberti
	Geronazzo (Rhone-Poulenc)
	Henkel
	Hoechst
	Marchon (Albright & Wilson)
	Zschimmer & Schwarz
Norway	Pronova
	Unger
Poland	Rokita
Portugal	Companhia Petroquimica Do Barreiro
Spain	ICI
	Kao
	Marchon (Albright & Wilson)
	Pulcra (Henkel)
	Unichema
	Union Derivan
Slovak Republic	Novaky
Sweden	Berol (Akzo Nobel)
Switzerland	Dr. Kolb
United Kingdom	Baker Performance Chemicals
	Croda
	Harcros
	Hickson Manro
	ICI
	Lanstar
	Marchon (Albright & Wilson)
	Shell
	Unichema
Africa	
South Africa	Croda
	Hoechst
	Manro
Americas	
Canada	ICI
United States	ADM

Table 2 Continued

	Air Products
	Amerchol (Union Carbide)
	American Ingredients
	Baker
	BASF
	Calgene
	Croda
	Du Pont
	Eastman
	Ethox
	Ecolab
	Grinsted
	Harcros
	Hodag
	Hoechst Celanese
	Henkel
	Heterene
	High Point (Kao)
	ICI
	Karlshamns
	Lipo
	Lonza
	McIntyre
	Mona
	Milliken
	Nalco
	Olin
	Piedmont
	PPG/Mazer
	Rhone-Poulenc
	Sandoz
	Scher
	Sip
	Stepan
	Shell
	Sherex (Witco)
	Texaco
	Union Carbide
	Vista
	Van Dyck
	Van den Bergh
	Witco
Mexico	Arancia Tensioactivos
	Canamex
	Christianson
	Henkel
	Hoechst
	Poliquimia Apizaco
	Poliquimia de Tlaxcala
	Polyurechemia
Brazil	Aquatec

Table 2 Continued

	Cliffs
	Henkel
	Hoechst
	Oxiteno
	Ultraquimica/Oxiteno
Colombia	Incolgasos
	Protechnio
Venezuela	Ethoxyl
Argentina	Dispersol San Luis
Asia	
India	Albright & Wilson
	Bayer India
	Colour Chemicals (Hoechst)
	Dai Ichi
	Globe Industries
	Henkel India
	Hico Products
	ICI
	India Glycols
	Laffans Petro
	Manali Petro
	Silvolical Industries
	S M Dye Chemicals
	Sunshield Chemicals
	Swastik
	United Pestichem
	Unitop Chemicals
	Vashuda Chemicals
	Venus Ethox
Japan	Asahi Denka
	Chuo Kasei
	DKS
	Henkel Hakusui
	Kao Corporation
	Kawaken Fine Chemicals
	Lion Corporation
	Marubishi Oil Chemicals
	Matsumoto
	Miyoshi Oil & Fat
	Mitsubishi Kasei
	New Japan Chemicals
	Nicca Chemicals
	Nikko Chemical Co.
	Nippon Nyukazai
	Nippon Oils & Fats
	Nippon Shokubai
	Nissan Chemical Ind.
	Nisso Yuka
	Riken Vitamin
	Sakamoto Yakuhin
	Sanyo Chemical Industries

Table 2 Continued

	Taiyo Chemical Industries
	Takeda Chemical
	Takemoto Oils and Fats
	Toho Chemical Industries
	Tokai Seiyu Ind.
	Yokkaici Gosei
	Yoshimura Oil Chemicals
Taiwan	Jintex
	Pan Asia Chemical
	Sino-Japan Chemical
	TWSC
People's Republic of China	Beijing Synthetic Chemicals Plant
	Jilin
	Nanying Zhongshou Chemical Plant
	Shanghai Auxiliary Plant
	Shanghai Gao Qiao Petrochemical Plant
	Shanghai Henkel Oleochemical Company
	Tianjin Auxiliary Plant
Korea	Dong Nam
	Dong Sung
	Han Nong
	Il Chil
	KOA
	Korea Polyol
Singapore	Croda
	EMPL
	Salim/A&W
	Shell/EGS
Thailand	Kao
Australia	Albright & Wilson
	Henkel/Nopco
	Harcros
	ICI

III. RAW MATERIALS

Nonionic surfactants are manufactured using raw materials derived from renewable resources as well as from fossil feedstock. Table 3 gives a summary of the natural/renewable component versus the petrochemical component in the major nonionic classes. It should be made clear that even those surfactants produced from completely natural and renewable raw materials require chemical processing and products, e.g., catalysts, for their manufacture and therefore have an environmental impact to a greater or lesser extent. Section VII deals in detail with the toxicological and environmental aspects of nonionic surfactants.

Figure 2 illustrates schematically the raw materials "road map" for commercial nonionic surfactants. The following subsections deal specifically with the individual raw materials.

Table 3 Natural/Renewable versus Petrochemical Component in the Major Nonionic Classes

Surfactants class	Hydrophile	Hydrophobe	Percent "renewable"
Fatty alcohols POE	P (POE)	P (SFA)	0
	P (POE)	N (NFA)	—[a]
Fatty alcohols POE/POP	P (POE/POP)	P (SFA)	0
	P (POE/POP)	N (NFA)	—[a]
Alkyl phenols POE	P (POE)	P (AP)	0
Alkylpolyglucosides	N (Glu)	N (NFA)[b]	100
Glycerol esters (incl. polyglycerol)	N (gly/PGL)[b]	N (NFAc)	100
Sorbitan esters	N (Sor)[b]	N (NFAc)	100
Sorbitan esters POE	P (POE)	N (NFAc)	—[a]
POE/POP copolymers	P (POE)	P (POP)	0
Sugar esters	N (Suc)	N (NFAc)	100
Polyglycerol ethers	N (PGL)[b]	N (NFA)[b]	100
PEG esters	P (POE)	N (NFAc)	—[a]
Amines POE and POE/POP	P (POE/POP)	N (NFAm)[b]	—[a]
	P (POE/POP)	P (SFAm)[c]	0
Amides POE	P (POE)	N (NFAm)	—[a]
Amine oxides	P (Oxy)	N (NFAm)	—[d]

Note: See Abbreviations at end of chapter.
[a]Depends on the HLB of the surfactant.
[b]Although the origin of the raw material is renewable, significant chemical processing is required.
[c]Uncommon.
[d]Depends on structure and molecular weight of the amine.

A. Ethylene Oxide

Ethylene oxide is one of the major building blocks for nonionic surfactants. It is a highly reactive three-member cyclic ether that gives nucleophilic addition reactions of the SN2 or SN1 type [18,19] with molecules containing active hydrogen atoms through cleavage of the carbon–oxygen bond. It was first prepared in 1859 by Wurtz [20] by dehydrohalogenation of 2-chloroethanol, and the initial industrial processes were based on this chemistry [21].

The direct oxidation of ethylene dates from 1931 [22,23]. Nowadays, the vast majority of ethylene oxide is produced by silver-catalyzed oxidation in air-based or oxygen-based processes. Other than for the manufacture of surfactants, it is used for the production of a variety of chemical intermediates, e.g., polyether polyols for polyurethanes, or polyethylene glycol for polyesters or products such as antifreeze glycols, glycol ether solvents, fire-resistant hydraulic fluids, quenchers, and lubricants.

It is estimated that the worldwide production capacity for EO in 1994 was 11.1 million metric tons split approximately 30% in Europe, 40% in the Americas, 20% in Asia and Australia, and 10% in Japan.

Much is known about the physicochemical properties of EO [19], but the nature of the chemical bond is not completely established. Figure 3 gives the structure, bond length, and bond angles as determined experimentally [24,25]. The carbon–oxygen–carbon angle is about 50° less than that in the dimethyl ether [26], and the length of the carbon–carbon bond (1.470 Å) is intermediate between that of a single bond and that of a double bond (1.55 and 1.35 Å, respectively). The CH_2 groups lie in a plane perpendicular to the plane of the ring [27].

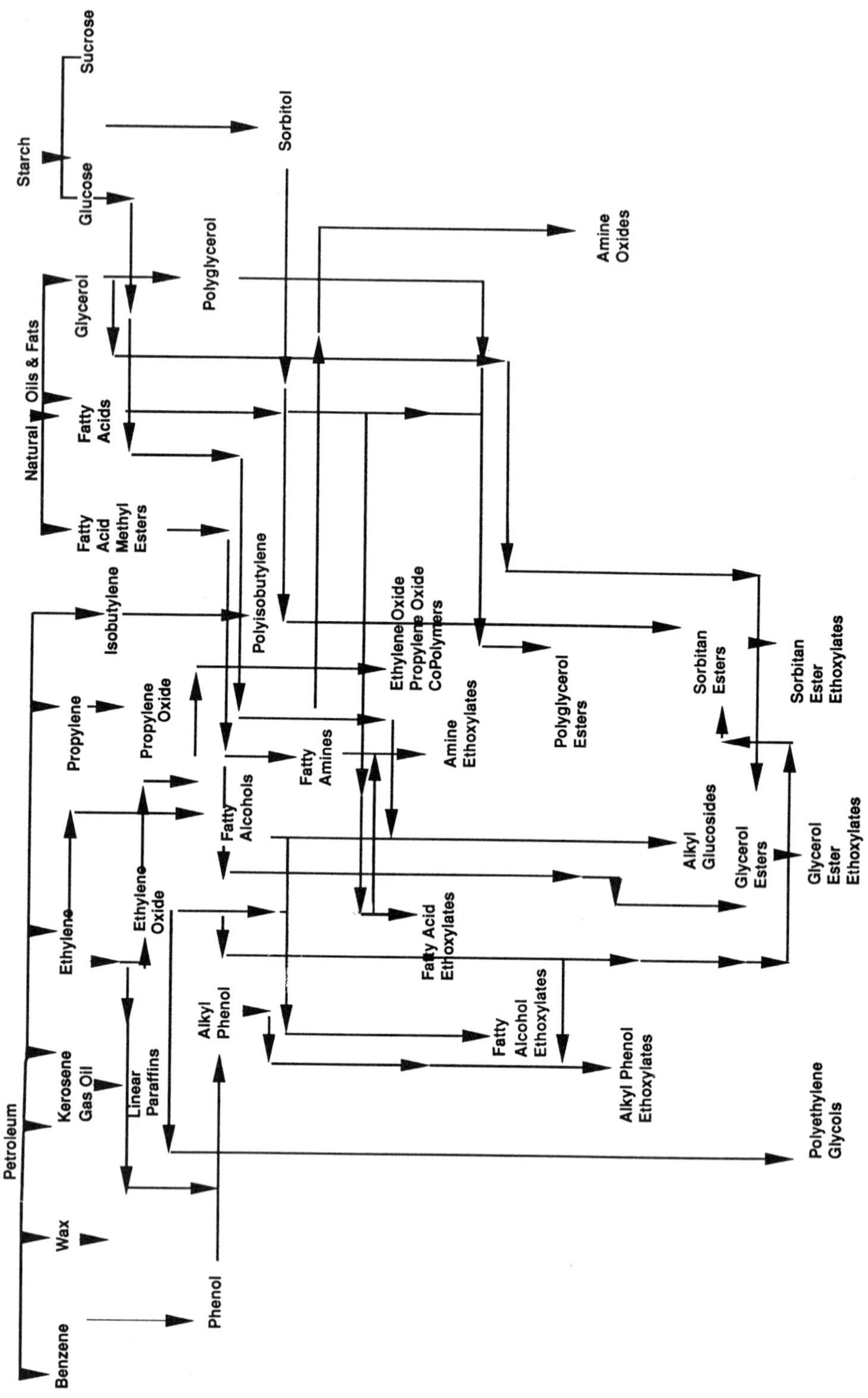

Fig. 2 Raw materials "road map" for commercial nonionic surfactants.

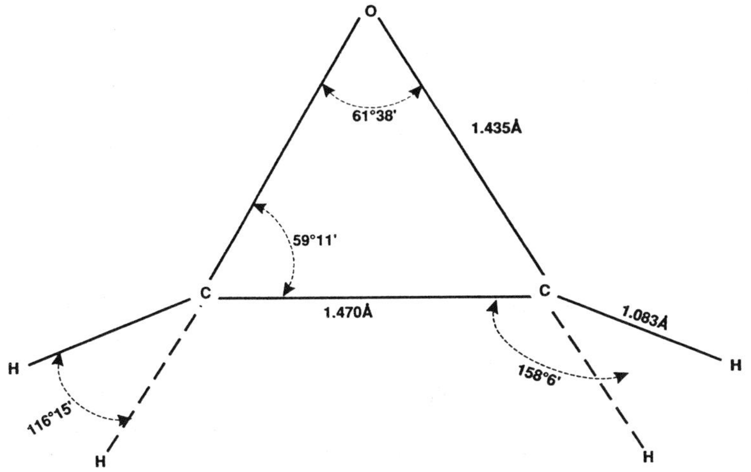

Fig. 3 Structure, bond lengths and bond angles for ethylene oxide. (From Ref. 199, p. 10.)

Because of its physicochemical properties and reactivity, EO must be handled with extreme care. It presents both physiological and explosive hazards, and its manufacturers continue to make considerable effort to improve the procedures and standards for its safe transport and handling [28].

B. Glycerol

Glycerol was discovered in 1779 by Scheele and was recognized as the ubiquitous component of fats and oils by Chevreul in 1813. It can be obtained either as a by-product in the conversion of fats and oils to fatty acid or fatty acid methyl esters (natural glycerol) or from synthetic processes.

Fats and oils are the main sources of natural glycerol, which is obtained from them by means of [29]

1. High pressure splitting. Water and fat are fed into a splitting column in a countercurrent fashion at 5–6 bar and 250–260°C.
2. Continuous methanolysis.
3. Saponification of natural oils with caustic soda.

Synthetic glycerol is produced from propene through any of the following intermediates:

1. Allyl chloride (IG Farben, 1943) [30–32]
2. Acrolein (Shell, 1958) [33]
3. Propene oxide [34]

The major uses of glycerol are in pharmaceuticals and cosmetics, carboxylic acid esters, alkyd resins, polyols, food, cellulose, and organic nitrates.

Polyglycerols are produced industrially by the base-catalyzed condensation of glycerol and as by-products of epichlorohydrin hydrolysis [35]. The worldwide consumption of glycerol is on the order of 600,000 tpa, with an approximate ratio of natural to synthetic of 3:1.

C. Sorbitol

Sorbitol (D-glucitol) occurs in nature in the fruits, leaves, and bark of many vegetable species. It is produced commercially by high pressure hydrogenation of glucose in the presence of heterogeneous catalysts, e.g., nickel phosphate [36], or homogeneous catalysts, e.g., ruthenium dichlorotriphenyl phosphine [37].

D. D-Glucose

D-Glucose (dextrose) is the most abundant sugar in nature, either as free monosaccharide or linked with other sugars (such as in sucrose) or as a polymer (as in starch or cellulose). Commercially, D-glucose is produced by the hydrolysis of a variety of starches. The high temperature acid-catalyzed process [38] has been replaced by the acid–enzyme or enzyme–enzyme process, in which the initial starch hydrolysis is carried out with acid [39] or enzyme (bacterial α-amylase [40,41]), respectively, whereas the subsequent hydrolysis is achieved with an enzyme (fungal glucoamylase [42]). Commercial products include the crystalline or monohydrate form as well as liquid solutions.

E. Sucrose

Sucrose (α-D-glucopyranosyl-β-D-fructofuranoside) occurs in practically every existing spermatophyte. It is extracted commercially from sugarcane, sugar beets, sorghum, and sugar maple. Hydrolysis of sucrose under mild acidic conditions releases the two component sugars, glucose and fructose. Hydrogenation with Raney nickel as catalyst produces a mixture of sorbitol and mannitol. Sucrose is used widely as a sweetener and in food processing as a preservative, bulking agent, flavor enhancer, and texturizer.

F. Propylene Oxide

Propylene oxide was also first prepared in Wurtz laboratory, by converting dibromopropane to propylene glycol, subsequently converting it to 1-chloro-2-propanol followed by dehydrochlorination with diluted potassium hydroxide. In the first industrial process (chlorohydrin process), the chlorohydrin is prepared by the reaction of propylene and chlorine in excess water followed by dehydrohalogenation. Because of the large volume of effluent produced in the process, the direct oxidation of propylene with hydroperoxides in the presence of a catalyst (oxirane process) is preferred [43–46].

The ring strain in propylene oxide is 29.57 kcal/mol [47], compared with 27.28 kcal/mol for ethylene oxide [48], and their reactions and chemistry are similar. In addition to its use in the manufacture of surfactants, propylene oxide is extensively used in polyether polyols, lubricants, and quenching fluids.

G. Fatty Acids

The fatty acids used in the manufacture of nonionic surface-active agents are virtually all of natural origin, an even-numbered series containing 8–22 carbon atoms [49,50]. The principal raw materials from which they are derived are tallow, coconut, palmkernel, soybean, rapeseed, and sunflower oils, marine oils, and crude tall oil. Except for those derived from tall oil, the acids are obtained from the triglycerides by hydrolysis with water or methanol:

$$\begin{array}{l} CH_2OCOR \\ | \\ CHOCOR \\ | \\ CH_2OCOR \end{array} + 3\,H_2O \longrightarrow 3\,RCOOH + \begin{array}{l} CH_2OH \\ | \\ CHOH \\ | \\ CH_2OH \end{array}$$

$$\begin{array}{l} CH_2OCOR \\ | \\ CHOCOR \\ | \\ CH_2OCOR \end{array} + 3\,CH_3OH \longrightarrow 3\,RCOOCH_3 + \begin{array}{l} CH_2OH \\ | \\ CHOH \\ | \\ CH_2OH \end{array}$$

High temperature and high pressure hydrolysis can be performed without a catalyst or with alkaline catalysts, e.g., lime or zinc and magnesium oxides. Hydrolysis at atmospheric pressure uses, preferentially, sulfuric or sulfonic acid catalysts.

Enzymes are used for hydrolytic processes operating at atmospheric pressure and ambient temperature [51]. The fatty acids obtained from hydrolysis are then purified by distillation or separated into their components by fractional distillation, pressing, or crystallization using solvents or, as in the Henkel SOT process, water–surfactant solutions.

Methanolysis gives methyl esters that are separated by fractional distillation. The grades suitable for detergent production are hydrogenated to alcohols; the others (usually the shorter chain lengths) are hydrolyzed to acids. Tall oil fatty acids are obtained by fractional distillation of crude tall oil.

The entire range of chain lengths, from C_8 to C_{22}, linear or branched, saturated or unsaturated, and substituted fatty acids are the starting raw materials to produce POE esters, glycerol, sorbitan, and sucrose esters and their alkoxylated derivatives. Fatty acids of vegetable origin are preferred by certain segments of the industry in response to the movement in public opinion toward the protection of animals. The C_{18} chains are most commonly used, because the surfactants thus obtained generally have good detergency and emulsification properties. Because of the ease of handling in its liquid form, oleic acid is preferred in many industrial processes. The C_{12}–C_{14} "cuts" give surfactants that are milder and therefore preferred in applications involving extensive skin contact, e.g., hair and body shampoos, creams and lotions for skin care, and washing-up liquid detergents.

Tall oil fatty acids from the sulfate or Kraft paper process are sometimes used to produce low cost esters. Isostearic acid, despite its high cost and limited availability, is used for specialty surfactants requiring a combination of the properties from the C_{18} chain length typical of stearic acid with the liquidity conferred by oleic acid but without the disadvantage of the unsaturation (which could lead to adverse effects, e.g., in polymerization reactions).

A building block that is most useful for the production of polymeric emulsifiers, described later, is 12-hydroxystearic acid.

It is estimated that, in 1990, about 1.8 million metric tons of fatty acids (including tall oil fatty acids, but excluding the fatty acid salts from continuous soapmaking processes) were produced in the United States, Europe, and Japan, in the approximate proportions 1:1:0.3.

H. Fatty Alcohols

Fatty alcohols are obtained either from natural sources (i.e., by hydrogenolysis of fatty acids, fats, or fatty esters) or synthetically by hydroformylation (oxo process) or by the Ziegler process. Hydrogenolysis is performed at high temperature (200–350°C) and high pressure (100–200 bar) [52] if a copper chromite catalyst is used. This process yields saturated alcohols. Unsaturated alcohols can be produced by using selective mixed oxide catalysts (Al_2O_3/CdO/Cr_2O_3)

$$RCOOH + 2H_2 \longrightarrow RCH_2OH + H_2O$$

The hydroformylation process involves the addition of carbon monoxide and hydrogen to an α-olefin in the presence of a catalyst to give an aldehyde that is then reduced to an alcohol containing one carbon atom more than the starting α-olefin [53]:

$$RCH{=}CH_2 \xrightarrow[H_2]{CO} RCH_2CH_2CHO \xrightarrow{H_2} RCH_2CH_2CH_2OH$$

$$RCH{=}CH_2 \xrightarrow[H_2]{CO} RCH(CH_3)CHO \xrightarrow{H_2} RCH(CH_3)CH_2OH$$

The alcohols are mixtures of linear and branched molecules, the composition being dependent on the α-olefin feedstock [54,55], the reaction conditions, and the catalyst [56,57]. The branching confers some degree of differentiation compared with the natural, or Ziegler, alcohols in terms of reactivity and physical properties.

The Ziegler process allows the production of straight-chain synthetic alcohols by reacting triethylaluminium and ethylene (58) (see Scheme 1). The alcohols synthesized by this route have carbon chains with an even number of carbon atoms, and their chain length composition follows a Poisson distribution. Branched chains are formed to a minor extent (below 5%) owing to side reactions, so the alcohols are predominantly linear.

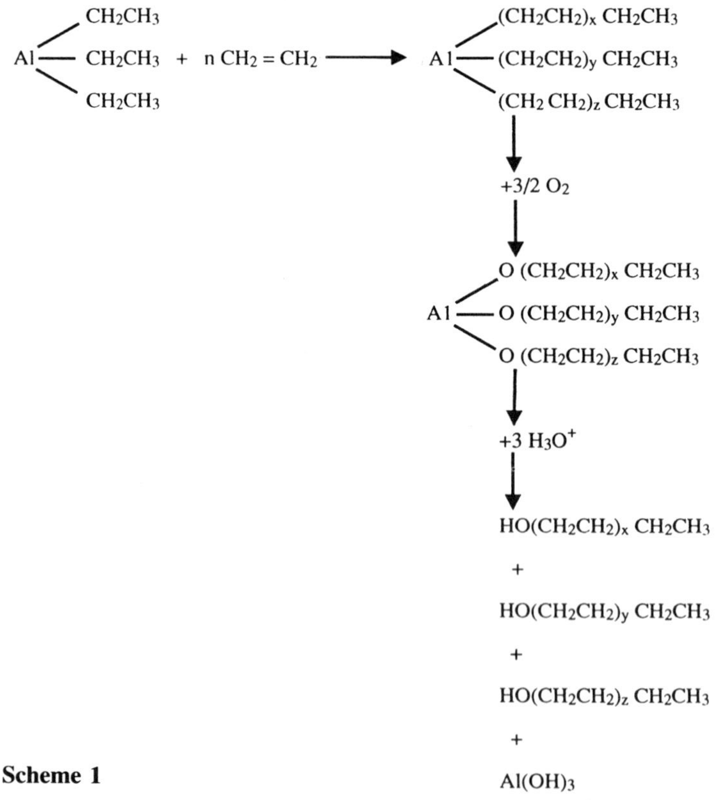

Scheme 1

A type of alcohol that is generating a growing interest is that in which a CH_2OH group is substituted at, or near, the middle of a hydrocarbon chain. Such molecules were first prepared by Guerbet in 1899 by heating shorter chain alcohols under alkaline conditions and are named after him. A key feature of the Guerbet alcohols is the much lower melting point compared with primary or secondary alcohols with a similar number of carbon atoms.

The oxidation of linear paraffins in the presence of boric acid yields secondary alcohols of the same chain length. The process is used on a larger scale in the former Soviet Union (FSU) and has gained significance in Japan. The oxidation is performed in the liquid phase at 150–170°C with a mixture of 3–4% oxygen in nitrogen. The alcohols produced have the secondary group distributed statistically along the whole carbon chain.

$$C_nH_{2n+2} + {}^1\!/\!{}_2O_2 \longrightarrow (C_nH_{2n+1})OH$$

Secondary alcohols can also be produced by hydration of α-olefins from the thermal cracking of petroleum waxes.

The total world capacity for detergent alcohols was in the region of 1.5 million tonnes in 1990, the "synthetics" being predominantly used in the United States and the "naturals" in Asia, with Europe having a balance between the two. Since then the expansion has drifted in favor of the natural materials, driven by environmental pressure, resource renewability considerations, production incentives, and the essential interchangeability of the two types in the large majority of detergents and other industrial applications. Economic considerations have not been of major relevance, as the prices of natural versus synthetic alcohols are fluctuating. The prices depend on the balance of supply and demand of the key raw materials (ethylene/α-olefins and oils and fats) more than the difference in the conversion costs for the various production routes.

J. Fatty Amines

Fatty amines were first introduced commercially in the mid-1930s by the Armour Industrial Chemical Company. They are produced by way of the hydrogenation of nitriles [59,60]:

$$RCOOH \xrightarrow{NH_3} RCOO^-NH_4^+ \longrightarrow RC{\equiv}N \xrightarrow{H_2} RCH_2NH_2$$

Nitriles can be produced by using the reaction of fatty acids with anhydrous ammonia in the liquid phase at about 300°C with catalysts such as alumina, titanium and zirconium alcoholates, and zinc and manganese oxides. Continuous processes can use fatty acids or triglycerides at temperatures between 300 and 400°C with aluminum or thorium oxide catalysts.

Nitrile reduction is performed at 100–150°C [61,62]; higher temperatures produce higher yields of secondary amines [63]. Catalysts of the Raney nickel type are preferred; metal oxides can be used, but they yield mixtures of primary, secondary, and tertiary amines.

N-Alkyl-1,3-propanediamines are prepared by the addition of acrylonitrile to *N*-alkylamines, followed by subsequent reduction of the resultant β-cyanoethylalkylamine:

$$RNH_2 + CH_2{=}CHCN \longrightarrow RNH(CH_2)_2CN \xrightarrow{H_2} RNH(CH_2)_3NH_2$$

K. Alkylphenols

Alkylphenols are produced industrially by catalytic alkylation of phenol with olefins [64]:

Diisobutene and di-*n*-butene yield octylphenols.
Propene trimer yields nonylphenol.
Propene tetramer and tri-*n*-butene yield dodecylphenol.

The preferred catalysts are ion-exchange resins, boron trifluoride, acid-activated clays, and synthetic aluminosilicates. The reaction temperature and the phenol/olefin ratio are dependent on the type of catalyst used and the end product specifications. They can be on the order of

50–85°C and 1.2:1 to 1.4:1, respectively, for the production of octylphenol with BF_3 catalyst [64]. Boron trifluoride catalyst has the advantage of having a high activity in relatively mild conditions and of yielding essentially para-substituted monoalkyl derivatives.

Higher alkylphenols are mainly used as building blocks for technical surfactants through alkoxylation processes. They are intermediates for the synthesis of antioxidants and light stabilizers. Dodecylphenols are the starting materials for a number of lubricating oil additives. It is estimated that the worldwide production of higher alkylphenols is on the order of 550,000 tpa, with western and eastern Europe accounting for about 50%, the United States about 40%, and Japan about 5%, the balance of 5% being supplied by the other regions.

IV. BASIC CHEMISTRY AND INDUSTRIAL PRODUCTION

The synthesis of nonionic surfactants is essentially based on well-known and long-exploited chemistry and centers on two fundamental types of reactions:

1. Condensation (i.e., water elimination through esterification, etherification, or amide formation)
2. Alkylene oxide addition

In the majority of cases only one type of reaction is involved. However, specialty surfactants may require both types, e.g., an esterification followed by etherification or alkylene oxide addition.

It follows that the industrial processes and equipment needed for production are comparatively simple and inexpensive. The major issues originate from the human safety and handling issues of ethylene oxide (and, to a lesser extent, propylene oxide), from the high temperatures associated with some esterification and etherification reactions, and from the finishing steps for the isolation, purification, and conditioning (if required) of the commercial products.

A. Esterification

The compounds described in this section are manufactured by way of esterification reactions.

1. Glycerol Esters

Glycerol esters can be produced by the direct esterification of fatty acids with glycerol, which yields mixtures of mono-, di-, and triglycerides depending on the temperature and time of reaction, the catalyst, and the relative proportions of the reagents [65]. The immiscibility of glycerol and fatty acids is a major disadvantage that can be overcome only by operating at high temperatures, typically in excess of 260°C.

The most important industrial method for the manufacturing of partial esters of fatty acids is the glycerolysis of fats. This is carried out at 180–250°C in the presence of alkaline catalysts, e.g., sodium hydroxide, sodium methylate, sodium and potassium soaps, potassium carbonate, trisodium phosphate, calcium oxide, or metallic zinc and aluminum. The resulting products are mixed mono-, di-, and triglycerides. The solubility of glycerol in fats (22% at 180°C and 40% at 250°C) limits the extent of glycerolysis [66,67]. Products with more than 90% monoglyceride can be obtained by molecular distillation of glyceride mixtures [68,69]. However, since the reaction between triglycerides and glycerol is reversible, monoglycerides can be subject to internal alcoholysis on heating or prolonged storage and disproportionate into glycerol and di- and triglycerides, especially if residual catalyst is present [67,70].

Transesterification (the interchange of acyl radicals between glycerides in the presence of appropriate catalysts) can be used to rearrange partial ester mixtures. By controlling the reaction conditions and the relative proportions of reagents, nearly pure mono- and diglycerides can be obtained [71].

Partial glycerol esters are hydrophobic surfactants that are suitable essentially for w/o types of emulsion. They can be converted to more hydrophilic species by the addition of ethylene oxide.

2. *Polyglycerol Esters*

Heating glycerol in the presence of alkaline or acid catalysts at 250–275°C results in condensation and the formation of polyglycerols [72]. Esterification of polyglycerol with fatty acids can be carried out at 190–220°C using the same catalysts as for the etherification reaction [73,74]. Polyglycerol esters can also be produced by the hydrolysis of triglycerides with polyglycerol.

3. *Tetritol and Pentitol Esters*

The synthesis of esters of erythritol and xylitol has been reported [75,76], but these products are of limited commercial significance. Pentaerythritol reacts readily to form tetraesters [75,77] that find use as lubricants and oil additives.

4. *Hexitol Esters*

The direct esterification of hexitol alcohols with fatty acids under industrial manufacturing conditions leads to the formation of internal ether linkages in the polyol moiety, so that the commercial products are, in reality, anhydrohexitol esters. The anhydrization is favored by heat and acid catalysts, e.g., sulfuric or *p*-toluene sulfonic or phosphoric acid [78], whereas alkaline catalysts such as sodium or potassium hydroxide or the alkaline salts of fatty acids promote the esterification reaction [79]. By an appropriate choice of catalyst and reaction conditions, particularly temperature, it is possible to produce esters of significantly different compositions, ranging from low polarity, heavily anhydrized diesters to the more hydrophilic, nonanhydrized monoesters. These processes, however, are hardly suitable for large-scale manufacture, which requires reasonably fast reaction kinetics, maximum yield, and product consistency combined with a minimal need for purification and postreaction treatments, e.g., to remove unreacted material or reduce color. Thus, statistically, the commercial products are heterogeneous mixtures containing free polyols and fatty acids and various sorbitan (1,4-,1,5-,2,5-sorbitan) and isosorbide mono-, di-, and triesters. Figure 4 illustrates some of the species present in a commercial sorbitan ester.

Hexitol esters can be prepared using the entire range of fatty acids [77–79], although the esters with lauric, palmitic, stearic, and oleic acid are those of major commercial significance. Sorbitol is the most common polyhydric alcohol; mannitol esters are equally available but are confined to niche applications.

Sorbitan esters are versatile o/w emulsifiers but are also the building blocks for the more hydrophilic ethoxylated species commonly referred to as polysorbates [80,81]. The products most successful from the commercial point of view are obtained by the addition of 4–5 mol of ethylene oxide to one theoretical mole of sorbitan monoester, or of 20 mol of ethylene oxide to one theoretical mole of mono- or triester.

Sorbitan esters and polysorbates have been in use for nearly half a century (they were originally introduced in the mid-1940s by Atlas Powder Co. under the trademarks SPAN and TWEEN, respectively) and have a remarkable record of safety in use, supported by extensive toxicological data, to the point that they are often used in toxicological studies as the bench-

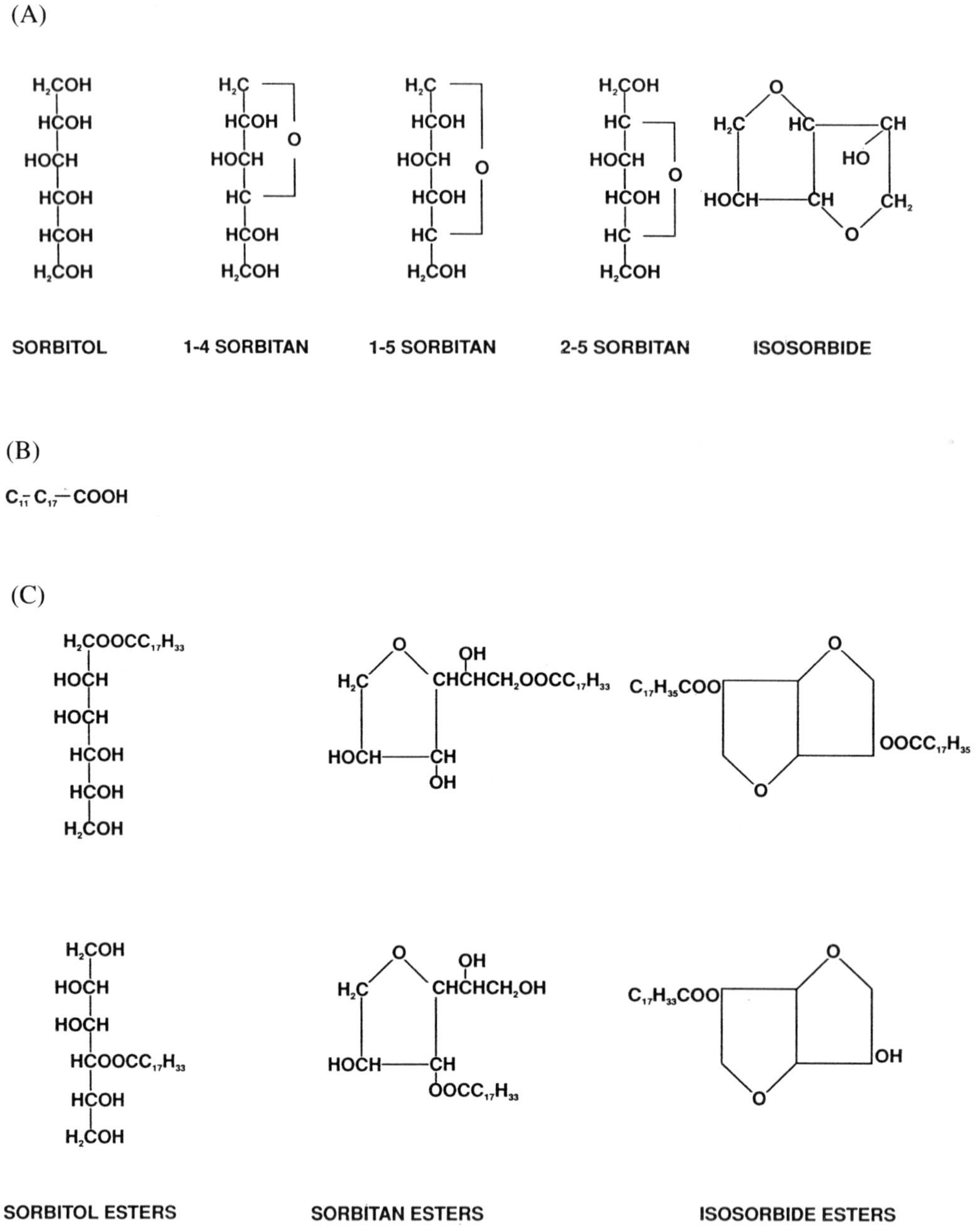

Fig. 4 Most common species present in a commercial sorbitan ester. (A) Free polyols; (B) free fatty acids; (C) sorbitol/sorbitan/isosorbide esters.

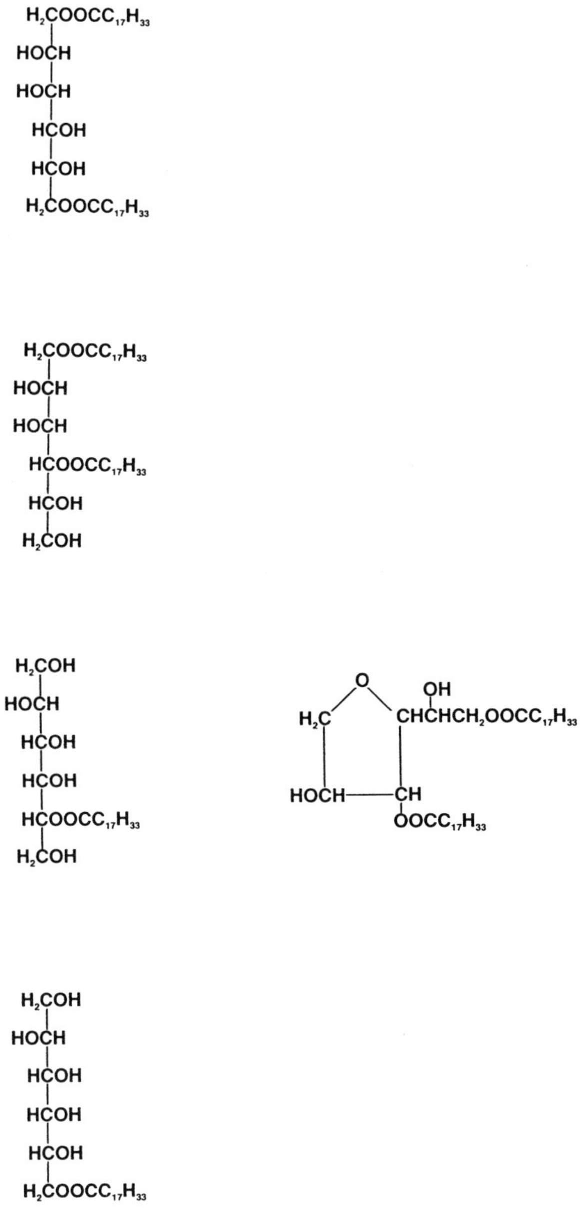

SORBITAN ESTERS

Fig. 4 Continued

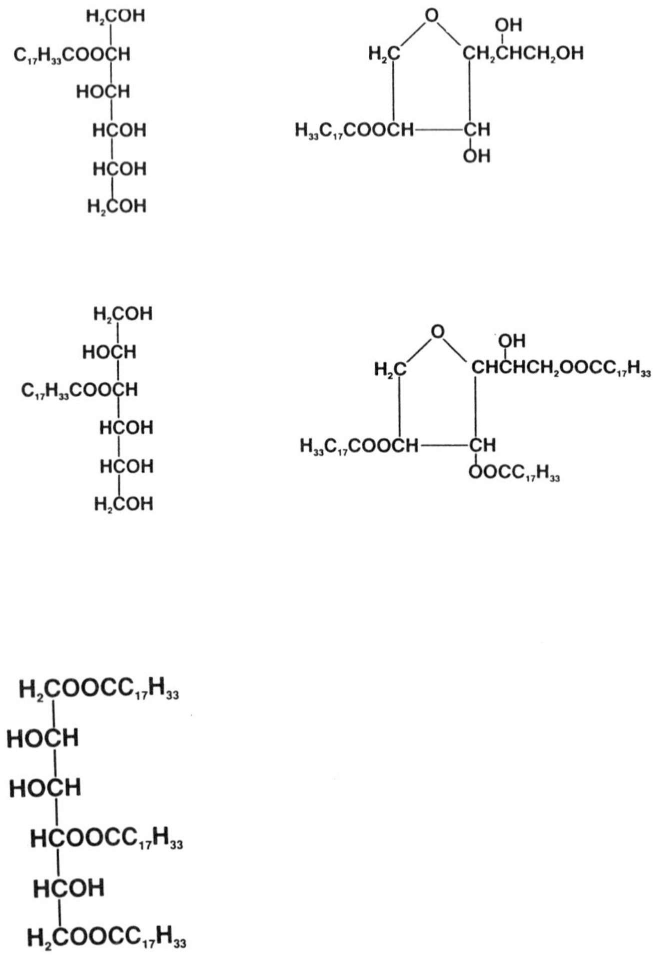

SORBITOL ESTERS SORBITAN ESTERS

Fig. 4 Continued

mark against which other surfactants are tested. They are reported in many pharmacopoeias [82–84] and are included in the positive lists of many regulatory bodies, e.g., FDA [85], BGA [86], and EU directives [87], as additives for polymers, adhesives, and paper products that come into contact with food and as direct additives for human food and animal feed.

Sorbitan esters and polysorbates also form a homogeneous series of molecules of different polarity and, as such, are particularly easy to formulate by applying, for example, the principles of the HLB (hydrophile–lipophile balance system) concept described in Section V. F. Because of the safety characteristics, the extensive regulatory approval, the good emulsification and dispersion properties, and the ease of formulation, these products are used in an amazing variety of applications, ranging from cosmetic creams and lotions through emulsion and suspension polymerization, pigment dispersion, industrial fluids, and food, to agricultural formulations, to name just a few.

A special category of hexitol esters is that in which the polyol is ethoxylated first and the resulting product is further esterified with fatty acids. The products with the best performance contain 6–7 mol of C_{18} fatty acids per theoretical mole of ethoxylated hexitol. Because of their structure, these molecules can function as steric stabilizers and are particularly useful in the formulation of crop protection chemicals and metalworking fluids.

5. *Sugar Esters*

The economic manufacture of fatty acid esters using mono- and oligosaccharides has been the subject of considerable research. This was—and still is—motivated by the drive to produce high HLB nonionic surfactants without using ethylene oxide. This is a target that is particularly attractive to the personal care and food/feed industries. It has already been shown that sorbitol esters are only a theoretical option, whereas polyglycerol esters are both expensive and, comparatively speaking, still too hydrophobic.

The difficulties associated with the industrial manufacture of sugar esters are the insolubility of the saccharides in the fatty acids and their derivatives, the charring of the sugar at the temperatures of esterification, the poor conversion rates, and the need for extensive purification of the finished product. Direct transesterification of sucrose with triglycerides, without further purification, yields complex mixtures that have found limited use as emulsifiers for animal feed, but they are too ill-defined and contain too many impurities to be suitable for other applications.

Manufacture of products with a broader range of applications involves the use of solvents such as dimethylformamide [88], dimethyl sulfoxide [89], or hydrocarbons [90] for the reaction or the subsequent purification. This adds significantly to the manufacturing costs and detracts from the "natural"/"green" image that these products could otherwise have. Recently, esters of α-methyl glucoside produced through enzymatic esterification have been proposed, and it is possible that, in the long term, enzymes will provide an economic route to the manufacture of other saccharide esters.

So far, apart from the sucroglycerides for animal feed, the few sucroesters commercially available have found only those niche applications that can tolerate their high cost or are used in specific foods in which the use of ethylene oxide derivatives is not allowed from a regulatory point of view. They continue, however, to receive considerable attention from the formulation chemists because of their "natural" connotation.

The ethoxylation of sucrose esters has been reported [91], but the products do not seem to have generated a commercial response. The derivatives obtained by propoxylation of sucrose, followed by esterification or transesterification, have some limited use as antifoam agents in the food and fermentation industries.

6. *Polyoxyalkylene Esters of Fatty Acids*

Polyoxyalkylene esters of fatty acids can be prepared by a variety of methods [92]. However, only the direct reaction of a fatty acid with an alkylene oxide olefin (usually ethylene oxide) and the esterification of a fatty acid with polyalkylene glycols are of industrial relevance.

Ethylene oxide reacts with fatty acids to give monoesters that, during the reaction, undergo a further transesterification, so that the final product is a mixture of polyoxyethylene glycol monoester, diester, and free polyoxyethylene glycol [93].

$$RCOOH + n(CH_2CH_2O) \longrightarrow RCOO(CH_2CH_2O)_nH$$

$$2RCOO(CH_2CH_2O)_nH \longrightarrow RCOO(CH_2CH_2O)_nOCR + HO(CH_2CH_2O)_nH$$

Similar mixtures are obtained by the direct esterification of fatty acids with polyoxyethylene glycols [94] if a 1:1 molar ratio of the reagents is used. A higher ratio of fatty acids favors the formation of diesters, whereas more monoester is produced with an excess of polyoxyethylene glycol. The esterification reaction is performed at temperatures of 100–250°C, with or without catalyst. The use of a catalyst (usually sulfuric or organic sulfonic acids) reduces the reaction temperature and shortens the reaction time. The reaction is reversible and is therefore never complete. In industrial processes the equilibrium is shifted either by removing the water of reaction as an azeotrope or by vacuum stripping or inert gas sparging.

B. Etherification

Apart from the production of ethers by ethylene oxide addition (discussed in detail later), the most important reactions of etherification by far are those leading to alkyl polyglycosides, in particular the glucose-derived products. Chemical glycosylation processes include acid-catalyzed reactions leading to complex mixtures of oligomers [95,96] and kinetically controlled, irreversible, mostly stereospecific substitution reactions with suitably activated carbohydrate substrates [97–100]. Enzymatic synthesis of alkyl glycosides [101] is not yet suitable for commercial scale application, and the industrial processes operated today are based on the Fisher process through different technical adaptations, depending on the molecular weight of the alcohol and the dextrose/alcohol ratio used to obtain the desired degree of polymerization. In addition, depending on the physical form and purity of the raw material, a direct acetalization or transacetalization may prove economically more attractive. These routes are illustrated schematically in Fig. 5.

The direct glucose conversion with fatty alcohols is a two-phase process, a disadvantage that can be minimized by using catalysts that do not accumulate in the polar phase, e.g., alkylsulfonic acids [102]. The molar ratio of glucose to fatty alcohol is typically 1:5, and the excess alcohol is removed (e.g., by using thin-film evaporators), purified, and recycled. Following dilution with water (fatty alcohol polyglucosides are highly hygroscopic and cannot be handled as such), a final purification may be required to remove colored species, by-products, or unreacted carbohydrate. During the synthesis, care must be exerted to prevent overreaction, which leads to excess polyglucosides with no surface activity, as well as underreaction, which leads to poor yield.

Industrial fatty alcohol polyglucosides are complex mixtures of isomeric mono-, di-, and triglucosides, with each higher oligomer formed in decreasing amounts and with an average degree of polymerization ranging from 1.3 to 1.8 units. The fatty alcohols more commonly used have chains of 8–10 carbon atoms if the alkylpolyglucoside is intended for wetting applications, 12–14 for cleaning/detergency, and 16–18 for emulsification. (In the latter case,

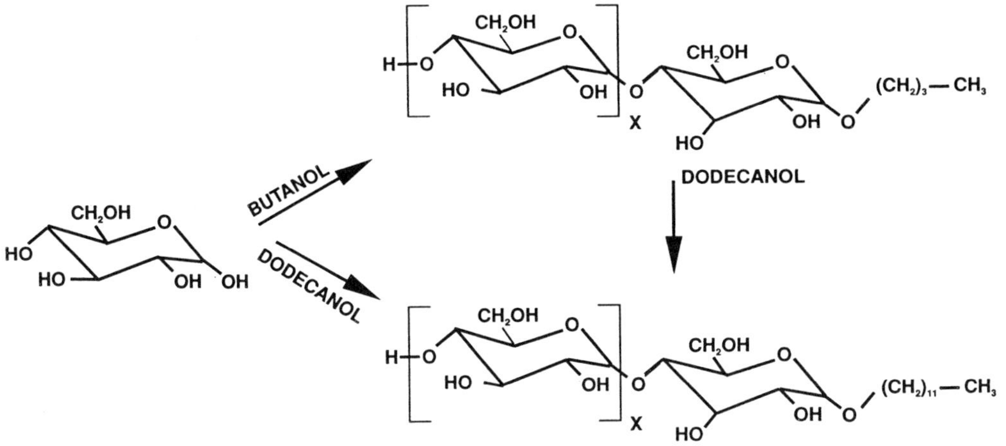

Fig. 5 Schematic route for the synthesis of alkylpolyglucosides.

however, large quantities of unreacted alcohol are present because of the impossibility of achieving its complete removal.)

The Williamson ether synthesis [103] is used to block the terminal OH of ethoxylated nonionics with alkyl groups such as methyl, benzyl, or allyl. Alkyl-terminated nonionics not only have reduced foaming and reduced water solubility but also are essential in the formulation of hydrolytically unstable actives, e.g., diketenes or alkenyl succinic anhydrides in paper sizing or crop protection chemicals, that would otherwise be attacked by the free hydroxyl groups.

Etherification reactions are also involved in the synthesis of small-volume specialty surfactants, e.g., polyglycerol ethers, that have a higher chemical stability than the corresponding esters. Synthetic processes based on alcoholates, epoxides, or chloroparaffin might be used.

C. Amide Formation

There are several routes for the synthesis of fatty acid amides [104], including reaction of fatty acid chlorides and anhydrides with ammonia, hydrolysis of nitriles, and aminolysis of glycerol esters. Industrially, amides are prepared by reacting fatty acids and ammonia, with the subsequent dehydration of the ammonium salt [60,105],

$$RCOOH + NH_3 \longrightarrow RCOO^-NH_4^+ \longrightarrow RCONH_2 + H_2O$$

Alkanoyl-*N*-methyl glucamides are a new class of nonionic surfactants that appear to be particularly suitable for the formulation of liquid dishwashing detergents because of their favorable performance and ecotoxicological properties. The synthesis involves reacting a tryglyceride [Eq. (1)] or a fatty acid ester of a lower alcohol [Eq. (2)] with *N*-methyl glucamine:

$$\begin{array}{l} CH_2OOCR \\ | \\ CHOOCR \\ | \\ CH_2OOCR \end{array} + 2\ CH_3NHCH_2(CHOH)_4CH_2OH \longrightarrow$$

$$\longrightarrow 2\ RCON(CH_3)CH_2(CHOH)_4CH_2OH + \begin{array}{l} CH_2OH \\ | \\ CHOH \\ | \\ CH_2OOCR \end{array} \qquad (1)$$

$$RCOOCH_3 + CH_3NHCH_2(CHOH)_4CH_2OH \longrightarrow$$

$$\longrightarrow RCON(CH_3)CH_2(CHOH)_4CH_2OH + CH_3OH \qquad (2)$$

The reaction temperature is kept within the range of 130–150°C, and there is no need for a catalyst. For practical applications, coconut fatty acids are preferred.

D. Alkylene Oxide Addition

The general reaction of alkylene oxide addition to compounds containing active hydrogen can be described as

$$RAH + n\ C_2H_4O \longrightarrow RA(C_2H_4O)_nH$$

The reaction can be viewed as a sequence of discrete steps that each add one alkylene oxide unit. This leads to a number of species with different alkylene oxide content, whose distribution is determined by the nature of the hydrophobe, the reaction rate, the catalyst, the equipment, and the reaction conditions:

$$RAH + n\ C_2H_4O \longrightarrow RA(C_2H_4O)H + (n-1)C_2H_4O \longrightarrow$$
$$RA(C_2H_4O)_2H + (n-2)C_2H_4O$$
$$RA(C_2H_4O)_iH + (n-i)C_2H_4O \longrightarrow RA(C_2H_4O)_{i+1}\ H + (n-i-1)C_2H_4O$$
$$RA(C_2H_4O)_{n-1}H + C_2H_4O \longrightarrow RA(C_2H_4O)_nH$$

The ring-opening reactions of ethylene oxide are nucleophilic substitutions [106,107]. With basic catalysts, there is first an attack of the nucleophilic at a ring carbon atom:

$$RA^- + C_2H_4O \longrightarrow RAC_2H_4O^-$$

which is then followed by proton exchange or further reaction with another alkylene oxide molecule as seen above:

$$RAH + RAC_2H_4O^- \longrightarrow RA^- + RAC_2H_4OH$$

or

$$RAC_2H_4O^- + C_2H_4O \longrightarrow RAC_2H_4OC_2H_4O^-$$

The rate-determining step is the first addition, which is dependent on both the concentration of the nucleophile and that of the alkylene oxide; the substitution is therefore of the SN2 type.

Acid catalysts can also be used but are less favored because of the formation of larger quantities of by-products such as polyoxyethylene glycols, dioxane, and 2-methyldioxolane. In the acid-catalyzed reaction, the slow stage is the formation of a carbocation following protonation and ring opening of the oxirane ring:

$$C_2H_4O + H^+ \longrightarrow C_2H_4\overset{+}{O}H$$
$$C_2H_4\overset{+}{O}H \longrightarrow \overset{+}{C}H_2CH_2OH$$
$$RAH + \overset{+}{C}H_2CH_2OH \longrightarrow RACH_2CH_2OH + H^+$$

In ideal conditions the reaction can be considered to be of the SN1 type. It has been suggested that in practice the reaction is intermediate between an SN1 and SN2 [107].

1. Polyoxyethylene Alkylphenols

Nonylphenol and ethylene oxide are, respectively, the most common hydrophobe and hydrophile used. A typical industrial manufacturing process involves oxyethylation in an autoclave at temperatures within the range 140–200°C, ethylene oxide pressure of 2–5 bar, with 0.1–0.5% of sodium or potassium hydroxide as catalyst. Other basic catalysts such as sodium acetate, potassium carbonate, and sodium methylate have also been described [108,109]. The products are statistical molecules, the composition of which approximates to a Poisson distribution with basic catalysts and a relatively low degree of ethoxylation (usually up to 10–15 mol of EO per mole of hydrophobe initiator [110]). Different distributions are possible, but the essential feature of a spread of molecules with different degrees of ethoxylation around a nominal average remains.

Octylphenol has a similar reaction pattern, but despite the nominal difference of only one carbon atom in the alkyl chains, the interfacial properties of its alkoxylated adducts differ markedly from those of the nonylphenol homologues owing to the structure of the alkyl chain. As such, octylphenol adducts are used in specialty niche applications, e.g., as emulsifiers in crop protection formulations. Dodecylphenol is a starting material for the synthesis of lubricating oil additives and antioxidants.

The polarity of the alkylbenzene ring confers to alkylphenols unmatched interfacial properties, but the limited industrial availability of hydrophobes restricts the range of possible HLB modifications. When a more pronounced hydrophobic character is needed, the water-insoluble moiety is extended by the addition of propylene oxide.

2. Polyoxyalkylene Fatty Alcohols

The oxyethylation of an alcohol was first reported in 1926, but it was the work of Wittmer and Schoeller in the early 1930s that produced the first derivatives of fatty alcohols. Industrial processes in current use favor base-catalyzed reactions with sodium or potassium hydroxide or sodium alcoholates at temperatures ranging from 120 to 200°C and pressures of 2–8 bar [109,111]. Mixing by impeller or loop circulation [112] must ensure adequate heat transfer throughout the reaction (which is highly exothermic) as well as maintaining intimate contact between the liquid phase and the ethylene oxide vapor phase. Primary alcohols give faster reaction rates than secondary or tertiary alcohols, while the rate of ethoxylation decreases as the carbon chain length of the alcohol increases. After the initial ethylene oxide addition, the differences in the rate of reaction diminish as the environment around the reacting center becomes similar following the growth of the polyoxyethylene oxide chains.

The product from the reaction is a mixture of alcohol ethers, with a distribution that is much wider for the secondary and tertiary alcohols. This distribution depends on the process parameters, in particular the catalyst and the mixing system used [113,114]. On completion of the reaction, the unreacted ethylene oxide is removed by vacuum stripping, and the catalyst is neutralized with organic or inorganic acid [115,116] and removed by filtration. The product may also be bleached to reduce color.

The propoxylation of fatty alcohols follows, in principle, the same procedure as ethoxylation [109].

3. *Polyoxyethylene Alkylamines*

The preparation of surface-active amines by the ethoxylation of fatty amines was first described by Schoeller and Wittwer. A large variety of addition products from primary, secondary, and tertiary amines as well as polyamines have been reported, although only the derivatives of the primary amines and of the *N*-alkyl-1,3-propane diamine are commercially important. In the reaction of ethylene oxide with primary amines, both reactive hydrogens are substituted prior to any further ethoxylation [110]:

$$R-NH_2 + 2\,(C_2H_4O) \longrightarrow R-\underset{\displaystyle |\atop\displaystyle C_2H_4OH}{\overset{\displaystyle C_2H_4OH\atop\displaystyle |}{N}}$$

$$R-\underset{\displaystyle |\atop\displaystyle C_2H_4OH}{\overset{\displaystyle C_2H_4OH\atop\displaystyle |}{N}} + n\,(C_2H_4O) \longrightarrow R-\underset{\displaystyle |\atop\displaystyle (C_2H_4O)_yH}{\overset{\displaystyle (C_2H_4O)_xH\atop\displaystyle |}{N}}$$

The reaction is initiated through ethylene oxide addition at about 100°C and is furthered by raising the temperature to 150°C or higher in the presence of base catalysts.

4. *Polyoxyethylene Alkylamides*

The polyoxyethylene alkylamides include the mono- and diethanolamide derivatives, which are the most important commercial products, together with the polyoxyethylene adducts.

Monoethanolamides can be prepared from fatty acids [117], from fatty esters [118], and by transamidification [119]. Diethanolamides from fatty acids have been prepared by reacting 1 mol of fatty acid with 2 mol of diethanolamine at 150–170°C [120] or by heating a fatty acid methyl ester with diethanolamine at about 115°C in the presence of sodium methoxide catalyst and distilling off the methanol released [121]. The reactions are schematically

$$RCOOH + NH_2CH_2CH_2OH \longrightarrow RCONHCH_2CH_2OH + H_2O$$

and

$$RCOOH + NH(CH_2CH_2OH)_2 \longrightarrow RCON(CH_2CH_2OH)_2 + H_2O$$

In the reaction of fatty acids with alkanolamides, ester amides are formed, which usually detracts from the foaming properties of the products. The formation of ester amides can be minimized by balancing the reaction parameters, in particular temperature and reaction time, and by using alkali metal catalysts [122].

Polyoxyethylene amides are prepared by reaction of amide or of a preformed alkanolamide with ethylene oxide in the presence of basic catalysts:

$$RCONH_2 + n(C_2H_4O) \longrightarrow RCONH(C_2H_4O)_nH$$
$$RCONHCH_2CH_2OH + n(C_2H_4O) \rightarrow RCONH(C_2H_4O)_{n+1}H$$
$$RCON(CH_2CH_2OH)_2 + n(C_2H_4O) \rightarrow RCON(C_2H_4O)_{n+2}H$$

5. *Polyalkylene Oxide Copolymers*

The polyalkylene oxide copolymers probably constitute the richest class of nonionic compounds. Several parameters—apart from the usual ones of catalyst type and reaction conditions—can be varied to achieve an amazing range of products and effects. In particular, the following variables can be taken into consideration.

1. Type of initiator. This can be mono- or polyfunctional (i.e., it may contain from one to several active hydrogen atoms). There can be further diversification depending on the nature of the carbon chain (a monofunctional alcohol initiator may have carbon chains containing from 1 to 18 atoms or more, a tetrafunctional diamine may have from 2 to 6 carbon atoms in the central bridge) and of the atoms or groups carrying the active hydrogen (usually oxygen or nitrogen, but possibly sulfur, amido, or carboxy).
2. Sequence of addition of alkylene oxide to the initiator, in particular
   a. A sequence of ethylene oxide units followed by one of propylene oxide units or by a mixture of ethylene oxide and propylene oxide.
   b. A sequence of propylene oxide units followed by one of ethylene oxide units or by a mixture of ethylene oxide and propylene oxide.
   c. A mixture of ethylene oxide and propylene oxide.
3. The total molecular weight of the adduct as well as the molecular weight of each specific sequence and, within a mixed sequence of ethylene oxide/propylene oxide, their relative proportions.
4. The type of alkylene oxide having the hydrophobe function (i.e., in a number of cases, butylene oxide has replaced propylene oxide units).

Polyalkylene oxide copolymers are peculiar in that, with a few exceptions, the amphipathic character of the molecule is attributable more to the balance of alkylene oxides in the molecule than to the type of initiator. It is impossible in this chapter to give a detailed discussion of the chemistry and manufacture of such a broad range of products. The most relevant commercially, apart from the glycerol, sorbitol, and sucrose polyols used as building blocks in polyurethane polymers, include derivatives from

1. Monofunctional alcohols. Alcohols of various hydrocarbon chain lengths are reacted in sequence with propylene oxide, followed by ethylene oxide addition [123]. Sodium hydroxide can be used as catalyst, the temperature for the propoxylation stage being about 100°C and for the ethoxylation stage about 140°C. The order of addition of the alkylene oxides can be reversed, yielding products with significantly different properties [124]. Tergitol XD, XH, and XJ from Union Carbide are produced starting from a C_1–C_8 alcohol, which is reacted first with a mixture of propylene oxide and ethylene oxide and subsequently with pure ethylene oxide [125].
2. Bifunctional alcohols. The Pluronic products from BASF are made by the sequential addition of propylene oxide and ethylene oxide to a propylene glycol initiator [126] using sodium hydroxide as catalyst and a reaction temperature of about 120°C. Block copolymers prepared by sequential condensation of butylene oxide and ethylene oxide have been reported [127].
3. Tetrafunctional polyamines. The Tetronic products from BASF have the structure

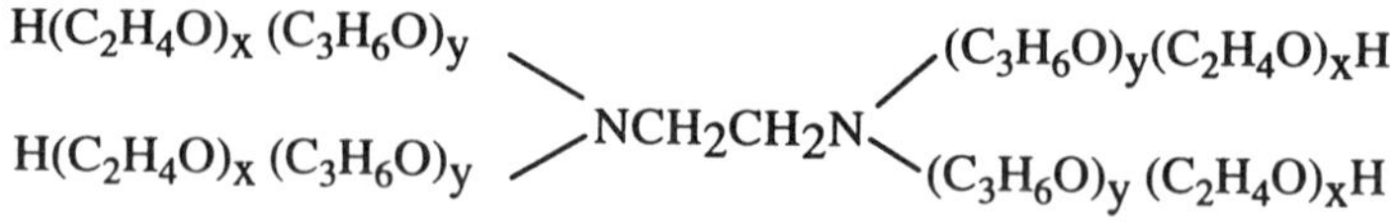

with each polyoxypropylene group containing about 4–110 propylene oxide units and ethylene oxide making up 20–90% of the total weight of the product.

E. Amine Oxides

Amine oxides are produced from the reaction of hydrogen peroxide on tertiary amines [128,129]:

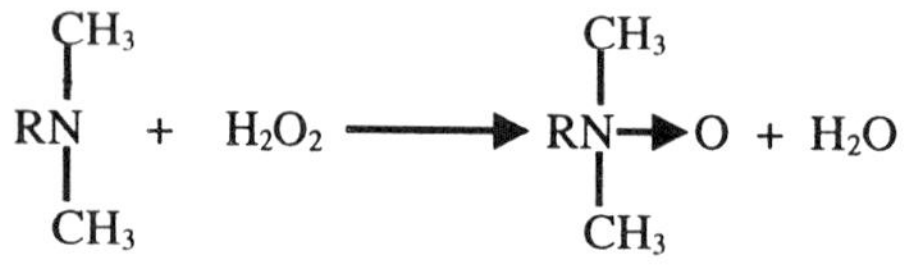

The tertiary amine (usually dimethyldodecylamine) is added to aqueous hydrogen peroxide with vigorous agitation, keeping the temperature in the region of 60–65°C. Initially, water is added in small amounts to prevent gel formation and to finally reach an active concentration of 30–35% while the temperature is raised to 70–75°C. Agitation is continued for another 2–3 h, and the excess peroxide is removed by addition of sodium sulfite.

F. Miscellaneous Nonionic Surfactants

Various nonionic compounds have been described, including sulfoxides [130], phosphine and arsine oxides [131,132], polyoxyethylene phosphonates and phosphonites [133,134], and acetylenic glycols and their ethoxylates. Of these, only the acetylenic derivatives, e.g., Surfynols from Air Products, have achieved commercial relevance as wetters, defoamers, pigment dispersants, and hydrotropes.

V. PHYSICOCHEMICAL PROPERTIES OF NONIONIC SURFACTANTS

The behavior of nonionic surfactants is determined by

1. The chemical class. This is at the root of many bulk and surface properties, e.g., physical form, surface and interfacial adsorption, micellar structure, and physical and chemical stability.
2. The distribution of molecular species. All the commercial products are statistical compounds, either polydispersed or heterodispersed [135]. In a polydispersed material, the variation of structure and properties of the components is gradual, and clusters form around a mean, as for example in ethoxylated alcohols or nonylphenols, alkoxylated amines or amides, and PEO/POP copolymers. In a heterodispersed material the variation is discontinuous and the individual constituent molecules have significantly different structures and properties as, for example, in the case of the sorbitan or glycerol esters.

These parameters impart to the different species distinct physicochemical properties that are the key to their performance in practical applications:

Critical micelle concentration
Foaming

Solubilization
Surface and interfacial adsorption
Wetting
Emulsification
Dispersion
Detergency

A. Critical Micelle Concentration

The critical micelle concentration (CMC) is the concentration at which the surface adsorption is complete to form an interfacial monomolecular layer. The surfactant then ceases to exist as an individual molecule in the bulk of the solution and begins to aggregate in clusters (micelles) of various shapes. The physicochemical properties of aqueous solutions of surfactants change abruptly at the CMC, and it is their value at concentrations higher than the CMC that determines the behavior and performance of a given surfactant.

Given the large variety of chemical classes and their significantly different properties, it is not possible to give more than broad generalizations here. Ethoxylated nonionic surfactants have been more widely investigated for the following reasons:

1. They are of major commercial interest.
2. They can be synthesized as well-characterized and pure molecules, using Williamson's condensation.
3. The polydispersed nature of the commercial products allows their use (instead of the pure products) for accurate and representative physicochemical studies.

Nonionic ethoxylates have CMC values that are significantly lower than those of ionic materials of comparable chain length and surface activity (see Table 4). In general, the CMC of nonionic surfactants

1. Decreases as the number of carbon atoms in the hydrophobic chain increases
2. Decreases with the temperature
3. Is hardly affected by electrolytes (except for hydrogen ions at very low pH)
4. Decreases for ethoxylated products as the number of ethylene oxide units decreases

Becher [136] produced an extensive review of the values of the CMC in aqueous solutions of homogeneous and heterogeneous ethylene oxide derivatives and of homogeneous surfac-

Table 4 Typical CMC Values of Ionic and Nonionic Surfactants

	Temp. (°C)	CMC (mol/L)	Percent (w/w)
Anionic			
$C_{12}H_{25}SO_4^-Na^+$	40	8.6×10^{-3}	0.2
$C_{12}H_{25}C_6H_4SO_3^-Na^+$	60	1.2×10^{-3}	0.04
Cationic			
$C_{12}H_{25}N^+(CH_3)_3Br^-$	25	2.0×10^{-2}	0.46
Amphoteric			
$C_{12}H_{25}N^+H_2CH_2COO^-$	27	1.3×10^{-3}	0.033
Nonionic			
$C_{12}H_{25}O(CH_2CH_2O)_4H$	25	4.0×10^{-5}	0.0014
$C_{12}H_{25}O(CH_2CH_2O)_7H$	25	5.0×10^{-5}	0.0025

tants with a hydrophile other than ethylene oxide. A more recent review, including CMC values in nonaqueus media, has been reported by Rosen [137].

B. Foaming

In aqueous media, nonionic surfactants generally produce less, and less stable, foam than ionic surfactants. For a given hydrocarbon chain length, the foam volume and stability reach a maximum at a particular POE chain length and then decrease.

In many industrial processes it is useful to have surfactants that exhibit surface or interfacial activity without producing foam. This can be achieved by structural changes that

1. Affect the diffusion characteristics, thus destroying the elasticity of the surface film. For this effect, large, straight-chain hydrophobes can be substituted with branched isomers, or the hydrophilic group can be positioned centrally in the hydrophobe chain.
2. Increase the area per molecule, thus forming loosely packed films. Suitable modifications involve putting two hydrophilic or hydrophobic groups onto the same molecule, using highly branched or unsaturated hydrophobes, or attaching two bulky hydrophobes on the same carbon atom. Of these the "capping" of terminal OH groups in polyoxyethylene nonionics with alkyl groups or propylene oxide is the most effective and preferred.

The structures of very low foaming nonionic surfactants are given in Table 5.

Table 5 Structures of Low Foaming Nonionic Surfactants

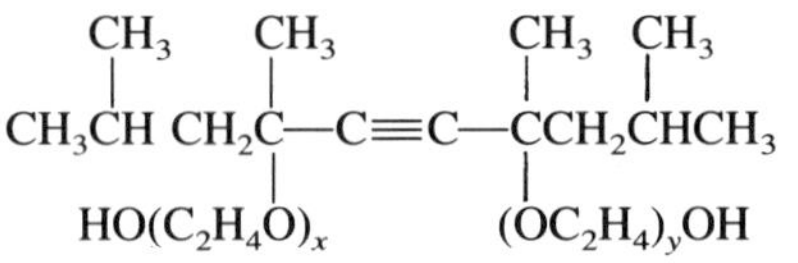	$x+y \leq 4$
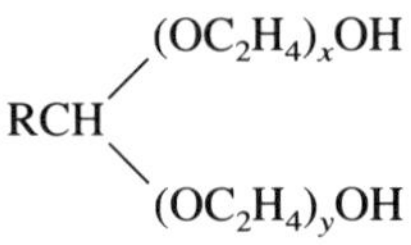	$R<C_{11}$; $x=y \leq 5$
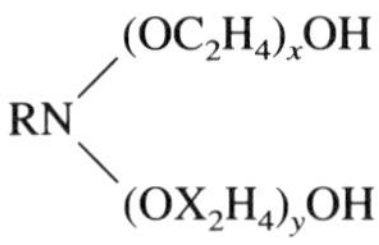	$R=C_{10}$; $x=y \leq 3$
$HO(C_2H_4O)_x(CH_2)_{12}(OC_2H_4)_yOH$	$x+y \leq 12$
$HO(C_2H_4O)_x(CH_2CH_2CH_2CH_2O)_y(C_2H_4O)_zH$	$y \leq 27$; $x+z \leq 82$
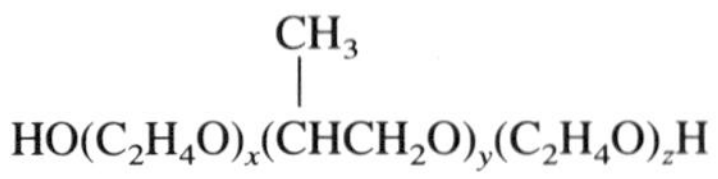	$y=35$; $x+z=45$
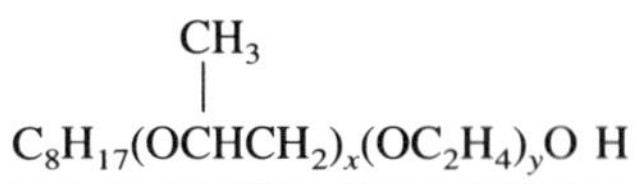	$x=y \cong 10$

Source: Ref. 137.

C. Solubilization

Solubilization is the spontaneous incorporation of a substance by reversible interaction with the micelles of a surfactant in a solvent, to form a thermodynamically stable isotropic solution. Solubilization into aqueous media is of major importance for processes such as cleaning and soil removal, micellar catalysis of organic reactions, emulsion polymerization, and enhanced oil recovery. Solubilization in nonaqueous media is the basis of dry-cleaning processes.

The solubilization by nonionic surfactants of polar molecules containing large hydrophobes and small hydrophiles, e.g., alcohols or amines, is believed to occur through comicellization with the surfactant's molecule (Fig. 6). For less linear molecules, e.g., aromatic and substituted aromatic hydrocarbons or branched esters, the solubilization may occur through inclusion in the interior of the micelles or through adsorption into their outer hydrophilic layer, depending on the polarity of the solubilisate (Fig. 7). In the case of polyoxyethylene nonionics it has been suggested that polar molecules may be solubilized into the outer region of the polyoxyethylene tails (Fig. 8). Because of the low critical micelle concentration, nonionic surfactants are better solubilizing agents than cationics and anionics of the same hydrophobic chain length.

The solubilization power seemingly attainable from the surfactant structure is influenced by the nature of the solubilisate, which determines the solubilization mechanism. In aqueous solutions, the solubility of aliphatic hydrocarbons increases with the chain length of the hydrophobic moiety and with a decrease in the number of ethylene oxide units. This reflects the increase in the aggregation number of the micelles produced by these changes and the probable dissolved state of the solute in the micellar core [138]. Ethylbenzene, which is more polar and probably solubilized both in the ethylene oxide chains and in the inner micellar core, shows an increase in solubility with an increase in the POE chain length [139].

Solubilization also depends on temperature and cloud point. In POE nonionic surfactants, the micellar weight increases and the CMC decreases with an increase in temperature [140–148]. Thus nonpolar materials that are solubilized in the inner core of the micelles show increased solubility as the temperature rises. However, as the increase in temperature causes

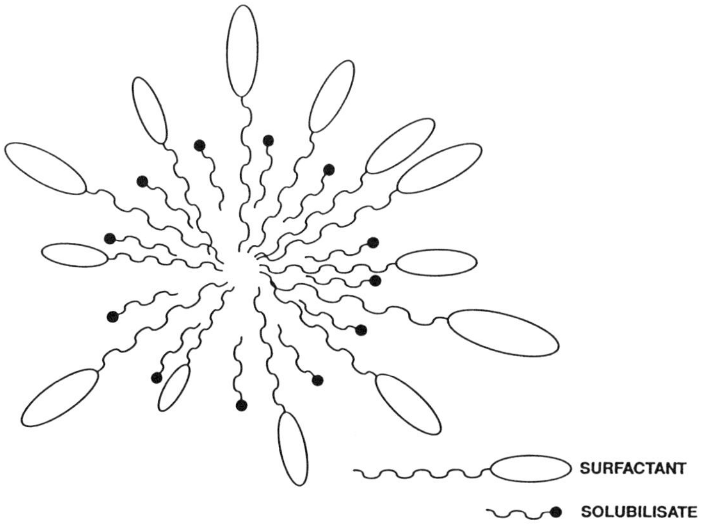

Fig. 6 Solubilization of polar molecules.

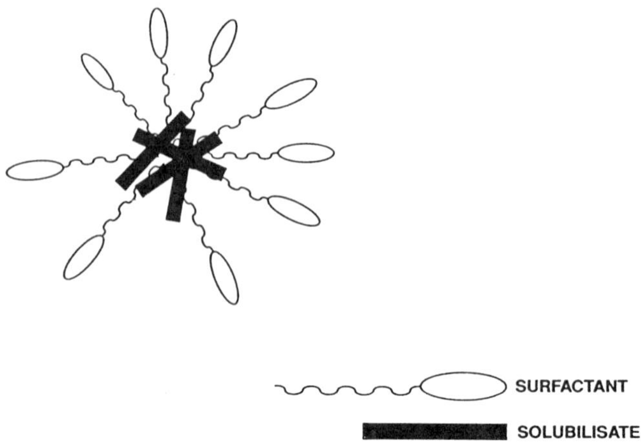

Fig. 7 Solubilization of low polarity molecules.

a partial dehydration of the ether oxygen atoms in the POE groups together with their tighter coiling, the solubilization of more polar species that may occur through surface adsorption or inclusion in the interfacial micellar layer is decreased.

D. Surface and Interfacial Adsorption

The key feature of surface-active agents is their adsorption at interfaces in an oriented fashion. The surface concentration and molecule orientation are critical in determining the performance of a surfactant with respect to many properties such as wetting, dispersion, emulsification, foaming, and detergency.

Adsorption may occur at liquid/gas, liquid/liquid, and liquid/solid interfaces. The adsorption at liquid/gas and liquid/liquid interfaces can be described by the Gibbs adsorption equation [149],

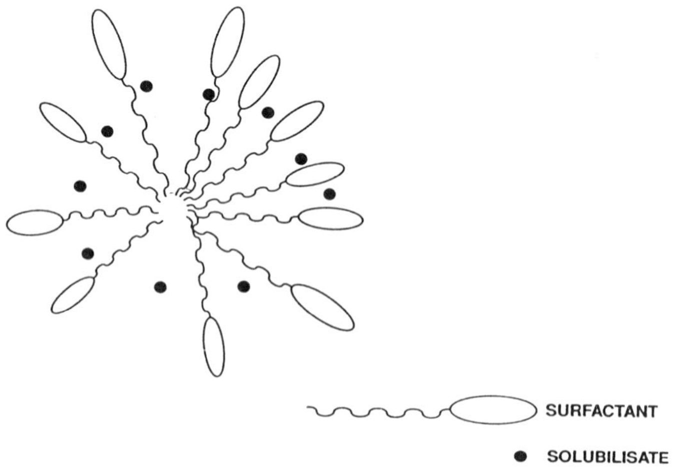

Fig. 8 Solubilization of polar molecules by polyoxyethylene nonionics.

$$d\gamma = -\sum_i \Gamma_i \, d\mu_i$$

where $d\gamma$ is the change in surface or interfacial tension of the solvent, Γ_i the surface excess concentration of the ith component of the system, and $d\mu_i$ the change in chemical potential of the ith component of the system. The surface excess concentration is defined as the excess per unit area of interface of any component over that present in a reference system of the same volume in which the bulk concentration in the two phases remains uniform up to a hypothetical dividing surface.

For dilute solutions of nondissociated surfactants, the above equation can be simplified to

$$d\gamma = -RT\Gamma \, d \ln C$$

where C is the molar concentration of the surfactant in the bulk of the solution. For surface-active agents, the surface excess concentration can be equated to the actual surface concentration; hence for solutions of nonionic surfactants containing a single species,

$$\Gamma = -\frac{1}{RT}\left[\frac{d\gamma}{d \ln C}\right]_{\mathrm{T}}$$

A typical plot of Γ versus ln C is given in Fig. 9. At low surfactant concentration, any addition of surfactant results in a decrease in γ, i.e., the surface is becoming progressively saturated with surface-active molecules. At high surfactant concentration, γ is practically constant, indicating that the surface is fully saturated. The break point at which the slope of the curve significantly changes corresponds to the critical micelle concentration (see Section V.A).

The surface concentration at surface saturation is a measure of the maximum value that saturation can attain and is consequently a useful measure of the effectiveness of adsorption. Table 6 gives the values for Γ and surface area per molecule for some common nonionic surfactants in comparison with ionic, cationic, and amphoteric reference compounds. Additional values are reported by Rosen [150].

The adsorption at solid/liquid interfaces can be calculated from the equation [158]

$$\frac{n_0 \, \Delta x_1}{m} = n^s{}_1 x_2 - n^s{}_2 x_1$$

where n_0 is the total number of moles in solution before adsorption; $\Delta x_1 = x_{1,0} - x_1, x_{1,0}$ is the mole fraction of component 1 before adsorption; x_1, x_2 are the mole fractions of components 1 and 2 at adsorption equilibrium; m is the mass of adsorbent in grams; and $n^s{}_1$ and $n^s{}_2$ are

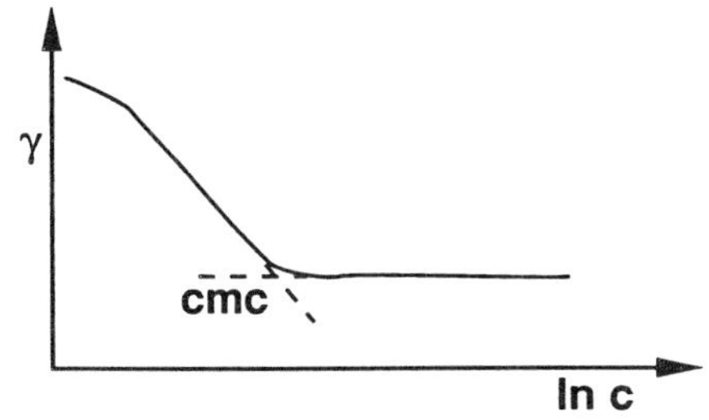

Fig. 9 Schematic representation of the variation of γ in function of surfactant concentration.

Table 6 Surface Excess Concentration (Γ) and Surface Area per Molecule for Some Common Nonionic Surfactants in Comparison with Anionic, Cationic, and Amphoteric References

Surfactant	Temp. (°C)	Γ_m (mol/cm^2 × 10^{10})	a^2_m (Å^2)	Ref.
$C_8H_{17}OCH\ (CHOH)$[a]	25	4.0	41	151
$C_{12}H_{25}(OC_2H_4)_4OH$	25	3.6	46	152
$C_{12}H_{25}(OC_2H_4)_6OH$	20	3.6	45	153
$C_{12}H_{25}(OC_2H_4)_8OH$	25	2.5	66	152
n-$C_{16}H_{33}(OC_2H_4)_6OH$	25	4.4	38	154
n-$C_{16}H_{33}(OC_2H_4)_9OH$	25	3.1	53	154
$C_8H_{17}OCH_2CH_2OH$	25	5.2	32	151
$C_{12}H_{25}N^+\ (CH_3)_2CH_2COO^-$	23	3.5	46	155
$C_{12}H_{25}N^+\ (CH_3)_3CH_3SO_4^-$	25	2.7	61	156
$C_{12}H_{25}SO_3^-Na^+$	25	2.9	57	157

[a] β·D-Octylglucoside.
Source: Ref. 137.

the numbers of moles of components 1 and 2 adsorbed per gram of adsorbent at adsorption equilibrium.

In the case of dilute solutions of surfactants the above equation can be simplified to

$$n^s{}_1 = \frac{\Delta C_1\ V}{m}$$

where $\Delta C_1 = C_{1,0} - C_1$, $C_{1,0}$ is the molar concentration of component 1 in the liquid phase before adsorption, C_1 is the molar concentration in the liquid phase of component 1 at adsorption equilibrium, and V is the volume (in liters) of the liquid phase.

The surface concentration ($C^s{}_1$, mol/cm^2) can be calculated from the surface area per unit mass of solid adsorbent (a_s, cm^2/g):

$$C^s{}_1 = \frac{\Delta C_1\ V}{a_s\ m}$$

and the surface area per adsorbate molecule (in Å^2) is

$$a^s{}_1 = \frac{10^{16}}{NC^s{}_1}$$

where N is Avogadro's number.

An adsorption pattern commonly observed in dilute solutions of surfactants is the Langmuir type isotherm [159], which can be expressed as

$$C^s{}_1 = \frac{C^s{}_m\ C_1}{C_1 + a}$$

where $C^s{}_m$ is the surface concentration of the surfactant (mol/m^2) at monolayer adsorption, C_1 is the concentration (mol/L) of the surfactant in the liquid phase at adsorption equilibrium, and a is a constant related to the absolute temperature and the free energy of adsorption at infinite dilution.

There are relatively few data on the adsorption of nonionic surfactants on different substrates. A review has been given by Von Rybinski and Schwuger [160]. Polyoxyethylene (POE) derivatives adsorb on silica surfaces following Langmuir type isotherms, possibly through hydrogen bonding between the SiOH groups on the surface and the oxygen atoms of the POE chains [161]. The interest in coal–water slurries in the mid 1980s prompted studies on the adsorption of alkoxylated alkylphenols on coal substrates [162]. The adsorption values decrease as the size of the POE chain increases, and the most likely orientation of the surfactant is with the hydrophobic chain adsorbing on the hydrophobic regions on the coal surface, leaving the POE chains to extend into the water phase.

E. Wetting

In a broad sense, wetting is the displacement from a surface of one fluid by another, although in practice it tends to be related to the displacement of air by water or aqueous solutions. Wetting is the basis of many natural and industrial processes. It is often the first necessary step for the further development of interfacial effects. The cleaning of oily, or particulate soil from garments, the dispersion of a pigment in an aqueous or a nonaqueous medium, the spreading of a crop protection formulation over the leaves of plants, and the transport of material across biological membranes are all dependent on wetting phenomena.

Wetting can be of three types [163]:

1. Spreading wetting: A liquid in contact with a substrate spreads over it and displaces another fluid, e.g., air. The work of spreading is then

   $$W_S = (\gamma_{SL} + \gamma_{LV}) - \gamma_{SV}$$

   where γ_{SV}, γ_{SL} and γ_{LV} are the interfacial tensions between substrate and vapor, substrate and liquid, and liquid and vapor, respectively. W_S is also referred to as the spreading coefficient. If it is negative, spreading can occur spontaneously.
2. Adhesional wetting: A liquid not originally in contact with a substrate makes contact with that substrate and adheres to it. The work of adhesion is

   $$W_A = \gamma_{SL} - (\gamma_{SV} + \gamma_{LV})$$

3. Immersional wetting: A substrate not previously in contact with a liquid is immersed completely in the liquid. The work of immersion is

   $$W_I = \gamma_{SL} - \gamma_{SV}$$

Introducing the contact angle θ from the Young equation,

$$\gamma_{SV} = \gamma_{SL} + \gamma_{LV} \cos \theta$$

the three equations above become

$$W_S = -\gamma_{LV} (\cos \theta - 1)$$
$$W_A = -\gamma_{LV} (\cos \theta + 1)$$

and

$$W_I = -4\gamma_{LV} \cos \theta$$

and

Spreading wetting is spontaneous when $\theta = 0$.
Adhesional wetting is spontaneous when $\theta < 180°$.
Immersional wetting is spontaneous when $\theta < 90°$.

Table 7 Examples of Contact Angles and Surface Tension for Nonionic Surfactants

Surfactant	Surface tension[a] (dyn/cm)	Contact angle[b] (°)		
		Glass	PET	Steel
POE (5) C_9C_{11} alcohol	27.9	10	6	32
POE (8) C_9C_{11} alcohol	33.3	8	17	34
POE (5) $C_{13}C_{15}$ alcohol	31.5	11	11	43
POE (9) $C_{13}C_{15}$ alcohol	33.2	9	16	41
POE (4) sorbitan monolaurate	37.0	23	48	63
POE (5) sorbitan monooleate	36.3	20	52	55
POE (20) sorbitan trioleate	40.1	15	59	70
POE (10) octyl phenol	37.7	12	22	40
POE (8.75) nonyl phenol	36.4	12	21	43
Alkyl polyglucoside[c]				
C_9–C_{10}–C_{11}, D.P. 1.4	28.7	13	29	39
C_9–C_{10}–C_{11}, D.P. 1.8	35.4	22	64	64

[a] Average of three measurements. 0.1% w/w, 18°C.
[b] Average of 12 measurements.
[c] D.P. = degree of polymerization.

The above considerations hold when the substrate to be wetted is a hard solid, so that the thermodynamics of the wetting process is the major controlling factor. In this situation, the effectiveness of a surfactant in modifying the wetting properties of a liquid can be assessed by means of surface tension and contact angle determinations. Zisman [164] has demonstrated that complete wetting can occur only when the surface tension of the wetting liquid falls below a critical value (critical wetting tension) characteristic of each substrate. The critical wetting tension can be determined by plotting cos θ versus γ_{LV} for liquids of different surface tensions and extrapolating for $\theta = 0$, i.e., for a zero spreading coefficient.

Table 8 Examples of Wetting Times for Nonionic Surfactants in Nonequilibrium Conditions

Surfactant	Conc. (%)	Wetting time in distilled water (s)	Ref.
$C_8H_{17}CHOHCH_2OH$	0.047	7	165
$C_8H_{17}(OC_2H_4)_2OH$	0.1	5	166
$C_{10}H_{21}(OC_2H_4)_2OH$	0.1	10	166
$C_{10}H_{21}(OC_2H_4)_4OH$	0.1	5	166
$C_{12}H_{25}(OC_2H_4)_4OH$	0.1	19	167
p-t-$C_8H_{17}C_6H_4(OC_2H_4)_4OH$	0.05	50	168
$C_8H_{17}CO(OC_2H_4)_2OH$	0.1	72	166
$C_{11}H_{23}CO(OC_2H_4)_4OH$	0.1	23	166
$C_{10}H_{19}(\Delta 10\text{–}11)CO(OC_2H_4)_4OH$	0.1	6	166

Source: Ref. 137.

The contact angles and surface tensions for a range of nonionic surfactants are given in Table 7.

In substrates with large surface areas, the kinetic effects are predominant over the thermodynamic equilibria. For nonionic surfactants at concentrations considerably above the CMC, the kinetics of wetting, as determined by the wetting time, depend on the diffusion constant, the molecular area of the surfactant at the air/solution interface, and the molecular weight of the surfactant. Under these conditions, high diffusion constants, large surface area, and low molecular weight will produce shorter wetting times. Table 8 gives some examples of wetting times for selected nonionic surfactants in nonequilibrium conditions.

F. Emulsification

Emulsification is the process of dispersing one immiscible liquid into another in the form of droplets with a particle size that is characteristic of the emulsion type and with a stability that is determined by the emulsion type and the intended use. Emulsion types are distinguished by particle size:

Emulsions (sometimes referred to as macroemulsions) have a particle size that is typically above 400 nm; they are opaque and thermodynamically unstable.

In contrast, *microemulsions* have a particle size below 100 nm; they are transparent and thermodynamically stable.

Miniemulsions can be described as systems with intermediate particle size and properties.

Emulsions can be of the water-in-oil (w/o) or oil-in-water (o/w) type depending on whether the water is the internal (disperse) or external (continuous) phase. Two less common categories are the multiple emulsions, in which the dispersed particles are themselves emulsions, and the bicontinuous microemulsions, which are of neither the w/o nor the o/w type, but in which the water and oil phases form an uninterrupted continuous phase.

Similarly to ionic surfactants, nonionics aid the emulsification process by reducing the interfacial tension between immiscible liquids and counter the increase in interfacial free energy from the increased surface area. In microemulsions the ultralow interfacial tension ensures that the emulsification occurs spontaneously. The nonionics, however, differ from the other surfactants in that the stabilization against coalescence of the dispersed droplets is the consequence of steric effects. In addition, nonionic surfactants are free of electrostatic repulsions between charged heads and, if properly formulated from a combination of different molecular species, can produce closely packed interfacial films of high interfacial elasticity. This further enhances the resistance to coalescence as the emulsion droplets collide, e.g., under the effect of Brownian motion. It is also an essential feature in the production and stabilization of w/o macroemulsions, since the water droplets in such emulsions have no significant electrical barrier against coalescence. Nonionic surfactants offer advantages over the ionic species in stabilizing emulsions in a colloidal sense but are not as effective as some anionics in giving small particle size and fast emulsion development (the "blooming" effect). For this reason synergistic combinations of anionic and nonionic surfactants are used, for example, for the formulation of crop protection actives, and in emulsion polymerization.

Both macro- and microemulsions are preferentially prepared by using combinations of surfactants rather than single ones. In macroemulsions, a mixed surfactant system gives a more dense interfacial packing and greater mechanical strength to the interfacial film. In microemulsions a mixture of surfactants and (usually) a cosurfactant are needed to give the ultralow interfacial tension and the proper balance between hydrophilic and lipophilic properties for the oil and water phases.

Table 9 Hydrophile–Lipophile Balance for Some Common Oil Phases

Oil-in-water emulsion	HLB (±1)
Acetophenone	14
Acetylated lanolin	14
Acid, dimer	14
Acid, isostearic	15–16
Acid, lauric	16
Acid, linoleic	16
Acid, oleic	17
Acid, ricinoleic	16
Alcohol, cetyl	15–16
Alcohol, decyl	15
Alcohol, hexadecyl	11–12
Alcohol, isodecyl	14
Alcohol, isohexadecyl	11–12
Alcohol, lauryl	14
Alcohol, oleyl	13–14
Alcohol, stearyl	15–16
Alcohol, tridecyl	14
Alcohols, C_{12}–C_{15}	13
Arachidyl propionate	7
Arlamol E	7
Arlamol F	11
Beeswax	9
Benzene	15
Benzonitrile	14
Benzophenone-3	7
Bromobenzene	13
Butyl stearate	11
Caprylic/capric triglyceride	5
Carbon tetrachloride	16
Carnauba wax	15
Castor oil	14
Castor oil (hydrogenated)	8
Ceresine wax	8
Cetyl palmitate	10
Chlorinated paraffin	12–14
Chlorobenzene	13
Cocoa butter	6
Coconut oil	5
Corn oil	6
Cottonseed oil	5–6
Cyclohexane	15
Decahydronaphthalene	15
Decyl acetate	11
Diethylaniline	14
Diisooctyl phthalate	13
Diisopropyl adipate	9
Diisopropylbenzene	15
Dimethyl silicone	9
d-Limonene (varies widely)	6–7

Table 9 Continued

Oil-in-water emulsion	HLB (±1)
Ethylaniline	13
Ethyl benzoate	13
2-Ethylhexyl oxystearate	10
2-Ethylhexyl-*p*-dimethylamino benzoate	10
2-Ethylhexyl stearate	7.5
Fenchone	12
Hydrogenated peanut oil	6–7
Hydrogenated polybutenes	11
Isodecyl isononanoate	9
Isoparaffinic hydrocarbon (MW ~ 190)	12.5
Isopropyl myristate	11–12
Isopropyl lanolate	14
Isopropyl palmitate	11–12
Isosorbide monolaurate (Arlamol ISML)	10
Isotridecyl isononanoate	8
Jojoba oil	6–7
Kerosene	12
Lanolin, anhydrous	9
Lanolin, liquid	9
Lard	5
Lauryl amine	12
Menhaden oil	12
Methyl phenyl silicone	7
Methyl salicylate	16
Methyl silicone	11
Mineral oil (light), naphthenic	11–12
Mineral oil, paraffinic	10
Mineral oil (light), paraffinic	10–11
Mineral oil (medium), paraffinic	9
Mineral spirits	14
Mink oil	5
Myristyl myristate	8.5
Nitrobenzene	13
N,N-Diethyl-*m*-toluamide	7–8
Nonyl phenol	14
Octocrylene	6–8
Octyl methoxycinnamate	8
Orthodichlorobenzene	13
Palm oil	10
Paraffin wax	10
Petrolatum	7–8
Petroleum naphtha	14
Pine oil	16
Polyethylene wax	15
Polyoxypropylene 11 stearyl ether	11
Polyoxypropylene 15 stearyl ether	7
Polyoxypropylene 30 cetyl ether	10–11
Propene, tetramer	14
Rapeseed oil	6

Table 9 Continued

Oil-in-water emulsion	HLB (±1)
Silicone oil	5
Silicone oil (volatile)	7–8
Soybean oil	6
Squalene	10
Styrene	15
Synthetic spermaceti (cetyl/stearyl palmitate)	10
Toluene	15
Trichlorotrifluoroethane	14
Tricresyl phosphate	17
Tridecyl stearate	6–9
Trimethylolpropane tri C_8–C_{10} ester	8
Vitamin A palmitate	6
Vitamin E	6
Xylene	14
Water-in-oil emulsions	**HLB (±1)**
Beeswax	5
Gasoline	7
Kerosene	6
Mineral oil, heavy	4
Mineral oil, light	4
Mineral oils, paraffinic and naphthenic	4–6
Paraffin	4
Petrolatum	4

Source: Ref. 169.

Empirical rules for the choice of emulsifying systems have been developed:

1. Surfactants with a broad molecular distribution or blends of surfactants are preferred over single species
2. There must be some solubility in the water and oil phase, but it must not be excessive, to allow migration to the interface.
3. Surfactants that are oil-soluble tend to form w/o emulsions, whereas o/w emulsions are usually produced with water-soluble surfactants.

To have a degree of consistency in the classification of nonionic surfactants and to save time in the formulation process, a systematic scheme for identifying the relatively few emulsifiers suitable for any given application was introduced in the late 1940s by Griffin [3]. This is called the hydrophile–lipophile balance, or HLB, system.

In the HLB system, each nonionic surfactant is assigned a number related to the proportion of hydrophilic and lipophilic moieties in the molecule. These numbers are calculated from the structure of the molecule [4] or determined experimentally. A similar range of numbers are assigned to oil phases. These are derived from experimental data. The HLB theory postulates that the emulsifying agents that will give the best performance for any specific oil are those whose HLB number more closely approaches that of the oil. To achieve this objective, usually a combination of surfactants has to be used.

Once the optimum HLB has been identified, there is still a need for further optimization to select the emulsifier type. This can be done by testing blends of emulsifiers of the same HLB but of different chemical structure.

Table 9 gives examples of the HLB value for some common oil phases. Table 10 gives the calculated HLB values for a number of nonionic emulsifiers.

Although the HLB system can assist in the process of the initial screening, it has numerous disadvantages, notably that it is not indicative of either the efficiency or the effectiveness of an emulsifier or of emulsifiers being used in combination. More recently Shinoda and Arai [5,7] proposed a selection method for nonionic emulsifiers based on the determination of the temperature at which an emulsion changes its nature (i.e., inverting from o/w to w/o or vice versa). At this phase inversion temperature (PIT) the lipophilic and hydrophilic properties of the surfactant are in balance for the particular system of oil and water phases, the interfacial tension is at a minimum, and the emulsion should have the finest particle size.

According to the PIT selection method, an emulsion is made with equal weights of oil and water phases and a given amount of emulsifier. The temperature at which the emulsion inverts is then determined. Suitable emulsifiers for o/w or w/o emulsions should give a PIT 20–60°C and 10–40°C, respectively, lower than the storage temperature of the emulsion. The PIT is affected and can be changed by a number of parameters [7,171]. These are summarized in Table 11.

The maximum solubilization method is based on the empirical observation [10,11] that a surfactant, or a blend of surfactants, producing the smallest particle size for a given water and oil combination are the ones that solubilize the largest amount of the water phase into the oil phase before inversion occurs.

G. Dispersion

The stabilization of dispersions in a colloidal sense can be achieved through electrostatic or steric mechanisms or a combination of the two. In the first instance, stabilization results from the repulsion of the electrical double layers surrounding the particles. This repulsion prevents them from approaching that critical distance below which the attractive forces become effective. Electrostatic stabilization is described by the DLVO theory [172,173] and relies on monomeric or oligomeric anionic surfactants, e.g., alkylbenzene sulfonates or naphthalene sulfonate-formaldehyde condensates.

In steric stabilization (reviewed in Ref. 174), amphipathic molecules are adsorbed onto the surface of the particles, which are prevented from reagglomerating by osmotic and entropic repulsions that originate as the layers of surfactants surrounding the particles overlap on short-range approach. This is the mechanism of stabilization of nonionic surfactants in aqueous media and generally for nonaqueous media, where the low dielectric constant of the liquid does not allow extensive electrostatic interactions to take place.

The steric/electrostatic stabilization relies on high molecular weight polyelectrolytes, e.g., lignin sulfonates, where the electrostatic repulsion from the ionic groups is superimposed on the steric hindrance created by the bulky molecules.

The stabilization of dispersions can be explained by considering that the attractive forces between particles (van der Waals forces) are balanced by electrostatic or steric repulsions. This allows the development of equations relating the attractive/repulsive potential to the elec-

Table 10 Calculated HLB Values for Nonionic Emulsifiers

Chemical composition	HLB
Ethoxylated sorbitol esters	
POE (40) sorbitol septaisostearate	9.0
POE (40) sorbitol hexaoleate	10.2
POE (40) sorbitol septoleate	9.0
POE (50) sorbitol hexaoleate	11.4
POE (40) sorbitol tall oil ester	9.7
Ethoxylated amines	
POE (18) fatty amine	15.5
POE (5) fatty amine	10.6
Mono- and diglycerides	
Glycerol mono- and distearate	3.2
Ethoxylated alcohols	
POE (4) lauryl alcohol	9.7
POE (23) lauryl alcohol	16.9
POE (2) cetyl alcohol	5.3
POE (10) cetyl alcohol	12.9
POE (20) cetyl alcohol	15.7
POE (2) stearyl alcohol	4.9
POE (2) oleyl alcohol	4.9
POE (2) oleyl alcohol, spec. quality	4.9
POE (10) oleyl alcohol	12.4
POE (10) oleyl alcohol, spec. quality	12.4
POE (20) oleyl alcohol	15.3
POE (20) oleyl alcohol, spec. quality	15.3
POE (10) stearyl alcohol	12.4
POE (20) stearyl alcohol	15.3
POE (21) stearyl alcohol	15.5
POE (100) stearyl alcohol	18.8
Ethoxylated fatty acids	
POE (8) stearate	11.1
POE (20) stearate	15.0
POE (30) stearate	16.0
POE (40) stearate	16.0
POE (50) stearate	17.9
POE (100) stearate	18.8
Sorbitan esters	
Sorbitan monolaurate	8.6
Sorbitan monopalmitate	6.7
Sorbitan monostearate	4.7
Sorbitan tristearate	2.1
Sorbitan monooleate	4.3
Sorbitan trioleate	1.8
Synthetic C_{13}/C_{15} alcohol ethoxylates	
POE (1) synthetic primary C_{13}/C_{15} alcohol	2.8
POE (2) synthetic primary C_{13}/C_{15} alcohol	5.9
POE (3) synthetic primary C_{13}/C_{15} alcohol	7.9
POE (4) synthetic primary C_{13}/C_{15} alcohol	9.1
POE (5) synthetic primary C_{13}/C_{15} alcohol	10.1
POE (7) synthetic primary C_{13}/C_{15} alcohol	12.2

Table 10 Continued

Chemical composition	HLB
POE (9) synthetic primary C_{13}/C_{15} alcohol	13.0
POE (11) synthetic primary C_{13}/C_{15} alcohol	13.9
POE (20) synthetic primary C_{13}/C_{15} alcohol	16.2
POE (50) synthetic primary C_{13}/C_{15} alcohol	18.3
Nonyl phenol ethoxylates	
POE (1) nonylphenol	3.3
POE (2) nonylphenol	5.7
POE (4) nonylphenol	8.9
POE (5) nonylphenol	10.5
POE (5.5) nonylphenol	10.7
POE (6) nonylphenol	10.9
POE (7) nonylphenol	11.7
POE (8) nonylphenol	12.3
POE (8.5) nonylphenol	12.6
POE (8.75) nonylphenol	12.7
POE (9) nonylphenol	12.8
POE (9.5) nonylphenol	13.0
POE (9.75) nonylphenol	13.2
POE (10) nonylphenol	13.3
POE (12) nonylphenol	13.9
POE (13) nonylphenol	14.4
POE (15) nonylphenol	15.0
POE (17) nonylphenol	15.4
POE (20) nonylphenol	16.0
Ethoxylated sorbitan esters	
POE (4) sorbitan monolaurate	13.3
POE (20) sorbitan monolaurate	16.7
POE (20) sorbitan monopalmitate	15.6
POE (4) sorbitan monostearate	9.6
POE (20) sorbitan monostearate	14.9
POE (20) sorbitan tristearate	10.5
POE (5) sorbitan monooleate	10.0
POE (20) sorbitan monooleate	15.0
POE (20) sorbitan trioleate	11.0

Source: Ref. 170.

trical double layer, or adsorbed layer thickness, as a function of the interparticle distance, yielding curves of the type schematically illustrated in Figs. 10a and 10b.

Of the two energy potential minima in the curve of Fig. 10a, the shallow one at large interparticle distance corresponds to a weak, reversible flocculation (i.e., particles cluster together but can be easily redispersed with a minimal input of mechanical energy). The energy maximum represents a barrier against particles approaching at close distances. However, if this maximum is overcome, e.g., because of Brownian motion, a large attractive energy develops and irreversible agglomeration occurs.

In Fig. 10b, apart from a possible shallow minimum at large distance, the interaction energy is always positive (i.e., there is a continuous interparticle repulsion, even at close distances). Because of this, dispersions produced with nonionic surfactants are more stable

Table 11 Parameters Affecting the Phase Inversion Temperature (PIT)

Parameter	PIT
Surfactant HLB	Increases with HLB.
Surfactant concentration	Decreases with concentration, more sharply at a low degree of ethoxylation. Reaches a constant value at a surfactant concentration of 3–5%.
POE distribution	Increases with broader POE distribution. Broader POE distribution gives greater emulsion stability.
Polarity of oil phase	Increases as polarity decreases.
Oil/water ratio	Increases as the oil/water ratio increases.
Mixed oils	Weighted average (by volume) of the PITs of the emulsions made from individual oils using the same emulsifying agent.

Source: Ref. 7, 171.

than those made with ionic molecules, even though they might not always allow the achievement of high dispersed phase ratios. An additional reason for the excellent stabilizing properties of nonionic surfactants is the significant extension of the steric barrier surrounding the particles, which can be adjusted by changing the degree of ethoxylation of the molecule and is further enhanced by the hydration of the POE chains. The minimum layer thickness for effective stabilization is on the order of 25 Å, which corresponds to approximately 20 POE units. In practice, optimum performance is obtained with 50–100 POE units, and the adsorption on the surface of the particles can be enhanced by modifying the chain length and the polarity of the hydrophobic moiety through the addition of POP units [162]. In addition, the layer of hydrated oxyethylene groups mimics the aqueous phase and significantly reduces the van der Waals attraction between particles.

A summary of suitable nonionic dispersing agents as a function of the dispersion medium, and of the polarity of the dispersed phase, is given in Table 12.

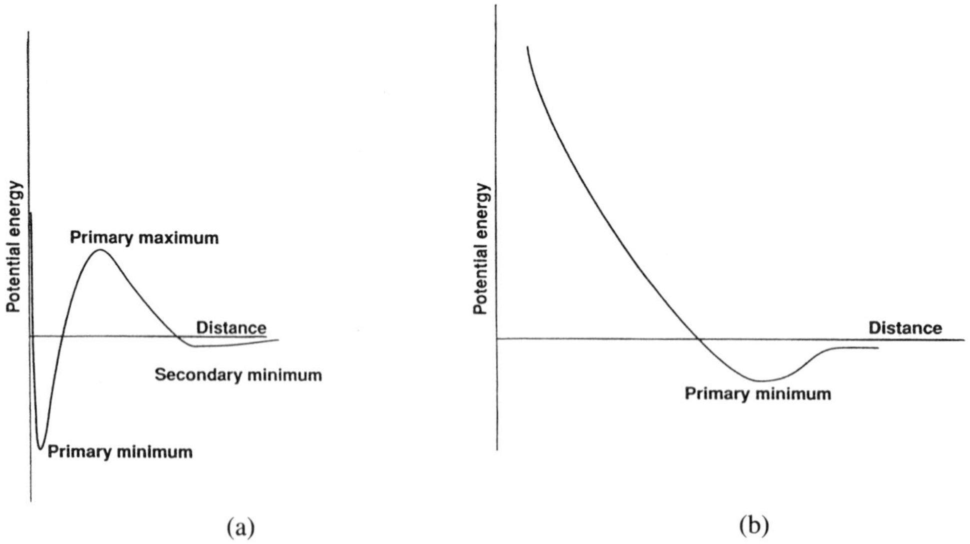

Fig. 10 Attractive/repulsive potential in function of interparticle distance for electrostatic (a) and steric (b) stabilization mechanisms.

Table 12 Suitable Nonionic Dispersing Agents as a Function of the Dispersion Medium and Polarity of the Dispersed Phase

Polarity of solid	Dispersion medium: Aqueous	Dispersion medium: Nonaqueous
Low	Long-chain alcohols, POE, and POE/POP	Low-polarity amphipathic molecules, e.g., alkylbenzenes
High	Alkylphenols, POE, and POP/POE, Unsaturated alcohols, POE, and POP/POE, POE/POP block copolymers, Di- and polyamines	Polyamines, High molecular weight polyesters, Low polarity POE alcohols and alkylphenols

H. Detergency

Detergency and cleaning represent by far the largest outlet for surface-active agents. There is extensive literature coverage of the physicochemical, formulation, and application aspects [175–177]; therefore these are not discussed here.

Polyoxyethylene nonionic surfactants are particularly efficient in removing oily type soil from hydrophobic substrates (probably because of the low CMC) and in preventing its redeposition (probably because of better particle coverage and steric stabilization). Oily type soil removal appears to be at a maximum when the cloud point of the surfactant is 15–30°C below the wash temperature; thus for a homologous series of POE alcohols maximum detergency occurs at increasing POE chain length as the washing temperature increases [178]. In addition to this, an increase in washing temperature demands an increase in the chain length of the hydrophobe [179], and the higher the polarity of the hydrophobic chain, the better the removal of oily type soil from nonpolar surfaces [180].

On cellulosic substrates there is evidence that POE nonionics adsorb through hydrogen bonding between the hydroxyl groups of the cellulose and the POE chains [181]. This reduces oily type soil removal and enhances redeposition.

Particulate soil is better removed with anionic surfactants; however, nonionics will help to reduce redeposition. Nonionics also solubilize those anionics that would otherwise be precipitated by multivalent ions in hard water, so they are powerful synergists in enhancing the performance of anionic surfactants.

VI. APPLICATIONS

Given the broad range of chemicals available, it is not surprising that nonionic surfactants are used in a very large number of formulations and manufacturing processes. The richness of product availability is the obvious consequence of the versatility of the alkylene oxide chemistry that allows the making of systematic, controlled changes in the polarity of the emulsifiers and consequently the production of tailor-made molecules fitting precisely the performance requirements in different, and often highly specific, applications. This capability is not matched by the chemistry of anionic, cationic, or amphoteric surfactants.

Another advantage of nonionic surfactants is their electrical neutrality. This ensures full compatibility with ionic species and the retention (if not the enhancement) of interfacial properties. But perhaps the most valuable consequence of electrical neutrality is that the interfacial

Table 13 Major Applications for Nonionic Surfactants

Application	Nonionic surfactant	Effect
Detergents and cleaners		
Heavy duty powder detergents	POE fatty alcohols	Soil removal
	(POE alkylphenols)[a]	Foam suppression
Liquid laundry detergents	POE fatty alcohols	Soil removal; wetting
Automatic dishwashing	Capped fatty alcohols	Low foam wetting
	POE/POP copolymers	
Hand dishwashing	Alkanolamides	Soil removal
	Amine oxides	Foam booster
	Alkyl polyglucosides	Foam booster; anti-irritant
Hard surface cleaners	POE fatty alcohols	Soil removal and wetting
	Amine oxides	Soil removal and wetting
Floor polishes	POE fatty alcohols	Wetting; Leveling
	POE alkylphenols	
Dry cleaning	Sorbitan esters; POE fatty alcohols; (POE nonylphenols)[a]	Water solubilization and soil removal
Industrial detergents	POE fatty alcohols; POE alkylphenols	Soil removal and wetting
Industrial dishwashers	Capped fatty alcohols; POE/POP copolymers	Low foam wetting
Metal cleaners and degreasers	POE fatty alcohols; POE nonylphenols	Cleaning and emulsification
	Proprietary blends	Electrolytic cleaning
Personal care		
Water-in-oil creams	Glycerol esters; sorbitan esters; specialty esters; proprietary blends	Emulsification
Oil-in-water creams	Polysorbates; POE fatty acid esters; sugar esters; polyglycerol esters; POE fatty alcohols; proprietary blends	Emulsification
Perfume solubilization	Polysorbates; POE fatty alcohols	Solubilization
Hair and body shampoos	Fatty acid alkanolamides	Foam booster; thickening
	Amine oxides	Foam booster
	POE fatty acid diesters	Thickening
	Polysorbates	Anti-irritant; solubilization
	Ethlene glycol/propylene glycol mono- and diesters	Opacifiers
	Undecylenic acid esters and amides	Antidandruff
	Sorbitan esters	Oiling and solubilization
Pharmaceutical		
Creams/gels	Sorbitan esters; polysorbates; glycerol esters	Emulsification
Vitamin/actives	Mannide esters; polysorbates; isosorbide ethers	Solubilization

Table 13 Continued

Application	Nonionic surfactant	Effect
Animal care		
Vitamin solubilization	Polysorbates; POE castor oils	Solubilization
Calves' milk replacers	Glycerol esters; sucroglycerides; polysorbates; POE fatty acid esters	Emulsification
Pigments, paints, inks		
Printing pastes	POE fatty alcohols; POE nonylphenols; sorbitan esters; polysorbates; proprietary blends	Dispersion and emulsification
	POE fatty acid diesters; proprietary blends	Thickening
Paints	POE fatty alcohols	Emulsification
	POE nonylphenols	Dispersion
	POE/POP copolymers	Antifoam
	Proprietary blends	Wetting
Printing inks	Sorbitan esters	Emulsification
	Polysorbates	Dispersion
Dyes/pigment dispersion	Proprietary copolymers	Stabilization
	POE fatty alcohols	Dispersion
	POE alkylphenols	Wetting
	Proprietary blends; POE alkylphenols	Dust suppression
Metalworking		
Rust prevention	Sorbitan, glycerol, pentaerithritol esters; alkanolamides; POE amines	Spreading/adhesion of protective oil; water scavengers
Cutting and grinding oils	POE alkylphenols; sorbitan esters; polysorbates; POE sorbitol fatty acid esters	Emulsification
	POE/POP copolymers	Antifoam
Rolling oils	POE fatty alcohols; POE fatty acid esters; POE alkylphenols; sorbitan esters; polysorbates	Emulsification
Hydraulic fluids	Polymeric surfactants; POE nonylphenols	Emulsification
Construction		
Cement asphalt mortar	Proprietary blends	Emulsification
Concrete water reducer and air entrainer	POE nonylphenols	Foaming; rheology modification
Energy		
Coal–water slurries	POE/POP high molecular polyamine and nonylphenol; polymeric surfactants	Dispersion and rheology modification
Oil-based drilling muds	Polymeric surfactants	Dispersion; emulsification
Crude oil demulsifiers	POE and POE/POP polymeric derivatives	Emulsion breaking
Fuel dehazing agents	POE/POP copolymers	Water scavengers
Fertilizers		
Anticaking agents	POE fatty acid esters; POE nonylphenols	Anticaking

Table 13 Continued

Application	Nonionic surfactant	Effect
Phosphate production	Sorbitan esters	Antifoam
Crop protection		
Emulsifiable concentrates	POE sorbitol fatty acid esters[b]; POE/POP copolymers[b]; POE alkylphenols[b]; POE castor oils[b]; sorbitan esters[b]; polysorbates	Emulsification
	POE/POP copolymers	Antifoam
Wettable powders	POE alkylphenols[b]; POE fatty alcohols[b]	Wetting, dispersion, and suspension stabilization
Flowables	POE alkylphenols; POE/POP copolymers	Dispersion
Adjuvants	Proprietary surfactants and blends; POE fatty amines; POE fatty alcohols; POE alkylphenols; sorbitan esters; polysorbates	Increase the biological activity of crop protection chemicals
Pulp and paper		
Pulp deresination	POE nonylphenols	Wetting; dispersion; emulsification
Paper deinking	POE nonylphenols[c]; POE fatty alcohols[c]	Dispersion, ink removal, ink collection, and emulsification
	POE/POP copolymers[c]	Low foam wetting; ink removal
Felt cleaning	POE fatty alcohols; POE nonylphenols	Cleaning
Paper coating	Sorbitan esters; polysorbates; POE castor oils; POE fatty alcohols; POE alkylphenols	Emulsification of silicone antifoams and wax/resin coatings
Textiles		
Staple spin finishing	POE fatty alcohols; POE nonylphenols; POE/POP copolymers; POE fatty acid esters; POE fatty amines; POE fatty amides	Emulsification, lubrication and antistatic
Scouring agents	POE fatty alcohols[d]; POE nonylphenols[d]	Wetting and cleaning
Dyeing auxiliaries	POE fatty alcohols	Dye leveling; wetting
	POE fatty amine	Dye leveling and retardant
	Proprietary blends	Dye dispersion; leveling
Textile printing	POE fatty alcohols	Wetting
	POE fatty acid diesters	Thickening
	Proprietary blends	Leveling
Resin finishing	Proprietary blends	Softening
Polymer compounding and fabrication		
Static electricity control	Glycerol esters; POE fatty amines; POE/POP copolymers	Static charge dissipation

Table 13 Continued

Application	Nonionic surfactant	Effect
Anticondensation (agricultural films)	Sorbitan esters; glycerol esters	Water vapor coalescence
Anticondensation (food wrapping)	Glycerol esters; polysorbates	Water vapor coalescence
Lubricants	Glycerol esters; polyol esters; fatty acid esters; proprietary blends	Reduce internal and external friction during polymer processing
Dye dispersants (liquid colorants)	Sorbitan esters; polysorbates; POE alkylphenols; glycerol esters	Dispersion
Plastisols	Polymeric surfactants	Rheology modifiers
Polymerization		
Suspension PVC	Sorbitan esters	Control of particle morphology
Emulsion PVC	POE nonylphenols; polysorbates	Rheology modification for finished polymer
Acrylic and vinyl polymers and copolymers	POE fatty alcohols	Emulsification; emulsion stabilization
Thermoset resins	Alkoxylated bisphenol A	Polymer building block
Acrylamide	Sorbitan esters; polymeric surfactants	Emulsification
	POE fatty alcohols; POE nonylphenols; POE/POP copolymers	Emulsion inversion
Mining		
Emulsion explosives	Sorbitan esters; polymeric surfactants	Emulsification
Food		
Margarine	Mono- and diglycerides	Emulsion stabilization; crystallization and rheology modification
	Polyglycerol esters	Emulsification; antispattering
	Sorbitan esters	Crystallization modifiers
Mold release	Polyglycerol esters; thermally oxidized soybean oils	Emulsification
Coffee whiteners	Monoglycerides; sorbitan esters; polysorbates	Emulsification and dispersion
Whipped toppings	Monoglycerides; sorbitan esters; polysorbates; polyglycerol esters; propylene glycol monostearate	Emulsification and dispersion
Ice creams	Mono- and diglycerides; sorbitan esters; polysorbates	Freeze stabilization; overrun; taste development, and melting profile
Bread	Mono- and diglycerides; polyglycerol esters; sucroglycerides; polysorbates; acetyl tartaric esters; mono- and diglycerides	Starch complexation and volume build-up

Table 13 Continued

Application	Nonionic surfactant	Effect
Cakes	Mono- and diglycerides; propylene glycol monostearate; sorbitan esters; polysorbates	Structure stabilization
Chocolate and chocolate toppings	Polyglycerol polyricinoleate; sucroglycerides	Rheology modification
	Monoglycerides	Dispersion
	Sorbitan esters	Antiblooming
	Polysorbates	Tempering aids
Synthetic creams	Mono- and diglycerides; polysorbates; polyglycerol esters	Emulsification and air entrainment
Soft drinks	Sorbitan esters; polysorbates	Essential oils and colors solubilization

[a]In only a limited number of countries and diminishing use.
[b]Used often in combination with anionic emulsifiers in proprietary blends.
[c]Mostly used in proprietary formulations.
[d]Used often in combination with anionic and amphoteric surfactants.

effects are the result of steric rather than electrostatic phenomena. This has advantages in terms of thermodynamic stability, while the molecular structure makes the molecules resilient to chemical destabilizers such as water hardness and multivalent ions. In this respect, nonionic surfactants have the advantageous feature of solubilizing the salts from sulfates and sulfonated anionics with multivalent ions. This allows the performance of these products to be enhanced in hard water conditions.

Nonionic surfactants are also much milder than their ionic antagonists, an advantage that is more and more valued in the current safety-conscious environment.

A. Industrial End Uses

Nonionic surfactants are used in such a broad spectrum of products and processes that it is impossible to give a full account of their industrial applications. Table 13 gives a summary (albeit incomplete) of the major outlets. It must be remembered that significant volumes of nonionic surfactants are consumed for the manufacture of intermediates (discussed later) or of other products used in further processes. Examples are silicone or wax emulsions for antifoam agents or paper coating, resin emulsions for leather finishing or paints, and emulsion polymers for adhesives. Many of these uses are not listed in the table, which essentially addresses those applications in which nonionic surfactants have a major identifiable role.

B. Nonionic Surfactants as Intermediates

Significant volumes of nonionics are consumed every year for the production of other ionic surfactants. These are discussed in the approximate order of decreasing commercial importance.

1. Sulfated Polyoxyethylene Aliphatic Alcohols

The major applications of sulfated polyoxyethylene aliphatic alcohols are in liquid hand dishwashing formulations and hair and body shampoos. In these, the three (or less commonly six) POE adducts of C_{12}–C_{14} (natural or synthetic) or C_{13}–C_{15} oxo alcohols are used.

2. *Sulfated Polyoxyethylene Alkylphenols*

Sulfamic acid is the common sulfating agent because it avoids ring sulfonation and the formation of inorganic salts. The original application in textile auxiliaries and hand dishwashing formulations used the low ethoxylated species. Nowadays, the domestic detergent applications have been replaced almost completely by other uses, e.g., emulsion polymerization where the 20–50 POE derivatives are used as steric/electrostatic stabilizers.

3. *Sulfosuccinates*

Sulfosuccinates are among the synthetic surfactants that emerged in commercial applications after World War II. They are produced by the esterification of maleic anhydride with 1 or 2 mol of a hydroxyl compound, e.g., on C_8–C_{18} fatty alcohol, an alkylphenol, or their ethoxylated derivatives, followed by sulfonation of the double bond with sodium metabisulfite.

The hemiesters are mild surfactants with good foaming and emulsification properties. The diesters are harsher, are poor foamers, and have low water solubility, but they are powerful wetters as low concentrations.

Sulfosuccinates find use in emulsion and suspension polymerization, crop protection formulations, textile dyeing formulations, oil spill dispersants, hair and body shampoo, dry-cleaning processes, concrete, and latex foamers.

4. *Phosphate Esters*

A large variety of phosphate esters are available, almost all directed toward specialized applications. They are stable in highly alkaline and high ionic strength environments. They are also effective as low temperature emulsifiers and have excellent metal-binding properties. All these attributes lead to their use as compatibilizing agents and hydrotropes, as emulsifiers in water-based hydraulic fluids and in personal care formulations as rust inhibitors in metalworking fluids, and as antiwear agents in lubricants.

5. *Ether Carboxylates*

Ether carboxylates were originally developed with a view to upgrading the performance of soaps by improving water solubility in general and in hard water in particular. Their interfacial properties are related to the pK_a of the carboxyl group, which is in the region of 5–6, and change significantly above and below this value. In alkaline medium, ether carboxylates function as excellent emulsion or dispersion stabilizers, whereas at lower pH they behave as emulsifiers. In this respect, ether carboxylates may be viewed as "fugitive" surfactants.

The products are comparatively mild and can be further processed to give a series of amides that can replace alkanolamides, for example, in shampoo formulations. The other major application of ether carboxylates is in metalworking fluids.

6. *Sulfonated Methyl Esters*

The major applications of sulfonated methyl esters are as dispersants, leveling agents, and retardants in textile auxiliaries formulations. They are also used in specific emulsion processes for polyvinyl chloride polymerization.

7. *Ether Sulfonates*

Ether sulfonates are an emerging class of anionic surfactants, first developed in the United States. They are meant to provide the features of a strongly ionized anionic group (hence surface and interfacial properties essentially independent of pH) without the hydrolytic instability inherent in the sulfate moiety. Manufacture is complex, and these surfactants are significantly more expensive than ether sulfates or carboxylates and are in more limited supply. Nevertheless, they are gaining increasing acceptance in high added value applications such as special emulsion polymerization processes.

8. *Sulfated Monoglycerides and Alkanolamides*

Nowadays, sulfated monoglycerides and alkanolamides have been almost completely displaced by other, more cost-effective surfactants.

C. Polymeric Surfactants

Polymeric surfactants are an emerging class of amphipathic molecules that exploit interfacial rather than surface effects [i.e., the molecules have strong interfacial activity with low (or nil) surface activity]. The implications are much broader than it might appear; by targeting a specific solid/liquid or liquid/liquid interface it is possible to develop the real chemicophysical effects wanted and avoid undesirable consequences from surface phenomena, e.g., foam, or from the ecotoxicological characteristics of monomeric species.

The chemistry of polymeric surfactants is not new. It involves mainly etherification, esterification, sulfonation, and addition reactions. Some products have been in the trade for a considerable number of years. What is new is the recognition of the unique opportunities that this technology can offer and its growing use in many industrial applications. Examples of anionic and nonionic macromolecules having strong interfacial effects include (1) polyacrylates, (2) naphthalene sulfonate/formaldehyde condensates, (3) lignin sulfonates, (4) polyvinyl alcohol and polyvinyl alcohol–acetate copolymers, (5) high molecular weight POE and block and random POE/POP glycols, (6) high molecular weight polypeptides and polysaccharides, and (7) POE sorbitol polyesters.

More recently, a number of high molecular weight polyesters and polyethers [182] have been produced and commercialized under the Hypermer trademark by ICI Surfactants. The versatility of the alkylene oxide chemistry allows the balancing of a number of parameters in the polymers, from relative hydrophilicity through solubility to molecular geometry, and the development of ad hoc structures for maximum interfacial effects.

Polymeric surfactants can be viewed as containing a number of anchoring groups (ensuring adsorption on the dispersed phase particles) and stabilizing chains that extend into the continuous phase. With copolymers, the anchoring is provided by the fraction of the molecule that has higher affinity for the dispersed phase. In the case of homopolymers the anchoring is driven by the free energy of binding polymer segments to the surface through van der Waals forces, hydrogen bonding, or solvent rejection.

Because of the many anchoring points, polymeric surfactants can adsorb strongly (sometimes almost irreversibly) at solid/liquid or liquid/liquid interfaces. Therefore, they provide a much more resistant interfacial coverage than simpler monomeric molecules that are easily displaced on particle collision. Furthermore, as the stabilization is the consequence of steric effects (entropic and osmotic), there is no real minimum in the potential energy curves (see Fig. 10b). Thus the emulsions, or dispersions, should be thermodynamically stable (i.e., the particles retain indefinitely their individuality, without any coagulation or coalescence). If phase separation occurs, it does so because of weak flocculation. The original properties of the system can be easily restored by a small input of mechanical energy.

Polymeric surfactants have proved their value in the realization of disperse systems for extreme conditions (i.e., systems characterized by unusually high conditions of dispersed/

continuous phase ratios, ionic strength, operational temperature, and pressure). These are encountered, for example, in [182] oil-based drilling muds, emulsion explosives, hydraulic fluids, rolling oils, nonaqueous dispersions, inverse-phase polymerization processes, and multiple emulsions. Under these conditions conventional surfactants would just not function, because of displacement from the surface, or salting-out, or precipitation.

VII. ECOTOXICOLOGICAL ASPECTS

The human safety and environmental impact of nonionic surfactants depends on the chemical class and structure.

Alcohol ethoxylates are the nonionic surfactants produced in the largest volume, with ethoxylates based on linear, or essentially linear, alcohols being the major type. These find use in household detergent products, institutional and industrial cleaning, cosmetics, agricultural formulations, and many industrial products and processes. The toxicity of alcohol ethoxylates has undergone extensive assessment in animal studies in view of the potential human exposure in consumer products. Alcohol ethoxylates exhibit a low order of acute toxicity by oral, dermal, and inhalation routes of exposure. The acute oral toxicity is low, with reported oral median lethal dose values in the range from 544 to greater than 25,000 mg/kg bodyweight [183]. Acute dermal toxicity in animal studies is equally low, with dermal median lethal dose levels typically exceeding 2000 mg/kg bodyweight [184]. Although undiluted and aqueous dilutions of certain alcohol ethoxylates can be skin irritants under exaggerated exposure conditions associated with animal studies, irritation is rarely encountered in humans under normal conditions of use and exposure [185]. Alcohol ethoxylates are reported to give negative responses in a variety of in vitro genotoxicity assays [186].

Polyethoxylated sorbitan esters (TWEEN, polysorbates) in acute and chronic toxicity studies have shown very low orders of effects and minimal potential for skin and eye irritation. Extensive clinical testing of TWEEN products has revealed little potential for human skin irritation or evidence of skin sensitization. This supports their application as safe-for-use in cosmetic products [187].

Polyethylene glycol esters of stearic acid (MYRJ, PEG stearates) are used extensively in cosmetics-related applications. The acute oral toxicity of PEG stearates is very low, with oral median lethal dose levels reported as exceeding 2 g/kg bodyweight. Skin and eye irritation in animal studies is reported to be minimal with mild to moderate ocular effects and mild irritation to the skin. The responses from clinical studies indicate no evidence of irritation or sensitization in preparations containing up to 25% PEG stearates [188].

The sorbitan fatty acid esters (e.g., SPAN) are generally recognized as having minimal skin and eye irritation, and nonsensitizing effects in animals. The results of oral toxicity studies on sorbitan fatty acid esters indicate very low orders of toxicity by ingestion, with typically the median lethal dose exceeding 30 g/kg bodyweight. Considerable longer term test data addressing general systemic toxicity, reproductive toxicology, and carcinogenicity have revealed no significant evidence of effects, supporting their extensive use in food packaging and cosmetics applications [189].

Nonylphenol ethoxylates have undergone a substantial degree of toxicity testing to establish their safety in industrial and consumer applications. The overall conclusion is that nonylphenol ethoxylates exhibit a low order of toxicity to mammalian species by oral, dermal, or inhalation exposure. Nonylphenol ethoxylates have not been found to cause reproductive, genotoxic, or carcinogenic effects. Skin irritation studies indicate responses ranging from nonirritating through to moderately irritating. The degree of ocular irritation appears to depend on the degree of ethoxylation, with generally the shorter chain variants showing more potential for causing eye irritation [183].

Fatty acid diethanolamides based on coco, lauric, and oleic acids are recognized as having a low order of oral toxicity, with the median lethal dose typically greater than 2 g/kg bodyweight. There is some variation in eye and skin irritation characteristics, with cocoamide DEA giving minimal eye irritation and moderate skin irritation in the rabbit, while lauramide DEA and linoleamide DEA were mild to moderate eye irritants and mild to severe skin irritants. Oleamide DEA is reported to be nonirritant to the rabbit eye, but a moderate skin irritant following single and cumulative application [190].

The environmental fate and effect of nonionic surfactants is increasingly recognized as a key determinant in their selection and application. The growing impact of legislative developments and the corresponding recognition of these developments has a major bearing on this process of selecting nonionic surfactants that match either regulatory driven or customer-led requirements.

Nonionic surfactants used in detergent applications in the European Union are required to comply with specific requirements set out in Council Directive 82/242/EEC, which prohibits the placing on the market, and the use of, a detergent if the level of biodegradability of the contained nonionic surfactant is less than 80% in accordance with specific test methods. The most common alkoxylated nonionic surfactants fully meet the conditions of this directive. Linear and branched alcohol ethoxylates, nonylphenol ethoxylates, and many alcohol alkoxylates all meet the minimum 80% primary biodegradation level in either the static or confirmatory test methods. The linear block copolymers of ethylene and propylene oxide, in contrast, do not meet the minimum primary biodegradation level, primarily through a combination of high molecular weight and the influence of the pendant methyl groupings on the oxidative-based biodegradation pathway [191].

The ultimate biodegradation (effectively the measure of whether complete mineralization of the organic compound to carbon dioxide, water, and biomass occurs) for nonionic surfactants has been extensively investigated. Linear and essentially linear fatty alcohol ethoxylates

Table 14 Overview of Environmental Releases from the Production of Specific Nonionic Surfactants[a]

	AE_3-PC	AE_3-PKO	AE_3-CNO	AE_7-PC	AE_7-PKO	AE_7-CNO	AE_{11}-PO	APG-PKO	APG-CNO
Atmospheric emissions									
Particulates	4.31	5.41	3.87	4.14	4.95	3.88	5.67	13	11.9
Nitrogen oxides	17.9	13.6	11.1	16.6	13.6	11.8	13.6	15.1	13.2
Hydrocarbons	36.2	25.2	24.2	36.8	29.8	29.1	30.5	16.4	15.7
Sulfur oxides	13.2	10.5	8.68	11.7	10	8.77	9.23	15.5	14.1
Carbon monoxide	1.41	1.48	0.89	1.3	1.36	0.95	1.47	2.22	1.78
Methane	0.012	14.5	0.015	0.015	10.1	0.018	15.7	10.9	0.0056
Waterborne emissions									
BOD	1.34	1.55	9.51	1.45	1.67	7.23	1.72	1.02	7.02
COD	2.7	3.72	10.7	2.73	3.51	8.41	3.85	2.81	8.08
Dissolved solids	1.94	4.38	25.5	1.51	3.18	17.9	4.41	5.37	21.3
Suspended solids	0.17	0.82	0.11	0.16	0.62	0.12	0.62	12.8	12.2
Solid waste (total)	66.97	74.1	57.9	64.04	69.9	58.3	62.9	147.8	135.4

[a]Values are expressed as kg/1000 kg surfactant.

Source: Ref. 196.

Table 15* Overview of the Resource Requirements for the Production of Specific Nonionic Surfactants[a,b]

Raw material (kg)	
Total AE_3-PC	2078.6
Total AE_3-PKO	3175
Total AE_3-CNO	3031
Total AE_7-PC	2143.1
Total AE_7-PKO	2939
Total AE_7-CNO	2859
Total AE_{11}-PO	2695
Total APG-PKO	2905.8
Total APG-CNO	2831.9

[a]See Abbreviations preceding References at end of chapter.
[b]Values are expressed as kg/1000 kg surfactant (1.3)
Source: Ref. 197.

are recognized as having a significant level of ultimate biodegradability, showing a BOD28: COD ratio greater than 60%. Alkylphenol ethoxylates, in contrast, although showing a high degree of primary biodegradability, reveal a lower extent of ultimate biodegradation. This lower degree of biodegradation is considered to reflect the influence of the highly branched alkyl chain and the steric effect of the branched phenyl entity on cleavage of the hydrophile from the hydrophobe, which is the accepted mechanism for biodegradation of essentially linear alcohol ethoxylates [192].

The effect of exposure of aquatic organisms to nonionic surfactants is generally well defined for individual types, but there is no one general trend relating toxicity with molecular structure. For alcohol ethoxylates, alteration of the hydrophilic moiety due to an increase in the degree of ethoxylation is responsible for a significant reduction in acute toxicity, for example, by at least two orders of magnitude by extending the EO number from 10 to 30 (LC_{50} increasing from 1 to 100 mg/L) [193].

With linear ethylene oxide/propylene oxide copolymers, typically the acute aquatic toxicity is greater than 100 mg/L, with reported 96 LC_{50} values for rainbow trout in the range of 200–1000 mg/L. By contrast, the alcohol alkoxylates introduced as alternatives to the block copolymers, effectively on the basis of improved biodegradation profile, have a higher order of toxicity in the range of 0.2–5 mg/L [194].

Ethoxylated fatty amines are used extensively as agrochemical surfactants, wetting agents, and emulsifiers. Their environmental characteristics are comparable with those of fatty alcohol ethoxylates, with acute toxicity to aquatic species generally decreasing with increasing ethoxylation, with reported EC_{50} values to *Daphnia magna* ranging from 1.3 to 30 mg/L for 2–25 ethylene oxide units [191]. The biodegradation of ethoxylated fatty amines is considered to be substantial and generally consistent with the criteria developed to distinguish substances as readily biodegradable [195].

Currently, the environmental aspects are receiving growing attention, prompted by the awareness of the environmental consequences of manufacturing and using chemicals in general, and surfactants in particular, and the need for a better understanding of the impact on the environment on a global basis. Thus life-cycle assessments are becoming common practice to help to understand the environmental implications of a product over its entire life span, from the extraction of raw materials through manufacture of the intermediate, to product use

Table 16 Overview of the Energy Requirements for the Production of Specific Nonionic Surfactants[a]

	Energy summary (GJ)			
	Process	Transport	EMR	Total
AE_3-PC	30.3	1.35	51.5	83.2
AE_3-PKO	22.2	1.79	46.4	70.4
AE_3-CNO	19.1	2.01	55	76.1
AE_7-PC	27.7	1.32	49.9	78.9
AE_7-PKO	21.8	1.71	46.6	70.1
AE_7-CNO	19.6	1.86	52.6	74.1
AE_{11}-PO	19.5	2.22	60.1	81.8
APG-PKO	26.6	1.99	32.1	60.7
APG-CNO	24.3	2.15	38.6	65.1

	Energy profile (GJ)							
	Natural gas	Crude oil	Coal	Hydropower	Nuclear	Biomass	Other	Total
AE_3-PC	43	32.2	4.93	0.33	2.8	0	0.0022	83.2
AE_3-PKO	18.1	17.9	4.1	0.74	1.69	27.9	0	70.4
AE_3-CNO	17.9	15.5	4.1	0.35	1.69	36.5	0	76.1
AE_7-PC	39.8	31	4.82	0.34	2.86	0	0.0016	78.9
AE_7-PKO	22.5	21.2	4.2	0.62	2.09	19.5	0	70.1
AE_7-CNO	22.4	19.6	4.2	0.34	2.09	25.5	0	74.1
AE_{11}-PO	23.1	22	4.1	0.23	2.22	30.2	0	81.8
APG-PKO	12.2	8.59	6.51	0.6	1.69	31.1	2.21E-05	60.7
APG-CNO	12.2	6.82	6.51	0.3	1.69	37.6	2.21E-05	65

[a] Values are expressed as GJ/1000 kg surfactant.
Note: EMR (energy of material resource) represents the energy equivalent that is diverted from other uses (e.g., fuel) to manufacture the specific surfactant.
Source: Ref. 196.

and final disposal. The first step in the assessment, the life-cycle inventory (LCI), has been recently completed for a range of large-volume products by the European Surfactants Study Group under the administrative support of the group ECOSOL of the European Chemical Industry council CEFIC. The group included 13 major manufacturers of surfactants and formulation detergents. It assessed the resources requirements and environmental releases during the production of detergent surfactants. Tables 14–16 summarize some of the key parameters for the nonionic surfactants included in the study [196–198].

As yet, there is no agreed-upon impact assessment methodology to enable a comparison on a qualitative basis of the global environmental effects arising from the production and use of products. This would involve aspects such as global warming, acid rain, surface water quality, and landfill implications. The data produced are, however, a powerful tool in understanding process issues on a much more complete basis than hitherto possible, and consequently for identifying and implementing improvements.

ACKNOWLEDGMENTS

I am indebted to Mr. L. Hughes and Dr. H. Thomas of ICI Surfactants for their contribution to the section on the ecotoxicological aspects of nonionic surfactants. I am also indebted to the works of M. J. Schick [*Nonionic Surfactants* (M. J. Schick, Ed.), Marcel Dekker, New York, 1966] and M. J. Rosen (*Surface and Interfacial Phenomena*, 2nd ed., Wiley, New York, 1989), which are milestones in the bibliography of nonionic surfactants and have provided ideas and material for the compilation of this chapter.

ABBREVIATIONS

AE_3; AE_7; AE_{11}	fatty alcohol (natural or synthetic origin) reacted with 3, 7, or 11 ethylene oxide units
AP	Alkylphenol
APG	Alkylpolyglucoside
BOD	Biochemical oxygen demand
CNO	Coconut oil
COD	Chemical oxygen demand
DEA	Diethanolamine
EG	ethylene glycol
EMR	Energy of material resource
Gla	Glucamine
Glu	Glucose
Gly	Glycerol
N	Natural/renewable origin
NFA	Natural fatty alcohol
NFAc	Natural fatty acid
NFAm	Natural fatty amine
Oxy	Oxygen
P (or PC)	Petrochemical origin
PET	Polyethylene terephthalate
PGL	Polyglycerol
PKO	Palmkernel oil
PO	Palm oil
POE	Polyoxyethylene
POP	Polyoxypropylene
SFA	Synthetic fatty alcohol
SFAm	Synthetic fatty amine
Sor	Sorbitol
Suc	Sucrose

REFERENCES

1. C. Schoeller and M. Wittwer, U.S. Patent 1, 970, 578 (Nov. 24, 1931).
2. A. Steindorff, G. Balle, K. Horst, and R. Michel, U.S. Patent 2, 213, 477 (Nov. 18, 1936).
3. W. C. Griffin, *J. Soc. Cosmet. Chem. 1*: 311 (1949).
4. W. C. Griffin, *J. Soc. Cosmet. Chem. 5*: 249 (1954).
5. K. Shinoda and H. Arai, *J. Phys. Chem. 68*: 3585 (1964).
6. K. Shinoda and H. Arai, *J. Colloid Sci. 20*: 93 (1965).
7. K. Shinoda, Proc. 5th Int. Congr. Detergency, September 1968, Barcelona, Vol. 2, p. 275.
8. K. Shinoda and H. Saito, *J. Colloid Interface Sci. 30*: 258 (1969).
9. H. Arai and K. Shinoda, *J. Colloid Interface Sci. 25*: 396 (1967).
10. T. J. Lin, H. Kurihara, and H. Ohta, *J. Soc. Cosmet. Chem. 28*: 457 (1977).
11. T. J. Lin, *J. Soc. Cosmet. Chem. 30*: 167 (1979).
12. J. T. Davies, *Interfacial Phenomena*, Academic, New York, 1963, pp. 371–383.

13. C. D. Moore and M. Bell, *Soap Perfumery Cosmet. 29*: 893 (1956).
14. H. L. Greenwald, G. L. Brown, and M. H. Fineman, *Anal. Chem. 28*: 1963 (1956).
15. L. Marszall, *Acta Polym. Pharm. 32*: 397 (1975).
16. L. Marszall, *Colloid Polym. Sci. 254*: 674 (1976).
17. L. Marszall, *Fette Seifen Anstrichm. 80*: 289 (1978).
18. C. K. Ingold, *Structure and Mechanism in Organic Chemistry*, Cornell Univ. Press, Ithaca, NY, 1953, p. 341.
19. R. E. Parker and N. S. Isaacs, *Chem. Rev. 59*: 737 (1959).
20. H. M. Stanley, in *Ethylene and Its Industrial Derivatives* (S. A. Miller, Ed.), Ernest Benn, London, 1969, pp. 16, 17, 527, 528.
21. J. N. Cawse, J. P. Henry, M. W. Swartzlander, and P. H. Wadia, in *Kirk-Othmer Encyclopedia of Chemical Technology*, 3rd ed., Vol. 9 (M. Grayson, Ed.), Wiley-Interscience, New York, 1980, p. 432.
22. S. Lenher, *J. Am. Chem. Soc. 53*: 3737 (1931).
23. S. Lenher, *J. Am. Chem. Soc. 53*: 3752 (1931).
24. G. L. Cunningham, A. W. Boyd, R. J. Myers, W. D. Gwinn, and W. I. Levan, *J. Chem. Phys. 19*: 676 (1951).
25. T. E. Turner and J. A. Howe, *J. Chem. Phys. 24*: 924 (1956).
26. L. Pauling, *The Nature of the Chemical Bond*, 3rd ed., Cornell Univ. Press, Ithaca, NY, 1960, p. 112.
27. N. Schachat and H. Greenwald, Nonionic surfactants, in *Surfactants Science*, Vol. 1 (M. J. Schick, Ed.), Marcel Dekker, New York, 1966, p. 10.
28. CEFIC, *Guidelines for the Distribution of Ethylene Oxide*, Revision 1, Brussels, 1992.
29. G. J. Jackobson, F. W. Kathagen, M. Klatt, and U. Steinberner, *Ullmann Encycl. Chem. Technol. A12*: 480 (1989).
30. Shell Development Co., U.S. Patent 2,605,293 (1952).
31. Shell Development Co., U.S. Patent 2,810,768 (1957).
32. Pittsburgh Plate Glass Co., GB Patent 926,804 (1963).
33. Shell Development Co., U.S. Patent 2,779,801 (1957).
34. G. Jackobson, F. W. Kathagen, M. Klatt, and U. Steinberner, *Ullmann Encycl. Chem. Technol. A12*: 485 (1989).
35. H. Dillenburg and H. Schufferus (to Deutsche Solvay Werke), DE-OS 3, 410,520 A1 (1984).
36. R. J. Capik and L. Wright (to Atlas Chemical Industries), U.S. Patent 3,588,019 (1970).
37. G. G. Boyers (to Engelhard Industries, Inc.), U.S. Patent 2,868,847 (1959).
38. R. W. Kerr (Ed.), *Chemistry and Industry of Starch*, 2nd ed., Academic, New York, 1950, Chapter 14.
39. A. D. Woolhouse, *Starch as Source of Carbohydrate Sweeteners*, Report CD 2237, Dept. of Scientific and Industrial Research, Aukland, New Zealand, 1976, p. 20.
40. E. R. Kool and F. C. Armbruster, in *Starch Chemistry and Technology*, Vol. 7 (R. L. Whistler and E. F. Paschall, Eds.), Academic, New York, p. 562.
41. R. V. MacAllister, in *Advances in Carbohydrate Chemistry and Biochemistry*, Vol. 36 (R. S. Tipson and D. Horton, Eds.), Academic, New York, 1979, p. 40.
42. Novo Industry A/S, *Dextrose and Starch Sugar*, Brochure B-007 g GB, Bagsvaerd, Denmark, 1977.
43. W. Kuhn, *Hydrocarbon Process 59*(10): 123 (1979).
44. R. A. Scheldon and J. A. Van Doorn, *J. Catal. 34*: 242 (1974).
45. R. A. Scheldon and J. A. Van Doorn, *J. Catal. 31*: 427 (1973).
46. K. H. Simmrock, *Hydrocarbon Process 57*(11): 105 (1978).
47. G. C. Sinke and D. L. Hildebrand, *J. Chem. Eng. Data 7*(1): 74 (1962).
48. A. S. Pell and G. Pilcher, *Trans. Faraday Soc. 61*: 71 (1965).
49. E. S. Pattison (Ed.), *Fatty Acids and Their Industrial Applications*, Marcel Dekker, New York, 1968.
50. D. Swern, *Bailey's Industrial Oil and Fat Products*, 4th ed., Wiley, New York, 1979.

51. Anon., Novel process for enzymatic fat splitting, *Chem. Week*, March 23, 1988, p. 28.
52. E. F. Hill, G. R. Wilson, and E. C. Steinle, *Ind. Eng. Chem. 46*(9): 1917 (1954).
53. A. Landgraf and O. Roelen, German Patent 931,405 (1955).
54. *Chemical Week*, August 1964, p. 19.
55. Continental Oil, Tech. Bull. AL-11, 1963.
56. I. Wender, S. Metun, S. Ergun, H. W. Sternberg, and H. Greenfield, *J. Am. Chem. Soc. 78*: 5401 (1956).
57. L. F. Hatch, *Higher Oxo Alcohols*, Wiley, New York, 1957.
58. K. Ziegler, Br. Patent 763,828 (1956).
59. A. W. Ralston, W. O. Pool, and H. J. Harwood (to Armour and Co.), U.S. Patent 2,061,314 (1936).
60. R. H. Potts (to Armour and Co.), U.S. Patent 2,546,521 (1950).
61. IG Farbenindustrie A. G., Fr. Patent 781,960 (1935).
62. A. W. Ralston, C. Hoerr, W. Pool, and H. Harwood, *J. Org. Chem. 9*: 102 (1944).
63. W. O. Pool and R. Potts (to Armour and Co.), U.S. Patent 2,358,030 (1944).
64. H. Fiege and H. W. Voges, *Ullmann Encycl. Chem. Technol. A19*: 314–347 (1991).
65. J. G. Baldinus (to Colgate Palmolive), U.S. Patent 2,910,491 (1959).
66. R. B. R. Choudhury, *J. Am. Oil Chem. Soc. 37*: 483 (1960).
67. R. O. Feuge and A. E. Bailey, *Oil Soap*, 23, 259 (1946).
68. N. H. Kuhrt, E. A. Welch, and F. J. Kovarik, *J. Am. Oil Chem. Soc. 27*: 310 (1950).
69. N. H. Kuhrt (to Eastman Kodak Co.), U.S. Patent 2,634,278 (1953).
70. A. Grun, *Chem. Umschau. 32*: 225 (1925).
71. E. W. Eckey and M. W. Formo, *J. Am. Oil Chem. Soc. 26*: 207 (1949).
72. C. S. Mimer and N. N. Dalton, *Glycerol*, Reinhold, New York, 1953, p. 366.
73. B. R. Harris, U.S. Patent 1,917,257 (1933).
74. B. R. Harris, U.S. Patent 2,023,388 (1935).
75. H. Burrell, *Ind. Eng. Chem. 37*: 86 (1945).
76. K. G. Mizuch, Z. A. Gorbunova, J. A. Roetermel, and V. A. Blinov, USSR Patent 125,572 (1960).
77. E. R. Mueller, P. L. Eness, and E. E. McSweeney, *Ind. Eng. Chem. 42*: 1537 (1950).
78. K. R. Brown (to Atlas Powder Co.), U.S. Patent 2,322,820 (1943).
79. W. L. Kubie, J. L. O'Donnel, H. M. Teeter, and J. C. Cowan, *J. Am. Oil Chem. Soc. 40*: 105 (1963).
80. W. C. Griffin (to Atlas Powder Co.), U.S. Patent 2,374, 931 (1945).
81. W. C. Griffin (to Atlas Powder Co.), U.S. Patent 2,380,166 (1945).
82. *U.S. Pharmacopoeia 23*: NF 18 (1995).
83. *British Pharmacopoeia*, London, Her Majesty's Stationery Office (1993).
84. *European Pharmacopoeia* published under the direction of the Council of Europe, European Treaty Series No. 50, Maisonneave S.A. Saint Ruffine, France. (July 1995).
85. *21st Code of Federal Regulations* (FDA 21 CFR).
86. *Empfehlung der Kunstoff-Kommision des Bundesgesundheitsamtes* (BGA), Kunstoffe im Lebensmittel verkehc, Karl Heymanns Verlag, Kölu, 1995.
87. European Parliament and Council Directive No 95/2/EC of Feb. 20, 1995; *Official Journal of the European Communities L 61/1*, March 18, 1995.
88. H. B. Hass, F. D. Snell, S. C. York, and L. I. Osipow (to Sugar Research Foundation, Inc.), U.S. Patent 2,893,990 (1959).
89. W. F. Huber and N. B. Tucker (to Procter and Gamble Co.), U.S. Patent 2,812,324 (1958).
90. L. Osipow, F. D. Snell, and A. Finchler, *J. Am. Oil Chem. Soc. 34*: 185 (1957).
91. S. Komori, M. Okahara, and K. Okamoto, *J. Am. Oil Chem. Soc. 33*: 44 (1956).
92. W. B. Satkowki, S. K. Huang, and R. L. Liss, in *Nonionic Surfactants* (M. J. Schick, Ed.), Marcel Dekker, New York, 1966, pp. 142–174.
93. K. Nagase and K. Sakaguchi, *Kogyo Kagaku Zasshi 64*: 1035 (1961).
94. J. D. Malkemus and J. I. Swan, *J. Am. Oil Chem. Soc. 34*: 342 (1957).

95. E. Fisher, *Ber. 26*: 2400 (1893).
96. J. Defaye and C. Peddersen, *Zuckerindustrie 116*(4): 271 (1991).
97. W. Konings and E. Knorr, *Ber. 34*: 957 (1901).
98. G. Zemplen, *Ber. 62*: 985 (1929).
99. J. Barnoub and D. R. Bundle, *Can. J. Chem. 57*: 2085 (1978).
100. R. J. Ferrier and N. Prasad, *J. Chem. Soc. (1969)*: 571.
101. K. G. I. Nilsson, *Trends Biotechnol. 6*: 256 (1988).
102. J. E. Davis and J. C. Letton, Eur. Patent 0,132,043 (1987).
103. W. D. Niederhauser and E. J. Smialkowski (to Rohm and Haas Co.), U.S. Patent 2,856,434 (1958).
104. N. O. V. Sonntag, in *Fatty Acids: Their Chemistry, Properties, Production and Uses*, Part 3, 2nd ed. (K. S. Markley, Ed.), Wiley-Interscience, New York, 1964, Chapter XVI.
105. R. H. Potts (to Armour and Co.), U.S. Patent 2,555,606 (1951).
106. C. K. Ingold, *Structure and Mechanism in Organic Chemistry*, Cornell Univ. Press, Ithaca, NY, 1953, p. 341.
107. R. E. Parker and N. S. Isaacs, *Chem. Rev. 59*: 737 (1959).
108. Jefferson Chemical Co., *Ethylene Oxide*, Tech. Brochure, Houston, TX, 1962.
109. J. D. Malkemus, *J. Am. Oil Chem. Soc. 33*: 571 (1956).
110. H. G. Nadean, D. M. Oaks, W. A. Nichols, and N. P. Carr, *Anal. Chem. 36*: 1914 (1964).
111. E. J. Mills and D. G. Leis, U.S. Patent 2,988,572 (1961).
112. R. W. Wilson (to Monsanto Co.), U.S. Patent 2,586,767 (1952).
113. G. J. Stockburger and J. D. Brandner, *J. Am. Oil Chem. Soc. 40*: 590 (1963).
114. A. N. Wrigley, A. S. Stirton, and E. Howard, Jr., *J. Org. Chem. 25*: 439 (1960).
115. W. P. Krause (to Jefferson Chemical Co.), U.S. Patent 2,983,763 (1961).
116. W. D. Beauchamp, R. E. Booth, and E. R. Degginger (to Allied Chemical Corp.), U.S. Patent 3,016,404 (1962).
117. R. Ernst (to Textilana Corp.), U.S. Patent 3,024,260 (1962).
118. G. C. Tessoro (to Onyx Oil and Chemical Co.), U.S. Patent 2,844,609 (1958).
119. Aktiebolaget Ferrosan, Br. Patent 803,769 (1958).
120. J. Kritchevsky, *J. Am. Oil Chem. Soc. 34*: 178 (1957).
121. V. Lamberti and G. J. McCrimslisk (to Unilever N.V.), Fr. Patent 1,321,790 (1963).
122. J. V. Schurman (to Colgate Palmolive Co.), U.S. Patent 2,877,246 (1959).
123. D. R. Jackson and L. G. Lundsted (to Wyandotte Chemicals Corp.), U.S. Patent 2,677,700 (1954).
124. D. R. Jackson and L. G. Lundsted (to Wyandotte Chemicals Corp.), U.S. Patent 3,036,130 (1962).
125. A. B. Steele, Jr., E. J. Mills, Jr., and D. G. Leis (to Union Carbide Corp.), U.S. Patent 3,078,315 (1963).
126. L. G. Lundsted (to Wyandotte Chemicals Corp.), U.S. Patent 2,674,619 (1954).
127. L. G. Lundsted (to Wyandotte Chemicals Corp.), Br. Patent 722,746 (1955).
128. D. B. Lake and G. L. K. Hoh, *J. Am. Oil Chem. Soc. 40*: 628 (1963).
129. W. Pilcher and S. L. Eton (to Procter and Gamble Co.), U.S. Patent 2,999,068 (1961).
130. I. Webb (to Union Oil Co.), U.S. Patent 2,787,595 (1957).
131. J. H. Wilson and J. Rudy (to Unilever Ltd.), Belg. Patent 604,156 (1951).
132. Procter and Gamble Co., Belg. Patent 641,766 (1964).
133. A. H. Lewis and R. D. Steiner (to California Research Co.), U.S. Patent 2,670,367 (1964).
134. W. H. Woodstock (to Victor Chemicals Works), U.S. Patent 2,586,897 (1952).
135. R. A. Gibbson, *Nature 200*: 665 (1963).
136. P. Becher, in *Nonionic Surfactants*, (M. J. Schick, Ed.), Marcel Dekker, New York, 1966, pp. 480–486.
137. M. J. Rosen, *Surfactants and Interfacial Phenomena*, Wiley, New York, 1989, pp. 70–80, 128–132, 149, 264–268, 296.
138. H. Saito and K. Shinoda, *J. Colloid Interface Sci. 24*: 10 (1967).

139. J. Xia and Z. Hu, in *Surfactants in Solution*, Vol. 5 (K. L. Mittal and P. Botherel, Eds.), Plenum, New York, 1986, pp. 1055–1065.
140. T. Nakagawa, H. Inue, K. Tori, and K. Kuryiama, *Nippon Kagaku Zasshi 79*: 1194 (1958).
141. K. Kuryiama, *Kolloid Z. 180*: 55 (1962).
142. K. Kuryiama, *Kolloid Z. 181*: 444 (1962).
143. K. Kuryiama, H. Inue, and T. Nakagawa, *Kolloid Z. 183*: 68 (1962).
144. R. R. Balmbra, J. S. Clunie, J. M. Corkill, and J. F. Goodman, *Trans. Faraday Soc. 58*: 1661 (1964).
145. P. H. Elworthy and C. B. Macfarlane, *J. Chem. Soc. 1963*: 907.
146. C. W. Dwiggins, Jr. and R. J. Bolen, *J. Phys. Chem. 65*: 1787 (1961).
147. M. J. Schick, *J. Phys. Chem. 67*: 1796 (1963).
148. J. M. Corkill, J. F. Goodman, and S. P. Harrold, *Trans. Faraday Soc. 60*: 202 (1964).
149. J. W. Gibbs, *The Collected Works of J. W. Gibbs*, Vol. 1, Longmans and Company, New York, 1928, p. 119.
150. M. J. Rosen, *Surface and Interfacial Phenomena*, 2nd ed., Wiley, New York, 1989, pp. 70–81.
151. K. Shinoda, T. Yamanaka, and K. Kinoshita, *J. Phys. Chem. 63*: 648 (1959).
152. M. J. Rosen, A. W. Cohen, M. Dahanayake, and X. Y. Hua, *J. Phys. Chem. 86*: 541 (1982).
153. J. E. Carless, R. A. Challis, and B. A. Mulley, *J. Colloid Sci. 19*: 201 (1964).
154. P. H. Elworthy and C. B. Macfarlane, *J. Pharm. Pharmacol. 14*: 100 (1962).
155. A. H. Beckett and R. J. Woodward, *J. Pharm. Pharmacol. 15*: 422 (1963).
156. H. Lange and M. J. Schwuger, *Kolloid Z. Polym. 243*: 120 (1971).
157. M. Dahanayake, A. W. Cohen, and M. J. Rosen, *J. Phys. Chem. 90*: 2413 (1986).
158. R. Aveyard and D. A. Hoydon, *An Introduction to the Principles of Surface Chemistry*, Cambridge Univ. Press, Cambridge, 1973 pp. 201, 212.
159. I. Langmuir, *J. Am. Chem. Soc. 40*: 1361 (1918).
160. W. Von Rybinski and M. J. Schwuger, in *Nonionic Surfactants, Physical Chemistry*, (M. J. Schick, Ed.), Marcel Dekker, New York, 1987, pp. 45–107.
161. J. R. Aston, D. N. Furlong, F. Grieser, P. J. Scales, and G. G. Warr, in *Adsorption at the Gas/Solid and Liquid/Solid Interface* (J. Rouquerol and K. S. W. Sing, Eds.), Elsevier, Amsterdam, 1982, pp. 97–102.
162. L. Trochet Mignard, P. Taylor, G. Bognolo, and Th. F. Tadros, *Colloids Surfaces A, Physicochem. Eng. Aspects 95*: 37–42 (1995).
163. H. J. Osterhof and F. E. Bartell, *J. Phys. Chem. 34*: 1399 (1930).
164. W. A. Zisman, *Advances in Chemistry*, No. 43, American Chemical Society, Washington, DC, 1964.
165. M. J. Rosen and C. C. Kwan, Novel Surfactants From d-Olefins, in *Surface-Active Agents*, Soc. Chem. Ind., London, 1979, pp. 99–105.
166. J. K. Weil, R. E. Koos, W. M. Linfield, and N. Parris, *J. Am. Oil Chem. Soc. 56*: 873 (1979).
167. A. W. Cohen and M. J. Rosen, *J. Am. Oil Chem. Soc. 58*: 1062 (1981).
168. E. H. Crook, D. B. Fordyce, and G. F. Trebbi, *J. Am. Oil Chem. Soc. 41*: 231 (1964).
169. ICI Surfactants, *The HLB System, A Time-Saving Guide to Emulsifier Selection*, Wilmington, DE, 1992.
170. ICI Surfactants, *Product Catalogue*, Brussels, 1993.
171. T. J. Mitsui, Y. Machida, and F. Harusawa, *Bull Chem. Soc. Jpn. 43*: 3044 (1970).
172. B. Derjaguin and L. Landau, *Acta Physicochim. 14*: 633 (1941).
173. E. Vervey and J. T. G. Overbeek, *Theory of the Stability of Lyophobic Colloids*, Elsevier, Amsterdam, 1948.
174. G. Bognolo, Proc. World Surfactants Congress, III C-3, pp. 185–199.
175. W. G. Cutler and E. Kissa (Eds.), *Detergency: Theory and Technology*, Marcel Dekker, New York, 1986.
176. A. M. Schwartz, in *Surface and Colloid Science* (E. Matijevic, Ed.), Wiley, New York, 1972, Vol. 5, pp. 195–244.

177. H. Lange, in *Solvent Properties of Surfactants Solutions* (K. Shinoda, Ed.), Marcel Dekker, New York, 1967, Chap. 4.
178. H. L. Benson, presented at 77th Annu. Am. Oil Chem. Soc. Meeting, Honolulu, 1986.
179. H. L. Benson, presented at 73rd Annu. Am. Oil Chem. Soc. Meet., Toronto, 1982.
180. K. W. Dillan, *J. Am. Oil Chem. Soc. 61*: 1278 (1984).
181. A. Waag, *Chim. Phys. Appl. Prat. Ag. Surface C.R. Congr. Int. Deterg. 5th*, Barcelona, 1968, III, p. 143.
182. G. Bognolo, Polymeric surfactants, paper presented at the Colloid and Surfaces Chemistry Group Meeting, Soc. Chem. Ind., Imperial College of Science and Technology, London, 12 Dec. 1990.
183. S. Talmage, *Environmental and Human Safety of Major Surfactants*, Soap and Detergent Association, Lewis Publishers, 1994, pp. 95–102.
184. G. M. Benke, M. N. Brown, M. J. Walsh, and R. B. Drotman, *Food Cosmet. Toxicol. 15*: 309–318 (1977).
185. R. Gingell and C. Lu, *J. Am. College Toxicol. 10*(4): 477–486 (1991).
186. J. Yam, K. A. Boorman, W. Broddle, J. Geiger, J. E. Heinze, Y. L. Lin, K. McCarthy, S. Reiss, V. Sawin, R. I. Sedlack, R. S. Slesinski, and G. A. Wright, Surfactants: a survey of short term genotoxicity testing, *Food Chem. Toxicol. 22*(9):761–769 (1984).
187. *J. Am. Coll. Toxicol. 3*(5): 1–82 (1984).
188. *J. Am. Coll. Toxicol. 7*(6): 881–910, (1988).
189. *J. Am. Coll. Toxicol. 4*(3): 65–121 (1985).
190. *J. Am. Coll. Toxicol. 5*(5): 415–454, (1986).
191. K. J. Bock, L. Huber, and P. Schorbel, *Tenside 25 Surfact. Deterg. 25*(2): 86–98 (1988).
192. L. Kravetz, J. P. Salanitro, P. B. Dorn, and K. F. Guin., *J. Am. Oil Chem. Soc. 68*: 610 (1991).
193. W. Guhl and P. Gode, *Tenside 26*(282): 3–8 (1989).
194. T. G. Blease, J. G. Evans, L. Hughes, and P. Loll, in *Defoaming* (P. Garnett, Ed.), Marcel Dekker, New York, 1993, p. 305.
195. C. G. Van Ginkel, C. A. Stroo, and A. G. M. Kroon, *Tenside 30*(3): 213–216 (1988).
196. M. Stalmans, H. Berenbold, J. L. Berna, L. Cavalli, A. Dillarstone, M. Franke, F. Hirsinger, D. Janzen, K. Kosswig, D. Postlethwaite, T. Rappert, C. Renta, D. Scharer, K. P. Schick, W. Schul, H. Thomas, and R. Van Sloten, European life-cycle inventory for detergent surfactants production, *Tenside 32*(2): 84–109 (1993).
197. W. Schul, F. Hirsinger, and K. P. Schick, *Tenside 32*(2): 171–192 (1995).
198. F. Hirsinger and K. P. Schick, *Tenside 32*(2): 193–199 (1995).
199. M. J. Schick (Ed.), *Nonionic Surfactants*, Marcel Dekker, New York, 1968.

26

Lipids

Their Use in Personal Care Products

Keith Coupland and Julie A. Nichols
Croda Universal Ltd., Hull, England

I. INTRODUCTION

The aim of this chapter is to outline the use of lipids in cosmetics and personal care products. It covers the types of lipids used, the reason for their use, and how they are formulated.

Cosmetics are defined as "something intended to adorn or beautify the body, especially the face" [1]. The common perception of the word is decorative, and it has an implied female bias. Since lipids are used in many products other than color cosmetics, the term "personal care product" is preferred.

The European Cosmetics Directive [2] provides a general description of personal care products that is the accepted European standard:

> Any substance or preparation intended to be placed in contact with the various external parts of the body (epidermis, hair system, nails, lips and external genital organs) or with the teeth and the mucous membranes of the oral cavity with a view exclusively or mainly to cleansing them, perfuming them, changing their appearance and for correcting body odors and/or protecting them or keeping them in good condition.

Increasingly, new products are appearing on the market that extend beyond these defined boundaries for personal care products into the verging area of pharmaceuticals. These products have been termed "cosmoceuticals."

Lipids or lipid derivatives are used in most, if not all, personal care products sold today. They are found in emulsions such as creams and lotions as carriers for other lipophilic materials and as emollients.

II. LIPIDS USED IN PERSONAL CARE PRODUCTS

The emollient most commonly used by the personal care industry is hydrocarbon or mineral oil. Its consumption is estimated to be around 700,000 tonnes per year (tpa) and is increasing by 2% per year [3]. Despite a long history of safe use, concern about trace contents of polycyclic aromatic hydrocarbons and a move toward renewable resources have resulted in the increasing use of natural lipids. There is a distinct preference for vegetable-derived oils, fats, and waxes.

Many personal care products contain ingredients with no real function other than to attract the customer to buy the product. However, if some benefit is claimed such as "antiwrinkle" or "moisturizer," it must be substantiated in accordance with the European Directive 6th Amendment [2] [Article 7a1(O)].

A. Triacylglycerols

Most plant lipids used in personal care products are triacylglycerols, i.e., esters of glycerol and carboxylic acids. The fatty acids are generally linear, even-numbered monocarboxylic acids in the range C_6–C_{24}. They can be saturated, monounsaturated, or polyunsaturated. In recent years it has become common to use a shorthand description of fatty acids indicating the number of carbon atoms, the degree of unsaturation, and the position of any double bond relative to the terminal methyl group. Using this scheme, γ-linolenic acid (all-*cis*-octadeca-6,9,12-trienoic acid) becomes 18:3 (*n*-6).

Table 1 lists five fatty acids contained in plant lipids and their plant sources. There are two other sources for triacylglycerol lipids used in personal care products: animal fats and marine oils. These are not considered as important as the plant oils both for reasons outlined earlier and because the polyunsaturation of many marine oils gives rise to oxidative instability.

The formulator has many good reasons for incorporating lipids into personal care products.

1. Lipids are emollients, or superfatting agents.
2. They have market appeal and are a renewable source.
3. They are mineral oil or hydrocarbon free.
4. Secondary benefits include ultraviolet absorbance.
5. Lipids serve as solubilizers or vehicles for lipophilic ingredients, e.g., botanical extracts.

Emolliency describes the smooth lubricant effect of applying an oil to the skin. This is particularly useful because dry skin is a common complaint caused by the environment, detergents, and disease. Removal of skin lipids by washing in detergent or soap is easily remedied by applying an emulsion containing a lipid, which has a "superfatting" action on the skin.

Market appeal is an obvious benefit, and a label claim stating, for example, "Contains Evening Primrose Oil," is often sufficient to attract the customer and sell the product.

The use of lipids as a replacement or alternative to mineral oil is of interest because it was oils, fats, and waxes that were used historically as cosmetics [4].

Several triacylglycerol lipids confer secondary benefits to the product such as ultraviolet absorbance. Unsaturated oils have a weak absorbance at 210 nm, but certain lipids such as shea butter contain ultraviolet absorbers in the unsaponifiable fraction of the oil. Shea butter absorbs strongly at 271 nm.

Table 1 Fatty Acids Contained in Plant Lipids

Fatty acid	Designation	Source
Myristic	14:0	Coconut oil, palm kernel oil
Oleic	18:1 (*n*-9)	Olive oil, hybrid sunflower seed oil
Linoleic	18:2 (*n*-6)	Safflower seed oil
γ-Linolenic	18:3 (*n*-6)	Evening primrose oil, borage oil
Stearidonic	18:4 (*n*-3)	Blackcurrant seed oil

Lipids may be used as solubilizers for other lipophilic components. Castor oil is a good solvent for bromoacid dyes in lipsticks. Vegetable oils are also employed as the extractant and carrier for plant or botanical extracts such as Calendula, Saint-John's-wort, Carrot, and Walnut.

In recent years it has become apparent that some triacylglycerol lipids are not simply passive lipophilic vehicles providing emolliency. Several oils containing polyunsaturated fatty acids of the *n*-3 and *n*-6 series are readily incorporated into skin lipids and are metabolized.

Dry skin is the result of a high rate of transepidermal water loss (TEWL). There are many causes for this symptom, including environment, diet, and essential fatty acid (EFA) deficiency [5]. EFA deficiency caused by poor diet is rare in developed countries, but it can be induced in disease conditions. A series of papers from the Unilever Laboratories demonstrated that the reduction of TEWL by topical application of EFA-containing lipids was not simply an occlusive effect but resulted from the repair of the impaired natural moisture barrier in the skin. EFA deficiency causes severely dry skin in patients with chronic fat malabsorbance. The cutaneous symptoms were corrected by applying an oil rich in linoleic acid [6]. This study was extended [7] to demonstrate that TEWL caused by detergent insult could also be corrected by cutaneous application of linoleic acid.

The repair of skin barrier function was also achieved by application of γ-linolenate [18:3 (*n*-6)] but not dihomo-γ-linolenate [20:3 (n-6)] [8]. Brod et al. [9] reported the use of an emulsion containing blackcurrant seed oil, which is rich in γ-linolenic acid containing triglycerides, to correct human dry skin deficiency.

B. Wax Esters

The second widely used group of lipids are known collectively as wax esters. They differ from the triacylglycerols by having fatty acid esterified to fatty alcohol. There are several sources in nature for both animal and vegetable wax esters. The best known of the wax esters

Table 2 Plant Lipids Used in Personal Care Formulations

Lipid	Plant source	Key fatty acids
Apricot kernel oil	*Prunus armeiaea*	18:1/18:2
Avocado oil	*Persea americana*	18:1
Babassu oil	*Oribgnya speciosa*	12:0/14:0
Blackcurrant seed oil	*Ribes niger*	18:2/18:3 (*n*-6)[a]
Borage oil	*Borago officianalis*	18:2/18:3 (*n*-6)
Castor oil	*Ricinus communis*	18:1 (OH)
Cocoa butter	*Theabroma cacao*	18:0/18:1
Coconut oil	*Cocus nucifera*	12:0/14:0
Evening primrose oil	*Oenothera biennis*	18:2/18:3 (*n*-6)
Macadamia nut oil	*Macadamia tetraphylla*	18:1/18:2/16:1
Meadowfoam oil	*Limnanthes alba*	20:1 (*n*-15)
Olive oil	*Olea europa*	18:1
Sesame oil	*Sesamum indica*	18:1/18:2
Shea butter	*Butyrospermum parkii*	18:0/18:1[b]
Sunflower oil	*Helianthus annus*	18:2/18:1
Sweet almond oil	*Prunus amygdalus*	18:1/18:2
Wheatgerm oil	*Triticum speciosa*	18:2/18:1[c]

[a] Also contains stearidonic acid 18:4 (*n*-3).
[b] Contains cinnamate ester.
[c] High vitamin E content.

are lanolin (purified wool wax derived from sheep's wool), jojoba oil (extracted from the jojoba plant), and orange roughy oil (from the deep sea fish *Hoplostethus atlanticus*).

The composition of lanolin is surprisingly complex; it has a range of around 200 fatty acids including linear acids (C_8–C_{38}), branched-chain acids (C_7–C_{41}), and hydroxy acids (C_{10}–C_{36}). These fatty acids are esterified or interesterified to a range of some 70 alcohols and sterols. Wool wax alcohols are characterized by their high cholesterol content (~30%).

The analysis and structural determination of lanolin are detailed in a series of papers by Fawaz and coworkers [10–14].

Lanolin is an excellent water-in-oil emulsifier, skin conditioner, and emollient. These effects have been quantified using skin image analysis, profilometry, dermal torque determination, and cohesography [15].

C. Minor Lipids

Several other classes of lipids are used in personal care products in relatively small amounts [16]. These include glycerol ethers, which are found in shark liver oil; phosphoglycerides; and sphingolipids, particularly the ceramides. This latter class has excited interest among cosmetic scientists following the discovery of the vital role played by ceramide in the permeability barrier of the mammalian epidermis [17]. Extraction of ceramide from tissue is expensive and gives variable results. The use of synthetic and semisynthetic products is expected to increase.

III. REFINING LIPIDS FOR PERSONAL CARE PRODUCTS

As with any raw material for personal care products, quality is of extreme importance. Trace impurities such as pesticide residues, oxidation products, thermal degradation products, pigments, and alkaloids may lead to problems with taste, odor, color, and irritation. The processes used to refine oils can themselves give rise to impurities. Although the use of chemical bleaching gives very pale colored lipids, it leaves behind autodecomposition products.

For very high purity lipids, process-scale absorption chromatography is used [18,19]. This has the advantage of removing polar lipids and pigments while leaving the substrate chemically unchanged. Peroxide values of less than 1 and complete removal of autoxidation products are possible. The resulting lipids are odorless, colorless, and tasteless.

There is a trend in the marketplace toward colorless oils and emulsions that are brilliant white. The use of oils that are pale yellow gives off-colored emulsions, particularly water-in-oil emulsions, though the slight coloration may be tolerated because it is deemed a natural artifact.

The expected shelf life of personal care products is usually 3 years. This includes the physical stability of emulsions, clouding, precipitation, and color and odor development. This entails that lipids with unsaturation must be adequately protected with antioxidants. Typical antioxidants include vitamin E, ascorbyl palmitate, butylated hydroxytoluene (BHT), butylated hydroxyanisole (BHA), and octyl gallate.

IV. FORMULATING PERSONAL CARE PRODUCTS

There are many considerations in formulating personal care products, and despite increased knowledge of the chemistry and physics of the systems formulation still remains a mixture of art and science. Modern formulations can contain as many as 20 components, and optimizing these systems is not trivial. Several examples of formulating with lipids are included in Appendix 1.

A. Emulsions

The majority of skincare products are in the form of emulsions. Formation of emulsions is facilitated by the incorporation of surface-active emulsifiers. Emulsions may be oil-in-water (o/w), in which the continuous phase is water, or water-in-oil, in which the continuous phase is oil. This allows a wide variety of emulsion characteristics to be obtained by varying water/oil ratios.

There are associated difficulties with the emulsification of vegetable oils, and formulation optimization is not always easy. One guideline technique often used is the HLB scale.

B. HLB Scale

The required HLB (hydrophilic–lipophilic balance) of a vegetable oil or emollient can help in identifying a suitable emulsifier, typically a blend of emulsifiers. However, the required HLB of the lipid does not mean that any surfactant of that HLB will produce the required emulsion. The chemical type or class of the surfactant used is at least equally as important, and a range around the required HLB will probably be needed. The lipophile of the surfactant needs to exhibit the greatest affinity for the oil-phase droplets. Similarly, the "like attracts like" concept may aid in surfactant choice; unsaturated lipophiles are attracted to unsaturated oils, and saturated lipophiles to saturated oils [20]. The surfactant type can also affect the texture, feel, viscosity, stability, and so on of an emulsion.

It should be noted that the HLB scale is limited to nonionic emulsions, and the use of anionic and cationic emulsifiers should not be ignored. A selection of required HLBs for o/w emulsions is shown in Table 3. In most cases other more complex factors have to be taken into consideration, but the HLB scaling provides a good starting point.

Other important points to consider are (as with other emulsions)

1. Ratio of oil phase to emulsifier
2. Ratio of oil phase to aqueous phase
3. Composition of the oil phase
4. Composition of the aqueous phase
5. Emulsifying method
6. Emulsifying equipment
7. Emulsifying temperature

All of these factors can affect the production and ultimately the stability of an emulsion.

Table 3 Required HLB

Material	Required HLB (o/w)
Mineral oil	10–12
Cottonseed oil	7–5
Castor oil	14
Cocoa butter	6
Corn oil	10
Jojoba oil	6–7
Palm oil	10
Rapeseed oil	6
Soybean oil	6

Source: Ref. 20.

C. Polarity Index

Another interesting consideration is the so-called polarity index [21], a measure of the interfacial tension of an oil against water. The more polar vegetable oils often afford better emolliency than nonpolar oils such as hydrocarbon or mineral oils. Polar oils offer improved skin surface distribution and solvency but are not the easiest materials to emulsify. The interfacial tension is affected by the emulsifiers used, which influence emulsification by their mobility, diffusion, and interfacial orientation.

The polarity index values for some common oils are listed in Table 4.

D. Mono-, Di-, and Triglycerides

Mixtures of mono-, di-, and triglycerides esterified with a variety of fatty acids allow esters to be obtained that function as emollient, stabilizer (viscosity modifier), and coemulsifier [22]. Varying the fatty acid composition and percentage of the three glyceride types can produce the desired end emulsion with the minimum number of components used in a formulation. This is the subject of a recent patent by Frank Johannsen [23].

E. Water-in-Oil Emulsions

Typically w/o emulsions are more difficult to stabilize than o/w emulsions. The stabilization of w/o emulsions is difficult with conventional mineral oil, and the same is true of water-in-triglyceride oil emulsions. In addition to optimization of the emulsifier in such w/o triglyceride emulsions, the quantity of fat crystals (solid triglycerides in the vegetable oil) is important. The lipid crystals can thicken the external oil phase and absorb at the water/oil interface, and they may form bridges between the discrete water-phase droplets. By varying the quantity and type of lipid crystals, stability or instability can be induced [24]. Other classical thickeners or rheology modifiers, e.g., clays/resins and particulate matter, may also be used.

F. Formulations

Natural lipids have a long history of use as emollients in personal care, originally for the simple oiling of skin and hair. Advances in technology allow their incorporation into a whole

Table 4 Polarity Indices[a]

Common oils (INCI)	Polarity index (mN/m)
Mineral oil	38.3
Cetearyl octanoate[b]	28.6
Caprylic/capric triglyceride	21.3
Peanut oil	20.5
Almond oil	20.3
Sunflower seed oil	19.3
Castor oil	16.9

[a] Substances below 30 mN/m are classified as polar; those above 30 mN/m are nonpolar.
[b] Cetearyl octanoate is a mixture of cetyl (16:0) and stearyl (18:0) octanoates.
Source: Ref. 21.

host of personal care formulations for a multitude of reasons as previously discussed. A selection of such formulations supplied by Croda Chemicals Limited is given in Appendix 1. Further details and information can be obtained from the company. Details of other suppliers are given in Appendix 2.

V. CONCLUSION

The use of lipids in personal care products is supported by many of the special or unique features they can offer the formulating chemist and consumer. They may be employed for functional benefits and provide an attractive marketing concept. Advances in technology and identification of novel lipid materials will ensure future interest in these materials.

APPENDIX 1. FORMULATIONS[1]

1. Moisturizing Evening Primrose Facial Oil C1610. This simple light facial moisturizing oil employs Evening Primrose oil, a material rich in γ-linoleic acid (GLA), to provide moisturizing effects and an attractive marketing angle. (See following data.)

A reconstituted vegetable oil, caprylic/capric triglyceride, is the base emollient vehicle and facilitates spreading the preparation over the skin.

Moisturizing Evening Primrose Facial Oil C1610

Ingredient	Supplier	Percent (w/w)
Crossential EPO (super-refined evening primrose oil)	Croda	1.00
Crodamol GTCC (caprylic/capric triglyceride)	Croda	to 100
Tocopheryl acetate	Roche	6.0
BHT	—	0.05
Perfume, preservatives, color	—	qs[a]

[a]qs = *quantum sufficit* (as much as will suffice to produce the desired effect).

METHOD OF MANUFACTURE. Incorporate BHT and preservative into Crodamol GTCC with gentle heating. Stir to cool, continue stirring, and add remaining ingredients.

2. Avocadin Cream C1776M2. This preparation employs Avocado oil unsaponifiables (a material rich in plant sterols; sitosterol, campesterol, and stigmasterol), to confer superior moisturization to the skin. This emollient o/w preparation uses nonionic ethoxylated fatty alcohols (Ceteareth 20) and sorbitan esters (Polysorbate 60), a small amount of anionic Triethanolamine Stearate stabilized with Glyceryl Stearate, and Cetearyl alcohol. (See Table 5.)

METHOD OF MANUFACTURE. Heat oil and water phases separately to 65–70°C. Add water phase to oil phase while stirring. Stir to cool, perfume at 40°C.

3. Carrot Oil Cream C1812. This light moisture cream employs Carrot Oil Extra to aid UV absorption and condition the skin. Peanut oil is used to extract the carrot. The flexible emulsification system of Sorbitan Stearate and Polysorbate 60 is used. (See Table 6.)

METHOD OF MANUFACTURE. Heat oil phase and water phase separately to 65–70°C. Add water phase to oil phase while stirring. Stir to cool, perfuming at 40–45°C. Fill off at 30°C.

[1]Available from Croda Chemicals Limited.

Table 5 Avocadin Cream C1776M2

Ingredient	Supplier	Percent (w/w)
Avocadin (avocado oil unsaponifiables)	Novarom	5.00
Crodamol IPP (isopropyl palmitate)	Croda	7.50
Crodamol CAP (cetearyl octanoate)	Croda	2.00
Silicone 200/100 cS (dimethicone)	Dow Corning	0.50
Crodacol CS90EP (cetearyl alcohol)	Croda	2.50
Crosterene SA4310 (stearic acid)	Croda	0.50
Mineral oil (25 cS at 25°C)	—	5.00
Cithrol GMS N/E GE0803 (glyceryl stearate)[a]	Croda	1.00
Cosmowax D (cetearyl alcohol and ceteareth 20)	Croda	2.10
Crillet 3 (Polysorbate 60)	Croda	2.50
Deionized water	—	to 100
Propylene glycol	—	2.00
Triethanolamine 99%	—	0.20
Perfume, preservative, color	—	qs

[a]GMS = glycerol monostearate.

4. *Highly Emollient Lanolin Cream C1440.* Ideal for rough, dry skin, this enriched moisturizing cream employs a combination of lanolin and hydrogenated lanolin (see Table 7). This cream affords a degree of occlusivity. Lanolin acts as an emulsifier and stabilizer.

METHOD OF MANUFACTURE. Heat oil and water phases separately to 70°C. Add water phase to oil phase with stirring. Stir to 30–35°C, then pass over triple roll mill. Fill off.

5. *Neck Cream C1267.* Formulation C1267 is a highly emollient-rich cream designed for use on the neck. The apricot kernel oil, squalene, and super sterol ester all contribute to its performance. (See Table 8.)

METHOD OF MANUFACTURE. For formula 1, predisperse Collasol in water (10% of total). Hydrate Carbopol in warm water, 60–65°C. Add remaining water-phase ingredients except triethanolamine, and heat to 65–70°C. Heat oil phase to 65–70°C and add to water phase while stirring. Allow to cool, adding predispersed Collasol phase for formula 1 at 30–35°C. Fill off.

Table 6 Carrot Oil Cream C1812

Ingredient	Supplier	Percent (w/w)
Crodamol GTCC (caprylic/capric triglyceride)	Croda	13.00
Crodamol OC (octyl cocoate)	Croda	3.00
Crodamol SS (cetyl esters)	Croda	3.00
Crodamol CS (cetearyl alcohol)	Croda	3.00
Carrot Oil Extra (peanut oil, carrot extract, and isopropyl myristate)	Novarom	4.00
Crill 3 (sorbitan stearate)	Croda	2.00
Crillet 3 (Polysorbate 60)	Croda	3.00
Deionized water	—	to 100
Croderol GA7000 (glycerine)	Croda	3.00
Perfume, preservative, color	—	qs

Table 7 Lanolin Cream C1440

Ingredient	Supplier	Percent (w/w)
White microcrystalline wax (No. 3520)	—	3.40
Satulan (hydrogenated lanolin)	Croda	2.50
Corona pure new lanolin (lanolin)	Croda	9.50
Light mineral oil (25 cS at 25°C)	—	16.00
White petroleum jelly, continental grade no. 3	—	2.50
BHT	—	0.01
Deionized water	—	to 100
Magnesium sulfate	—	0.70
Croderol GA7000 (glycerine)	Croda	3.50
Perfume, preservative, color	—	qs

6. *Night Cream C1520.* Rich skincare properties are given by the inclusion of orange roughy oil (wax ester). The emulsification system for this o/w cream is PEG-24 stearate, an ethoxylated fatty acid, stabilized with glyceryl stearate and cetearyl alcohol. (See Table 9.)

METHOD OF MANUFACTURE. Predissolve Collasol in water (10%). Heat oil and water phases separately to 65–70°C. Add water phase to oil phase while stirring. Stir to cool, perfuming at 45°C and adding Collasol solution at 30–35°C.

7. *Physical Block Sunscreen Cream C1643.* A highly protective water-resistant sunscreen that uses microfine titanium dioxide as the physical UV blocker. Caprylic/capric triglyceride, a reconstituted vegetable oil derived in combination with shea butter, provides

Table 8 Neck Cream C1267

		Percent (w/w)	
Ingredient	Supplier	Formula 1 (with Squalene)	Formula 2
Polawax GP200 (nonionic emulsifying wax)	Croda	4.00	4.00
Super-refined apricot kernel oil	Croda	5.00	15.00
Squalene	—	10.00	—
Crodamol MM (myristyl myristate)	Croda	2.00	2.00
Super sterol ester (C10–30 cholesterol/lanosterol esters)	Croda	2.00	2.00
Crosterene SA4310 (stearic acid)	Croda	3.00	3.00
BHA	—	0.05	0.05
Dow Corning 580 Wax (stearoxytrimethylsilane and stearyl alcohol)	Dow Corning	2.00	2.00
Deionized water	—	to 100	to 100
Croderol GA7000 (glycerine)	Croda	5.00	5.00
Triethanolamine	—	1.50	1.50
Carbopol 941 (Carbomer 941)	BF Goodrich	0.30	0.30
Deionized water	—	10.00	—
Collasol (1%)	Croda	1.00	—
Perfume, preservative, color	—	qs	qs

Table 9 Night Cream C1520

Ingredient	Supplier	Percent (w/w)
Super-refined orange roughy oil (orange roughy oil)	Croda	8.00
Silicone fluid 200/300cS (dimethicone)	—	1.00
Crodalan LA (cetyl acetate and acetylated lanolin alcohol)	Croda	1.00
Crodamol MM (myristyl myristate)	Croda	3.00
Crodet S24 (PEG-24 stearate)	Croda	3.00
Crodacol CS90 EP (cetearyl alcohol)	Croda	2.00
Cithrol GMS N/E GE0803 (glyceryl stearate)	Croda	7.00
Propylene glycol	—	3.00
Collasol 3% (soluble collagen and water)	Croda	1.00
Deionized water (for collasol)	—	10.00
Deionized water	—	to 100
Perfume, preservative	—	qs

emollience. This o/w system uses an emulsification system of a nonionic emulsifying wax and triethanolamine stearate (soap) stabilized with glyceryl stearate. (See Table 10.)

METHOD OF MANUFACTURE. Heat oil phase (including Titanium Dioxide MT-100T) and water phase separately to 80–85°C. Add water phase to oil phase with vigorous agitation. Add triethanolamine. Continue stirring until temperature reaches approximately 55°C, then pass through suitable homogenizing equipment.

8. *W/O Cream with Zinc Oxide C545.* This protective w/o baby cream uses lanolin alcohols as the low HLB emulsifier. Lanolin acts as a stabilizer as well as a skincare emollient. (See following data.)

W/O Cream with Zinc Oxide C545

Ingredient	Supplier	Percent (w/w)
Super Hartolan (lanolin alcohols)	Croda	2.00
Corona Pure New Lanolin (lanolin)	Croda	4.50
Mineral oil (70 cS at 25°C)	—	17.00
White petroleum jelly, continental grade No. 3	—	13.30
BHT	—	0.01
Croderol GA7000 (glycerine)	Croda	5.00
Deionized water	—	to 100
Zinc oxide	—	7.00
Perfume, preservative, color	—	qs

METHOD OF MANUFACTURE. Heat water and oil phases to 70°C. Add water phase to oil phase under continual stirring. A planetary mixer with side scraping arms is preferred. Add the zinc oxide at 60°C and continue stirring. Add perfume at 45°C. Homogenize at 30–35°C using suitable equipment, e.g., colloid or roller mill.

9. *Avocadin Facial Scrub Cream C1817.* This traditional detergent cleanser contains a combination of almond oil and Avocadin to aid moisturization and improve skin feel (Table

Table 10 Sunscreen Cream C1643

Ingredient	Supplier	Percent (w/w)
Crodamol MM (myristyl myristate)	Croda	1.5
Silicone 200/200 cS (dimethicone)	Dow Corning	7.0
Cithrol GMS N/E (glyceryl stearate)	Croda	2.5
Crosterene SA4310 (stearic acid)	Croda	3.0
Antaron V220 (PVP eicosene copolymer)	ISP	0.5
Polawax GP200 (nonionic emulsifying wax)	Croda	2.0
Crodamol OHS (octyl hydroxystearate)	Croda	5.0
Super-refined shea butter	Croda	5.0
Crodamol GTCC (caprylic/capric triglycerides)	Croda	3.0
Deionized water	—	to 100
Croderol GA7000 (glycerine)	Croda	3.0
Triethanolamine 99%	—	0.9
Titanium Dioxide MT-100T (titanium dioxide)	Mitsubishi	10.0
Perfume, preservatives, color	—	qs

11.) It may be modified into a facial scrub by addition of exfoliating agents such as polymer granules or crushed seeds or stones.

METHOD OF MANUFACTURE. Hydrate the Veegum in water. When fully hydrated, add the remaining water components (except Hydromond) and heat to 65–70°C. Heat oil phase to 65–70°C. Add water phase to oil phase while stirring. Stir to cool, adding perfume and Hydromond at ~40°C.

10. Two-Layer Conditioning Shampoo with Jojoba Oil C1569M1. This unusual shampoo (Table 12) employs jojoba oil in the upper opaque/cream layer for extra conditioning and market appeal.

Table 11 Avocadin Facial Scrub Cream C1817

Ingredient	Supplier	Percent (w/w)
Cithrol GMS A/S ES0743 (glyceryl stearate and PEG-100 stearate)	Croda	6.00
Dehydag Wax E (sodium cetearyl sulfate)	Henkel	7.50
Polawax GP200 (nonionic emulsifying wax)	Croda	1.00
Crovol M40 (PEG-20 corn glycerides)	Croda	2.00
Super-refined sweet almond oil	Croda	1.00
Volpo L3 Special (C_{12}–C_{13} Pareth-3)	Croda	2.50
Avocadin (avocado oil unsaponifiables)	Novarom	2.00
Deionized water	—	to 100
Veegum Regular (magnesium aluminum silicate)	Vanderbilt	1.00
Crodasinic LS35 (sodium lauroyl sarcosinate)	Croda	5.00
Empicol LZ (sodium lauryl sulfate)	Albright & Wilson	7.50
Hydromond (hydrolyzed almond protein)	Croda	0.50
Perfume, preservative, color	—	qs

Table 12 Two-Layer Conditioning Shampoo C1569M1

Ingredient	Supplier	Percent (w/w)
Empicol ESB3 (sodium laureth sulfate)	Albright & Wilson	30.00
Crodasinic LS35 (sodium lauroyl sarcosinate and water)	Croda	8.00
Crodafos SG (PPG-5-ceteth-10 phosphate)	Croda	1.00
Crotein Q (hydroxypropyl trimonium hydrolyzed collagen)	Croda	0.50
Incromine Oxide C (cocamidopropylamine oxide and water)	Croda	0.50
Incronam 30 (cocamidopropyl betaine)	Croda	5.00
Solan X (PEG-150 lanolin)	Croda	2.00
Empilan CDE (cocamide DEA)	—	3.00
Deionized water	—	to 100
Jojoba oil	—	1.00
Crodamol OSU (dioctyl succinate)	Croda	4.00
Perfume, preservative, color	—	qs

METHOD OF MANUFACTURE. Dissolve Crotein in water at 60°C. Add Crodafos while stirring; warm if necessary. Add remaining water-soluble ingredients, stirring until dissolved; then add emollient oils. Stir to disperse. Fill off while stirring.

11. Wheat Germ Shampoo C1535. Incorporation of wheat germ oil into this clear shampoo system provides a suitable treatment for dry hair and an attractive marketing concept. Crillet 1 and Crodafos SG are used in combination as the solubilizing agents for the oil, as the detergent base alone is not sufficient when used at a level of 1%. (See Table 13.)

METHOD OF MANUFACTURE. Solubilize wheat germ oil and perfume in Crillet and Crodafos. Combine the rest of the ingredients, and when homogeneous add Crillet–Crodafos premix. Stir until clear and fill off.

Table 13 Wheat Germ Shampoo C1535

Ingredient	Supplier	Percent (w/w)
Empicol ESB3 (sodium laureth sulfate)	Albright & Wilson	50.00
Incronam 30 (cocoamidopropyl betaine)	Croda	5.00
Super-refined wheat germ oil	Croda	1.00
Crillet 1 (Polysorbate 20)	Croda	3.00
Crodafos SG (PPG-5-ceteth-10 phosphate)	Croda	3.00
Triethanolamine 99%	—	to pH 6.5–7.00
Deionized water	—	to 100
Sodium chloride	—	—[a]
Perfume, preservatives, color	—	qs

[a]To required viscosity.

Table 14 Lipstick C1496

Ingredient	Supplier	Percent (w/w)
Fluilan (lanolin oil)	Croda	20.00
Novol (oleyl alcohol)	Croda	15.00
Candelilla wax	—	10.00
Carnauba wax	—	10.00
BHT	—	0.05
Polytrap 603 (acrylates copolymer)	Witco	0.25–0.50
Castor oil	—	44.60
Perfume, preservatives, color	—	qs
Pigment		
Ariabel Ruby 300503	Warner Jenkinson	1.75
Ariabel Saffron 300508	Warner Jenkinson	0.75
Ariabel Magenta 300522	Warner Jenkinson	2.25
Ariabel Pearl White 300330	Warner Jenkinson	10.00
Titanium Dioxide 300309	Warner Jenkinson	1.25

12. Lipstick C1496. Lipsticks contain a wide variety of oils, fats, and waxes that impart properties such as gloss, moisturization, and ease of application. This formulation (Table 14) uses castor oil as the primary emollient, blended with lanolin oil and oleyl alcohol to reduce drag and improve application. Wax combinations impart structure to the stick, preventing bleeding, sweating, and blooming if correctly formulated.

METHOD OF MANUFACTURE. Mill pigments in castor oil, add waxes, and combine with stirring over heat at 80°C. When homogeneous, stir slowly to deaerate. Stir to cool, perfuming and filling into molds at 55°C.

APPENDIX 2. OTHER SUPPLIERS

1. F Hoffman - La Roche,
   CH-4002, Basle, Switzerland.
   (Tel 00 4161 688 1111;
   Fax 00 4161 688 9391)
2. Novarom c/o Croda
   Chemicals Ltd, Cowick Hall,
   Snaith, Goole, North Humberside,
   DN14 8NF UK.
   (Tel 01405 860551;
   Fax 01405 86025)
3. Dow Corning Europe,
   62 Rue General de Gaulle,
   1310 La Hulpe, Brussels, Belgium.
   (Tel 00 322 655 2111;
   Fax 00 322 655 2001)
4. BF Goodrich Chemical Europe,
   742 Rue de Verdun,
   B-1130 Brussels, Belgium.
   (Tel 00 32 2 247 1911;
   Fax 00 32 2 247 1991)
5. ISP Europe Ltd,
   40 Alan Turing Road,
   Surrey Research Park,
   Guildford, Surrey GU2 5YF, UK.
   (Tel 01483 301757;
   Fax 01483 302175)
6. Mitsubishi Corp (UK) Ltd,
   Bow Bells House, Bread Street,
   London EC4M 9BQ, UK.
   (Tel 0171 822 0192;
   Fax 0171 822 0082)

7. Henkel KGA,
Postfach 1100,
D-4000 Dusseldorf 1, Germany.
(Tel 00 49 211 7970;
Fax 00 49 211 7987696)
8. R. T. Vanderbilt Company Inc,
30 Winfield Street (P.O. Box 5150),
Norwalk, CT 06856-5150.
(Tel 203 853 1400;
Fax 203 853 1452)
9. Albright & Wilson Ltd
(Detergents Group),
Whitehaven Works,
P.O. Box 15, Cumbria CA28 9QQ
UK.
(Tel 01946 693131;
Fax 01946 68181)
10. Witco Corporation,
10 Rue Cambaceres,
75008 Paris, France.
(Tel 00 33 142 65 9903;
Fax 00 33 142 65 6761)
11. Warner Jenkinson Europe Ltd,
Old Meadow Road,
Hardwick Industrial Estate,
Kings Lynn,
Norfolk PE30 4JJ, UK.
(Tel 01553 770550;
Fax 01553 770707)

REFERENCES

1. Oxford Dictionary.
2. *EEC Cosmetics Directive*, updated version, August 1994, Dir. 7617681EEC.
3. Anon., Natural selection, *Soap*, *Perfumery*, and *Cosmetics*, November 1994, pp. 45–47.
4. G. De Navarre, Oils and fats the historical cosmetics, *J. Am. Oil. Chem. Soc. 55*: 435 (1978).
5. G. O. Burr and M. M. Burr, On the nature and role of the fatty acids essential in nutrition, *J. Biol. Chem. 82*: 345 (1929).
6. C. Prottey, P. J. Hartop, and M. Press, Correction of the cutaneous manifestations of essential fatty acid deficiency in man by application of sunflower seed oil to the skin, *J. Invest. Dermatol. 64*: 228 (1975).
7. C. Prottey, P. J. Hartop, J. G. Black, and J. I. McCormack, The repair of impaired epidermal barrier function in rats by the cutaneous application of linoleic acid, *Br. J. Dermatol. 94*: 13 (1976).
8. P. J. Hartop and C. Prottey, Changes in transepidermal water loss and the composition of epidermal lecithin after applications of pure fatty acid triglycerides to the skin of essential fatty acid–deficient rats, *Br. J. Dermatol. 95*: 255 (1976).
9. J. Brod, H. Traitler, A. Studer, and O. De Lacharriere, Evolution of lipid composition in skin treated with blackcurrant seed oil, *Int. J. Cosmet. Sci. 10*: 149 (1988).
10. F. Fawaz, C. Miet, and F. Puisieux, Analyse des pommades, des huiles et des cires. XII. Etude de la composition chimique de la lanoline. 1. Examen preliminaire et fractionnement de la l'enchantillon etudie, *Ann. Pharm. Fr. 31*: 63 (1973).
11. F. Fawaz, M. Chaigneau, and F. Puisieux, Analyse des pommades, des huiles et des cires. XIII. Composition de la lanoline. 2. Etude des acides non hydroxyles de la lanoline totale et des differentes fractions, *Ann. Pharm. Fr. 31*: 217 (1973).
12. F. Fawaz, C. Miet, and F. Puisieux, Analyse des pommades, des huiles et des ceres. XIV. Composition de la lanoline. 3. Etude des acides hydroxyles de la lanoline totale des differentes fractions, *Ann. Pharm. Fr. 32*: 59 (1974).
13. F. Fawaz, M. Chaigneau, and F. Puisieux, Analyse des pommades, des huiles et des cires. XV. Composition chimique de la lanoline. 4. Etude des alcools aliphathiques de la lanoline totale et des ses differentes fractions, *Ann. Pharm. Fr. 32*: 215 (1974).

14. F. Fawaz, M. Chaigneau, and F. Puisieux, Analyse des pommades, des huiles et des cires. XVI. Composition chimiques de la lanoline. 5. Etude des sterols et des alcools triterpeniques de la lanoline totale et des differentes fractions, *Ann. Pharm. Fr. 32*: 301 (1974).
15. J. Thewlis, The emollience of lanolin, *Fragr. J. Jpn. 9*: 108 (1992).
16. K. Coupland, Natural lipids—valuable raw materials in cosmetics, Croda, Internal report.
17. D. T. Downing, Lipid and protein structures in the permeability barrier of mammalian epidermis, *J. Lipid. Res. 33*: 301 (1992).
18. K. Coupland, Super-refining natural lipids for pharmaceutical adjuvants, *Pharm. Manuf. Rev.*, March 1990.
19. K. Coupland and N. A. Langley, Modifying natural oils by adsorption chromatography, in *Oils and Fats in the Nineties* (V. K. S. Shukla and F. D. Gunstone, Eds.), Denmark, 1992, p. 176.
20. ICI Americas, Inc. *The HLB System*, ICI publication.
21. N. Lawson and G. Dahms, Formulating better cosmetics, *Manuf. Chem.*, April 1985.
22. H. Traitler and H. Winter, UK Patent GB 2 118 567 (1984).
23. F. Johannsen, Fat compositions and their use in cosmetic and pharmaceutical emulsion products, Int. Pub. No. WO 92/16184, October 1992.
24. D. Johansson, B. Bergenstahl, and E. Lundgren, Water-in-triglyceride oil emulsions. Effect of fat crystals on stability, *J. Am. Oil Chem. Soc.*, 72(8): 939 (1995).

27

The Use of Oils and Fatty Acids in Paints and Surface Coatings

John Bentley*
Consultant, Buckinghamshire, England

I. INTRODUCTION

This chapter describes the use of oils and fatty acids in what are generally referred to as surface coatings, and in this context, because they use essentially the same range of materials, we include inks. Surface coatings are generally seen as those materials applied to substrates ranging from wood and paper to a variety of metals, plastic, and many composite assemblies. They generally have a dual role, which is to protect and to decorate, the latter including an ability to disguise. The protective role is that of shielding the substrate from such environmental agents as radiation, moisture, and oxygen and possibility from more aggressive attackers such as atmospheric SO_2 in industrial environments. In addition to its role in decoration, paint may disguise inferior construction materials or even prevent recognition of the object, as with camouflage and IR reflective paints. Inside this spectrum of uses, inks may be seen as a subset of products, characterized as being more highly pigmented and used as thin coatings laid down in a precise manner, where properties of color and hiding power are paramount.

Paints consist of three principal components: binder, pigment, and solvent [1]. The first two are the permanent constituents, with the binder providing the adhesion and cohesion, keeping the pigment within the coating and ensuring that the paint remains attached to the substrate. Pigments provide color and opacity. Solvents are present to aid in manufacture and application. Varnishes are nonpigmented coatings. Binders have evolved through the ages from the initial range of naturally available gums and resins such as casein, tree, and fossil resins, through to a vast range of carefully designed synthetically based polymeric materials. Of all the natural substances used, oils have played a dominant role over at least the last 200 years, with various blending and enhancing modifications. Latterly, they have been chemically combined into a range of polymer compositions, as will be described. Although the overall use of oils has decreased in recent years, a number of reasons make prediction of their imminent demise premature, as will be discussed later.

*Previously with ICI Paints, Slough, England.

This chapter gives an overview of the size and breakdown of the paint industry worldwide and its use of oils and fatty acids, as far as can be deduced. After an overview, the use of oils in modern surface coatings and the chemistries involved are discussed in some depth. This includes the impact of legislation and environmental concerns as referred to above and attempts a brief look into the future. It is not possible here to cover the full chemistry or composition of all surface coatings, and for that the reader is referred elsewhere [2,3]. Older compositions are also not covered in as much depth where their use has significantly diminished or ceased, but again references give more information for those interested. Hence, the reactions, processes, and products discussed are principally those connected with the use of oils or fatty acids in coatings or resins in current or recent manufacture [4].

II. OVERVIEW AND ECONOMICS OF THE COATINGS INDUSTRY

The world coatings industry is split by volume on a geographical basis [5] as follows:

North America	22%
Western Europe	24%
Eastern Europe	20%
Asia	13%
Japan	10%
South America	8%
Rest of world	3%

Separately sourced figures give the total western European coatings market for 1993 as 5,854,000 tonnes, with a value of ECU 15 billion and a global market share of 31%, employing 110,000 people [6]. European printing ink production for 1993 is reported to have been 679,000 t with a value of ECU 2,547,000. The total U.S. paint and coatings market for 1993 has been put at 1100 million gallons US (~4,200,000 t) [7]. The Japanese market for that year was 1,956,000 t including thinners [8]. The total world surface coatings output can hence be estimated at around 20 million tpa (tonnes per annum).

Of the western European market above, the western part of Germany contributed just over one-third of the total, followed by Italy, France, and the United Kingdom with 18.5%, 16.6%, and 13%, respectively. To add perspective to these figures, in the United Kingdom, an industrial size comparison, taking the size of the total paint, varnish, and printing ink industry at a level of 100, has plastic packaging at 115 and glass and glassware at 110 [9]. The stationery market follows at 95, and perfume, cosmetics, and toiletries at 90.

The markets for paint are variously subdivided, with major sectors being "architectural," meaning the interior and exterior of land-based structures; product or original equipment manufacturer (OEM), embracing the largest industrial sectors, including automotive, coil, can, and appliance manufacturers; and special-purpose (depending on which statistics are used), which includes car refinishing, powder coatings, road marking paints, and inorganic coatings. Since the industry still uses, and also blends for resale, large quantities of organic solvents, solvent sales are significant as both visible and hidden parts of the industry totals, not always separable from the figures given.

In the United States the market split is 50% architectural, 30% product (OEM), and 20% special-purpose [7]. The European decorative (architectural) market has similarly been estimated to be 50% of the total. It should be noted that some areas, notably automotive and can, are fast becoming world markets, with a few manufacturers taking large quantities of material to a global specification, also from a limited range of suppliers. By contrast, the large architectural/decorative market, though increasingly dominated by large suppliers, sells in small pack sizes to users who may often be individual homeowners, with product tailored both to local conditions and to fashions that may be shaped by advertising.

The polymer or resin systems now used are wide and varied, and the relative importance of the different types continues to shift. Aqueous addition polymer dispersions made by emulsion or dispersion polymerization dominate in volume terms and find application in all markets. Nevertheless, oil-modified alkyd resins, discussed in detail in this chapter, are the next highest in volume. These two resin types have this position from their use in architectural/decorative paints. Other mixed aqueous polymer dispersions, partly or principally epoxy-based, have achieved significance in automotive primer and can coatings. More traditional industrial coatings now include a range of solvent-borne polymers including oil-modified alkyds, polyesters, acrylics, epoxies (including epoxy esters), and phenolics, with additional cross-linking resin including melamine formaldehyde and urethane polymers. Non-oil-containing resins are not discussed further in this chapter.

Oil use figures for the ink and coatings industries are not directly available but may be estimated from partial breakdowns of resin use by type and from general knowledge of these coatings. While trends in decorative paints continue toward water-borne emulsion paint, alkyd paint still accounts for 25–35% of this sector, varying with country. Thus 16% of total European paint and ink output is estimated still to be based on alkyd resins, of which the predominating decorative alkyd typically has an oil content of around 70% based on the solid resin; in Japan, 18% of resin production is oil-based. A recent survey in the United Kingdom [10] estimated that 80,000 t of alkyd are produced annually, requiring 48,000 t of oil. This would imply a European use of around 400,000 t for coatings. The European demand for fatty acids for the manufacture of alkyds and derivatives is estimated at around 120,000–160,000 tonnes as part of the above total. In Europe, consumption of vegetable oils for printing inks is currently 12,000–15,000 tpa, predicted to rise to possibly as much as 25,000 tpa by 1997 [11].

III. OIL USE IN PAINTS

Table 1 shows the principal oils now used in surface coatings, in probable order of their use, and also their fatty acid content. Of those oils or fatty acids listed, linseed, soybean, castor, and coconut oils are all derived from seeds or nuts; tall oil (in fact available only as fatty acid and not combined with glycerol) is a by-product of paper manufacture and derives its name from the Swedish name for pine. Dehydrated castor oil, included in order to give its fatty acid content here for comparison, is derived from castor oil and is described in a later section.

Oils used in the coatings industry will always have been refined, by processes described elsewhere in this book, to remove waxes, free fatty acids, lecithins, and other natural contaminants, because the presence of those substances would in almost all cases have deleterious effects on the derived coating resins. Tall oil in particular is now used almost exclusively in distilled form, in order to free it from any rosin content [12,13]. While all except tall oil occur naturally as a triglyceride, free fatty acids are available from almost all oils as alternative reactants to extend formulating flexibility. The use of both oils and fatty acids in coating resins is described later.

Oils are classified as drying, semidrying, or nondrying, depending on whether the oil on its own has sufficient unsaturation in its total fatty acid chains to autoxidize fully to a dry film. Linseed oil, with over 60% of its fatty acids doubly or triply unsaturated, is classed as

Table 1 Oils Used in Coatings and Their Fatty Acid Composition

	Component, wt %				
Oil	Saturated acid	Oleic acid	9,12-Linoleic acid	9,12,15-Linolenic acid	Other acid
Soybean oil	14	22–28	52–55	5–9	0
Tall fatty acid	3	30–35	35–40	2–5	10–15[a]
Linseed oil	10	20–24	14–19	48–54	0
Castor oil	2–4	6–8	3–6	0	85–87[b]
Castor oil, dehydrated	2–4	6–8	48–50	0	40–42[c]
Coconut oil	89–94	6–8	0–2	0	0

[a]Conjugated and pinolenic (octadeca-5,9,12-trienoic) acids, amounts dependent on source and method of refinement.
[b]Principally ricinoleic (octadeca-12-hydroxy-9-enoic) acid.
[c]Conjugated 9,11-linoleic acid.

a drying oil, whereas soybean, tall, and dehydrated castor oil are generally classified as semi-drying. Castor oil and coconut oil are nondrying. The concepts of drying and autoxidation are discussed later in this chapter. In all of these oils, with one exception, the saturated fatty acids are mainly palmitic and stearic acids; the exception is coconut oil, which contains very high amounts of lauric acid.

Many other oils have been, and in various parts of the world still are, used in coatings. Historically tung oil (or China wood oil) was used extensively in oleoresinous materials. However, it has not been generally available in good supply for a number of years; in any case it is less favored because of its highly conjugated unsaturation and hence tendency to yellow in derived resins. Sunflower and safflower oils have significant use and are similar to soybean oil in application. Other oils are mentioned in the following sections.

IV. TERMS USED IN THE COATINGS INDUSTRY AND FORMULATING PRINCIPLES

Since this chapter uses terms specific to the industry, it is essential to define some of these now to avoid confusion. The term "surface coating" could be very widely embracing, but in our context it refers to essentially organic paints or finishes. Varnishes are unpigmented coatings, so the substrate will at least partly show through. They are used on wood and over pigmented coatings to provide additional protection. The term "lacquer" may confuse, but to the technologist it normally refers to those coatings that form a film only by solvent evaporation; lacquers may be clear or pigmented. "Drying" covers all mechanisms of film formation, which is the transformation of applied liquid paint into a solid film, and may involve physical change with polymer particle coalescence and chemical change, not just the evaporation of solvent. "Curing" most usually applies to processes in which applied agents such as heat and light are used to effect drying. Drying and curing are generally polymerization processes.

The terms "resin," "vehicle," and "binder" are all used for the film-forming part of the paint, which is nowadays a polymer (wholly or partly synthetic), though in early paints it may have been simply oil or gum. Resins are made in kettles or reactors and thinned in churns. The heat polymerization processing of oils is frequently referred to as "bodying." Both paints and resins are characterized by their nonvolatile or solids content, their viscosity, and frequently their acidity, expressed as acid value (mg KOH/g resin). For resins containing oil, the terms "long oil," "medium oil," and "short oil" are used to categorize the oil or fatty acid content of the total resin. This concept guides both the flexibility of the film and, if the film contains drying oil, the rate and extent of oxidative drying to be expected. Oleoresinous vehicles containing less than 65% oil are termed "short oil"; those with more than 65% oil, "long oil." For alkyds, "long oil" contains more than 60% oil, "medium oil" 40–60%, and "short oil" less than 40%. The terms "drying," "semidrying," and "nondrying" have already been referred to.

The age-old balance required with paint formulation and hence directed to the resin preparation is that of adequately achieving final properties, particularly mechanical strength, getting surface coverage and build, and being able to apply the paint evenly, all at acceptable cost. In molecular terms, good mechanical properties require solid polymeric binders of high or infinite molecular weight; most application techniques demand low viscosities, and hence paint compositions always contain some solvent or diluent to make the product fluid for application. If the binder is of high molecular weight, a proportionately higher quantity of solvent will be required for application than for a low molecular weight binder; the latter, however, has inferior mechanical properties. The effect of this can be seen from spirit varnishes and nitrocellulose paints (lacquers), which may contain 80–90% solvent as they are applied. The well-recognized solution to this is to employ a curing or drying reaction so that the binder will increase in molecular weight and its properties improve following application.

Paint systems can be applied, for example, by brush, spray, roller coating, and dipping. Reaction can be assisted by component or catalyst addition just prior to application and by the use of heat or other radiation after application. Utilizing the drying of oils through autoxidative cross-linking has provided one of the most enduring examples of this type of system.

Solvent or diluent, as already stated, is an essential component. Solvents of choice have moved on from turpentines to white spirit; short oil alkyds and acrylic and epoxy resins require the use of the full range of organic solvents to achieve full solubility. Organic solvent content is expressed as the volatile organics content (VOC) of the coating, and users expect VOC to fall within certain limits for different coating types. These limits are being enforced by legislation. Water is now the preferred diluent, with polymers being used in dispersions rather than solutions.

V. CHEMICAL PROCESSES IN OIL AND FATTY ACID USAGE

A. Chemistry of Oil and Fatty Acid Reactions

Apart from hydrolysis and esterification reactions involving oils and fatty acids and their esters, a range of other reactions are possible when one or more double bonds, or attached groups such as hydroxyl, are present on the fatty acid chain. Reactions carried out at the unsaturation site include dimerization, maleinization, epoxidation, vinylation, dehydropolymerization, and hydrogenation. Oxidation reactions are described in the next section.

Epoxidized oils find little use in coatings, and their manufacture and use are discussed in another chapter in this book. However, naturally occuring vernonia oil has recently been advocated as a reactive diluent for high solids alkyds [14].

Heat bodying dimerization of unsaturated oils is an integral part of the process for producing oleoresinous and alkyd resins (see later). This heat bodying may also be carried out as a separate process to form what are referred to as stand oils. This process, carried out at temperatures around 300°C (lower with catalysis), consists predominantly of bimolecular additions or combinations resulting in so-called dimer fatty acid structures in the oil. The strongest contributor is probably the Diels–Alder reaction, as shown in Fig. 1, which involves conju-

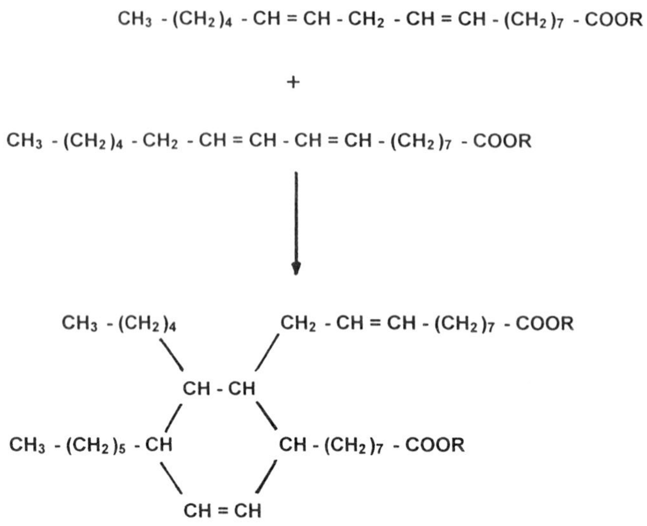

Fig. 1 Thermal dimerization of fatty acids.

gated pairs of double bonds on one of the fatty acid moieties (formed in situ if not already present) [15]. The separate dimer fatty acids are used both in alkyd resin manufacture and in the preparation of polyamide resins, which are used as thixotropic modifiers for alkyd resins.

Maleinization of oils and fatty acids is carried out both separately and in situ in alkyd resin preparation to provide a flexible, higher functionality acid component of formulations. In the stand-alone process, typically linseed oil fatty acid (LOFA) is reacted with maleic anhydride at 200°C, and addition occurs principally by the ene reaction (Fig. 2a), because the bonds involved are nonconjugated [16]. Where conjugated double bonds are present, a Diels–Alder reaction is possible (Fig. 2b), which proceeds rapidly and exothermically. At higher maleinization levels of LOFA, since the ene reaction moves the double bond from the nonconjugated

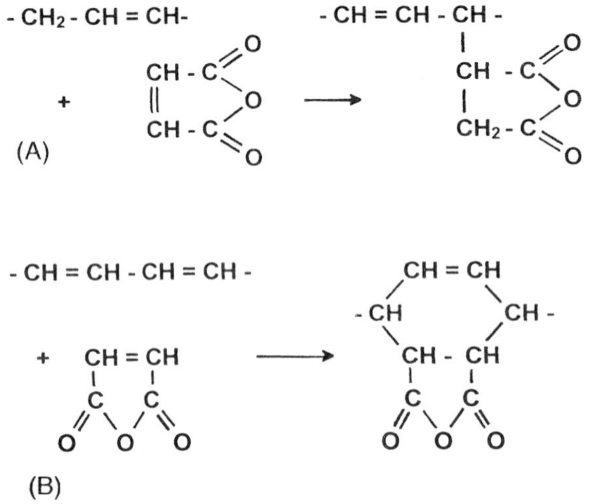

Fig. 2 (a) Ene reaction with maleic anhydride. (b) Diels–Alder reaction with maleic anhydride.

to the conjugated configuration, it is possible that a sluggish ene reaction is followed by a faster Diels–Alder reaction on the same fatty acid.

Unsurprisingly to the polymer chemist, with the potential availability of the double bond for addition reaction, vinylation with styrene or vinylbenzene, either as a preliminary or final stage of manufacture of an alkyd is practiced to obtain polymers that combine polyester and chain addition polymer structures in order to exploit the properties of both [17]. The process is designed to graft the two structures together to overcome the consequences of incompatibility, which would otherwise be very apparent. In practice, the reactivity of double bonds to addition reaction is not high, and some grafting is also achieved by the use of peroxidic initiators such as di-tert-butyl peroxide, which almost certainly adds grafting sites by abstraction of hydrogen atoms from positions adjacent to double bonds. Graft sites to the main polymer in alkyds have also been provided by the reaction of small amounts of methacrylic acid into the polymer prior to vinylation.

A novel reaction demonstrating the susceptibility of unsaturated fatty acids to hydrogen abstraction is that of "dehydropolymerization" [18]. With this, treatment of a range of oils, including soybean oil and tall oil, with di-tert-butyl peroxide at 145°C was found to generate dimer structures superficially similar to those described for heat bodying but believed to have a single -C-C- link between chains.

Dehydration of castor oil, composed mainly of glyceryl triricinoleate, is carried out to generate a second double bond, and hence semidrying oil status, by removing the hydroxyl group on the carbon in the 12-position along with the hydrogen atom on one of the adjacent carbons [19]. This process is carried out at temperatures of 280–300°C in the presence of strong acids such as sulfuric acid and is illustrated in Fig. 3. Since oil bodying as described earlier can then occur, skill in temperature control, choice of catalyst, and use of high vacuum is necessary to produce low viscosity, good colored products. As will be evident from the structure, both conjugated and nonconjugated linoleic acids may be produced (see Table 1). As a semidrying oil, both dehydrated castor oil (DCO) and the fatty acids (DCOFA) find application in alkyds and epoxy esters. However, owing to a residual characteristic surface "nip" when DCO/DCOFA is used, its use is restricted to industrial and mainly stoving applications.

Castor oil fatty acids may be hydrogenated to produce 12-hydroxystearic acid, which can be self-condensed to a limiting molecular weight by virtue of its residual stearic acid content [20]. Poly-12-hydroxystearic acid is the essential aliphatic-soluble oligomeric component of a range of graft copolymers (see later). Castor oil may also be sulfonated (sulfated), and sulfated castor oil, or Turkey red oil, is used as a pigment dispersant.

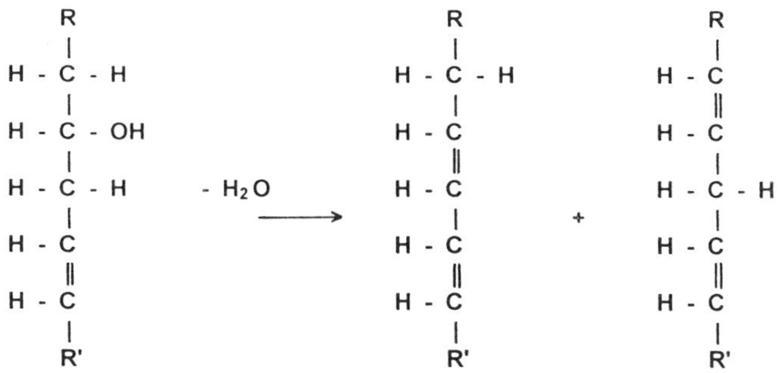

Fig. 3 Dehydration of ricinoleic acid to form conjugated and unconjugated forms of linoleic acids. [$R = C_5H_{11}$ and $R^1(CH_2)_7COOH$]

With the different reactivities shown by conjugated and nonconjugated fatty acids and also some differences in reactivity of their isomeric forms, isomerization is a useful process to enhance properties, principally drying. As an example of the possibilities, linoleic acid can be isomerized to the conjugated form and further modified from the cis,trans to the trans,trans configuration [21]. A number of patented processes exist, and isomerized fatty acids are used particularly in medium oil air-drying alkyds. They improve the initial drying, color, and resistance properties of the alkyds.

B. Drying Mechanisms

Drying or polymerization of fatty acid moieties is the most widely used of all hardening mechanisms for surface coatings. It is an oxidative process, and early evidence (though not then necessarily fully understood) includes Priestley's use of linseed oil to remove the oxygen that had been sustaining rats enclosed in a bell jar. The process, generally referred to as autoxidation, has drawbacks that encourage continued examination of its mechanisms to further control and overcome them. Foremost, it is a cross-linking mechanism that does not stop at a convenient point when mechanical properties are achieved but continues, ultimately causing embrittlement and degradation. Second, control of drying speed, through-drying, and drying in adverse conditions depends on the use of combinations of metal driers, whose cost, balance, availability, and toxicity may all cause concern. Furthermore, side reactions cause production of unpleasant volatile materials, and yellowing occurs to some degree in most environments; both of these latter are features that paint chemists would like to control.

In the past, commercial pressures have triggered spurts in investigation. More recently, moves to water-based paints and legislation requiring lead removal have led to changes in how driers are used. However, it is notable that the availability of new investigative techniques has over time enabled step changes in the quality of reported investigations, so that the literature shows application of tracer techniques [22], much IR investigation [23], including time lapse [24] and recently FTIR spectroscopy, ^{13}C-NMR spectroscopy, and most recently SIMS [25], EXAFS [24], and SAXS [26].

Although parts of the process are incompletely understood, the overall process clearly involves initial formation of hydroperoxides followed by their breakdown, induced by cobalt in modern systems, and subsequently by free radical linking and scission processes to form cross-links. Practical complications in investigation arise through the range of fatty acids present in typical oils and the number of reactive positions, the practice of adding a blend of driers to a typical system, the fact that only the surface is available for oxygen ingress and the escape of volatile products, and the presence of other components in the paint that may affect drying.

For a preliminary overview, the reader is urged to consult the excellent review by Wexler [27], which gives much background on oxidation and radical reactions and affords insight into formulation practice in the context of drier use.

The first stage in drying, often preceded by an induction period, is the formation of hydroperoxides, with attack occurring predominantly at allylic hydrogens, whose position is activated by double bonds. The reaction is enhanced when the double bonds are multiple and conjugated. Khan [28], using a variety of techniques, showed with nonconjugated linoleic acid the clear-cut formation of 9- and 13-*cis*,*trans*-hydroperoxides with a shift both to conjugation and to the *trans* configuration at the position adjacent to the hydroperoxide. Figure 4 illustrates this. A free radical mechanism is now thought most unlikely.

Much investigation has been carried out on purified esters, and this has led to the preparation of high purity crystalline linoleate hydroperoxide [29].

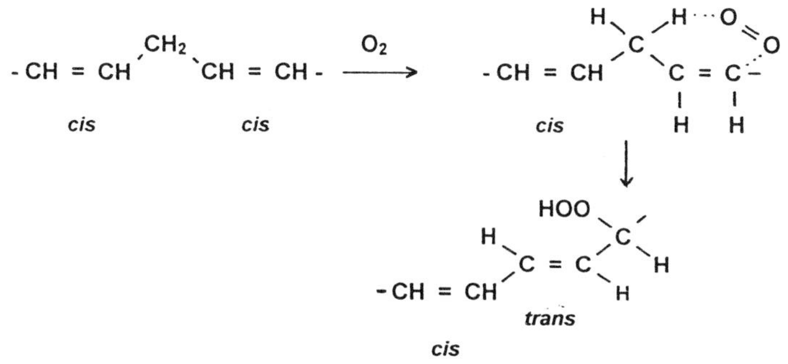

Fig. 4 Hydroperoxide formation and rearrangement. (From Ref. 28.)

Part of the subsequent reaction involves not just useful polymerization but also scission and breakdown products. These are responsible for the characteristic odor of paint drying, which can be acrid and lachrymatory in conditions of poor ventilation. A number of examinations have been published, of which some of note are those by Loury and Forney [30,31], Frankel [32], and the recent studies by Leeves [33,34]. All studies identified a multiplicity of alcohols, aldehydes, ketones, and esters. Of these, the aldehydes propanal and hexanal and acids such as formic are major contributors to the unpleasantness of drying alkyd fumes. Thus, for example, Loury and Forney's examination of methyl linolenate oxidation under mild conditions at 20°C found 29 volatile compounds [30]. They also produced a complete mechanistic scheme showing how by oxygen fixation at all possible positions followed by scission, almost all of these compounds could be formed. Other esters examined were methyl linoleate and methyl oleate [31]. Frankel identified 27 compounds using similar conditions [32].

In addition to examining the same esters as Loury and Forney, Leeves extended his investigation to alkyds using typical drying oils (soybean, linseed, DCO, and tall oils), an isomerized acid, and purified linoleic acids [33,34]. He used two typical drier combinations and investigated less common catalyst systems. Differences were found, but interpretation was difficult.

Cobalt, as has been stated, is now the major drier used, with supplementary driers including calcium, zirconium, and aluminum; lead is no longer in general use because of environmental and toxicity concerns. These driers are now used as octoates (octanoates), except for aluminum, which is used as the ethyl acetoacetate substituted alkoxide. Manganese and barium have been used, and zinc may also be considered partially active as a drier in paints, since zinc oxide when used in pigmentation may partly dissolve as soap formed through reaction with any residual acid.

The cobalt redox system is shown in Fig. 5, which illustrates why only a low concentration of cobalt is required and the free radical nature of decomposition. The nature of the cobalt drier complex [35] and the kinetics and mechanism of the complete process have been studied by Waters [36]. Film weight studies show that linseed oil initially takes up about 40% by

$$\mathrm{ROOH} + \mathrm{Co}^{2+} \longrightarrow \mathrm{Co}^{3+} + \mathrm{RO}\cdot + \mathrm{OH}\cdot$$

$$\mathrm{ROOH} + \mathrm{Co}^{3+} \longrightarrow \mathrm{Co}^{2+} + \mathrm{ROO}\cdot + \mathrm{H}^{+}$$

Fig. 5 Cobalt–catalyzed hydroperoxide decomposition.

ROOH		⟶	RO· + ·OH	Scission
RH +	RO·	⟶	ROH + R·	Abstraction
R· +	R·	⟶	R - R	Oxidation and Dimerisation
R· +	RO·	⟶	R - O - R	
R· +	O_2	⟶	ROO·	
R· +	ROO·	⟶	R - OO - R	

Fig. 6 Dimerization reactions occurring with hydroperoxide decomposition.

weight of oxygen, around half of which is ultimately lost in the form of volatile scission products. It has been concluded that polymerization, which forms both -C-O- and -C-C- cross-links, primarily involves the addition of peroxy and alkoxy radicals to a conjugated double bond, with ether and peroxy links predominating. Other recent work using NMR and SIMS confirms this conclusion [25], showing the links to result essentially from dimerization reactions. Figure 6 shows many of the possible reactions, where R indicates attachments to fatty acid chains. This and other studies indicate that oleates can have at least some participation in the later stages of cross-linking. The final film polymer structure will be a cross-linked three-dimensional network, swellable but not soluble unless bond breakage occurs.

Various views have been expressed on the roles of the ancillary driers, but although their effectiveness has been fully demonstrated, the mechanisms can only be postulated at present. Nevertheless, formulators will be aware of the way different combinations can be chosen to be effective with different pigmentation systems.

The auxiliary metal driers are known to enhance the catalytic activity of cobalt, thus improving through-drying and low temperature curing. It is also probable that driers such as lead react preferentially with some of the autoxidation by products, which would otherwise build up in the film, interfering with the catalytic effect of the cobalt. Aluminum and zirconium, in particular, are capable of reacting with functional groups in the medium, thus yielding a cross-linked structure; both function by forming coordination and covalent cross-links [37,38].

Investigation continues into the drying mechanism in new situations. For example, a recent paper [39] suggests that in high solids systems, behavior is different to that found in conventional alkyds, with singlet oxygen believed to be involved in the reaction.

C. Other Drying Processes

Other processes are available for the curing of finishes containing drying oil. The processes described above occur under ambient conditions in air. However, under stoving conditions with industrial paints, clearly the thermal polymerization reactions described earlier also play a part. In addition, paints that contain a strong acid catalyst, primarily to promote cross-linking reactions with added urea formaldehyde resin, for example, may also have autoxidation promoted by these acids.

Strong acid catalysts such as alkyl- and arylsulfonic acids have been shown to accelerate alkyd resin drying at room temperature [40]. To be most effective, semidrying oil fatty acid must be present in an alkyd structure formulated with a significant excess hydroxyl content. With 1–6% of catalyst, drying is comparable to that with conventional driers, and the mechanism postulated is the acid-catalyzed decomposition of peroxy groups formed in the initial stage of other autoxidative mechanisms.

Photoinitiated autoxidation has also been demonstrated, with autoxidation initiated either directly by UV radiation or with the aid of photoinitiators. In other applications, photoinitiation of unsaturated polyester reaction by straightforward radical mechanisms is an effective process, using the combination of a ketone such as benzil and a reducing agent capable of reducing the photoexcited state of the ketone. In translating this to autoxidative systems, it has been shown that the drying of alkyd resin can be initiated by the use of benzil alone in daylight [41,42]. It is believed that benzil in its excited state abstracts a methylenic hydrogen from unsaturated fatty acid, which produces radicals and results in cross-linking. In this case, although hydroperoxide is present during drying, it is doubtful that these peroxides contribute significantly to the drying process.

Enzymatically catalyzed autoxidation has been reported, with cytochromes, heme, and similar metal-containing enzymes cited as being active. Whether this is principally metal-catalyzed autoxidation or involves other mechanisms is unclear, and much of this area has yet to be explored.

D. Yellowing

Yellowing is an intrinsic problem with autoxidative drying, its effect being more noticeable with films kept in the dark and in the atmospheres found in kitchens and bathrooms. The effect of ammoniacal compounds has long been recognized, and accelerated testing of yellowing is frequently carried out in an ammonia cabinet. It should be noted, however, that bleaching processes also occur, and yellowing–bleaching cycles can be demonstrated, so that the process is to some extent reversible.

Yellowing is a process linked with residual unsaturation, as is drying ability, and drying and yellowing proceed together. In general, chromophoric groups include unsaturated and aromatic ring structures, -N=C- and >N=O, and those always present in oils, including -C=C- and >C=O, and already involved in the drying process. Mechanisms causing the formation of -NR_2, -OH, etc. as auxochromes can readily be postulated.

The process of yellowing can be seen as proceeding stepwise, involving the production of colorless precursors followed by further reaction to colored compounds [43]. Yellowing of aldehydes, themselves produced in the process, can be demonstrated, with unsaturated aldehydes yellowing most severely. Davison [44] showed that hexadienal yellows the worst of those unsaturated aldehydes she tested in the dark at 70°C. Other authors have identified benzoquinone structures in films [45].

E. Degradation and Weathering Mechanisms

In many respects, degradation mechanisms can be seen as extensions or reversals of the polymerization and drying processes. It has long been recognized, particularly with respect to the autoxidation mechanism, that to get an initial cure that allows handleability and hardness to develop in reasonable time for the coating to withstand normal wear and tear, the system requires high concentrations of unsaturation. Hence a continuation of this process is likely, which leads to a progressive increase in cross-link density and the loss of the pendant groups plasticizing the structure, producing embrittlement. This may be followed by degradative processes with bond breakage causing polymer structure failure. Some very detailed studies of the processes that occur have been published [46]. Unsaturation is both reacted and regenerated in the drying process, and residual unsaturation provides target sites for a host of degradative mechanisms. Depending on the use of the painted article, oxygen, UV, and moisture are all likely to be aggressors in various ways.

Many of the polymer structures used are ester-based, and fatty acid moieties are also linked into the structure by ester groups. Hydrolysis is therefore a factor in film breakdown when films are in contact with moisture, especially if soluble salts are present. A factor in facilitating this process is the number and location of the more polar hydroxyl and carboxyl groups on the polymer, as their presence allows water to enter and swell the film, thus enabling attack. The drying and scission processes result in increases in the concentrations of both of these groups, thus accelerating the process. In damp conditions, fungal and bacterial agents contribute to breakdown.

Pigmentation and additives are well recognized as having significant effects. Different grades of titanium dioxide can, for example, either shield the resin from radiation or, alternatively, photocatalyze breakdown. Inhibitors and UV absorbers are frequent additives, though they must be carefully chosen to avoid interference with basic curing.

It should be said, however, that the breakdown processes are relatively slow, so a properly applied decorative alkyd paint system, with the use of primer, undercoat, and the optimum thickness topcoat, can give 3–5 years of full protection before some partial failure at, for example, edge joints necessitates retreatment. Normally, the first apparent failure mode is that of surface dulling and loss of gloss, and possibly some chalking (powdery loss of the surface layer); in early stages this will be seen as an aesthetic rather than mechanical failure. Alkyd-based industrial paints, for which reactions other than autoxidation provide the major curing following application, generally retain a satisfactory appearance and other properties for the service life of the product (car, appliance, etc.).

VI. SURFACE COATING RESIN SYSTEMS

A. Oleoresinous Systems Including Treated Oils

In the United Kingdom, linseed oil probably came into use in the 1400s. Certainly by the 1600s, white and red lead were used in what we would now identify as oleoresinous binders. These comprised linseed oil and resins such as rosin and shellac. Thus we can trace over 400 years of use of indigenous linseed oil, which was used with imported tung oil along with rosin and turpentine as naval stores, and with white and red lead included as protective pigments, also providing catalysis for autoxidation.

Oleoresinous media are made by heating together oils and chosen solid resins until compatibility and a predetermined viscosity are reached [47]. Solvent and driers are added to give an unpigmented varnish composition. For paints, the solid components, which are now classified as pigments if they are colored or scatter light significantly and otherwise are classified as extenders, must be dispersed in the resin. Only drying oils (with possibly a limited amount of semidrying oil) can be used in this class of paint vehicle. This type of vehicle has use now only in varnishes, in some undercoats and primers, and in printing inks.

The major oils used in oleoresinous systems have always been linseed oil and tung oil, with the former dominant, now overwhelmingly so as there are very limited supplies of tung oil. Oiticica oil has been a satisfactory replacement for the latter, although a range of oils blended with linseed oil have been used. These include fish oils, dehydrated castor oil, soybean oil, and tall oil esters with glycerol or pentaerythritol. The choice of oil blend will always be such that in order to keep up the drying potential, proportions of highly unsaturated drying oil, drying oil, and semidrying oil need careful control, with the amounts of soybean oil or tall oil esters strictly limited. Clearly, in various parts of the world, indigenous oils have been used, and a long list of those useable includes safflower, sunflower seed, rubber seed, and tobacco seed oils.

Natural resins used have been the copal fossil resins found in a number of locations and rosins obtained either as gum rosin from the live tree or wood rosin from stumps. As required with oils, preheat treatments for these resins evolved. Thus the copals were normally "run" to effect partial decomposition and improve oil solubility. Rosins were distilled, maleinized, and partly polymerized by a variety of techniques.

Synthetic resins have been used since they became available; these include phenol formaldehyde (PF), coumarone indene, and most recently epoxy resins. PF resins may be modified with rosin to aid oil solubility and reduce cost. Unmodified and modified resole type PFs [3] have become the most used resins in oleoresinous vehicles. Epoxy resins, which are used to impart hydrolysis and chemical resistance, are more expensive.

Oleoresinous vehicles are manufactured by heating oils and solid resin together at temperatures up to 260°C. Since the heat polymerization of oils, particularly tung oil, can lead to gelation, skilled techniques are required to ensure sufficient reaction that the resin is "solubilized" while avoiding gelation. Methods to stop reaction involve cooling back either with oil held out of the formulation or with thinning solvent. Monitoring techniques have often been primitive, involving judging incipient gelation by the tendency of samples taken from the process to "string" and judging the attainment of compatibility by the clarity of a "pill" of material cooled on a metal plate. The potential for fire during processing is great but can be reduced by an inert gas sparge and by heating reaction vessels indirectly. These processes are undesirable in that considerable fumes are produced from decomposition of the oils and resins used, and this requires containment for environmental reasons.

While tung oil heat-bodies and gels rapidly and hence can be used only directly in vehicles, it is advantageous to use linseed and other oils prebodied as "stand oils"; this is still true of alkyd resins. This heat treatment, previously done in directly heated open pots, is now more safely carried out in indirectly heated stirred vessels under inert gas. All drying and semidrying oils may be treated in this way, the process improving the drying, elasticity, and durability of the derived coatings, possibly because polymerization is through direct -C-C- links, reducing residual unsaturation. The unsaturation may also have undergone some rearrangement (to conjugated form or cis–trans transformation), and there will be some early attendant loss of volatiles, mimicking the effects of partial drying.

A simple wood treatment involves the use of so-called boiled linseed oil, which was once prepared by dissolving metal oxides in the "boiling" oil but is nowadays made by blending soluble driers with linseed stand oil.

Blown oils have also been manufactured; the heated oil is preoxidized by an air sparge, the temperature being kept to around 130°C. Blown castor oil was an ingredient used in the manufacture of wrinkle finishes.

B. Alkyd Resins

1. *Alkyd Overview*

Alkyd resins, sometimes known as glyptals or oil-modified alkyds, are polyester resins whose structures incorporate a significant amount of combined oil-derived fatty acids. This may be achieved through processing via the parent oil, and in these cases glycerol will constitute at least part of the alkyd's polyol content. Alkyd resins may be made equally as well by using preseparated fatty acid and the fatty acid process, as described later. The name "alkyd" resins possibly derives from their manufacture from *al*cohol and an a*cid*, with the "cid" changed to "kyd." Alkyds have been subclassified as oil-modified and oil-free. "Alkyd resin" is now generally used as the generic name for the former type; the term "oil-free" is less usually applied to polyester resins, normally formulated from polyhydric alcohols and di- and tribasic acids but sometimes containing short non-oil-derived monobasic acids as minor components.

Alkyd resins were one of the first large-scale applications of a synthetic polymer in surface coatings and have endured because of the variety of ingredients available, which make it possible to use alkyds in virtually all areas of coatings application [48]. As a resin class, they are cost-effective and can be made with relatively simple equipment.

Alkyd resins have a polyester backbone structure, and the acid most commonly used in the backbone is orthophthalic acid, always as the anhydride (PA). A range of di- or polyhydric alcohols are used. The polymer-forming reaction is hence esterification, with water the reaction by-product. The process can be single-stage with all ingredients charged together if fatty acid is used (the so-called fatty acid process), but it is more economical, and also more sensible if glycerol is to be part of the formulation, to incorporate oil directly. However, this does not occur readily, since oils are triglycerides, and a preliminary stage known as a monoglyceride stage is normally carried out to make the oil reactive. As discussed below, two functional groups are required to allow molecules to react to form a polymer structure, and hence each oil molecule must shed two fatty acid groups to make available two of the three hydroxyl groups of the glycerol for subsequent reaction.

Although any polyol (or polyol blend) may be used, the monoglyceride process is illustrated in Fig. 7 with glycerol, which is used in simpler alkyds at all levels of oil content. This process is typically carried out at around 240°C in the presence of a catalyst such as sodium hydroxide and requires around 30 min. The process does not go to the completion implied above but approaches an equilibrium mixture consisting of some unreacted oil and glycerol, and also the diglycerides, in addition to the expected and predominant monoglycerides. It is essential to ensure that sufficient randomization has occurred, and, depending on the polyol, this may be tested by a simple alcohol tolerance test. In some circumstances, where a polyol such as PE is used, it will be necessary to perform a small-scale trial alkyd preparation, taking the monoglyceride sample and additional polyol and phthalic anhydride (PA) as appropriate, heated in a wide-necked tube or small flask. The penalty for incomplete equilibration can be the preparation of an alkyd that is cloudy or separates or is even prone to gelation. Equally, where the oil is prone to heat bodying, it is essential not to prolong this stage, and the best practice is to standardize both the heat-up rate and the time of reaction for this stage so that subsequent stages will be predictable and reproducible.

The resinification or polyesterification stage is carried out by heating the reaction mixture at temperatures between 180 and 260°C. Catalysis is not essential in conventional alkyd preparation. High temperatures are used for long oil alkyds, where oil bodying during the process is wanted to increase viscosity. The lower temperatures may be required with short oil alkyds formulated to higher functionality with a strong gelling potential, so that they are more controllable. Water of reaction is removed at this stage, though this is not practically a reliable indicator of extent of reaction. If a monoglyceride stage has been carried out, the charge will have been cooled so that additional polyol and dibasic acid can be added safely before reheating. This stage, using monoglyceride, is shown schematically in Fig. 8.

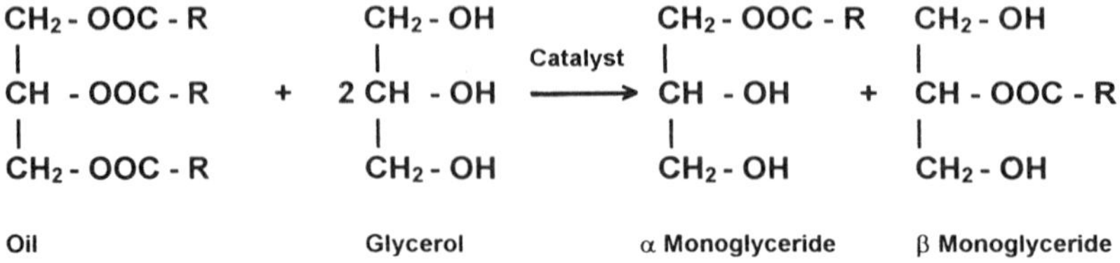

Fig. 7 Monoglyceride formation. (—OOC—R is a fatty acid moiety.)

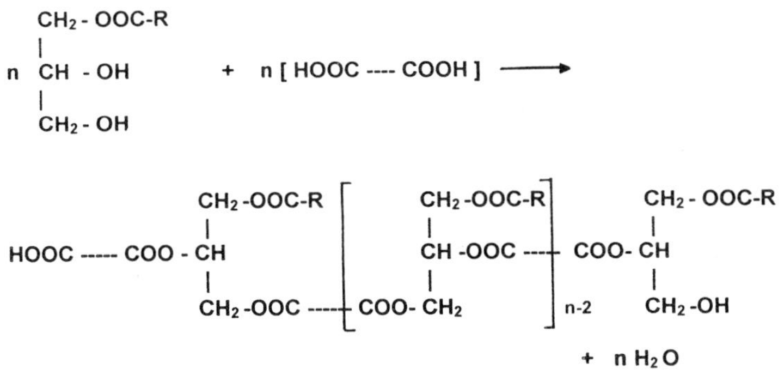

Fig. 8 Alkyd formation with monoglyceride and dibasic acid. (—OOC—R is a fatty acid moiety; HOOC—COOH is a dibasic acid.)

The progress of reaction will be followed by observing the increase in viscosity, which is related to an increase in molecular weight, and also the decrease in the acidity of the mixture, measured as acid value. The rate of viscosity increase generally increases as reaction proceeds, and sampling intervals decrease during preparation. As stated above, some preparations are prone to gelation, and in this case operators need be ready to stop the reaction by cooling and, after the initial cooling by adding solvent. When a preparation runs out of control and gelation occurs, this may often be reversed and the reactor contents recovered if prompt action is taken in adding fatty acid or glycerol; this is further indication that a process of randomization (by transesterification or ester interchange) is always occurring to some degree during processing.

Where an oil such as tung oil is concerned, it is not generally possible to carry out a monoglyceride process because of the very high rates of oil bodying. Nevertheless, a proportion of any oil, including tung oil, may be incorporated directly by using the fatty acid–oil process. Here, fatty acid and oil are charged together, reacting directly with polyols and phthalic anhydride in a single stage during which sufficient transesterification can take place to make a satisfactory product. Another case where oil can be esterified directly is when unmodified castor oil is used, which in formulation terms is considered a triol because all three hydroxyl groups are reactive.

It will be clear from the foregoing that normal processing is either single-stage or, if a monoglyceride stage is carried out, two-stage. However, stepwise addition of components has been claimed to give narrower molecular weight and functional group distributions and lower viscosity. The original claims concerned better drying from tall oil alkyds [49–51], and the concept is potentially capable of being of benefit with high solids alkyds (see later).

An alternative process for incorporating oil into alkyd resins is the acidolysis process, in which oil is equilibrated with phthalic anhydride and the polyol is added during a second stage. This process is not widely practiced.

It has been claimed that some of the properties of alkyds are derived from the presence of "microgel" in the structure [52]. A physical explanation of gelation is that it occurs through separation of microgel particles, which at high concentrations coalesce to form a "macrogel." Otherwise, gelation is thought of as occurring when molecular weight becomes infinite, and continued reaction results in the development of a three-dimensional structure in the system. Practically, a gel is characterized by a significant presence of insoluble material.

2. *Alkyd Formulation and Manufacture*

As has already been implied, the basic requirement for the formation of an alkyd or polyester polymer structure is for a blend of molecules with two hydroxyl groups and a dibasic acid to be reacted. Clearly in a system with equimolar amounts of glycol and difunctional acid present, complete reaction would be capable of producing a polymer of very high molecular weight, limited in extent only by system viscosity and the ability to remove the water of reaction in manufacture.

Practically, mixtures of components are used, and very high molecular weight is not required. Inspection reveals that, taking all molecules into account, there will be a finite presence of polymer ends, and hence if we average the reacted number of all groups per molecule charged we will find that the value always falls short of 2 by a small amount. This concept of *mean functionality* of the system is the basis of formulating with mixtures of components [53,54], where in practical systems monoreactive acids such as benzoic acid, in particular, may be used satisfactorily, as may polyols of functionality 3 or 4. Since we have also implied that reaction overall may not be complete, a further consideration in both formulating and actual preparation is the extent of reaction. In typical formulations, there may be an excess of hydroxyl groups present, such that while the extent of acid group reaction may approach unity, that of the hydroxyls may be much less. Again, if reactants with more than two functional groups are present, and if the average functionality of all groups present is greater than 2, then if the extent of reaction is taken to a high level, gelation can very possibly occur.

Practical formulating will involve consideration of both stoichiometry and oil length and calculation of the extent of reaction at gelation using equations derived from Flory–Stockmeyer theory, of the type

$$p^2 = \frac{(\Sigma gB)^2}{(\Sigma f^2A - \Sigma fA)(\Sigma g^2B - \Sigma gB)}$$

where p is the degree of reaction of acid groups at gelation; f, g are functionalities of carboxy and hydroxy moieties, respectively; and A, B are the number of moles of carboxy and hydroxy moieties present.

The effectiveness of these equations has been tested [55]. It should be noted that it will normally be necessary to consider phthalic anhydride as having a functionality of less than 2 because of its nonideal behavior. This includes a tendency to form oligomeric ring compounds and also to exhibit some reversibility in its reaction, and theory has developed to take this into account.

Those formulating alkyds are likely to use computer programs to carry out the necessary calculations. Programs are available for purchase [56], and from raw material suppliers [57].

Alkyd resin manufacture requires high temperatures, and while esterification processes are only mildly endothermic, more significant heat input is required to remove water of reaction in the polymerization stage. Also, heat-up rates need to be reasonably fast for consistency in manufacture, and for this reason heating systems need to be quite powerful. Alkyd resins have been made in primitive equipment, venting water and fumes to containment systems; in this case, formulations need to allow for a loss of around 5% of such volatile ingredients as phthalic anhydride.

Better control is achieved using equipment with provision for using xylene as azeotroping solvent, where xylene boiled from the batch is condensed, water is separated, and cold xylene is returned to the batch. Good design is necessary to prevent overheads from being blocked by phthalic anhydride sublimate, and plants variously have fractionating columns or bubble cap scrubbers to overcome this.

Polymerization stages need control with frequent sampling, testing for viscosity rise and acidity fall (measured by acid value). Endpoint prediction may be employed by graphing results, possibly using guide lines from previous batches. Shorter oil alkyds, which are particularly prone to gelation, need the most frequent sampling.

At the endpoint, resins require cooling to stop further reaction and will be thinned to maintain handleability, using the appropriate solvent.

3. *Modified Alkyds*

Although some resin additives have already been discussed, some further alkyd modifications merit mention. These are the ones recognized as significant in terms of imparting special or unique properties.

The use of a thixotropic additive is a useful enhancement to brush-applied paints to give nondrip and improved film build properties, and alkyds may be made thixotropic by modification with polyamide resin. The resins used are simple linear structures with diamine and dimerized fatty acids. It is normal practice to prepare a long oil thixotropic alkyd with a higher level of polyamide modification (5%) for blending with other resins to give the required properties. In practice, great care is required in selecting the polyamide resin as well as skill in formulation and preparation [58,59]. It may be difficult to characterize the resin in rheological terms, as much subjective judgment is required in properties such as brush loading and brush mark flow-out. Modification is carried out after alkyd condensation, and strict time, temperature, and endpoint control are necessary to gain optimum clarity, viscosity, and gel strength, with the most critical thixotropic properties peaking and then falling off during preparation.

Urethanation involves the replacement of some of the difunctional acid with a diisocyanate, with toluene diisocyanate (TDI) the most common. In resin terminology, a urethane oil is a resin formulated with complete replacement of PA by TDI, whereas a urethane alkyd is made with partial replacement. The latter is the more usual. In this case, a two-stage (or strictly three-stage if monoglyceride) process is used in which an alkyd is first prepared to low molecular weight and acid value but with a significant excess of hydroxyl groups. Reaction with diisocyanate is then a further stage, carried out at lower temperature. As it is essential to react isocyanate groups to completion, the reaction is principally controlled at the alkyd-making stage. If a monoglyceride stage is included, care is needed to avoid using a catalyst that will cause side reactions of isocyanate groups in the urethanation stage. These resins are harder and tougher than unmodified alkyds, and for this reason they are blended with other alkyds to improve these properties.

Styrenation (vinylation) of oils has already been described. It is used to improve "lacquer dry" and hardness in medium oil alkyds for maintenance paints and primers. The process is as described earlier. Although drying is improved, durability may be adversely affected. This modification of principally DCO alkyds is carried out on the thinned resin after the alkyd preparation stage is complete.

Silicone resin modification may be used to improve heat resistance and weatherability in both alkyds and other resin types such as polyesters. Although it is an expensive ingredient, silicone resin will give improvement in direct proportion to the amount used [60]. These modified resins are either hydroxy or methoxy functional and are generally partly reacted into the alkyd after its own condensation is complete.

4. *Use of Alkyds in Paint*

Long oil drying oil alkyds remain the major vehicles used in glossy exterior paints and varnishes, and in volume terms they are by far the largest volume oil-containing resins now made. These alkyds were developed with around 70% linseed oil, with PE and PA as the other components. With the use of white paints increasing, partly due to the development of nonchalking grades of rutile titanium dioxide, formulation has changed to the use of lower yellowing semidrying oils, notably soybean oil. Soybean oil alkyds are generally formulated with a significant proportion of soybean stand oil in the composition. The use of linseed oil alkyds is now reserved for undercoats and varnishes. Other semidrying oils such as sunflower and safflower may also be used.

Soybean oil is widely available, with its major use in food, but naturally suffers some seasonal variation in yield and therefore in price. Because of this variation, distilled tall oil is now used as an alternative to soybean oil. For applications in decorative alkyds particularly, where tight cost control is necessary, manufacturers may have formulations with tall oil (TOFA) available as alternatives to soybean oil resins so they can switch as price varies. Since tall oil does not contribute any glycerol to the formulation, long oil tall alkyds may be made with all PE and by the fatty acid process. Higher temperatures are then used.

Long oil alkyds of this type are of relatively low molecular weight and viscosity, used at high solids, and diluted with an aliphatic solvent such as white spirit. Decorative paint is pigmented with titanium dioxide for white paints, with inorganic pigments rather than the more expensive organic pigments used where possible to provide color. The most expensive high durability organic pigments are generally too costly for use in decorative paints.

Decorative paints are generally formulated with a blend of alkyds and include urethane alkyd to toughen, thixotropic alkyd to improve application rheology, and possibly another modified alkyd to provide, for example, easy brush cleaning. These have been described previously.

These paints are generally applied by brushing, mainly to wood substrates (doors, windows, trim) in interior and exterior building decoration, but also to metal where appropriate. Long oil alkyd base paints of this type are also used in commercial transport paints and in maintenance of marine structures.

Medium and short oil alkyds have for many years been the workhorses of almost all industrial coatings markets, and they remain so despite being displaced by polyesters and acrylics for spray-applied top coats and by epoxies for more corrosion resistant primer and undercoat vehicles. The range of oils used is extremely broad, all available polyols and acids are to be found, and the range of modifiers is extensive. To illustrate this, some typical applications are described below.

Medium oil semidrying oil alkyds are still used for spray-applied vehicle refinishing as an alternative to lacquers where two-pack finishes cannot be used. Since rapid drying to a good hardness is required, the formulations may include isomerized fatty acids and chain-stopping acids such as benzoic or *para-tert*-butylbenzoic acids (ptBBA). These will be partly aliphatic-soluble; both the use of ptBBA and trimethylol propane (TMP) may be used to increase this solubility by making the polyester part of the structure more aliphatic in character. The overall alkyd properties are greater hardness and durability and faster drying speed than long oil alkyds.

Medium oil alkyds made from a range of semidrying oils have been used in a wide range of industrial stoving paints. For example, light fittings and the full range of domestic appliances (refrigerators, etc.) may use stoving paints with soybean oil, dehydrated castor oil (DCO), and tall oil alkyd, along with urea formaldehyde (UF) resin or melamine formaldehyde (MF) for higher performance. Alkyd resins of this type can be made less expensive by incorporating rosin, which will increase drying speed and hardness but decrease durability. Performance can be enhanced by using isophthalic acid in place of PA and by careful choice of polyol. The range of polyols includes glycerol, PE, TMP, neopentyl glycol, and other glycols. Nowadays this kind of composition competes with thermosetting acrylics, which (admittedly with the use of more expensive and specialized monomers) have better performance regarding detergent resistance, for example.

Blending or partial reaction with phenolic or epoxy resins can be used to improve chemical resistance, though these will reduce both yellowing and weathering performance.

Short oil coconut oil alkyds were used with MF resins to provide factory-applied stoving automotive top coats and for some appliances. For automotive use, these are now fully supplanted by acrylic resins, which can give the rheology control necessary in application for metallic finishes, but other industrial uses remain.

Castor oil alkyds have been used in industrial stoving appliance finishes and, formulated to higher hydroxyl content, still have particular use as plasticizing resins in nitrocellulose lacquers.

5. *Waterborne Alkyds*

Although the ability to modify alkyds to confer partial or total water solubility has been well used over many years, this has been explored further with recent trends to overall solvent reduction, where the change to water as diluent achieves this aim. Alkyds can be made water-dispersable by the incorporation of the hydrophilic polyethylene glycol, as a polyol, into the alkyd structure [61]. This has been used to produce alkyds that confer water brush washability in decorative paints and to produce alkyds with surfactant properties for a variety of industrial cleaning purposes [62].

To attain higher levels of dispersability or complete water solubility, highly polar groups, which will in practice need to be ionizable, will have to be incorporated in the resin, and these are typically acid groups. Hence the alkyd is formulated to a much higher acid value than normal, using, for example, maleinized fatty acid or using trimellitic anhydride in place of phthalic anhydride. In the latter case, a technique to get high acidity involves a two-stage process whereby terminal hydroxyl groups are reacted with trimellitic anhydride in a final "ring-opening" stage, where the anhydride ring is reacted with one hydroxyl to give a single ester group and two acid groups.

These resins require preparation in the absence of water and will be dissolved or dispersed subsequent to polymerization. This is of advantage in circumstances where another resin such as a cross-linking resin needs be coemulsified with the solubilized resin.

The presence of ionizable groups is not in itself sufficient to confer water solubility, and in the case of acid groups, partial or full neutralization is necessary to confer full hydrophilic properties. Ammonia, amines [63], or alkalis may be used, depending on the application. The former clearly will be released and hence help to lower water sensitivity of the final film but may present toxicity problems. In the case of alkali neutralization, since they will remain in the film, cross-linking mechanisms need to be sufficiently effective to make the film fully insoluble. Their use reduces the resin's VOC and avoids problems caused by amine interfering with autoxidative drying.

Dimethylol propionic acid (DMPA) is one material specially exploited in making this type of resin, where the hydroxyl groups react into the resin structure but the acid groups, normally reluctant to react, remain for subsequent salt formation. Recent use has been made of the alkali metal salts of sulfo-isophthalic acid, claimed to be very effective at imparting solubility and dispersability [64]. These two structures are shown in Fig. 9.

While acid groups have been described here as the groups to confer solubility, with other resin types amine groups are an alternative to give cationic rather than anionic resins. These are not, however, easy to obtain in an alkyd structure, and their use has been with acrylic and epoxy resins.

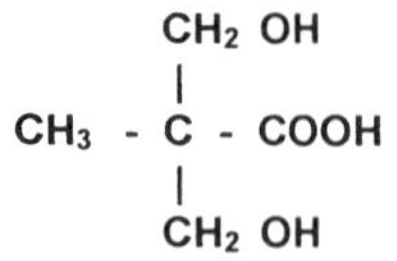

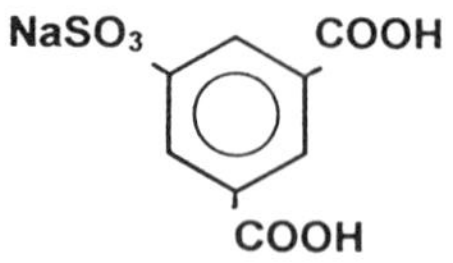

Fig. 9 Structures of dimethylol propionic acid and a sodium salt of sulfo-isophthalic acid.

A drawback to using alkyds as waterborne vehicles is that the alkyd backbone is prone to hydrolysis, and for this reason, to aid storage stability, water dispersability rather than increased water solubility is often more practical. Hence alkyd emulsions are being explored as solvent-free paints. While these may be prepared by blending surfactant with the alkyd prior to emulsification, the use of reactive surfactant reacted into the alkyd structure is more satisfactory [65]; the resin may include ionizable groups as described above. The phase inversion temperature method is the most efficient for emulsification [66].

Although there are problems in the use of these water-soluble resins, with hydrolyzability and with the inability to get the highest gloss finishes, it should be noted that alkyd emulsions have been used successfully in a number of European countries. Novel techniques have evolved for their preparation, and microemulsion techniques have been examined [67]. Certain industrial products have been made with water-based technology, and for many years alkyd-based waterborne primers, especially dip primers, were used on larger structures such as car bodies.

6. *High Solids Alkyds*

As an alternative to full removal of solvent and the use of water to attain approval for certain environmental labeling requirements (e.g., Blue Angel), some current formulations are supplied at a higher solids content with smaller amounts of solvent required to achieve application properties. "High solids" is generally taken to mean >85% solids in a brush-applied paint.

To get high solids in alkyds, it is necessary to formulate to higher oil length, since viscosity constraints require lower molecular weight, which in turn requires a greater extent of reaction to achieve cross-linking. To this end, compositions may contain higher functionality synthetic oils, for example, full fatty acid esters of polyols such as di- and tripentaerythritol [68,69]. Considerable attention must be given to technique as well as design, since a narrower molecular weight distribution is very desirable; it is claimed that this is achieved by replacing phthalic anhydride by isophthalic acid [70].

An additional technique for formulating high solids alkyds is to add "involatile monomer," exploiting the ability of the free radical nature of cross-linking to copolymerize other unsaturated compounds into the structure [71]. Although this is not a recent discovery, a number of materials have been used in this modern application. In addition to vinyl monomers, allyl ether compounds have been used as diluents. The allyl ether group has autoxidative ability and has been extensively investigated as a synthetic replacement for the drying oil moiety [72]. As a reactive diluent an oil-like structure, pentaerythritol diallyl ether di-TOFA ester, has been used.

Finally, with considerable effort being made to control polymer architecture in other polymer systems, such as group transfer polymerization (Du Pont), and in making dendrimers and hyperbranched structures [73], dendritic structures have been synthesized as high solids alkyds [74] based on a PE/DMPA core, with the many hydroxyl groups of the final product esterified with fatty acid.

C. Epoxy Esters

The third major class of coatings, which have been made in significant volume in one particular application, is epoxy esters, formed by the reaction of fatty acids with epoxy resins [75]. Epoxy resins are made by the reaction of epichlorohydrin with diphenylol propane (also known as Bisphenol A) and have a structure with both in-chain hydroxyl groups and terminal epoxy groups. Although their preparation is difficult and hazardous owing to the toxicity of epichlorohydrin, a range of these resins are available from the "monomeric" form with $n=0.5$ to grades with $n=12$ or more. A generalized form of this structure is shown in Fig. 10. It is common practice now for larger users to "chain extend" liquid grades with diphenylol propane to higher molecular weight.

Both the terminal epoxy groups and the secondary hydroxyl groups can be reacted with fatty acid, and the resin with $n=4$ is typically used. The reaction is carried out at 240–260°C under reflux to an acid value and viscosity endpoint in a manner similar to that used with an alkyd.

It is possible to use the full range of fatty acids, though linseed and DCO are most common. Although a nomenclature similar to that used for other oil-containing resins may be used (high, medium, short oil length), there is another specific code in which, for example, "L8 resin" refers to a 4-type epoxy resin modified with 0.8 equivalent of linseed oil fatty acids to 1.0 epoxy. These "straight" epoxy esters have been used for primer formulations and a range of industrial finishes in which the higher chemical resistance of the epoxy backbone is of benefit.

Medium and long oil esters of drying oil fatty acids have been used in air-drying finishes, while short oil drying and nondrying fatty acid esters are used in industrial stoving primers and finishes. Stoving finishes frequently include formaldehyde resins and phenolic resins to provide cross-linking and further modify properties.

Water-thinnable epoxy esters have been prepared by a number of routes. An epoxy resin ester emulsion for use in water-thinned primers can be made by an inversion emulsification process in which water is added with stirring to a D4 (DCO-containing) epoxy ester that has been preblended with amine and linseed oil fatty acids. An alternative method is to prepare an epoxy ester using maleinized linseed oil fatty acid, and this kind of resin has been made in high volume for use in dip primers, particularly automotive electrocoat primers [76]. In this latter process the anionic resin, which will have been partly neutralized again with alkali or amine, is deposited onto a metallic object by passing a current through a bath containing the paint. In this case the object (car body) is made the anode of the electrical system and resin is deposited onto it and becomes insolubilized; as film resistance increases deposition slows and film thickness is relatively even. This resistance increase also promotes "throw," whereby paint is deposited inside box sections of chassis (where spray or brush processes cannot reach). These resins are stoved to harden them completely and are pigmented with anticorrosive pigments. In the main, this process has been superseded by the cathodic process using non-fatty acid–containing resins.

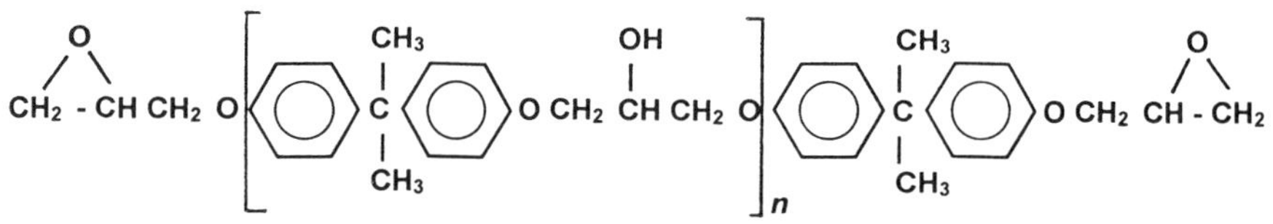

Fig. 10 Bisphenol epoxy resin structure.

D. Other Oil Applications in Coatings

Castor oil, by virtue of its hydroxyl content, has also been used as a polyol. Hence it has served as a component of polyurethane resin, where it is reacted with an isocyanate such as toluene diisocyanate.

As mentioned earlier, 12-hydroxystearic acid can be self-condensed to a limiting molecular weight by virtue of its residual stearic acid content, and this polymer is the essential aliphatic-soluble oligomeric component of a range of graft copolymers [20]. These have been the key to the manufacture of nonaqueous dispersion (NAD) polymers [77]. This versatile, though now lesser exploited, technique has been employed to prepare both addition and condensation polymers, used as rheology modifiers in automotive paints and currently in woodstain formulations [78]. The graft copolymers have also found separate application in industrial detergents and cleaners [60].

A number of attempts have been made to incorporate drying oil fatty acids into addition polymer acrylic resin backbones to prepare an analogue of the drying oil alkyd resin, but the results have not had significant commercial success. Saturated fatty acid derived monomers are, however, used in significant quantities in acrylics, notably lauryl acrylate and methacrylate and to a much lesser extent stearyl methacrylate.

VII. CURRENT STATUS AND FUTURE TRENDS

Current environmental concerns and legislation, principally targeted to reduce and ultimately eliminate solvent emissions, expressed as VOC, have enabled oil-based polymers to continue to find use through some reformulation to higher solids and in water-emulsified form.

As already described, drying oils remain nearly unique in their ability to autoxidize to "convert" a liquid polymer to a solid cross-linked durable film at room temperature, despite some adverse features of the process. Only the allyl grouping, incorporated as either the allyl ether or the vinyl cyclic acetal, has shown useful autoxidative ability [72], and research into other synthetic alternatives may still be warranted. Although there are side effects from the use of coatings containing drying oil in their contribution to "true" VOC via emissions from their autoxidative curing, they are currently tolerated because of their history of consumer use over many years. Again, there is currently no alternative to the use of drying oil alkyds in the formulation of high gloss trim paints, so that while consumer demand for these continues, their use seems ensured.

With environmental awareness now a permanent force in shaping both consumer acceptance and industrial practice, and with developments being steadily enforced by legislation, the movement to high solids and particularly waterborne coatings [79] looks set to continue. There is now consumer pressure to use renewable rather than petroleum-based resources. On cost grounds, oils compete well with the petrochemically derived ingredients normally used to form polymers. In certain cases, pressures are triggering some new applications where there is demand for ecologically responsible products; one is the replacement of mineral oil in inks with rapeseed oil; in the United Kingdom, 1000–2000 t is already used annually [10]. This type of substitution will undoubtedly continue.

One further factor in materials use is the demand for recycling of products, where "cradle-to-grave" life cycle studies are being carried out [80, 81]. These studies can include energy requirement calculations for all components used in manufacture. Some goods are being redesigned so they can be fully recycled, and legislation is beginning to enforce this. It might be added that the role of paints is often "green" in greatly prolonging the life of products, preventing their premature scrapping and hence the energy and environmental burden of providing new structures and scrapping the old ones. The use of a few micrometers of paint properly applied externally plus appropriate inner surface treatments enormously extend the lifetimes of motor bodies, for example. It is difficult to imagine this paint coating itself being recyclable, but certainly changes in ingredient use, already shown by restricting use of lead and chromates, will continue. Recycling does, however, affect paint waste and used containers. "Cradle-to-grave" studies on paint itself are inspiring ecolabeling that will embrace performance requirements, toxicity of ingredients, and limitations on the release of volatiles as VOC.

Where oils are concerned, with crop yields already improved by selection and by farming methods, genetic engineering looks set to change fatty acid compositions and produce new varieties. [82,83] The effects of this will become evident in future years.

ABBREVIATIONS

DCO	Dehydrated castor oil
DCOFA	Dehydrated castor oil fatty acid
DMPA	Dimethylol propionic acid
EXAFS	Extended X-ray fine structure analysis
FTIR	Fourier transform infrared spectroscopy
LOFA	Linseed oil fatty acid
MF	Melamine formaldehyde
NAD	Nonaqueous dispersion
NMR	Nuclear magnetic resonance
OEM	Original equipment manufacture
PA	Phthalic anhydride
PE	Pentaerythritol
PF	Phenol formaldehyde
ptBBA	*para-tertiary*-Butylbenzoic acid
SAXS	Small-angle X-ray scattering
SIMS	Secondary ion mass spectroscopy
TDI	Toluene diisocyanate
TMP	Trimethylol propane
TOFA	Tall oil fatty acid
UF	Urea formaldehyde
VOC	Volatile organic content

REFERENCES

1. R. Lambourne (Ed.), *Paint and Surface Coatings*, Ellis Horwood, London, 1987.
2. D. H. Solomon, *Chemistry of Film Forming Polymers*, Krieger, 1977.
3. J. Bentley, Organic film formers, in *Paint and Surface Coatings* (R. Lambourne, Ed.), Ellis Horwood, Chichester, England, 1987, Chap. 2.
4. Surface Coatings Association of Australia, *Surface Coatings*, Vol. 1, *Raw Materials and Their Usage*, 3rd ed., Chapman and Hall, London, 1993, Chaps. 4 and 5.
5. A. Kimber, Recovery hampered by raw material hikes, *Eur. Chem. News*, 14 Nov. 1994, pp. 20–21; data from Akzo Nobel.
6. *CEPE Annual Statistical Review 1994*, presented at CEPE Annual Conference 1994.
7. M. S. Reisch, Paints and coatings, *Chem. Eng. News*, 18 Oct. 1993, p. 34.
8. Japanese Paint Manufacturers Association. (Data from *Coatings COMET*, No. 3, 1994, p. 65).
9. British Coatings Federation, *U.K. Coatings Industry Facts & Figures 1994*.

10. *Industrial Markets for U.K. Produced Oilseeds*, Home Grown Cereals Authority, U.K., Research Review No. OS9, January 1995.
11. J. Mispreuve, *Oil Fats Int. 10*: 3 (1994).
12. K. S. Ennar and J. Oxley, Composition and fractionation of tall oil, *J. Oil Colour Chem. Assoc. 50*: 557 (1967).
13. A. Hase and S. Pajakkale, Tall oil as a fatty acid source, *Lipid Technol. 6*: 110 (1994).
14. P. Muturi, Epoxidised vegetable oils as reactive diluents 1. Comparison of vernonia, epoxdised soya and linseed oils, Prog. Org. Coatings 25(1): 85 (1994).
15. L. F. Byrne, The chemistry, manufacture, and uses of dimerised fatty acids, *Paint Technol. 26*(3): 28 (1962).
16. H. R. H. Hoffmann, The ene reaction, *Angew. Chem. Int. Ed. 8*: 556 (1969).
17. H. W. Chatfield, *Varnish Constituents*, Leonard Hill, London 1953, p. 76.
18. R. Lambourne, The dehydropolymerisation and copolymerisation of drying oils and other reactive compounds exhibiting α-methylenic activity, *J. Oil Colour Chem. Assoc. 51*: 310 (1968).
19. H. W. Chatfield, *Varnish Constituents*, Leonard Hill, London, 1953, p. 47.
20. F. A. Waite, The development of graft copolymers as macromolecular surfactants for use in non-aqueous media, *J. Oil Colour Chem. Assoc. 54*: 342 (1971).
21. P. L. Nichols, S. F. Herb, and R. W. Reimenschneider, Isomers of conjugated fatty acids. I. Alkali-isomerised ricinoleic acid, *J. Am. Chem. Soc. 73*: 247 (1951).
22. N. A. Khan, Application of tracer techniques to studies on autoxidation reactions, *J. Am. Oil Chem. Soc. 30*: 273 (1953).
23. T. L. Helmon and M. Möller, Förändringar i Klarlackers Kemiska och Fysikaliska Egenskaper vid Utomhus-och Laboratorieprovning 5b. IR-spektroskopiska studier av härdnings-och nedbrytningsförlopp, *Färg Lack 13*: 155 (1967) (in Swedish).
24. J. H. Hartshorn, Time-lapse infrared spectroscopic investigation of alkyd and linseed oil cure, *J. Paint Technol. 54* (687): 53 (1982).
25. W. Z. Muizebelt, J. C. Hubert, and R. A. M. Venderbosch, Mechanistic study of drying of alkyd resins using ethyl linoleate as a model substance, *Prog. Org. Coatings 24*: 263 (1994).
26. C. Carr, I. S. Dring, and N. A. R. Falla, Studies of the curing of soya oil/alkyd films containing metal driers, XVI Int. Conf. Org. Coat. Sci. Technol., Athens, July 1990, p. 99–109.
27. H. Wexler, Polymerisation of drying oils, *Chem. Rev. 64*(6): 591 (1964).
28. N. A. Khan, Some specific features in autoxidation of methyl linoleate, *Can. J. Chem. 37*: 1029 (1959).
29. W. J. Bailey and G. L. Barlow, Preparation of crystalline linoleate hydroperoxide for drying oil research, *J. Paint Technol. 42*(544): 287 (1970).
30. M. Loury and M. Forney, Étude de l'autoxidation de l'acide linolénique, *Rev. Corps Gras 16*: 167 (1969).
31. M. Loury and M. Forney, Étude de l'autoxidation de l'acide linoléique, *Rev. Corps Gras 15*: 663 (1968).
32. E. N. Frankel, *Lipids 16* (5): 279 (1981).
33. R. A. Hancock, N. J. Leeves, and P. F. Nicks, Studies of autoxidation. Part I. The volatile by-products resulting from the autoxidation of unsaturated fatty acid methyl esters, *Prog. Org. Coat. 17*: 321 (1989).
34. N. J. Leeves, Autoxidative degradation of unsaturated fatty acid esters, Thesis, Royal Holloway College, Univ. London, 1985.
35. L. A. O'Neill and N. R. Falla, Structure of the cobalt drier complex, *Br. Polym. J. 5*: 115 (1973).
36. W. A. Waters, The kinetics and mechanism of metal-catalysed autoxidation, *J. Am. Oil Chem. Soc. 48*: 427 (1971).
37. J. H. W. Turner and P. Womersley, Aluminum compounds as major components of surface coatings, *J. Oil Colour Chem. Assoc. 62*: 419 (1979).
38. J. H. W. Turner and P. Womersley, Coordination drying: an alternative to catalysed autoxidative polymerisation for paints, *Chem. Ind.*, March 1975, p. 203.
39. K. H. Zabel, J. C. Hubert, W. J. Muizebelt, and R. A. M. Venderbosch, Singlet oxygen cross-linking in high solids alkyd paints, *ACS Polym. Mater. Sci. Eng. 70*: 64 (1994).

40. ICI, Eurp. Patent Spec. 0 009 356 (1980).
41. ICI, Germ. Off. 2,350,646 (1974).
42. R. A. Hancock, N. J. Leeves, and P. F. Nicks, Studies in autoxidation, Part II. An evaluation of the by-products formed by the autoxidation of fatty acid modified alkyd resins under the influence of different promoters, *Prog. Org. Coat. 17*:337 (1989).
43. O. S. Privett, M. L. Blank, J. B. Covell, and W. O. Lundberg, Yellowing of oil films, *J. Am. Oil Chem. Soc. 38*: 22 (1961).
44. S. E. L. K. Davison, Chemical processes accompanying the autoxidation of paint films, Thesis, Royal Holloway and Bedford College, Univ. London, 1989.
45. H. G. McAdie and R. V. L. Nicolls, in *Film Formation and Film Degradation* (C. R. Braydon, Ed.), Interscience, New York, 1958.
46. T. Helmen and M. Möller, series of 15 papers entitled Förändringar i Klarlackers Kemiska och Fysikaliska Egenskaper vid Utomhus-och Laboratorieprovning, with the last in *Färg Lack 15*: 155 (1969) (in Swedish).
47. H. W. Chatfield, *Varnish Constituents*, Leonard Hill, London 1953, Chap. 4.
48. J. F. Rooney, Practical application of alkyds, *Off. Dig.*, August 1964, Part 2, p. 32.
49. W. M. Kraft, High polymer alkyd technique, *Am. Paint J. 41*(28): 6 (1957).
50. H. Tobita and Y. Ohtani, Control of molecular weight distribution in step growth polymerisation: the intermediate monomer feed method, *Polymer 33*:801 (1992).
51. H. Tobita and Y. Ohtani, Control of molecular weight distribution in step growth polymerisation by an intermediate monomer feed method: effect of interchange reactions, *Polymer 33*: 2194 (1992).
52. E. G. Bobalek, E. R. Moore, S. S. Levy, and C. C. Lee, Some implications of the gel point concept to the chemistry of alkyd resins, *J. Appl. Polym. Sci. 8*: 625 (1964).
53. L. A. Tysall, *Calculating Techniques in the Formulation of Alkyd and Related Resins*, Paint Research Association, Teddington, 1982.
54. T. C. Patton, *Alkyd Resin Technology: Formulating Techniques and Alkyd Calculations*, Interscience, New York, 1962.
55. J. J. Bernardo and P. Bruins, Gelation prediction and related concepts, *J. Paint Technol. 40*: 558 (1968).
56. P. J. C. Nelen, Recon 36X, a computer program for the design of polycondensation resins, 17th FATIPEC Congress 1984, v.4, p. 283–298.
57. AMOCA Chemicals Company, "Polycalc II," polyester characterisation computer program, Tech. Rep. No. GTSR48; *Perstorp Polyols, Guidelines to Perstorp Alkyd Calculations—Percalc* 3.0, Tech. Info. Leaflet No. 0005.
58. W. E. Shackleford and D. W. Glaser, Thixotropic modification of alkyd resin paints with polyamides, *J. Paint Technol. 38*: 293 (1966).
59. I. Garbett, Long life air drying paints, P.R.A. 2nd Middle East Conference, Bahrein 1995, Paper 3.
60. J. C. Martin, Silicone copolymers, *Paint Polym. Colour J. 177*(4182): 20 (1987).
61. ICI, U.K. Patent 1,370,914 (1972).
62. A. S. Baker, Applications of polymeric surfactants, *Paint Resin Manuf.*, Mar/Apr. 1982, p. 37.
63. C. A. Hare, Anatomy of paint—alkyd resins, *J. Prot. Coatings Linings*, February 1994, p. 97.
64. Eastman Chemicals, U.S. Patent 5,378,757 (1995).
65. K. Holmberg, Polymerisable surfactants, *Prog. Org. Coatings 20*: 325 (1992).
66. D. Fenstel, Emulsifier systems for modern surface coatings, *Surface Coatings Australia*, January/February 1994, p. 6.
67. G. Österberg, M. Huldén, B. Bergenståhl, and K. Holmberg, Alkyd emulsions, *Prog. Org. Coatings 24*: 281 (1994).
68. BASF, Eur. Patent 87/00589 (1987).
69. ICI, Ger. Offen. 2,818, 102.
70. L. W. Hill and Z. W. Wicks, Design considerations for high solids coatings, *Prog. Org. Coatings 10*: 55 (1982).
71. K. E. J. Barrett and R. Lambourne, Air drying alkyd paints, *J. Oil Colour Chem. Assoc. 49*: 443 (1966).

72. K. B. Gilbes and T. Hunt, New oils and fatty acids for surface coating resin manufacture, *J. Oil Colour Chem. Assoc. 51*: 399 (1968).
73. R. Engel, Cascade molecules, *Polym. News 17*: 301 (1992).
74. B. Pettersson and K. Sorenson, High-solid alkyds based on hyperbranched (dendritic) polymers—a new concept with new opportunities, Proc. Waterborne, High Solids and Powder Coatings Symposium, New Orleans, 1994, p. 1–13.
75. Shell Chemicals, *Epikote Resins for Paints*, publication Res: 64:4, Shell Chemicals, London.
76. E. I. DuPont de Nemours, U.S. Patent 4,042,478 (1977).
77. K. E. J. Barrett (Ed.), *Dispersion Polymerisation in Organic Media* Wiley, New York, 1975.
78. C. W. A. Bromley, Non-aqueous dispersion polymer microgels as film formers in architectural paints, *J. Coatings Technol. 61*(768): 39 (1989).
79. P. K. Nielsen and J. H. Hanson, Paint and pollution—a question of solids, *Färg Lack Scand. 6*:113 (1992).
80. I. Särvimäki, The implications of life cycle assessment for R&D in the paint industry, *Surf. Coatings Int. 1994*(8): 339.
81. P. J. A. Geurink and E. L. J. Bancken, Life cycle assessments of decorative paints, PRA 12th Int. Conf, Waterborne Coatings, Milan, 1992, Paper 12.
82. E. M. S. van Hamersueld, F. P. Cuperus and J. T. P. Derksen, Innovation in paint technology using new vegetable oils and novel concepts, PRA 15th Int. Conf, The Future of Industrial Coatings, Brussels, 1995, Paper 7a.
83. J. T. P. Derksen, F. P. Cuperus, and P. Kolster, Renewable resources in coatings technology: a review, *Prog. Org. Coatings 27*: 45 (1996).

28

Lubricants

Theo Mang
Fuchs Petrolub AG Oel + Chemie, Mannheim, Germany

I. INTRODUCTION

Environmental awareness, the desire to protect resources, and the aim of agricultural producers to use available capacities for the cultivation of technical raw materials—harvestable raw materials—have led to a new generation of lubricants and functional fluids. However, it is increasing environmental awareness that has been most instrumental in accelerating the development of rapidly biodegradable lubricants and greases.

Although lubricants are not especially hazardous from an environmental point of view, their innumerable applications make them omnipresent. Mineral oil, representing the largest single component in lubricants (>90%), and some chemical additives are environmentally undesirable because of their poor biodegradability. This is particularly true from the point of view of protecting soil and groundwater.

Mineral oil products were not always used as the base substance during the long history of lubrication technology, or tribology, as the scientific study of friction, wear, and lubrication is called today. Until the second half of the 19th century, nearly all lubricants were based on natural vegetable or animal oils and fats. Only after its industrial production and refining was mineral oil used as a base oil for lubricants. Refined mineral oils displayed much better thermal and oxidative stability than natural fatty oils. With the development of modern engineering practices and the help of chemical additives, it became possible for these "inert" mineral oils, which alone are not particularly good lubricants, to deliver the desired properties. The additive technology developed since the early 1950s has now gained a thoroughly "high tech" character. It appeared that mineral oil was seen as a solvent for the chemical additives and that the disadvantages of mineral oil compared to natural fatty oils could be more than balanced out.

Fortunately, lubricant developers never forgot the benefits of natural fats and their derivatives; some are still used in the form of refined oils and chemically modified substances as additives because to their good lubricity. In the area of metalworking oils in particular, the use of natural raw materials never went out of favor. As metalworking, wire drawing, and press oils form a particular focal point in some specialized laboratories, the use of rapeseed oil, oxidized rapeseed oil, rapeseed oil esters, and other natural additives was never abandoned.

As long ago as the late 1970s, the lube industry began developing environmentally harmless and non-workshop-polluting metalworking oils. This work was successfully continued in the 1980s within the scope of large-scale projects sponsored by the Federal Ministry for Research and Technology in Germany, for example.

At the beginning of the 1970s, at the time of the oil and energy crisis, scientists throughout the world began investigating the possibilities of using harvestable "alternative" raw materials in order to conserve resources. Much of the work done at this time was put aside when oil prices fell. Natural fatty base oils for lubricant development regained favor in the early 1980s, and this fit well into development programs for environmentally harmless and non-workshop-polluting products.

II. THE SEARCH FOR ENVIRONMENTALLY FRIENDLY LUBRICANTS—THE DRIVING FORCE BEHIND THE USE OF VEGETABLE OILS AND THEIR DERIVATIVES

Apart from the efforts made in agriculture to find new areas of application in non-foodstuff sectors, the search for environmentally harmless lubricants is the main reason vegetable oils and their derivatives are used as base oils. To understand the development of lubricants based on natural fatty oils and their derivatives, one must examine the criteria used to define the environmental compatibility of lubricants. It then becomes clear that vegetable oils and their oleochemical derivatives have significantly better chances than petrochemical alternatives even though industrially manufactured synthetic carboxylic acids have produced good results in lubricants. In terms of cost also, natural oils and their derivatives are thoroughly competitive.

More than one-third of lubricants sold end up polluting the environment either via total-loss applications, spillages, evaporation, or in other ways (Fig. 1). In western Europe, 1.5–

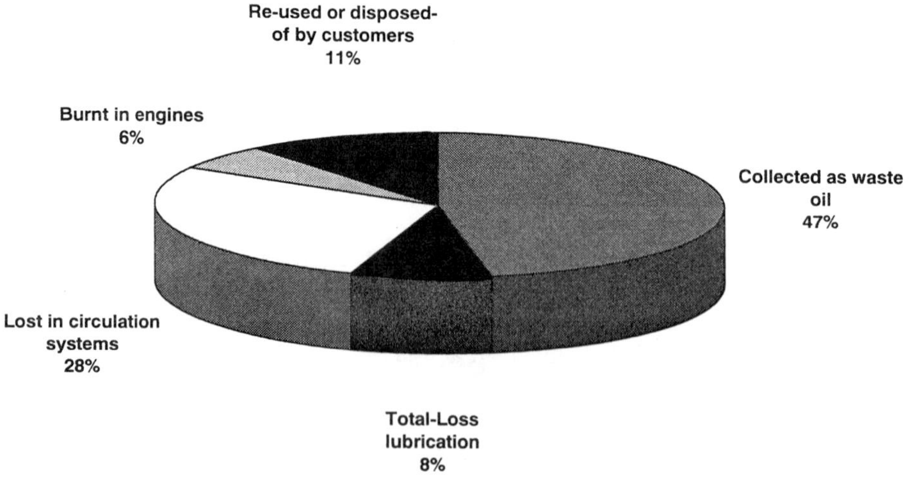

Fig. 1 Lubricant consumption and lubricants in the environment in Western Europe.

2.0 million tonnes per annum (tpa) pollute the environment. This fact and ever-increasing environmental awareness led to the development of environmentally harmless lubricants at the beginning of the 1980s. "Environmentally harmless" means in particular, rapidly biodegradable and ecologically nontoxic. A significant increase of such lubricants is expected in the next few years, helped to some extent by the agricultural/political situation in Europe. Agriculture will get the opportunity to produce technical raw materials.

Rapidly biodegradable two-stroke outboard marine engine oils were developed as long ago as the late 1970s for ecological reasons. But the use of rapidly biodegradable base oils for other lubricants only began in the early 1980s. Predominant among these were total-loss lubricants based on vegetable oils, especially rapeseed oil. Since then, the additive industry has supported these developments with environmentally harmless lube additive packages.

Lubricant manufacturers and users are now in a continuing process of incorporating current know-how into further product improvements and adapting machine designs to this new generation of lubricants [1–4].

A. Biodegradability and Additional Environmental Requirements

The meaning of "environmentally friendly" or "ecologically compatible" has been subject to much debate over the last few years, with existing conventional lubricants used as the yardstick. With regard to ecology, lubricants are not a particularly critical group of substances, but they are omnipresent, and mineral oil as well as some chemical additives are not desired in the environment. "Bio-oils" are also a good idea if they are used in biologically inactive areas, thus providing additional groundwater protection.

B. Biodegradability

Rapid biodegradability is a generally desirable feature for substances that may eventually enter the soil and waters and also for biological purification plants.

Evaluations are based on laboratory procedures because the numerous systems presently used have greatly differing microbiological degradation conditions. An important aspect of natural degradation is the distribution of the substance to be degraded. Biodegradation is more rapid if the substance is finely distributed in the soil or water and there is sufficient oxygen. If a substance reaches the biologically inactive (deeper) reaches of the ground because it degrades too slowly or because the ground is too porous, it can become a long-term source of water contamination.

The following laboratory biodegradability tests are generally used in Germany and in other European countries (test duration in parentheses): closed bottle test OECD 301 D (28 days); AFNOR test OECD 301 A (28 days); MITI test OECD 301 C (14 days); OECD screening test OECD 301 E (19 days); Sturm test EC 79/831 (28 days); CEC test CEC-L-33 (21 days); Zahn–Wellens test OECD 302 B (28 days). These international tests were developed mainly for water-soluble substances. Only the CEC test was developed for lubricating oils (two-stroke outboard engines).

In general, the tests use oxygen consumption, carbon dioxide production, the drop in DOC (dissolved organic carbon), or the reduction in CH valency fluctuations in the infrared spectrum (CEC test) to measure biodegradation. The duration of the tests is 14–28 days; depending on the test, rapid biodegradability means degradation of (at least) 60–80%. The Zahn–Wellens test defines a potential degradation of >20% and is of significance for the evaluation of some additives. Even though considerable variations can occur between different

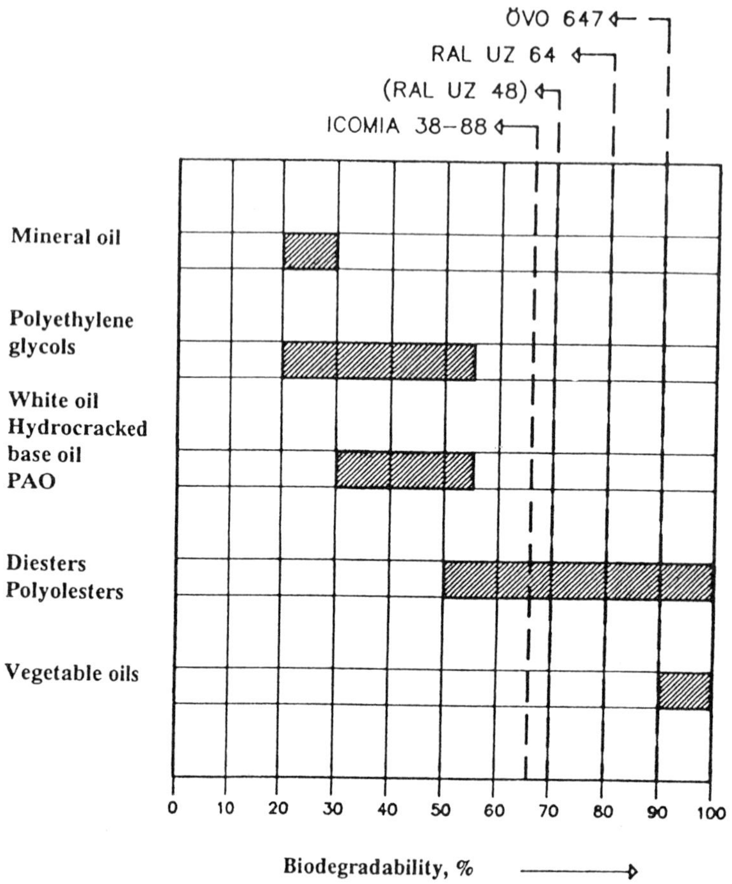

Fig. 2 Biodegradability of base oils. Övo, Austrian law, May 1, 1992; RAL, German authority for labeling (ecolabel for lubricants, 1988, 1991); ICOMIA, International Council of Marine Industry Association (two-stroke oils), 1988.

test procedures on some substances, the basic statement on a substance's degradation remains valid, and as a rule the relative order stays the same in a comparison of tests of a given series of substances.

The biodegradability of used lubricants can be altered by contamination. Measurements have shown that a deterioration of up to 15% can occur. This means that a lubricant that is 90% degradable when fresh may only be 75% degradable when used.

Figure 2 shows the CEC biodegradability of various base oils and the threshold values contained in European specifications for rapidly biodegradable lubricants.

C. Toxicity Values

To develop environmentally compatible lubricants, both biodegradability and ecological/toxicological criteria must be taken into consideration. The aim is to protect life in various areas, especially in waters (aquatic areas) and in nonaquatic areas (terrestial areas). The following ecological/toxicological test procedures are of significance:

Bacteria toxicity according to DIN 38 412, part 8; this determines cell multiplication inhibition (EC_{10} and EC_{50} values). The *Pseudomonas* type used for the test is found in wastewater and in the soil.

The bacteria toxicity test according to ISO 8192 determines acute toxicity by the inhibition of oxygen consumption; results of this test are EC_{50} values.

The algae toxicity test according to DIN 38 412, part 9 is yet another test for aquatic systems (measurement of chlorophyll fluorescence and determination of EC_{10} and EC_{50} values).

One of the most important test procedures in German legislation concerning aquatic areas is a test on small living organisms (*Daphnia magna* Straus, water flea, small crustacean) called the Daphnia test, according to DIN 38 412, part 11 or OECD guideline 202. In the aquatic area, fish toxicity according to DIN 38 412, part 15 performed on a special fish (*Leuciscus idus*) is of importance. Test results are the LC_0, LC_{50}, and LC_{100} values. The German environmental seal also incorporates fish toxicity according to OECD guideline 203. Possible pollution of the nonaquatic terrestial area, i.e., soil and plants, is evaluated by the plant growth test according to OECD guideline 208 (e.g., testing of wheat, cress, and rape seeds).

Naturally, toxicity testing for environmental protection purposes must also include mammalian and human toxicity, all the more so if safety-at-work and environmental protection concerns are combined. Toxicity terminology and toxicity classes according to national laws and EC guidelines need to be considered. The lethal dose LD_{50} is an important measure of toxicity.

D. Mobility

If one includes groundwater in environmental protection, the mobility of lubricants in the ground is a further measure of environmental compatibility. Rapidly degradable but easily soluble substances can drain through the soil and reach underground waterways. This fact has led to criticism of the use of water-soluble, rapidly degradable glycols in lubricants, especially hydraulic fluids.

III. ENVIRONMENTAL LEGISLATION

German legislation does not prohibit the use of mineral oil based lubricants for any application. However, a series of environmental protection provisions and orders encourage the use of rapidly biodegradable lubricants.

A. Water Protection

German water legislation defines the water pollution potential of substances. The arithmetic average of three toxicity values (mammalian toxicity, fish toxicity, and bacterial toxicity) forms the basis of a water pollution number (WGZ). Water pollution numbers 0–6 form the basis for four water pollution categories: WGK 0 (WPN 0–1.9), generally non-water-polluting; WGK 1 (WPN 2.0–3.9), slightly water polluting; WGK 2 (WPN 4.0–5.9), water polluting; and WGK 3 (WPN 5.9–6.0), highly water polluting. Merit ratings, e.g., for biodegradability and bioaccumulations, are included in water pollution categories along with water pollution numbers. Ratings within a list are decided by an Environment Ministry commission. At present, the following ratings apply for lubricants:

Volume in m³	WGK 0	WGK 1	WGK 2	WGK 3
< 0.1	stage A	stage A	stage A	stage A
0.1 - 1	stage A	stage A	stage A	stage C
1.0 - 10	stage A	stage A	stage B	stage D
10 - 100	stage A	stage A	stage B	stage D
100 - 1000	stage A	stage B	stage D	stage D
> 1000	stage A	stage C	stage D	stage D

Fig. 3 Definition of water pollution potential according to four hazard stages.

WGK 0 Some rapidly biodegradable lubricants, most of the natural fatty oils
WGK 1 Mineral oils without additives
WGK 2 Mineral oils with additives
WGK 3 Water-miscible mineral oils with additives

EC guideline 91/325/EWG, 1991 specifies the criteria contributing to the description "hazardous to the environment" for aquatic and terrestial areas. Step by step this is being adopted into national environmental legislation and will lead to the classification of numerous lubricants.

Water pollution categories for water-polluting substances have a major influence on the storage and handling of these substances. Considerably greater care must be taken with highly water polluting substances than for less polluting substances. The hazard potential based on water pollution categories and quantity is divided into four levels when manufacturing, processing, and application plants for water-polluting substances are planned. The highest hazard level is achieved when only 1 m³ of cutting fluid (WGK 3) is stored (Fig. 3).

Rapidly biodegradable lubricants can attain WGK 0 if suitable additives are used, and this has become a general aim of development work.

The German wastewater levy provides a further motivation to develop rapidly biodegradable lubricants. Industrial customers have to pay additional charges if nondegraded organic substances (CSB) are disposed of into waters. This logically raises the level of interest in easily degradable substances.

Special requirements apply to certain protected water zones. In some areas, prohibitions exist for the transport of water-polluting substances, the use of grass mowers with internal combustion engines, or DIY engine oil changing.

B. Soil Protection

A certain degree of soil protection is offered by water protection legislation. A specific law for the protection of soil is currently being drafted in Germany and will be influenced by the previously mentioned EC guideline.

In Europe, the "Holland list" applies for the clean-up of soil contaminated with mineral oils. This states that a clean-up is necessary if the contamination is greater than 500 mg/kg. At present clean-up costs in Germany are about DM 1500/m^3. This can be reduced to DM 150m^3 if the lubricant is rapidly biodegradable.

Austria was the first country to impose a ban on mineral oil based chain saw oils (effective May 1, 1992). This ban requires this type of total-loss lubricant to be at least 90% biodegradable (CEC test) along with fulfilling the other technical specifications. In addition, the products must not be water-soluble, which eliminates glycols as base fluids and promotes the use of natural fatty oils and some derivatives.

C. Safety at Work, Law on Chemicals

The development of rapidly biodegradable lubricants must also take into account, as previously mentioned, other ecological criteria, particularly the selection of nontoxic raw materials. Such products are especially interesting in terms of safety at work and thus their effect on workshop personnel. The German Law on Chemicals, which contains numerous European guidelines, categorizes substances according to their hazard potential and establishes classification regulations. The German environmental seal for rapidly biodegradable lubricants severely limits the use of hazardous substances as defined by the Law on Chemicals.

D. Keeping the Atmosphere Clean

The evaporation losses of rapidly biodegradable lubricants are generally much lower than those of conventional oils. This can assist in meeting emission threshold requirements. An important example is the significant reduction of diesel particulate emissions by engine oils that display low evaporation losses (Fig. 4).

IV. ENVIRONMENTAL AWARDS FOR LUBRICANTS

More and more countries are awarding environmental seals for products that are environmentally friendly. In Germany, the first environmental seal (Blue Angel) was awarded in 1988

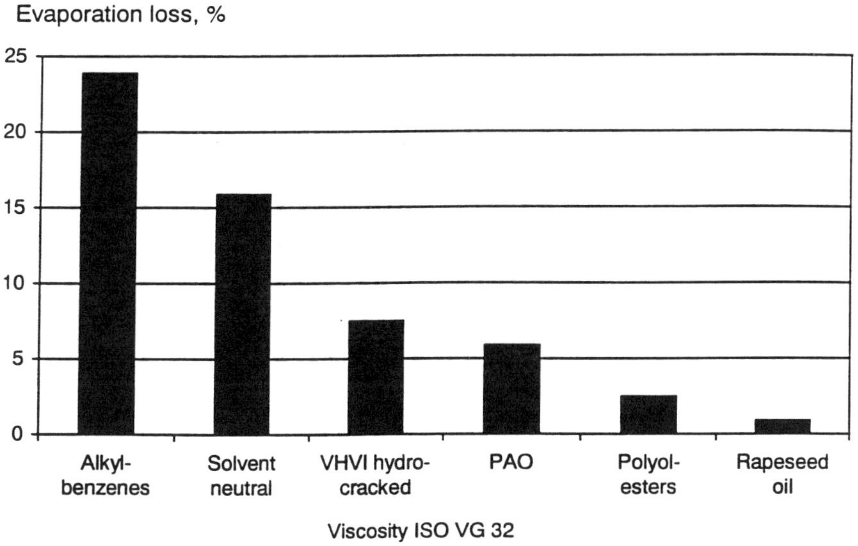

Fig. 4 Evaporation loss of lubricant base oils (Noack evaporation, 250°C, 1 h).

for rapidly biodegradable chain saw oils. In 1991, the scope of this award was extended to cover rapidly biodegradable lubricants and mold release oils. A third award was being prepared for hydraulic oils in 1995.

The German "Blue Angel" is a quasi-official seal with a considerable public profile. It is awarded by a panel of experts that includes members of the Federal Environment Ministry.

If the term "environmentally friendly" is defined in line with the two German environmental seals for lubricants, the following requirements result:

1. The products must not contain substances that need to be classified according to the law on hazardous substances (either under the threshold concentration that leads to classification or absolutely none).
2. None of the components must be in water pollution categories 2 or 3.
3. The products must not contain any organic chlorine bonds or nitrite.
4. The products must not be ecologically or toxicologically problematic. They must pass OECD guidelines 202 (Daphnia test), 203 (acute fish toxicity test), and 208 (growth on higher forms of plants test).
5. Each of the fluids must be at least 70% biodegradable according to OECD guideline 301 A-E and EGC 3–7. Biodegradability according to CEC-L-33-T-82 is also a possible test procedure (for which the RAL environmental seal UZ 64 requires at least 80% degradation).
6. The additive content must not exceed 5%. The additives must be potentially degradable (Zahn–Wellens test).
7. Ecological and toxicological threshold values apply to the additives.

In December 1991, the EC environment ministers decided to introduce a European environmental seal. Lubricants are not yet included in the first 12 product groups.

V. BASE OILS FOR RAPIDLY BIODEGRADABLE LUBRICANTS

In line with the previously described ecological/toxicological requirements, the selection of base oils concentrates mainly on ester substances. A wealth of know-how has been gathered over the years for this group of substances in the lubricant area even though their use as base oils has been limited to a few specialized applications.

A. Natural Fatty Oils

Natural fatty oils such as castor oil, palm oil, rapeseed oil, neatsfoot oil, lard, and degras or sperm oil have been used in lubricants for years. Over the past decade, these oils have been used mainly as additives. Since the early 1980s, rapeseed oil has gained increasing acceptance as a base fluid because it represents a good compromise in terms of availability, price, low temperature properties, viscosity, and thermal/oxidative behavior when suitable additives are used for some applications in the temperature range between −25 and +80°C.

Overall, one can say that as the concentration of saturated long-chain fatty acids in the triglyceride molecule (e.g., palmitic acid, stearic acid) increases; the low temperature behavior worsens. As the polyunsaturated fatty acids (e.g., linolenic acid, linoleic acid) increase, the thermal/oxidative resistance worsens. Even very long monounsaturated fatty acids (e.g., erucic acid, 22:1) worsen the low temperature behavior. Today's low erucic acid rapeseed oil has considerably better low temperature characteristics than the rapeseed oil of the 1970s, which contained erucic acid [8,9].

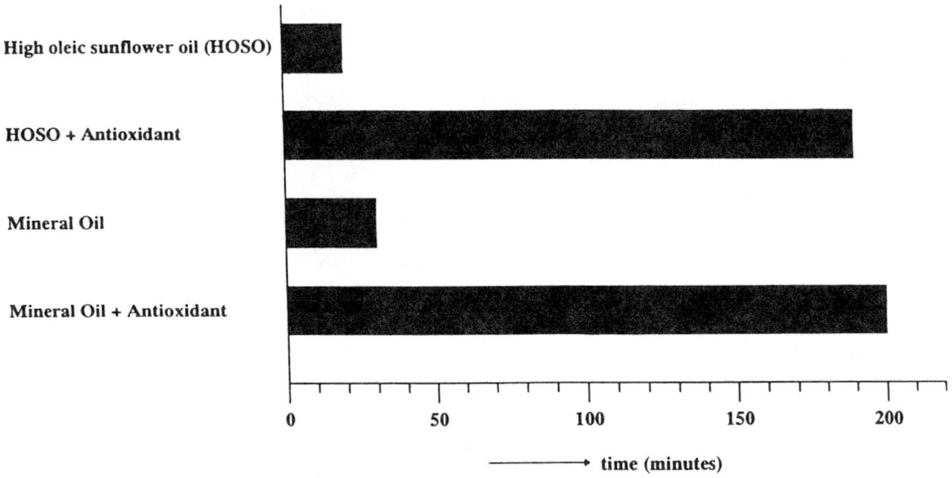

Fig. 5 Improvement of oxidation sensitivity of high oleic sunflower oil (rotary bomb test). (From Ref. 10).

B. Cultural Improvements to Natural Fatty Oils for Use in Lubricants

From an application point of view, the objectives of the cultivation of "technical rapeseed" is to achieve a different fatty acid pattern with a higher content of monounsaturated oleic acid and less double and triple unsaturated (linoleic and linolenic) acids. A high oleic acid sunflower oil (~ 90% oleic acid) is presently undergoing technical trials. High oleic acid rapeseed oil (70% oleic acid) is already available in limited quantities. Apart from these objectives of obtaining high oleic acid products by cultural improvements, efforts are also being made to produce linoleic acid–free rapeseed oil.

Figure 5 shows how the sensitivity of a new high oleic acid sunflower oil to oxidation can be improved by addition of an antioxidant [10].

The weakness of vegetable oil based lubricants can best be described by the structural elements of the double bond and the β-CH group (they cause thermal and oxidative instability). The ester group can hydrolyze, i.e., it can be split by water [11].

Lubricant applications generally employ fully refined rapeseed oil, i.e., foodstuff quality oil that has undergone all refining stages such as dewatering, desliming, neutralization, bleaching, and steaming (water vapor). These, however, still do not guarantee optimum lubricant suitability.

C. Chemical Modification of Natural Fatty Oils

The agricultural improvement of natural fatty oils, in particular rapeseed and sunflower oils, is a slow process. Moreover, their potential for use in lubricants is not sufficient to optimize the objectives for lubricants when new plants are cultivated or genetic alterations are made. The question, therefore, is Can fundamental processes such as chemical modification lead to more rapid success? An interesting example from lubrication technology is the manufacture of 12-hydroxystearic acid from castor oil. 12-Hydroxystearic acid as an alkaline or earth alkaline soap (mainly lithium soap) is the most important thickener for greases and is obtained by hydrogenation.

Selective hydrogenation is a process to remove polyunsaturated fatty acids and thus improve the oxidative stability of vegetable oils. In this regard, the work of Behr et al. [12] should be consulted. In this work, polyunsaturated fatty acids in rapeseed oil and rapeseed fatty acid methyl esters were reduced to 1% without significantly increasing the stearic acid content. The latter would negatively affect cold flow properties. Also important is the fact that the *cis-/trans* 18:1 ratio remains very high because large amounts of 18:1 in the trans form also detrimentally affect cold flow properties.

Further modifications to the multiple double bonds to improve thermal oxidative stability are possible, and it is conceivable that more cost effective processes will be developed to increase the use of natural fatty oils that have been chemically modified.

D. Synthetic Esters

The term "synthetic esters" covers a broad range of chemicals with differing qualities and prices. Their base components (alcohols and carboxylic acids) may have a natural or synthetic origin (0–90% content of naturally grown components). They can be more or less biodegradable, depending on their structure; their thermal/oxidative stability and cold flow properties can be much better than those of triglycerides. The toxicological properties are in many cases comparable to those of naturally grown fatty oils.

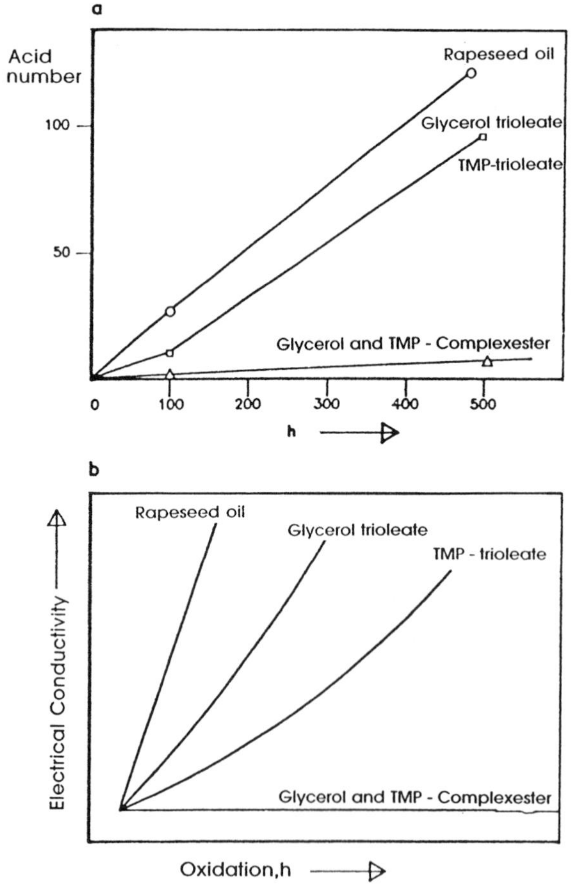

Fig. 6 Hydrolysis and oxidation of various base oils according to H. G. Schmidt. (From Ref. 11.)

Table 1 Time Line for the Development of Synthetic Esters as Lubricant Base Fluids

1937–1944	Production and research on 3500 ester-based oils. In Germany, basic acids from the coal chemistry (from phenols and cresols).
1940–	Esters from neopentyl polyols (neopentyl glycols, trimethylolpropane, pentaerythritol). Produced by IG Farben.
1960–	Great importance of aviation jet engines with medium-chain carboxylic acids.
1970–	Importance of other synthetic lubricants.
1985–	Importance of biodegradable lubricants.
2000–	Increasing importance as lube base oils.

Chemistry offers a wide variety of possibilities in the area of synthetic esters. A major development objective is to find inexpensive esters among all the possible substances that meet the specific requirements of a particular application.

Before ecological aspects became part of lubricant development, ester oils were used in special lubricants for technical reasons, e.g., as base fluids for aviation turbine oils and components for fuel economy oils.

The most important groups of esters include monoesters, diesters, polyol esters, and complex esters. At present, polyol esters such as trimethylolpropane esters (TMP esters) or pentaerythritol esters dominate. The basis of these are mainly alcohols from the petroleum and chemical industries and fatty acids derived from natural oils.

As for hydrolytic stability, "normal" polyol esters (trimethylolpropane or glycerol trioleate) differ only slightly from rapeseed oil; the difference in oxidative resistance is much greater. Both characteristics are significantly improved with complex esters. Normally an improvement in hydrolytic stability worsens the substance's biodegradability. However, there are complex esters [medium chain length saturated fatty acids on trimethylolpropane (TMP) or glycerol] that combine excellent thermal/oxidative characteristics with good hydrolytic resistance and good biodegradability. Figure 6 compares the oxidative stability and hydrolytic stability of rapeseed oil to those of glycerol trioleate and TMP trioleate and a complex ester.

With regard to the important application criterion of oil viscosity, the combination of dicarboxylic acids (e.g., azelaic acid and sebacic acid) with fatty acids is an excellent instrument for adjusting viscosity.

Table 1 shows the historical development of esters as lubricant base fluids.

Table 2 lists esters that are presently offered by the oleochemical industry for use in lubricants. The table is arranged by product viscosity at 40°C. This is one of the most important values used to define the possible applications of a product. Low temperature behavior is described as the pour point.

E. Polyalkylene Glycols

As with esters, polyalkylene glycols (aliphatic polyethers) are highly diverse and display greatly differing characteristics (e.g., water-soluble and non-water-soluble products). Water-miscible ethylene glycols are well known as lubricant base fluids and have achieved considerable significance in fire-resistant HFC hydraulic fluids. They are also easily biodegradable. Compared to esters, they suffer from the disadvantage that they are highly mobile in soil and

Table 2 Oleochemical Ester Base Fluids for Lubricants

Viscosity at 40°C (mm^2/s)	Viscosity index	Pour point cold flow behavior (°C)	Chemical name
4,7	313	−15	Methyl oleate
4,7	207	16	Methyl tallowate
5,2	173	12	Isopropyl palmitate
5,4	205	−12	Methyl colzate
5,8	237	−18	Methyl erucate
7,7	148	<−53	Di-(2-ethylhexyl) adipate
8,0	196	−30	2-Ethylhexyl colzate
8,6	196	−4	2-Ethylhexyl palmitate
8,6	193	−21	2-Ethylhexyl oleate
8,9	179	−21	2-Ethylhexyl oleate
8,9	179	−6	2-Ethylhexyl tallowate
9,2	187	3	Isotridecyl stearate
10	246	−10	*n*-Decyl oleate
14	146	<−60	Diisodecyl adipate
16	151	−27	Capryl-caprinic acid ester
27	148	<−60	Diisotridecyl adipate
30	192	−33	Neopentylglycol dioleate
45	198	−45	TMP trioleate
48	180	−42	TMP trioleate
50	189	−39	TMP trioleate
53	163	−21	TMP trioleate
60	186	−45	TMP ester with long-chain fatty acids
86	157	−9	TMP ester with long-chain fatty acids
88	146	−27	TMP ester with long-chain fatty acids
102	146	−3	TMP ester with long-chain fatty acids
148	140	−24	PE tetraisostearate

thus have a greater water pollution potential. In applications where water solubility is an advantage (e.g., in the foodstuff and beverage industries), ethylene glycols remain an important group of base fluids.

The chemical combination of the above-mentioned esters and ethers can lead to suitable base oils from both technical and environmental points of view.

VI. ENVIRONMENTALLY FRIENDLY ADDITIVES

In applications that use natural and synthetic esters as base fluids, a series of mineral oil additives would be suitable. However, a selection should be made according to ecological criteria, especially toxicological values, water pollution categories, biodegradability, and waste management considerations. The German environmental seal defines the selection criteria more precisely. Apart from such selections, there is still a need to develop new additives whose technical performance and ecological characteristics are optimized.

The German environmental seal for rapidly biodegradable lubricants only approves of additives in water pollution category 1. This raises the question of whether, for example, 0.5% of a water pollution category 1 oxidation inhibitor is a greater hazard to water than 0.1% of a

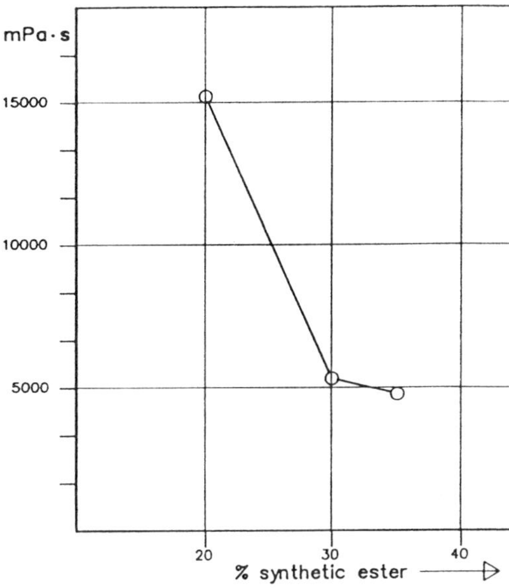

Fig. 7 Brookfield viscosity for high oleic sunflower.

water pollution category 2 substance. Such points need clarification. On the other hand, some additives are unsuitable for rapidly biodegradable, environmentally friendly lubricants; this applies especially to substances containing heavy metals.

A. Improving Low Temperature Properties

Many synthetic biodegradable esters have excellent cold flow characteristics. The use of environmentally friendly pour point depressants can give vegetable oils such as rapeseed oil adequate long-term cold flow characteristics (rapeseed oil, −30°C). Mixing in easily biodegradable synthetic esters can improve the pour point, long-term low temperature behavior, and low temperature viscosity. Figure 7 illustrates this for a high oleic acid sunflower oil [10]. New laboratory procedures have been developed for vegetable oils that define cold flow characteristics over time.

B. Extreme Pressure and Antiwear Additives

Most vegetable oils and synthetic esters used in lubricants display better lubricity in boundary lubrication conditions than mineral oils because of their polarity. This has been proved by a series of tests [13,14]. Recently a twin-disk test apparatus showed that the coefficient of friction of esters and vegetable oils is half that of mineral oils.

Table 3 shows Michaelis and Höhn's results [14] for rapeseed oil, various synthetic esters, polyglycols, and mineral oils. The table lists products with viscosities between 32 and 220 mm^2/s. The tests were performed on base fluids without additives. The generally used solvent raffinates were used for the mineral oil tests [14].

Sulfurized esters have established themselves as environmentally friendly extreme pressure (EP) additives, and the additive sulfur in vegetable oils and esters also has an antiwear (AW) effect. Sulfur carriers with 15% total sulfur and 5% additive sulfur have proven themselves particularly effective in rapidly biodegradable esters and vegetable oils [15].

Table 3 Coefficient of Friction of Rapidly Biodegradable Lube Base Oils

Type of lubricant base oil	ISO viscosity	Coefficient of friction μ
Rapeseed oil	VG 32	0,023
	VG 46	0,022
	VG 220	0,078
Synthetic esters	VG 46	0,022
	VG 46	0,022
	VG 46	0,045
	VG 68	0,020
	VG 68	0,079
	VG 220	0,022
	VG 220	0,078
Polyglycols	VG 32	0,020
	VG 68	0,020
Refined mineral oils	VG 32	0,050
	VG 46	0,050
	VG 68	0,057
	VG 220	0,040

Source: Ref. 14.

Table 4 shows a comparison of antiwear values and friction coefficients such as are used in numerous lube laboratories, for example, for a rapeseed oil raffinate and for an equiviscous VG 32 mineral oil raffinate. Here again, the values show that rapeseed oil has much better lubricity than mineral oils without additives. One can assume that the favorable friction and wear behavior of natural fatty acids and most esters leads to a corresponding reduction in additive content. From an environmental point of view, this reduction in chemical additives can play an important role in the move toward rapidly biodegradable products.

Table 4 Wear Protection and Friction Behavior for Rapeseed Oil and Solvent-Refined Mineral Oil (Solvent Neutral VG-32)

Test	Rapeseed oil	Solvent neutral VG 32
VKA test according to DIN 51 350 T2 welding load (N)	1300–1400	1000–1100
FZG test according to DIN 51 354 T 2		
Breakdown load stage	9	6
Spec. wear (mg/kWh)	0.16	0.28
Timken test according to IP 240		
OK load (lb)	12	10
Tannert test according to DIN 51 387 slide index (N)	2070	160

C. Antioxidants

The focus is on phenolic products more than amine products as antioxidants. The best known product, BHT (2,6-di-*tert*-butyl-1-4-methylphenol), is effective at lower temperatures in esters and vegetable oils. It has the advantage of being in water pollution category 1 and having FDA foodstuff approval. Another phenolic product with these advantages and high temperature suitability is named [15]. Mixtures based on amine components have been used successfully in biodegradable greases.

D. Viscosity Index Improver

The viscosity index of lubricants shows how much viscosity drops when the temperature increases. The paraffinic mineral oil raffinates generally used in lubricants today have a viscosity index of about 100. For a number of lubricant applications, polymer viscosity index improvers are used to improve viscosity–temperature behavior. The disadvantage of these polymer compounds is that they can be mechanically destroyed when subjected to high shearing forces. This in turn lessens or eliminates the desired effect, and chemical reactions with the molecular fragments can negatively influence the lubricant (e.g., sludge formation, deposits on machine elements).

Natural fatty oils have much better viscosity indices than mineral oils, mostly over 180, so that no viscosity index improvers are necessary. As no polymer compounds are present in the lubricant, it is not affected when shearing forces increase. Many synthetic esters have viscosity indices over 150, with some over 200.

VII. PLANT OIL AND ESTER-BASED LUBRICANTS IN THE MARKET

Biodegradable two-stroke oils have been available since 1976. In Germany, this development was helped by inland boating regulations. The CEC-L-33-T-82 biodegradability test developed for two-stroke oils was agreed on in 1982. Biodegradable chain saw oils have been available since 1984, and biodegradable hydraulic oils for almost as long. In 1996, about 35,000 t of rapidly biodegradable lubricants were sold in Germany, about 4% of total lubricant sales [4].

A. Total-Loss Lubricants

Total-loss lubrication is a technology in which the lubricant performs its function for only a relatively short period of time before it is lost into the general environment. This type of lubricant does not create any used products that have to be disposed of because it is designed to enter the environment. Table 5 shows some typical total-loss applications of rapidly biodegradable lubricants. These make up about 7–8% of total lubricant sales. In terms of volume, chain saw oils (0.3–0.6 L/m^3 wood) and some greases are of particular significance [16].

As long-term characteristics of total-loss lubricants, in particular their thermal/oxidative behavior, are of limited significance, rapidly biodegradable products are relatively simple to formulate. The first German environmental lubricant award was dedicated to chain saw oils in 1988 (RAL UZ 48); in 1991 a second environmental award was issued for total-loss lubricants and mold release oils (RAL UZ 64).

Table 5 Total-Loss Lubricants and Their Applications

Lubricant	Application
Two-stroke oil	Air and water cooled two-stroke engines
Chain saw oils	Chain saws
Tacky oils	Exposed lubrication points on industrial machinery
Greases	Exposed and central lubrication systems in commercial vehicles and industrial machinery
Railroad switch greases	Railroad switches
Wheel flange lubricants	Railroad wheel flanges
Spray lubricants	Lubrication and corrosion protection of machinery
Mold release oils	Construction

B. Hydraulic Oils

The development of rapidly biodegradable hydraulic oils attracted attention early on because of their volume potential (11–14% of lubricant sales) and their ecological relevance in automotive hydraulic applications. Lubricant manufacturers, additive producers, and universities have performed basic research, which is still in progress [17–19]. Of the oils sold in Germany, considerably more than half are based on rapeseed oil (~ 40% are based on synthetic esters). The problems that arose with vegetable oils and esters, especially with respect to elastomer compatibility [20], nonferrous metal corrosion caused by hydrolysis [21], and viscosity increase, could all be clearly eased.

The draft of the German norm for hydraulic oils based on natural triglycerides with HETG or synthetic esters with HEES or polyglycols are characterized with HEPG [22]. An ISO standard of international importance is being drafted. New threshold values need to be established for used oil evaluation; e.g., the common measurement of the neutralization number, as with mineral oils, is no longer relevant. Figure 8 details the results of a three-year long-term test in two stationary hydraulic systems of deep-drawing presses at a car maker. In spite of a neutralization number (NZ) of 7 and a viscosity increase of about 40–70, no problems occurred (Figure 8a).

C. Hydraulic Oils for Mobile Equipment

Good results have been achieved with biodegradable hydraulic fluids in agricultural forestry and off-road construction. A broad range of applications under mild conditions may be covered using products based on vegetable oils (in Europe, mainly rapeseed oil). More severe demands concerning high temperature stability or cold flow properties, or both, may be satisfied by the use of synthetic ester base fluids. Similar aspects should be considered for gear oils or oils for combined gear and hydraulic systems. They are often used in farm tractors with mounted agricultural equipment [28].

Special problems are involved with integrated wet brakes, because their function is dependent on the friction coefficient. This is different for mineral oils and natural or synthetic esters or other biodegradable base fluids. Apart from this, we believe that mineral oils can be replaced by biodegradable products less harmful to the environment in a considerable number of hydraulic fluids and gear oils in mobile equipment.

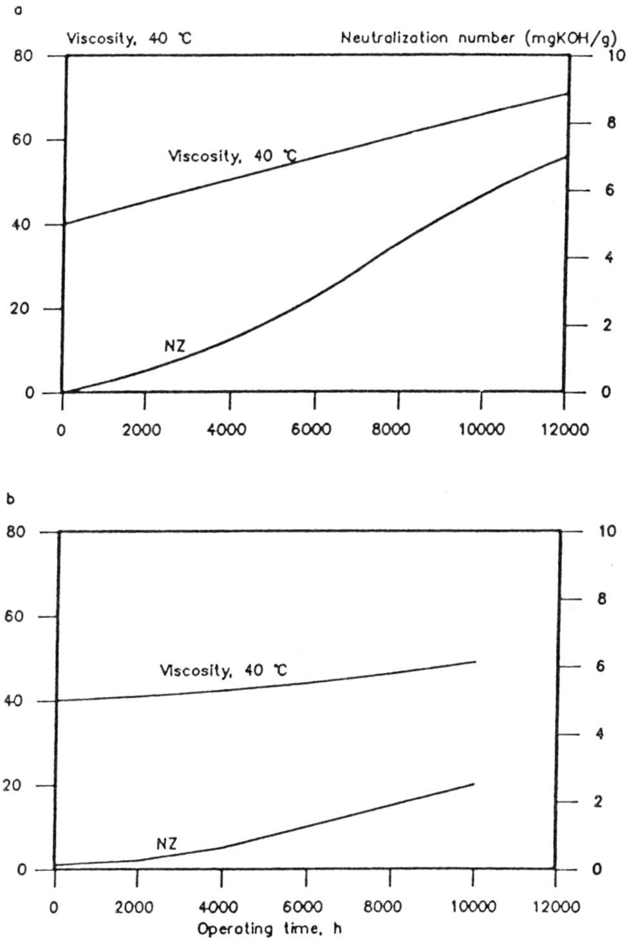

Fig. 8 Long-term trial of a rapeseed oil based hydraulic fluid in two stationary hydraulic systems.

D. Ester-Based Machine Tool Fluid Family

The development of a new family of fluids based on rapidly biodegradable esters in water pollution category 0 offers not only considerable ecological but also economic advantages. This involves the replacement of water-miscible cutting fluids by low viscosity esters. Hydraulic oils, gear oils, and other machine lubricants are also formulated on biodegradable esters that are compatible with the cutting fluids. Even though such products are more expensive than conventional mineral oil based lubricants and functional fluids, the total process cost analysis still shows advantages. Along with the general ecological considerations regarding the handling of water-polluting fluids and disposal costs, the improved workshop conditions offered by "natural" raw materials play an important role. This model has been successfully realized in a large diesel engine manufacturing plant for a couple of years now [23,24,27].

Figure 9 illustrates the complexity of the "biofluid" family circuit based on oleochemical esters. This completely new concept closes the material circle with losses due to only evaporation and residual oil on chips.

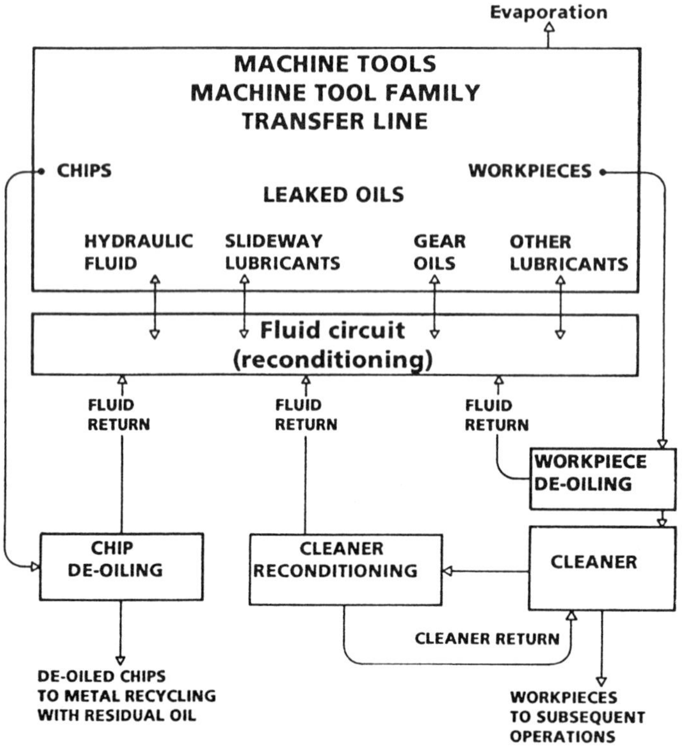

Fig. 9 Closed fluid circuit in chip-forming machining for an ester-based machine tool fluid family. (From Refs. 23 and 24.)

Table 6 shows the viscosities of the most important products in the machine tool fluid family, and Table 7 comprehensively lists the characteristics of the special biohydraulic oil.

E. Engine Oils

Premixed rapidly biodegradable two-stroke engine oils based on esters have been available since 1976 and are of particular relevance to ecological demands and endangered waters. In Germany, Mercedes-Benz was the first company to perform ecological evaluations of engine oils [25]. These put a certain emphasis on biodegradability. A project initiated by the German Federal Ministry for Agriculture [26] foresees the development of biodegradable engine oils based on vegetable oils and their derivatives. At the end of 1993, a leading European manu-

Table 6 Biofluid Family for Machine Tools

Biofluid	Viscosity
Cutting oil	8 mm^2/s
Hydraulic oil	32 mm^2/s
Gear oil	220 and 320 mm^2/s
Way lubricant	68 and 220 mm^2/s
Other machine tool oils	68 mm^2/s
Grease	NLGI 000

Table 7 Specification for Ester-Based Machine Tool Hydraulic Fluids

Characteristic	Value	Test method
Density at 15°C	914 kg/m^3	DIN 51 757
Viscosity at 40°C	32 mm^2/s	DIN 51 562
Viscosity at 100°C	7.3 mm^2/s	DIN 51 562
Viscosity index	198	DIN ISO 2909
FZG mechanical gear rig test	> stage 12	DIN ISO 51 354-2
Rotary vane pump test	< 120 mg	DIN 51 389-1
Ring vanes	< 30 mg	DIN 51 389-1
Steel corrosion	degree 0–A	DIN 51 585
Copper corrosion	degree 1–1000 A 3	DIN 51 759
Air release	< 5 min	DIN 51 381
Demulsification	20 min	DIN 51 599
Biodegradability	> 95%	CEC-L-33-T-82
Water pollution category	WGK 0	

facturer of diesel engines issued a first approval for an ester-based oil whose biodegradability is greater than 80% (Fuchs Plantomot 5W-40).

At present, the use of chemically unchanged vegetable oils with traditional fatty acid patterns as base oils for normally lubricated four-stroke internal combustion engines does not appear to be possible. The previously mentioned research project also foresees a development in which the continuous burning of some of the engine oil should eliminate aging products. This would make less thermally stable oils suitable for engine lubrication. A rapeseed oil based engine oil has been run for about 15,000 miles in the diesel engine of a Volkswagen passenger car. Figure 10 shows the lubrication model flowchart. A prerequisite is that a "low

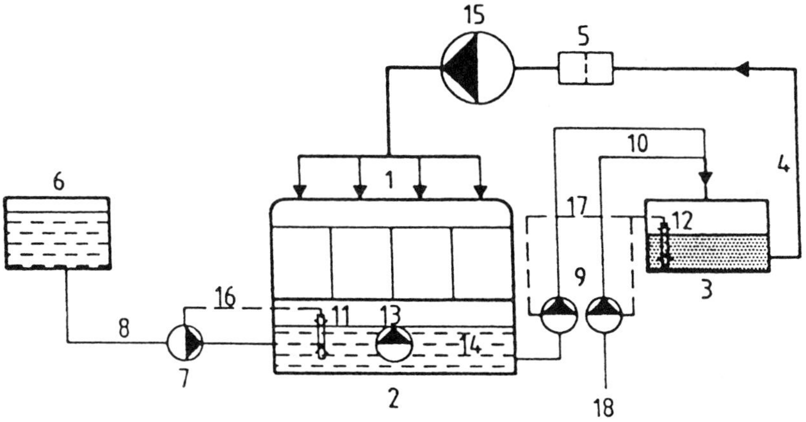

Fig. 10 Vegetable oil based engine oil with partial "total-loss lubrication" in a diesel engine flowchart for environmentally compatible diesel engine lubrication. 1, Engine; 2 oil sump; 3, tank for fuel–used oil mixture; 4, fuel–used oil mixture feed to injection; 5, fuel filter; 6, fresh oil tank; 7, fresh oil dosing pump; 8, fresh oil feed; 9, coupled pumps for used oil and fuel; 10, connecting pipes; 11, control device for oil volume/level; 12, control device for fuel–used oil mixture; 13, pump for internal motor oil circuit; 14, motor oil; 15, fuel injection pump; 16, 17, control lines; 18, fuel feed from main tank.

chemical" oil is burned with few emissions. In spite of high oil consumption, the injection of 2% used oil into the fuel is an economically acceptable solution.

VIII. SUMMARY AND OUTLOOK

Increasing environmental awareness has intensified the development of lubricants and has generated new evaluation criteria. The new generation of rapidly biodegradable lubricants was in its infancy in the 1980s but continues to gain importance in the 1990s. Considerably more development work needs to be done, and consumers must be informed about the benefits of lower environmental risks because the products are usually significantly more expensive than conventional mineral oil based products.

Some of the applications of rapidly biodegradable lubricants and functional fluids mentioned in this chapter have shown that this area of work can best be promoted by not just focusing on lubricant developments but also by machine tool manufacturers and their suppliers demanding new technical solutions for environmentally friendlier lubrication technologies.

Figure 11 shows the possibility of substituting 90% of mineral oil based lubricants with

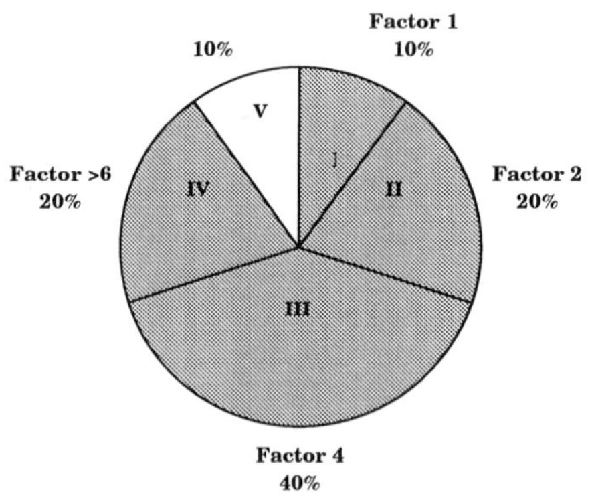

Cost factor 1
- I Vegetable oils incl. HO-types and other adapted cultivation products

Cost factor 2
- II TMP oleates, other low-price polyol esters, chemically modified plant oils, some dicarbonic acid esters

Cost factor 4
- III TMP or glycerol complex esters with saturated fatty acids, some dicarbonic acid esters

Cost factor > 6
- IV Complex esters of various structures, other saturated special esters

- V Not biodegradable

Fig. 11 Technical realization of rapid biodegradability for 90% of total lubricant volume (hypothesis)

rapidly biodegradable products. The rapidly biodegradable base oils are divided into four groups (I–IV) according to cost factors. Group V is the 10% nonbiodegradable that would remain in use. By the year 2000, it is forecast that 10% of German lubricant demand could be met by such products. Base oils will be mainly regular vegetable oils (rapeseed), new vegetable oils, and synthetic esters of an oleochemical nature. Economically interesting alternatives for used oils could be treatment and incineration or recycling if larger quantities are involved.

REFERENCES

1. T. Mang, *Schmierstoffe und Funktionsflüssigkeiten auf Pflanzenölbasis, Erfahrungen eines Herstellers*, Schriftenreihe des Bundesministers für Ernährung, Landwirtschaft und Forsten, Reihe A, Heft 391, 1990.
2. T. Mang, Umweltschonende und arbeitsplatzfreundliche Schmierstoffe, *Tribol. Schmierungstech. 38* (4): 231–236 (1991).
3. H. Ihrig, Umweltverträgliche Schmierstoffe in den 90er Jahren, *Tribol. Schmierungstech. 39* (3): 121–125 (1992).
4. T. Mang, Schmierstoffe und Umwelt—die Schmierstoffentwicklung im Einfluss der Umweltgesetzgebung, 9th Int. Kolloq. Tribologie, Esslingen, Jan. 11–13, 1994, pp. 4.1.1–4.1.10.
5. T. Mang, Lubricants and legislation in the Federal Republic of Germany, *Erdöl Kohle 42*(10): 400–407 (1989).
6. T. Mang, Legislative influences on the development, manufacture, sale and application of lubricants in the Federal Republic of Germany, CEC Symp., Paris, Apr. 19–21, 1990.
7. T. Mang, Environmentally harmless lubricants, current status and relevant German environmental legislation, *NLGI Spokesman 57* (6) 9-233–15-239 (1993).
8. U. J. Möller, with T. Mang and P. Studt, Gesellschaft für Tribologie, GfT-Arbeitsblatt Pflanzenöle als Schmierstoffe, Moers, 1993.
9. A. Hubmann, Additivierung pflanzlicher Schmierstoffe, in *Biologisch schnell abbaubare Schmierstoffe und Arbeitsflüssigkeiten* (W. J. Bartz, Ed.) Expert-Verlag, Ehingen, 1993.
10. K. Lal and V. Carrick, Performance testing of lubricants based on high oleic vegetable oils. 9, Int. Kolloq. Tribologie, Esslingen, Jan. 11–13, 1994, pp. 2.9.1–2.9.14.
11. H.-G. Schmidt, Komplexester aus pflanzlichen Ölen, 9th Int. Kolloq. Tribologie, Esslingen, Jan. 11–13, 1994, pp. 2.2.1–2.2.9.
12. A. Behr, N. Döring, S. Durowicz-Heil, B. Ellenberg, C. Kozik, C. Lohr, and H. Schmidke, Selektive Härtung mehrfach ungesättigter Fettsäuren in der Flüssigphase, Henkel-Referate 30/1994, pp. 21–26.
13. S. Odi-Owei, Tribological properties of some vegetable oils and fats, *Lub. Eng. 45* (11): 685–690 (1988).
14. K. Michaelis and B.-R. Höhn, Reibungsverhalten biologisch leicht abbaubarer Schmierstoffe, 9th Int. Kolloq. Tribologie, Esslingen, Jan. 11–13, 1994, pp. 8.1.1–8.1.8.
15. A. Fessenbecker and J. Korff, Additive für ökologisch unbedenkliche Schmierstoffe, 9th Int. Kolloq. Tribologie, Esslingen, Jan. 11–13, 1994, pp. 11.12.1–11.14.14.
16. W. Dresel, Schmierfette auf pflanzlicher Basis, in *Biologisch schnell abbaubare Schmierstoffe und Arbeitsflüssigkeiten* (W. J. Bartz, Ed.), Expert-Verlag, Ehingen, 1993.
17. W. Backé, Bestandsaufnahme und Trends der Fluidtechnik, *Ölhydraul. Pneum. 36* (1): 16–19 (1992).
18. C. Busch and W. Backé, Biologisch schnell abbaubare Hydraulikflüssigkeiten, *Tribol. Schmierungstech. 41* (1): 17–23 (1994).
19. T. Mang, Umweltbedingte Alternativen zu Mineralölen als Grundkomponenten in Hydraulikflüssigkeiten, 10th Aachener Fluidtech. Kolloq., Mar. 17–19, 1992, pp. 11–36.
20. K. Nagdi, Dichtungswerkstoffe für umweltfreundliche Flüssigkeiten, *Ölhydraul. Pneum. 34* (1): 42–50 (1990).
21. H. F. Eichenberger, Biodegradable hydraulic lubricant—an overview of current development in

central Europe, SAE Tech. Paper Ser., 42nd Earthmoving Ind. Conf., Peoria, IL, Apr. 9–10, 1991.

22. E. Pelzer, Normung und Vergaberichtlinien des Umweltzeichens von umweltschonenden Hydraulikflüssigkeiten, in *Biologisch schnell abbaubare Schmierstoffe und Arbeitsflüssigkeiten* (W. J. Bartz, Ed.), Expert-Verlag, Ehingen, 1993, pp. 117–135.
23. T. Mang, C. Freiler, H. Hanagarth, and E. Strobel, Niedrigviskose Esteröle als Kühlschmierstoff—Grundlagen und Anwendung in der Zerspanung eines Nutzfahrzeugmotorenwerkes, IWT-Seminar, Bremen, October 1995.
24. T. Mang, C. Freiler, M. Spilker, W. Göttert, H. Hanagarth, and E. Strobel, Hydrauliköle für die Werkzeugmaschine—Bestandteil einer neuen Fluidfamilie für die spanende Fertigung, *Ölhydraul. Pneum. 10*: 759–762 (1995).
25. Fuchs Mannheim and Motoren-Werke Mannheim AG, R&D project sponsored by the German Ministry of Agriculture and Forestry, Umweltfreundliche Lösungen zur Schmierung von Dieselmotoren, 1993.
26. Fuchs Mannheim and Mercedes-Benz AG, Stuttgart, R&D project sponsored by the German Ministry of Research and Technology, Entwicklung emissionsarmer Schmierstoffe unter Verwendung nachwachsender Rohstoffe, Published by Bundesumweltamt in UFOKAT '92, p. 276.
27. Mercedes-Benz AG, Stuttgart, Tech. Paper 1992.
28. K. Höhn, Bio-Hy-Gard in John Deere Traktoren, Mannheimer Morgen, July 30, 1993.

29

Epoxidized Oils

Frank D. Gunstone
Mylnefield Research Services Ltd., Scottish Crop Research Institute, Invergowrie, Dundee, Scotland

I. INTRODUCTION

Epoxidised oils are made from unsaturated acids or from their glycerol or alkyl esters such as soybean oil, linseed oil, rapeseed oil, tall oil (fatty acids), oleic acid, or alkyl oleates. For the most part, they are produced by reaction with hydrogen peroxide in some appropriate form; most often this is a peroxy acid (RCO_3H).

In conducting epoxidation reactions it should always be remembered that the reaction is exothermic and that *high concentrations of peroxy acid should be avoided*. Whatever procedure is employed, it is advisable to add the epoxidizing agent slowly or at an appropriate rate to the olefin. The reverse procedure can be hazardous. Further, it should be noted that the epoxides themselves are reactive compounds, especially in acidic solution, and efforts must be made to ensure that the epoxides are not converted to other products following the process of epoxidation.

Other reviews of this topic are available elsewhere [1,2].

II. NATURAL EPOXY ACIDS

Since the discovery of vernolic (*cis*-12,13-epoxyoleic) acid in *Vernonia anthelmintica* seed oil some 40 years ago [3], several other natural epoxy acids have been recognized (Table 1). Epoxy acids are among the monomers of cutin and the phytoaxelins (toxic antimicrobials in plants) [5].

Vernolic acid is still the most readily available natural epoxy acid and continues to be the most widely studied. It attains high levels in several seed oils (Table 2), and interest has been expressed in developing some of these as new crops of potential industrial value. This applies particularly to *Vernonia galamensis* (in the United States) and *Euphorbia lagascae* (in Europe).

These vernolic oils can be epoxidized to produce material rich in diepoxystearate such as epoxidized soybean oil, but it seems to me that this is not the best approach. Such material

Table 1 Natural Epoxy Acids

Acid[a]	Chain length and unsaturation				
9,10-Epoxy	18:0	18:1 (12*c*)[b]	18:1 (12*a*)	18:2 (12*c*15*c*)	18:2 (3*t*12*c*)
12,13-Epoxy		18:1 (9*c*)[c,d]		18:2 (6*c*9*c*)	18:2 (6*t*9*c*)
15,16-Epoxy				18:2 (9*c*12*c*)	

[a] All these acids are *cis* epoxides except for the 18:0 acid, which occurs in both *cis* and *trans* forms.
[b] Coronaric acid.
[c] Vernolic acid.
[d] C_{20} homologue—alchornoic acid.
Source: Ref. 4.

would probably be more expensive than epoxidized soybean oil, and, more significantly, it is wasteful of the multifunctionality present in vernolic acid. It would be better to exploit the fact that vernolic acid contains one epoxide group and one unsaturated center with differing reactivities and to use the acid as a source of specialty chemicals (i.e., low volume, high value products). This is discussed further in Section VII.

Useful studies have already been carried out on *Vernonia galamensis* seed oil. There were reports in 1990 that this plant was being cultivated in Kenya (~200,000 hectares) and in Zimbabwe (20,000–30,000 ha) [6], and procedures for extraction of the oil (40–42%) [7] and its refining [8] have been described. Typical fatty acid and triacylglycerol analyses are given in Table 3. It is clear that with around 80% vernolic acid this oil contains substantial proportions of trivernolin (50–60%) and divernolins (20–30%). It is readily biodegradable [12] by *Acinetobacter* and *Pseudomonas* species.

In Europe attempts are being made to develop *Euphorbia lagascae* as a new commercial oilseed crop. The seed contains about 45–50% oil, and the oil yield in the Netherlands is at present 500–1000 kg/ha. This is expected to rise to about 1200 kg/ha or to even higher values [13]. The fatty acid composition of this vernolic-rich oil is given in Table 4. The oil is reported to be about 41% trivernolin and 18% divernolic triacylglycerols.

The extracted oil is bright yellow and has a low level of phosphorus (<1 ppm) and a high specific gravity. Though more viscous than soybean or rapeseed oil, it is less viscous than epoxidized soybean oil or epoxidized linseed oil. It can be used, like these epoxidized oils, as a stabilizer for PVC. But it can also be used as a diluent in paints, avoiding the necessity for volatile organic solvents. It has been estimated that the addition of a vernolic oil to paint at around the 10% level would reduce the amount of volatile organic compounds by 160 million pounds in the United States alone [15].

Table 2 Natural Occurrence of Vernolic and Coronaric Acids in Seed Oils

Vernolic acid (12,13-epoxyoleic), wt %			
Vernonia anthelmintica	72	*Euphorbia lagascae*	57–62
V. galamensis	73–78	*Erlangea tomentosa*	52
Stokes aster	65–79	*Crepis aurea*	52–54
Cephalocroton cordofanus	62	*Crepis biennis*	68
Coronaric acid (9,10-epoxy-12-octadecenoic), wt %			
Chrysanthemum coronarium	14	*Helichrysum bracteatum*	14
Xeranthemum annuum	8		

Source: Ref. 4.

Table 3 Composition of *Vernonia galamensis* Seed Oil

Fatty acids, wt %			
Vernolic	79–81	Stearic	2–3
Linoleic	11–12	Palmitic	2–4
Oleic	4–6		

Triacylglycerols,[a] wt %				
V_3	V_2X	VX_2	X_3	Ref.
59.2	28.1	9.5	3.2	9[b]
50	21	20	9	10[c]
52				11[d]

[a] V = vernolic acid; X = acids other than vernolic.
[b] Gas chromatography.
[c] Reverse phase HPLC.
[d] Mass spectrometry.

III. THE QUALITATIVE AND QUANTITATIVE ANALYSIS OF EPOXY OILS

Total epoxide content can be measured by hydrogen bromide in acetic acid (Durbetaki reaction [16]), by gas chromatography, and by a colorimetric procedure based on reaction with picric acid. This last has also been used as a specific spray reagent for epoxides in thin layer chromatography. Epoxides show characteristic absorption bands in the infrared spectrum at 848 and 826 cm^{-1}.

The position of an epoxide function in a long chain is easily determined by mass spectrometry or by chain fission with periodic acid [3]. Epoxide configuration can be derived from the ^1H NMR spectrum (Section VI) or from the thin layer chromatographic behavior of epoxide-derived diols on silica impregnated with boric acid [17].

The stereochemical configuration of epoxy acids has been determined by gas chromatography of appropriate derivatives [18, 19]. If unsaturated, the epoxide is first subjected to mild catalytic hydrogenation, and the resulting saturated epoxide is converted to an allylic hydroxy compound by reaction with selenophenoxide anion and hydrogen peroxide. This is converted to its menthoxycarbonyl derivative, oxidized by ozonolysis, methylated with diazomethane, and analyzed by gas chromatography using appropriate reference compounds.

Hammond [20] has suggested HPLC examination of unsaturated oils after epoxidation as an alternative to silver ion chromatography. The oil is epoxidized with *m*-chloroperoxybenzoic acid, and the separation is effected on a silica column. Each double bond is replaced by an

Table 4 Composition of *Euphorbia lagascaea* Seed Oil

	Fatty acid						
	14:0	16:0	18:0	18:1	18:2	Vernolic Acid	Other
Whole oil	1.3	3.8	1.5	19.1	8.6	64.3	1.4
2-MAG[a]	—	—	—	10.2	13.8	75.0	1.0

[a] 2-Monoacylglycerol remaining after lipolysis.
Source: Ref. 14.

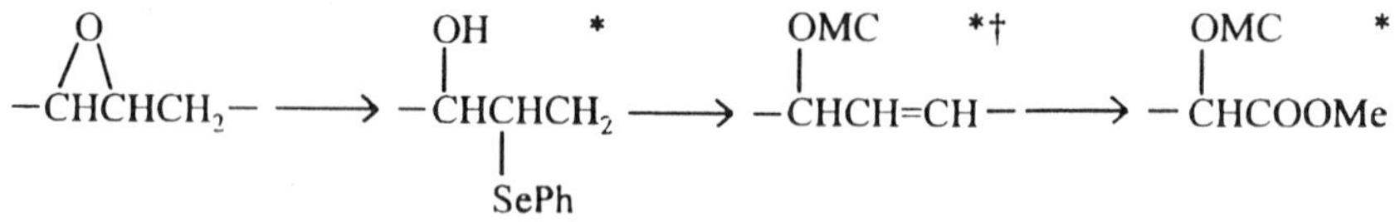

epoxide function, with the possibility of stereoisomerism when there is more than one such group. Though developed as a way of examining unsaturated oils, it can be applied directly to epoxy oils. The method is less satisfactory at high epoxide levels because of the complications arising from stereochemistry. It would therefore be more suitable for epoxidized palm oil than for epoxidized soybean or linseed oil. See also Reference 21.

Epoxy esters have long been examined by mass spectrometry, either as epoxides or as derivatives based on the epoxide function [22]. Marx and Classen [23] describe the mass spectra of the dimethyloxazoline derivatives prepared by reaction of the epoxy acid with amino isobutanol [$Me_2C(NH_2)CH_2OH$] alone at 180° for 1 h or in the presence of dicyclohexylcarbodimide at ambient temperatures for 4 h. Many fragments are observed in the mass spectrum, but the most significant result from cleavage α or β to this epoxide group. This is applied to compounds with up to three epoxide groups, which may be saturated or unsaturated.

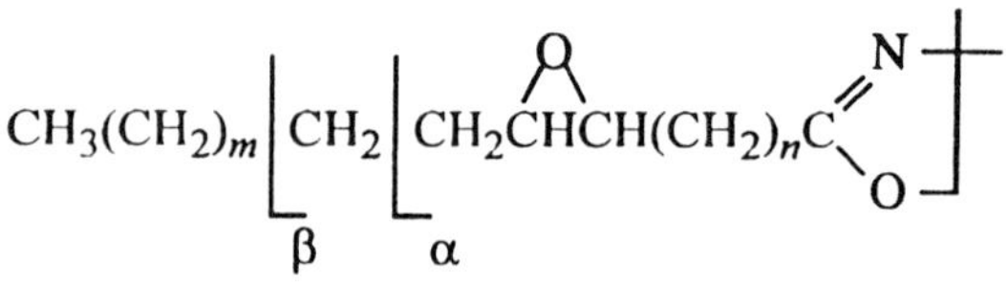

Gunstone [24] has used ^{13}C NMR to indicate the composition of epoxidized oils. Because of the presence of saturated acyl chains and chains with one to three epoxide groups, some of which exist in stereoisomeric forms, the spectra are fairly complex. Nevertheless it is possible to distinguish between and to determine the levels of triepoxides from α-linolenic acid, diepoxides from linoleic acid, monoepoxides from oleic acid, and nonepoxy compounds from saturated acids.

IV. THE EPOXIDATION PROCESS

Before considering examples of epoxidation in Section V, some general comments about the epoxidation process are in order.

- When carrying out epoxidation reactions it must always be remembered that the reaction is exothermic, that high concentrations of peroxy acid should be avoided, and that epoxides are acid-labile.
- The epoxidation reaction is second-order, and reaction rate depends on the concentrations of alkene and peroxy acid. The situation is more complex if the reaction is biphasic.
- The rate of epoxidation increases with the degree of substitution as shown in the relative rates for the following alkenes:

$CH_2{=}CH_2$	$CH_3CH{=}CH_2$	$CH_3CH{=}CHCH_3$	$(CH_3)_2C{=}CHCH_3$
1	25	600	6000

Oleic and linoleic acid are disubstituted alkenes and therefore react quite readily. *cis*-Alkenes are more reactive than their *trans* isomers.

*Mixtures of two compounds, only one structure shown; †mc = menthoxy carbonyl.

- The reactivity of peroxy acids is related to their pK_a values, and the order of reactivity of the more common peroxy acids is given by the listing

Peroxytrifluoroacetic > monoperoxymaleic > monoperoxyphthalic > *p*-nitroperoxybenzoic > *m*-chloroperoxybenzoic > peroxyformic > peroxybenzoic > peroxyacetic

These are not pure compounds, and the level of peroxy acid should always be determined before use.

- Reaction rates are influenced by solvent, and epoxidation is facilitated particularly by halogenated solvents. Many laboratory-scale epoxidations are carried out in dichloromethane solution.
- Epoxidation is a *cis* addition process, so *cis* alkenes give *cis* epoxides and *trans* alkenes give *trans* epoxides. A *cis*, *cis* diene, such as linoleic acid, will give two racemic diastereoisomeric bis epoxides. Bartlett's proposal for the mechanism of this reaction has been criticized but remains the most widely accepted

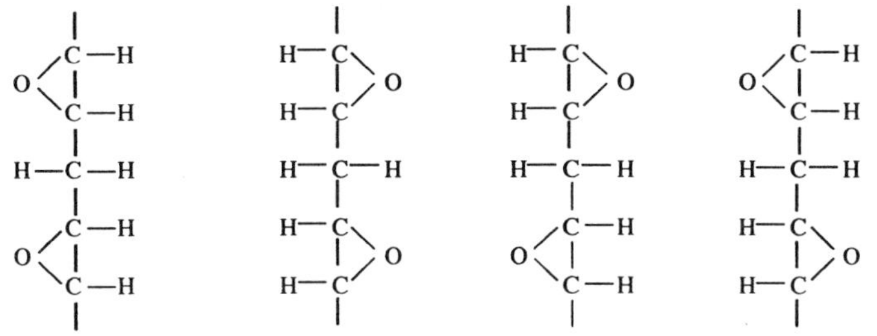

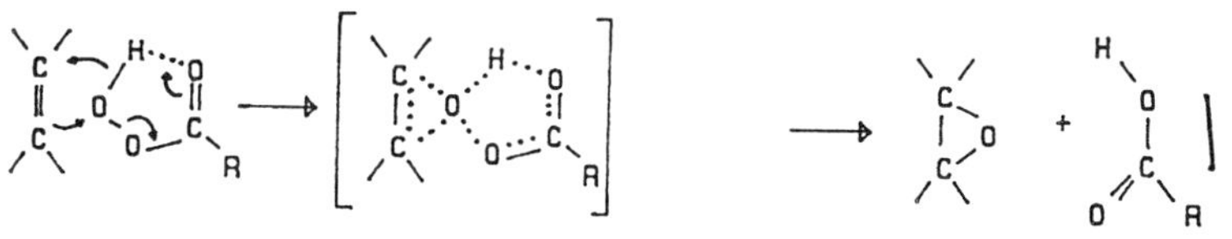

- The epoxides resulting from reaction between alkene and peroxy acid are reactive compounds, especially in acidic solution (Section VII). They are readily converted to monoacylated diols, and this is the major cause for reduced yields. With the more reactive carboxylic acids it may be necessary to buffer the reaction mixture to inhibit this further reaction.

$$-\mathrm{CH{=}CH}- \xrightarrow{RCO_3H} -\overset{\overset{O}{/\ \backslash}}{\mathrm{CHCH}}- \xrightarrow{RCO_2H} -\mathrm{CH(OH)CH(OCOR)}-$$

The kinetics of oxirane cleavage in epoxidized soybean oil by acetic acid at temperatures between 60 and 90° have been reported [25].

- The most important alkenes used as substrates for the purpose of this chapter include
  1. Glycerol esters containing olefinic acids, especially soybean oil and linseed oil, but also palm oil and some others
  2. Alkyl esters of olefinic acids such as oleic acid or tall oil fatty acids
  3. α-Olefins ($RCH{=}CH_2$) produced by oligomerization of ethene.
- The reagents most used on a commercial scale for the epoxidation of the materials listed above are peroxyacetic acid in various forms or peroxyformic acid. Other peroxy acids or mixtures of hydrogen peroxide and catalyst are used mainly for smaller scale or laboratory operations.

- Peroxyacetic acid can be made from acetic acid and hydrogen peroxide or by atmospheric oxidation of acetaldehyde (ethanal). It is available from a major supplier in a number of forms:
  1. Peroxyacetic acid in excess of acetic acid prepared in situ, i.e., in the presence of the alkene to be epoxidized.

$$CH_3CO_2H + H_2O_2 \rightleftharpoons CH_3CO_3H + H_2O \xrightarrow{\text{alkene}} \text{epoxide}$$

  2. Preformed peroxyacetic acid. One such preparation (Proxitane 4001) contains peroxyacetic acid (36–40%), acetic acid (44–46%), hydrogen peroxide (3–4%), water (13–17%), and sulfuric acid (0.5–1.0%). This is often used along with sodium acetate as a buffer.
  3. Distilled aqueous peroxyacetic acid (Proxilate) is made by collecting the peroxyacetic acid–water azeotrope from the distillation of a dilute aqueous peroxyacetic acid equilibrium mixture. It contains peroxyacetic acid (up to 35%) and water (60–65%) with only a little acetic acid (less than 5%). Its advantages are that it contains no mineral acid, has a low level of acetic acid, and has a high content of water. Its main disadvantage is its relatively high reversion rate to hydrogen peroxide and acetic acid. This makes it less suitable for transportation, and it is more desirable to produce it on site. The epoxidation reaction will usually be biphasic.
  4. Higher grades of peroxyacetic acid can be obtained from Proxilate by extraction into organic esters such as ethyl acetate (Proxisolv I) or isopropyl acetate (Proxisolv II). These contain 25–30% peroxy acid in ester solution. Epoxidation occurs in a one-phase system.

V. EXAMPLES OF EPOXIDATION

A. Epoxidation on a Technical Scale

Soybean oil is the most extensively epoxidized oil, with annual world production of around 100,000 t. This is generally produced with peroxyformic or peroxyacetic acid generated in situ from hydrogen peroxide and the appropriate alkanoic acid (Table 5). The final oxirane level at 6.5–7.1% is lower than might have been expected (Table 6). It depends on the level of unsaturation in the starting oil and the final product. The major side reaction is epoxide ring-opening, which leads to a lower oxirane value. Sometimes the reaction is carried out in the presence of a solvent such as hexane, heptane, or toluene.

Table 5 Typical Epoxidation Processes for Soybean Oil with (A) Peroxyformic Acid or (B) Peroxyacetic Acid Prepared in situ

	A	B
Soybean oil (units)	1	1
Formic acid (units)	0.14	—
Acetic acid (units)	—	0.2
Hydrogen peroxide (70% w/w) (units)	0.3	0.3
Sulfuric acid (units)	—	0.01
Temperature (°C)	60	60
Reaction time (h)	10–12	10–15

Source: Epoxidation (Interox, publishers).

Table 6 Oils Used for Epoxidation

Oil	Iodine value	Theoretical epoxide value[a]
Soybean	130	8.2
Linseed	180	11.3
Safflower	145	9.1
Sunflower	125	7.8
Rapeseed	99	6.2
Palm	54	3.4

[a]Based on the iodine value; the values actually attained are slightly lower than these.

Other vegetable oils can be epoxidized in a similar way, but only soybean and linseed oil are reacted on a commercial scale. Epoxidized palm oil is being exploited in Malaysia.

Epoxidized soybean oil, epoxidized linseed oil, and epoxidized octyl oleate are used mainly as plasticizers and stabilizers in PVC. They add flexibility to the polymer and scavenge any hydrogen chloride that is liberated by heat or light. Epoxidized palm oil with its lower oxirane value is less effective in this respect, but it is a satisfactory plasticizer.

B. Epoxidation on a Laboratory Scale

The same procedures can be applied on a laboratory scale though it is then more convenient to use a preformed peroxy acid such as m-chloroperoxybenzoic acid in dichloromethane solution. Hammond [20] suggests that *m*chloroperoxybenzoic acid be used at the level of 1.3 mol equiv (when adjusted for purity) for 1.25 h at 25–30°C. Some examples are listed in Table 7. The reaction is reported to be quicker if conducted in a microwave oven [35].

VI. SPECTROSCOPIC PROPERTIES

Reference has already been made to the use of infrared, *NMR*, and mass spectrometry in the recognition and identification of epoxy esters (Section III). Some additional information on NMR shifts (^{1}H and ^{13}C) is collected in Table 8. With more than one epoxide function the

Table 7 Epoxides Obtained from Unsaturated Oils, Acids, and Methyl Esters

Epoxides from	Ref.
18:1 isomers	26
18:2 isomers	27,28
Linoleate	29,30
Linolenate	31
Eicosatetraenoate	32
Eicosapentaenoate	33
Lesquerella oil	34
Meadowfoam oil	34

Table 8 Chemical Shifts (^1H and ^{13}C) in *cis* Epoxy Esters

Epoxy ester	^1H and ^{13}C shifts				
	a	*b*	*c*	*d*	*e*
$-CH_2CH(O)CHCH_2-$ (epoxide O over CHCH); *a b b a*	1.43	2.70*	—	—	—
$-CH_2CH(O)CHCH_2CH(O)CHCH_2-$ (two epoxides); *a b c d c b a*	1.45	2.79	2.91	1.60	—
$-CH_2CH(O)CHCHCH{=}CHCH_2-$ (epoxide O over CHCH); *a b b c d d e*	1.45	2.73	2.1–2.3	5.41	2.03
$-CH_2CH_2CH_2CH(O)CHCHCH_2CH_2-$ (epoxide O over CHCH); *d c b a a b c d*	56.67	27.48	26.25	28.85	*(cis)*
	58.84	31.89	25.80	28.98	*(trans)*

[a]2.45 in *trans* isomer.
Source: Refs. 24, 26–28.

^{13}C NMR spectra become rather complex, though, as already indicated, some useful information can be obtained [24].

VII. CHEMICAL REACTIONS

The epoxide function is very reactive, especially in acid solution, and undergoes a wide range of reactions.

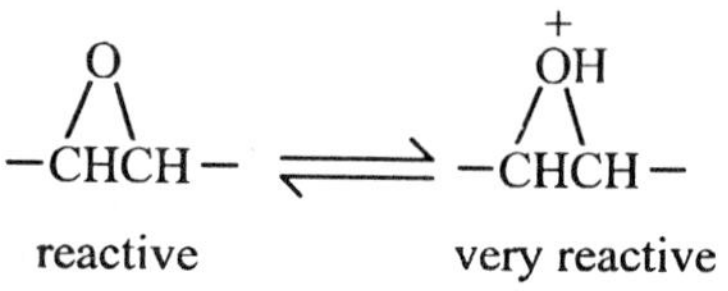

Addition of molecules designated HZ, usually in the presence of acid, will give a mixture of two products if the epoxide is unsymmetrical:

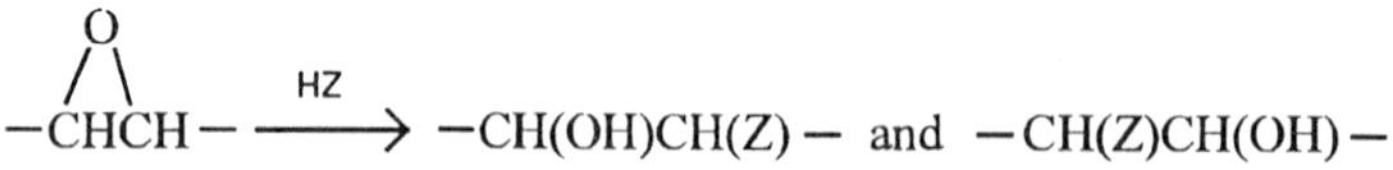

The reaction with hydrogen chloride to give a chlorohydrin is the basis of the use of epoxides as stabilizer for PVC, which can liberate hydrogen chloride under the action of light or heat. Other examples, with the product type indicated in parentheses are water (diol), alcohol (hydroxy ethers), carboxylic acid (hydroxy esters that are readily hydrolyzed to diols), hydrogen sulfide (hydroxy thiols), sulfites (hydroxy sulfonates), thiourea or thiocyanates (thioepoxides), hydrogen cyanide (hydroxy nitriles), alkyl carboxylates (lactones), ammonia, or primary or secondary amines (hydroxyamines), carbon dioxide (carbonate), and others.

The acid-catalyzed alcoholysis of methyl vernolate with methanol, ethanol, propanol, and isopropanol at reflux temperature takes 4, 48, 72, and 264 h, respectively. The product is mainly the 12-hydroxy-13-alkoxy derivative rather than its isomer (80:20) [36]. These are mainly derivatives of ricinoleic acid, and it would be interesting to study them in the typical reactions of this acid. Reaction with diols gives polymeric material with free hydroxyl groups that can be converted to polyurethanes [37].

Reaction between epoxidized palm oil and acrylic acid gives a vinyl derivative that can be cured by radiation:

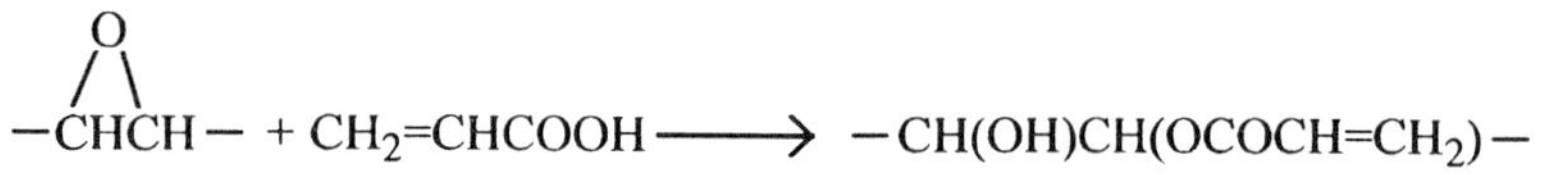

Rearrangement of epoxides to ketones occurs with sodium iodide in polyethylene glycol, with 9,10-epoxystearate being converted to the 9- and 10-oxostearates [38].

Reaction with glycine (in the presence of aluminum trichloride and dimethylformamide) gives a fatty morpholide that shows activity against bacterial and fungal groups [39].

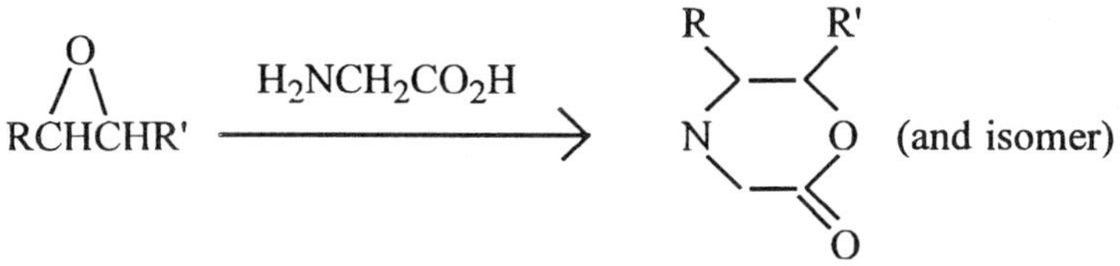

Other reactions of fatty epoxides are related to the acid/ester function and (when present) to the double bond. For example, epoxy oils can be converted to amides with C_4–C_6 amines at room temperature (reaction is 80% complete after several days), at reflux temperatures (1–5 h), or in boiling hexane solution (2–5 days) [40]. Compounds of this type have been used as antirust derivatives in water-based cutting fluids [41].

Several dibasic acids have been prepared from vernonia oil. Oxidation with nitric acid (18 h, 90–100°C) gives a mixture of C_5–C_{11} dibasic acids from which 95% pure suberic (C_8) and azelaic (C_9) acids have been isolated in yields of 15 and 11%, respectively [42]. After saponification and hydrogenation, oxidation of the 12,13-epoxystearic acid with chromic acid gives dodecanedioic acid (C_{12}) in 4–7 days at room temperature [43].

The oil reacts with dibasic acids such as suberic to give a polyester that produces a toughened elastomer with styrene and with divinylbenzene [44].

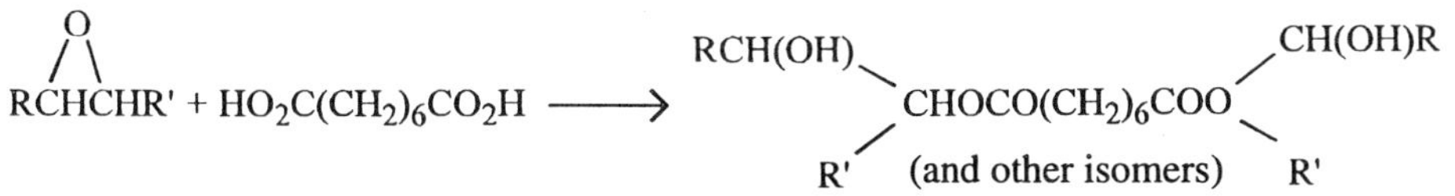

In attempts to produce high value products from new oilseed crops, Derkson and Cuperus [45] converted vernolic acid from Euphorbia lagascae to compounds such as bombykol, bombykal, traumatic acid, and some lactones.

Bombykol (C_{16})	$CH_3(CH_2)_2CH{=}CHCH{=}CH(CH_2)_8CH_2OH$
Traumatic acid (C_{12})	$HOOCCH{=}CH(CH_2)_8COOH$

REFERENCES

1. D. Swern, Organic peroxy acids as oxidizing agents. Epoxidation, in *Organic Peroxides* (D. Swern, Ed.), Wiley Interscience, New York, 1971, p. 355.

2. F. D. Gunstone, The formation and reactions of long-chain epoxy acids, in *Fatty Acids* (E. H. Pryde, Ed.), AOCS, Champaign, IL, 1979, p. 379.
3. F. D. Gunstone, The nature of the oxygenated acid present in *Vernonia anthelmintica* seed oil, *J. Chem. Soc. (Lond.)*: 1611 (1954).
4. F. D. Gunstone, J. T. Harwood, and F. B. Padley (Eds.), Lipid Handbook, 2nd, Chapman and Hall, London, 1994.
5. E. Blée, Oxygenated fatty acids and plant defenses, *INFORM 6*: 852 (1995).
6. Editor, Natural plant material, *Oils Fats Int.* (2): 9 (1990).
7. F. O. Ayorinde, K. D. Carlson, R. P. Paulik, and J. McVety, Pilot plant extraction of oil from *Vernonia galamensis* seed, *J. Am. Oil. Chem. Soc. 67*: 512 (1990).
8. F. O. Ayorinde, B. D. Butler, and M. T. Clayton, *Vernonia galamensis*: a rich source of epoxy acid, *J. Am. Oil. Chem. Soc. 67*: 844 (1990).
9. K. D. Carlson and S. P. Chang, Chemical epoxidation of a natural unsaturated epoxy seed oil from *Vernonia galamensis* and a look at epoxy oil markets, *J. Am. Oil. Chem. Soc. 62*: 934 (1985).
10. W. E. Neff, R. O. Adlof, H. Konishi, and D. Weisleder, HPLC of the triacylglycerols of *Vernonia galamensis* and *Crepis alpina* seed oils, *J. Am. Oil. Chem. Soc. 70*: 449 (1993).
11. M. A. Anderson, L. Collier, R. Dilliplane, and F. D. Ayorinde, Mass spectrometric characterisation of *Vernonia galamensis* oil, *J. Am. Oil Chem. Soc. 70*: 905 (1993).
12. E.-S. Yalley, F. D. Ayorinde, and B. E. Eribo, Biodegradation of *Vernonia galamensis* seed oil by *Acinetobacter* and *Pseudomonas* sp., *J. Am. Oil. Chem. Soc. 69*: 1046 (1992).
13. L. J. M. van Soest and F. Mulder, Potential new oilseed crops for industrial use, *Lipid Technol. 5*: 60 (1993).
14. B. G. Muuse, F. D. Cuperus, and J. T. P. Derksen, *Ind. Crops Products 1*: 57 (1992).
15. J. T. P. Derksen, F. P. Cuperus, and P. Kolster, Paints and coatings from renewable resources, *Ind. Crops and Products 3*: 225 (1995).
16. American Oil Chemists' Society, Oxirane oxygen, AOCS method Cd 9-57 (89), AOCS, Champaign, IL.
17. L. J. Morris, Separation of long-chain polyhydroxy acids by thin layer chromatography, *J. Chromatog. 12*: 321 (1963).
18. M. Hamberg, A method for determination of the absolute stereochemistry of $\alpha\beta$-epoxy alcohols derived from fatty acid hydroperoxides, *Lipids 27*: 1042 (1992).
19. P. Fahlstadius and M. Hamberg, A gas-liquid chromatographic method for steric analysis of epoxy acids, *Chem. Phys. Lipids 51*: 15 (1989).
20. E. W. Hammond, *Chromatography for the Analysis of Lipids*, CRC Press, Boca Raton, FL, 1993, pp. 147–152.
21. H. C. Gérard, R. A. Moreau, W. F. Fett, and S. F. Osman, Separation and quantitation of hydroxy and epoxy fatty acids by HPLC with an evaporative light-scattering detector, *J. Am. Oil Chem. Soc. 69*: 301 (1992).
22. F. D. Gunstone, *Fatty Acid and Lipid Chemistry,* Blackie, London, 1996, p. 150.
23. F. Marx and E. Classen, Analysis of epoxy fatty acids by GC-MS of their 4,4-dimethyloxazoline derivatives, *Fat Sci. Technol. 96*: 207 (1994).
24. F. D. Gunstone, The study of natural epoxy oils and epoxidised vegetable oils by ^{13}C NMR spectroscopy, *J. Am. Oil Chem. Soc. 70*: 1139 (1993).
25. F. A. Zaher, M. H. El-Mallah, and M. M. El-Hefnawy, Kinetics of oxirane cleavage in epoxidised soybean oil, *J. Am. Oil Chem. Soc. 66*: 698 (1989).
26. F. D. Gunstone and F. R. Jacobsberg, The preparation and properties of the complete series of methyl epoxyoctadecanoates, *Chem. Phys. Lipids 9*: 26 (1972).
27. F. D. Gunstone and H. R. Schuler, Epoxyoctadecenoates, dihydroxyoctadecenoates, and diepoxyoctadecanoates, *Chem. Phys. Lipids 15*: 174 (1975).
28. F. D. Gunstone and H. R. Schuler, PMR spectra of several epoxyoctadecenoic, epoxyoctadecynoic, and diepoxyoctadecanoic esters, *Chem. Phys. Lipids 15*: 189 (1975).
29. G. Maerker, E. T. Haeberer, and S. F. Herb, Epoxidation of methyl linoleate. II. The two isomers of methyl 9,10;12,13-diepoxystearate, *J. Am. Oil Chem. Soc. 43*: 505 (1966).

30. A. Debal, G. Rafaralahitsimba, A. Bonfand, and E. Ucciano, Catalytic epoxidation of methyl linoleate—cyclisation products of the epoxy acid esters, *Fat Sci. Technol. 97*: 269 (1995).
31. M. Alaiz, M. P. Maza, R. Zamora, F. J. Hidalgo, F. Millán, and E. Vioque, Epoxidation of ethyl (Z)-9-(Z)-12-(Z-15)-octadecatrienoate with *m*-chloroperbenzoic acid, *Chem. Phys. Lipids 49*: 221 (1989).
32. E. J. Corey, H. Niwa, and J. R. Falck, Selective epoxidation of eicosa-5,8,11,14-tetraenoic (arachidonic) acid and eicosa-8,11,14-trienoic acid, *J. Am. Chem. Soc. 101*: 1586 (1979).
33. M. Van Rollins, Synthesis and characterisation of cytochome P-450 epoxygenase metabolites of eicosapentaenoic acid, *Lipids 25*: 481 (1990).
34. K. D. Carlson, R. Kleiman, and M. O. Bagby, Epoxidation of *Lesquerella* and *Linanathes* (meadowfoam) oils, *J. Am. Oil Chem. Soc. 71*: 175 (1994).
35. M. S. Lie Ken Jie and C. Yan-Kit, The use of a microwave oven in the chemical transformation of long-chain fatty acid esters, *Lipids 23*: 367 (1988).
36. K. A. A. Bryant, C. P. Nwaonicha, M. A. Anderson, and F. O. Ayorinde, Acid-catalyzed alcoholysis of *Vernonia galamensis* oil, *J. Am. Oil Chem. Soc. 69*: 1023 (1992).
37. B. Dahlke, S. Helbart, M. Paetow, and W. H. Zech, Polyhydroxy fatty acids and their derivatives from plant oils, *J. Am. Oil Chem. Soc. 72*: 349 (1995).
38. G. Stoll and K. Worschech, Keto fatty acids—a simple procedure for their preparation, *Fat Sci. Technol. 94*: 332 (1992).
39. R. Agarwal, M. H. Ansari, M. W. Y. Khan, M. Ahmad, and K. D. Sharma, Synthesis and antimicrobial activity of fatty 2-morpholinones prepared from epoxy fatty acid methyl esters, *J. Am. Oil. Chem. Soc. 6*: 825 (1989).
40. K. A. A. Bryant, C. P. Nwaoicha, M. Hassan, M. A. Anderson, and F. O. Ayorinde, Synthesis and isolation of epoxy secondary amides via direct amidation of *Vernonia galamensis* seed oil, *J. Am. Oil Chem. Soc. 70*: 457 (1993).
41. S. Watanabe, T. Fujita, and M. Sakamoto, Epoxides of unsaturated fatty acids as anti-rust additives in water-based cutting fluids, *J. Am. Oil Chem. Soc. 65*: 1311 (1988).
42. F. O. Ayorinde, G. Osman, R. L. Sheppard, and F. T. Powers, Synthesis of azelaic acid and suberic acid from *Vernonia galamensis* seed oil, *J. Am. Oil Chem. Soc. 65*: 1774 (1985).
43. F. O. Ayorinde, F. T. Powers, L. D. Streete, R. L. Shepard, and D. N. Tabi, Synthesis of dodecanedioic acid from *Vernonia galamensis* oil, *J. Am. Oil Chem. Soc. 66*: 690 (1989).
44. O. A. Ofolabi, M. E. Aluko, G. C. Wang, W. A. Anderson, and F. O. Ayorinde, Synthesis of toughened elastomer from *Vernonia galamensis* seed oil, *J. Am. Oil Chem. Soc. 66*: 983 (1989).
45. J. T. P. Derkson and F. P. Cuperus, High value-added applications from unusual fatty acids, Lecture presented to ISF Congress, The Hague, October 1995.

30

Biofuels

Y. M. Choo and A. N. Ma
Palm Oil Research Institute of Malaysia (PORIM), Kajang, Selangor Darul Ehsan, Malaysia

A. S. H. Ong
Malaysian Palm Oil Promotion Council, Kuala Lumpur, Malaysia

I. INTRODUCTION

The energy crisis in the mid-1970s coupled with the fast diminishing energy reserves, greater environmental awareness, and increasing energy consumption have led to an intensified global search for viable alternative sources of energy. In recent years a great deal of attention has been directed to plant-based sources for fuels, particularly fuels for diesel engines. Some plants produce oil directly. Vegetable oils that are renewable have the greatest potential as alternative fuel sources. The choice of vegetable oil depends on the cost of production and reliability of supply. At present, annual worldwide vegetable oil production is around 77 million tonnes, with 14.3 million tonnes of soybean oil, 10 million tonnes of palm oil, and 7.7 million tonnes of rapeseed and sunflower oil [1]. Of all the vegetable oils available, palm oil, being the highest oil yielding crop would be the preferred choice (Table 1). Since factors other than yield must be considered, biofuels are made from palm (Malaysia), coconut (Philippines), rapeseed (Europe), and soy and tallow (United States), as well as from used frying oil (Japan and other countries), i.e., in general from oils in greatest supply.

Many researchers have investigated the possibility of using vegetable oils straight or blended as diesel substitutes. In fact, the use of vegetable oil as fuel for compression-ignition engines is not new. The inventor of the diesel engine, Rudolf Diesel, displayed an engine fueled with peanut oil at the Paris Exposition of 1900 [3]. The use of palm oil as a diesel fuel was reported as early as 1920 [4]. Since then many studies have been conducted in various parts of the world, particularly in the period 1920–1940 [3,5–8]. The 1973 oil embargo and the ensuing energy crisis reactivated the search for alternative sources of fuel. A good account of these attempts is given by Goering et al. [9] and in the 1983 AOCS symposium on vegetable oils as diesel fuels [10–13]. The AOCS symposium revealed that vegetable oils have good potential as alternative fuels if the following problems could be overcome satisfactorily: high viscosity, low volatility, and the reactivity (polymerization) of the unsaturated hydrocarbon chains when the oils are highly unsaturated. These give rise to cooking on the injectors,

Table 1 Average Yield of Oil-Producing Crops

Oil	Yield [kg/(ha·yr)]
Palm	3636
Coconut	1272
Peanut	909
Soybean	432
Rapeseed	999
Sunflower	801

Source: Ref. 2.

carbon deposits, oil ring sticking, and thickening and gelling of the lubricating oil as a result of contamination with vegetable oils.

It is possible to reduce the viscosity of the vegetable oils by incorporating a heating device in the diesel engine as has been successfully demonstrated by the Elsbett engine manufacturer [14]. Other factors that may have long-term effects on the engine are free fatty acids and gummy substances found in the crude vegetable oils. The incomplete combustion residues may contribute to undesirable deposits on the engine components. The gummy substances cause filter plugging. This calls for more regular and frequent servicing and maintenance of the engine. The long-term effects of the oil on the engine have to be monitored.

"Biofuel" is the term usually used to describe biodiesel. According to Reed [15], the term "biodiesel" refers to methyl or ethyl esters derived from vegetable oils. It can also refer to pyrolysis products, diesel–vegetable oil blends, microemulsions of alcohols and water in vegetable oils, and fermentation butanol. This chapter, however, deals exclusively with alkyl esters derived from vegetable oils. Esters are used because they have lower viscosity and better fuel properties than vegetable oils. Vegetable oil esters are used on a commercial scale in many cities of the world where there are air pollution problems. Engines run on vegetable oil esters are known to emit less dark smoke and CO and no SO_2. Though vegetable oil esters cannot meet total diesel demand in the transport sector, they can help reduce air pollution if they are used in heavy vehicles such as buses and trucks.

II. PRODUCTION TECHNOLOGY OF ALKYL ESTERS FROM VEGETABLE OILS AND FATS

Commercially, methyl esters of fatty acids can be manufactured either by esterification of fatty acids or by transesterification of fat triglycerides. The predominant process for the manufacture of methyl esters is the transesterification of fats and oils with methanol. The ester interchange, i.e., the replacement of the alcohol component glycerol by methanol, takes place quite easily at low temperatures, 50–70°C, and under atmospheric pressure with an excess of methanol and in the presence of an alkaline catalyst such as sodium hydroxide [16–19].

These mild reaction conditions, however, require an oil neutralized by means of alkaline refining, steam distillation, or preesterification of free fatty acids.

Esterification may be carried out batchwise at 200–250°C under pressure. In order to obtain a high yield, the water produced during the reaction has to be removed continuously. Esterification can also be carried out continuously in a countercurrent reaction column using superheated methanol [17]. Esterification is the preferred method for ester preparation from specific fatty acid cuts such as palm stearin and olein.

Vegetable oil + (Triglyceride) 1 tonne	Alcohol 0.1 tonne	---------->	Esters 1 tonne	+	Glycerol 0.1 tonne

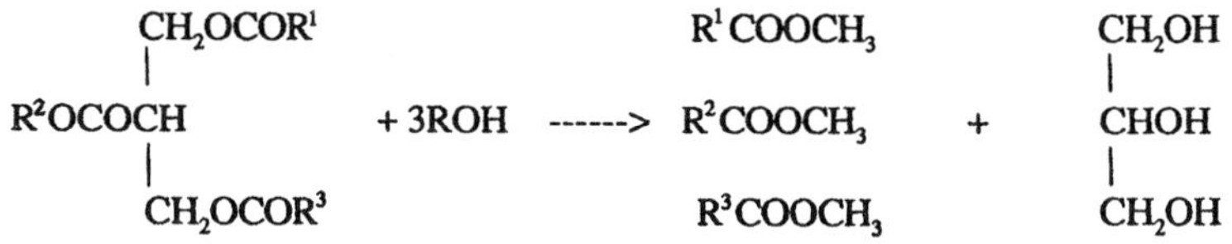

Some of the available ester production technology for biodiesel application is shown in Table 2. It is also possible to convert vegetable oils such as crude palm oil to esters with varying amounts of free fatty acids in a continuous process by combining the esterification and transesterification processes as developed by PORIM. This has been successfully demonstrated in a 3000 tpa pilot plant [21].

The novel aspect of this patented process is the use of solid acid catalysts for the esterification [21–23]. The reaction is carried out below 100°C at atmospheric pressure in a column of solid catalysts. The resultant reaction mixture, which is neutral, is then transesterified in the presence of an alkali catalyst. The conventional washing stage or neutralization step after the esterification process is obviated, and this is an economic advantage.

III. EVALUATION OF VEGETABLE OIL ALKYL ESTERS AS BIOFUEL

The use of vegetable oils in the form of esters as diesel substitute has been quite extensively reviewed [1,24,25]. Vegetable oils such as soybean, rapeseed, sunflower, canola, coconut,

Table 2 Available Biodiesel Production Technologies

	Reaction conditions			
Company	*P* (atm)	*T* (°C)	Catalyst	Mode of operation
Transesterification				
Comprimo/Vogel and Nott	1	Ambient	KOH	Batch
University of Idaho	1	Ambient	KOH	Batch
Novamont/Technimont	1	>Ambient	Organic	Batch
Conneman/Feld and Hahn	1	60–70	NaOH	Continuous
Lurgi	1	60–70	Alkaline	Continuous
IFP/sofiproteal	1	50–130	Alkaline/acidic	Batch
Gratech	3.5	95	?	Continuous
Desmet	50	200	Nonalkaline	Continuous
PORIM	1	<100	Acidic/alkaline	Continuous
Others: Oleofina,				
Procter & Gamble, Me KFT		360–380		
Hydrogenation	40–100		Cobalt/molybdenum	Continuous
Cammit/Arbokem				

Source: Refs. 1 and 2.

Table 3 Fatty Acid Composition of Vegetable Oils and Tallow

Fatty acid	Coconut oil	Babassu oil	Castor oil	Cotton-seed oil	Karanja oil	Lin-seed oil	Maize oil	Palm oil	Palmkernel oil
6:0	0.0–0.6	nd	—	nd	—	—	nd	ns	0.0–0.8
8:0	4.6–9.4	2.6–7.3	—	nd	—	—	nd	ns	2.4–6.2
10:0	5.5–7.8	1.2–7.6	—	nd	—	—	nd	ns	2.6–5.0
12:0	45.1–50.3	40.0–55.0	—	0.0–0.2	—	—	0.0–0.3	0.0–0.4	41.0–55.0
14:0	16.8–20.6	11.0–27.0	—	0.6–1.0	—	—	0.0–0.3	0.5–2.0	14.0–18.0
16:0	7.7–10.2	5.2–11.0	1	21.4–26.4	10	8	8.6–16.5	40.1–47.5	6.5–10.0
16:1	nd	nd	—	0.0–1.2	—	—	0.0–0.4	0.0–0.6	ns
17:0	nd	nd	—	nd	—	—	nd	ns	ns
17:1	nd	nd	—	nd	—	—	nd	ns	ns
18:0	2.3–3.5	1.8–7.4	1[a]	2.1–3.3	7	5	1.0–3.3	3.5–6.0	1.3–3.0
18:1	5.4–8.1	9.0–20.0	3	14.7–21.7	49	21	20.0–42.2	36.0–44.0	12.0–19.0
18:2	1.0–2.1	1.4–6.6	4	46.7–58.2	19	64	39.4–62.5	6.5–12.0	1.0–3.5
18:3	0.0–0.2	nd	tr	0.0–0.4	—	2	0.5–1.5	0.0–0.5	—
20:0	0.0–0.2	nd	—	0.2–0.5	4	—	0.3–0.6	0.0–1.0	—
20:1	0.0–0.2	nd	—	0.0–0.1	—	—	0.2–0.4	ns	}
20:2	nd	nd	—	0.0–0.1	—	—	0.0–0.1	ns	}
22:0	nd	nd	—	0.0–0.6	5	—	0.0–0.5	ns	}
22:1	nd	nd	—	0.0–0.3	}	—	0.0–0.1	ns	} 0.0–1.0%
22:2	nd	nd	—	0.0–0.1	} 6	—	nd	ns	}
24:0	nd	nd	—	0.0–0.1	}	—	0.0–0.4	ns	}
24:1	nd	nd	—	nd	}	—	nd	ns	
			90[b]						

corn, safflower, cottonseed, peanut [26], palm [2], buffalo gourd, and Chinese vegetable tallow [27] have been studied and reported to be potential feedstocks for biodiesel production. Animal fats such as tallow [28] and used cooking oils of all types [29] have also been considered for use as fuels.

Table 3 shows the fatty acid composition of the vegetable oils used for biodiesel productions [30]. Table 4 shows the fuel characteristics of the methyl esters of these oils [25], and Table 5 summarizes engine testing of alkyl esters from various vegetable oils and tallow [31–50].

A. Fuel Characteristics

The fuel properties of vegetable oil esters (Table 4) clearly show that they can be attractive as diesel substitutes. The specific gravities of all the vegetable oil esters are slightly higher than that of diesel. Though their viscosities are somewhat higher, they flow under warm weather conditions. The fluidity of the esters depends on the fatty acid composition (FAC) or the iodine values (IV) (unsaturation) of the vegetable oils. Obviously, oils with high IV will produce esters with better fluidity and will not solidify at low temperature. For example, palm oil has an IV of 52, and its esters, which have a pour point of 15°C, will solidify at that temperature. Rapeseed oil, which has an IV of >90, will solidify at a much lower temperatures, about 4°C. However, the high pour point can be lowered by adding pour point depressing agents or by blending with solvents [25].

Palm stearin	Peanut oil	Rapeseed oil	Rapeseed (low erucic acid)	Ricebran oil	Soybean oil	Sunflower oil	Premier and edible tallow
nd	—		nd	—	nd	nd	
nd	—	0.1	nd	—	nd	nd	
nd	—		nd	—	nd	nd	<2.5
0.1–0.4	0–0.1		nd	—	0.0–0.1	0.0–0.1	
1.1–1.8	0–0.1	0.2	0.0–0.2	0.4	0.0–0.2	0.0–0.2	
48.4–73.8	8.3–14.0	1.5–6.0	3.3–6.0	16.4	8.0–13.3	5.6–7.6	1.4–78
0.05–0.2	0–0.2	0.0–3.0	0.1–0.6	0.1	0.0–0.2	0.0–0.3	17–37
nd	nd	nd	0.0–0.3	—	nd	nd	0.7–8.8
nd	nd	nd	0.0–0.3	—	nd	nd	0.5–2.0
3.9–5.6	1.9–4.4	0.5–3.1	1.1–2.5	1.8	2.4–5.4	2.7–6.5	<1.0
15.6–36.0	36.4–67.1	8–60	52.0–66.9	42.3	17.7–26.1	14.0–39.4	6.0–40
3.2–9.8	14.0–43.0	11–23	16.1–24.8	37.1	49.8–57.1	48.3–74.0	26–50
0.1–0.6	0–0.1	5–13	6.4–14.1	1.3	5.5–9.5	0.0–0.2	0.5–5.0
0.3–0.6	1.1–1.7	0.0–3.0	0.2–0.8	0.4	0.1–0.6	0.2–0.4	<2.5
nd	0.7–1.7	3–15	0.1–3.4	0.2	0.0–0.3	0.0–0.2	<0.5
nd	—	0.0–1.0	0.0–0.1	—	0.0–0.1	nd	<0.5
nd	2.1–4.4	0.0–2.0	0.0–0.5	—	0.3–0.7	0.5–1.3	
nd	0–0.3	5–60	0.0–2.0	—	0.0–0.3	0.0–0.2	
nd	—	0.0–2.0	0.0–0.1	—	nd	0.0–0.3	
nd	1.1–2.2	0.0–2.0	0.0–0.2	—	0.0–0.4	0.2–0.3	
nd	0–0.3	0.0–3.0	0.0–0.4	—	nd	nd	

nd = not detected; ns = not specified.
[a]Dihydroxystearic acid.
[b]Ricinoleic acid 18:1 (OH).
Source: Ref. 30.

All vegetable oil esters have higher cetane numbers and flash points than petroleum diesel. The higher cetane numbers indicate shorter ignition delay characteristics, and the higher flash points represent greater safety in storage and transportation.

B. Engine Testing

Engine testing of alternative fuels usually comprises two phases: (1) bench endurance tests (single-cylinder or multicylinder engine tests) and (2) vehicle field tests.

1. Bench Endurance Tests

Many researchers have carried out bench endurance tests with biodiesel from various vegetable oils as feedstocks [31–54]. Table 5 summarizes the results reported by them. As the tests were carried out under different conditions with different engine makes, it is expected that the results will vary accordingly.

However, in general, biodiesel shows improved exhaust emission compared to diesel in terms of smoke opacity, unburned hydrocarbons (HC), and carbon monoxide (CO). Though sulfur dioxide (SO_2) was not monitored except for palm oil methyl esters, it is expected to be much lower as all the vegetable oils (with the possible exception of poorly refined rapeseed

Table 4 Fuel Characteristic of Methyl Esters of Various Vegetable Oils

	Vegetable oil										
Test conducted	Coconut	Palmkernel	Palm	Peanut	Cotton-seed	Rapeseed (erucic)	Rapeseed (non-erucic)	Sun-flower	Soybean	Linseed	Typical diesel fuel
Iodine number	10.0	15.0	52.0	94.0	105.0	98.0	114.0	129.0	131.0	183.0	—
Density at 15°C (g/L)	872.0		872.0–877.0				882.0	885.0		891.0	830.0–840.0
Lower heating value											
MJ/kg	35.3	5.5	37.0	37.2	37.0	37.7	37.2	37.1	37.1	37.0	42.7
MJ/L	30.8		32.4				32.8	32.8		33.0	35.5
Viscosity at 40°C (mm^2/s)	2.7	4.3–4.5	4.3–4.5				4.2	4.0		3.7	2.0–3.5
Cetane number	62.7		64.3–70.0				51.0–59.7	61.2	56.0	52.5	51.0

Source: Ref. 25.

oil) contain negligible amounts of sulfur. There are conflicting results on the nitrogen oxides (NO_x) in the exhaust emissions. Most reported an increase in NO_x with biodiesel fuel.

All the tests experienced a drop in engine power output and increased fuel consumption with biodiesel fuel compared to diesel fuel. This is expected to be marginal as the energy content of biodiesel is about 10% less than that of diesel. Most of the differences reported, especially in the combustion process, were mainly attributed to differences in the ignition delay and load conditions between the two fuels.

Some material compatibility problems with biodiesel have been observed and reported [2,35,38,55,56]. Paints, various plastic and rubber construction materials, and various adhesives experience degradation through hardening and/or deterioration. These observations suggest that fuel lines and other components should be replaced with a more compatible material.

2. *Vehicle Field Tests*

There is limited published information available on the exhaustive field testing of biodiesel [1,2,25,55–58]. In France, 5–100% blends of biodiesel (from rapeseed oil) and diesel are used experimentally in buses and trucks in about 30 cities [1]. In Italy, it was reported that 19 cities used Iveco engines in 72 buses running on biodiesel starting in the latter part of 1991 [1] in a field test involving 520,000 km of operation. Results showed that the power output and fuel costs were comparable to those of diesel fuel. There was marked reduction in smoke emissions, unburned hydrocarbons, and particulate matter and little dilution of lubricating. In Germany, following the favorable results obtained with both single-cylinder and multicylinder engine bench tests, Mercedes-Benz AG conducted exhaustive field tests on the rapeseed oil methyl esters (RME) in trucks, buses, and taxis from January 1991 to October 1992 [25]. Very encouraging results were obtained: Fuel consumption and lubricating oil consumption were comparable to those obtained during diesel fuel operation and the drivers were satisfied with the vehicles' behavior; the slightly lower rated output due to the lower caloric value of the RME only rarely led to complaints. However, the unpleasant odor of the exhaust gas (smell like French fries) was noted! This is very subjective. Also there were some problems with the auxiliary heaters installed in the vehicle, which must be operated with RME. Similarly, in the United States, blends of biodiesel from soybean oil and diesel in a 20 : 80 or 30 : 70 ratio were used to run 1200 buses, trailer trucks, and vans in 400 cities in 1992–1993 in operational tests over a total of 4.8 million km [1]. Observations similar to those in tests conducted in Europe were obtained. It was also found that biodiesel involves no technical problems and has many energy and environmental advantages, particularly in urban areas. Field trial results have so far confirmed that biodiesel and diesel behave in a similar manner and that the use of biodiesel requires no specific engine modification.

Perhaps the most systematic and thorough evaluation of vegetable oil methyl esters as diesel substitutes was conducted in Malaysia. Over the past ten years, palm oil methyl esters have been extensively evaluated as diesel substitutes by evaluating their fuel characteristics, followed by bench endurance tests and finally exhaustive vehicle field trials involving a wide range of diesel engines such as stationary engines, passenger cars, lorries, trucks, taxis, tractors, generators, land cruisers, portable generators, water pumps, and vans running in Malaysia [2]. There were two fleets of a total of 30 buses mounted with Mercedes engines operating in Kuala Lumpur, the capital city of Malaysia using crude palm oil methyl esters as fuel from 1992 to 1995. Direct injection engines were used in these fleet tests. Each bus traveled 300,000 km, the exhaustive field trial distance recommended by the engine manufacturer. Ten buses were run on 100% palm oil methyl esters, 10 on 50 : 50 blends of palm oil methyl esters and petroleum diesel, and 10 on 100% petroleum diesel as control. Another fleet of six

Table 5 Engine Testing

Biofuel	Engine test	Engine investigated	Observations and results compared to petroleum diesel	Ref.
1. Soybean oil methyl Esters (SBME)	Five endurance tests at 200 h each	4C, D1	Lube oil viscosity decreased.	31
	EMA cycle 200 h. (Tests include soybean oil ethyl esters and soybean oil butyl esters)	4C, DI, TC	NOx and fuel consumption increased. Power output decreased.	32
	Engine Durability Test 578 h.	6C, DI, NA	Fuel consumption increased. Lube oil viscosity decreased.	33
	Various load conditions; 20–100% rated output.	1C, DI	Smoke increased. Thermal efficiency and ignition delay decreased.	34
	Power tests and emission tests.	6C, DI, TC, 1C	Fuel consumption and NOx increased. Power output, smoke, CO, hydrocarbons, and particulates decreased.	35
	On-road vehicle test 550 h; power tests and emission tests.	6C, DI, 1C	Power output increased. CO, CO_2, and hydrocarbon decreased.	35
	Various load conditions.	6C, DI, TC	Fuel consumption and NOx increased. CO, hydrocarbons, power output, and particulates decreased.	36
	Field tests	4C, DI 6C, DI	White smoke increased (reduced by advance in injection timing). Lube oil viscosity and black smoke decreased.	31
	Various loads BMEP: 50–600 kPa at 1800 rpm. Retard injection timing 5 deg.	4C, DI, NA	Hydrocarbons, smoke, and CO decreased. Hydrocarbon increased NOx decreased.	37
	On-road vehicle test 825 h and power tests.	6C, DI, TC	Power output increased.	38
	On-road vehicle test 535 h and power tests.	DI, TC, IC	Power output decreased.	38

	ISO 8178 C1 (Tests include blending with petroleum diesel)	4C, DI, TC 6C, DI, TC 6C, DI, TC and IC 4C, DI, NA 4C, DI, TC	NOx increased as concn. of SBME increased. Power output, smoke, CO, and hydrocarbon decreased as concn. of SBME increased.	39
	On-road vehicle test. Durability test, 1418 h. Power test and emission test (Tests include blending with petroleum oil).	4C, IDI	Fuel consumption and volumetric specific consumption increased. Volumetric calorific value, smoke emission, and CO decreased. Power and torque differ marginally. No lube oil dilution.	40
	Bench test 100 h and mixed driving test (Incorporation of methyl esters of babassu oil). Lube oil dilution tests.	DI 4C, IDI	Lube oil dilution increased with increase in double bonds of ME in DI engines. No lube oil dilution with IDI engines. Lube oil normal at 7500 km, the drain period.	41
2. Rapeseed oil methyl esters (RME)	Various load conditions; 20–100% rate output.	1C, DI	Smoke increased. Thermal efficiency and ignition delay decreased.	34
	EMA cycle 200 (h), advance injection timing 2 deg.	3C, DI, NA	Smoke, exhaust gas, and temperature decreased.	42
	Advance injection timing 2 deg.		Smoke, peak pressure, and ignition delay decreased. Power output increased.	42
	Short-term and engine durability tests 300 (h) (Tests includes ethyl esters of rapeseed oil).	4C, DI, TC 1C, PC	Fuel consumption and injection coking (ME) increased. Output power (EE lowest), torque (EE lowest), smoke, and exhaust temp. decreased.	43
	On-road vehicle test 364 (h) (Test includes blending with petroleum diesel).	6C, DI, TC, IC	Power output, fuel consumption, and smoke decreased.	44
	On-road vehicle test 364 (h) (Test includes blending with petroleum diesel).	8C, PC, NA	Power output and smoke decreased.	44

Table 5 (Continued)

Biofuel	Engine test	Engine investigated	Observations and results compared to petroleum diesel	Ref.
	US-FTP 75 (Test includes raw and half-refined rapeseed oil methyl esters).	4C, DI, TC, EGR	Fuel consumption, NOx, and formaldehydes increased. Hydrocarbons decreased.	45
	EMA cycle 200 h.	3C, DI, NA	Lube oil viscosity decreased.	46
	EMA cycle 1000 h.	3C, DI, NA	Fuel consumption increased.	47
	EMA cycle 1000 h (Test includes blending with petroleum diesel).		Lube oil viscosity, coking, and fuel consumption decreased.	
	Various load conditions	1C, MC	Power output and smoke number decreased. BMEP higher; NOx and CO increased.	25
3. Sunflower oil methyl esters (SFME)	1/3, 2/3, and full rack @ 2400 rpm	1C, DI, NA	Thermal efficiency, NOx, aldehydes, and formaldehydes increased. Particulates decreased.	48
	Various load conditions; 20–100% rated output	1C, DI	Smoke increased. Thermal efficiency and ignition delay decreased.	34
	Perkins 4.236 engines on a cyclic load; endurance test; on-road vehicle test.		Brake thermal efficiency increased. Fuel consumption increased marginally. Smoke decreased. Power output: comparable.	49
4. Palm oil methyl esters (PME)	Various load conditions; pure and blending with diesel.	1C, MC	Power output and smoke number decreased. BMEP higher. NOx increased. HC and CO decreased.	25
	Pure and blends, various loads.	1C, 4-stroke	BMEP and torque increased.	50
5. Cottonseed oil methyl esters (CSME)	1/2, 2/3, and full rack @ 2400 rpm.	1C, DI, NA	Thermal efficiency, NOx,aldehydes, and formaldehydes increased. Particulates decreased.	48

		Perkins 4.236 engines on a cyclic load; endurance test.		Brake thermal efficiency increased. Fuel consumption slight increased. Smoke decreased. Power output: comparable.	49
6.	Peanut oil methyl esters (PEME)	Endurance test 100 h.	4C, DI	Lube oil viscosity decreased.	31
		Perkins 4.236 engines on a cyclic load; endurance test		Brake thermal efficiency increased. Fuel consumption slight increased. Smoke decreased. Power output: comparable.	49
7.	Ricebran oil methyl esters (RBME)	Various load conditions; 20–100% rated output	1C, DI	Smoke increased. Ignition delay decreased.	34
8.	Karanja oil methyl esters (KAME)	Various load conditions; 20–100% rated output	1C, DI	Smoke increased. Thermal efficiency and ignition delay decreased.	34
9.	Tallow methyl esters (TAME)	SAE J 1003	6C, DI, TC, AC	NOx and fuel consumption increased. Hydrocarbons, CO, and particulates decreased.	28
10.	Maize oil methyl esters (MME) and ethyl esters (MEE)	Perkins 4.236 engines on a cyclic load. Endurance test.		Brake thermal efficiency increased. Fuel consumption slight increased. Smoke decreased. Power output: comparable.	49
11.	Castor oil methyl esters (CME) and castor oil ethyl ester (CEE)	Perkins 4.236 engines on a cyclic load. Endurance test.		Fuel consumption, brake thermal efficiency, and ignition delay increased. Smoke value decreased but increased at high speed, low load. Power output: comparable.	49

AC, air-cooled; BMEP, brake mean effective pressure; DI, direct injection; EGR, exhaust gas recirculation; IC, intercooled; IDI, indirect injection; MC, multicylinder; NA, naturally aspirated; PC, precombustion chamber; TC, turbocharged; *n*C, *n* cylinders.

buses, three running on 100% palm oil methyl esters and three on 100% petroleum diesel, were tested at about the same time. A different lubricating oil was tested with these six buses. Each bus traveled between 200,000 and 250,000 km. The field trial has been completed, and detailed analyses of the results are being carried out. However, throughout the field trial period, no technical problem was reported as long as the engines were maintained according to engine service manuals. The operators were not able to detect any difference in terms of general engine performance. The conclusions derived from the above evaluation are as follows:

1. No modification of the diesel engines is required.
2. The performance of the engines is generally good. The engines are easy to start, with no knocking and with smooth running. There is a shorter ignition lag.
3. The exhaust gas emission of the engines is much cleaner with reduction of hydrocarbon, NO_x, CO, and SO_2 content; therefore palm oil methyl esters are more environmentally friendly. These results are based on steady-state emission.
4. The engine oil is still usable at the recommended mileage as no engine oil dilution was observed either by infrared spectroscopy or by gas chromatography.
5. Palm oil methyl esters do not produce explosive air–fuel vapor. They also offer enhanced safety characteristics with a higher flash point (174°C compared to 96°C of petroleum diesel).
6. The cetane number of palm oil methyl esters is 64, which is higher than that of petroleum diesel, 30–40. Thus palm oil methyl esters can act as a cetane number improver.
7. Carbon build-up on the injector nozzles was normal except that the nature of the carbon was different.
8. The fuel consumption of the engines on palm oil methyl esters is comparable to that with petroleum diesel (e.g., 3–4 km/L for the buses under trial).
9. Palm oil methyl esters attack low grade plastic and rubber products such as hoses and seals and also react with the binding material in cement floors.

IV. GENERAL SUMMARY OF VEGETABLE OIL ALKYL ESTERS AS BIOFUEL

The highlights of the findings are as follows.

1. Methyl esters of vegetable oils are better diesel substitutes than the original oils.
2. Simple alkyl esters, in contrast to triglycerides, improve viscosity, but volatility remains poor. This may cause a problem in direct injection engines.
3. Introduction of a double bond resulted in increased efficiency; thus ethyl oleate had the highest thermal efficiency.
4. Among the methyl esters of saturated acids, thermal efficiency is inversely related to the chain length of the fatty acid.
5. A study of the efficiency of various esters of fatty acid as diesel fuel suggests that a vegetable oil having a high content of oleic acid or short-chain saturated acid transesterified to produce ethyl esters should produce a good test material as an alternative diesel fuel. Hence, the above observations indicate that alkyl esters of palm oil have the potential of being good diesel fuel substitutes.

V. CONCLUSION

Various technologies have been developed for the production of alkyl esters from vegetable oils containing varying amounts of free fatty acids. Many studies have demonstrated that vegetable oil esters (or biodiesel) can be used as fuel to run standard diesel engines without modification. The diesel engines are able to run on pure biodiesel or a blend with petroleum diesel. Biodiesels have higher cetane numbers than diesel, and they can be used as diesel improvers. All the studies from bench engine tests to exhaustive field trials have demonstrated that there was no discernible difference in terms of engine performance and fuel consumption in field trials. However, the exhaust emission was found to be cleaner. It emits no dark smoke and less CO and SO_2. Thus it is a more environmentally benign fuel. Nevertheless, the high cost of biodiesel compared to petroleum diesel at the present time is a major obstacle in the commercialization of biodiesel. It is believed that biodiesel has a strategic role to play in the near future.

ACKNOWLEDGMENT

We thank the Director General of PORIM for his permission to publish this chapter.

REFERENCES

1. T. Murayama. Evaluating vegetable oils as a diesel fuel, *INFORM 5*: 1138 (1994).
2. Y. M. Choo, A. N. Ma, and Y. Basiron, Palm oil methyl esters as diesel substitute, Presented at An Alternative Energy Conference—Liquid Fuels, Lubricants and Addition from Biomass, June 15–18, 1994, Kansas City, MO, 1994.
3. R. W. Nitske and C. M. Wilson, *Rudolf Diesel, Pioneer of the Power Age*, Univ. Oklahoma Press, Norman, OK, 1965.
4. E. G. Shay, Diesel fuel from vegetable oils: status and opportunities, *Biomass Bioenergy 4*: 227 (1993).
5. J. Walton, The fuel possibilities of vegetable oils, *Gas Oil Power*, July 1938, p. 167.
6. R. H. Seddon, Vegetable oils in commercial vehicles, *Gas Oil Power*, August 1942, p. 136.
7. A. W. Baker and R. L. Sweiget, A comparison of various vegetable oils as fuels for compression-ignition engines, *Proceedings of the 19th Oil and Gas Power Conference*, American Society of Mechanical Engineers, Cleveland, OH, 1947, pp. 40–48.
8. J. S. Aggarwal, H. D. Chowdury, S. N. Mukerji, and L. C. Vermoh, Indian vegetable oils as fuels for diesel engines, *Bull. Indian Ind. Res. 19* (1952).
9. C. E. Goering, M. D. Schrock, K. R. Kaufman, M. A. Hanna, F. D. Harris, and S. J. Marley, Evaluation of vegetable oil fuels in engines, ASAE Paper No. 87-1586, ASAE, St. Joseph, MI, 1987.
10. C. Adams, J. F. Peters, R. J. Rand, B. J. Schroer, and M. C. Ziemke, Investigation of soybean oil as a diesel fuel extender: endurance tests II, *J. Am. Oil Chem. Soc. 60*: 1574 (1983).
11. W. K. Klopfenstein and S. Walker, Efficiencies of various esters of fatty acids as diesel fuels, *J. Am. Oil Chem. Soc. 60*: 1596 (1983).
12. E. H. Pryde, Vegetable oils as diesel fuels: overview, *J. Am. Oil Chem. Soc. 60*: 1557 (1983).
13. R. C. Strayer, J. A. Blake, and W. K. Craig, Canola and high erucic rapeseed oil as substitutes for diesel fuel: preliminary tests, *J. Am. Oil Chem. Soc. 60*: 1587 (1983).
14. Y. Basiron and A. Hitam, *Cost Effectiveness of the CPO Fuel in the Mercedes Elsbett Engine Car*, PORIM Inf. Ser. No. 4 (Published by PORIM), July 1992.
15. T. B. Reed, An overview of the current status of biodiesel, Proceedings of the First Biomass Conference of the Americas, Vol. II, 1993, p. 797.
16. N. O. V. Sonntag, Fat splitting, esterification and interesterification, in *Bailey's Industrial Oil and Fat Products*, 4th ed., Vol. 2 (D. Swern, Ed.), Wiley, New York, 1982, pp. 97–173.
17. U. R. Kreutzer, Manufacture of fatty alcohols based on natural fats and oils, *J. Am. Oil Chem. Soc. 61*: 343 (1984).
18. B. Freedman, E. H. Pryde, and T. L. Mounts, Variables affecting the yields of fatty esters from transesterified vegetable oils, *J. Am. Oil Chem. Soc. 61*: 1638 (1984).

19. B. Freedman, R. O. Butterfield, and E. H. Pryde, Transesterification kinetics of soybean oil, *J. Am. Oil Chem. Soc. 63*: 1375–1380 (1986).
20. R. D. Farries, Methyl esters in the fatty acid industry, *J. Am. Oil Chem. Soc. 56*: 770A (1979).
21. Y. M. Choo, A. S. H. Ong, K. Y. Cheah, and A. Bakar, Production of alkyl esters from oils and fats, Australian Patent 43519/89 (1992).
22. Y. M. Choo and S. H. Goh, Esterification of carboxylic acids/glyceride mixtures, U.K. Patent 214 8897 (1987).
23. Y. M. Choo and A. S. H. Ong, Carboxylic acid esterification, U.K. Patent 216 1809 (1989).
24. S. C. Borgelt, T. S. Kolb, and L. G. Schumacher, Biodiesel: world status, in *Proceedings of an Alternative Energy Conference* (B. E. Dale, Ed.), ASAE, St. Joseph, MI, 1994, pp. 67–76.
25. A. Schafer, The use of biofuel in modern diesel engines, Presented at 1995 PORIM International Biofuel Conference, Jan. 17–18, 1995, Langkawi, Malaysia.
26. C. L. Peterson, M. E. Casada, L. M. Saflley, Jr., J. D. Broder, L. Perkins, and D. L. Auld, Agriculturally produced fuels, Paper No. 905532, Am. Soc. Agric. Engineers, St. Joseph, MI, 1990.
27. L. J. Icerman and E. B. Shultz, Buffalo gourd as a source of diesel fuel and ethanol on arid lands: preliminary economic analysis, *Proceedings of Seminar III on Vegetable Oil As Diesel Fuel*, Agric. Rev. and Manuals, USDA Agricultural Research and Service, Peoria, IL, 1983, pp. 21–22.
28. W. F. Marshall, Effects of methyl esters of tallow and grease on exhaust emissions and performance of a Cummins L10 engine, IIT Research Institute, Nat. Inst. for Petroleum and Energy Research, Bartlesville, OK. (Prepared for Fats and Proteins Research Foundation, Inc., Fort Myers Beach, FL), 1993.)
29. M. Mittelbach and P. Tritthart, Diesel fuel derived from vegetable oils, III. Emission tests using methyl esters of used frying oil, *J. Am. Oil Chem. Soc. 65*: 1185 (1988).
30. F. D. Gunstone, J. L. Harwood, and F. Padley (Eds.), *The Lipid Handbook*, 2nd ed., Chapman and Hall, London, 1994, pp. 72, 77, 92, 116, 118–119.
31. L. M. Ventura, A. C. Nascimento, and W. Bandel, First results with Mercedes-Benz Di diesel engines, Int. Conf. on plant and Vegetable Oils as Fuels, Fargo, ND, August 1982.
32. L. E. Wagner, S. J. Clark, M. D. Schrock, and P. O. Pienaar, Soybean oil esters as alternative diesel fuels, SAE Paper No. 841385, SAE, Warrendale, PA, 1984.
33. J. Einfalt and C. E. Goering, Methyl soyate as a fuel in a diesel tractor, *Trans. ASE 28*: 70 (1985).
34. P. Srinivasa Rao and K. V. Gopalakrishnan, Vegetable oils and their methyl esters as fuels for diesel engines, *Indian J. Technol. 29*: 292 (1991).
35. L. G. Schumacher, W. G. Hires, and S. C. Norgelt, Fueling a diesel engine with methyl-ester soybean oil, ASAE Proceedings of an Alternative Energy Conference, Nashville, TN, Dec. 14–15, 1992.
36. D. Fosseen and W. Goetz, Evaluation of methyl soyate/diesel blend in a DDC 6V-92TA engine: optimization of NOx emissions, Fosseen Manufacturing and Development, Radcliffe, IA. (Report prepared for National SoyDiesel Development Board, Jefferson City, MO), 1993.
37. K. W. Scholl and S. C. Sorenson, Combustion of soybean oil methyl ester in a direct injection diesel engine, SAE Paper No. 930934, SAE, Warrendale, PA, 1993.
38. L. G. Schumacher, S. C. Borgelt, W. G. Hires, and J. K. Humphrey, Biodiesel on the road—a report from Missouri, ASAE Paper No. 93-5017, ASAE, St. Joseph, MI, 1993.
39. L. G. Schumacher, S. C. Borgelt, and W. G. Hires, Soydiesel/petroleum blend research, ASAE Paper No. 93-6523, ASAE, St. Joseph, MI, 1993.
40. G. H. Pischinger, A. M. Falcon, R. W. Siekmann, and F. R. Fernades, Methyl esters of plant oils as diesel fuels, either straight or in blends, *Vegetable Oil Fuels* (Proc. Int. Conf. on Plant and Vegetable Oil as Fuels), Am. Soc. Agric. Eng., St. Joseph, MI, 1982, pp. 198–208.
41. R. W. Siekmann, G. H. Pischinger, D. Blackman, and L. D. Carvalho, The influence of lubricant contamination by methyl esters of plant oils on oxidation stability and life, *Vegetable Oil Fuels* (Proc. Int. Conf. on Plant and Vegetable Oil as Fuels), ASAE, St. Joseph, MI, 1982, pp. 209–217.
42. M. E. Feldman and C. L. Peterson, Fuel injector timing and pressure optimization on a DI diesel

engine for operation on biodiesel, in *Liquid Fuels from Renewable Resources—Proceedings of an Alternative Energy Conference*, Nashville, TN, 1992.
43. C. L. Peterson, D. L. Reece, R. Cruz, and J. Thompson, A comparison of ethyl and methyl esters of vegetable oil as diesel fuel substitutes, ASAE Alternative Energy Conference, Nashville, TN, Dec. 14–15, 1992.
44. D. L. Reece and C. L. Peterson, A report on the Idaho on-road vehicle test with RME and neat rapeseed oil as an alternative to diesel fuel, ASAE Paper No. 93-5018, ASAE, St. Joseph, MI, 1993.
45. M. Mittelbach, P. Tritthart, and H. Junek, Diesel fuel derived from vegetable oils. I. Emission tests using rape oil methyl ester, *Energy Agric. 4*: 208 (1985).
46. G. Zhang, M. Feldman, and C. L. Peterson, Diesel engine durability when fueled with methyl ester of winter rapeseed oil, ASAE Paper No. 88-1562, ASAE, St. Joseph, MI, 1988.
47. L. A. Perkins, C. L. Peterson, and D. L. Auld, Durability testing of transesterified winter rape oil (*Brassica napus* 1) as fuel in small bore, multi-cylinder, Di, Ci engines, SAE Paper No. 911764, SAE, Warrendale, PA, 1991.
48. S. M. Geyer, M. J. Jacobus, and S. S. Lestz, Comparison of diesel engine performance and emissions from neat and transesterified vegetable oils, *Trans. ASAE 27*: 375 (1984).
49. C. S. Hawkins and J. Fuls, Comparative combustion studies on various plant oil esters and the long term ignition engine, *Vegetable Oil Fuels* (Proc. Int. Conf. on Plant and Vegetable Oil as Fuels), ASAE, St. Joseph, MI, 1982, pp. 184–197.
50. H. Masjuki and M. Sohif, Performance evaluation of palm oil diesel blends on small engine, *J. Energy Heat Mass Transfer 13*: 125 (1991).
51. D. Fosseen and W. Goetz, Performance of a 6V-92 engine using methyl ester soybean oil, Fosseen Manufacturing and Development, Radcliffe, IA. (Report prepared for National SoyDiesel Development Board, Jefferson City, MO, 1993.)
52. W. F. Marshall, Status of biodiesel work at the National Institute of Petroleum Energy Research (NIPER), Presented to the research committee of the National SoyDiesel Development Board, Dec. 10, 1993.
53. J. R. Needham and D. M. Doyle, The combustion and ignition quality of alternative fuels in light duty diesels, SAE Paper 852101, SAE, Warrendale, PA, 1985.
54. C. L. Peterson, Vegetable oil as a diesel fuel: status and research priorities, *Trans. ASAE 29*: 1413 (1986).
55. J. Fuls, C. S. Hawkins, and F. J. C. Hugo, Tractor engine performance on sunflower oil fuel, *J. Agric. Eng. Res. 30*: 29–35 (1986).
56. C. L. Peterson, Vegetable oil as a diesel fuel: status and research priorities, *Trans. ASAE 29*: 1413 (1984).
57. J. J. Bruwer, B. Van D. Boshoff, F. J. C. Hugo, L. M. du Plesis, J. Fuls, C. Hawkins, A. N. van der Walt, and A. Engelbrecht, Sunflower seed oil as an extender for diesel fuel in agriculture tractors, Symposium of the South African Institute of Agricultural Engineers, June 11, 1980.
58. C. R. Engler, L. A. Johnson, W. A. LePori, and C. M. Yarbrough, *Biomass Energy*, A Monograph, Texas A&M University, 1985, pp. 174–212.

31

Products from Castor Oil

Past, Present, and Future

Henri-Jean Caupin
Elf Atochem S-A, Paris, France

I. INTRODUCTION

Castor oil has a long history and special status relative to other vegetable oils. Castor oil seeds have been found in Egyptian graves of 4000 BC. The seed was used in medicine (strong purgative action), and the hollow stalks of the plant were used to store cosmetics (especially kohl). Known as huile de carapathes, it was widely used in Africa for care of the skin and hair. In India, oral tradition described 59 illnesses cured by internal and external use of castor oil. The oil is clearly different from other vegetable oils, and it is soon apparent that it can only be ingested in a controlled fashion.

II. COMPOSITION

Castor oil is characterized by the presence of ricinoleic acid (12-hydroxyoleic acid) to the extent of almost 90%. Its components are listed in Table 1.

Ricinoleic acid has the molecular formula $CH_3(CH_2)_5CH(OH)CH_2CH{=}CH(CH_2)_7COOH$. The presence of a midchain hydroxyl group influences both the physical and chemical properties of this acid and its esters. These substances show many interesting chemical reactions, some of which are exploited commercially. For this reason the oil should be expressed in a clear and neutral form, as bleaching and other refining processes may result in its deterioration. Many of the important uses of castor oil depend on the presence of the β-hydroxyalkene unit $—CH(OH)CH_2CH{=}CH—$.

The more sophisticated downstream products are more valuable than the long-established and local uses of the oil. There are reports of castor oil being used as a fuel when other fuels are in short supply.

Table 1 Fatty Acid Composition of Castor Oil (wt %)

Acid	Wt %
Ricinoleic	89.5
9,10-Dihydroxystearic	0.7
Palmitic	1.0
Stearic	1.0
Oleic	3.0
Linoleic	4.2
Linolenic	0.3
Arachidic	0.3

Source: Frank Naughton of Cambrex.

III. MAJOR USES OF CASTOR OIL AND RICINOLEIC ACID ESTERS

Castor oil has a very wide range of uses [1,2]. Some are simple and traditional and exist only in the producer countries; others are more advanced and involve chemical modifications carried out only in the developed world.

The oil has well-recognized medicinal and cosmetic effects, and significant amounts are applied to the skin and hair. Almost 60 such uses are recognized in the oral traditions of India, and some of these uses are reflected in European cosmetic products.

A. Sulfation (Turkey Red Oil)

The first industrial modification of castor oil came in 1874 with the production of Turkey red oil. Reaction with sulfuric acid and sulfur trioxide results in sulfation of the hydroxyl group and further modification of the double bond:

$$\text{—CH(OH)—} \rightarrow \text{—CH(OSO}_2\text{OH)—}$$

This yields a product with improved surfactant properties. Apart from the soaps known from ancient times, Turkey red oil was the first anionic surfactant. It is used in textile processing, inks, industrial detergents, and leather treatment and in lubricant additives for cutting oils and hydraulic fluids. The sulfated hydrogenated oil has the consistency of an ointment and gives adjustable viscosity to water-based formulations with excellent skin compatibility.

B. Atmospheric Oxidation

All highly unsaturated oils such as linseed undergo atmospheric oxidation when blown, i.e., heated in air. This change occurs much more readily with castor oil, which has free hydroxyl groups. Oils of different viscosities result depending of the condition of oxidation, and they have different uses. The products are highly compatible with other resins, polymers, and waxes. They have low volatility, and there is no migration. They are used extensively in nitrocellulose lacquers.

C. Dehydration

Dehydration of castor oil (DCO) and of castor acids (DCOFA) converts castor oil to a mixture of diene esters and acids that are partly conjugated:

12—OH 18:1 (9*c*) → 18:2 (9*c*11*t*, 9*c*12*t*, and other isomers)

They have been used as a valuable alternative to drying oils such as tung oil (rich in eleostearic acid), which are not always available in the quantity required. DCO is used in paints and coatings in traditional nonaqueous systems and in more recently introduced aqueous systems.

Nonaqueous coatings require drying oils, and DCO has been much used for this purpose. Other drying oils present problems with smell, availability, and price stability. DCO and DCOFA are used in the now-preferred aqueous systems and can be incorporated into acrylic paints and into polyacrylate and epoxy ester systems.

D. Hydrogenation

Hydrogenated castor oil and hydrogenated castor acids have higher melting points than the nonhydrogenated materials. The products still contain the hydroxyl group, and they are used in cosmetics as well as in paints, coatings, and greases.

Greases prepared from tallow had many shortcomings, and their quality was dramatically improved when derivatives of hydrogenated castor oil were incorporated. Lithium salts increase the hardness and raise the melting point of the products, permitting their use under more severe operating conditions and higher temperatures.

Hydrogenated castor products are also used in melamine resins, where they increase flexibility; in acrylics; and in paints, where their thixotropic properties permit control of product viscosity.

E. Reactions with Ethylene Oxide and Propylene Oxide

Reaction of the hydroxyl group in castor oil with ethylene or propylene oxide gives rise to a nonionic surfactant. By suitable adjustment of the reactants, products of diverse water solubility and of oil/water compatibility can be obtained. Although the products are of high quality, they are in competition with cheaper products, and supplies of castor oil have been too uncertain to warrant the large investment necessary for surfactant markets.

F. Reactions with Isocyanates

Like other hydroxyl-containing molecules, castor oil reacts with appropriate isocyanates to produce polyurethanes, which find many uses in the developed world. In America polyurethane coatings for houses made of wood are in great demand, but this is not true in Europe, where wooden buildings are less common. Polyurethanes based on castor oil have also been developed for use as encapsulating materials, especially in the electrical industry. Developments in this area are now concerned more with formulations than with new scientific issues (see Ref. 5) [3].

G. Splitting with Caustic Soda (C_8 and C_{10} Products)

Sodium hydroxide splits the C_{18} ricinoleic acid molecule into C_8 and C_{10} products, the exact nature of which depends on the reaction conditions. At 180–200°C with a caustic/castor ratio of 1:1, the major products are 2-octanol and 10-hydroxydecanoic acid. At 250–275°C, with a caustic/castor ratio of 2:1, the products are mainly 2-octanol and sebacic (decanedioic) acid. Of these the most valuable product is the dibasic acid (sebacic). This has been used along with the C_6 diamine to produce polyamide 6,10, which has better solvent and water resistance than the shorter chain polyamides. The C_{10} dibasic acid has also been combined with a range of alcohols to give diesters that serve as efficient lubricants at both low and high temperatures.

H. Splitting with Steam (C_7 and C_{11} Products)

The historical procedure for splitting the C_{18} ricinoleic acid chain into C_7 and C_{11} fragments suffered from low yields and high environmental pollution. It was carried out mainly to give C_7 and C_{11} aldehydes, which are used as perfumes and fragrances. Heptanal is still condensed with benzaldehyde to give amyl cinnamic aldehyde ($PhCH{=}C(C_5C_{11})CHO$). Undecenal, a key component of high quality perfumes and a quality enhancer of large volume products, is still made from undecenoic acid. The older uses are only a small part of the current applications of these materials, to be discussed in detail later.

The antifungal properties of the zinc and calcium salts of undecenoic acid were recognized in the early 1940s, and these compounds are still used extensively for this purpose, particularly for the care of adult feet and baby bottoms.

The C_7/C_{11} splitting process has been much improved by the development of a continuous steam cracking process. The production streams are managed on the basis of the sales profile, and in recent years there have been many new developments in the use of these C_7 and C_{11} compounds. These are discussed in the following section.

IV. NEW USES OF CASTOR OIL DERIVATIVES WITH SPECIAL REFERENCE TO C_7 AND C_{11} COMPOUNDS

A. C_7 Compounds

The C_7 compounds are available as aldehyde, acid, and alcohol. These compounds are of interest in that they have borderline properties between short- and long-chain compounds.

1. *Corrosion Control*

Heptanoate salts act as stable anticorrosion agents even in hard water. The sodium salt gives efficient corrosion control at concentrations between 0.5 and 4.0 g/L. The pK of calcium heptanoate at these levels is such that there is no precipitation of the metallic soap. This is also true for the C_6 acid but not for derivatives of the C_8, C_9, and C_{10} acids.

The protective coatings of heptanoic acid obtained on copper, iron, aluminum, and zinc are probably molecular layers of organized crystals. Further investigation is needed to find the optimum levels and formulation conditions for corrosion control. It is known that the C_6 acid is too short to provide protection and that whereas the C_8, C_9, and C_{10} acids are protective their efficiency is handicapped by hard water precipitation at the concentration required for corrosion control. It appears that the C_7 acid is unique in combining corrosion control with hard water compatibility [4].

2. *Low Temperature Properties*

The boiling and freezing points of heptanoic acid derivatives are generally such that these compounds are not lost in normal high temperature use, and they remain liquid at low temperatures. For example, they make good plasticizers for polymers to be used at very low temperatures.

Among the many esters examined as potential lubricants for jet engines, automobiles, and stationary engines, those based on heptanoic acid continue to find a place because of their outstanding properties.

B. C_{11} Compounds

In addition to its antifungal properties, undecenoic acid finds other uses both actual and potential based on its differing functional groups (double bond and carbonyl) at either end of the molecule. Some recent studies are described in patents [2, 5–10].

1. Polyamide 11

Currently the major use of undecenoic acid is the production of 11-aminoundecanoic acid and its polymerization to give polyamide 11. This has superior water resistance and solvent compatibility compared with polyamides 6 and 66. Under dry conditions its mechanical properties are weaker than those of the shorter chain polyamides, but they are much better maintained under long-term use and wet conditions than those of the C_6 polyamides.

The physical mixture of polyamide 11 and silk has been known for some time, since polyamide 11 is a better quality improver for silk than the shorter chain polyamides. This is due to a very meaningful and highly surprising technical feature: Silk takes in about 10% water, but the dimensional changes are sideways—the length of the filament remains constant, as does that of polyamide 11.

2. New Uses for C_{11} Compounds

Other uses now developed for C_{11} compounds depend on the previously established antifungal, antismell, and insecticidal properties of the various derivatives, of undecenoic acid.

Biostability and Basic Antifungal Properties. Formulators of performance products are much concerned with the preparation of water–oil emulsions that maintain physical stability over the life of the product. This holds for a wide range of water-based products such as paints, cosmetics, cutting oils, and other types of emulsified compounds. Once the necessary physical stability has been secured, biological stability must also be guaranteed from production through to consumption.

Currently, the necessary biocides are added late in the production process, and they must be selected to have no deleterious effect on emulsion stability. Appropriate rules and regulations have to be applied to comply with EPA and FIFRA documents. A compound can be described as a biocide only if it decreases the existing germs in a water-based formulation by a factor of 100–1000 in a given very short time.

Some compounds combine natural and limited bioresistance as well as surfactant activity. Fats present in hair that exercise a protective function include significant levels of C_{11} acids, no or very little C_{12} acid, and a mix of C_{14}, C_{16}, and C_{18} acids. This is important when the high fermentation situation around the hair is recognized. With a constant mild temperature and permanent humidity at root level, it is necessary to have an effective bioresistant and renewed mixture. This must allow for constant and healthy control of microorganisms and for permanent monitoring of the scalp lice population [11,12].

The situation for many organic–water emulsions is similar to that of the hair and scalp, and the problem is how to introduce an approved C_{11} compound without destroying the physical stability of the emulsion. Many trials over the last 10 years have failed because the C_{11} compounds differ from conventional surfactants and tend to destroy the emulsions. However, a number of specialized C_{11} derivatives are now available that can be used in the preparation of emulsions and are introduced into the formulation early rather than at a later stage [13].

The C_{11} derivatives mentioned are the counterparts of the even-chain fatty acid derivatives, already long known by any cosmetics formulator.

Economic and regulatory constraints encourage formulations with controlled levels of approved biocides. This can be achieved through fine-tuning of combinations of naturally derived bioresistant materials. C_{11} compounds at 10–20% levels are already allowed in OTC formulations, and the addition of such compounds at levels of 0.1–5.0% could assist in reducing biocide levels.

Animal Waste Smells. C_{11} esters decrease the malodorousness of pig and other animal wastes to acceptable levels. Efficient addition rates are below 100 g/t (i.e., <100 ppm). Smells associated with water treatment stations and with industrial effluents can be reduced by C_{11} esters in the same way.

Antilouse Activity of Selected Derivatives. Compounds similar to those just described have also been used for the control of lice in human situations. Details are given in Ref. 7.

Boron-Free Cutting Oils Technology. C_7 and C_{11} monoethanolamides are used in cutting oils where they replace boron compounds and reduce the environmental impacts of the effluent. The combined efficiencies allow optimization of the corrosion control as well as of the natural bioresistance of the emulsions.

Evidence of Specific Properties Linked to Acetylenic Bifunctionality in the C_{11} Linear Chain. The C_{11} products are unique in that they contain different functional groups (one carboxyl group and one unsaturated center) at the ends of the molecule. The common unsaturated acids have midchain unsaturation. The α,ω difunctional aliphatic compounds are either short-chain (C_3, C_4) or neutral (α-olefins) or contain two identical functional groups (dibasic acids, diols), so differentiation becomes difficult.

There is increasing information about the surface effects of C_{11} compounds that appear to be spontaneously organized.

Typical "self-assembled biolipid tubular templates" are shown in Ref. 14, and new mesogenic diacetylenic monomers and polymers in Ref. 15. Perfluoroalkylated dienes can be obtained from perfluoroalkylated alkynes, and these lead to fluorinated analogues of *Lepidoptera* pheromones [16]. An important pheromone from *Leucoptere scitella* can be obtained from conjugated diyne diols [17, 18].

Molecular assemblies of functionalized diynes are described in Refs. 19 and 20 along with polymerization of supramolecular assemblies requiring a C_{11} monoyne and conjugated C_{22} diynes. Rare dienoic acids of marine origin can be obtained starting from undecynoic acid [21]:

$$CH_2{=}CH(CH_2)_8COOH \rightarrow CH{\equiv}(CH_2)_8COOH \rightarrow HOOC(CH_2)_8C{\equiv}C{-}C{\equiv}C(CH_2)_8COOH$$

Receptor–ligand interactions have been identified using diyne compounds as part of the lipid system. Direct colorimetric detection of the poly-ynes can be exploited for diagnostic applications in various pharmaceutical fields [22]. Glycerides combined with C_{22} diyne structures derived from the C_{11} acid inhibit the influenza virus in vitro [23].

In addition there are many abstracts making reference to diynes based on the medium-chain C_{11} material. A partial list of applications over the period 1980–1994 is given in Table 2.

It is recognized that the cost of undecynoic acid and its derivatives is very high and there is no industrial source of these at reasonable and reliable cost. Laboratory-prepared material is reported to be available at $100 per gram for 99% pure grade and around $20 per gram for a 95% pure grade.

Undecynoic acid, its alcohol, and other derivatives have been known and used for experimental purposes since the beginning of the century. It is clear from the list in Table 2 that there is a potential demand for acetylenic compounds of this chain length and the materials

Table 2 Applications of Diynes Based on Undecynoic Acid—A Time Line

1980	Photographic element
1985	Nonlinear optical elements
1986	Electrically conductive polydiacetylenes
1986	Semiconductor device comprising an organic material
1987	Recording medium for optical memory
1987	Urethane-substituted diacetylene monomers
1987	Thermochromic polydiacetylenes as temperature indicators
1987	Polymerized diacetylene Langmuir–Blodgett films as optical modulators
1988	Laser recording
1989	Monosubstituted diacetylene compounds for electrically conductive polymers and resist materials
1989	Polarized inner-shell absorption spectroscopy of organic molecular and polymeric solids
1989	Preparation of polymerizable fatty acids
1990	Reaction of diynoic acid chlorides with para-substituted phenols
1990	Coating for increasing sensitivity of a radiation imageable polyacetylenic film
1990	Polydiacetylene derivative-containing thin film for electric conductor
1990	Process for sensitizing polyacetylenic imaging dispersion
1991	Organic film materials
1991	Langmuir–Blodgett organic films and preparation thereof
1991	Recording medium of improved stability
1991	Immunolesioning: selective destruction of neurons using immunotoxin to rat NFG receptor
1991	Optical recording medium and method
1991	Preparation of heteroarylaminoalkanols as drugs
1992	Preparation of PLA_2 inhibitors as anti-inflammatories
1992	Process for the preparation of 2-alkyl-4,6-*tert*-butylphenol compounds
1992	Manufacture of liquid crystal orienting film
1992	Visually imageable polyacetylene salt dyes
1992	Recovery of imageable polyacetylene compounds from binder containing aqueous dispersions
1992	Gas-encapsulated or gas precursor encapsulated microbubbles for ultrasound contrast agents, ADN preparation, and polymerization of amphiphiles therefor
1992	Process for acylation of 2-alkyl-6-*tert*-butylphenol
1993	Preparation of 2,3-epoxy derivatives as anti retroviral pests
1993	Contrast agents consisting of carbohydrate particles and amphiphilic carboxylic acids
1993	Spicamycin derivatives and the use thereof
1993	Phospholipids for modulation of microstructure morphology
1993	Preparation of cis olefins
1994	Photochemical labeling of nucleic acids with europium-chelating reagents and their use in gene probe systems
1994	Molecular composites

derived from them. This could be a significant and valuable development of castor oil, because there is no convenient source of these undecynoic acid derivatives other than the steam cracking of this oil.

V. FUTURE TRENDS FOR CASTOR OIL AND ITS DERIVATIVES

Castor oil is the most convenient source of C_7 and C_{11} compounds and materials made from them. This arises from the unique structure of ricinoleic acid (12-hydroxyoleic acid) and from the fact that these products are easily available only from castor oil. If the demand for C_7/C_{11}

chemicals increases in the next 5–10 years, then the pressure to find a bioengineered solution will be strengthened. Attempts are already being made to transfer the required gene(s) to the rape plant so that it generates ricinoleic acid. Such a modified rape could be grown in nontropical areas in North America and northern Europe.

The most recent Berkeley publications report on delicately organized "nanochemical" layers for many applications. The basic requirement appears to be a difunctional C_{11} chain with terminal ethylenic or acetylenic unsaturation that would provide layers of 1 nm (1×10^{-9} m) thick. We are still a long way from that, with layers of 1 μm thickness, i.e., 1000 times thicker. There is a need for creative thinking by new generations of chemists and technologists to develop this nanochemistry, and ricinoleic acid, as the most convenient source of C_{11} compounds, will have a significant role in this.

The existing uses of castor oil will continue to be significant, and the oil of *Ricinus communis* will continue to be the biggest source of ricinoleic acid for many years. New developments can only add to the value of this material.

ACKNOWLEDGMENTS

This review is based on information in the public domain and on information coming from Elf Atochem relating to that company's products and to its castor oil cracking process. I also thank Frank Naughton and Frank Vignolo, both of whom played a significant role in ICOA, itself part of the American Oil Chemists' Society. Frank Gunstone of St Andrews, a joint editor of this book, made useful suggestions based on his knowledge of oleochemistry. He also assisted me in the production of the final version of this chapter. I also thank many others who have published in this field and whose work is cited.

Five key companies making extensive use of castor oil issue booklets and data sheets on their range of products: Alberdingk Boley GmbH, Germany; Caschem Inc, Elizabeth, NJ (former Baker Castor Oil Co); Elf Atochem, Paris, France; Itoh Oil Manufacturing Co, Yokkaichi, Japan; and Thai Castor Oil Industries Co, Bangkok, Thailand.

REFERENCES

1. Kirk-Othmer Encyclopedia, Interscience, New York, 1994, pp. 301–320, includes 120 references.
2. F. C. Naughton and Robert L. Vignolo, *The Chemistry of Castor Oil Derivatives and Their Applications*, Int. Castor Oil Assoc., Westfield, NJ; includes 391 references.
3. Elf Atochem S-A, Undecylenate deodorants for animal manures, U.S. Patent 006050 173 (Dec. 19, 1990).
4. Elf Atochem S-A, Corrosion inhibitory formulation based on carboxylic acid and corrosion inhibition activity for ferrous and nonferrous metals, Fr. Patent 92 14233 (Mar. 24, 1995).
5. Elf Atochem S-A, Deodorizing the muds and sludges from water treatment stations using C_{11} ester, Eur. Patent 0434 522 B1 (Mar. 1, 1995); U.S. Patent 5.338.511 (Aug. 16, 1994).
6. Elf Atochem S-A, Animal nutrients deodorizing agents using undecylenic acid ester, Eur. Patent 0 434 524 B1 (June 15, 1994).
7. Elf Atochem S-A, Lice killing formulation using undecylenic acid derivatives, Fr. Patent 92 06 149 (May 20, 1992).
8. Elf Atochem S-A, Deodorizing paper mill effluents with undecylenic acid esters, Fr. Patent 92 09 457 (July 30, 1992).
9. Elf Atochem S-A, New polymer resin loaded with undecylenic ester, Fr. Patent 92 13 369 (Nov. 6, 1992).
10. Elf Atochem S-A, Animal dejecta deodorizing formulation and full process treatment, Fr. Patent 93 07 631 (June 23, 1993).

11. L. Leon, L. D. Gershbein, and B. S. Metcalfe, *J. Invest.* Dermatol. *46*(5): 477–479, (1966).
12. W. Stodt Meister, H. P. Nissen, C. Schirren, and H. W. Kreysel, Fettsaüreverteilung der Haarlipide bei der Psoriasis vulgaris, *Arch. Dermatol. Res. 264*: 399–343, (1979).
13. A. Domsch, Chemistry and application of undecylenic acid and derivatives thereof for cosmetics, *SÖFW J. 6*: 120 (1994).
14. S. L. Burkett and S. Mann, Spatial organization and patterning of gold nanoparticles on self-assembled biolipid tubular templates, *Chem. Commun. 1996*: 321.
15. P. T. Hammond and M. F. Rubner, Synthesis and characterization of new mesogenic diacetylene monomers and their polymers, *Macromolecules, 28*(4) (1995).
16. F. Saremi, E. Maasen, and B. Ticke, Self-assembled alternating multilayers built up from diacetylene and bolaamphiphiles and poly(allylamine hydrochloride): polymerization properties, structure, and morphology, *Langmuir 11*: 1068–1071 (1995).
17. Z. Wang and X. Lu, Stereoselective synthesis of perfluoroalkylated (*E*, *E*) dienes from perfluoroalkylated alkynes: The synthesis of fluorinated analogs of *Lepidoptera* pheromones, *Tetrahedron 51*: 11765–11774 (1995).
18. B. Crousse, O. Pichon, and J. Soulié, Synthese totale du (10Z, 13Z, 16Z)-nonadeca-10,13,16,18-tetraen-1-ol, Elsevier, Paris, 1993.
19. W. Spevak, J. O. Nagy, and D. H. Charych, Molecular assemblies of functionalized polydiacetylenes, in VCH, Weinheim, 1995.
20. D. F. O'Brien, C. S. Marvel Laboratories, Polymerization of supramolecular assemblies, *Trip 2*(6): June (1994).
21. B. A. Kulkarni, A. Chattopadhyay, and V. R. Mammdapur, A facile synthesis of (10Z, 15Z) cicosa-10,15-dienoic acid, a marine natural product, *J. Nat. Prod. 57*(4): 537–538 (1994).
22. D. H. Charych, J. O. Nagy, W. Spevak, and M. D. Bednarski, Direct colorimetric detection of a recepto-ligand interaction by a polymerized bilayer assembly, Center for Advanced Materials, Lawrence Berkeley Laboratory, Berkeley, CA.
23. J. C. Paulson, Polymerized liposomes containing C-glycosides of sialic acid: potent inhibitors of influenza virus in vitro infectivity, in *The Receptors* (M. Conn, Ed.), *Academic*, New York.
24. Elf Atochem S-A, Filtration process equipment for a mud and specific equipment, Fr. Patent 94 03 814 (March 31, 1994).

Index

Numbers in italic refer to figures; "t" refers to tables.

THE UNIVERSITY OF READING LIBRARY

Reading University Library